Welding Metallurgy
Volume I, Fundamentals

Dedicated to
George Michael Linnert
1943-1952

Welding Metallurgy
Carbon and Alloy Steels

Volume I
Fundamentals

George E. Linnert
GML Publications
Hilton Head Island, South Carolina, USA

Fourth Edition

Published by the American Welding Society
Miami, Florida, USA

Library of Congress Catalog Card Number: 94-074489
International Standard Book Number: 0-87171-457-4
Copyright © 1994 by the American Welding Society
550 N.W. LeJeune Road; Miami, Florida; USA

Fourth Edition 1994

All rights reserved. No part of this book may be reproduced in any form or by any means — electronic or mechanical — including photocopying, recording, or any other information storage or retrieval system, without permission in writing from the publisher.

The American Welding Society, Inc. assumes no responsibility for the information contained in this publication. An independent, substantiating investigation should be made prior to reliance on or use of such information.

Text edited, designed and formatted by Alexander M. Saitta

Printed in the United States of America

Contents

Chapter One:
Background to Welding Metallurgy 1
MILESTONES IN WELDING HISTORY 1
THE FUTURE OF WELDING 4
WHAT IS WELDING METALLURGY? 6
PUTTING WELDING METALLURGY TO USE 12
WELDING TECHNOLOGY RESOURCES 12
SUGGESTED READING 15

Chapter Two:
The Structure of Metals 18
ATOMS 18
Elementary Particles 20
Electrons 22
Positrons 26
Atomic Nuclei 26
Protons 27
Neutrons 28
Atom Construction 32
Isotopes of Elements 33
Isobars 34
Atomic Weight 34
Atomic Mass 34
Atom Valency 35
Ionization 36
Radioactivity 37
Atom Size or Diameter 38
THE ELEMENTS 39
AGGREGATES OF ATOMS 41
The Solid State 45
The Crystalline Solids 45
Amorphous Solids 47
The Liquid State 48
The Gaseous State 49
FUNDAMENTALS OF CRYSTALS 50
Identification of Planes and Directions in Crystals 56
Basic Types of Crystals 56

        Inert Gas Crystals 58
        Ionic Crystals 58
        Covalent Crystals 59
        Metallic Crystals 59
    THE CRYSTALLINE STRUCTURE OF METALS **61**
        How Does a Crystal Grow from the Melt? 64
        The Formation of Dendrites 66
        The Formation of Grains 68
            The Shape of Grains 71
            The Size of Grains 72
        Undercooling 72
    THE IMPORTANCE OF A CRYSTALLINE STRUCTURE **74**
        Allotropic Transformation 75
        Solubility in the Solid State 76
        Plasticity in Metallic Crystals 77
            Slip in Crystalline Structures 77
            Slip and Lattice Orientation 78
            Slip in Polycrystalline Metals 79
            Observing Slip 80
            Twinning in Crystalline Structures 81
        Lattice Imperfections: Dislocations 84
            Point Defects 85
            Edge Dislocations 86
            Screw Dislocations 88
            Stacking Faults 88
            Other Lattice Imperfections 88
        Cold-Working Metals 88
    EXAMINATION OF METAL STRUCTURES **91**
        Fracture Appearance Assessment 92
        Metallography 92
            Metallography Using Optical Microscopy 92
            Quantitative Metallography 105
            Metallography Using the Electron Microscope 106
            Metallography Using Ion Microscopy 115
            Tunnel-Effect Microscopy 116
            Chemical Analysis of Microstructural Constituents ... 116
        Newer Techniques in Metallurgy 121
    FRACTOGRAPHY **123**
    SUGGESTED READING **131**

Chapter Three:
The Properties of Metals 133

STRUCTURE SENSITIVITY OF PROPERTIES **133**
DIRECTIONALITY IN PROPERTIES **135**

MECHANICAL PROPERTIES 136
Elastic Behavior of Metals 137
Young's Modulus of Elasticity 139
Poisson's Ratio 139
Limits of Elasticity and Proportionality 141
Plastic Yielding in Metals 142
Yield Strength 143
Breaking Strength of Metals 143
Tensile Strength 144
True Stress and True Strain 144
Notched Tensile Strength 147
Ductility 147
Elongation 148
Reduction of Area 149
Ductility Indications from Special Tests 149
Hardness 149
Static Indentation Hardness Testing 150
Microhardness Testing 151
Dynamic Hardness Testing 152
Scratch Hardness Testing 153
Conversion of Hardness Numbers 153
Toughness 154
Introduction of Impact Testing 155

FRACTURE IN METALS 156
Ductile Fracture 158
Brittle Fracture 159
Intergranular Fracture 162
Conditions Affecting Fracture Toughness 164
Effect of Temperature 164
Effect of Stress Axiality 166
Stress Gradient 169
Stress Multiaxiality 170
Effect of Rate of Strain 171
Effect of Cyclic Stress (Fatigue) 174
Fatigue Crack Initiation 176
Fatigue Crack Propagation 176
Fatigue Crack Failure 176
Cyclic Stress Limits to Avoid Fatigue Failure 181
Cyclic Stress Conditions 183
Variable Loading and Cumulative Fatigue Damage 185

FRACTURE MECHANICS: ASSESSMENT OF FRACTURE TOUGHNESS 188
Brittle Fracture Test Parameters 190
Section Dimensions 192
Plotting Coordinates 192
Crack Surface Displacement Mode 193
Plane-Strain 195

- Plane-Stress 195
- Stress Distribution..196
- Procedures for Evaluating Propensity for Brittle Fracture 197
  - Use of Linear-Elastic Fracture Mechanics...198
- Development of Elastic-Plastic Fracture Mechanics 201
  - Crack Tip Opening Displacement Testing...205
  - The J-Integral Test Method 211
- Fatigue Cracking Assessment by Fracture Mechanics 215
- Mechanical Properties at Low Temperature...218
  - Strength at Low Temperature..219
  - Impact Toughness at Low Temperature 220
  - Test Methods for Toughness Evaluation...226
  - Correlation of Results from Fracture Toughness Tests 236
- Improved Mechanical Properties for Low-Temperature Service 238
- Mechanical Properties at Elevated and High Temperatures....................238
  - Short-Time Elevated Temperature Testing 241
  - Long-Time Elevated Temperature Testing..242
- Mechanical Properties After Plastic Work..247
  - Hot Work 248
  - Cold Work 248
  - Peening 248
  - Irradiation 249

PHYSICAL PROPERTIES 258
- Density 258
- Thermal Properties..259
  - Specific Heat..260
  - Thermal Conductivity 261
  - Melting Point or Melting Range 264
  - Heat of Fusion 264
  - Viscosity and Surface Tension of Molten Metals 264
  - Boiling Point and Heat of Vaporization..266
  - Thermal Expansion and Contraction..266
  - Thermionic Work Function..268
- Electrical Properties 268
- Magnetic Properties 270
  - Evaluation of Magnetization 272
  - Summary of Magnetic Behavior 273
  - Involvement of Magnetization in Welding..274

CHEMICAL PROPERTIES 274
- Corrosion of Metals..275
  - Corrosion in Aqueous Solutions 276
  - Corrosion in Hot Gases...282
  - Corrosion in Molten Metals...282
  - Corrosion in Molten Salt 283
- Forms of Corrosion Pertinent to Weldments 283
  - Stress Corrosion Cracking (SCC) 284

SUGGESTED READING 293

Chapter Four:
Effects of Alloying Elements 295

ALLOYING 295
Alloys in the Liquid State 296
Phase Diagrams 299
Binary Phase Diagrams 304
Ternary Phase Diagrams 307
Phase Diagrams for Multi-Element Alloys 308
Alloys in the Solid State 310
Factors Influencing Solid Solubility 310
Formation of Intermediate Phases and Compounds 316
Mechanisms and General Effects of Alloying 316
Role of Crystalline Structure 317
Role of Microstructure 319
Mechanisms for Altering Mechanical Properties 319

ALLOYING ELEMENTS IN IRON 326
Carbon 329
Analysis of the Iron-Iron Carbide Diagram 331
Manganese 337
Phosphorus 338
Sulfur and Selenium 340
Silicon 342
Copper 343
Chromium 345
Nickel 346
Molybdenum 346
Niobium (Columbium) 347
Vanadium 348
Aluminum 348
Nitrogen 349
Titanium 352
Boron 354
Cobalt 355
Tungsten 355
Lead 355
Other Alloying Elements 356

BENEFIT OF REVERSING THE ALLOYING TREND 357
Residual Elements 358

SUGGESTED READING 359

Chapter Five:
Types of Steel and Their Manufacture 361

GENERAL CATEGORIES OF IRON AND STEEL 361
IRON PRODUCTION BY ORE REDUCTION 362
Blast Furnace 362

Direct Reduction Processes 365
CAST IRON 366
WROUGHT IRON 367
POWDER METALLURGY 367
STEELMAKING PROCESSES 368
Significance of Acid and Basic Steelmaking 368
Bessemer Converter 369
Open Hearth Furnace 370
    Rimmed Steel 372
    Capped Steel 373
    Killed Steel 373
    Semikilled Steel 379
    Vacuum Deoxidized Steel 379
Oxygen Steelmaking 380
Basic Oxygen Steelmaking 380
    L-D Process 381
    Kaldo Process 383
    Off-Gas BOF 383
    Q-BOP Process 383
    Lance-Bubbling-Equilibrium 385
Ladle Refining 385
    Slag Removal 385
    Mixing Capability 386
    Alloying Additions 386
    Vacuum Treatment 386
    Temperature Adjustment 387
    Desulfurization 388
Electric-Arc Furnace 389
Electric-Induction Furnace 389
Electroslag Remelting 391
SPECIAL MELTING PROCESSES 392
Vacuum Induction Melting 392
Vacuum Consumable-Electrode Remelting 393
Electron-Beam Melting 395
Argon-Oxygen Decarburization (AOD) 396
FOUNDRY AND STEEL MILL OPERATIONS 397
Ingot Steelmaking Practice 398
Continuous Casting of Steel 400
HOT WORKING OPERATIONS 404
Thermo-Mechanical Control Process (T-MCP) 406
COLD FINISHING 407
HEAT TREATMENT 408
CONTINUOUS COATING OF STRIP STEEL IN COILS 409
TYPES OF STEEL 410
Carbon Steels 410

Alloy Steels 411
    Construction 412
    Automotive, Aircraft, and Machinery 412
    Low-Temperature Service 413
    Elevated Temperature Service 413
High-Alloy Steels 413
    Austenitic Manganese Steel 413
    Stainless Steels 414
    Heat-Resisting Steels 415
Tool Steels 415

STANDARDS AND SPECIFICATIONS FOR STEELS 416
Unified Numbering System 416
AISI-SAE System of Standard Carbon and Alloy Steels 418
ASTM Standards 420
API Specifications 424
Aerospace Material Specifications 426
ASME Material Specifications 426
AWS Specifications, Codes and Rules 429

CARBON AND ALLOY STEEL USED IN WELDED CONSTRUCTION 430
Qualities of Steel Important to Welding 431
Factors Affecting the Weldability of Steel 432
    Chemical Composition 432
    Mechanical Properties 434
    Metallurgical Structure 435
    Internal Soundness 435
    Cleanliness 436

THE FUTURE OF STEELS AND THEIR WELDABILITY 437
New Steels and Product Forms 437
Dissimilar-Metal Welding 437
Repair Welding — The Ultimate Challenge 439

SUGGESTED READING 439

Chapter Six:
Welding Methods and Processes 444

SOLID-STATE WELDING (SSW) 444
FUSION WELDING 445
BRAZING AND SOLDERING 445
HEAT SOURCES FOR WELDING AND CUTTING 448
Electrical Heat Generation 448
    Electric Arc 448
    Electron Beam 459
Electric Resistance 461
Electromagnetic Radiation 461
    Laser Beams 463

Chemical Heat Generation..465
Mechanical Heat Generation 466

THE WELDING AND CUTTING PROCESSES **467**
Arc Welding Process..467
Power Sources for Arc Welding 467
Auxiliary Equipment for Arc Welding 470
Basic Forms of Arc Welding 471
Shielded Metal Arc Welding (SMAW)..472
Stud Arc Welding (SW)...477
Gas Tungsten Arc Welding (GTAW) 478
Gas Metal Arc Welding (GMAW)...489
Flux Cored Arc Welding (FCAW) 501
Submerged Arc Welding (SAW)..505
Plasma Arc Welding (PAW)..511
Percussion Welding (PEW) 514
Magnetically Impelled Arc Welding 515
Welding Arc Technology...517
Resistance Welding Processes 520
Resistance Spot Welding (RSW) 522
Resistance Seam Welding (RSEW) 531
Projection Welding (PW) 532
Upset Welding (UW)..533
Flash Welding (FW) 537
Electrical Metal-Explosion Welding Process 541
Induction Welding (IW) 542
Electroslag Welding (ESW) 542
Electron Beam Welding (EBW)..548
Welding in a High Vacuum (EBW-HV)...550
Welding in a Medium Vacuum (EBW-MV) 565
Nonvacuum Electron Beam Welding (EBW-NV)..565
Tracking Joints During Electron Beam Welding...565
Laser Beam Welding (LBW) 568
Nature of Laser Beams and Plasma Generation 569
Basic Techniques in Laser Welding 571
Attributes of Laser Welding 571
Laser Welding Difficulties and Defects..572
Shielding Gas Effects in Laser Welding 575
Filler Wire Feeding in Laser Welding 576
Oxyfuel Welding (OFW) 582
Oxyacetylene Welding (OAW)..582
Thermite Welding (TW) 585

SOLID-STATE WELDING PROCESSES..**587**
Hot Pressure Welding (HPW) 589
Induction Welding (IW)..592
Friction Welding (FRW)...593
Inertia-Drive Friction Welding 594
Direct-Drive Friction Welding..595

Contents **xiii**

    Materials Suited for Friction Welding...595
    Mechanical Properties of Friction Welds..598
    Other Forms of Friction Welding 600
  Explosion Welding (EXW)..607
  Diffusion Welding (DFW) 612
  Ultrasonic Welding (USW) 613
  Cold Welding (CW) 616
  Electrostatic Bonding 617
  Electrodeposition Welding...619

BRAZING AND SOLDERING PROCESSES 620
  Brazing Processes (B) 620
  Soldering (S) 624

SURFACING BY WELDING AND THERMAL SPRAY 628
  Buildup 628
  Buttering 628
  Hardfacing 629
  Overlaying and Cladding...632

THERMAL CUTTING PROCESSES 633
  Metallurgical Effects of Thermal Cutting..633
  Oxygen Cutting Processes (OC) 634
    Oxyfuel Gas Cutting (OFC) 635
    Chemical Flux Cutting (FOC) 636
    Metal Powder Cutting (POC) 636
  Electric Arc Cutting Processes (AC) 637
    Air Carbon Arc Cutting (CAC-A)...639
    Plasma Arc Cutting (PAC)...640
  Electron Beam Cutting (EBC) 643
  Laser Beam Cutting (LBC)...647

SUGGESTED READING 651

Chapter Seven:
Temperature Changes in Welding 653

TERMS AND DEFINITIONS 653
  Heat...653
  Temperature...654
  Thermal Flow..654
  Conduction, Convection, and Radiation..655
  Enthalpy and Entropy 656

TEMPERATURE AND TIME IN WELDING 656
  Heat Flow Equations..657
  Heat Source Characterization 658
  Rate of Heating 659
  Heating Potential of Energy Sources 661
    OAW Heating Potential...662
    AW Heating Potential..662

xiv   Welding Metallurgy

  FRW Heating Potential 662
  EBW Heating Potential 663
  LBW Heating Potential 663
  Electrical Resistance Heating Potential 663
 Peak Temperatures 664
  Defining the Weld Zone 665
  Numerical Modeling of Temperatures 666
 Temperature Distributions 668
  Effects of Temperature Distribution on Cooling Rate 670
  Special Considerations Regarding Temperature Distribution 672
 Time at Temperature 674
  Furnace Heating for Welding Simulation 675
  Temper Color as an Indicant 675
  Resistance Heating for Welding Simulation 676
 Cooling Rate of Heated Zones 677
  Correllation with Heat Input 677
  Workpiece Pre-Weld Temperature 681
  Instantaneous Cooling Rate 683
 Influence of Travel Speed on Weld Zone Size 686
 Hardness/Microstructure/CoolingRate Relationship 693
  Cooling End Points 696

CONTROL OF TEMPERATURE IN FUSION WELDING **698**
 Predictive Diagrams for Fusion Welding Parameters 699
 Mathematical Modeling of Fusion Welding 699
 Sensory Systems for Adaptive Control of Fusion Welding 701
 Adaptive Control of Solid-State Welding Processes 703

SUGGESTED READING **705**

Chapter Eight:
Fluxes, Slags, and Gases for Shielding 708

OXIDATION OF IRON **708**
OXIDATION OF STEEL **710**
 Carbon/Oxygen Reaction in Molten Steel 710
 Oxidation of Solid Steel 713

PREVENTING OXIDATION DURING WELDING **714**
 Shielding Slags 714
 Fluxes 716
 Controlled Atmospheres 717
 Vacuum 718
 Technique 720
 Deoxidizers 721
 Protective Surface Alloys 721
 Liquid Blankets 722

SHIELDING THE JOINING PROCESSES FROM AIR **722**
 Carbon Arc Welding 722

Metal Arc Welding 723
    Covered Electrodes 724
    Generic Electrode Coverings and Typical Formulas 731
    Flux Cored Electrodes 738
Submerged Arc Welding (SAW)..744
    Types of SAW Fluxes and Their Classification...............................744
    Methods of Manufacturing SAW Fluxes 748
    Physical Chemistry of Fluxes in SAW Process 750
    Transfer of Elements Between SAW Flux/Slag and Weld Metal..........758
Electroslag Welding...765
Gas Shielded Arc Welding 766
    Argon..768
    Helium..768
    Carbon Dioxide 769
    Propane...770
    Nitrogen...770
    Hydrogen...770
    Miscellaneous Gas Additives 771
Gas Tungsten Arc Welding 771
Gas Metal Arc Welding 772
    Globular Transfer in GMAW 775
    Repelled Transfer in GMAW...775
    Projected Transfer in GMAW..776
    Streaming or Axial-Spray Transfer in GMAW................................776
    Pulsed Spray Transfer in GMAW-P..777
    Rotating Droplet (Kinking) Transfer in GMAW 778
    Explosive Drop Transfer in GMAW..778
    Short Circuiting Transfer in GMAW-S 778
Flux Cored Arc Welding (FCAW) 779
Plasma Arc Welding (PAW) 780
Electrogas Welding (EGW) 780
Other Welding Processes Using Gas Shielding 781
    Laser Beam Welding (LBW) 781
    Non-Vacuum Electron Beam Welding (EBW-NV) 781
Protecting Brazing Processes From Air 782
Soldering — Ways to Achieve Bonding 784
    Fluxes for Soldering...784
    Mechanical Means to Accomplish Solder Bonding.....................784
SUGGESTED READING **785**

Chapter Nine:
Simple Welds in Iron and Steel 787

FUSION WELDS **787**
    Solidification of Weld Metal..788
        Modes of Primary Solidification Structure................................793
        The Weld Zone...797

    The Unmixed Zone......799
    The Partially Melted Zone......800
    The Heat-Affected Zone......801
    Unaffected Base Metal......802

SOLID-STATE WELDS......**802**

MICROSTRUCTURAL TRANSFORMATIONS IN SOLID IRON AND STEEL......**802**
    Phase Changes in Steel......803
        Ferrite......804
        Austenite......804
        Cementite......806
        Pearlite......807
        Widmanstätten Pattern......809
    Microstructural Changes in Steel During Heating......810
    Microstructures Formed in Steel During Cooling......816
        Martensitic Microstructures......818
    Isothermal Transformation of Austenite......824
        Pearlite Formation Isothermally......826
        Bainite Formation Isothermally......828
        Martensite Formation......828
    Reappraisal of Microstructures Formed in Steel......831
        Upper Bainite......832
        Lower Bainite......833
    Importance of Critical Cooling Rate......834
    Importance of Delay-Time Before Austenite Transformation......836
    Martensite: Implications in Welding......836
        Temperature Range for Martensite Formation......837
        Quantitative Prediction of Martensite Formation......839
        Martensite Hardness Rationale......841
        Martensite Formation Monitoring by AE Signals......842
    IT Diagrams: Summation of Usefulness......843
    Transformation of Austenite During Continuous Cooling......848

PREDICTION OF MICROSTRUCTURES IN THE HEAT-AFFECTED ZONES OF WELDS......**851**
    Jominy Method of Predicting HAZ Microstructure......852
    Mathematical Approach to Prejudging HAZ Suitability......856

TRANSFORMATIONS IN WELD METAL......**857**
    Continuous Cooling Transformation Diagrams for Weld Metal......867
        Importance of Weld Metal Composition......870
        Role of Grain Size in Weld Metal......872
        Influence of Nonmetallic Inclusions in Weld Metal......872

STUDY OF A TYPICAL FUSION WELD IN STEEL......**876**
    Making Welds with Good Toughness......883
    The Challenge of Optimizing Welding Procedures......888
        Base Metal......889
        Weld Metal......890
        Welding Process and Procedure......890

    Weldment Property Testing 890
    Nondestructive Examination 890
  SUGGESTED READING 891

Appendixes 893
  I. ACRONYMS FOR ORGANIZATIONS 893
  II. STANDARD TERMINOLOGY REFERENCES..............................895
  III. SYMBOLS USED IN TEXT AND TABLES 896
  IV. ALPHABETS USED IN SCIENTIFIC NOTATION 896
  V. ABBREVIATIONS & ALPHABETICAL DESIGNATIONS........................897
  VI. SI BASE UNITS..901
  VII. STRESS CONVERSION: MPa ⇔ ksi..902
  VIII. TEMPERATURE CONVERSION: CELSIUS ⇔ FAHRENHEIT 905
  IX. THE ELEMENTS: SYMBOLS & PROPERTIES............................907
  X. THE ELEMENTS: ELECTRON CONFIGURATIONS 913
  XI. ELECTRONIC DATABASES & COMPUTER PROGRAMS 919

Index 923

NOTE: To assist the reader in finding specific information on a particular subject, the contents of each chapter has been demarcated in a separate listing located on the first verso page preceding that chapter. In addition, the contents of each chapter, including the Technical Briefs and the tables in the Appendixes, have been extensively cross-catalogued in the index located on pages 923 to 940.

Technical Briefs

Brief No.	Title	Page No.
1.	Transition Joints Between Different Steels	4
2.	What is an Angstrom?	18
3.	Thermonuclear Fusion	32
4.	Hydrogen - The Building Block of the Elements	35
5.	Miller Indices	57
6.	Burgers Vectors	90
7.	Laser Scanning Microscopy	96
8.	Specimen Preparation for TEM	109
9.	Dislocation Movement in Body-Centered-Cubic Metals	162
10.	Crack Arrest	164
11.	Flow Strength Versus Cohesive Strength	174
12.	Griffith's Theory of Fracture Mechanics	189
13.	Fracture Mechanics Testing of Ductile Metals	202
14.	Developing an Analytic Procedure for Elastic-Plastic Fracture Mechanics	206
15.	Improving the Value of the Charpy Impact Test	227
16.	The Larson-Miller Parameter	247
17.	Quantifying the Severity of Exposure to Neutron Irradiation	252
18.	Thermoelectric Effects as Related to Metallurgy	271
19.	Lead as an Immiscible Alloy in Iron and Steel	297
20.	Application of the Lever Law to Solidification of Metals	302
21.	The Gibbs Phase Rule	310
22.	Hume-Rothery's Classification of Elements	313
23.	Correlation Between Electron Configuration and Crystal Structure	317
24.	Coherency of Atoms in Precipitation Hardening	324
25.	Rationale for Letter Designations in the Iron-Iron Carbide Phase Diagram	332
26.	Effects of Manganese Sulfide Inclusions in Steel	343
27.	Nitrogen Retention in Steel	351
28.	Manufacture of Fine-Grain Steel Using an Aluminum Additive	377
29.	Continuous Casting of Thin Steel Strip	404
30.	The AISI System for Generic Designation of Sheet Steels	423
31.	Ionization	452
32.	Initiating a Welding Arc	455
33.	Effects of Weld Pool Circulation on Penetration Depth in GTAW	486
34.	Hot Cracking Susceptibility in Submerged Arc Welds	507
35.	Problems in Spot Welding Zinc-Coated Steel	529

Brief No.	Title	Page No.
36.	Preventing Arcing in Electron Beam Welding	557
37.	Friction Welding in Undersea Applications	599
38.	Effects of Alloying Elements in Steel on Oxygen Cutting	637
39.	Controlling HAZ Cooling Rate via Arc Energy Output	678
40.	Development of the SMAW Electrode	727
41.	Safeguards for Handling Low-Hydrogen SMAW Electrodes	736
42.	The Wall Neutrality Number	747
43.	Assessing Basicity of SAW Fluxes	753
44.	Solidification Structures in Steel Fusion Welds	792
45.	The Solidification Mechanics of Weld Surface Patterns	797
46.	Austenite: The Mother of Microstructures	806
47.	Pearlitic Microstructures	817
48.	Martensitic Microstructures	840
49.	Bainitic Microstructures	845
50.	Acicular Ferrite in Weld Microstructures	885

Tables

Table No.	Title	Page No.
2.1	Elementary Particles	20
2.2	Ionization Potential of Gases and Vapors	37
2.3	Diameters of Atoms	39
2.4	The Elements	42
2.5	Features of the Seven Crystal Systems	56
2.6	Densities of Some Pure Metals and Their Change	66
2.7	Crystal Structures of Some Pure Metals	75
2.8	Crystal Structures of Some Pure Metals Capable of Allotropic Transformation	76
2.9	Relationships Between ASTM Grain Size Numbers	99
2.10	Comparison of Methods for Localized Analysis of Surfaces	122
3.1	The Properties of Metals	134
3.2	Typical Elastic Constants for Various Materials at Room Temperature	140
3.3	Influence of Crystallographic Orientation on Elastic Modulus at Room Temperature	141
3.4	Strain Rates Common to Various Exposures, Services, and Mechanical Tests	173
3.5	Temperature Conversion Scales and Fixed Points of Interest	218
3.6	Notched - Bar Impact Values for Metals Tested at Low Temperatures	221
3.7	Effect of Cold Work on Mechanical Properties of Low-Carbon Steel Sheet	249
3.8	Thermal Neutron Cross Section for Iron and Other Elements Found in Steel	251
3.9	Density and Strength-Weight Ratio of Metals, Alloys and Nonmetals	259
3.10	Specific Heat of Some Metals and Nonmetals	260
3.11	Thermal Conductivity of Metals, Alloys, and Nonmetals	262
3.12	Melting Points or Ranges of Metals, Alloys, and Nonmetals	263
3.13	Melting Point, Latent Heat of Fusion, Boiling Point and Heat of Vaporization of Metals	265
3.14	Coefficient of Linear Thermal Expansion for Some Metals, Alloys, and Nonmetals	267
3.15	Electron Thermionic Work Functions of Metals and Nonmetals	268
3.16	Electrical Properties of Metals, Alloys and Nonmetals	270
3.17	Forms of Corrosion and Instigative Conditions	275
3.18	Galvanic Series, and Electromotive Force Values for Metals	278
3.19	Commonly Used Materials and Environments in Which Failure by Stress-Corrosion Cracking Has Been Experienced in Industry or in Tests	287
4.1	Elements Used for Alloying Irons and Steels, and Their Influence on Crystalline Structure of the Alloy	319

Table No.	Title	Page No.
4.2	Changes in Pure Iron During Cooling and Temperatures of Occurrence	321
4.3	Interstitial-Free I-F Steel	358
5.1	Comparison of Rimmed, Capped, Semikilled and Killed Steels	379
5.2	Gas Content of Alloy Steel Ingots	380
5.3	Typical Gas Contents of Alloy Steel Ingots Melted by Four Electric Furnace Processes	394
5.4	Common Terms for Hot-Rolled Mill Products	405
5.5	Classification and Usage of Carbon Steels	411
5.6	SAE-ASTM Unified Numbering System	417
5.7	AISI-SAE System for Designation of Carbon and Alloy Steels	419
5.8	AISI-SAE System for Characterization of Carbon and Alloy Steels	420
5.9	AISI and SAE Systems for Classification of High-Strength Carbon and Low Alloy Steel Sheet	421
5.10	ASTM Annual Book of Standards - 1993 Edition	422
5.11	API Specifications Governing Materials	425
5.12	ASME Boiler and Pressure Vessel Code	427
5.13	ASME B31; Code for Pressure Piping	428
5.14	AWS Codes, Specifications and Standards	429
6.1	Characteristics of Argon and Helium Gases When Shielding the GTAW Process	483
6.2	Typical Chemical Compositions of Weld Metal Deposited by Flux Cored Arc Welding Electrodes	504
6.3	Thickness and Weight of Zinc Coatings on Steel	528
7.1	Temper Colors Formed on Iron and Carbon Steel	676
8.1	Degree of Vacuum Used to Protect Metal During Various Processes	719
8.2	Materials Used in Coverings on Steel Electrodes for SMAW Arc Welding Process	726
8.3	Recovery of Elements From Coverings of Electrodes Deposited by SMAW Process	729
8.4	Typical Covering Formulas for Steel SMAW Electrodes	732
8.5	Typical Flux Core Formulas for Steel FCAW Electrodes	740
8.6	IIW Classification of Fluxes for Submerged Arc Welding of Steel	754
8.7	Typical Flux Compositions for Submerged Arc Welding of Steel	755
8.8	Typical Flux for Electroslag Welding of Steel	766
8.9	Gases Used in Shielding Welding Arcs	767
8.10	Controlled Atmospheres Used in Furnace Brazing Carbon and Low-Alloy Steels	782
9.1	Influence of Alloying Elements in Steel on Martensite Formation	837
9.2	Microstructures Found in Steel Weld Metal	858
9.3	Microstructures Found in Cast and Wrought Steels; Including their Weld Heat-Affected Zones	877

Preface

Readers of Welding Metallurgy in their comments about previous Volumes 1 and 2 of the Third Edition have offered helpful suggestions toward betterment of future editions. All offerings from students, engineers, experienced welding personnel, and educators have been carefully weighed by the author in planning this present revision. A principal aim in carrying out this work was to acquire updated information wherever available, but input from readers prompted several changes. One change will be found in the sequence of chapters, which now provides more logical progression from the structures and properties of metals, through the mechanics and effects of heating for welding, and finally to metallurgical description of typical welds in steel. Also, the remodeled text now includes additional material of a fundamental nature to assist in developing broader, analytical explanations for the chemical and physical events that arise in welding. The need for this assistance had been voiced by readers who are confronted with the complex technology of welding, but who have not had exposure to chemistry and physics of the depth needed to rationalize the phenomena encountered. Relationships between fundamental features of metals and their behavior in welding are pointed out to encourage development of more complete understanding of each particular happening in welding. This kind of awareness of the factors involved and the mechanics entailed aids greatly when a decision must be made regarding the most effective measure for dealing with a given troublesome situation.

Several features of the new text deserve comment. Metric practice has been employed as the principal system for dimensions and quantities because of expanding usage world-wide; however, U.S. customary units are in parentheses following SI units throughout the text and are included in most tables and figures. The text does not include references to particular articles in periodicals, reports, and books that served as resources because of the great number from which information was extracted. Instead, keywords from the International Welding Thesaurus are included in the text to guide readers to helpful literature for additional details whenever needed. Widely available computerized search services cognizant of these keywords can produce lists of relevant references, including those published henceforth. Each chapter concludes with a short list of selected publications under the heading "Suggested Reading." These are recent publications that provide broader range of coverage of the area being reviewed in the chapter.

Colleagues and organizations that assisted the author in collecting and preparing material for publication are so large in number that individual recognition here is not practical. The author is keenly aware of the great importance of this support. In the course of time, sincere appreciation will be expressed to contributors individually. Two organizations provided protracted help with their staffs and their facilities. The American Welding Society gave encouragement and assistance at every possible turn, and waited patiently for completion of the author's manuscript. Mr. Robert L. O'Brien, Director of Technical Services, for AWS at the time, arranged peer reviews to verify the manuscript's technical content. Mrs. Debbie Givens, Publishing Coordinator for AWS, deserves a note of appreciation for her diligence in arranging reproduction of the many line drawings of figures and graphs.

A bright spot in the lengthy publishing procedure was the engaging of Mr. Alexander M. Saitta by AWS as Production Editor to handle editing and formatting of this substantially larger edition of Volume I. Alex, a newcomer to the welding field, quickly demonstrated ability to fuse his language knowledge with the idiom of welding technology. In providing assistance on shortening the manuscript, Alex proposed the inclusion of "Technical Briefs" to avoid deletion of useful fundamental information — an effective format measure for which this author is very appreciative.

The Welding Institute, at Abington, England, with which the author was affiliated for many years as head of its North American Office, was extremely generous with staff assistance, and provided many photographs and photomicrographs for new illustrations. This new edition would have been very difficult to complete without the support received from TWI.

Hopefully, a revision of Volume II will receive similar support from the many colleagues and organizations during its preparation, and can be issued in a fourth edition without undue delay.

George E. Linnert
Hilton Head Island, South Carolina
December, 1994

Chapter 1

In This Chapter...

MILESTONES IN WELDING HISTORY 1
THE FUTURE OF WELDING 4
WHAT IS WELDING METALLURGY? 6
PUTTING WELDING METALLURGY TO USE 12
WELDING TECHNOLOGY RESOURCES 12
SUGGESTED READING 15

Background to Welding Metallurgy

Throughout industry, welding is the most widely used, cost-effective means for joining sections of metal to produce an assembly that will perform as if cut or formed from solid material. Proper application of welding technology by the user ensures the making of welds that are "fit for purpose" in virtually all kinds of service. Even in manufacturing operations which do not produce welded articles, production facilities would be difficult to construct without the use of welding. While most people would agree that nuclear power plants, spacecraft, and deep-diving submarines must make use of welding, they might not realize that many vital components, such as high-pressure piping systems, transducers, vacuum tubes, and a multitude of other articles must also use welding in practical production. Without welding, we would be hard-pressed to build computers, television systems, jet planes, and the many devices that support our present mode of living.

MILESTONES IN WELDING HISTORY

A brief historical review of welding will identify the driving forces that spurred its remarkable growth, and will help explain engineering constraints that have been placed on welding from time to time in past years. Although man began to pressure-weld noble metals such as gold more than 2000 years ago, and learned to forge-weld iron about 1000 years later, modern implementation of welding did not begin until a little more than 100 years ago. In the early 1880s, the acetylene torch and the electric arc were introduced as tools capable of fusion welding. Elementary forms of electric resistance welding processes were demonstrated about the same time. Therefore, in a single century just passed, welding has grown from virtual inception to its present indispensable role in industry.

Initially, proponents of riveting, bolting, screwed joints, folded seams, and other forms of mechanical fastening contrived to place obstacles in the path of welding. Admittedly, the early welding equipment and processes were crude. Welding techniques were essentially exploratory, and the results were not always satisfactory. Nevertheless, the potential of welding was recognized by many innovators, and they worked to improve the art and science of this new method for joining metals.

In 1917, a dramatic situation provided an ideal proving ground for welding. As the United States prepared to join the Allies in World War I, the crews of German ships in certain ports realized that their vessels would soon be seized. The crews acted to quickly put the vessels out of commission. Boilers and heat exchangers were purposely damaged by overheating, and propulsion systems were broken by jamming gears. Overall damage was so extensive that more than a year of repair work using customary methods would have been required to put these badly needed vessels in commission and to have them operate within the Allied merchant marine. However, at the urging of welding advocates, the Emergency Fleet Corporation organized a wartime welding committee which was given free rein to plan this massive repair project and to make maximum use of oxygen cutting and welding. The repairs were completed in less than half the estimated time, and the welding effort contributed greatly to the improved Allied shipping position in 1918.

After the Armistice, the Wartime Welding Committee members acted in 1919 to organize a not-for-profit technical body of interested persons and companies dedicated to further the advancement of welding and its allied processes. This organization grew to become the present American Welding Society (AWS).[1] At the 1919 milestone, however, welding processes were regarded by industry merely as means to repair metal articles and structures. Welding was not yet recognized as a production operation.

In a 20-year growing period (1920-1940), proponents of welding began to gain acceptance of its processes for production operations in addition to repair work. Industry was slow to recognize the advantages of welding, but the various processes steadily drew the attention of builders of steel-framed buildings, steel bridges, earth-moving machines, mining machinery, etc. Limited processes and consumables were available for production operations during this early period, but welding proved to be efficient and economical, and it gradually emerged as a major endeavor worldwide. Companies large and small became engaged in the manufacture of welding equipment, product research and development, and marketing. Research in welding was taken up by university laboratories, not-for-profit institutions, technical societies, and metal producers.

In 1935, the Engineering Foundation in New York City established the Welding Research Council (WRC), and since that time WRC and AWS have cooperated to encourage research in welding and the allied fields. By 1940, a budding welding industry had responded with an impressive array of new processes and improved filler metals and fluxes. A need was recognized during this early growth period to disseminate authoritative, up-to-date information on the rapidly developing processes. Comprehensive welding encyclopedias and handbooks emerged shortly after 1920, and periodic editions of the AWS *Welding Handbook* were prepared with open-end planning for master charts of the processes, standard definitions, and welding symbols. Efforts were also being directed to the establishment of guidelines for welding specific metals and alloys. Yet, this early bright picture was not without an occasional dark cloud. Although welding equipment was relatively inexpensive, processes and procedures were not always clearly defined and welders were not always adequately trained. It was not long before weld failures began to tarnish the industry's record. Because so many welding operations were being carried out without due regard for sound welding engineering and welder training, it became necessary for proponents of welding to launch a substantial educational effort to encourage and assist proper planning and performance of welding operations.

---

[1] Acronyms are used throughout this book for technical organizations. Full names and locations are provided in Appendix I - Acronyms.

At the start of the next 20-year time frame (1940-1960), the welding industry was again confronted with a major challenge. Even before World War II, shipbuilders all over the world had been steadily moving toward welded construction to take advantage of cost savings over riveted construction. When hostilities became worldwide in 1941, a tremendous need arose for ships to transport materiel for the Allied Forces. Shipbuilders met this challenge by selecting a limited number of ship designs for production and specifying maximum use of welding for rapid hull construction. A large fleet was built and rushed into service. The ships sailed in ocean waters that ranged from warm to frigid.

The pressures of this international emergency forced an "art-before-science" approach to the shipbuilding task. Approximately 5000 ships were built for military transport service (mostly merchant ships and tankers of the *Liberty* and *Victory* design). Of these, about 20 percent developed cracking in the hulls and main decks in or near welds in steel plate. Some cracks occurred while the ships were at sea, and these cracks rapidly progressed to fractures sufficiently large to cause catastrophic breakup. Occasionally, cracking or major fractures occurred while these ships were in port, particularly during a sharp drop in ambient temperature, or during uneven loading or ballasting. After thorough investigation by a number of concerned companies and federal maritime safety agencies, ship fractures were found attributable to three principal causes — briefly,

(1) steel that had poor toughness, especially at low temperatures,
(2) hull and deck construction techniques that had not been refined to avoid points of severe stress concentration, and
(3) some questionable workmanship done under the intense production pressure of the wartime emergency.

Toward the end of the 1940-1960 period, two important endeavors would require the best welds that could be made. One was the fabrication of pressure vessels for the nuclear power plants in submarines, ships, and utility stations; the second was the building of solid- and liquid-propellant rocket motors for aerospace vehicles. Rocket motors required large thin-wall casings and pressure vessels. The components had to be light in weight despite the loads to be carried; therefore, high-strength alloys (some with barely adequate toughness) were used for construction. A few of the early welded rocket motor casings failed at welds during pressure testing because (1) welds contained small physical imperfections, (2) weld metal required unusual modifications in chemical composition to achieve mechanical properties comparable to those of the base metal, and (3) certain welding processes exerted influences on weld metal properties that were not anticipated or completely understood. These findings motivated research on the welding of high-strength steels and other alloys used in thin-wall cylinders and tanks that are expected to withstand proof testing and service loading which approaches yield strength. As a result of this work, welding is now used without reservation as the preferred joining method for these important constructions.

In the nuclear power field, the performance record of welding is virtually free of serious complaint. A few small problems have occurred which represent unavoidable consequences of modern welding processes, but these problems were not entirely unexpected (e.g., the effects of residual stresses at joints in the as-welded condition).

Three reasons can be cited for the highly successful use of welding in building nuclear power facilities. First, the metals and alloys used for fabricating welded components are thoroughly tested and governed by detailed standards covering property and weldability requirements. Second, the planning of welding procedures has benefited from extensive experience in building petrochemical plants and other facilities using these same materials. Finally, codes and detailed specifications have been established to cover qualification of welders, permissible processes, procedures, inspection, etc. In general, nuclear facility construction makes use of proven materials, processes, and procedures. Nothing is left to chance.

Many other examples can be selected from technical literature to illustrate ways in which welding has improved and expanded to meet demands of industry, including such widely diversified applications as (1) thick-wall pressure vessels for use in coal-gasification plants, (2) large offshore drilling platforms, and (3) miniature capsules and circuitry for microelectronic devices, which undoubtedly contain the smallest welds made by man. The welding industry has mounted an impressive record of successful applications, but weld difficulties are encountered from time to time, and many are of a metallurgical nature. A good example is described in Technical Brief 1, shown below.

THE FUTURE OF WELDING

American industry must achieve significant cost reductions and production efficiency over the next decade to maintain a healthy position in the world manufacturing market. The trend to use existing welding processes for appropriate applications has been recognized by perceptive industrialists as a means of appreciably reducing manufacturing costs. Production experience will verify that welding is the most efficient joining method, and the savings can be documented; but there is another trend in the USA which is having a counterproductive influence. Although a highly impressive record of welding process invention and equipment development can be credited to U.S. companies and laboratories over past years, budgeting in this area has been drastically cut. As a result, dedicated research and development in the USA on welding has decreased substantially. In time, this policy will allow other countries with greater ded-

---

TECHNICAL BRIEF 1:
Transition Joints Between Different Steels

When electric power utilities began to build steam plants that would operate at higher temperatures and pressures, stainless steel tubing had to be used in the high-temperature operating portions of certain components. However, cost considerations dictated that low alloy steel tubing be used in the lower temperature operating portions; consequently, transition joints had to be made in the pressure-containing tubular systems between these two markedly different kinds of steel. Welding engineers responded with appropriate filler metals and special procedures for welding these joints, and many thousands of such welds have been made and placed in service. While operating experience with these dissimilar-metal welds has been acceptable, cracking sometimes develops at the weld after long-term service, and remaking the joint becomes necessary. This complex, ongoing metallurgical problem will be treated in detail in Volume II.

ication to welding research and development to originate and control the use of new processes. Results of this trend have already appeared in the welding equipment market, and supportive research in the metallurgical study of welding is also likely to decrease as the trend continues.

An encouraging trend in U.S. industry, however, is a strong movement toward greater mechanization of welding to reduce manufacturing costs. Welding is a labor-intensive operation, and manual welding requires skilled workers. The mechanization of welding has long been an aim of its promoters and users, and a wave of effort has swept through the welding industry in the direction of automation. This effort started with a surge of specially built units designed to perform a prescribed welding process and procedure on a specific assembly. Riding the crest of this wave is the robot, a programmable device intended to supplant the human welder. Purpose-built welding units and welding robots are being brought into manufacturing plants in increasing numbers; and, judging from future plans of welding equipment builders and the programs of technical societies regarding robotics and automated welding, efforts to automate welding will continue to grow. Profound changes will take place in manufacturing plants that incorporate welding in production operations. One aim of this book is to explain the importance of mastering the metallurgical behavior of metals during automated welding.

Welding automation should be more specifically defined because of encumbrances that it will place on the metallurgical behavior of the metals being welded. The term *complete automation* has been used to describe a completely controlled welding function. Complete automation infers that the operation is not only performed without the manipulative aid of a human welding operator, but that if conditions vary during welding, compensating adjustment is automatically made to assure an acceptable weld. This immediate adjustment to obtain the desired weld is called *adaptive control*. It is sometimes called *feedback control* or *closed-loop control* in the literature.

Some planners of complete automation require that the weld be nondestructively tested or examined during or immediately after its formation and before the welded article moves to the next stage of production. Completely automatic welding and inspection would reduce costs by decreasing the manpower required to make the weld and ensure that it is of acceptable quality.

All segments of industry, not only welding, will be pursuing complete automation because of the need for cost reduction. A trend in this direction is well underway in forging, machining, plating, and other industrial operations and processes; and mechanization of these processes in manufacturing has been amply demonstrated. However, the goal ahead is to fit many different operations into an articulated system to form an automated factory staffed by a minimal force of supervisors and trouble-shooting technicians. A number of forward-looking companies in the USA are planning the construction of completely automated factories. Welding processes must be readied to fit into these highly coordinated manufacturing facilities if welding is to maintain its position as the favored means of joining. Automated welding processes are described in Chapter 6 of this volume, along with facets of metallurgical behavior of steels during welding that requires consideration to maintain consistent, acceptable weld quality.

Microelectronics is a technology that will assist in the complete automation of welding. Computers and microprocessors store data, analyze input, and issue output signals to controllers. A vital component of a complete system for automatic regulation is the sensory linkage between the analyzer/control unit and metallurgical reactions or changes occurring in the weld area during welding. Each process has different features related to weld quality that can be measured continuously to generate input signals for analysis. Signals can be very rapidly analyzed in a microprocessor, and any adjustment in operating parameters required to maintain weld quality can be made almost immediately through output commands from the microprocessor. This type of monitoring and response is termed *in-process* or *real time control*. Some welding processes generate easily detected electrical phenomena that can be continuously monitored for particular deviations which in turn can guide corrective measures. Other processes require more obtrusive features to be measured or appraised continuously as the weld is being made in order to feed corrective input to the microprocessor. A variety of sensory devices have been developed to serve as linkages between the microprocessor and the welding operation. These include:

(1) instrumentation for instantaneous measurement of electric current and dynamic electrical resistance,
(2) gauges for precisely measuring localized dimensional changes that occur during thermal expansion and contraction,
(3) coaxial optical systems for weld image viewing and machine vision analysis, and
(4) fiber-optic transmitters for thermographic monitoring of infrared radiation emitted from a point on a weld surface.

An important aspect of a sensory device is its ability to assess changes that stem from metallurgical peculiarities of the materials being welded. Painstaking development work is required for each sensory device to achieve a reliable, complete automation system that will meet the demands of industry.

WHAT IS WELDING METALLURGY?

What is welding metallurgy? And how does a knowledge of it contribute to the future of welding, especially to the effort of achieving complete automation of its processes? Discussion of these questions helps define the information included in this book, and how the information should be weighed and interpreted before being put to use.

Welding, in the broadest sense, is described as the art and science of joining metals by using the intrinsic adhesive and cohesive forces of attraction that exist within metals. This description encompasses metallic bonds as observed in various welded, brazed, and soldered joints; and it rules out joints accomplished with purely mechanical design, mechanical fasteners, or with nonmetallic adhesives. This description acknowledges that a certain amount of artistic skill is necessary in making a weld — skill that is acquired by practice rather than by scientific rationalization. The need for artistic skill in welding decreases, of course, as scientific and engineering knowledge increases. Therefore, to retain its position in the high-technology future as the principal method of joining metal components, welding processes must continue to improve and depend less on art by making maximum use of science and engineering.

Metallurgy, simply defined, is the science and technology of metals, and consists of two main parts:

(1) *process metallurgy*, which involves the reduction of ores, refining of metals, alloying, casting, and the working and shaping of metal to semi-finished and finished products; and

(2) *physical metallurgy*, which includes heat treatment, mechanical testing, metallography, and numerous other subjects dealing with the application, design, testing and inspection of semi-finished and finished products.

At first thought, welding might seem to be merely a part of process metallurgy; that is, an operation for shaping metal into a finished product. However, welding is more than a constituent of process metallurgy or of physical metallurgy. Welding encompasses the entire scope of metallurgy, but requires a specialized viewpoint to comprehend the peculiarities of its processes. It is necessary, therefore, to concentrate on the study of *welding metallurgy*.

Both process metallurgy and physical metallurgy are involved in welding. For example, as nickel oxide is added to the thermit welding mixture of iron oxide and aluminum, the nickel oxide will be reduced to nickel during the thermite reaction, and it will be alloyed with the molten metal discharged from the crucible. Here, knowledge of process metallurgy is used, because the operation is similar to ore reduction in a furnace. Process metallurgy is involved in the formulation of flux coverings and core filling in arc welding electrodes. Many of the materials used for making slags in connection with the melting of metals in a furnace are also used in electrode fluxes for welding; limestone, fluorspar and rutile as used in making steel are three good examples. After minerals and other suitable materials are compounded to make a flux with the proper acidity (or basicity) and viscosity at welding temperature, process metallurgy techniques can be put to further use; the alloy composition of the weld metal can be altered by adding materials to the flux covering on the electrode, or the flux in its core.

The same alloying materials used in metal production are also used in powdered form in welding fluxes. For example, a steel electrode core wire devoid of alloying elements can be covered with a flux containing an addition of ferrochromium to form a chromium-containing alloy steel weld metal. The transfer of chromium from the flux to the weld metal during deposition of the electrode is highly efficient. Most flux-bearing arc welding electrodes, both flux covered and flux cored, contain additions of metals to function as deoxidizers, or to alloy with the weld metal. However, certain metallurgical precautions must be taken to ensure proper transfer of alloying elements, and to guard against pickup of unwanted minor elements that adversely affect weld properties.

Welding can be compared to a series of metallurgical operations involved in metal production, like steel making, but welding is performed on a very small scale with pertinent steps carried out in rapid succession. During most welding processes, a volume of molten metal (melt or bath) is formed (cast) within the confines of the solid base metal (mold). Weld metal initiates its solidification in a unique manner, unlike the molten metal cast in a mold (See Chapter 9). Yet, weld metal also has a susceptibility to blowholes and internal porosity caused by the evolution of gases, the same as experienced with ingots and castings (Figure 1.1).

The base metal of a weld can be preheated to retard cooling rate and solidification just as molds are often preheated to delay cooling of the contents. After solidification, the weld metal or nugget (ingot) can be placed in service in the as-welded (as-cast) condition, or it can be peened or worked (wrought) to obtain a specified shape or particular mechanical properties.

One striking difference between welding and other metal-producing operations is the contrast in the mass of metal being heated and worked, and the effect of mass on physical and metallurgical changes. Metal production operations usually deal with large masses that are subjected to relatively slow rates of heating and cooling. Welding, on the other hand, usually involves comparatively small masses that are heated very rapidly by intense heat sources and that cool rapidly because of intimate contact with a larger surrounding mass of colder base metal.

The structure and properties of metals are most readily observed and compared when they have been heated and cooled under equilibrium conditions, entailing extremely slow rates. Equilibrium conditions are seldom seen in conventional metalworking operations, and welding usually represents a further departure from equilibrium conditions. Consequently, it can be expected that weld zones are prone to display unusual structures and properties.

The hot and cold peening of weld metal is similar to hot forging or cold finishing mill operations, and the principles of process metallurgy apply. There are pro and con statements in the literature concerning the advisability of using peening on a weld. It is true that peening has resulted in unexpected difficulties in a number of instances. In many of these cases, however, peening was seized as an easily-applied remedy for a particular condition, perhaps to eliminate cracking, and little or no forethought was given to the eventual consequences of the operation (Figure 1.2).

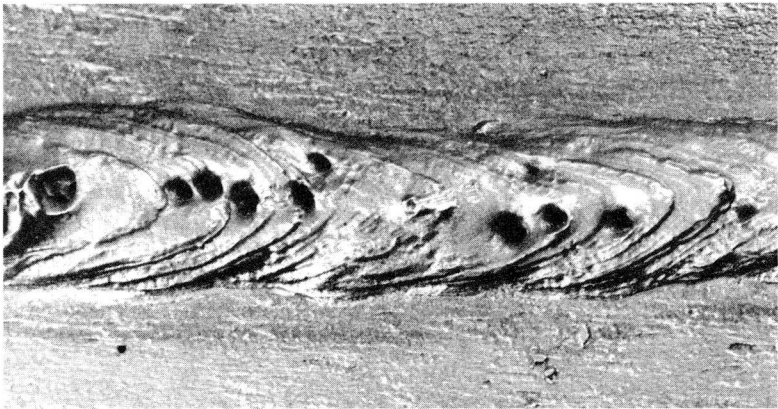

Figure 1.1 — Blowholes in a weld
Gases can erupt from the weld leaving blowholes not unlike the gas pockets on the surface of a casting or an ingot. Blowholes can be avoided most effectively through a knowledge of welding metallurgy.

Figure 1.2 — Cracking in an arc welded corner joint as revealed by dye penetrant examination
A knowledge of welding metallurgy will assist in preventing the kinds of cracking that possibly can occur in or near welds. (Courtesy of The Welding Institute, England)

Peening is a useful operation if it is performed in accordance with sound metallurgical reasoning. Guided by a knowledge of process metallurgy, welds can be hot or cold peened to relieve shrinkage stresses. Cold peening the weld will increase the strength of the metal. In some instances, components made of cast or wrought steel can be welded into an assembly which then is hot formed or forged to produce a complicated shape. This shape might be difficult, if not impossible, to produce by casting or by forging alone. A knowledge of process metallurgy is essential to success in multiprocess production of this nature.

Many aspects of physical metallurgy are involved in welding. After a weld is made, postweld heat treatment may be required to alter the unusual microstructure and properties produced by rapid cooling. The treatment can be one designed to soften (anneal) the weld zone, or it can be a complete hardening and tempering treatment designed to obtain properties in the weld zone comparable to those of the base metal. Sometimes this postweld heat treatment is integral to the welding operation. For example, it is common practice to accomplish localized heat treatment of a resistance spot weld, when required, in the machine before the newly made weld is removed from between the electrode tips. Again, this postweld heat treatment can be a softening operation, or a more complex treatment. Immediately following formation of the spot weld nugget, a precise program of heating and cooling can be performed to carry the nugget and a calculated portion of the heat-affected zone through a bona fide heat-treating operation. The weld area can be allowed time to cool nearly to room temperature, whereupon a controlled surge of electrical current heats the metal locally to proper temperature for hardening. The withdrawal of heat by surrounding base metal

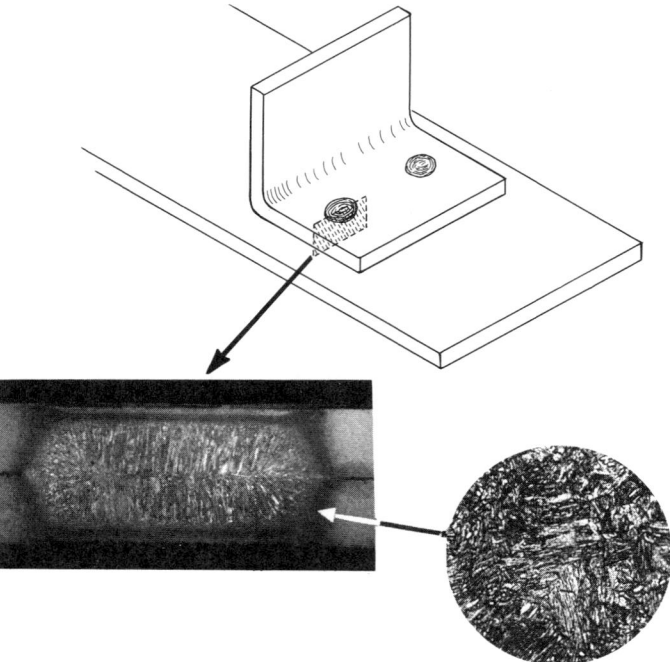

Figure 1.3 — Metallography is often used to examine microstructures of welds and their heat-affected zones in the base metal

This resistance spot weld between two alloy steel sheets has a hardened microstructure which will detract from the toughness of the weld.

and the cool copper electrodes is equivalent to a quenching operation. After another short cooling period, a final smaller surge of current is applied to temper the previously hardened metal. In some cases the entire operation, consisting of spot welding, hardening and tempering, may be carried out within a total time of two seconds (Figure 1.3).

Mechanical testing and metallography are important parts of welding metallurgy. Work is constantly underway to gather data on the strength and toughness of weld metal and welded joints so that new structures can be fabricated with confidence that the welded joints will be reliable during the service of the assembly. The metallographic microscope is the "workhorse" of welding metallurgy because it provides an inside view of the microstructural changes within the metal caused by the thermal effects of welding. By examining microscopic features, the metallographer can provide valuable data for diagnosing the fundamental causes of defects and can assist in the interpretation and classification of these defects. Progress in metallography has been truly spectacular as newer tools, such as the electron microscope and the electron probe microanalyzer, have been applied in welding studies (Figure 1.4).

Metallurgists have drawn freely from the classic sciences, particularly from physics and inorganic chemistry, for basic knowledge. Recent technological demands involving the selection and the fabrication of steels have required broader studies. Other scientific fields, such as solid-state physics and nuclear physics, have required study because they hold explanations for some of the anomalous behavior encountered in new applications. The metallurgy of welding seems to have no limits as to the scope of science and engineering that must be probed to gather useful information. The many materials used in welding operations — and the wide variety of methods employed for heating, surface protection, application of force, etc. — extend the search for vital data to a diverse number of engineering fields, including ceramics, electronics, mineralogy, optics, thermodynamics, and ballistics.

Mechanical design is also an important concern. Welding, because of its versatility, makes it possible to assemble articles of almost any configuration with little difficulty. Yet, these assembled articles are expected to perform in service as if they were crafted from a solid piece of metal. Many structural failures have been blamed on a weld, but actually have been caused by inadequacies in design. This was proven in a number of instances by fashioning a similar article from a solid piece of metal and subjecting it to the same service conditions; the unwelded article failed in essentially the same manner as the weldment. These demonstrations have made design engineering an obligatory facet of welding metallurgy (Figure 1.5).

Figure 1.4 — **Electron probe microanalyzer used on constituents in the microstructure of metals**

(Courtesy of TI Research, England)

Figure 1.5 — Failure by fatigue in a weldment
Welding metallurgy relates joint design and endurance limit under dynamic loading to avoid fatigue cracking like that extending horizontally just below the fillet welds in this rugged-appearing weldment.

PUTTING WELDING METALLURGY TO USE

The aim of *Welding Metallurgy, Fourth Edition*, as of previous editions, is to present a tutorial text that will provide the reader with rationale for many of the important conditions, problems, and phenomena encountered in welding. This text is not intended as rote which is to be followed in a given welding application, or when dealing with a welding problem. Experience shows that each welding situation is unique when all influential factors and conditions are considered. Therefore, decisions regarding welding procedure must be based on circumstances in each application, and the required results "in toto." When a new application is studied in preparation for planning the welding procedure, there is usually good reason to review information from various sources. This preliminary exercise can help avoid problems like those that may have been experienced earlier, elsewhere, and can also capitalize on reported advances in welding technology. The intricacy of welding not only calls for individual study of each situation, but the technology that often must be put to use to ensure success is so vast that no single volume can hold all of the information that deserves review.

WELDING TECHNOLOGY RESOURCES

The multiple volume AWS *Welding Handbook* is a good example of a general reference, but specialized sources often must be consulted when seeking the details needed for complete planning of a welding procedure. This volume of *Welding Metallurgy* is tailored to serve as an additional specialized source; but attention can be directed to additional useful resources, some of which may not be appreciated until now because of the sheer bulk of welding literature that confronts the welding procedure planner.

Each year approximately 5000 specialized articles, papers, documents, books, and cinematographic film and tape are added to the world's works on welding. A large part of the accumulated material continues to have value for aiding modern welding practices, but effective utilization of this vast resource requires good familiarity with its multimedia.

Books are good resources because most of them undergo peer review and intensive editing prior to publication, and this screening contributes to the reliability of the information therein.

Technical journals often contain articles and papers on welding because of widespread interest by the manufacturing sector, but only a limited number of the journals devoted mainly to welding give the subject meticulous coverage.

Technical papers on welding which focus on a particular aspect can be an excellent source of information. However, the profuseness of this portion of the literature can require a tiresome search unless a reference system is skillfully employed to pinpoint pertinent, applicable information.

Codes, standards and recommended practices covering welding and welded construction are issued by regulatory agencies, branches of the military, and by technical organizations. These documents embody much welding knowledge, but most of them do not include the technical rationale that would explain the need for a particular provision or requirement.

Patents are a unique resource for information on newest welding technology, but patents can be difficult to interpret because they may present only a concept that requires additional research and engineering before the invention can be utilized in practice.

Slides, motion picture films, and video tapes are available covering virtually the entire scope of welding technology from research endeavors to proper process application. Welding processes are shown in actual operation in films and video, and by magnification of the image, minute details are revealed that cannot be seen with the unaided eye. Playback can be in real time, or in slow motion to assist observation of details that occur in rapid sequence. High-speed cine photography has been employed to examine very short-lived phenomena, such as the modes of metal deposition from consumable electrodes during arc welding.

Fortunately, this tremendous mass of welding information has been digested electronically and organized into databases and programs to permit rapid searching and selective retrieval.[2] Computer software to a large extent consists of programs loaded with data on a specific subject or the requirements of a code or specification. The programming then assists in such steps as; calculations, conversions, interpretation of requirements, material choices, and selection of welding process and operating parameters.

An information service listed in Appendix XI, Weldasearch, deserves comment because of its focus on welding and many allied subjects, and the scope of its coverage. Weldasearch was started by The Welding Institute (TWI) in 1969 using key words

---

[2]Examples of electronic databases and software available at the present time are given in Appendix XI - Electronic Databases and Computer Programs.

Figure 1.6 — Interactive dialogue between searcher at computer terminal and server at remote database can quickly assemble information on proper welding and inspection procedures

(Courtesy of The Welding Institute, England)

for indexing and cross-referencing. Professional staff reviewed each item before entry and prepared an abstract for cataloged storage. The key word system proved so effective that in 1974 the International Institute of Welding (IIW) published it as the *International Welding Thesaurus*, which presently in its third edition contains more than 1000 keywords for controlled language indexing. Today, Weldasearch contains more than 100 000 items. Both AWS and WRC advocate use of this database because of its completeness and the efficacy of its retrieval procedure. The files of Weldasearch can be searched for any defined topic by all means of communication, including on-line searching, which now is accessed via Orbit Search Service (Figure 1.6).

Databases, information services, and computer programs covering welding technology were introduced a relatively short time ago, but these proprietary resources are growing at an impressive rate. They allow rapid searching of extensive areas of technology, retrieval of pertinent information on a selected subject, and rapid transmission of findings to any point in the world via computer screen or hard-copy FAX. Every aspect of joining metals by welding, brazing, and soldering is covered to some extent among these multimedia; and they are usually quite current. However, it should be recognized that these resources do not generate new information; that is, they are not the "point of origin." Instead, they have recorded facts and data first presented elsewhere, and have staged this material in electronic format for convenient, rapid, selective retrieval.

SUGGESTED READING

Introductory Welding Metallurgy, from AWS Educational Activities Committee, American Welding Society, Miami, FL, Book Code IWM-T, 1968.

Introduction to the Physical Metallurgy of Welding, by Kenneth Easterling, Butterworths Monographs in Metals, Butterworths & Company Limited, London, England, 1983, ISBN 0-408-01351-6.

Modern Welding Technology 2nd Edition, by Howard B. Cary, Prentice-Hall, Incorporated, Englewood Cliffs, NJ, 1989, Library of Congress No. 78-2966, ISBN: 0-13-599290-7.

Maintenance Welding, by Edgar Graham, Prentice-Hall, Incorporated, Englewood Cliffs, NJ, 1985, Library of Congress No. 84-11439.

Welding Technology, Volume 1, Welding Handbook, Eighth Edition, American Welding Society, Miami, FL, 1987, Library of Congress No. 87-071919, ISBN: 0-87171-281-4.

Fundamentals of Welding Metallurgy, by H. P. Granjon, AP Abington Publishing, Abington, Cambridge, England, 1991, ISBN: 1-85573-019-7.

Chapter 2

In This Chapter...

ATOMS **18**
   Elementary Particles 20
      Electrons 22
      Positrons 26
   Atomic Nuclei 26
      Protons 27
      Neutrons 28
   Atom Construction 32
      Isotopes of Elements 33
      Isobars 34
      Atomic Weight 34
      Atomic Mass 34
      Atom Valency 35
      Ionization 36
      Radioactivity 37
      Atom Size or Diameter 38
THE ELEMENTS **39**
AGGREGATES OF ATOMS **41**
   The Solid State 45
      The Crystalline Solids 45
      Amorphous Solids 47
   The Liquid State 48
   The Gaseous State 49
FUNDAMENTALS OF CRYSTALS **50**
   Identification of Planes & Directions in Crystals 56
   Basic Types of Crystals 56
      Inert Gas Crystals 58
      Ionic Crystals 58
      Covalent Crystals 59
      Metallic Crystals 59
THE CRYSTALLINE STRUCTURE OF METALS **61**
   How Does a Crystal Grow from the Melt? 64
   The Formation of Dendrites 66

   The Formation of Grains 68
      The Shape of Grains 71
      The Size of Grains 72
   Undercooling 72
THE IMPORTANCE OF A CRYSTALLINE STRUCTURE **74**
   Allotropic Transformation 75
   Solubility in the Solid State 76
   Plasticity in Metallic Crystals 77
      Slip in Crystalline Structures 77
      Slip and Lattice Orientation 78
      Slip in Polycrystalline Metals 79
      Observing Slip 80
      Twinning in Crystalline Structures 81
   Lattice Imperfections: Dislocations 84
      Point Defects 85
      Edge Dislocations 86
      Screw Dislocations 88
      Stacking Faults 88
      Other Lattice Imperfections 88
   Cold-Working Metals 88
EXAMINATION OF METAL STRUCTURES **91**
   Fracture Appearance Assessment 91
   Metallography 92
      Metallography Using Optical Microscopy 92
      Quantitative Metallography 105
      Metallography Using the Electron Microscope 106
      Metallography Using Ion Microscopy 115
      Tunnel-Effect Microscopy 116
      Chem. Analysis of Microstructural Constituents ... 116
   Newer Techniques in Metallurgy 121
FRACTOGRAPHY **123**
SUGGESTED READING **131**

The Structure of Metals

Examination of metal structure in the vicinity of a weld is an important part of welding research and has helped gain a general understanding of welded and brazed joints. As metals and alloys are subjected to manufacturing steps, fabricating operations, and heat treatment, changes in structure which influence their properties and behavior are methodically documented. This information can be used to make predictions about strength, toughness, corrosion resistance, and other properties of welds, and often provides explanation for troublesome problems such as loss of strength, embrittlement, and cracking. Study of metal structure also assists in the development of improved construction alloys.

The thermal cycles involved in most welding procedures are quite unusual when compared with conventional hot working operations and heat treatments, so it is common practice to determine the structural changes produced by actual welding. The term *microstructure* is used to describe the structure of metals.

Various tools and techniques are used to study microstructure. A simple visual examination of etched metal surfaces and fractures will reveal some configurations in etched patterns that are related to structure, but magnification of minute details will yield considerably more information. Four widely used tools for examination of structures in metals are listed below:

(1) The low-power *magnifying glass*, when applied to etched metal surfaces, reveals gross details in microstructure. Also, telltale facets can be seen in the morphology of fractures.

(2) The *metallograph*, an optical microscope, is usually fitted with an inverted stage for convenience in scanning the flat face of prepared specimens. Metallographic examination usually requires magnification in the range of about 50 to 1500X. Because of the wave length of visible light, there is an upper limit of about 2000X for magnification with an optical microscope.

(3) The *x-ray diffraction camera* permits more positive identification of the basic crystallographic structures in metals by analyzing the diffraction of regulated x-rays as they are passed through a specimen and emerge to register as spots on a precisely positioned photographic film.

(4) The *electron microscope* is capable of magnifying at least 200 000X with remarkably good depth of focus and resolution. Images of metal microstructure are obtained either by electron beams transmitted through a specimen, or by beams that are reflected and emitted.

Other microscopes and analytical devices used in special metallographic studies are reviewed in this chapter. For example, the need frequently arises to know the chemical composition of minute phases and particles observed in a microstructure. The *electron microprobe analyzer* can analyze a precisely selected area as small as one micron (0.000 0394 in.) in diameter on the surface of a metallographic specimen.

Pursuit of theoretical physics enabled mankind to utilize nuclear energy, and while theoretical physics, thermodynamics, and chemical phase analysis can be applied to the study of metal structure, these methods are complex and difficult to verify by experimentation. No approach to the study of metals has been so broadly rewarding as the "atomistic" approach; that is to study metals and their structure from the perspective of the atoms of which the material is comprised. An atomistic framework is the basic perspective used in this text.

ATOMS

The nature of atoms and their role in the structure of matter, including metals, had been clearly hypothesized in early research, but only recently has a microscope become available that is capable of viewing an atom directly. This was first accomplished in 1960, using a *field ion microscope* which projected relatively crude registrations of atoms in the tip of a needle-like metal specimen. An atom is the smallest unit of matter that behaves like an element. Visualization of atoms can be started by picturing them as tiny solid spheres. Atoms exert attractive forces in all directions, and this achieves cohesion, or bonding, in a quantity of the element which often results in a solid mass of matter.

Atoms cannot be seen directly with commonly available magnifying instruments because the atom's diameter is only in the order of one angstrom (See Technical Brief 2). Atoms can be indirectly located by various examinations, using x-rays, electrons, and ions, which produce registrations on screens or photographic film that are related to the positions of the atoms in a specimen of the material. The x-ray diffraction camera produces films for interpretation called Laue photographs. The ionic technique used in the field emission microscope produces a registry of atoms on a screen at great

---

TECHNICAL BRIEF 2:
What Is An Angstrom?

An angstrom (Å) is one ten-billionth of a meter, often expressed as 1×10^{-8} cm. Because the angstrom is not a coherent SI unit, the nanometer (nm), one billionth of a meter, has been recommended in the metric system for indicating dimensions in this order of magnitude. However, the angstrom is widely used as a unit of measure in crystallography and related sciences because of its "close fit" to atom size. The angstrom unit, which is equivalent to 0.1 nm, will be used throughout this text.

One angstrom (Å), the approximate diameter of an atom, is equal to 0.000 000 003 94 in., or about four-billionths of an inch. The human eye cannot resolve an object smaller than about one hundred microns (0.00394 in., or about four-thousandths of an inch); it would be necessary to magnify an atom approximately one million times to see it as a mere speck. A magnification of about ten million times would allow the human eye to compare sizes of different kinds of atoms, and even greater magnification would be needed to perceive any details of atom construction.

magnification and shows the positions of the atoms. Although these techniques have produced only "shadows" of the atoms in the specimen, or reflections from them, much can be surmised about atoms in the structure of metals. The atomistic view of structure, despite being somewhat superficial, has proven to be a functional concept, useful to all who are intent on efficient processing, fabrication, and utilization of metals.

In a simplistic view, atoms are pictured as tiny solid spheres in close proximity making up a mass of matter. Each individual atom is believed to consist of a central nucleus around which one or more electrons revolve at distances up to approximately one angstrom in diameter. The nucleus of the atom, consisting of a cluster of protons and neutrons, is very small, measuring only about 10^{-12} cm across (0.3937×10^{-12} in.). The electrons revolving around the nucleus are believed to be distributed on orbital layers, or *shells*, that have rather firmly fixed numbers of electrons on each shell. Figure 2.1 schematically illustrates the arrangement of nucleus and electrons in an atom of the metal zinc.

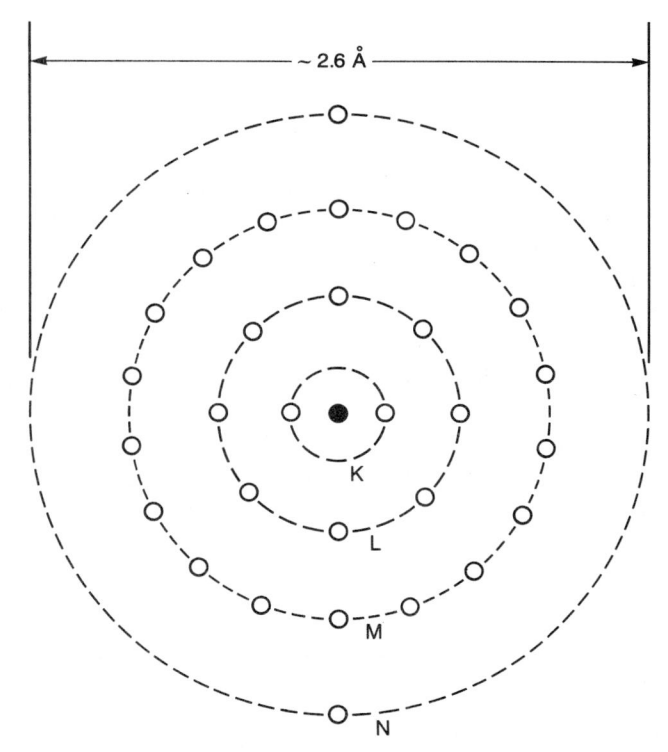

● SOLID CIRCLE AT CENTER IS NUCLEUS OF ATOM
○ OPEN CIRCLES IN ORBITS ARE ELECTRONS
ORBITAL "SHELLS" ARE LETTERED K, L, M AND N STARTING WITH INNER SHELL

Figure 2.1 — Schematic arrangement of nucleus and electrons in an atom of the metal zinc ($_{30}Zn^{65}$)

This concept of the atom is generally accepted, but there are factors that seem to refute certain aspects of this simple picture of the atom. As an example, previous mention that atom diameter is only about one angstrom is true of the smallest atom, that of hydrogen. The zinc atom in Figure 2.1 is indicated to be approximately 2½ times larger. This seems logical in view of the relative weight or mass of hydrogen as compared with that of zinc; however, as reasons arise to discuss various elements and their atom size some unexpected variations will be encountered.

Elementary Particles

The three electrically-structured constituents of atoms, the electron, proton, and neutron, are called *elementary particles* because they were the first to be identified, evaluated, and named. In early studies, these appeared to be the basic particles of which all atoms were constructed. These particles differ in ways that profoundly influence their role and behavior, both in the intact atom as well as a "free" particle. As shown in Table 2.1, the electron and the proton have equal, but opposite, electrical charges; the electron is negative, and the proton is positive. However, they are vastly different in mass. The electron has a mass only $1/1837$ that of a proton. The neutron is very slightly heavier than a proton, and displays no electrical charge. Despite their greater mass, the proton and the neutron are confined to a much smaller volume than electrons when they are grouped together to form the nucleus of an atom. In fact, the mass of an atom is contained essentially in the nucleus, where its protons and neutrons are held together by powerful forces.

No complete explanation has yet been developed for the forces that hold protons and neutrons together in the nucleus of an atom. The stability of nuclei suggests that attractive forces are the principal ones in operation. These attractive forces must be of short range because when techniques are applied to separate the elementary particles in the nucleus by a very small distance, perhaps in the order of one nucleus diameter, the binding force becomes negligible.

Atom construction should be examined beyond the three elementary particles which have been described because observations concerning radioactivity of certain elements can be helpful in understanding nondestructive methods of examination which apply various forms of radiation to weldments. When gamma radiation from elements such as radium and uranium was detected and the emission mechanism studied, researchers realized that atoms in these unstable elements were undergoing spontaneous, permanent intrinsic change. In addition to the radiation emitted, a stream of

Table 2.1 - Elementary Particles

Particle Name	Common Symbol	Charge	Electrical State esu*	Mass gram
Electron	e	Negative	4.802×10^{-10}	9.107×10^{-28}
Proton	p	Positive	4.802×10^{-10}	1.672×10^{-24}
Neutron	n	None	0	1.675×10^{-24}

*Electrostatic Units

particles, consisting of positively-charged alpha particles (actually helium atoms with two electrons removed), was being discharged. Also, atoms that disintegrated in this manner were undergoing transmutation to new chemical elements.

Continued research in atomic structure has isolated many more sub-atomic particles of an elementary nature. The *positron* was an early discovery made during analysis of cosmic rays. The positron was found to have the same small mass as an electron, but it carried a positive charge; therefore, the positron was essentially a "positive electron". A particle given the name *meson* has a mass several hundred times that of an electron, and can carry either a positive or a negative charge. Mesons exist in the nucleus of atoms, where they play a role in establishing the short-range forces that bind the elementary particles together.

The advent of high-energy particle generators in the 1930s allowed man to smash the nuclei of certain atoms and to analyze the fragments. The cyclotron and the linear accelerator produced an exciting number of new elementary particles; in fact, almost two hundred particles have been identified and named. Study of these particles is important to anyone seeking a complete understanding of the atom (e.g., the nuclear physicist), but details of the nature and possible function of the particles is not pertinent to welding metallurgy. The existence of these many elementary particles should be recognized, but it must be noted that most of them are the product of forcibly disintegrated nuclei, and whether these sub-atomic particles actually exist in the atom is unclear. These particles may be transitory forms of matter that facilitate the escape or transfer of energy from the disturbed nucleus.

It is not necessary to consider the *leptons* and *hadrons* that make up the majority of sub-atomic particles, but there is good reason to examine more closely the three elementary particles listed in Table 2.1. The proton and the neutron are indeed basic particles that form the nuclei of atoms, and electrons orbit each nucleus as illustrated in Figure 2.1; however, a number of important details must be added to this rudimentary picture of an atom.

A fundamental detail is the "spin" of particles. This spinning motion has not been observed directly, but physicists have postulated that all three elementary particles must have an axis of symmetry about which rotation takes place. Every manner of mathematical treatment applied to atomic structure (including the quantum theory) supports the concept of spinning elementary particles. This theoretical deduction provides a rational explanation for a number of properties of matter that are important to metallurgy, for example, magnetism. A preliminary study of magnetism will help illustrate the spin of particles and will emphasize the far-reaching significance of this feature.

To understand the significance of particle spin, a pertinent reminder can be offered that flowing electric current always generates a magnetic field. When the electron is examined, this corpuscle of negative electricity will be found to be spinning on an axis like a toy top. The spinning motion is analogous to current flow (i.e., electron movement or flow) and this imparts a negative magnetic moment. Next, the orbital path of the electron around the nucleus must be considered. This orbital motion is also analogous to current flow, and therefore a broader magnetic field is generated.

The spinning electron, on experiencing the magnetic moment from orbiting, will strive to align its magnetic axis parallel to the imposed field. This change in axial orientation is called *Larmor precession*. Interaction between the magnetic field from electron spin and the magnetic field from electron orbit can result in increased angular momentum. This feature is the product of moment of inertia about the axis of rotation and the spin angular velocity, just as linear momentum of a body moving along a straight line is equal to mass times velocity. Angular momentum, therefore, is the vector sum of the electron spin and orbital motion. As angular momentum increases, a stronger magnetic field is generated.

The axis of the spinning electron during Larmor precession undergoes slow gyration so that each of its poles describe a cone. Random orientation of particle axes from one atom to another finds them counterbalancing the magnetic moment of each other in substances. Consequently, no overall magnetic field is exhibited by the substance. When a magnetic field from an outside source is imposed, the axes are brought into more common orientation and the substance becomes more strongly magnetized.

The proton also spins on an axis, and its motion should be considered next. A proton carries a positive charge and its spin exerts a positive magnetic moment. For this reason the magnetic field of the proton is parallel to its spinning motion instead of being oppositely directed as in the case of the electron. Also, the proton has a larger mass and spins slower than the electron. Consequently, the magnetic moment of a proton (even of a whole nucleus) is only one-thousandth as strong as the magnetic moment generated by the electron.

Neutrons also spin, but they generate a weak negative magnetic moment. This magnetic effect is opposite to that anticipated from the direction of neutron spin, and is surprising in view of the absence of electrical charge. This finding led physicists to hypothesize that the neutron is made up of positively and negatively charged particles in equal strength to balance the static electrical charge. However, they believe the negative particles are further from the axis of rotation in the neutron and this causes some negative electrical influence to be exerted during spinning.

The fundamental concept of magnetism gained from studying the spin of elementary particles is that the magnetic properties exhibited by all substances are mainly effects of the motion of electrons within the atomic structure. The magnetic moments of nuclei and their constituent particles, on the other hand, are quite small and in general can be disregarded. The magnetic moment of atoms, therefore, appears to have three main sources;

(1) spin of electrons,
(2) orbital motion of electrons around the nuclei, and
(3) orientation change in the axes of the spinning electrons induced by an applied magnetic field.

Individual examination of the three principal elementary particles — the electron, proton, and neutron — will help gain an understanding of the atom as a whole. All of these particles play key roles in specific aspects of welding technology. Therefore, familiarization with each of the particles, per se, is pertinent.

Electrons

An electron has never been viewed directly as a tangible object. Knowledge of the electron is mostly hypothetical, but is supported by innumerable experiments in which the observed effects of electrons present good indirect evidence that their character is much like the reported consensus. Yet, scientists have debated at length whether an electron should be regarded as a point particle, or as a corpuscle of electromagnetic radiation. One view of electrons in terms of radiation recognizes that a stream of electrons exhibits wave length, and strong similarity exists between optical systems handling light and magnetic fields acting on electrons. Also, a cloud of electrons can exist in several states, and transition from one state to another usually results in the emission of photons, which are elementary quanta of radiant electromagnetic energy. Nevertheless, electrons have mass, momentum, and kinetic energy, and one way to understand the role of the electron in the atom is to picture an electron as a minuscule, indivisible particle with a strong negative electrical charge, as outlined in Table 2.1.

All atomic matter normally contains electrons, and the number and distribution of electrons in the atom determine to a large extent the chemical and physical properties of the element. The electrical charge carried by an electron is amazingly high, and this charge, along with energetic spin, enables the electron to exert strong effects despite its small mass. The principal properties of an electron are electrical charge, mass, spin, and wave length.

The wave length of electrons is a property that deserves further explanation. In Figure 2.1, electrons in the structure of an atom are depicted as spinning particles orbiting the nucleus on shell-like layers. This concept was proposed in 1913 by Bohr, who provided a model that appeared to fit the many working equations which were subsequently developed for calculating such features as orbit radii, orbital velocity, angular momentum, etc. Recent findings on atomic structure suggested to some scientists that the deployment of electrons in an atom is a more complex arrangement. They reasoned that electrons in the structure of the atom must have a wave length because electrons emitted as a stream, or beam, display wave length. Wave-apperceptive proponents visualize the atom as a nucleus surrounded by electron waves. Although there is some basis for this concept, description of the wave-like arrangement of electrons in the atomic structure is complex and lengthy. Fortunately, in final analysis, the model of an atom incorporating electron waves also has layers, or shells, of "charge density" surrounding the nucleus. This similarity between Bohr's concept and the wave theory permits continued use of the convenient, easily-visualized atom model with spinning electrons orbiting on shells around the nucleus as shown in Figure 2.1.

Electrons orbit the nucleus of the atom on shells that ordinarily remain stable because the inward centripetal force (from electrostatic attraction between the positively-charged nucleus and the negatively-charged electron) is counterbalanced by the outward centrifugal force, which is the angular momentum determined by orbital velocity. Bohr also provided a quantum hypothesis to explain why definite orbital shells are fixed for the electrons. The innermost shell is calculated to have a radius of 0.5 Å, and only two electrons can be accommodated on this first shell. The second shell is calculated to be four times larger; the third shell is nine times larger, and so on. Obviously, electrons must travel at different speeds on each shell in order to maintain discrete orbits. Electrons orbiting on shells closer to the nucleus must travel at higher velocities. Those electrons on the innermost shell have a velocity that is $1/37$ the speed of light.

Electrons on the next shell travel only one-half as fast as those on the first shell. The velocities of electrons orbiting on the remaining shells are reduced accordingly.

Both chemical and metallurgical studies give attention to specific electron orbital shells. As shown in Figure 2.1, the shells are commonly identified by alphabetical designations, starting with the letter "K" for the innermost shell immediately surrounding the nucleus, and continuing with the letters L, M, N, O, P and Q to identify a total of seven shells (as found in the heavier atoms). Some scientists prefer to use numerical designations for the seven principal shells, numbering the innermost shell with "1", and applying numbers 2 to 7 consecutively to shells of increasing diameter.

Each orbital shell of an atom is considered to bear no more than a specific number of electrons, with two electrons on the K-shell (No. 1), eight electrons on the L-shell (No. 2), 18 on the M-shell (No. 3), 32 on the N-shell (No. 4), 32 on the O-shell (No. 5), and more complex distributions on the P-shell (No. 6) and the Q-shell (No. 7).

The total number of electrons orbiting the nucleus increases as the mass of the atom representing a given element becomes larger. Each added electron is usually placed in orbit in the outermost shell unless this addition would exceed the limiting number for the particular shell, and when this is the case a new shell would be started. This statement regarding the addition and deployment of electrons is only partly true. It is not unusual for a new shell to be started before the completion of the predecessor shell. These variations in electron population on the shells affect the chemical and metallurgical behavior of the atom.

Application of quantum mechanics to the electron shell theory made it clear that the main shells have *subshells*. The orbits of these subshells are characterized by a principal quantum number (n), an angular momentum number (l), and a magnetic magnitude or quantum spin number (m). Full description of this area of electron theory can be found in any introductory book on atomic physics, however, a brief notation is included in this text to characterize four different states of motion by electrons in the subshells. Using quantum mechanics, a principal quantum number was obtained to distinguish the subshell orbits of each main shell. The orbital angular momentum (l) of 0, 1, 2 and 3 (in units $\hbar$) are identified by the letters s, p, d and f, respectively. A table is provided as Appendix X listing the electron configuration of each of the elements by giving the numbers of electrons normally present on each shell and subshell. This table will be useful as the role of electrons becomes evident in a number of important areas, including the chemical behavior of elements, bonding between the atoms in an aggregate, control of crystal structure, and phase prediction in alloying metals.

The concept of electrons as they are incorporated in an atom is important in welding metallurgy because electrons in orbit around the nucleus are not as tightly bound as are the protons and neutrons in the nucleus. Relatively little energy change is involved when an electron is moved from one orbit to another, or an electron is stripped free of an atom. The amount of energy possessed by an individual electron varies with the specific shell and subshell in which it is orbiting.

Electrons are bound in a given shell by energy measured in electron volts (eV). This unit is the energy change experienced by a particle having one electronic charge (e = 1.602×10^{-19} coulomb) in passing through a potential difference of one volt. One electron volt (eV) is equal to 1.6×10^{-19} joules. As applied to electron emittance (ionization), the energy required to strip free an outermost electron ranges from about 5 to 25 eV.

There can be a shifting or "jumping" of electrons from one shell to an adjacent shell under various circumstances, but this change in orbital location must be complete and the electrons cannot stop in between shells. Consequently, these jumps must be accomplished by the introduction of a definite minimum amount of energy if the movement is outward to a larger diameter shell, or entails complete removal of the electron. If the orbital change is an inward movement to a location on a smaller diameter shell, there will be a release of energy as photons of electromagnetic radiation. Further details concerning the electron in the structure of an atom, particularly the movement of electrons in real materials, and especially in metals, are provided later in this chapter as aggregates of atoms are examined.

Free electrons can be generated rather easily in a variety of devices, and emitted in quantity in various ways, but since these particles carry a negative charge, their path, or trajectory, is easily controlled by applying magnetic fields along the beam pathway. If a cathode and an anode are arranged in space and sufficient electrical potential is applied, electrons will leave the cathode and flow to the anode. Of course, it helps to heat the cathode to aid electron emission, and to have a vacuum in the intervening cathode-anode space to avoid collision of the electrons with the atoms and molecules in air.

The *cathode ray tube* (CRT) is a classic illustration of a device which produces a controlled flow of electrons, however, the flow of electrons as free particles can be made to occur in a myriad of ways. When emitted as a free particle, the electron normally travels in a straight path. The velocity of the electron is increased as higher electrical potential (voltage) is applied between the emitting cathode and the anode. Electrons usually travel about ⅕ the velocity of light, but they can be accelerated to approach the speed of light. As higher velocities are attained, the mass of an electron increases by a small, but measurable amount.

A beam of electrons acts somewhat like x-rays because it also has wave form. In fact, in early research the wave form and wave length of electrons were established from the pattern of electron diffraction from the face of a solid crystal of nickel metal. The wave length of an electron is inversely proportional to its velocity, that is, as velocity increases the wave length decreases. Electrons with an applied voltage of 50 kV would be expected to attain a speed of about 100 000 kilometers per second (60 000 miles per second) and would exhibit a wave length of approximately 0.05 Å. Electron velocity near the speed of light (299 774 kmps, or 186 280 miles per second) would be achieved with applied potential of one million volts. Electrons with this velocity would exhibit a wave length in the order of 0.01 Å.

Some important and useful phenomena occur as electrons bombard the surface of a substance. The electron beam will not penetrate as deeply as the x-ray because the negative charge on the electrons causes a strong interaction with other particles in the atom structure, especially those particles carrying a positive charge. The electron beam is highly effective in eliminating dissolved gases, if the substance being bombarded is a metal. An electron beam impinging on a metal will produce fluorescence, phosphorescence, and radiation. Secondary electrons will be emitted from the surface area beneath the beam as well. If the electron beam has sufficient energy, the substance being bombarded will be melted and will possibly be vaporized (as occurs in electron beam welding and cutting processes). Electron beams also exhibit behavior for which scientists have yet to find explanation. This is especially true in the field of microelectronics, where electron flow through special conductors and the blockage of flow at insulating barriers are all-important in operating circuitry. For example, electrons have the ability to tunnel through the film-like layers of various metals that have been built up, sandwich style, and then treated in various ways by "doping". This *resonant tunneling effect* allows passage of electrons through the metal "sandwich" at several distinct energy levels. Utilization of this electron behavior could allow a single transistor to serve several functions in an integrated circuit.

Positrons

The positron, the electron with a positive charge and one of the many elementary particles, was first detected among the cosmic rays that come from outer space and concentrate in the earth's stratosphere. Cosmic rays, on entering the earth's atmosphere, cause unusual fragmentation of the nitrogen, oxygen, and carbon atoms encountered by the rays. The atomic debris produced by this cosmic radiation consists mostly of positively charged particles (protons), and a smaller amount of heavier nuclei. Positrons were not easily detected among the debris because of their very short life. Electrons (negatively charged) are in abundance almost everywhere and the strong attraction between dissimilar charges brings the electrons and positrons forcibly together, whereupon annihilation occurs with an outburst of gamma radiation. These positively charged particles, positrons, can be produced in high-energy accelerators, and they are also emitted in small amounts by certain elements in radioactive form (e.g., Na, Co, Cu, Ge). These sources can be used to carry out a recently developed high-resolution NDE method that holds great promise for examination of metals. The *positron annihilation* (PA) technique determines certain lifetime measurements of the photons that comprise the gamma rays emitted during the annihilation process. Analyses of the lifetime data enables determinations to be made of minute material changes, and the detection of defects that are considerably smaller than can be resolved by most other NDE methods.

Atomic Nuclei

A review of an intact nucleus of an atom will help illustrate the individual properties of protons and neutrons. The term *nucleon* is often used to describe protons and neutrons when they are combined in the nucleus of an atom. Nucleons are pictured as being ball-like, with a diameter of approximately 2.5×10^{-13} cm. Nucleons are held to the confines of a spherical field by extremely high short-range attractive forces between the elementary particles. Despite their close packing, nucleons are believed to be moving around inside the nucleus and spinning, because all elementary particles maintain this motion whether they are contained in the atomic structure or are in free flight.

The size of the spherical configuration which results when the nucleons form a bona fide nucleus provides a surprise, because the diameter or volume is not a simple arithmetical function of the number of nucleons present. Nuclei are much smaller in size than would be expected if related to the total number of component nucleons. For example, a nucleus containing a couple of hundred nucleons (as would be present in a rather heavy atom) might have a diameter only six times that of a single proton. Yet, there is a relationship between the number of nucleons present and the diameter of the nucleus, and all atomic nuclei appear to have approximately the same density. This has been estimated at the tremendously high value of 100 000 tons per cubic millimeter (about 1.75 million tons per cubic inch). This value, in turn, suggests that the nucleus of an atom occupies only about two parts in 10^{12} of the volume of an atom.

In view of the constancy of density in the nuclei of the elements, it would seem reasonable that the spherical nuclei would increase in volume proportionately with the number of particles contained, but this is not the case either for the atomic nuclei or for the overall size of atoms. There are many unanswered questions about the "size" of atomic particles. Unexpected variations are found in the physical dimensions of atoms as well as in certain phenomena that would be expected to correlate closely with atom size. These aberrations are discussed under such topics as atom size and *neutron-capture cross-section*.

Protons

The proton is the positively-charged elementary particle that serves as the basic building block for the nucleus of an atom. When compared with the electron, the proton has an equal, but opposite electrical charge. From the standpoint of mass, the proton is 1837 times greater than the electron, yet the proton is much smaller in size. Consequently, the proton has a tremendous density commensurate with that of the nucleus of an atom.

The characteristics of protons can be studied rather easily because the particle can be generated in quantity and the motion of the protons can be controlled by virtue of their positive charge. The hydrogen atom, when stripped of its lone electron, is simply a proton. No other element has a single proton as its nucleus. There are, however, two variations of the hydrogen atom (isotopes) that include neutrons in the nucleus along with the single proton, and these will be detailed because of their great importance. When the hydrogen atom (having only a proton in its nucleus) is stripped of its orbital electron, the remaining particle carries a single positive charge. Other elements can be stripped of their electrons to form positively-charged atoms or nuclei, and then these particles can be manipulated in experiments with the aid of applied electrical or magnetic fields.

Atoms will carry one positive charge for each electron stripped away. The helium atom is a favorite for this kind of experimentation. When helium is stripped of its two orbital electrons, the nucleus that remains consists of two protons and two neutrons and it carries a double positive charge. A stream of helium nuclei is called an *alpha ray*. Alpha particles are emitted during certain phenomena in nature, such as the radioactive disintegration of radium.

Protons, as fast-moving particles, are utilized in many ways. Of course, they play a major role in electric arc operations, but this comes about quite naturally as the counter-flow to the passage of electrons. Although many applications utilizing protons actually employ atomic nuclei (deuterons, alpha particles, and ions), nevertheless, simple protons (hydrogen nuclei) produced in the cyclotron are frequently used as high-energy sub-microscopic bullets in atomic research. When a substance is bombarded by fast-moving protons, these particles travel only a very short distance into the target substance (perhaps about 0.005 cm, or about 0.002 in.) because of the abundance of oppositely charged particles (electrons) present. The strong attraction of dissimilar charges causes direct collisions of protons and atoms of the target substance, and results in a condition termed ionization.

Neutrons

A neutron, as shown in Table 2.1, is very slightly heavier than a proton, and is the only elementary particle that exhibits no electrical charge. Even though the neutron was recognized as a nuclear particle in early research, scientists have yet to arrive at a consensus understanding of its nature. Despite uncertainties about the neutron, a working hypothesis has developed which enables man to accomplish much with this particle. The foremost accomplishment, of course, is the advent of commercial nuclear power generation. Nuclear weapons must also be mentioned, but attention should be directed to unique uses of neutrons in analytical and metal examination procedures pertinent to welding metallurgy; such as:

(1) composition profiling,
(2) crystal structure identification,
(3) radiography,
(4) magnetic property determination, and
(5) residual stress measurement.

Study of the neutron has presented a special challenge, because although it can be emitted as a free particle, its electrical neutrality does not allow acceleration and manipulation with applied electrical or magnetic fields as allowed by the electron, proton and other particles that carry a charge. The neutron does have a very weak magnetic moment, and this is one reason why scientists have speculated that the neutron possibly is a proton combined with an electron which is an arrangement that would account for electrical neutrality. This view is supported by experiments with neutrons that found the particles to be quite unstable when removed from nuclei. Free neutrons tend to transform into a proton by releasing an electron plus a *neutrino*, and to exhibit radioactivity. Half-life of free neutrons appears to be approximately 13 minutes.

The magnetic moment of the neutron is -1.93 μ_N which is only sufficient to have these particles interact with unpaired electrons in magnetic atoms. The cause of this weak magnetic moment is suspected to arise from slightly off-center distribution of the electron with respect to the rotational axis of the spinning neutron.

Absence of an electrical charge on the neutron is very important because it enables the neutron to approach an atomic nucleus without being repelled by the positive charge of the nucleus. Consequently, a neutron, when colliding with an atom, can enter the nucleus; whereupon any number of remarkable actions may take place. The frequency of collision between a neutron in free flight and the atoms of a target material is relatively low because atomic nuclei are extremely small and this provides ample open space within the atoms for the free-moving neutrons to pass within the interatomic spaces without colliding with any particles. Neutrons are not particularly prone to react with the electrons present, but instead are destined for nuclear interaction. This is quite different from the attenuation of x-rays which expend their energy principally on the electron shells. The electrical neutrality of the neutron enables a beam of these particles to penetrate to a remarkable depth when it strikes a target material. The penetration can amount to as much as five centimeters (about two inches) in steel, and to almost 30 cm (12 in.) in aluminum. This penetration capability is utilized in neutron radiography where the beam is attenuated to a lesser extent than are x-rays by heavier atoms, and where unique selectivity is exhibited as the beam expends its energy on lighter atoms.

Neutrons are not ordinarily found in the free state outside the nuclei of atoms, but they can be emitted or liberated from nuclei in the following ways:

(1) bombardment of a suitable target material by charged particles in an accelerator,

(2) irradiation of a substance with gamma rays, and

(3) release of a suitable kind of atom during nuclear fission.

The velocity of neutrons emitted from nuclei depends on the production technique employed, but once in flight nothing can be done to increase the speed of neutrons because they are unaffected by an applied electrical or a magnetic field. Considerable energy is needed to free a neutron from the very high short-range attractive forces that bind nucleons together. The minimum energy required for neutron emittance is probably in the vicinity of five MeV, but this minimum depends on a number of conditions, for example, the ratio that exists between the numbers of protons and neutrons present in the nucleus. When energy above this level is applied, perhaps by particles from an accelerator striking target nuclei, the neutrons emitted can have a velocity close to the speed of light. These particles are referred to as *fast neutrons*. Although the neutrons emitted in free flight travel in a straight line, they spin and exhibit wave form.

Neutrons cannot be accelerated once in free flight, but their velocity can be reduced in a controlled manner. Control is accomplished by directing a beam of fast neutrons through a "moderator", which is a permeable barrier made of relatively light elements such as a thin sheet of aluminum, or a block of paraffin or wax. Slower-moving neutrons produced by this technique are called *thermal neutrons* because their speed is approximately equal to the movement of a hydrogen molecule in a quantity of this gas at room temperature and atmospheric pressure. Thermal neutrons have a velocity of only 2200 m/sec (7218 ft/sec), and at this velocity they offer a kinetic energy of 0.0253 eV. Also, thermal neutrons exhibit a wave length of 1.8 Å, which has a significance related to the crystallographic structure of metals. Because this value is comparable to the center-to-center spacing of atoms in a solid, it offers possibilities for measurement of lattice parameters by diffraction.

The two general kinds of neutrons, as differentiated by velocity, are important to keep in mind, because while fast neutrons are easier to generate or eject from nuclei, thermal neutrons display much greater effectiveness when a stream of them is directed at a solid substance. Comparison of the behavior and effects of fast neutrons versus thermal neutrons requires examination of the mechanics of passage of particles through the atomic structure of the target substance.

High-flux reactors are often used as a source of neutrons, and modern reactors can produce a beam of controlled velocity, wave length, and alignment. Precise beam control allows a number of important metallurgical determinations to be made; usually without damage to the specimen, and including measurement of crystallographic parameters, and levels of residual stress that exist in a section of metal at localized points or areas.

When a freely-moving neutron does collide with the nucleus of an atom, a number of different reactions can take place and the outcome depends largely on the kind of atom struck and the velocity of the neutron. The most important of these are the following:

(1) diffraction of the neutron to a new trajectory (i.e., scattering),
(2) displacement of the struck atom to an adjacent site,
(3) ejection of one or more neutrons from the nucleus,
(4) capture of the neutron by the nucleus to form a heavier atom, and
(5) fission of the nucleus to form lighter atoms.

Nuclear scientists have found these reactions, particularly neutron capture and fission, more readily accomplished by thermal neutrons than by fast neutrons. Since the neutron does not experience a repulsive force, as protons do, it appears that the slower moving thermal neutron has a greater opportunity to enter the nucleus, remain captured (at least momentarily), and then expend its kinetic energy.

Controlling the velocity of neutrons is a critical part of utilizing their chargeless kinetic energy to obtain a desired reaction. When neutron-capture results in the formation of a heavier atom, the new entity is usually radioactive. When fission occurs it usually produces two nuclei of slightly unequal mass and is accompanied by the ejection of from one to five free neutrons. A certain amount of gamma radiation is also emitted as fission occurs. Those neutrons emitted almost immediately after fission are called *prompt neutrons*, or *secondary neutrons*. More neutrons can emerge seconds later as unstable fragments decay and become stable nuclei. These are called *delayed neutrons*.

Neutron capture intrigued not only nuclear scientists, but also practical metallurgists and welding engineers because of the vital role it played in nuclear power technology. In addition to the selectivity exercised by the nuclei of atoms for thermal neutrons, the atoms of various elements displayed great differences in the number of interactions, or captures, that occurred between their nuclei and incident neutrons. Naturally, the degree of change that took place among the atoms of a given element and the amount of radioactivity generated were dependent on the number of interactions and captures that occurred between neutrons in flight and the atomic nuclei of the element.

To measure the probability of an interaction of a given kind taking place between a nucleus and an incident neutron, a special concept of geometrical cross-section for atoms and their nuclei was devised. An arbitrary unit of area measurement was adopted and termed the *barn* (b). The barn was set at 1×10^{-24} cm^2, and is defined as the area in which the number of interactions occurring in a specimen exposed to a beam of neutrons is equal to the product of the cross-section, the number of nuclei per unit volume in the specimen, and the number of neutrons in the beam that would enter the specimen if their trajectories were perpendicular to it. In Chapter 3, when the influence of irradiation on mechanical properties is reviewed, neutron-capture cross section is described from the standpoint of the number of barns ascribed to various elemental metals which are present in steels as residual or as alloying elements. The quantity of neutrons passing through a given area is quantified as *neutron fluence rate*.

The elementary particle, the neutron, plays a vital role in controlling energy within the atom. The short-range forces that bind the nucleons together are tremendous, and removal of a proton from the nucleus requires energy a million times greater than the energy of any chemical reaction involving similar masses. However, any incident that puts a sufficient burst of energy into a nucleus, such as the impact and entry of a high-speed particle, may upset the balance of internal nuclear forces and exceed the minimum energy needed to initiate disintegration. When a neutron strikes the nucleus, a very important aspect is whether energy is required to carry the reaction to completion or, on the contrary, whether energy is released in sufficient quantity during the reaction to cause continued reaction in nearby nuclei. The stability of a nucleus, that is, its resistance to disintegration, depends on the relative numbers of protons and neutrons present. Atoms of lesser mass have nuclei with approximately equal numbers of protons and neutrons.

Heavier elements, for example, uranium, have a greater number of neutrons relative to the protons in their nuclei. Protons in large number in a nucleus are difficult to hold, so the binding force of excess neutrons serves to insure stability. However, if the nucleus contains a large number of protons with barely enough neutrons to bind the nucleus together, the nucleus will be more likely to split (fission) when bombarded by a high speed particle. When a neutron strikes an atom that is susceptible to fission, the impacting neutron is captured and causes the nucleus to split and release one or more additional neutrons. These neutrons are quickly absorbed by other nearby atoms, which in turn split and simultaneously liberate more neutrons. Obviously, this can be the onset of an uncontrolled reaction. When fission does occur under these conditions, all offspring nuclei possess a positive charge. The electrostatic force of like-charges drives the smaller nuclei apart until they are separated at least beyond the short-range forces. This separation may occur, of course, with explosive violence.

The most effective way of accomplishing nuclear fission, to release some of the energy within the atom, is by thermal neutron bombardment. Application of this technique to selected elements (e.g., uranium-235, and plutonium-239) has allowed man to accomplish fission with closely controlled release of energy (as in nuclear power generation), or with runaway chain reaction (as in an atomic bomb). A more direct means of utilizing nuclear forces as energy is to cause fusion of free nuclear elementary particles. The amount of energy available when protons and neutrons are forced to combine into a nucleon can be estimated from the following:

(1) Neutron mass, $_0n^1$ .... = **1.00898 amu***

(2) Proton mass, $_1H^1$ .... = **1.00814 amu**
    Sum .... = **2.01712 amu**

(3) Mass found when the proton and the neutron are combined to form a nucleon (deuteron), $_1H^2$ ...... = **2.01474 amu**
    Difference .... = **0.00238 amu**

    Atomic mass unit

> **TECHNICAL BRIEF 3:**
> **Thermonuclear Fusion**
>
> Thermonuclear fusion has been the object of intensive research for nuclear power generation because it holds great promise for utilizing nuclear forces as energy. Lighter elements enter into the fusion process more readily than the heavy elements. Hydrogen is the prime example of a low-mass element that can be made to undergo fusion to produce heavier elements and to release a tremendous amount of energy through the process. Iron appears to have the limiting array of nuclear particles for exothermic nuclear fusion behavior. Elements heavier than iron resist undergoing fusion for reasons not entirely understood. When the atoms of heavy elements are somehow compelled to fuse together, the combined mass of all new atoms produced will be slightly greater than the combined mass of the original atoms. Obviously, energy must be added to carry out the fusion process with heavy elements. Scientists believe that the heavier elements that make up part of planet Earth were formed by nuclear fusion that took place inside stars (e.g., the Sun) where ample energy was available for endothermic fusion.

The difference in mass after the proton and the neutron have combined to form a nucleon is the *annihilation energy* expended during the binding action of *thermonuclear fusion* (see Technical Brief 3). An atomic mass unit of 0.00238 is equal to 2.2 MeV.

Atoms of many elements favor particular values of *neutron velocity* (i.e., kinetic energy) at which they capture more-readily impinging neutrons and then react in some characteristic manner. An ultrasensitive analytical method called *neutron activation analysis* (NAA) makes use of this mechanism and can determine quantitatively the presence of certain elements in extremely small amounts. The NAA method is based on a resonance phenomenon and requires a high-intensity source of neutrons. NAA can be a very useful tool in welding research because of its capability to determine the location and the dynamics of elements present in trace amounts; for example, as little as 15 µg of hydrogen per gram of metal.

Atom Construction

One hundred and nine kinds of atoms (elements) have been identified and differentiated by the manner in which elementary particles are apportioned and deployed in their construction, and several additional elements are under study. The scheme of electron and nucleus portrayal in Figure 2.1 will continue to be used as other kinds of atoms are reviewed.

In all atoms, the total number of protons and neutrons in the nucleus is approximately equal to the atomic weight, that is, the mass of the atom as determined principally by its dense nucleons. The additional mass contributed by its electrons is virtually insignificant. The atomic weight reported in most textbooks for an element is often the average weight of one atom of the element as compared with an arbitrarily selected reference element or standard.

Many years ago, hydrogen, the only element that does not contain a neutron in the nucleus of its atom, was used as the standard because of its unique simplicity. Because the hydrogen atom consists of only one proton and one electron, its atomic weight was arbitrarily set at one (1.000). However, hydrogen proved to be a poor choice as a standard because of the extreme lightness of the atom. Also, hydrogen is a constituent of a limited number of compounds, and this handicapped the verification of atomic and molecular weights of other elements and compounds.

For a long period of time oxygen was used as the reference atom, with its atomic weight arbitrarily set at 16, but again this choice led to problems with verification. Finally, in 1961, an international agreement was reached on a precise standard: all atomic masses would be related to the mass of a specific carbon atom, which has been assigned 12.000 00 atomic mass units (amu). When "atomic weight" is reported as a simple, whole number for an element, the value is probably an approximation.

All atoms of a given element contain the same number of protons in its nucleus, and this number is known as the atomic number, usually denoted by the letter "Z." The atomic number determines to a very large extent the chemical properties of the element. The number of orbital electrons is normally equal to Z (the number of protons in the nucleus). Appendix IX is a tabulation of all known elements in order of their atomic number.

If the number of protons in the nucleus is equal to the electrons in orbit, the atom as a whole is electrically neutral. If this state of electrostatic neutrality is disturbed by removing or adding electrons, the atom becomes electrically charged, and thereupon is considered an *ion*. *Ionized* atoms cause important changes in atom behavior when this state is assumed.

The number of neutrons in the nucleus of an atom is commonly identified with the letter "N" The sum of protons (Z) and neutrons (N) present in the nucleus is known as the *mass number*, and is identified by the letter "A." As previously mentioned, one function of the neutron is to bind with the protons present to form a stable nucleus. While the lighter elements have approximately equal numbers of protons and neutrons in their nuclei, the heavier elements have a greater number of neutrons than protons. In fact, the heaviest atoms have almost twice as many neutrons as protons. A second important observation in the structure of atoms is that the number of neutrons is not rigidly fixed for a given element as is the proton count. Variation in the number of neutrons is the reason that the atomic weight reported is likely to be an approximation.

Isotopes of Elements

Atoms of an element, all of which have the same number of protons, Z, in their nuclei, but which differ in atomic weight (by virtue of having different numbers of neutrons, N, in the nuclei) are called *isotopes* of that element. Approximately 25 of the elements have only a single atomic weight, that is, Z and N are each firmly fixed at a particular number. The remaining elements have from two to 11 isotopes. The variation in the number of neutrons, N, in the nuclei of these isotopes can be more than 10 percent of the total present.

For reasons not yet understood, elements of an odd atomic number have only one or two kinds of atoms or isotopes. Elements with an even atomic number have several isotopes, and in some cases may have as many as 11. Manganese, for example, with the odd atomic number of 25 has only a single kind of atom with the atomic weight of 55; whereas iron (atomic number 26) has ten isotopes, four are stable and six are unstable, or radioactive. Most textbook tables list iron as having an atomic weight of

55.85. Yet, the ten isotopes of iron have atomic mass numbers ranging from 52 to 61, so it must be understood that the value of 55.85 is the average weight of all ten isotopes weighed according to their relative natural abundance.

While 109 distinctive atoms or elements have been discovered, more than 350 isotopes have been identified as stemming from these elements, differentiated by variations in neutron number, N, in their nuclei. Tables are readily available in chemical handbooks listing all known isotopes of the elements and their natural occurrence or abundance, lifetime, and real atomic mass.

Small differences in chemical and physical properties can be observed between isotopes of a given element, although for most purposes, isotopes are considered to represent the same element. However, some of these differences can be significant. Melting points and boiling points can differ by several degrees, and diffusion rates for gaseous elements can differ by small but measurable amounts among the isotopes. The cryogenic temperature at which certain metals become electrically superconductive depends on isotopic composition. For reasons related to welding metallurgy, perhaps the most important difference among the isotopes is the occurrence of radioactive isotopes. Radioactivity may present a problem, or it may provide a useful tool. For example, the element cobalt has a highly radioactive isotope, cobalt-60, and serves as an excellent source for radiographic examination of weldments.

Isobars

Because of the variations in the mass of atoms (A) as related to neutron number (N), it is possible to have atoms of different elements with the same atomic weight. These atoms are referred to as *isobars*. Even though they have the same mass (A), isobars are likely to be quite dissimilar in properties, because it is the number of protons in the nucleus (Z) that establishes chemical individuality. Therefore, a comparison of two atoms that are isobars will show that they have the same number of nucleons (Z + N), but one will have a greater number of protons than the other.

Atomic Weight

The practice of reporting the atomic weight of an element as a simplified, or average, number is widespread, and is sometimes carried further by expressing the weight of an element in kilograms. This value is obtained by multiplying the atomic weight by 1.66×10^{-27}, which is the unit atomic mass of the hydrogen atom. These average values serve well for many computations in chemistry and material handling, particularly if the element is in a form that does not exclude any of its naturally occurring isotopes.

Atomic Mass

A more precise way to designate a particular kind of atom, and its weight or mass, consists of writing the symbol for the element with the atomic number (Z) as a pre-subscript, and the atomic mass (A) as a post-superscript. This notational system was applied when the zinc atom, illustrated schematically in Figure 2.1, was captioned as $_{30}Zn^{65}$. Although the terms atomic weight and atomic mass essentially are synonymous, atomic mass is a more precise description of the property of interest. Technical Brief 4, which presents background information on hydrogen, uses this exacting symbolism for the isotopes of hydrogen. As further demonstration of this symbol system

> **TECHNICAL BRIEF 4:**
> **Hydrogen - The Building Block of the Elements**
>
> Hydrogen has the smallest mass of all elements, and its atom is constructed with the simplest possible arrangement of elementary particles. The nucleus consists of a lone proton. A single electron orbiting the nucleus provides electrostatic neutrality ($_1H^1$). Despite its odd atomic number, hydrogen has two additional isotopes, and they are so important that each has acquired a name. Deuterium, $_1H^2$, has an atomic mass of approximately two because a neutron is included in its nucleus along with the proton. Naturally, a single orbiting electron is sufficient to maintain electrostatic neutrality. The special nucleus of $_1H^2$ is called a deuteron. Deuterium atoms are a natural constituent of water, where they are present to a small extent, usually only about 0.015 percent. Deuterium can be separated from ordinary hydrogen. When two deuterium atoms are combined with an oxygen atom to form water (H_2O), the product is known as heavy water, a vital component of certain systems for nuclear power generation.
>
> The second important isotope of hydrogen, tritium, contains the usual single proton in its nucleus, but also has two neutrons present ($_1H^3$). Tritium also occurs in nature to a very small extent, but can be manufactured by a special process. This isotope of hydrogen is highly radioactive and is a key element in certain kinds of weaponry.
>
> From this overall perspective of hydrogen and its two isotopes, it is obvious that atoms are constructed in orderly fashion using three elementary particles – the proton, the neutron, and the electron. This theory of atom construction can be applied to all of the elements and has enabled man to make some very astute observations about the fundamental nature of matter. This simple concept of atoms can assist in gaining a keen insight into chemical reactions among flux and slag constituents, the mechanics of alloying metals, and many other facets of welding metallurgy.

and its helpfulness, iron will be examined for isotopes, since this element is at the heart of this text. The most abundant isotope of iron is $_{26}Fe^{56}$. This notation indicates that the particular atom is 26th in the numerical progression of elements, the nucleus contains 26 protons and 30 neutrons, and the atomic mass expressed as the whole number nearest to the actual mass is 56. When a general reference is made to the atomic weight of iron, the value ordinarily includes all of the naturally occurring isotopes in their usual percentages. When there is reason to single out a specific isotopic form of iron atom, this symboling system, based on the international standard carbon atom $_6C^{12}$, allows a more a more precise atomic mass to be indicated. As a final example, $_{26}Fe^{61}$ is an unusual isotope of iron inasmuch as it is highly radioactive.

Atom Valency

The atomic number of an element determines the number of orbital shells normally filled, or partly filled by electrons. The electrons on the outermost shell of any atom are called the valence electrons, and are of great importance in determining chemical properties of the atom. Elements that act to release electrons from an incomplete outermost shell and thus attempt to form a stable, or completely filled, outer shell are considered electropositive; those elements that accept electrons for the purpose of achieving a properly filled outer shell are regarded as electronegative. Metals are electropositive elements because they possess only a few valency electrons, and it is easier for them to lose electrons than to gain a sufficient number to form a completely filled, stable outer shell. This electron trait is often regarded as the distinguishing feature of a metal. However, distribution of electrons on the orbital shells must also be considered in terms of the energy levels of the electrons. This is a complex matter that requires quantum mechanics analysis to provide a more complete explanation. As applied to welding metallurgy, it is sufficient to note that distribution of electrons on the shells surrounding the nucleus of the atom, along with the electron-to-atom ratio, helps to do the following:

(1) It indicates the probable chemical behavior.

(2) It provides insight into important physical properties.

(3) It identifies elements that are metallic.

(4) It indicates a tendency to alloy and to form intermetallic compounds with other elements.

Ionization

A single isolated atom normally exists as an electrically neutral structure; that is, the electrostatic positive charge attributable to the protons in its nucleus is balanced by the negative charge of the electrons in its shells. To alter atom construction, it is possible to remove electrons and to add extra ones, and this process of changing the electrostatic balance is called ionization. Removal of electrons leaves the atom with a positive charge, and it is then regarded as a positive ion. The addition of electrons to the normal atom produces a negative ion. The atom's readiness to become ionized is a characteristic of each element, and the conditions or energy required to attain this state vary considerably among the elements.

The potential for ionization (expressed as electron volts, eV) of a number of gaseous elements, and elements in the vapor state are shown in Table 2.2. Note that helium, because its only outer K-shell is filled perfectly with two electrons, requires a very large amount of energy, 24.9 eV, for ionization. In contrast, cesium, with the atomic number 55, requires only 4.0 eV to remove a lone electron from its outermost P-shell. Ionization potential is an important characteristic of the materials involved in arc welding because the complex atmosphere in which the arc forms consists of shielding gases, metals in the vapor phase, and elements used as fluxing agents. The readiness with which these materials become ionized determines the ease of arc initiation, the capacity of the arc to carry electrical current, and the stability of the arc.

The shells formed by electrons orbiting the nucleus are the chemically and optically active part of an atom. The introduction of additional energy to an electron in orbit can alter this orbital system. Under some circumstances, energy received by an electron on an inner shell (i.e., one other than the outer valency shell), may cause the electron to be dislodged and to jump outward to find an available vacancy in one of the outer shells. When an electron is forced from its ground state into an outer orbit, this atom is said to be excited (not the ionized state). This is an unstable position for the electron and the positive attraction of the nucleus will attempt to return the orbital electron system to normalcy by pulling the wayward electron into its regular shell location in the ground state. This corrective action may occur in one or more inward jumps, but each change that the electron makes when moving to a smaller shell involves release of energy in the form of photon quanta.

Radioactivity

Radioactivity, the spontaneous disintegration of the nucleus of the atom, is an important phenomenon related to atom construction. The inclination to undergo this disintegration varies among the elements; in fact, different isotopes of a given element can show great differences in radioactivity.

Radioactive disintegration is a very complex mechanism which results in transmutation of the atom into another kind of element. During disintegration, the atom may pass through several stages, and in the process can emit alpha particles, beta rays, and gamma radiation. Alpha particles have been described as helium atoms that have been stripped of their two orbital electrons, causing them to behave as doubly ionized helium nuclei carrying a positive charge because of the two protons contained therein. Beta rays are the streams of electrons ejected from the disintegrating atoms. Gamma rays are electromagnetic waves, or photons, representing another form of energy given up by the atom. Gamma rays have a wave length somewhat shorter than x-rays, and they carry no electrical charge. For these reasons, gamma rays have a much greater ability to penetrate matter than alpha or beta rays.

The rate of disintegration of an atom exhibiting radioactivity is measured by its half-life, which is the time required for half of a given quantity of that element to disintegrate into a new element. The half-life of the various radioactive elements or their isotopes ranges from an extremely short time (milliseconds) to very, very long periods measured in years. The intensity of radiation emitted by the disintegrating atom depends on the particular elemental isotope and the rate of decay.

In welding metallurgy the phenomenon of radioactivity can be a problem or can be very useful. As examples, the embrittlement of metals subjected to irradiation in weldments used in nuclear power generation is a major engineering consideration

Table 2.2 - Ionization Potential of Gases and Vapors*

Element	Symbol	eV	Element	Symbol	eV
Argon	A	15.7	Iron	Fe	7.9
Aluminum	Al	6.0	Lanthanum	La	5.6
Barium	Ba	5.2	Lithium	Li	5.4
Boron	B	8.3	Magnesium	Mg	7.6
Carbon	C	11.3	Manganese	Mn	7.4
Carbon Dioxide	CO_2	14.4	Molybdenum	Mo	7.4
Carbon Monoxide	CO	14.1	Nickel	Ni	7.6
Calcium	Ca	6.1	Nitrogen	N	14.5
Cerium	Ce	7.0	Oxygen	O	13.6
Cesium	Cs	4.0	Phosphorus	P	10.6
Chlorine	Cl	13.0	Potassium	K	4.3
Chromium	Cr	6.8	Silicon	Si	8.1
Cobalt	Co	7.9	Sodium	Na	5.1
Copper	Cu	7.7	Sulfur	S	10.4
Helium	He	24.9	Titanium	Ti	6.8
Hydrogen	H	15.6	Tungsten	W	8.1

*Total energy in electron-volts required to extract the least tightly bound electron from the valency shell of the atom or molecule and place it at rest at an infinite distance.

which requires costly monitoring, while the use of radioactive isotopes in the radiographic examination of welds is an effective tool.

Atom Size or Diameter

Since atoms are constructed from the elementary particles, it might be expected that the size or diameter of the individual atom could be determined simply by atomic mass, the number of particles involved. Since the atomic number and atomic weight fix the number of protons in the nucleus and the orbital electrons required to establish electrostatic neutrality, it would appear that atomic mass should directly dictate the atom's overall diameter. This is not quite the case, and this aspect of atom construction deserves examination because atom size is an important factor in alloying metals, and in determining the distribution of a particular element in the structure of an alloy.

Because of the difficulty in actually "seeing" an atom to measure its diameter, the assumption has been made that in a solid the atoms are virtually touching each other; at least it has been demonstrated that this must be the case in certain kinds of solids. By using x-ray diffraction methods, the center-to-center distance of atoms in a solid can be measured to deduce the diameter of the atoms involved.

Unfortunately there is no straightforward relationship between apparent atom diameter and atomic mass. There is less variation in sizes or diameters of different atoms than in their masses. Despite atomic weight ranging from approximately one (for hydrogen) to more than 200, all of the elements appear to have atom diameters of approximately one to six angstroms.

Heavier elements do not necessarily develop larger diameters with their given mass than the lighter elements. For example, the largest atoms, in terms of diameter/mass ratio, are those of the alkali metals. As a specific example, sodium, with an atomic weight of 23, has an atom diameter of 3.72 angstroms. In contrast, silver, with an atomic weight of 108 (almost five times the atomic mass of sodium), has a smaller atom diameter of 3.04 angstroms.

The forces that bind the electron shells around the nucleus appear to hold the explanation for this inconsistent relationship between atomic mass and atom size. As atomic mass increases and additional protons appear in the nucleus and electrons are placed on the orbital shells, the nuclear charge grows stronger and this tends to pull the shells closer to the nucleus. However, a good understanding has yet to be gained of the mechanism that controls the shape of the shells, and the distances between the shells and the nucleus.

The matter of atom size or diameter becomes even more complex when a departure is made from an aggregate of the same atoms and x-ray diffraction measurements are made of atoms in a mixture of elements, or in a chemical compound. It appears that the nuclear charge of an atom and its influence on the radii of the electron shells is not determined entirely within the spherical boundary of the atom. The size of the atom also depends upon whether it is free, in an aggregate of atoms, in an alloy with other kinds of atoms, or in a chemical compound. Table 2.3 lists estimated atom diameters for a number of elements, and gives a general perspective of this basic dimension.

THE ELEMENTS

To date, 109 kinds of atoms or elements have been isolated through discovery or by nuclear transmutation. An alphabetical list of these elements can be found in Appendix IX. Approximately three-quarters of the elements are metals; at least, they are metallic as defined by the chemist. Some of the metals can appear to be non-metals, for example, tin at a temperature below 13°C (56°F). Some non-metals may have a metallic appearance, for example, silicon.

A very informative way to list the elements is to arrange them in order of their atomic number, starting with hydrogen, which has an atomic number of 1, and ordinarily has an atomic weight of essentially one (1.008). The second element listed in this arrangement would be helium with its atomic number 2, and an atomic weight of 4.003. The list would continue with elements of increasing atomic number, next showing lithium, then beryllium, and so on. However, on encountering neon (atomic number 10), which also is an inert gas, the element neon would be placed under helium to start forming a column. As this list continued, argon, another inert gas, with an atomic number of 18, would be listed below neon in the vertical column.

This method of tabulation was first suggested by Dimitri I. Mendeleyev in 1869, who pointed out that a remarkable periodic progression could be seen in the properties of the elements when arranged in this manner. Mendeleyev's Periodic Table has been used by scientists and adapted through the years to develop tables or systems that pay greater attention to the atomic number of the elements, and to the disposition of electrons on the shells surrounding the nucleus of the atom. His original periodic table was instrumental in the discovery of a number of elements, because at the time of its introduction man had not yet discovered such elements as erbium, hafnium and ruthenium. The periodic scheme not only revealed where elements of a particular atomic number and nature were awaiting discovery, but it also indicated that some serious errors had been made in determinations of the atomic weight of certain elements.

Table 2.3 - Diameters of Atoms*

Element	Atomic Weight	Atom Diameter, Å	Element	Atomic Weight	Atom Diameter, Å
Aluminum	26.98	2.52	Nickel	58.71	2.49
Beryllium	9.01	1.96	Niobium	92.91	2.85
Calcium	40.08	3.93	Nitrogen	14.01	1.40
Carbon	12.01	1.54	Oxygen	15.99	1.32
Cesium	132.91	5.24	Phosphorus	30.97	2.20
Chromium	51.99	2.50	Plutonium	239.00	6.00
Copper	63.54	2.70	Selenium	78.96	2.28
Hydrogen	1.01	1.00	Silicon	28.09	2.35
Iron	55.85	2.58	Silver	107.87	3.04
Lead	207.19	3.48	Sodium	22.99	3.72
Lithium	6.94	3.03	Sulfur	32.06	2.08
Magnesium	24.31	2.80	Titanium	47.90	2.89
Manganese	54.94	2.24	Tungsten	183.85	2.73
Mercury	200.59	2.96	Zinc	65.37	2.62
Molybdenum	95.94	2.72	Zirconium	91.22	3.16

*Determined by measuring the center-to-center spacing between nearest atoms in an aggregate of the pure elemental material by means of x-ray diffraction.

Table 2.4 is one of the columnar arrangements of Mendeleyev's Periodic Table, illustrating the remarkable periodic change in the properties of elements with increasing atomic number. Hydrogen precedes the columns because its unique nuclear structure (single proton) and its lone electron appears to serve as the basic building block for all other elements. The names of the elements and their properties of general importance are given in Appendix IX.

A study of Table 2.4 provides a basic understanding of the elemental matter encountered in welding. It can also provide an appreciation of the problems that scientists face in extending knowledge of the elements and in accomplishing transmutation.

The electron shells or sub-orbits of the *inert gases* in Column 1 of Table 2.4 are completely filled in a manner that prevents them from combining with other elements to form compounds. Disposition of the electrons in the orbits of each kind of atom is given in a table of electron configuration in Appendix X. The inert gases undergo liquefaction only at extremely low temperatures.

Elements listed in Column 2, the *alkali metals*, all have a single electron in their outermost shell, and this electron disposition induces destructive chemical characteristics. These metals are electropositive, and are quite soft. They are very active chemically, and therefore they corrode easily.

The *alkaline earth metals* in Column 3 have two electrons in their outermost shell. They display less tendency to corrode and are slightly harder than those metals in the preceding column. The *rare earth-like metals* in Column 4 are a little less susceptible to corrosion and are a little harder.

The elements in Columns 5 through 18, the *rare earth metals*, oxidize very easily, and possess moderate hardness. All elements of this group are similar in chemical behavior, which causes considerable difficulty in separating one from another. Some published tables of the elements arranged in a periodic system set apart these elements (having atomic numbers 58 through 71) as a separate series and usually identify them as the *lanthanides*. Also, the heavier elements in the same Columns numbered 5 through 18, but having atomic numbers 90 through 103, are sometimes designated as the *actinides* and are paired with the lanthanides.

Carrying this sorting technique a step further, the elements with atomic numbers 104, 105, and 106 are designated as *transactinides*.

Rutherfordium (104) has been placed in Column 19 of Table 2.4 because its chemical properties were found to be very similar to those of hafnium (72), which is listed immediately above. No adequate chemical studies have been reported for hahnium (105), or for element 106, since these are extremely short lived through radioactive decay.

Elements heavier than uranium (92) have not been found in nature in significant quantity. Yet, scientists have had some success in devising ways of increasing the number of elementary particles in the atomic structure of existing elements to produce new elements. For example, neptunium (93) was first synthesized in 1940 by bombarding uranium with deuterons. Element 106 was produced in 1974 by bombarding californium (98) with an oxygen isotope. In 1976, element 107 was synthesized by bombarding bismuth (83), in the form of one of its isotopes having an atomic weight of 204, with heavy nuclei of chromium (24) having an atomic weight of 54. However, only traces of element 107 were produced on a spinning cylinder and they existed for only a couple thousandths of a second.

Efforts to produce element 108 have been unsuccessful to date. In 1982, element 109 was synthesized by bombarding bismuth (83) with the nuclei of iron atoms (26) having an atomic weight of 54 (i.e., an isobar of chromium). A week of bombardment was required to produce a single fused nucleus of element 109. This new nucleon existed for only five msec and then began to decay by emission of an alpha particle to element 107. This transmutation was followed immediately by emission of another alpha particle to become hahnium (105), but then an electron was captured in the nucleus to convert a proton into a neutron, and this action formed rutherfordium (104). As this research on man-made elements continues, attention is directed to the production of element 116 by bombarding an isotope of cerium (58) with the nuclei of an isotope of calcium (20). Element 116 would be expected to undergo a down-stepping decay through a series of new nuclides.

In general, the very heavy elements that are produced synthetically tend to be very unstable and they undergo transmutation reactions to produce other atoms of lesser atomic number. The half-life of the synthetically produced atoms can be extremely short, but exceptional cases are anticipated. For example, calculations indicate that if element 110 can be successfully produced as a homolog of platinum (78), this new element would have a half-life as long as 100 million years. Research on synthesizing new elements is very important to extending knowledge of the structure of matter, and there is speculation that elements as heavy as atomic number 127 can probably be synthesized. It is impossible, of course, to predict the properties and usefulness of these new elements.

Elements in Columns 19 through 25 are called the *transition metals*. These have been a great boon to mankind because of their plentiful supply and their suitability for construction purposes. Rhenium, in Column 22, is a strong, ductile metal. It is quite expensive because of scarcity. Nevertheless, rhenium is used commercially as an addition to tungsten and molybdenum where the so-called "rhenium-effect" markedly improves strength, toughness, plasticity, and weldability. Column 26 contains the *noble metals*, but only copper, silver and gold are listed in this column. The metals palladium (46) and platinum (78), listed in Column 25, are sometimes regarded as noble metals. In Column 27, zinc (30) and cadmium (48) are soft, and these two elements have somewhat similar chemical properties. Mercury (80) is unique because it is a metal that is liquid at room temperature. The elements in Column 28 are soft, with the exception of boron, which is hard. The elements in Columns 29 through 32, with some exceptions (nitrogen, oxygen, lead and tin), are either semiconductors or semi-metals.

Note that the periodic arrangement in Table 2.4 shows that most of the electropositive elements are on the left side of the table, with electronegative elements on the right. The significance of this observation will become clear when bonding between atoms and formation of compounds between unlike atoms are explored.

AGGREGATES OF ATOMS

When a large quantity of atoms of the same element are brought together, the general nature of this aggregate is determined primarily by the interatomic forces characteristic of that element, but the exact state of the aggregate also depends on prevailing

Table 2.4 — The Elements

The Elements — Grouped in vertical columns according to similarity of properties and increasing atomic weight.

	1	2	3	4	5	6	7	8	9	10	11	12	13	14	15	16
	Inert Gases	Alkali Metals	Alkaline Earths	Rare Earth-like	←——————————— Rare Earth Metals ———————————→											
1	**H** 1.008 g															
	2 **He** 4.003 g	3 **Li** 6.940 m bcc	4 **Be** 9.01 m hcp													
	10 **Ne** 20.18 g	11 **Na** 22.99 m bcc	12 **Mg** 24.31 m hcp													
	18 **Ar** 39.95 g	19 **K** 39.10 m bcc	20 **Ca** 40.08 m fcc	21 **Sc** 44.96 m hcp												
	36 **Kr** 83.80 g	37 **Rb** 85.47 m bcc	38 **Sr** 87.62 m fcc	39 **Y** 88.91 m hcp												
	54 **Xe** 131.30 g	55 **Cs** 132.91 m bcc	56 **Ba** 137.34 m bcc	57 **La** 138.91 m hcp	58 **Ce** 140.12 m fcc	59 **Pr** 140.91 m hcp	60 **Nd** 144.24 m hcp	61 **Pm** 145 m ?	62 **Sm** 150.35 m hcp	63 **Eu** 151.96 m bcc	64 **Gd** 157.25 m hcp	65 **Tb** 158.92 m hcp	66 **Dy** 162.50 m hcp	67 **Ho** 164.93 m hcp	68 **Er** 167.26 m hcp	69 **Tm** 168.93 m hcp
	86 **Rn** 222 g	87 **Fr** 223 m bcc	88 **Ra** 226.05 m bcc	89 **Ac** 227 m fcc	90 **Th** 232.04 m fcc	91 **Pa** 231 m ⊗	92 **U** 238.03 m ⊗	93 **Np** (237) m ⊗	94 **Pu** (239) m ⊗	95 **Am** (243) m ⊗	96 **Cm** (245) m ⊗	97 **Bk** (249) m ⊗	98 **Cf** (252) m ⊗	99 **Es** (254) m ⊗	100 **Fm** (257) m ⊗	101 **Md** (258) m ⊗

KEY
- Atomic Number
- Element Symbol
- Atomic Weight*
- State at RT**
- Crystalline Form***

* Values in parentheses represent atomic weight of most stable isotope of that element if element does not occur naturally.

** State at Room Temperature
- g – Gas
- m – Metal
- l – Liquid
- s – Semiconductor
- sm – Semimetal

*** Crystalline Form
- bcc – Body-Centered Cubic
- fcc – Face-Centered Cubic
- hcp – Hexagonal Close-Packed
- dia – Diamond
- tet – Tetragonal
- ⊗ – Crystallographic Form Not Simple

Table 2.4 — The Elements (continued)

17	18	19	20	21	22	23	24	25	26	27	28	29	30	31	32
Rare Earth Metals →		←—————————— Transition Metals ——————————→							Noble Metals	Group IIb	Group III Elements	Semiconductors (Except Pb)	Semimetals (Except N and O)		Halogens
															9 **F** 18.99 g
											5 **B** 10.81 ⊗ m	6 **C** 12.01 s dia	7 **N** 14.01 g	8 **O** 15.99 g	17 **Cl** 35.45 g
											13 **Al** 26.98 m fcc	14 **Si** 28.09 s dia	15 **P** 30.97 ⊗ sm	16 **S** 32.06 ⊗ sm	
		22 **Ti** 47.90 m hcp	23 **V** 50.94 m bcc	24 **Cr** 51.99 m bcc	25 **Mn** 54.94 ⊗ m	26 **Fe** 55.85 m bcc	27 **Co** 58.93 m hcp	28 **Ni** 58.71 m fcc	29 **Cu** 63.54 m fcc	30 **Zn** 65.37 m hcp	31 **Ga** 69.72 ⊗ m	32 **Ge** 72.59 s dia	33 **As** 74.92 ⊗ sm	34 **Se** 78.96 ⊗ sm	35 **Br** 79.91 g
		40 **Zr** 91.22 m hcp	41 **Nb** 92.91 m bcc	42 **Mo** 95.94 m bcc	43 **Tc** 99 m hcp	44 **Ru** 101.07 m hcp	45 **Rh** 102.91 m fcc	46 **Pd** 106.40 m fcc	47 **Ag** 107.87 m fcc	48 **Cd** 112.40 m hcp	49 **In** 114.82 ⊗ m	50 **Sn** 118.69 m tet	51 **Sb** 121.75 ⊗ sm	52 **Te** 127.60 ⊗ sm	53 **I** 126.90 ⊗
70 **Yb** 173.04 m fcc	71 **Lu** 174.97 m hcp	72 **Hf** 178.49 m hcp	73 **Ta** 180.95 m bcc	74 **W** 183.85 m bcc	75 **Re** 186.20 m hcp	76 **Os** 190.20 m hcp	77 **Ir** 192.20 m fcc	78 **Pt** 195.09 m fcc	79 **Au** 196.98 m fcc	80 **Hg** 200.59 lm ⊗	81 **Tl** 204.37 m hcp	82 **Pb** 207.19 m fcc	83 **Bi** 208.98 ⊗ sm	84 **Po** 210 ⊗ sm	85 **At** 211 s
102 **No** (259)	103 **Lr** (262)	104 **Rf** (261)	105 **Ha** (262)	106 ⊘ (263)	107 ⊘ (262)		109 ⊘							116 ⊘	?

⊘ Element symbol not yet established.

teristic of that element, but the exact state of the aggregate also depends on prevailing temperature and pressure.

As two atoms of the same element are brought together, starting from remote positions, they experience virtually no attractive force until they are within a few angstroms of each other. At this distance, the attractive force becomes stronger, but as closer approach takes place, repulsive force develops. This repulsion is considered to be a short range force because it is weak at large distances and becomes stronger at small distances. Eventually, at a characteristic distance for the particular kind of atoms, the attractive and repulsive forces become equal and this establishes the equilibrium spacing.

The nature of the interatomic forces and resulting spacing determines approximately whether the aggregate of atoms is a gas, a liquid, or a solid; the latter two (because of their much closer interatomic spacing) are regarded as condensed states. Although materials are commonly classified as gases, liquids and solids, actual classification depends on the physical condition of a substance at the time of observation.

It is a convenient practice to classify substances by what is found at room temperature and the pressure of one atmosphere. Under these familiar conditions, as observed from the Table of Elements in Appendix IX, only two of the 109 elements are indicated to be liquids (bromine and mercury). Eleven of the elements are shown as gases, and the remaining elements are listed as solids. However, these are the states of aggregates of the elements merely when examined under ambient conditions. If aggregates of the elements were examined at temperatures ranging from absolute zero (-273°C, or -460°F) to very high temperature, perhaps in the order of six thousand degrees Celsius (about 11 000° Fahrenheit), all elements would be found to be capable of existing in any one of the three stable states at certain levels of temperature and pressure. Obviously, the state assumed by an aggregate depends on the kind of atoms and prevailing temperature and pressure. Table 2.4, the periodic table of elements, shows that the properties of elements appear in recurring themes, and this has helped scientists anticipate how an element of a particular atomic number is likely to appear as an aggregate. Additional research is needed, however, to understand the relationship between elementary particles that make up atom construction and the final properties of a substance in aggregate form.

Since the terms gas, liquid, and solid are sometimes used rather loosely, these preliminary observations will help identify the states of aggregates of the elements. If a body maintains a definite size and shape, it is apt to be considered a solid. A liquid also has a definite size because it will fill a container to a particular level and occupy a specific space or volume, but a liquid does not have a definite shape of its own. A gas has neither a definite shape nor a definite size or volume, but a gas will fill any container no matter how small a quantity of atoms of the gas are put inside. Another way of distinguishing gases and liquids from a solid is to consider the property of fluidity in gases and liquids, and to take note of the innate capability of a solid to return to its original shape after forces that have been imposed upon it are removed. Consequently, a true solid is expected to be elastic to some degree.

In the construction of an atom, whether it is natural or man-made, the electrons present in the structure that maintain an electrostatic balance swarm in clouds around the nucleus in a behavioral pattern that is by no means random. As each electron is added to an orbital pattern, it interacts with all others present to establish a system which controls momentum and energy of these negatively-charged particles and holds them in orbit. When numbers of a particular kind of atom are organized into a condensed body, either solid or liquid, the electrons "sense" the system in which they will operate and arrange their disposition and behavior in a way that is most energy efficient. The overall system will determine whether the aggregate is a gas, a liquid, a solid, a metal, a semiconductor, or an insulator.

Atoms in an aggregate also have motion and movement as individual bodies. Each atom possesses kinetic energy which is manifested as vibratory motion. The vigor of this motion is temperature dependent. As temperature of the aggregate is raised, the atoms gain thermal energy and their vibratory motion increases. With increased temperature, the vibrational frequency of each atom remains nearly constant, but the atom oscillates with greater amplitude within an instantaneous spacing. Also, some atoms will jump from one position to another. Further increase in temperature causes larger numbers of atoms to jump from one position to another, and the range of these jumps, or excursions, becomes greater.

If the temperature of the aggregate were lowered to absolute zero, the substance would be devoid of all thermal energy and the atoms be would be expected to remain locked in position and motionless. This is not the case, because all solids appear to have an inherent residual energy of vibration. It is extremely difficult to cool a substance to absolute zero and remove all residual energy. Some metals that have been refrigerated to a temperature as low as 0.05 °K continue to display minute effects of retained kinetic energy.

Atoms can be gathered together as an aggregate through a number of different processes, including condensation, pressurization, chemical reaction, electrodeposition, and melting. The process usually determines, at least initially, whether the aggregate will be a gas, a liquid, or a solid. The state of an aggregate may change as its temperature or the ambient pressure is altered. The three states in which an aggregate can exist will be examined to note the general behavior of atoms in each state, and to explore the mechanisms by which aggregates of atoms change from one state to another.

The Solid State

Melting is the process most often used to form an aggregate of atoms. When the temperature of a melt is lowered to permit a particular kind of atom to form a solid, the liquid will form either a crystalline solid or an amorphous solid. Welding technology involves both of these solid aggregates, so basic knowledge of these will help explain the behavior of welded metals, as well as the nonmetallic materials that are sometimes involved in welding processes.

The Crystalline Solids

When a liquid crystallizes as a result of cooling below T_m (melting temperature), the solid that forms has an equilibrium spacing between its atoms that is characteristic of the aggregate, and also has a specific form of atom positioning in space. The periodicity of atoms arranged in this manner represents a three-dimensional array. This structural aspect of the crystalline solid is the feature that allows diffraction of x-rays, permits positive identification of the crystalline form, and even allows dimensional measurement of structural details in the arrangement (e.g., center-to-center spacing of atoms). Crystallinity in an aggregate is a very important feature because it determines to a large extent the physical and mechanical properties of the solid. Metals will ordinarily solidify from the liquid state into a crystalline solid, and much of the behavior of metals and alloys during welding is related to their crystallographic forms. Yet, there can be notable exceptions to the usual inclination of metals to solidify as a crystalline solid, and *amorphous metals* are a reality.

When a liquid is cooled to form a crystalline solid, it should transform into the solid state as soon as it reaches its freezing point. This does not actually happen, because to get the solidification mechanism started the melt must be undercooled to a small extent below the freezing point. Undercooling is required to achieve a controlled difference in free energy between the newly formed solid and the remaining liquid. During solidification to a crystalline form, the viscosity (η), volume (V), and internal energy (H) change discontinuously because of the complex mechanics taking place as the solid forms in the liquid, and because of differences in the properties of the solid and the liquid phases that exist at given temperatures and times. Solidification normally proceeds by heterogeneous nucleation followed by competitive growth of the nuclei as the latent heat of solidification is dissipated.

The freezing of a pure substance as a crystalline solid is characterized by the constancy of temperature during the process. The constancy of the freezing point is so typical that change of freezing point is one of the most convenient tests for the presence of an impurity in the substance.

Some crystalline solids, including metals and alloys, undergo change in crystallographic form as their temperatures are changed. This change is termed *allotropic transformation*. A transformation of this kind that occurs on cooling will often be reversible on heating. The particular allotropic forms present in crystalline solids have a very strong tendency to allow the most stable crystalline form to exist for the prevailing circumstances, that is, the crystallographic arrangement should have the lowest free energy for the substance at the given temperature and under the ambient conditions. Very small energy differences can call for an allotropic transformation. Sometimes allotropic transformation will occur with a change in pressure, during mechanical deformation of the solid, or under the influence of magnetic forces.

Some elements have more than one crystalline form, and may even have a non-crystalline form, which is a property termed *polymorphism*. For example, at room temperature carbon can exist as graphite (a crystalline solid), as diamond (another form of crystalline solid), and as lampblack (a non-crystalline solid), all of which are solid forms of carbon.

To summarize, the "atom aggregate" story is an exciting one, showing that an atom is a lively bundle of energy with its elementary particles busily spinning and the whole atom vibrating vigorously as dictated by temperature level. Even when atoms reach a condensed state (i.e., liquid, or solid), they are still able to assume the atom-to-atom distance (equilibrium spacing) for the aggregate under the prevailing circumstances. Also, the atoms in a crystalline solid are able to take the fixed position-centers that form a three dimensional pattern in space. Even in the crystalline solid, however, the atoms continue to have some capability for jumping from one position to another, especially short-range excursions from one position to an immediately adjacent position where an atom happens to be missing.

When a crystalline solid is heated and reaches its melting temperature, and sufficient additional thermal energy is added, the vibrational forces break the interatomic forces that hold the atoms at equilibrium spacing and in crystalline array; consequently, the aggregate becomes a liquid. The additional energy required to accomplish this conversion from a solid to a liquid at the melting point is called the *latent heat of fusion*. Although the long-range order of crystallinity is destroyed as the solid changes to a liquid, a residue of local order persists in the liquid. This residue performs a nucleation role on re-solidification during cooling.

Amorphous Solids

A seemingly solid substance with its atoms held apart at an equilibrium spacing, but with no long-range periodicity in atom location in its structure is an *amorphous solid*. This form of solid is similar to a liquid with respect to atom location, and in certain ways it behaves like a highly viscous liquid. The term *undercooled liquid* is sometimes applied.

Ordinary glass is a classic example of an amorphous solid, and definition of the word *glass* connotes absence of crystallization in the material. Many other solid-appearing substances fall into this non-crystalline category, including pitch, wax, and various organic materials. However, the term "glass" is usually applied to relatively hard substances like isinglass, plexiglass, and metallic glass. These amorphous solids share a number of common traits. They tend to flow progressively under constantly applied stress. The rate of flow among these solids differs greatly, but ordinarily increases with higher temperature. In certain amorphous solids there are deviations from this typical behavior caused by stress level, loading rate, and total time under stress.

Amorphous solids can be produced in a number of different processes, but, from a welding standpoint, production by cooling a liquid yields the most useful information. Unlike the solidification of a crystalline solid, formation of an amorphous solid during cooling of a liquid takes place gradually as a homogeneous process over a range of temperatures. Amorphous solids do not exhibit a freezing or solidification point, and their viscosity (η), volume (V), and internal energy (H), change continuously in the liquid as temperature decreases. Eventually, the substance reaches a stage where the viscosity is so high that it appears to be solid. Many of the molten slags and fluxes used in welding become amorphous solids when cooled to room temperature.

Some substances that usually form crystalline solids on cooling from the liquid state can be undercooled in a manner that bypasses the crystallization process. The technique for avoiding crystallization during cooling of the liquid requires exceeding a critical quench rate for the particular substance. In fact, study of *rapid solidification technology*, RST, has revealed that nearly all liquids will form an amorphous solid if cooled to a sufficiently low temperature and at a rate sufficiently rapid to avoid crystallization. As a liquid cools to the amorphous solid state, there is a salient point on cooling that is designated the *glass transition temperature*, T_g. This temperature is defined by a sharp upward inflection in a plotted curve of rising specific heat for a given substance with increasing temperature. This transition temperature (T_g) is not a constant of the substance, but it is a function of conditions surrounding its determination. Above T_g the amorphous aggregate is considered a supercooled liquid, but below T_g it is considered a glass. T_g correlates with various changes in the properties of amorphous solids, and this is especially true of the metallic glasses.

Certain amorphous solids, when reheated just above their T_g, can develop a degree of short-range, highly ordered atomic structure, termed microcrystallinity. In this case, the aggregate does not possess the long-range, complete periodicity of a crystalline solid, but instead has micro-regions of short-range periodicity that measure from about 10 to 20 Å in diameter. The microcrystals exist without correlated orientation, and they are very difficult to detect and assess.

Another structural configuration that is believed to lie between that of an amorphous solid and one with microcrystallinity has been termed the continuous random model. Unambiguous differentiation between fully amorphous and microcrystalline atomic configurations is yet to be achieved. Precise assessment of atomic structure is a matter of considerable importance in the development of metallic glasses, which comprise a new kind of material for manufacturing.

The Liquid State

Atoms in a liquid are packed almost as close together as in a solid, but they do not locate themselves with the periodicity found in a crystalline solid. Liquids, as ordinarily viewed, yield to the slightest pressure because the atoms move freely among themselves; yet they maintain a fixed equilibrium spacing. Despite the fluidity of a liquid, the relatively close, fixed spacing allows very little compressibility. In a few instances, atom packing in a liquid can be closer than in the solid form (i.e., just below the solidification temperature) as evidenced by the expansion that occurs when water freezes and becomes ice.

Knowledge of the liquid state is somewhat lacking, because comparatively more research effort has been directed to solids and gases. In fact, the liquid state has been neglected as the intermediate form which matter assumes in changing from a solid to a gas. Solids are characterized by low kinetic energies relative to the powerful cohesive forces that bind atoms in a very orderly crystalline pattern. Gases, in contrast, have high kinetic energies relative to the weak forces that allow random particle distribution. A liquid, however, with its intermediate kinetic energies relative to cohesive forces, has neither the orderliness of a crystal, nor the randomness of a gas.

As the temperature of a liquid is lowered, the interatomic spacing becomes very slightly smaller and thermal agitation and vibration of individual atoms lessens, therefore, the viscosity of the liquid usually increases as the temperature drops. As temperature continues to decrease, the liquid may appear to have solidified, that is, the aggregate of atoms may seem to exhibit some of the physical characteristics of a solid. Yet, on closer examination and study of its properties, this aggregate can remain noncrystalline, showing it to be nothing more than an undercooled liquid. Reasons for this view are easily summarized. The undercooled liquid does not display a solidification point, or even a distinct range of solidification. During cooling, the viscosity of the liquid gradually increases until the aggregate seems to be solid. There is no marked change in the various properties of the aggregate at any particular temperature. Therefore, the difference between the liquid and the solid-like undercooled form is one of degree rather than kind.

It is well known that some liquids tend to evaporate, and many circumstances arise in welding that require an understanding of this phenomenon. Each substance possesses a critical temperature and pressure state above which there is only one fluid (non-solid) phase; that is, above this critical state of temperature and pressure there is actually no distinction between liquid and gas. At each temperature below the critical state, a pressure called vapor pressure exists; this is the point at which the two fluid phases are capable of coexistence. The more dense phase is designated as the liquid, and the less dense phase is the vapor, or gas. At pressures exceeding the vapor pressure, the liquid is stable. When a liquid reaches a temperature at which the vapor pressure is equal to the external pressure, it reaches its *boiling point*. Any additional heat supplied at this point acts to convert the substance from the dense phase (liquid) to the less-dense phase (gas). The energy required to accomplish this changed state of the aggregate of atoms or molecules is the *latent heat of vaporization*.

The Gaseous State

Gases consist of an aggregate in which the atoms (or molecules) are too energetic for their attractive forces to overcome. Consequently, the spacing between atoms in a gas is relatively large. In many instances, gas atoms bond with only one or two neighbors to form a molecule of particular gas.

Three familiar gases, oxygen, nitrogen and hydrogen normally exist in a molecular form at room temperature. Each molecule of these gases contains two atoms, expressed in the symbols for these gases, O_2, N_2, and H_2, respectively, when they are in stable molecular form. Molecular bonds between gas atoms are relatively weak, and there are various ways of dissociating molecular gas to bring about the atomic or nascent form.

When gases are involved in certain metallurgical situations in welding, it is important to know whether the gases are in the atomic form or in the molecular form, because there are notable differences in these two forms that affect the chemical activity and solubility in a liquid or a solid metal.

Although a still gas may appear to be motionless, its atoms or molecules are in constant motion from the kinetic energy that each holds. In addition to the vibratory motion of the atoms, the molecules which may be present go through rotational gyrations. The outward trajectory of these particles is stopped only when they collide with each other or with the confining walls of the gas container.

Kinetic theory suggests that the gas particles travel at a very high speed, perhaps in the order of 300 meters per second (1000 ft/s), but they travel only a minute distance before colliding with another particle. The impact of atoms or molecules on the confining surfaces of the gas container causes gas pressure. The motion of these particles becomes more energetic as the temperature is raised, and therefore a proportionate increase in pressure occurs (*Charles' law*).

When a gas is subjected to increasing pressure, the atoms or molecules are moved closer together, thus reducing the inter-particle spacing. At a constant temperature there is an inverse but proportionate relationship between the volume of a gas and its pressure

(*Boyle's law*). As the volume of the confining chamber is decreased, the pressure of the gas contained inside will increase. Of course, the energy put into this action to bring the bombarding particles closer together results in a rise in the temperature of the gas.

Change from the gaseous state to the liquid state is accomplished by condensing liquid from the gas by applying pressure or lowering the gas temperature, or both. However, each gas has a critical temperature above which no amount of pressure will accomplish liquefaction.

FUNDAMENTALS OF CRYSTALS

It has been established that solid aggregates of atoms often have a crystalline structure, and that this is commonly true of metals. Examination of crystallinity in the solid state can lead to a better understanding of the behavior of the metals and alloys handled in welding. This examination must address several basic questions: Why do atoms of a solid assume a crystalline array? What is a crystal? What holds the atoms so firmly together? Highly detailed answers to these questions can be found in the voluminous literature on solid state physics, but the welding engineer will concentrate on crystals of metals to find a useful rationale concerning metal structures and properties.

Why do atoms in a solid assume a crystalline array? Atoms forming a solid assume a crystalline array because the laws of thermodynamics dictate that the aggregate will represent the lowest energy-stable state for the ambient conditions. Crystals develop because the forces exerted by the aggregate on each new atom (or molecule) that joins the aggregate is a repetition of those forces acting on earlier atoms, and therefore each atom occupies a similar position in concourse with its neighbors. Although most elements and compounds always form the same array or crystalline pattern, a few may take on different forms. This versatility is explained by the fact that the particular crystalline forms for the aggregate all have virtually the same level of free energy, and therefore ambient conditions control the initiation, or nucleation, of a particular array, which then proceeds to completion. When a solid undergoes an allotropic transformation in crystalline form, again the change is in compliance with the energy difference between the two crystalline forms, even though the difference is small, and even though it is a change in intrinsic energy source.

What is a crystal? A widely accepted definition describes a crystal as a homogeneous body bounded by smooth plane surfaces that are an external expression of an orderly internal atomic arrangement. A crystal can also be described as a solid composed of atoms, ions, or molecules arranged in a repetitive, three dimensional pattern. A specific requirement of this solid is that it must be able to diffract x-rays into an orderly beam pattern. To exhibit this capability, the solid must have consistent spacing in the three-dimensional arrangement of particle repetition, therefore, the distances between all particles making up the unit cell must be characteristic of the crystals of the particular material. The simplest crystal can contain as few as four atoms, but if examination is extended to cover organic compounds, such as the protein crystals, many thousands of atoms are contained in the molecules that form these crystals.

Methodical procedures are required to analyze the structure of crystalline solids. In the case of metals, the factors to be considered are the geometry of the external crystal faces or planes, the positions of atoms at internal sites, and even the size and shape of interstices, although they may be a relatively small part of the crystal volume. The science of crystallography, as applied to metal crystals, is used extensively by the welding engineer.

To facilitate study of basic structures, several notation techniques have been developed for describing crystals. As a first step, lines can be drawn through centers of the atoms to develop a space lattice. The intersections of these lines then provide a set of imaginary points in space, which are the atom sites. Figure 2.2 (A) illustrates that the lattice is made up of cells which are square in shape when viewed on this single plane. If the smallest possible three-dimensional unit from this lattice is singled out, as illustrated in Figure 2.2 (B), a unit cell is isolated. Usually there is one kind of unit cell that is most easily visualized in any crystalline structure, and the unit cell serves as the repeatedly-used building block in locating atoms in the lattice, as the crystal increases in size by accumulating atoms in the aggregate.

Although the unit cell in Figure 2.2 (B) is drawn as a solid cube, the lines indicate only imaginary edges of the crystal cell to assist visualization. Atoms would be positioned at the eight corners of the cube. The imaginary six faces (as delineated by lines along the edges) are called planes. Of course, a crystal can be more complex than the simple cube or isometric form illustrated in Figure 2.2 (B). In addition to atoms located at the intersections of lines outlining the form of the cell, a crystal can have atoms located at particular sites within the cell, or even on the planes. The notation techniques of the crystallographer are used to identify both the external geometry of the crystal cell and the positions of every atom in the crystalline structure.

The crystallographer often draws lines to indicate the axes of a crystal's unit cell. An axis is a line drawn through the center of the crystal to serve as an axis of reference. For all kinds of crystals except one (hexagonal), three axes are drawn and identified, as shown in Figure 2.3 (A). One axis, a, is horizontal and perpendicular to our view; a second axis, b, is horizontal and parallel to our view; and a third axis, c, is vertical. The ends of the axes are identified by a plus or a minus sign, with the front end of a, the right end of b, and the upper end of c labeled with a plus sign.

The angles between the axes, as shown in Figure 2.3 (B), and given the identifications alpha (α), beta (β) and gamma (γ) respectively, may not necessarily be perpendicular, since this depends on the form of the crystal. However, this graphic system facilitates analysis of the geometry of crystal structure.

Sometimes an axis is drawn as a symmetry axis, meaning that the crystal can be revolved on that axis and will repeat itself in appearance two or more times during a complete revolution. A plane of symmetry can sometimes be drawn within the body of the crystal to divide it into halves that are mirror images of one another.

Reference axes can also be drawn along three edges of a unit cell to intersect at one corner, as illustrated in Figure 2.4. These axes serve as cartesian coordinates for any crystal structure, and any point in the unit cell can be designated by specifying the fraction of the length a, b, or c in the direction of the coordinates, with their origin located at the corner of the cell as shown.

Despite the multitude of crystal structures found in solids, there are only 14 types of lattices into which the unit cell of any solid can be fitted. These 14 types are called the *Bravais lattices*, and they are illustrated in Figure 2.5, complete with crystal edges drawn to outline their forms, and with atoms located at their normal sites. Where additional atoms are situated on planes, or within the unit cell, light diagonal lines have been drawn to indicate exact locations.

Another way of depicting forms of crystals involves analysis of the lattice parameters and interaxial angles. For example, the crystal in Figure 2.4 is shown with equal-length lattice parameters along three (edge) axes, with these axes mutually perpendicular. The character of a cell fitting this lattice can be written as $\alpha = \beta = \gamma = 90°$. This summary of the unit cell identifies it as plain, cubic or isometric in form. When the lattices presented in Figure 2.5 are analyzed in the same way, the 14 Bravais lattices can be categorized in seven crystal systems and their structural features expressed as a formula. Incidentally, the hexagonal lattice requires four axes for reference: three horizontal axes of equal length that intersect at an angle of 60 degrees, and a fourth axis which is perpendicular to the plane of the other three and which often is of different length.

Table 2.5 lists the seven crystal systems and their typifying features to indicate the respective parameters along their lattice axes, and the interaxial angles. Another feature of a crystal structure sometimes noted is the coordination number, which is the number of nearest neighboring atoms to any particular atom located in the space lattice. The coordination number for the body-centered cubic crystal structure is 8; while both the face-centered cubic and the hexagonal structures are 12.

Crystallographers strive to use the simplest possible concept of the various crystal structures and they sometimes rely on the coordination number for absolute ratio-

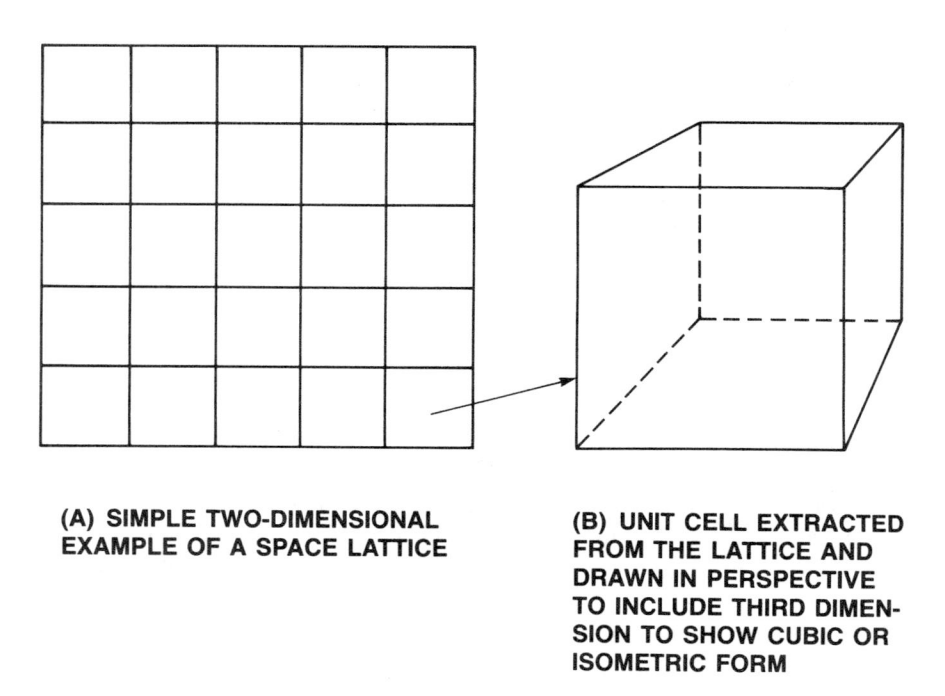

(A) SIMPLE TWO-DIMENSIONAL EXAMPLE OF A SPACE LATTICE

(B) UNIT CELL EXTRACTED FROM THE LATTICE AND DRAWN IN PERSPECTIVE TO INCLUDE THIRD DIMENSION TO SHOW CUBIC OR ISOMETRIC FORM

Figure 2.2 — Diagrammatic sketch of a crystal space lattice (A, left) and a single crystal (B, right)

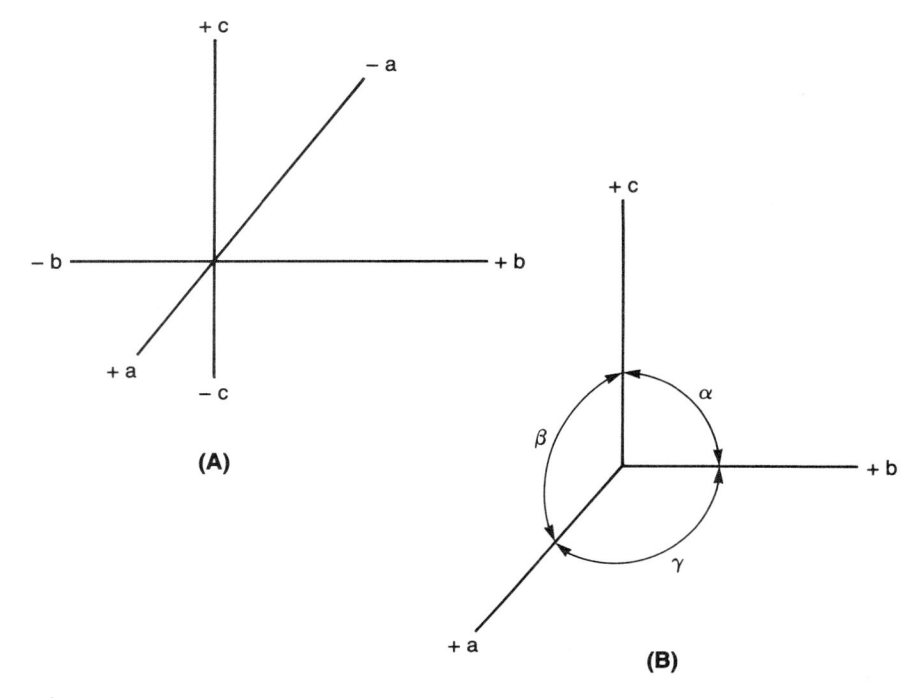

Figure 2.3 — Method of identifying crystal axes (A, left) and their relative angles (B, right) in any form of crystal; except those in the hexagonal system which use four axes

nalization. For example, if asked for the minimum number of atoms comprising the "cell" of the body-centered cubic structure, they would respond with the number two, because it is only the arrangement of the atom in the center of the cube and its relationship with an atom at any of the eight adjacent corners that is unique. All structural arrangement beyond this relationship is merely repetition.

Metallurgists, however, have found the crystal structures of metals more easily visualized by using the cell that best shows complete symmetry (as exemplified in Figure 2.5). Consequently, the metallurgist pictures the body-centered cubic crystal structure as having a structure cell of nine atoms; even though the eight corner atoms are shared by eight adjacent cubes. It will be helpful to be familiar with the geometric features of these systems and to note the variations produced by additional atoms that take positions on particular planes and within the cell. When attention is concentrated on metallic crystals, only a few of the systems deserve continued study because, in general, solid metals, especially iron and its alloys, adopt only a limited number of crystalline forms. Nevertheless, the form that exists is very important in establishing the properties of the metal or alloy.

The American Society for Testing and Materials (ASTM) provided a standard method of assigning designations for both crystal structure and its composition in metallic systems in ASTM E 157. This document suggests a two-part designation, with the first part consisting of chemical symbols and possibly formulae proportions, (in parentheses), and the second part consisting of an arabic numeral indicating the

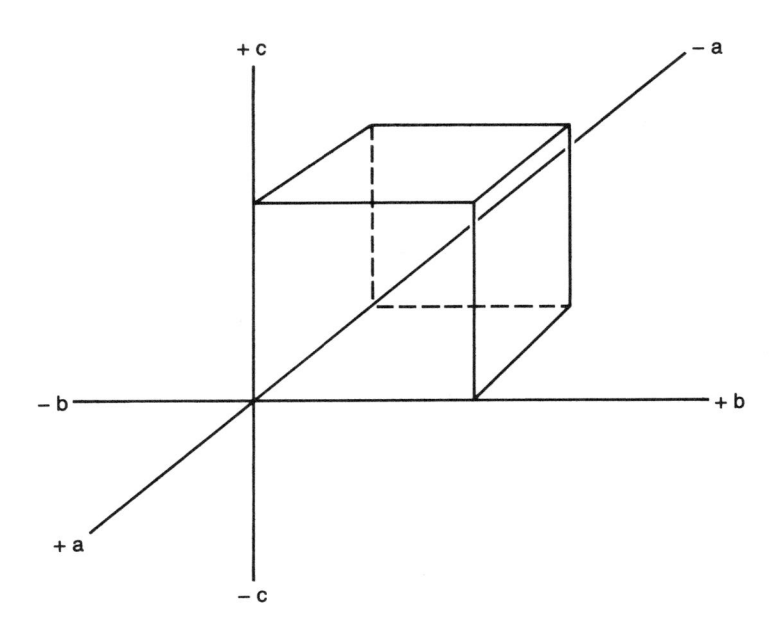

Figure 2.4 — Axes arranged to coincide with three intersecting crystal edges

When dealing with a cubic crystal, the three axes are mutually perpendicular and are ordinarily oriented to have the *a* axis horizontal and perpendicular to the observer; the *b* axis is horizontal and parallel to the observer; and the *c* axis is vertical. The ends of each axis is designated by either a plus or a minus sign, and ordinarily the positive sign is placed at the front-end of *a*, at the right-hand end of *b*, and at the upper end of *c*. The opposite ends of the axes, of course, are negative.

number of atoms in a unit cell, followed by a capital letter (in italic form) showing the Bravais lattice according to the following arrangement:

Letter	Bravais Lattice
C	Primitive cubic (plain)
B	Cubic, body-centered
F	Cubic, face-centered
T	Primitive tetragonal (plain)
U	Centered tetragonal (body-centered)
R	Rhombohedral (trigonal)
H	Hexagonal
O	Primitive orthorhombic (plain)
P	Body-centered orthorhombic
Q	Base-centered orthorhombic
S	Face-centered orthorhombic
M	Primitive monoclinic (plain)
N	Centered monoclinic (base-centered)
Z	Triclinic (plain)

An example of the ASTM method written for an aluminum-zinc alloy, is (Al,Zn) 4F. Although these alphabetical designations are not commonly used in the literature, the welding engineer should be familiar with them.

CUBIC, PLAIN   CUBIC, BODY-CENTERED   CUBIC, FACE-CENTERED

TETRAGONAL, PLAIN   TETRAGONAL, BODY-CENTERED

ORTHORHOMBIC, PLAIN   ORTHORHOMBIC, BASE-CENTERED   ORTHORHOMBIC, BODY-CENTERED   ORTHORHOMBIC, FACE-CENTERED

MONOCLINIC, PLAIN   MONOCLINIC, BASE-CENTERED   TRICLINIC, PLAIN

RHOMBOHEDRAL (TRIGONAL)   HEXAGONAL

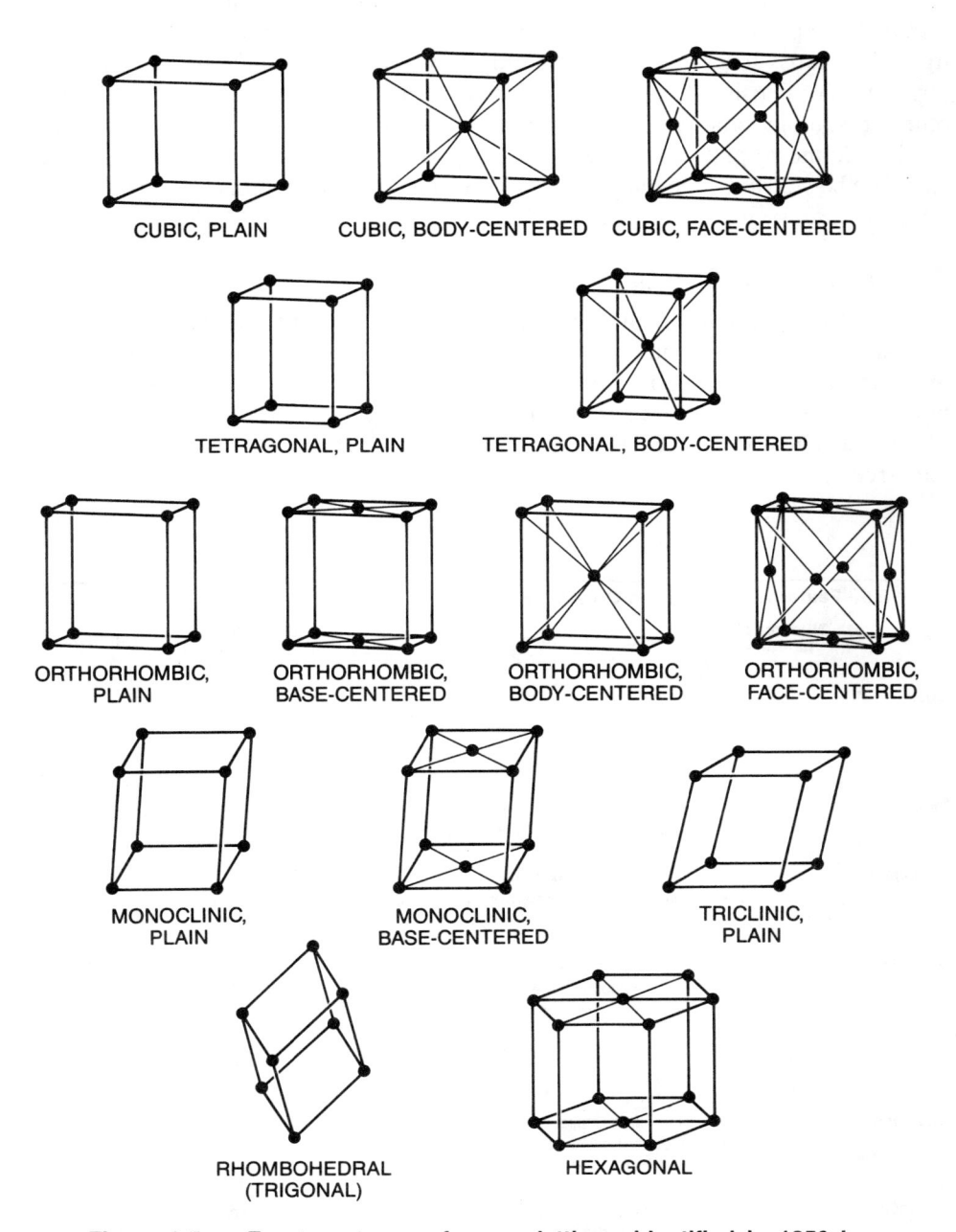

Figure 2.5 — Fourteen types of space lattices, identified in 1850 by Auguste Bravais

Identification of Planes and Directions in Crystals

Certain properties can vary with changes in orientation in the crystal, so a system of notations was developed to identify crystal planes and directions in the crystal structure. The visualization of crystal faces by outlining edges with drawn lines proved so effective that this technique was extended to outlining significant internal planes. Yet, it must be kept in mind that these easily-sighted outlines (with the planes often highlighted by cross-hatching) are really only rows of atoms, which for reasons of atom population or density along the plane (usually referred to as packing density), or some other crystallographic feature, constitute a significant "boundary" in the crystal. The need to delineate particular faces, planes, boundaries, directions, etc. in a crystalline structure has led to the development and use of a notation system called *Miller Indices*, which is discussed in Technical Brief 5.

Basic Types of Crystals

What holds the atoms in a crystal so firmly together? The bonding of atoms in a crystal is a result of a number of forces acting in concert. These forces can include gravitation effect, magnetic attraction, stabilization energy, and electrostatic interaction. However, electrostatic attraction is chiefly responsible for cohesion in the crystal of a solid. The cohesive energy of a crystal is the difference in total energy between the "free atom" and the lower "crystal energy."

Table 2.5 - Features of the Seven Crystal Systems

System	Number of Lattices	General Description	Unit Cell Edge Lengths	Interaxial
Cubic	3	Sometimes called isometric or regular. Three equal axes mutually perpendicular. Plain, body-centered, and face-centered unit cells.	$a = b = c$	$\alpha = \beta = \gamma = 90°$
Tetragonal	2	Two equal axes perpendicular to each other and to a third (unequal) axis.	$a = b \neq c$	$\alpha = \beta = \gamma = 90°$
Orthorhombic	4	Three unequal axes mutually perpendicular. Plain, base-centered, body-centered, and face-centered.	$a \neq b \neq c$	$\alpha = \beta = \gamma = 90°$
Monoclinic	2	Three unequal axes, only one of which is perpendicular to the other two. Plain, and base-centered.	$a \neq b \neq c$	$\alpha = \gamma = 90° \neq \beta$
Triclinic	1	Three unequal axes, no two perpendicular. Plain only.	$a \neq b \neq c$	$\alpha \neq \beta \neq \gamma$
Rhombohedral	1	Sometimes called trigonal. Three equal axes mutually inclined at an angle not equal to 90°. Plain only.	$a = b = c$	$\alpha = \beta = \gamma < 120°, \neq 90°$
Hexagonal	1	Three equal axes inclined at an angle of 120° and a fourth (unequal) axis perpendicular to their plane.	$a = b \neq c$ $a = b = c \neq d$	$\alpha = \beta = 90°$ $\gamma = 120°$

The cohesive energy of solids is determined by measuring the various work-units of energy required to form separated neutral atoms from the crystalline solid. This can be defined as the energy required to remove an atom from the surface of a crystal to infinity, that is, the *heat of sublimation*. Values for various elemental solids are not included in this text, but are reported in technical handbooks.

Four basic types of crystals are differentiated by the behavior of electrons in their structure. The deportment of these elementary particles can directly affect the cohesive strength in each crystal type. Electron distribution in the crystal structure is equally important and is the key to understanding many physical properties of solids, such as hardness, brittleness, electrical resistivity, thermal conductivity, etc. These four types of crystals are: (1) crystals of the inert gases, (2) ionic crystals, (3) covalent crystals, and (4) metal crystals.

TECHNICAL BRIEF 5:
Miller Indices

Miller Indices consist of a series of digits, which are always given in a sequence that refers to the major axes of a crystal (e.g., axes a, b, and c of Figure 2.4). The numbers indicate points on these major axes where a particular plane intersects. Usually, three digits suffice for an index, but four digits are necessary for noting planes in the hexagonal crystal system, and here the notation is a Miller-Bravais indice.

Indices are established by first obtaining the intercepts by the plane on the major axes in terms of multiples or submultiples of one lattice parameter. Then the reciprocals of these numbers are noted and reduced to the smallest integers having the same ratio. The numbers are written to indicate the intercepts on the a, b, and c axes, in that order; the ratio sign is eliminated for brevity, and the index is enclosed in parenthetic brackets ( ). Figure 2.6 is a rudimentary illustration of Miller Indices for internal planes in a cubic crystal. Note that when a plane is parallel to an axis, the intercept is assumed to occur at infinity and is given a zero index digit. Additional notations of Miller Indices can be extended to define the features of the plane more closely. When the planes intercept axes on the negative side, a bar is placed over the particular digits indicating the axes involved. A variety of brackets are used to signify specific planes. For example, if all planes, regardless of orientation, of a given crystallographic type are to be indicated, the indices might be enclosed in curley brackets {}.

Directions in a crystal structure are indicated with different insignia than those used for planes. Directions can be regarded as motions along the major axes of the crystal, and the magnitudes of these motions are indicated in terms of distances. By drawing a line from a point of origin along a given direction, points can be located along the line using cell edges as coordinates and as units of length. The indices of direction are the smallest integers proportional to these coordinates. Direction indices are enclosed in square brackets [ ] to distinguish them from indices for planes. In the case of cubic cells, their symmetry is such that indices may be the same numerical combinations for directions and for planes that are perpendicular to each other; for example, the [111] direction is perpendicular to the (111) plane in a cubic crystal. To indicate a class or family of directions in a particular crystallographic type, regardless of orientation, the direction index is enclosed by carets < >. Again, negative axes intercepts are noted with over-bars.

58 Chapter 2

Inert Gas Crystals

The simplest form of crystal is one made up of neutral atoms with the outer electron shell completely filled; that is, the outer shell has neither valency electrons, nor vacancies for electrons. Atoms of the inert gases are of this type. When an aggregate of inert gas atoms is caused to solidify at extremely low temperature, the atoms are held in the structural configuration of the crystal only by forces arising from fluctuation in charge distribution, known as *Van Der Waals forces*. Consequently, the crystals of an inert gas are weakly bonded.

Ionic Crystals

Ionic binding takes place when an aggregate of unlike atoms with incomplete outer shells are presented to each other and a transfer of electrons takes place. As a result of electron transfer, the atoms become oppositely charged. The electrostatic

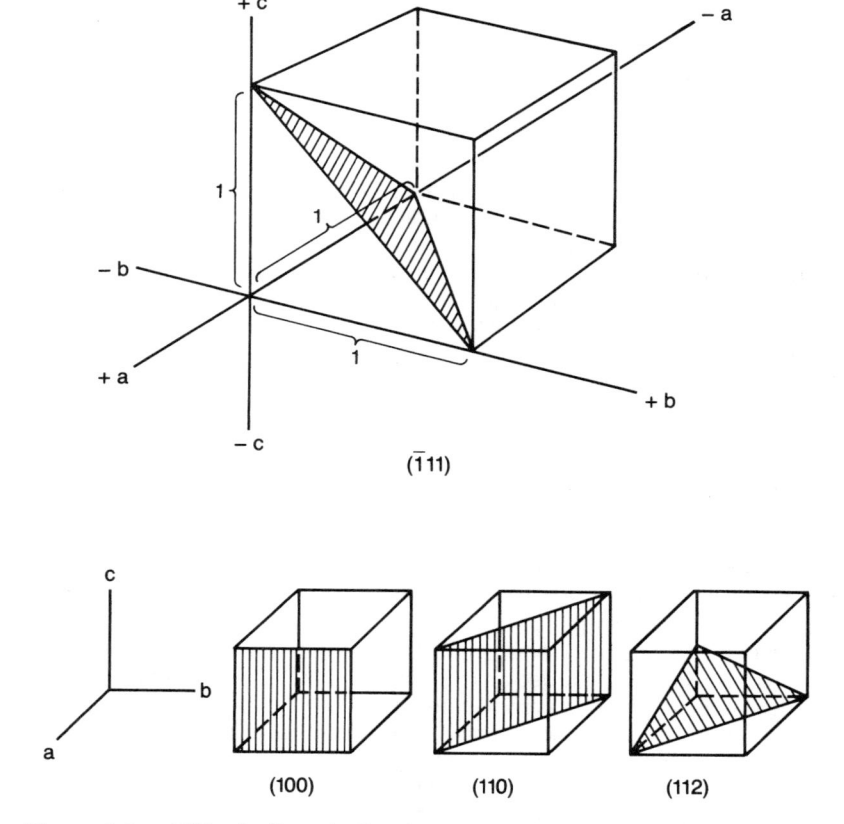

Figure 2.6 — Miller indices indicating the location and orientation of particular external and internal crystallographic planes (shaded areas) in cubic crystals

In the upper illustration, the plane is designated ($\bar{1}$11) because it intercepts the a, b, and c axes each at one unit of lattice parameter, and the intercept on the a axis is indicated to have occurred on its negative side.

attraction of opposite charges then strongly pulls the atoms together. When they reach a center-to-center distance between the atoms of about three angstroms, repulsive forces act to establish the interatomic spacing characteristic of the particular solid. Also, the atoms arrange themselves in whatever crystalline pattern forms when the strongest attractive interaction is compatible with repulsive interaction.

The principal binding force in an ionic crystal is electrostatic attraction between ions of opposite charge; this is called *Madelung energy* in recognition of one of the foremost investigators of ionic valence forces. Ionic crystals, sometimes referred to as salt crystals, display greater strength than inert gas crystals.

Covalent Crystals

A covalent bond is established in a crystal when a small number of valency electrons exist in the outermost shell of the atoms, and adjacent neutral atoms act to share their electrons with them. Cohesion, or binding, is attributable to the attraction of the positive ions to the negative valency electrons passing between them. Most often, the bond is formed by only two electrons, one from each adjacent atom. The binding of two hydrogen atoms to form molecular hydrogen (H_2) is the simplest example of a covalent bond. The neutral hydrogen atom, because it has only one electron, should normally form a covalent bond with only one other hydrogen atom.

There is a strong bond between atoms in a covalent crystal, and it is strongest when the spins of the two electrons participating in the binding are not parallel. The covalent bond strength can be highly directional. The diamond is an example of a crystalline mineral that utilizes the covalent bond, and its directional property is well known.

There is a continuous range of crystals between the ionic bond and the covalent bond types. Whether the binding between atoms is ionic or covalent depends primarily on the electron distribution in the outer valency shells of the bonded atoms and the extent of electron (ionic) transfer, or electron (covalent) sharing that develops. For this reason, covalent bonded crystals are sometimes called valency crystals.

A detailed examination of crystal binding reveals that sometimes atoms initiate "ionic" binding, but when one of the atoms supplies both needed electrons, the final crystal is covalent bonded. Silicates, for example, possess properties resembling both salts (ionic) and valence crystals. Even the simple hydrogen molecule, under certain exceptional circumstances, can exhibit an unexpected kind of binding. In this extraordinary binding, three hydrogen atoms approach and one loses its electron to another. The bare proton (nucleus) then acts to bind the two intact hydrogen atoms together as a molecule. This is a unique, important form of binding which has been termed the *hydrogen bond*.

Metallic Crystals

The form of crystal binding found in metals is the fundamental feature that controls the mechanical and physical properties of metals and alloys and makes them so useful. The bonds in a metal crystal, while closely related to a covalent bond, are different enough to endow the crystalline structure with a remarkable array of properties. Each atom in the structure is attracted equally and indiscriminately to one another.

The distance between atoms in a metal crystal can be precisely measured by means of x-rays, and these dimensions vary with the composition of the crystal and its temperature. The closest approach of atoms (center-to-center) is about three to five angstroms (about 1.0×10^{-8} to 2.0×10^{-8} in.). The closest approach results from the balance of attractive and repulsive forces between the atoms. Thus, beneath the surface of a crystal the field surrounding an atom is entirely in equilibrium with the neighboring atoms, and these forces tend to prevent movement.

The principal cohesion of atoms in a metal is ascribed to the overlapping of atomic orbits that allows valence electrons to seek bonding orbits, causing the electrons to interact with two or more nuclei and hold them together. In this manner, electrons serve as a universal "glue" for materials. Of course, if the outer shell of the atoms is complete with electrons, as is the case with the inert gases, the atoms will tend to repel one another and the overall bond strength will be minimal.

For effective metallic bonding it is necessary to have certain electron configurations to make effective use of the valence electrons. Extensive studies have found that this aspect of bonding is influenced by (1) number of electrons available, (2) number of electron pairs used in each bond, and (3) types of orbits in which the electrons reside.

Even though electrons do not have a free choice in their distribution around the nuclei of atoms in a metal crystal, the electrons in the outer valency shells have a remarkable facility to shuttle back and forth by exchanging positions from atom to atom. In fact, this ability is so evident that the valence electrons have been likened to a "gas" surrounding an aggregate of atoms, which in turn might be regarded as an aggregate of metal cations (positively charged ions). Despite the paradox of an electron "gas" that can "glue" atoms together to provide the bond strengths found in metal crystals, there is additional support for these unique views. For example, the valency electrons have much lower kinetic energy in the atoms of a crystal (cohesive energy) than valency electrons of a free atom. There is also evidence that even the electrons in the secondary shells contribute to the strong bonding between atoms in metallic crystals. As a consequence of the strong electrical and magnetic forces that bond metal atoms together, very close-packed crystalline structures are formed in which the atoms occupy minimum volume.

The strength of the valence bond between atoms in metal crystals has been studied, and according to best calculations, the alkali metals have a valency of one (the weakest bonding exhibited by metal crystals). The alkali-earth metals have a value of 2. The valence bond of metals with atomic numbers from 20 through 30, ranges from 3 for scandium to about 6 for chromium, and then decreases for succeeding elements such as iron, cobalt, nickel, copper and zinc. Iron has a valence bond of about 5, which accounts for the good basic strength and hardness of iron crystals.

The concept of gas-like valence electron freedom in a metal crystal provides explanation for many of the properties of metals. The energetic motion of the valency electrons accounts for the high thermal conductivity of metals; that is, the heat energy is transmitted by atomic agitation through the medium of the electron gas. Also, under an applied electrical potential, or an imposed magnetic field, the electrons (usually one or two per atom, called conduction electrons) "flow" very efficiently by progressive displacement along the pathway formed by the metal, hence the good electrical conductivity of metals.

The electron gas theory finds support, too, in the observation that at certain low temperatures, the ratio of thermal conductivity to electrical conductivity is directly proportional to the temperature, with the value of the constant of proportionality independent of the particular metal. Yet, the free electron cloud that is envisioned as surrounding the atoms does not fully resemble a gas. For instance, the free electrons tend to repulse one another because they possess like charges. Nevertheless, each atom appears to abide by the laws of attraction and repulsion for electrically charged bodies and the electron shells maintain a normal apparent atomic volume for each atom.

Despite the incompleteness of the free electron model of metals, this concept explains much about the physical properties of metals, metallic compounds, and even some non-metals. An understanding of this concept is improved by recent research, which suggests that electrons are arranged in energy bands separated by regions in energy for which no wavelike electron orbits exist. These regions are called energy gaps or band gaps, and they occur because of the interaction of the conduction electron waves with the ionic nuclei of the crystal. The crystal acts as an insulator if the allowed energy bands are either completely filled or empty, because under these circumstances no electron can move through. The crystal behaves as a metal if one or more bands are only partly (10-90 percent) filled. The crystal is a semiconductor or a semi-metal if all bands are filled except for one or two, which are slightly filled or slightly empty. If valence electrons exactly fill one or more energy bands, leaving others empty, the crystal will be an insulator, and application of an electrical potential will not allow current to flow.

The influence of electrons in determining crystalline structure of alloys and their properties is re-examined in Chapter 4 because these negative elementary particles in the atom play a key role in alloying of metals.

THE CRYSTALLINE STRUCTURE OF METALS

Metals usually solidify as a "true solid"; that is, they have a structure that is crystalline in every aspect of the term. There is an area of exception to this norm: when a liquid metal is cooled through the solidification process at a tremendously high rate, for example, at about ten million °C per second (approximately 18 million degrees F per second), perhaps by "splat cooling," an amorphous solid metal is produced. When first produced in a laboratory in about 1960, solid metals with a non-crystalline structure, for obvious reasons, were called metallic glasses. Several techniques have since been devised for producing amorphous metals or metallic glasses in the form of fine wire or thin ribbon. (These unusual but useful metal products are detailed in Chapters 3 and 14). It is also possible to melt the surface of a solid piece of metal superficially and to have the very rapidly solidified surface "glaze" assume an amorphous structure. However, unless highly unusual steps are taken to cool solidifying metal at an extremely fast rate, solid metal will have a crystalline structure. The cooling rate commonly used during casting and welding processes is not fast enough to preserve an amorphous arrangement of atoms, and metals and alloys ordinarily have a crystalline structure when in solid form.

The majority of metals develop a crystalline structure during solidification which is characteristic of the particular elemental metal, and this remains the permanent crystalline form while it stays solid. Some metals, after solidifying and on further cooling, will undergo the previously mentioned allotropic transformation at specific levels of temperature. (This phenomenon is reversible and is quite important). However, approximately three-fourths of the elemental metals that do not undergo further crystallographic transformation after solidifying have crystal structures that are face-centered cubic, body-centered cubic, or hexagonal close-packed. These crystal structures are detailed in Figure 2.5, but this illustration shows only the simple base-centered hexagonal structure. The hexagonal close-packed arrangement, which incorporates an intermediate layer of atoms between the top and bottom layers, is actually the more common form whenever a metal adopts the hexagonal crystal structure. Figure 2.7 is a more detailed illustration of the three crystal structures favored by metals, designated by the abbreviations bcc, fcc and hcp. Other crystal structures found in solid metals include simple cubic, tetragonal, rhombohedral, orthorhombic, and a number of variations of these forms. Note that Figure 2.7 presents dual sketches: The crystal unit cells on the left emphasize the atom sites schematically, and those on the right show the atoms nearly touching, as is believed to exist in an actual solid metal.

The use of solid spheres to illustrate the unit cells on the right in Figure 2.7 may be a bit misleading because an atom is mainly empty space. The nuclei and the orbiting electrons are so small that a fast-moving particle like a neutron can be shot through a million atoms without hitting anything. Even the positions on the atom sites in the lattice represent a small exaggeration, because the atoms are normally moving about or vibrating with thermally-induced kinetic energy. Possibly the only accurate features of the unit cells in Figure 2.7 are the lattice parameters and the number of atoms in a single unit. Nevertheless, these admittedly crude impressions of crystal structures are very helpful in understanding the behavior of metals under many conditions. In fact, the prevailing crystal form determines in a fundamental way many of the properties and characteristics of the metal, so its crystalline nature should probably be considered its most important feature.

There are subtle differences among the three crystal structures, body-centered cubic (bcc), face-centered cubic (fcc), and hexagonal close-packed (hcp), illustrated in Figure 2.7. Geometric analysis shows that the most efficient way of packing spheres together to achieve maximum density is to utilize either the fcc or the hcp arrangements. These two arrangements have planes of spheres packed together as closely as possible so that each sphere is in contact with six others lying on the close-packed plane, and three others in each of the two adjacent planes. In both the fcc and the hcp arrangements, the total volume occupied by the spheres is 74 percent. The bcc arrangement achieves only 68 percent. Consequently, transformation of a metal from bcc crystalline structure to fcc at a given temperature will result in a contraction in volume.

Studies of alloys, particularly steel, have found that the location, shape and size of the vacant areas among the spheres or atoms are also important. Even with its efficient packing, the fcc crystal structure offers a larger opening or void at a particular place among its atoms than does the more loosely packed bcc structure, or the hcp structure, thus permitting easier insertion of a small alloy atom. A very useful void in the fcc structure is located in the center of the unit cell (See detailed illustration in Chapter 9).

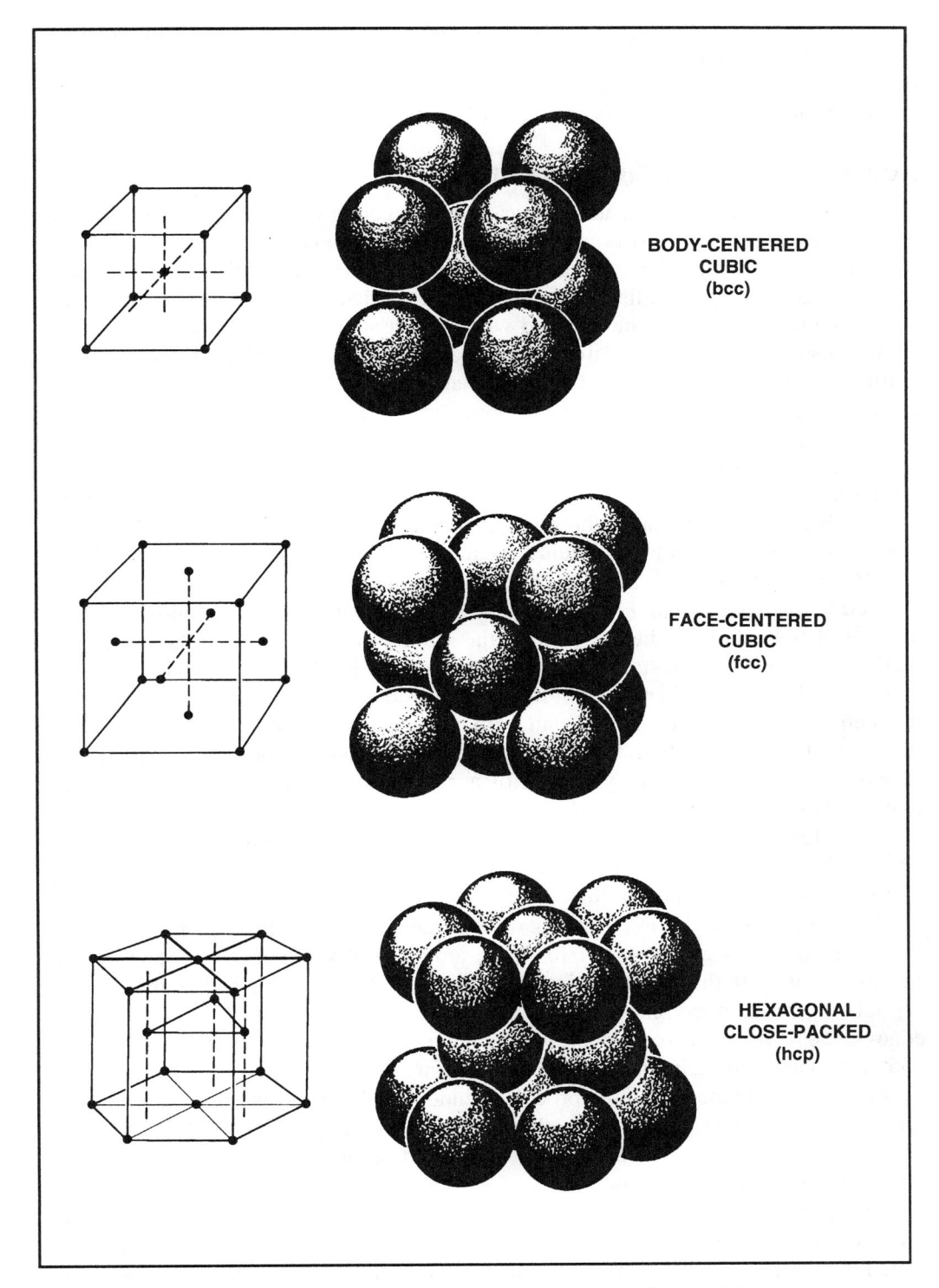

Figure 2.7 — Three kinds of crystal structure favored by metals

To analyze the metallurgical behavior of the various crystal structures it is sometimes necessary to picture them as multiple layers of many atoms rather than to single out one unit cell. This is especially true when considering the movement of atoms during allotropic transformation, twinning, and plastic deformation.

How Does a Crystal Grow from the Melt?

When liquid metal cools, atoms transfer thermal energy to any cooler surrounding material. Each degree of lowered temperature is accompanied by a consistent release of energy per unit mass. This energy can be measured, and the resulting measurement corresponds to the specific heat of the metal. At a particular temperature for pure metals, a quantity of heat will be released without a continued decrease in temperature. This temperature level is the solidification point, and the heat released per unit mass at this point is the latent heat of fusion. The atomic latent heat of fusion is the difference between internal energy per mole possessed by atoms in the solid state and the liquid state. When the liquid is converting to the solid crystalline state, this energy is liberated by the atoms in the form of heat. This heat must be dissipated in order for crystallization (solidification) to proceed. As might be expected, the same quantity of heat is absorbed by a metal without any perceptible temperature rise as the metal is converted from the solid to the liquid state at the melting point.

When liquid metal solidifies, the crystalline structure forms in two distinct stages —*crystal nucleation* and *crystal growth*. The nucleation stage requires a small amount of free energy to be available in the molten metal to assist in the initial formation of a complete crystal unit. The exact manner in which the atoms of the first solid crystal come together is one of the unexplained occurrences in nature. However, it is believed that foreign embryos can form and dissipate in the liquid just above and below the freezing point, and that these embryos possess a semblance of crystalline structure. These embryos may consist of a near-crystalline cluster of two or three hundred atoms. As the temperature falls, the embryos multiply and attract the freely moving atoms in their vicinity. Some investigators of solidification phenomena believe that almost any foreign particle in the liquid metal can supply the free energy to help form the initial crystals.

Regardless of the driving force, as soon as the first crystal units are formed in the melt, these crystals act as nuclei for more atoms to attach themselves, and crystal growth begins with the addition of atoms to sites on planes established by the nucleus crystal. There is good experimental evidence that crystals increase in size by edge-wise growth of the basal planes of the unit cell. The atoms coming from the melt continue to add three-dimensionally to the existing crystal structure to form additional unit cells. While the nucleation process is endothermic (requires energy to proceed), the process of crystal growth is exothermic (releases energy). As solidification proceeds by crystal growth in a pure metal, the evolution of heat acts to arrest temperature drop, as illustrated in Figure 2.8, under very slow cooling conditions.

In the case of an alloy, the energy released by the atoms moving into fixed crystal positions occurs over a range of temperature. The upper temperature of this range is called the liquidus and the lower temperature is called the solidus. Below the solidus, the metal consists entirely of solid crystals, and so the only energy that can be released by the atoms present is the thermal energy corresponding to the specific heat of the particular alloy.

Examination of the three cooling curves in Figure 2.8 shows that the plot for pure iron develops a horizontal step at 1535°C (2795°F), which is the solidification temperature for this metal in pure form. The plot for pure copper develops a step at 1083°C (1981°F), which identifies its solidification point. The cooling plot for a specimen melt composed of 80 percent iron and 20 percent nickel displays a marked evolution of heat starting at 1485°C (2705°F), which retards the rate of cooling, but allows the temperature to drop very slowly until it reaches the level of about 1470°C (2678°F), whereupon the regular rate of cooling is resumed. Obviously, 1485°C is the liquidus, and 1470°C is the solidus for this particular alloy composition.

Study of the mechanisms involved in the solidification of metal into a crystalline structure has contributed to the many accomplishments in welding and other manufacturing processes. The need for further knowledge in this area continues. For example, the shrinkage of metal upon solidification is a curious action. It is understandable that as a metal freezes and the heat of fusion is liberated, the release of energy would allow the individual atoms to assume a more intimate fixed bond (crystallinity).

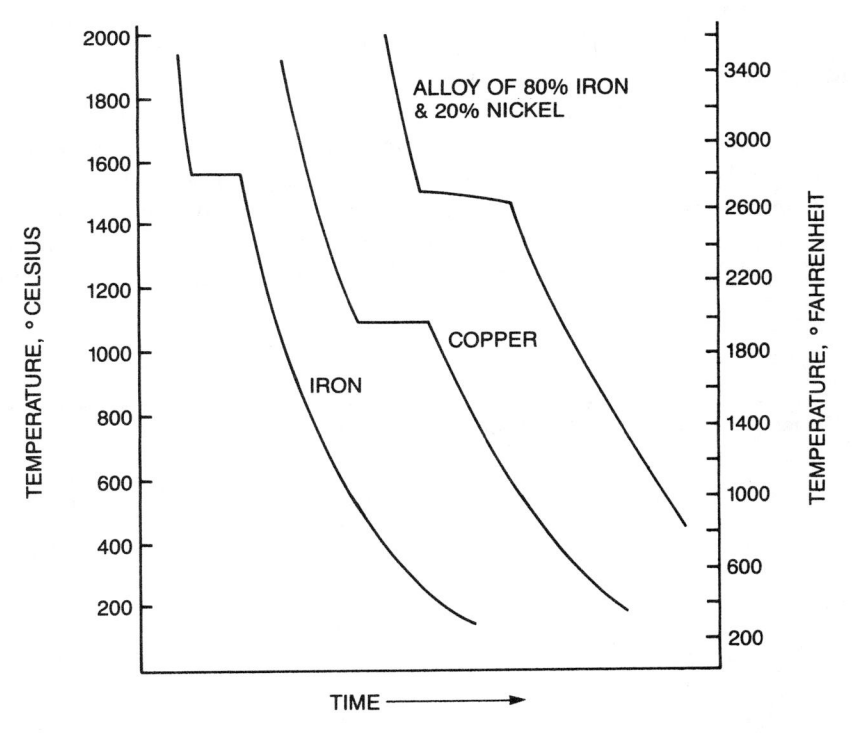

Figure 2.8 — Cooling curves depicting solidification point exhibited by pure metals, as contrasted with solidification range of alloys

During very slow cooling, the temperature of a pure metal remains constant at the solidification point until all of the metal has frozen, whereas an alloy undergoes a reduced rate of cooling between the liquidus and the solidus and therefore displays a solidification range.

Naturally, this more intimate bonding would be expected to result in a volume decrease, (shrinkage occurring on solidification) and this is the case with most metals. Yet, five of the pure metals (Ga, Si, Ge, Sb and Bi) and a number of their alloys increase in volume on solidification. The behavior of these elements is attributed to the formation of crystalline structures in which a fewer number of atoms occupy a unit volume in the solid metal than in the liquid. Table 2.6 provides data on the density of a number of pure solid metals and the volume change that occurs on solidification.

Another feature of solidification is the endothermic nature of the nucleation process. Increased knowledge of this aspect has provided a basis for the development of techniques to speed up or to retard crystallization in the solidifying metal. Some of these techniques are used in making castings and weld metal.

The Formation of Dendrites

An atom at the surface of a crystal is not surrounded on all sides by neighboring atoms. Because of this, the crystal possesses available energy that is directed outwardly from its surfaces. The crystal, with this surface energy can attract other atoms to itself, and if the environment is suitable it can grow. Different capacities for growth under the driving force of free surface energy depends on the types of crystal planes exposed at the surfaces. For example, a cubic crystal usually has greatest capacity for growth on its faces, identified as the {100} planes.

Figure 2.9 shows how a cubic crystal nucleus can grow in a melt in three directions at right angles, corresponding to the six faces of the cube. For clarity of illustration, Figure 2.9 shows the continued solidification from the six faces as simple lines instead of unit cells. The pattern of crystal growth in many materials resembles a pine tree and its branches, thus the term dendritic, which simply means "branching." Unlike salt and other substances that often crystallize from water solutions as compact, well-shaped crystals, metal crystals grown in the solidifying melt have a pine-tree, or dendritic appearance.

Table 2.6 - Densities of Some Pure Metals and Their Change

Element	Symbol	Density (grams per cc at room temperature)	% Change in volume upon solidification
Aluminum	Al	2.70	-6.4
Antimony	Sb	6.62	+1.0
Bismuth	Bi	9.80	+3.4
Copper	Cu	8.96	-4.2
Gallium	Ga	5.91	+3.2
Germanium	Ge	5.36	+12.0
Gold	Au	19.32	-5.2
Iron	Fe	7.87	-2.8
Lead	Pb	11.34	-3.5
Lithium	Li	0.53	-1.6
Magnesium	Mg	1.74	-4.1
Manganese	Mn	7.43	-1.7
Silicon	Si	2.33	+12.0
Silver	Ag	10.49	-5.0
Tin	Sn	7.30	-2.8
Zinc	Zn	7.13	-4.2

As detailed schematically in Figure 2.10, if one of the solid crystalline arms growing perpendicularly from the face of the nucleating crystal is examined in greater detail, it will be seen that as early solidification proceeds in the <100> growth directions, the primary arm soon forms secondary arms, and eventually tertiary arms emerge. In the final stage of solidification when many arms have grown in the melt and they come into contact, the open spaces between them become filled with unit cells of solid metal.

This basic knowledge of solidification mechanics can be very helpful in solving problems with weld metal, and with the high-temperature portion of a heat-affected zone. If impurities having a lower solidus are present in the metal, they will be concentrated in the areas last to freeze, that is, in the spaces between the dendrite arms. Naturally, slower cooling results in coarser dendritic structure and gives low-melting elements greater opportunity to segregate into the areas of final solidification. The presence of this lower-solidus material can cause interdendritic cracking to occur. Also, a higher concentration of an impurity element can affect the mechanical properties of the solid metal.

It is frequently necessary to analyze welding problems from a solidification standpoint. Although solidification by growth of dendrites in the melt has been described as the normal mode, the occurrence of this mechanism can be altered, or even completely circumvented by a number of conditions often encountered with welds. These include the relatively fast cooling rates experienced with welds, direction of heat flow, nature of the nucleating crystals presented by the base metal to the molten weld metal,

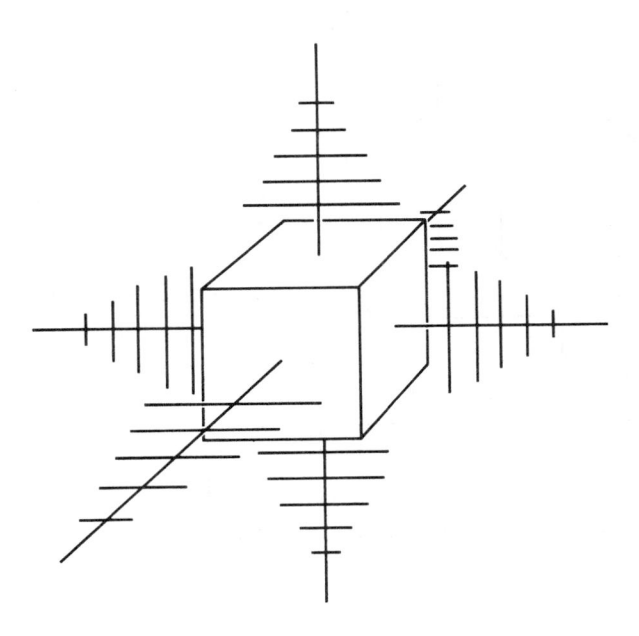

Figure 2.9 — Pattern of dendritic growth from a crystal during solidification

and effects of alloying in the weld metal. More information on the manner in which welds solidify is contained in Chapters 6, 7 and 9.

The Formation of Grains

If molten metal were poured into an ingot mold and allowed to solidify, and if the crystal structure were to nucleate only at one point and continue to grow without interruption of any kind, then the entire mass would form a single crystal. Although this crystal would have an external shape the same as the inside of the mold, the internal structure of the crystal would have a repeated pattern of unit cells in all directions, and an undeviating orientation of planes from one side of the crystal to the other. The vol-

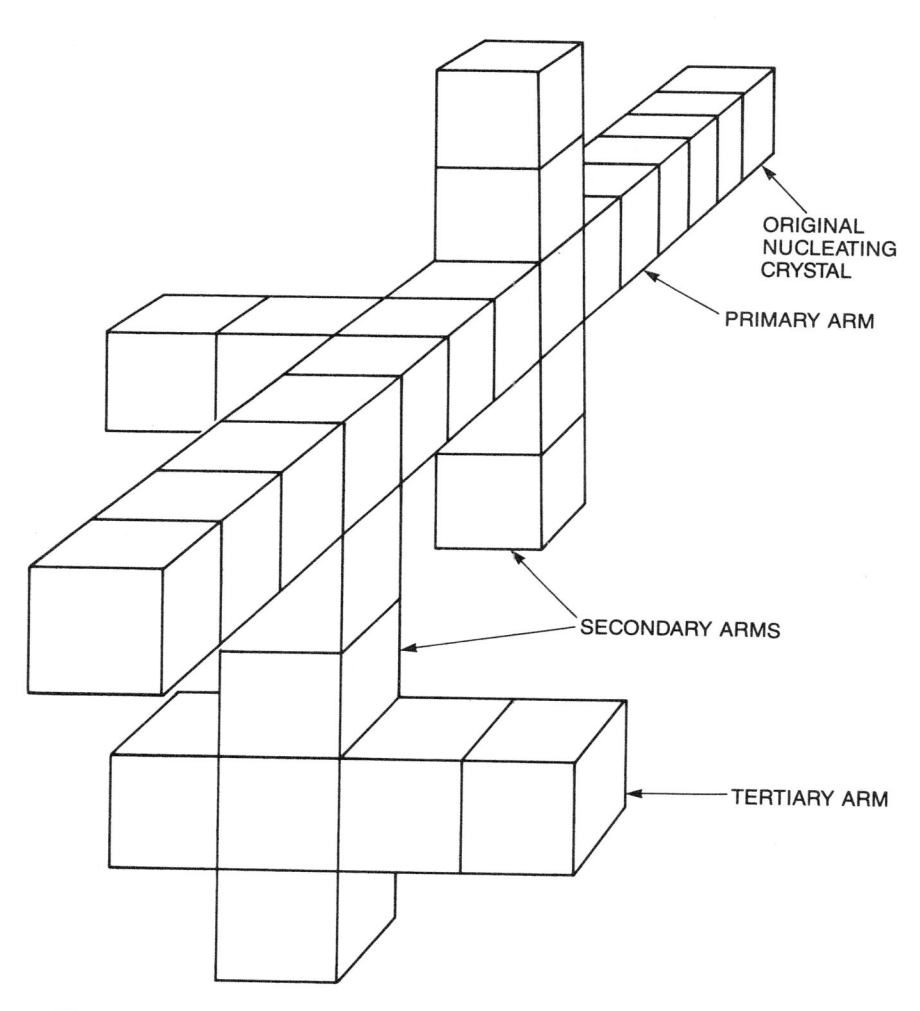

Figure 2.10 — Schematic of a dendrite in a cubic metal during early stage of solidification

ume of solid crystal would be slightly smaller than that of the liquid for those metals that shrink during solidification. This single crystal would exhibit some unusual properties and characteristics. Single crystals are difficult to grow and are uncommon in most metal and alloy articles or products. Polycrystallinity is the result of the usual multiple nucleation, which interferes with the simple, orderly process of crystallization during solidification.

The development of some highly specialized techniques for making metal sections that are a single crystal has been based on knowledge of polycrystallinity. These single-crystal sections include ribbons of metal, rods, small bars and small shapes with unusual and very useful physical and mechanical properties. The manufacturing techniques usually involve highly directional cooling of a solid shape as the section is slowly withdrawn from the melt.

To understand why the great majority of metal sections handled in industry are not comprised of a single crystal, an instructive demonstration can be observed using a glass of a cold carbonated drink. As the liquid in the glass grows warmer, tiny bubbles of carbon dioxide gas evolve from the water first at the walls of the glass, from which they become detached by buoyant force and rise to the surface. The gas evolves at the wall for two reasons:

(1) the glass is warmer than the liquid because its outer surface is exposed to ambient air, and

(2) there are minute peaks and valleys on the wall of the glass that have high surface energy, and therefore they favor physical and chemical changes that require a little free energy to set them in motion.

A somewhat similar situation is found when molten metal freezes in a metal ingot mold. Heat is abstracted by the mold through its walls, causing the liquid metal at the mold wall to be the first to drop in temperature to the nucleation point. Consequently, the first crystal nucleus will form in the liquid metal immediately adjacent to the wall or bottom of the mold. Does only one crystal form and grow? No, a number of conditions favor the formation of many nuclei in the liquid metal when the temperature has dropped to the solidification point. Although a small amount of energy is required to initiate formation of a crystal nucleus, this energy can be found at the surface of the mold. This nucleation energy is also found on the still-solid face of a weld groove in base metal when it acts as a mold for the liquid weld metal. Therefore, many crystals will form and grow to cover the wall of a metal mold or a weld groove in base metal. Incidentally, in the case of a weld groove, the intimacy of contact between the molten weld metal and the solid base metal also has an influence on the initial layer of unit cells that forms whenever composition similarity permits.

Because the crystal structure of the solid base metal is in intimate contact with the liquid weld metal, the atoms in the liquid metal that are ready to join a solid crystal unit cell are pulled by free energy sources of the crystalline base metal into the same unit cell arrangement. The term *epitaxial solidification* is used to describe this process, which is frequently observed in welds in iron and steel. Epitaxial growth, in addition to dictating crystal form and orientation of the initial weld metal, also influences the size of initial segments of crystal possessing a particular orientation that form along the weld interface.

The periodicity, or repeated pattern, of atom arrangement, along with undeviating orientation of planes in the single crystal, must originate from a single nucleus when molten metal cools to the solidification point. Each of the many tiny nuclei grows independently by allowing atoms in the melt to join the particular crystallographic arrangement three-dimensionally as soon as each atom can lose energy equivalent to the heat of fusion. The arrangement of atoms and the orientation of the planes are essentially precise in each crystal, but as two crystals approach each other by growing larger in the melt and finally impinging along a common front, their planes are often not in alignment.

As illustrated in Figure 2.11, three crystals with differing orientation have met with the atoms at the juncture accepting the best possible compromise positions. Along these fronts where the crystals meet, the lattice is distorted and the atoms at or near the juncture accept slightly abnormal positions. The distance over which the atoms gradually change from the orientation of one crystal to that of the neighbor has not been firmly established, but scientists have speculated that at least two rows of atoms must be affected.

Despite the facility that atoms have for accommodating mismatch in orientation, the juncture between crystals of differing orientation exhibit a number of unusual properties because of lattice abnormalities. For example, it is very easy to determine where these junctures exist using various methods of examination. Etching polished sections of metal with acid is particularly effective, because the highly localized chemical attack reveals the juncture as a very narrow, line-like boundary. The individual segments or areas having their own uniform crystal structure and orientation, called *grains*, can be clearly detected. The junctures between them are called *grain boundaries*. When neighboring crystals are mismatched in orientation about 20 degrees or more, the grain boundaries are further defined as *large-angle boundaries*.

This description of grains in metals is based on the ideal crystal. In reality, a number of complications often arise during formation of the actual crystalline structure in metals, and these cause small departures from the ideal atomic arrangement. Many kinds of small irregularities develop in the crystalline structure during solidification and during various processing operations conducted on solid metal. The imperfections that occur are well documented in the literature; because, despite their minuteness, they have significant influence on the properties of the metal. These crystal imperfections have become known as *point defects, lattice vacancies, dislocations, stacking faults, impurity atoms*, etc.

Grains in metals often display *subgrains*. In fact, a grain that is delineated by a regular large-angle boundary may appear to be comprised of a mosaic of *crystallites*, which have very small mismatches in orientation from one crystallite to another. The size of crystallites may be in the approximate range of 1000 to 10 000 atomic diameters. The networks of veins between crystallites are called *low-angle boundaries* because of the very small angular difference in orientation of adjoining crystalline structures. These boundaries are not easily observed when the mismatch in orientation is less than one degree.

The substructure in the grains of polycrystalline metal and in a single crystal of metal has been intensively studied to identify specific types of crystal imperfections and determine what causes them. Heat treatments and other metal processing steps are used to reduce the number of imperfections in the crystalline structure. An abbreviated list of categories of crystal imperfections and the terms describing them follow:

(1) *veining* – the sub-boundaries that are either low-angle boundaries or boundaries that have acquired a greater than average concentration of solute atoms or precipitates,

(2) *lineage structure* – a mosaic of crystallites that originated during the solidification process,

(3) *cellular structure* – usually a substructure that results from cold work,

(4) *impurity substructure* – low-angle boundaries or arrays of dislocations which have been "decorated" with foreign solute atoms, and

(5) *polygonized structure* – a mosaic or a substructure resulting from cold work followed by heat treatment.

The Shape of Grains

Grains have irregular curved boundaries instead of straight boundaries with sharp corners because the numerous branches of a dendrite do not grow at equal rates. The ends of branches in a given crystal consequently reach ends of branches in other crys-

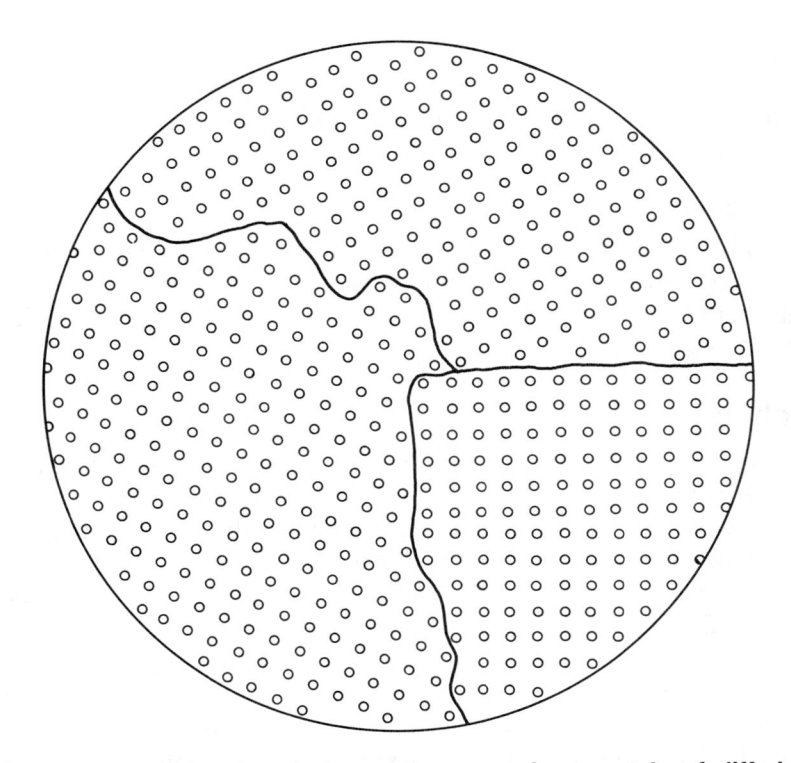

Figure 2.11 — Juncture between three growing crystals of differing crystallographic orientation to form grain boundaries

Fine lines have been drawn simply to delineate the mismatch in orientation that comprise grain boundary conditions.

tals at varying distances from the axis of the crystal. This crystal growth in the dendrites form grains of varying shape and size, as shown in Figure 2.12. The grain boundaries drawn in this illustration are relatively smooth, flowing lines, and this is the appearance of the boundaries in a metal specimen when revealed by etching and observed by metallographic examination. In reality, however, the boundaries represent the junction of disoriented crystal structures, as shown in Figure 2.11.

A factor that strongly influences the shape of grains during solidification is the direction of heat flow. As an example, when molten metal is cast into a cold mold, crystals are quickly nucleated in the metal at the mold wall. These crystals grow into grains covering the mold wall. The remaining liquid is, in effect, contained in a mold consisting of its own grains. Heat continues to flow from the liquid metal through the solid grains and into the mold wall. The flow of heat is mainly perpendicular to the mold wall. This strongly directional abstraction of heat from the liquid encourages atoms to join the crystal structure on the extremities of the grains, that is, on the crystal faces through which the heat is flowing. Consequently, the grains grow more rapidly in a direction perpendicular to the mold wall and the final shape they assume is noticeably elongated or columnar. If the abstraction of heat is less directional, the grain shape tends to be equiaxed. As a result, the metal that solidifies last in the center of the mold will have an equiaxed grain shape as its heat flows in many directions toward surrounding mold walls.

Welding usually involves intense localized heating of relatively cool metal, followed by highly directional heat flow from the weld into the base metal. Consequently, columniation is often seen in the shape of grains that are formed during solidification and cooling of the weld.

The Size of Grains

The size of individual grains that grow in metal while it solidifies may change during cooling to room temperature, but grain size of newly solidified metal generally has an important bearing on its subsequent behavior. A number of conditions influence the initial grain size.

Undercooling

Undercooling is a phenomenon in solidification which can significantly affect crystallization. If, on solidification, a liquid substance that normally forms a crystalline solid is cooled very slowly and without agitation, the temperature of the liquid may drop many degrees below the freezing point before the crystallization mechanism initiates. Molten metals, however, will ordinarily undercool only to a very small extent before solidification and crystallization proceeds.

Undercooling of metal to any appreciable degree requires very special conditions and techniques. Metals have a very strong inclination to crystallize on solidification, and the slightest disturbance in liquid metal at the solidification point can provide the energy needed to generate nuclei upon which the crystals will grow. Merely dropping a tiny particle of solid material into the undercooled liquid metal can trigger the crystallization mechanism. Although appreciable undercooling is not ordinarily observed in the solidification and crystallization mechanism of metals, crystal structure trans-

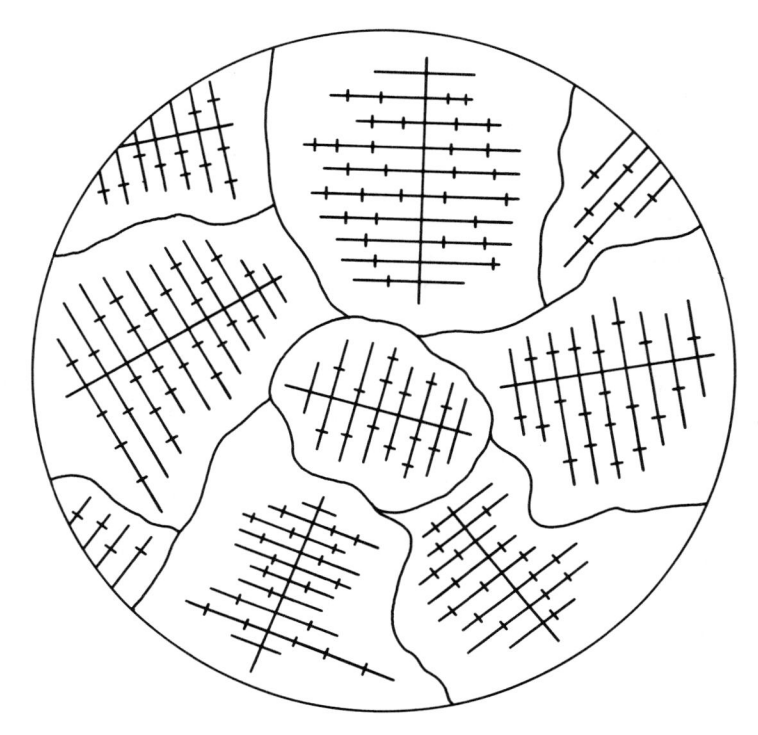

Figure 2.12 — Grain size and shape in solidified metal are determined by the manner in which the branches of dendrites meet

formations in the solid state as affected by undercooling should be studied. Chapter 9 contains information on the influence of cooling rate on crystalline transformations in solid steel at low temperatures. As a general rule, if liquid metal is cooled very rapidly, more nuclei tend to form and consequently a greater number of smaller grains appear in the solid metal. Another general rule produces the opposite: the higher the temperature from which a liquid metal cools, the larger the size of grains formed in the solid metal. Two conditions exert this influence:

(1) overheating lowers the rate of cooling, which tends to reduce the number of nuclei and results in a corresponding increase in grain size, and
(2) overheating dissolves foreign particles which would be available to serve as nuclei if left undissolved in the melt.

Artificial conditions can be created to promote the formation of small grains on solidification. If tiny particles of solid material, even nonmetallic, are distributed throughout the melt, these particles can act as nuclei for crystallization. Adding small quantities (as little as 0.1 percent) of certain favorable elements to the metal can be effective. Insoluble nonmetallic material, such as aluminum oxide added to iron, will function in this manner. Vibrating or stirring the metal immediately before it undergoes solidification will act to produce smaller grains, and several techniques can be used to accomplish this. Mechanical vibrators attached to a mold represent a simple application of this principle. High frequency vibration, such as ultrasonic vibratory energy produced by a magnetostrictive transducer, is said to be very effective in creating small grains. Stirring the metal can also be accomplished by applying an alternating electromagnetic field.

While these general rules provide guidance on solidification and the shape and size of grains, weld metal presents unique conditions that influence its solidification grain structure. Although the weld metal reaches a rather high temperature in most processes, the solidified weld metal is usually relatively fine-grained because of the rapid cooling rate that ordinarily prevails. Liquid weld metal never undercools to any extent, because it is always in contact with metal crystals; these have the crystalline structure of the still-solid base metal. Consequently, no need exists for initial nucleation by embryos in the liquid metal, and therefore crystals of weld metal form at the very start of solidification on the faces of the weld groove. As previously mentioned, this aspect of weld metal solidification is termed epitaxial solidification.

Because the crystalline structure of the weld metal originates at its interface with the base metal, the grains of weld metal are extensions of the orientation and the size of individual base metal grains at the weld interface. Epitaxial grain growth may not persist very far into the weld metal, since the rapid cooling rate and other factors may exert sufficient influence to cause considerable changes in solidification grain shape and size.

Directional heat flow frequently affects the structure of weld metal, and the resulting columnar grains are particularly noticeable when chill bars are included in the tooling or welding fixture, or when water-cooled copper shoes, as used in electroslag welding, confine and contact the weld metal. Structures found in welds are illustrated and discussed in detail in Chapter 9.

THE IMPORTANCE OF A CRYSTALLINE STRUCTURE

To a great extent the particular crystalline structure of a metal or an alloy determines the properties of the metal, including mechanical properties, corrosion behavior, magnetic characteristics, and even light reflection qualities. Significant changes can be made in the properties of a metal by simply altering or transforming its crystal structure without altering its chemical composition. Depending on the metal or alloy involved, the change may be quite unavoidable as the material cools from an elevated temperature, or the change can be made to occur in a particular manner by heat treatment. With certain highly alloyed materials, crystalline change will sometimes occur by exposure to very low (sub-zero) temperatures.

Mechanical properties, in particular, are controlled by crystal structure. Body-centered cubic (bcc) crystals, as shown in Figure 2.7, are found in the metals iron and tungsten and are characterized by a high yield strength and limited cold workability. Face-centered cubic (fcc) metals, as shown in Figure 2.7, include copper and nickel, and are characterized by lower yield strength but a greater capacity for cold work. Consequently, when a new construction metal or alloy is given an initial properties appraisal, knowing its crystalline structure can help identify its mechanical properties.

The majority of metals have a characteristic space lattice or crystalline structure which forms on solidification and is a permanent arrangement despite varying conditions of temperature, strain, etc., below the solidification point. A number of commonly used metals that have a single, characteristic crystal structure are listed in Table 2.7

Allotropic Transformation

Some metals have the propensity to transform their crystalline structure with changes in temperature. The driving force for this transformation is the innate tendency for the metal to adopt the structural form or arrangement having the lowest level of free energy under the prevailing conditions. No change in chemical composition occurs during these crystallographic transformations, but the density, mechanical properties, and other physical characteristics of the metal may change significantly because of the new form of atomic array that makes up the crystalline structure. The transformation is usually reversible with reversal of temperature change; however, the rate of heating or cooling can have a profound influence on whether the transformation initiates, and whether it proceeds to completion. Table 2.8 lists a number of metals that exhibit this property of crystalline structure allotropy. One important benefit of allotropic transformation is grain refinement, but not all transformations produce favorable metal properties. The properties or characteristics that change and the degree of change depends on the particular metal, and whether results are favorable or unfavorable must be appraised on an individual basis. For example, at room temperature the metal tin, is white, soft, ductile, corrosion resistant, and has a body-centered tetragonal crystalline structure (identified as ß phase). By cooling tin below 13°C (56°F), allotropic transformation of the crystalline structure to the face-centered cubic form (α phase) can take place, causing the metal to disintegrate into a coarse, gray powder. Disintegration is the result of a marked volume increase that accompanies the ß-to-α transformation. This transformation in tin is difficult to initiate, and it progresses very slowly. Also, the presence of impurities in the tin will greatly inhibit the transformation. In the case of iron, the allotropic transformation from body-centered cubic (α phase) to face-centered cubic (γ phase) on heating through 912°C (1674°F) results in a number of changes that give this plentiful metal its versatility. These changes involve atomic spacing, magnetic properties, mechanical properties, and a number of physical characteristics. The allotropic transformation in iron is reversible, and is easily regu-

Table 2.7 - Crystal Structures of Some Pure Metals*

Metal	Symbol	Form of Cubic Structure
Aluminum	Al	Face-centered
Copper	Cu	Face-centered
Molybdenum	Mo	Body-centered
Nickel	Ni	Face-centered
Niobium (Columbium)	Nb	Body-centered
Silver	Ag	Face-centered
Tantalum	Ta	Body-centered
Tungsten	W	Body-centered
Vanadium	V	Body-centered

*Listed are some of the metals that have a single characteristic cubic crystallographic structure which remains unchanged at all temperatures while in the solid state.

lated through cooling rate and by alloying. This feature of iron is fundamental and its crystallographic behavior makes it a truly remarkable and useful metal.

Although titanium undergoes allotropic transformation at a rather convenient heat treating temperature (882°C, or 1620°F), the change that occurs in crystalline structure cannot be utilized as effectively as the transformation in iron.

Solubility in the Solid State

Crystal structure becomes even more important when foreign atoms are added to a metal. The foreign atoms may represent another metal added for alloying purposes, or possibly an element that is retained in the metal as an impurity. If two or more elements are present in solid metal, they can exist as independent crystals. For example, crystals of body-centered cubic iron can exist simultaneously with crystals of face-centered cubic lead in an iron-lead alloy. Under these circumstances, the two metals are insoluble in the solid state.

When the mixture of two or more elements crystallizes to form only one type of crystal, the elements involved exist in some type of combination, (if the resulting crystal is different in type from the lattice of any of the elements involved). An example of combination is iron and carbon, which form iron carbide, a compound found in many steels. Elements are soluble in each other if the resulting crystal is of the same type as one of the elements. Solubility can be complete, or can be partial to any degree. An example of a pair of elements that are soluble in each other is copper and nickel. Virtually complete insolubility is exemplified by solid iron and lead.

Table 2.8 - Crystal Structures of Some Pure Metals Capable of Allotropic Transformation

Metal	Symbol	Phase	Crystal Structure	Temperature (t) of Existence — Celsius Scale	Temperature (t) of Existence — Fahrenheit Scale
Beryllium	Be	α	Hexagonal close-packed	t ≤ 1250°C	t ≤ 2282°F
		β	Body-centered cubic	1250°C > t ≥ mp	2282°F > t ≥ mp
Cobalt	Co	α	Hexagonal close-packed	t ≤ 418°C	t ≤ 785°F
		β	Face-centered cubic	418°C > t ≥ mp	785°F > t ≥ mp
Iron	Fe	α	Body-centered cubic	t ≤ 912°C	t ≤ 1674°F
		γ	Face-centered cubic	912° > t ≥ 1394°C	1674° > t ≥ 2541°F
		δ	Body-centered cubic	1394°C > t ≥ mp	2541°F > t ≥ mp
Manganese	Mn	α	Cubic	t ≤ 710°C	t ≤ 1310°F
		β	Cubic	710° > t ≥ 1079°C	1310° > t ≥ 1975°F
		γ	Tetragonal	1079° > t ≥ 1143°C	1975° > t ≥ 2089°F
		δ	Face-centered cubic	1143°C > t ≥ mp	2089°F > t ≥ mp
Tin	Sn	α	Face-centered cubic	t ≤ 13°C	t ≤ 56°F
		β	Body-centered tetragonal	13°C > t ≥ mp	56°F > t ≥ mp
Titanium	Ti	α	Hexagonal Close-packed	t ≤ 882°C	t ≤ 1620°F
		β	Body-centered cubic	882°C > t ≥ mp	1620°F > t ≥ mp
Zirconium	Zr	α	Hexagonal close-packed	t ≤ 862°C	t ≤ 1584°F
		β	Body-centered cubiic	862°C > t ≥ mp	1584°F > t ≥ mp

Foreign atoms can be held in solution in the crystal structure in two different ways. In the first method, a *substitutional solution* is formed when the foreign atoms (solute atoms) replace metal atoms (solvent atoms) at their regular lattice sites. The foreign atoms are usually scattered at random among the metal atoms. In some cases, foreign atoms may enter the solvent lattice on a substitutional basis, but instead of assuming a random distribution, the foreign atoms may have a strong preference to occupy certain lattice sites. Various patterns, or ordering, may be formed by the foreign atoms within the solvent metal lattice, and these ordered solutions are called *superlattices*. Improved mechanical properties, such as greater strength at high temperature, can be achieved in certain alloys when manufactured and treated to develop a superlattice.

The second distinct way in which foreign atoms can be held in solution in solid crystalline material is by forming an interstitial solution. In this type of solid solution, the solute atoms are interspersed among the solvent atoms in the "holes" or interstices that exist in the lattice structure. The interstices will vary in size, depending on the particular lattice structure and the size of the solvent atoms, but in general will be quite small. Therefore, the foreign atoms must be quite small to fit in them and form an *interstitial solution*. Sometimes the lattice structure will allow a small amount of distortion, or straining, in crystal units to accommodate a foreign atom interstitially. For many years it was believed that small foreign atoms capable of interstitial solution would be dispersed uniformly in a lattice structure, but there is evidence that the randomness can be altered by certain conditions. (See Chapter 4, Alloying).

Plasticity in Metallic Crystals

The property of metallic crystals to sustain extensive plastic deformation without fracturing is the quality that makes metal a highly useful construction material. Ionic or covalent bonded crystals do not have this plastic property and tend to fracture. Metals that can undergo plastic deformation for fabrication can be expected to tolerate localized plastic strain without failure as excessive stresses are re-distributed, and can be entrusted with high loads in service. As explained in Chapter 3, this valuable plastic property of metal is not so unfailing that it can be taken for granted. When welding various metals, many special circumstances arise that require an understanding of the deformation process.

A basic understanding of plasticity and other mechanical properties involves further study of crystalline structure to determine (1) what controls the strength of metal, (2) what happens in the crystalline structure as deformation occurs, and (3) how deformation affects the final strength and plasticity of metal.

Slip in Crystalline Structures

A single crystal serves as the best starting point for study of plasticity because it illustrates how the individual grains of a polycrystalline metal will probably behave when stressed. Single metallic crystals are readily deformed because they present a uniformly oriented system of crystallographic planes to facilitate deformation. In some low-melting metals, a load of only about 20 kPa (about three psi) is sufficient to cause permanent deformation, when in the form of a single crystal, and specimens must be carefully handled to avoid plastic strain in the lattice. Fortunately, not all metals are so delicate, and single crystals frequently serve useful purposes, even under rugged mechanical conditions.

The markings that appear on the surface of a single crystal after deformation are very informative. In early studies, when only the optical microscope was available, deformation of a metal crystal was observed to be accompanied by *slip* along certain crystallographic planes and in certain crystallographic directions. Single crystal specimens were given a flat polished surface prior to application of tensile force, and when examined after stressing, a number of parallel slip lines were found on the polished surface as illustrated schematically in Figure 2.13.

Slip and Lattice Orientation

By relating direction of slip lines with lattice orientation in the crystal, it was established that slip occurred more readily between planes most closely packed with atoms and most widely spaced. In the bcc crystal structure, as illustrated in Figure 2.14, the dodecahedral planes, (011) or {110} family, are most susceptive to slip, but the (112) and (123) planes are also amenable to slip.

Because the fcc structure has close-packed planes building outward in layers from the eight corners of the cube, these octahedral planes {111} provide four planes that are highly amenable to slip in the <110> family of directions (12 directions in all, because each close-packed plane has three close-packed directions). Since plastic deformation involves the slipping of close-packed planes over one another in the direction of a close-packed row, the number of close-packed planes and close-packed directions available in a crystal structure determines the degree of plasticity that probably will be exhibited. In a fundamental way, this explains the generally greater plasticity of metals having a fcc structure than those with a bcc structure.

The hexagonal close-packed (hcp) structure is similar in a number of aspects to the fcc structure, but the hcp structure has only one set of parallel close-packed planes on which slip can be readily accepted. Therefore, metals with a hcp structure, in general, do not exhibit as much plasticity as metals with a fcc structure. These early studies of slip are described here mainly to emphasize the value of metallographic examination at magnification greater than that offered by the optical microscope.

Briefly, when slip lines on specimens were first viewed, they were assumed to represent planes where entire rows of atoms had slipped past the neighboring rows en masse, and when the gliding motion stopped the atoms were believed to have instantly re-established regular interatomic binding forces without abnormal separation or fissuring. Although slip could initiate at relatively low load, the metal gained strength as deformation proceeded. This was termed *strain-hardening*. Consequently, when strengthening, or some other condition brought slip to a halt on one plane, the continued application of load would bring slip into play on the next susceptible plane. Continued study of the slip mechanism indicated that slip would occur first on the most susceptible plane when a component of force exceeded a critical value and provided a critical value of stress. This became known as the *critical resolved shear stress law*, and the concept was used to predict the approximate yield stress for particular slip systems in various crystallographic structures.

Slip in Polycrystalline Metals

As research was extended to polycrystalline metal, slip lines appeared to develop by the same mechanism found in single crystals; that is, planes at a particular angle to the direction of tensile force displayed greatest readiness to undergo slip. This angle, usually identified by the symbol ø, is usually close to 45°. As the stress direction becomes perpendicular to a plane of potential slip, a much larger force is required to cause slip to appear. Also, different metals varied significantly in their tendency to strengthen (to strain-harden) as deformation by slipping progressed. In general, metals and alloys with a fcc crystalline structure exhibited the greatest ability to strain-harden or to work-harden, before fracture, and this was credited to the larger number of intersecting slip systems in the fcc crystalline structure.

Slip in polycrystalline metals appeared much the same as observed in a single crystal except that the planes susceptible to this movement had orientations at many different angles to the direction of stress. Therefore, slip lines were found in varying amounts over a scattering of grains where planes by happenstance met the criteria for initiation. During slip line development in polycrystalline metal there were many more obscure occurrences to halt the slipping motion.

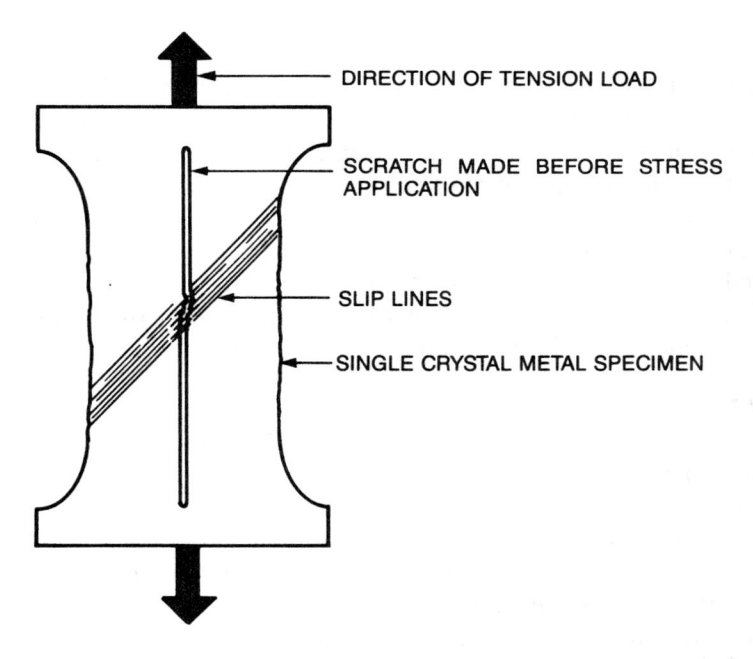

Figure 2.13 — Schematic of slip lines as seen at about 50X on a polished single crystal metal specimen after being stressed in tension to cause slight plastic elongation

A vertical scratch made on the polished surface prior to stress application now displays slight sideways displacements where step-like slipped planes have developed. Plastic displacement at each visible slip line would be approximately 300 Å.

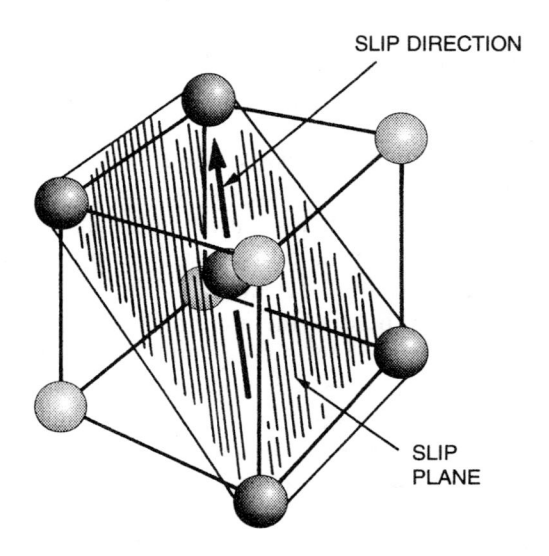

Figure 2.14 — In a body-centered cubic unit cell, closely packed dodecahedral planes (shaded) are most susceptive to slip. Direction of slip is indicated by the arrow.

As illustrated in Figure 2.15, grain boundaries present obstructions to slip because the crystal lattices in adjacent grains are tilted with respect to each other. The angle between the lattices of adjacent grains determines how effective a barrier the grain boundary will be to the slipping motion. If lattices of adjacent grains have almost the same orientation, their boundaries may have almost negligible influence in stopping slip. However, the majority of grain boundaries are positive barriers to slip. Fine-grained metal with a larger amount of boundary region has greater slip-stopping capacity than coarse-grained metal. This helps explain why fine-grained metal customarily has higher strength than coarse-grained metal.

Observing Slip

The planes on which slip occurs are clearly visible under the optical microscope if the polished specimen is examined after plastic deformation (Figures 2.13 and 2.15), but if the specimen is repolished the slip lines, or steps, almost disappear. However, the metal in the immediate vicinity of slip is strain hardened, and often these regions will appear in slight relief on a repolished specimen. Etching usually attacks the slip planes more deeply than unslipped metal, and therefore they are revealed. While examples of slip lines in Figures 2.13 and 2.15 are shown as straight, parallel lines (obviously following a particular system of planes in each grain), this is not always their appearance. In iron, for example, when stressed at room temperature, slip may occur on several systems of planes, depending on which are in a favorable position. Slip lines that appear on a polished surface of the iron specimen often are forked, irregular, or wavy, and apparently are not closely controlled by any particular planes in the systems available.

In early studies of slip lines, the slip process was envisioned to involve several thousand unit cells on a plane gliding across its neighboring plane so as to displace one section of the crystal or grain relative to the other. When the slipped plane came to a halt, registry or binding immediately was resumed between its atoms and those on the adjacent plane. However, continued examination of slip disclosed that crystals preferred to deform by partial sliding rather than complete sliding of a plane. Slip lines usually could be seen on a number of parallel planes about one micron apart, but there was some evidence that slipping was not continuous. Instead, the slip line seemed to be the end result of a more complex mechanism in which the planes moved in discrete steps.

Twinning in Crystalline Structures

Another distinct mechanism of deformation that can be observed by optical metallography is known as *twinning*. This mechanism does not involve slip, but it develops in certain forms of crystal structure by a unique process of homogeneous shear. Twinning was first observed in mineral crystals, but was later recognized in those metals and alloys with crystal structures which did not favor deformation by slip.

Twinning derives its name from the deformation twins that develop when a localized portion of a crystal or a grain rotates under stress to produce a narrow band in

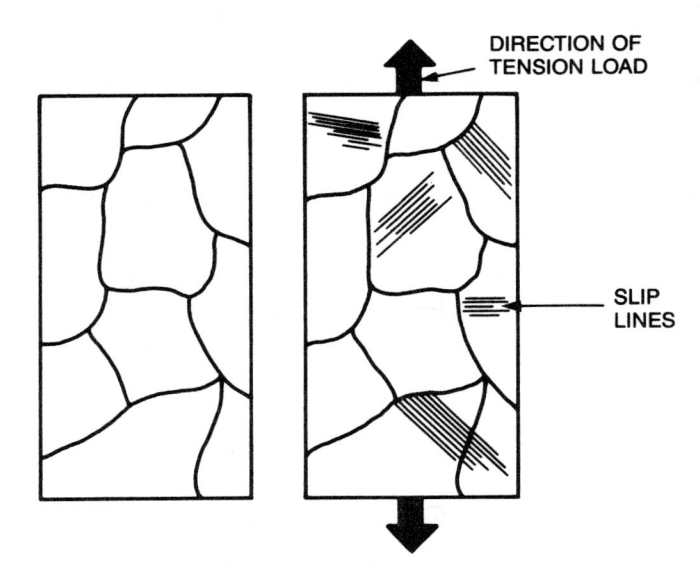

Figure 2.15 — Schematic of slip lines in polycrystalline metal

On left; before stressing, polished and etched microstructure at about 100X displays no lip lines. On right; after application of tension, grains with most favorably oriented planes are first to develop slip. Note the obstruction to continuation of slip presented by grain boundaries.

which the lattice is given new orientation. This particular rotation involves layers of atoms that undergo the same amount of shear, one by one, to generate a newly oriented lattice, as illustrated in Figure 2.16. The atoms in each layer move only a fractional distance, and the new configuration is a mirror image of the parent's crystal lattice, as viewed by reflection from the twinning plane. Twins are seen during metallographic examination because differing orientation between the original lattice and the twinned portions affects their etching rate. However, if the twinned bands are extremely narrow, as illustrated schematically in Figure 2.17, identification by metallography alone may be difficult. x-ray diffraction usually provides positive differentiation between twinned bands and slip lines.

Twins form very rapidly, and can be expected to occur under lower applied loads than those required to produce slip. Some metals emit audible sounds when twinning occurs, and "tin cry" is a classic example. Some metals are difficult to mechanically polish for metallographic examination without twinning occurring on the polished surface. There is evidence that twinning takes place to facilitate subsequent deformation by slip. The rotation in the twinned portion places planes at a more favorable orientation for slip to occur. The formation of twins under an applied stress is a common mode of deformation in metals having a hcp crystalline structure, and also in metals having a bcc structure, when the stress is applied at low temperature.

As metallographers continued to report observations of slip lines and twinning in metals, they speculated on the mechanisms that probably governed these occurrences during deformation. Scientists conducting fundamental studies of the strength of metals began to find significant discrepancies in previous conclusions. For example, actual measured values for tensile strength and yield strength were not in agreement with

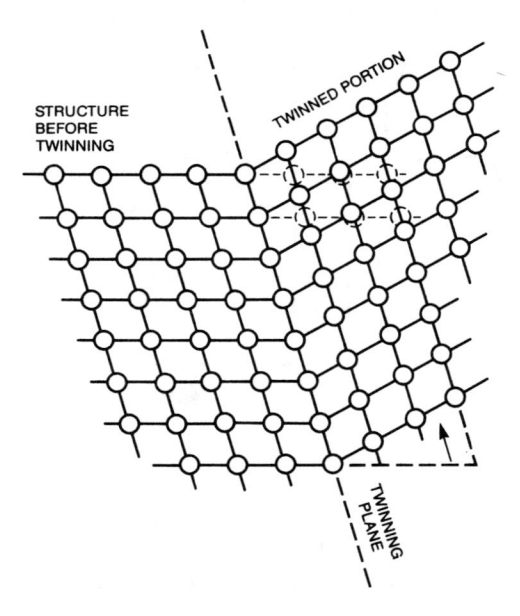

Figure 2.16 — Schematic showing shift or rotation of atom positions during twinning in crystal structure

(A) GRAINS BEFORE APPLICATION OF TENSION STRESS

(B) SAME GRAINS AFTER APPLICATION OF TENSION STRESS

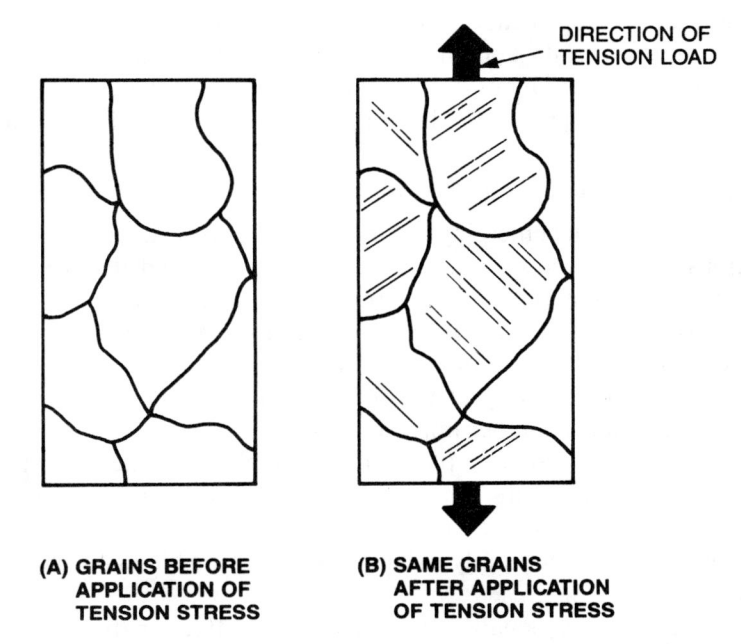

Figure 2.17 — Schematic of twins at about 100X in microstructure of polycrystalline metal

On left; before stressing, grains display no twinning. On right; after subjecting metal to stress, twin bands have developed in grains that offered favorable crystallographic orientation for this deformation mechanism.

values calculated from laws governing interatom binding forces. Calculations indicated that simultaneous slip of an entire layer of atoms (pictured as a slip line by the metallographer) would occur when the shear stress in the plane was about 690×10^3 MPa (approximately 100 000 ksi). Measured values in experiments were several orders of magnitude lower. While metals were being praised by some for being so strong, the materials scientist was concerned because they were so weak! Indeed, metals exhibited strengths that were 10 to 1000 times lower than judged from interatom binding forces.

In about 1920, metallurgical investigators began to center attention on the difference between predicted and measured strength of metals. One of the first tentative explanations for this difference was that metal crystals contained submicroscopic defects, such as minute cracks. These defects were suspected to detract from the calculated strength in a straightforward mechanical way, but there was no consistent evidence to support this belief. However, after about ten years of additional research, scientists became convinced that real crystals and polycrystalline metals must contain some form of imperfection that exerts a controlling influence on the mechanical behavior of metals as they are deformed under load. Although lacking the means to actually view the atoms making up the crystalline lattice, they predicted that on an atom scale metals contain a variety of imperfections. These were described as "elastic defects", "disclinations" and "dispirations."

Still without microscopes capable of resolving atoms forming the crystalline lattice, scientists became firmly convinced of the probable nature of lattice imperfections, and the term *dislocation* emerged in the mid-1930s as the generic name for the lattice imperfections that appeared to be at the root of the strength problem.

Lattice Imperfections: Dislocations

An important part of the emerging theory of dislocations concerned the movement of imperfections through the crystal lattice. It was postulated that when stress was applied to a crystalline material, certain forms of dislocations would move quite easily by consecutive displacement of single atoms. In this way, a comparatively low level of stress could set a dislocation in motion along a plane of atoms. It was believed that two moving dislocations meeting at the intersection of planes in the lattice would sometimes merge and form a larger imperfection in the lattice. In other instances, two dislocations moving on intersecting planes might "pin" each other and start a pile-up of dislocations. Depending on the number and distribution of dislocations which coalesced during the application of stress to a metal, the lattice might develop a microvoid, a plane of slip, or simply exhaust itself of possible dislocation generation and movement.

Because of the extremely small size of dislocations affecting the crystalline lattice, estimated to cover a distance less than 100 Å, it was impossible for investigators to observe this abnormality with optical microscopes. They concentrated on indirect means of rationalization, and applied mathematical analysis with some measure of success. The realization grew that an entire plane of atoms need not slide rigidly across an adjacent plane during deformation. Instead, slip lines could be an end result of copious dislocation movement.

As researchers proceeded in the study of dislocations in metal, there was considerable enthusiasm for the premise that a basic mechanism of deformation had been conceived which would provide answers for many questions concerning strength, plasticity, work hardening, and other phenomena of metal properties. In this early period of dislocation study, it was observed that a carefully prepared, nearly perfect single crystal of a metal (one that probably had few dislocations) possessed tensile strength or critical shear strength of a very high order. Iron whiskers, for example, which were virtually dislocation-free, displayed a yield stress in the vicinity of 12×10^3 MPa (about 1700 ksi). However, immediately on introducing dislocations into the crystalline structure, the actual strength was considerably reduced. This phenomenon was explained in terms of movement of the smaller dislocations present in the structure by relatively low applied stress. It was further observed that as the number of dislocations or imperfections grew larger, the yield strength began to increase, but the strength did not reach the very high level achieved by the dislocation-free metal whisker specimens. When plastic deformation of a specimen was carried beyond yielding, new imperfections seemed to form in the crystalline structure and the original ones appeared to be annihilated.

The overall effect that can be clearly seen from cold working of metals is to increase the stress required to cause further deformation. At first thought, it may seem incongruous that dislocations can move in a crystalline lattice by having atoms jump from one lattice site to an adjacent one so that a void, or an atom vacancy, will be shifted and the crystal or grain will still remain structurally sound. However, it must be noted that the direction of the bond between atoms is not fixed. Atoms are held flexibly, rather than rigidly coupled together, and a dislocation can progress across the lattice of a crystal like a wrinkle moves across a carpet on a floor; that is, only a small portion of the carpet is displaced at a time, despite the fact that the wrinkle moves a considerable distance. Similarly, a dislocation can move through the crystal lattice over relatively large distances without permanent parting of the interatomic bonds that would allow a rift or gap to form. In the earliest stage of deformation under stress, movement is accomplished by the gliding of some form of extremely small dislocations along planes. The number of dislocations usually available to initiate this process is very large, perhaps in the order of one million per square centimeter. Of course, this dislocation population is only a small fraction of the lattice where atoms occupy all of the sites normal for the particular crystal form.

Shortly after the electron microscope was introduced in about 1950, the remarkable magnification capability of this instrument provided direct confirmation that dislocations actually exist. With transmission electron microscopy (described in this chapter), and ingenious techniques for "decorating" the lattice imperfections with fine precipitates of foreign atoms to make them more discernible, an appraisal could be made of the various forms of dislocations. This breakthrough greatly aided studies of not only the formation of dislocations, but also the influence of these lattice imperfections on the properties of metals.

A tremendous amount of information appears in the literature regarding dislocations, and much of it is pertinent to welding metallurgy. The microstructural conditions that develop in the vicinity of welded joints involve unique arrays of dislocations that can affect the behavior of weld metal and base metal heat-affected zones. Examination of a few of the many forms of dislocations that exist in the crystal lattice of metals will provide general familiarity with their origin and size, the manner in which they move through a crystal lattice during deformation, and the ways in which they can affect the properties and behavior of metals.

Point Defects

Point defects are vacant atom sites which are believed to exist in the crystal lattice of all metals. Electron microscopy capable of resolving this single unoccupied atom position verified the existence of point defects in about 1955. Because this task requires a magnification of nearly one million times, not many laboratories are equipped to accomplish it.

When dealing with the very small atoms that are normally held in interstitial solid solution, it is possible to have unoccupied interstitial sites constitute a point defect. Although point defects are numerous in most sections of metal, they obviously represent the minimal dislocation, per se, and therefore they play a minor role in the mechanics of deformation. Under some circumstances, point defects may migrate together to form slightly larger dislocations termed a divacancy (a pair of adjacent vacant atom sites), or a trivacancy (three unoccupied neighboring sites). As conditions encourage increasing numbers of point defects, they may tend to gather in clusters; the

86 Chapter 2

formation of clusters will depend on the distribution of internal stress. Despite the minor role of point defects in the deformation process, their numbers will increase exponentially as the temperature of metal is raised. The movement of these single vacancies in the lattice is an important part of the mechanics of elemental diffusion in metals. Point defect movement is also important to the mechanics of the diffusion welding process.

Edge Dislocations

Edge dislocations, sometimes referred to as *line defects*, commonly exist in all crystals, and they are particularly abundant in polycrystalline metals along the grain boundaries. An edge dislocation is the edge of an incomplete plane of atoms as shown in Figure 2.18, a schematic representation normal to the cross-sectioned plane of atoms in question. In this illustration, the plane extends down from the top of the crystal and terminates about midway through its depth. The symbol ⊥ is commonly used to indicate the lower edge of the dislocation. This imperfection is also considered to be a "line defect" because the area partially void of atoms exists for some distance above and below the intersecting plane of the illustration. Naturally the abnormally structured lattice along the terminus (⊥) of the edge dislocation is an area of locally increased stress and free energy.

The usual stress pattern associated with an edge dislocation, as illustrated in Figure 2.18, is compressive force on the plane above the edge, and tension force between a number of planes below. These acting forces produce a vector that is perpen-

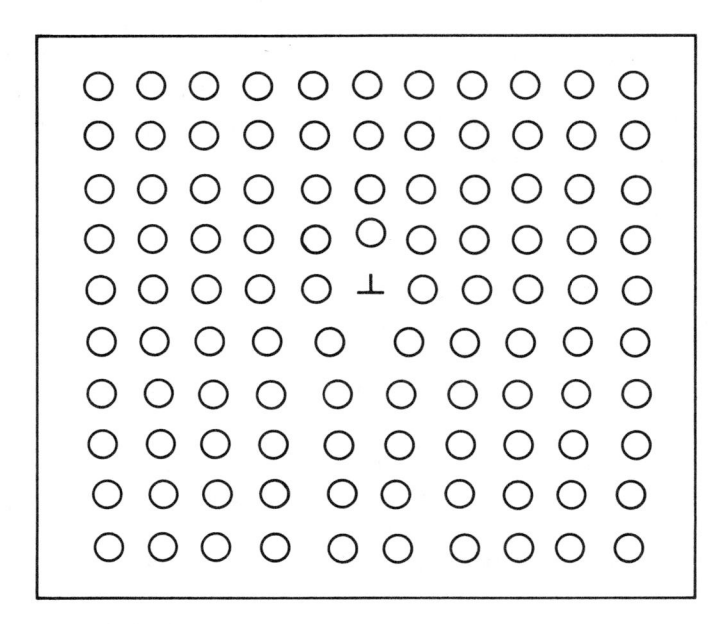

Figure 2.18 — Edge dislocation or line defect in crystal lattice of a metal as viewed in two dimensions. The third dimension (length) is perpendicular to plane of the schematic sketch.

dicular to the dislocation line, but normally the forces are in balance, and no lateral movement of the dislocation occurs. However, external application of only a small load can result in an imbalance, and the dislocation will be set into motion to achieve equilibrium. This explains the relatively low true yield strength of crystalline materials.

Grain boundaries are believed to be made of edge dislocations, as schematically represented in Figure 2.19. The angle of disorientation between the lattices of adjacent grains has a significant effect on the array of dislocations along the common boundary. Large-angle grain boundaries contain regions of good and bad fit, and overall they offer relatively high free energy. Low-angle boundaries are those that have less than about 15° of disorientation, and they are made up of dislocation arrays with much lower energies than those in large-angle boundaries. The low-angle boundaries also have been called *tilt boundaries*. Arrays of dislocations can gather within grains of metal and divide the grains into the subgrains. The subdivided areas have virtually no difference in lattice orientation, and yet electron microscopy shows the subgrains to be faintly outlined by boundary-like arrays of dislocations. Grain boundaries are a very important feature of the microstructure of metals because they often play a leading role in metallurgical phenomena. In general, grain boundaries can be viewed as "internal

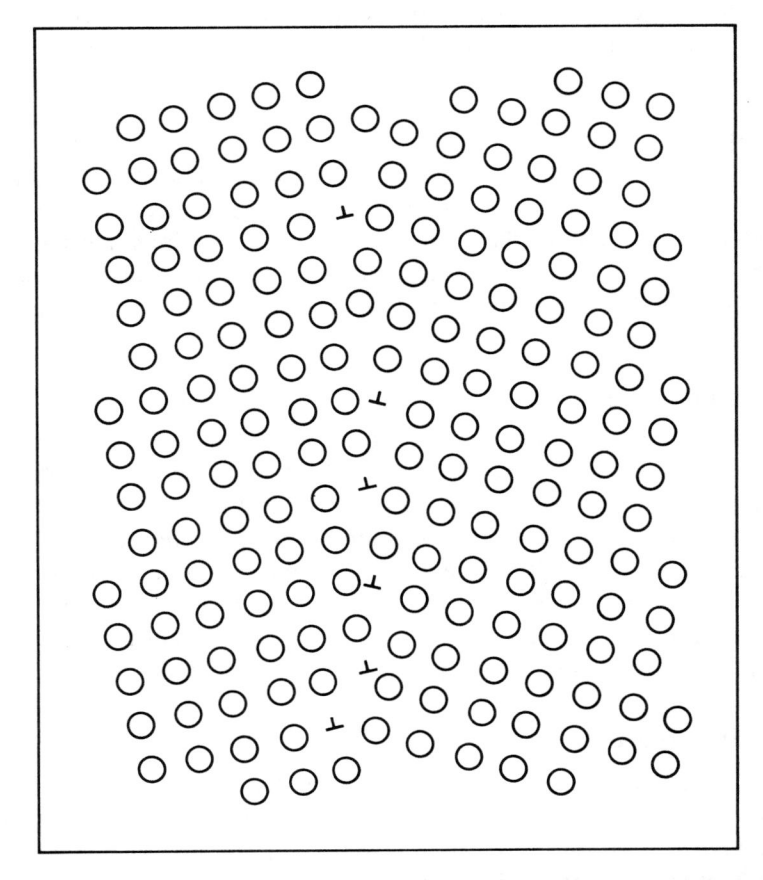

Figure 2.19 — Schematic of boundary between grains of differing crystallographic orientation showing multiple edge dislocations

surfaces" offering various levels of free energy. The amount of energy available is obviously related to the kinds of dislocations present and their population along the boundary. The manifestation of dislocations in the arrays is often the key to understanding larger-scale structural features and the properties associated with them.

Screw Dislocations

Screw dislocations exist when a portion of a crystal has a continuous plane in the form of a helicoid through its lattice. This unique form of lattice imperfection develops when an edge dislocation skips over a region and causes displacement in the lattice in a parallel direction. As a result of this slippage, a portion of the crystal does not now have parallel planes of atoms, but instead has a single plane in the form of a helical ramp or screw. The center of the screw dislocation is fixed, and the step rotates around that point.

The nature of a screw dislocation is that an incomplete face (or step) is always present on the top face of the crystal, as illustrated in Figure 2.20. Screw dislocations are believed to play an important role in the solidification of metals, inasmuch as spiral growth can often be observed on the surface of a crystal taken from solidifying melts. As can be seen in Figure 2.20, when all of the screw dislocation has emerged from the lattice, the crystal is again perfect, but the portion on the left in the illustration is displaced upward by one unit cell from that on the right.

Stacking Faults

Stacking faults are two-dimensional imperfections in a crystalline lattice consisting of planes that are in error by less than a full lattice parameter. They can be found in the growth of the crystal, or they can result from dislocations that move only partially through the lattice. Unlike a full dislocation that produces displacement equal to the distance between lattice points, a stacking fault may amount to no more than the bending of the lattice planes.

Other Lattice Imperfections

Additional lattice imperfections, or dislocations, have been identified and designated as dipoles, multipoles, nodes, tangles and pile-ups. The nature of these lattice abnormalities, their mechanism of development, and their significance in terms of mechanical properties is extensively documented in literature on the subject. These resources, continually growing in coverage and complexity, document the behavior of metals during fabricating operations, including welding, and during exposure of the weldments to particular service conditions. The many different types of dislocations are defined in Burgers vectors, which are discussed in Technical Brief 6.

Cold-Working Metals

The cold-working of metals and the resulting changes in strength, ductility, and toughness are important areas from a dislocation viewpoint. The number of dislocations is greatly increased during plastic deformation; although it might seem that as deformation continues, all of the dislocations in a crystal might be used (by moving to the surface) so that slip would have to stop. However, experiments show that new dis-

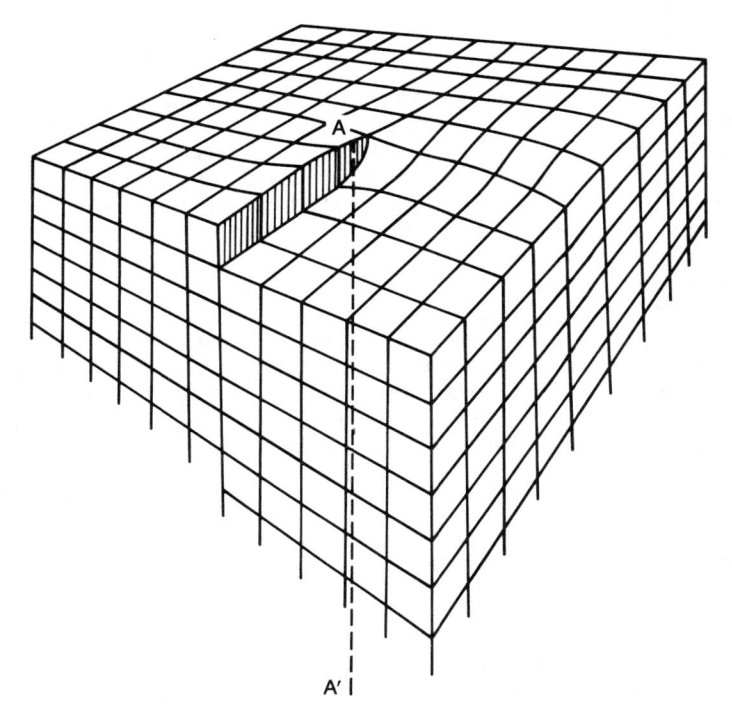

Figure 2.20 — Screw dislocation (A-A') as shown schematically on the face of a multi-cell crystal

Instead of showing atoms on their sites, the individual cubic unit cells are outlined to portray clearly the advance of the rotating screw plane.

locations are continually generated during the deformation process itself from Frank-Read sources, so that slip on any given plane can continue indefinitely.

Work-hardening or strengthening of a metal or alloy occurs when obstacles to the usual easy passage of dislocations through the lattice are encountered. Fundamentally, any lattice inhomogeneity that has an associated stress field can be an obstacle to the easy glide of a dislocation striving to migrate under imposed conditions. These obstacles may already be present in the lattice, or they may be generated during straining and deformation. Two examples of obstacles are (1) grain boundaries (with their usual array of dislocations) cutting across the plane on which dislocation movement is favored to proceed, and (2) the presence of solute atoms of different size that can cause localized distortion in the lattice. In fact, the strain induced by dislocations in the lattice can cause impurity atoms to be attracted to these stressed sites, and these odd-size atoms form an aggregate in the strained region. These aggregates of impurity atoms are termed *Cottrell atmospheres*, and are believed to play a part in causing etch pits, which are used by materials scientists as a means of identifying dislocations on a polished surface being examined by optical metallography. The effectiveness of an obstacle to block dislocation movement depends on (1) its strength — that is, resistance of

> **TECHNICAL BRIEF 6:**
> **Burgers Vectors**
>
> Burgers vectors are unique to each type of dislocation. The Burgers vector defines the displacement that occurs when a dislocation passes a given point in the lattice. When viewed rearward from the aftermath of dislocation movement, the Burgers vector can be described as the exact reverse translation which would restore normality to the particular crystal lattice. This method of tracking dislocation movement in the lattice provides the opportunity to study the relationships between metallurgical phenomena and specific dislocation movement when a crystalline material is strained. For example, motion of a dislocation perpendicular to its slip plane is known as climb, and the Burgers vector indicates that climb occurs by mass transport. This transport is induced either by time and temperature acting in the diffusion regime for the metal, or by high external load that rearranges atoms by sheer force. As another example, when two non-parallel dislocations cut through one another, the result is a jog, which is important in analyzing dislocation mechanics. The jog is very difficult to visualize or to illustrate schematically by line-drawing, but the use of Burgers vectors allows exact tracking of the dislocation movements involved.

the dislocation to passage through itself — and (2) the spacing between obstacles. Of course, dislocations approaching each other interact because each is surrounded by its own localized stress field.

The science of predicting the kinds of dislocations generated in a particular metal by a certain processing operation is virtually in its infancy, but discovery of the dislocation in the crystal lattice of metals has provided the key for revealing the secrets of mechanical properties, and much more.

Crystals of material other than metals contain dislocations, but they do not exhibit the plasticity or the work-hardening characteristics of metals. Covalent crystals, for example, have directional bonds between the atoms, and the dislocations present are difficult to move. The ease with which a dislocation can be set in motion in a metal is determined mainly by dislocation width; that is, the distance along a slip plane where atoms are not in their normal sites (even if only slightly out of line). The wider the dislocation zone, the lower the stress needed to keep it moving. Calculations confirm that the required force to put a dislocation in motion is at a minimum when the projected plane is close-packed and movement is in the close-packed direction.

The *meshlength theory* is receiving much attention as a guiding principal in understanding work-hardening of metals. This theory has been written in equation form, and it encompasses a number of factors, parameters, and constants that require considerable description. Briefly, however, this theory proposes three basic stages of work hardening. In Stage I, when the initial dislocation density is low, single dislocations from isolated sources glide along free paths until they interact in a pile-up to form boundaries of small cells within the grains. These cells are only about one to three microns in diameter (0.04 to 0.12×10^{-3} in.). As deformation is increased and a greater number of dislocations move through the lattice, Stage II occurs as the average path available to permit dislocation movement shrinks inversely with the root of dislocation density. In Stage III, the free dislocation path is constant, and "forest" dislocations pile up and become stabilized. The pile-ups represent centers of intense internal stress fields, but they continue to have mobility under stress. The meshlength theory holds that a momentary flow (or yield) stress for a crystal can be computed specifically as that stress necessary to bow out dislocation links and thereby effect plastic deformation.

Cold-worked metal is usually annealed to reduce hardness and to increase ductility. Dislocations in the lattice of the metal play a specific role in the annealing process. For many years, metallurgists were perplexed about the reduction in strength that occurred in cold-worked metal at the start of annealing, because no apparent change in the microstructure could be seen by optical metallography. Electron microscopy however, has established that the first changes to occur in the lattice of the cold-worked crystalline structure during annealing are (1) redistribution of dislocations, (2) a decrease in the number of dislocations, and (3) outlining of subgrains by dislocations. These changes in dislocation population and distribution are the early stages of the annealing process and require the high magnification capability of the electron microscope for detection and assessment.

When applied to weldments, knowledge of dislocations will assist in developing rationale for a number of the phenomena encountered in welded steel. For example, untreated weld metal of steel has unusually high yield strength, and this is related to the very large number of dislocations that normally appear in the lattice of the newly solidified weld metal. Weld strength after reheating by multi-pass welding or by post-weld heat treatment will depend on the disposition of these dislocations during exposure to temperature, time and stress.

EXAMINATION OF METAL STRUCTURES

In the preceding portion of this chapter, elementary particles of matter, atoms, molecules, crystal lattices, and even dislocations in lattices were described, despite the fact that the descriptions have been based largely on indirect evidence. Nevertheless, these indirect methods of obtaining submicroscopic information about metal structure have been spectacularly successful. Neutrons, protons and electrons can be studied with the aid of the cyclotron, cloud chamber, cathode-ray tube, and a multitude of other instruments. Atoms can be located in crystalline materials, even in opaque material like metal, by use of x-rays, electron, or ion diffraction techniques. In addition to fundamental knowledge secured using very high magnification microscopy, much information of a more overt nature has been gathered by workers who deal with the practical problems that occur during production and fabrication of metals. Consequently, there is reason to look at the entire array of methods employed to examine metal structures.

Fracture Appearance Assessment

Undoubtedly the first evaluation of metal structure took place when the size of crystal-like facets on the fractured surfaces of broken sections of metal were closely inspected by the unaided eye and noted. This visual examination, or one made with a low-power magnifying glass, can provide preliminary determinations of the possible cause of a problem. Programs were established to assess and record these observations; they involved a series of fracture-appearance specimens which were prepared to serve as standards for metals when tested by a "nick-break" procedure. Some of these fracture evaluation programs were successful, but most have been set aside as more discerning methods of fracture examination and metal evaluation were developed.

Prior to the use of metallography, the texture of a fracture was believed to be indicative of the grain size of the metal. This belief was not always correct. Fracture texture can be used to make a rough approximation of grain size, but the method is reliable only when a number of variables which affect fracture texture, such as chemical composition, temperature, and loading method for fracturing the specimens, are closely controlled. Consequently, the search continued for a better way of determining conditions in the metal that affected both the propensity to fracture, and the texture of the remnant surface. Greater magnification was clearly indicated. Although this now may seem an easy route, early microscopes that employed a magnificaton in the range of 5 to 100X were not very suitable for examining the rough topography of a fractured metal surface mainly because of limited depth of focus. The real rewards of fracture examination are secured at very high magnification, where facets, dimples and other revealing markings can be distinguished, indicating both internal structure and fracture development and causes. These studies, made possible by certain advances in microscopy, are termed fractography, and are described near the end of this chapter. It is worthwhile to note that "double-teaming" fractographic examination with metallographic examination will maximize findings.

Metallography

The term metallography originally covered the microscopic study of metals under substantial magnification, and the recording of microstructural details by photography. Initially all work was done with an ordinary visible-light microscope, and the photographs made of details observed at various magnifications were called photomicrographs. In about 1960, the electron microscope was put to use for examining metals. Although examinations with the electron microscope represented an extension of metallography, the total activity was divided into two principal branches, optical metallography (using the visible-light microscope), and electron metallography. Photographs of magnified images obtained by electron microscopy are called electron micrographs.

Metallography Using Optical Microscopy

Much of the practical knowledge of the structure of metals was obtained using the optical metallograph. This is a special microscope with an inverted stage that allows a flat specimen to be placed face-down on it, so that portions of interest on the specimen can be scanned. The metallograph usually has an integral camera, and can have a number of features for changing specimen illumination, and for measuring details observed on a specimen (see Figure 2.21).

Metallographic examination requires a metal specimen (usually not over 25 mm square, or 1-inch square) that is cut to provide a flat surface for preparation and survey. This surface may represent an arbitrarily chosen plane to determine the general microstructure present, or it might be directed to intersect an area of specific interest, such as a weld, or a heat-affected zone containing a defect. This flat surface is ground and polished by a specific procedure until it is as scratch-free as possible. The metal should not be allowed to heat significantly during cutting and grinding because heat could change the microstructure. Grinding cannot be done vigorously, because this can produce flowed metal on the specimen surface and this distorted metal can mask cracks, and will possibly alter the microstructure. A complete procedure for preparing metallographic specimens can be found in ASTM Standard E 3, *Standard Methods for Preparation of Metallographic Specimens.*

A polished specimen surface, when examined with the metallograph at a magnification in the range of 100 to 500X, is uniformly reflective and featureless unless there are cracks, porosity, or nonmetallic inclusions in the metal. Pits and scratches caused by polishing should not be confused with the structure of the metal, or mistaken for porosity. Nonmetallic inclusions often require a very careful polishing technique so they will remain fixed, thus permitting assessment of their origin and significance. Occasionally, a harder metallic constituent will polish in raised relief, and will be visible in the unetched specimen.

Etching to Reveal Microstructure

Etching the polished surface of the metallographic specimen to reveal microstructure can be done in a number of different ways, depending on the metal or alloy, and conditions such as whether the material is cast, wrought, weld metal, etc. Many ferrous specimens can be etched merely by dipping for a few seconds in a solution of one to five percent nitric acid in alcohol (commonly, two percent nital). Metals and alloys that are resistant to acid attack, like nickel or stainless steel, can be etched electrolyt-

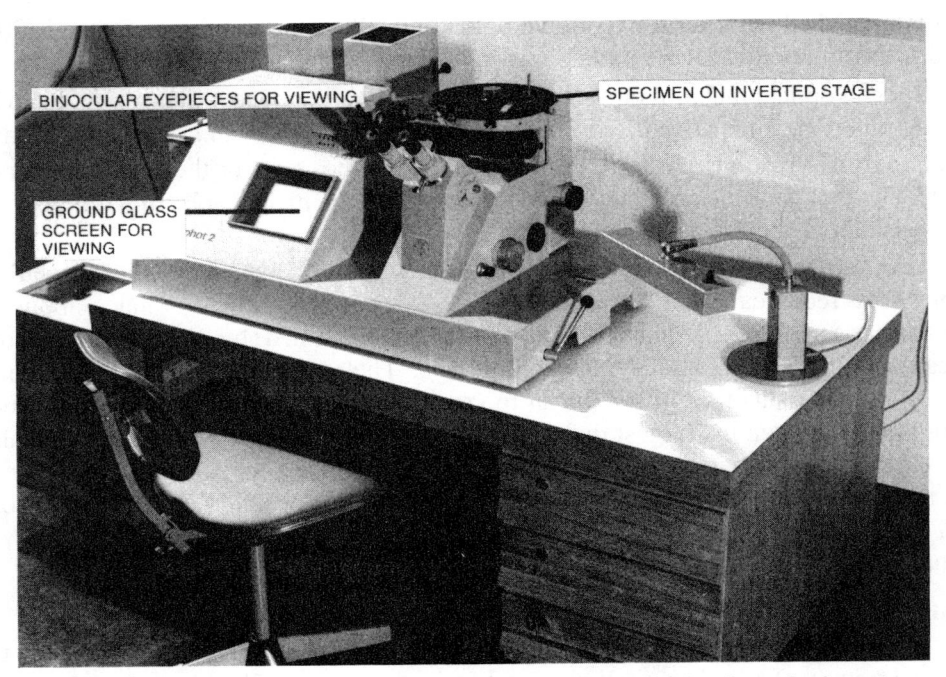

Figure 2.21 — Metallographic optical microscope for examination of microstructure in metals

(Courtesy of The Welding Institute, England)

ically by having the specimen serve as the anode in a small cell containing a suitable dilute acid solution, or by using an acid solution of much higher concentration for immersion etching. The entire procedure for etching the polished surface to reveal the microstructure must be carefully controlled. Even rinsing and drying is important, because stains left behind can mask or confuse the appearance of microstructural details. ASTM E 407 provides *Standard Methods for Microetching Metals and Alloys*.

Specimens containing a weld often present a challenge to the metallographer because of marked differences in etching rate between base metal and weld metal, especially when working with dissimilar metal welded joints. Sometimes specimens incorporating dissimilar metals may require electrolytic etching of the more resistant metal (during which the less resistant metal is electro-polished), and after the electrolytic etching, the less resistant metal is etched by immersion in a dilute acid which does not affect the more resistant metal. Proper selection and application of etchants to metallographic specimens has become highly refined, and many microstructural constituents can be identified by their appearance in the etched specimen.

Ultramicroscopic Examination Methods

The capability limits of the visible-light microscope must be heeded when stretching optical metallography to its uppermost magnification. Resolving power is an important consideration. This is the ability to make two closely spaced points visible as separate entities. For example, a resolving power of one micron (1 µm) (0.000 0394 in.) means that two point-shaped particles located one micron apart will appear just separated. With optimum conditions and best technique, the smallest particle that can be distinguished with the conventional optical metallograph is approximately 1750 Å or 0.175 µm (about 7×10^{-6} in., or seven-millionths of an inch).

Special optical metallographs equipped with a quartz optical system and ultra-violet illumination can resolve particles as small as 800 Å (3×10^{-6} in., or three-millionths of an inch). To resolve questions concerning the composition, or the crystalline structure of constituents, examination can be extended to include electron microprobe analysis and x-ray diffraction methods.

Laser scanning microscopy (LSM) is the newest procedure for optical metallography (see Technical Brief 7), but even with its capabilities, there is much about the crystalline structures of metals that cannot be examined by optical microscopy because of its intrinsic limit on magnification. The unit cell, which is about three angstroms across its face in iron, is far too small to be revealed under the optical metallograph, and even the largest dislocations, which are perhaps 100 Å overall, cannot be resolved. Nevertheless, very useful information can be secured by examination in the range of about 100 to 500X. Also, the limited magnification of optical microscopy can be aided by other capabilities or special techniques, such as the use of polarized light, or dark-field illumination to distinguish and to identify certain constituents.

Grain Size Assessment

The first feature usually noted by a metallographer during examination of the microstructure of a polycrystalline metal is its grain size. The size of grains often exerts a profound effect on the properties of a metal, particularly mechanical properties. When a specimen of solid metal is prepared for metallographic examination, and etched to delineate its grain boundaries, some difference in the size of grains will always be observed. The normal variation in size is explained in Figure 2.22, in which a mass of grains in a metal section has been cut by plane abcd. When this plane is examined under the metallograph, some grains appear larger because they have been intersected by the plane at approximately the mid-section; whereas the smaller grains represent intersections at various distances away from the mid-section. The smallest grains obviously represent the extreme tips or nodes cut by the intersecting plane. Generally, when grains of a metal are of normal uniform size, the field of a metallographic specimen will have about three-fourths of the grains within one size category.

Sometimes a considerable variation in the size of grains will be observed in metallographic examination; therefore the range of sizes present is carefully measured and noted. The most prevalent grain size is estimated. Both the range of grain size, and the distribution of sizes can have an influence on metal properties.

Grains in metals and alloys can differ greatly in size, from extremely small grains barely distinguishable under the optical metallograph at maximum magnification, to

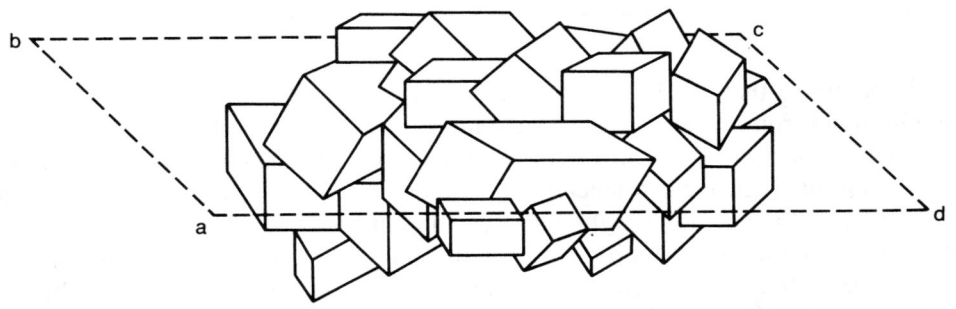

(A) A SAMPLE OF SOLID POLYCRYSTALLINE METAL BEFORE BEING CUT FOR METALLOGRAPHIC EXAMINATION

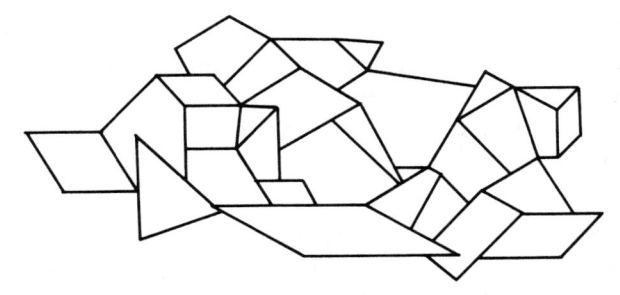

(B) VIEW FROM ABOVE BY METALLOGRAPHY ON PLANE abcd

Figure 2.22 — Grains of varying size are observed metallographically (B, lower sketch) because the sectioned plane (A, shown as abcd in upper sketch) intersects bodies of grains at various positions

The grains have been drawn as cubic and rectangular solid bodies in upper sketch only for clarity instead of the polygonal or rounded shapes observed commonly in the microstructure of metals.

> **TECHNICAL BRIEF 7:**
> **Laser Scanning Microscopy**
>
> Laser scanning microscopy (LSM) is one of the latest advances in optical metallography. LSM uses a laser for illumination that provides a beam of selected wavelength, suitable for the material being examined, to produce the kind of reflective response desired. LSM presently utilizes gas lasers producing argon blue with λ at 488 nm, helium-neon red with λ at 633 nm, or with IR at λ 1152 nm. The laser beam is focused to an extremely small spot and is made to rapidly scan a selected field. A spatial filter frames reflection from the spot to eliminate any unwanted reflection from light rings surrounding the spot's Airy disk or from planes other than the one targeted at the focal point for scanning. The image of the field is collected on-the-fly and is recorded by a detector. A number of field images are made from multiple planes by changing the levels of beam focus, perhaps by increments of one micron. All reflected illumination signals are processed electronically into digital information and stored in a computer as superimposed images. The optical-depth sectioning performed plane-by-plane provides the remarkable depth contrast when the whole field image is displayed on a CRT screen or is printed as a photomicrograph. The stored digital information can be transmitted directly over a network and thereby avoid any degradation in resolution as commonly occurs when a photomicrograph is scanned and transmitted via FAX.

very large grains that can be approximately finger-nail size visible with the unaided eye. Grains in iron and steel castings commonly have a mean diameter of about 0.1 mm (approximately 0.004 in.), whereas wrought products usually display a finer grain size, perhaps having a mean diameter in the order of 0.05 mm (about 0.002 in.). A photomicrograph of a typical wrought polycrystalline microstructure at 100X is shown in Figure 2.23. The highly magnified images in Figure 2.23 might be a misleading perspective; to realize a true concept of grain size, it should be noted that a cubic millimeter of this metal contains approximately ten thousand of these tiny, intimately bonded solid bodies (grains), or about 180,000,000 grains in a cubic inch.

In most metals and alloys either grain growth or grain size reduction can be accomplished through working, heat treatment or both. Although processing steps are frequently used to produce grains of a size that will assist in obtaining the metal properties desired, care must be exercised that some oversight in handling the metal does not result in a highly unfavorable grain size. This can happen in any one of a number of ways, and for this reason grain size is the first microstructural feature that should be measured and recorded.

Measuring Grain Size

In measuring grain size, the precise magnification of the view under the metallograph must be considered. The specimen of iron, illustrated in Figure 2.23, has a mean grain diameter of about 0.05 mm (0.002 in.), and this allows only about 25 grains to be seen in a 25 mm square (1 in.) area under the metallograph. If the examination is conducted under less magnification, for example 75X, then approximately 45 grains would be seen in the same area.

A basic problem in developing a method of measurement is that the observer is not looking at the three-dimensional crystals in a metallographic specimen, but only at a two-dimensional pattern of grain boundaries. As explained in Figure 2.22, the surface prepared for examination will be a plane that may cut through one grain through its middle, which will be seen as a large grain, whereas another may be cut through one corner and will be seen as a small grain. On the average, what is seen in a specimen is not the maximum cross-section of the grains, but only a fraction of their cross-

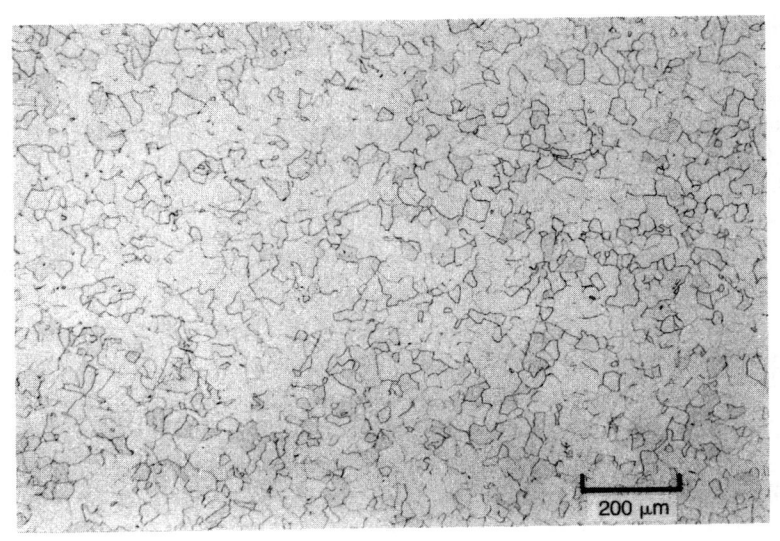

ORIGINAL MAGNIFICATION 100X

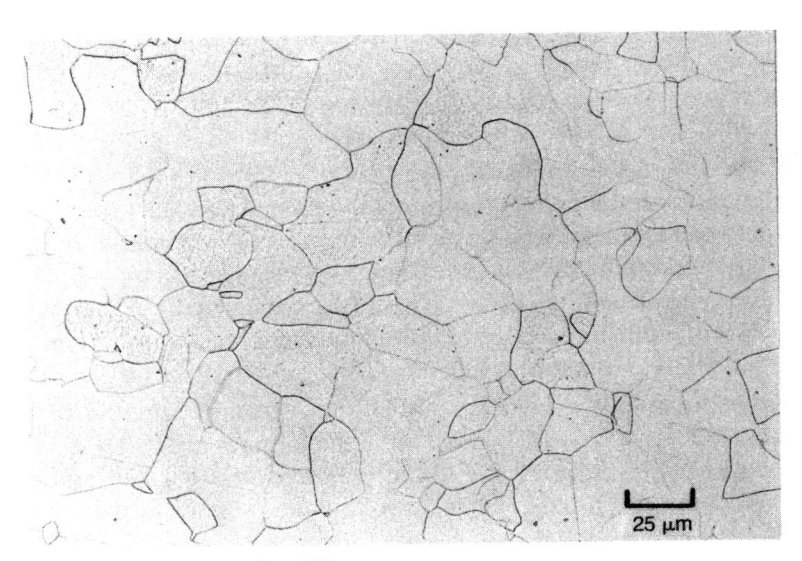

ORIGINAL MAGNIFICATION 500X

Figure 2.23 — **Microstructure of iron consisting of polygonal ferrite grains (alpha phase) that are virtually equiaxed and rated ASTM micrograin size no. 6**
Nital etchant; magnifications as indicated (Courtesy of The Welding Institute, England)

sections. The observed average area is the best that can be accomplished under the circumstances and this has been termed the *grain cross-section*. It was recognized that metallographers would need a standardized method of measuring grain size to permit evaluation of investigations of metal properties, specification, and control.

Three methods of grain size determination have evolved over the years, and all are described in ASTM E 112, *Standard Methods for Determining Average Grain Size*. These methods are: comparison procedure, intercept (or *Heyn*) procedure, and planimetric (or *Jeffries*) procedure. This ASTM Standard provides a scale of micro-grain size numbers and relates these to grains of graduated sizes. A convenient table lists the calculated diameter of an average grain for each grain size number, and contains other information on the number of grains in a unit volume, etc. The original scale assigned No. 1 for a very coarse grain (diameter 0.25 mm, or 0.010 in.) and No. 14 for the finest grain (diameter 0.0025 mm, or 0.000 1 in.). However, the coarse end of the scale proved to be too limiting and therefore was extended by adding Nos. 0.5, 0, and 00. Grains of 00 size have an average diameter of 0.5 mm (about 0.02 in.).

The comparison procedure for grain size determination facilitates rating of the size of equiaxed grains, that is, grains not elongated by directional solidification, or by working. In this procedure, a representative field of the specimen is magnified (usually 100X) and projected on a screen or recorded in photomicrograph. The image from the specimen is matched against charts in the Standard that represent a series of microstructures of different grain sizes. Each structure in the series is identified with an ASTM Grain Size Number. If necessary, when matching specimen image against the ASTM standards, an interpolation can be made to secure a number that most closely approximates the Micro-Grain Size Relationships tabulated in ASTM E 112. A digest of these relationships is given in Table 2.9. The information in this table can also be related to a photomicrograph inscribed with one square inch circles or rectangles. The number of equivalent whole grains within one of the inscribed areas is counted, and the observed number of grains is then converted to ASTM Grain Size Number.

When an unusually wide range of sizes is found among the grains in metal, the metallographer can use a measuring eyepiece in the metallograph to make note of the actual size of abnormally large or small grains present, and their approximate percentage in the microstructure. If the microstructure is made up of grains that are not equiaxed, then either the intercept procedure, or the planimetric procedure is preferred to establish an ASTM Grain Size Number.

The intercept procedure involves drawing one or more straight lines across the microstructure (most conveniently done on a photomicrograph at 100X), taking care to have lines parallel to, and normal to the long axis of elongated grains. Each line must be sufficiently long to intercept at least 50 grains. Grains touched by the ends of the line count as half grains. The number of millimeters measured in the length of each line is divided by the number of grains intercepted. Several such lines are usually employed to establish an average intercept distance, even for microstructures composed of equiaxed grains. When grains are not equiaxed, intercept lines are ruled on longitudinal and transverse sections along axes that lie in all three principal directions of the specimen. When the average intercept distance is secured, this figure is related to an ASTM Grain Size Number as given in Table 2.9.

In the planimetric procedure, a circle or a rectangle of specific area (usually 5000 sq mm, or 7.75 sq in.) is scribed or overlayed on the microstructure, using a field magnification that will show at least 50 grains in the area to be counted. When the grains are not equiaxed, a grain count is made on three mutually perpendicular planes determined by the longitudinal, transverse, and normal directions. ASTM Standard E 112 gives guidelines for counting grains, establishing the number of grains in a unit volume, and equating this information to an ASTM Grain Size Number. Another publication, ASTM Standard E 1181, *Methods of Characterizing Duplex Grain Sizes*, covers duplex grain size.

Often the metallographer will examine the microstructure of a metal to determine if grains have grown after undergoing heating and cooling. The grains of a metal may increase in size under a variety of conditions, but the fundamental explanation for their behavior lies in thermodynamics. On heating a metal, larger grains will grow at the expense of smaller grains. The driving force for this growth is the free energy that is available in the structure at any condition that represents a surface or an interface. A grain boundary, for example, represents an area with some available free energy. Since the metal structure is aspiring to gain a thermodynamically stable form, it would find this state of minimum free energy when in the form of a single crystal or grain with minimal surface area. Therefore, any polycrystalline material is capable of grain growth. All that the metal requires is an increase in temperature and a little free energy within the structure. If grains could not grow, annealing would be far less effective in softening a metal, and welded steel would have not nearly its characteristic ductility.

The grain boundaries are rather easily moved by a diffusion process. The rate of grain boundary movement, that is, the extent of grain growth, depends on the metal, its prior condition, and the environment to which it is exposed. Several common circum-

Table 2.9 - Relationships Between ASTM Grain Size Numbers*

ASTM Grain Size Number	Total Number of Grains Observed in One Square Inch at 100X	Total Number of Grains Calculated to Exist in One Square Millimeter Not Magnified	Actual Size of Grains (Average Nominal Diameter) Microns (µm)	Actual Size of Grains (Average Nominal Diameter) Mils (in. × 10⁻³)
00	¼	4	510	20.1
0	½	8	360	14.2
0.5	¾	11	300	11.8
1	1	16	250	9.8
2	2	31	180	7.1
3	4	62	125	4.9
4	8	124	90	3.5
5	16	248	65	2.6
6	32	496	45	1.8
7	64	992	32	1.3
8	128	1 980	22	0.9
9	256	3 970	16	0.6
10	512	7 940	11	0.4
11	1 024	15 870	8	0.3
12	2 048	31 700	6	0.2
13	4 096	63 500	4	0.16
14	8 192	127 000	3	0.12

*The relationships tabulated here are rounded-off for simplification. ASTM Standard E 112 contains Table 2, which provides more precise numbers and detailed information.

stances which strongly encourage grain growth are (1) heating cold-worked material, (2) effecting an allotropic transformation, and (3) precipitating a new phase from a solid solution alloy. The forces that promote growth of grains are probably different in each of these circumstances. In the case of cold worked grains, the difference in strain energy between two contiguous grains will encourage one to grow at the expense of the other. Where an allotropic transformation is involved, there can be a difference in the energy of formation of the allotropic forms. In the case of precipitation, chemical energy may be released. Regardless of the differences in the source of the energy, the ultimate effect accompanying grain growth is a reduction in the free energy of the system.

Recrystalization of Cold-Worked Metal

Because weldments may benefit or may suffer from grain growth in cold-worked metal, depending on the particular situation, the mechanics of recrystallation must be considered. Grains in a metal or alloy are altered in shape during cold-working in about the same manner as the shape of the workpiece is altered. Heavy reductions produce very elongated grains. The grain boundaries persist despite severe grain deformation. When temperature is increased, grains that have been permanently distorted or strained by plastic deformation (cold work) recrystallize to undistorted equiaxed grains. The temperature at which the distorted grains are replaced by equiaxed grains is called the *recrystallization temperature*.

The greater the application of cold work, such as by hammering, drawing, or rolling at room temperature, the lower the recrystallization temperature will be. If grains are distorted while above their recrystallization temperature, they recrystallize immediately. This type of deformation is defined as hot work, as when steel is forged or rolled at a red heat.

Two very important factors concerning the recrystallization of cold-worked metals have an effect on the final grain size of the metal. In general, the smaller the amount of cold work and the higher the temperature to which the metal is heated, the larger the resulting grain size will be. A small but specific amount of cold work is necessary to promote recrystallization. When this amount of *critical cold work* is applied, maximum coarsening of the grains is obtained for any recrystallization temperature. In fact, the grain growth may be excessive for most purposes.

The amount of deformation that represents a critical reduction varies with different metals. With iron, a cold reduction in the vicinity of five percent is critical. Alloys acquire critical cold work at somewhat higher levels; perhaps in the vicinity of 10 percent reduction. General practice is to cold-reduce material 20 to 30 percent, if considerable grain coarsening is not desired. Some pure metals undergo excessive grain growth if they are severely worked and recrystallized at a very high temperature. This phenomenon appears to differ from the formation of abnormally large grains on heating critically cold-worked metal.

The recrystallization temperature varies with the metal or alloy, its nonmetallic content, prior grain size, amount of cold work, and other factors. The recrystallization temperatures of some metals are below room temperature; lead, tin and cadmium are examples. Iron, on the other hand, has a *recrystallization limit* (the lowest temperature that produces complete recrystallization of cold-worked grains) of approximately 425°C (about 800°F). As an example of iron's recrystallization characteristics, the commercially pure iron in Figure 2.24 (A) has an equiaxed ferrite grain structure. When this metal is cold worked to reduce the cross-section by about 60 percent, Figure 2.24 (B), the grains are severely distorted and their general shape clearly indicates compression in a vertical plane and elongation in a horizontal plane. After heating to 720°C (1328°F), which is below any temperature for allotropic transformation in crystalline structure, and holding for 20 hours, complete recrystallization has taken place, as can be seen from the equiaxed grain structure in Figure 2.24 (C).

Grain Growth at High Temperature

Temperature can be effective in causing grain growth, even when additional free energy has not been made available in the structure by cold work, transformations, or precipitation phenomena. When the metal is heated to a very high temperature, little free energy is required to stimulate grain growth, and this small amount of energy is always present somewhere in the structure. Both temperature and time determine the grain size. Consequently, if a metal is heated to an excessively high temperature for purposes of annealing, forging, bending, or welding, it may display abnormally large grains, particularly when exposure at the high temperature is of long duration. The temperature at which this grain coarsening becomes significant depends a great deal on the metal or alloy. Some steels undergo exaggerated grain growth at temperatures above about 925°C (about 1700°F), while others are quite resistant to grain growth at temperatures up to approximately 1150°C (2100°F). While the grain coarsening characteristics of a metal at high temperature are important, it is equally important to know if the grains established at the high temperature will persist during cooling to room temperature. Many metals, fortunately, have metallurgical characteristics that bring about refinement in grain size during cooling.

Grain Orientation

A condition that can develop in varying degrees during the growth of grains is *preferred orientation*. This simply means that the lattice structures of an unusually high number of grains are arranged in the same manner, that is, the major crystallographic planes of these grains have assumed a common orientation with respect to the body of metal. The amount of preferred orientation exhibited may range from a bare majority of the grains up to an almost completely preferred orientation of grains. Also, the grains may have only two axes located in preferred directions, or in the case of certain body-centered and face-centered cubic metals, may have all three major axes in a common arrangement (cubic oriented). Special steels for electromagnetic applications are often processed to obtain preferred orientation of grains.

The development of preferred orientation is actually initiated prior to any heating to obtain recrystallization or grain growth. During cold working, the slip planes and slip directions in the grains gradually approach the section axis or direction of tension in the metal. Even though the original grains had random orientations, the cold-deformed grains tend to assume a common orientation. The extent to which the grains are able to gain this preferred orientation depends on many conditions. However, drastic working of a specific kind and special reheating practices can produce grains with a very high degree of preferred orientation. Industrial cold-working and annealing practices comonly followed will usually prevent preferred orientation from developing in the grains of a metal.

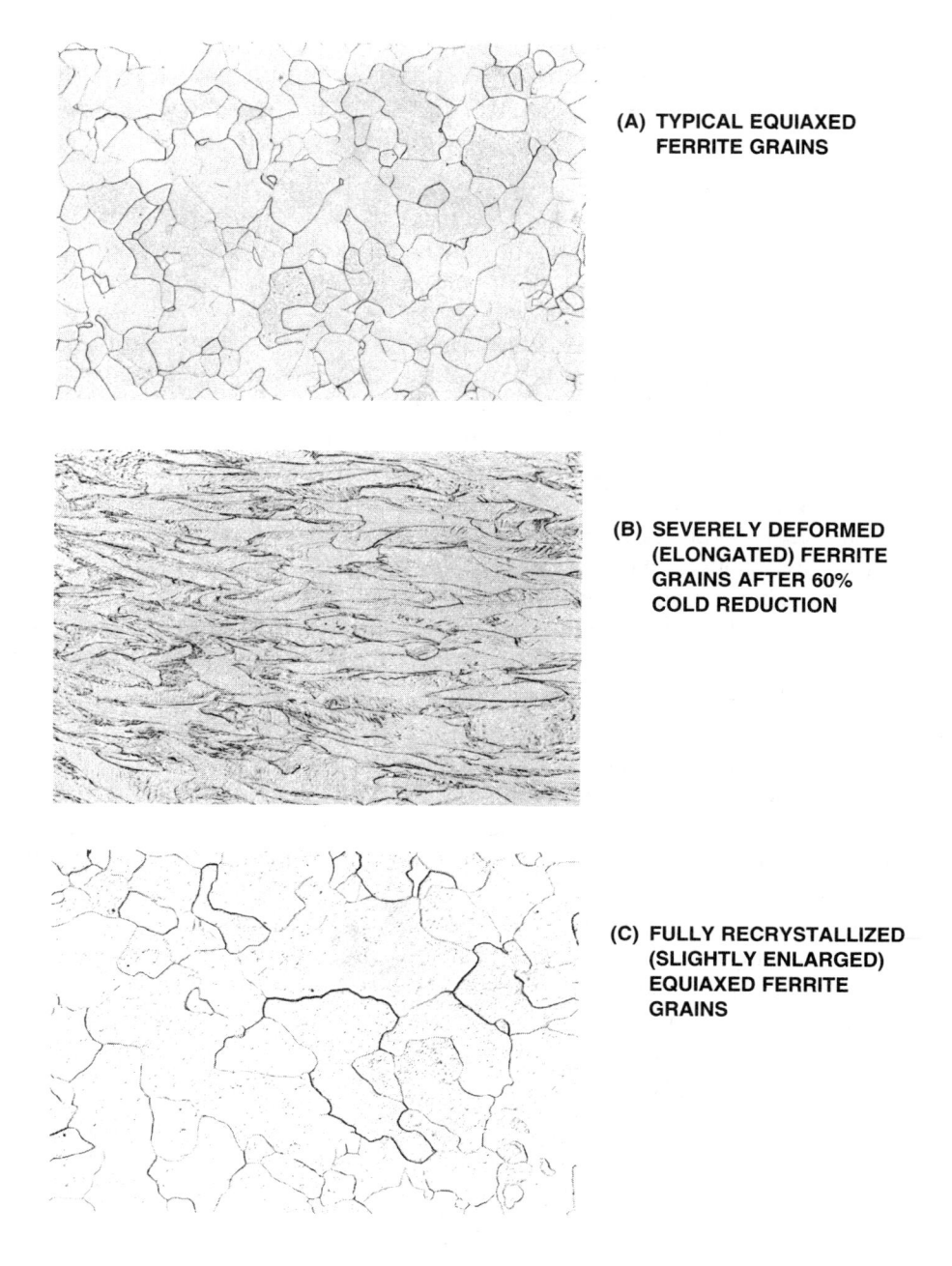

Figure 2.24 — Microstructure of iron, (A) annealed, (B) after 60% cold reduction of section, and (C) after recrystallization heat treatment

Nital etchant; original magnification: 320X. (Courtesy of The Welding Institute, England)

Grain Refinement

Sometimes it is necessary to examine the structure of a metal to see if its grain size has been reduced. For most purposes and in most metals and alloys, a fine grain size improves mechanical properties, particularly toughness, and it is not uncommon to apply working or treatments to refine the grains. If metal has been inadvertently heated to a high temperature and has grown very large grains, grain size can be reduced by one of several methods, depending on the nature of the article and the metallurgical characteristics of the metal. The ASTM numerical system of rating grain size is used to compare the structures of metals before and after these grain refinement treatments.

Cold working will lead to grain refinement if the nature of the article will permit some form of rolling, drawing, or swaging. Metals in the form of sheet and wire are often scheduled to receive a specific amount of cold reduction during manufacture to take advantage of the grain refinement that can be accomplished after cold working. It should be understood that the grains do not become smaller during cold working. Even though they may become quite deformed (elongated when rolling or drawing), the original boundaries can be seen among the slip planes that develop during working. Grain refinement is accomplished by heating or annealing the cold-worked metal to recrystallize a new generation of grains whose sizes are smaller than the original grains. This is accomplished in most metals by cold working beyond the critical reduction and by subsequent annealing above the recrystallization temperature, and limiting the peak temperature.

Hot working the metal also can refine the grain size. The difference between hot working and cold working is that the hot deformed grains immediately recrystallize into a new set of grains. This immediate development of new grains occurs because the hot working is carried out at a temperature above the recrystallization point. This is, in fact, the criterion for deciding whether an operation is hot working or cold working. To accomplish grain refinement by hot working, the amount of reduction and the temperatures at which hot working is started and completed are important. Generally, the greater the extent of hot deformation, the smaller the grain size will be. Temperature, however, is very important. The grains in the metal can coarsen on heating to excessively high temperatures, and even though they will be recrystallized during the working, a high temperature at completion of the working operation will permit these new grains to grow. Consequently, *finishing temperature* is carefully controlled in hot working. Grain size control has led to the development of controlled rolled steels, which offer high strength and better toughness by virtue of the fine grain size steelmakers have been able to achieve. This highly developed method of processing steel plate, termed Thermo-Mechanical Control Processing (TMCP), is explained in Chapter 5.

Transformation of the crystalline structure within grains provides a very efficient method of refining grain size because it can be done without working or reducing the section of the metal. Metals that undergo allotropic change at an elevated or moderately high temperature are best suited for grain refinement through this phenomenon; iron is an excellent example.

Iron undergoes allotropic transformation at 912°C (1674°F), the point at which the crystalline structure changes during heating from the body-centered cubic lattice to face-centered cubic. This transformation occurs in the grains of iron, on heating, by the formation of minute new grains, which usually form at many points along the boundaries of the original grains. However, any discontinuity in the structure which makes some free energy available can nucleate a grain. The new grains continue to grow as transformation progresses during the continued introduction of heat. The temperature of the metal does not rise during transformation, because the change is endothermic. When sufficient thermal energy has been supplied to transform the metal entirely into new grains of face-centered cubic structure, the continued application of heat produces a temperature rise. The new grains of face-centered cubic crystalline structure will be smaller than the original grains. Of course, if heating is continued to higher temperature levels these grains will commence to grow. On cooling, however, the iron goes through a similar grain refinement because the allotropic transformation is reversible. When the temperature falls to 912°C (1674°F), new grains of body-centered cubic structure nucleate at many places on the boundaries and in the high-temperature grains. These new grains grow, as heat is evolved (the change from fcc to bcc crystalline structure being exothermic). When transformation to the new grain structure is complete, the temperature continues its descent. Again the new grains are smaller than the grains from which they transformed. Because the temperature is dropping, the new grains are not likely to grow in size. As a matter of fact, if cooling is accelerated by some means prior to and during the transformation, the number of nuclei for new grains will be increased and the extent of their growth will be restricted. When cooling is accelerated, the transformation may sweep across the grain structure in a massive action, and in some instances even across a grain boundary. Therefore, the crystallographic orientation of the new grains may not necessarily be coherent with the original grains from which they formed. Rapid cooling also results in undercooling, which amounts to transformation at a temperature lower than the normal level for the change.

The conditions that accompany a transformation are analogous to those found during the freezing of a metal. The critical point for the transformation corresponds to the freezing point. At the transformation temperature, nuclei form in a solid crystalline structure instead of forming in a melt. There are no crucible walls, but the original grains have edges and corners that can supply the energy to start the formation of new grains. If there are foreign particles in the metal, such as alumina in steel, these particles may also act as nuclei during a transformation.

Undercooling, while more accentuated in a transformation than in freezing, also leads to more complex situations than freezing. In the absence of undercooling, the change in temperature during cooling through a transformation has the same appearance as a cooling curve plotted during the freezing of a metal. If impurities or alloying elements are added to a metal they affect the cooling curves during melting and freezing in ways that obey the same laws as those that affect transformations.

Transformation is different from recrystallization in that the latter can occur only when cold-worked metal is heated to its recrystallization temperature or above, or the metal is hot worked. Transformations, on the contrary, should occur every time a metal capable of allotropic transformation is heated or cooled through its transformation temperature or critical point. The product of recrystallization is a crystal of the same type as the original undistorted crystalline structure. On the other hand, the type of crystal changes as a result of a transformation. Transformation and recrystallization do have two elements in common: (1) The nuclei are extremely small, and (2) the transformation and recrystallization may be controlled so that the final grain size is finer than the original grain size.

The effects exerted by grain size greatly depend on the particular metal or alloy. As previously mentioned, the crystalline structure of the metal and the dislocation population of the lattice must be kept firmly in mind. These aspects are often the basis for determining the influence of grain size. In general, coarse grain size can detract from toughness at room temperature and lower, while fine grain favors toughness, particularly at low temperature. At elevated temperature, fine grain may lower the strength of metals. Therefore, the requirement of coarse grain for steels designed for high-temperature service will be mentioned often. Obviously, grain size is a feature of microstructure that deserves close scrutiny in the examination of metal structures.

The use of optical metallography to study the microstructure of metals has dealt only with the shape and size of grains, and means of effecting changes in metal. As other elements are added to form alloys, many other micro-constituents will appear within the grains and in the grain boundaries. In many instances, the nature and distribution of these constituents, or secondary phases, will exert greater influence on properties than grain size or shape. This is particularly true when carbon is added to iron to make steel. Metallographers recognized that the influence on the properties of the alloyed metal of any secondary phases or particles would depend on the volume, size and distribution of these secondary phases, or particles. Quantitative determination of these features, however, was a time consuming task that usually involved point counting or area measurements of the microstructural image. The need for rapid, precise quantitative determination of microstructural features resulted in the evolvement of a new branch of metallography, termed *quantitative metallography*.

Quantitative Metallography

This branch of metallography deals with mathematical analyses of morphology observed in the structure of metals. Morphology, in this instance, is external form exhibited in the microstructure as related to internal crystalline structure. Quantitative metallography provides numerical values for many microstructural features of metals which influence properties or behavior. Microstructural features often evaluated are:

(1) texture,

(2) total volumes of phases or particles present,

(3) shape and size of grains, secondary phases, or particles,

(4) distribution of secondary phases or particles within the grains or in grain boundaries, and

(5) homogeneity.

Quantitative metallography is used in such specific functions as defining optimum microstructure for a particular application, setting specification limits for acceptable material, exercising quality control, guiding the development of new alloys that have structure-sensitive properties, and making detailed comparisons of different lots of metal that have similar chemical composition, but dissimilar properties or fabrication behavior. Electron microscopy can be also employed in quantitative metallography.

Auxiliary units are now available that will perform quantitative image analysis automatically. Those designed to be coupled with the optical metallograph (or the electron microscope) can carry out a number of measuring and analyzing functions on the image of the microstructure of a metallographic specimen. In applying this new form of metallography, great care must be taken that the specimen to be examined and compared is truly representative, and that the entire preparation of the specimen followed a closely regulated procedure. Replicate specimens and cross-checking is often advisable. The microstructural image from the metallograph is presented to the analyzer unit, which consists of three principal components as outlined below and illustrated in Figure 2.25:

(1) a video camera that transports the image to a special fluorescent screen where it is precisely adjusted for grey-level of all microstructural features,

(2) a detector for scanning the electronic image on the screen, allowing the metallographer to discriminate and count by selecting grey-level sensitivity, and

(3) logical counting circuits for processing detection signals, and peripheral equipment for recording and analyzing.

Numerous features of the microstructural image can be measured, counted, classified, etc., by the automatic analyzer. It is possible to scan the image and count all particles that are larger, or smaller than a prescribed size, or that are oriented in a particular way, or that have a specific shape. In addition to measurement and comparison analyses of phases in the microstructure that have been revealed by etching, it is possible to conduct quantitative image analyses of nonmetallic inclusions in the metal. Ordinarily, examination for inclusions and their analyses are conducted with the specimen in the as-polished (unetched) condition. Putting microstructural examination on a quantitative basis adds significantly to the value of metallography.

Metallography Using the Electron Microscope

The advent of the electron microscope in about 1930 generated much enthusiasm among metallographers, because the magnification capability of this instrument (see Figure 2.26) was so much greater than the optical metallograph. Even the earliest electron microscopes achieved magnification as high as 300 000X. This capability depended on the very short wave length of the electron beam, which could be calculated from the following equation:

$$\lambda = \sqrt{150/V} \qquad \text{(Eq. 2.1)}$$

where
λ is wave length in angstroms, and V is accelerating potential in volts.

The first generation of electron microscopes applied to the study of metals used accelerating voltage in the range of 100 to 200 kV. Consequently, the electron beam wave length usually was in the order of 0.04 Å, which is one hundred-thousandth as long as the wave length of visible light.

While resolving power theoretically was determined by wave length and should have been spectacularly good, resolving power in early electron microscopes was restricted by spherical aberrations. Other problems also arose as the metallographer attempted to adapt the electron microscope, and almost 25 years of painstaking effort was required to achieve the desired results in metal examination. Today, electron microscopy is a "wonderland" of microscopes of different kinds, with accessories for orienting specimens, hot stages, video recorders, stereographic image printers, and analyzers. The information available on the construction and operation of this equipment is so voluminous that only those instruments proven most useful in studying welded metals will be introduced in this chapter, with an explanation of how they operate and what results can be obtained.

Electron microscopy has substantiated many of the theories and hypotheses drawn from laborious studies by optical microscopy, and has identified previously undiscovered microstructural features that affect the properties of metals. A comparison of optical and electron metallography in terms of operating components and wave-paths shown in Figure 2.27 illustrate the substantial gain in capability provided by the electron microscope. Of course, the metallographer will be satisfied only with a direct view of the crystalline structure that clearly shows how the atoms form the unit cells, which allows first-hand examination of the nature of dislocations in the lattice of grains and at the grain boundaries, and which permits determination of size, distribution, and composition of secondary particles within the crystalline structure.

Figure 2.25 — Quantitative image analyzer
Image on CRT screen can be provided by either optical microscopy, or by electron microscopy. (Courtesy of The Welding Institute, England)

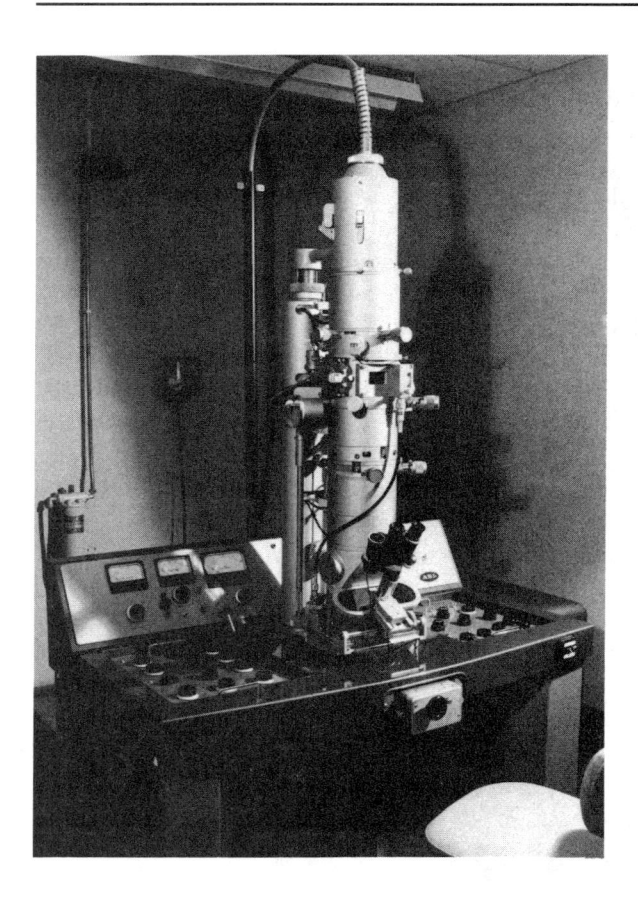

Figure 2.26 — Electron microscope used for metallography

The electron gun is located at the top of the vertical column and electrons are ejaculated downward in vacuo through magnetic focusing lenses to the metal specimen. Electrons from the specimen form an image on a fluorescent screen that can be viewed directly, or examined through binocular eyepieces. (Courtesy of The Welding Institute, England)

Transmission Electron Microscopy (TEM)

The electron microscope was initially used as a transmission instrument; that is, the electron beam was passed through the specimen before magnification of the beam image by electromagnetic or electrostatic lenses and registry on a fluorescent screen. This technique was quite satisfactory for translucent materials, and for tiny opaque particles that could be suspended in a transparent film to permit shadow enlargement. However, to examine the etched microstructure on the surface of a solid metal specimen, it was necessary for the metallographer to make a replica of the surface using formvar, polystyrene, or a carbon-platinum coating. Despite elaborate techniques to accurately reproduce the etched configuration on the specimen surface, the results left much to be desired. Fortunately, methods for directly viewing solid metal specimens under the electron microscope were developed, and the tedious preparation of surface replicas is no longer necessary.

Because the primary beam of early transmission electron microscopes could not pass through a metal specimen thicker than about 5000 Å (approximately 0.000 02 in.), examination of the usual solid metal specimen was not feasible. Furthermore, a beam passing through a very thin specimen presents a confused image, because of the many microstructural features encountered during its passage. To avoid these problems, electron microscope technicians devised unique procedures for specimen preparation, some of which are discussed in Technical Brief 8. With these prepared specimens, TEM could resolve details as small as 50 Å (0.000 000 2 in., or two-tenths of a millionth of an inch). This level of resolution made it possible to view directly the larger dislocations present in the crystalline structure of metals. Of course, a "point defect" (single missing atom) and other highly localized lattice imperfections could not be seen, because of their very small size. Yet, it was reassuring to observe directly for the first time dislocations such as line defects, illustrated in Figure 2.28. Here, approximately a dozen line defects have developed in the crystalline lattice of a three percent silicon electrical steel. The propensity of the dislocations to coalesce on a particular plane with favorable orientation, as related to direction of applied stress, is evident in the TEM micrograph.

Continued development of electron microscopes led to the use of higher accelerating voltage at the electron gun, and this brought considerable improvement in capabilities, particularly with potential in the range of 300 to 400 kV. Point-to-point resolution of approximately 2 Å has been achieved with some microscopes employed in TEM, and this makes it possible to view individual atoms in thin crystalline specimens, revealing the real nature of most kinds of dislocations.

Interpretation of the image from TEM that appears on the fluorescent screen, or in an electron micrograph, requires great care. The final image of a specimen is composed of two superimposed images, that of the upper surface (on which the beam impinges) and that of the lower (beam exiting) surface. The loss of definition of image from the upper surface, and similar loss in intermediate levels provides means of singling out the meaningful lower-surface image. In addition to providing a highly magnified image of the structure of the specimen, the diffraction of electrons permits identification of the crystalline structure of the metal through which the beam passes. This examination is termed diffraction analysis.

Very small areas can be selected for electron diffraction analysis by focusing the beam to an extremely small spot, perhaps as small as 500 Å in diameter if the beam

---

TECHNICAL BRIEF 8:
Specimen Preparation for TEM

In the early years of transmission electron microscopy, one popular procedure for specimen preparation was to form a hole through a metal section of wafer-like, but manageable thickness in the area of interest. Then, a special electropolishing operation was applied to produce tapered electro-thinning at the edge of the hole. There, the hole would be razor-sharp, and the specimen gradually would increase in thickness to its original wafer thickness. An area very close to the edge of the hole, where the electropolished metal was only about 1000 Å thick (about 0.000 004 in.) was scanned by TEM to produce a screen image or an electron micrograph.

Other procedures have employed electropolishing until final thinning reached the point where perforation or a "window" indicated that a thickness had been achieved suitable for TEM. Regardless of the preparation method, careful interpretation of the image secured by TEM was required, because the microstructural features encountered during beam passage might be superimposed upon each other.

has a high electron acceleration. A number of "problems" encountered in the course of applying electron microscopy have led to useful knowledge. For example, when an array of diffraction spots is generated in examining a metal foil, a network of white and black lines appears along with the spots. These lines are called *Kikuchi lines*, and it is now appreciated that they will pass through the centers of the diffraction spots whenever the Bragg conditions of reflection become ideal. "Kikuchi cards" are available for various crystalline structures to assist in their identification and to determine lattice orientation.

The capabilities of 300-400 kV electron microscopes did not completely satisfy research metallographers. They encouraged the building of ultra-high-resolution instru-

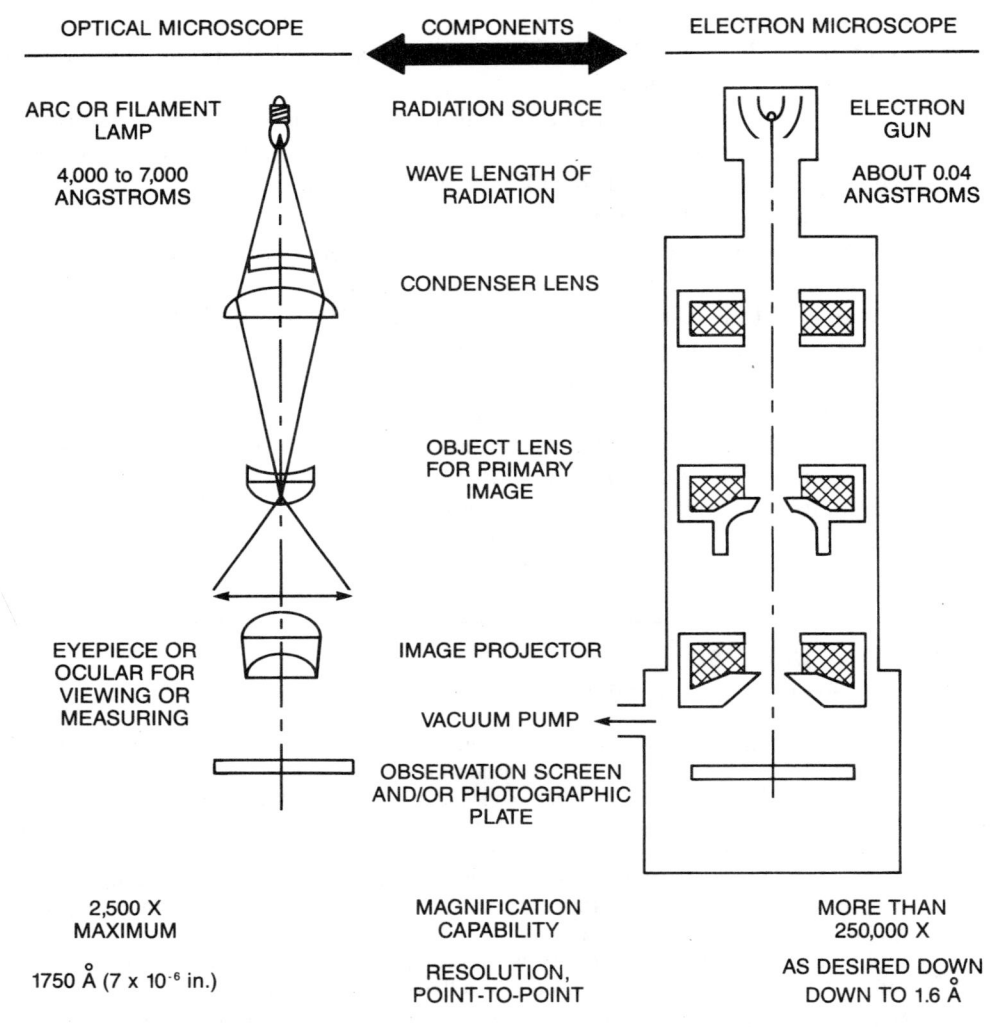

OPTICAL MICROSCOPE	COMPONENTS	ELECTRON MICROSCOPE
ARC OR FILAMENT LAMP	RADIATION SOURCE	ELECTRON GUN
4,000 to 7,000 ANGSTROMS	WAVE LENGTH OF RADIATION	ABOUT 0.04 ANGSTROMS
	CONDENSER LENS	
	OBJECT LENS FOR PRIMARY IMAGE	
EYEPIECE OR OCULAR FOR VIEWING OR MEASURING	IMAGE PROJECTOR	
	VACUUM PUMP	
	OBSERVATION SCREEN AND/OR PHOTOGRAPHIC PLATE	
2,500 X MAXIMUM	MAGNIFICATION CAPABILITY	MORE THAN 250,000 X
1750 Å (7 x 10^{-6} in.)	RESOLUTION, POINT-TO-POINT	AS DESIRED DOWN DOWN TO 1.6 Å

Figure 2.27 — Comparison of components, wave-paths, and capabilities of optical (light) microscope and electron microscope used in metallography

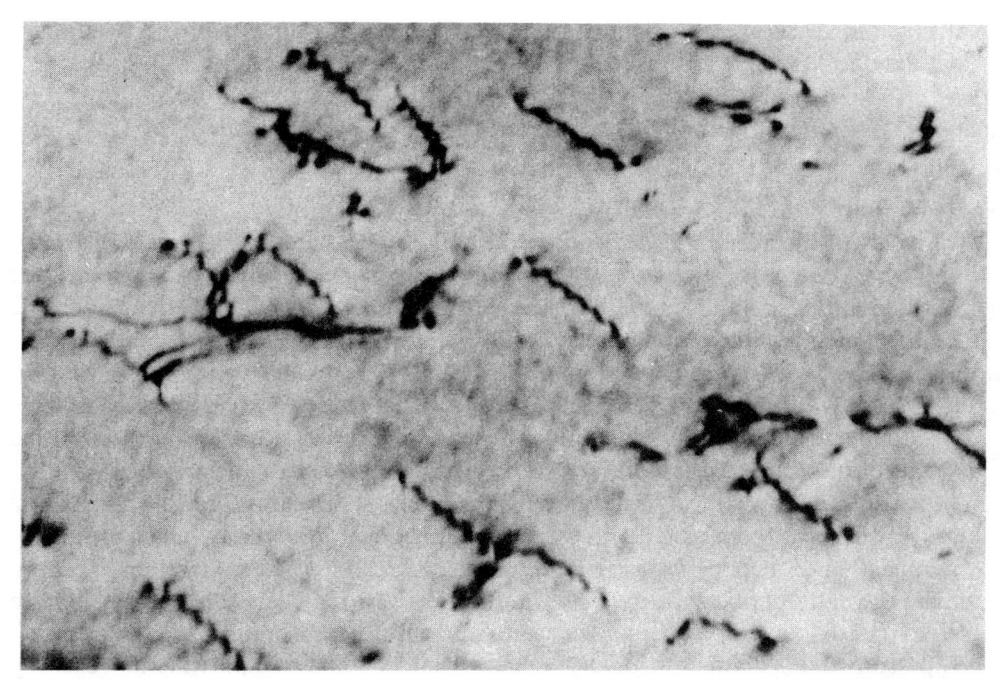

Figure 2.28 — Transmission electron micrograph (TEM) of "line type" dislocations in 3% silicon electrical steel

Thin foil prepared by Bollman technique using 10% perchloric acid and 90% acetic acid. Original magnification: 70 000X. (Courtesy of Armco Inc. - Middletown, OH)

ments with increased acceleration potential. A limited number of special electron microscopes have been built that use accelerating voltage in the range of 600 kV to 1.5 MV. The beam produced by these ultra-high-voltage microscopes can pass through much thicker metal specimens, perhaps ten to 40 times as thick as the 1000 Å thick "hole-edge" mentioned earlier. Maximum thickness permitted in a specimen is determined by (1) the penetrating power of the beam, (2) the proportion of electron scattering by the metal being examined, and (3) the amount of aberration of the lens system.

The high-voltage accelerated beam, with its shorter wave length, is able to pass through the metal with less chromatic aberration, and with better resolving power. Passage of the electrons through the metal and transfer of some beam energy results in a number of inescapable effects, including the following:

(1) a slight temperature increase,
(2) partial ionization,
(3) breakage of some interatomic bonds,
(4) significant ejection of atoms,
(5) diffraction of electrons,
(6) emission of visible and ultraviolet light, and
(7) emission of x-rays and secondary electrons.

These effects should be noted, because some are used to good advantage in other forms of electron microscopy. Of course, the primary objective of the ultra-high-voltage electron microscope is to achieve better resolution. Instruments of this kind operating at the one MV accelerating voltage level have demonstrated point-to-point resolution in the order of 1.6 Å. Despite this availability of increased beam-penetration capability, there are advantages to using the thinnest possible specimen for TEM studies of metals.

Scanning Electron Microscopy (SEM)

A significant advance in electron metallography was introduced in about 1950 with the development of the scanning electron microscope. This microscope uses the electron beam, under an accelerating potential of 1 to 50 kV, which impinges on the surface being examined to produce secondary electrons, back-scattered electrons, and photon emission (including characteristic X-radiation), any of which can be used for developing an image of the specimen's surface. The entire field of view cannot be bathed with electrons to obtain the image. Instead, the SEM sweeps a sharply focused primary electron beam (perhaps 100 to 200 Å in diameter) across the area being examined in a raster, either a linear raster like that used on a television screen, or perhaps a pattern of squares. Electrons emitted by the specimen are attracted by an electron collector consisting of a concentrator, detector, and a scintillator. These auxiliary components are optically connected to a photomultiplier. The signals from this system are amplified and displayed on a fluorescent screen, as shown in Figure 2.29. The primary beam scans the specimen in precise synchronism with the display signal reaching the screen. The signal is constantly regulated in accord with the number of electrons being emitted or reflected from each minute detail on the specimen's surface. The image is formed on the screen point-by-point as the electron probe beam scans the surface. SEM cannot follow fast-changing events that might appear on the surface of a specimen undergoing a treatment, such as those taking place on a hot stage, because the usual time for scan-and-registry on the screen requires from a tenth of a second to perhaps as long as four seconds.

The SEM can be operated in various modes to obtain an image of the specimen's surface on the viewing screen. Three favored modes and reasons behind their selection are:

(1) Secondary electron emission provides the highest resolution because of the energy of electrons emitted from the specimen. The image is easily interpreted because of its similarity to what is normally seen by the human eye.

(2) Back-scattered electrons come from a greater depth in the specimen and in larger volume, and therefore the image is adversely affected to some degree. Because elements having high atomic numbers backscatter a larger proportion of the primary electrons than elements having low atomic numbers, the screened image can show composition differences that might not have been revealed by the metallographic etchant.

(3) X-ray emission is the mode used when composition determinations are the principal kind of information needed for both qualitative and quantitative analyses. The primary beam can be located on a single point so that all elements (within detectable limits) will be analyzed, or the

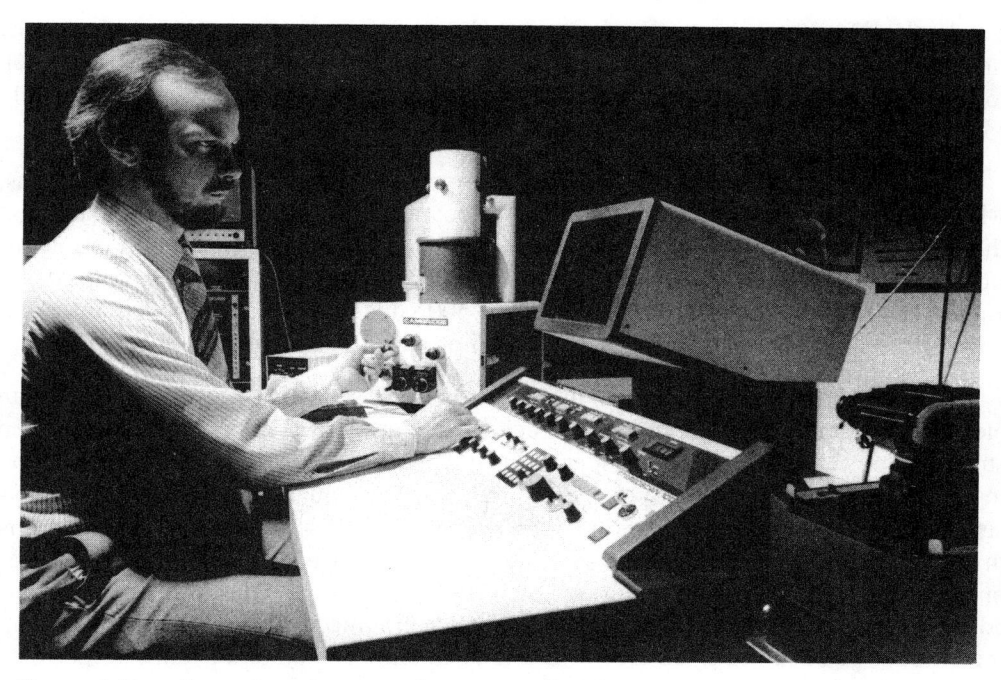

Figure 2.29 — Scanning electron microscope (SEM)
Scanning is inserted in vacuum chamber and moved on mechanical stage under focused electron beam. Reflected electrons and photon emissions form image on fluorescent screen. Image can be highly magnified view of very small selected area, or can be relatively large area at lower magnification but with remarkable depth of focus. (Courtesy of The Welding Institute, England)

beam can be traversed across the specimen so that any variation in the concentration of a single element along this path can be indicated. In some cases, channeling of X-radiation emission can be used to identify atom location, such as pin-pointing the crystallographic position of residual or trace elements, even when present in such minute levels as only 0.1 atom percent.

Resolution in the image produced by a SEM is determined by the fineness of the primary probe beam, and the mode of operation. Current microscopes achieve resolution in the order of 100 Å; which is not as good as with TEM, but is quite adequate for many microstructural studies. When better resolution is needed, a number of precautions can be taken to control stray magnetic fields and other factors that tend to degrade image quality, yielding a point-to-point resolution of 30 Å with SEM. The resolution of the SEM is at least 20 times better than the best optical metallography.

SEM is the most widely used form of electron metallography because it can directly view and magnify the usual polished-and-etched specimen, and can accomplish equally dramatic results with unprepared surfaces that have been fractured, corroded, worn, etc. The depth of field held in sharp focus when viewing with SEM can be several hundred times better than in optical microscopy. At a magnification of 10 000X, the depth of focus can be a full micron (ten-thousand angstroms); while at a very low magnification (e.g., 10X) the depth of focus can be as much as several millimeters (up to 0.12 in.). This capability makes SEM very useful in fractography, when relatively rough, unprepared surfaces are examined.

This excellent depth-of-field produces impressive single-exposure micrographs, which seemingly have a three-dimensional appearance. Examination can be started on the SEM at a magnification as low as 5X, and then continuously increased in magnification to as high as 300 000X (depending on the particular microscope employed). However, magnifications ranging from about 500X to 20 000X have proven most productive because of good resolution and depth of focus in this range.

A further advantage of the scanning electron microscope is the ease with which an electron diffraction camera can be included in the system to identify the crystalline structure of selected areas in the specimen. It can also include an x-ray spectrograph to analyze emission from the specimen surface to secure quantitative elemental analysis. This multi-purpose system is possible for SEM because as high-energy electrons in the primary beam strike the surface, a portion of their energy is converted to secondary electrons, and to X-radiation, which is characteristic of the atoms being excited by electron impact. By applying sufficient potential to the primary electron beam, the amount of back-scattered electrons and X-radiation is quite adequate to perform crystallographic and chemical analyses.

When higher magnification, beyond 300 000X, is needed the metallographer usually turns to TEM (or to scanning transmission electron microscopy, an advanced technology).

Scanning Transmission Electron Microscopy (STEM)

TEM microscopes operating with 660 kV to 1.5 MV potential have resolving power of 1.6 Å under optimum conditions. There is a current effort to develop STEM, a scanning transmission electron microscope which ultimately will be capable of achieving a resolution of 0.5 Å, which would allow a direct examination of atoms and the unit cells in the crystalline structure of metals. STEM-type microscopes with a claimed capability of magnifications to ten million times are installed in a few laboratories. When the new STEM has proven resolution capability better than 1 Å, it can be used immediately in many areas of research.

Essentially, the new STEM makes use of a field emission type electron source designed to emit a very bright, tightly-focused probe beam. This probe impinges on a suitably thin specimen following a prescribed raster. Detectors located beneath the specimen will record electron transmission. The electrons emerging from the back of the specimen are categorized as (1) unscattered, (2) large-angle elastically scattered, and (3) small-angle elastically scattered. The detection system must separate and differentiate between these electron types to achieve the desirable high resolution. In the final stage, the output of detectors is recorded synchronously on a greatly magnified version of the raster in a manner similar to the present-generation scanning electron microscope.

Metallography Using Ion Microscopy

The first microscope that enabled man to indirectly "see" individual atoms in crystalline array in solid metal was the Mueller field emission instrument introduced in about 1960. Instead of electrons, as in electron microscopy, the field emission microscope uses ions in the transmission and magnification of the image. Because of profound differences in the interactions that occur when ions collide with matter (as compared with impingement by a beam of electrons), ion microscopy can reveal unique information about the structure of metals. The earliest of these instruments had restricted application, but did provide views of atoms in the crystalline lattice in a very small volume of solid metal. This ion microscope consists of a highly evacuated tube (at 10^{-9} torr), in which a metal specimen in the form of an extremely sharp needle is mounted opposite a fluorescent screen. A very high electrical potential is impressed on the needle which, because of its sharp tip, has a field strength of 0.3 to 4 million volts per square millimeter on its point. The needle is used either as the cathode (as in the electron field emission microscopy), or as the anode. If used as the cathode, the needle tip does not have to be heated to emit electrons. Under high negative potential, a large number of electrons will stream radially away and flow by tunnel effect of wave mechanics until they strike the screen.

When the needle serves as the anode (ion field emission microscopy), a trace of gas (helium or neon) is admitted into the previously evacuated tube to generate ions. The gas atoms assume a positive charge and follow the straight lines of electrostatic force, which run from the tip of the needle-like specimen to the screen. On the screen, the ions (or the electrons, as mentioned previously) can produce a pattern that represents the nature of the atom array on the surface of the needle. The imprints of individual atoms are not always completely sharp because particles hurled from the tip have a small tangential velocity. The magnification of the pattern is determined by the ratio between the needle tip-to-screen distance and the radius of the needle tip. A magnification of 2 000 000X can be produced, but the resolving power is in the order of 3 to 20 Å.

Although the field emission microscope has great power for magnifying and resolving ultrafine microstructures, there are limits to the information if can divulge. Images are distorted by strain in the surface layer of the specimen. At present, only the strongly bonded metals and alloys, like tungsten, molybdenum and platinum are easily examined with the field emission microscope. Metals with melting points below about 1250°C (2285°F) evaporate when subjected to very high electrical potential (represented by the four million volts per square millimeter of the ion emission instrument), so this limits the application of field emission microscopy. Nevertheless, the patterns observed on the screen of a field emission instrument are the closest approach to "seeing" individual atoms in a crystalline metal.

Ion metallography now has new instruments with scanning capability, with brighter ion sources that broaden their application. In one type of microscope, a tungsten needle protrudes through a heated filament and a liquid metal of choice is applied to the tip. The liquid metal forms a sharp cusp, from which ion emission takes place. This beam can be focused on a spot as small as 100 Å in diameter for use as a probe, and for scanning. These instruments are equipped with the latest auxiliary devices and analyzers, such as mass spectrometer capable of secondary ion analysis. Dramatic results can be achieved with these microscopes. A preliminary sweep can be made with the ion beam to effectively remove a surface layer to a controlled depth before performing crystallographic examination, or chemical analysis. Ion microscopy is very sensitive to crystal orientation, and quickly reveals twinning in grains. Not to be overlooked in the capability of ion scanning is isotopic analysis. This is an important metallurgical feature of microelectronic devices, where microjoining is an especially critical step in manufacturing.

Tunnel-Effect Microscopy

Operation of the scanning tunneling microscope (STM) is based on the tunnel effect of wave mechanics, like that experienced with the field-emission ion microscope. The STM scans a specimen by passing a small sharply pointed probe back and forth at only a few atom-diameters distance above the surface. An electric current flows across the very small gap between the probe tip and the specimen surface, and tunnel effect from the atoms on the surface acts to change the electron flow as the probe moves across the site of each individual atom. This minute constricted flow of electrons is a phenomenon known as *quantum tunneling*.

The STM can achieve a variety of metallographic objectives. For example, as the flow of electrons from the material's surface changes, the probe-to-surface vertical distance can be immediately adjusted to maintain a constant current. By recording the vertical movement to the probe, a profile is traced as the probe moves along its path. By making many closely spaced parallel profiles, the current reaction of the specimen surface is traced in three dimensions. The performance of the STM is very promising, because the profile emitted by the surface can be plotted as a photographic image with vertical resolution believed to be 0.1 Å, and a horizontal resolution of 2 Å. Although the map of the surface is a two-dimensional array of bumps across a plane, these projections are interpreted as the electron cloud surrounding individual atoms, revealing the atomic structure of the surface. Also, analysis of the images of individual atoms, which appear as a variety of shapes (other than spheres), apparently indicate bonds between the atoms. Scientists using the STM on thin wafers of silicon incorporated as semiconductor crystals in microcircuitry have identified various kinds of chemical bonding between atoms making up the surface layer, and the specific directions in which bonds extend to second-layer atoms. Imperfections in the lattice structure of the crystal can also be seen with the STM.

Chemical Analysis of Microstructural Constituents

Electron metallography has the capability to provide information on both the crystallographic structure and the chemical composition of the metal being examined. These functions have proven helpful in understanding the relationships between microstructure and the properties of metals and alloys. (See text in Chapter 4 dealing with the heterogeneous distribution of elements on a microscopic scale in steels and other alloys).

Frequently, there is a need to know the chemical composition of minute entities observed on the surface of a metallographic specimen. As an example, analysis of the composition of slag and other nonmetallic inclusions found in steel provides the first clue to their origin, and probably will suggest ways of avoiding their entrapment in the metal. Also, surface coatings, films and platings may require chemical analysis. A recourse to these analytical challenges is provided by instruments that utilize characteristic emissions from the metal's surface to identify and determine the quantity of elements (atoms) present in a highly localized area. No single instrument serves all needs. Each instrument has useful capabilities, but some have limitations that must be kept in mind when a selection is made to analyze the chemical composition of a microconstituent, or a superficial surface material.

Electron-Probe Microanalyzer

The electron-probe microanalyzer, often called the *microprobe*, was introduced in the 1950-60 period, and it quickly became a favored analytical tool of metallographers because of the information on microstructural constituents that could be revealed. It is similar to an electron microscope, but usually includes an optical microscope for ease in locating the specific area to be analyzed. Because one or more forms of analyzers are incorporated for measuring electron and x-ray emission from the surface of the specimen, it is not possible to outline a single, detailed procedure for analyzing a specimen. A spectrometer can be coupled to an electron microscope, and the assembly will function in a manner quite similar to the microprobe.

A metallographic study (including use of the microprobe) might start with the preparation of a specimen, following a rigorous procedure to avoid contamination by polishing abrasive that could affect analytical results. A photomicrograph or an electron micrograph can be made to record appearance of the matrix and the size and shape of secondary constituents distributed in it. If these secondary constituents stand out prominently, stereo-scan micrographs can be prepared for viewing the microstructure in three-dimensional relief. The primary electron beam of the microprobe is usually focused on the smallest spot that still provides adequate emission for analyses. With some instruments, this focused spot can be as small as 0.5 micron in diameter (0.000 02 in.). The beam energy, determined by a number of analytical considerations, will probably be in the range of one to ten electron kilovolts. An important aspect of incident or primary beam strength to keep in mind is that beam energy determines the depth into the material being "probed" and analyzed. Therefore, if minute particles in the matrix structure are to be analyzed, the beam should be focused to a commensurately small spot and the power carefully regulated to probe no deeper than necessary. Typically, the electron beam striking the specimen surface will initiate analysis of all metal approximately within a one-micron radius. If a larger area is to be examined for possible composition changes, or gradients, or segregation of other elements, the probe beam is scanned along a prescribed pathway and is recorded on the photomicrograph. A raster can be programmed to analyze an area and plot concentrations of elements over this same area on a micrograph. It is also possible, with certain microprobe units, to display an image of the microstructure on a fluorescent screen. With these units, scanning of the probe beam is synchronized with the output signal from the spectrometer, as it retraces the screened image. Localized brightness of the image from the probe signal will vary in proportion to the concentration of a designated element. This type of analytical display usually has a resolution of approximately one micron.

Early electron-probe microanalyzers had a detection limit of about 100 ppm for most elements, and could not produce quantitative determinations for elements having an atomic number less than 11. Consequently, some important lighter elements often present in metals, like carbon, nitrogen and oxygen, could not be analyzed. However, present microprobes have been improved by a ten-fold increase in detection, and can analyze a specimen for any of the elements except hydrogen and helium. The degree of analytical accuracy is within about one percent.

Figure 2.30 is an example of the capabilities of the electron microprobe analyzer. A specimen of ASTM A 516 steel of very low sulfur content, aluminum-killed and calcium-treated, contained very small complex inclusions as shown in the SEM micrograph (upper left). To determine qualitatively the elements contained in these inclusions and their distribution, analytical scanning with the microprobe was conducted for the elements oxygen (O), aluminum (Al), magnesium (Mg), calcium (Ca) and sulfur (S), and the five responses (arranged in counterclockwise order in the illustration) were presented to the viewer. The nonmetallic particles were clearly indicated to be sulfide inclusions of a duplex Type III MnS variety. Approximately half of each inclusion is made up of an aluminate phase. There are traces of magnesium present, which probably came from furnace slag and refractories. Calcium is shown to be present in both the aluminate and sulfide phases.

Findings like this are helpful in understanding the behavior of nonmetallic inclusions in hot rolling. Calcium modification of sulfide inclusions makes them very hard at rolling temperatures, thus they resist deformation. Quantitative analyses of these inclusions could be obtained with the microprobe, but knowing the shape and composition distribution often is sufficient, when investigating metallurgical problems, such as hydrogen cracking and lamellar tearing.

Ion-Probe Microanalyzer

Because the electron microprobe is troubled by background signals that affect its sensitivity, and because it is not capable of analyzing elements of low atomic number (Z), efforts have been directed toward developing an ion-probe microanalyzer, sometimes called the atom-probe microanalyzer. The first successful ion microprobes appeared about 1970, offering the combined attributes of ion microscopy and mass spectrometry. One type of instrument consists of a field ion microscope (FIM) coupled with a time-of-flight mass spectrometer (TOFMS). The specimen to be analyzed is shaped to form a needle and the tip is electropolished to a radius of about 500 Å. The specimen is placed in a high-vacuum chamber and the tip is bombarded with a primary beam of ions having an energy level of about 10 keV. Particles emitted from the surface of the tip are ionized and constitute secondary ions that characterize the atoms (isotopes) present in the specimen's surface. Electrostatic optics incorporated in the instrument accelerate and focus the secondary ions into a beam, which carries an image. The ion beam can be analyzed in a number of ways, and then the beam can be fed to an image converter to be translated into an electron image which can be displayed on a fluorescent screen.

The ion-probe microanalyzer offers a number of advantages over its electron-operated counterpart. There is no background noise, so sensitivity is very good. An image of the surface and its microstructural constituents can be provided on which lateral resolution is about one micron and depth resolution is in the order of 100 Å. All elements regardless of atomic number can be analyzed quantitatively with about the same degree of accuracy. Specific elements can be determined at the specimen surface within a spatial distribution of 100 Å, and is some cases, 10 Å spatial resolution has been achieved. Techniques are being developed to use the ion-probe microanalyzer to conduct quantitative composition analysis on an atomic plane-by-plane basis. From field ion images and ion analyses, it has been possible to identify the size, crystalline structure, and composition of discrete particles as small as 300 Å in diameter in the microstructure of complex alloys.

A number of devices which are not necessarily part of the metallographic investigative tools are available to perform chemical or crystallographic analyses of metal specimens. However, these devices offer unique capabilities that could serve the welding engineer very well under certain circumstances. These devices have been developed by the physical chemist by taking advantage of characteristic emission from a surface as a result of bombardment with elementary particles, and with specific kinds of photons. Some of the devices do not require the usual metallographic specimen, but can accept an unprepared specimen from service. This can be a distinct advantage, if

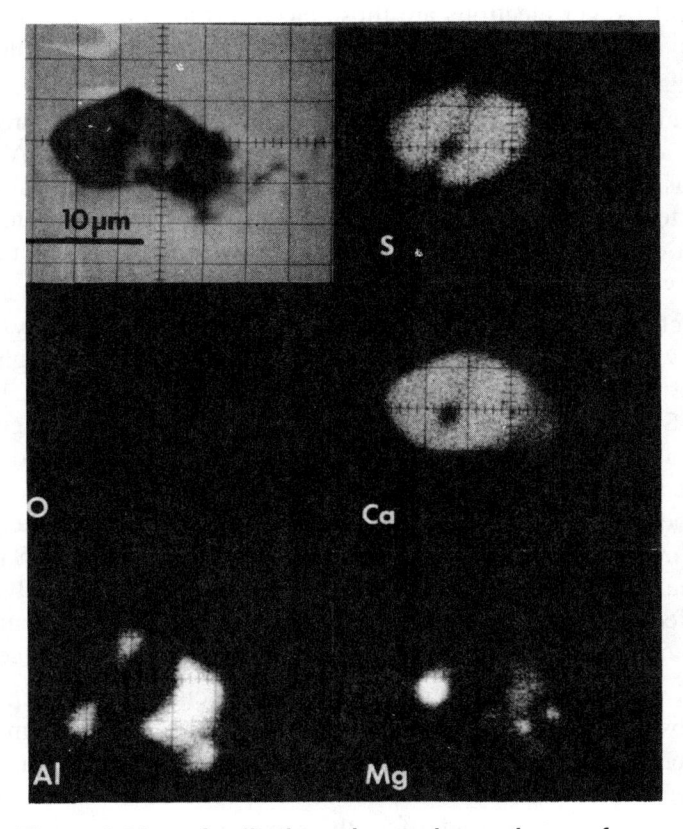

Figure 2.30 — Qualitative microprobe analyses of nonmetallic duplex Type III manganese sulfide inclusion in ASTM A516 steel

(Courtesy of Lukens Steel Company - Coatesville, PA)

the aim of the analytical work is to identify and quantify elements present only in a superficial surface layer, such as found in a plating, a coating, a layer of corrosion product, or oxide.

Auger Electron Spectroscopy

Auger electron spectroscopy (AES) is a method well suited to analyze the first few atomic layers of a metal surface. The Auger effect, (named after its principal investigator, *P. Auger*, pronounced "oh'zhay'") is determined when an atomic structure that has had energy introduced by particle or photon impingement ejects an electron from an inner shell. This atom, therefore, is ionized because of the absence of one electron. In this condition, there is a tendency for an electron in an outer shell to drop down into the smaller orbit and to fill the vacancy. Through this mechanism, the atom passes to a lower energy level, and the difference in energy can be radiated in the form of a photon with a wavelength characteristic of the particular kind of atom, or the atom may elect to emit another of its electrons, in which case the excess energy is divided between the work of electron emission and imparting kinetic energy to the freed electron. This energy division, or balance, follows an established equation, which depicts the Auger effect. Auger electrons are those ejected from the atom during AES that have a low-level kinetic energy, which is characteristic of the particular element (according to its atomic number).

Auger electron spectroscopy is performed with monokinetic electrons in the primary beam at a selected energy level in the range of one to five keV. The emitted Auger electrons have energy in the range of 50 to 500 eV. These electrons are guided (in a vacuum) to an electron spectroscope to determine Auger peaks and interpret the atomic number of the parent atom. All elements can be analyzed except hydrogen and helium, because they cannot exhibit Auger effect.

Auger electron spectroscopy offers distinct advantages in analyzing surfaces, because the low-energy Auger electrons have difficulty escaping through the lattice of the metal, and their emission is confined to depths of only about 10 Å or less. Therefore, AES is a method well suited for surface analysis. Although the primary beam in AES also generates x-ray fluorescence, and this emission could be readily analyzed using a crystal spectrometer, the x-ray photons move more readily through the lattice of the metal. Therefore the photon emission could be received from depths as great as several thousand layers of atoms. Care must be taken, if AES is selected for quantitative analysis of a surface, because a number of conditions involved in the procedure can affect analytical accuracy. For example, a polished specimen produces stronger emission to the detector-spectrometer than does a rough surface.

Some AES units include an ion gun in the vacuum chamber to clean the specimen surface, and possibly to erode it in a controlled manner to obtain chemical composition results at prescribed distances below the original surface. When an element to be analyzed by AES is chemically bonded as a compound in a surface layer, this can modify its electronic structure and affect the shape and position of its Auger peaks. Also, the atomic numbers of the elements being analyzed, that is, whether light elements with an atomic number less than ten, or heavier elements, and the levels of their concentration can require adjustments in analytical procedure and check analyses assisted by selected standards. Nevertheless, the abilities of AES to perform superficial surface analyses should not be overlooked.

Photo-Electron Spectroscopy

Photo-Electron Spectroscopy is often called ESCA, an acronym for electron spectroscopy for chemical analysis. This analytical method differs from AES, because the specimen is irradiated with mono-chromatic x-rays, to cause emission of electrons that are suitable for spectroscopic analysis. The emission comes from a limited number of atom layers at the surface, and although Auger electrons may be included among the secondary electrons, the sought-after photo-electrons are easily distinguished in the spectra. The photo-electrons are passed through a system consisting of a detector, amplifier, integrator and counter, which accurately performs the analytical determinations. All elements except hydrogen may be determined qualitatively, however, quantitative determinations have some restrictions, governed by the kind of specimen available, and the particular elements present and their levels. In addition to relatively shallow penetration, ESCA offers opportunities to measure directly the energy levels of electrons, to study complex surface phenomena, and to deal with molecular compounds. Restricting analysis to a highly localized area, using a polished specimen or an unprepared specimen, is often important in conducting an investigation of phenomena that occur in welding. Table 2.10 gives approximations of detection and sampling limitations to serve as guidelines.

Newer Techniques in Metallurgy

Significant advances continue to be made in studying the structure of metals with the use of new techniques and new kinds of instruments. Increased magnification is, of course, one of the new strengths; however, improved resolution is virtually of equal importance. A promising avenue of development in microscopy is the introduction of instruments that allow magnified observation of objects or areas of interest even when they are contained within the metal specimen and thus are buried from surface view.

The *thermal wave microscope* scans the surface of a specimen with a focused beam of energy, which can be directed from a laser, or an electron gun. The beam's energy is absorbed at or near the surface, and the material responds by producing thermal waves. These waves are imaged, revealing features on the specimen's surface, or immediately below the surface, which are indicative of metal soundness and microstructure. The specimen need not be one prepared for metallographic examination by polishing and etching. The thermal wave microscope can scan unpolished sections of metal and reveal such features as grain size, secondary phases and particles, voids, heat-affected microstructure, and joints made by welding, brazing or soldering.

Thermal wave microscopy is based on photoacoustic methodology established more than a century ago, when it was discovered that optical or periodic energy absorbed by a material specimen will thermally generate acoustic signals. Only recently has it been put to use in microscopy; the incentive came from the microelectronics industry, which has a great need to survey the character of circuitry for variations in lattice structure.

The probe beam for thermal wave microscopy can be secured as the energy beam in a scanning electron microscope, but the beam must be at an intensity-modulated frequency in the range of 10 Hz to 10 MHz. This is accomplished by beam-blanking, that is, 100 percent modulation by off-axis electrostatic deflection. As the probe beam scans the surface of a specimen, periodic heating at the modulation frequency results in an output of thermal waves similar to eddy current waves, or other critically damped wave phenomena that travel only one or two wavelengths before decaying to an unmeasurable level. However, the thermal waves interact with thermal features and undergo scattering and reflection in a manner similar to other propagating waves.

Detection and imaging of the thermal wave can be accomplished with several systems, such as (1) photoacoustic microscopy using a gas microphone, (2) photothermal imaging using infrared detection, and (3) mirage effect using optical beam deflection. All of these systems can detect the scattering of thermal waves caused by local structural features or physical properties on or near the surface of the specimen. Some of the features that can be detected and imaged in thermal waves are density, specific heat, thermal conductivity, significant dislocations in the lattice, and abrupt localized changes in composition such as implantation of foreign ions. These systems produce thermal wave image magnification to at least 1000X with good resolution because of controlled thermal wave length and small probe beam spot size when scanning surfaces. Also, the wave image is not degraded by depth of penetration, which could diffuse the image. Thermal wave microscopy is useful in studies of welded specimens.

The concept of *acoustic microscopy*, examination for defects by ultrasonic testing (UT) and by acoustic emission is not new. Well known and widely used, the potential of acoustic emission for microscopy was recognized as early as 1936, when it was discovered that the ultrasonic energy directed through a specimen could be used to produce magnified images. However, an additional 35 years of research were required to make the first acoustic micrographs.

Acoustic microscopy uses high-frequency sound waves instead of light of to view and measure microscopic features in a material. The beam of energy directed into the material should be referred to as a stress wave, rather than a sound wave, because it is not necessarily audible to the human ear. Several kinds of equipment are available that provide a system for directing a periodic stress wave (usually in the range of 10 to 500 MHz) into the surface of material and penetrating into the interior. The acoustic microscope then analyzes the point-by-point reactions of the material and translates the signals into an image display. Even though the material may be opaque in optical metallography, acoustic microscopy can "view" subsurface conditions that alter microelastic features. Typical conditions detectable include cracks, inclusions, and unwelded areas.

Table 2.10 - Comparison of Methods for Localized Analysis of Surfaces

Method	Minimum In-Depth Penetration	Minimum Lateral Sampling. Diameter	Approximate Limit of Detection
Electron - Microprobe	1 micron	1 micron	0.05%
Ion - Probe Microanalyzer	10 Å	1 micron	10 ppm
Auger Electron Spectrometry	10 Å	30 microns	10^{-10} gr
Photo - Electron Spectrometry	50 Å	—	10^{-10} gr

In ultrasonic testing, a specimen is placed before the acoustic microscope, where it is provided with a fluid coupling of oil or water to facilitate ultrasonic energy transmission. Sometimes the specimen is scanned with optical (perhaps using a laser) and sonic equipment, and the two images produced by these different systems are superimposed to display both external form and internal conditions. These scans are made synchronously and displayed directly on a CRT screen at 30 images per second. This system of optical-sonic dual examination employs a *scanning laser acoustic microscope* labeled a SLAM.

Small-angle neutron-scattering (SANS) examination is a new technique analogous to x-ray scattering as employed with the x-ray diffraction camera for determination of crystallographic structure. SANS makes use of the neutron's unique properties to accomplish a number of difficult evaluations beyond the identification of crystallographic structure. SANS requires a source for a neutron beam, usually a neutron reactor, and a highly specialized system for control of neutron velocity to secure a particular wave length to suit the kind of examination. The greater penetration capability of neutrons relative to x-rays (about three orders of magnitude greater than x-rays of the same wave length) allows SANS to be applied to metal components and have the beam probe to an appreciable depth rather than undergo near-surface diffraction as occurs with x-rays.

The SANS interactions that occur between projected neutrons of precisely controlled velocity and wave length during collision with the nuclei of atoms in the specimen enables this method to discriminate between neighboring elements in the atomic number series (e.g., iron and manganese) during composition determinations, and in some cases even the isotopes of the same element can be identified (e.g., hydrogen and deuterium).

When extremely small microstructural particles must be resolved and analyzed for composition and crystalline structure, SANS is a very efficient tool. For example, precipitated particles in the microstructure of an alloy in the size range of about 0.5 to 500 nm (5 to 5000 Å) can be identified with respect to composition, crystallographic structure and its orientation, coherency with the matrix structure, and inter-particle spacing. For certain kinds of examination, "cold" neutrons are favored to obtain a beam probe of smaller spot size and to produce better resolution. Cold neutrons have an energy level less than 0.005 eV; consequently they have longer wave length which usually is about 0.4 nm (4 Å). Although facilities for conducting examination by SANS are few in number, this method offers extraordinary capabilities and its use undoubtedly will increase. A national facility is available to all U.S. researchers at NIST.

FRACTOGRAPHY

Fracture study is described by the terms *fractography* and *microfractography*. The appearance and texture of a fracture and how the fracture initiated holds much information about the microstructure of the metal through which the fracture propagated. However, simple visual examination or low-level magnification reveals very little, and optical microscopy at higher magnification is handicapped by shallow depth of focus. Fractography progressed slowly over the 1940-50 period with the best available deep-focus microscopes. As improved equipment became available to magnify fractured surfaces to perhaps 300 X and above, tantalizing bits of evidence appeared. Highly magnified images of fracture topography permitted study and interpretation of details.

Significant contributions then began to appear in the literature, which presented photographs of profiles and other details on fracture surfaces at magnifications ranging from 5 to 500 X. These fractographs enabled investigators to identify specific features, and to interpret them in terms of underlying microstructure and fracture mechanism. Distinguishing features that were clearly related to fracture initiation and fracture propagation were recognized. Observed zones were defined with such descriptive terminology as cleavage, fibrous, woody, dimpled, shear lipped, radial, herringbone patterned, fatigue striated, and healed cavities. These features are illustrated and discussed in later chapters.

In about 1950, the electron microscope was used to study fractures, and this almost immediately made fractography a more rewarding endeavor. Initially, transmission electron microscopy (TEM) was applied. Early TEM techniques required replication of the fracture surfaces, of course, and this was carried out using procedures similar to those employed on polished-and-etched metallographic specimens intended for TEM. Replicas of fractures often required "shadowing" by directional vapor deposition of a heavy metal (electron opaque), to secure contrast and to allow the topography to be viewed. Naturally the rough fracture surface was likely to cause artifacts to appear on the replica. TEM allowed the replicated surface to be magnified through a range of about 200 to 300 000 X, but the most informative working magnifications appeared to be in the range of 3000 to 30 000 X. If a "single-stage" replica of a fractured surface were made, this would represent a "negative" of the topography — that is, low areas filled out with more replicating material and therefore thicker on the stripped-off film. Consequently, TEM displayed these low areas in a lighter color in contrast to the darker portions. Sometimes "two-stage" replication would achieve a "positive" reproduction of the fracture surface. The preparation, shadowing and handling of surface replicas required great care, and in some cases, the extremely fine textures of very small particles were lost in replication.

In about 1955, when the scanning electron microscope (SEM) became available, it was used to study fractured metal surfaces because of its capabilities for directly viewing and analyzing a specimen that consisted of a section of the actual surface. SEM and all of its auxiliary equipment and techniques were brought to bear in the pursuit of fracture knowledge, including special stages for heating and cooling the specimen, stereography for height and depth measurements of topography, and spectrometers for chemical analysis using Auger electron, secondary electron, and photon detection methods.

Analysis of fractured surfaces became especially important when, with increasing frequency, evidence was found that foreign elements could apparently segregate in the microstructure of the metal and the fracture tended to propagate through these segregated areas. Embrittling elements, such as antimony, arsenic, bismuth and tin, were detected on the facets of fractured surfaces. Although SEM offered magnification from 5 to 250 000 X, magnifications ranging from 500 to 20 000 X were found to be most informative.

The use of the transmission electron microscope (TEM) was not discontinued completely in favor of the scanning electron microscope (SEM) for fractography examination because TEM offered greater magnification and better resolution. When suitably thinned specimens could be prepared from the fractured surfaces, extremely minute details could be examined. Also, the metal adjacent to the fracture could be examined for dislocations (provided they exceeded approximately 100 Å) to ascertain whether these lattice imperfections were providing a fracture path.

A few of the ways in which fractography can assist a study of metal failure by fracture can be reviewed through the complement of examination methods applied in a typical case. First, specimens of metal, including selected areas of fractured surface, are prepared to determine microstructural features adjoining the fracture, and to see if there are any embryo cracks that would help explain initiation of the fracture. Prior to etching polished metallographic specimens, a preliminary examination is made to determine the nature, number and size of nonmetallic inclusions in the metal. Because inclusions can serve as initiation sites for fracture, they can significantly influence fracture toughness and fatigue performance. When conducting fractographic examination by SEM, the inclusions can usually be seen in various arrays of distribution on the fracture surface.

Interpretation of topographical details in fractographics and correlation with the observed microstructure in metallographics can be very rewarding. Fractography can reveal textures which indicate embrittlement of the metal by hydrogen before its fracture. Hydrogen embrittlement in steel is a condition that cannot be detected with metallographics unless perhaps sufficiently large micro-cracks have developed in the embrittled metal. If the steel was hydrogen embrittled prior to fracture, a revealing pattern can be rather easily detected by fractography.

In general, formation of a fracture surface in a metal takes place by complex deformation and micro-cracking processes that occur at a crack tip region, where triaxial stresses develop. The microstructure, the medium in which the crack tip progresses, is usually composed of a variety of sub-units that must deform in concert immediately prior to formation of micro-cracks or a continuing fracture surface. Also present are dislocations, nonmetallic inclusions and other possible forms of weakness, such as cracks and interdendritic shrinkage voids. The interfaces between these imperfections and the metal provide ample sites for crack initiation. Therefore, under triaxial stress conditions, this heterogeneous microstructure will act in some manner to determine (1) at what points fracture will initiate, (2) the path of fracture, and (3) the amount of energy required for initiation and for propagation. Fractographic studies have provided significant information about the mechanism of fracturing in metal, and the influence of microstructural details.

When the surface of a brittle fracture in metal is examined by SEM, as illustrated in Figure 2.31, the cleavage failure will display facets that are closely related to the grains in the metal. In the case of iron or steel, the favored plane of fracture propagation is one of the {100} planes of the body-centered cubic lattice. Naturally, the angle and size of the facets on the fracture surface will change from grain to grain as dictated by orientation of lattice and size of each grain. Fractography provides a positive

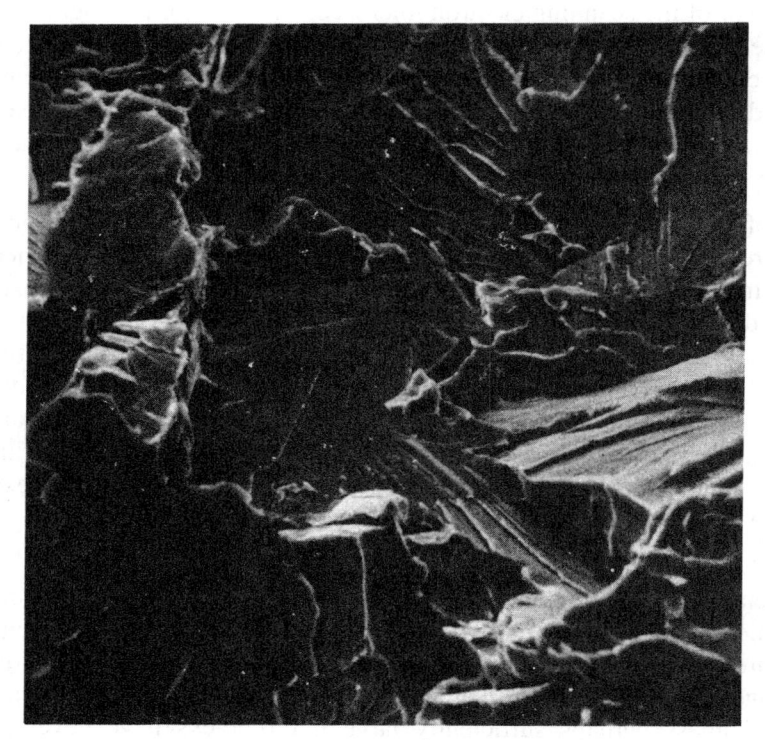

Figure 2.31 — Brittle fracture surface in a plain steel
SEM fractograph at original magnification of 2200X shows flat facets from cleavage through {100} planes across grains with various orientations. (Courtesy of The Welding Institute, England)

means of identifying a low-energy-absorbing (brittle) fracture through examination of fracture topography.

Tough, ductile fracture is easily recognized in fractographic examination by its fibrous texture. The rough surface, viewed at very high magnification, displays many small "dimples". The mechanism of ductile fracture is believed to entail the initiation of microvoids in the metal ahead of the crack tip, where triaxial stress conditions develop. As previously mentioned, many initiation sites can enter into the mechanism as determined from the various available interfaces between inclusions, secondary microstructural constituents, etc. The microvoids expand as stress conditions increase in intensity until the microvoids coalesce and contribute to the formation of a through-fracture. A face of this broken surface will contain hemispherical cavities or dimples as illustrated in Figure 2.32. Fractography allows quantitative evaluation of the shape and size of dimples, as related to the conditions of applied stress, and as related to the fracture toughness of the metal.

Failure by fatigue cracking is an important concern because it is a major problem in metal articles of many kinds, especially those containing welds. Although much is known about the fatigue problem through mechanical testing (aided by specimen examination at relatively low magnification), high-cycle dynamic stress produces fatigue fractures on which the characteristic wave-like pattern of crack development occurs in extremely small increments, and these are most easily seen by fractography at very high magnification.

Figure 2.33 shows the fracture surface of metal failed by fatigue, which would appear smooth and unblemished at low magnification. Yet, at very high magnification, a fractograph clearly shows the striation of incremental crack growth. The spacing of the striae are indicative of the frequency rate of cyclic stress application, and the number of striae corresponds to the number of applied loading cycles. Striation counting is a practical method for assessing conditions contributing to fatigue crack growth. Techniques have been developed to reduce the time and effort for this method of assessment, using either a SEM fractograph (usable up to about 30 000X), or a fracture surface replica as prepared for TEM examination. A computer digitally displays the fracture image, and the data can then be used to produce a Fourier transform, which in

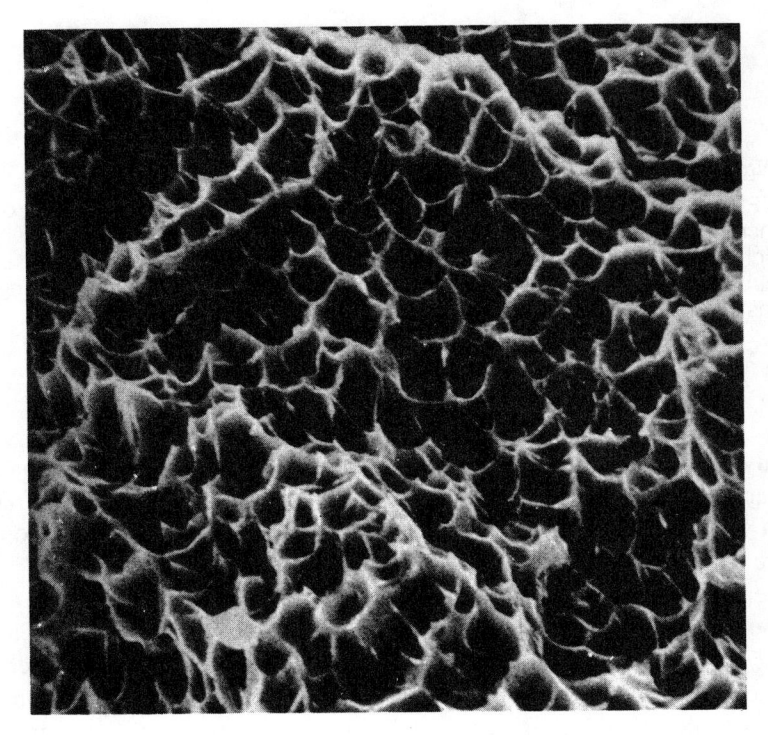

Figure 2.32 — Ductile fracture surface in a C-Mn steel
SEM fractograph at original magnification of 2800X shows microvoids that have expanded during deformation. When through-fracture occurred, the microvoids coalesced to produce a completely dimpled texture. (Courtesy of The Welding Institute, England)

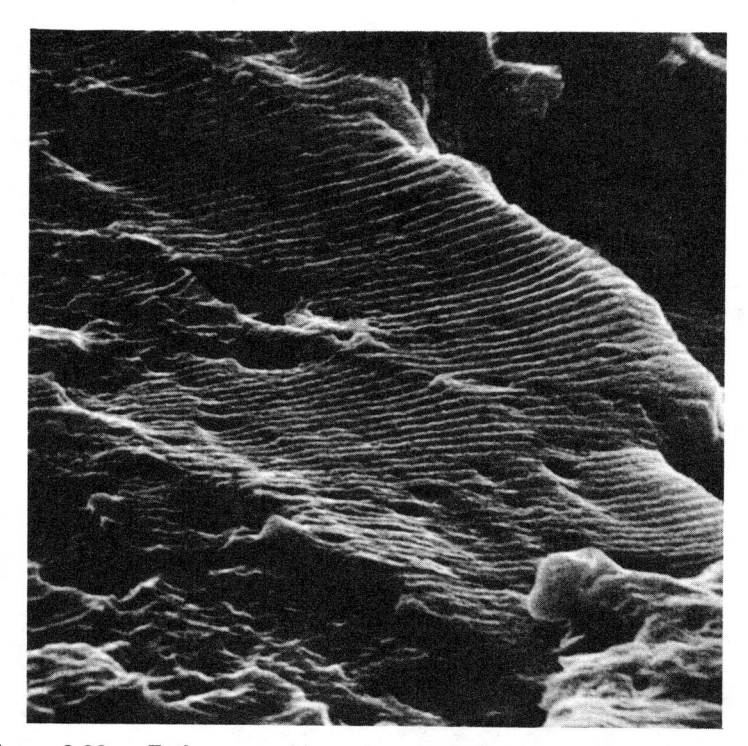

Figure 2.33 — Fatigue cracking on a steel surface
SEM fractograph at original magnification of 7200X shows striations of incremental crack growth. Direction of crack propagation is from lower-right corner toward upper-left. Spacing of striae from peak to peak is approximately 200 nm, or about 8×10^{-6} in. (Courtesy of The Welding Institute, England)

turn is analyzed to determine the spatial frequency of its characteristic projected spots. The Fourier transform is very sensitive to repeat patterns (just like a diffraction grating), and it will extract repeat-feature frequency information, despite the complexity of the fracture pattern.

Another concept for characterizing fatigue fracture-surface profiles, and relating fracture roughness to fatigue threshold in a given metal, is a mathematical treatment termed *fractals*. The fractal-dimension characterization (D_F) provides a measure of the roughness of the fracture surface, with which mechanical property parameters for a material can be correlated. A larger fractal dimension corresponds to a greater surface roughness. The fractal dimension, (D_F), has also been found to correlate with the toughness of metals. Fractographic studies have established that fatigue crack propagation can take place by ductile tearing or by brittle cleavage. This depends on whether the repeated cycles of stress are being applied to the metal while it is above, or is below a set of ductile-brittle transition conditions, with temperature an important factor. (See Chapter 3 for more information on this transition in fracturing mechanism.

Using SEM or TEM, examination of details best observed and measured at very high magnification can help identify the character of the fracture, and assist in determining the stress and other conditions that produced the fracture.

Only an elementary view of fractography has been given on the preceding pages by the several examples illustrating typical textures or topographies of brittle, ductile, and fatigue fractures. Actual fractures through metal sections often display what appears at first glance to be a potpourri of various textures. Nevertheless, careful examination and classification of areas differing in appearance can usually produce an account of (1) where the fracture initiated, providing this area is present on the specimen, (2) the direction of fracture propagation, (3) the modes of fracture as it progressed through the metal, and (4) specific details in the metal that are exerting an influence on the fracture mechanism and are influencing the topography.

As an example of the searching capability of fractography, particularly by SEM, Figure 2.34 is a fractograph of a C-Mn steel specimen, which displays an overall pattern of dimpled texture. This now should be recognized as typical of ductile fracture.

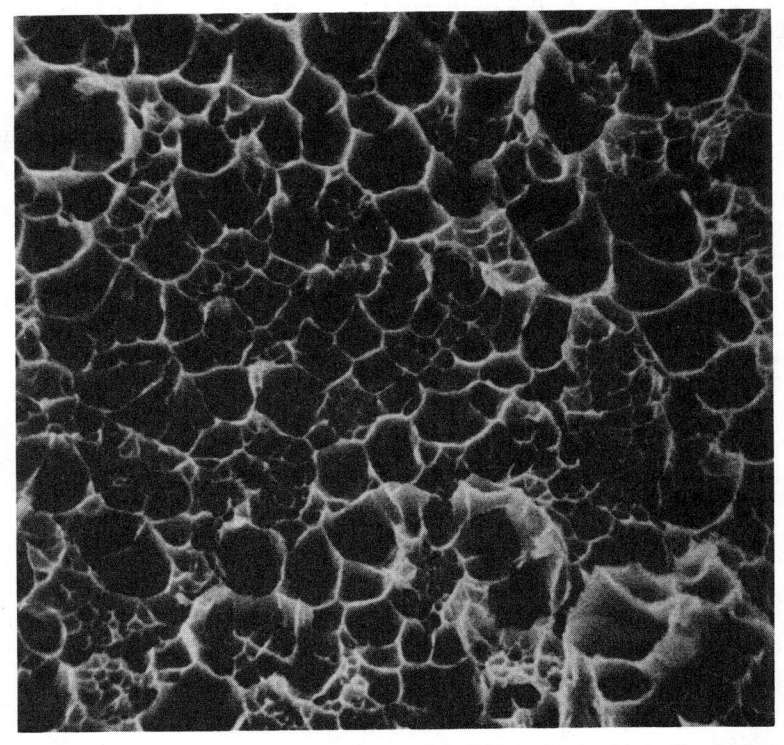

Figure 2.34 — Ductile fracture surface on a C-Mn steel which displays topographical features that assist appraisal of toughness

SEM fractograph at Original magnification: of 2400X. (Courtesy of The Welding Institute, England)

However, Figure 2.34 has distinct areas of larger dimples, with areas of much smaller dimples. The conclusion reached in interpreting this fractograph is that the steel fractured with a finely-dimpled texture, except where coarser microvoids were initiated at second-phase particles. Note that at the bottom of each larger dimple, a nonmetallic inclusion in the shape of a small spheroid can usually be clearly seen. These particles were analyzed and found to be manganese sulfide inclusions. The presence of additional phases in the microstructure of steel can readily influence fracture mechanism; this is especially the case if a brittle constituent, or a friable and easily disrupted type of nonmetallic inclusion is present. If the second phase aids progress of the fracture, discrete areas of brittle (cleavage) fracture can appear in the fracture topography, which is otherwise dimpled and ductile. The manganese sulfide inclusions in Figure 2.34 appear essentially innocuous because they did not interfere with the formation of microvoids prior to ductile fracturing. The larger dimples suggest that if the inclusions were present in greater number or size, the steel would probably exhibit lower toughness.

Laser scanning microscopy (LSM), mentioned earlier as a new tool for metallographic examination of microstructures found on polished-and-etched specimens, has been used with success in fractography. LSM offers capability to scan rough fracture surfaces both horizontally and vertically, and the examinations are performed sans vacuo. In fractographic work, reflected light from the specimen's surface is detected by an acousto-optic reflector during horizontal scanning, and by a galvano mirror during vertical scanning. The output image is focused to allow a CCD to sense the image as it develops and to produce a digital record. High-contrast, high-resolution computerized images are recorded for immediate playback, for digital analysis, and for tape or diskette storage. Work on relatively smooth surfaces achieved a resolution of 0.2 microns (8×10^{-6} in.) at magnifications up to 6000X.

SUGGESTED READING

Microstructural Characterization of Metals and Alloys, P. E. J. Flewitt and R. K. Wild, The Institute of Metals, London, England, 1985, IM Book No. 367, ISBN: 0-904357-76-7

Metallography and Microstructures, Metals Handbook, 9th Edition, Volume 9, American Society for Metals, now ASM International, Materials Park, Ohio, 1985, Library of Congress No. 78-14934, ISBN: 087170-015-8

Fractography, ibid, Volume 12, 1987, ISBN: 087170-018-2

Structure of Metals Through Optical Microscopy, by A. Tomer, ibid, 1990, ISBN: 087170-410-2

Optical Microscopy of Carbon Steels, by L.E. Samuels; ibid, 1980; ISBN: 0-87170-082-4

Chapter 3

In This Chapter...

STRUCTURE SENSITIVITY OF PROPERTIES 133
DIRECTIONALITY IN PROPERTIES 135
MECHANICAL PROPERTIES 136
   Elastic Behavior of Metals.....................................137
      Young's Modulus of Elasticity 139
      Poisson's Ratio 139
      Limits of Elasticity and Proportionality 141
   Plastic Yielding in Metals 142
      Yield Strength..143
   Breaking Strength of Metals................................143
      Tensile Strength..144
      True Stress and True Strain 144
      Notched Tensile Strength 147
   Ductility..147
      Elongation...148
      Reduction of Area 149
      Ductility Indications from Special Tests 149
   Hardness..149
      Static Indentation Hardness Testing..................150
      Microhardness Testing......................................151
      Dynamic Hardness Testing...............................152
      Scratch Hardness Testing.................................153
      Conversion of Hardness Numbers 153
   Toughness...154
      Introduction of Impact Testing...........................155
FRACTURE IN METALS 156
   Ductile Fracture...158
   Brittle Fracture...159
   Intergranular Fracture 162
   Conditions Affecting Fracture Toughness............164
      Effect of Temperature 164
      Effect of Stress Axiality.....................................166
      Stress Gradient...169
      Stress Multiaxiality 170
      Effect of Rate of Strain......................................171
   Effect of Cyclic Stress (Fatigue)...........................174
      Fatigue Crack Initiation 176
      Fatigue Crack Propagation 176
      Fatigue Crack Failure..176
      Cyclic Stress Limits to Avoid Fatigue Failure 181
      Cyclic Stress Conditions 183
      Variable Loading/Cumulative Fatigue Damage ...185
FRACTURE MECHANICS: ASSESSMENT
OF FRACTURE TOUGHNESS...............................188
   Brittle Fracture Test Parameters..........................190
      Section Dimensions..192
      Plotting Coordinates..192
      Crack Surface Displacement Mode 193

      Plane-Strain..195
      Plane-Stress...195
      Stress Distribution 196
   Procedures for Evaluating Propensity for Brittle
Fracture 197
      Use of Linear-Elastic Fracture Mechanics...........198
   Development of Elastic-Plastic Fracture Mech. 201
      Crack Tip Opening Displacement Testing...........205
      The J-Integral Test Method..............................211
   Fatigue Cracking Assessment by Fracture Mech...215
   Mechanical Properties at Low Temperature 218
      Strength at Low Temperature 219
      Impact Toughness at Low Temperature............220
      Test Methods for Toughness Evaluation 226
      Correlation of Results; Frac.-Toughness Tests ...236
   Improved Mech. Properties for Low-Temp. Svs. 238
   Mech. Properties at Elevated and High Temp.........238
      Short-Time Elevated Temperature Testing..........241
      Long-Time Elevated Temperature Testing 242
   Mechanical Properties After Plastic Work............247
      Hot Work 248
      Cold Work...248
      Peening 248
      Irradiation...249
PHYSICAL PROPERTIES 258
   Density...258
   Thermal Properties..259
      Specific Heat 260
      Thermal Conductivity..261
      Melting Point or Melting Range.........................264
      Heat of Fusion..264
      Viscosity & Surface Tension of Molten Metals 264
      Boiling Point and Heat of Vaporization 266
      Thermal Expansion and Contraction..................266
      Thermionic Work Function 268
   Electrical Properties...268
   Magnetic Properties 270
      Evaluation of Magnetization..............................272
      Summary of Magnetic Behavior........................273
      Involvement of Magnetization in Welding 274
CHEMICAL PROPERTIES 274
   Corrosion of Metals 275
      Corrosion in Aqueous Solutions........................276
      Corrosion in Hot Gases.....................................282
      Corrosion in Molten Metals 282
      Corrosion in Molten Salt....................................283
   Forms of Corrosion Pertinent to Weldments.........283
      Stress Corrosion Cracking (SCC) 284
SUGGESTED READING 293

The Properties of Metals

The term *properties* has different meanings to the many people who use or work with metals. The mechanical engineer is usually occupied with strength, hardness, and toughness. The electrical engineer thinks in terms of electrical conductivity, resistivity, and magnetic permeability. The manufacturing planner usually looks for malleability, machinability, and other characteristics that affect fabrication. The corrosion engineer is concerned with chemical reactivity, electromotive force, oxidation resistance, and other aspects of behavior under corrosive conditions. Finally, the purchasing agent is mainly interested in documented properties of materials that satisfy the requirements of an imposed code or specification.

It is not possible to point out particular metal properties that are more important than others, because the use of each metal determines the relative importance of each property. Also, in the production of the metal itself or during its fabrication, any of the known properties can control a key manufacturing step. Therefore, it is important for the welding engineer to be familiar with virtually all metal properties, because welding requires knowledge of most of them.

Terminology is an important part of a discussion of metal properties. For example, strength, hardness, and ductility are often incorrectly referred to as "physical properties." These are *mechanical properties*. They are determined by the application of force to the metal to test its suitability for mechanical service. *Physical properties* of metal are characteristics such as density, electrical conductivity, thermal expansion, and specific heat, which are inherent to a particular composition and crystallographic structure. A number of references that provide terminology for particular properties, as well as many other terms used in metallurgy and in welding, can be found in Appendix II.

In this chapter, the properties of metals are divided into three general groups: (1) mechanical properties, (2) physical properties, and (3) chemical properties. Other groups, such as optical and nuclear properties, are not included. However, additional properties are described as necessary. Table 3.1 provides a list of metal properties divided into structure-insensitive and structure-sensitive categories. This separation of properties emphasizes the consideration that should be given to property values reported in handbooks and other references.

STRUCTURE SENSITIVITY OF PROPERTIES

Structure-insensitive properties are well defined and do not vary from one metal sample to another of the same kind. This is true for most engineering purposes and is usually verified by test data. Structure-insensitive properties can sometimes be calculated from the chemical composition of a metal. These properties (Table 3.1) are commonly termed *constants*. The size or condition of the specimen on which a property determination was conducted is seldom mentioned, since composition is assumed to be the controlling feature of the metal.

Structure-sensitive properties, on the other hand, are dependent not only on chemical composition, but also on the manufacturing and processing history of the metal. The size of the specimen can influence the value obtained for a structure-sensitive property.

Table 3.1 shows that all mechanical properties are structure sensitive, with the possible exception of elastic moduli. This view suggests that mechanical properties reported for a particular metal or alloy should be accepted with some reservation. Table 3.1 also indicates that physical and chemical properties, with few exceptions, are structure insensitive. These properties are often re-evaluated, however, because the "constants" reported in handbooks may not be as well-founded as has been assumed.

The metals most used in industry are *polycrystalline aggregates*. These aggregates may display properties that do not represent an average of those properties exhibited by single-grain structures. Properties of multi-grained metals are generally determined by their chemical composition, but they are also strongly influenced by variations in microstructural features. These features include the size and shape of the grains, the presence of secondary phases and constituents, and the amounts and micro-distribution of these secondary components.

When evaluating metal properties, the analysis should begin by noting the method of manufacture to determine if the article is a casting or a wrought material. Heat treatment applied to cast or wrought products can have a profound influence on their properties. Metal thickness is another consideration; heat treatment applied to achieve property improvement may be less effective on very thick metal than on thinner metal. This is especially true for many kinds of steel.

Table 3.1 - The Properties of Metals

	Structure-Insensitive Properties		Structure-Sensitive Properties	
Mechanical Properties	Elastic Moduli		Tensile Strength True Breaking Strength Yield Strength Elastic Limit Proportional Limit Creep-Rupture Strength Strain Hardening Rate Tensile Reduction of Area	Fracture Strength Fatigue Strength Impact Strength Hardness Damping Capacity Tensile Elongation Creep Strength
Physical Properties	Thermal Expansion Thermal Conductivity Melting Point Specific Heat Emissivity	Vapor Pressure Density Thermal Evaporation Rate Thermoelectric Properties Thermionic Emission	Magnetic Properties Electrical Conductivity	
Chemical Properties	Electrochemical Potential Oxidation Resistance Catalytic Effects			

The testing of metals for physical, mechanical, and corrosion properties is well covered in Section 3 of the *ASTM Book of Standards*. These volumes provide excellent guidance on testing procedures and methods of data analysis. This text concentrates on metallurgical conditions that cause metal property variations in different product forms and in different portions of a given product.

DIRECTIONALITY IN PROPERTIES

In addition to gross structural variations, which make the properties of one metal different from those of another despite similarity in composition, a metal can also exhibit different properties with changes in microstructural direction. This characteristic is termed *anisotropy*. Conversely, *isotropy* is the quality of exhibiting identical properties in all directions, a desirable quality not easily achieved in commercially produced metals. Anisotropy, with respect to a mechanical or physical property, is quite common in metal products; fortunately, the property differences are usually relatively small. Most applications do not require close attention to this detail. Nevertheless, circumstances may be encountered in welding where care must be exercised to avoid problems that can stem from anisotropy. Stresses resulting from welding can be generated in many directions in the joint members during fabrication and during service, and any one of these may require adequate load-carrying behavior by the metal.

Anisotropic behavior can be caused by (1) directionality in microstructure, (2) heterogeneity in composition, and (3) preferred crystallographic orientation. The mechanisms by which each of these induce anisotropy is discussed in later chapters, but the following comments provide a general understanding of the topic.

Microstructure can be a strong controlling factor in establishing the properties of metals on a micro scale. Many circumstances determine whether a metal has a homogeneous microstructure, and thereby offers isotropic properties, or whether manufacturing conditions will produce a heterogeneous microstructure that induces anisotropy. Microstructural conditions that cause anisotropy include banding, columniation, and changes from the surface to center of a metal member.

Composition heterogeneity is often the root cause of anisotropy, but the compositional differences that exist may not be readily apparent from observable changes in microstructure. Composition differences can range from slight to gross; nevertheless, some properties can be affected by minute composition differences. Chemical analyses, perhaps with an electron microprobe analyzer, may be required to determine whether a relationship exists between composition heterogeneity and anisotropy. In some metals, composition heterogeneity may be indicated by random distribution of nonmetallic inclusions in the material rather than by an intrinsic feature of the microstructure.

Ordinarily, the random orientation of crystallographic structure in polycrystalline metals causes them to behave isotropically because they exhibit average properties, even though the individual grains are anisotropic. Metals can be processed to achieve a high degree of preferred orientation in a polycrystalline microstructure, and as a result, the metal properties may be strongly directional. This anisotropy benefits a number of commercial applications; as an example, grain-oriented electrical steels offer highly directional and desirable magnetic properties.

In welding, however, strongly directional heat flow and progressive solidification can promote subtle degrees of preferred orientation in the crystallographic structure of grains. The anisotropy in properties that results from this condition may be less pronounced than that induced by banding, columniation, etc., but it must be considered as a factor that can add to the total anisotropy exhibited by a welded structure. The microstructure of weld metal, because of the solidification mechanism during directional heat flow into the colder base metal, will often display preferred orientation. This has been found to cause small but measurable directionality in mechanical properties.

Both cast and wrought metals can exhibit anisotropy in properties. In castings, solidification is closely related to heat flow and cooling rate, and characteristics such as dendritic growth, columniation, and segregation can promote anisotropy. The influence of these factors depends on their degree of development in a particular casting. Wrought metal varies in properties with respect to the principal axis of working. Various forms of heterogeneity in the ingot or in direct-cast semifinished metal can become elongated along the working or rolling axis, and this imparts directionality. Even the nonmetallic inclusions are altered in shape and distribution by working. For this reason, when considering wrought metal — and particularly when using flat-rolled products such as plate or strip — it is important to maintain identity of (1) the longitudinal direction, which is parallel to the principal direction of working; (2) the transverse-to-longitudinal direction, in the same horizontal plane; and (3) the through-thickness direction of the metal (commonly referred to as the "z" direction). The anisotropic behavior of wrought metal can be sufficiently pronounced to require precautions when designing welded joints, in order to avoid problems such as lamellar tearing.

MECHANICAL PROPERTIES

Metals are highly favored as construction materials because they offer combinations of mechanical properties not found among nonmetallic materials. Metals are generally strong, and many metals can be loaded or stressed to very high levels before breaking. Also, the fact that metals exhibit *elastic* behavior in the early portion of their load-carrying capacity is of great importance. Elasticity is a vital factor in the success of many metal applications. Indeed, some uses — detents and springs, for example — depend primarily on consistent elastic behavior. Finally, when loaded beyond their elastic range, many metals are *ductile* and *tough*. These qualities permit cold forming, and they facilitate the successful performance of metals in rigorous service. If a metal structure is subjected to unanticipated overload, sudden or catastrophic failure is often forestalled by absorption of mechanical energy as members of the structure undergo permanent deformation.

The earliest engineering approach to selection of a metal for structural use was based on breaking strength. Values for this strength were obtained from simple tension tests, and were reported in handbooks as "tensile strength." Working stresses allowed in structures then were commonly set at a fraction of the tensile strength. The remaining or surplusage fraction of strength was regarded as a "safety factor" to guard against breakage from accidental overloading. Allowable stresses were arbitrarily selected by code-writing bodies, and one-fourth of tensile strength appeared in many specifications as the imposed fractional limit.

As years of metal usage went by, particularly when employing steels, engineers began to realize that the practice of basing allowable stresses on tensile (i.e., breaking) strength had become archaic for two reasons. First, most structures are expected to resist permanent or plastic deformation under normal working loads, and therefore the end-point of elastic behavior by the particular metal used in construction should be the proper determinant for selecting an allowable stress. Second, much progress had been made in obtaining higher yield strengths in steels; more specifically, the yield/tensile strength ratio of a number of new steels had been substantially increased. This added strength in the elastic regime, if not used to advantage, would detract from the economic utilization of the material of construction. These sharpened perspectives led code-writing bodies to shift their attention from the tensile strength of constructional metals to the stress level above which non-elastic yielding or plastic deformation occurred. This decisive yield stress level proved not simple to delineate; consequently, several different expressions of yield point or yield strength began to appear in engineering handbooks.

Elastic Behavior of Metals

A clear understanding of the various values published for the strength of metals can be gained by examining the behavior of metals over the entire range of loading — that is, from initial application of a light load; to a higher level beyond which the metal ceases to behave elastically; and finally to an ultimate load that causes the metal to break.

From a fundamental viewpoint, metals are regarded as being quite strong because of their ability to support considerable loads without appreciable deformation. Also, the small amount of deformation that does take place may not remain after unloading. For instance, a 25 mm (1 in.) cube of tool steel is compressed only 0.0254 mm (0.001 in.) under a stress of 207 MPa (30 000 psi). When this stress is removed, the cube regains its original dimensions. Liquids also can support loads. For example, liquid mercury can float a piece of iron, but not without considerable displacement. Other materials, such as rubber, may also be considered strong, but they likely will stretch more than a metal under a given load. To appreciate why metals are "strong," further use can be made of the schematic view of atoms in the crystalline structure of metals as illustrated earlier in Chapter 2. The effects of applying certain levels of tensile or compressive stress can be envisioned with respect to position of the atoms and the lattice parameters of the crystalline structure. This rationalization provides a basis for the engineering stress diagrams and discussion that follow.

The changes in the interatomic spacing that occur in the crystalline structure of the metal when initially loaded are accompanied by similar changes in external dimensions of the metal body. As illustrated in Figure 3.1, a cube of metal is assumed to have a simple, cubic crystallographic structure, with a typical atom-to-atom distance of about 3Å (0.000 000 012 in.). When a tensile stress of 69 MPa (10 000 psi) is applied, the cube is extended 0.008 89 mm (0.000 35 in.) in the direction of the tensile force; this is an extension of 0.035 percent. Also, the cube contracts 0.003 02 mm (0.000 12 in.) normal to the direction of the force. When the load is removed, the metal returns

Chapter 3

SI (Metric) Values	US Customary Values
UNSTRESSED A 25.4 mm cube of metal contains atoms spaced on centers 3 Å apart.	**UNSTRESSED** A one-inch cube of metal contains atoms spaced on centers 0.000000012-inch, apart.
UNDER TENSION STRESS OF 68.95 MPa (A LOAD OF 44 480 NEWTONS)	**UNDER TENSION STRESS OF 10,000 PSI (A LOAD OF 10,000 POUNDS)**
Cube of metal has extended to 25.40889 mm in the direction of stress and has contracted in lateral direction to 25.39698 mm. Spacing of atoms has increased in vertical direction and has decreased in horizontal direction in same proportions.	Cube of metal has extended to 1.00035-inches in the direction of stress and has contracted in lateral direction to 0.99988-in. Spacing of atoms has increased in vertical direction and has decreased in horizontal direction in same proportions.
UNDER TENSION STRESS OF 137.9 MPa (A LOAD OF 88 965 NEWTONS)	**UNDER TENSION STRESS OF 20,000 PSI (A LOAD OF 20,000 POUNDS)**
Cube of metal has extended to 25.41778 mm in the direction of stress and has contracted in lateral direction to 25.39396 mm. Interatom spacing is further increased vertically and has decreased horizontally in similar proportions.	Cube of metal has extended to 1.0007-inches in the direction of stress and has contracted in lateral direction to 0.99976-in. Interatom spacing is further increased vertically and has decreased horizontally in similar proportions.
UNDER COMPRESSION STRESS OF 68.95 MPa (A LOAD OF 44 480 NEWTONS)	**UNDER COMPRESSION STRESS OF 10,000 PSI (A LOAD OF 10,000 POUNDS)**
Cube of metal has been compressed to 25.39111 mm in the direction of stress and has expanded laterally to 25.40302 mm. Interatom spacing has decreased vertically and has increased horizontally in proportion to dimensional changes.	Cube of metal has been compressed to 0.99965-inch in the direction of stress and has expanded laterally to 1.00012-inches. Interatom spacing has decreased vertically and has increased horizontally in proportion to dimensional changes.

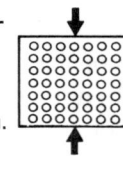

Figure 3.1 — **Dimensional changes and inter-atom spacing in a metal cube as tensile and compressive stresses are applied**

A one-inch cube is used in this figure as a compromise to illustrate changes both in SI and U.S. customary units. Inter-atom spacing and shape of the metal section in the figure are exaggerated for clarity.

to its original cubic form. If the tensile force is doubled, the dimensional changes are doubled. When the load is changed from tensile to compressive, the reversal of forces produces dimensional changes that are the opposite of those from tensile force.

If the cube of metal in Figure 3.1 is subjected to various stresses between zero and 140 MPa (20 ksi), a linear proportionality would be found between the stress and dimensional change in each instance. This firm relationship was expressed in 1660 as *Hooke's law*, which states that in an elastic body, the ratio of the stress applied to the strain produced is constant. However, the proportionality between stress and strain is not the same in all elastic materials. Different metals display marked differences in the amount of elastic elongation that accompanies a tensile load, or in the reduction in length under a compressive load.

Young's Modulus of Elasticity

A convenient way of appraising the ability of a material to resist elastic stretching or compression under load is to define the ratio between the stress and the corresponding strain. For example, if this ratio is calculated between the stresses applied to the metal cube in Figure 3.1 and the resulting changes in length in the direction of the force, a ratio of approximately 197 GPa (28 million psi) is obtained (unit stress divided by unit strain). This ratio is known as *Young's modulus* or the *modulus of elasticity*. It commonly is expressed in the following equation and terms:

$$E = \sigma/\varepsilon \qquad (Eq.\ 3.1)$$

where
    E is the modulus of elasticity,
    σ is the unit stress, and
    ε is the unit elastic strain.

Young's modulus is a constant characteristic of a polycrystalline metal measured in tension, compression, bending, or other mechanical tests in which stress and strain are correlated. Fundamentally, elastic moduli for metals are established by the interatomic binding forces present in each metal, and these forces are not altered unless the basic nature of the metal is changed. Consequently, the elastic modulus is one of the most structure-insensitive mechanical properties. It is virtually unaffected by grain size, cleanliness, or heat treatment. In fact, the modulus of elasticity often remains unchanged even after substantial alloying additions have been made. Table 3.2 lists Young's modulus for a number of metals and for a few nonmetals for comparison. Note that iron, steel, and stainless steel have almost the same modulus, despite marked differences in chemical composition. Some metals, like aluminum, copper, lead, and titanium, undergo a greater elastic strain than iron for a given stress, while others (e.g., beryllium and tungsten) show lower strains. However, non-metals such as glass, plastic, and rubber undergo much greater elastic strain for a given stress than most metals. A standard test method for determining Young's modulus is provided in ASTM E 111, *Young's Modulus at Room Temperature*.

Poisson's Ratio

The dimensional change in the direction of stress, as explained above, is accompanied by a lesser change that occurs *transverse* to the stress direction. In Figure 3.1, when the cube of metal was loaded with a tensile stress of 69 MPa (10 ksi), the elastic elongation of 0.008 89 mm (0.000 35 in.) in the direction of stress was accompanied by a transverse contraction of 0.003 02 mm (0.000 12 in.). This proportionality between strain (ε) in the direction of applied stress (σ) and the lateral contraction (expansion in the case of applied compressive stress) is called *Poisson's ratio*; this is usually identified by the symbol "μ". Poisson's ratio for many elastic materials, as shown in the right-hand column of Table 3.2, ranges between ¼ and ½. For the metal in Figure 3.1, Poisson's ratio is 0.34. Like Young's modulus, Poisson's ratio is considered to be structure-insensitive. ASTM E 132, *Poisson's Ratio at Room Temperature*, gives a standard method for determining this elastic property.

Both Young's modulus and Poisson's ratio are temperature dependent. Increasing temperature results in a greater elastic strain for a given stress. Although Young's modulus for steel is approximately 200 GPa (29×10^6 psi) at room temperature, the elastic modulus decreases by about 10 percent for each 200 °C (400 °F) temperature increase.

Grain structure in the metal that is strongly oriented can produce deviations in the expected elastic modulus. Columnar microstructure, such as often found in weld metal, will exhibit directional deviations when precise measurements are made. Specimens with pronounced crystallographic orientation have been studied to determine changes in elastic moduli. Table 3.3 contains values for several metals, and these data confirm that crystallographic orientation can have a substantial influence on elastic modulus.

Changes in moduli are seldom observed with regular (isotropic) polycrystalline metals and alloys, because random orientation of their grains produces average values. Regardless of the conditions that affect the amount of strain, elastic moduli are very useful in practical metal applications. Uses include computing deflection of beams and

Table 3.2 - Typical Elastic Constants for Various Materials at Room Temperature*

Material	Young's Modulus of Elasticity (E) GPa	10^6 psi	Poisson's Ratio (μ)
Aluminum	69	10	0.33
Beryllium	290	42	0.43
Copper	110	16	0.33
Iron	193	28	0.29
Lead	14	2	0.43
Nickel	207	30	0.26
Niobium	103	15	0.38
Titanium	117	17	0.35
Tungsten	407	59	0.28
Steel, Carbon and Alloy	200	29	0.28
Stainless Steel	193	28	0.30
Rubber, Vulcanized	14×10^{-3}	2×10^{-3}	0.50
Plastic, Phenolic	7	1	
Glass	69	10	
Human Muscle	0.90	0.013	
Human Tendon	14	2	
Human Bone	221	32	

*Approximately 25 °C (about 75 °F)

other stressed components and determining stresses by measuring the amount of strain incurred by loading.

Limits of Elasticity and Proportionality

An effective way to understand the behavior of metal under tensile loading is to plot stress versus strain. Figure 3.2 considers the same metal used in Figure 3.1, with loading beyond the limit of elastic behavior. Starting from no load and no strain at the lower left corner of the diagram, the straight solid portion of line A - A' traces the specimen elongation over a 25.4 mm (1 in.) gauge length with increasing stress. This linear proportionality between stress and strain is in accordance with Hooke's law, and it represents Young's modulus for this metal — namely, 197 GPa (28.6×10^6 psi). If the load on this tension specimen is removed at any point along the straight portion of A - A', then the specimen length will return to its original dimension; thus absolute elasticity is demonstrated by the metal.

Next, it should be noted in Figure 3.2 that when the tensile stress on the specimen rises above approximately 193 MPa (28 ksi), the strain becomes greater for each unit of stress applied, and the stress-strain plot gradually becomes a curve that depicts increasing strain for each unit of stress applied. The exact level of stress at which deviation from a straight line occurs is highly dependent on the precision of measurement and plotting, and the clarity of the lines. Nevertheless, the point of initial deviation from the straight modulus line is termed the *proportional limit*. Although this mechanical property has been listed in handbooks for many years, its use is not favored because of the practical difficulties in obtaining a reproducible value. It has already been made clear that if the stress is reduced or removed at any point along the *straight* portion of the stress-strain plot, the strain decreases in strict compliance with Young's modulus. Consequently, the strain disappears at zero load, and the original dimensions of the metal body are regained. Yet, mention must be made that if the stress is raised by just a small amount above the proportion limit, there is a very limited range where removal of stress still allows the metal to respond in an elastic manner and regain its original dimensions.

Now it is very important to note that at some relatively small increment of stress above the proportional limit, an *elastic limit* is encountered. This can be defined as the highest stress the metal can withstand without measurable permanent strain remaining after removal of stress. The elastic limit is very dependent on the accuracy of strain measurement. Uncertainty about this accuracy, along with tedious incremental loading

Table 3.3 — Influence of Crystallographic Orientation on Elastic Modulus at Room Temperature

Metal	Crystal Orientation				Ratio Between Planes ($E_{111} : E_{100}$)
	E_{111}		E_{100}		
	GPa	10^6 psi	GPa	10^6 psi	
Aluminum (fcc)	75.84	11.0	63.43	9.2	1.19
Copper (fcc)	190.98	27.7	66.88	9.7	2.87
Gold (fcc)	116.52	16.9	42.75	6.2	2.72
Iron (bcc)	273.03	39.6	124.80	18.1	2.18
Tungsten (bcc)	384.73	55.8	384.73	55.8	1.00

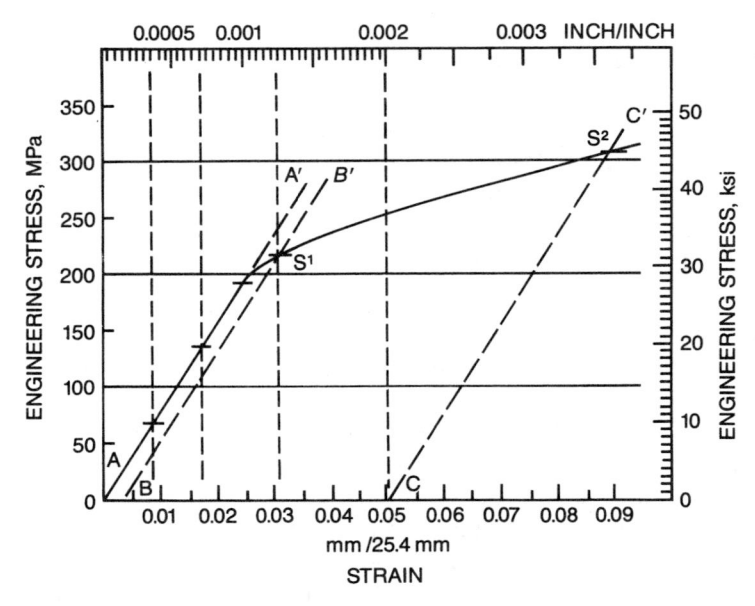

Figure 3.2 — Stress/strain diagram for a metal loaded in tension beyond its limit of elastic behavior

and unloading test procedure, has discouraged use of the elastic limit as an engineering mechanical property.

Plastic Yielding in Metals

When a metal is stressed beyond its somewhat elusive "elastic limit," plastic yielding is produced that is permanent, and that can be readily measured by an *extensometer* clamped to reference marks on the tension specimen. Despite the minuteness of the stretching that takes place in the specimen as load is increased, the plotting of stress versus the growing increments of strain is easily carried out as illustrated by the curve portion S^1 to S^2 included in Figure 3.2. Brief mention was made earlier of the need to know the stress level at which a metal would commence significant plastic yielding. This piece of information is valuable because the majority of structures and components are designed to avoid plastic deformation under anticipated service loads.

Note that when the metal is stressed to the level designated S^1 on the stress-strain curve in Figure 3.2 — specifically, 221 MPa (32 ksi), removal of the stress at this point causes the metal to recover elastically along the dashed line B - B′ (parallel to the linear elastic loading line). However, on arrival at zero stress, the 25 mm (1 in.) length of metal would have permanently elongated by 0.003 81 mm (0.000 15 in.). From this plot the general observation can be made that multiplying unit strain by Young's modulus gives a product that is unit stress only when the stress is below the proportional limit and no plastic strain has occurred.

The mechanics of yielding under tensile stress during the early stages of plastic deformation may not plot a simple, smooth curve for stress-strain as shown in Figure 3.2. Some metals, including certain very ductile steels, exhibit a marked propensity to yield plastically immediately after being stressed above the proportional limit. In fact, once yielding has started, the plastic flow continues (somewhat heterogeneously) even at a slightly lower level of stress. In this case, the stress-strain curve will display an upper *yield point* followed by appreciable yield-point elongation, albeit at a slightly lower level of stress, until work-hardening causes resumption of rising stress. Stress-strain curves for various metals can have a variety of configurations starting from a stress just above the elastic limit, but the majority of metals, especially most of the steels, will produce a smooth curve similar to that shown in Figure 3.2. Standardized methods are described in ASTM E 8, *Tension Testing of Metallic Materials*, for determination of a yield point, and as is more often the case, a *yield strength*. This latter term requires further definition to serve for engineering purposes, and it has proved a practical, reliable index for an important feature of metal strength.

Yield Strength

The procedure used most frequently to establish the yield strength of a metal is to determine the stress required to produce a specified amount of plastic strain. Although no universal agreement exists on one particular amount of plastic strain (yielding) that should take place to fix the *offset yield strength*, offset (plastic deformation) amounts of 0.1, 0.2, and 0.5 percent are being used in various sectors of the world-wide engineering communities. Figure 3.2 illustrates the procedure for determining the 0.2 percent offset yield strength, which is the value most often specified and published in the USA. ASTM recommends that the offset value used should be included in parentheses after the term — thus, yield strength (offset = 0.2 percent) — but this seldom is done. When Figure 3.2 is examined, Point C on the abscissa will be noted to represent 0.2 percent strain in the tension specimen's gage length; precisely 0.0508 mm (0.002 in.). A line is drawn from Point C parallel to the elastic modulus (shown by lines A - A' and B - B') extending above the stress level where appreciable plastic yielding occurs. This line, C - C', represents a strain axis of 0.0508 mm/25.4 mm (0.002 in./in.) The point S^2, where the 0.2 percent offset line (C - C') intersects the stress-strain curve, is the stress required to produce 0.2 percent plastic flow or yielding — namely, 310 MPa (45 ksi). Although this value is called the *0.2 percent yield strength* in the USA, and is presented as such in handbooks, it is referred to as *0.2 percent proof stress* in the UK. Care must be exercised when accepting a "yield strength" to ascertain the percent offset used in fixing the value inasmuch as a different offset would be likely to affect the value significantly.

Breaking Strength of Metals

Metal will break when sufficiently overstressed by any of the three basic means of loading — tension, compression, and shear. When these externally applied forces exceed the internal resisting stresses, the metal breaks. Torque force (twisting) sometimes is applied as a load, but torsion amounts to shear stress in the metal. The tension test ordinarily is employed for determining breaking strength; hence frequent use of "tensile strength" to loosely indicate the load-carrying capacity of a metal. However, even this term must be supplemented to clearly define the property value being provided. While the modulus of elasticity and yield strength values for most metals are approximately equal in tension and compression, the shape of stress-strain diagrams for compression tests is strongly influenced by the ductility of the metal and the size and shape of the compression specimen. In compression tests of brittle metals, the specimens usually break with a shearing type of fracture. When testing ductile metals, compression is accompanied by lateral expansion to produce a flattened disk, and fracture of the metal often does not occur.

Tensile Strength

When the tension specimen of Figure 3.2 undergoes continued loading until it breaks, a new stress-strain diagram must be plotted to accommodate the substantial range of plastic strain that occurs. Therefore, Figure 3.3 covers the complete stress-strain history of this ductile metal from the start of loading to the breaking point. The abscissa of this diagram does not have a dimension scale, because there is no further need for precise incremental elongation values (after the offset yield strength has been established). Figure 3.3 indicates the relatively small amount of elastic strain compared to the extensive plastic strain during the complete testing procedure. The usual practice in tension testing requires the extensometer to be removed from the specimen after passing the offset yield strength level (to avoid damage to the instrument), and only the stress is recorded until the specimen breaks. Metal plastically deformed beyond the elastic limit continues to develop progressively higher strength until its *ultimate tensile strength* is reached. This is the value reported as "tensile strength" among the engineering mechanical properties of a metal. This maximum recorded stress is also the approximate stage where localized "necking" commences in the specimen, and this acts to fix the location of final fracture. The pronounced reduction in specimen cross-section that occurs during necking normally causes a decrease in the load-carrying capability of the specimen as the test proceeds toward the final fracture. Consequently, the stress-strain curve slowly descends from the ultimate tensile strength until it terminates at the point of breaking stress.

ASTM E 8 covers the tension testing of any form of material to determine elastic limit, yield point, offset yield strength, ultimate tensile strength, elongation, and reduction of area while at room temperature (considered to be 10 to 38 °C, or 50 to 100 °F). An important requirement of the standardized testing method is the speed of testing, because the rate of plastic straining can affect test values for certain metals. Standard specimen designs are suggested for different metals in various product forms.

True Stress and True Strain

To this point, most material properties have been presented in terms of load per unit area (e.g., MPa or ksi), based on the initial cross-sectional area of the specimen. This was done despite the fact that, upon application of load, the specimen's cross-section decreases in accordance with Poisson's ratio, and as the specimen elongates during plastic yielding the cross-sectional area decreases uniformly and rapidly along its gauge length. However, the strain hardening in the metal during plastic flow more than compensates for the decrease in area, and the stress recorded continues to rise. The start of the necking stage is a signal that the decrease in specimen cross section is more significant than the increase in strength from strain hardening, and the load falls off quickly to the fracture load.

Obviously, the value reported as ultimate tensile strength is not the true stress-supporting capacity of the metal. The true strength of the metal just prior to fracture is substantially higher. While tensile strength has been used for many years to control allowable design loads (which are reduced by a factor of safety), the present trend is to have permissible loads based upon a specified fraction of the offset yield strength, since this is widely regarded as a more meaningful criterion.

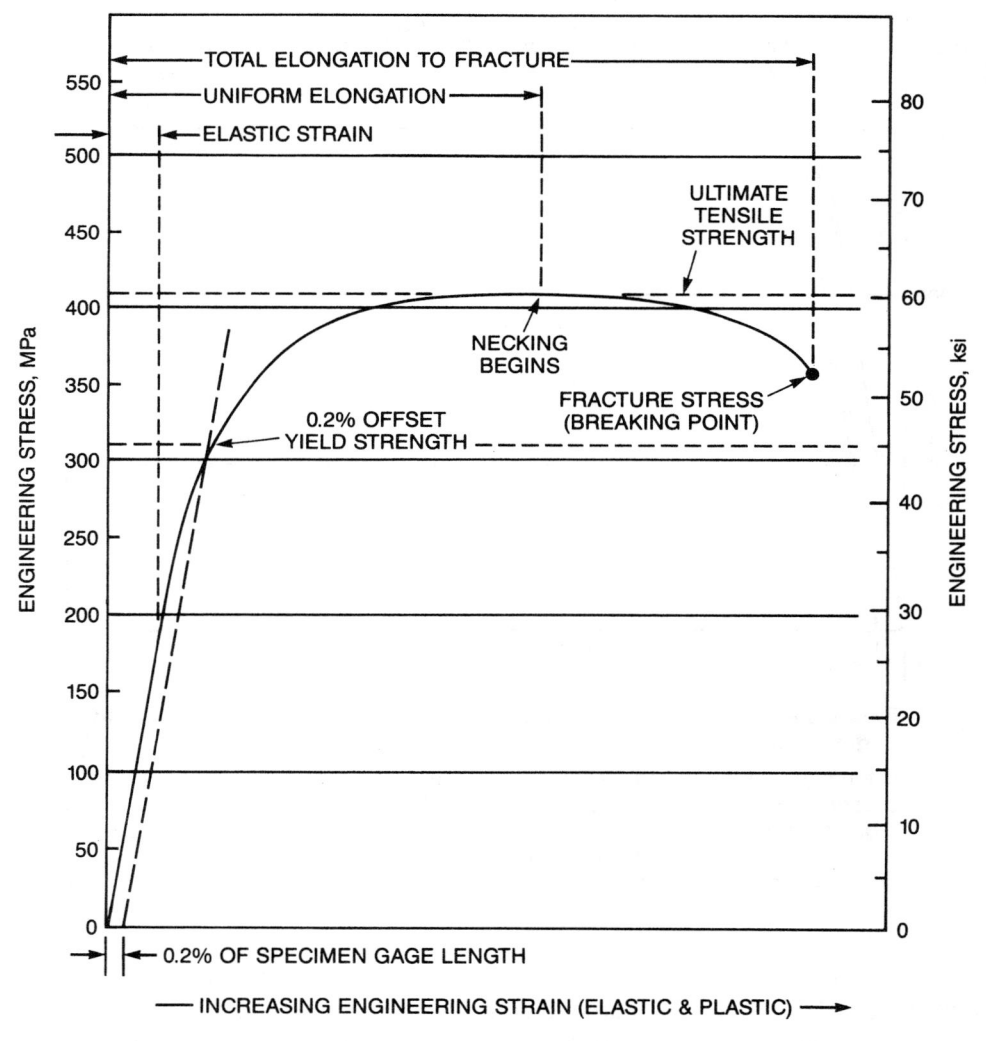

Figure 3.3 — Stress/strain diagram for complete history of a metal tension test specimen from the start of loading and carried to the breaking point

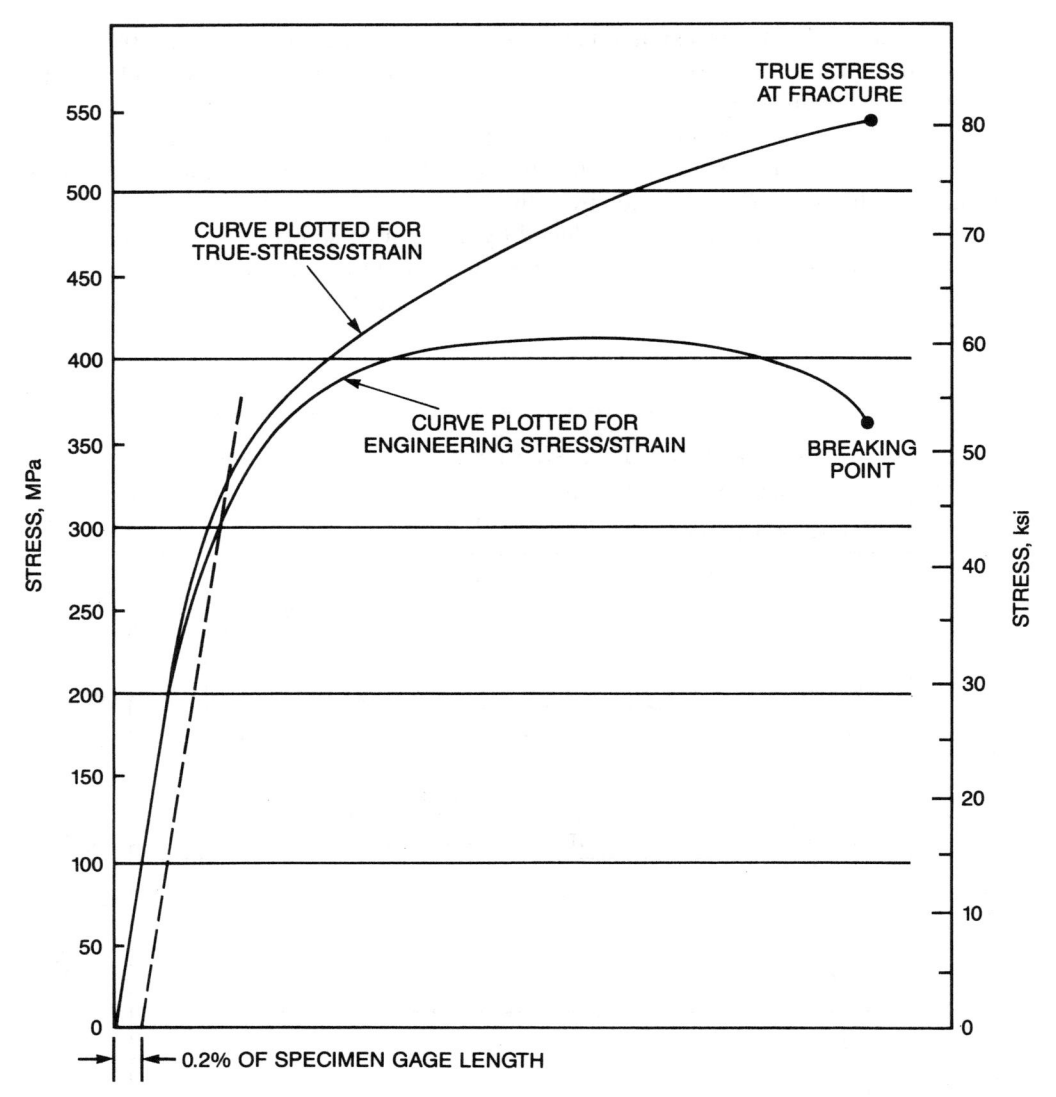

Figure 3.4 — Comparison of a curve for true-stress/strain with curve for engineering-stress/strain

During course of tension testing and computing true-stress/strain, the cross-section of the specimen is continuously measured to obtain the actual net cross-section at each unit of stress. Computed true-stress/strain also is dependent to some extent on the degree of constraint to necking in the specimen just prior to breaking.

To reveal the unit area strength of metals as they are strained to the breaking point, Figure 3.4 has been prepared to compare the stress-strain curve plotted earlier with a curve that would likely result if the cross section were continuously monitored during loading to permit calculation of the true stress. From zero load up to the elastic limit, the straight-line (modulus) portions are very close to one another, but beyond the limit of elasticity, the reduction in cross section that accompanies plastic elongation produces higher computed stress values than those indicated by the original curve. The true stress/strain curve shows tensile strength steadily increasing up to the moment of fracture. However, there is an additional factor acting on this curve that should be identified.

Fracture mechanics has shown that when metal is constrained from contracting or decreasing in thickness transversely to a longitudinal load, the transverse stresses that develop through this area of constraint increase the longitudinal stress required to cause plastic flow or elongation. Formation of the neck in the tension specimen during the final stage of loading introduces complex triaxial stresses in the constricted portion, equivalent to those produced by a notch in the metal. The effect of triaxial stresses is to increase the stress required to produce further plastic deformation before the fracture stress is exceeded.

Notched Tensile Strength

The influence of triaxial stresses on the strength of a metal and on plastic deformation under load has been sufficiently important to justify special mechanical tests for analyzing *notch sensitivity*. Tension testing is only one means of evaluating notch sensitivity. The tension specimens are of special design containing a notch of precise form, sharpness, and size within the gauge length. The maximum load sustained by the notched specimen (usually at the moment of fracture) is divided by the initial cross-sectional area at the base of the notch. Because of constraint surrounding the notched area, the stress calculated for the notched specimen is ordinarily higher than the ultimate tensile strength determined with a smooth tension specimen.

The practice for reporting the notch sensitivity of a metal is to give the ratio of notched to unnotched tensile strength. If the ratio is less than unity (notched strength less than value with no notch), it is indicative of *notch brittleness*. A standard testing method can be found in ASTM E 338, *Sharp-Notch Tension Testing of High-Strength Sheet Materials*. Although notched tension specimens are appropriate for evaluating notch sensitivity in some metals, they are seldom used to evaluate carbon and low-alloy steels. As will be explained shortly, better discrimination is possible in these steels with notched-bar impact tests or precracked fracture tests.

Ductility

The amount of plastic deformation that a metal undergoes in a mechanical test to fracture is considered its *ductility*. Value for ductility secured from various mechanical tests does not measure a fundamental characteristic but merely provides a relative value for comparing metals subjected to identical test conditions. The plasticity exhibited by a specimen is simply the deformation that took place during the yielding process. At the point where fracture releases the stress, the yielding process is terminated, and no further deformation occurs. Ductility, regardless of the method of evaluation, is a structure-sensitive property. It is affected by a number of compositional and microstructural features in the metal, as well as many of the conditions involved in conducting the test. The size and shape of the specimen, the ambient temperature, and the rate of straining are examples of testing conditions that influence the amount of plastic deformation prior to fracture.

Ductility data obtained from common mechanical tests cannot be used directly in design or in judging the ability of a metal to withstand forming that involves considerable plastic deformation. Most structures are designed to operate at stresses below the yield strength, and any significant deformation usually makes the component unfit for service. Nevertheless, ductility may approximate the capacity of a metal to withstand the cold forming involved in brake bending, deep drawing, cold heading, and cold straightening. Ductility may also indicate the reserve of plasticity available to resist sudden shattering fracture under unforeseen localized overloading.

Elongation

Elongation of a tension specimen after breaking is one indicator of ductility. The total extension of the gauge length, after placing the fractured ends tightly together, is derived from two contributing functions. One is the uniform plastic extension that occurs up to the point of necking, and the other is the localized extension that takes place during necking and up to the point of breaking. Total elongation is expressed as the percent increase in a gauge length. This value is not only dependent on the capacity of the metal to yield plastically before breaking, but it is also affected by the size and shape of the specimen, particularly the cross-sectional shape, and the gauge length. Gauge length is quite important, because localized elongation during necking can significantly add to the total elongation. Consequently, a shorter gauge length results in higher elongation. For reasons related to tension specimen size, a gauge length of 25 mm (1 in.) is used for most subsize specimens, while 50 mm (2 in.) is used on the widely standardized 12.5 mm (0.500 in.) diameter round specimen; a 200 mm (8 in.) length is used for many rectangular cross section specimens prepared from plate. ASTM Standard E 8 explains proper specimen preparation and testing procedures to ensure obtaining representative elongation values.

Much research has been done to relate a metal's elongation to the degree of ductility necessary for satisfactorily cold forming, or to ensure adequate plastic behavior in a structure that is loaded far beyond the elastic limit. Despite efforts in this area, relationships observed between elongation in tension testing and performance in an operation involving plastic deformation are mainly empirical. Nonetheless, most material standards include minimum elongation among their tension test requirements. Depending upon the metal and product form, minimum elongation values can range from five percent (for hard drawn steel wire) to 40 percent (for soft annealed low-carbon steel). The minimum elongations set for various metals are arbitrarily selected from accumulated tension test data.

In considering elongation as a measure of ductility, perhaps an upper limit should not be established. Metal components are frequently overstressed in localized areas, and the ability of a metal to deform plastically without fracture is the mechanism by which these areas limit the overstressing. The ductility must be accompanied by strength during deformation. Later, this dual capability will be more clearly defined as *toughness*, and mechanical tests for measuring toughness will be described.

Reduction of Area

A second indicator of ductility obtainable from broken tension test specimens is the *reduction of area* at the point of fracture in the necked portion. Percent reduction of area is most accurately determined from round tension test specimens. With square or rectangular specimens, considerable restraint to plastic deformation occurs at the corners, and the fracture location usually necks down nonuniformly. Therefore, the final area is more difficult to measure. For this reason, material standards do not ordinarily include reduction-of-area minima where nonround tension specimens are used. Reduction of area, like elongation, cannot be taken as a true or absolute property of metal. During necking, the stress state that develops in this region is highly dependent upon specimen geometry, and this can significantly affect deformation behavior. Nevertheless, reduction of area at the break is highly structure sensitive, and values obtained can be very useful in assessing a metal's capacity to deform prior to fracture. In addition to having minimum reduction of area in the tension test requirements of many material standards, this test value can be used in evaluating anisotropy in tensile properties and in determining susceptibility to lamellar tearing.

Ductility Indications from Special Tests

Certain fabricating operations, particularly the deep drawing of sheet metal, rely so heavily upon ductility that a number of specialized mechanical tests have been developed to measure this capability. Detailed descriptions of simulative tests that involve bending, stretching, punching, drawing, and cupping appear in the literature. As an example, the *Olsen ductility test* uses a round specimen of sheet metal that is restrained around the edge by a hold down ring while a cup is deep drawn in the center by a flat-bottomed cylindrical punch. The depth of the Olsen cup at specimen rupture is one measure of the drawability. Other methods of testing entail determination of the maximum ratio of specimen diameter to punch diameter (limiting draw ratio), limiting specimen diameter, and critical wrinkling strain. These specialized ductility tests for sheet metal can be found in the literature through the key words Erichsen, Olsen, Swift, and Yoshida. ASTM E 643, *Conducting a Ball Punch Deformation Test for Metallic Sheet Material*, provides a standard ductility testing method.

A novel method has been developed to determine the largest amount of localized plastic strain that a metal experiences during the drawing or stretching in these tests. A grid of small circles is imprinted on the surface of the specimen, and after drawing or punching, the elliptical distortion of the circles on the outer surface of the cup or dome is analyzed for localized plastic strain, particularly the extent of major strain.

Hardness

The hardness of metal is commonly regarded as resistance to deformation, usually by indentation. Several general types of mechanical testing can be used to measure hardness. Metallurgists usually rely on *static indentation testing*, because they find resistance to deformation in tests of this kind (i.e., hardness) to be closely related to strength. Indeed, tables giving approximate correlations between hardness and ultimate tensile strength are widely published. A materials engineer, on the other hand, may be interested in resistance to wear, and he might prefer a *dynamic hardness test* in which a standard tup or ball is dropped on the surface of the material; the height of rebound is a measure of hardness. Finally, the mineralogist may rely on a scratch test for measuring hardness.

Static Indentation Hardness Testing

A number of different instruments are widely used for hardness testing; they slowly indent the metal surface with a ball, cone, or pyramid. The area or depth of the indentation resulting from a given load is converted to a hardness number on a scale developed for each type of test. Indentation hardness testing is a complex measurement because of the different degrees of work hardening that occur in metals and the influence of the indenter used. Although indentation testing is the most common hardness indicator for metals, the choice of indentation equipment requires consideration of specimen size, form, and ultimate purpose of the hardness measurement. Sometimes the type of specimen or the equipment available dictates the kind of test. Circumstances may demand a portable tester. This equipment requires additional precautions, because it usually does not use dead weights for applying force to the indenter. For more details see ASTM E 110, *Indentation Hardness of Metallic Materials by Portable Hardness Testers*.

Brinell Hardness (HB)

Brinell hardness is a test with a long history of application. See ASTM E 10, *Brinell Hardness of Metallic Materials*, and ASTM E 370, *Mechanical Testing of Steel Products*. The Brinell test is widely used to monitor mechanical properties in metal articles of substantial size (bars, beams, plates, etc.). The test usually involves the application of a 3000 kg load for 30 seconds with a very hard ball, 10 mm in diameter, resting on the smooth surface of the metal. However, the load can be varied to accommodate a particular specimen or location on the metal surface. The 30-sec test time ensures that plastic flow of the metal surrounding the indentation has ceased. A standard procedure is used to measure the diameter of the indentation and to compute the *Brinell hardness number* (HB), using an equation that relates load applied, ball diameter, and indentation diameter to the hardness number. Computation is seldom needed, since most test results are available in tabular form.

Brinell hardness numbers can range from 15, for very soft metals such as lead-tin solder, to 940, for very hard steel. Much of the steel fabricated by welding ranges from about 150 HB for low-carbon steel to 325 HB for high-strength steel.

The Brinell test is not widely used on welded products, simply because of specimen size and shape requirements and the large size of the test indentation. Nevertheless, familiarity with the Brinell test is useful because of its wide usage and because of the close relationship between Brinell hardness number and ultimate tensile strength. The Brinell number, when multiplied by 3.5, indicates the approximate tensile strength in MPa; when multiplied by 500, the strength is indicated in psi.

Vickers Hardness (HV)

Vickers hardness uses a diamond pyramid indenter instead of a ball. A standard method is available in ASTM E 92, *Vickers Hardness of Metallic Materials*. Further discussion of Vickers hardness testing may be found in this text under "Microhardness."

Rockwell Hardness (HR)

Rockwell hardness has become the most widely used method for determining hardness, because the test can accommodate a wide variety of specimen shapes and sizes and kinds of metal. ASTM E 18, *Rockwell Hardness and Rockwell Superficial Hardness of Metallic Materials* provides standard test methods. The Rockwell hardness test is simple to perform, the hardness number is easily read directly on the testing machine, and the testing can be automated if desired. The procedure involves initial application of a minor seating load to the indenter to establish a zero datum position. A diamond-tipped indenter with a sphero-conical shape is used for hard metals, and a small hardened steel ball of prescribed size is used for softer metals. Both the minor load and the major load can be selected to best suit the specimen, and more than a dozen scales of hardness numbers have been tabulated; each is designated by a letter of the alphabet. These basic scales are supplemented by additional scales that provide modified conditions to compensate for specimen form (e.g., curvature) and approximate level of hardness.

Three Rockwell scales are most popular for measuring the hardness of steels: (1) the *C Scale*, which makes use of a sphero-conical indenter and applies a 150 kg major load, (2) the *B Scale*, which uses a ball indenter [usually 1.588 mm (¹⁄₁₆ in.) diameter] and a major load of 100 kg (these conditions can be changed and compensated for by an established correction factor), and (3) the *N Scale*, which encompasses many established conditions for superficial hardness testing. Rockwell hardness numbers should always be quoted with a scale symbol, because this clearly indicates the kind of indenter, major load, and other testing conditions. For example, 62 HRC represents a hardness of 62 on the C Scale, whereas 100 HRB represents a relatively low hardness of 100 on the B Scale. Sometimes the scale symbol includes digits to indicate the scale and test conditions used. For example, 81 HR30N represents Rockwell superficial hardness number 81, determined by conditions outlined in the 30N Scale. The ranges of Rockwell hardness scales overlap to some extent, and comparative tables are available.

Microhardness Testing

Microhardness tests can be performed with a number of instruments that use a very small indenter and a very light, precise load to make an indentation in a polished surface. The resulting indentation is measured with a microscope. A polished and etched metallographic specimen is frequently used to allow hardness determinations on individual phases or constituents in the microstructure. By using an indenter with a test load in the range of one to 1000 gf, the indentation can be confined to a single grain in the microstructure. ASTM E 384, *Microhardness of Materials*, covers microhardness testing using the Knoop and the Vickers instruments, which are the most widely used.

The Knoop indenter is a highly polished, pointed, rhombic-based pyramidal diamond that produces a diamond-shaped indentation with a large ratio between the long and short diagonals (7:1), and the depth is only about one-thirtieth of its length. The Vickers indenter is a pointed, square-based pyramid diamond with face angles of 136°. The indentation made by the Vickers instrument has a depth about one-seventh the length of its diagonal. Because of the relatively shallow depth of the Knoop impression, this microhardness test is more sensitive to superficial surface conditions than is the Vickers test, but the long diagonal of the Knoop indenter is less convenient for localized measurements. Both the Knoop and the Vickers indenters can be used in any microhardness testing machine capable of holding the indenter and applying the precise required load. A microscope with a filar eyepiece is used to measure the lengths of the diagonals, and their average dimension is converted to a hardness number. Although depth is not measured with either Knoop or Vickers tests, the indentation in each case does not ordinarily exceed about 20 microns (about 0.0 001 in.).

The *ultrasonic microhardness tester* is a recently developed instrument that features a lightly loaded indenter affixed to the end of a magnetostrictive metal rod. The diamond-tipped rod is excited to its natural frequency and is applied to the surface of the metal with 800 gf. The resonant frequency of the rod changes as contact is made between indenter tip and metal surface. By calibrating the instrument for the modulus of elasticity of the metal, the area of contact between the indenter tip and metal surface can be calculated from the measured resonant frequency. The contact area is inversely proportional to the hardness of the metal, and the resonant frequency also correlates with metal hardness. The test is rapid (usually taking less than 15 seconds), and the indentation is so superficial that the test is virtually nondestructive. This method of hardness testing can also be conducted with a heavier test load. A hand-held unit employs a Vickers diamond tip on the end of the oscillating rod. This spring-loaded indenter is pressed against the metal surface to be tested until a load of precisely five kgf is reached. At this instant, and for the next six milliseconds, the resonant frequency of the oscillating rod is monitored and the frequency shift (as affected by the depth of indenter penetration) is evaluated in terms of hardness. By operator choice, hardness can be displayed as a Rockwell, Vickers, or a Brinell digital value.

Dynamic Hardness Testing

The first device for determining hardness by measuring the rebound of a small hardened ball, tip, or hammer dropped from a fixed height onto the surface of a metal was invented by Shore in 1907, primarily to test hardened steels. This tester was named the *scleroscope*, and it consists of a precision-bore glass tube that has a graduated scale engraved along its length. This tube is supported in a vertical position against the surface of the metal to be tested, and a diamond-tipped hammer is dropped from a precisely set height through the length of the tube. The rebound height of the hammer is dependent on the surface hardness of the metal.

The scale of hardness tested with the scleroscope (HS) was established by determining the rebound from fully-hardened, untempered tool steel (68 HRC) and dividing this height into 100 units. The scale is continued upward to 140 units to accommodate harder materials. Relatively soft steel (e.g., 80 HRB) will exhibit approximately 23 HS. The scleroscope is available in two models. Model C requires visual reading, whereas a Model D offers analog or digital displays.

Proper testing with the scleroscope requires great care in specimen selection, surface finish, location,and alignment. Recommended practice is contained in ASTM E 448, *Scleroscope Hardness Testing of Metallic Materials*. This method of hardness testing is restricted by specimen requirements, but if offers advantages in speed and virtual absence of surface damage. The latter advantage explains its popularity for testing hardened rolls on their finished surface.

Scratch Hardness Testing

Scratch hardness testing is the oldest kind of hardness evaluation. The *Mohs scale* (circa 1822) was developed by mineralogists, and it is based upon the capability of a harder material to scratch the surface of a softer material. The ten minerals listed below were selected to form the Mohs scale, and each makes a permanent scratch (one that cannot be rubbed off) on all minerals listed beneath it.

MOHS SCALE OF HARDNESS

Material	Position	Material	Position
Diamond	10	Apatite	5
Corundum	9	Fluorite	4
Topaz	8	Calcite	3
Quartz	7	Gypsum	2
Feldspar (orthoclase)	6	Talc	1

There is a very large hardness difference between diamond and corundum compared to the relatively small differences between the softer minerals. A human fingernail can scratch gypsum. A rough hardness approximation of a polished metal surface can be obtained by determining which pair of adjacent minerals scratch and do not scratch the sample. Hardened tool steel is between 8 and 9 on the Mohs scale. Low-carbon steel is usually between 3 and 4, while annealed copper is between 2 and 3. Although the Mohs scratch test is not ordinarily applied to metals because of its semiquantitative nature, a uniformly applied scratch with the tip of a sharp needle across the face of a polished and etched metallographic specimen can give useful information on the metal's microstructural constituents. Careful observation, and perhaps measurement of the scratch width along its length, can provide an indication of the relative hardness of the microstructural constituents.

The *file hardness test* is another early form of scratch testing that can be useful. A machinist's steel file, hardened and sharp, can determine if a metal is "file hard" (about 60 HRC). If the material is softer the file will cut the metal surface, inflicting pronounced scratches and possibly removing metal. This simple test also can determine if soft spots exist on the surface of a metal that has been heat treated to the file-hard condition. Also, by using test blocks of known hardness and special files treated to various hardnesses, one can approximate hardness in a metal part of almost any size and shape.

Conversion of Hardness Numbers

Relationships between scales for the various hardness tests are given in standard conversion tables in ASTM E 140, *Standard Hardness Conversion Tables for Metals*. The tables provide data for Brinell, Knoop, Rockwell, Scleroscope, and Vickers testing. The ASTM document emphasizes that each type of hardness test is subject to errors, and differences in accuracy exist among the testing instruments. The relationships between the scales are mainly empirical. Because indentation hardness determinations are influenced to some degree by the modulus of elasticity of the metal being tested, separate tables are provided in ASTM E 140 for a number of commonly used ferrous and nonferrous alloys. When a reported hardness number has been secured by conversion from a different type of test, the original determination should be indicated in parentheses, e.g., 400 HRB (43 HRC).

In concluding this discussion of hardness, strength, and ductility of metals, one must recognize the overly-simplified, wide usage of *uniaxial tension test values*, despite the fact that these values are influenced by conditions of testing and by structure. Theoretically, the strength of a metal should be directly proportional to the square root of the modulus of elasticity and the surface energy, and inversely proportional to the square root of the lattice parameter. A rough estimate on this basis would predict a strength of approximately one-tenth the modulus of elasticity. Strength of this order has never been found in real metals. Actual strength, as mentioned in Chapter 2, is generally 10 to 1000 times less than theoretical strength, because dislocations in the metal's crystalline structure are able to move in the lattice and trigger the deformation and fracture mechanisms that control actual strength.

Tension test procedures are being standardized to minimize the influence of specimen size, shape, orientation, rate of loading, and many other aspects. For example, see ASTM E 8, *Tension Testing of Metallic Materials*. The chemical and metallurgical conditions in metals that make their mechanical properties structure sensitive include grain size, microstructure and its degree of directionality, nonmetallic inclusions, and residual elements. Although strength and ductility are reported for a given metal, these limited data from uniaxial tension tests cannot be expected to predict the load-bearing capability of a structure.

Engineering evaluation of a structure requires analysis of full-scale members exposed to the service stresses. The stresses from loading are almost always biaxial and are frequently triaxial. The complex stresses that may arise allow the metal to carry somewhat higher loads than those measured under uniaxial testing, but these stresses reduce the capacity for localized plastic deformation in the structure, because fracture terminates the deformation process.

Toughness

The ability of a metal to deform plastically and to absorb energy in the process before fracturing is termed *toughness*; this is an important property that must be understood and carefully considered when planning the fabrication of structures, especially those joined by welding.

Tension and bend tests for metals were adopted shortly after 1900, but as experience was gained with these tests, it became clear that the tests could not fully predict the mechanical behavior of metals, particularly under the conditions encountered in service. Even the notched tension test specimens did not consistently select metals that might behave in a brittle manner; these specimens poorly defined the conditions that accentuated brittleness. The frequency of brittle failure indicated a serious problem in predicting a metal's ability to perform in a tough, dependable manner under varying conditions.

Despite a tremendous amount of research, complete understanding of ductile/brittle behavior has yet to be achieved. The features of metals that affect mechanical behavior have been isolated and special tests to evaluate toughness have been developed, but application of this knowledge to building structures is only beginning.

Only one assessment of toughness can be made with reasonable certainty from ordinary tension and bend test results; a metal that displays very little ductility is not likely to behave in a ductile manner in any other test carried to fracture. However, a metal that displays good ductility in the tension or bend test will not necessarily behave in a ductile manner in other kinds of mechanical tests. Investigations of brittle fractures that occurred in service have supported these conclusions. These early cases of unexpected non-ductile behavior made clear that failure might occur at an abnormally low stress and with less than expected energy absorption.

Studies of brittle fracture in testing and brittle failures in structures revealed several factors that affected toughness. The *temperature* of the metal was found to have a profound influence. As temperature was lowered, the propensity for brittle fracture increased. Incidents of catastrophic brittle failure in iron and steel bridges, storage tanks, etc., showed a marked increase when ambient temperature was low, particularly when it was below freezing. A second significant factor was the rate of loading, or the *strain rate*. Although failure could occur in structures carrying static loads, impact loading noticeably increased the occurrence of brittle fracture. A third factor, termed *notch effect,* was also suspected of contributing to brittle failure, because in many cases the fracture appeared to have initiated at a notch-like location.

Stress distribution, or *multiaxiality of stresses*, surrounding a notch proved to be a contributing factor. A metal might display good toughness when the applied stress was uniaxial, but under multiaxial stresses (as would develop in the vicinity of a notch), it might not withstand the simultaneous elastic and plastic deformations in the various directions, as demanded by the stresses. Another factor in brittle fracture was the steel itself. Steels produced during this early period were without the refinements now deemed vital to ensure good toughness. As a result, there was a strong incentive to develop reliable methods of evaluating toughness in metals and to determine required toughness ranges for various applications.

Introduction of Impact Testing

Recognition of the influence of temperature, strain rate, and distribution of stresses on the toughness of metals encouraged investigators to make use of drop-weight and pendulum machines for obtaining higher strain rate. Much early impact testing was performed at room temperature, but compelling reasons arose for conducting tests at closely controlled temperature levels over a considerable range of temperature. To further increase the sensitivity of impact tests, notches of various shapes were cut into the specimens to simulate notch effects sometimes found in structures. The shape of the specimen notches had to be readily reproducible. For reasons of cost, convenience, and ease of material comparison, early tests used small specimens cut from the metal used in a structure. However, this course of action produced results that failed to satisfy the ultimate aim of toughness testing.

From 1900 to 1905, a number of notched-bar impact tests were developed: some were named after their originators, for example, Charpy, Izod, and Mesnager. Two types of notched-bar impact specimens emerged from early work to become part of standard methods used in the United States. The Izod design is a small bar of round or square cross section that is held as a cantilevered beam in the gripping anvil of a pendulum machine. The specimen is broken by a single overload of the swinging pendulum, and the energy absorbed in breaking the specimen is recorded by a stop pointer moved by the pendulum. The Izod specimen can be tested as an unnotched bar, or it can be prepared with a 45° V-notch in the face struck by the pendulum.

The Charpy design is the second standard specimen type, and the majority of work on impact testing of steel uses the Charpy design. Although the Charpy specimen can have one of three different shape notches, labeled keyhole, U-notch, and 45° V-notch, the Charpy V-notch design has been used almost exclusively in research and quality control testing. The Charpy specimen is placed in a pendulum machine as a simple beam in the horizontal position and is supported at each end by anvils. The specimen is struck in the middle of its length by an edge affixed to the swinging pendulum which contacts the specimen side opposite the notch. A single overload breaks the specimen, and the energy absorbed is indicated by the stop pointer. The Charpy specimen is favored over the Izod specimen because it can be chilled or heated to a prescribed temperature; and in less than five seconds the specimen can be centered on the anvils and broken by the pendulum. The very short setup time minimizes temperature change in the specimen after removal from a temperature-controlled bath or oven.

The energy absorbed in breaking Charpy or Izod specimens is now reported in joules, but for many years, impact energy was reported in foot-pounds (one joule equals 0.737 ft-lb). While some handbooks list the results of Charpy and Izod testing under the heading "impact strength," this is a misnomer. Impact energy determinations are quantitative values from a particular test conducted under specific conditions, and these values cannot be used in engineering design calculations to establish the strength or stress-resisting capacity of structural members. Impact energy values should only be used as an index to compare the toughness of different metals and to monitor toughness in quality control testing.

Standard methods for Charpy and Izod impact testing can be found in ASTM E 23, *Notched Bar Impact Testing of Metallic Materials.* In addition to providing pertinent details on specimen design, sizes, testing apparatus, and method of energy measurement, this document outlines additional determinations that can be made on the tested specimens, i.e., fracture appearance and lateral expansion. To understand why these additional determinations have been suggested, and to appreciate the significance of the quantitative values that can be obtained from an impact specimen, research findings regarding toughness and fracturing in metals are reviewed here. Also, recent work has revealed certain shortcomings of notched-bar impact tests related to specimen size, rate of load application, and distribution of strain. These aspects of toughness testing will be discussed in detail shortly.

FRACTURE IN METALS

The term *fracture* is used in a number of ways when discussing breakage in metals. Strictly defined; a fracture is the irregular surface that forms when a metal is broken into separate parts. If the fracture has progressed only part way through, and the metal remains in one piece, somewhere there is a crack with at least one end buried within the metal. A crack can be defined as two coincident free surfaces in a metal that join along a common front called the crack tip, which is usually very sharp. As knowledge of cracking and fracturing has broadened, many new terms have been proposed. ASTM has adopted as standard the terminology contained in ASTM E616, *Standard Terminology Relating to Fracture Testing*. This text will make use of these terms and definitions as much as possible. Generally speaking, fracture and rupture have the same meaning: the separation of metal under stress into two or more parts. However, the term fracture may be used when metal separation occurs at relatively low temperature and toughness performance is the chief topic; the term rupture is more often found in discussing metal separation at elevated temperatures (e.g., see rupture strength).

Basically, there are two kinds of fracture that occur in metals: *ductile fracture* and *brittle fracture*. These two modes are easily recognized when they occur in exclusive form, but fractures in metal often have a mixed morphology, called a *mixed mode*. Several distinct mechanisms that result in fracture have been identified in metals, and frequently more than one of these mechanisms can be initiated. The mechanisms are termed (1) *shear* fracture, (2) *cleavage* fracture, and (3) *intergranular* fracture. Only the shear mechanism produces a ductile fracture. These fracture mechanisms can operate in competition with each other to produce mixed-mode fracture, either on a macro-scale or on a micro-scale. For example, the ductile shear mechanism may be initiated at the start of fracture, but the fracture may change to the brittle cleavage mode as it progresses through the metal. The area in the cross section where fracture mode changes from ductile to brittle can usually be identified, and this can be useful information.

On a micro-scale, the overall morphology of a fracture may appear to be that from the ductile shear mechanism, but closer examination may find tiny intermixed areas where some brittle cleavage was instigated by a micro-feature of the metal. Avoidance of brittle fracture is a prime objective, but this goal should realistically include minimizing the amount of brittle fracture during development of a mixed-mode fracture. Before attempting to analyze mixed-mode fractures, simple ductile and brittle fractures will be examined to gain an understanding of their fundamental features.

Whether the fracture is ductile or brittle, the fracture process is viewed as having two principal steps: the first is crack *initiation* and the second is crack *propagation*. This view of the process is helpful, because there is often a noticeable difference in the amount of energy required to execute the two steps. The relative levels of energy required for initiation and for propagation determine the course of events which will occur when the metal is subjected to stress. For example, if the stress required to initiate a fracture crack is greater than the stress required to propagate it, the metal will probably behave in a brittle manner, because a crack, once initiated, will find ample strain energy in the system for continued propagation. Conversely, a crack can be initiated at a specific energy, but the metal may have considerable capacity for plastic flow, and thus a large amount of energy is needed to deform the metal ahead of the crack tip and to create the two new fracture surfaces. In this case, crack propagation is restricted, and the metal will display some degree of toughness.

The plastic properties of metal are important because of their influence on crack propagation. Small cracks or sharp discontinuities are almost always present, and a primary concern is the possibility of brittle fracture propagation from a discontinuity that is already sited in the metal.

Studies of the fracturing mechanism have revealed that crack initiation is often a repetitive step, and that propagation can take place by progressive cracking. Sometimes, fracturing results from the impermanent unloading of a metal structure. Further propagation may be halted because the crack front has entered a stress field wherein too little energy is available to continue propagation in the particular metal under the prevailing circumstances. Nevertheless, these stationary cracks continue to be a threat, because they act as stress raisers. Any change in conditions may cause one of these cracks to serve as an initiator, whereby fracture propagation can resume.

The mechanism of fracture was more clearly understood after examination of the frontal edge of an initial crack and the localized stress/strain field that surrounded the crack tip. ASTM has accepted the term *ideal-crack-tip stress field*, but for brevity in this text, this important area of stress/strain is referred to as the *fracture process zone*. The stress that arises at the tip of the crack is concentrated in a localized zone and produces a commensurate amount of strain. Initial strain is elastic, but if the stress is sufficiently high, then plastic strain will result. When stress at the crack tip demands plastic deformation, this is the instant that the properties of the metal and certain prevailing conditions determine whether the metal will experience *stable crack growth*, which is a limited increment of ductile fracture, or *running cracks*, which are often damaging brittle fractures.

A number of revealing signs can often be observed at a fracture which are helpful in analyzing the nature and cause of the fracture, namely:

(1) profile of the overall sections on each side of the fracture,
(2) amount of plastic deformation intimately associated with the fractured ends of members,
(3) general topography of the fracture surface,
(4) growth markings in the texture of the fracture surface, and
(5) micro-details in the morphology of the fracture surface.

Details concerning the development of these telltale signs during the fracturing process and their interpretation are included in the following sections, which describe the two basic fracture mechanisms.

Ductile Fracture

A ductile fracture is characterized by substantial plastic deformation in and near the fracture, and by considerable absorption of energy in accomplishing the deformation. In general, a ductile fracture is usually not sudden or catastrophic. The gradually developing deformation often signals impending failure, and the energy absorbed usually lessens its consequences. Also, there is a greater likelihood of crack arrest if prevailing conditions favor a ductile fracture.

A fully ductile fracture is often a *shear fracture*, because much of its surface is at an angle of approximately 45° in relation to the major stress axis. In simplest form, the entire broken end of a specimen with a fully ductile shear fracture can exhibit a 45° angle, as illustrated in Figure 3.5 (A), and this form is sometimes referred to as "slant shear." More often, only portions of the fracture are near the 45° angle, and these portions appear as slanting, fibrous protuberances and cavities on the fracture surfaces. The slanting regions are more apt to develop at the edges of the broken specimen, where the stresses tend to be biaxial (rather than triaxial), and the final portion of the fracture occurs under shear strain as seen in Figure 3.5 (B). However, note that the central portion of the fracture is flat and perpendicular to the axis of the major tensile stress. This "flat face" form of shear fracture occurs because of the triaxiality of stresses localized in the central area of the specimen.

A ductile fracture in a round tension specimen often occurs as a "cup-and-cone" type, as illustrated in Figure 3.6, where a slanting fibrous portion encircles the outer edge of the break, and the central portion is flat-face shear fracture. The profiles of fractures through plate frequently have slanting fibrous regions localized at the surfaces of the plate, and these regions are called *shear lips*. Even relatively brittle fractures in plate or bar may have some shear lips along their surfaces, where reduced multiaxiality of stresses allows localized ductile shear fracture to occur.

In addition to substantial plastic deformation at the breaking point and possibly a profile that includes slanting fibrous regions, a ductile fracture in steel usually has a gray, rough-looking morphology. This topography is due to the ductile fracture mechanism. Much of the incremental, stable growth cracking initiates at nonmetallic inclusions, second-phase particles in the microstructure, grain boundaries, and any other microstructural feature that permits dislocations to pile up (and plastic deformation to concentrate). This initiates microvoids in the metal, and each of these immediately produces a stress-strain field. Plastic flow and formation of the microvoid momentarily relieve stress; however, if stress application continues, the microvoids enlarge, and some of them connect (usually referred to as *microvoid coalescence*) to form larger voids in the metal.

When capacity to undergo plastic flow is virtually exhausted, ductile fracturing continues by *tearing* the solid membranes of metal that lie between the microvoids. This causes the relatively rough morphology of the ductile fracture. At high magnification, the appearance of the final fracture can be like that shown through SEM fractography (Figure 2.32). The enlarged microvoids appear in the fracture surface as cupules, or *dimples*, with surrounding tear ridges. The ductile fracture propagates at a relatively slow rate that is commensurate with loading, and it proceeds only as long as energy is supplied by external loading to exceed the plastic flow capacity. When deformation is stopped by limiting the imposed energy (stress), propagation of the ductile fracture stops.

The dimpled surfaces of ductile fractures have been studied extensively using TEM and SEM fractography, and much has been learned from the dimple size, shape, orientation, population, etc. These studies have called attention to additional details in the ductile fracture morphology, for example, ripples, serpentine glide planes, and stretched zones. These details are helpful in determining exactly how the ductile fracture progressed through the metal.

Brittle Fracture

A completely brittle fracture has little or no plastic deformation, as shown in Figure 3.5 (C) and the upper specimen in Figure 3.6. When the fractured surface exhibits many

(A) DUCTILE SHEAR FRACTURE IN ALUMINUM **(B) DUCTILE FRACTURE IN STEEL WITH SHEAR PORTIONS IN FRACTURE AT EDGES** **(C) BRITTLE FRACTURE IN STEEL**

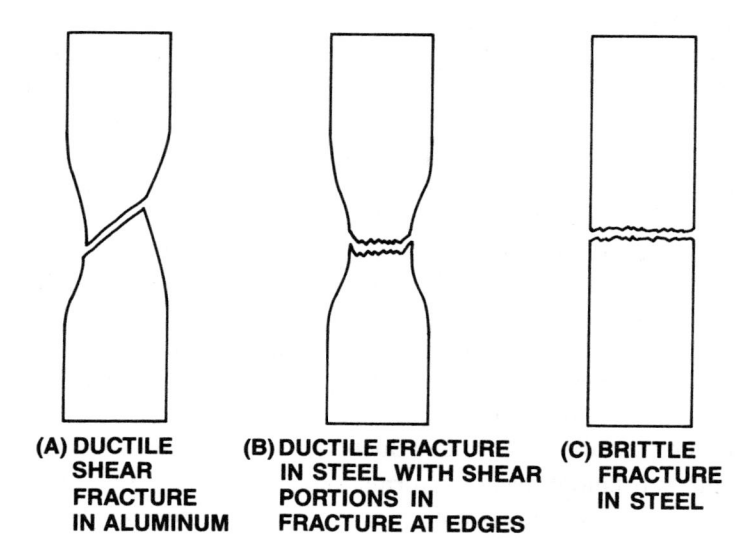

Figure 3.5 — Simple schematic of types of fractures observed in metals
Profiles represented flat strips of metal broken in uniaxial tensile loading. Despite specimen simplicity a number of telltale features are embodied in the illustrations that can assist in the interpretation of real fractures.

bright reflecting facets in its texture, the term *cleavage fracture* is applied. These facets are the remnants of the grains that experienced *crystallographic cleavage*, i.e., separation of planes that have the largest interplanar spacing and that exhibit little or no ability to undergo plastic flow by cross slip. In polycrystalline metals, the cleavage facet size is often controlled by the metal grain size. In iron and low-carbon steel, the ferrite grain size is usually the controlling feature, but when the microstructure is multiphase, the prior austenite grain size may exert a stronger influence.

When grain size is very small, a somewhat larger amount of energy is required to propagate a cleavage fracture because of the many-faceted crack path. Nevertheless, this fracture mechanism enables a crack to initiate more readily and to propagate more rapidly than in the ductile type of fracture. The incremental voids and tears that develop during stable ductile fracture grow slowly, perhaps only at about six meters per second (20 ft/s). Unstable brittle fracture propagates very rapidly, and crack propagation rates of 1000 m/s (about 3000 ft/s) have been measured. Furthermore, once initiated, the fast brittle fracture can progress at this high speed completely through a member without the need for additional energy from externally applied stress. The internal elastic stress present in the metal can be an adequate driving force for the brittle fracture.

Metallography and fractography have revealed details in brittle fracture morphology that not only identify occurrence of this kind of fracture but also allow mapping of the fracture's origin and its propagation paths. Crystalline structure and microstructure play important roles in brittle fracture, and the carbon and low-alloy steels and other body-centered cubic (bcc) metals have a proclivity for brittle fracture. This susceptibility is due to the manner of dislocation movement in the bcc crystalline structure when stress is applied. It is explained in Technical Brief 9.

The face-centered cubic (fcc) crystalline structure is not susceptible to this form of brittle fracture, and therefore metals like aluminum, copper, nickel, and the austenitic high-alloy steels characteristically display good toughness. However, if their structure contains a significant amount of second phase or if the matrix structure changes from fcc to bcc (as can occur in the austenitic high-alloy steels under certain circumstances), then the susceptibility to brittle fracture increases.

Fractography has shown that the faceted cleavage fracture has additional characteristics. The facets usually display a *river pattern*, which is surface marking arising from the mechanics of microcrack growth in each grain. As the cracks initiate on parallel planes and grow wider, thin ligaments of solid metal are found between them. These ligaments are eventually fractured, and the cleavage-crack segments join, but this leaves step-like markings (a river pattern) on the facets, as shown in Figure 2.31. River patterns are informative because, as segments of the pattern run across the grain, they broaden and approach each other. Consequently, the river-like step lines on the cleavage facets converge in the direction of crack propagation in a particular grain (or in a cluster of grains having closely aligned orientation). This is not necessarily the direction of crack propagation for the main brittle fracture. However, the propagation direction of local cleavage can be helpful in locating the fracture origin.

Another feature of brittle fracture, *feather marks*, can appear on the cleavage facets between the step lines of a river pattern. These are faint chevron-shaped decorations, and their apex usually points back to the initiation point of local cleavage. Other features sometimes observed in brittle fracture texture include *cleavage tongues*, which are deviations in the cleavage plane caused by traversing a deformation twin, and a *herringbone pattern*, which arises when the {110} and {112} planes are involved in microcracking, in addition to the {100} planes, because of interaction with deformation twins.

Figure 3.6 — Brittle fracture (upper specimen) and ductile fracture (lower specimen) in standard ASTM round tension test specimens

Lower (ductile) specimen displays "cup-and-cone" break and has undergone appreciable elongation and reduction of area during tension testing as seen in side view and fractured end. Upper (brittle) specimen has cleavage fracture and has developed very little plastic deformation during testing.

> **TECHNICAL BRIEF 9:**
> **Dislocation Movement in Body-Centered-Cubic Metals**
>
> Where two planes of the {110} type intersect in the crystalline structure, migrating dislocations occurring as pairs on these planes will combine to form a new larger dislocation, which in turn may move onto adjoining {100}-type planes. The {100}-type planes are not slip planes in the bcc crystallographic system, and the mobility of dislocations along these planes is impeded. This favors dislocation coalescence on the {100} planes and early initiation of microcracks. These microcracks cleave the grains along their {100} planes, thus creating the grain-like flat facets shown in Figure 2.31. The microcracks actually initiate in the microstructure a short distance in advance of a larger fracture, and if the metal has little capability to undergo slip in the areas surrounding the microcracks, these defects quickly develop into a running crack.

Fractures that are deemed brittle can occur because of early breakage with little plastic deformation, and the fracture texture may display only a small number of cleavage facets. Furthermore, a certain amount of dimpling or microvoid coalescence may be found in the texture. If there are no discernible boundaries between the cleavage facets and the dimpled rupture areas, the term *quasicleavage* is applied. The term *mixed mode*, mentioned previously, is used when distinct areas that fractured by different mechanisms can be seen. For example, if a brittle second phase succumbed to cleavage separately from a matrix that fractured by microvoid coalescence, then the term mixed mode would apply.

When an overload initiates an unstable fast fracture and the crack must travel through a portion that has an appreciable width-to-thickness ratio (e.g., plate), there is an inherent tendency for the rapidly propagating crack to take the shortest path to free surfaces. This inclination produces another characteristic macro-pattern in the fracture texture, which is called a *chevron pattern* (sometimes incorrectly referred to as herringbone pattern). Figure 3.7 illustrates a chevron pattern in the texture of a fast fracture in thick steel plate. Chevron patterns can often be seen with the unaided eye. A chevron pattern present in the fracture texture can be very helpful in determining fast fracture propagation direction, because the apex of each chevron points back toward the point of crack origin.

The subject of *crack arrest* has received much attention because this aspect of the fracture process has potential for forestalling complete failure of a metal structure. Crack arrest is summarized in Technical Brief 10, and it will be discussed further at appropriate points — both from a fundamental standpoint and in the context of it practical role in safeguarding the integrity of weldments.

Intergranular Fracture

Intergranular fractures have an appearance that makes them easy to distinguish from fractures developed by other mechanisms. However, intergranular fractures that appear similar may have resulted from different causes, and establishing the nature of the cause can be a challenge. Among the causes of intergranular fracture are:

(1) the presence of grain-boundary phases that are weak or brittle;

(2) environmental factors such as stress corrosion, hydrogen damage, and heat damage; and

(3) mechanical conditions such as extreme triaxiality of stresses.

Many intergranular fractures — such as those caused by hydrogen embrittlement, stress corrosion, brittle grain boundary phases, and overheating — often have a "rock candy" appearance. This is because the grain boundaries separate almost catastrophically; and plastic flow, which would normally deface the crystalline structure as observed in the texture of the fracture surfaces, does not occur. The texture of intergranular fractures is not always the rock candy morphology. This crystalline appearance develops when the extent of grain boundary damage, or degree of boundary weakness, is sufficiently advanced that catastrophic intergranular separation occurs completely upon the initial application of stress.

Many circumstances arise, however, where mixed mechanisms of fracture operate and the resulting fracture texture is mixed mode. For example, the segregation of a brittle constituent in the grain boundaries may vary, and if the matrix of grains has some toughness, the intergranular fracture will probably display some dimpled areas of ductile fracture distributed on a micro-scale among the smooth, separated grain facets of the intergranular fracture. These facets do not display a river pattern, which helps to differentiate them from the cleavage facets of brittle fracture.

The toughness of a metal that is susceptible to intergranular fracture cannot be estimated unless all information regarding the nature and extent of conditions affecting the grain boundaries is available. Some conditions can be so deleterious (e.g., extensive

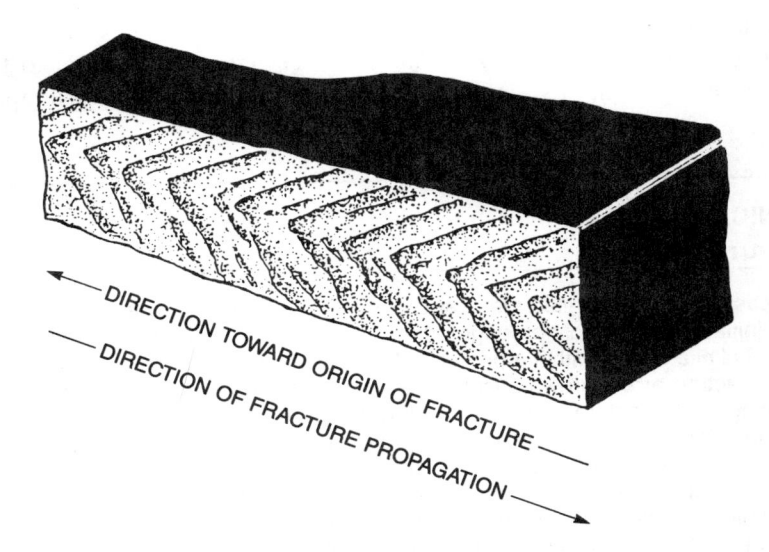

Figure 3.7 — Sketch of chevron pattern sometimes observed on fracture surfaces

Chevron pattern is prone to appear on the surface of fast fractures that have occurred in heavy steel sections, such as thick plate. The apex of each chevron points back toward the point of fracture origin, which is helpful information when searching for the initiator of cracking.

intergranular corrosion) that the metal may lose its customary "metallic ring," and it will have virtually no strength or ductility. More often, a mixed-mode fracture is found, even though an intergranular condition was the cause. For example, hydrogen embrittlement may have occurred to a lesser extent than that which would cause complete intergranular fracture. Consequently, the metal initially fractures intergranularly; but at a later stage, some tearing develops and produces indications of ductile fracture. Fractography can detect these details and assist in a full diagnosis of the fracture.

Conditions Affecting Fracture Toughness

Two mechanisms have been described as the bases for ductile fracture and brittle fracture in metals, but the interplay between these mechanisms in the course of crack development is highly sensitive to conditions outlined in the following sections. Much research has been directed to ways of quantifying the influence of these conditions, and this information is very useful when appraising a weldment for its fitness-for-service.

Effect of Temperature

Temperature is highly influential among the factors that govern the mechanical and physical properties of metals. Iron and steel, because of their low cost and wide availability, are used over a very wide range of temperature. In general, inadequate toughness usually limits low-temperature usage, whereas loss of strength limits high-temperature service. There is a need to determine the service temperature range for each application and to note the temperatures at which specific properties of a given material deserve consideration.

At least a half-dozen arbitrary temperature ranges in which iron and steel see service have acquired commonly used designations. However, no specific ranges have been widely adopted or recognized. *Sub-zero service* usually refers to temperature below the freezing point of water, i.e., 0°C (32°F). *Cryogenic service* temperatures are

---

TECHNICAL BRIEF 10:
Crack Arrest

Occasionally, when failed metal sections were examined, brittle fractures were found to have initiated and rapidly propagated to a damaging extent, but the fractures then terminated as finite cracks within a portion of the section of metal. These infrequent incidents of brittle fracture arrest drew considerable interest because of a possible safeguard that would bring brittle fracture to an early stop and thus keep it from running rampant through a structure. For example, brittle fracture in steel pipelines sometimes ran for more than a kilometer (about 0.6 mile) with costly consequences. Naturally, a pipeline failure of this magnitude spurred thinking about the installation of some kind of intermittent "crack stopper" to limit the extent of a brittle fracture, such as the inclusion of a short length of very tough metal, albeit expensive, at intervals in the pipeline.

A number of crack arrest test methods have been developed and used in studies on steels (Robertson crack arrest test, Esso test, Japanese double tension test, drop weight test, wide plate test, etc.). Yet, the evaluation of crack arrest capability continues to be a challenging problem because of the difficulty of reproducing the stiffness and other fracture-promoting conditions that exist in a large structure in a laboratory test specimen.

below about -100°C (-150°F). Although metals are used to contain liquified gases down to almost absolute zero [-273°C (-459°F)], the carbon and low-alloy steels are not ordinarily used in cryogenic service because of inadequate toughness. Atmospheric service implies temperatures in the range of about 0 to 25°C (32 to 75°F). Room temperature is commonly understood as approximately 25°C (75°F).

Elevated temperature service for iron and steel has no established limits, because the minimum and maximum temperatures depend on the properties being considered. From the standpoint of strength in iron and steel, temperature must be above about 200°C (400°F) before a significant effect is seen; however, dramatic improvement in toughness is frequently seen in service above room temperature.

High-temperature service also has no specific limits, because it depends on the metal being considered. Iron and carbon steels suffer loss of strength rather rapidly as temperature exceeds about 400°C (750°F), but small additions of certain alloying elements when making low-alloy steels can extend the strength to about 600°C (1100°F). Generally, high-temperature service for the carbon and low-alloy steels implies temperatures above about 260°C (500°F), and marked changes in toughness can occur above this temperature. Although these designations for service (temperature) regimes are useful for reference, the toughness of iron and steel is so temperature sensitive that the actual service temperature must be stipulated whenever toughness is considered.

In evaluating the influence of temperature on the fracture, behavior at elevated temperatures should be considered first. Fracture that occurs in a metal at a high temperature is often intergranular, and little or no deformation may have occurred. A transgranular fracture, on the other hand, is nearly always accompanied by distorted grains, and the fracture propagates along slip planes rather than grain boundaries. This difference in microstructural path usually permits determining whether a high temperature fracture has occurred, and this can be important when conducting a failure analysis. The surface of a high temperature fracture in iron or steel that extends to the component surface and allows air to enter the crack is usually brown or blue because of the formation of temper colors.

Recrystallization Temperature

As their temperature is increased above room temperature, most metals become more ductile and toughness increases; this is also generally true of steel. While elevated temperatures permit the crystal planes to slip on each other, there is usually a change in the type of fracture at the *recrystallization temperature*. Below this temperature, the path of fracture is through the grains (transgranular). Above this temperature, the fracture path is along the grain boundaries. Sometimes the recrystallization temperature is called the *equicohesive temperature*, because it is the temperature at which the grain boundaries appear to have the same strength as the grains.

For a given metal, the recrystallization temperature is not constant in the sense that melting point is a constant. As explained in Chapter 2, increases in cold work and duration of heating result in lower recrystallization temperatures. The composition of the metal or alloy also has a profound influence on recrystallization temperature. Figure 3.8 shows the role of recrystallization temperature in controlling fracture in a single-phase microstructure, as would be found in iron. Figure 3.8 (A) illustrates the intergranular fracture that develops when the iron is fractured above the recrystallization temperature. Under these circumstances, any slip that occurs causes immediate recrystallization, and the fracture crack proceeds along the relatively weak grain boundaries. Below the recrystallization temperature, the planes have slipped, and the grains have become deformed (elongated) by the applied stress [see Figure 3.8 (B)]. The fracture crack has proceeded transgranularly by following the weakest of the slipped planes, and this gives the fracture path an irregular, saw-tooth appearance.

Brittle fracture below the recrystallization temperature and in the low-toughness regime is illustrated in Figure 3.8 (C). A running transgranular fracture has separated fracture-susceptible planes by the cleavage mechanism, and the grains display little or no deformation. It is important to remember that the grain boundaries are stronger than the grains themselves below the recrystallization temperature, but the boundaries are weaker above this temperature. As discussed later, this relationship is one of the bases for making steels intended for high-temperature service.

Near and below room temperature, the fracture toughness of some metals can seriously degrade. As previously mentioned when describing the features of cleavage fracture, metals with a bcc crystalline structure have a tendency for brittle fracture. This is especially true of iron, carbon steel, and the low-alloy steels. These steels are used extensively in a wide range of temperatures, and their susceptibility to brittle fracture is a continuing problem. The bcc metals invariably display a transition from ductile to brittle behavior over a narrow range as temperature is decreased. The approximate temperature at which this transition occurs depends on the particular metal, its condition, and the nature of imposed stress. The transition may occur at as high as 200°C (400°F) or as low as almost absolute zero, (-273°C or -459°F), but many bcc metals undergo this transition in the range of 200°C to 5°C (400°F to 40°F).

Metals that have an fcc crystalline structure, and those that are hexagonal close-packed (hcp), do not show a marked transition over a narrow range of temperature, but their fracture toughness also decreases gradually with temperature. Figure 3.9 illustrates how the three common crystalline metal structures compare with respect to fracture behavior over a wide range of temperature. This marked transition in the behavior of bcc metals is examined more closely, and typical quantitative data are provided, when notched-bar impact testing is discussed again.

Effect of Stress Axiality

The stress imposed on a metal is a source of energy in the fracturing process. Applied stress may provide energy that can be stored in the metal as elastic strain. Beyond the elastic limit, excess energy should be dissipated if it induces plastic deformation through slip or flow. The greater the fraction of energy that can be dissipated through plastic flow, the smaller is the possibility of reaching the critical energy level needed to start separation (running crack formation). Since the creation of a crack also consumes energy, movement must take place in metal at any point where the atoms cannot maintain a state of energy equilibrium under the imposed load. The the metal must undergo plastic flow to dissipate energy or new surfaces must be created by fracturing.

The brittle form of fracture has been estimated to require only about 0.001 joule per square millimeter (one foot-pound per square inch) of energy to sustain the growth of a fractured surface in metal. This amount of energy can easily be found in the metal as elastic strain. Therefore, a brittle fracture has little difficulty propagating through any thickness of metal, as long as the imposed stress is above the cohesive strength of the metal and conditions do not permit plastic flow or slip. However, tension testing indicates that the force required to supply the separation energy depends on whether the applied stress is uniaxial or multiaxial.

According to Hooke's law, if a tensile stress below the elastic limit is applied to a metal, it should elongate a certain amount, and the cross section should diminish in

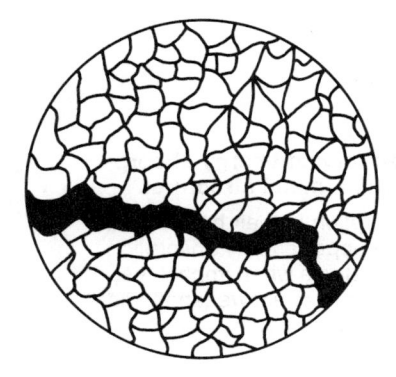

(A) INTERGRANULAR FRACTURE
THERE IS NO DEFORMATION OF GRAINS BECAUSE SEPARATION HAS OCCURRED ALONG THE WEAK GRAIN BOUNDARIES ABOVE THE RECRYSTALLIZATION TEMPERATURE.

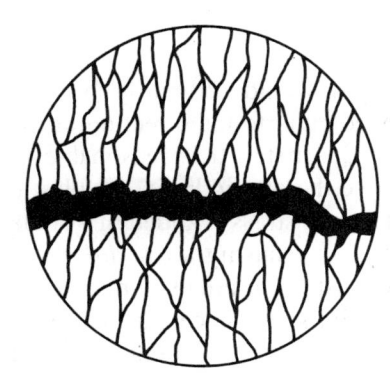

(B) DUCTILE TRANSGRANULAR FRACTURE
THE GRAINS HAVE UNDERGONE DEFORMATION BY SLIP AND THE FRACTURE HAS PROCEEDED THROUGH THE WEAKEST OF THE SLIP PLANES AT A TEMPERATURE BELOW THE RECRYSTALLIZATION LEVEL.

(C) BRITTLE CLEAVAGE FRACTURE
TRANSGRANULAR SEPARATION OF GRAINS ALONG PARTICULARLY FRACTURE-SUSCEPTIBLE PLANES WITH LITTLE OR NO DEFORMATION OF THE GRAINS.

Figure 3.8 — Schematic metallographic view of three types of fracture in metal controlled by temperature at the time of propagation

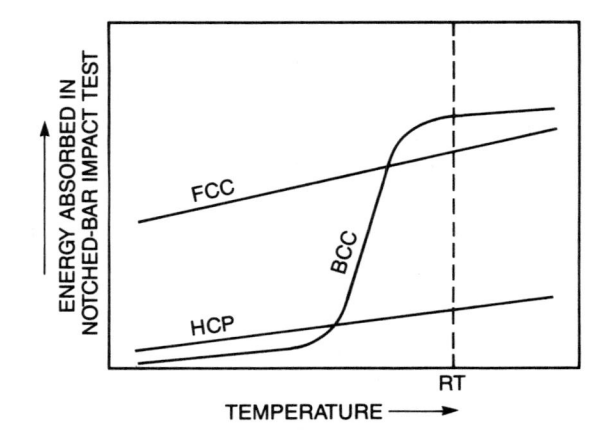

Figure 3.9 — Approximate relationship between impact energy absorption and temperature for three important crystalline metal structures

The bcc metals have a lower "shelf" of energy absorption in their behavior at low temperatures, and they exhibit a transition to an upper energy absorption "shelf" at higher temperatures. Carbon and low-alloy steels when impact tested over a range of temperature often produce this pattern of a lower shelf and an upper shelf when their energy absorption values are plotted.

accord with Poisson's ratio. Both dimensional changes would ordinarily be predictable. If a second tensile stress is also applied perpendicular to the direction of the initial stress, *biaxial stress* conditions are imposed on the metal. It is reasonable to expect that if the biaxial tensile stresses are raised above the elastic limit, the plastic deformation would be significantly impeded. If a third tensile stress is added in a plane perpendicular to the plane formed by the first two stresses, causing a *triaxial stress state*, then the circumstances under which plastic flow must occur are even more constrictive. In fact, this *plastic constraint* increases the axial stress necessary to cause yielding. Although the triaxial stress state can arise by deliberately loading the metal as described, the most prevalent producers of triaxial stresses are either notches in the stressed specimen or the presence of cracks.

Notches, cracks and other forms of flaws deserve study to understand how they create multiaxial stresses when a uniaxial load is applied. A notch may be a feature of the specimen, such as a hole or an inside corner; or it may be an unintentional flaw, such as a nonmetallic inclusion with sharp extremities, or a crack. Any notch affects stress distribution in several ways:

(1) Stress in a highly localized area near the root of the notch is concentrated and increased above the nominal or average level of stress present in the gross cross section.

(2) A stress gradient exists from the point of greatest concentration to the nominal level in the specimen.

(3) A state of multiaxial stresses exists in the metal at the root of the notch or at the tip of a crack.

Stress-Intensity Representation

Increased stress in the root area of the notch is often referred to as *stress concentration*, and the notch is referred to as a *stress raiser*. The stress in the root area can easily be several times higher than the nominal stress. The actual stress depends on the shape of the notch; sharper notches promote higher stresses. Methods have been developed for calculating and measuring the concentrated stress, and the symbol K_t has been widely adopted to represent a *theoretical stress-intensity factor*. This is the ratio of greatest stress in the root region of the notch to the nominal stress in the gross cross section. K_t is calculated according to various theories, and it may make use of elastic constants, shear lag equations, finite element analyses, etc. Sometimes a qualitative estimate of stress distribution and intensity is made with a transparent plastic model and a photoelastic technique, or an indication of surface stress distribution can be obtained with brittle lacquer applied to the metal surface prior to loading. These efforts to evaluate the distribution and amount of stress associated with notches have shown that blunt notches have a moderate effect; they elevate the nominal stress by a factor of two or three.

To point out a familiar notch as reference, consider the V-notch used in the standard 10 mm (0.394 in.) square Charpy impact test specimen. It has a 45° angle, a radius of 0.25 mm (0.010 in.) at the root, and a depth of two mm (0.079 in.) in the face of the specimen opposite to that being struck. This Charpy V-notch is also considered to be relatively blunt and shallow. Most test specimens that have machined or ground notches usually have a K_t value of about six. A fresh crack in a brittle material presents the sharpest kind of notch at its tip, where it is atomically sharp. Values of K_t as high as 18 have been calculated for cracks. Welded joints are recognized to have stress concentration associated with their shape, weld toe angle, degree of root penetration, and other features. Stress-intensity factors for fillet welds increase as the toe angle increases, and K_t values may reach five when the weld toe angle approaches about 60°.

The ratio K_t is often a calculated or a theoretical value. The actual stress-intensity factor is usually somewhat different. Nevertheless, K_t provides an approximation of the higher level of stress imposed on the metal in a localized area. Although the stress factor provided by K_t may indicate localized stress above the yield strength, and perhaps even above the ultimate tensile strength, it is important to know whether plastic flow has occurred. Plastic constraint created by stress multiaxiality can raise the yield strength in the localized area by as much as a factor of three. If the concentrated stress produces only *elastic* strain, the stress concentration factor can be quite high. However, if the artificially high yield strength is exceeded and plastic flow occurs, stress concentration at the root of the notch will be reduced. In a general way, plastic flow tends to blunt the notch.

Stress Gradient

Stress gradient must be considered when studying notch effect because of the increased strain rate over this localized area as the metal is loaded. Even when a metal is loaded elastically, stress and strain increase more rapidly over the stress gradient area. Consequently, plastic flow is required earlier and to a greater extent than that occurring elsewhere in the nominally stressed metal. The metal should be capable of plastic flow when stress exceeds the yield strength in the gradient area.

Stress Multiaxiality

Stress multiaxiality is very important in analyzing notch effect, and the mechanics by which a notch creates a multiaxial stress state can be quickly outlined. Assume a substantial length of metal, as shown in Figure 3.10. This length, for example, could be within a long tensile tie-bar in a highway bridge. Next, examine the notch in the surface marked "S." This notch has a V-shape, and it penetrates to a relatively shallow depth. When the length is loaded in tension, stress cannot be transmitted across the notch gap, and therefore the longitudinal axial stress near the notch must follow the deviated path indicated by the stress line marked "L." The flow of stress around the notch tends to be concentrated in the root area, as shown in the enlarged drawing on the right. This pattern of stress distribution can be verified with strain gauges or with photoelastic modeling.

As should be expected, the longitudinal stress (L) and the strain in the member of Figure 3.10 require constriction in the direction normal to the principal stress axis, per Poisson's ratio. Therefore, this would be a constriction across the width of the member. However, the walls of the notch are unstressed and therefore have no cause for constriction, consequently, they act as a restraint against the lateral contraction that should be occurring in the metal surrounding the notch. This restraint causes a lateral tension stress (W), which is especially important since it arises in the root area and produces a biaxial stress state. Naturally, the existence of longitudinal stress (L) and lateral stress (W), and their accompanying strain, requires a constriction in the third direction, that is, through-the-thickness of the section. With the appreciable thickness of the member in Figure 3.10, the mass of metal resists constriction, and a third stress (T) develops in the thickness direction (sometimes called the *z direction*). As a final consequence, a localized area of metal under the apex of the notch is now in a state of triaxial stress. If the member were not so thick, perhaps only of sheet thickness, there would be no significant restraint in the z direction, and therefore little or no stress would be developed in this direction. The final state of stress would then be biaxial.

Design engineers have long dealt with metal structures that are loaded in more than one direction simultaneously (multiaxially). One or more of the five basic types of stress (tensile, compressive, bending, torsion, and shear) may be present and must be accommodated. Design formulas are ordinarily available (with terms for calculating stress, member size and kind, and strain or deformation) to cope with combined primary stresses from anticipated loading. Yet, it must be appreciated that secondary stresses can also arise from self-constraint against strain, and these stresses may be in different planes than those of the primary stresses that produced them. Additional stresses can come from other extrinsic causes, such as dynamic stress from rapid changes in service loads, thermal stresses from rapid heating or from fast cool-down, wind, earthquakes, etc.

The effects of multiaxial combined stresses are not new, but the detrimental nature of notches and the demands that their highly localized multiaxial stresses make on the

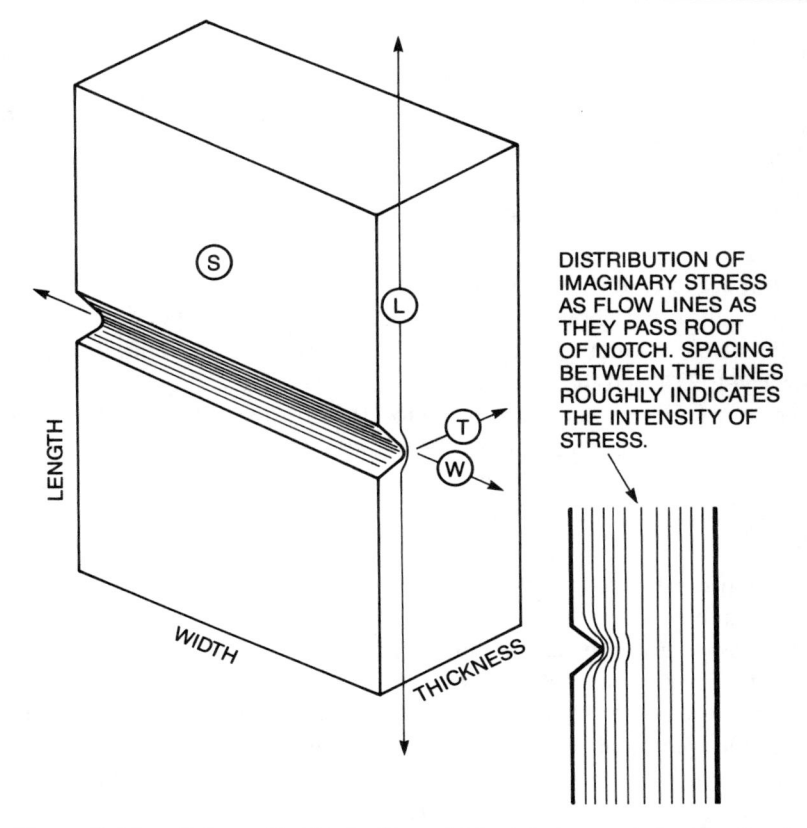

Figure 3.10 — Schematic illustration of triaxial stress state development in thick metal member with a notched surface when subjected to longitudinal tension

Line (L) in drawing of metal member at left-hand illustrates general direction and path of longitudinal stress from applied tensile load. Line (W) indicates lateral stress in biaxial direction that develops from constriction across the width of the member due to restraint of unstressed walls of notch. Line (T) indicates through-the-thickness stress (triaxial, or z direction) that arises from the restraint of the thick member to the constriction demands of longitudinal and lateral stresses.

toughness of metal too often are overlooked. The threat of notch effect receives considerable attention throughout this text.

Effect of Rate of Strain

Although the rate of strain has long been recognized as a principal factor influencing the toughness of metals, its effect is less clear than the effect of temperature or multiaxial stresses. The load on a metal component can be an unchanging dead weight, i.e., a static load, or it can increase rapidly. Slowly applied loads often are termed *quasi-static*, which signifies that their effects are essentially the same as those produced by static loads. The highest loading rates are those produced by explosions. Between the two extremes of static load and loading by explosive action, operations are encountered that involve rapid loading by mechanical, hydraulic, electromagnetic, ballistic, and other systems. In early experience with these actions, the general term *impact* was applied to indicate that loading was rapid and that increase of stress and strain was expected to be equally rapid. *Dynamic fracture toughness* was another general term applied to impact testing to signify that toughness evaluation involved very rapid loading as opposed to quasi-static loading.

At first, little attention was given to the actual rate of loading. Even the notched-bar impact test, during its early development, included no prescribed loading rate, except that the load should be delivered by a swinging pendulum or a falling hammer with sufficient kinetic energy to break the specimen. Eventually, the test procedure in ASTM E 23 fixed the velocity of the hammer delivering the breaking impact blow at three to six meters per second (10 to 20 ft/s) to standardize the loading rate. This velocity range was selected mainly because it could be achieved by existing impact test machines. Yet, when compared with the loading rates imposed on many metal structures and components in service, the velocity of the standard impact test is relatively low.

Table 3.4 outlines loading rates in four general regimes of service or exposure. As indicated in this table, many structures and components are subjected to rapid straining because of fast loading in service; this is particularly true of earthmoving machinery and heavy military armament. Some large structures, such as steel building frames and bridges, are expected to survive occasional very high loading rates resulting from accidents or other unavoidable circumstances, e.g., earthquake shock. Safety codes in some areas require that construction materials and design anticipate these unusual demands on toughness and structural integrity.

Metals loaded very rapidly usually respond with somewhat higher strength than if loaded slowly, but they can exhibit a marked decrease in toughness. The reduction in toughness is likely to occur regardless of the nature of the applied stress — that is, whether it is tensile, compressive, bending, shear, or torsion. Rapidly applied shear stress seems most significant in reducing toughness. In fact, a change in the shear failure mechanism appears as the strain rate becomes very high. At medium strain rates, toughness may increase, because flow stress rises faster than ductility decreases. Shear failure at the lower strain rates involves the familiar uniform deformation and void coalescence; whereas at very high strain rates, plastic instability can occur, and localized flow can terminate in failure along adiabatic shear bands. In this manner, shear instability prevents the normal shear mechanism from contributing toughness.

Special testing equipment and techniques have been developed for *high strain-rate testing*, and information on the effects of very rapid loading can be found in the literature. As loading velocity increases, the transmission and distribution of stress and strain through a metal member are more complex than with static loading. As indicated in Table 3.4, when the strain rate exceeds approximately 200 s^{-1}, stress is transmitted in waves that propagate at the speed of sound in varying form and direction through the metal, sometimes reverberating until equilibration occurs. With extremely rapid short-pulse loading, perhaps greater than 1×10^4 s^{-1}, as generated by detonation of an explosive charge, a shock front is transmitted. This is followed by wave propagation, which must overcome inertial constraints.

Under these stress and strain conditions, it appears that fracture can occur by the near-simultaneous nucleation of microvoids and the subsequent development of interconnecting cracks to produce a continuous fracture. This mechanism of metal failure slows the rate of toughness decrease associated with increasing loading rate; however this slowing occurs at a much lower general level of toughness than if the normal microvoid initiation-and-coalescence process were operating. Despite the complexity of conditions that must be fulfilled within a few microseconds, investigators have devised ways of measuring the intensity of stress and strain in test specimens, and a better understanding is being gained about the behavior of metal under these unusual dynamic conditions.

The type of metal plays a major role in determining the amount of toughness reduction caused by rapid loading. The bcc metals are more adversely affected, and the carbon and low-alloy steels usually start to display significant effects as the rate of loading exceeds approximately 10 m/s (33 ft/s). When the loading rate exceeds 100 m/s (325 ft/s), many steels with a bcc structure exhibit a marked drop in toughness. Of course, the onset of toughness reduction also depends on metal temperature and the axiality of stress. If these factors are unfavorable, low toughness can be experienced at a lower loading rate, such as that of the Charpy impact test. This change in toughness associated with high strain rate has prompted intensive research to develop steels

Table 3.4 - Strain Rates Common to Various Exposures, Services, and Mechanical Tests

Approximate Range of Strain Rate ($\dot{\varepsilon}$) *	Characteristic Time	Typical Exposures and Services	Applicable Testing Techniques	Remarks
Low ($<0.1 s^{-1}$)	>1s	Majority of structural applications	Conventional standard mechanical tests in hydraulic or screw-actuated testing machines	Quasi-static conditions
Medium (0.1 to 200 s^{-1})	1s to 1 ms	Metal forming, impacts in operating machinery and structures	Pendulum impact, drop-weight, special servo-hydraulic loading	So-called dynamic or impact conditions
High (200 to 10^5 s^{-1})	1 ms to 1μs	Earthquakes, terminal ballistics	Hopkinson bar, Taylor rod, Kolsky bar, ballistic testing	Elastic-plastic wave transmission. Stress and strain possibly not distributed uniformly
Very High (>10^5 s^{-1})	<1μs	Explosive forming	Flyer plate impact, explosion testing	Inertia effects, shock fronts, and wave propagation occurs

*$\dot{\varepsilon}$ is rate of change of strain (ε) with time (t). Units of strain rate are inverse time, or s^{-1}. As a benchmark, an ordinary tension test usually is performed at a strain rate of about 10^{-3} s^{-1}, which produces a strain of 0.5 in 500 s; but conventional mechanical test equipment easily can achieve strain rates as high as 0.1 s^{-1}.

which will meet the demands of military applications such as gun barrels and armor plate. One such investigation is described in Technical Brief 11. Findings from these efforts have been extended to making tough steels for other applications.

Effect of Cyclic Stress (Fatigue)

Cyclic stress imposed on a metal member can produce progressive cracking, and ultimately will cause the form of failure commonly termed *fatigue*. This is a serious problem, because more service failures are produced by fatigue than any other cause. The behavior of a metal in a structure or a component under cyclic stress is fraught with so many influential factors that fatigue failure requires implementation of technology from many disciplines. Design, mechanical engineering, and metallurgy are a few of the disciplines that contribute to preventing fatigue. Fracture mechanics (to be described next) offers new analytical procedures for dealing with the fatigue problem. Yet, fatigue failures continue to occur with disturbing frequency in many metal structures. Weldments commonly have features that make them especially vulnerable to

---

TECHNICAL BRIEF 11
Flow Strength Versus Cohesive Strength

In seeking a fundamental explanation for the reduced ability of bcc metals to undergo plastic flow, or *slip* (and thus alleviate brittle fracture), under certain conditions of temperature, stress, and strain rate, investigators postulated that fracture behavior may be controlled by *two* strengths: *flow strength*, which is the resistance to plastic flow; and *cohesive strength*, which is the resistance to cleavage fracture.

When these two strengths were plotted against temperature, it was found that as the temperature was lowered the normal yield strength increased more rapidly than the cohesive strength. In fact, the yield strength doubled as the temperature was reduced from room temperature to about -255 °C (-425 °F). Cohesive strength was found to increase at a lesser rate; it rose steadily as a straight-line function of temperature as the temperature was lowered. Researchers theorized that although the yield strength is generally reached first in mechanical tests performed near room temperature, yield strength could conceivably be raised above cohesive strength at a very low temperature.

In pursuing this concept, the stress required for plastic flow in the presence of a notch (triaxial stress state) was also given consideration. Figure 3.11 illustrates the relationships between yield strength, brittle strength, and plastic flow stress under a triaxial stress state, as the temperature ranges from above room temperature to a very low temperature. This figure predicts that metal at a temperature of T_1 and lower will invariably fail in a brittle manner when stressed to fracture, because it cannot support a stress where yielding will occur. Furthermore, if a notch is present to create a triaxial stress state, the stress required for plastic flow increases very quickly with decreasing temperature. Again referring to Figure 3.11, note that a metal containing a notch fails in a brittle manner when stressed to fracture at temperatures as high as room temperature. The brittle temperature (point at which flow strength exceeds cleavage strength) depends on a number of factors; these certainly include notch acuity and rate of strain. The brittle temperature for metals becomes successively lower with (1) a test with a very high strain rate (e.g., explosion bulge test), (2) a high strain rate test (a Kolsky bar), (3) a medium strain rate test (the drop-weight tear test), and (4) a regular (quasi-static) tension test.

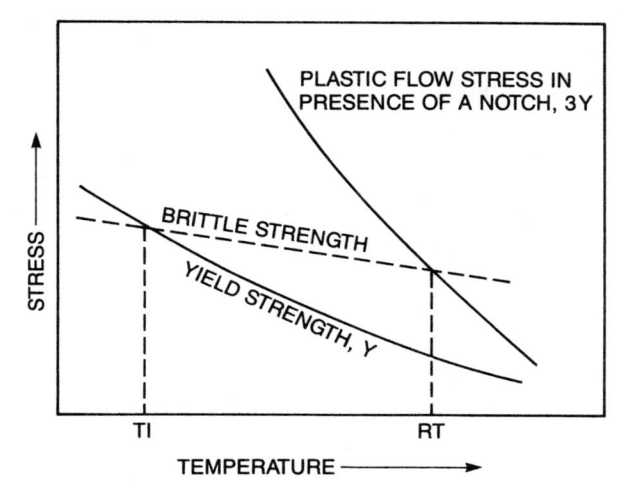

Figure 3.11 — Approximate relationship between yield strength, brittle strength, and plastic flow stress under a triaxial stress state
Below temperature T1 the metal is fully brittle. Above temperature RT the metal is tough under all conditions. Between temperatures T1 and RT a transition range exists wherein uniaxial stressed metal will fracture in a ductile manner, but metal containing a notch is likely to encounter brittle strength level and succumb to cleavage (brittle) fracture (data according to Orowan).

this problem. While available technology can ensure avoidance of fatigue failure, the need to apply this information is often overlooked.

In earlier discussion of strength and fracture characteristics of metals, only *monotonic loading* was considered — that is, whether the load was static or applied rapidly, it was increased continuously from zero to the ultimate level at which the specimen broke. Study of fatigue requires examination of the behavior of metals when they are subjected to variable (repeated or fluctuated) loading. Stresses in metals under these loading conditions are commonly called *cyclic stresses*. The *stress cycle* often involves complete reversal of stress, and therefore tension and compression are equally applied by alternating loads. Damaging cyclic stress can have a wide range of patterns and frequencies, and neither tensile nor compressive stresses are necessarily involved.

Under cyclical loading, fracture can occur at a stress substantially below a metal's yield strength. Since more metal members fail by fatigue than any other cause, the effort directed to the study of fatigue and the number of tests for fatigue strength data are greater than those for any other mechanical property. Because of the complexity of the subject and the many variables that affect the performance of metal under cyclical loading, the literature is voluminous. A comprehensive presentation on the subject here is not possible because of space limitations. Instead, the fatigue mechanism by which metals can fail is described, and the fundamental aspects of the behavior of metals as they are subjected to cyclic (fatigue) stresses are reviewed. In Volume II, Chapter 13, the special attention required to avoid fatigue failures in weldments will be discussed.

Fatigue Crack Initiation

Fatigue in metal initiates under the action of cyclic stress by formation of microcracks. The microcracking starts in areas where dislocations in the crystalline structure are inclined to migrate as stress is applied. These areas include twin boundaries, slip bands, and grain boundaries. As dislocations pile up in these micro-locations and coalesce under the driving force of repeated stress, a critical condition develops in the sound but dislocation-riddled area; a microcrack forms as a result. Usually, more than one microcrack develops in a particular area, because the metal is not isotropic, and dislocations are moved by relatively low stress to the crystallographic sites that are the most receptive. As the microcracks grow during further application of cyclic stress, they ordinarily follow a transgranular path. This feature can be helpful in identifying fatigue fractures during metallographic examination. As multiple cracks grow, they often link by the shearing of intervening membrane-like walls. This produces a revealing feature, called a *ratchet mark* that can easily be seen on the face of the fatigue fracture.

Logically, fatigue cracks should initiate in a member where the applied load produces the highest stress or where the metal is the weakest, but fatigue resistance exhibited by a metal member is much more dependent upon the *localized state* of stress than upon the static strength of the metal. For this reason, fatigue cracking often initiates at the surface of metal members, where the stresses are usually higher. Furthermore, when a *stress raiser* (notch, crack, or sharp change in thickness) is present, it may accelerate the start of fatigue failure by serving as a crack or by raising the localized stress. Thus the defect or design error may act as a fatigue crack initiator.

The stress concentration produced by notch effect was discussed earlier in terms of a simple V-notch cut into one surface of a member (Figure 3.10), and the significance of K_t was explained. To show how easily a notch effect can arise in a welded article, Figure 3.12 provides a sketch of a typical welded assembly of bracket and plate. In addition to the sharp corners formed by the two components, the start of the vertical weld has melted irregularly at the top, and it thus forms a small notch at the upper corner that is subjected to loading in tension. As indicated by the stress distribution diagram just above the schematic, stress at this upper corner is certain to rise to a level considerably above that calculated by conventional cantilever beam formulas.

Fatigue Crack Propagation

Fatigue microcracks grow during the tensile portion of cyclic stress, and they maintain their size when compression loading is being applied. Although a cyclic stress created purely by compression loading does not initiate fatigue cracks or aid in their propagation, compressive loading *can* produce and abet the fatigue mechanism through tensile stresses that are inclined to develop in certain areas as a compressive load is released. As multiple microcracks link to form a main fatigue crack, the main crack continues to propagate through the remaining metal that carries the highest stress; thus it becomes an increasing threat to the integrity of the member.

Fatigue Crack Failure

When the uncracked portion of a member is no longer able to carry the applied load, the member fails. If the final fracture takes place with some plastic deformation and absorption of energy, this final ductile fracture may give some warning of the

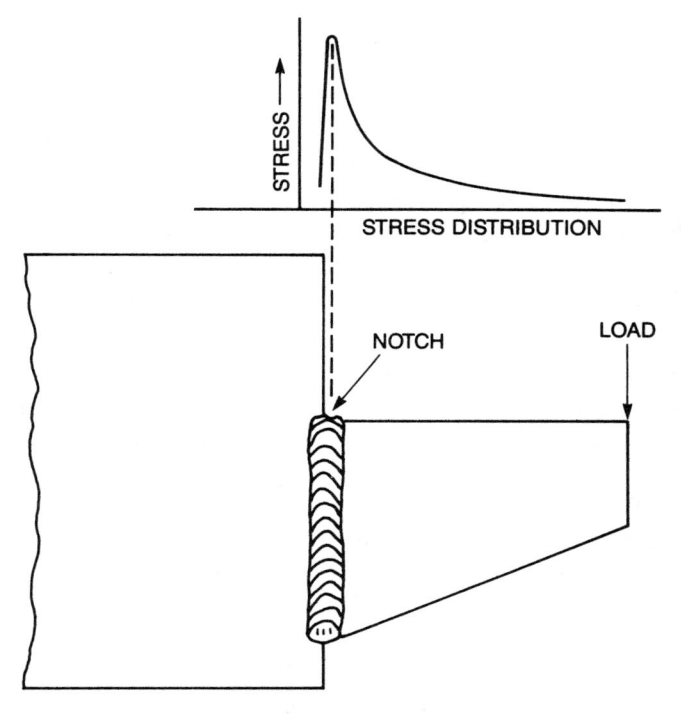

Figure 3.12 — Stress distribution in a weldment under load
Sketch of typical plate-and-bracket component with load applied to outboard end of cantilevered bracket. Notch in corner at starting point of weld causes significant stress concentration.

impending failure. On the other hand, many failures occur suddenly by brittle fracture, without prior significant deflection or distortion of the member to signal that it has undergone serious crack propagation by the fatigue mechanism. A fatigue crack can constitute a very severe notch and initiate a sudden brittle fracture.

To summarize the above, the fatigue process is viewed as having three general stages:
 (1) *initiation*, wherein cyclic stresses lead to the formation of microcracks (although an existing sharp defect also can serve as an initiator);
 (2) *propagation*, wherein progressive growth of the fatigue cracks takes place with each stress cycle, until the remaining uncracked cross-section is inadequate to support the imposed load; and
 (3) *fatigue failure*, during which the remaining uncracked metal succumbs to complete breakage by ductile tearing, or by brittle fracture, or by a succession of different fracture modes.

In sound, uniform metal, the initiation of microcracks by the fatigue mechanism is significant. Indeed, much research has been conducted to identify the start of this stage and to determine how its inception could be delayed. For example, temporarily stopping the application of damaging stress cycles provides a small benefit. Annealing, which decreases the number of accumulated dislocations and eliminates the pile-ups that presage microcracking, is a better treatment. The formation of a microcrack represents permanent damage in the metal that cannot be removed by resting or annealing; however, new techniques are being studied that involve the application of a periodic static overload stress to blunt the tips of the microcracks and thus decrease their propagation rate. Although microcrack initiation deserves continued attention as a fundamental step in fatigue, this initial stage is often supplanted in structures and components by a defect that acts like a fully developed crack. The presence of this existing initiator allows propagation to begin virtually immediately at the start of cyclic loading. This observation is particularly significant in the case of weldments, and the defects in a welded structure that can behave as initiators will be discussed in Volume II, Chapter 13.

Appearance of Fatigue Failures

Fatigue fractures are easily identified when the faces are parted for examination. Although fatigue cracks usually initiate at a surface because of the higher stress ordinarily present at this location, stress raisers within the metal can also cause internal crack initiation. Figure 3.13 portrays a typical example from a welded structure that failed by fatigue. One face of a broken member of round cross section has been photographed, and the schematic indicates where revealing information appears on the fractured face.

The area at the top of the round member was subjected to cyclic loading during service, and stresses were concentrated along the toe of a fillet weld (not seen in the photograph because of its position above the opposing face). Cyclic stresses that concentrated on this area of the surface initiated several fatigue microcracks. The faint *beach marks* (also called *fatigue striations*) radiating downward from the surface at the top of the member show the incremental growth of the cracks.

The vertical rifts that extend a short distance from the surface near the top of the member are ratchet marks, which represent sheared walls where cracks became linked by rupturing thin intervening membranes of metal. Below the termini of the ratchet marks, the beach marks show continued propagation of a main crack across the entire cross section of the round member. When the main fatigue crack advanced about halfway through the member, the remaining (uncracked) lower half was unable to carry the imposed load, and failure occurred.

Closer examination of the fracture face in Figure 3.13 with the aid of a magnifying glass reveals several regions of different fracture morphology. The region of microcrack initiation at the top of the round member has only faint beach marks, and the fracture texture is smooth and silky. The smoothness at this location is caused by a very slight rubbing action between the two opposing fracture faces, which occurs after the crack has propagated some distance into the metal. This rubbing tends to obliterate the beach marks. Deeper in the member, however, the fracture faces become rougher and the beach marks are plainly seen. The increasing prominence of the beach marks results from less rubbing and from increasing stress on the decreasing (uncracked) cross-section, which produces a larger increment of crack propagation during each load cycle.

The rough fracture morphology of the lower half of the round member suggests that the final break was a ductile fracture that entailed some plastic flow and energy absorption during failure. This example of a fatigue failure, illustrated in Figure 3.13,

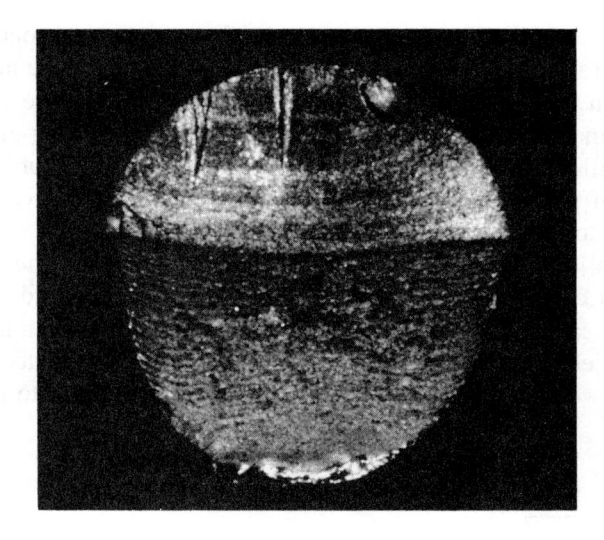

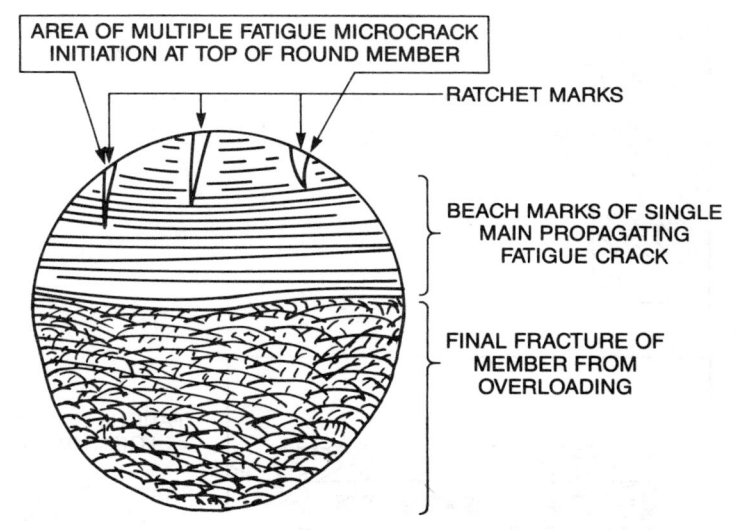

Figure 3.13 — End-face of a round steel member (upper photograph) that fractured in service which displays features characteristic of fatigue failure

Areas of significance in the photograph are explained by labeling on the lower drawing of the fractured end-face.

is clear and easily analyzed, but many other conditions are often involved in metal structures that influence the appearance of fracture. Sometimes an existing crack that serves as a fatigue initiator may be due to an entirely different cause; therefore, the initial portion of the fracture may be devoid of fatigue striae or beach marks.

Other revealing features may indicate the cause of the initiator. If the loading on the member is low-cycle or if the metal has a high work-hardening rate, then the stress required to cause fatigue may be above the static yield strength (but below the ultimate tensile strength). Low-cycle and high-stress loading produces broad beach marks on the fatigue fracture faces. Conversely, if the loading is high-cycle and low-stress, the crack propagation increments may be so small that the fracture faces may appear devoid of beach marks, even when examined under a magnifying glass or low-power microscope. As illustrated in Chapter 2, Figure 2.33, fractographic examination by scanning electron microscopy may be necessary to detect the very closely spaced fatigue striae produced by high-cycle, low-stress loading.

In weldments joined by a process that forms beads of weld metal, typical convex fillet welds often concentrate stresses at their toes. To better understand the fatigue cracking at this location, refer to Figure 3.14. This figure shows an arc-welded T-joint that has been subjected to cyclic stress. Fatigue cracks have initiated at the toes of the upper and lower weld beads on the butting member. In addition to the stress concen-

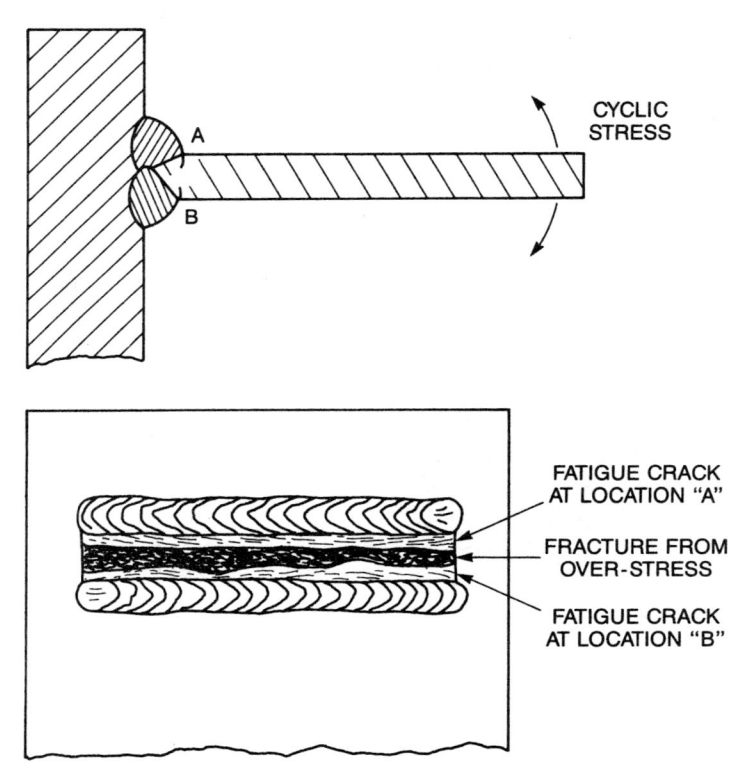

Figure 3.14 — Fatigue fracture in a T-joint where the thinner cantilevered member was subjected to alternating stresses

In the lower sketch of the heavy base plate, the fatigue cracks can be seen to have initiated at the most highly stressed locations A and B which are the toes of the fillet welds deposited against the thinner member. The dashed lines indicate gradual inward extension of the fatigue cracks. The central area of the fracture may exhibit crystalline facets as often seen in the fractures of tension test specimens. This is the last portion of the component to fracture under load, sometimes in a brittle manner by a running crack.

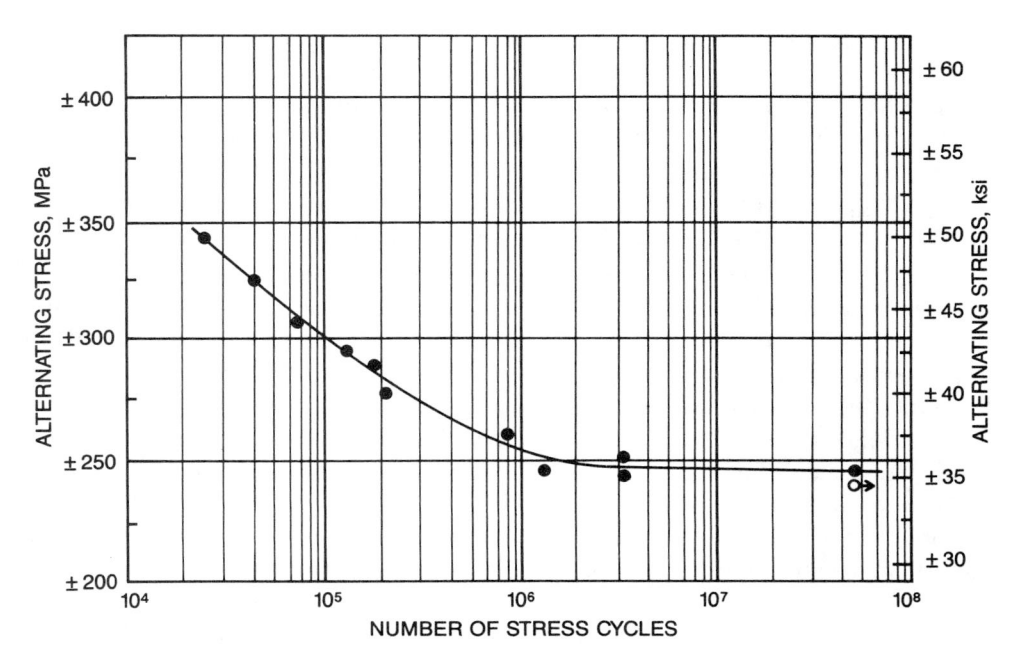

Figure 3.15 — S-N diagram which correlates stress level and number of stress cycles to produce fatigue failure in steel

Open-circle point plotted on diagram in lower-right represents specimen which is unbroken after approximately 45 million cycles of reversing tension-compression loading. This performance suggests an endurance limit of about 240 MPa (35 ksi)

tration along these toes, there can also be a metallurgical condition (such as a slag inclusion) at this location, as will be described in Chapter 13 of Volume II. This condition, if present, assists in the initiation of fatigue cracking.

Cyclic Stress Limits to Avoid Fatigue Failure

In general, the highest stress that can be resisted by a metal without fracture will decrease with the number of times the stress is repeated. The maximum stress so determined for a stated number of cycles is the *fatigue limit*. With lower stress, the number of stress cycles that the metal can withstand without fracture is increased. The tolerable number of cycles increases exponentially with decreasing stress, until a stress is reached in some metals where apparently no amount of repeated loading produces failure fatigue. This limiting stress is considered the inherent "fatigue strength" and is called the *endurance limit*.

To illustrate resistance to fatigue over a wide range of loading, the S-N diagram is often used, and a typical example of this diagram for a low-carbon steel is shown in Figure 3.15. To produce this diagram, a dozen smooth round-bar specimens of the steel were tested as rotating cantilever beams at various cyclic stresses, ranging from a high of 345 MPa (50 ksi) to a low of 240 MPa (35 ksi). The applied stress involved complete reversal from tensile to compressive during each cycle of rotation. Because of the large number of stress cycles needed to produce fatigue failure, especially at low stress, the number of cycles is regularly indicated on the abscissa of the S-N diagram using a logarithmic scale. The stress may also be expressed logarithmically on the ordinate, in which case the curve becomes an almost-straight line.

The S-N diagram in Figure 3.15 shows that three specimens were tested at a stress of approximately 245 MPa (36 ksi), and the number of cycles to failure were approximately 1.5, 3, and 45 million, respectively. Substantially extended fatigue life by some specimens is not unusual as imposed stress is reduced to a level near the endurance limit. This scatter in specimen failure points arises from the wide range of cycles over which microcracking initiates at low stress levels. Once this initial cracking has occurred, the propagation stage usually progresses in a more predictable manner. Note that another specimen tested at a stress slightly below 240 MPa (35 ksi) remained intact after 45 million cycles and perhaps could have withstood this level of cyclic stress indefinitely. This performance suggests that the steel being tested has an endurance limit approaching 240 MPa (35 ksi), but this supposition is based on the performance of only a single specimen. The very long testing time required to execute more than 40 million cycles, as plotted for the two specimens, discourages replicate testing above and below this estimated endurance limit to confirm its validity. Fatigue testing commonly involves multiple specimens, and scatter of data is regularly seen as results are plotted on the S-N diagram. Some diagrams are drawn with a "scatter band" to indicate the range of results secured from tests. A single curve represents the median number of cycles expected at a given level of stresses.

Many metals, particularly nonferrous metals, do not display an endurance limit in fatigue testing. Instead, they simply continue to fail at an extended number of cycles as lower levels of imposed stress are applied for extended tests. Because this behavior is widespread among metals other than steels, there is some skepticism that an endurance limit actually exists in any metal. While this matter is of scientific interest, the endurance limit data plotted in Figure 3.15 — gathered from tests with small, smooth specimens — have little engineering value in designing structures or components that must withstand cyclic loading. Fatigue test performance, as shown in Figure 3.15, serves mainly as a reference in further studies of the steel to determine the influence of changes in composition, processing, heat treatment, etc. To provide information for design purposes, fatigue testing must closely simulate features anticipated in the fabricated structure or component, and it must be conducted under conditions expected to be encountered in service. These degrading features and service conditions are discussed briefly in fundamental aspects of fatigue strength, but a more detailed examination from the standpoint of weldments will be made in Volume II, Chapter 13.

The shape of the curve plotted on the S-N diagram in Figure 3.15 indicates that fatigue cracks initiate quite early, when the stress is high. Perhaps only one to 10 percent of the fatigue life has passed when irrecoverable damage (microcracking) occurs, and complete fatigue failure of the specimen occurs within a million cycles. This set of circumstances is regarded as the *low-cycle* fatigue region. Because the applied stress in this region ordinarily exceeds the yield strength of the metal, the plastic deformation that occurs tends to produce unstable conditions in a specimen. It is not unusual to have considerable scatter in the fatigue test results. For this reason, much fatigue testing in the low-cycle region is conducted with strain rather than stress as the controlled condition.

At lower cyclic stress (that is, below the yield strength of the metal) is the *high-cycle* region, where fatigue life exceeds a million cycles and a valid endurance limit may be indicated by the S-N curve. At these lower levels of cyclic stress, the movement of dislocations in the crystallographic structure of the metal is more orderly, and pile-ups and slip do not develop as quickly. With alternating stress, an unfavorable dislocation grouping may form during tensile loading and then disperse during compressive loading. In this manner, a longer time is required at the low alternating stress to develop a crack. In fact, in the high-cycle region, a greater number of stress cycles is required to initiate a fatigue crack than to propagate the crack to a complete fracture. This observation emphasizes the concern regarding existing defects that can substitute for the microcracking initiation stage.

Fatigue Ratio

Fatigue ratio is another overly-simplified figure calculated with the aid of the S-N diagram, and this value is often listed in handbooks on metal properties. This is the ratio of estimated endurance limit to the ultimate tensile strength. Steels, for example, commonly have a fatigue ratio of about 0.5. Although helpful when making a very coarse comparison of the fatigue resistance of different metals, the fatigue ratio is of no practical use in determining allowable cyclic operating stress for structures or components.

Cyclic Stress Conditions

In the S-N diagram in Figure 3.15, the stress applied to the test specimens was a repeated, uniformly alternating, tensile-compressive stress. This pattern has the greatest potential for fatigue damage, and it is therefore favored for fatigue testing. A simple rotating round-bar specimen facilitates rapid accumulation of applied alternating stress cycles, reducing the time required for a fatigue test. Also, the repeated tensile-compressive cycle is simple to count for a rotating bar, and the effect of a precise sinusoidal-wave form of stress on fatigue life is easily recorded. It must be recognized, however, that a cantilever bar rotating with a dead-weight load on the outboard end will subject only the metal at the surface of the specimen to maximum stress. Only a relatively small volume of metal is actually under the test conditions. Consequently, fatigue test results obtained with this type of specimen are usually higher than those secured from axially-loaded specimens of the same cross section.

As previously stated, cyclic stresses do not always involve complete stress reversal. Indeed, the loading imposed on metals by many kinds of service often develops complex stress patterns. Stresses in a metal member may result from loading in axial tension, axial compression, bending, torsion, or shear. Furthermore, the load and resultant stresses can take one or more of the following forms:

(1) varying from zero to a maximum level,
(2) varying from a nonzero minimum to a maximum,
(3) varying proportions of alternating tension and compression,
(4) varying combinations of the five basic kinds of stress,
(5) varying stress over a number of cycles, and
(6) varying cycle rate.

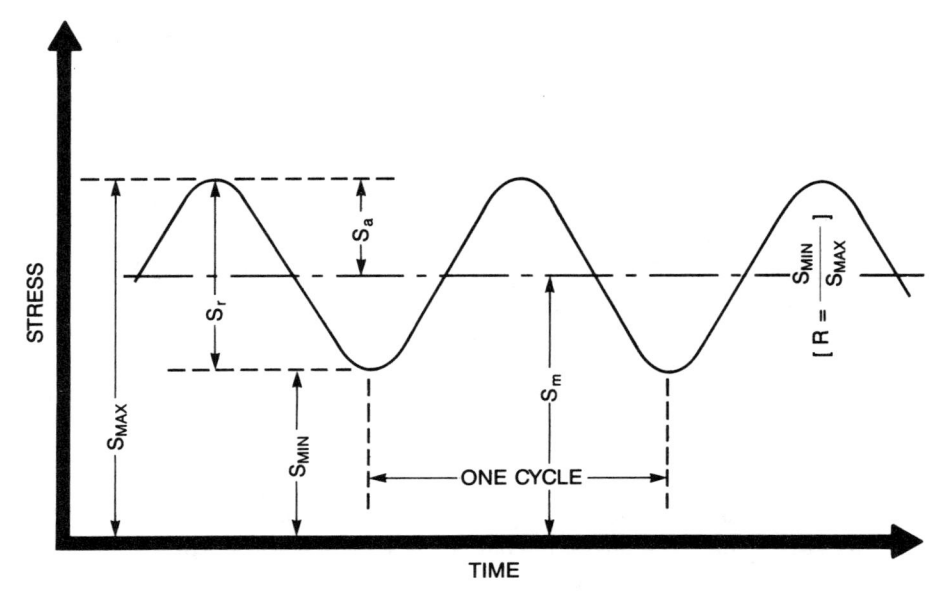

Figure 3.16 — Parameters used in characterizing cyclic stress: as noted on a schematic stress-time trace for pulsating tension-tension stress cycles

Because of the many stress patterns encountered in service and duplicated in fatigue testing, attention was directed to characterizing a given stress spectrum as its fatigue propensity was recorded. Despite the many patterns of stress, a stress cycle can be taken as the smallest section of a stress-time function that is repeated periodically and identically. The following four basic parameters are regularly used in characterizing a stress cycle; it is generally necessary to determine two of them:

(1) minimum stress in the cycle [S_{min}],
(2) maximum stress in the cycle [S_{max}],
(3) mean stress [$S_m = \frac{1}{2}(S_{min} + S_{max})$], and
(4) stress range [$S_r = S_{max} - S_{min}$].

Stress Amplitude

Stress amplitude, S_a, is sometimes mentioned in the literature; for an alternating stress, this is one-half of the stress range. Also, *stress ratio*, R, may be used to indicate the ratio of the minimum stress to the maximum stress in one cycle (R = S_{min}/S_{max}). These conditions are illustrated in Figure 3.16.

The above conditions were selected to define any form of stress cycle, but it was also necessary to plot data to illustrate the effects of different stress cycles or patterns. One early effort to portray the fatigue strengths in this manner was the *Goodman diagram*. Originally, the Goodman diagram (Figure 3.17) depicted endurance limit over an array of cyclic stress patterns — progressing from complete stress reversal (equal tension and compression), to a stress cycle that ranges from zero to maximum tensile stress, to a tension-tension cyclic stress that pulsates between minimum and maximum stress in the cycle. As fatigue testing experience broadened, the Goodman diagram was modified to include fatigue strengths at given numbers of cycles. Presently, the Goodman-type diagram is referred to as a *constant lifetime fatigue diagram*, and an example is shown in Figure 3.18. Lines drawn through data points on this diagram indicate the combinations of mean stress and stress amplitude that are expected to produce a stated lifetime (number of cycles). Although fatigue limits are defined by mean stress and stress range, the latter is the dominant factor.

Variable Loading and Cumulative Fatigue Damage

Structures and components are subjected to many kinds of cyclic and variable loading in service. Because any fluctuation in loading has the potential to cause fatigue failure, investigators have studied the stress spectra produced by the service loading of bridges, cranes, earthmoving machines, aircraft, etc. Strain gauges were installed on main metal members, and all changes in stress were continuously recording over a significant time. In nearly all cases, the stress traces showed that main members are subjected to high stresses only a limited number of times during service. Most of each spectrum consisted of lower variable stresses. Furthermore, the spectra showed little

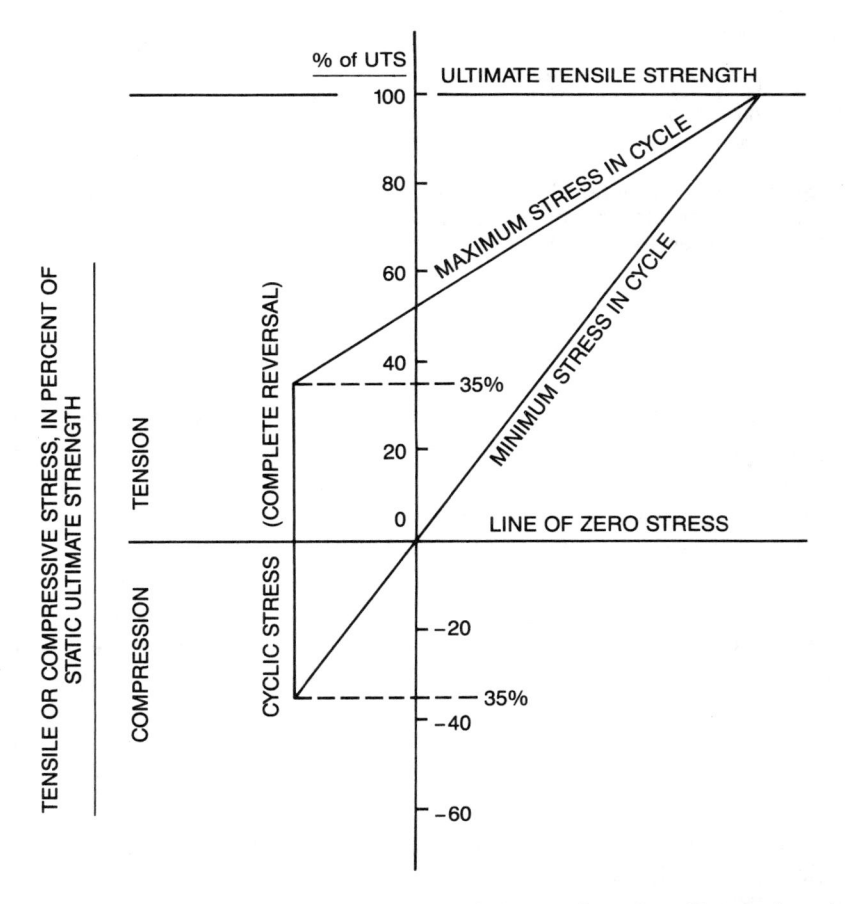

Figure 3.17 — Goodman diagram (original form) illustrating the effects of various patterns of cyclic stress on endurance limit

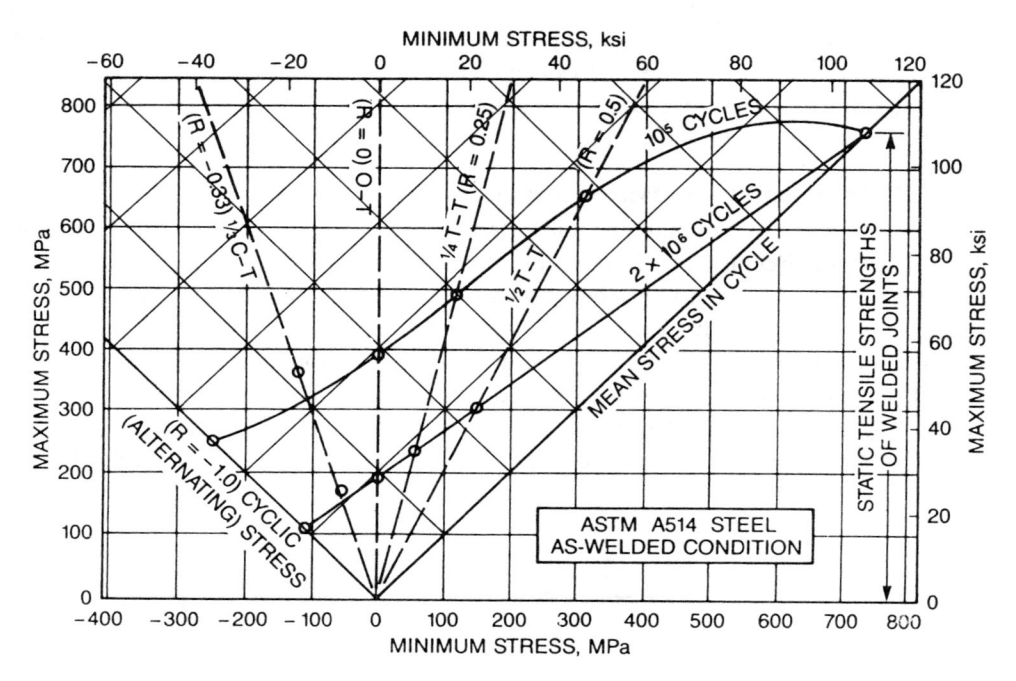

Figure 3.18 — Constant-lifetime fatigue diagram showing interrelationships between stress amplitude and mean stress with fatigue strengths at one-hundred thousand cycles, and at two million cycles

repeated cycling of a particular stress, and the levels of stress varied randomly. The stresses encountered by rotating or reciprocating machinery were considerably more cyclic, but peak stresses from unanticipated loadings were sometimes superimposed on the normal cyclic operational stresses.

These observations regarding service stress spectra presented a formidable challenge to investigators. It was difficult to use the data from fatigue testing, which had been conducted with the usual constant amplitude loading, to estimate the real-world fatigue life of metal subjected to the variable loading. A method of addressing this problem on a specialized basis was proposed as early as 1924 (Palmgren's work on ball bearings), but it was not until 1945 that a more generally applicable method emerged from work by Miner. Gurney, a noted fatigue researcher, explained three features of a stress spectrum, regardless of its complexity, that offer opportunity for counting and assessment. These features are:

(1) the number of peaks and troughs of each stress level,
(2) the number of stress ranges of each of several magnitudes, and
(3) the number of level crossings, that is, the number of times the stress-time curve crosses each of several defined stress levels.

Peak and Trough Counting

Peak and trough counting has been practiced with two different techniques. One method counts every peak or trough where the slope of the stress-time curve changes in sign. The second method requires a count only if the stress has passed through zero since the preceding peak or trough. Limited use has been made of peak-and-trough counting due to the poor resolution which often accompanies a stress-time spectrum.

Range Counting

Range counting has also seen limited use — because of uncertainty over what constitutes a stress range that should be counted, and because of difficulty in evaluating the ranges found in the curve. A range can be defined as that portion of a curve occurring between two adjacent points of stress reversal. Therefore, a salient range can be either an ascending or a descending stress. A number of measuring techniques have been explored to assess range units. To effectively apply range counting, an evaluation procedure must be devised for each case.

Rainflow Counting

Rainflow counting is named for the analogous flow of rain drops down a pagoda roof. It is an important variation method of range counting which can be used to determine cumulative stresses that would cause fatigue damage. To conduct a rainflow count on a service stress spectrum, the trace is positioned vertically, extending downward. The stresses are considered to follow a flow path through cycles and to reach a peak in a manner similar to the rainwater on a pagoda roof. Application of rainflow counting requires detailed study of this method; results obtained can be very informative.

Level-Crossing Counting

Level-crossing counting requires selection of three or more stresses, and a count is recorded each time the stress rises above any one of the designated stresses. The total count for each stress level allows construction of a cumulative frequency diagram. This method does not consider the sequence of stress amplitudes, and its application is better suited to traces where the mean stress is approximately constant.

Miner's Rule

Miner's rule, the most widely used cumulative damage assessment method, is based on the work done during each loading cycle. The life under each random loading is related to the constant-amplitude S-N diagram for the particular metal, and a measure of fatigue life is assumed to be exhausted by a selected stress. More specifically, the fatigue damage at each stress is assumed to be proportional to the number of cycles of that stress applied, and the damage accumulates linearly until failure occurs. Stresses below the normal constant-amplitude S-N endurance limit curve must be taken into consideration. These stresses, although insufficient by themselves to initiate a fatigue crack, can contribute to fatigue damage by propagating a crack initiated earlier by occasional higher stresses.

In Miner's analytical procedure, the order of application of the various stresses is assumed not to affect fatigue life; however, this may not be true. For Miner's rule to give a useful approximation of a metal structure's performance under fatigue loading, care must be exercised in counting stress cycles and in forecasting the sequence of high and low stresses expected in the service stress spectrum.

Fracture Mechanics

Fracture mechanics, used as an analytical engineering tool to predict the fracture behavior of a cracked metal member, was also applied by fatigue investigators to calculate the propagation rate of a fatigue crack during cyclic loading. They reasoned that, in many structures and components, an existing condition probably already provided a small crack-like defect with a stress-strain field at its terminus. This is the principal element needed for fracture mechanics computations. Furthermore, the stress-strain field at the terminus of this initial crack would likely be quite small. This would allow calculations using *linear elastic fracture mechanics* (LEFM), with the objective of predicting the incremental crack extension from each stress cycle or tensile stress application.

LEFM is not as difficult to use as *elastic-plastic fracture mechanics* (EPFM), which must be used when the crack is larger and when nonlinear or plastic strain exists at the crack front. Fracture mechanics holds much promise for assessing cumulative fatigue damage, because a structure or component can have considerable life remaining even if a crack already exists or after a fatigue crack initiates and propagates. LEFM and EPFM can forecast the growth of subcritical cracks and thus be used to prevent a crack from reaching a critical size that invites brittle fracture or from becoming a through-thickness crack (e.g., in a pressure vessel).

FRACTURE MECHANICS: ASSESSMENT OF FRACTURE TOUGHNESS

Fracture mechanics was previously pointed out to be a rapidly developing means of toughness evaluation. Also, fracture mechanics methods are accepted as valid means of assessment in certain codes and standards. Fracture mechanics is a collection of experimental techniques which, along with associated mathematical analytical methods, are used to avoid premature fracture in structures and components subjected to stress.

Fracture mechanics originated as a discipline for understanding *brittle* fractures, but its scope has been extended to cover several other kinds of fracturing. This discipline used stress analysis or energy release rate criteria to analyze stresses and strains near the tips of *sharp flaws* (crack-like discontinuities) from which brittle fractures usually initiate. The flaws were presumed to be already present in the metal, or they were believed to appear in the metal through mechanisms which serve as fracture initiators. The discipline of fracture mechanics held the following basic assumptions:

(1) Fractures initiate under stress at flaws.
(2) Metals always contain flaws.
(3) Sharp flaws can be treated mathematically as cracks with zero radii of curvature at their termini.
(4) Linear elastic stress analysis can be used to calculate stress intensity at the tip of a crack when the nominal stress is less than the yield stress.

When this approach was applied, the results clearly showed that fracture toughness of metals could be clearly determined by testing special specimens in a prescribed manner. Furthermore, the fracture susceptibility of a structure or component under load could be evaluated to predict service reliability. Also, when considering fracture by fatigue and the difficult problem of appraising cumulative damage from variable cyclic stresses, fracture mechanics appeared to indicate fatigue properties accurately by predicting fatigue crack growth rates.

The basic concept of fracture mechanics, as proposed by Griffith, is explained in Technical Brief 12. Over the years following Griffith's work, until about 1940, relatively little was done to apply his theory to the brittle fracture problem in metals. Attention was being centered on notched-bar impact testing during this period, and the results apparently were serving most needs related to toughness evaluation of metals. Then World War II began, bringing new and greater material demands. The War soon revealed that the notched-bar impact test was not sufficiently discerning to guide toughness evaluations in applications such as hull steel for ships, armor plate, and other structures in which a brittle fracture could have serious consequences. This awakening to the shortcomings of the impact test resulted in the development of tests that would provide a more discriminating evaluation of toughness.

New test methods began to appear in the literature which purportedly would better satisfy the requirements of various applications. Some noteworthy tests pertinent to welded construction will be described in Volume II, Chapter 16. Even as this toughness test development continued, the notched-bar impact test remained most frequently used, mainly because of its widespread entrenchment in codes and specifications. Yet, the limitations of this test continued to unfold, especially as toughness problems arose in new applications. Two examples are (1) embrittlement of steel reactor vessels in nuclear service, and (2) brittle fracture of thin-wall cylinders for large rocket motors.

Fortunately, at the same time that the need arose for a better toughness test, those working with fracture mechanics also began to re-examine the problem from a fundamental perspective. Hollomon and Zener, in 1944, studied Griffith's work and decided

---

TECHNICAL BRIEF 12:
Griffith's Theory of Fracture Mechanics

The basic concept of fracture mechanics was proposed by Griffith, who advanced a theory in 1920 to explain the change in conditions in brittle solids (like glass) whereby an existing flaw would suddenly become a running crack. He believed that two types of energy governed the engineering strength of materials: (1) the strain energy of the material system and (2) the surface energy of a crack within the material.

Strain energy was proposed as the driving force for crack growth. Griffith postulated that microcracks were invariably present in the material, and because of stress concentration at the crack tips, accelerated crack extension should be expected at some critical point of applied stress which was dependent on the size of the existing crack. He assumed an energy balance relationship between the applied stress and crack size, in which the crack would become unstable and rapidly propagate if the release of elastic strain energy was at least equal to the surface energy of the newly formed crack surfaces. The symbol, G, was used to denote the strain energy released for each unit area of fracture. The condition of *criticality*, i.e., the start of unstable crack propagation, was given the symbol G_c. Griffith proved the validity of his theory in work predicting the strength of glass rods and plates. He showed that the residual strength of glass members is inversely proportional to the square root of crack length. This work established an energy release rate criterion for brittle materials.

that the Griffith concept applied equally as well to metals. They proposed that the microcracks envisioned by Griffith in glass need not be inherently present in metals. Rather, the microcracks could easily form by fracture of brittle constituents, or at interfaces between phases in the metal's microstructure. Although they found that Griffith's theory fit the brittle fracture behavior of metal reasonably well, it was in 1945 that Orowan discovered by x-ray diffraction of fracture surfaces that a minute amount of plastic flow was involved even during brittle crack formation in a metal. This contrasted with the almost purely brittle fracture of glass encountered in Griffith's analyses, which did not require adjustments for any energy consumed in plastic deformation.

This recognition of the true character of brittle crack formation in metal allowed Orowan — and Irwin, working independently — to extend Griffith's concept by making an allowance (albeit very small) for the added work of plastic deformation. Their efforts established an energy term for the work associated with brittle crack growth in metals. Again, the symbol G was used for the strain energy release rate, while G_c was used to identify the critical event at which the release of strain energy exceeds the surface energy of a new crack surface, and unstable crack propagation occurs.

Brittle Fracture Test Parameters

At this point, fracture mechanics people struggled to find a parameter that was a constant related to fracture propensity; the ideal parameter would have a value for a given metal that would not be affected by shape of specimen, size, or kind of metal. Naturally, they sought a value that could be employed in calculations to predict fracture toughness. An important breakthrough occurred when Irwin proceeded to equate the strain energy established earlier by Griffith with a *stress intensity* approach. Irwin envisioned a crack as an extremely sharp opening with a front extending along a line, and with the stress field at the crack front being annular and two-dimensional. Although he recognized that some plastic strain existed at the tip of the crack, he believed this could be initially ignored if relatively small in size.

Using *linear elastic* treatment, Irwin developed a mathematical expression of the elastic stress field generated by a sharp crack, and extracted from the computations a *stress intensity factor*, which is a material constant that expresses how the stress increases as the crack front is approached. Irwin's stress intensity factor commonly is identified by the symbol K (As a cautionary remark, Irwin's K is not the same as K_t, described earlier as the theoretical stress intensity factor for notches and sharp corners). Considerable mathematical development surrounded Irwin's stress intensity approach, but the final equation established for determining K is

$$K = \sqrt{E \times G} \qquad \text{(Eq. 3.2)}$$

In the above equation, the stress intensity factor, K, is equal to the square root of Young's modulus of elasticity, E, times Griffith's energy release rate, G.

Irwin next observed that under almost completely elastic strain conditions, the extension of a sharp crack occurred suddenly, and repeatedly at a particular stress at the crack front; that is, at a constant stress at the crack front, as predicted by K. This observation led to a method of testing metals using specimens that contained a small crack, and of carefully engineered size and shape to satisfy certain linear-elastic requirements. The specimen was loaded in tension, or by bending, until the *critical stress intensity* for the particular metal was established by reaching a point of significant crack extension or growth. This critical stress intensity factor is identified as K_c, and is that value of stress intensity factor, K, at which the start of crack growth occurs.

When specimen size and other conditions associated with the crack allowed only minimum plastic deformation, the start of crack growth would occur by the cleavage mechanism, and a running brittle fracture would result. In general, the greater the K_c value, the greater will be the innate resistance of a given metal or alloy to brittle fracture. Equations that evolved from Irwin's work for relating critical stress intensity, K_c, to known values are given below:

$$K_c = \sqrt{EG_c} \qquad \text{(Eq. 3.3)}$$
$$K_c = \sigma \sqrt{\pi a} \qquad \text{(Eq. 3.4)}$$

Equation 3.3 is similar to 3.2, but it introduces Griffith's critical strain energy release rate, G_c, as a factor instead of G. Equation 3.4 states that a determinate level of stress at the crack front (σ), times the square root of pi, times one-half the crack length (a) will exceed tolerable stress intensity and prove to be the critical stress intensity K_c, at which crack growth occurs. Crack length (a) in Equation 3.4 is the length

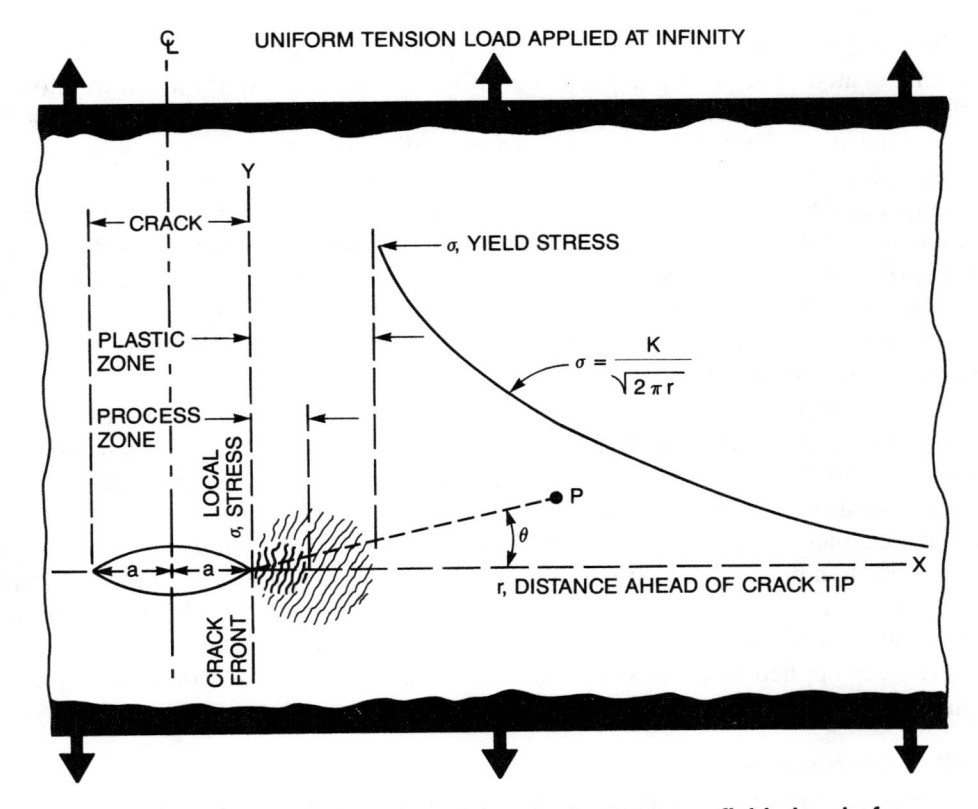

Figure 3.19 — Stress distribution in the elastic stress field ahead of a through-thickness crack as characterized by Irwin's stress intensity factor, K

of an internal or buried crack — that is, a crack length divided in half. For a surface crack, however, the total crack length is used in calculating K_c. The K_c factor, therefore, is a compound parameter involving stress and the square root of crack size. If a crack is small, and if the metal is not "crack sensitive," the application of a typical service load will probably cause only slight yielding of the metal at the terminus of the crack. In this way, the potential of the crack for initiating a brittle fracture is decreased. On the other hand, when dealing with a sharp flaw of critical crack size, there is a good possibility of having brittle fracture initiate from the existing flaw or crack as soon as the stress intensity at the tip reaches the K_c level.

Figure 3.19 is an elementary sketch illustrating conditions of stress and strain that exist in a metal and extend from the tip of a sharp crack when load is applied. In addition to showing discrete zones of plastic strain, the sketch illustrates how Irwin's stress intensity factor, K, characterizes the elastic stress-strain field — that is, the relationships between load applied, crack length, and stress intensity at any point ahead of the crack front in the elastic field. A number of key elements and notations are included in this schematic figure, and the following explanations are offered regarding their function and importance.

Section Dimensions

Figure 3.19 represents a section of thick plate within which a *through-thickness* crack is located. The plate is considered infinitely large, and quite thick, to allow a valid fracture mechanics analysis. The plate edges must not be near the crack, because the stress field along the crack front, particularly at its tip, would be unduly influenced. A principal aim is to have the plastic zone at the crack tip be negligible in comparison to the plate dimensions. Of course, in structures, members are not of infinite size; and cracks easily can be located on or near plate edges. Also, the crack, instead of extending through the thickness, possibly will be at one surface only; or the crack may be buried and not extend to any surface. Corrective terms can be entered into final fracture mechanics computations to adjust for location of the crack and other dimensional effects.

When fracture mechanics analysis is applied, it is essential to have the smallest possible plastic zone. This small zone is achieved in specimens through their size and design, and by the length of the crack. The extent to which this original small plastic zone is able to expand under stress during testing will depend on the strength, ductility and toughness of the metal, and on the degree of constraint in the vicinity of the crack front. In a plate with a through-thickness center crack, loaded uniformly in tension as shown in Figure 3.19, when the half-crack length is less than 10 percent of the plate width, Irwin's stress intensity factor (K) is a valid representation of the relationship between applied local stress (σ) and half-length crack (a), as would be expected from an infinitely wide plate.

Plotting Coordinates

Coordinates for analysis and plotting, presented as dashed lines in Figure 3.19, are labeled x on the abscissa, and y on the ordinate. These present the distance ahead of the crack tip (r), and local stress (σ) arising from load applied to the plate at infinity.

The third direction, perpendicular to both the x and y coordinates, is identified as the z direction (although this is not shown in Figure 3.19). For a plate, z is the thickness direction. The z direction is an important third coordinate, because with it a determination is made whether a condition of *plane-strain*, or of *plane-stress* exists. The nature of these two conditions and their significance will be explained shortly under their respective headings.

The *crack plane*, as related to the axis of principal stress, is an important feature, and in the case illustrated in Figure 3.19, the plane of the crack (abscissa, x) is perpendicular to the tensile stress imposed across the crack front. When the tensile stress is applied, the surfaces of the crack will move apart, and this motion is called a "mode" of crack deformation. It is one of three basic modes, to be illustrated shortly. During extension of the crack under stress, the crack will proceed along a plane that tends to remain perpendicular to the largest stress, and the plane will remain in a straight line if the direction of largest tensile stress does not change.

The *process zone* was mentioned in earlier discussion of fracture as the highly localized region of stress and strain immediately surrounding the crack tip. This is the zone where cracking initiates, either as limited increments of stable cracking when the metal behaves in a tough manner, or as a brittle running crack when a critical stress is exceeded and cleavage fracture occurs. Because stress concentration at the tip of the crack is extremely high, the metal in this immediate vicinity yields, and plastic strain occurs in a very small region. Consequently, the initiation site for the start of crack extension is in a zone of plastically deformed metal, which suggests that an analysis of the fracture process zone probably should be approached via plasticity considerations. However, this entails very complex computations, and proponents of this course of action have not been able to agree upon analytical treatment.

The *plastic zone*, as shown schematically in Figure 3.19, is a region enveloping the process zone at the crack tip, and this somewhat larger plastic zone sometimes is called the *crack front plastic zone*. The plastic zone, nevertheless, is relatively small in size, and it encompasses nonlinear strains of somewhat less severity than those in the process zone. When the plastic zone is small, as compared with the elastic stress-strain field surrounding it, the assumption is made that both process zone and plastic zone can be ignored, because the nonlinear strains therein will be virtually controlled by the surrounding elastic field. This assumption, essentially, is the basis of the current *linear-elastic fracture mechanics* approach.

Crack Surface Displacement Mode

The mode of crack surface displacement during crack extension must be considered when analyzing stress patterns and stress intensity. The crack in Figure 3.19 was described as lying in a plane perpendicular to imposed tensile stress. This would result in a direct opening of the crack and further propagation along the same plane. This is termed the *opening mode*, and is only one of three basic classes of crack (surface) displacement adjacent to the crack tip.

Figure 3.20 illustrates all three basic modes and their common designations — namely direct opening (Mode 1), edge sliding (Mode 2), and screw sliding (Mode 3). Sometimes these modes are considered from a stress standpoint as tensile, shear, and tearing modes, respectively. Arabic numerals are positioned as post-subscripts (e.g., K_1, K_2, K_3) to indicate the mode involved for stress-intensity-factor determination. Roman numerals are used when the mode is executed under plane-strain crack front conditions, which are described next. In most applications of linear-elastic fracture mechanics (LEFM), the direct opening of a crack (Mode I) is the applicable mode.

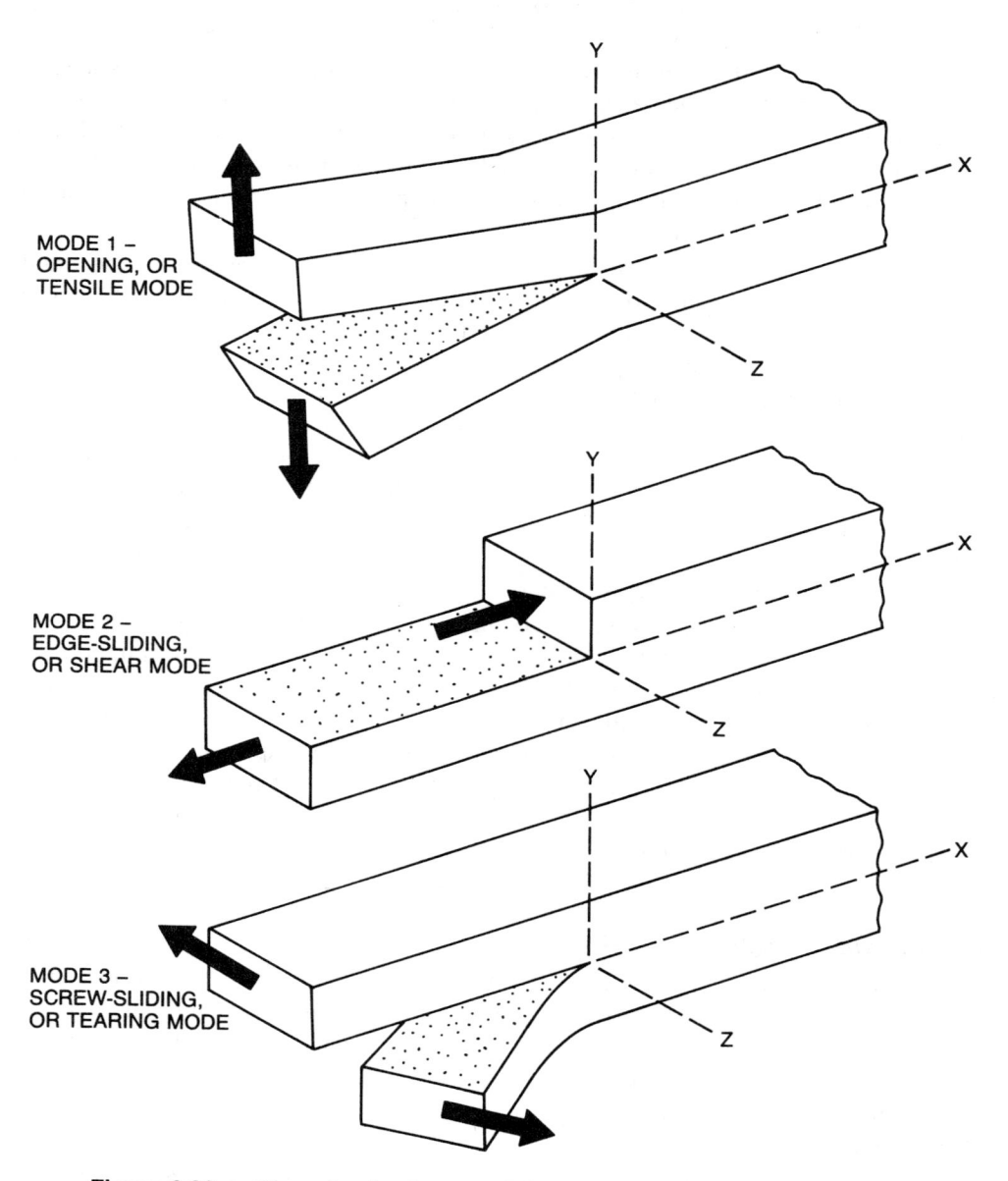

Figure 3.20 — Three basic classes of crack deformation or crack (surface) displacements associated with stress-strain fields around the crack tip

Plane-Strain

Plane-strain is the term used to indicate that a condition prevails ahead of the crack tip, where there is zero strain in the direction perpendicular to both the axis of principal stress (y), and the axis of crack growth (x). In other words, there is no strain in the z direction. This condition arises when thick members are loaded in a direction parallel to their surface. In plane-strain conditions, the plane of fracture instability is perpendicular to the direction of principal normal stress. When loaded, a crack located as shown in Figure 3.19 will develop both high stress and plastic strain at its tip. However, from earlier discussion of Hooke's law, Poisson's ratio, and triaxial stress generation, we know that a small cylindrical zone of metal lying parallel to the crack front (extending, therefore, through the thickness of the plate) is restrained from contraction by surrounding metal. This constraint produces tensile stress in the z direction, which is the longitudinal axis of the cylindrical zone along the crack front. However, the amount of strain in this same direction is essentially nil.

When plane-strain conditions prevail, the crack tip conditions are of maximum severity, and crack growth occurs at the lowest possible level of K. Consequently, the fracture toughness index that ordinarily serves as the lowest value for the particular testing conditions (temperature, loading rate, etc.) is K_{Ic}. It should be noted that Roman numeral "I" has been used in the post-subscript. This is to confirm that the critical stress intensity being conveyed has been determined under plane-strain conditions, with crack surface displacement occurring in the direct opening or tensile mode. Under these plane-strain conditions, the principal stress near the crack tip can become greater than the uniaxial ultimate tensile strength. Of course, the actual tensile and yield strengths that exist in the plastic zone are not known. This is because the plastic zone is under triaxial constraint, a state that increases tensile and yield strengths. Nevertheless, when the stress developed at the crack tip increases to a critical level, a running cleavage fracture usually occurs. It is important to recognize that at the moment of cleavage fracture initiation, the gross stresses from loading of the member might be well below the metal's uniaxial yield strength; but as pointed out earlier, the stress parameter with much information at the moment of fracture is the stress intensity at the crack tip.

Fracture under plane-strain conditions provides a relatively simple example for analysis using fracture mechanics. Fracture is initiated directly from the existing crack tip, and when the plastic zone is ruptured, crack extension follows. Under plane-strain conditions, there is no period of complex, growing resistance to crack extension prior to the occurrence of unstable fracture. Linear-elastic analysis requires only a minor mathematical treatment to adjust the existing crack length for the presence of the plastic zone at the crack tip.

Plane-Stress

Plane-stress fracture toughness conditions exist when thickness (z direction) is not sufficient to maintain complete constraint along the crack front. In general, for through-thickness cracks in plate, the amount of constraint depends upon the length-to-diameter ratio of the cylindrical zone mentioned above. If the zone is long, as in a thick plate, *plane-strain* conditions likely will exist. If the zone is short, as in sheet, constraint is less, and *plane-stress* conditions probably will exist. In that case, the critical stress intensity factor is written as K_c. Plane stress can be briefly defined as a condition occurring when the plane of fracture instability is inclined 45° to the direction of principal normal stress. If constraint is relatively small, so that either plane stress or some intermediate state of strain exists, then the fracture toughness (K_c) will be greater than what is indicated by K_{Ic}.

The question often arises as to how thick a plate or a member must be to have plane-strain conditions at a crack tip, or conversely, to what maximum thickness will plane-stress conditions exist. Dimensions are dependent upon the kind of metal and its strength. A high-strength steel plate 12 mm (½ in.) thick can be enough to produce plane-strain conditions, but for a low strength aluminum alloy even 50 mm (2 in.) may not be sufficient. When a crack extends only partially through a member, the state of conditions at the crack tip will be plane strain, unless the crack is almost through the thickness. Therefore, cases of *part-through cracks* are treated in terms of K_{Ic} rather than K_c, regardless of thickness. This practice will be demonstrated later, when the fracture mechanics of fatigue cracking is reviewed.

Stress Distribution

Stress distribution in the elastic field ahead of a crack is shown graphically in Figure 3.19, along with some mathematical treatment included in the figure. Although space limitations will not permit a detailed review of the mathematics involved in completing a fracture mechanics analysis, it is helpful to understand the derivation of salient factors employed in practical application of this analytical tool. The solid curved line (σ) in Figure 3.19 represents the stress levels found along the crack plane (abscissa, x) when a tensile load is applied to the plate in the vertical direction (ordinate, y). Because the crack cannot support any of the load, the stress is transferred to adjacent metal. As should be expected, this transfer does not result in a uniform increase in net stress; instead, a concentration of stress occurs near the crack, with the highest stress near the tip as shown. Note that the level of stress is not indicated immediately adjacent to the crack tip. This is because yielding has occurred in the plastic zone, and the stress will have reached a plateau at an indeterminate multiaxial yield strength. This level undoubtedly is somewhat higher than the uniaxial tension test yield strength for this metal. Stress (σ) along the crack plane (x) decreases as shown by the solid curved line, and the values for plotting are computed with the following equation:

$$\sigma = K / \sqrt{2\pi r} \qquad \text{(Eq. 3.5)}$$

Stress at any point away from the crack plane, but still in the elastic field generated by the presence of the crack, can be calculated, using r and θ as polar coordinates. As shown by the example in Figure 3.19, a line drawn from a point labeled P to the crack tip establishes an angle, which commonly is identified as θ. The abscissa or crack plane usually is taken as zero base. The distance along the line from point P to the crack tip is the value for r. The stress at point P can be calculated using the following equation:

$$\sigma = K/\sqrt{2\pi r} \cos \theta/2 \; (1 + \sin \theta/2 \sin 3\theta/2) \qquad \text{(Eq. 3.6)}$$

Equation 3.6 reveals in a broader way how the K factor indicates stress intensification as the crack tip is approached from any point in the elastic stress field generated by the crack. It also reveals how the loading and geometry factors are related to this single parameter, K. A simple relationship can exist between K and applied stress and with crack length, as implied earlier by Equation 3.4, but often the relationship is of greater complexity in structures. Handbooks are available that provide correction factors for stress intensity and elastic compliance.

Procedures for Evaluating Propensity for Brittle Fracture

Determination of toughness by a fracture mechanics test is more complex than most other established tests for a mechanical property. The many details that must be considered require much study beyond the rudimentary information given here. The science of fracture mechanics is progressing so rapidly that the latest issues of testing specifications, along with the findings from research in current literature, must be applied to ensure that the best possible test procedure is followed and that results are correctly interpreted. One standard test method for determining plane-strain fracture toughness (i.e., K_{Ic}, the critical stress intensity factor) is presented in ASTM E 399, *Plane-Strain Fracture Toughness of Metallic Materials*. This document details several configurations of tension and three-point bend specimens. All of these specimens are single edge-notched to provide a locus for cracking by a closely prescribed fatigue procedure. This important crack is of controlled length and has a very sharp terminus. The size of the specimen must meet or exceed specific minimum requirements for thickness, width, and uncracked ligament ahead of the crack. The specified dimensions are related to the 0.2 percent tensile yield strength of the metal, and an estimate of K_{Ic}. In some cases, Young's modulus can be used instead of an estimated K_{Ic}.

The aim of specimen size and design is to establish suitable crack front conditions in a tri-tensile stress field, with a minimum of plastic strain zone at the crack tip. In other words, plane-strain conditions are to exist. When specimens are tested, an autographic record is made of load versus notch-mouth opening. This record will initially show a straight line relationship; but when the load results in a two-percent crack extension, as determined by deviation from the linear portion of the record, then the critical stress intensity has been reached. At this point, K_{Ic} is calculated from the load, using established equations.

In addition to strict adherence to the prescribed procedure in ASTM E 399, the results of a test must be examined closely to make certain of a valid K_{Ic}. This includes examination of the specimen after fracture to ensure that the initial crack front (cracked by fatigue) conformed to requirements of the ASTM Standard, as well as a number of features of the crack extension. When K_{Ic} is determined by the ASTM E 399 procedure, it is related only to a small initial amount of crack extension, which sometimes can be stable crack growth rather than brittle fracture. Close examination of the load-deflection line may reveal a phenomenon called pop-in, which is a discontinuous short burst of cracking during loading. The burst can be readily detected by various acoustical devices, and the incident may cause enough deviation in the line to indicate the two-percent increment of crack extension that marks the K_{Ic} threshold.

When a very small increment of stable crack extension occurs, it sometimes is considered *subcritical crack extension*. The extension can be involved in the growth of a crack toward a size that is detrimental to structural integrity; that is, the critical crack size. Fortunately, subcritical crack extension advances in short increments at relatively slow speed through the metal, and if loads are enough to promote this kind of growth, the progress of cracking can be predicted and monitored. If the start of unstable brittle cracking must be clearly established, a larger value of K may be required to initiate this kind of cracking, and to arrive at a K_{Ic} that fits the circumstances of the particular application.

In reviewing the details of ASTM E 399, it will be noted that the procedure involves slow loading of specimens at room temperature. As would be expected from earlier observations of fracture in metals, temperature and rate of loading can have a significant influence on the critical stress intensity. A lower testing temperature can cause a marked decrease in K_{Ic}; but, the decrease will be dependent upon the kind of metal. Carbon and low-alloy steels are among the metals most adversely affected by decreasing temperature. The conditions of temperature and loading rate in ASTM E 399 were selected for convenience and consistency in laboratory testing, and they are not usually comparable to the service conditions. Consequently, plane-strain fracture toughness may have to be re-appraised under modified testing conditions to secure a more useful K_{Ic}. Temperature adjustment to make testing comparable to service is a relatively straightforward matter.

Tests at higher rates of loading have been conducted by a number of investigators, because of their concern with service involving impact. To simplify testing, one investigator loaded fracture mechanics specimens by dropping a large weight from a small height above the specimen, and providing a soft-metal cushion beneath to prevent elastic stress fluctuations. The fracture toughness values reported were identified as K_{Id}. As expected, K_{Id} was lower than the corresponding K_{Ic}. The British Standards Institution (BSI) is planning to issue BS 6729 for dynamic fracture testing. Slow loading (ASTM E 399) requires several minutes to establish K_{Ic}. When loading requires only one or two seconds, investigators often identify the critical stress intensity as $K_{I(t)}$. Impact loading usually takes only 0.01 sec or less, and K_{Id} is applied.

The environment in which the metal will be exposed under load may require consideration from a fracture toughness standpoint. The usual laboratory testing in air establishes K_{Ic} for a neutral environment. Yet, the metal possibly will not exhibit this level of critical stress intensity if the service environment is aggressive; ambient conditions may be of a nature that promotes embrittlement or attack. Because fracture mechanics considers crack growth, and does not consider the initiation phase, investigators have made good use of fracture mechanics to study crack-growth kinetics, when the environment is aggressive and induces some form of cracking. Data concerning the stress-corrosion cracking critical stress intensity factor (K_{Iscc}) can be found in the literature.

Use of Linear-Elastic Fracture Mechanics (LEFM)

Linear-elastic fracture mechanics provided the initial break-through to a rewarding, multi-use analytical procedure for toughness evaluation. Unfortunately, LEFM is limited by the kind of metal sections on which meaningful determinations can be made. Nevertheless, LEFM can be helpful in the fabrication of certain weldments, and it is also a basis for understanding other procedures developed to broaden the application of fracture mechanics to metal sections — both welded and unwelded. The areas of LEFM application to weldments that are most appropriate and potentially rewarding include design, stress analysis, metal selection, nondestructive examination, and failure analysis. The mathematics of LEFM treatment are sound, but even the best procedure often involves approximations and inescapable small errors in factors of the equations. Also, the reliability of K_{Ic} determined through LEFM depends heavily on close adherence to plane-strain conditions at the crack front.

In general, little difficulty is encountered in applying LEFM to high-strength limited-ductility metal of substantial thickness. Plane-strain conditions usually prevail; and, judging from reports in the literature, the K_{Ic} obtained has been useful. Because of the three basic quantities embodied in LEFM — applied stress (σ), crack length (a), and stress intensity (K) — it is possible, after K_{Ic} has been determined, to compute the following:

(1) the maximum stress that can be applied without causing an existing crack or sharp flaw of specific length to commence unstable growth,
(2) the longest crack or sharp flaw that can be tolerated in a given member under a specific load without danger of unstable crack growth, and
(3) the minimum fracture toughness required of the metal to avoid brittle failure, with known maximum length of crack or sharp flaw (as ascertained by NDT), and with a specified maximum load.

These values deserve further comment. Although users of LEFM strive to obtain such values by test — and to calculate the most precise values possible — it is important to realize that in actual practice LEFM analyses are only approximations. Even so, this approximate character of the LEFM value does not nullify its usefulness.

In addition to the usual differences that exist between test specimens and structures, fracture toughness values per se are regularly quite variable. While test procedures and testing apparatus have been examined closely for causes of this variability, the principal cause seems to lie in the test itself. Only a relatively small volume of metal, located at the tip of a sharp crack, is directly involved in a K_{Ic} test. If any kind of defect is present in this tested volume, such as a nonmetallic inclusion, it could influence the test result. If the metal has been welded, differences in the performance of a specimen versus the structure usually become greater.

Along with the issue of duplicating welding effects in specimens, there is the problem that the distribution of stress in a real structure ordinarily is more complex. This complexity arises because of interactions between weld metal, heat-affected zones, and base metal, as well as the development of residual stresses.

The distribution of load in real structures, even those that are not welded, is not as straightforward as in a test specimen. The stress intensity factor, K, which is the precursor of K_{Ic} and is essential to establishing the three values listed in the numbered paragraphs above, ordinarily is viewed as having simple relationships with applied load and with crack length. Because of the more complex distribution of load in a structure, relationships originating from K have been determined both experimentally and by advanced analytical methods for many patterns of loading, member geometry, and form of crack-like defects. These findings and improved equations have been published in handbooks to assist the application of LEFM.

When results from specimens are projected to assist in structural assessment, it must be remembered that test specimens frequently perform better than the actual structure. This is because less metal is involved at the crack tip in a specimen, and there is lower constraint from surrounding metal. On the other hand, the crack or sharp flaw in a structure may have a blunted terminus that resulted from some form of exposure or deliberate treatment, such as warm stressing. Consequently, the K_{Ic} produced by a specimen might be overly conservative. When all of the factors that are involved in testing and for structural assessment are considered, it seems almost impossible to gain accurate knowledge of the fracture toughness of a structure without simply loading it to destruction. Despite its shortcomings, fracture mechanics still provides a more realistic appraisal of the fracture toughness expected in a structure than does information obtained from any other kind of mechanical property test, such as the Charpy notched-bar impact test.

When using the numerical values for factors involved in LEFM equations, the stress intensity factor (K) is expressed in a manner unlike tensile and yield strength. Instead of familiar measures of force per unit cross-section (e.g., MPa, or ksi) K is expressed as units of stress-intensity, which is the product of unit applied stress and the square root of half-crack length. The single parameter K, therefore, includes both the effect of applied stress and the effect of a crack of given size on the amount of energy available at the crack tip to initiate extension. For example, a steel that has an ultimate tensile strength (UTS) of 700 MPa, and a 0.2 percent offset YS of 600 MPa, may be found to have a K_{Ic} of 55 MPa × m$^{1/2}$. Corresponding U. S. customary units for this steel are 100 ksi UTS, 87 ksi YS, and 50 ksi in.$^{1/2}$. The worthiness of the K_{Ic} is in its indication that unstable crack extension could occur in this steel under plane-strain conditions at a stress level substantially below its yield strength. More to the point, a structure built of this steel, when loaded to a nominal stress of only one-half its yield strength, i.e., 300 MPa (44 ksi), could suffer unstable crack extension from a through-thickness crack of only 19 mm (¾ in.) length, since this would represent critical crack size. An elementary equation for the relationship that allows estimation of critical crack size was presented earlier as Equation 3.4; to wit,

$$K_c = \sigma \sqrt{\pi a} \qquad \text{(Eq. 3.4)}$$

As a general trend, the fracture toughness of steels under plane-strain conditions tends to decrease as the yield strength increases. However, there is a considerable spread in the range of K_{Ic} exhibited by different steels, and the changes in fracture toughness performance with changes in temperature and rate of loading are not the same for all steels. Consequently, researchers continue to search for a more discriminating and precise method of fracture toughness evaluation, and for ways of increasing the fracture toughness of steels at all levels of strength without significant increase in cost.

The three values described above, determined with the aid of LEFM, are essential information that should be available for a structure to ensure its fitness. To have this LEFM technology available ordinarily would be welcome. However, this is only half of the fracture mechanics story. The remaining half is prefaced with the statement that LEFM cannot be successfully applied with the majority of commercial metals. The reason is that the linear-elastic mathematical treatment, despite all compensatory measures, fails to predict the higher fracture toughness ordinarily exhibited when crack front conditions change from plane-strain toward plane-stress. This improved fracture toughness sometimes can be due to thin metal, which fails to constrain the plane-strain condition. However, more often it is attributable to the lower strength and greater ductility of the metal, which allow a much larger plastic zone to develop at the crack front.

These facts are not intended to convey that with commercial metals there is little need for a toughness evaluation method as found in LEFM. On the contrary, even though plane-strain conditions arise only infrequently (mostly with very thick metal), inadequate toughness can be a problem under plane-stress conditions. Consequently, it was realized that fracture toughness also had to be assessed on an *elastic-plastic* fracture mechanics basis, and this realization encouraged renewed attack on the problem.

Development of Elastic-Plastic Fracture Mechanics (EPFM)

There is no clear boundary between plane-strain and plane-stress conditions to indicate where linear-elastic fracture mechanics analysis is no longer applicable. When a larger plastic zone develops at the crack tip, especially one of moderate size relative to the thickness of the metal (z direction), then LEFM becomes less discriminating as a means of evaluating the factors involved. Consequently, predictions by LEFM analyses about fracture toughness may not be of value — or worse, they can be misleading. Because of the analytical complexities that arise when more extensive plastic deformation occurs at the crack front, the task of developing a test method to evaluate fracture toughness under plane-stress conditions has proved fully as difficult as expected.

The importance of considering a larger plastic zone is the reason why testing and evaluation in this area is called *elastic-plastic fracture mechanics* (EPFM). The term retains the word "elastic" because, even though significant plasticity develops at the crack front, the nominal stress is in the elastic range below the uniaxial yield strength. As metals develop a plastic zone substantially larger in size at the crack front, and as the nominal stress exceeds the uniaxial yield strength, there is an inclination to use the term *general yielding fracture mechanics*. There also can be an extreme set of circumstances where a structure constructed of relatively low strength, high toughness metal containing a crack is able to accept application of maximum nominal load and undergo commensurate yielding *before* crack extension or tearing initiates. In this case, the expressions *plastic instability* or *plastic collapse* are often used to indicate the kind of failure. The majority of metals used in industry (this is especially true of the structural steels) have mechanical properties for which EPFM is the appropriate form of fracture-mechanics analysis in most cases. As mentioned earlier, if fracture should happen to occur, even in these metals under a condition that clearly is plane strain, then the specific case would be analyzed by LEFM, since this procedure likely would obtain the most meaningful results.

Development of a fracture mechanics test method for assessing the toughness of very ductile metals, such as the structural steels, actually began shortly after findings with LEFM first appeared in the literature during the mid-1940s. The history of this development is presented in Technical Brief 13. The need for an EPFM method drew attention because of the puzzling cases of brittle fracture in steel bridges, ships, cranes, storage tanks and other structures — and because the widely used notched-bar impact test often did not provide an adequate explanation for fracture occurrence.

TECHNICAL BRIEF 13:
Fracture Mechanics Testing of Ductile Metals

New EPFM tests introduced during the 1940s consisted of either small specimens with machined notches as a crack starter, or larger structural prototype specimens. A fundamental objective that confronted the originators of these early tests was to achieve brittle fracture in a specimen at a nominal stress below the yield strength, because this is the manner in which many failures occurred in real structures in service. A number of features were tried in these early tests as a means of measurably provoking brittle fracture; these included the use of (1) precracked weld beads; (2) high loading rate achieved by pendulum or falling-weight impact, or by an explosive charge; (3) lower temperatures; and (4) increased restraint by larger mass, or rigid design of specimen.

The *Wells wide plate test* was one of the early tests designed to closely simulate the conditions that appeared to cause brittle fracture susceptibility, even in a very ductile structural steel. The specimen consisted of a very large plate, perhaps measuring one meter (39 in.) square. The specimen sometimes had a welded butt joint across its middle and perpendicular to the axis of applied tensile load. A flaw of controlled size and shape usually was made in order to localize fracture initiation and to simulate the kind of defects commonly encountered in structures. The plate specimen was mounted between powerful hydraulic jacks, which applied a uniform tensile load while strain gauges recorded localized stress at many selected points, particularly those points in close proximity to the flaw. The specimen sometimes was heated or cooled prior to load application.

The Wells wide plate test proved very informative as a research tool, and produced results that closely paralleled the behavior of steel structures in which brittle fracture had been experienced. However, the test specimen was too large and the test was too costly to routinely serve the many fracture toughness evaluation needs that continued to arise. Nevertheless, data and information collected from Wells wide plate tests have seen continued use as references, although tests with smaller and less costly specimens have been made. Also, the realness of the test has led investigators to sometimes use biaxial applied loads. Testing fixtures presently available in laboratories can accommodate plate as thick as 150 mm (6 in.), and can apply loads as great as 4000 metric tons. Consequently, the wide plate test can be a very close simulation of real structure circumstances.

Many fracture toughness tests have been introduced over the past 40 years, and a number of these make use of specimens that are designed to be evaluated by EPFM analysis. Many of the specimens have a common feature — namely, a machined notch which is extended as a fatigue crack using a prescribed procedure. Obviously, a crack of controlled length and sharpness has proved to be the best kind of initiator to preplace in a specimen when susceptibility to further crack extension is to be determined. Differences between the various EPFM test methods are seldom in the design of their specimens, but are more likely to be found in the procedures for analyzing specimen behavior during loading. In fact, many of the specimens currently used for EPFM testing are virtually identical to those intended for LEFM testing under ASTM E 399.

ASTM standards for fracture testing include systems for coding specimen configuration, method of loading, crack-plane orientation, and expected direction of crack propagation during testing. These systems can be found in ASTM E 616 - *Standard Terminology Relating to Fracture Testing*. Despite the well planned arrangements of the ASTM code systems, writers often take liberties with symbols for specimens in their reports. Because fracture toughness is affected by anisotropy in wrought metal, close attention should be paid to crack-plane orientation, and to crack propagation direction.

When EPFM specimens are loaded, either in tension or in bending, the nominal stress is recorded autographically along with the crack surface displacement. The latter usually is measured by a double-cantilever displacement gauge of the electrical resistance strain type, which is clipped into the mouth of the machined notch from which a fatigue crack extends. By this means, the increase in mouth dimension or crack opening is measured and recorded during load application. Loading may be either continuous or incremental, depending on the particular test and procedure. Because much of the EPFM testing is with the structural steels — and these are metals that have a strain rate sensitivity — the standards for EPFM testing prescribe a specific range for rate of load application to minimize any strain-rate influence. Most testing is performed at a relatively slow rate, but increasing attention is being given to special tests conducted at higher loading rates to secure fracture toughness indicative of service at such loading rates. Much standardized testing is performed on specimens at room temperature, but temperature also is recognized as highly influential. Therefore, a test temperature (often lower than RT) might be selected for the expected service temperature.

No EPFM test proposed to date for fracture toughness evaluation of the more ductile metals serves all needs. Consequently, the number of available tests and the variety of procedures often is intimidating. Literature describing EPFM tests is too voluminous to review in detail here, but general observations can be made on methods used for analyzing specimen behavior during loading. The crux of every EPFM test method for determining fracture toughness is found in (1) the appraisal of the crack tip and its associated plastic zone, and (2) the analysis of non-linear yielding that develops in the immediate vicinity as crack extension occurs.

Before starting this overview of EPFM testing to make general observations, it would be well to recall Figure 3.9, where the impact toughness curve for bcc metals showed a lower shelf of poor energy absorption at low temperatures, and an upper shelf of much better energy absorption at higher temperatures. Naturally, intermediate temperatures produced a gradual transition from poor energy absorption to better absorption as temperature increased. Because the bcc, or ferritic carbon and alloy steels were the principal metals used in EPFM test development, an examination should be made first of how this general trend in energy absorption or toughness limits the applicability of EPFM testing on ferritic steels.

A properly designed steel specimen tested at a low temperature, somewhere along the lower toughness shelf, probably would fracture under elastic conditions. Therefore, linear-elastic (LEFM) analysis, in accordance with ASTM E 399, is applicable for obtaining K_{Ic}. With increasing test temperature, the fracture toughness would increase; and fracture probably would occur only after some yielding. This is typical behavior for a metal specimen in its transition range, and here is where the need arises for some form of yielding fracture mechanics analysis. When temperature is increased beyond the point where toughness is elevated to the upper shelf, specimens likely will not fracture; rather, they will fail by plastic instability. Inasmuch as struc-

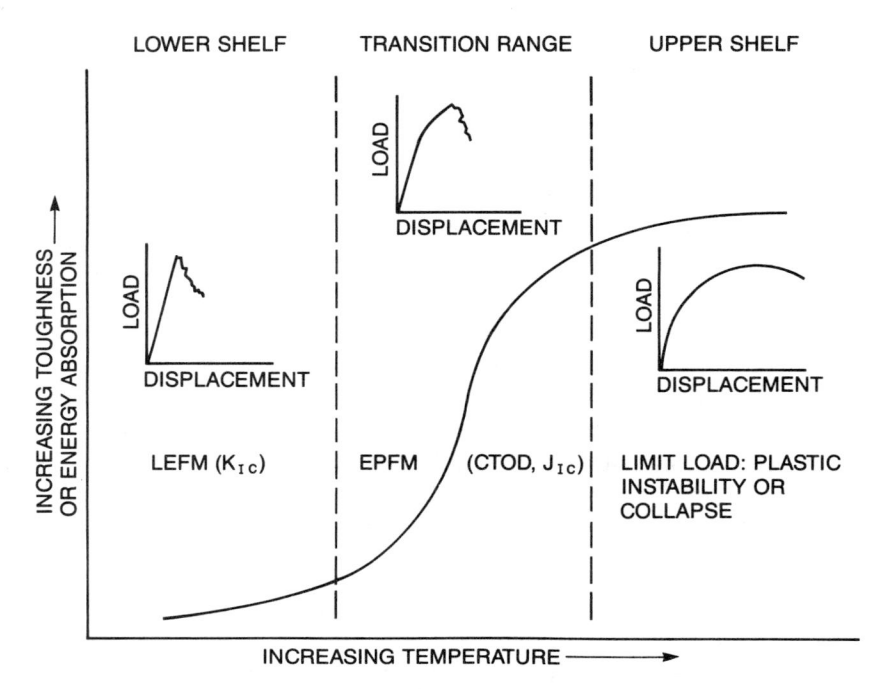

Figure 3.21 — Schematic toughness-temperature curve plotted for a bcc metal illustrating transition from lower-shelf energy absorption to upper-shelf

Also shown in the diagram are the three general regimes where typical load/displacement records for fracture toughness tests would ordinarily be analyzed on the bases of LEFM, EPFM and limit load.

tures subject to this failure mechanism easily can be assessed on a "limit load" basis, there is no reason to attempt an EPFM analysis.

Figure 3.21 is a schematic illustration that includes typical autographic records of applied load/crack surface displacement. Such records are obtained when fracture toughness tests are performed on a bcc metal over its entire range of behavior, as affected by temperature. The load/displacement records represent three distinct behaviors: brittle fracture in the lower shelf region, where LEFM analysis can provide a K_{Ic} value; elastic-plastic or yielding fracture over the transition range; and plastic instability or collapse in the upper shelf region. This schematic illustration will be a useful reference later, as various fracture toughness tests and analysis methods are discussed and their parameters are described (e.g., CTOD and J_{Ic}).

Unfortunately, many steels have a toughness transition range that includes room temperature; hence there is the frequent need for a test method that is able to decipher the elastic-plastic conditions that evolve when fracture toughness in the transition temperature range is determined. Of course, special steels are produced for low-temperature service which maintain upper-shelf toughness at lower temperatures. However, the economics of construction and manufacturing are such that many steels are used at temperatures in their toughness transition range. Therefore, the need will continue for an EPFM test that is easily executed in order to obtain a fracture toughness measure that can be translated into critical crack size, allowable stress, and other pertinent limiting quantities — just as can be accomplished with LEFM and its K_{Ic}.

A discussion of the issues concerning development of an EPFM analytical procedure is presented in Technical Brief 14. ASTM's E-24 Committee on Fracture Testing has been intensively studying EPFM methods in order to establish their merits. Several ASTM standards have been issued, and additional ones are under consideration. The British Standards Institute (BSI) also has published documents that deal with EPFM, and these will be discussed later. Again, space limitations will not permit sufficient description to guide the selection and conduct of a particular EPFM test. Here the aim will be to discuss several noteworthy EPFM tests, to explain how fracture toughness determinations are made with each, and to indicate how results can be employed when welded steel structures are discussed in later chapters.

Crack Tip Opening Displacement Testing (CTOD)

Crack tip opening displacement as a method of testing and assessment, originated in the United Kingdom around 1960, when Cottrell and Wells, working independently, concentrated on the need for a parameter that would characterize fracture toughness under elastic-plastic conditions. Both investigators aimed to accomplish this objective with a relatively small, simple test specimen. They both observed that when a sharply-notched bend specimen was loaded in three-point bending beyond the limit of elastic deformation at the root of the notch, the notch tip blunted and the opposing faces moved apart. It was hypothesized that when stress caused the faces to open beyond a critical dimension that was commensurate with the innate toughness of the metal, a crack would initiate at the terminus of the preplaced notch. If unstable cracking ensued, this usually would not allow calculation of K_{Ic} by LEFM analysis. More often, with the lower-strength very ductile steels of interest, the stable or unstable cracking that initiated called for appraisal by some form of EPFM.

In early tests, the machined notch was sharpened with a jeweler's saw or a slitting wheel, and measurement of opening action during loading of the bend specimen was made by a paddlemeter located in the mouth of the notch. For this reason, the test originally was called crack opening displacement (COD) method. Indeed, the "COD" designation continues to persist in some current documents. However, as the test underwent further refinements, it became apparent that the single edge-notched bend specimen should have its machined notch extended by a fatigue crack to match the acuity of flaws that often were encountered in structures.

Soon after the change to the precracked bend specimen, an innovative change was made in the method of appraising crack surface displacement under load. Although measurements were made at the mouth of the machined notch with the common double-cantilever clip gauge, crack tip opening displacements were ascertained and employed in establishing a toughness value. At this point, the test designation was changed from COD to CTOD. Several methods were found feasible for ascertaining crack tip opening at given loads, but CTOD usually is inferred from clip-gauge measurements using formulae developed for this function. These formulae are based on the finding that a relationship exists between notch mouth opening and CTOD, because of specimen rotation about hinge positions at specific distances along the still-solid ligament ahead of the crack tip.

British Standard 5762

The British Standards Institution issued BS 5762 in 1979 to provide methods for conducting the CTOD test. The test is conducted with a single-edge-notched, fatigue-cracked bend specimen which has been slowly loaded via three-point bending. Equivalent results can be obtained with a compact tension specimen.

The CTOD test is geometry dependent; therefore, a decrease in specimen thickness will reduce constraint and increase CTOD. For this reason, the test specimen thickness should be the full thickness of the material in order to obtain results that are indicative of the material or member of interest. For brevity, further description will be confined to testing with a bend specimen of full-thickness, since this is regarded as the most practical type for determining CTOD.

For a loaded CTOD bend specimen, BS 5762-1979 identifies four possible points on load/displacement records that can be useful, and selection as to the significant cri-

---

TECHNICAL BRIEF 14:
Developing an Analytical Procedure for Elastic-Plastic Fracture Mechanics

Development of an EPFM analytical procedure has been exceedingly difficult, because attempting to extend an existing crack under conditions involving some yielding or plastic flow is a complex undertaking. It is not merely a matter of creating new surfaces by displacement — that is, simply opening the crack tip. The extreme tip of the existing crack, albeit freshly formed by fatigue, can be curved, rounded, or wedge-shaped when observed on a microscopic scale. Briefly, various views are held on how to measure the crack tip and secure numbers as a basis for analysis. Also, there is disagreement among investigators on the point to use as the terminus of the crack in calculations. The problem is whether to use the exact position of the original crack tip, or to use an *effective crack size*, which for calculation purposes is regarded as being located a small distance ahead of the original crack terminus by a dimension called the *plastic adjustment factor*.

One view concerning the crack tip and its initial forward movement has been prompted by an observation of a *plastic strip zone* (also called the *stretch zone*), which can be seen by fractography. This concept views the absolute initiation of crack tip opening as a displacement by nonlinear yielding concentrated within a very narrow zone or strip, perhaps only 25 microns (0.001 in.) in width, extending directly ahead of the crack tip. Movement across this zone is believed to be caused by Poisson's ratio contraction rather than by one of the familiar fracture mechanisms. Evidence of this unusual form of crack extension appears as an amorphous fracture texture on the region between the original preplaced fatigue crack tip and the line of demarcation, where crack growth obviously is proceeding by cleavage or by microvoid coalescence. Figure 3.22 is an SEM fractograph that clearly shows the presence of an intermediate stretch zone located between the fatigue crack tip and the crack growth by the cleavage fracture mechanism. There are additional differences in views held by investigators regarding assessment of crack tip location, the start of crack extension, and the nonlinear deformation which accompanies crack growth. Although these differences may seem of small importance, they influence numerical analysis to a significant degree.

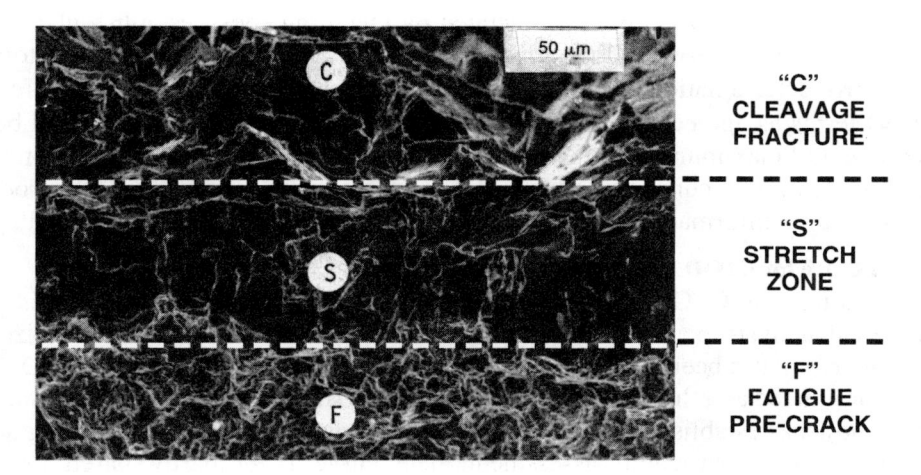

Figure 3.22 — Stretch zone formed in fracture just prior to cleavage propagation as revealed in SEM fractograph
C-Mn-Nb microalloyed steel bend test specimen 10 mm or 0.394 in. thick, pre-cracked by fatigue and tested per ASTM E 399 or E 616 for specimen of SE (B), (L-T) type. (Unetched specimen; original magnification: 650X; provided courtesy of The Welding Institute, England)

terion of the test is determined by the parties involved. The significant values of CTOD correspond to elastic, plane-strain conditions, and to measurable amounts of plastic deformation. The symbols used in BS 5762 and explanation of these significant events in load/displacement records are:

δ = An undefined value of CTOD,

δ_c = CTOD, either at brittle fracture initiation, or at the onset of pop-in without prior slow (ductile) crack growth,

δ_i = CTOD at point of slow crack growth initiation,

δ_m = CTOD at first attainment of maximum force level, and

δ_u = CTOD at point of brittle fracture (either unstable cracking or pop-in), but preceded by some slow (ductile) crack growth. Both δ_i and δ_u to be reported.

CTOD fracture toughness values are calculated using the test specimen dimensions along with particular values of the force and displacement on the load/displacement plots. CTOD is intended to be a test of fracture initiation, as is K_{Ic}; but CTOD is designed to produce a quantitative assessment of initiation under conditions of greater plastic strain. BS 5762 acknowledges this by recommending R-curves (CTOD versus slow crack growth) using multiple specimens, and then extrapolating the crack growth curve back to zero, which should represent the threshold of initiation. This would estimate CTOD at the original crack tip prior to stretch-zone development — before any ductile tearing initiated. The entire range of fracture initiation circumstances can be surveyed by the test; that is, from completely brittle fracture to fully plastic collapse.

If conditions, especially those dictated by metal thickness, establish plane strain, then the test is analyzed and reported in K_{Ic} terms. If thickness or other factors preclude a K_{Ic} determination, then the fracture toughness is reported in one or more of the CTOD terms as agreed upon by the concerned parties. CTOD is not applied beyond attainment of maximum load, because this is the threshold for application of limit load analysis. However, current work is being conducted on a modified CTOD procedure using R-curve information to predict maximum nominal stress and plastic collapse.

Design Curve for CTOD Assessment

Although the CTOD approach to structural assessment can be accomplished by numerical analysis to arrive at limiting values for applied load, critical crack size, etc., a *design curve* has been devised to simplify the task of CTOD application. The design curve originated as a theoretical model, but much testing and structural appraisal experience now has established it as a workable, conservative means of predicting quantities involved in structural assessment. The curve is primarily based on small through-thickness cracks in a tensile stress field. Of course, many cracks in structures are buried, but analyses imply that the tolerable size of a buried crack is more than twice that of one at the surface.

At stresses up to one-half of yield strength, the design curve follows linear-elastic fracture mechanics analysis with a safety factor of two. Above one-half yield strength, the curve is based on an empirical correlation between CTOD tests and wide-plate tests. Critical CTOD results from small specimens have been correlated with failure strain and crack size in wide-plate tests. Although the factor of safety varies when stress exceeds one-half yield strength, the design curve conservatively predicts the largest sharp-terminus flaw or crack that can be safely tolerated, rather than citing a critical size. An early equation that led the way to the design curve concept is as follows:

$$\Phi = \delta E / 2\pi \sigma_{ys} \bar{a} \qquad (Eq. 3.7)$$

where

Φ	=	non-dimensional CTOD
δ	=	applied CTOD
σ_{ys}	=	0.2 percent offset yield strength at testing temperature
E	=	Young's modulus at the testing temperature
$\bar{a}$	=	half-length of a through-thickness crack

This equation is merely an elementary step into the computations that are regularly made when assessing the significance of crack-like defects in a structure. Although most of the work that confirmed the validity of the design equations and curves was conducted on ferritic steels, design curves and supporting equations for a variety of metals are shown in Figure 3.23. The curve in this figure identified as (C) is the original CTOD "design curve."

Application of the CTOD Test

CTOD testing was studied by the ASTM E-24 Committee and described in ASTM STP 668. In 1991, BSI issued BS 7448, Part One, which included coverage of K_{Ic} and J testing. In 1989, ASTM issued E 1290 - *Standard Test Method for Crack Tip Opening Displacement (CTOD) Fracture Toughness Measurement*, which was updated in 1993.

CTOD is now widely used in structural and component design, and in fitness assessment, where the acceptability of crack extension must be defined, and a judg-

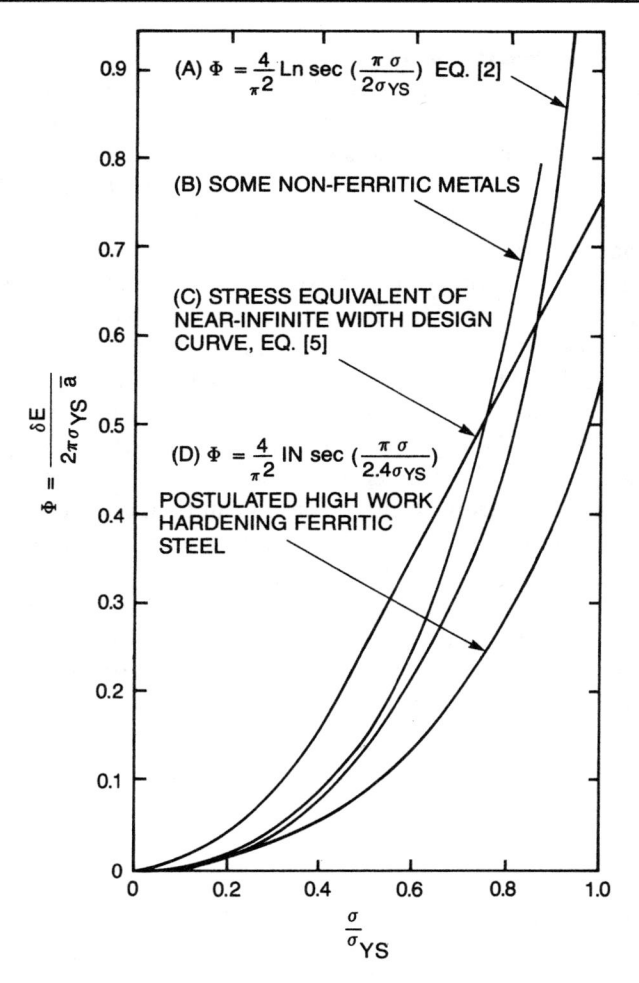

Figure 3.23 — Theoretical relationships between Φ and σ/σ_{ys} for flat plate materials in uniaxial tension
(Courtesy of The Welding Institute, England)

Curves shown:
- (A) $\Phi = \dfrac{4}{\pi^2} \operatorname{Ln\,sec}\left(\dfrac{\pi\sigma}{2\sigma_{YS}}\right)$ EQ. [2]
- (B) SOME NON-FERRITIC METALS
- (C) STRESS EQUIVALENT OF NEAR-INFINITE WIDTH DESIGN CURVE, EQ. [5]
- (D) $\Phi = \dfrac{4}{\pi^2} \operatorname{ln\,sec}\left(\dfrac{\pi\sigma}{2.4\sigma_{YS}}\right)$ POSTULATED HIGH WORK HARDENING FERRITIC STEEL

Vertical axis: $\Phi = \dfrac{\delta E}{2\pi\sigma_{YS}\,\bar{a}}$

ment made regarding allowable applied loads, tolerable flaw size, and minimum metal fracture toughness. The application of CTOD testing and analysis to weldments, however, is a venturous extension of the art. This is due to the multitude of influential conditions involving weld metal — heat-affected zones, diffusible hydrogen, residual stresses, etc. — which must considered. Despite the complexities of a weldment, CTOD is quite successful, and guidance on its application is currently available in the literature. Particularly, there is BSI PD 6493-1991, *Guidance on Some Methods for the Derivation of Acceptance Levels for Defects in Fusion Welded Joints*. Weldment assessment by fracture-mechanics methods, including CTOD, will be discussed further in Chapters 13 and 16.

Resistance Curve Test Method (R-Curve)

The resistance-curve (R-curve) method of testing, given brief mention earlier when describing the CTOD test method, has been long recognized as a means of fracture toughness evaluation, and it warrants further explanation. ASTM STP 527, issued in 1973, describes fully the R-curve method. A more current standard test procedure can be found in ASTM E 561, *R-Curve Determination*. This method predicts the stable growth of crack-like defects in metals during small-scale yielding. The R-curve itself is a graphical representation of the resistance of a particular metal to fracture during slow, stable, incremental cracking from a sharp notch that is enveloped at its tip by a growing plastic zone. This resistance to cracking is expressed as a stress intensity factor, K_R, which is related to applied load and crack length.

When *extensive* yielding occurs during testing, the R-curve is more readily expressed in terms of CTOD, or the "J-integral", which will be described next. The stress intensity factor, K, is determined by linear-elastic treatment, even though the plastic zone that develops during R-curve testing can be large enough to initiate general yielding. This treatment is justified by making a *plastic zone adjustment* (identified as *ry* in E 561), which is an addition to the physical crack size that compensates for the plastic, crack-tip deformation encompassed by the linear-elastic stress field.

Two test methods are included in ASTM E 561. One makes use of multiple specimens that are individually loaded at rising step-like levels to secure increasing amounts of notch-mouth opening. The specimens are heat tinted and broken apart to measure the

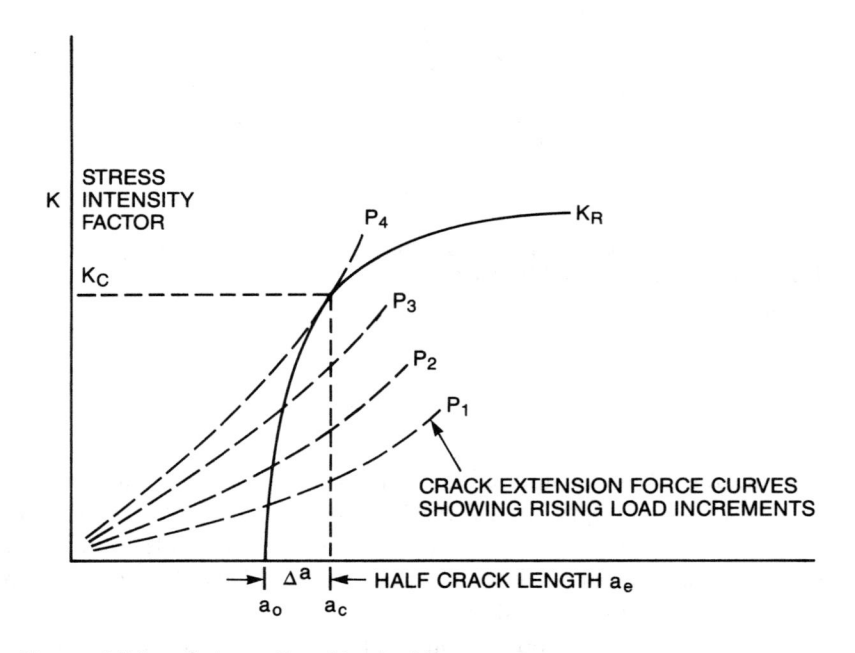

Figure 3.24 — Schematic of typical R-curve

Also included are crack-extension force curves (broken lines showing rising load increments, and labeled P1 to P4, inclusive) superimposed on the R-curve. From ASTM E 561. (Reprinted, with permission, copyright American Society for Testing and Materials, Philadelphia, PA).

extension of cracking that occurred. This is the most reliable method of securing the data needed to establish the R-curve, but metal to prepare at least five specimens may not be readily available. Therefore, a second method that employs a single specimen has been developed. This method makes use of an *unloading compliance* technique to determine the increments of cracking caused by incrementally higher loads.

During testing to develop an R-curve, the resistance to slow, stable crack extension is plotted as shown by the solid curve (K_R) in Figure 3.24. Equations similar to those employed for LEFM analysis are used for calculating K. However, an adjustment is made to account for effective extension of the crack length, because of the larger crack front plastic zone. In addition to the R-curve, dashed-line curves (labeled P_1 to P_4 inclusive) have been superimposed to illustrate the calculated crack extension force available to cause extension of the crack at a given load.

Force curve "P_4" is unique inasmuch as its tangency with the K_R curve defines the critical load or stress that exceeds the crack extension resistance of the metal and causes the start of unstable fracturing. This stress intensity is identified as K_c, and is cited as the plane-strain fracture toughness. R-curves for most steels are developed under test conditions in which the start of unstable (brittle) fracture is unlikely. The R-curve and K_c are highly dependent on metal properties, temperature, loading rate, thickness, and crack length; therefore, test conditions must be carefully adjusted to match any structural application for which the R-curve will serve in making engineering judgments. For this reason, most R-curves are developed to assist in specific applications, and have limited use as general reference.

The J-Integral Test Method

The J-integral method of fracture toughness evaluation was proposed in 1968 by Rice, et al; and it is described in ASTM STP 536. The J-integral, a mathematical expression, is a line or a surface integral that encloses the crack front from one surface to the other. It characterizes the stress-strain field which exists around the crack front but remote from the crack tip. The J-integral expression for a two-dimensional crack lying in the x-z plane, with the crack front parallel to the z-axis, is the line integral. This line integral is employed to appraise the toughness of metal at or near the initiation of slow, stable crack growth from an existing sharp, crack-like flaw. The ultimate objective of this test method is to establish the J-integral at the very threshold of crack extension. Specifically, "J" represents the *resistance* of the metal to stable cracking, just before any extension occurs, at which point the designation J_{Ic} is applied.

Although this test method has become widely recognized and praised for its potential usefulness in structural design and integrity assessment, J-integral testing is both complex and expensive. Consequently its adoption by industry has been somewhat slow. A standard method is available in ASTM E 813, J_{Ic}, *A Measure of Fracture Toughness*. Like K_{Ic} and CTOD, J_{Ic} can be employed to establish the suitability of a metal for a specific application, in which stress conditions are prescribed and flaw size will be monitored by NDT.

Specimens used for J-integral testing are usually either fatigue-cracked bend or compact-tension types, and they are designed somewhat like those included in ASTM E 399. The three-point-bend, single-edge-notched specimen is preferred. For J-integral testing, specimens must comply with a number of requirements concerning design, dimensions, and length of notch-plus-fatigue crack. Also, there is a special requirement that the specimen must permit measurement of displacement *at the load point*. This ensures that the area under the plotted load-displacement curve is the true total of combined elastic and plastic strain energy — and therefore is considered numerically equal to the potential energy available for crack extension. Thus, in an elastic solid, the J-integral is equal to the crack extension force, G. The displacement measuring point on the single edge-notched bend specimen is the middle of the front face, while a compact tension specimen must be modified, in accordance with E813, to allow a clip gauge to be attached on the load line.

When testing a J-integral specimen, the loading rate is slow, and it is either at room temperature, or some other suitable temperature. Using the three-point bend specimen, a transducer is placed on the load line to measure displacement and permit calculation of J; but, a COD gauge can be applied to the notch mouth to determine crack length using an elastic compliance technique. As load is applied to the specimen, load versus displacement is recorded autographically. The area under this curve (A) is taken as the total elastic and plastic strain energy. The J-integral usually is established with replicate specimens to produce at least four data points that must lie within specified limits for crack extension, and within minimum and maximum offset exclusion limits. Originally, J was defined by a complex analytical expression, but now J is calculated with the following equation:

$$J = 2A / B(W-a) \tag{Eq. 3.8}$$

where

B = specimen thickness
W = specimen depth (width)
a = length of notch-plus-fatigue crack
A = area under load-versus-displacement curve (or total energy absorbed by specimen)
J = J-integral expressed as K Joules/m² (in. lb/in.²)

The value of J calculated for each specimen with the above equation is plotted against crack extension (Δa), as shown by the diagram in Figure 3.25. Standard ASTM procedure requires a power law curve through the data points. This curve then is extended leftward as a regression curve (J_R).

As stress is applied to the specimen at the start of the test, the first effect prior to any crack extension is *crack tip blunting*. Also plotted in Figure 3.25 is a *blunting line*, which is a line that approximates apparent crack advance due to crack tip blunting in the absence of stable, slow crack growth. The blunting line is based on the assumption that a pseudo-crack advance (Δa_B) should be considered, which is equal to one-half of the crack-tip opening displacement. Data for plotting this line are calculated with the following equation, which considers the effective yield strength of the metal:

$$J = 2 \Delta a \; \sigma y \tag{Eq. 3.9}$$

where

Δa = increase in crack length
σy = effective yield strength (an assumed value of uniaxial yield strength, as modified by E 813)

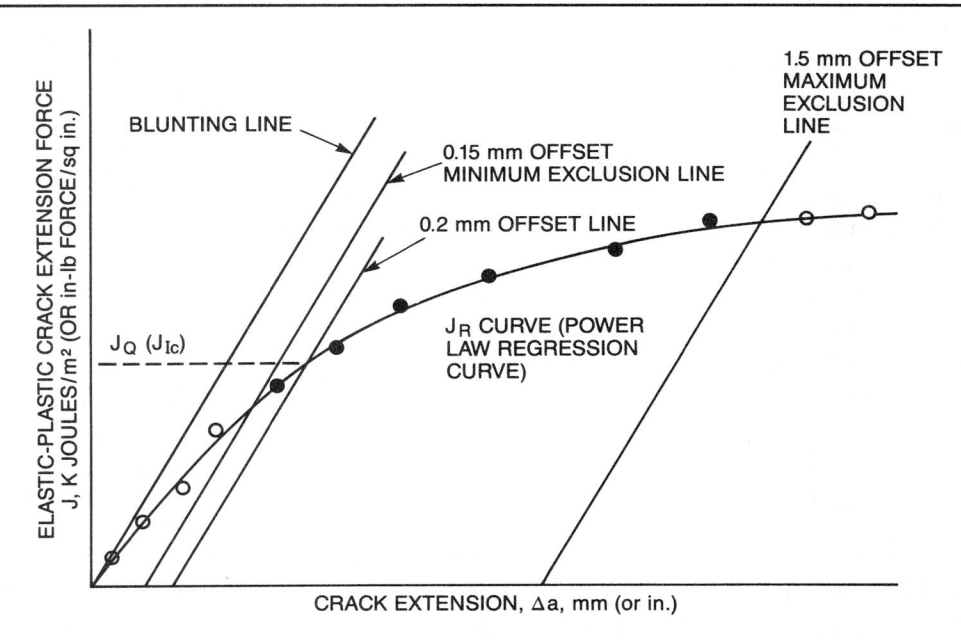

Figure 3.25 — Typical results obtained from J-integral testing, but plotted without all refinements and validation requirements of ASTM E 813

The crack-tip blunting line indicates the initiation of stable, slow-growth cracking or tearing at the point where it intercepts the regression curve, J_R. The standardized J-integral procedure in E813 requires an offset line parallel to the blunting line to be set at 0.2 mm (0.008 in.), as shown in Figure 3.25. This small amount of stable, slow crack extension is used to establish J_{Ic}. Actually, a series of steps must be closely followed, using provisional or qualifying J values such as J_Q. Only after the validity of all data becomes confirmed is a J_{Ic} accepted.

The J values employed in this test method probably should consider the effects of crack extension, but because the amount of extension allowed is relatively small, the inaccuracies in J caused by crack growth also are small. If a more accurate J_{Ic} must be determined, an equation for refining the crack growth resistance curve is available in ASTM E 813.

A distinct advantage of the J-integral test method is its capability of producing valid results while accomodating a certain amount of crack-tip blunting and plastic deformation in the specimen. When the amount of plastic deformation is quite small, J_{Ic} is identical with G_{Ic}. Therefore J_{Ic} can be used to estimate the plane-strain toughness of a metal where circumstances hinder testing by ASTM E 399 to determine K_{Ic}. This assumes that a valid K_{Ic} for the given metal exists. A valid J_{Ic} can be converted to an estimated K_{Ic} using the following equation:

$$K_{Ic}^2 = J_{Ic} E / (1-v^2) \qquad \text{(Eq. 3.10)}$$

where,
  E = Young's modulus
  v = Poisson's ratio

In essence, the J_{Ic} may be established by applying a nonlinear-elastic form of analysis rather than laboring through the complexities of nonlinear plastic analysis. To some investigators, the J_{Ic} analytical procedure appears quite justified. This is because the amount of plastic deformation is only a little more than normally accommodated in LEFM analysis, and it is less than the ductile crack extension that usually occurs in CTOD testing. Therefore, investigators reason that despite the relatively small size of the specimen and the utilization of a small crack extension to establish J_{Ic}, it is a valid fracture-mechanics determination and is acceptable for conversion to K_{Ic}. Other investigators do not agree that restriction of the J_{Ic} determination to a small, stable crack extension in the elastic-plastic region validates the J_{Ic} analysis. They have pointed out that the ductile metals for which J_{Ic} ordinarily might be determined are likely to receive a very conservative estimated K_{Ic}, using the aforementioned estimation procedure. Nevertheless, the J_{Ic} test method is an approach to an estimated K_{Ic} that allows use of a much smaller specimen. This alone can be a valuable advantage.

A single test specimen can be used to determine J_{Ic}, using the unloading compliance technique mentioned earlier under R-curve testing. A suggested method is outlined in ASTM E 813. This technique is receiving increased attention because of the speed with which computerized unloading compliance testing can be accomplished.

Determining J-R Curves

The determination of a *J-R Curve* is a relatively new application of the J-integral concept. Although it is a logical extension, it probably was prompted by the debate over the allowance of some small, stable crack extension in J-integral testing. Determination of J-R curves simply amounts to extending the J-integral test to conditions of larger crack extensions. The J-R curve is intended to be applicable to the stable, slow crack growth resistance, by a manner of testing that is not affected by geometry. Higher values of J_{Ic} can be measured on the R-curve. An increase in R with increased extension of the elastic-plastic crack, defined in J_{Ic} terms, is considered a metal property; that is, a measure of fracture toughness. Steels of high R (upper-shelf) are expected to develop larger increases in J_{Ic} with crack extension, while steels of low R (lower-shelf) are likely to develop relatively small increases in J_{Ic} with crack extension.

One standard method — ASTM E 1152, *Determining J-R Curves* — designates the J-R curve as a plot of the far-field J-integral versus crack extension. However, it is recognized that the far-field value for J may not represent the local stress-strain field of a growing crack. A detailed procedure for determining the J-R curve is provided by E1152 using a single test specimen along with the unloading (elastic) compliance technique. The J-R curve is reported to be useful as an index of metal toughness, and when established with the three-point bend specimen, the curve defines the lower bound estimate of J-capacity as a function of stable crack extension. Thus the J-R curve can be used to assess the stability of cracks in structures. As these rather complex determinations and assessments are made, there must be an awareness of the differences that can exist between laboratory test results and the conditions in a structure. The practical application of J-R curves to weldments will be reviewed in Chapters 13 and 16.

Fatigue Cracking Assessment by Fracture Mechanics

When metal failure by fatigue was discussed earlier, and the complexities of predicting crack growth by fatigue were explained, fracture mechanics was mentioned as a potential tool for this difficult problem. Following review of LEFM and EPFM testing and analytical methods for determining propensity to various kinds of cracking under sustained load, procedures for calculating *fatigue crack growth rate* now can be outlined. In general, fatigue is concerned with structural members that already contain crack-like defects, and the extension of these by cyclic stresses.

Paris and Erdogan first suggested that the crack-tip stress-intensity factor, K, as defined by linear elasticity, provides a geometry-independent value that represents the driving force for fatigue crack growth. Therefore, the stress intensity concept could be used as a basis for quantitative determination of the fatigue crack propagation rate. To repeat a pertinent point, fatigue cracks usually can be assessed on a linear-elastic basis This is because most existing flaws and their fatigue-cracked extensions initially are small, and the plastic zone surrounding their tip likely will be of a size that does not preclude the use of K_{Ic} from LEFM analysis.

During the earlier preliminary discussion of fatigue, emphasis was placed on the fact that fatigue cracking could initiate and propagate at nominal stresses that were well below the yield strength of the metal. Now, in the context of appraising stress intensity at the tip of a sharp flaw or crack, the inclination for crack growth at seemingly low levels of nominal stress is understandable. In fact, crack growth can occur under cyclic loading at a stress intensity (K), substantially below the critical stress intensity (K_c). Some alloy steels display fatigue crack propagation at stress intensities as low as five percent of their K_{Ic}.

After some study of cyclic loading from the standpoint of stress intensity factor, it was established that the relevant condition for use in assessing fatigue crack growth is the range of stress intensity, and this was assigned the designation ΔK. Therefore,

$$\Delta K = K_{max} - K_{min} \qquad (Eq.\ 3.11)$$

where K_{max} and K_{min} are the values of K calculated for a crack of a given size at the upper and lower limit of stress during cyclic loading.

A standard test method for determining fatigue crack growth at medium to high rates (i.e. above 10^{-8} m/cycle or 3.9×10^{-7} in./cycle) is found in ASTM E 647, *Constant-Load-Amplitude Fatigue Crack Growth Rates Above 10^{-8} m/cycle*. This ASTM method covers the determination of fatigue crack growth rate by linear-elastic analyses, and can be used for applications where fatigue lives up to a million cycles are anticipated. Longer fatigue lives usually involve slower growth rate. Thus, it may be necessary to employ a special fatigue testing procedure because there is a stress intensity threshold for cracking below which cracks will not grow. Testing in accordance with E 647 is performed at room temperature in an air atmosphere. If fatigue crack growth rate must be ascertained for a another gaseous, or a liquid environment, and at a temperature other than RT, then the testing conditions can be adjusted. Indeed, this often is done.

Although any suitable fatigue-cracked specimen may be used, regardless of design and size, E 647 proposes the standard use of either a compact-type tension specimen or a center-cracked tension specimen. A specimen may be of any thickness and size, providing the test specimen contains a through-thickness crack and remains predominantly elastic during testing.

While LEFM analysis of fatigue cracking was stated to be geometry-independent, uncertainty has existed for some time on the influence of thickness on crack growth rate, since mixed results had been reported. Recent studies show that substantially thicker structural steel members will have significantly higher crack growth rates (i.e. lower fatigue strength, or shorter fatigue life). Therefore, the potential influence of thickness on performance under fatigue conditions may require an investigation designed for a particular application involving thick members.

Conducting a fatigue test in accordance with E 647 requires constant amplitude loading, with an optimal load ratio (R) selected to provide results that conform to the application. The loading can represent any desired stress ratio, but the specimen must be able to accommodate the imposed cycle without buckling. Consequently, compression loading is not often included, unless the specimen has adequate cross-section to display a uniformly distributed elastic response. E 647 recommends that each specimen be tested at a constant load range (P) and at fixed loading conditions. Other details concerning test specimens and testing procedure are included in E 647. An important detail during the fatigue test is measurement of fatigue crack growth rate, so that a precise value is obtained as a function of cycles and the stress intensity factor range. Measurement can be done either visually (usually with a microscope) or by a nonvisual technique, such as using electric potential difference. Methods for measurement of fatigue crack length and shape can be found in ASTM STP 738.

Because of the variability which is regularly encountered in fatigue testing, an appropriate number of replicate tests are suggested to substantiate values and to indicate the width of a scatterband. If replicate testing is not feasible, E 647 suggests that tests be conducted to obtain data from overlapping regions of cracking-rate versus stress-intensity. As crack extension measurements, designated as "a," are made at selected stress cycles, "N," these data are analyzed by a multi-step procedure and reduced to the format shown below using any one of several available numerical methods:

$$\log da/dN \text{ versus } \log f(\Delta K) \qquad (Eq. 3.12)$$

here, da/dN is crack extension during n^{th} stress cycles, and $f(\Delta K)$ is the range of stress intensity factor.

E 647 recommends use of either the secant *(point-to-point) method* or the *incremental polynomial method*, both of which are detailed in the Standard's appendix. Briefly, the secant method is simply the slope of a straight line connecting two adjacent data points on the "a" versus "N" curve, with the rate computed in terms of da/dN. The average crack length normally is used to calculate ΔK. The incremental polynomial method for computing da/dN involves use of basic regression equations and fits a second order polynomial expression (parabola) to approximately six adjacent data points. The slope of this expression is the growth rate. These two methods provide approximately equivalent results, except that scatter of individual da/dN values is likely to be somewhat greater when using the secant method. Plotting is done with the independent variable ΔK on the abscissa, and the dependent variable da/dN on the ordinate. Log-log coordinates are normally used.

Figure 3.26 shows typical fatigue crack growth behavior for a structural steel, when determined in accordance with E 647 and tested in air at room temperature. The da/dN versus ΔK log-log plot has a somewhat sigmoidal shape that is characteristic of structural steels in the medium strength range. However, the crack growth rates at certain levels of stress intensity would depend upon the metal. The width of the scatterband would depend not only on the metal, but also the precision of the testing procedure.

The plotted curve in Figure 3.26 has three distinct regions. Region 1 includes an approximate range of stress-intensity factors, below which fatigue cracks do not prop-

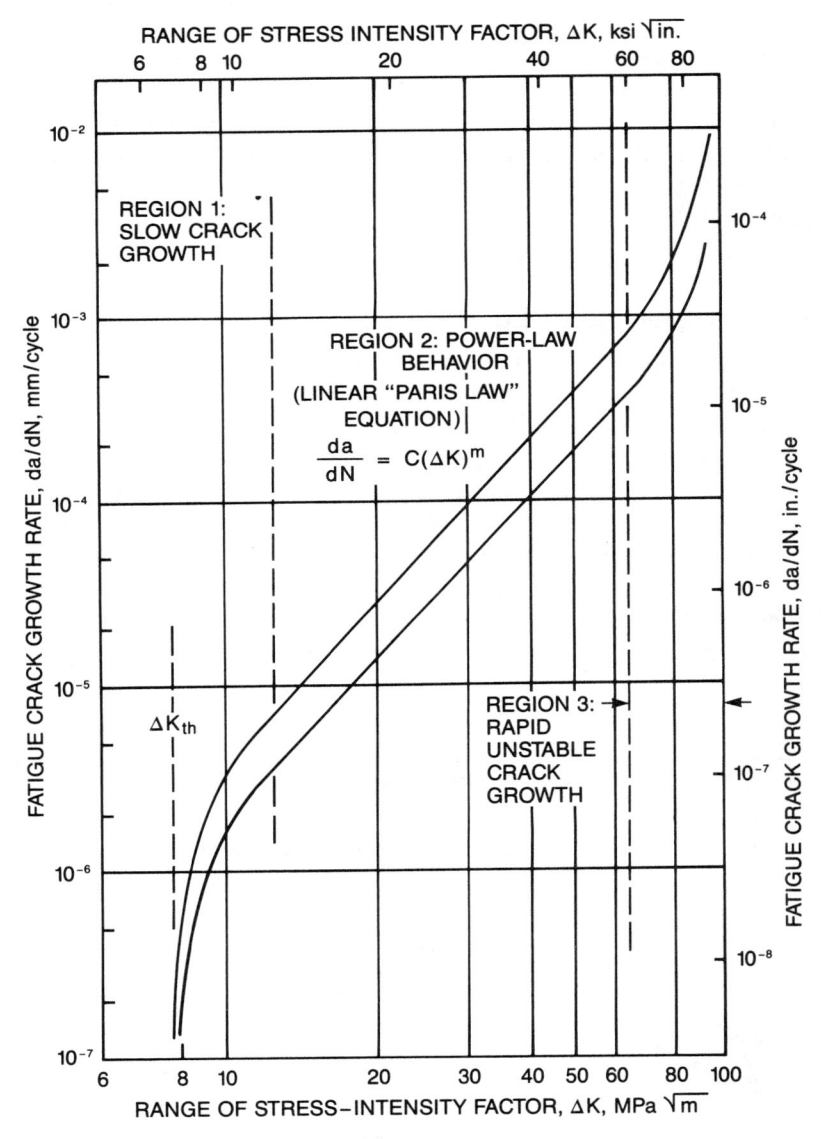

Figure 3.26 — Typical fatigue crack growth rate behavior for steel showing expected scatter band for results

agate. This is designated the threshold value (ΔK_{th}). Just above this threshold, cracks grow slowly until the intermediate Region 2 is reached. Here, cracks grow in conformance with a power law, which often is referred to as the Paris Law equation:

$$\frac{da}{dN} = C(\Delta K)^m \qquad \text{(Eq. 3.13)}$$

where C and m are metal constants — although not in the strictest sense of the term. Values for C and m usually must be defined experimentally because they vary with a number of principal conditions including environment and stress range. Finally, the upper end of the crack growth rate curve enters Region 3, wherein a marked acceleration in growth rate occurs. This may be either because K_c for the metal has been exceeded, or because gross plastic deformation is occurring.

An obvious limitation of fracture mechanics for assessing fatigue crack growth is the restriction to the intermediate region where the power law applies. There is good reason to avoid extending fracture mechanics analysis to higher values of $\frac{da}{dN}$ and ΔK, because fatigue life could be grossly overestimated. However, for low-strength metals it appears likely that gross yielding would warn of invalid use of fracture mechanics at excessively high levels of ΔK. Likewise, the occurrence of plane-strain (K_{Ic}) fracture would be the warning sign with high strength metals.

Mechanical Properties at Low Temperature

Lowering the temperature of metal profoundly affects fracture behavior, particularly metals with the body-centered cubic crystalline structure. Strength, ductility, toughness, and other properties are changed in all metals when they are exposed to temperatures near absolute zero. The properties of metals at very low temperatures are of more than casual interest, because welded pressure vessels and other equipment are expected to operate satisfactorily at levels far below room temperature. Moderate sub-zero temperatures are imposed on equipment for dewaxing petroleum and for the storage of liquified fuel gases. Much lower temperatures are involved in cryogenic service, which entails the storage and use of liquified industrial gases such as oxygen and nitrogen. Down near the very bottom of the temperature scale, there is a real challenge

Table 3.5 - Temperature Conversion Scales and Fixed Points of Interest

Temperature (Degrees)			
°F	°K	°C	Physical Changes in Matter
-459	0	-273	Absolute zero: All motion of atoms ceases (-459.69° F)
-452	4	-269	Helium boils
-423	20	-253	Hydrogen boils
-321	77	-196	Nitrogen boils
-305	86	-187	Fluorine boils
-297	90	-183	Oxygen boils
-108	195	-78	Dry ice (CO_2) sublimates
-36	235	-38	Mercury freezes
32	273	0	Water freezes
75	297	24	Room temperature
212	373	100	Water boils
449	505	232	Tin melts
621	600	327	Lead melts
675	630	357	Mercury boils

for metals that are used in equipment for producing and containing liquid hydrogen and liquid helium, because these elements in liquified form are increasingly important in new technologies. Liquified helium is only slightly above *absolute zero*, which is -273.16° C (-459.69° F).

Absolute zero is the theoretical temperature at which matter has no kinetic energy, and atoms no longer exhibit motion. Man has yet to cool any material to absolute zero, so it is not known how metals will behave when cooled to this boundary condition. Absolute zero has been approached within a millionth of a degree, with the aid of special magnetic techniques, but absolute zero continues to be just beyond reach. Nevertheless, some amazing new phenomena have been found with metals at temperatures just above absolute zero. It is likely that welded equipment will continue to be used for service at this extremely low temperature. While the solid state physicist continues to struggle to remove the last bit of thermal energy, the metallurgist must contend with problems caused by changes in metal properties under cryogenic conditions.

This realm of cold temperature for welded structures will seem more familiar if the points of special interest listed in Table 3.5 are noted. Throughout this text both the Celsius and the Fahrenheit scales have been used to indicate temperature, because of their use by metallurgists and welding engineers. Discussion of subjects related to cryogenics frequently involve the *absolute temperature scale*, which is graduated in degrees Kelvin (°K), and which corresponds with the Celsius temperature scale in the following manner:

$$°K = 273.16 + °C \qquad \text{(Eq. 3.14)}$$

Table 3.5 shows how the three temperature scales compare and provides practical information that mark changes in the state of familiar materials.

Strength at Low Temperature

As temperature is lowered over the range indicated in Table 3.5, the atoms of an element move closer together by dimensions easily computed from the coefficient of thermal expansion. A number of understandable changes occur as a result of this smaller lattice parameter. The elastic modulus, for example, increases. In general, the tensile and yield strengths of all metals increase as the temperature is lowered. Some exhibit a small increase, while others show marked increases. For instance, if the temperature of low-carbon steel is lowered from room temperature to that of liquid helium, a five-fold increase in yield strength can occur. The increased strength of metals at lower temperatures is attributed to an increase in resistance to plastic flow. The actual increase, therefore, would depend on the manner in which lowered temperature opposes the movement of dislocations in the crystalline structures and hampers slip and twinning motions. Since plastic flow by these mechanisms is strongly dependent upon the nature of the crystalline structure, it would be logical to expect metals with the same kind of structure to react in a similar manner; and this is, in general, the case.

Mechanical property testing of metals at low temperatures is more time-consuming and costly than testing conducted at room temperature. Consequently, far less data on low-temperature properties are available in the literature. Testing for tensile and yield strengths is conducted mainly to obtain minimum values for design purposes in a particular application. Figure 3.27 gives a general perspective of how strength is

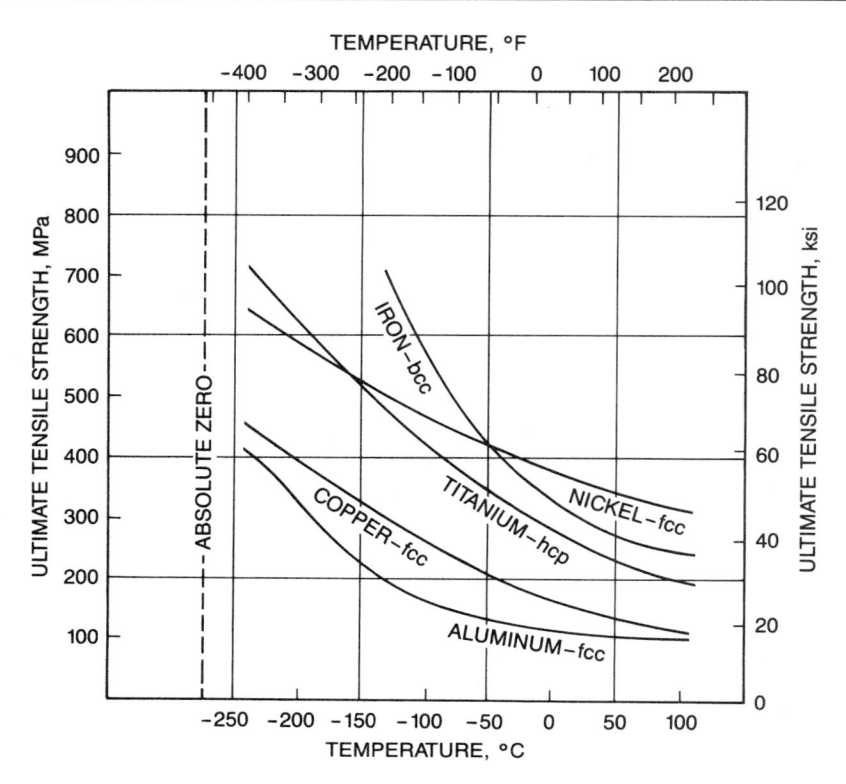

Figure 3.27 — Effect of low temperatures on the tensile strength of five metals representing three basic kinds of crystalline structures

altered in five familiar metals as temperature is lowered. The curves for tensile strength show that construction metals having a bcc or a hcp crystalline structure undergo a more rapid increase in strength with decreasing temperature than found in the fcc metals. This is to be expected, because the bcc and the hcp structures with their limited deformation systems have a greater temperature dependency.

The data in Figure 3.27 have been obtained from tension tests made with smooth specimens. Unfortunately, the attractive high strengths that the bcc and hcp metals exhibit at very low temperatures seldom can be utilized in structures, particularly in the form of weldments, because of fracture-toughness deterioration. The addition of alloying elements to the bcc and hcp metals can help offset this brittle fracture tendency, but the use of bcc or a hcp metals still requires careful selection and toughness testing. Fortunately, metals with an fcc crystalline structure (including the austenitic high-alloy steels) retain good toughness to very low temperatures. Although possibly at a disadvantage from a strength standpoint, they serve very well in cryogenic applications.

Impact Toughness at Low Temperature

Because toughness tends to decrease as temperature is lowered, especially in the ferritic (bcc) steels, much testing is conducted in industry to measure and to monitor this property. The most frequently used test specimen is the notched-bar impact specimen, despite the shortcomings mentioned in earlier discussion. The popularity of impact testing may be attributed in part to its long-established position in standards. Another reason is the convenience of small test specimens, which can be cooled easily to the test temperature. Also, the popular Charpy V-notch specimen may be quickly inserted into the test machine and broken to obtain a result for the specified temperature.

Table 3.6 contains representative notched-bar impact energy absorption for the five metals used above to illustrate tensile strength. Note that the fcc metals actually improve as temperature decreases — at least, down to -196 °C (-320 °F). The bcc and hcp metals, on the other hand, suffer a marked reduction. Even though impact test values cannot be used in engineering computations concerning the design or load-carrying capability of a particular structure, per se, they can serve well when selecting a tough metal from among a group of available candidates. They are also useful when testing to show compliance with a code or specification that includes impact test requirements, or when monitoring toughness with a quality control program.

A reason for further discussion here of the notched-bar impact test is to point out additional determinations that can be made with these comparatively simple specimens, which are regarded by some investigators as providing better criteria for judging toughness than does energy absorption. Also, mention should be made that Charpy specimens can be reduced from the familiar 10×10 mm (0.394 in.) square bar to relatively thin metal. Energy absorption acceptance values have been published for V-notch specimens of ¾, ⅔, ½, ⅓, and ¼ of the 10 mm thickness, with the width at the usual 10 mm (0.394 in.).

Energy Absorption in Impact Testing

Energy absorption has been analyzed in a number of different ways to evaluate toughness and to gain unanimity in decisions regarding acceptance. While minimum energy absorption often is specified, it must be recognized that typical values differ quite widely for different metals. For example, nickel was shown in Table 3.6 to have a much higher C_v value at RT than does aluminum, copper, and titanium. Despite the high energy absorption of nickel, we should not disallow the use of other metals with

Table 3.6 - Notched - Bar Impact Values for Metals Tested at Low Temperatures*

Temperature		Aluminum[1]		Copper[1]		Nickel[1]		Iron[2]		Titanium[3]	
°C	°F	joules	ft. lbf	joules	ft. lbf	joules	ft. lbf	joules	ft. lbf	joules	ft. lbf
25	75	27	20	54	40	122	90	102	75	20	15
-18	0	27	20	57	42	125	92	41	30	18	13
-73	-100	30	22	60	44	126	93	3	2	15	11
-129	-200	33	24	63	46	128	94	1	1	12	9
-196	-320	37	27	68	50	129	95	1	1	10	7

Notes:
*Charpy V-notch specimens tested in accordance with ASTM E 23 to secure energy absorption values
1. Face-centered cubic crystalline structure.
2. Body-centered cubic crystalline structure.
3. Hexagonal close-packed crystalline structure.

lower toughness. Each useful metal deserves study to ascertain its energy absorption capacity and to judge whether it is adequate for intended applications. If not, ways possibly can be found to improve toughness.

When analyzing the C_v performance of structural steel plate used in ships, investigators found that steels which had a C_v below 14 J (10 ft-lbf) were quite susceptible to brittle fracture initiation. They also observed that when C_v exceeded approximately 28 J (20 ft-lbf), the plate usually possessed sufficient toughness to arrest a running crack. These findings, along with other supportive data, prompted the inclusion of an impact test requirement in some steel plate specifications calling for a minimum C_v of 20 J (15 ft-lbf), as an assurance against brittle fracture initiation. Energy absorption for these steels was profoundly influenced by temperature, and therefore the specifications included the temperature at which a stated minimum C_v was required. For example, if steel for a storage tank was likely to be exposed to temperature as low as -8°C (20°F), the specification then might require a minimum C_v of 20 J (15 ft-lbf) at that temperature. For some steel pressure vessels, where "leak-before-break" was to be assured, the steel plate specification likely would require a higher minimum C_v at a stated temperature that was commensurate with the anticipated low service temperature.

Transition Temperature for Energy Absorption

Transition temperature provides somewhat similar criteria for analyzing C_v test results. This method requires C_v tests over a range of temperatures — from a relatively high level at which the metal exhibits its best toughness, down to a low temperature at which cleavage can initiate. As illustrated earlier in Figure 3.9, when absorption energy is plotted against temperature, metals with a bcc crystalline structure undergo a precipitous drop in energy over a relatively narrow mid-range span of temperature. This drop in energy absorption starts when some cleavage occurs during fracture — an occurance which usually can be confirmed by the appearance of some crystalline texture in the fractured faces of the broken specimen.

References already have been made to the upper shelf and the lower shelf of this typical energy-temperature transition curve of bcc metals, as well as the fracture character in ferritic steel at these two levels. Much attention has been given to analyzing this toughness transition behavior, because the transition usually is not sharp and some accepted procedure must be followed. A common method for defining the transition temperature is to find the temperature at which an arbitrarily selected amount of fracture energy is exhibited. For example, a C_v of 20 J (15 ft-lbf) may be selected as a deciding criterion. This value may be required at -20°C (-4°F), or below, which is an arbitrary selection. Nevertheless, criteria are thereby clearly set for judging C_v test performance.

The selection of energy-temperature criteria that intersect on the transitional slope of the energy absorption curve is a speculative undertaking, because the curve should be bordered above and below by scatterband limits. The scatterband width ordinarily is greatest along the transition slope. Sometimes the symbol $T_è$ is used in the literature to define a temperature that intersects the transition slope, but there is no general agreement on exactly what the symbol means. Of the various methods for defining $T_è$, the one most often used is to compute the mean energy absorption between the upper shelf and the lower shelf. Thus $T_è$ can be the average energy absorption between the values found for completely ductile and completely brittle fracture. Other symbols appear in the literature which use the letter "T" for transition temperature and a subscript to indicate a particular means of definition. Care must be exercised in interpreting these symbols.

There is another method of evaluating C_v test performance to reflect the possible behavior of a structure of the same metal under comparable conditions of temperature, dynamic loading, and notch-effect. This method calls for selecting an energy absorp-

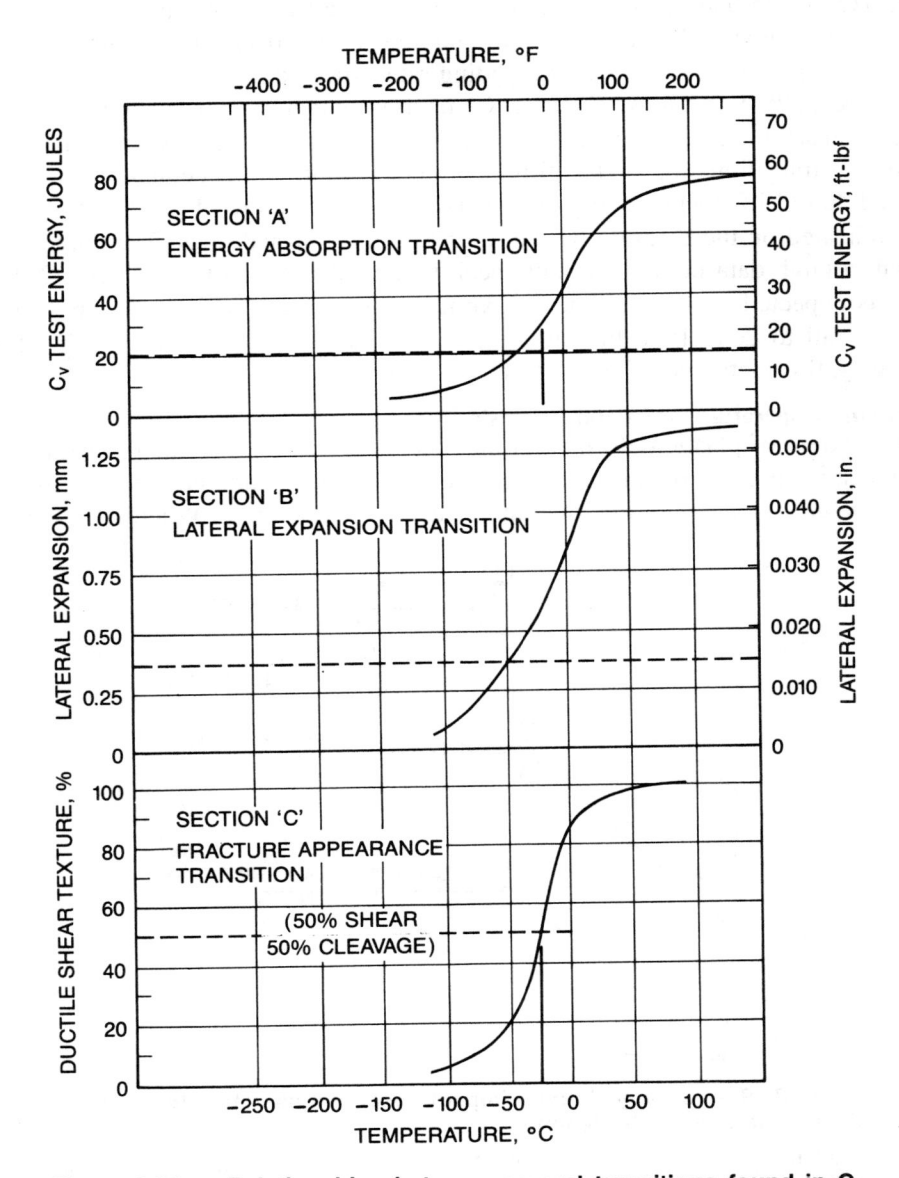

Figure 3.28 — Relationships between several transitions found in C_v impact test parameters with changes in temperature for a low-carbon steel

tion minimum and a temperature maximum that lie on the upper shelf. Although a few specifications have been written that require upper-shelf C_v test performance to a specified maximum temperature, most specifications require firmly defined energy-temperature limits. For example, American Association of State Highway and Transportation Officials (AASHTO) requires steels used in the primary tension members of bridges to meet toughness requirements expressed as minimum C_v energy absorption at specified testing temperatures. A lengthy tabulation is provided by AASHTO, which lists steels according to their strengths and outlines three ranges of service temperatures. The minimum C_v required is based on the fracture toughness which corresponds to the maximum loading rate expected in service.

Figure 3.28, Section A, shows a C_v energy-temperature transition curve for a low-carbon steel that has typical upper and lower energy shelves, along with the interconnecting transition slope. The aforementioned example of acceptance criteria — a required 20 J (15 ft-lbf) C_v energy at a maximum temperature of -20°C (-4°F) — has been indicated on the diagram by the intersection of dashed lines. The energy curve (drawn through data points that have been removed) shows a C_v of 30 J (22 ft-lbf), which is expected at the stipulated maximum temperature. In fact, the predicted C_v does not fall to 20 J (15 ft-lbf) until the temperature drops to about -45°C (-50°F). Obviously, this metal would be judged to have an acceptable C_v.

Transition Temperature for Lateral Expansion

The extent of plastic deformation that occurs in the C_v specimen's cross-section during testing also is a quantifiable value, and this feature undergoes a marked transition in the bcc metals with the lowering of testing temperature. When a C_v specimen

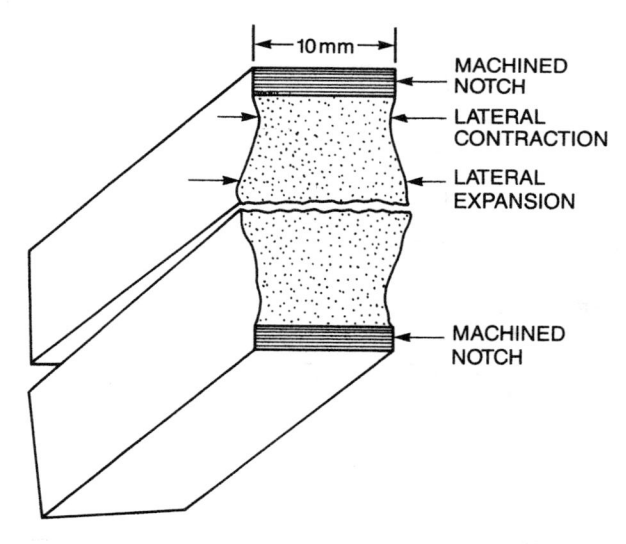

Figure 3.29 — Charpy V-notch impact specimen sketched to illustrate significant details after testing

Fractured surfaces are positioned back-to-back to show lateral locations where contraction and expansion have taken place during testing. Measurement of these features can supplement energy-absorption values when evaluating notch-toughness.

is broken, a small amount of lateral contraction ordinarily occurs across the width, close to and parallel with the root of the notch, as shown in Figure 3.29. Conversely, expansion should occur across the width opposite the notch. Both changes in dimension from the original 10 mm (0.394 in.) width of the specimen are easily measured, and both dimensional changes are indicators of ductility in the presence of a notch. The extent of lateral expansion opposite the notch is the value presently favored for appraising the capacity of metal to flow plastically during fracture under impact loading. This is considered an acceptable means of quantitative toughness evaluation.

A useful document that includes a requirement for lateral expansion is ASTM A 20, *General Requirements for Steel Plate for Pressure Vessels*. This specification covers a variety of carbon and low-alloy steels, and provides general guidelines on microstructural features and mechanical properties, including many aspects of C_v impact testing that affect test results. It outlines acceptance criteria for C_v testing in six classes of steels differing in tensile strength, or deoxidation during steel making, or both. In addition to providing a minimum C_v for a number of steels in five of the classes at various test temperatures, ASTM A 20 recommends a minimum lateral expansion of 0.38 mm (0.015 in.) at a specified test temperature, when this feature of the C_v specimen is to be used as an acceptance criterion for steels in Class VI. This same lateral expansion applies as a minimum to all sub-size specimens, since they all have the 10 mm (0.394 in.) width parallel to the notch.

To illustrate how lateral expansion undergoes a transition with decreasing temperature, note that the low-carbon steel in Figure 3.28, Section B, is able to display C_v lateral expansion values as high as 1.35 mm (0.05 in.) when tested at a temperature above 50°C (122°F). The curve at lower temperatures shows that the minimum lateral expansion of 0.38 mm (0.015) would not be likely to occur until the test temperature decreased to about -50°C (-58°F). If this feature were not being employed as the acceptance criterion, at least the lateral expansion record would confirm the reliability of energy absorption, as illustrated in Figure 3.28, Section A.

Fracture Appearance Transition Temperature

Another feature of the broken C_v specimen that is regarded as a valid toughness indicator is the relative proportion of ductile fracture versus brittle fracture observed on the broken surfaces. This feature changes, as shown in Figure 3.28, Section C, and has been assessed in various ways. The transition can represent the percentage of ductile shear or fibrous texture that appears on the fractured end of the specimen, and a minimum percentage can be specified. On the other hand, the percentage of brittle cleavage or crystalline texture can be plotted against temperature, and a maximum percentage can be specified. In Figure 3.28, Section C, the temperature at which a tested specimen will display half-brittle and half-ductile fracture texture denotes the *fracture-appearance transition*. Although a record often is made of the brittle and ductile fracture, the percentage seldom is used as an acceptance criterion.

The fracture-appearance transition in C_v specimens of the low-carbon steel tested for Figure 3.28, Section C occurred at a test temperature of approximately -30°C (-22°F), which is in reasonable agreement with the transition temperatures established by setting values for energy absorption and for lateral expansion. However, many structural steels, when C_v tested at low temperatures usually display a fracture-appearance transition that is slightly higher than established by energy absorption or deformation criteria.

Mechanics of Fracture Initiation and Propagation

There are reasons why the fracture-appearance transition may not be coincident with the energy transition for a given steel. Indeed, the transition from a ductile shear fracture to a brittle cleavage fracture may not occur at all, despite the occurrence of an energy transition as testing temperature is lowered. Explanation for differences in temperature of these two transitions is found in the mechanics of fracture when breaking an impact specimen.

Energy transition is controlled by the *initiation* of a cleavage fracture, whereas the fracture-appearance transition is determined by the resistance of the steel to the *propagation* of a cleavage fracture. Confirmation of this reasoning comes from tests with precracked Charpy impact specimens, which show that the major portion of the total energy employed in breaking the specimen is absorbed in accomplishing fracture initiation. Thus a specimen can fracture initially through shear, with considerable energy absorption, but then allow the fracture to propagate easily by cleavage through the greater portion of the specimen. Therefore, the shear-to-cleavage transition will become apparent before a significant decrease in energy absorption occurs.

In discussing alloy steels in Volume II, Chapter 15, an unusual form of fracture behavior will be addressed which occurs in some heat-treated steels. At subzero temperatures, these steels are shown to be susceptible to a *low-energy shear* or *low-energy tear* form of fracture. In other words, a transition occurs in energy absorption, without a change from the ductile-appearing shear-type fracture.

Much effort has been directed toward improving the Charpy impact test to secure more discerning information while retaining the convenience of small, low-cost specimens (see Technical Brief 15). Yet, despite the described limitations of the C_v impact test, and the severe restrictions on even the instrumented cracked C_v specimen to provide K_{Ic}, the regular (ASTM standardized) C_v test is widely employed in industry, research, and in failure analyses because of extensive correlation with actual service performance. The mass of data and information available in the literature has given the C_v test a good degree of credibility. This will become more apparent in later chapters, as data from C_v testing are related to structures.

Test Methods for Toughness Evaluation

Several noteworthy fracture toughness tests were developed ibetween 1950 and 1960, when dissatisfaction with the Charpy impact test was high and fracture mechanics tests had not yet matured. Some of these toughness tests drew enough attention to become ASTM standards. Although presently receiving less usage, the findings from these tests are prominent in diagrammatic schemes that appear in the literature. To assist in interpreting some of the more popular diagrams, a brief description follows of various alternative toughness tests. These tests have contributed useful values for toughness evaluation, and their abbreviated designations still serve as labels for their respective values.

TECHNICAL BRIEF 15:
Improving the Value of the Charpy Impact Test

In the search for an improved Charpy impact test, instrumented Charpy testing has drawn considerable attention as a research tool. By applying an electrical strain gauge to the striker on the pendulum, the load-time behavior of the specimen can be recorded with a high-speed storage oscilloscope. When load-time response is plotted, the various stages in the fracturing process can be separated and identified — namely (1) general yielding, (2) arrest after some fast fracture, (3) maximum load, and (4) energy absorbed in accomplishing any of these functions.

Instrumentation also can include an optical velocity measuring system, and a signal analyzer for interpretation of inertial oscillations that often arise when the striker contacts the specimen. Sometimes electrical strain gauges are applied to the surfaces of the specimen at locations near the crack tip. The instrumented C_v test has not been used in industry for quality control. However, it has been successfully applied in research to gain specific objectives, such as determination of critical stress intensity factor for the dynamic condition, K_{Id}.

The precracked Charpy specimen is a further step in the continuing effort to improve on the test by removing at least one of its limitations, that of the mechanical notch with its rounded root. To achieve greater notch acuity, some early precracked C_v specimens employed a "low-blow" technique to initiate a short crack extension from the root of the mechanical notch, before loading to complete fracture. Eventually, the fatigue technique was adopted to obtain a crack of controlled size and indisputable crack acuity. One obvious advantage of the fatigue-cracked specimen is the opportunity to apply the analytical methods of fracture mechanics to results — and thus attempt a correlation between the small C_v specimens and the much larger LEFM, EPFM and CTOD specimens. Efforts to correlate precracked C_v specimen data with fracture mechanics tests have achieved only limited success, apparently because of basic differences that exist between the two kinds of tests and their specimens. With the precracked C_v specimen, the loading rate is somewhere in the range of three to six m/s (10 to 20 ft/s), while in the usual fracture mechanics bend or tension test, the loading is considered quasi-static.

The size of the fracture mechanics test specimen [as compared to the 10 mm (0.394 in.) square C_v specimen] undoubtedly affects results, as do the degrees of restraint imposed on the process zones at the crack tips. Optimum correlation in results is found when both kinds of specimens fail by the same micromechanism of fracture — that is, by brittle cleavage or by ductile microvoid coalescence. When fracture occurs before general yielding in both kinds of specimens, linear-elastic analytical methods can be applied in each case. The K_{Ic} computed for the instrumented cracked C_v test usually is in agreement with a valid K_{Ic} obtained by a test conducted in accordance with ASTM E 399. However, when general yielding precedes fracture in one or both kinds of specimens, an energy-based J-integral must be used. The procedure followed in analyzing the resulting ductile fracture requires assumptions that tend to derange correlation between the two kinds of specimens.

Explosion Bulge Testing

The explosion bulge test originated at NRL about 1950 to study brittle fracture susceptibility in structural steels, such as plate steels used in ship hulls. The test was performed both on plate specimens containing a welded butt joint and on unwelded metal. As illustrated in Figure 3.30, two different approaches were employed in evaluation. In the "crack starter test plate," a weld bead of very hard, brittle weld metal was made by short circuit gas metal arc welding (GMAW-S). The bead was located in the center of one face of the plate, and a sharp notch was ground into the bead to serve

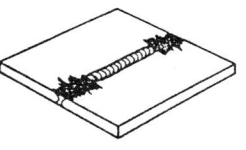

CRACK STARTER TEST PLATE
BEAD OF BRITTLE WELD METAL ATOP WELDED JOINT IS NOTCHED TO PROVIDE SMALL FLAW. WELD REINFORCEMENT ON FACE OF WELDED JOINT IS GROUND FLUSH AT ENDS TO PERMIT INTIMATE CONTACT WITH SUPPORTING DIE. TWO SHOTS ARE APPLIED TO PLATE

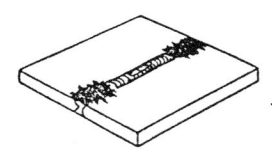

TEST CONDITIONS ALTERED WITH CHANGES IN PLATE THICKNESS

EXPLOSION BULGE TEST PLATE
WELDED (AND UNWELDED) TEST PLATES ARE SUBJECTED TO REPEATED SHOTS UNTIL A CERTAIN MINIMUM BULGE DEPTH IS DEVELOPED WITHOUT VISIBLE FAILURE OCCURRING. REDUCTION OF PLATE THICKNESS IN APEX OF BULGE ALSO IS MEASURED

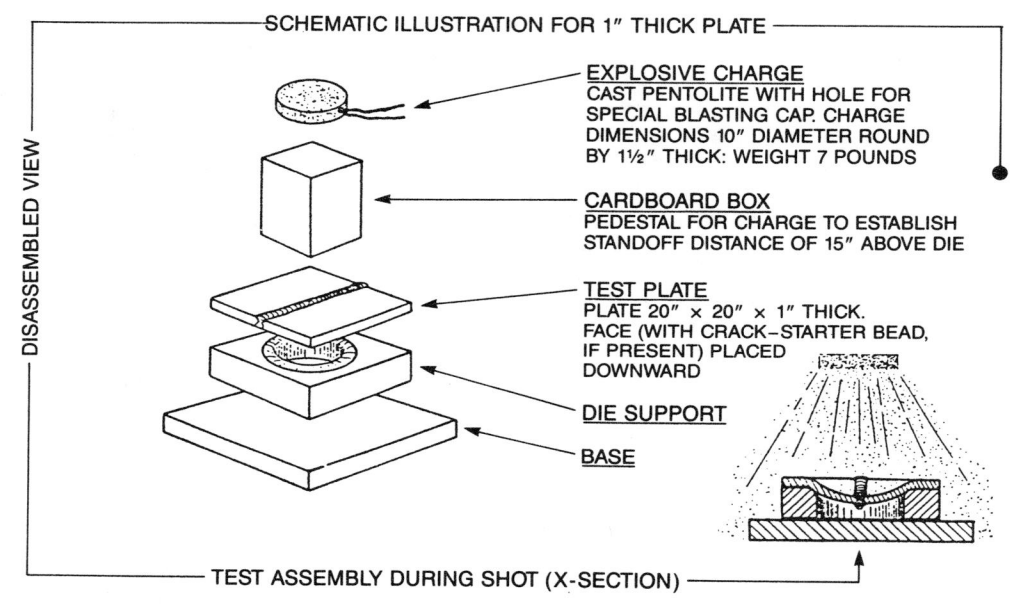

SCHEMATIC ILLUSTRATION FOR 1" THICK PLATE

EXPLOSIVE CHARGE
CAST PENTOLITE WITH HOLE FOR SPECIAL BLASTING CAP. CHARGE DIMENSIONS 10" DIAMETER ROUND BY 1½" THICK: WEIGHT 7 POUNDS

CARDBOARD BOX
PEDESTAL FOR CHARGE TO ESTABLISH STANDOFF DISTANCE OF 15" ABOVE DIE

TEST PLATE
PLATE 20" × 20" × 1" THICK. FACE (WITH CRACK-STARTER BEAD, IF PRESENT) PLACED DOWNWARD

DIE SUPPORT

BASE

DISASSEMBLED VIEW

TEST ASSEMBLY DURING SHOT (X-SECTION)

DETERMINATION OF THESE THREE CRITICAL FRACTURE TRANSITION TEMPERATURES FROM FAILURE MODE IN CRACK-STARTER TEST PLATES

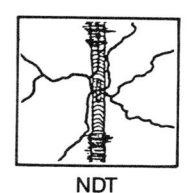

NDT
NIL DUCTILITY TEMPERATURE, FLAT BREAK, BRITTLE FRACTURES EXTEND TO EDGE OF PLATE

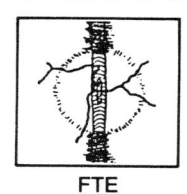

FTE
FRACTURE TRANSITION ELASTIC. FRACTURES ARE ARRESTED IN ELASTICALLY LOADED DIE-SUPPORTED REGION

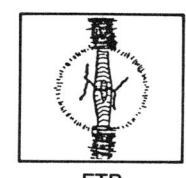

FTP
FRACTURE TRANSITION PLASTIC. FRACTURES ARE ARRESTED IN PLASTICALLY LOADED (BULGED) REGION

Figure 3.30 — Explosion bulge test as originally developed at the U.S. Naval Research Laboratory

Test conditions for the test are often modified to suit plate thickness, material strength, and other influential circumstances, such as test plate temperature at time of detonation.

as a cleavage crack initiator. This face of the test specimen was placed downward on a circular supporting die. Explosion bulge test plates, without the crack-starter bead, were located on the die in a similar manner. A charge of explosive was detonated at a fixed position above the test specimen.

The use of explosives limited the test to laboratories that had a remote, guarded test site and personnel licensed to use explosives. Consequently, this test was not widely used in industry. Nevertheless, the test produced very useful results and led the way to new tests suitable for the ordinary mechanical testing laboratory. Furthermore, the explosion bulge test continues to be used occasionally on metals that must demonstrate resistance to brittle fracture at continuously applied high strain rates, as outlined earlier in Table 3.4, and at a low temperature.

Figure 3.30 provides a rudimentary perspective of the explosion bulge test as performed on steel plate 25 mm (1 in.) thick. The weld in the butt joint, shown across the middle of the two test plates, may not be present. The original conditions for this test were given in a Federal specification, NAVSHIPS 0900-500-5000; however, various alterations are made when the test is employed in industry, in order to accommodate a particular metal. The usual purpose of the explosion bulge test is to establish fracture-transition behavior. This requires a series of replicate specimens to be heated or cooled to progressive temperatures over a range that encompasses the steel's expected transition. The test conditions are quite severe, and the range from ductile to brittle fracture ordinarily is quite narrow. At temperatures below the transition range, the test specimens show virtually no deformation. This is because the notch causes a brittle, running fracture, and the plate is unable to resist its propagation by cleavage fracture. Above the transition temperature range, the fracture is arrested in the plate, and the specimen is deformed by bulging of the central area into the die cavity. Any cracking that does occur would be the ductile, shear-fracture type, which does not extend beyond the bulged area.

As summarized by three illustrations of tested specimens in the lower portion of Figure 3.30, termination of cleavage fractures in the outer edges of the plate over the supporting die, where stresses are lower, can be a sign that the ductile-to-brittle fracture transition temperature is near. The occurrence of a shear lip along the plate surface, on otherwise cleavage fractures, can predict fracture transition to ductile failure. In general, the fracture-transition temperature range, as determined by the explosion bulge test, can be used to estimate a temperature above which brittle fractures are highly unlikely to propagate in a structure, including its weldments.

Drop Weight Testing

The drop-weight test was conceived by Pellini and Puzak at NRL in 1952. Because brittle fracture had been found to occur in weldments with no more energy available than the elastic strain in the structure, these investigators shifted their attention from high-energy fracture analysis (explosion bulge) to a relatively low-energy fracture study. This approach was more consistent with the residual stresses found in weldments and with the service loading conditions imposed upon conventional structures. Pellini and Puzak elected to use a specimen containing a crack, on the premise that many weldments contain small flaws that may act as a crack. They decided to load the specimen by means of a falling weight, to simulate the velocity associated with impact loading and to limit the imposed stress to the yield stress. This represents the approximate amount of elastic energy often available in the vicinity of a weld.

In view of the strong dependence of fracture mechanism on temperature, the NRL investigators decided to arbitrarily fix factors such as flaw size and imposed stress, and vary the temperature to determine the level at which a steel would lose its ability to develop even a minute amount of plastic deformation in the presence of a crack-like notch. This temperature, at which the NRL drop-weight specimen first initiates a major cleavage or running fracture in the metal, is called the *nil-ductility transition temperature* — or "NDT" temperature, as it is now widely known.

Figure 3.31 provides essentials of the specimen with the crack-starter weld, and the procedure employed to test a series of specimens over a range of temperatures to determine the NDT for a steel base metal, or for a steel weld. NDT is defined as the maximum temperature at which a specimen will break by fracturing to either or both edges on the tension surface, and where at least two specimens do not break at a temperature 5 °C (10 °F) above NDT. The drop-weight test was devised for testing relatively thick plate and bar, and is not recommended for base metal less than 16 mm (⅝ in.) thick. A standardized method of conducting the drop-weight test is given in ASTM E 208, *Conducting Drop-Weight Test to Determine Nil-Ductility Transition Temperature of Ferritic Steels*. Many additional references can be found in the literature on the drop-weight test, with extensive results. As an abbreviated identification of this test, *DWT-NDT* will be used in this text to avoid confusion with another drop-weight test to be described next.

The DWT-NDT test is surprisingly reproducible, and it requires only six to eight specimens to establish the temperature below which a ferritic steel is susceptible to brittle fracture. Of course, a steel used below its NDT does not inevitably fail. The NDT simply indicates that catastrophic cracking is a possibility if circumstances require the steel to dissipate energy by plastic flow at the tip of a crack. Below the NDT, the steel is likely to permit the crack front to propagate at high speed by cleavage, with little or no dissipation of energy. The NDT concept has been utilized in a number of procedures for estimating the brittle-fracture safety of weldments in structural steels, and this work has produced fracture-analysis diagrams that appear in the literature. These diagrams — which illustrate the relationships that exist between fracture toughness, temperature, flaw size, etc. — will be discussed further in Chapter 16 of Volume II, when tests of welds are reviewed.

Drop-Weight Tear Testing

The drop-weight tear test (DWTT) emerged as another method for evaluating fracture behavior of carbon and low-alloy steels shortly after NRL's NDT test. The DWTT was developed at BMI to test steels used for pipelines in thicknesses of 3.18 to 19.1 mm (0.125 to 0.750 in.). The minimum thickness requirement of 16 mm (⅝ in.), specified in ASTM E 208, often would preclude use of the NDT test. Primary aims during development of DWTT were (1) to use an easily prepared and inexpensive specimen for testing, either by a vertical drop-weight or by a pendulum machine; and (2) to employ a simple method of evaluation. As illustrated in Figure 3.32, the DWTT specimen consists of a specimen cut from the plate thickness. A notch of 45° is pressed into one edge at the center with a sharp chisel to a depth of 5.1 mm (0.20 in.). This notch is not cracked before impact loading, but the acuity of the notch root is controlled by using a carefully sharpened chisel that produces a very small radius. A standard

CONDUCTING DROP-WEIGHT TEST TO DETERMINE THE NIL-DUCTILITY TRANSITION TEMPERATURE (NDT) OF FERRITIC STEELS

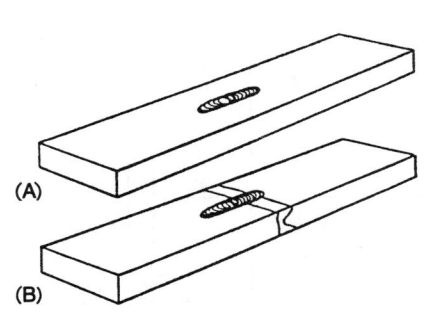

DROP WEIGHT SPECIMENS (A) & (B)
STANDARD DIMENSIONS ARE GIVEN IN ASTM E 208.

EACH SPECIMEN IS PROVIDED WITH A BRITTLE WELD BEAD IN WHICH A NOTCH INITIATES AN EARLY RUNNING FRACTURE. ALTHOUGH INTENDED PRIMARILY FOR EVALUATING PLATE, A WELDED SPECIMEN CAN HAVE THE CRACK-STARTER NOTCH LOCATED OVER THE WELD METAL, OR OVER THE BASE METAL HEAT-AFFECTED ZONE AS ILLUSTRATED IN SPECIMEN (B)

AT LEAST SIX REPLICATE SPECIMENS ARE REQUIRED TO DETERMINE NDT

DETAILS OF NOTCHED WELD BEAD
AS SHOWN IN (C), WELD BEAD FROM A SPECIAL ELECTRODE IS MANUALLY DEPOSITED BY SMAW PROCESS IN TWO INCREMENTS STARTING AT ENDS AND OVERLAPPING IN AREA TO BE NOTCHED BY ABRASIVE DISK

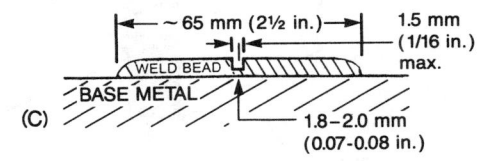

ANVIL FOR DROP-WEIGHT TESTING
ANVIL WITH GROOVED STOP BLOCK TO RECEIVE WELD BEAD SUPPORTS SPECIMEN DURING LOADING 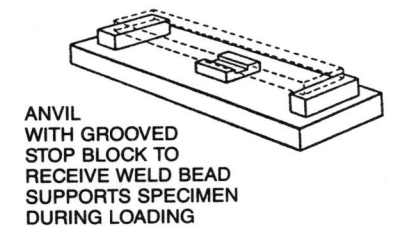

CONDITIONS FOR DROP-WEIGHT LOADING
YIELD STRENGTH OF STEEL DETERMINES DROP HEIGHT

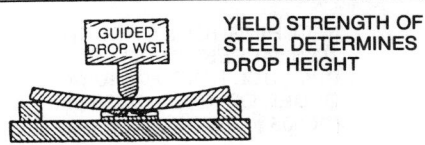

YIELD POINT LOADING IN PRESENCE OF SMALL CRACK IS TERMINATED BY CONTACT WITH STOP BLOCK ON ANVIL

DETERMINATION OF NIL-DUCTILITY TEMPERATURE FROM TESTED SPECIMENS
A SPECIMEN IS CONSIDERED BROKEN IF FRACTURE ON TENSION SURFACE EXTENDS TO ONE OR BOTH EDGES. DUPLICATE NO-BREAK PERFORMANCE IS REQUIRED 5 °C (10 °F) ABOVE NDT. NDT FOR STEEL IN SPECIMENS BELOW IS − 18 °C (0 °F).

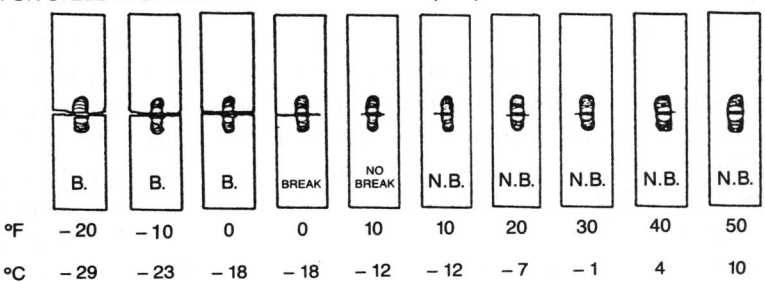

B.	B.	B.	BREAK	NO BREAK	N.B.	N.B.	N.B.	N.B.	N.B.
°F	−20	−10	0	0	10	10	20	30	40	50
°C	−29	−23	−18	−18	−12	−12	−7	−1	4	10

Figure 3.31 — Drop-weight test for determination of nil-ductility transition (NDT) temperature (See ASTM E 208)

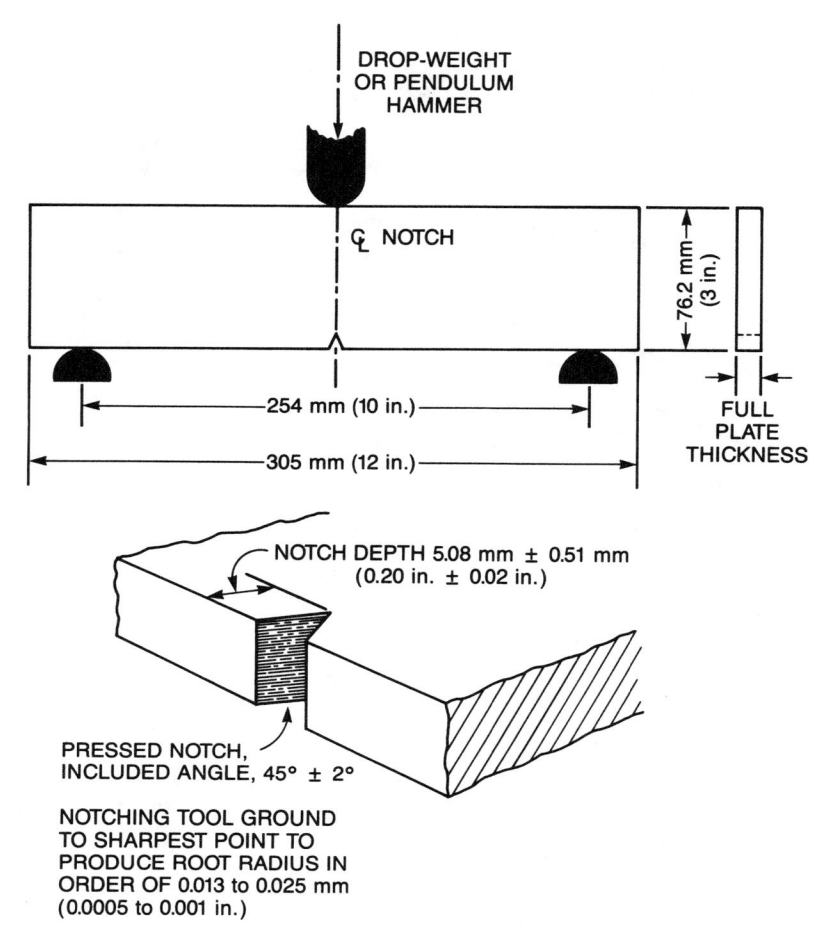

Figure 3.32 — Drop-weight tear test (DWTT) specimen and three-point impact loading procedure (See ASTM E 436)

method for conducting the DWTT test is contained in ASTM E 436, *Drop-Weight Tear Tests of Ferritic Steels*.

The DWTT specimen is broken via three-point impact loading using a drop-weight tup or pendulum hammer having a minimum velocity of five meters per second (16 ft/s). Specimens heated or cooled in a liquid bath to a progressive series of temperatures must be broken within 10 seconds after removal from the bath. Depending on the strength and toughness of the steel, impact energy up to 2700 J (2000 ft-lbf) may be required to break a specimen in a single blow. No attempt is made to measure absorbed energy; evaluation of the broken specimen is based entirely on textures found on the fractured ends. Fracture surfaces having a dull gray, silky appearance, and inclined to the specimen (plate) surfaces, are considered ductile shear fracture. Fracture surfaces that are bright, crystalline, and perpendicular to the plate surfaces are considered brittle cleavage fracture. Appraisal of performance is based on the percentage of ductile shear fracture. A number of measuring procedures are

DWTT TESTING TEMPERATURE	PHOTOGRAPHS OF MATCHING FRACTURE FACES OF SPECIMENS	PER CENT SHEAR IN FRACTURE	PER CENT SHEAR AREA AS DETERMINED ON SPECIMEN BY THE CENTERLINE METHOD
°C / °F			
66 / 150		100	
51 / 125		93	
39 / 100		75	
29 / 85		40	
24 / 75		20	
-18 / 0		2	
-46 / -50		0	

$$\frac{A+C}{A+B+C} \times 100 = \% \text{ SHEAR}$$

CENTERLINE SHEAR AREA METHOD

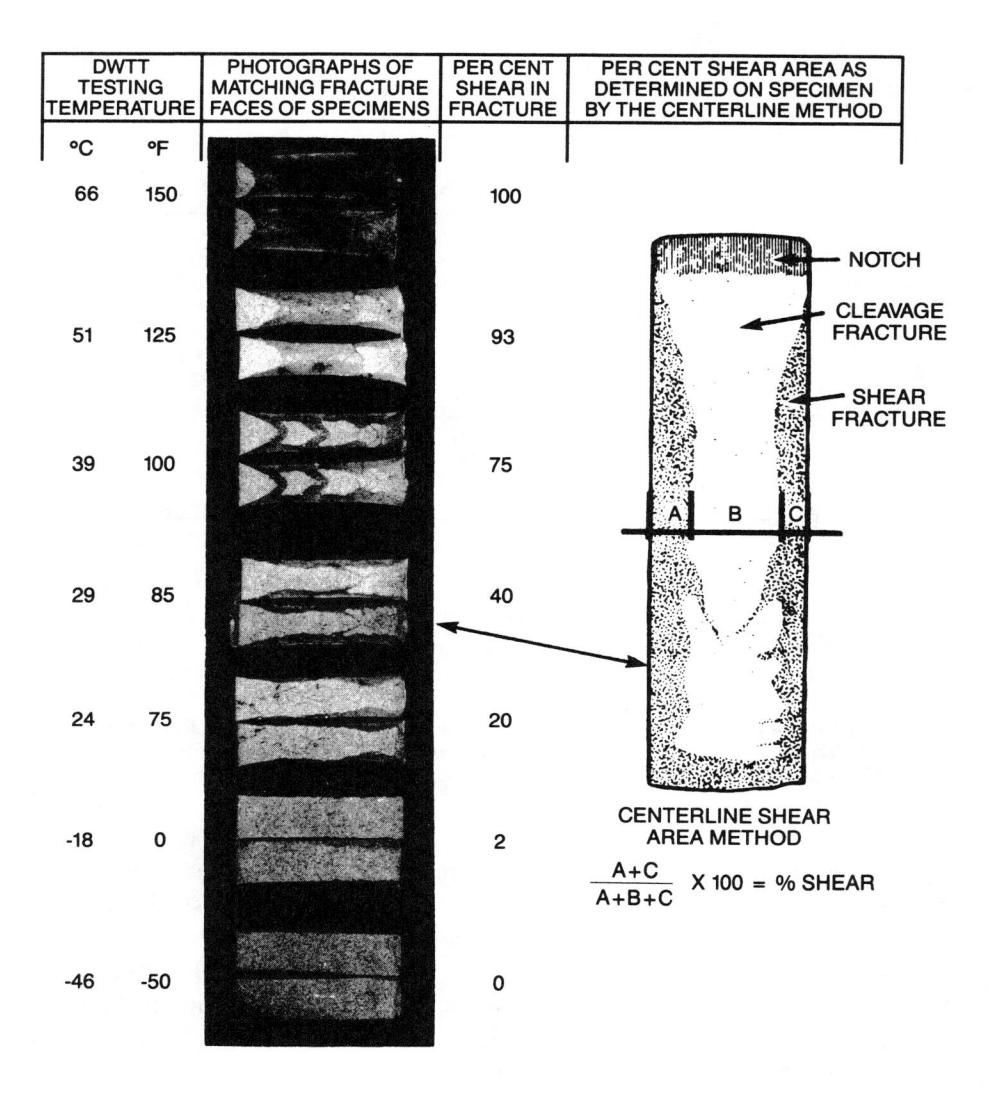

Figure 3.33 — Fracture faces of drop-weight tear test (DWTT) specimens of ASTM A 36 steel broken over a wide temperature range

Percent ductile shear texture in each specimen's fracture has been determined by the centerline shear area method in ASTM E 436.

detailed in ASTM E 436 for determining the percent shear area. The procedure selected should depend on the configurations in the fracture texture and the degree of precision desired. Examination of a series of DWTT specimens of a particular steel broken over a wide range of test temperatures will show that although a cleavage fracture initiates from the sharp notch in the specimen's edge, the mode of fracture changes over a relatively narrow range of temperature from cleavage to shear as test-

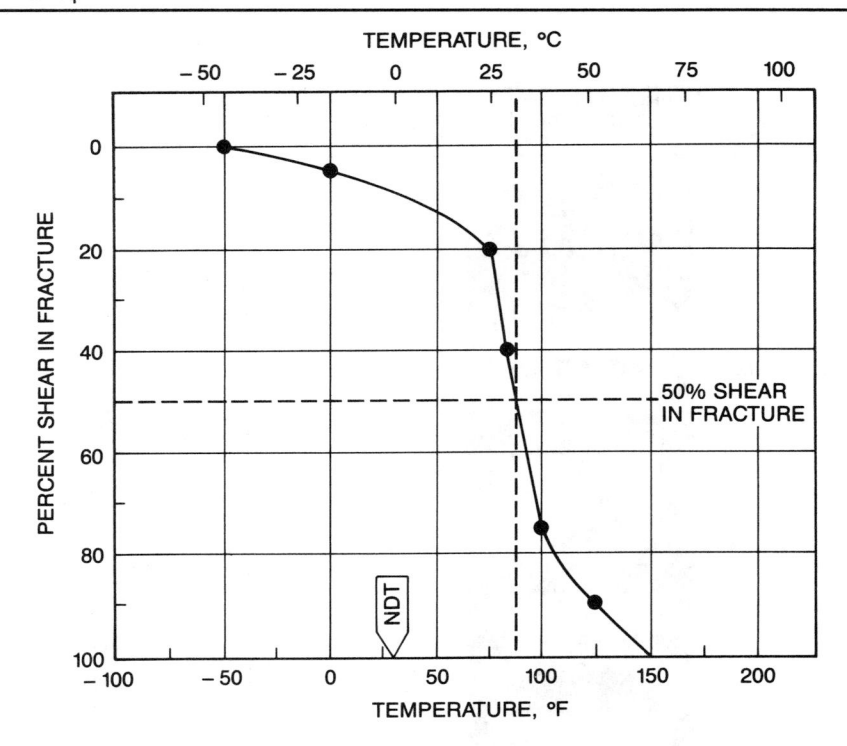

Figure 3.34 — Curve plotted from drop-weight tear test (DWTT) data for an ASTM A 36 steel depicting toughness transition as temperature is lowered

ing progresses to higher temperatures. Figure 3.33 shows a typical group of broken DWTT specimens, cut from 19 mm (¾ in.) plate of ASTM A 36 steel. The fractured faces of each half of each specimen have been examined to determine the amount of shear fracture. As can be seen in the photographs of the fractured ends of the specimens, the regions of shear and cleavage texture form rather complex patterns.

To determine the approximate amount of ductile shear fracture in each specimen, the "centerline shear area" method has been applied. Starting at the lowest temperature, -46°C (-50°F), the fracture was 100 percent cleavage. As the testing temperature was increased, a small amount of shear lip appeared along the edges of the fractured face at the plate surfaces. The amount of shear lip grew slowly as testing temperature increased, until a very rapid change from cleavage to shear occurred over the temperature range of approximately 24 to 38°C (75 to 100°F). When the amounts of shear fracture are plotted against temperature, a familiar curve is obtained. As shown in Figure 3.34, the curve consists of a lower and upper shelf, and an intermediate steeply sloping transition. Because of the rapid fracture transition, the temperature at which 50 percent shear fracture occurs is a convenient way of expressing the performance of a steel in the DWTT test. The 50 percent-shear temperature for the A 36 steel represented in Figures 3.33 and 3.34 is 32°C (90°F), as noted in Figure 3.34. This same A36 steel was tested by the DWT-NDT test, and the NDT established was -1°C (30°F), as also indicated in Figure 3.34.

The DWTT test has been used mostly as a research tool in studying the fracture problem in large diameter welded steel pipe. Pipe has been tested by initiating fractures under various conditions of stress, temperature, and defect acuity, and the results were compared with results from DWTT testing of the same steel. Fracture appearance in the failed pipe (removed from the initiation area) occurred at the same temperature as the transition in the DWTT tests. Thus, the DWTT 50 percent-shear transition appears to define a fracture propagation transition temperature.

Dynamic Tear Testing

The *dynamic tear test* (DT) was developed at NRL about the same time as the DWTT test originated at BMI (circa 1960). However, the DT test incorporated several features in specimen design and testing procedure that produced results amenable to analysis by more discerning methods, including fracture mechanics. For a number of years following its origin, the DT test was required by various U. S. Federal procurement documents for steels to be used in applications involving low temperatures, and where toughness had to be assured. After the DT test gained recognition, a standard test method was issued in 1983 by ASTM as E 604, *Dynamic Tear Testing of Metallic Materials*. This ASTM standard is intended for metal in the thickness range of five to 16 mm (3/16 to 5/8 in.). When metal exceeds the maximum, a standard specimen is prepared by reducing the thickness to 16 mm (5/8 in.).

The DT specimen is single-edge notched, and is impact loaded via three-point bending in a manner similar to the DWTT specimen illustrated earlier in Figure 3.32. However, E 604 has additional requirements. A V-notch is machined into the edge as a preliminary step, followed by the pressed-chisel technique to obtain a sharp root with a radius of 0.02 mm (0.001 in.). This two-step procedure reduces cold-work effects associated with a notch made entirely by pressing; but, the notch in the DT specimen still does not have the tip acuity of a real crack. Nevertheless, the dimensions and shape of the sharpened notch specimen have been selected to provide high constraint, in an attempt to equal the conditions in a cracked K_{Ic} specimen. The DT specimen also can be fractured in a drop-weight, or a pendulum testing machine, but E 604 requires the loading velocity to be in the range of 4.0 to 8.5 m/s (13 to 28 ft/s). This range corresponds to a weight being dropped from heights of 0.8 to 3.7 m (32 in. to 12 ft). The potential energy of the load must be able to fracture the specimen completely in one blow.

When a DT specimen is tested, total energy absorption must be recorded, and E 604 provides suggestions for ascertaining this value both with pendulum, and with drop-weight testing machines. Also, the percentages of shear and of cleavage textures in the fracture should be recorded for nonaustenitic steels. Minimum requirements for the DT test are agreed upon between parties involved, and the metal has a large influence on the values selected. Ordinarily, replicate specimens are tested as a progressive series over a range of temperatures, perhaps from about 100°C (212°F) down to below the NDT temperature.

In the case of ferritic steels subjected to the DT test, when impact energy absorption and fracture appearance are plotted against temperature, the plotted points usually form two transition curves, each with an upper and a lower shelf and a steeply sloping transition between the two shelves. At the NDT temperature and below, the

DT specimen fractures in a flat-brittle mode, which is indicative of plane-strain conditions. Shear lip is not developed, and lateral contraction at the notch is virtually zero. Energy absorption is so low that it amounts almost entirely to the kinetic energy involved in ejecting the broken halves of the specimen from the testing machine. This has been substantiated by installing the broken halves in the machine (by holding them together with paper tape) and retesting. The taped specimen will absorb the same energy as did the original unbroken specimen. This experiment gives an impressive indication of the severe brittleness that corresponds with dynamic plane-strain fracture. Because of the small size of the DT specimen, its plane-strain capability is limited, and as testing temperature is raised above the NDT, energy absorption increases sharply. A double-pendulum testing machine has been used for DT testing in which both the hammer and the anvil-supported specimen swing and meet at mid-point to achieve a "shockless" impact.

Drop-weight DT tests have been conducted on ferritic steel plate with specimens as thick as 300 mm (12 in.). These tests required a very tall tower for dropping a large weight to attain the necessary overload impact energy. Although use has been made of the results of this DT testing — for example, in helping establish transitions under plane-strain conditions in K_{Ic}, $K_{I(t)}$ and K_{Id} testing — the DT test lately has been replaced with tests using fatigue-cracked specimens that are amenable to fracture mechanics analysis.

Correlation of Results from Fracture-Toughness Tests

There was a period (approximately 1955-70) when much attention was directed to the many proposed toughness tests in order to establish correlations between results. These activities are described in the literature, and they were encouraged by the following objectives. First, there was interest in how the various tests compared with regard to precision, degree of discrimination, and reproducibility. Next came the problem of knowing whether results from one kind of toughness test could be accepted in lieu of results from a different test, when all available data was reviewed. Then came the matter of cost for testing, along with availability of testing equipment and test material. Mention was made earlier of the Charpy V-notch specimen's low cost and small amount of metal required, as well as the widespread availability of pendulum machines for testing. It was reasoned that if Charpy V-notch test results could be reliably and closely correlated with results from more complex and expensive toughness tests, then obviously the Charpy test would be employed whenever permissible.

Another purpose of correlation studies was to ascertain if any basic information on toughness was revealed during analyses of the numerous test data. If significant fundamental criteria could be identified, perhaps they could be incorporated in new tests that would indicate the probable behavior of engineering structures exposed to low-temperature service.

One early organized study by the IIW is summarized in Figure 3.35. The figure gives transition temperatures from five methods of testing that were popular at the time, and for ten different steels. The values vary for the different steels, and the results from all the tests must be considered to see the general trend, with the toughest steels on the left and the less tough steels on the right (i.e., lowest to highest transition temperature). No single test provided results that rated the steels in the same order, although DWT-NDT results were in best agreement with the general trend.

Despite intensive work to correlate the results of the early toughness tests, the aforementioned objectives were not satisfied. Although effort was made to normalize conditions (for example, by fixing the testing temperature), there remained too many differences in notch acuity, rate of loading, degree of constraint, etc. All of these influenced the results to some degree. For the most part, the result from each test was an appraisal of toughness measured in arbitrary terms that could not be readily equated because no basic parameters were being handled. This impasse persisted until the advent of fracture-mechanics analyses for metals, whereupon new tests were developed for such fundamental values as K_{Ic}, J_{Ic}, and CTOD.

Efforts are continuing to correlate toughness tests by utilizing as much as possible the more fundamental values from fracture-mechanics tests. For example, empirical formulas have been proposed for the calculation of K_{Id} curves from C_v data. The formulas were developed by correlation of K_{Id} curves with C_v energy absorption curves over the same range of temperature. Considerable scatter was encountered in the data, which formed a curved scatterband more so than a correlation curve. In 1970, ASTM issued STP 466, *Impact Testing of Metals*, which reported on efforts to correlate C_v energy absorption with K_{Ic}, since this correlation would allow quantitative assessment of critical flaw size and permissible stresses. Some success was achieved with an equation that expressed the relationship between K_{Ic} and the lower portion of the C_v

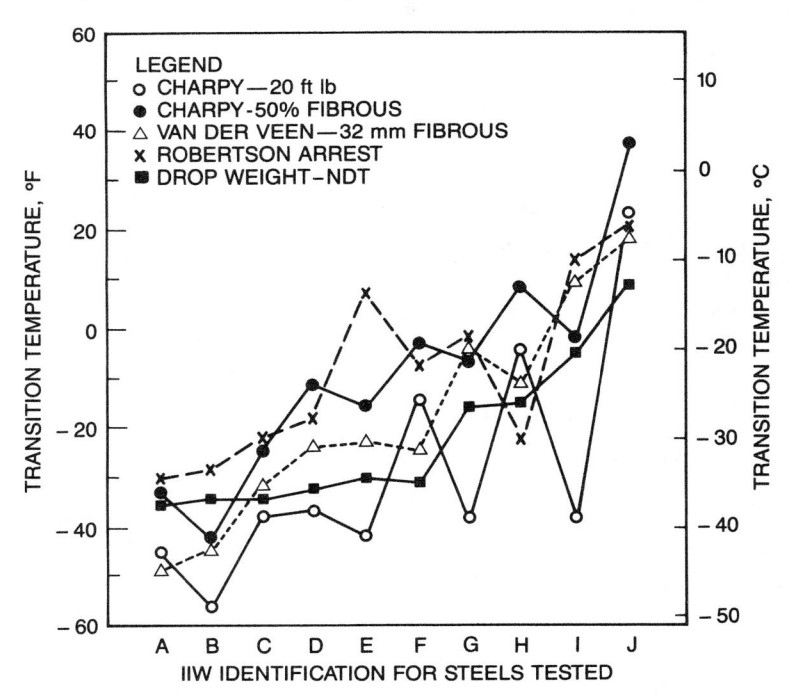

Figure 3.35 — Relative behavior of ten structural steels when tested by four different kinds of toughness tests

(Courtesy of the International Institute of Welding, Document No. IX-114-58)

energy transition curve, but another equation appeared necessary for calculation of K_{Ic} from C_v upper-shelf energy values.

While a number of detailed correlation studies can be found in current literature, correlative procedures proposed to date for converting results from a small, inexpensive test (e.g., Charpy V-notch impact test) to the more adaptable, meaningful fracture mechanics results are useful only for estimating fracture toughness and securing approximate appraisals of various criteria.

Improved Mechanical Properties for Low-Temperature Service

Even before the turn of the century, there was a growing awareness of a serious toughness problem with ferritic steels when they are exposed to low service temperatures. Dramatic and catastrophic failures had been experienced with bridges, storage tanks, and other large structures. One would expect that this early awareness, and the great amount of metallurgical research conducted since then, should have brought the problem under control. Unfortunately, this is not the case, and premature failures from inadequate toughness still occur occasionally.

There are several general reasons why problems with brittle fracture at low temperatures will continue. Firstly, steels that offer good strength have a ferritic structure, which is innately susceptible to metallurgical conditions that have a deleterious influence on toughness. Secondly, although low-temperature toughness can be improved by reduction of some residual elements, addition of alloying elements, and special processing or heat treatment, all of these increase the cost. Thirdly, economics plays a key role in determining how much toughness will be specified to assure fracture-safe performance at the lowest expected service temperature.

If austenitic steels and other metals with a fcc crystalline lattice offered better strength and were available at an acceptable cost, they would virtually eliminate the problem with toughness at low temperatures. However, commercial considerations commonly compel use of the ordinary ferritic structural steels. In Chapters 14 and 15, alloying, special processing, and heat treatment to improve toughness will be discussed. Because the low-temperature toughness problem is even more critical with weld metal, Chapter 10 of Volume II will consider the role of filler metal.

Mechanical Properties at Elevated and High Temperatures

Elevated temperatures are assumed to be somewhere above room temperature [i.e., 25°C (75°F)] but not exceeding about 260°C (500°F). *High temperatures* are those exceeding about 260°C (500°F), and often there is reason to discuss temperatures up to the melting point. Mention was made that carbon and low-alloy steels offer useful mechanical properties only up to approximately 600°C (1100°F). Beyond this, high-alloy steels, superalloys, and heat-resisting alloys ordinarily would be employed. Time is a significant factor in determining behavior of metals at elevated and high temperatures, and this makes necessary the study of *short-time* and *long-time* properties. Different methods of testing are employed for determining strength and other properties at elevated and high temperatures, depending on the time, and standardized or widely-used test methods will be discussed.

The Properties of Metals   **239**

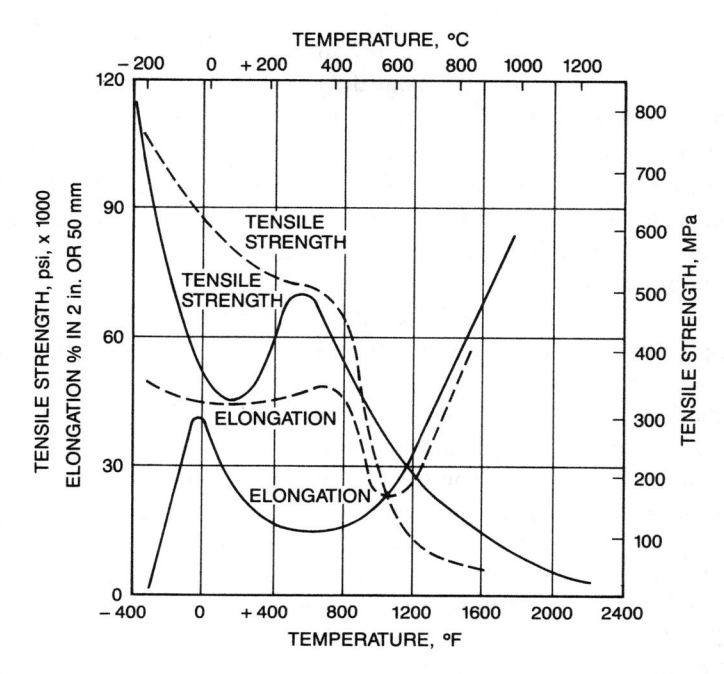

Figure 3.36 — Change in short-time tensile strength and elongation with temperature found in iron (solid curves) and in Ni-Cu alloy (dashed line curves)

Both materials are in the annealed condition prior to testing. The Ni-Cu alloy contains 68% Ni, 29% Cu, 1.5% Fe and 1.5% Mn. Results are approximate because of influence of speed of testing.

In general, the strength of metals decreases as temperature is raised, and the atoms move more freely. Consequently, the elastic modulus is reduced. A method for determining Young's modulus at low and elevated temperatures was issued as ASTM E 231, but it was discontinued in 1986. As temperature is increased, plastic deformation mechanisms operate more freely. This would suggest that plastic flow occurs more readily, and perhaps would be extended to greater limits. Metals differ in the way their strength and other properties change with temperature, and the change does not necessarily occur gradually. In fact, there are many unusual patterns of change, as determined by composition, microstructure, and other conditions.

As a simple example of this individualistic behavior, Figure 3.36 presents the tensile strength and ductility of iron compared with those of Monel (68 percent Ni, 29 percent Cu, 1.5 percent Fe, 1.5 percent Mn) over a wide range of temperature. The curves for iron show an unexpected increase in strength, and a decrease in ductility over a narrow temperature range that is centered at approximately 260°C (500°F). The Ni-Cu alloy does not show these pronounced aberrations in mechanical properties.

These changes in strength and ductility of iron are caused by an embrittling phenomenon called *blue brittleness*. This name is applied because the embrittlement occurs during tempering over a temperature range in which a blue color appears on the surface of clean steel. Blue brittleness is caused by precipitation hardening that develops in iron and some steels over the temperature range of about 200 to 425°C (400 to 800°F). The severity of the embrittlement depends on the strain present in the metal prior to heating, and on the time spent in the blue brittleness temperature range.

Blue brittleness is only one of a number of forms of embrittlement that can develop in iron and steels when they are exposed to elevated and high temperatures during heat treatment, heating for hot working, welding and brazing, and in service. Other forms are strain-age embrittlement, quench-age embrittlement, temper embrittlement, liquid-metal embrittlement, and liquation embrittlement. To avoid these forms of embrittlement, each must be studied individually. First, the type of metal or general chemical composition that is susceptible to each form of embrittlement must be established. Sometimes, secondary elements or impurities are responsible for the embrittlement, and sometimes these elements can be restricted. The range of temperature over which the embrittlement develops, and the amount of time spent at influential temperatures must be determined. While some forms of embrittlement develop during heating, and adversely affect mechanical properties in the elevated temperature range, most forms affect properties after cooling to room temperature and below. The effect often is measured by the upward shift in toughness transition temperature. Embrittlement phenomena will be reviewed in Volume II, Chapter 13, to explain the causes of various forms, their severity, and remedies. Significant forms of embrittlement in specific types of steels will be discussed in Chapters 14 and 15 of Volume II.

Many steel members employed in weldments are produced by rolling, forging, or extrusion at high temperatures — in the range of 980 to 1200°C (1800 to 2200°F). If the steel does not possess suitable properties for hot working at these high temperatures, defects can be created that may later interfere with the making of sound welds. Some steels exhibit a susceptibility to intergranular cracking or tearing over certain ranges at high temperature. An example is high-sulfur steel at about 1100°C (2000°F). Lack of ductility or plasticity at a red heat commonly is called *red shortness*. The usual cause is the melting of a small portion of the metal. Often, the molten portion develops in the grain boundaries (liquation) because of lower-melting segregates that tend to concentrate there. Consequently, the metal tends to crack or tear intergranularly in brittle fashion. The attack of oxidizing gases along grain boundaries, and intergranular penetration by external molten metals can cause similar embrittlement defects.

Sometimes, to avoid the problems that occur at high temperatures, only moderately elevated temperatures are used for simple forming. If the temperature is sufficient to lower the elastic limit, then the metal bends more easily, and spring-back is lessened.

Heat treatment to improve properties, to eliminate the effects of cold work, or to relieve internal stresses also requires a knowledge of mechanical properties to judge whether objectives have been accomplished. Welding and brazing almost always require heating, albeit usually only locally and for a relatively short time. Nevertheless, in order to predict behavior under the stresses that develop, information often is needed on the mechanical properties of metals as the temperature changes.

A commonly encountered reason for detailed information on the properties of metals at elevated and high temperatures is that the fabricated article will be heated during service. The conditions encountered with welded process equipment range from steady-state to cyclic or variable temperature. Design of this equipment and the selection of metals regularly require much data to verify that load-carrying members possess adequate strength at the temperatures to which they will be subjected. Assurance must be gained that no significant changes will occur in any of the metal properties during long-time service that would jeopardize the performance of the equipment.

Degradation of fracture toughness is a possible change that deserves close scrutiny. Short-time testing for properties of metals at elevated and high temperatures provides useful data, but more often there is need for long-time testing — such as when one must determine a metal's suitability for processing equipment intended for long-time service at elevated temperatures. When equipment intended for elevated-temperature service is studied, at least six different types of tests are required in order to cover commonly encountered service conditions — namely (1) short-time tension testing, (2) long-time tension testing, (3) long-time exposure to elevated temperature, followed by short-time and long-time tension testing, (4) thermal shock testing, (5) thermal fatigue testing, and (6) time-dependent fatigue testing. Designers sometimes require metal qualification tests under simulated service conditions.

Short-Time Elevated Temperature Testing

Standard methods for tension testing are available in ASTM E 8M (E 8), but these documents include tests only at room temperature. For short-time tension testing at elevated and at high temperatures, ASTM E 21 - *Elevated Temperature Tension Tests of Metallic Materials* provides standard recommended practice. Because tensile properties are usually more affected by rate of strain at elevated temperatures, the recommended procedures of ASTM E 21 include specified strain rates for the portion of the tension test in which yield strength is determined. After yield strength has been determined, a ten-fold increase in rate of strain is allowed in further loading of the specimen. This ASTM document does not include tension tests with rapid heating or high strain rates. For these conditions, ASTM E 151 was available for about 20 years; but it was canceled in 1984, because rapid heating and high strain rate testing usually requires special test conditions. There is available, however, a current ASTM standard practice, E 209, for compression tests at elevated temperatures, with conventional or rapid heating rates and strain rates.

Elevated temperature tension testing requires much more attention and care than testing at room temperature. For example, uniform axial loading of the specimen is very important, because bending strains can lower the strength.

Short-time tension test data for elevated temperatures are published for many metals, including the carbon and low-alloy steels. However, the data have limited engineering application, because the effect of extended time is not included. For service temperatures up to approximately 340°C (650°F), computation of allowable design stress for a steel structure may be based on the short-time yield or tensile strength at the maximum expected service temperature. The steel members undergo elastic deformation immediately upon load application, but no significant further deformation occurs with time. At temperatures above approximately 340°C (650°F), however, members again deform elastically as load is applied; but, more importantly, deformation continues slowly with time. This deformation with time at service temperature is called *creep*.

The creep phenomenon was discovered at temperatures and loads where deformation was expected to be determined by elastic behavior. Despite the innocuous term applied, creep can proceed to a serious extent and cause major difficulty, particularly in equipment where dimensions and clearances at elevated temperature are expected to remain within prescribed limits. The temperature of 340°C (650°F) is only a very rough approximation. The limiting temperature that applies in a given situation is dependent on the chemical composition and the microstructure of the steel, the service temperature, and the time. Carbon steels may be prone to creep at a lower temperature — particularly if they are low in carbon, or if the microstructure lacks a favorable grain size and adequate distribution of phases, which ordinarily serve to strengthen the steel. Objectionable creep can be avoided by limiting the applied load. However, this is seldom practical, because usually either the low stress will restrict loading too severely, or the cross-section of the members must be increased too much. The more effective solution is to use a steel offering greater resistance to creep at elevated temperature.

Low-alloy steels can contain additions of elements, such as molybdenum, that effectively increase resistance to deformation at elevated temperatures. Some of these steels can serve at temperatures somewhat above 340°C (650°F) without significant creep. Since higher service temperatures must be accommodated, creep must be recognized as unavoidable and limited to tolerable amounts during the life of a structure. This requires property data from long-time elevated-temperature tests. Short-time tension tests at elevated temperatures provide information that is useful only up to the temperature at which creep becomes significant, and this limiting temperature must be ascertained through examination of metal, service temperature, time, stress, and acceptable deformation.

Long-Time Elevated Temperature Testing

In addition to determination of creep rate at elevated temperatures, long-time testing includes measurement of time for fracture, when sufficient load is applied, and measurement of stress relaxation by creep. These additional tests are called *creep-rupture*, *stress-rupture*, *notched-bar rupture*, and *relaxation tests*. Standard practices for most of these long-time tests can be found in ASTM documents — for example, ASTM E 139, *Conducting Creep, Creep-Rupture, and Stress Rupture Tests of Metallic Materials*. These tests are conducted with a relatively slow heating rate, usually with a radiation furnace. The procedure for conducting creep and creep-rupture tension tests with rapid heating and short times was available in ASTM E 150, but this standard was discontinued in 1984 for the same reasons explained above for E151.

Creep Testing

Creep testing at a little above 340°C (650°F) will show that a reaction often occurs which defers creep until a higher temperature. Strain-hardening occurs in steel that is cold worked and then heated to temperatures up to about 425°C (800°F); and since creep is low at this temperature, many steels will not exhibit measurable

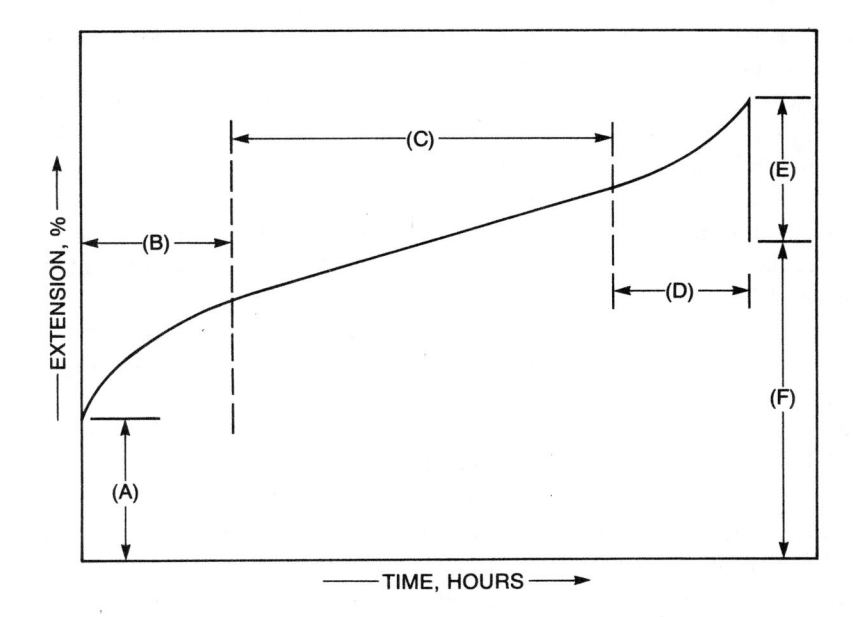

(A) **IMMEDIATE ELASTIC STRAIN UPON APPLICATION OF GIVEN STRESS**
(B) **FIRST STAGE (PRIMARY, OR DECREASING RATE) CREEP**
(C) **SECOND STAGE (SECONDARY, OR CONSTANT RATE) CREEP**
(D) **THIRD STAGE (TERTIARY, OR INCREASING RATE) CREEP**
(E) **IMMEDIATE ELASTIC CONTRACTION UPON REMOVAL OF APPLIED STRESS**
(F) **PERMANENT CHANGE IN SPECIMEN LENGTH CAUSED BY COMBINED EFFECT OF ALL THREE STAGES OF CREEP**

Figure 3.37 — Typical creep curve showing six stages in which a change occurs in test specimen length

creep. As the temperature is increased, strain-hardening becomes negligible, and creep will commence.

A creep curve can be obtained by plotting change in length of the specimen against time; a typical curve is shown in Figure 3.37. The initial constant load produces an immediate elastic extension (A). The specimen then stretches gradually with time at a decreasing rate during *first-stage creep* (B), which also is called *primary creep*. Next, the rate becomes virtually constant, and this is called either *second-stage creep* or *secondary creep*. The slope of the curve in this second stage (C) is the rate commonly used in design. If the temperature is sufficiently high and time is long enough, the rate will increase to *third-stage* or *tertiary creep* (D), and the accelerating deformation will terminate in complete rupture, which would be *creep rupture*. However, if the load is removed before rupture occurs, the resulting elastic contraction (E) will be equal to the elastic extension (A) caused by the load applied at the start of the test. This indicates that metals creeping under stress at elevated temperature possess both elastic and plastic properties.

Primary creep should be considered transient creep because it represents an early adjustment in the crystalline structure. This adjustment starts when thermally activated plastic strain facilitates the movement of dislocations into more stable arrays and more stable positions in the lattice. The dislocation movement then decreases until it reaches a level of activity that is commensurate with stress and temperature. This is the threshold of secondary creep, which is near a steady-state process representing equilibrium between work hardening and recovery. The almost constant rate of creep is generally referred to as the *minimum creep rate* (MCR). An increase in either stress or temperature accelerates creep, and the minimum creep rate is increased. Even though secondary creep is a transition between primary and tertiary creep, secondary creep occupies the major portion of time. There is no distinct beginning of tertiary creep, because it is simply the start of an increasing rate of elongation that is followed by rupture.

To determine creep strength, tension-type specimens are subjected to appropriate selected loads and held at selected elevated temperatures. Often, the testing temperature is a little above the maximum service temperature expected, to be certain that an unexpectedly high creep rate will not result from a service temperature overshoot. The temperature must be held within very narrow limits during the entire testing time. The limits are usually ±½°C (1°F), because even a momentary temperature deviation outside these limits could cause enough thermal expansion or contraction to introduce unacceptable measurement error.

The rate at which metals creep increases rapidly with increasing temperature. Consequently, the time required for a metal to deform too much to be usable can vary — from many years at a slightly elevated temperature to a few minutes at a temperature near the melting point. Metals differ considerably in their creep rates. At room temperature, only metals with low melting points — such as lead, tin, and zinc — will creep. However, at sufficiently high temperatures, all metals will creep. The slope and length of the curve during secondary creep will depend on the metal, the stress, and the temperature. The shape of the curve for secondary creep ordinarily is almost linear; but, if a change should occur in the microstructure as a result of temperature and time exposure, then the resulting metallurgical instability could cause a change in the slope of the curve and thus alter the minimum creep rate. Most changes of this kind result in an increased creep rate.

Creep tests ordinarily are conducted for at least 1000 hours, but the duration may be extended to several thousand hours or more if a metallurgical or microstructural change is likely to occur that will alter the creep rate. Stress and temperature conditions are selected that will establish minimum creep rates (MCR) ranging between 0.0001 and 0.00001 percent/h. When the data are plotted on a log stress versus log creep rate for several temperatures, a group of nearly parallel straight lines ordinarily are formed. Common practice is to extrapolate the results and determine the stress that produces one percent creep in 10 000 h (a little longer than a year), and the stress that produces one percent creep in 100 000 h (about 11 years), i.e., a creep rate of 0.00001 percent/h.

Caution must be exercised in extrapolating creep results. For example, if a metallurgical instability is observed during secondary creep at high temperature, then the possibility exists that a similar instability or slope change can occur at lower temperature at a longer time. Obviously, a thorough understanding of the metallurgical behavior at various temperatures for very long durations would be useful when deciding to what extent a reliable extrapolation can be made. The elevated-temperature mechanical properties of metal are highly dependent on chemical composition and microstructure. Grain size, impurities, cold-working history, and heat treatment are just a few of the many factors that influence properties at elevated temperatures.

Creep can be predicted remarkably well from a minimum amount of high-temperature test data. By taking into consideration crystalline structure, melting temperature, valence for pure metals, and elastic modulus, an estimate of creep rate can be made for metals at particular temperatures. The crystalline structure with greatest creep resistance is the diamond lattice. Close-packed structures, as found in fcc and hcp lattices, are second-best, and the bcc structure is third. While creep resistance is better in metals that have high melting temperatures, none of the commonly used metals have useful strength at temperatures exceeding about 65 percent of their melting points, expressed in °K. Also, higher valency and elastic modulus results in higher resistance to creep.

The mechanical properties of weld metal and heat-affected zones frequently deserve individual attention because of the variations in their microstructure. At elevated temperatures, the creep behavior of these zones requires even greater attention, particularly for low-alloy steels that are rolled or heat treated to achieve good properties. Elevated-temperature testing of welds may require special specimens to ensure that the weld metal and the heat-affected zones have acceptable creep strength and ductility.

Low-ductility failure by *creep brittle response* is a difficulty encountered in the heat-affected zones and weld metals of low-alloy steels. In this creep failure, microscopic intergranular cavities develop. As these cavities accumulate, either during elevated-temperature service or during postweld stress-relief heat treatment, the damage becomes a small crack, and eventually develops into a large crack. The problem appears to originate when the high temperature of welding dissolves the fine precipitates that strengthen the steel. These precipitate-free zones tend to be more prevalent at prior austenite grain boundaries, and their lower strength allows shear deformation to concentrate in and adjacent to the zones. The presence of sulfide inclusions in the zones adds to cavitation development, since the sulfide particles serve as non-wetting initiators. Detailed studies have been reported on cavity initiation and growth, damage accumulation, and weldment service life as affected by creep brittle response.

Stress-Rupture Testing

Stress-rupture testing is performed to establish time-temperature limits that would avoid fracture in the metal. Although time-versus-extension data can be obtained during the stress-rupture test and can be used for a creep curve, often only the stress, temperature, time to fracture, elongation, reduction of area, and character of the fracture are recorded. The stress-rupture strength usually is expressed as the stress that produces failure at a particular temperature in a stipulated period of time (often a 100 h period is used). The stress-rupture strength in the relatively short-time test will be greater than for the conventional minimum creep rate (MCR) for the same temperature. *Ductile ruptures* at elevated temperatures are transgranular, and an appreciable amount of elongation and necking occurs. *Brittle ruptures* occur intergranularly, with very little elongation or necking. The kind of rupture and the ductility not only depend on the properties of the metal, but also can be influenced by temperature and strain rate. In general, ruptures with substantial ductility are considered desirable when metals are evaluated by stress-rupture testing. Premature failures sometimes occur when tertiary creep is present and the ductility is low.

Notched-bar stress-rupture tests reveal metals which are notch-sensitive at elevated temperatures, and which do not sustain the loads for temperatures and times as predicted by tests with smooth specimens. Because structures in elevated-temperature service usually are subjected to complex stresses from design details and defects, there often is good reason to conduct stress-rupture tests with both notched and smooth specimens to determine their stress-rupture strength ratio. Notch sensitivity also can be based on ductility, as measured by reduction of area. For smooth specimens, metals with a reduction of area exceeding five percent usually are not notch sensitive. If a notched specimen has a reduction of area below three percent, the metal may be notch sensitive.

Relaxation Testing

Relaxation testing — sometimes called *stepdown relaxation testing* — is a special form of creep testing. In relaxation testing, a specimen is loaded at room temperature to produce a certain stress, and the increase in length caused by elastic strain is recorded. This extended length is held constant at an elevated testing temperature by reducing the load as the metal attempts to creep. This stress reduction to prevent creep is continued for a given time.

The objective of relaxation testing is to determine the load-carrying ability of a specimen at a given temperature for a significant period of time without permanent distortion (creep). Information from this test is very useful in structural applications. For example, when bolts are used for a flanged joint, they are under tensile stress from tightening. When this bolted joint is used at elevated temperature, the metal will relax rather than undergo creep (elongation), resulting in a loss of bolt tightness due to tension relief.

Methods for converting results from one form of elevated temperature property test to another, and for predicting stress-rupture and creep performance, have been proposed in the literature by a number of investigators. The most widely used system is the Larson-Miller parameter, presented in Technical Brief 16.

Thermal Shock

Thermal shock is the term applied to a very rapid change in temperature. In addition to the strains and stresses that arise from expansion and the lack of uniformity during heating, there can be additional stress from very rapid heating. A wide variety of thermal shock conditions are encountered in aircraft, spacecraft, and many kinds of industrial process equipment. The purpose of thermal shock testing is to determine the ability of the metal to accommodate the strains that occur as a result of a single very rapid change in temperature, or from repeated rapid changes. To simulate severe thermal shock, a metal can be heated in less than a minute to temperatures as high as 1650°C (3000°F) by resistance heating; or, using the capacitor discharge technique, a specimen can be heated to test temperature within 200 milliseconds. Under these very

TECHNICAL BRIEF 16:
The Larson-Miller Parameter

The Larson-Miller parameter converts time and temperature by the following equation:

$$P = 1.8T (C + \log t) \times 10^{-3} \qquad \text{(Eq. 3.15)}$$

where

  P = Larson-Miller parameter
  T = temperature in °K
  C = a constant
  t = time to rupture in hours

The constant "C" depends on the metal and varies with the stress. It is established by plotting a graph using all available data to correlate stress with temperature and rupture time. The constant may range from about 15 up to 45, but approximately 30 is a good average for most metals. Many low-alloy steels favored for service at elevated temperature have a constant of approximately 20.

If the reader regularly uses the Fahrenheit scale, the Larson-Miller equation can employ temperature in degrees Rankine as follows:

$$P = T (C + \log t) \times 10^{-3} \qquad \text{(Eq. 3.15a)}$$

The relationship expressed by the Larson-Miller parameter is that, at a given creep stress, a longer time at a lower temperature causes as much deformation as a shorter time at a higher temperature. Also, the higher the temperature, the more striking is the difference in times required at different temperatures to cause the same amount of creep. This relationship holds true for such rate processes as recovery, recrystallization, creep, and stress rupture. Of course, if precipitation, aging, or grain growth occurs, use of the parameter would not result in an accurate prediction. Where the conditions permit creep or stress rupture as a rate process, a few data points interpreted as Larson-Miller parameters can be used to develop a master rupture graph at minimum cost and time.

---

rapid heating rates, metals exhibit greater strength than with regular elevated-temperature tests. Therefore, thermal shocks of very short duration possibly can be withstood by a metal that might ordinarily have inadequate strength for the load and temperature expected.

The rate of cooling after reaching the high temperature also deserves consideration, because equally rigorous strains and stresses can be caused by non-uniform contraction. Heated metal might be quenched in a liquid bath, as often is the case during heat treating; or cold liquid may impinge upon the heated surface, as can occur in process vessels and piping. These temperature changes and gradients can impose strains that the metal cannot accommodate, resulting in cracking. If the thermal changes are cyclic, the repeated strains and stresses can cause fatigue cracking. Low-cycle thermal fatigue failures have been experienced in equipment with only a few thousand starts and stops. Thermal cycling not only produces cyclic stresses that cause fatigue cracking; it also reduces fatigue strength itself at elevated temperatures.

Mechanical Properties After Plastic Work

Plastic deformation of metals commonly is divided into *hot work* and *cold work*. This differentiation is made because of the marked differences in mechanical properties after each kind of working. When a metal is cold worked, the crystalline structure is distorted. Cold work also results in strain hardening — that is, an increase in resistance to further deformation. The strength increases and the ductility decreases. A more subtle form of plastic work is the effect of radiation by neutrons and gamma rays. The changes to the crystalline structure produced by irradiation are similar to those induced by cold work, but irradiation effects are important in today's technology, and will be given specific attention.

Hot Work

Hot work does not cause permanent deformation of the crystalline lattice. All lattice distortion from working is almost immediately removed through recrystallization to form equiaxed grains. The recrystallization temperature of cold-worked low-carbon steel is about 510°C (950°F), but since a transformation occurs during heating low-carbon steel through the temperature range of about 727 to 910°C (1340 to 1670°F, the exact range depends on the carbon content), hot working is commonly done above this temperature range.

The higher the temperature at which hot work is completed, the coarser will be the grain size. As a general rule, grain size increases as the temperature increases above the transformation or recrystallization temperature. Yet, when steel cools through the transformation range of 910 to 727°C (1670 to 1340°F), new smaller grains will be formed. Notwithstanding, the coarse grain established by a high temperature can influence the finer grains that result from cooling through the transformation range. The high initial temperature can form a *Widmanstätten pattern* in the microstructure (see Chapter 9), which is not a ductile condition. For finest grain size, therefore, hot work should be completed at as low a temperature as possible without cold working the metal.

Cold Work

Cold work is deformation below the recrystallization temperature. The effects of this plastic work will be closely related to the extent of the permanent distortion left in the crystalline lattice, and other responses of the particular metal, such as age hardening. Examples of cold working include hammering, drawing, and rolling. Machining methods such as turning, milling, and drilling will cold work the metal chips being removed; but ordinarily machining does not cold work the metal remaining, with the possible exception of a superficial surface layer.

The extent of cold work can be very roughly estimated by the amount of distortion seen in the grains. Figure 2.24 illustrated the distortion that can be seen in the microstructure of iron after a reduction of 60 percent by cold working. Distortion of the grains increases strength, but decreases ductility and toughness. When a metal is subjected to cold working (e.g., drawing or rolling) a close relationship exists between increase in strength and reduction in thickness. The effects of cold rolling on the properties of low-carbon steel sheet, and the remedial effects of recrystallization by annealing are shown in the tabulation of tension test results and hardness in Table 3.7.

Peening

Peening is plastic working that long has been used (and abused) supplementary to welding. The effects of peening on mechanical properties can be quite significant, but results usually are difficult to predict. This is because of uncertainty about whether the metal is being hot worked or cold worked, and about the extent of deformation. Welds are sometimes peened to control dimensions and distortion; and for this reason, peening will be discussed more extensively in Volume II, Chapter 11. Also, welders sometimes resort to peening in attempts to prevent welds from cracking. This is a dubious remedy, however, as will be explained in Chapter 13 of Volume II, where the subject is covered in detail.

Irradiation

Neutron radiation causes concern because of its effects on the mechanical properties of metals. Other radiation forms, such as infrared, and strong gamma rays, might impart enough heat to affect mechanical properties, but these effects are easily evaluated. Neutron radiation must be considered in all areas of nuclear power generation, including commercial electrical power generation and marine propulsion. Irradiation by neutrons, especially high-energy (fast) neutrons, can produce significant changes in strength and toughness, and also can cause other forms of radiation damage. The changes in mechanical properties are somewhat like those from cold working — that is, strength is increased and ductility and toughness are reduced. The mechanics involved in changes by neutron irradiation are more complex than in cold working, and the results are unique.

Much metallurgical research has been conducted on neutron-induced changes and damage. Findings are extensively documented in the literature because of the serious consequences from unexpected failure in nuclear equipment. Since welding and brazing are used extensively for assembly of this equipment, it is important to know how specific areas of a weldment are affected by neutron irradiation. Weld metal, braze metal, and heat-affected zones cannot be assumed to be affected by irradiation to the same degree as the base metal. The pressure vessels used in nuclear reactors present challenges to the metallurgist and the welding engineer, because they are weldments subjected to irradiation by neutrons during service. Studies of metal specimens from structural members, tanks, and pressure vessels that have been under irradiation require remote sampling, handling, machining, and testing. Until recently, only limited studies of this kind had been made, but now that obsolete equipment is being retired from nuclear service, there is greater opportunity to test actual components.

Table 3.7 - Effect of Cold Work on Mechanical Properties of Low - Carbon Steel Sheet

Condition	Ultimate Tensile Strength MPa	ksi	Yield Strength MPa	ksi	Elongation in 2 in.	Rockwell Hardness
Hot rolled	415	60	310	45	25	65 HRB
Hot rolled, then cold reduced 20% by rolling to "half-hard temper"	515	75	450	65	20	80 HRB
Hot rolled, then cold reduced 40% by rolling to "full-hard temper"	590	85	550	80	5	90 HRB
Hot rolled, then cold rolled 40%, then process annealed at 760°C (1400°F) to produce "dead soft" condition	380	55	275	40	30	60 HRB

Changes in the properties of steels during neutron irradiation appear to be the result of more than one mechanism acting on crystalline structure and microstructure. The mechanisms are (1) generation of dislocations, (2) cavity nucleation, (3) boron-to-helium transmutation, and (4) precipitation of unfavorable compounds or phases. The latter mechanism, radiation-induced precipitation, has the most damaging influence on the toughness of carbon and low-alloy steels, but fortunately this can be controlled through chemical composition. Precipitation phenomena will be reviewed in detail shortly to single out the elements in steel that are involved, and to describe their effects on mechanical properties as a result of radiation-induced precipitation.

Irradiation by high-energy neutrons can displace metal atoms from their normal lattice sites by impact, generating various kinds of dislocations. An abnormal number of point defects often develop in the crystalline lattice. These simple lattice defects then migrate and cluster under operating stress and temperature conditions to form larger dislocations. In fact, the higher-energy neutrons (above one MeV) cause multiple point defects by sequential displacement, and thereby generate a greater number of abnormalities in the crystalline structure.

Neutrons at very high energy, upon impact with a metal atom, also can force the atom into an interstitial position in the lattice. This very unusual position causes a very high short-range strain and stress — a condition which fosters increased strength, but which markedly reduces ductility and toughness. Microscopic cavities can form under sustained exposure to these high-energy neutrons if the irradiation temperature is sufficiently high. These cavities cause a dimensional change, which usually results in localized swelling. If the steel contains boron, thermal neutrons can cause transmutation of the boron into lithium and helium gas. Because helium has no solubility in the solid steel, the gas exerts several effects:

(1) Helium appears to nucleate the formation of cavities, both in the matrix and at precipitate-matrix interfaces where greater frequency of cavity nucleation and growth is found.

(2) Helium accumulates in the cavities and develops considerable pressure, which increases the amount of swelling.

(3) Helium, because of its influence on metallurgical reactions at the interface of matrix and precipitates, can affect the type and distribution of precipitates that form during irradiation. When a high-energy neutron strikes a metal atom, a "thermal spike" occurs and kinetic energy is transferred to a very small region. Thermal spikes produce highly localized temperatures that may reach 5 000°C (9 000°F) for an extremely short time. From a macroscopic viewpoint, effects of thermal spikes are manifested as highly localized hardening and embrittlement.

The radioactivity induced by neutron irradiation brings on a variety of safety problems in testing irradiated metal, and in the repair or disposal of used equipment from nuclear service. The neutron-capture cross-section is a mass property, and it is directly proportional to the amount of each element contained in the metal. The value can be determined from the sum of the weight fractions of each element multiplied by its cross-section per gram. The value obtained is called the *microscopic cross-section*, and has the dimension of cm^2/g. The cross-section per gram, when multiplied by its density, gives the conventional *macroscopic cross-section*, which has the dimension cm^{-1}. The absorption cross-sections for thermal neutrons of various elements found in steels are presented in Table 3.8.

Although iron has a cross-section that is not especially high, under neutron irradiation it generates its radioactive isotope $_{26}Fe^{55}$, which has a half-life of almost three years. Consequently, the repair of irradiated steel equipment has required the development of extraordinary remote-controlled welding processes and procedures. Also, certain elements in steels, which may be present either for alloying or as impurities, have cross-sections higher than that of iron. Furthermore, they may generate their own radioactive isotopes. Cobalt is a good example of an element with a relatively high cross-section; it is the element that generates the radioactive isotope $_{27}Co^{60}$. Although cobalt is not ordinarily found in significant quantities in carbon and low-alloy steels, this property of cobalt becomes quite important in highly alloyed metals.

Boron deserves further comment, because its neutron-capture cross-section is a little higher than cobalt and more than 1000 times greater than iron. As background for discussion of neutron capture, free neutrons can be described as not very stable, and regardless of their velocity — fast or slow — they are readily captured by nuclei. Yet, reduced velocity (i.e., energy) enhances capture. When a thermal neutron impacts a boron atom, the reaction produces lithium and helium, and the capture and transmutation is complete in one ten-billionth of a second. This extremely rapid and gross capture of neutrons by boron makes this element a very useful addition to the special high-alloy materials that serve in control rods for nuclear reactors. However, as a minor alloying addition or as a residual element in other steels, boron is likely to play an unfavorable role.

Most elements and compounds capture neutrons in lesser quantity and more slowly, usually in the order of one ten-thousandth of a second. Specific substances to keep in mind for their neutron capture rate are graphite (used as a moderator) at 0.01 seconds, air at about 0.1 seconds, and hydrogen at many seconds.

Before describing changes in mechanical properties from neutron irradiation, the problem of quantifying the severity of exposure should be reviewed; this issue is addressed in Technical Brief 17. Practices for surveying and reporting the effects of

Table 3.8 - Thermal Neutron Cross Section for Iron and Other Elements Found in Steel

Element	Cross Section, cm^2/g
Iron	0.028
Carbon	0.0002
Phosphorus	0.0037
Sulfur	0.009
Manganese	0.146
Silicon	0.0034
Chromium	0.036
Nickel	0.047
Molybdenum	0.017
Niobium	0.007
Tantalum	0.070
Cobalt	0.380
Copper	0.036
Boron	42.500

neutron irradiation on nuclear power reactor vessels, and for determining changes in mechanical properties of metals, can be found in ASTM standards E 185 and E 184, respectively. These ASTM standards, as well as regulatory guides of the Nuclear Regulatory Commission (NRC), recommend the Charpy V-notch impact test to assess decrease in toughness from irradiation. The Charpy impact test is used not only because of its convenient size, but also because irradiated specimens are radioactive and must be handled by remote-operated testing equipment. The requirements for conducting C_v tests on irradiated specimens and the procedure for evaluating results are not universal, and they will likely be changed from time to time. However, the following paragraph is intended to provide some familiarity with evaluation procedure.

TECHNICAL BRIEF 17:
Quantifying the Severity of Exposure to Neutron Irradiation

The neutral electrical state of the neutron particle makes dosimetry more difficult than measurement of other elementary particles, or of radiation from radioactive elements. Whereas a "neutron gun" can provide a stream of neutrons having kinetic energy within a very narrow band, the neutrons commonly encountered in nuclear operations have a very broad spectrum of energies. Because the need frequently arises to indicate the approximate energy of neutrons, the spectrum commonly is described as having three broad classes of neutrons: (1) thermal neutrons with an energy of 0.5 eV or less, (2) epithermal or intermediate neutrons with more than 0.5 eV up to 100 eV, and (3) fast neutrons of more than 100 eV. Fast neutrons have up to several MeV, and often are described as high-energy neutrons.

The number of neutrons at any energy across the spectrum varies considerably from one reactor to another. Consequently, this entire broad, irregular neutron spectrum is not easily used as a quantitative severity factor. The severity of neutron irradiation might require that the *neutron flux density* be considered, which is the number of neutrons passing through a defined space per unit of time, usually expressed as either neutrons per square meter-second or neutrons per square centimeter-second.

Some use has been made of an activation technique to determine severity of exposure. This technique involves exposing a piece of foil or wire of a suitable metal to the neutron flux, thus giving all neutrons opportunity to generate radioactive isotopes that can be measured by radioanalysis. When nickel is used, neutrons with energies about three MeV and greater will generate radioactive isotopes of cobalt, and the isotope population created is measured by their gamma ray radiation to gain an assessment of the neutron flux. ASTM E 261 provides standard practices for measuring neutron flux by radioactive techniques.

Since fast neutrons represent the portion of the neutron spectrum that causes most of the changes in mechanical properties, a practice has evolved using a dosimeter to count neutrons with energies greater than 1 MeV. Equations for integrating a neutron spectrum and the density of energies across the spectrum into flux are quite complicated, but the usual practice is to report severity of neutron irradiation as *fluence*, which is an approximation of the number of fast neutrons (E >1 MeV) passing through a defined space; typically an area of one square centimeter. Sometimes neutron irradiation severity will be found reported as NVT, which means "neutron-velocity-time" for total integrated flux. As an example of fluence values (n/cm^2 >1 MeV), the pressure vessel of a pressurized-water nuclear reactor for commercial electrical power generation might have a yearly integrated fast neutron dosage at the beltline (the region of maximum flux) of approximately 1×10^{17}. Because the design lifetime for a reactor vessel is in the order of 30 to 40 years, the estimated cumulative dosage at EOL (end of life after service at effective full power) will be a fluence of about 1×10^{19}.

The relationship of C_v absorbed energy to test temperature usually is required for both unirradiated and irradiated specimens. Two C_v measures might be determined: any increase in the 41 J (30 ft-lbf) transition temperature, and any decrease in upper-shelf energy absorption. The transition temperature in the unirradiated condition is considered a *reference temperature*, and accordingly it has been given the designation RT_{NDT}. This selected point on the C_v energy transition curve (41 J, or 30 ft-lbf) should not be confused with the "NDT" temperature from NRL's drop-weight test, described earlier. In the irradiated condition, the 41 J (30 ft-lbf) transition commonly occurs at higher temperature, and the upward transition temperature shift (also called the adjustment of reference temperature) is given designation ΔRT_{NDT}. Note that ΔRT_{NDT} is the difference in number of degrees between the 41 J (30 ft-lbf) index temperatures measured before and after irradiation.

Upper-shelf energy absorption, the second criterion for evaluation, is defined as the average value for all C_v specimens (usually three) whose test temperature is above the upper end of the transition region. Average upper-shelf energies for unirradiated and irradiated specimens usually do not differ as much as those found in the transition region, especially the 41 J (30 ft-lbf) transition. Nevertheless, the irradiated specimens may display lowered upper-shelf energy absorption. A C_v upper-shelf energy of 68 J (50 ft-lbf), along with a lateral expansion value of 0.89 mm (0.035 in.), is cited in ASTM E 185 as marginal. If lower values are projected in the beltline region of a reactor vessel, then additional fracture toughness testing is advised.

The practice presented in ASTM E 185 for conducting surveillance testing was developed for light-water cooled nuclear-power reactor vessels for which the predicted maximum neutron fluence (E >1 MeV) at the end of the design lifetime would exceed 1×10^{17} n/cm^2 at the inside surface of the reactor vessel. The decrease in steel toughness from neutron irradiation is a matter of interest wherever nuclear power is generated, and it is understandable that views in various countries differ on the best choice of mechanical tests for maintaining surveillance. A worldwide review of this question suggests that many of the mechanical property tests commonly used for evaluating toughness can give useful quantitative measurements for the change of the ductile-to-brittle transition temperature, and that the specimen need not impose plane-strain conditions. Consequently, in addition to fracture mechanics tests that provide K_{Ic}, other acceptable tests for monitoring toughness under neutron irradiation would include Charpy V-notch impact tests, large slow-bend tests, and wide plate crack-arrest tests. Because of irradiation space limitations in the reactor vessel, large specimens ordinarily are not included in surveillance programs. ASTM standard E 636 identifies specimen types that can be considered for supplementing C_v tests and tensile tests.

The temperature of the metal has a strong influence on the extent of neutron damage that occurs. Thermal gradients often arise when nuclear radiation is superimposed onto operating temperature, particularly if specimens are large and are being exposed to strong gamma radiation. Consequently, temperature measurement requires attention to possible localized temperature increases. Higher metal temperatures, especially those above approximately 232°C (450°F), tend to reduce the harmful effects of irradiation. Many reactor vessels operate at higher temperatures; therefore, their normal service temperature should lessen the effects of irradiation. For most steels, the greatest amount of damage occurs when irradiation takes place in the temperature range of about 50 to 150 °C (125 to 302 °F). The effects of irradiation are cumulative, particularly in the case of toughness. If the irradiated metal is heated to more than 300°C (580°F), annealing of radiation damage can be significant in many cases because of annihilation of the point defects in the lattice and other metallurgical processes (to be discussed). Successively higher temperatures increase annealing in continuous, accelerating fashion; however, post-irradiation heating is not a reliable solution for degraded fracture toughness. The recovery processes necessary in some steels to restore toughness can require a temperature of 425°C (800°F) or higher.

Irradiation increases the strength and hardness of the ferritic carbon and low-alloy steels by modest amounts. Yield strength usually is increased proportionately more than tensile strength. No attempt is made to utilize this increase in strength, however, because of concern with the accompanying decrease in toughness. The various types of ferritic steels, as well as their weld metals, suffer different degrees of toughness degradation when subjected to near-saturation irradiation, which usually entails a fluence in the order of 1×10^{17} n/cm^2 of >1 MeV, or more. The ΔRT_{NDT} after this irradiation has been reported to be as much as 260 Celsius degrees (470 degrees on the Fahrenheit scale). A ΔRT_{NDT} of this magnitude would have the ductile-to-brittle transition of many steels raised above room temperature. Along with the higher adjusted reference temperature, the upper shelf energy level may be decreased by as much as 25 percent. Every effort is made in selecting steel for a nuclear reactor to secure metal with excellent toughness, i.e., the lowest practical RT_{NDT}, and the highest practical C_v upper-shelf energy absorption.

Chemical composition and microstructure of steels used for reactor vessels are carefully controlled to minimize susceptibility to neutron damage and toughness loss. Fine grain size has been found to lessen susceptibility to irradiation embrittlement. In addition to the known composition control that favors good toughness, certain elements have been found to increase neutron radiation damage. By exercising precautions, it appears possible to limit a ΔRT_{NDT} by the steel in a reactor vessel. Thus, instead of experiencing a ΔRT_{NDT} increase by as much as 260 Celsius degrees (470 Fahrenheit degrees) at the end of operating life (EOL), shift can be held to less than approximately 80 Celsius degrees (about 145 Fahrenheit degrees). Also, every possible effort is made to have the initial reference temperature at the lowest practical level, which probably is a subzero temperature. Nevertheless, great care is exercised after nuclear reactor vessels have been in service to avoid placing the vessel under stress while the metal temperature is below the temperature for crack arrest. This temperature is determined by conducting a crack-arrest type of fracture toughness test. An alternative measure inslused in ASTM E 185 to ensure crack arrest capability in the irradiated condition is to estimate the number of degrees that should be added to the reference temperature (RT_{NDT} + ΔRT_{NDT}) to provide an *adjusted reference temperature* at which to index the C_v 41-J (30 ft-lbf) impact test requirement.

When examining the ferritic carbon and low-alloy steels to determine the mechanisms that induce irradiation embrittlement, it appears that their macroscopic neutron-capture cross-section should be indicative of their susceptibility. Also, it seems that the microscopic neutron-capture cross-section of the individual elements might be tabulated to determine susceptibility to irradiation damage.

Neutron-capture cross-section is a complex property, and the first important fact to note is that the barn values can range from very low figures, such as 0.2 b for zirconium, to very high values, such as 61 000 b for gadolinium. This is one reason for favoring zirconium alloy tubing to encapsulate nuclear fuel. Although there is an approximate trend for heavier atoms to exhibit higher barn values, this relationship has been reported to have many deviations. Aluminum, as a pure metal, has a b value of only about 3, and for this reason it is successfully used in foil form as a moderator for slowing fast neutrons to thermal neutrons. Barn values reported for alloying elements commonly used in steels — such as chromium, nickel, copper, and manganese — are not much greater than that of aluminum, and yet certain of these elements can have a profound influence on a steel's irradiation embrittlement. Obviously, the behavior of steels with respect to mechanical properties after neutron irradiation cannot be explained simply by macroscopic neutron-capture cross-section.

Since an understanding of the behavior of ferritic steels during neutron irradiation would require painstaking, extensive testing and examination, a number of comprehensive programs of research were initiated in the USA and overseas. These programs were established to study not only specially prepared laboratory steels, but also metals removed from both old and new power-producing nuclear plants. These studies made use of all applicable analytical and examination methods, and the results have been very informative.

First, it is important to distinguish between rolled plate, forgings, castings, or weld metal, because these forms differ in their susceptibility (with weld metal being the worst). Second, chemical composition exerts a major influence on the radiation sensitivity of ferritic steels, and chemistry factors (CF) have been derived for projecting radiation embrittlement (see NRC Regulatory Guide 1.99). Third, perhaps the most rewarding results have come from examination of the microstructures of irradiated steels using highly advanced metallographic methods — including scanning electron microscopy (SEM), transmission electron microscopy (TEM), Auger electron spectroscopy (AES), and microprobe analysis.

Work in this area revealed that radiation-induced precipitation in the microstructure is to a large degree the mechanism by which strength and hardness increases and, more importantly, by which toughness is lowered. The nature and distribution of the precipitates are dependent on the level of certain elements in the steel's composition, as described in the following paragraphs. It is important for the welding engineer to note that at least two particular elements commonly present in steel in residual amounts have been selected as principal contributors to neutron irradiation sensitivity. These two elements, phosphorus and copper, can increase to an intolerable level in weld metal if proper precautions are not exercised in selection of filler metal and fluxes. Nickel, in general, is a favored alloying element for weld metals; yet, the role of nickel in the radiation embrittlement problem is not completely clear. Consequently, this element deserves close scrutiny.

Influence of Copper

Copper is the most potent offending element in causing irradiation embrittlement in ferritic steels, and although ordinarily present at low levels in most steels merely as a residual element, its influence is sufficiently strong to warrant exercising strict control of copper content. The preferred copper maximum is approximately one-tenth of one percent, but somewhat more can be tolerated if other offending elements are low. The National Regulatory Commission has provided guidance on the appraisal of copper (and other elements) in its regulatory documents, and ASTM has assisted by adding appendices to specifications for materials used in nuclear reactors with advice on impurities (e.g., ASTM A 533/A 533M, Appendix XI). Knowledge regarding the influence of copper is of particular importance to the welding engineer, because of the care that must be exercised in limiting the copper content of weld metal. Sometimes bare filler metal is given a very thin copper plating for rust protection, and even this small amount of copper may not be tolerable in some cases.

The strong influence of copper on irradiation embrittlement in ferritic steels has been confirmed by a number of investigators, and their work shows that radiation-induced precipitation is indeed the mechanism by which copper contributes to the problem. The precipitates formed in the microstructure were not pure copper, but were a complex of copper and other alloying elements, mostly manganese and nickel. The precipitated particles were found to have either sphere or disc shapes, and tended to form coherent copper-rich clusters. Microprobe analyses found some evidence of copper depletion in areas surrounding the clusters of precipitated particles. When the copper content was sufficiently high to form precipitates during irradiation, any phosphorus available entered into copper phosphide formation. Recovery of toughness by annealing has been studied via changes observed in precipitate size, number, and composition, but more research remains to be done.

Influence of Phosphorus

Phosphorus had been reported in some early studies to be of secondary importance with regard to irradiation embrittlement, but more recent work shows this impurity is deserving of close attention. Phosphorus long has been known to adversely influence the toughness of steel, especially the higher-strength types. Therefore, good reason often existed to hold this element to a low level whenever toughness was important. Presently, phosphorus is receiving increasing attention, since radiation-induced precipitation of phosphide particles has been discovered in the microstructure of steel that has suffered radiation embrittlement. When sufficient copper is present to form the copper-rich phosphide, phosphorus preferentially migrates to the copper clusters. Consequently, the influence of phosphorus and the precise mechanism will be dependent upon the amount of copper. Nevertheless, enough evidence has been gathered to suggest that phosphorus is a contributor to irradiation embrittlement, and that reducing the level of this residual element is a worthwhile objective.

Influence of Nickel

Nickel, alone, does not appear to cause irradiation embrittlement, but nickel does seem to reinforce the adverse effect of copper. When the copper content is as low as 0.10 percent, nickel up to about one percent is likely to increase irradiation damage only by a small amount. However, with more copper, perhaps about 0.40 percent, nickel becomes a stronger factor in promoting irradiation embrittlement. Steel weld metal follows these same trends, except the effects are more pronounced.

Influence of Sulfur

Sulfur is another element in steel that affects toughness, especially as measured by upper-shelf energy-absorption in the C_v impact test. Consequently, control of this impurity, either in amount, or in the morphology of nonmetallic inclusions, is advisable to maximize steel's ability to resist neutron damage and have good C_v upper-shelf energy absorption remaining after irradiation.

Influence of Vanadium

Vanadium is another element reported to be of secondary importance in promoting irradiation embrittlement. When vanadium is present as an impurity, the amount ordinarily is quite low. A maximum of 0.05 percent is the aim for steel intended for service in the beltline location of a nuclear reactor vessel.

Influence of Nitrogen

Nitrogen appears to increase irradiation embrittlement when in solution in steel. Presumably, this propensity arises because the nitrogen atoms are easily driven farther into interstitial lattice positions.

The steelmaker has reasonable guidance as given above for producing ferritic steels with sensitivity to irradiation embrittlement reduced to an acceptable level. However, the welding engineer's task is more difficult. Although the same chemical composition information appears to apply to weld metal, in general, the initial toughness of weld metal often is lower than wrought base metal. In addition, weld metal after irradiation is inclined to display greater toughness damage. Most importantly, the final composition of the weld metal, including the maximum levels of residual elements that may be of critical significance, is the responsibility of the welding engineer; and great care is required to assure consistent compliance with specified limits. Filler metal and flux, dilution, and pick-up of unfavorable elements must be controlled to prevent intolerable composition deviations. Furthermore, the welding engineer often must deal with weldment designs in which the most rigorous combinations of localized stresses and stress multiaxiality occur in or near the welds. Consequently, weld metal must have adequate toughness to prevent the propagation of crack-like flaws, and thus avoid brittle fracture.

Metals with a face-centered cubic crystalline structure, such as aluminum and the austenitic stainless steels, usually exhibit admirable toughness. However, they can decrease in C_v energy absorption with only modest neutron irradiation exposure if the metal temperature is in the range of approximately 25 to 150 °C (75 to 300°F). At higher temperatures, as with the ferritic steels, property changes are progressively less. In nuclear service, the use of metals with a fcc crystalline structure in lieu of the ferritic steels often is restricted by strength, cost, or other considerations — hence the effort to reduce irradiation embrittlement in the ferritic steels, and continue employing them in weldments subject to neutron irradiation.

PHYSICAL PROPERTIES

The physical properties of metals seldom receive the searching attention regularly given to mechanical properties in a general treatise on welding. Nevertheless, physical properties are a very important part of a metal's characteristics, and often the welding engineer (and the welder) may be unaware that the success of a particular joining operation depends heavily on a certain physical property. The physical properties of regular polycrystalline metals are not as structure-sensitive as the mechanical properties. Constant values are published for metals and alloys, and these serve satisfactorily for most engineering purposes. Only those physical properties which may require consideration in designing, fabricating, and treating a weldment are discussed here.

Density

Density is the concentration of matter, measured as mass per unit volume. It is important when the weight of a structure or component must be kept as low as possible. Specific gravity is the weight of a volume of the metal compared with the weight of an equal volume of water at 4°C (39.2°F). In the metric system, since one cubic centimeter of water weighs one gram, the numerical value of specific gravity is equal to that of metric density when expressed as g/cm^3. In the U. S., the customary unit for density is in pounds of one cubic inch of a metal. Of course, the specific gravity, which has no units, has the same value for both metric and U.S. Customary style for a given material.

The density of a metal or alloy is proportional primarily to the atomic weight of the elements present, and secondarily to the metal's crystalline structure. The metal with the highest density, osmium, is about 40 times as dense as lithium, which is the lightest metal. Heating a metal will usually decrease its density because it expands. Significant changes in density also can occur with allotropic transformations, because the packing of the atoms may differ in each crystallographic structure. For example, iron will decrease in density at a given temperature if the metal can be transformed from the fcc (close-packed) crystalline form to the bcc form.

If two mutually insoluble molten metals (like aluminum and lead, or lead and zinc) are mixed together, the lighter metal will rise to the top, and there will be a sharp line of demarcation between the two liquid layers. Dissimilar metals that have no mutual solubility in the liquid state generally are unweldable by fusion processes.

Density is also important in fusion welding from the standpoint of flux and slag. If the flux or slag has a higher density than the molten metal, these nonmetallic molten materials will collect mainly at the bottom of the weld melt. Under such circumstances surface tension usually forces a film of flux or slag to cover the surface of the metal. The lower the density of the flux or slag, the faster it will rise to the surface of the molten metal. While gases have a very low density, they may remain dissolved in metals. When the gas is insoluble, bubbles of the gas attempt to rise quickly through the molten metal because of the great difference in density, and the bubbles will escape at the surface if the surface tension is not too great. Table 3.9 lists the densities of a number of familiar metals, alloys, nonmetals, and gases.

Density is not the sole criterion when determining the suitability of a metal for minimum weight equipment. Strength is equally important. For this reason, the strength-to-weight ratio is used to compare the load-carrying ability of metals on an equal weight basis, as indicated in Table 3.9 in the right-hand column. Efforts have been directed to making lighter, stronger alloys by adding elements of much lower atomic weight to decrease density — for example, adding lithium to aluminum. While it is possible to reduce density as much as 10 percent by alloying aluminum with lithium, changes in strength are equally important and depend on the crystalline matrix of the Al-Li alloys and the effects of phases and precipitates formed on the properties. In the case of iron, none of the very light elements (e.g., aluminum, lithium, magnesium, titanium) are metallurgically-suitable for alloying in sufficient amount to make a steel of lower density.

All types of steel have a specific gravity of about 7.8 (a density of about 0.281 lb/in.^{3}). Consequently, to reduce the weight of components, the strength of steel must be increased to permit smaller sections. Steels that can be hardened by heat treatment usually provide the greatest strength per pound. The strongest alloys of beryllium, titanium, and iron all compare closely on a strength-to-weight basis as indicated in the right-hand column of Table 3.9.

Thermal Properties

Metals exhibit a number of characteristics which are related in some manner to temperature change. These characteristics comprise the *thermal properties of metals*. Thermal properties include specific heat, thermal conductivity, expansivity, melting point, thermomagnetic effects, and others. Most of the thermal constants vary with the temperature at which they are measured, so this must be kept in mind

Table 3.9 - Density and Strength - Weight Ratio of Metals, Alloys and Nonmetals

Material	Density SI Metric (g/cm^3)	Density U.S. Customary (lb/in.^{3})	Approximate Strength-Weight Ratio, UTS/Density ($\frac{psi}{lb/in.^3}$)
Aluminum	2.7	0.098	105 000
Beryllium	1.8	0.067	650 000
Copper	8.9	0.321	87 000
Magnesium	1.7	0.064	545 000
Iron	7.9	0.284	125 000
Lithium	0.53	0.019	
Nickel	8.9	0.322	140 000
Osmium	22.6	0.818	
Titanium	4.5	0.163	180 000
Tungsten	19.3	0.697	309 000
Uranium	18.7	0.677	
Zinc	7.1	0.257	
Zirconium	6.4	0.232	125 000
Aluminum Alloy (530 MPa; 77 ksi)	2.6	0.096	800 000
Steel (2 068 MPa; 300 ksi)	7.8	0.281	1 100 000
Titanium Alloy (1 207 MPa; 175 ksi)	4.6	0.168	1 100 000
Beryllium Alloy (510 MPa; 74 ksi)	1.8	0.067	1 100 000
Carbon	2.0	0.072	
Silica	2.3	0.083	
Alumina	4.0	0.145	
Hydrogen	0.000089	0.0000032	
Oxygen	0.001429	0.000052	

when making quantitative calculations. These properties are not as structure-sensitive as the mechanical properties. Most thermal properties of metals are given proper consideration in welding, particularly where heating is essential to success of the joining operation.

Specific Heat

The heat or thermal capacity of a substance is measured as the number of calories required to raise one gram of that substance 1°C (or the number of Btu to raise one pound by one degree Fahrenheit). Specific heat is expressed as the ratio between the thermal or heat capacity of a substance and the thermal capacity of water at 15°C (59°F). Table 3.10 shows the surprising differences that exist between the specific heats of familiar materials. Water, it will be noted, requires more thermal energy per weight to raise its temperature than most of the materials commonly encountered in welding. Helium is unique among the gases, and its high specific heat (and thermal conductivity) explains its use as a coolant in sealed generating equipment for electricity. Very little change can be made in the specific heat of metals by alloying.

Heat in a substance was described in Chapter 2 as the kinetic energy in the vibratory motion of its atoms or molecules. In metal crystals, the atoms vibrate continuously about their equilibrium positions. As the temperature is raised, the amplitude of vibration steadily becomes greater, increasing the kinetic and potential energies of the atoms and causing thermal expansion. The thermal or heat capacity of a substance also increases with increasing temperature, requiring even more heat to raise its temperature. Consequently, comparison of the specific heats of different substances requires determining their values at a given temperature. The specific heat values in Table 3.10

Table 3.10 - Specific Heat of Some Metals and Nonmetals

Substance	Specific Heat* (Determined at 25°C, or 77°F)
Aluminum	0.215
Copper	0.092
Iron	0.108
Nickel	0.106
Titanium	0.125
Tungsten	0.032
Zinc	0.093
Zirconium	0.067
Potassium	0.180
Sodium	0.293
Carbon (graphite)	0.170
Water	1.00
Steam	0.50
Ice	0.50
Glycerine	0.60
Argon	0.124
Helium	1.240
Nitrogen	0.249

*Ratio between the thermal or heat capacity of substance and the thermal capacity of water, as defined by the number of calories required to raise one gram of substance through 1°C, or by the number of Btu required to raise one pound by 1°F.

show that various materials require different amounts of thermal energy to raise a fixed weight of the substance by the selected unit of temperature.

As a practical illustration, consider a piece of aluminum and a piece of tungsten each weighing the same. Although tungsten melts at a temperature 2750°C above that of aluminum, more heat is required to raise the piece of aluminum to its melting point than to heat tungsten to its melting point. This will be the case because the specific heat value of aluminum is about seven times that of tungsten. However, aluminum is only one-seventh as heavy as an equal volume of tungsten. Therefore, when welding a joint in a component made of aluminum, a much smaller quantity of metal by weight must be heated than if the same component was made of tungsten. Also, it is necessary to consider the rate at which this energy is conducted away from the joint, because conductivity could outweigh the specific heat of the metal as an influential property.

Thermal Conductivity

There are three mechanisms significant to welding by which heat can be transmitted from a heat source to a material: conduction, convection, and radiation. Of these three, conduction plays a significant role in welding more often than does convection or radiation. Metals conduct thermal energy mainly by electron transfer and by lattice vibration. Electron transfer can carry heat as well as electricity. Consequently, metals that are good conductors of electricity also have high thermal conductivity. When heat is applied locally to the metal, atoms in the lattice are excited to larger amplitudes of vibration. This increased vibration is transmitted to less energetic neighboring atoms through the interatomic bonding that couples the atoms together in the solid. However, the propagation of atom vibrations through the crystal lattice is impeded by impurities and lattice dislocations; so the role played by the lattice in thermal conductivity can be altered by alloying or by any other condition or treatment that significantly changes the number and type of lattice imperfections.

The amount of heat being conducted through a body of matter is proportional to the cross-sectional area and the difference in temperature or gradient between the measuring points; and it is inversely proportional to the distance or length between the measuring points. These factors can be arranged in an equation with a proportionality constant for thermal conductivity. In the literature on thermal conductivity, at least half a dozen different combinations are used to picture the quantity of heat that in one second passes through a cube when two opposite faces are maintained at a given difference in temperature. This thermal transmittance rate usually is identified as k, and it can be used to calculate the amount of heat flowing through any sized body of the same substance.

Thermal conductivity values for a number of metals and other materials used in welding are listed in Table 3.11. As can be seen from the values in Table 3.11, metals differ in their thermal conductivity; but, in the main, the metals are much better heat conductors than are the nonmetals. Copper is an excellent conductor, which accounts for the difficulty in welding copper using a relatively low-temperature heat source, like an oxyacetylene flame. On the other hand, the good conductivity of copper explains its efficiency as a "heat sink" when employed as a hold-down fixture or as a backing bar. Iron is a relatively poor conductor as metals go, which partly accounts for the ease with which steel can be welded and thermally cut. The addition of alloying elements to a pure metal always lowers the thermal conductivity because of (1) the greater number of dislocations and other lattice defects that exist in an alloy, (2) the effects of mixed atoms in the crystalline structure, and (3) the possible presence of mixed phases in the microstructure. All of these conditions shorten the *mean free path* in the metal along which thermal energy is transmitted. The mean free path is defined as the average distance traveled by an electron between collisions, and this dimension increases rapidly with decreasing temperature. In general, the thermal conductivity of a metal is directly proportional to the absolute temperature and to the mean free path for electrons in its crystallographic structure. Since the mean free path tends to diminish with increasing temperature because of thermal agitation, thermal conductivity usually tends to decrease with increasing temperature.

Convection

Convection as a transmittance mechanism for heat can take place only in a gas or a liquid and involves mechanical motion of the substance. Various circumstances can set into motion the gas or liquid that transports energy-bearing atoms or molecules to cooler regions, where their higher level of energy is given up until the temperature difference disappears. The common cause of convection currents is that heated gas or liquid is less dense, and therefore this portion tends to rise. Upon transfer of some of its heat, the cooled substance tends to descend. The rise and fall of the convection current attempts to establish a circulatory heat transfer system. Although convection current transfer is widely employed in building heating systems and in various industrial

Table 3.11 - Thermal Conductivity of Metals, Alloys, and Nonmetals

Substance	W/m/°K*	cal/cm²/cm/°C/s **
Aluminum	238.5	0.57
Copper	393.3	0.94
Iron	75.3	0.18
Magnesium	154.8	0.37
Mercury	8.43	0.02
Nickel	92.1	0.22
Silver	418.4	1.00
Titanium	221.8	0.53
Tungsten	167.4	0.40
Zirconium	225.9	0.54
Steel, low-carbon	71.1	0.17
Steel, high-carbon	66.9	0.16
Stainless Steel, 18-8	15.5	0.037
Carbon (graphite)	25.1	0.060
Glass	1.05	0.0025
Water	0.58	0.0014
Paper	0.12	0.0003
Argon	1.80×10^{-2}	0.043×10^{-3}
Carbon Dioxide	1.67×10^{-2}	0.040×10^{-3}
Helium	15.06×10^{-2}	0.360×10^{-3}
Nitrogen	2.59×10^{-2}	0.062×10^{-3}
Oxygen	2.64×10^{-2}	0.063×10^{-3}

* To convert values in W/m/°K to Btu/sq ft/ft/°F/hr, multiply by 0.577789.
** To convert values in cal/cm²/cm/°C/s to Btu/sq ft/ft/°F/hr, multiply by 242.08.

processes, it never constitutes the principal heat exchange mechanism in a welding process because it is relatively slow.

Radiation

Radiation as a heat transfer mechanism makes use of electromagnetic waves emitted from a source. While those rays which are in the visible light range (4000 to 7000 Å) characterize many of the welding processes, it is the invisible infrared rays which accomplish most of the actual heating. Despite the intense radiation emitted by the electric arc in welding, this is not the principal transfer mechanism that provides the heat needed to melt the joint and the filler metal. In fact, none of the welding processes for joining metals accomplishes its goal through radiation heating. Only limited use is made of radiant heat lamps in brazing. However, this method of heating is more effective for soldering, where lower temperatures are required.

The effectiveness of radiation as a heat transfer mechanism depends much more on its absolute temperature, the surface color, and the surface texture of the particular metal than on any inherent metal property. For instance, a blackened, dull surface is an excellent emitter of radiant heat as well as a good absorber; while a bright, polished metal surface will absorb less than 10 percent of incident energy reaching it, reflecting the remainder.

The ability to radiate heat is called *emittance*, and the numerical values used to express this capability are termed *emissivity*. Different metals will have unequal emissivity values. The intrinsic emissivity of different metals varies sufficiently with specific radiation wave length to warrant including this as a factor in equations when defining radiant exitance. Temperature measuring devices that register radiation from

Table 3.12 - Melting Points or Ranges of Metals, Alloys, and Nonmetals

Material	Melting Point, or Melting Range °C	°F
Aluminum	660	1220
Copper	1083	1981
Iron	1535	2798
Magnesium	650	1202
Nickel	1453	2647
Silver	961	1761
Titanium	1668	3035
Tungsten	3410	6170
Zirconium	1852	3366
Steel, 0.2% Carbon	1490 - 1520	2720 - 2770
Steel, 0.8% Carbon	1380 - 1490	2520 - 2710
Stainless Steel, 18-8	1400 - 1450	2550 - 2650
Cast Iron, 3.5% Carbon	1130 - 1200	2065 - 2200
Alumina (Al_2O_3)	2072	3760
Calcium Oxide, Lime	2614	4740
Iron Oxide, FeO	1369	2500
Iron Oxide, Fe_2O_3	1594	2900
Iron Oxide, Fe_3O_4	1565	2850
Magnesia, MgO	2800	5070
Silica, SiO_2	1720	3140
Titania, TiO_2	1840	3344
Zirconia, ZrO_2	2715	4920
Water	0	32

heated surfaces are usually designed to operate on a selected wave band, and they ordinarily require adjustment for the emissivity of the particular metal being scanned.

Melting Point or Melting Range

The melting point is defined as the temperature at which a substance under normal atmospheric pressure changes from the solid state to the liquid state, or vice versa. Elements and compounds have melting points; that is, they melt at a single or fixed temperature. Unlike compounds, mixtures of elements melt over a range of temperatures, unless they are eutectics or have other special characteristics. The melting points or melting ranges of a number of materials of interest are shown in Table 3.12.

The melting temperature is an approximate indication of the strength of bonds between atoms. A strongly bonded atom must be supplied with more thermal energy to dislodge it from a fixed position in a crystal. The higher the melting point or range, the larger must be the amount of heat supplied to melt a given volume of metal. Welding two metals of dissimilar composition becomes increasingly difficult as the difference in melting range is widened. It is easy to weld tin to lead, but not to tungsten. However, two pieces of a material, such as iron, may be joined with a metal of lower melting point, such as bronze if it wets and adheres to the iron body to which it has been applied. The iron is not melted during this braze welding.

Heat of Fusion

The heat added to a solid to bring it to its melting point is termed *sensible heat* because its input can be sensed by temperature rise. Upon reaching the melting point, however, an additional quantity of heat is required to convert the solid into a liquid, and this addition occurs without a corresponding rise in temperature. The quantity of heat required to complete the transition of a substance to its molten state is a finite amount characteristic of that substance, called the *latent heat of fusion*. The heat of fusion is a quantity less than the total sensible heat required to raise the temperature from ambient to the melting point, but it is greater than the endothermic and exothermic energies involved during allotropic transformations in a crystalline structure. Table 3.13 lists a number of metals to illustrate their differences in latent heat of fusion. These same quantities of heat are given up by each particular substance upon solidification; that is, when the liquid form cools to the solidification point, no further drop in temperature occurs until these latent heat quantities are liberated. Water (also listed) provides an interesting comparison, since the solid form (ice) has a latent heat of fusion that is comparable to most metals.

Viscosity and Surface Tension of Molten Metals

In the molten state, metal behaves as a liquid; and viscosity, surface tension, and interfacial tension are properties that can require consideration in welding, brazing, and soldering. Viscosity is a measure of the resistance to relative motion of two adjacent portions of the liquid, such as the resistance to instantaneous change of shape due to internal friction. As the layers of liquid move relative to each other, there will be a transfer of some of the momentum from one layer to the other. To overcome this fluid friction, which is termed *absolute viscosity*, force must be applied to maintain the relative laminar flow motion of the layers. The *coefficient of viscosity* is the shearing force required to maintain a unit difference of velocity between the adjacent layers. Therefore, viscosity has been given the dimensions of dynes-seconds/cm², and this standard unit is called a *poise*.

Water (at 20 °C, or 68 °F) has a viscosity of one centipoise. Many molten metals and alloys at a temperature just above their melting points have a viscosity of about two centipoise, which is not much different from that of water. Oils at room temperature commonly have viscosity values ranging from 100 to 1000 centipoise. The viscosity of molten metals and alloys is determined to some extent by their composition, but temperature is a much more influential factor, viscosity falls rapidly with rise in temperature. Conversely, the viscosity of gases increases with temperature rise because the higher kinetic energy of the gas atoms or molecules increases their severity of impact as they collide when moving in opposite direction.

Fluidity in a liquid, which is the reciprocal of viscosity, is measured in units called the *rhe*. This explains use of the term *rheology*, which is the science that deals with the intrinsic flow of materials.

Viscosity must not be confused with *surface tension*, which causes liquids in contact with their own vapors to reduce to minimum area. In the absence of gravity or other external forces, the stable form of a mass of liquid is spherical, because this formation has the smallest surface area for a given volume. The surface of a liquid acts at all times as if it is covered by an invisible membrane which is stretched under tension and is striving to contract. This membrane-like effect is attributable to forces that arise across the surface of the liquid because the atoms (or molecules) at the exposed surface are subject to interatomic forces only from within the liquid. Surface tension is measured in units of dynes/cm, and several methods are available for making determinations. These include sessile drop, bubble pressure, and pendant drop.

The degree to which surface tension is exhibited by a liquid depends on the strength of binding forces between its atoms. However, the temperature of the liquid and the presence of minor elements are also controlling factors. To avoid the effects

Table 3.13 - Melting Point, Latent Heat of Fusion, Boiling Point and Heat of Vaporization of Metals

Substance	Melting Point (°C)	Latent Heat of Fusion, (cal/g)	Boiling Point (°C)	Heat of Vaporization, (cal/g)
Aluminum	660	95	2467	2575
Copper	1083	50	2567	1130
Iron	1535	58	2750	1675
Magnesium	650	90	1090	1335
Manganese	1244	65	2095	976
Nickel	1453	72	2732	
Silver	961	25	2212	628
Sodium	98	27	883	951
Potassium	63	14	760	470
Tin	232	14	2770	641
Titanium	1668	105	3287	2347
Tungsten	3410	45	5660	
Zinc	420	24	907	425
Zirconium	1857	60	4377	1557
Water	0	80	100	540

Note: To convert the above values in cal/g.°C to J/kg. °K multiply by 4.187×10^3

of surface films of oxides or nitrides, surface tension is usually determined in vacuo or under an inert atmosphere. Water at room temperature under an air atmosphere has a surface tension of about 75 dynes/cm. Molten iron under a non-oxidizing atmosphere at a temperature of 1650°C (3000°F) has a surface tension of about 1700 dynes/cm, which is only slightly higher than that of molten copper (about 1300 dynes/cm at 1150°C, or 2100°F). Some minor elements in molten iron cause a decrease in surface tension. Carbon additions up to two percent appear to have little influence, but sulfur in amounts exceeding about 0.01 percent cause a marked decrease in surface tension.

When a liquid is in contact with a solid, or another immiscible liquid, this contact is characterized by *interfacial tension*, which determines whether the liquid in question will wet or spread on the other material. This more often is the set of circumstances dealt with in welding metallurgy — that is, to have the molten metal in contact with a molten flux or slag. Usually the interfacial tension for a specific combination of molten metal or alloy and a nonmetallic material must be determined experimentally.

Boiling Point and Heat of Vaporization

The boiling point is defined as the temperature at which the vapor pressure of a liquid equals the normal atmospheric pressure imposed on it. These temperatures also are listed in Table 3.13; and, like the process of melting, a characteristic amount of heat is required to convert a substance from liquid to vapor, and this quantity is termed the *heat of vaporization*. As this thermal energy is added to the substance at its boiling point, the temperature does not rise. The added heat is not retained by the liquid, but is carried away in the vapor. When the vapor is allowed to cool and return to the liquid state, this same amount of heat is given up by the vapor during the condensation process. The temperature does not fall below the boiling point until this heat is fully liberated. It must be recognized, however, that some evaporation or volatilization of a liquid, including metals, can occur at temperatures below the boiling point (witness the evaporation of water at room temperature in a dry room). If the ambient pressure over a liquid is increased, the boiling point will be raised. Conversely, if the pressure is decreased, the boiling point is lowered.

Accurate boiling point temperatures and heats of vaporization seldom are needed in welding. For this reason, the values in Table 3.13 are those which are most often presented in the literature. In applying these values, it should be noted that if a material of low boiling point dissolves in a material of high boiling point, the solution generally will have an intermediate boiling point. Phosphorus boils at 280 °C (536°F); but, when 0.75 percent P is dissolved in steel, it remains therein even at temperatures a little above the melting range of steel without boiling out. In metal arc welding, the temperature between the melt and the end of the electrode usually exceeds the boiling point of the metal. Consequently, the arc atmosphere contains a substantial amount of metal vapor.

Thermal Expansion and Contraction

Most metals increase in volume when they are heated, and they contract by the same volume when they are cooled. Common practice is to measure changes only in one dimension; hence, values are provided for linear expansion. Linear thermal expansion regularly is expressed as the ratio of the change in length per degree change in temperature to the initial length. This coefficient can vary substantially with the temperature at which it is determined. Therefore, it is important when calculating dimensional changes to be expected in a section of a particular material to use the appropriate coefficients that are applicable to specific ranges within the total anticipated temperature change.

The coefficient of thermal expansion of a metal is closely associated with its crystalline structure — unlike thermal conductivity, which is more dependent upon chemical composition. In fact, in a crystalline solid, the expansion may differ appreciably along the various axes of each crystal. This can be an important consideration with a metal section composed of a single crystal. Of course, in the usual polycrystalline metals, (assuming that no preferred orientation exists), the expansion rate will be an average of any variations associated with the crystallographic axes. Table 3.14 lists the coefficient of linear thermal expansion for a number of familiar metals and nonmetals as determined at room temperature. This tabulation provides an interesting comparison of materials, but if calculations are to be made for dimensional change in a section of any of these materials caused by temperature excursion, a handbook must be consulted to secure the proper coefficients to employ over specific ranges of the excursion.

Crystalline structure plays an important role in the thermal expansion of metals. The importance of crystalline structure is emphasized by the behavior of iron. At room temperature, the coefficient of thermal expansion for alpha iron with a bcc crystalline structure is 11.7 µm/m/°K. Gamma iron, with a more close-packed fcc structure, has a coefficient of about 23 µm/m/°K. Of course, this fcc phase exists above 912°C (1674°F) in iron. The room temperature coefficient of the gamma phase is seen in Table 3.14 as the value for 18-8 stainless steel, which has a fcc crystalline structure at room temperature. This value is about 50 percent higher than that of the carbon steel and of the three plain-chromium stainless steels listed, all of which have a bcc crystalline structure at room temperature.

Table 3.14 - Coefficient of Linear Thermal Expansion for Some Metals, Alloys, and Nonmetals *

Material	µ m/m/°K	10⁻⁶ in./in./°F
Aluminum	23.6	13.1
Copper	16.7	9.3
Iron	11.7	6.5
Lead	29.3	16.3
Magnesium	25.2	14.0
Nickel	13.3	7.4
Titanium	8.4	4.7
Tungsten	4.5	2.5
Zinc	39.7	22.0
Zirconium	5.8	3.2
Steel (AISI 1020)	12.1	6.7
Stainless Steel (18-8)	17.0	9.6
Stainless Steel (12% Cr)	11.7	6.5
Stainless Steel (17% Cr)	10.4	5.8
Stainless Steel (27% Cr)	10.4	5.8
Alumina	7.2	4.0
Silica (fused quartz)	0	0
Silica (ganister)	10.8	6.0

* Coefficient of linear thermal expansion and contraction determined at approximately room temperature.

Thermionic Work Function

When a metal or a metallic oxide is heated, electrons are emitted from the surface. The rate at which these elementary particles are ejected depends on the condition of the surface, temperature, ambient pressure, and intrinsic work function of the material — that is, the amount of energy measured in electron-volts required to free the electron from atoms present at the surface. An electron-volt (eV) is the energy acquired by any charged particle carrying unit electronic charge when it falls through a potential of one volt (1 eV equals 1.6020×10^{-19} J). Thermionic work functions of a number of metals and some nonmetals are shown in Table 3.15. A low work function means that electrons are more easily removed from the atom. Perhaps more important is the temperature to which a metal can be heated as a cathode, since the emissivity of electrons increases exponentially with temperature. This is why tungsten, with its high work function, is an effective electrode in the GTAW process. This property of metals and oxides plays an important role in arc welding, and its practical implications will be discussed in Chapter 6. Electrons released to the plasma of an arc can be used to carry electric current and to maintain a stable arc.

Electrical Properties

The electrical properties of metals and other materials play a major role not only in the joining operations, but also in post-weld treatments and in the testing of weldments. The property of conducting an electric current is common to all states of matter, although ability to pass current varies enormously among the materials. Metals are the most efficient conductors, and the dielectrics are the least effective inasmuch as they will transmit a charge only under extreme conditions. Metals are good conductors of both heat and electricity because of the unique freedom of their valence electrons. When a voltage difference is applied, there is a net flow of electrons in the direction of the field, which is from negative to positive.

Table 3.15 - Electron Thermionic Work Functions of Metals and Nonmetals

Material	Range of Measured Values, Electron-Volts (eV) Minimum	Maximum
Aluminum	3.8	4.3
Boron	-	4.4
Cerium	1.7	2.6
Cesium	1.0	1.6
Chromium	4.4	5.1
Copper	1.1	1.7
Gold	4.2	4.7
Iron	3.5	4.0
Lanthanum	3.3	3.7
Magnesium	3.1	3.7
Nickel	2.9	3.5
Silver	2.4	3.0
Strontium	2.1	2.7
Titanium	3.8	4.5
Tungsten	4.1	4.4
Zirconium	3.9	4.2
Carbon	-	5.0
Barium Oxide	4.9	5.2

Once the electron current is accelerated in a metal conductor, the electrons would continue to flow if it were not for the *resistance* to flow that is offered by the field of the positive ions in the metal lattice. An increase in the temperature of the metal increases the vibrational activity of the ions, which in turn interferes with electron flow; consequently, *conductivity* decreases. The addition of alloying elements to a metal invariably leads to decreased conductivity, because the presence of the solute atoms disturbs the regularity of the lattice. This derangement of the crystalline structure causes scattering of the conduction electrons. Conductivity would be infinitely greater if all of the atoms were completely at rest and the crystalline lattice were perfect. Conductivity is finite, however, because the free electrons that act as conductors for electrical and thermal energy are crowded together only about 2 Å apart. At room temperature, these electrons have a mean free path of about 10 Å, but when the temperature is lowered to 1 °K, the mfp is longer than 10 cm. This effect of near-absolute-zero temperature on mfp suggested long ago that electrons could flow through the lattice of a material and encounter virtually no resistance, a state now termed *superconductivity*.

The addition of only a small number of foreign atoms to a pure metal has a more potent effect on resistivity than is experienced when relatively large changes are made in the composition of alloys. This is understandable, because the initial minor addition to a pure metal causes distortion in an almost perfect crystalline structure; whereas, in a heavily alloyed material, the crystalline perfection has been substantially deranged, and further alloying does not make it much worse. Cold working a metal, as would be expected, decreases its conductivity. At the surface of a conducting material, electrons are subject to forces that prevent them from leaving the lattice.

One general law applies to all true conductors — namely *Ohm's law*, which states that the current transmitted by a material is directly proportional to the applied potential difference. Common terms are widely used for expressing the fundamental electrical properties governed by Ohm's law, but symbols for denoting these properties in equations vary to some degree in the literature. Electrical current commonly is denoted by I, and the potential (voltage) by V. The constant ratio V/I is the resistivity (ρ) of the material. The reciprocal of the resistivity is conductivity, sometimes denoted by the Greek letter "σ". Conductivity is the term used when discussing the capacity of a material for carrying electrical current; in fact, a standard unit that has been accepted internationally is footnoted in Table 3.16 for stating the volumetric measure of conductivity. Nevertheless, electrical capacity is more often measured and discussed quantitatively from the standpoint of ohmic resistance of a unit volume of the material. The measure of resistivity most widely used is microhm-cm, as shown in Table 3.16. Also listed in this table is the *temperature coefficient of resistivity*, which is defined as the increase in resistivity per degree divided by the resistivity at 0 °C.

Among the common metals, copper has excellent conductivity and low resistivity; only silver offers better values. This explains the wide usage of copper for conductor wires and buss bars, and the use of silver (even if only a plating) for smaller components where good conductivity and low surface resistance are all-important. Iron and its alloys, as can be seen from Table 3.16, have fairly high resistivity, which precludes their use as efficient conductors.

The technology of electricity includes a number of phenomena that may, at first, seem of secondary importance; but they often perform a crucial role in welding. Among these phenomena are *thermoelectric effects*, which are discussed in Technical Brief 18.

Magnetic Properties

All substances, whether in the form of a gas, liquid, or solid, will respond in some manner to an applied magnetic field, although to varying degrees. The magnetic field may be produced by an electric current, or it may be the flux from either a permanent magnet or an electromagnet. Five basic kinds of magnetic properties are recognized in the differing responses — namely (1) ferromagnetism, (2) paramagnetism, (3) ferrimagnetism, (4) antiferromagnetism, and (5) diamagnetism.

Ferromagnetism is the magnetic property of greatest interest in the context of welding metallurgy, because this particular magnetic behavior is frequently involved in welding operations. Ferromagnetism is the familiar property of some materials that allows them to be attracted by a magnet, and enables them to become magnetized themselves by proper application of a permanent magnet, or by an electromagnetic field. A distinction is made between "soft" and "hard" ferromagnetic materials. Soft ferromagnetic materials are easy to magnetize, but they retain little or none of the induced magnetism when the magnetizing force is removed. Hard ferromagnetic materials are difficult to magnetize, but they retain a significant degree of magnetization

Table 3.16 - Electrical Properties of Metals, Alloys and Nonmetals

Material	Conductivity, Volumetric, IACS, % *	Resistivity Microhm-cm	Resistivity Temperature Coefficient, per °C
Aluminum	64.9	2.7	0.0047
Copper	103.1	1.7	0.0043
Gold	73.4	2.2	0.0040
Iron	17.6	9.7	0.0066
Lead	-	19.3	0.0034
Magnesium	38.6	4.5	0.0178
Nickel	25.2	6.8	0.0067
Silver	108.4	1.6	0.0041
Tin	15.6	11.0	0.0046
Titanium	-	42.0	0.0048
Tungsten	-	5.6	0.0045
Zirconium	4.1	40.5	0.0040
Steel (Plain carbon)		10 to 50	
Stainless Steel (Cr-Ni)		72 to 85	
Stainless Steel (Plain Cr)		50 to 65	
Aluminum Alloys		2.7 to 7	
Copper Alloys		2 to 15	
Lead-Tin Solder (50-50)		15.0	
Titanium Alloys		50 to 80	
Zirconium Alloys		40 to 75	
Carbon (Graphite)		1375.0	

*Values listed are percent electrical conductivity as compared with that of tough pitch copper (99.92-99.96% purity copper with 0.03% oxygen) established to have a resistance of 0.15328 ohm for length of one meter weighing one gram at 20 °C (68 °F). This International Annealed Copper Standard (I.A.C.S.) has been universally adopted for industrial purposes.

> **TECHNICAL BRIEF 18:**
> **Thermoelectric Effects as Related to Metallurgy**
>
> Thermoelectric effects are utilized in various pieces of equipment, especially temperature measurement instrumentation. The *Seebeck effect* is experienced when two different metals or alloys are brought into electrical contact at two points, forming a circuit, and the two junctions are at different temperatures. Because of the difference in voltage that arises between two junctions, a small but consistent current will flow through this circuit. The Seebeck effect is the basis of precise temperature measurement by thermocouples. The *Peltier effect* is an inversion of the Seebeck effect; that is, when two different metals are brought into electrical contact and a current is passed across the junction, this contact point will be heated or cooled depending on the direction of the current flow and the nature of the dissimilar metals. The *Thompson effect* is somewhat similar, inasmuch as a dissimilar metal circuit with its junctures at different temperatures will undergo a transport of thermal energy from one junction to the other when an impressed current is passed through the circuit.
>
> One other thermoelectric effect worthy of mention is that of *superconductivity*. More than 75 years have passed since Onnes discovered that lowering temperature of certain metals to nearly absolute zero would virtually eliminate resistivity, but only recently have the tools and know-how reached the stage for successful development and construction of superconducting electrical equipment. Great forces accompany the very strong magnetic fields produced by the concentrated electric current in superconducting coils, and these components would "blow up" if they were not firmly restrained. Steels of many types are used in this equipment, and welding plays a key role in its construction.

after the force is removed. Of the 109 elements (listed in Table 2.4) only three are ferromagnetic at room temperature: iron, cobalt, and nickel. However, ferromagnetic alloys can be formulated using various metallic elements which individually are not ferromagnetic. Alnico is an example of an Al-Ni-Co-Cu-Fe alloy used to make permanent magnets that can lift hundreds of times their own weight.

Magnetization in a substance arises fundamentally from the magnetic moment generated by electrons surrounding the nuclei of its atoms (as briefly described earlier in Chapter 2). When the axes of the spinning and orbiting electrons become aligned through precession, a "coupling" is effected between the orbital and spin moments, and this produces magnetization. Whether magnetization will occur in a material depends on how its atomic moments are coupled.

Magnetization in a body of material is not a uniform, homogeneous state. Looking at a section as a whole, particularly in the case of a polycrystalline ferromagnetic metal, small regions called *domains* are seen which are already magnetized to saturation even before a magnetizing force is applied. However, the section as a whole does not appear to be magnetized because the direction of magnetization in each domain is randomly oriented, giving the section of material an overall net magnetization of zero. The domains are quite small and can be observed by optical metallography using various outlining techniques with magnetic powder. When a magnetizing force is applied to a ferromagnetic material, magnetization occurs by realignment of domains to achieve parallelism.

There are certain crystallographic axes along which magnetization is more easily achieved. This directionality in magnetization is termed *magnetocrystalline anisotropy*. When a magnetizing force of modest strength is applied, the domains are aligned by movement of Bloch walls to allow the overall section to perform as a magnet. This first occurs by growth of the domains having favorable crystal orientation. At higher field strengths, domains are coerced into directional alignment, and magnetization occurs quite readily along the favorably oriented crystallographic axes. Dislocations in the crystalline lattice retard magnetization. In fact, the presence of foreign atoms or compounds positioned interstitially, or appearing as a precipitate, can exert a dynamic effect upon magnetic properties termed *magnetic after-effect*. This refers to increasing magnetization of a material after a magnetizing force has been applied and removed. The time-dependent magnetization is attributed to redistribution of the interstitial atoms as encouraged by the magnetizing force applied earlier.

Magnetization is not a simple reversible process. When a magnetizing force on a material is reduced to zero, a certain amount of residual magnetism usually remains. This reluctance of the material to return to its earlier unmagnetized state is termed *hysteresis*, and the residual induction is termed *remanence*. Residual magnetism therefore can be looked upon as "magnetic inertia." Of course, the residual magnetism can be reduced to zero by applying a precise magnetizing force of opposite polarity or sign. The value of the force required to remove remanence is called *coercive force*, and this is an important quality both from basic and practical standpoints. When magnetization is depicted by plotting induction (B) against field strength (H) in one direction to determine the extent of remanence and coercive force, the magnetization and demagnetization curves form a *hysteresis loop*, which by its area indicates energy loss as heat during traversal of the loop.

Evaluation of Magnetization

A number of physical quantities are used in evaluating magnetic behavior. *Magnetization* (M) is the magnetic moment per unit volume of material. This, of course, provides a measure of the intensity of magnetization. When all of the dipoles in a material are completely aligned parallel to the applied field, the intensity of magnetization produced is called *saturation magnetization*. *Magnetizing force*, or field strength, usually is assigned the identifying letter H. Magnetic field force is measured in the oersted, which is a unit proportional to ampere-turns per centimeter. *Induction*, which is the flux density developed by a material from an applied magnetic field, is identified by the letter B. So-called lines of flux are measured at a right-angle to the direction of the flux. The flux of a magnetic field is the total number of lines passing through a surface, and this can be measured with a search coil or Hall probe. Flux is expressed as *webers* (Wb). One weber (one line) per m^2 is often called a *tesla* (T), and one tesla is equal to 10^4 gausses (because a gauss is equal to one line of induction per cm^2). When welding problems that are related to magnetic fields associated with workpieces are discussed, the *gauss* is the usual unit of measurement employed to express the strength of the magnetic field. This custom has popularized the term *degaussing* for the demagnetizing operation.

To compare the response of different materials to an applied magnetic field, reference often is made to *permeability*, which usually is denoted by the Greek letter "μ". This is the ratio of induction (B) in the material to the magnetizing field (H), and therefore μ is a measure of the ability of the material to accept magnetization. However, permeability is not always a constant. The permeability of a vacuum is unity. Nonmagnetic materials have a μ value of about one, and they display little variation with changes in the strength of the applied field. For example, Cr-Ni stainless steel of the nonmagnetic austenitic type ordinarily will display a μ value of about 1.003 in the annealed condition. Cold working this stainless steel results, of course, in the transformation of some of the austenite in the microstructure to martensite, and this can increase the μ value to perhaps as high as 15.0. Ferromagnetic materials have a permeability much greater than one, and their μ value varies in nonlinear fashion with the strength of the applied magnetic field. Iron at room temperature, depending upon the applied field, can display a μ value ranging from two thousand to possibly as high as ten thousand. Numerical figures for permeability are used mainly as a convenient criterion for comparison of materials.

The *Curie point* is a characteristic temperature exhibited by all ferromagnetic and ferrimagnetic materials above which enough thermal energy is available to remove spontaneous magnetization. Therefore, the Curie point is a transition temperature above which a ferromagnetic or a ferrimagnetic material is changed to a paramagnetic material. Iron encounters a Curie point at approximately 770°C (1420°F). Although loss of magnetism does not occur at a single temperature, it occurs over a very narrow range of temperature.

Summary of Magnetic Behavior

The two important kinds of magnetic properties listed at the start of this review are covered in the following capsule descriptions.

Ferromagnetism is the familiar strong attraction of one magnetized body to another, and it is displayed by a material when its atoms tend to have the magnetic moment of electrons in parallel alignment even in the absence of an applied field. The principal requirements for a material to exhibit ferromagnetism are (1) that the material's magnetic moments must be in parallel, and (2) that they should be strongly maintained in this position by interatomic force. Bulk material will have many small magnetic domains which exert a net magnetization that is dependent upon composition and processing history. In general, all ferromagnetic materials are solids; they usually contain iron, cobalt, or other elements of either the first transition series (titanium through nickel) or of the later transition series (rare earths). They have permeabilities of considerable magnitude that vary with the field, they exhibit hysteresis, and they lose those properties fairly abruptly at the Curie point. Magnetism vanishes above the Curie point because the ordered ferromagnetic character of the material becomes disordered. The material then is paramagnetic. However, upon cooling below the Curie point, the material again becomes ferromagnetic, but it is unlikely to exhibit magnetic remanence.

Ferrimagnetism is a magnetic behavior somewhat like ferromagnetism. Although the atoms in the material tend to assume an ordered arrangement of magnetic moment, a certain amount of non-parallelism persists among the domains. Consequently, the spontaneously induced magnetism of a ferrimagnetic material is not as complete as in the base of a ferromagnetic material. Ferrimagnetism is observed in certain compounds (often referred to as "ferrites" because they consist of a metal oxide in chemical combination with iron oxide) and in certain alloys. The term ferrimagnetic was coined to describe the "ferrite-type" magnetic spin order and scheme of coupling. Many ferrimagnetic materials are poor conductors of electricity, and this can be an advantage in certain devices. When ferrimagnetic materials are subjected to rising temperature, they pass a Curie point and become paramagnetic.

Involvement of Magnetization in Welding

The role played by magnetization in welding is quite broad. In addition to the normal anticipated functions of magnetism in generating and transforming electrical energy, magnetic forces that produce unwanted effects must be dealt with. One such unwanted effect is "arc blow," which sometimes arises in an arc-welding operation. Many welding processes such as electron-beam welding are particularly sensitive to the magnetic properties of the workpieces even though the process itself is basically dependent upon magnetization technology.

Magnetization of a metal, particularly ferromagnetization, can affect a number of the metal's mechanical and physical properties. Although the effects of magnetization are comparatively minor, the changes can touch upon such physical properties as thermal expansion, specific heat, and electrical resistivity. Consequently, if the affected property plays a key role in the performance of a welding process, some consideration or adjustment in parameters may be required to ensure normal operation. For example, ferromagnetic metals, in general, have relatively low electrical resistivity; however, magnetization usually causes a small but measurable increase in resistivity — sometimes as much as five percent. Also, the presence of ferromagnetic workpieces within the magnetic field of a resistance welding machine can sap electrical energy and cause lack-of-fusion because of inadequate power to make a proper weld.

The side effects of magnetization, although slight, sometimes can be used to good advantage, and they can lead the way to new welding processes and methods of weld examination. For example, *magnetostriction* is a physical change in the dimensions of a body of metal produced by magnetization, and this phenomenon is utilized in the heart of ultrasonic welding machines. When a rod-like member of ferromagnetic metal is magnetized, a small but reproducible change in dimensions occurs. If a nickel rod is magnetized, it will contract slightly in length (about 0.004 percent). This dimensional change can be instantly nullified by cutting off the magnetic field force. By applying a suitable alternating magnetic field, the rod can be caused to vibrate lengthwise. In this manner, magnetostriction energizes the transducer or "driver" of an ultrasonic welding machine. Iron and other ferromagnetic alloys undergo magnetostriction that usually is negative (contraction), but sometimes is positive under the influence of an applied magnetic field.

CHEMICAL PROPERTIES

Metals were the first elemental substances discovered, separated, and named by man, and the documenting of their chemical properties started long before the

Christian era. Although the chemical behavior of certain metals is pointed out in this text from time to time as various welding procedures are described, *corrosion* is the multiform chemical activity of metals that gives chief cause for concern about the performance of some weldments in service. Man pays a tremendous price for the ravages of corrosion on metal structures, vehicles, tools, and equipment of many kinds. Because metals and alloys differ greatly in their propensity to corrode under various conditions, this subject has been methodically studied, and the literature holds much information concerning both *corrosion resistance* as a desirable quality and *corrosive attack* in its many forms. With the exception of Noble metals such as gold, nearly all metals are found as compounds in the earth's crust. They exist as compounds because the formation of compounds decreases free energy. This same driving force to decrease free energy is the reason why man-made metallic articles will revert to compound form whenever the environment can act on the metal. Iron, for example, when exposed to air under certain common circumstances, will slowly combine with oxygen to form red rust similar to the familiar hemantitic ore. Corrosion is a problem with most metals because as products of corrosion, their compound forms are more stable thermodynamically.

Corrosion of Metals

Corrosion occurs under many environmental conditions, the attack manifests in many different forms, and many kinds of metals and alloys are subject to this problem. Table 3.17 lists forms of corrosion encountered in metals to illustrate the wide variety of attack, and it proposes a separation of these corrosion forms into five categories to give an indication of factors that can be highly influential in instigating attack. Note

Table 3.17 - Forms of Corrosion and Instigative Conditions

I. General Attack, or Uniform Metal Loss

Aqueous Solution Corrosion	Microbiological Corrosion
Atmospheric Corrosion	Molten-Salt Corrosion
Corrosion Fatigue	Molten-Metal Corrosion
High-Temperature Oxidation	Sulfidation
Hot Gas Corrosion	

II. Localized Attack

Biological Corrosion	Galvanic Corrosion
Crevice Corrosion	Pitting Corrosion
Filiform Corrosion	Stray-Current Corrosion

III. Mechanically Assisted Degradation

Erosion Corrosion	Cavitation Corrosion
Fretting Corrosion	Impingement Corrosion
Fretting Fatigue	Corrosion Fatigue

IV. Metallurgically Influenced Corrosion

De-alloying Corrosion	Nonmetallic Inclusion Attack
Intergranular Corrosion	Exfoliation Corrosion

V. Environmentally Induced Attack

Caustic Embrittlement	Liquid-Metal Embrittlement
Hydrogen Damage	Stress-Corrosion Cracking

that the forms listed are not always instigated through the circumstances outlined by the five categories. In fact, it is not unusual for corrosion to initiate as one of the forms and then, after progressing to some degree, translate its attack into another form. Perhaps the most common example of this changeability occurs when corrosion initiates as pitting, and later continues as stress-corrosion cracking. The nature of various forms of corrosion is best understood, and the reasons for change from one form to another rationalized, when the fundamentals involved in each form have been clearly established. Certain reactions and mechanisms are common to a number of the corrosion forms listed in Table 3.17. This discussion will center on these fundamental aspects and will emphasize the behavior of iron, along with the carbon and low-alloy steels. The corrosion of other metals and alloys will be given attention only when their behavior adds to needed understanding.

Welding introduces unusual conditions into a metal structure, many of which can become unfavorable factors in a corrosion process. Under corrosive conditions differences may be observed between the weld metal and the base metal, and sometimes between the heat-affected zones of the base metal and the unaffected metal. Even subtle surface conditions left by welding — such as thin oxide films, residues of deliquescent flux particles, ripples on the weld surface, or the toe condition of an as-deposited fusion weld — can become factors in the attack by a corrosive environment.

The local atmosphere, the most common environment to which metals are exposed, determines a metal's *atmospheric corrosion*. A greater tonnage of ferrous alloys is lost to corrosion in the atmosphere than in any other environment. Yet, this seemingly simple medium can be as variable and as complex as a chemical process. The most important variable is moisture in the atmosphere. Pollutants in air can also greatly accelerate corrosion. Depending on the local atmosphere to which it is exposed, iron and steel can either (1) remain essentially free of corrosion (rusting) when air is very dry (i.e. desert-like conditions); (2) succumb to rust over a period of time when exposed to dampness; or (3) be short-lived if repeatedly wetted, especially if certain pollutants also are present. Despite widespread encounter with atmospheric corrosion, attention must be directed to forms of attack that occur very frequently in four particular media: aqueous solutions, hot gases, molten metals, and molten salts.

Corrosion in Aqueous Solutions

Fresh natural water, rain, sea water, contaminated water, and many man-made aqueous solutions represent the most frequently encountered environments that can pose a corrosion threat to metals. Although the reactions in aqueous solutions are quite complicated, a simplified description should give a helpful perspective. Because aqueous solutions are capable of ionic conductivity, they are *electrolytes*. Corrosion therefore, is an electrochemical process. Metal exposed to electrolyte tends to dissolve and form ions which migrate from anodic areas on the surface into the electrolyte, making the metal an electrode. A proportionate quantity of hydrogen ions are released from the electrolyte and deposited on adjacent metal surfaces which function as cathodes. The rate at which the metal loses ions to the electrolyte depends on the rate at which hydrogen is removed from the cathodic areas. Since the hydrogen usually escapes by evolution or is oxidized to water by oxygen available in the electrolyte, the electrochemical reaction proceeds at the expense of the anode. The final products of this electrochemical activity can dissolve in the electrolyte, or be precipitated as corrosion products. If dissolved in the electrolyte, its conductivity will increase, and the corrosion rate probably will increase accordingly.

Environments that function as aqueous electrolytes usually are classified according to their acidity, or their basicity (alkalinity). This is done with the aid of the "pH" scale, a system for expressing hydrogen-ion concentration and electrochemical potential. The number 7 has been selected to indicate a neutral solution. Decreasing numbers below seven indicate increasing hydrogen-ion concentration and stronger acidity. Increasing numbers above seven indicate increasing alkalinity. Acid solutions with low pH values (pH <5) tend to be very corrosive to iron and steel.

Acids also can be classified as being either *reducing* or *oxidizing* in character. Hydrochloric acid and dilute sulfuric acid are reducing-type acids. Nitric acid is oxidizing in character, as evidenced by its ability to form a protective oxide film on some alloys that can make the metal passive and resistant to attack. Near-neutral pH conditions (approximately pH 5 to 9) greatly reduces the rate of attack as compared with the strong acidic conditions. Above pH 9, however, new corrosion characteristics are likely to be encountered, particularly with steels. First, there will be a tendency for a small increase in rate of uniform corrosion as compared with near-neutral corrosion rate. Some metals — like aluminum, zinc and lead — undergo accelerated uniform corrosion as alkaline aqueous solutions are increased in pH value. Secondly, at very high pH, and especially at somewhat high temperatures, steels can suffer stress-corrosion cracking, a form of attack to be described shortly.

To estimate the direction and rates of electrochemical reaction between a substance and an electrolyte, a method has been developed using reference electrodes to measure potential difference across their interface. The hydrogen electrode has been selected as the international standard for determining the electro-potential of metals. This allowed extensive plotting of diagrams to show the thermodynamic conditions as a function of electrode potential and concentration of hydrogen ions. These plots of E - pH (equilibrium potential versus pH) are called *Pourbaix diagrams*, and they relate to the electrochemical and corrosion behavior of metals in aqueous solutions. They do not provide information on rates of reaction, but they do give guidance on how corrosion can be minimized or avoided.

An electrochemical cell is produced when two different metals or alloys are in electrical contact in an electrolyte, and accelerated attack will occur on one of the materials depending upon their relative electromotive potentials. The less noble (anodic) metal is attacked to a greater degree than if it were exposed alone. Likewise, the more noble (cathodic) metal is attacked to a lesser degree than if it were exposed alone, if indeed it is attacked at all. This behavior is known as *galvanic corrosion*. Often it can be recognized by markedly more severe attack on the anodic metal's side near the junction than has occurred elsewhere on the surfaces of the two dissimilar metals.

The corrosion behavior of metals and alloys can be compared with a list called the *galvanic series*, which arranges them according to their corrosion when coupled in sea water (as the electrolyte). At the top of the galvanic series are the least noble or anodic metals — those having greatest potential for providing protection through self-sacrificial action to metals listed at lower positions. Toward the bottom of the series are the most-noble metals, which tend to behave as cathodes. These metals do not corrode. The larger the potential difference between two metals, the greater will be the probability of galvanic attack on the metal occupying the higher position in the series, and the greater will be the likelihood of galvanic protection for the metal in the lower position on the list.

Table 3.18 lists a number of metals and alloys in the sequence of the galvanic series, including the potential for each elemental metal as measured with the standard hydrogen electrode (SHE). Although the potential determined with the SHE is strictly a laboratory reference value, the sign and the progression of the values support the long-standing galvanic series established in sea water. It will be noted that titanium is listed in both the upper and lower parts of the series. The first listing indicates corrosion because of anodic attack, and the second listing indicates resistance to corrosion and the metal's tendency to perform as the cathode in a coupling with metals positioned above it. This dual listing is explained by the fact that titanium remains chemically active until a thin oxide film forms on its surface whereupon it becomes resistant to most corrosive media. Coupling titanium to other metals does not cause sacrificial corrosion of the titanium except in reducing environments where the surface of the titanium cannot remain passive (oxide protected). In fact, titanium usually is the cathodic member of any galvanic couple in an aqueous solution; and when full passivity is maintained, its standard potential is slightly more favorable than that of platinum.

Table 3.18 - Galvanic Series, and Electromotive Force Values for Metals

	Metal or Alloy	Reaction			Standard Potential (E°,V) *
Corroded End (Anodic or Least Noble)	Lithium	$Li^+ + e^-$	$\Rightarrow$	Li	-3.04
	Magnesium	$Mg^{2+} + 2e^-$	$\Rightarrow$	Mg	-2.37
	Beryllium	$Be^{2+} + 2e^-$	$\Rightarrow$	Be	-1.85
	Aluminum	$Al^{3+} + 3e^-$	$\Rightarrow$	Al	-1.66
	Titanium (when active)	$Ti^{2+} + 2e^-$	$\Rightarrow$	Ti	-1.63
	Zirconium (when active)	$Zr^{4+} + 4e^-$	$\Rightarrow$	Zr	-1.53
	Zinc	$Zn^{2+} + 2e^-$	$\Rightarrow$	Zn	-0.76
	Iron	$Fe^{2+} + 2e^-$	$\Rightarrow$	Fe	-0.44
	Cadmium	$Cd^{2+} + 2e^-$	$\Rightarrow$	Cd	-0.43
	Low-carbon steel	—	$\Rightarrow$	—	—
	Nickel (when active)	$Ni^{2+} + 2e^-$	$\Rightarrow$	Ni	-0.26
	Lead-Tin Solder 50-50	—	$\Rightarrow$	—	—
	Tin	$Sn^{2+} + 2e^-$	$\Rightarrow$	Sn	-1.38
		$Sn^{4+} + 4e^-$	$\Rightarrow$	Sn^{2+}	-0.15
	Lead	$Pb^{2+} + 2e^-$	$\Rightarrow$	Pb	-0.13
	Hydrogen (Reference)	$2H^+ + 2e^-$	$\Rightarrow$	H_2	0.0000
	Copper	$Cu^+ + e^-$	$\Rightarrow$	Cu	0.52
		$Cu^{2+} + 2e^-$	$\Rightarrow$	Cu	0.34
Protected End (Cathodic or Most Noble)	Nickel and Nickel Alloys (passive)	—	$\Rightarrow$	—	—
	Stainless Steels (passive)	—	$\Rightarrow$	—	—
	Silver	$Ag^+ + e^-$	$\Rightarrow$	Ag	0.80
	Platinum	$Pt^{2+} + 2e^-$	$\Rightarrow$	Pt	1.12
	Titanium (passive)	$Ti^{3+} + 3e^-$	$\Rightarrow$	Ti^+	1.25
	Gold	$Au^{3+} + 3e^-$	$\Rightarrow$	Au	1.50

* Standard potential (E°,V) measured at 25°C or 77°F, determined with SHE.

A number of metals and alloys display this dual active/passive character — as examples, the metal zirconium, and the high-chromium alloys (e.g., the stainless steels).

Another aspect of galvanic corrosion that affects the rate of attack on the anodic metal is the cathode/anode area ratio. The larger the cathode area that is electrically coupled to the anode area, the more rapid will be the dissolution of the anode. When service conditions foster galvanic corrosion, accelerated anodic attack promoted by a large cathode and a small anode is common. This rate-controlling phenomenon has been put to practical use to slowly sever the legs of offshore platforms by wrapping a wide, thin copper band (the cathode) around the heavy-wall tubular steel leg leaving it electrically coupled it to the surface. A circumferential slit about 0.13 mm (0.005 in.) wide through the copper band exposes the steel, allowing concentrated anodic attack to take place. This arrangement provides a cathode-to-anode area ratio exceeding 10 000 to 1, and corrosion in sea water can penetrate the steel leg at a rate of approximately 200 mm (8 in.) per year.

Uniform Corrosion

Uniform corrosion in an aqueous solution defines what appears to be an equal rate over the entire surface of a metal. However, it still involves an electrochemical reaction of anode dissolution and cathodic protection in which the anodic and cathodic areas are quite small. They are created by the micro-heterogeneity in chemical composition and microstructure of the metal. Indeed, as the reaction and attack progresses, there is a constant changing of anodic and cathodic areas as dictated by the changes in micro-distribution of chemical elements and microstructure. Consequently, the attack progresses at an average rate that appears uniform on the metal section. While this form of corrosion accounts for great loss of metal articles, it is not a kind of attack to be feared.

Uniform corrosion usually is anticipated; if not, its occurrence can be detected in the early stage. Often, the rate of metal loss through uniform attack can be predicted, and the useful life of the article can be estimated with reasonable accuracy by use of a *potentiodynamic polarization resistance measurement* technique (see ASTM G 59).

Corrosion rates also can be determined directly by exposing specimens to actual or artificially accelerated attack, and then removing the corrosion products and measuring the loss in specimen weight. The weight-loss data often are converted to loss-of-thickness, and they are usually expressed in terms of penetration per unit of time — as examples, mm/yr or mils/yr. If corrosive attack on a test specimen is not uniform, then a penetration rate obtained by conversion of weight-loss data can be misleading. Specimens containing welds must be inspected carefully after testing to determine whether accelerated attack occurred to any significant degree in the vicinity of the weld.

Pitting

Pitting in an aqueous solution is a more serious form of corrosive attack that makes many kinds of metal articles unfit for further service. Often, metal in the pitted areas is impractical to repair, and must be replaced. Pitting also is the result of an electrochemical action, but it occurs in very localized spots; often the pits appear initially in scattered areas. These very small pits increase in number, diameter, and depth with exposure time. Pitting is particularly damaging in thin plate or sheet weldments because the diameter of the individual pits does not necessarily indicate their depth. A vessel or container can be made unfit for service by perforation from pitting, even though the metal surface shows little evidence of serious corrosive attack.

Pitting often occurs when a metal possesses just borderline resistance to a corrosive environment. Localized breakdown of passivity on the surface allows miniature electrochemical cells to function and form a pit. The corrosive attack continues at the bottom of the pit and is accelerated by the depletion of oxygen in the confined area. Pitting also may start at impurities or at defects exposed at the surface of the metal. Practically all metals and alloys are subject to pitting corrosion because the environment can be a major factor in promoting this form of attack. For example, if small particles of foreign solid material are deposited at random on a metal surface which is exposed to an aqueous solution, these particles may cause severe pitting of the metal just beneath them. Among the possible actions which may occur here are (1) a depletion of oxygen in the electrolyte between the particle and the metal surface, and (2) an increase in the concentration of ions which contribute to the strength of the electrolyte. Weight-loss determinations are of no practical value when studying pitting corrosion. Even if the depth of pits are probed and measured, there seldom is good reproducibility between test specimens and actual articles in service. Selection of material with resistance to pitting is best guided by tests in the actual service environment rather than in accelerated laboratory tests. Nevertheless, a number of relatively new potentiodynamic and potentiostatic polarization testing techniques have been developed that give helpful guidance.

Crevice Corrosion

Crevice corrosion is another form of localized attack in an aqueous solution which also is electrochemical. This attack often is concealed from view because of its occurrence beneath gaskets, in seams, cracks, abutting (unwelded) edges, and many other kinds of narrow openings or gaps that as a group are called crevices. Corrosion in the crevice, although undetectable on the surface, may cause severe damage before corrosion products are exuded and recognized. Crevices enable aqueous solutions to instigate electrochemical cell activity in a shorter period of time and to concentrate the anodic dissolution in a localized area because of the oxygen differential that develops. Within the crevice, as oxygen is consumed, its replacement is impeded by restricted circulation into the confined area. Because the surrounding exposed surface has ready access to oxygen, this area becomes cathodic relative to the crevice area. The passivity of the exposed cathodic surfaces is maintained by available dissolved oxygen, but the oxygen-starved solution in the crevice continues to concentrate ions and become a more potent electrolyte. As a practical example, sea water typically has a pH of about 8, but when sea water has opportunity to perform as an electrolyte in a crevice, this medium can attain an acidic pH of 1, whereupon electromotive force between anodic areas in the crevice and cathodic areas on the exposed surface quickly activates vigorous electrochemical cell action.

Intergranular Corrosion

Intergranular corrosion takes place when the corrosion resistance of the grain-boundary areas of an alloy is less than that of the grain interior. This attack is electrochemical in nature, and it is promoted by any microstructural differences that produce either a zone of reduced corrosion resistance or a zone having a markedly negative standard potential along the grain boundaries. This zone may be depleted of an alloying element helpful to the corrosion resistance. This depletion can occur through precipitation of the element in the form of a phase or a compound in the boundary.

Galvanic difference between the metal in the interior of the grains and an intergranular phase can promote intergranular attack under corrosive conditions. As grains of metal are loosened by corrosion through the grain boundaries, they are dislodged from the surface. This results in extensive localized penetration. The surface may take on a crystalline appearance, and feel "sandy" to the touch. Sometimes the intergranular attack penetrates deeply before grains are lost at the surface. Consequently, thin sections can display leakage before serious section loss is observed. Carbon and low-alloy steels are not especially prone to intergranular corrosion. This form of attack is more likely to be observed under certain conditions with the stainless steels, and with alloys of aluminum, copper, and nickel.

Biological and Microbiological Corrosion

Biological and microbiological corrosion are two forms of attack that are only slightly different in overall appearance, and they are somewhat similar in electrochemical nature. *Biological corrosion* usually refers to localized crevice attack that occurs under relatively large organisms that do not form a continuous film on an exposed metal surface; the most common example is the accumulation of barnacles on a metal structure or boat hull in sea water. The area covered by the attached barnacle is shielded from dissolved oxygen in the sea water and thus becomes the anode in an electrochemical cell.

Microbiological corrosion is less localized because it is initiated by microscopic organisms deposited as films or coatings. However, conditions sometimes cause irregular areas to be coated leaving, the remainder of the surface essentially bare. Microbiological corrosion has been encountered with a number of aqueous solutions, emulsions of oils, sewerage effluents, and water-bearing soils — all of which present favorable conditions of temperature, etc., for growth of the organisms having unique biological activity. The oxygen-cell effect created by a coating of these micro-organisms is only part of their corrosion threat. In some cases, the organisms themselves enter into the attack mechanism. Although their functions are not completely understood in every case, a number of biological activities carried out by various kinds of bacteria and fungi have been studied and identified. The following is an example of their step-like progressive action:

(1) colonization of iron bacteria (*Gallionella*) and iron/manganese bacteria (*Siderocapsa*) on the exposed metal surface;
(2) microbiological concentration of iron and manganese compounds, primarily as chlorides when this anion is available;
(3) microbiological oxidation to corresponding ferric and manganic chlorides, which, in addition to forming these severe pitting corrodants, creates an acidic chloride solution; and
(4) penetration of the protective oxide film on the metal surface, which likely is already weakened by oxygen depletion under the biodeposits.

Microbiological corrosion has been experienced with a wide variety of metals (aluminum, copper, iron) and alloys (aluminum alloys, copper-nickel alloys, and stainless steels). Often the attack is in the form of minute pits at the start, but these will increase in size and number as the attack progresses. In some environments, small knobby protuberances of corrosion product will rise above the corroded surface, and this form has been called *tuberculation*. When this form occurs on the inside of a tube, it can progress to fill the entire bore and close off flow.

Corrosion in Hot Gases

Heated gas to be held in a welded pressure vessel or piping system may be simply air or steam, or it may be a more aggressive gas. Regardless, oxygen will be widely encountered, and its rate of attack will be highly dependent upon whether the metal or alloy normally forms a protective type of surface oxide. Therefore, the reaction in oxidation by the hot gas is not electrochemical; rather, it is a diffusion-controlled process. When metal is exposed to oxygen at an elevated temperature, the first step in the oxidation process is the absorption of atomic oxygen. Oxides then nucleate at favorable sites on the surface, and grow laterally to form a complete, very thin film. As this oxide layer thickens, its physical nature and its compatibility with the metal beneath determines whether it will afford protection against further conversion of the metal to an oxide. If the layer is porous or flaky oxidation likely will continue until the metal is completely destroyed.

The progressive scaling of iron or plain steel at a high temperature, possibly 1000°C (roughly 1800°F), in air is a good example of attack which continues unimpeded by the products of the attack on the surface. On the other hand, if the metal forms an impervious, tight film of corrosion product on the surface, this layer may serve as a barrier and effectively prevent further attack. The addition of substantial amounts of chromium, silicon, or aluminum to iron can change the scaling characteristics of the alloy so that a protective type of oxide film will form. Yet, this protective film must be fully compatible with the base metal upon which it forms. In fact, the base metal acts as a substrate, and the oxide that grows upon it often tends to align its crystalline structure in some registry with that of the base metal. This can be an unfavorable epitaxial tendency since the structural form of the oxide thusly formed may be more-or-less unstable. Differences with the base metal in thermal expansivity, or expansions due to phase transformations in the base metal may cause cracks in the oxide, leaving the base metal vulnerable to localized oxidation.

Attack by hot gases containing oxides of chlorine or sulfur usually are more damaging than oxidation alone. The corrosion products seldom form an impervious protective surface layer. In fact, under some conditions involving sulfur, this element can diffuse into the base metal, often progressing to an appreciable depth by intergranular penetration. Furthermore, the sulfides formed may be driven to even greater depth as finger-like protrusions through a unique mechanism of preferential oxidation and continued diffusion. When the protrusions act as stress-raisers, this form of attack can adversely affect the mechanical properties of the metal section.

Corrosion in Molten Metals

Molten or liquid metals often are handled in vessels and piping systems, and they often must be endured for extensive lengths of time without unacceptable attack. Examples of molten metal containment range from large iron pots holding molten zinc for galvanizing operations to sodium-cooled fast-breeder nuclear reactors, in which stainless steel piping is used in the heat-transfer area and refractory metal alloys must be used for the reactor. The metal or alloy selection for a welded containment system in some applications may be based upon long-standing empirical experience, but in new situations the material selection may have to be guided by accelerated laboratory testing.

Prediction of corrosion rate and the performance of welds in the molten-metal containment system is fully as complex as predictions with aqueous solutions. Accelerated attack by molten metals occurs in several distinctly different forms that bear little resemblance to those encountered with aqueous solutions. There can be (1) uniform direct dissolution of the solid surface of the metal container; (2) surface degradation, whereby only certain elements in an alloy container or in the molten alloy enter into a reaction; and (3) intergranular penetration by the molten metal into the solid metal of the container. The latter form of attack commonly is labeled *liquid metal cracking* (LMC), and this attack can be greatly accelerated under certain circumstances by the presence of molten cracking-agents. In the case of steel, the offending agents include cadmium, copper, lead, tin and zinc. Ways in which these elements can be encountered during the fabrication of weldments and cause LMC difficulty will be reviewed in Volume II, Chapter 13.

Corrosion in Molten Salt

Molten salts held in welded metal containers and piping systems can cause two kinds of corrosion. The first is metal dissolution caused by the solubility of the metal in the molten salt. The second occurs more frequently, and consists of oxidation of the metal to ions. This mechanism is analogous to corrosion in an aqueous solution; because molten salts usually have good ionic conductivity and this makes the molten-salt/metal surface interface electrochemically similar to the aqueous solution/metal surface interface. In fact, all forms of attack experienced with aqueous solutions have been observed in fused-salt handling equipment — galvanic attack, stress-assisted corrosion, etc. Furthermore, because molten salt systems operate at high temperatures, forms of attack not seen with aqueous solutions are possible. Also, thermal gradients in a molten salt system can cause more rapid dissolution of metal at hot spots. Because the salts used in the molten state may consist of fluorides, chlorides, nitrates, sulfates, hydroxides, carbonates, etc., the reactions on metals vary greatly. For example, molten fluorides, because of their fluxing powers, do not allow protective films to form on metal surfaces. They also can carry out selective removal and oxidation of elements from alloys. Chromium is especially vulnerable to this kind of attack. Welding is used extensively in fabricating the equipment to contain molten salts. Weld metal and heat-affected zones require special attention to ensure that these areas withstand attack as well as the unaffected base metal.

Forms of Corrosion Pertinent to Weldments

Welding introduces so many chemical, mechanical and physical conditions in a fabricated article that it is entirely possible for a type of steel which ordinarily would be considered safe in a particular environment to suffer some form of aggressive attack as a weldment. Of course, attention must be given to extraneous conditions often present in welds such as adhering slag particles, oxide films, irregular surfaces, etc. Certain features in design also may make the weldment more susceptible to corrosion — sharp corners that accumulate foreign deposits, fluid-harboring crevices between overlapping flat surfaces, and dissimilar metals joined in a combination that poses a galvanic corrosion threat. Furthermore, there are attendant conditions and circumstances of a metallurgical nature in weldments that can lead to disabling corrosion.

Four general features common to many weldments that sometimes instigate corrosive attack are outlined below:

(1) Weld metal, even when its chemical composition is essentially the same as wrought base metal, has an as-welded microstructure that is markedly different and perhaps not as corrosion-resistant as the base metal.

(2) Weld metal, particularly when secured mainly from filler metal, can be sufficiently dissimilar from the base metal in chemical composition to generate a potential difference in some environments; thus the emf can trigger galvanic attack.

(3) The heat-affected zones of base metal (as-welded) can have a markedly different microstructure than the unaffected base metal, and for this reason the HAZ may display localized corrosion susceptibility.

(4) Residual stresses — both short-range and long-range, both compressive and tensile — will be present in the weldment (as-welded), and the magnitude of these stresses will be high enough to figure prominently in stress-corrosion phenomena.

These general features of steel weldments should be kept in mind, because they can act as the root-cause of corrosive attack. Although a particular feature of a weldment may trigger corrosion, actual aggressive attack often is compounded by multiple adversarial factors in the weldment and in the service environment. General guidelines on guarding against unanticipated loss of metal section by corrosion are difficult to formulate, because a single change in fabricating procedure sometimes will result in a complete reversal of results in service. Consider, for example, the matter of dissimilarity in chemical composition between base and weld metals as outlined in the second feature described above. If the weld metal is only slightly different in composition, then galvanic effect may cause increased loss by corrosion of the weld metal section. On the other hand, if the weld composition is significantly richer in alloy content and has better corrosion resistance, then galvanic attack may be directed to the base metal. Sometimes the attack is highly localized on the base metal heat-affected zones because of lowered corrosion resistance in these areas; such an attack is called "knife line" corrosion.

Stress-Corrosion Cracking (SCC)

Stress-corrosion cracking is the most frequently encountered form of disabling attack in steel weldments. Sometimes the SCC is completely unexpected, but most often it is anticipated because of previous experience, and it may be viewed as a virtually inescapable problem in certain industrial processes or applications. Despite much industrial experience and research, SCC continues to be a major corrosion problem, particularly as industries become involved with new technologies. To minimize the problem, a good understanding is needed of the circumstances that are likely to foster attack by SCC.

SCC Mechanics

The essential mechanics of SCC involve a delayed metal failure that results from a synergistic interaction between mechanical stresses in the metal and a corrosion reaction at its surface. Cracking initiates at the surface exposed to an aqueous solution, and when viewed metallographically, the progress of attack can be either *intergranular* as illustrated in Figure 3.38 (A) or it can be *intragranular* as illustrated in Figure 3.38 (B). These two patterns of cracking seldom occur concurrently, and a prominent feature of SCC that aids in its identification is the multiple-branching pattern shown. Applied independently in alternating cycles, neither the stress nor the corrosive exposure would result in the same kind of failure. Failure by SCC proceeds as time-dependent subcritical crack propagation. Ordinarily, there is little visible evidence of corrosion on the metal's surface when SCC takes place, as might be indicated by weight loss or by the presence of corrosion products on the surface (i.e., rust on iron or steel). Nevertheless, SCC eventually can penetrate a metal section and provide a pathway for leakage. A fortunate inclination of tanks and vessels that operate with pressurization is that the early stage of oncoming failure by SCC usually involves leakage, and this signals serious pending difficulty. Under some circumstances, however, failure of a pressurized container can occur catastrophically. This is because the extent of SCC penetration, although progressing undetected, has exceeded tolerable defect size.

Although the mechanism by which SCC progresses depends upon the kind of material exposed, in general the intergranular form of SCC is developed by an anodic mechanism, whereas the intragranular form involves a cathodic mechanism. Separation of the observed mechanisms into two general categories has been quite helpful in developing a basic understanding of each form of SCC attack.

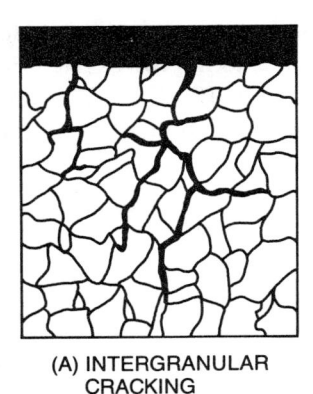

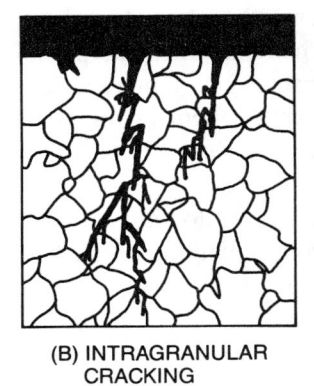

(A) INTERGRANULAR CRACKING

(B) INTRAGRANULAR CRACKING

Figure 3.38 — Two forms of stress-corrosion cracking (SCC) that occur in metals; simply illustrated

Both forms of SCC penetrate the metal by multiple-branching, but the intergranular pattern (A) develops by an anodic mechanism, whereas the intragranular form (B) involves a cathodic mechanism. These features of SCC assist in identifying the problem and determining the cause.

Anodic SCC

Anodic SCC mechanisms involve extremely localized dissolution of metal to form very narrow, corroded, crack-like pathways via the grain boundaries which progress through the metal section. Anodic dissolution involves passive film rupture, which exposes bare metal at the crack tip and then corrodes very rapidly to extend the fissure-like pathway. Although metal at the crack tip may repassivate, stress concentration at the tip appears to cause emergence of slip-steps which rupture the passive film, bringing on further rapid penetration by anodic dissolution. The intergranular pattern of SCC also is dependent upon grain-boundary conditions that foster electrochemical potential and give the boundary area a negative standard potential. Intergranular SCC is more likely to be the form of attack when potential is in the active-to-passive transition.

Cathodic SCC

Cathodic SCC mechanisms involve four steps in sequence: (1) the evolution of nascent, atomic hydrogen at the cathode, just as would be expected during the chemical corrosion reaction; (2) adsorption and absorption of this hydrogen into the metal; (3) diffusion of hydrogen in the metal to highly localized areas under stress which, in the main, are crack tip zones; and (4) embrittlement of these localized zones to the degree that hydrogen-induced subcritical cracking initiates. As more hydrogen is liberated at the cathodic surfaces of the crack, the steps just summarized are repeated, and crack growth takes place in micro-increments. With this mechanism of SCC, it is understandable that cracking would be influenced by crystallographic orientation of the grains and therefore would proceed along intragranular planes. This hypothesis of the cathodic mechanism appears valid, because the behavior of steels undergoing the intragranular form of SCC is in close accord with established knowledge of hydrogen-induced cracking of steel by hydrogen. For example, high yield-strength steels are more susceptible to intragranular SCC. Furthermore, the maximum growth rate of these cracks has been found to occur near room temperature (20°C or 70°F), while either higher and lower temperatures decrease the growth rate — exactly the effect with hydrogen-induced cracking brought on by straightforward gaseous introduction. On the other hand, anodic (intergranular) SCC increases in growth rate as temperature is raised [especially as temperatures of approximately 100°C (212°F) are exceeded] because the chemical corrosion reaction is accelerated at higher temperatures.

Three particular aspects of a weldment act as the determinants of occurrence of SCC, namely:

(1) the metal or alloy employed and its microstructural condition,
(2) the environment to which the weldment is exposed and the duration of exposure, and
(3) the level and distribution of tensile stresses.

If any of the above can be removed or effectively altered to meet prescribed criteria, then SCC will not occur. Although many other secondary factors will influence the onset of SCC and its rate of growth, the three features listed above are the principal deciding factors with respect to susceptibility of the weldment to SCC.

SCC-Susceptible Materials

Materials susceptible to SCC include most of the commonly used metals and alloys, but propensity to fail by this form of corrosion is very much dependent on the particular environment to which the given metal is exposed. Briefly, while most metals can fail by the SCC mechanism, the susceptibility or immunity of each metal to SCC must be related to behavior in a defined environment. Table 3.19 lists a number of commonly used metals and alloys that have been selected to illustrate in a preliminary way the recognized combinations of metals and environments that portend the threat of SCC. Later, all three determinates of SCC — metal, environment, and stress — will be examined for their individualized contribution to this corrosion process, but for the sake of space this review must be confined to information concerning the carbon and low-alloy steels.

Carbon and low-alloy steels vary in their susceptibility to SCC; but, as yet, there is no mechanistic model by which a steel's behavior can be rationalized or predicted. While carbon steels display somewhat greater susceptibility to SCC than the low-alloy steels, no particular alloying addition appears to significantly assist the resistance of steels to SCC (other than major alloying as found in the stainless steels). In fact, low-alloy steels, because of their higher yield strengths, often have lower resistance to SCC. The microstructure of steels, especially grain size, has a measurable influence both on SCC initiation and crack-growth rate. Steels with finer ferritic grain size tend to have a lower crack-growth or penetration rate during the SCC process. A quenched-and-tempered microstructure will have a good resistance to SCC even in the more potent environments, providing the tempering temperature is high enough to eliminate untempered martensite, and the final microstructure is not very hard. High hardness usually is accompanied by appreciable internal, short-range residual stresses. In general, weld metal with an as-welded microstructure is more susceptible to SCC than wrought steel of similar composition.

Nonmetallic inclusions in steel can be detrimental to SCC resistance because they provide crack-initiation sites. Sulfide inclusions are especially undesirable for this reason, and therefore elevated levels of sulfur in steel will tend to reduce resistance to SCC. The same can be said for high residual phosphorus content; although the mechanism by which phosphorus aids crack initiation and growth in the SCC process is not as evident as in the case of sulfide inclusions.

Table 3.19 - Commonly Used Materials and Environments in Which Failure by Stress-Corrosion Cracking Has Been Experienced in Industry or in Tests

General Classification of Material	Broad Nature of Environment
Aluminum alloys	Aqueous chloride solutions, and tropical marine conditions
Brass	Solutions of ammonia, and mercurous nitrate
Copper alloys	Aqueous solutions of ammonia, ammonia salts, and moist SO_2 gas
Magnesium alloys	Aqueous solutions of chlorides
Nickel alloys	Aqueous solutions containing lead ions
Steels, carbon and low-alloy	Hot caustic, acidified cyanide solutions, aqueous solutions of nitrates, phosphates, and sulfates; especially if hydrogen sulfide is present
Stainless steels (austenitic)	Hot aqueous solutions containing chlorides
Titanium alloys	Red fuming nitric acid, and chlorinated hydrocarbons
Zirconium alloys	Aqueous solutions containing chlorides, and gaseous iodine

SCC-Supporting Environments

Environments reported to support SCC also are included in Table 3.19, but those listed are described only in very broad terms because actual service media vary so widely in chemical species. The listings in Table 3.19 are a small sample of the service media reported to have supported SCC. The potency of media employed in industry to promote SCC differs greatly according to their make-up of chemical components, including minor elements and compounds that enter as tramp constituents of raw materials. Most of the threatening environments reported for carbon and low-alloy steels are aqueous solutions. These can be bulk liquid in contact with the steel's surface; but more threatening is a condensed film of an aqueous solution on the surface, especially if alternating cycles of condensing and drying cause an offensive solute to increase in concentration in the condensate. However, the chemical constituent that promotes SCC need not be present in a large amount. A number of elements and compounds are reported to act as a "poison" with respect to SCC, and surprisingly low levels of these offensive elements in the environment can greatly increase the likelihood of SCC initiation as well as increase the SCC growth rate. They include antimony, arsenic, carbonates, phosphates, selenium, sulfur, nitrates, and hydrogen sulfide. Hydrogen sulfide (H_2S), is very aggressive in promoting SCC by the cathodic mechanism, and H_2S frequently is encountered as a contaminant. Some poisonous agents can enter the environment from materials used for lubrication, caulking, oxygen-scavaging, buffering, and other purposes incidental to the principal operation. Consequently, analyses performed on the actual operating environment often are needed to be certain that agents with adverse influence are not present at a level that is likely to be threatening.

Environments designed to form a protective film of some kind on the surface of the exposed metal can promote SCC if the film coverage is imperfect. If the metal surface becomes exposed at pinholes or at rifts in the film, highly localized dissolution of metal is likely to occur at the bare areas. Pitting can be a damaging form of corrosion in itself; but seemingly innocuous pits can provide micro-points of stress concentration and oxygen scarcity, and in this way they can trigger the initiation of SCC.

Irradiation as occurs in nuclear power generation must be included among the environmental conditions that promote SCC. As neutrons and other high-energy particles penetrate the metal, they can produce abnormal lattice vacancies in the crystalline structure which accelerate the SCC process.

Sulfide stress cracking (SSC) and sulfide stress-corrosion cracking (SSCC) are terms that appear frequently in the literature to call attention to the very strong propensity to SCC that hydrogen sulfide (H_2S) imparts to environments. Hydrogen sulfide sometimes is almost inescapable as a contaminant in industrial operations. Wet H_2S-bearing environments are among the most aggressive in promoting hydrogen entry into metal. A substantial portion of crude oil, feedstocks, natural gases, mineral slurries, and water from deep wells handled by industry can be characterized as "sour" because they contain wet H_2S to a significant extent. Also, effluents from process streams and scrubbing operations that contain organic materials also can undergo decomposition reactions that generate H_2S.

When H_2S dissolves in water, the solution is weakly acidic; but it is sufficient to react with iron and steel and generate hydrogen ions at the cathodic areas of attack. Furthermore, H_2S also acts as a catalyst to promote absorption of the hydrogen by the metal. The presence of additional acidifying agents such as acetic acid can make the potency of a sulfide-containing environment much greater. An aqueous solution of H_2S is so effective in promoting hydrogen pick-up by steel that, in addition to causing SCC, other types of hydrogen damage can occur. All told, one or more of the following five difficulties can be encountered with an environment that includes H_2S:

(1) stress-corrosion cracking (SCC),
(2) hydrogen embrittlement (HE),
(3) hydrogen-induced cracking (HIC),
(4) blistering and delamination, and/or
(5) Hydrogen stress cracking (HSC).

The terms used above for difficulties stemming from hydrogen, along with their acronyms, appear throughout the literature and can guide a search for data on each particular form of damage. Further description of these forms of failure and their mechanisms will be put off for reason of space until Chapter 13 of Volume II. However, the presence of H_2S is especially threatening to steels of higher strength and hardness, and some helpful guidelines have come from technical groups for avoiding these difficulties.

There is a direct relationship between H_2S concentration in the environment, the strength and hardness of the steel, and the stress threshold for the initiation of cracking. NACE has recommended in their MR-01-75 that steel components which contact corrosive fluids containing H_2S should have their hardness limited to 22 HRC (approximately 248 HB) to avoid difficulty with sulfide stress-corrosion cracking (SSCC). Although this hardness limit can be met without difficulty when selecting a steel base metal, close attention to welding procedure may be required to ensure that the hardness of the weld metal and the HAZ conform to the requirement. When stronger steels (i.e., above approximately 620 MPa, or 90 ksi) fail in a H_2S-containing environment, the damage often is in the form of HSC. Under these circumstances, the cracks travel through the microstructure as intragranular fissures, with little or none of the secondary branching that is characteristic of SCC.

Stress and SCC

The level of stress needed to promote SCC is highly dependent upon the material being exposed and the nature of the environment. Although stress level is the third decisive factor in determining whether SCC will initiate, it appears that the matter of stress is of lesser importance than material or environment. This view is held because stress that initiates SCC in an unfavorable situation can be as low as 15 MPa (about two ksi). Consequently, the causative stress can be residual tensile stresses remaining from any cold working of the material, from heat treating operations that may have been performed to harden the material, or from welding of the material. If attention is given only to stress from applied load, residual stresses possibly would not be recognized and considered during computations. Failure to consider the influence of residual stresses could result in unexpected failure by SCC. This oversight is not uncommon.

Compressive stresses do not cause SCC; in fact, they can suppress SCC in an environment that ordinarily would initiate failure. If the stresses are cyclic, then any envi-

290   Chapter 3

ronmentally induced cracking is termed "corrosion fatigue," and the crack-like penetration into the metal is unlikely to display the multi-branched appearance that is typical of anodic and cathodic SCC as illustrated in Figure 3.38. There is general agreement that SCC has three recognized regions of crack-propagation rate versus stress level — namely:

- Region I - *Initiation of cracking* at a stress threshold followed immediately by increasing crack growth rate as stress intensity rises at the frontal tips of the cracks;
- Region II - *Steady-state crack growth* as stress at the crack tips tends to stabilize at intermediate intensities and growth rate tends to plateau; and
- Region III - *Rapid crack propagation* as the unattacked load-carrying section becomes much smaller and stress at the crack tips sharply increases. Fracture of the entire section, or intolerable penetration then becomes a final stage of SCC.

Efforts have been made to calculate stress intensity (K) either by LEFM or by EPFM at the crack tips during SCC. This determination for the threshold of crack initiation (Region I) has been labeled K_{ISCC}. Theoretically, no crack initiation should be encountered with a given metal or alloy in a particular environment at a stress below K_{ISCC}. As cracking develops under stress in tests, the velocity of crack growth, da/dt, is plotted against K. In this second stage of crack growth (Region II), a steady applied load increases the crack length gradually and raises the stress intensity (K) at the crack tip accordingly.

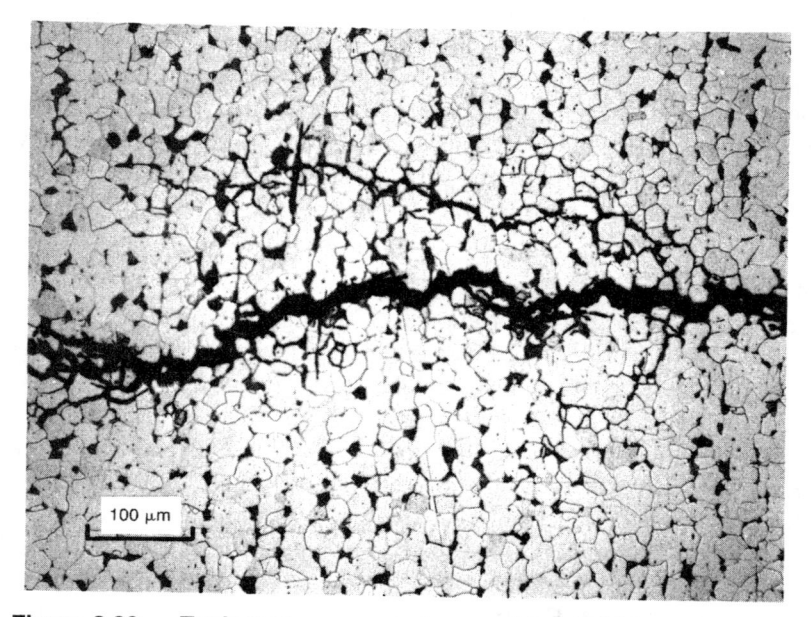

Figure 3.39 — Typical stress-corrosion cracking (SCC) of the intergranular form produced by the anodic mechanism in the microstructure of plain steel (Nital etchant; original magnification: 150X)

(Courtesy of The Welding Institute, England)

Finally, in the third stage (Region III), the rate of SCC growth rapidly increases as the crack lengths grow and the stress intensity at the crack tips approaches the critical stress intensity for brittle fracture.

Other SCC Factors

The secondary factors influencing SCC, particularly those associated with environment, are numerous. They include temperature, pressure, hydrogen-ion concentration (pH, acidity/basicity), cyclic exposure, fluctuating humidity, circulation and viscosity of contacting fluids, and degree of aeration. Because weldments are used in very diverse

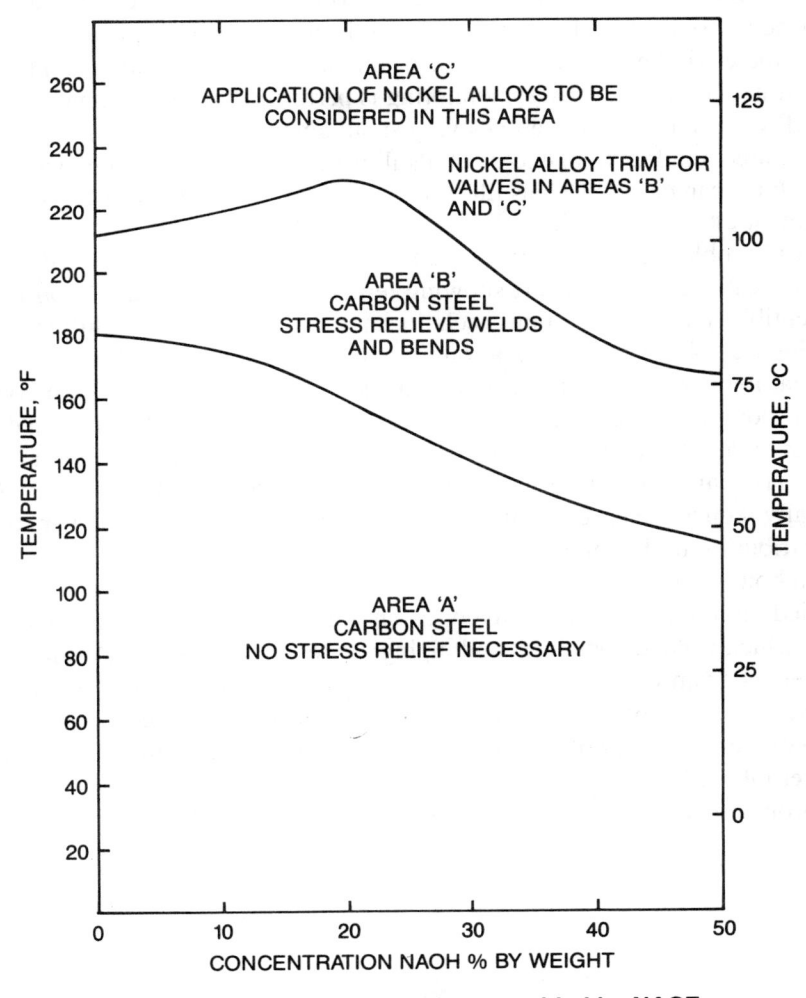

Figure 3.40 — Caustic soda service chart provided by NACE

Corrosion data survey, Metals Section, National Association of Corrosion Engineers, 6th Edition, 1985.

service environments, assessing the threat of SCC should be based on case-by-case reports of similar applications in which problems have occurred.

Literature on *caustic embrittlement* in steel is an example of a form of SCC that has confronted operators of boiler plants for many years. Caustic embrittlement was a very serious problem, because it resulted in a number of dramatic boiler explosions. However, the steel does not become embrittled; its still-solid grains remain ductile. Instead, intergranular attack penetrates the steel, weakening the steel section and reducing its capacity to act in a ductile manner. This intergranular attack is recognized as SCC that has developed by the anodic mechanism. Unfortunately, failures occasionally occur in steel boilers, piping and valves when responsible personnel are unaware of the environmental conditions that must be monitored to avoid this form of SCC.

Briefly, the aim in hot water and steam boiler operations is to avoid handling acidic water (pH < 7) because of metal loss by uniform corrosion. Therefore, small amounts of alkali (caustic soda) are added to the water periodically to keep the pH at about 10.5, at which level there will be practically no corrosion on plain steel. To maintain this slightly alkaline water condition, regular monitoring is performed and small amounts of alkali are added when required. Sometimes very small amounts of other inhibitors, such as nitrates, also are included. If an excess of alkali is added, such water at elevated operating temperatures can cause SCC, as illustrated in Figure 3.39. This is a classic example of intergranular attack by anodic dissolution in plain steel.

The SCC problem also can arise when an operational fault allows salts to accumulate at a localized area. A chart showing temperature and concentration limits for SCC susceptibility of plain carbon steels in aqueous caustic soda solutions, such as shown in Figure 3.40, can serve as a rough guide to safe operating limits. Additional metallurgical items should be taken into consideration, such as the carbon content of the steel. Although iron and decarburized steel are resistant to SCC, the propensity of carbon steel to develop SCC increases when carbon is brought below about 0.10 percent, with maximum susceptibility at approximately 0.05 percent carbon. Fine ferrite grains usually reduce the susceptibility to SCC, as do strong carbide-formers that alter carbide distribution in the microstructure, particularly those which minimize carbides in the grain boundaries.

Liquified ammonia, used extensively as an agricultural fertilizer, is another environment to which welded containers and piping are regularly exposed. Ammonia tanks, fabricated either from carbon or from high-strength low-alloy steels, have developed SCC despite the relatively low service temperature at which liquefied ammonia is held. It is caused by trace amounts of free oxygen and nitrogen in the ammonia. A small addition of water (about 0.2 percent) to the ammonia greatly reduces susceptibility to SCC in this environment, a practice now used in the manufacture of liquefied ammonia.

SUGGESTED READING

Metals Handbook, Corrosion, 9th Ed., vol. 13.; ASM International, Materials Park, OH; 1987; ISBN: 0-87170-019-0.

Mechanical Testing, ibid, Volume 8; 1985; ISBN: 0-87170-014.

Standard Methods for Mechanical Testing of Welds, ANSI/AWS B4.0-92; American Welding Society, Miami, FL; 1992.

Stress-Corrosion Cracking, Materials Performance and Evaluation, Jones, R.H. ed.; ASM International, Materials Park, OH; 1992; ISBN: 0-87170-441-2.

Designing for Corrosion Control, Landrum, R.J.; National Association of Corrosion Engineers, Houston, TX; 1989; ISBN: 0-915567-34-2.

Guidelines for Fracture-Safe and Fatigue-Reliable Design of Welded Structures, Pellini, W.S.; The Welding Institute, Abington, Cambridge (UK); 1983; ISBN: 0-85300-166-9.

Fatigue of Welded Structures, Gurney, T.R; Cambridge University Press, Cambridge, England; 1979; ISBN: 0-521-225582.

Chapter 4

In This Chapter...

ALLOYING **295**
   Alloys in the Liquid State 296
   Phase Diagrams 299
      Binary Phase Diagrams 304
      Ternary Phase Diagrams 307
      Phase Diagrams for Multi-Element Alloys 308
   Alloys in the Solid State 310
      Factors Influencing Solid Solubility 310
      Formation of Intermediate Phases/Compounds .. 316
   Mechanisms and General Effects of Alloying 316
      Role of Crystalline Structure 317
      Role of Microstructure 319
      Mechanisms for Altering Mech. Properties 319

ALLOYING ELEMENTS IN IRON **326**
   Carbon 329
      Analysis of the Iron-Iron Carbide Diagram 331
   Manganese 337
   Phosphorus 338
   Sulfur and Selenium 340

   Silicon 342
   Copper 343
   Chromium 345
   Nickel 346
   Molybdenum 346
   Niobium (Columbium) 347
   Vanadium 348
   Aluminum 348
   Nitrogen 349
   Titanium 352
   Boron 354
   Cobalt 355
   Tungsten 355
   Lead 355
   Other Alloying Elements 356

BENEFIT OF REVERSING THE ALLOYING
TREND **357**
   Residual Elements 358

SUGGESTED READING **359**

Effects of Alloying Elements

A material to be welded is seldom a pure or elemental metal, since the properties of metals for constructional purposes can be markedly improved by the addition of other elements. Therefore, the great majority of materials subjected to welding are alloys of two or more metals. Occasionally, so-called pure metals are handled in special joining operations — for example, in the electronics field, or the nuclear industry. Even then, the metals are not pure, because it is extremely difficult to avoid or remove the last traces of foreign atoms from an elemental metal. Therefore, labels are applied to metals, signifying them as being "commercially pure," or "chemically pure." Often a list is provided comprising residual impurities that are still present.

The majority of materials handled in weldment fabrication contain not only the residual elements that are difficult or costly to remove, but also deliberate minor additions of elements that facilitate the production of the material in the form required. Nickel, for example, may contain a small addition of manganese to minimize hot shortness induced by residual sulfur. Because such minor additions do not change the chemical behavior and characteristic properties, the metal is not regarded as an "alloy," but is looked upon as a commercially pure metal.

There are many reasons for preparing alloys of metals. The most common reason, of course, is to improve mechanical properties such as hardness, tensile strength, or ductility and toughness. Other reasons for alloying could be to improve corrosion or scaling resistance, to alter electrical or magnetic properties, and perhaps to improve longevity in nuclear service. Many years ago, alloys were formulated mainly to improve the service performance of the material. Gradually, fabricators and manufacturers who converted the alloys into useful products called for improvement in the capability of the alloy to be cold formed, hot forged, machined, polished, and shaped by other specific operations. Joining by welding, however, was given little attention for many years. Therefore, many of the currently available alloys may or may not have any real capability for welding. As welding became more widely used as a method for joining, the weldability of alloys grew in importance. Producers of metals and alloys now regularly give consideration to all aspects of weldability when developing a new material for the commercial market. Even the older alloys which display unsatisfactory welding qualities are adjusted or modified in composition to make them more amenable to modern joining processes.

ALLOYING

When alloying is studied from the standpoint of welding, the entire gamut of this subject should be pursued because any of its facets can be pertinent to the production or to the utilization of a weldment. An alloy, by common definition, is a metallic substance that contains more than one element. The real interest lies in commercial alloys that contain a number of deliberately added elements. However, before consideration of these multi-element additions and their effects, initial focus should be on binary alloys. Binary alloys are comprised of only two elements — a base metal and one added element that may be a metal, a metalloid, or a nonmetal (perhaps even a gaseous element). Later, it will be necessary to deal with ternary alloys (three elements), quaternary alloys (four elements), and more.

Alloys in the Liquid State

Because welding often makes use of melting in its processes, there is good reason to focus discussion of alloying on that which originates in the liquid state. The usual starting point for a binary alloy is to prepare a molten bath of the base metal and to add either another kind of metal or a nonmetal. This addition can be in solid form if its melting point is below that of the molten bath; but if not, the addition may require pre-melting to ensure proper alloying. Melting is not the only way of preparing an alloy; new techniques involving powder metallurgy and ion implantation also can form an alloy of controlled composition. Nevertheless, the majority of material supplied commercially for constructing weldments is produced by one of the melting processes to be described in Chapter 5.

When a second element is added to the molten bath of base metal, the atoms of the added element can enter into several different states. The atoms can either (1) remain more or less undissolved and thus retain their identity, (2) enter fully into solution in the molten base metal, or (3) form a compound with the base metal. When the added element does not dissolve in the base melt, this is called an *immiscible alloy*. Two distinct layers of molten metal will form in the bath when the percentage of each metal exceeds the solubility limit in the other. Lead and zinc provide a good example of almost complete immiscibility, much in the same manner as water and oil. Immiscible alloy systems are not uncommon; at least 500 immiscible alloy combinations have been identified, one classic example of which is described in Technical Brief 19.

When the element added to molten base metal enters into solution, the combination of metals may be completely soluble in each other, and a homogeneous liquid alloy is formed. This may be true regardless of the proportions of each element involved. As examples, any proportions of molten iron and nickel, tin and lead, or aluminum and copper will commingle and form homogeneous liquid solutions. On the other hand, some combinations of metal and an added element may show only partial solubility, and two immiscible layers will appear in the molten bath when the percentage of one metal exceeds its solubility limit in the other. Chromium in quantities up to about 35 percent by weight can be dissolved in copper; however, any chromium added beyond this amount forms a second separate chromium-rich liquid layer. When the element added to a melt enters into solution with the base metal, at least in part, a significant effect occurs with regard to the freezing and melting of the alloy. Homogeneous binary alloys usually freeze and melt over a temperature range rather than at a single temperature as found with pure metals. Therefore, with an alloy, a *liquidus* is found, which is the temperature at which various compositions of the system begin to freeze on cooling, or complete their melting on heating, and a *solidus* is found at which temperature freezing is complete upon cooling, or at which melting begins in

> **TECHNICAL BRIEF 19:**
> **Lead as an Immiscible Alloy in Iron and Steel**
>
> A classic example of an immiscible alloy combination — and one that deserves special attention — is the metallic element, lead, added to iron or steel. Lead has an extremely low solubility in molten iron, and it is not oxidized in the presence of molten iron. Consequently, virtually all lead added to iron will collect as pools of nearly pure lead and settle under the pull of gravity because of its higher density. Ordinarily, when two liquid phases of an immiscible system form stratified layers based upon density difference, the heterogeneous liquid freezes as stratified solid layers. If the metals involved are completely immiscible, the separate layers freeze and melt at the same temperatures as their respective pure metals. Little use has been found for such stratified materials in the solid form. In the absence of the earth's gravity, the density difference is inconsequential and the two immiscible metals can be melted and mechanically mixed or allowed to achieve homogeneity by diffusion. This is a promising new direction that metallurgical research presently is taking on NASA shuttle flights. Experimental results will show whether unique, homogeneous alloys of elements ordinarily immiscible can be produced under microgravity conditions.

the solid alloy on heating. The span of temperature between the liquidus and the solidus is an important piece of information about an alloy because it can be indicative of susceptibility of the particular composition to cracking during solidification. A large difference between liquidus and solidus favors conditions that promote hot cracking.

When a compound is formed by the element added, the nature of the elements in the binary alloy determines the kind of compound and its distribution in the melt. If the added element is a strongly electronegative nonmetal (e.g., sulfur, or oxygen), a true chemical compound usually is formed and this phase does not possess the metallic properties of an alloy. When the added element is another metal, the two metallic components of the binary alloy may combine and form an *intermetallic compound*. Metals do not obey the usual rules of valence chemistry when they combine, and the composition of the intermetallic compound often is a simple ratio when expressed as atom fractions; for example, in the magnesium-tin binary alloy system the intermetallic compound Mg_2Sn can form. When the percentage of tin in the alloy reaches about 71 percent, the material consists entirely of the intermetallic compound. The compound Mg_2Sn has a higher melting point than either magnesium or tin, and it is hard, brittle, and virtually devoid of metallic properties. Of course, if the tin content is lower, or if a third alloying element is added, the intermetallic compound plays a lesser role in the microstructure, and a substantial change can occur in the mechanical and physical properties.

When a molten alloy is allowed to cool and solidify, it is important to know the temperature range over which it solidifies, and exactly what crystalline structure or structures will appear in the solid cast alloy. The solid state assumes a structure with stable thermodynamic equilibrium only when cooling is extremely slow; and even then some alloys persist in solidifying with an unstable crystalline structure. In the latter case, reheating to a temperature just below the solidus for a period of time may be

298 Chapter 4

needed to bring about an atomic arrangement that truly represents the equilibrium state. Commercial operations that produce castings, ingots, slabs, etc. normally entail cooling rates that are much too fast to allow thermodynamically stable crystalline structures to form. Nevertheless, to provide basic information about each alloy that can serve initially as a convenient guide, the metallurgist uses diagrams to display facts about alloys that have been determined under very slow cooling conditions. In early literature, these were called *equilibrium diagrams*. As appreciation grew that these diagrams often did not represent complete thermodynamic equilibria, the term

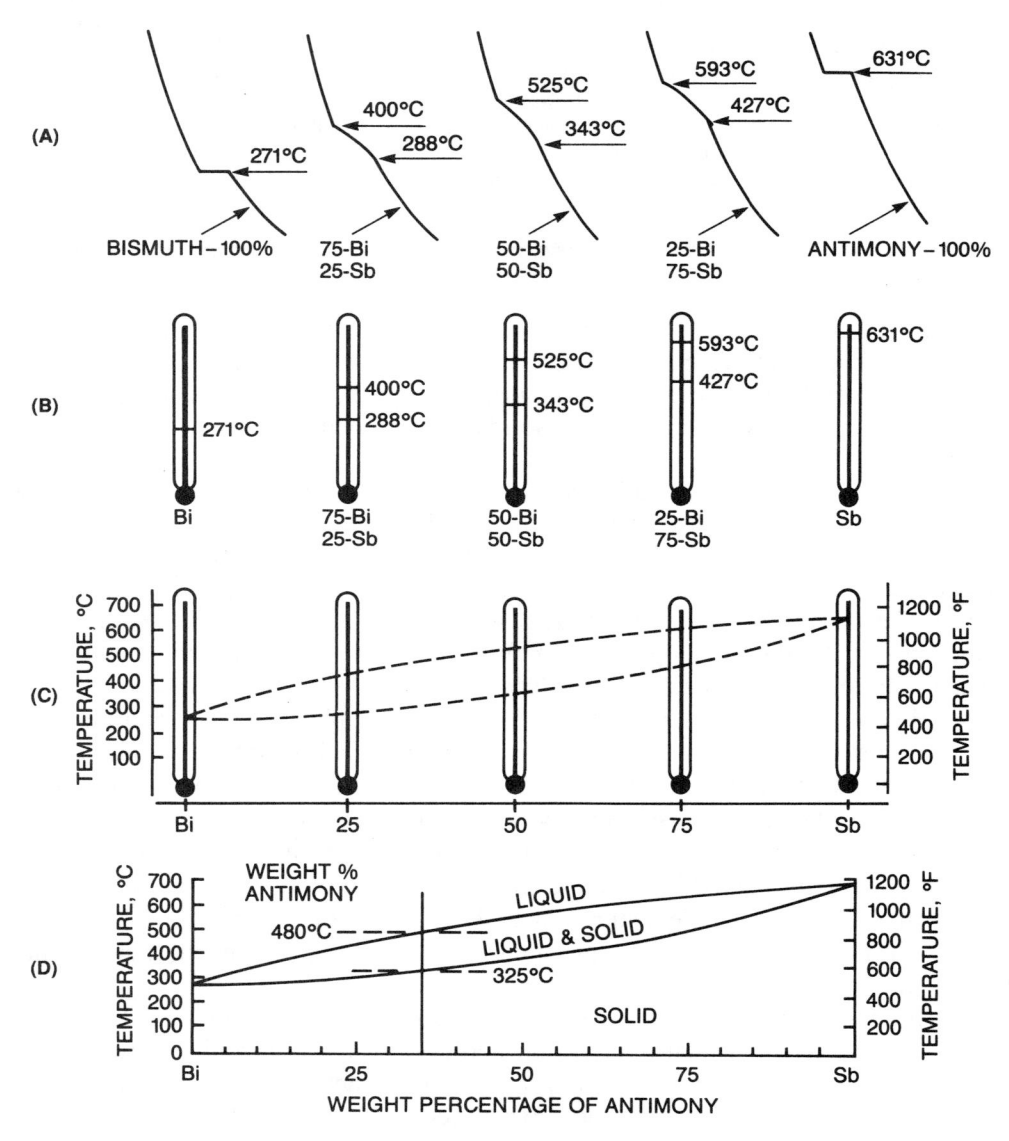

Figure 4.1 — Steps in the development of a phase diagram for the binary alloy system of bismuth and antimony

constitution diagram was applied. Although this term made no claim to represent the stable state of constitution for the alloy involved, it too was set aside in favor of applying the term *phase diagram*. A phase diagram now is understood to portray the temperature and composition limits of phase fields found in alloy systems as determined under specific conditions of heating or cooling, and pressure. However, most phase diagrams are developed by procedures that involve very slow cooling of the molten material under ambient pressure to room temperature. Therefore, a phase diagram may be an equilibrium diagram for a particular alloy system, or it can be a constitution diagram that depicts the metastable phases developed under stated conditions.

Phase Diagrams

The value of phase diagrams as basic guides to liquidus and solidus temperatures, and to the crystalline structures to be found in alloys, has encouraged their continued preparation. ASM has seen to the extensive publication of diagrams for binary alloys, as well as for ternary alloys. As mentioned above, sometimes the information obtained for an alloy does not represent the stable equilibrium state, but instead represents a *metastable* or nonequilibrium state. Usually, differences in temperatures and crystalline structures found under true stable equilibrium conditions, as compared with the metastability that arises when cooling is not entirely slow enough to achieve thermal equilibria, are far less than will be found after the cooling conditions of commercial casting or fusion welding operations. Phase diagrams serve as the first stepping stone in proceeding to develop a complete understanding of a particular alloy when cast and processed under the conditions employed in commercial production. ASTM has provided a standard practice for presentation of phase diagrams in ASTM E 391. Both binary and ternary type diagrams are covered by this document, and the recommendations therein range from the scale for plotting data, labeling of phases, to other details that should be considered prior to publication of a diagram.

Because recorded phase diagrams are voluminous, attention here will be confined to the method of obtaining data required to plot the diagrams, and to their utilization. In the four drawings of Figure 4.1 — labeled A, B, C and D — procedures for constructing a diagram for a binary alloy are outlined. The diagram is two-dimensional, with scales placed horizontally and vertically for the two principal variables — namely chemical composition and temperature. The composition is on the abscissa or horizontal base, and the unit markings usually indicate the percentage by weight of one element in the other; although atom percentages sometimes are shown also. Temperature is indicated on the ordinates or vertical sides. A first step in establishing a diagram for a binary alloy is to determine the freezing temperature and any transformation temperatures for the two pure metals involved. Next, a suitable number of intermediate alloy compositions are melted, and similar determinations are carried out. As a rule, substances possess only one liquid form; the only elemental exception is found in helium, which can assume either of two different liquid forms. The freezing points for the two pure metals and the freezing ranges for their alloys are determined by recording a time-temperature graph or *cooling curve* as a quantity of each molten metal and alloy cools very slowly to room temperature through the solidification and transformation points or ranges.

In Diagram A of Figure 4.1, a simple binary alloy system, that of bismuth and antimony, has been selected for the purpose of illustration. Cooling curves are shown from the molten state for pure bismuth and pure antimony, and for three Bi-Sb alloys of selected composition. Note the sharp breaks in the curves for the two pure metals. These horizontal steps accurately mark the freezing points for these metals. During the entire time each metal is undergoing solidification, the latent heat of fusion (as described in Chapter 3) is being liberated. The amount of latent thermal energy given up is sufficient to halt temperature drop of the partly frozen sample. When solidification is complete, liberation of latent heat stops, and release of sensible heat continues. Thus the cooling curve has indicated the freezing *points* for the two pure metals. Next examine the cooling curves for the three Bi-Sb alloys. The hump in each of their curves due to slowed cooling is a result of latent heat of fusion being liberated over a range of temperatures; this indicates that freezing is taking place over a range of temperatures. Therefore, these cooling curves provide the freezing *range* for each of the three alloys, and the temperatures of occurrence have been noted.

In Diagram B, a set of five schematic thermometers have been marked with the temperatures secured from the cooling curves shown in Diagram A. These thermometers provide valuable data, but if the freezing temperature range of a Bi-Sb alloy composed of 65 percent Bi and 35 percent Sb is wanted, a rough estimate would have to be made, or another cooling curve for this particular composition would be required.

In Diagram C, a base line has been laid down to indicate composition. The line is divided into units representing percentage by weight. On the extreme left, 100 percent bismuth is found, while on the extreme right is 100 percent antimony. As a traverse is made across the line from left to right, increasing amounts of antimony are added to bismuth. At the proper locations above this composition line, the thermometers from Diagram B have been positioned. Dotted lines have been drawn through the temperature levels previously marked on the thermometers.

Diagram D is the phase diagram for the bismuth-antimony alloy system. The dotted lines shown previously on Diagram C now are drawn solid. This final diagram gives the freezing range for any Bi-Sb alloy composition. While the pure metals melt and freeze at a point on the temperature scale, alloys undergo these changes over a temperature range — unless, as will be illustrated shortly, the particular alloy is of eutectic composition, or it forms an intermetallic compound. The alloy of 65 percent Bi and 35 percent Sb composition, for which no thermometer had been available earlier, can have its freezing range ascertained by drawing a vertical line from the 35 percent Sb point on the base of the diagram and taking note where this (vertical) line intersects the liquidus and the solidus. These temperatures have been noted on Diagram D as 480°C (896°F) for the liquidus, and 325°C (617°F) for the solidus. The final bits of information added to Diagram D are the liquid and solid phases present in the metals and alloys at any level of temperature. The shape of the liquidus and solidus curves and the simplicity of phase distribution indicates that the two metals are soluble in each other over the entire range of composition in both the liquid and the solid states. The phase boundaries of the diagram in Figure 4.1 essentially depict free energy curves. The crystalline structure of the solid phase must be determined by x-ray diffraction examination of specimens of the solids. Both bismuth and antimony are known to have a hexagonal (rhombohedral equivalent) crystalline structure in the solid form. Therefore, it comes as no surprise that all Bi-Sb alloys acquire this selfsame hexagonal crystalline structure in the solid state.

A *lever law* can be applied to phase diagrams for alloy systems to extract useful information concerning the process of solidification. Application of the lever law to solidification is discussed in Technical Brief 20. Later, this same law will be applied to the process of crystallographic transformation in a solid phase. The lever law, applied at different temperature levels in the solidification range, will show that the composition of the solid phase changes during the interval of solidification over a falling temperature. Therefore, when cooling slowly enough to achieve equilibrium in the solidifying alloy, the early dendrites are richer in the higher-melting-point metal than in the solid formed at lower temperature. Unless diffusion occurs to remove this composition difference, in a state of so-called equilibrium the material may actually exist in *heterogeneous equilibrium*. In fact, equilibrium is practically never achieved in commercial casting or fusion welding operations, and there will be differences in both temperatures and compositions as indicated by a phase diagram, and as actually found in the alloy.

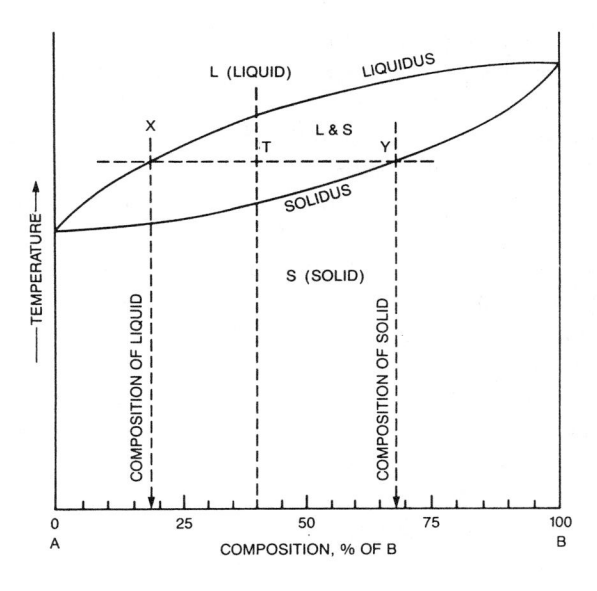

Figure 4.2 — Application of the lever law to a phase diagram

At a selected temperature, T, within the solidification range for this hypothetical binary alloy, the lever law has indicated the proportions of liquid and solid phases existing in the alloy and their average compositions.

> **TECHNICAL BRIEF 20:**
> **Application of the Lever Law to Solidification of Metals**
>
> To illustrate application of the lever law to solidification, picture the simplest form of phase diagram [such as that shown earlier in Figure 4.1(D)] where two metals are mutually soluble in both the liquid and the solid states. For the hypothetical alloy system illustrated in the phase diagram of Figure 4.2, the metals will be labeled A and B. On this diagram, a vertical dotted line has been drawn from the 60% A / 40% B composition location upward through the liquidus and the solidus. The lever law allows determination of (1) the proportions of the alloy in the solid and in the liquid states, and (2) the composition of each of the phases existing at a given temperature level during the solidification process.
>
> When the 60% A / 40% B liquid alloy cools to the point designated as T on the dotted vertical line, the lever law is implemented by drawing a horizontal line (also dotted) through T and extending to the liquidus on the left, and the solidus on the right. From the points at which the horizontal line intersects the solidus and the liquidus, vertical lines (also dotted) are dropped to the composition scale on the abscissa. At the temperature level T, the amount of solid alloy formed and the liquid alloy still remaining is indicated by the relative parts of the line XY, with the percentage of solid indicated by the segment XT [i.e., percent solid = XT/XY (100)]. The composition of the solid forming at temperature level T can be found by noting where the vertical dotted line from Y intersects the abscissa (about 32% A / 68% B). The remaining liquid is indicated by a similar line from X to consist of approximately 82% A / 18% B.

Likely, there will be *undercooling* of both the liquidus and the solidus, especially the latter. The more rapid the freezing, the greater the undercooling that increases the amount of liquid still present. This may explain why the solidus is difficult to determine precisely upon cooling. Experience shows the solidus is more-easily pin-pointed by heating the homogeneous solid alloy to successively higher temperatures until incipient melting is identified metallographically, or by mechanical testing.

Even in a binary alloy system involving metals that are completely soluble in both the liquid and the solid states, the first crystals to solidify from the liquid are richer in the higher-melting-point element. Naturally, the remaining liquid surrounding these early solid crystals increases in its content of the lower-melting-point element (the extent being indicated by the lever law). As freezing progresses in the surrounding material, the crystals that form later are richer in the lower-melting-point element. Unless cooling is exceedingly slow, there likely will be insufficient time for equalization of composition by diffusion of elements between the first core crystals and the last layers of crystalline material that solidify. The resulting difference in composition between the center and the outer portions of a crystal or a grain is called coring. As should be expected, the core generally is richer in the higher-melting-point element than the outer layers. Actually, the coring appears in a dendritic configuration because of the directionality involved in the solidification mechanism (illustrated earlier in Figure 2.10). The primary arms of the dendrites are richer than the metal that solidifies last between the arms. The degree of coring that occurs in a casting or in weld metal will vary considerably with the various alloys. For the most part, the amount of coring will be determined by (1) compositional spread between the liquidus and solidus, which controls the composition difference between the phases during solidification, and (2) the deviation from equilibrium conditions during the formation of solid crystals from the liquid alloy during solidification. Little evidence of coring appears in carbon steel weld metal, but alloy steel weld metals are apt to display this feature when examined by particular metallographic techniques. Annealing may be

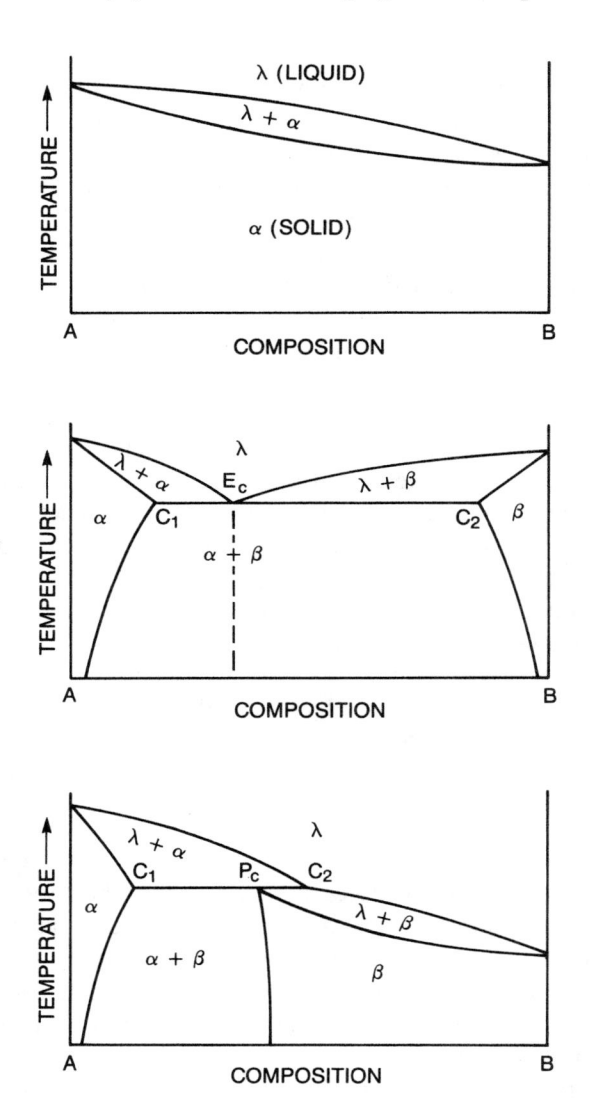

Figure 4.3 — Three types of phase diagrams commonly found for binary alloy systems

Top Diagram: Complete solid solubility by components A and B. Middle Diagram: Partial solid solubility with eutectic reaction when amount of B alloyed in molten base A is indicated by composition Ec. Lower Diagram: Partial solid solubility with peritectic reaction when composition of alloy falls between C1 and C2.

effective in removing coring, that is, restoring composition homogeneity even on an atomic scale. Coring in alloy castings and weld deposits may, or may not, exert a noticeable effect upon the properties of the metal. The significance of coring will be pointed out from time to time in discussing alloy steel weld metals.

Binary Phase Diagrams

Binary alloy phase diagrams become more complex as the metals melted together differ in such features as atom size, crystalline structure, valency, electronegativity, etc. These features will be examined for their effects shortly, but several common types of phase boundary configuration can be pointed out to serve as bases for discussion of diagrams. Figure 4.3 consists of three schematic phase diagrams that are shaped by the degree of solid solubility between the two components of a binary alloy system. When two metals are of comparable atom size, and have similar chemical characteristics, and also have the same crystalline structure, they will likely be completely soluble in one another in the solid state. A phase diagram for a binary alloy of this nature is shown as the top schematic in Figure 4.3. Despite its simplicity, this kind of diagram can provide much useful information as demonstrated earlier with Figure 4.2. Identification of the phases on the diagrams of Figure 4.3 follows the practice started many years ago of using letters of the Greek alphabet for phases within their boundaries of composition and temperature. Usually, the practice uses lower case characters and assigns them to phases in order of their occurrence during heating, or during cooling. However, many exceptions appear in diagrams in the literature and each diagram must be examined closely to note the different phases that exist in various compositions over specific ranges of temperature. The particular identification letters applied to phases give no clue as to their structure. As mentioned earlier, crystalline structure must be ascertained by x-ray diffraction examination. If a diagram is not appended with this information, the structure possibly can be found in the extensive tables of crystalline structure of phases of binary alloy systems published in metal handbooks.

Eutectic-Type Diagrams

Eutectic-type diagrams, as illustrated by the middle schematic in Figure 4.3, can be produced when two components of a binary alloy system have partial solubility. The term eutectic comes from the Greek meaning "easily fused." The eutectic reaction is an isothermal reversible mechanism in which a liquid solution is converted into two or more intimately mixed phases during solidification. The number of phases formed equals the number of components in the alloy system. Note in the middle diagram of Figure 4.3 that the alloy consisting of approximately one-third component B in the base metal A has a solidification point at a temperature much lower than either of the two components. The alloy that solidifies at the point marked E_c is called the *eutectic composition*, and it represents an instance in this system where an alloy has a freezing and melting point rather than a range. Incidentally, intermetallic compounds also exhibit fixed melting points. In the hypothetical alloy system portrayed by the middle diagram of Figure 4.3, the components are fully soluble in the liquid state, but are sparingly soluble in each other in the solid state. As indicated by the solid solubility boundary on the left-hand side of the diagram, base metal A can retain only a small percentage of B in solid solution in its α phase at room temperature. As soon as this solubility limit is exceeded, the α phase commences to include some ß phase in the alloy's microstructure. Alloys having compositions located to the left of the eutectic point upon freezing first will form crystals of α, and as this phase continues to solidify the percentage of B is enriched in the yet unfrozen liquid (λ). This mechanism of freezing and enrichment continues until the temperature drops to a level just above the solidus line where the remaining liquid has been enriched with component B until its composition has been shifted to the eutectic point (E_c). Distribution of the α and ß phases in the eutectic microstructure is often a complex array that might resemble fibers, rods, nodules, or lamellae, and the form that develops depends upon the nature and amounts of the phases and the conditions of solidification. In an alloy on the right-hand side of the eutectic point (hypereutectic compositions), the first crystals to form will be ß phase. Consequently, the liquid remaining during solidification of the ß phase will become enriched with the metal A. The mechanism of freezing and enrichment, however, is exactly the same as described for the hypoeutectic compositions. The horizontal line extending from C_1 to C_2 in the diagram is an important isothermal boundary that marks for all alloy compositions of this particular eutectic system the lowest temperature at which liquid λ phase can exist. Many alloy systems produce phase diagrams of the eutectic-type because the number of systems that show complete solubility is limited. The majority of alloys are mixtures of two or more phases. The phase diagrams for lead-antimony alloys, and for silver-copper alloys are good examples of a simple eutectic-type diagram. The diagram for iron-carbon alloys (e.g., steel), to be described in detail later, includes the eutectic reaction, but has additional features that increase its complexity.

Peritectic-Type Diagrams

The peritectic-type diagram shown in the lower part of Figure 4.3 is produced by alloy systems having partial solubility and in which there is a wide difference in the melting points of the two components. The peritectic type of diagram is similar to the eutectic-type, but a unique reaction takes place during freezing that deserves description because of its occurrence in some compositions of steel. The term peritectic comes from the Greek meaning "around," and it applies to an isothermal reversible reaction in which a solid phase reacts with a liquid phase to produce another solid phase. During this mechanism, the first phase to solidify will be surrounded by a second phase, which in turn will be surrounded by liquid. In the lower diagram of Figure 4.3, point P_c is the *peritectic composition*. The isothermal boundary C_1 - C_2 is the *peritectic temperature* above which no ß phase can form in any of the alloys. In this alloy system, two separate solid solutions are formed during solidification, α phase and ß phase. The free energy curve of the ß phase is much higher than that of the α phase, and this produces an arrangement of phase boundaries that essentially are the inverse of those found in a diagram of the eutectic-type. When an alloy has a composition between point C_1 and point P_c, the primary crystals of solid solution α phase are formed first. Then, at the peritectic temperature, the crystals of a phase react with the remaining liquid λ phase to give a mixture of α and ß phases. If the alloy composition lies between point P_c and C_2 the solid crystals of α phase are converted entirely to ß phase during the peritectic reaction. Time at the peritectic temperature must be sufficient for diffusion between the solid phase and the liquid phase to carry the reaction to completion. The phase diagram for the silver-platinum alloy system is a good simple example of the peritectic type. The peritectic reaction in steel will be shown by the iron-carbon diagram to occur over a composition range of approximately 0.09 to 0.53 percent carbon.

Most phase diagrams for alloy systems are more complex than the examples in Figure 4.3, even for binary alloy systems. Complications arise mainly because of phase transformations and the development of intermediate phases in the solid state during cooling following solidification. Allotropic transformation in the crystalline structure of phases was described in Chapter 2, but now the subject must be examined in greater detail because of frequent occurrence of allotropic transformation in alloys.

Polymorphism is the term used to indicate ability by a solid to exist in more than one form. In metals and alloys, this means ability to exist in two or more crystalline structures, or in an amorphous state and at least one crystalline structure. The polymorphic transformations of iron are a very important aspect of the metallurgical behavior of this metal, and they play a major role in the far-reaching successful use of steels.

Figure 4.4 is the phase diagram for a hypothetical binary alloy system in which crystallographic change occurs in the solid state through an allotropic transformation. This simple diagram has been tailored as a harbinger for later discussion of the phase

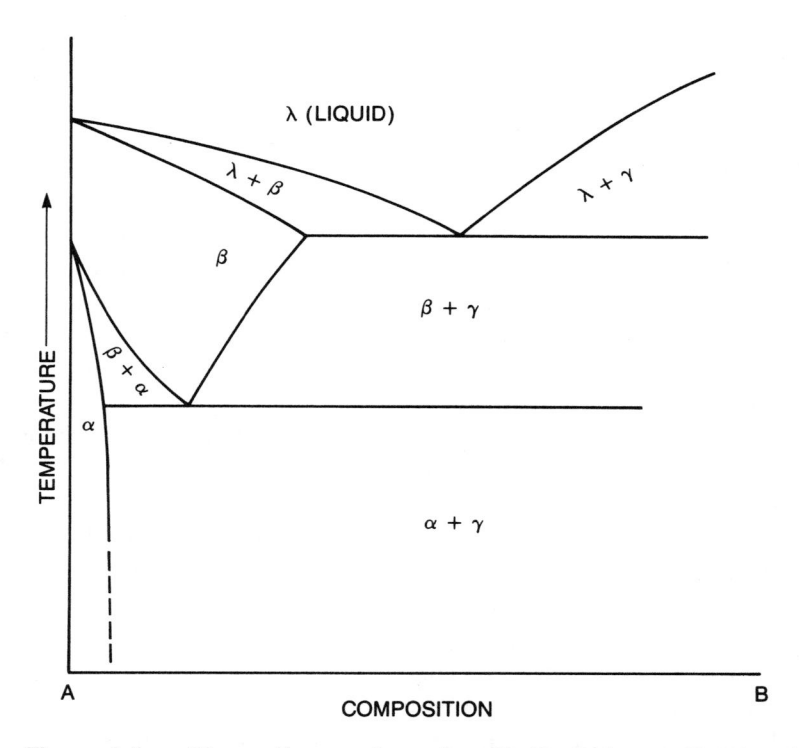

Figure 4.4 — Phase diagram for a hypothetical binary alloy in which solid beta phase β undergoes allotropic transformation of its crystalline structure to alpha phase α during cooling

diagram for alloys of iron and carbon which provides valuable insight into the metallurgical behavior of steels. The configuration of liquidus and solidus curves in Figure 4.4 for the alloys made up of metals A and B indicate that these components have limited solubility in each other, and at a particular content of B (the solute) in base metal A, a eutectic of ß and γ phases is formed during solidification. If attention is directed to the left-hand ordinate of the diagram, the ß phase of metal A will be seen to undergo transformation at a point during further cooling, and will form a phase which is retained as the metal is cooled to room temperature. As small amounts of component B are added to base metal A, the ß to α transformation in the resulting alloys occurs over a range of temperature as indicated by the downward sloping boundaries of the ß + α field. Because of the metastability of the ß phase, continued cooling of these alloys encounters a *eutectoid* reaction at the isothermal (horizontal) boundary where the ß phase decomposes into α phase and γ phase. These eutectoid transformation products often are distributed as a lamellar structure, although some nonferrous alloys sometimes develop a granular distribution. Phase changes in the solid state produce the same general configurations of boundaries as those marking the liquid-solid changes, that is, the eutectoid observed in the solid state is analogous to the eutectic, whereas a peritectoid, which is observed in some alloy systems in the solid state, is analogous to the peritectic illustrated in the lower diagram of Figure 4.3. Determination of the temperature limits for the start and completion of allotropic solid-state transformations can be carried out in the same manner as described earlier for Figure 4.1 when establishing liquidus and solidus temperatures. Allotropic transformation ranges for solid alloys are indicated by humps in the cooling curves caused by gradual liberation of latent heat of transformation. Of course, when the alloy is of eutectoid composition, a sharp step is produced in the cooling curve by release of latent heat of transformation at the isothermal boundary for the eutectoid reaction. Transformation in the solid state can be analyzed on the phase diagram by applying their lever law in the same way as explained earlier in Figure 4.2 for the process of solidification in alloys.

Only a relatively small number of alloy systems involve allotropic transformations of their solid phases. Examples can be found in phase diagrams for alloys of beryllium-copper, copper-iron, copper-tin, and copper-antimony. However, this configuration in a phase diagram is particularly pertinent because of its frequent occurrence in alloys of iron; this includes iron-carbon alloys, which are the bases of the carbon and low-alloy steels that will be studied intently. Phase diagrams can be subjected to searching scrutiny for correctness by using the Gibbs phase rule, which is presented in Technical Brief 21.

Ternary Phase Diagrams

When three components are present in an alloy, the composition usually is plotted within a triangle as shown in Figure 4.5. Each side of the triangle serves as the composition abscissa for a binary system; namely, A-B, B-C and C-A. The composition for the alloy at point X on the diagram is 20% A, 40% B and 40% C. To indicate the role of temperature and phases present at given temperature levels on a ternary diagram, it is necessary to move onto the third dimension. This can be done in several different ways. Diagrammatically, a binary phase diagram can be drawn from each of the abscissa on the sides of the triangle as shown in Figure 4.6 for a eutectic alloy system.

Although an approximation can be made as to where the eutectic liquidus for each of the three binary alloys would intersect to produce the lowest-possible liquidus point, the ternary phase diagram does not facilitate the designation of temperature limits for phase fields. It is possible, of course, to construct one or more pseudo-binary diagrams that would represent sections through the ternary system at fixed contents of one of the components. Also, a single ternary alloy system can be illustrated in detail by constructing (triangular) isothermal sections at selected temperatures of interest. Another entirely different method, which is most-easily visualized but rather costly, is to build up a solid prismatic model of transparent pieces using different colors to portray the various phase fields and their temperature limits. These problems in picturing ternary alloy systems help explain the greater use of binary alloy phase diagrams.

Phase Diagrams for Multi-Element Alloys

When four or more elements are contained in an alloy, a single diagram or model cannot be constructed to portray phases and temperatures, because a space of more than three dimensions would be required. The practical recourse in dealing with such a multi-element alloy is to use a series of quasi-binary diagrams showing the phases

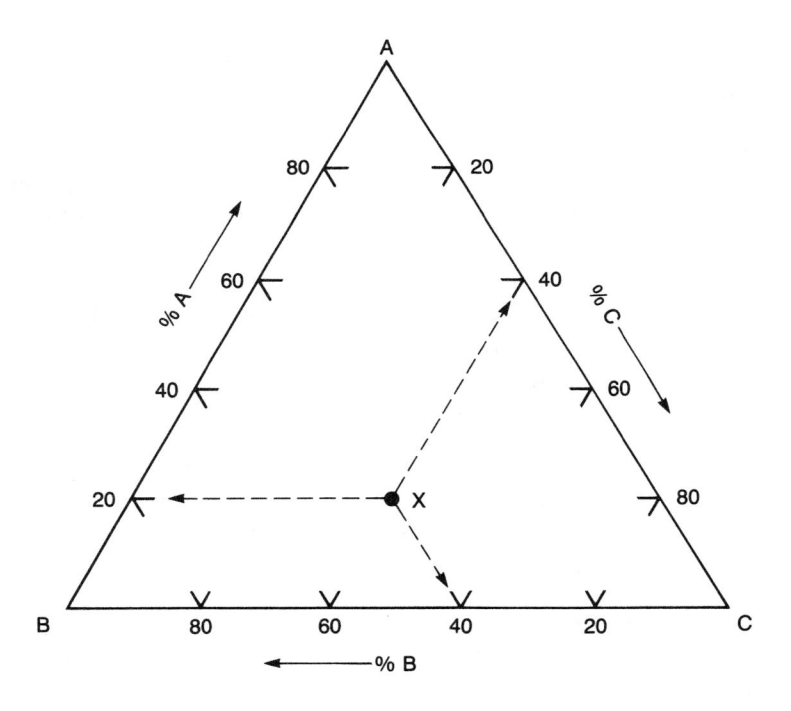

Figure 4.5 — Triangular diagram for portraying compositions in a ternary alloy system

Point labeled "X" represents composition consisting of 20% A, 40% B and 40% C. The phase in existence for this particular composition requires further description on the diagram in terms of temperature and nature.

present as the base element and all alloying additions are held constant except one which is varied over a range of interest.

Phase diagrams are not yet available for all alloy systems of interest because these diagrams are costly to prepare. Several sample alloy compositions must be produced, subjected to lengthy testing and examination, and then charted merely to construct a simple binary phase diagram. At least 40 of the elements have been employed, either as a base metal or as an addition to a base metal, in metallurgical exploration of alloy systems to create useful combinations for constructional purposes. If alloys for special applications, such as solid-state electronic devices, are included in this list, the total of elements employed probably would exceed 75. Coverage of only the promising binary alloys already has required hundreds of phase diagrams. If ternary systems were to be documented in similar manner, thousands of diagrams would be needed. Even with a phase diagram to lead the way toward a new alloy, much cut-and-try is involved in adjusting the final composition to secure the desired combination of structure, properties and characteristics. Consequently, there long has been strong incentive to seek

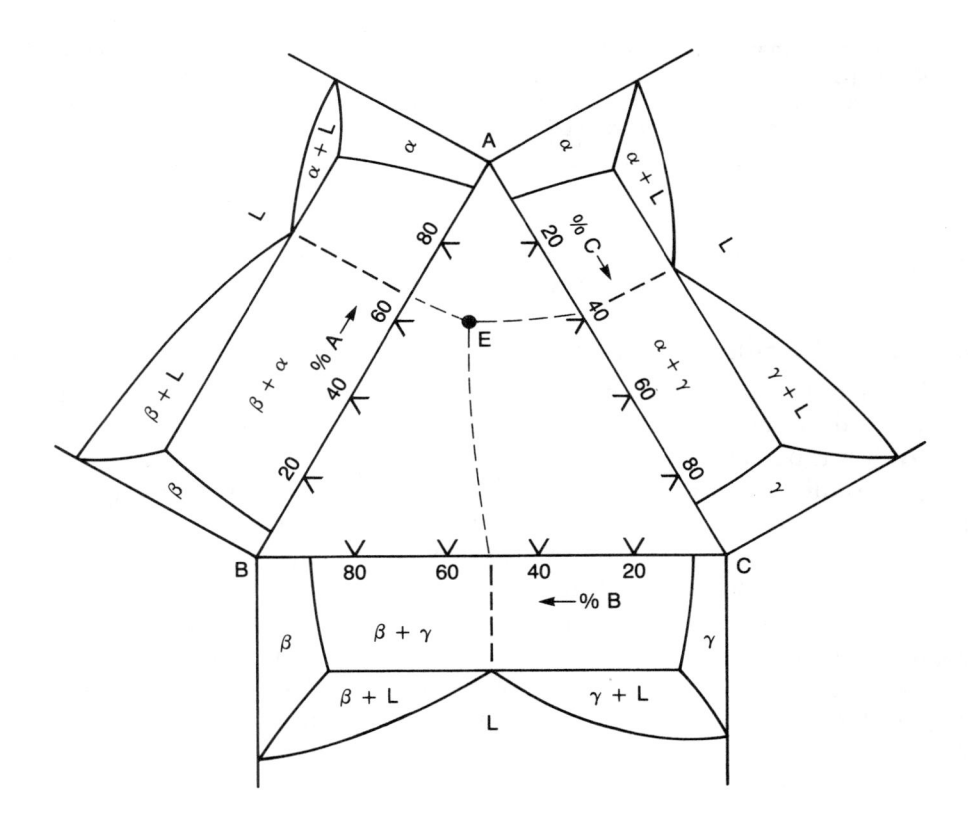

Figure 4.6 — Phase diagram for eutectic ternary alloy system augmented by binary diagrams for each pair of components

The lowest-melting eutectic composition in this ternary alloy system is the composition at the three-way intersect labeled "E".

> **TECHNICAL BRIEF 21:**
> **The Gibbs Phase Rule**
>
> The Gibbs phase rule is based upon thermodynamic reasoning enunciated by Gibbs in 1878, and it was hailed as one of the most notable contributions ever made to physical chemistry. The phase rule states a relationship that must hold between (1) the degrees of freedom (F) of the system; that is, the number of independent variables, (2) the number of components (C), and (3) the number of phases (P). The Gibbs phase rule tells us that the number of degrees of freedom of any system is equal to the number of its components plus two, minus the number of phases present: Hence, the following equation applies:
>
> $$F = C - P + 2 \qquad (Eq.\ 4.1)$$
>
> Application of the phase rule requires detailed consideration of external forces (e.g., pressure, magnetic forces, etc.), temperature, and composition in the theorem governing construction of a phase diagram. Because pressure can be regarded as a constant (simply atmospheric pressure) rather than a variable in the formation of alloys, this reduces by one the possible number of degrees of freedom: Therefore, in the case of alloys the phase rule can be expressed as;
>
> $$F = C - P + 1 \qquad (Eq.\ 4.2)$$
>
> thus signifying that the number of degrees of freedom (F) is equal to the number of components (C) minus the number of phases (P), plus one. A complete understanding of the phase rule and its application is a challenging task, and ordinarily the originators of a phase diagram will have tested their work by this rule.

fundamental principles that govern the end result of alloying in order to save time and reduce the cost of alloy development.

Alloys in the Solid State

Mention was made in Chapter 2, under *Solubility in the Solid State*, of possible modes of distribution of foreign elements and alloying additions in solid metal. Briefly, it was pointed out that even in solid metal there are three general modes that are quite similar to those outlined for molten alloys — namely (1) *solubility* of elements in each other; (2) *insolubility*, where different elements retain their own identity; and (3) *compound formation*, where elements combine to form a new material, which may or may not be soluble in the solid crystalline structure. The distribution of atoms of the added element can have a profound influence not only on the crystallographic form of the alloy, but also on its microstructure and its properties.

Solubility is the first topic examined because this mode of distribution in alloys provides the broadest variety of useful effects. As outlined in Chapter 2, atoms of a second element (the solute) can be held in solid solution in the crystalline structure of a base metal (the solvent) in two different ways — by *substitutional* solution or by *interstitial* solution.

Factors Influencing Solid Solubility

Hume-Rothery was a leading investigator of factors that influence the solid solubility of added elements in an alloy (circa 1930). His contribution of a means for identifying these factors is described in Technical Brief 22. Hume-Rothery's work led to the identification of three features of atoms that strongly influence the extent of solid solubility: atom size, electronegativity, and valence.

Atom Size

The size of atoms often must be considered when dealing with alloying, and some preliminary information was given in Chapter 2 under the subheading, *Atom Size or Diameter*. Cautionary remarks were offered about the vagarious size-behavior of atoms as found in different structures, or in a pure aggregate. Nevertheless, even a best-estimate of atom size can be helpful when anticipating the likely distribution of a solute element in an alloy. For this reason, atom diameters of a number of elements in aggregate form were listed in Table 2.3. Information contained in the footnote of Table 2.3 can be augmented by stating that diameter is best measured when atoms are in contact along a specific direction in a unit cell of the crystalline structure. This is the close-packed direction set by nearest neighboring atoms. The atoms are regarded as solid spheres and are assumed to be in contact along this direction. Therefore, the lattice constant along this axis by x-ray measurement is taken as equal to the atom diameter.

Substitutional solid solubility is favored when atoms of the solute element and the solvent base metal are nearly equal in size, and a solute atom can replace one of the solvent atoms without significant strain developing in the crystalline lattice because of this substitution. However, if atoms of the two elements differ in size by more than 15 percent, substitutional solid solubility is unfavorable. This is because the elastic strain developed around each solute atom represents strain energy, and this energy may be sufficient to make the solution unstable. While there is no such thing as one metal being completely insoluble in another, a large number of solute elements added to a solvent metal are virtually insoluble because of the mismatch of atom diameters. To provide a helpful measuring stick, Figure 4.7 has a vertical ruler that is graduated in angstrom units. On the left-hand side, a number of familiar elements have been listed according to their reported respective atom diameters. On the right-hand side, dotted horizontal lines have been drawn to indicate probable solid solubility of these elements in iron. The criterion mentioned above for substitutional solid solution (size within 15 percent of that of iron) suggests that elements with an atom diameter of about 3.00 Å and larger are too big to have significant solid solubility in iron. Indeed this is true for lead (3.48 Å), which has been reported to have no solid solubility in iron. Zirconium (3.16 Å) is reported to have only a very small solid solubility in iron — perhaps less than 0.10 percent. Lithium (3.00 Å) and silver (3.04 Å) are reported to be virtually insoluble in solid iron.

As Figure 4.7 also indicates, elements with atoms ranging from about 2.2 to 2.9 Å diameter have a good chance of solid solubility in iron by substituting at regular lattice sites. This does not portend that the solute atoms of favorable size will be permitted complete or unlimited solubility in the solvent metal. Complete solid state substitutional solubility is possible only when the metals involved have the same crystallographic structure, are nearly equal in size, and are compatible from an electronic or valency standpoint. A number of elements satisfy the 15 percent size factor criterion with regard to iron, and this accounts for the use of such elements as chromium, manganese, molybdenum, nickel, silicon, etc., as alloying additions to steel. However, even when a solute element is within the favorable atom-diameter range, this does not mean it will readily enter into substitutional solid solution because of possible limitations imposed by valency and electronegativity. Also, when these two features are favorable, there still can be anomalous exceptions.

Elements with atom diameter (in parentheses) reported for pure aggregate	Atom diameter in angstroms	Criteria reported to hold for solid solution solubility feasibility
LEAD (3.48)	3.50	
		ATOMS TOO LARGE FOR SUBSTITUTIONAL SOLID SOLUTION IN IRON
ZIRCONIUM (3.16)		
SILVER (3.04)		
LITHIUM (3.03)	3.00	
TITANIUM (2.89)		
MAGNESIUM (2.80)		ATOMS WITH DIAMETERS SUITABLE FOR ENTERING INTO SUBSTITUTIONAL SOLID SOLUTION IN IRON
COPPER (2.70)		
IRON (2.58)		
ALUMINUM (2.52)	2.50	
NICKEL (2.49)		
SILICON (2.35)		
PHOSPHORUS (2.20)		
SULFUR (2.08)		
BERYLLIUM (1.96)	2.00	ATOMS TOO SMALL FOR SUBSTITUTIONAL SOLID SOLUTION, AND TOO LARGE FOR INTERSTITIAL SOLID SOLUTION IN IRON
BORON (1.90)		
CARBON (1.54)	1.50	
NITROGEN (1.40)		ATOMS SMALL ENOUGH IN DIAMETER TO ENTER INTO INTERSTITIAL SOLID SOLUTION IN IRON
OXYGEN (1.32)		
HYDROGEN (1.00)	1.00	

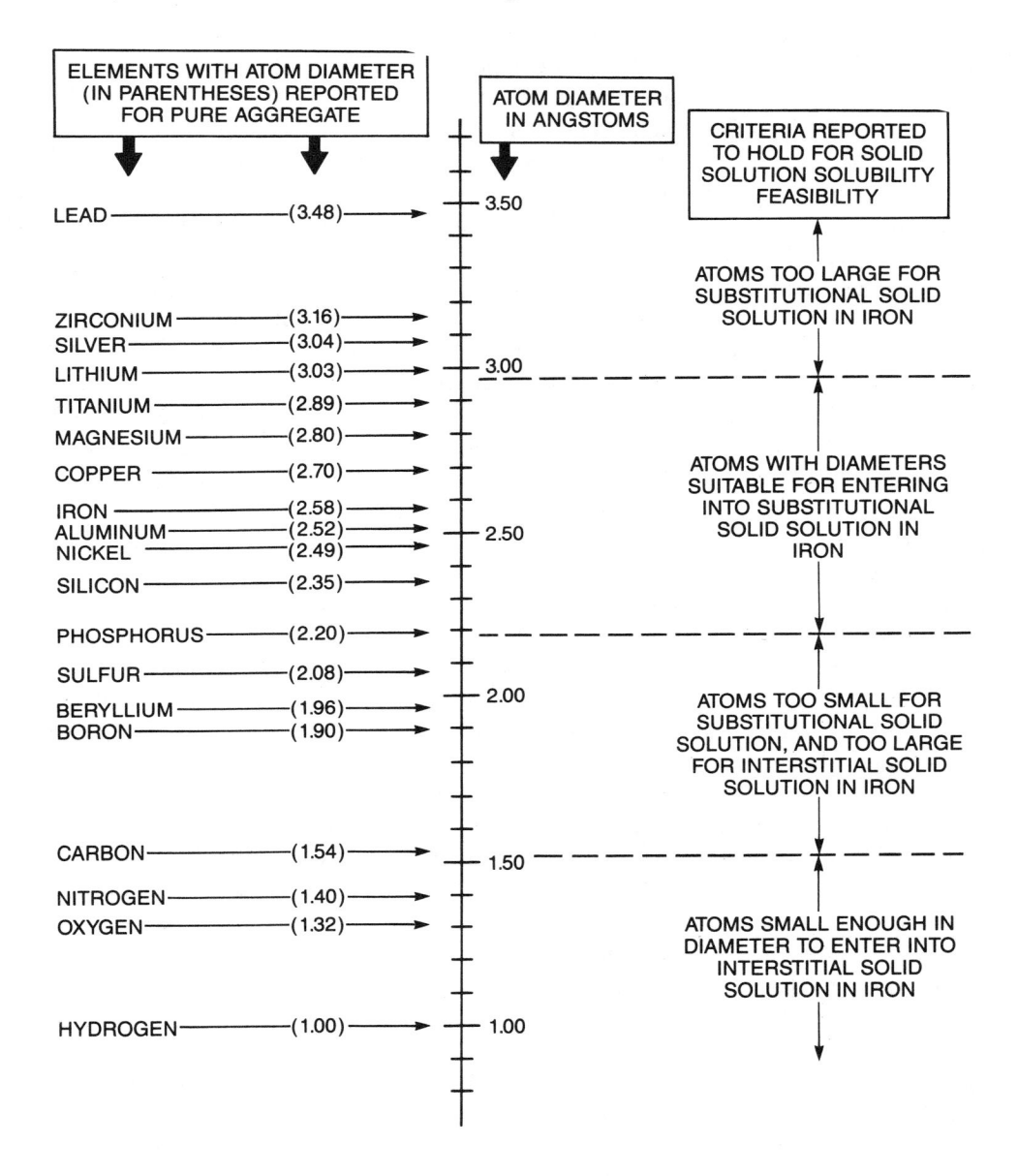

Figure 4.7 — Role of atom size in controlling entry of elements into either substitutional solid solution, or into interstitial solid solution in iron-base alloys

Also see Table 2.3 which lists diameters of atoms as determined by measuring the center-to-center spacing between nearest atoms in an aggregate of the pure elemental material by means of x-ray diffraction.

> **TECHNICAL BRIEF 22:**
> **Hume-Rothery's Classification of Elements**
>
> Hume-Rothery defined principles still followed today as guides to the formulation of new alloys. He proposed that the elements — with the exception of hydrogen, fluorine, nitrogen, oxygen and the inert gases — be divided into three groups labeled Class I, Metals; Class II, Intermediate Elements; and Class III, Non-Metals. These classifications were based upon the crystalline structure and the coordination number of atoms for each element; the latter has been defined in Chapter 2 as "the number of nearest neighboring atoms to any particular atom in its crystalline structure."
>
> Of the metallic elements in Class I, those with a body-centered cubic (bcc) structure have the closest distance between the central atom and the corner atoms. Consequently, there are eight nearest neighbors to the center atom, and the *coordination number* for bcc is 8. For the face-centered cubic (fcc) and the hexagonal close-packed (hcp) crystalline structures, the coordination number of each is 12. In the case of the fcc structure, the *close-packed direction* is along the diagonals of the cube faces. The non-metallic elements of Class III have coordination numbers ranging from one to four, while the Class II intermediate elements crystallize in forms that are similar partly to Class I and partly to Class III.

Interstitial solid solubility is shown by Figure 4.7 as being favored only when solute atoms are very much smaller than that of the solvent metal. Hägg found that two conditions must be satisfied for an element to be distributed interstitially in the crystalline structure of an alloy: first, the interacting metals must be close together in the electromotive series of elements (see Table 3.18); and second, the diameter of the solute atom must be less than about 0.59 times the diameter of the solvent atom. Elements that clearly satisfy this size criterion when added to iron are hydrogen, nitrogen, oxygen and carbon. Further comment can be offered regarding boron and carbon. Although boron is reported to have an atom diameter of 1.90 Å, which gives a calculated boron-iron atom diameter ratio of 0.74, this element appears to have a small interstitial solid solubility in iron before an excess appears out-of-solution in the microstructure as an iron-boron intermetallic compound. Possibly the boron atom is not quite as large as the 1.90 Å dimension when it is surrounded by atoms of iron. Even though the amount of boron that enters into interstitial solid solution is very limited (possibly less than 0.01 percent), this element when distributed in this manner plays a significant role in the hardening of steel (as will be discussed in Volume II, Chapter 15). Carbon, with its atom diameter of 1.54 Å (a carbon-to-iron atom diameter ratio of 0.60), lies just above the borderline of interstitial solid solubility in iron as defined by Hägg, and it is very fortunate that carbon can enter into interstitial solution to some extent in all of the crystalline forms of iron. If this were not so — that is, if the carbon atom were somewhat larger, or if the 0.59 ratio were an absolute maximum — then mankind would have been denied the properties offered by the most useful of all alloys, namely, heat-treatable steel.

Attention must be called to the atom diameter range of about 1.5 to 2.1 Å in Figure 4.7 where additional elements are listed (sulfur and beryllium). Over this atom size range such elements added to iron encounter the 0.59 maximum ratio propounded by Hägg for interstitial entry into solid solution, and the 0.85 minimum ratio that

Hume-Rothery had in mind for substitutional solution. Consequently, both sulfur and beryllium appear virtually insoluble in solid iron for valid reasons. Yet, lack of solubility cannot be ascribed entirely to unadaptable atom diameter, inasmuch as electronegativity or valency may play a part in the insolubility of a particular element. Also, remarks made here deal with solubility determined in essentially pure iron. Much of alloying is done with multiple-element material as the base, and the presence of other elements in the solvent metal can easily alter circumstances. For example, whereas sulfur is virtually insoluble in solid iron (certainly less than about 0.005 percent can be contained in substitutional solution), seldom does the opportunity arise to witness this limitation. This is because manganese is almost always present in irons and steels, and the stronger affinity of manganese for sulfur invariably ties up the sulfur as manganese sulfide, a compound that is insoluble in the steel matrix. To close this discussion of the size factor in binary alloys, it can be pointed out that a special kind of intermediate phase can be formed when one atom is smaller than the other by approximately 20 to 30 percent. In fact, when the participating atoms differ in size by about 22.5 percent, they can pack together in a crystal structure of higher coordination number than can be achieved by atoms of nearly equal size. These are called *Laves phases*, and they commonly have the composition AB_2. Examples are $MgCu_2$, $TiFe_2$ and $MgNi_2$. Generally, a combination of geometric and electronic circumstances result in the formation of a Laves phase type of crystalline structure. This alloying phenomenon is not likely to be encountered in the carbon and low-alloy steels under study here, but are more often found in nonferrous alloys, or in high-alloy materials.

Electronegativity

Electronegativity pertains to the difference between two elements from the standpoint of valency electrons in their outermost shells. It is the electrons in the s and p states of the outer shell that mainly determine chemical behavior, and when the negativity difference is large there is a strong likelihood that a compound of definite composition will be formed rather than a solid solution. Also, a large electronegativity difference between two elements usually limits the solubility of one element in the other. For example, magnesium and tin have favorable atom diameters for substitutional solid solution (2.80 and 3.02 Å, respectively), but the solubility of tin in magnesium is less than one percent at room temperature because electronegativity difference. Brief mention was made in Chapter 2 that most of the elements can be placed in two groups; *electropositive* ones that are inclined to give away electrons in order to form a stable outer shell, and electronegative ones that readily accept electrons for the same reason.

Metals are electropositive elements because they usually have only a few electrons in their outermost valency shell at most and they are more likely to give up this limited number than to gain a larger number of electrons to complete the outer shell. When there are a half-dozen or more valency electrons in the outermost shell, the element will display electronegative behavior, such as found with non-metals like oxygen ($2s^2\ 2p^4$), sulfur ($3s^2\ 3p^4$) and chlorine ($3s^2\ 3p^5$). Many metals have only one valency electron, and therefore most are inclined to be chemically active electropositively. As the number of valence electrons increases, as is the case with aluminum ($3s^2\ 3p^1$), tin ($5s^2\ 5p^2$) and bismuth ($6s^2\ 6p^3$), the atoms vacillate on whether to donate, or to accept electrons upon entering into a chemical reaction. Consequently, difference in electronegativity plays an important role as these high-valency metals encounter another element during alloying.

Valence

Valence requires further discussion as a factor in alloying, because this feature of atoms influences a number of conditions in the resultant material. These conditions include its crystalline structure; its solid solubility for particular elements; and the emergence of intermediate phases, intermetallic compounds, and chemical compounds in its microstructure. Valence, in general, is the combining capacity of an element; but, more precisely, it is that property measured by the number of atoms of hydrogen (or its equivalent) that one atom of the given element can hold in combination if electronegative, or can displace in a chemical reaction if the given element is electropositive. However, metals frequently disregard the usual rules of valence chemistry and they can combine in unique ways over wide ranges of composition that bear no relation to the recognized combining proportions in chemical compounds. The nature of valence forces operating during alloying has been studied starting in the late 1920s, but even today the valency factor has not been completely rationalized. Nevertheless, analyses have been made of the behavior of valence electrons with the aid of quantum mechanics. Hypotheses drawn from this effort have provided fundamental assistance for explaining some very puzzling matters, such as the color of alloys, and the transparency of certain alloys to ultraviolet radiation. The work of Born and Oppenheimer has been very helpful in gaining a basic understanding of cohesion, electrical conductivity, magnetic properties, optical properties, etc.

Valence electrons have been described in Chapter 2 as "gaseous glue" that binds aggregates of metal cations together, and yet these electrons were claimed to have considerable freedom of movement through the solid crystalline structure. While this simplified overview remains appropriate, the valence electrons do not have free choice of distribution around the nucleus of an atom, and their movement in space is restricted to certain orbits until transported by external force or energy to another orbit: hence the importance of the table in Appendix II giving the main shell and subshell electron configurations for all of the elements. This table probably holds the key to atomic behavior in alloying, but a full understanding has yet to be achieved.

Originally, Hume-Rothery discovered a correlation — first in pure metals, and then in alloys —between electron configuration and crystal structure. This correlation is described in Technical Brief 23. The relationships between e/a ratio and crystalline structure hold true for many materials, but application to the transition elements is not as straightforward. The transition elements now number twenty-five (see Table 2.4), and they include some of our most useful industrial materials, iron being a prime example. All transition metals are characterized by a particular feature of electron configuration; all have incompletely-filled shells immediately within their outermost valency shell. Initially, investigators felt that the electrons in the next-to-last incomplete shell (often having unfilled d-subshell) need not be considered in the e/a computation because their orbitals are localized as compared to those of the s-subshell electrons in the outermost valency shell. The s-state electrons were believed to be capable of ranging far out into the metal and to exercise control over the long-range order of the crystalline structure. The d-subshell electrons were viewed as adding primarily to the stability of the solid but were not likely to directly influence the type of crystalline structure formed. More recently, the view has been expressed that with a given transition metal, the unfilled shell just beneath the outermost valency shell indeed can affect valence factor during alloying. Investigators feel that available valency electrons can be drained off to some extent into the unfilled subshells of the adjacent inner shell, and for this reason the electrons will not be able to participate in the e/a relationship effect.

Formation of Intermediate Phases and Compounds

Phases can form in solid alloys which have structures that are crystallographically different from the matrix of the solvent metal. These intermediate phases can be bonded by electron sharing, or by covalent bonding as described in Chapter 2. Such phases are called *intermetallic compounds*. Although intermetallic compounds are characterized by a sharp melting point and their own distinctive microstructural appearance, they seldom have the precise composition of a chemical compound. This is because, in the majority of instances, factors other than chemical bond formation are dominant in the composition of the intermetallic compound — even though the composition may constitute a simple ratio between the atoms of its component elements. In general, there is a tendency for metals that are strongly electropositive to form stable intermetallic compounds with other metals and metalloids that are prone to act electronegatively. Thus, the valencies of the elements entering into the compound are satisfied, even though chemical bonding does not develop. Intermetallic compounds can have properties quite unlike a metal. Often they are brittle, and they are poor conductors of electricity.

There is a class of *intermetallic structures* that cannot be easily related to ordinary chemistry. These structures tend to form when the electrochemical characteristics of the components do not differ greatly. Their compositions often represent simple atomic ratios, but the ratios do not bear any relation to the recognized valencies of the components. These intermetallic structures usually display properties typical of a metal, including good electrical conductivity. Here again, relationships between electrons in the valence shells and among the atoms appear to govern the crystallographic structure and the properties of these intermetallic structures; therefore they often are referred to as *electron compounds*. In the main, their occurrence can be attributed to the tendency of a metallic structure to assume an arrangement wherein the relatively free (i.e., non-ionic) electrons possess the lowest possible energy. The arrangement of atoms in an electron compound can be completely random, or it can be highly ordered.

Mechanisms and General Effects of Alloying

To this point, Chapter 4 has covered the various ways in which an added element can exist in an alloy, and the features of atoms that determine the mode of distribution of the solute element in the solvent metal. Knowing the mode or state of distribution of the solute in the crystalline structure of the solvent is the first step in estimating likely effects on the mechanical and physical properties of the alloy. Now an examination will be made of the mechanisms that are put into place by the particular distributions

> **TECHNICAL BRIEF 23:**
> **Correlation Between Electron Configuration and Crystal Structure**
>
> Hume-Rothery observed that many pure metals, alloys, and intermetallic compounds with the same number of valence electrons per atom had the same crystalline structure — in other words, the ratio of valence electrons to atoms (e/a) generally fell into the same range. His findings required some alteration in accord with the work of Pauling, and that of Engel, but the e/a ratio concept led to useful correlations. When the e/a ratio is approximately 1.5 or somewhat less, the bcc crystal structure will be formed: When the e/a ratio is 1.7 to 2.1 the hcp structure forms: When e/a is 2.5 to 3 the fcc structure forms; and for e/a of 4 or greater the diamond structure forms. An example of this correlation in a simple binary alloy, is beta brass, which is composed of 50% Cu / 50% Zn. Copper (fcc structure) provides one valence electron ($4s^1$), while zinc (hcp) provides two ($4s^2$): Therefore, *e* totals 3 and *a* totals 2, and the e/a ratio is 1.5. As expected from the relative valency effect, the crystalline structure of beta brass indeed is bcc (despite the dissimilar crystal structures of the component metals). Also, the bcc crystals of this 50% Cu / 50% Zn alloy form an *ordered structure*, more specifically, the copper atoms are segregated at the corners of the unit cubes and the zinc atoms are located at the center of the cube, or vice versa. This ordered arrangement of the two kinds of atoms takes place upon cooling through a temperature level of about 468°C (875°F). The ordering occurs so rapidly that it cannot be prevented or even retarded by quenching.

of solutes, and the manner in which each mechanism operates to influence or to control certain properties.

Role of Crystalline Structure

The addition of an alloying element can affect the crystalline form of the solvent metal in several different ways. Mention was made earlier that if the solvent metal displays an allotropic transformation at some temperature point as a pure metal, this transformation will occur over a range of temperatures when the metals are alloyed together. As a second effect, the level of temperature at which the transformation takes place may be raised, or it may be lowered in the alloy. Third, the alloying element may markedly affect the nature of the crystalline change, either by suppressing the formation of a particular crystalline structure, or by generating the formation of an entirely new structure. In the latter case, it may not be simply a matter of the alloying element forcing its own crystalline form on the solvent metal, but a new, unexpected crystalline structure may appear as influenced by the valency factor described earlier. Another factor to be considered when dealing with alloys that undergo allotropic transformation of crystalline structure is that a rapid rate of cooling through the temperature range over which transformation normally would occur (as indicated by a phase diagram) may be depressed to a significantly lower temperature, or may be completely suppressed in some cases. Rapid cooling also may force a crystalline structure to hold in solid solution a larger percentage of solute atoms than would be held under slow cooling conditions. Consequently, the rapidly cooled alloy would represent to some extent a supersaturated solid solution.

When the crystalline structure takes foreign atoms into solid solution, either substitutionally or interstitially, a certain amount of distortion occurs in the surrounding lattice. This distortion can change lattice parameters to some extent. If a substitutional solution is formed, the amount of distortion will be determined by the difference in size between the solute and the solvent atoms, as well as the percentage of solute atoms added. The relationship between change in lattice parameter and atomic percent usually is linear (either increasing or decreasing depending on the relative sizes of atoms involved). But sometimes unexpected deviations occur; if the solute atoms are suitably small and an interstitial solid solution is formed, the lattice parameters will be expanded. Often this expansion occurs to a greater extent than can be explained on the basis of the number of solute atoms present. It may be necessary to ascertain exactly where in the interstices of the crystalline structure of the solvent metal the solute atoms are able to position themselves. Commonly, a simple hard-sphere model of the crystalline structure is used, and geometric analysis is applied to determine the largest sphere (i.e., foreign atom) that can be sited in specific interstices. It is useful to study

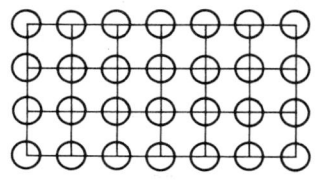

(A) CRYSTALLINE STRUCTURE AND LATTICE OF PURE BASE METAL.

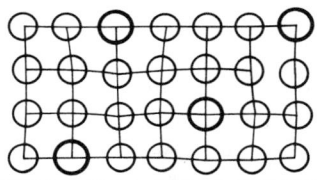

(B) SUBSTITUTIONAL SOLID SOLUTION. BASE METAL LATTICE IS DISTORTED BY PRESENCE OF FOREIGN ATOMS.

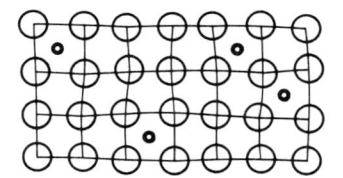

(C) INTERSTITIAL SOLID SOLUTION. FOREIGN ATOMS LOCATE IN INTERSTICES OF BASE METAL CRYSTALLINE STRUCTURE AND CAUSE DISTORTION IN LATTICE.

Figure 4.8 — Solid-solution hardening
Entry of solute atoms into crystalline lattice of solvent metal (A), either by substitutional solid solution (B), or by interstitial solid solution (C), distorts lattice of alloyed metal and causes strain hardening.

the interstices of the bcc and the fcc crystalline structures to see exactly how they serve with carbon in the principal hardening mechanism for steel. In the final stages of this hardening process additional changes occur in lattice parameters; these must be recognized to understand why steel hardens so effectively, and why significant volumetric changes occur which sometimes cause cracking.

Role of Microstructure

Microstructure can be viewed in a general way as alterations that are produced in the basic framework of the crystalline lattice. Alloying not only allows control over basic crystalline structure, but it is even more effective in manipulating the atomistic nature of the microstructure. When iron is alloyed to form steel, the effects involve the dissolution of phases and compounds in the microstructure by heating above a particular high temperature. This effect is followed by either (1) retention of the elements in solid solution via cooling at some predetermined rate, or (2) if not held in this manner, their dispersal in some form in the microstructure. A final important point concerning microstructure is that *rate of cooling* from the solution-treating high temperature can have a profound influence on many microstructural features — such as grain size; formation of phases; and the dispersal or distribution of phases, compounds, and dislocations in the matrix structure.

Mechanisms for Altering Mechanical Properties

For brevity, mechanical properties will be discussed simply in terms of hardness, ductility and toughness. Usually, strength is commensurate with hardness; but, where a close correlation ceases, a distinction will be made and an explanation given. Ductility and toughness will be treated as two distinct properties in view of the findings that were fully covered in Chapter 3.

At least five different mechanisms may act in an alloy to increase its hardness (not including cold work hardening as described in Chapter 3). These five mechanisms are (1) solid-solution hardening, (2) order-disorder hardening, (3) precipitation hardening, (4) dispersion hardening, and (5) transformation hardening. All five can be utilized in

Table 4.1 — Elements Used for Alloying Irons and Steels, and Their Influence on the Crystalline Structure of the Alloy*

Ferrite Formers	Austenite Formers
Elements that favor the formation of alpha phase in iron, i.e., ferritic (bcc) crystalline form in steel.	Elements that favor the formation of gamma phase in iron, i.e., austenitic (FCC) crystalline form in steel.
Aluminum Chromium Molybdenum Niobium Silicon Tungsten Vanadium	Carbon Cobalt Manganese Manganese Nickel Nitrogen

*The crystalline forms of the alloying elements themselves can be found in Tables 2.7 and 2.8, but their intrinsic structures do not necessarily bear any relationship to their influence when alloyed in iron.

ferrous alloys. Transformation hardening is an especially important mechanism in many kinds of steel.

Solid-Solution Hardening

Solid-solution hardening is achieved by adding a quantity of an alloying element that is soluble in the host metal. This usually produces a mild increase in hardness, with very little loss of ductility or toughness. The presence of foreign atoms as a solute, whether in substitutional or interstitial solid solution, results in a strained lattice as illustrated schematically in Figure 4.8. This condition increases hardness, because localized strain around each foreign atom retards the movement of dislocations and the formation of slip planes.

Foreign solute atoms interact with dislocations and obstruct their movement in several ways. First, distortion in the crystalline lattice alters the size of regular atom sites and interstices; this causes unnatural fit of the solute atoms, thereby increasing the number of dislocations and raising the resistance to dislocation movement. Second, an atmosphere of solute atoms tends to collect in those parts of the lattice being strained and distorted; and this atom concentration, although sub-microscopic, produces more distortion, more dislocations, and more resistance to normal dislocation movement and plastic flow. Third, an increased force is needed to break the atom's bond between dislocations and its atmosphere of foreign atoms; but, once this required stress is applied, a sharp yield point occurs. After this yielding is underway and dislocations have moved away from their atmosphere, the large break-away force is not required. Dislocation theory allows an estimate to be made of the conditions under

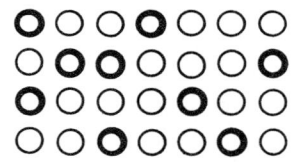

(A) DISORDERED CRYSTALLINE STRUCTURE OF ALLOY IN UNHARDENED STATE.

(B) ORDERED CRYSTALLINE STRUCTURE OF ALLOY AFTER HARDENING TREATMENT.

Figure 4.9 — Order-disorder hardening

Binary alloy of substitutional solid solution type usually has solute atoms in disordered or random distribution as illustrated in condition (A). By treatment, a highly ordered (superlattice) distribution of solute atoms can be attained as illustrated in (B). Real alloys in which this mechanism is employed may be completely ordered in their crystalline structure, or only partly ordered.

which maximum solid-solution hardening can be achieved. There is a critical stage where internal forces acting on dislocations do not cancel each other, and where solute atoms and their dislocations are sufficiently close together to prevent a dislocation line from taking a by-pass route of movement. The solute atoms must be maintained in solid solution to exert hardening effect. If solubility limit is exceeded, the foreign atoms will likely contribute little to the strength and hardness of the alloy, unless their nature and number result in the formation of a highly dispersed phase in the matrix structure. Under these circumstances, a second phase may increase hardness.

Order-Disorder Hardening

Order-disorder hardening is a mechanism that can operate automatically in certain alloys by forces already present in the alloy system, or this hardening mechanism can be made to occur in certain alloys by special heat treatment. Fundamentally, the extent to which order-disorder hardening is accomplished in an alloy is determined by the arrangement of solute and solvent atoms in the substitutional solid solution of a crystalline structure. As illustrated schematically in Figure 4.9 (A) foreign atoms in substitutional solid solution normally are distributed at random lattice sites. The mild hardening that comes from the "solid-solution" effects has been described above. Certain alloys, however, can produce appreciable ordering in their crystalline structure, whereby the foreign atoms occupy particular sites in the lattice as shown in Figure 4.9 (B).

The ordered structure shown in Figure 4.9 (B) is only one possible arrangement. An ordered distribution of the atoms in an alloy sometimes is called a "superlattice," and it will cause a further mild increase in hardness beyond that attributable to solid-solution hardening. Ductility and toughness are usually not adversely affected to any significant extent. Although hardness cannot be increased appreciably by ordering in the crystalline structure, the high-temperature strength of alloys having a superlattice can be substantially better than the same alloy with a disordered structure. This strengthening technique is sometimes used in commercial alloys for high-temperature service.

Table 4.2 — Changes in Pure Iron During Cooling and Temperatures of Occurence

Alphabetical Designation*	Nature of Change	°C	°F
A	Liquid iron freezes to form solid delta iron with bcc crystalline structure	1538	2800
N	Transformation of delta (bcc) phase to gamma iron with fcc crystalline structure	1394	2541
G	Transformation of gamma (fcc) phase to alpha iron with bcc crystalline structure	912	1674
M	Approximate temperature level where alpha iron (bcc) becomes ferromagnetic (the Curie point) on cooling	770	1418
Q	"Room temperature" approximation	25	77

*Alphabetical designations shown as used in Figure 4.12, "Iron-Iron Carbide Phase Diagram."

Ordering in crystalline structures has been long recognized (mention was made earlier of ordering in beta brass when discussing the influence of valency). Although this mechanism is not used as a principal means of hardening in steels, considerable research has been directed to the phenomenon to learn why an alloy formed at a high temperature with a disordered structure will tend to shift to an ordered structure during cooling to a certain temperature. Equations are available in the literature that are based upon entropy, the concentration of solute atoms, and other factors to predict whether a given system will react to achieve the ordered structure.

Precipitation Hardening

Precipitation hardening is a principal hardening mechanism in many commercial alloys, including some steels, and it often operates as a secondary hardening process. Precipation hardening is based on a fundamental characteristic of solutions — the solubility of an alloying element in a solvent metal may decrease as the temperature is lowered. A simple analogy is that of salt being dissolved in hot water. A considerable quantity of salt may enter solution when the water is hot; but when the water cools, some of the salt is rejected from the solution as salt crystals. Obviously, the salt was present in the water beyond its maximum solubility at the lower temperature, and crystals precipitated to relieve the supersaturation. The same performance may take place in an alloy of the proper type; as the solid solution is cooled, a portion of the alloying element may crystallize out of solution in the matrix structure.

STEP 1 - SOLUTION TREATMENT

(A) ALLOY IS HEATED TO ELEVATED TEMPERATURE TO INCREASE SOLUBILITY FOR SOLUTE ELEMENT WHEREUPON DISSOLUTION OF FREE PARTICLES OCCURS IN MATRIX OF SOLVENT METAL.

(B) ALLOY IS RAPIDLY COOLED FROM ELEVATED TEMERATURE TO RETAIN SOLUTE ELEMENT IN SOLID SOLUTION. MATRIX OF ALLOY IS THEN IN A STATE OF SUPERSATURATION.

STEP 2 - PRECIPITATION HARDENING

SOLUTE ELEMENT IS INDUCED TO MOVE TOWARD PRECIPITATION STAGE IN MATRIX BY (a) AGING AT ROOM TEMPERATURE, (b) HEATING TO A SLIGHTLY ELEVATED TEMPERATURE, OR (c) REFRIGERATING TO A LOW TEMPERATURE.

WHEN CORRECTLY HARDENED, PRECIPITATED PARTICLES, AS ILLUSTRATED IN THE GRAINS OF THE MATRIX STRUCTURE IN THE LOWER SCHEMATIC SKETCH, USUALLY ARE NOT VISIBLE BY OPTICAL METALLOGRAPHY. DISTINCT PARTICLES APPEAR WHEN THE PRECIPITATION TREATMENT IS CARRIED INTO THE "OVERAGING" STAGE.

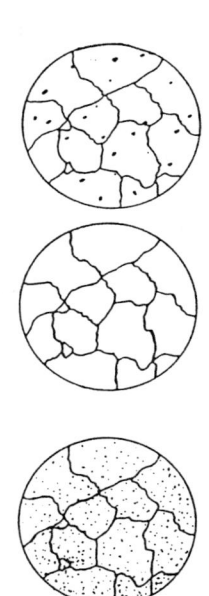

Figure 4.10 — Precipitation hardening: illustration of two essential steps

The behavior of an alloy that is capable of precipation hardening can be made to vary by using certain treatments of heating and cooling as illustrated in Figure 4.10. If the solid solution is cooled rapidly (by quenching), then the dissolved element may be retained in the solvent metal as a supersaturated solid solution. A mild increase in hardness can be expected in the supersaturated condition, but a much greater increase in hardness can be obtained by inducing the solute element to move toward a stage where very fine, uniformly dispersed particles are ready to precipitate out of the solvent metal. This mechanism can be put into motion in several possible ways: (a) by aging the alloy at room temperature, (b) by raising the temperature to a level where precipitation starts, or (c) by refrigerating the alloy to a low temperature where the condition of supersaturation is more intense and precipitation is forced to occur. If precipitation is made to take place at a lower temperature than it normally would during slow cooling, a finer dispersion of particles is secured and the hardening effect of the precipitate is increased.

The efficacy of the solute atoms to produce precipitation hardening is dependent upon the state of the solute-element atoms in the matrix structure of the alloy. In fact, maximum hardening effect is achieved at a stage just before actual particles appear in the microstructure of the matrix metal. This stage involves a condition termed *coherency*, which is explained in Technical Brief 24. The aim in treating a precipitation hardening alloy is to secure a uniform dispersion of the smallest possible coherent particle-like zones. If maximum hardness is desired, this condition achieves the objective. However, if better ductility or toughness must be obtained and some hardness can be sacrificed, the coherent zones will be encouraged by treatment to form precipitated particles and to grow in size until an acceptable combination of mechanical properties is obtained. Good dispersion in a precipitation hardening alloy often is secured whenever the lattice is in a state of mild internal strain or stress. This state may exist in the alloy by virtue of the rapid cooling from the solution treating temperature, but another effective measure is to cold work the alloy in the solution treated condition prior to aging or precipitation hardening. The energy induced in the structure by the cold work increases the diffusion rate and greatly speeds up the precipitation hardening reaction. The size of the precipitated particles usually is controlled by the temperature level employed when reheating the solution treated alloy. *Overaging* is the term applied to any form of treatment that involves further growth of particles with a gradual reduction of strength and hardness, and a beneficial increase in ductility and toughness. Optimum treatment for the various compositions of precipitation hardening alloys varies with the particular alloy system employed.

Dispersion Hardening

Dispersion hardening is not widely used in strengthening and hardening alloys because of limitations of the mechanism that are difficult to avoid. Briefly, extremely fine particles of an insoluble material are added to the base metal, and these particles appear in the matrix as incoherent *dispersoids*. The material that forms the dispersoid can be added to a melt, which then must be agitated just prior to solidification to maintain dispersal in the matrix. Also, the dispersoid can be incorporated in a powder-metallurgy mixture prior to sintering. A rapid solidification operation is an effective way of maintaining dispersal of the particles, but only limited-size sections can be pro-

> **TECHNICAL BRIEF 24:**
> **Coherency of Atoms in Precipitation Hardening**
>
> Coherency requires that a sufficient number of the solute atoms migrate to extremely minute regions of the crystalline structure, where they can concentrate to form the composition of the phase or compound about to be precipitated. Yet, the "particles", of precipitate a composition do not develop an interface between themselves and the surrounding matrix. Instead, each tiny particle-like area represents a sharp difference in composition and a change in crystalline structure, although the orientations of the two structures may coincide. These areas of minute heterogeneity in the matrix are called zones; they are believed to be in the order of 100 Å, and they are without any particular shape. This characteristic of the particle-like zones generates severe elastic distortion in the surrounding structure, which in turn means a large energy term in the total free energy of the zone. The surface energy term of the zone or coherent particle will be nil, because there is no interface between it and the matrix.
>
> Each coherent particle will affect a relatively large surrounding region in the matrix structure, and thus will increase the likelihood of obstructing dislocation and slip-plane movement. Hence a maximum hardening effect takes place that cannot be distinguished metallographically. In this highly hardened condition, ductility and toughness may be less than desired. When treatment fostering precipitation from the matrix structure is carried further, actual microscopic particles (observable metallographically) of the phase or compound are precipitated. Each particle is outlined by an interface or boundary with the surrounding matrix. Initially, the small precipitated particles are coherent with the matrix structure; but with growth in size, as encouraged by whatever precipitation treatment is applied, they eventually become incoherent with the matrix structure.

duced. Another technique sometimes employed is to incorporate an easily oxidized element such as zirconium as a solute in the alloy composition, thus forming a substitutional solid solution. At an appropriate stage, the alloy is subjected to highly oxygenating conditions in order to convert as much of the solute as possible into oxide particles.

Alloys that are hardened by the presence of a dispersoid in the matrix usually maintain their mechanical properties over a wide range of temperature, because the particles act as keys on the slip-planes and are not easily dissolved or coalesced. Metals, oxides, and other compounds are selected as the dispersoid particles that are chemically and physically stable in the matrix. To be effective, the particles must be very small in size, large in number, and uniformly dispersed. A typical alloy which satisfies these requirements would contain particles ranging from 0.5 to 10 microns in diameter, numbering approximately two million per square millimeter, and resulting in an interparticle spacing of about 0.3 micron.

Dispersoids also act to pin grain boundaries and subgrain boundaries, and this tends to stabilize the substructure of an alloy and inhibit recrystallization and grain coarsening. Dispersion strengthening is used in a few commercial high-alloy materials intended for high-temperature service, because it is an effective means of increasing creep resistance and stress-rupture life. This mechanism of hardening is unlikely to be encountered in any of the carbon and low-alloy steels; although the use of oxide particles formed in situ during steelmaking sometimes is used to inhibit grain coarsening in the heat-affected zones of welded low-alloy steels.

Transformation Hardening

Transformation hardening is outlined by the simple schematic illustration in Figure 4.11. In some respects, this mechanism closely resembles precipitation hardening; however, one additional detail is involved in transformation hardening that greatly improves its efficacy. The planned sequence of the mechanism is to first select a solvent metal that undergoes an allotropic transformation of crystalline form at some convenient temperature level. Next, an alloying element is added that is soluble in the high-temperature crystalline form to an appreciable extent, but is soluble in the low-temperature form only to a limited extent. An alloy containing a suitable percentage of the solute element is heated above the transformation range so that its high-temperature crystalline form exists and is able to take the solute into solid solution. Upon cooling this alloy to a temperature where a reversal of the allotropic transformation occurs, the solute element suddenly feels the full effect of limited solubility — in other words, the solvent metal suddenly has become a supersaturated solid solution. The question that immediately arises is whether the solute can escape from the supersaturated lattice of the solvent metal. Circumstances at this point in the mechanism are highly dependent upon the particular elements in the alloy, as well as the temperature at which the reversal of crystallographic transformation occurred. If the transforma-

STEP 1 - SOLUTION TREATMENT

ALLOY IS HEATED TO ELEVATED TEMPERATURE TO DISSOLVE ELEMENTS OR COMPOUNDS IN A HIGH-TEMPERATURE PHASE.

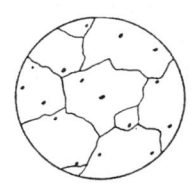

STEP 2 - QUENCH HARDENING

ALLOY IS RAPIDLY COOLED TO RETAIN SOLUTE ATOMS OR MOLECULES IN SOLID SOLUTION, AND TO OBTAIN TRANSFORMATION TO FINER-GRAINED, LOW-TEMPERATURE PHASE.

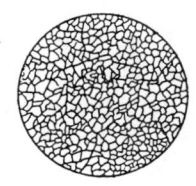

STEP 3 - TEMPERING

ALLOY IS REHEATED TO A SLIGHTLY ELEVATED TEMPERATURE TO ALLOW SOLUTE ATOMS OR MOLECULES TO ESCAPE FROM SUPERSATURATED SOLID SOLUTION.

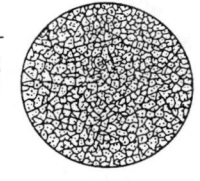

Figure 4.11 — Transformation hardening: including tempering treatment

Essential steps consist of (1) solution of solute atoms in crystalline lattice of solvent metal after allotropic transformation at elevated temperature, (2) rapid cooling to temperature below that for reversal of allotropic transformation and to retain solute atoms in supersaturated solid solution, and (3) tempering to cause precipitation of solute atoms as extremely small particles from the crystalline lattice of the solvent metal.

tion temperature was sufficiently high to permit some atom mobility, then the solute element likely would form some kind of precipitate in the microstructure of the transformed matrix. However, if the reversal of transformation occurred at a low temperature, the solute might be virtually unable to escape from the lattice — unless, of course, the alloy was reheated to a somewhat higher temperature after the reversal of transformation was completed. The reheating (tempering) temperature would be selected to allow the solute to precipitate and form particles of desired size. The rate of cooling from the solution treating temperature often affects the transformation reversal; that is, rapid cooling usually tends to delay the onset of transformation until a lower-than-normal temperature is reached. Sometimes the cooling rate can be regulated so that immediately after transformation, automatically, a fine dispersion of precipitated particles is produced throughout the matrix. Naturally, this microstructural condition results in a marked increase in hardness, but, in many cases, unfortunately, a marked decrease in ductility and toughness takes place. Of course, the precipitated particles can be increased in size to reduce their hardening effect by heating to a selected intermediate temperature (overaging, or increased tempering), and this, in turn, will improve ductility and toughness. Transformation hardening is the principal mechanism employed for increasing the strength and hardness of carbon and alloy steels.

ALLOYING ELEMENTS IN IRON

Because iron is so abundant, is handled in such vast quantities in our steel industry, and is consumed so rapidly by a multitude of users, man often fails to appreciate the remarkable, fortuitous properties of this element. Iron has an atomic weight of approximately 56; and, at 7.9 g/cm^3, it is a metal of moderate density. It has a fairly high melting point 1538°C (2800°F), which in many applications is desirable. The most remarkable and commodious characteristic of this element, however, is its ability to undergo reversible changes in crystallographic structure (allotropic transformations) at two specific temperature levels during heating and cooling. At room temperature, the iron atoms are crystallized in the body-centered cubic (bcc) structure. When heated beyond 912°C (1674°F), the bcc structure (alpha iron) transforms to the face-centered cubic (fcc) structure (gamma iron). If heating is continued past 1394°C (2541°F), the fcc structure reverts to the bcc form, which is called delta iron to distinguish it from alpha iron. Delta iron (bcc) persists to the melting point. On cooling, the transformations just described are repeated in the reverse order, providing that cooling rate is very slow (i.e., equilibrium conditions).

In the course of these transformations during cooling, heat is evolved during the change of one form of crystalline structure to another. These transformations also involve a volume change. When iron in the bcc form transforms at 912°C (1674°F) to the fcc form, a contraction in volume occurs. On these two facts are based methods for noting the temperature levels for transformations, namely, thermal and dilatometric analyses. By thermal analysis, that is, by measuring the temperature of the metal during the passage of time, evolution of heat is determined and matched with transformations. Similarly, by observing any sudden change in the length of a specimen during heating or cooling, a determination can be made of the temperature at which the transformation occurs. Transformations ascertained from (a) interruptions in the normal uniform rate of heating and cooling, or from (b) sudden changes in expansion or contraction of a specimen, are known as *critical points*. Iron also undergoes a loss of ferromagnetism upon heating through the Curie temperature 770°C (1418°F). Long ago, iron heated beyond the Curie point but still in the temperature range of 770 to 912°C (1418 to 1674°F) was called *beta iron*. However, the Curie temperature no longer is recognized as a critical point, because the loss of magnetism takes place over a very narrow range of temperatures and is not associated with a crystallographic transformation.

There are shortcomings of iron that can be mitigated by adding alloying elements to the iron in order to form steel and other ferrous alloys. For instance, iron is progressively oxidized at a fairly rapid rate by rusting and scaling, even in mildly corrosive media; but the addition of elements like chromium and copper can make a substantial improvement in the inherent corrosion resistance of iron. Likewise, pure iron has tensile strength and hardness too low for general construction purposes; but the mechanical properties of iron are improved by adding carbon for increased hardness, and alloying elements like chromium, nickel, or molybdenum for increased hardenability. Much of the success in altering the mechanical properties of iron can be attributed to the reversible transformation of alpha iron to gamma iron and the wide difference in the solubility of carbon in each of these crystalline forms. Among the effects of alloying elements, it will be seen, are alterations in electrical resistance, magnetic properties, and many other properties and characteristics. The addition of alloying elements to iron to accomplish a desired improvement in properties may require only a few tenths of one percent of the element. In fact, most alloy additions are less than three percent. Sometimes improved properties are gained by stringent reduction of a harmful element that may have been tolerated as a residual element. This is a relatively new approach in alloying technology which can be credited to improved steelmaking methods.

Following is a review of alloying elements added to iron for property improvement, and the corresponding residual elements that impart harmful effects. Fundamental information on each element is presented first (atomic weight, melting point, crystalline structure at RT, strong inclinations, etc.) to convey familiarity. Finally, a summary is given of effects imparted when quantities of the element are alloyed in iron and steel. Many of these elements are characterized by either of two new terms; they might be labeled as either a *ferrite former* or an *austenite former*. These classifications indicate the influence of the given alloying element on the crystallographic inclination of iron in the matrix structure of the alloy. It will be recalled from Table 2.7 that a number of metals have a single characteristic crystalline structure whenever in the pure solid state. Aluminum, copper and nickel, as examples, always crystallize in the fcc form. In Table 2.8, metals that are capable of allotropic transformation in crystalline structure were listed; and in addition to iron, the behavior in this respect of elemental metals like manganese and titanium was recorded. When added to iron, certain elements will favor retention of the bcc crystalline structure (alpha iron, or ferrite in steel) and in doing so will restrain the inclination for the allotropic transformation to take place to the fcc structure (gamma iron, or austenite in steel). A sufficient quantity of one of these ferrite-formers (chromium, for example) will suppress the fcc crystalline form completely, and the alloy will retain the bcc structure at all levels of temperature up to melting. Conversely, the addition of a sufficient amount of an austenite-former (nickel, for example) can stabilize the fcc crystalline form.

The effects of alloying elements on the crystallographic behavior of the iron matrix is much more complex than outlined above, because these effects involve not only the expansion or contraction of the phase fields for the respective crystalline structures, but also the alloying element's valency interactions with iron. Nevertheless, the simple explanation given for ferrite-formers and for austenite-formers will serve the purpose providing a very important fact is kept in mind — the intrinsic crystalline structure of the alloying element in pure form (as shown in Tables 2.7 and 2.8) is not necessarily the structure that is induced in iron. For example, aluminum has been stated to have an fcc crystalline structure, but it is a powerful ferrite-former when added to iron. In fact, the addition of little more than one percent aluminum to iron will completely prevent the allotropic transformation of alpha iron to gamma iron. Silicon, a metalloid which in pure form has a diamond cubic structure, is such a profound ferrite-former that a stable or permanent bcc structure is formed when this element is added to iron in an amount approaching about three percent. Whether a significant ferrite-forming or austenite-forming effect will be observed in iron alloyed with multiple elements will depend on the balance between the totals of the two classes of alloying elements. Because multiple elements are commonly made to steels for various reasons, no significant change in ferrite-austenite balance often is the net result. However, there are irons and steels that have a preponderance of elements of one type and this is reflected in the crystalline structure. The steels used in electrical transformers (silicon irons) are a good example. Also, circumstances can arise, and this is true of welding, where pick-up of a substantial amount of a particular element can result in a dramatic effect on the crystalline structure of the newly formed material. Table 4.1 lists a number of elements commonly used to alloy steels, and it indicates whether they are ferrite formers or austenite formers.

A review follows of principal alloying elements used in irons and steels, and residual elements that deserve scrutiny because of their common occurence in commercial products. Interactions are known to take place in ferrous alloys between elements when more than one is present; in fact, significant synergism occurs with certain combinations of elements. Detailed accounts will be provided in Chapters 14 and 15 on effective combinations when multiple elements are used for alloying, but at this point attention will be directed to individual elements as a preliminary step toward explaining the various metallurgical mechanisms by which particular elements affect specific properties. Most of the elements reviewed here are added for alloying purposes, but a few are commonly present as residuals. Despite their low levels, these residual elements are nontheless quite influential; and some of them can have profound adverse effects on weldability.

While the steelmaker or the foundryman has handled the task of properly apportioning elements in the base metals provided for weldments, the composition of the weld metal is largely in the hands of welding personnel and is controlled by the process and procedure applied. Therefore, welding personnel should be aware of the ranges into which alloying elements in the weld metal are expected to fall, and the maxima for residual elements that can be tolerated in most cases. This overview of weld composition should be augmented with available information on quantitative interdependence between given elements and the properties which they influence. This total knowledge can be put to use in judging whether deviations in composition from target values are acceptable.

Carbon

A greater variety of changes can be made in iron by alloying with carbon than with any other element. Extremely small amounts of carbon, as little as one-tenth percent, can cause significant changes in metallurgical behavior and mechanical properties. Although this alloying element is "dirt cheap," it requires more precision in controlling the amount present and accuracy in quantitative analysis than any other regularly used alloying element. Alloys of iron and carbon which contain up to approximately two percent carbon are called *steels*; above about two percent carbon, they are called *cast irons*.

Carbon is an element with an atomic weight of 12, and it can exist in three allotropic forms: amorphous carbon (e.g., lampblack), graphite, and the diamond. Carbon has an indeterminate melting point which is somewhere above 3300°C (about 6000°F); nevertheless, solid carbon is readily taken up by molten iron to form a liquid solution. In solid iron, however, the state or mode of carbon as an alloying element is quite complex. Carbon is an austenite former (see Table 4.1) because it extends the temperature range over which the fcc crystalline structure is stable. An appreciable amount of carbon can dissolve in gamma iron (the fcc form), where it forms an interstitial solid solution. However, the solubility of carbon in alpha iron (bcc form) is very much restricted, and the maximum that can be held in interstitial solution is 0.0218 percent at 727°C (1341°F). At room temperature, solid solubility is only about 0.008 percent. The carbon that cannot remain in solid solution in the alpha iron or ferrite will form a compound for the most part rather than exist as some free form of carbon. This compound is *iron carbide* — Fe_3C, also called cementite — and it is a material of very high hardness (6.5 on the Mohs scale). Because of this trait, the phase diagram for alloys of iron and carbon is called the *iron-iron carbide diagram*. Carbon can exist in iron alloys as free carbon (usually in the form of graphite), but this form is generally found in very high-carbon-content materials such as gray cast iron, which contains approximately three percent carbon.

A fact not to be overlooked, however, is that experimental work has shown the true equilibrium diagram for the iron-carbon system should have iron and free carbon as the final stable phases. Indeed, a number of cases have been reported concerning low-carbon steels which contained a minimum of other alloying elements, and which were held in a service temperature range of about 650 to 700°C (about 1200 to 1300°F) for a couple of years or longer. The hard iron carbides in these steels decomposed to soft graphite in a ferritic matrix in the microstructure. These experiences are interpreted as further evidence that iron carbide is really a metastable phase, and that under true equilibrium conditions the stable form of carbon would be one of its free unbonded forms, most likely graphite. Fortunately, it is relatively easy to alloy the steel with any of a number of elements so that the carbon is more tightly bonded as a carbide;

330   Chapter 4

whereupon, the carbon will remain bonded as such under all conditions of heating, holding, and cooling at temperatures below the transformation range. In this way, the hard carbide particles will continue to serve as a microstructural constituent that maintains strength and hardness. This matter of carbon in steel seeking to assume its stable

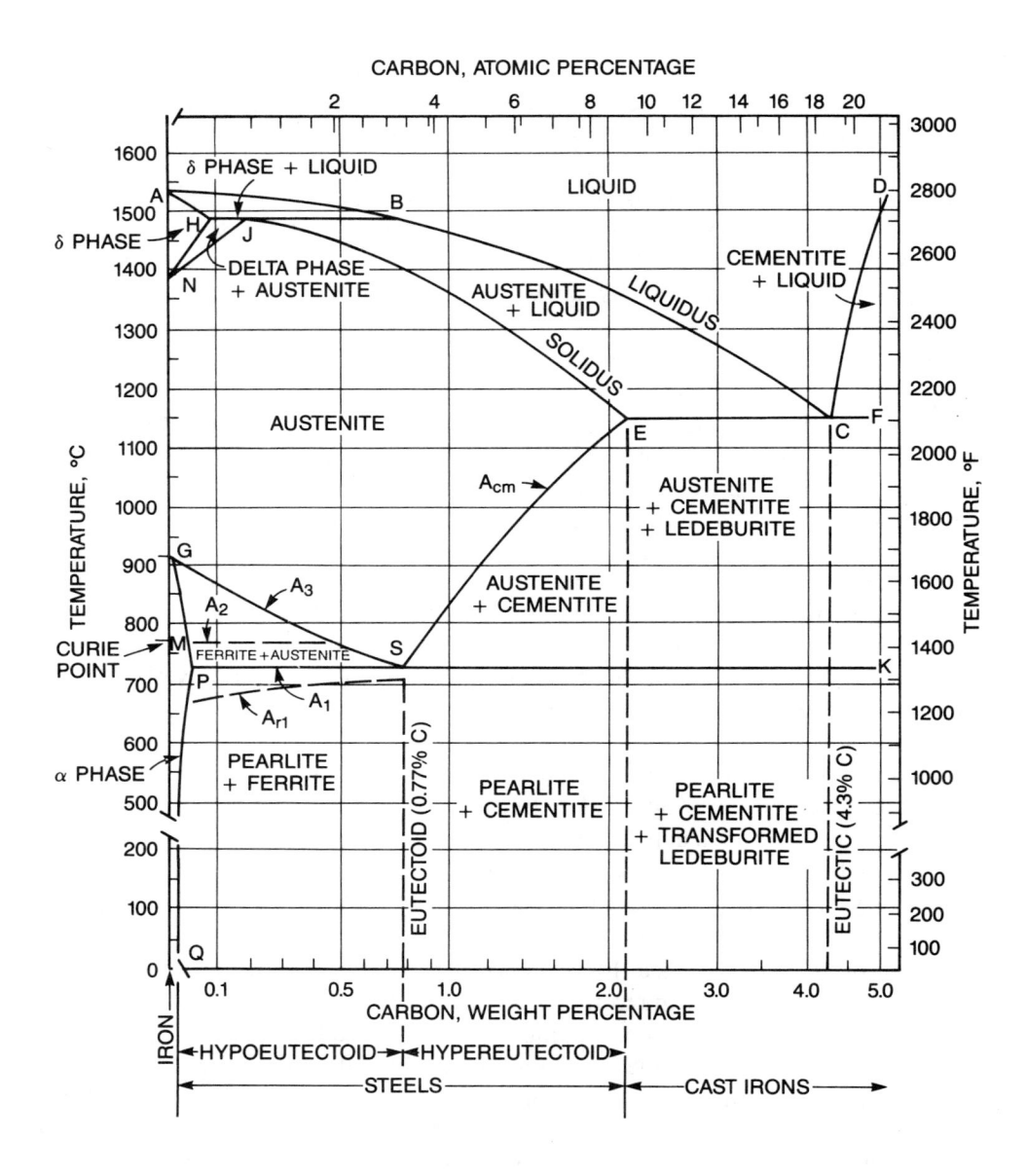

Figure 4.12 — iron-iron carbide phase diagram
Carbon content is plotted logarithmically on the diagram to magnify and make clear the important effects of small percentages of this potent alloying element in iron.

form is not a trivial metallurgical quirk. As will be pointed out in Chapter 13 of Volume II, plain-carbon steels employed in long-time service within the 450 to 600°C (about 850 to 1100°F) temperature range have been known to suffer breakdown of carbides and to form small patches of graphite in the microstructure. This microstructural change seriously reduced strength. Steels alloyed with chromium and/or molybdenum have provided a practical remedy for the problem.

Analysis of the Iron-Iron Carbide Diagram

A phase diagram for the iron-iron carbide system is shown in Figure 4.12. On this diagram, composition is correlated with temperature at normal atmospheric pressure. Temperature is plotted in the vertical direction with the Celsius scale on the left and the Fahrenheit scale on the right. On the abscissa (lower horizontal scale), composition in *weight percent* carbon is plotted on a logarithmic scale. This magnifies the units representing the lower carbon contents, and emphasizes the potency of this alloying element to cause changes. Logarithmic plotting also compresses the area of cast irons where precise carbon content is of less importance. On the upper horizontal scale, carbon is given in *atomic percent*. The number of atoms of this relatively light element among the heavier atoms of iron indicates why carbon is so potent even at low weight-percentages. On a supplementary vertical ordinate at the left side of the diagram, changes in pure iron (i.e., free of carbon) during cooling from the molten state are marked by alphabetical designations positioned along the temperature scale. These designations are described in Table 4.2. Two of these designations deserve further explanation — namely, A_2 and M — and this explanation is provided in Technical Brief 25.

According to Figure 4.12, the iron-iron carbide alloy system has a eutectic reaction like that illustrated earlier in the middle diagram of Figure 4.3, and a peritectic reaction like that shown in the lower diagram. The addition of carbon lowers the freezing temperature of alloys containing up to about five percent carbon and causes solidification to take place over a range of temperature, a typical effect of alloying. As carbon is added to iron, the liquidus steadily decreases along the curve *ABC* until a eutectic composition is reached at about 4.3 percent carbon. Further increase in carbon content causes the liquidus to rise along the curve *CD*. In general, the liquidus of steels is lowered over a relatively small temperature range as carbon content is increased up to the maximum incorporated in high-carbon steels. Cast irons, however, have a substantially lower liquidus as compared with steels. This difference is noticeable when fusion welding the two kinds of iron alloys.

Most steels fabricated by welding usually contain less than about 0.5 percent carbon, so there is reason to give closer attention to this portion of the phase diagram. In a steel melt containing less than 0.53 percent carbon, the first crystals of solid to appear below the liquidus *AB* are δ phase (bcc). Their compositions are represented by the line *AH*, which is part of the solidus *AHJEF*. The terms *delta iron* and *delta ferrite* are used to designate either iron that is free of carbon, or a solid solution of iron and carbon which possesses the bcc crystalline structure just below its solidus. Those alloys having carbon contents between 0.09 and 0.53 percent encounter the peritectic boundary at a temperature level of 1495°C (2723°F).

TECHNICAL BRIEF 25:
Rationale for Letter Designations in the Iron-Iron Carbide Phase Diagram

Two of the alphabetical designations in Figure 4.12 deserve further explanation — the *M* designation, which marks the change in magnetic properties of iron at 770°C or 1418°F (i.e., the Curie point); and the A_2 designation, which indicates this same change in steel. The importance of *critical temperatures* in dealing with steels has fostered the practice of marking not only the occurrence of these changes in the solid state, but also their sequence of occurrence and the conditions under which the temperature was determined. Extensive use is made in the literature of these designations for phase transformation temperatures in steel. Since determination of a critical temperature commonly is made by plotting temperature change during heating or cooling and noting an "arrest" in the resulting curve, the letter *A* is employed as the base of the designation, and this comes from the French *arrêt* meaning arrest. Transformations in the solid state can involve diffusion and thereby require more time than in the liquid state. Consequently, it is important to signify the conditions under which the critical temperature was determined. When a phase change is determined during a slow rate of heating, the critical temperature would be labeled *Ac* (with the letter *c* from the French *chauffage* or *chauffant* meaning "heating"). If the determination was made during slow cooling, the designation would be *Ar* (with the letter *r* from the French *refroidissement* or *refroidissant* meaning cooling). When a critical temperature for phase transformation has been established under equilibrium conditions the designation is *Ae*. The sequence of phase transformations in steel is viewed from their occurrence during heating, and therefore the lowest critical temperature is marked Ac_1. The next higher critical temperature is Ac_2, and this is followed by Ac_3. The "A_2" in Figure 4.12 commonly is included on the iron-iron carbide phase diagram, but this change in magnetic properties is not regarded as a true transformation point. Another extra notation often included on the phase diagram is A_{cm}, which marks the curve that indicates start of austenite transformation in hypereutectoid steels (> 0.77 percent carbon) during cooling where cementite forms.

When the carbon content is above 0.53 percent, the crystals that form as the temperature reaches the liquidus *BC* are called *austenite*. This phase is an interstitial solid solution of carbon in γ iron (fcc). The compositions of these primary austenite crystals are shown by the line *JE*. Austenite that forms from the δ phase along the line *HN* during cooling is called *secondary austenite*. A summation of important phases that form in steels during the solidification process, the phase compositions, and relevant levels of temperature is given below:

- *H* — Delta phase (δ, bcc) containing 0.09 percent carbon
- *J* — Gamma iron (γ, fcc) containing 0.17 percent carbon (peritectic composition)
- *B* — Peritectic liquid containing 0.53 percent carbon
- *HJB* — Peritectic temperature 1495°C (2723°F)
- *NJE* — Austenite (γ, fcc) with carbon in interstitial solid solution

Little attention is paid to the peritectic reaction during the solidification of carbon and low-alloy steels, because the outcome of the reaction is not seen in the distribution of phases in the final microstructure of these steels. In the high-alloy steels (e.g., the stainless steels), the peritectic reaction has more significance when it does occur;

this is because the reaction often does not proceed to completion, and small amounts of δ phase (bcc) may be retained in a matrix that is of a different structure.

Iron-carbon melts containing more than about two percent carbon and up to 4.3 percent carbon during cooling below the liquidus *BC* first form austenite. The carbon content of the austenite can be ascertained by applying the lever law to the solidus *JE*. This primary solid will be lower in carbon than the overall alloy composition, and when the temperature reaches the eutectic line *EC* the remaining liquid will have been enriched to the eutectic carbon content of 4.3 percent. Final solidification at 1154°C or 2109°F — the eutectic line — proceeds by formation of a eutectic consisting of metastable austenite and cementite; this structure sometimes is referred to as *ledeburite*. On continued cooling below the eutectoid boundary *PSK* (727°C or 1341°F), the

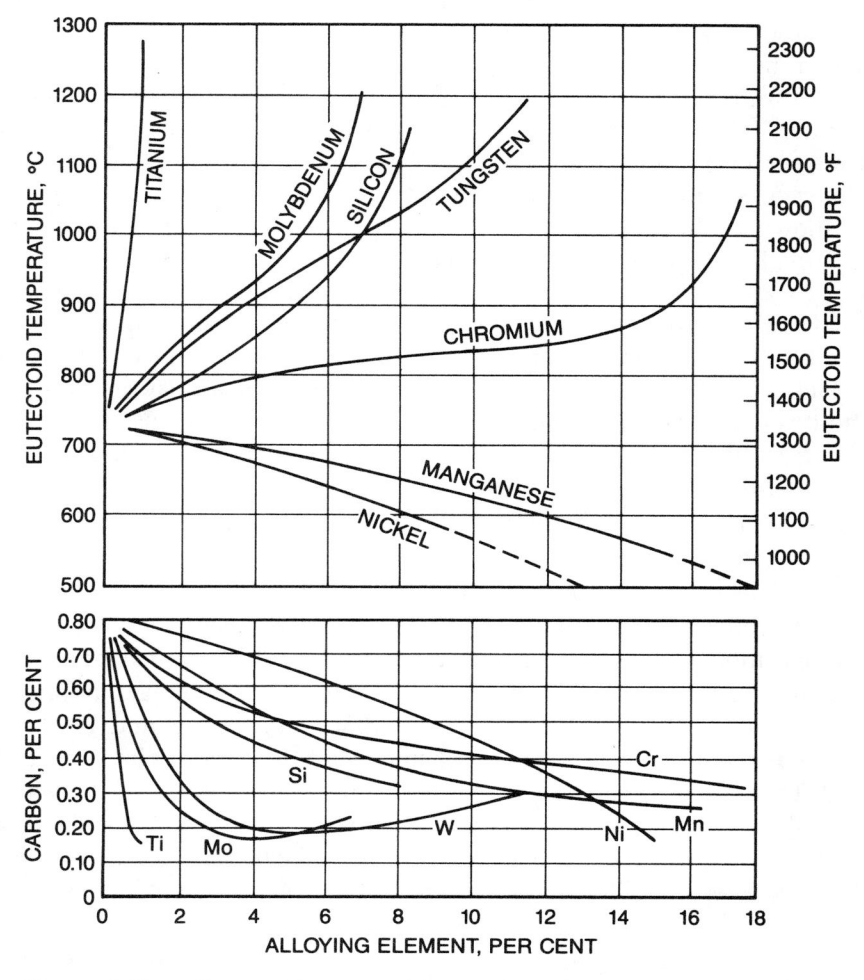

Figure 4.13 — Eutectoid composition and its temperature in steels as influenced by alloying elements

(After E. C. Bain)

metastable austenite undergoes a eutectoid reaction and decomposes into ferrite and cementite (a structure known as *pearlite*).

In melts containing more than 4.3 percent carbon, the first crystals to appear when the temperature reaches the liquidus *CD* during cooling are crystals of cementite — the iron-carbon compound, Fe_3C, which contains 6.67 percent carbon. Pertinent details in the eutectic portion of the phase diagram are:

ECF — Part of the eutectic line, which extends to 6.67 percent carbon
E — Eutectic austenite, 2.1 percent carbon
C — Eutectic liquid, 4.3 percent carbon

Along *ECF*, the three components — austenite, *E*; eutectic liquid, *C*; and cementite — are in equilibrium. If heat is applied, austenite *E* and cementite react to form a liquid containing 4.3 percent carbon at constant temperature 1154°C (2109°F). If heat is removed, the eutectic liquid *C* discharges austenite *E* and cementite in finely divided eutectic form. Liquid is not found below the solidus *AHJEF*.

Aside from the unimportant lines *NH* and *NJ*, there are only five lines in the phase diagram below the solidus, and only two other points, *P* and *S*, which remain to be discussed in detail. Lines *GS* and *SE* are analogous to curves *AC* and *CD*; molten alloy (liquid solution) is analogous to austenite (solid solution). When the solid solution of austenite reaches *GS* (the Ar_3) on cooling, crystals of alpha iron appear. Their composition is shown by the boundary *GP*. The alpha phase in alloys of iron and carbon is called *ferrite*. Ferrite alone, without cementite or without austenite, can exist only in the area to the left of *GPQ*. For example, at an arbitrarily selected temperature of 760°C (1400°F) a ferrite crystal cannot contain an appreciable amount of carbon. Ferrite at this particular temperature can contain no more than about 0.015 percent carbon. In fact, the maximum solubility of carbon in alpha iron appears to be 0.0218 percent, and this occurs at a temperature of 727°C (1341°F). When austenite reaches *SE* (the A_{cm}) on cooling, crystals of cementite start to appear. Pertinent details in the eutectoid portion of the phase diagram are:

PSK — Part of the eutectoid line, 727°C (1341°F)
P — Eutectoid ferrite, containing 0.0218 percent carbon
S — The eutectoid, containing 0.77 percent carbon

The right end of the eutectoid line extends to 6.67 percent carbon, corresponding to the composition of cementite, Fe_3C. Along the line *PSK*, the materials *P*, *S*, and cementite are in equilibrium. If heat is applied, *P* and cementite react to form austenite of composition *S*, containing 0.77 percent carbon, at the temperature level of 727°C (1341°F). If heat is removed, on the other hand, then *S* decomposes into ferrite (*P*) and cementite. The crystals of ferrite and cementite arrange themselves in a lamellar structure with a thin plate of cementite separating every pair of ferrite plates. This structure of alternate ferrite and cementite plates is called *pearlite* because of its resemblance to natural mother-of-pearl. Photomicrographs of this microstructural constituent will appear in figures later.

Transformation temperatures identified as A_1 and A_3 for iron-carbon alloys can vary from those shown by the boundaries drawn in Figure 4.12. If either the heating rate or the cooling rate is faster than the very slow rates required to establish near-equilibrium conditions, the transformation will occur at somewhat different temperatures.

For example, if an alloy in the austenitic condition is allowed to cool in air through the A_1-A_3 transformation range to form ferrite and pearlite, the A_1 temperature might be lowered by approximately 20 to 40°C (40 to 80°F) as indicated by the dashed boundary labeled Ar_1 in Figure 4.12. If an alloy is heated through this transformation range at a comparatively fast rate, as often is done in industrial operations, the Ac_1 and Ac_3 temperatures are shifted to slightly higher levels.

Effects of Alloying Elements Interacting with Carbon

The presence of additional alloying elements, as commonly found in steels, has a profound influence on the critical temperatures, even when heating or cooling through the transformation range is performed at the very slow rates employed for establishing equilibrium temperature levels. This influence of a number of alloying elements commonly used in steels has been studied to determine their effect on the eutectoid temperature (Ae_1) and the findings are shown in the upper graph of Figure 4.13. Manganese and nickel additions shift the Ae_1 temperature downward from the familiar 727°C (1341°F) level, while the elements chromium, molybdenum, silicon, titanium, and tungsten raise the Ae_1 temperature.

The presence of other alloying elements also can cause a significant shift in the eutectoid carbon content, which normally is found at 0.77 percent carbon in alloys consisting of only iron and carbon. As shown by the lower graph in Figure 4.13, the addition of two percent chromium to an iron-carbon alloy would lower the eutectoid composition to approximately 0.60 percent carbon. Additions of manganese, nickel, molybdenum, silicon, titanium, and tungsten act in similar fashion and will shift the eutectoid composition to lower carbon contents. Naturally, this change in the location of the eutectoid point also alters temperatures for the Ae_3 boundary. When heat treatments are applied to steel, it is important to know the Ac_1 and Ac_3 temperatures for the specific composition or type of steel, because structural changes are manipulated largely by control of temperatures to which the material is heated. These critical temperatures are given in handbooks for standard types of steel. Often the cooling rate from the various heat-treating temperatures is of equal importance. Attention will be directed to this metallurgical aspect of steel behavior shortly.

The usefulness of a phase diagram is limited, because alloys during actual metalworking operations practically never are subjected to the very slow heating and cooling rates used in laboratory development of a phase diagram. This is particularly true in welding, where very fast rates of heating and cooling are the rule. Nevertheless, the phase diagram is a first-step guide to temperatures and phases in specific alloys, and in some respects the diagram can be very close to reality and can help make practical metalworking decisions. For example, prediction of solidus temperatures by the diagram will be very close, even for weld metal, regardless of the rate of cooling. This is because molten metal will tolerate very little undercooling before the initiation of crystallization and solidification. It is important to recognize, however, that transformations in the solid state — such as the change from austenite to ferrite and some configuration with cementite, which takes place during cooling through the Ar_3 - Ar_1 critical range — can be delayed by rapid cooling; and they may even be suppressed when substantial amounts of certain alloying elements are present. The delay of austenite transformation until a lower temperature is reached is a fact of considerable importance. If the Ar_3 - Ar_1 transformation range is lowered and there is sluggishness on the part of austenite to transform, then the formation of lamellar pearlite must take place quickly, or not at all. When the transformation of austenite is made to occur more rapidly and at a lower temperature, one of two effects will result: either finer lamellae will be produced in the pearlite, or the cementite will not find an opportunity to appear in this form at all.

From this point onward, frequent use will be made of the word *hardenability*. Hardenability is the inclination of a steel to undergo transformation in a markedly different fashion when it is quickly cooled from the solution-treated or austenitic condition. The microstructural product of this new kind of transformation is called *martensite*, and it is considerably harder than the pearlitic structure. Whereas the pearlite structure forms by a nucleation and growth process, martensite forms through a very rapid shear-type transformation process. Chapter 9 includes information about the influence of cooling rate on the mode of cementite distribution in steels containing various amounts of carbon and modest amounts of other alloying elements, and, in turn, how hardness and other properties are affected.

Increasing the carbon content, in general, will raise the level of hardness in steels. Their hardness increases mainly because a greater amount of cementite is formed in their ferritic matrix. Cast irons do not adhere as closely to this trend because a portion of the carbon that they contain may appear in their microstructure as soft graphite, and this form does not contribute to the hardening mechanism. Earlier it was stated that the mode of distribution for carbon in solid iron or steel is quite complex, and certainly the iron-iron carbide diagram has substantiated this assertion by revealing the numerous transformations in phases that occur at progressively changing temperatures in this binary alloy system. The description of the effects of carbon provided thus far is only a small part of the full account of carbon.

There are many additional ways in which other alloying elements interact with the carbon to produce unique effects. For example, carbon, rather than exist as the free (unbonded) element when out-of-solution, will combine with iron to form the compound, Fe_3C (cementite). Yet, in steels where other alloying elements ordinarily are present, the cementite usually is not simply iron carbide: instead, it is a complex metal(s) carbide formed by additional elements present which have an affinity for carbon that is equal to or greater than that of iron. Manganese, chromium, molybdenum and vanadium are important alloying elements that have strong carbide-forming propensity. When these elements are present, only a small portion of the cementite particles formed in the steel may be iron carbide whereas others will be of a more complex metal-carbide composition. This is a significant fact, and the value of having chromium and/or molybdenum present in steel for sustained service at a high temperature was touched upon earlier.

Furthermore, the more complex multi-metal carbides, in addition to being more stable in the ferritic matrix, have a number of characteristics that improve the mechanical properties and general performance of the steel. To reap maximum benefits from the cementite phase, the metallurgy of this constituent must be thoroughly understood. Sometimes an appraisal must be made of the particular composition of cementite that appears in the microstructure of a given alloy steel, because the nature of the particles can affect at least four characteristics of the steel:

(1) the temperature and time required for dissolution when heating to austenitize the steel,
(2) the tempering temperature needed to gain a particular level of hardness,
(3) the occurrence of *secondary hardening* during tempering, during stress-relief heat treatment, or during elevated temperature service, and
(4) the toughness of the steel which can be affected by the way carbide particles are dispersed in the microstructure and the level of temperature that was employed to produce the dispersion.

The effects of carbon as an alloying element will be detailed further as other aspects of alloying steel provide background for discussion.

Manganese

Manganese is a helpful additive in iron alloys, therefore it is commonly found in some quantity in most steels. It is not as costly as most other alloying elements; and, in general, it is utilized as much as permissible in gaining desired effects. Consequently, the manganese content ordinarily is posted next to carbon whenever the analysis of a steel is reported. Manganese is a grayish-white metallic element which resembles iron in many ways. It has an atomic weight of 54.9 and a melting point of 1245°C (2268°F). Raw material for making a manganese addition is available as electrolytic manganese, and as ferromanganese alloys with various manganese (and carbon) contents. Like iron, manganese exists in different allotropic crystalline forms depending on the level of temperature. From room temperature to 710°C (1310°F) the crystalline form is simple-cubic, and it is designated as alpha manganese. Heating above this temperature will produce several crystalline forms, as indicated earlier in Table 2.8, until the delta form (fcc) melts. The diameter of the manganese atom ordinarily is 2.24 Å, as compared with iron at 2.58 Å, which certainly favors solid solution by substitution. Furthermore, the electron configuration of both elements is electropositive, and each has two electrons in its valency shell (N, or 4) where they are located in the s-subshell. These aspects of atom size, valency, and crystalline structure suggest good compatibility.

The chemical properties and metallurgical characteristics of manganese enable it to perform several helpful functions when alloyed in iron. As compared with iron, manganese has a stronger affinity for oxygen, sulfur and carbon. When added to molten iron, manganese reacts with oxygen contained in the melt to form manganese oxide (MnO). This reaction does not go to completion to take up all of the oxygen present, but proceeds to a point of balance or equilibrium. Therefore, although manganese is a deoxidizer, it is not as powerful in this respect as some other elements, like aluminum and silicon. The manganese in the melt also will combine preferentially with any sulfur present to form manganese sulfide (MnS), a compound which has only limited solubility in the molten iron and rises to escape in the slag if conditions permit. Manganese sulfide also is virtually insoluble in the solid metal, where it appears as non-metallic inclusions. Any manganese, beyond the amount required to combine with all of the sulfur present, will combine with the carbon in the steel to form manganese carbide, Mn$_3$C, as the steel cools through the Ar$_1$ critical point. The properties of manganese carbide and the appearance of particles of Mn$_3$C in the microstructure are indistinguishable from iron carbide.

Manganese is commonly found as an alloying addition in all types of carbon and low-alloy-steel base metals and filler metals for welding, save perhaps Armco ingot iron. The purpose of manganese generally is threefold: (1) combine with oxygen in the molten steel and thus assist in its deoxidation, (2) tie up any sulfur that may be present to avoid the formation of iron-sulfide inclusions that cause hot cracking, and (3) promote greater strength by increasing the hardenability of the steel. A bonus-effect of these actions by manganese is that the fracture toughness is usually improved.

When considering influence on the crystalline structure of iron, manganese is found to be an unusual alloying element. The iron-manganese phase diagram shows that increasing amounts of manganese expands the gamma field, mainly by markedly lowering the gamma-to-alpha transformation which normally occurs in iron at 912 °C (1674 °F). Therefore, manganese has been long regarded as an austenite former in steel. In fact, a very large addition of manganese to iron, perhaps about 25 percent, lowers the gamma-to-alpha transformation so effectively that it does not take place when the alloy is cooled to room temperature — that is, the structure remains fcc gamma phase. However, in certain high-alloy steels, particularly where a large amount of chromium also is present, evidence has been uncovered that manganese in large amounts (perhaps greater than six percent) will lower the delta-to-gamma transformation range. Therefore, a very high manganese content acting in concert with chromium can induce some bcc delta phase to remain in the crystalline structure. Manganese is employed in carbon steels in amounts up to about 1.5 percent. If this element is above the 1.5 percent level, this would place the steel in the low-alloy category. Manganese in amounts exceeding about 2.5 percent is likely to be found only in the high-alloy steels. There is, however, one notable exception. An austenitic steel known as *Hadfield's manganese steel* was developed as a work-hardening material many years ago; although it contains only 10 to 14 percent manganese, an austenitic structure at room temperature is achieved by including 1.0 to 1.5 percent carbon as a supplementary austenite-forming element.

Phosphorus

Phosphorus is an enigma as an alloying element in iron. While regarded as an undesirable residual in many steels, phosphorus is also used very effectively (at little cost) to secure added strength and to improve corrosion resistance in a number of proprietary and standard types of steels. Many years of testing and experience were required to achieve a firm understanding of how to make appropriate use of phosphorus in alloying steels.

Phosphorus has an atomic weight of 31, and in pure form is a very soft, waxy metal with a cubic crystalline structure that melts at quite a low temperature, 45°C (112°F). Pure phosphorus has such a strong affinity for oxygen that it must be protected from the air and from moisture to keep it from reacting with the oxygen in these media. When added to steel as an alloying element, it usually is handled as ferrophosphorus alloy, which contains about 20 percent phosphorus. In molten iron, phosphorus is soluble up to at least 30 percent, a level far beyond the amounts safely usable in iron alloys and steels.

In the solid state, phosphorus forms a compound with iron, Fe_3P, which in turn dissolves in the iron to form a solid solution. Alpha iron, or ferrite in steel, can hold only about 0.1 percent phosphorus in solution at room temperature. Phosphorus in excess of this amount exists as Fe_3P particles embedded in the matrix structure of the alloy. Phosphorus is a weak ferrite former; but, because of the limited amounts used in alloying, this effect can be disregarded.

Phosphorous can be troublesome in steelmaking because it has a marked tendency to segregate during solidification of a melt. During production of steel by the ingot method, phosphorous-rich areas can form in the last metal to solidify. These areas then create an objectionable microstructural condition by rejecting carbon into the surrounding metal. This gives rise to *ghost bands* in the microstructure, which now are recognized as rolled-out areas that contain larger amounts of phosphorus and smaller amounts of carbon. Therefore less cementite and more ferrite appears in the lighter-etching bands.

In steels of compositions designed to be strengthened or hardened by heat treatment, phosphorus tends to decrease fracture toughness. Therefore, in many carbon and alloy steels designed to be used in the heat-treated condition, the phosphorus content is held as low as practicable; certainly no higher than 0.04 percent. In some special steels where fracture toughness is especially important — and where other metallurgical phenomena must be controlled, such as temper embrittlement, and neutron irradiation embrittlement — the phosphorus content is held to a very low limit, in some cases as low as 0.005 percent. This very low maximum for residual phosphorus

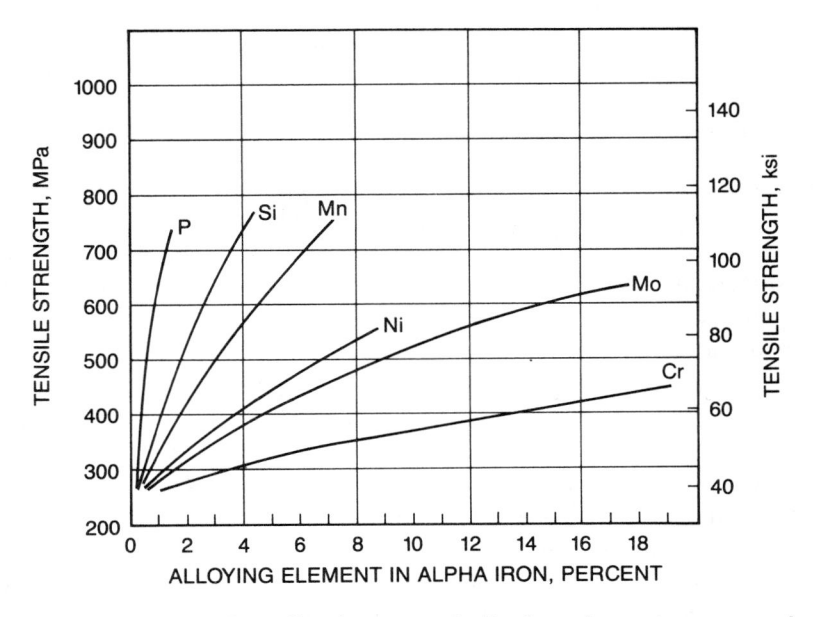

Figure 4.14 — Relative effectiveness of alloying elements commonly used in steels to strengthen ferrite, the alpha phase (bcc) of iron

requires special steps in steelmaking. It is unfortunate that phosphorus has an embrittling influence in the hardened steels because, as will be shown later, phosphorus is very potent in promoting hardenability. In this respect, phosphorus can be helpful, but its contribution to hardenability is limited by the relatively small amount of phosphorus permitted in most steels. Phosphorus probably is the most potent alloying element available for steels to strengthen ferrite by substitutional solid solution, as shown in Figure 4.14. Here again, the amount that can be utilized for this purpose is limited for reasons of segregation, embrittlement, etc. Nevertheless, a small number of carbon steels and high-strength low-alloy steels safely use small additions of phosphorus for increased strength.

The embrittling influence of phosphorus has been found to be dependent on the carbon content in steel, with the higher-carbon types being more adversely affected. For this reason, a number of types of low-carbon steel are made with phosphorus contents somewhere in the range of 0.04 to 0.15 percent. Bessemer steels (popular many years ago, but produced only in limited quantity today) have an elevated phosphorus content that stems from the nature of the steelmelting process; but this residual phosphorus imparts useful characteristics, such as good machinability in certain metal-cutting operations. Consequently, some low-carbon steels produced today by more-modern steelmaking processes also employ elevated phosphorus to improve their machinability. A number of high-strength low-alloy steels (which also are of low carbon content) include a high phosphorus content as part of their alloy design to gain strength and corrosion resistance. To help offset any embrittling influence by the phosphorus level, these steels may employ aluminum as an agent in the deoxidation of the steel inasmuch as this benefit has been observed through experience.

Corrosion resistance of steels in atmospheric exposure can be improved by elevating the phosphorus content, especially if a small amount of copper is present. Phosphorus and copper together provide a greater benefit corrosion-wise than is produced by either of these elements when added individually at the same levels.

Sulfur and Selenium

Sulfur and selenium are mentioned together because their behavior in iron and steel is practically identical. Of course, sulfur often is pointed out as an objectionable residual element in many types of steel, and considerable research has been devoted to less costly steelmaking procedures for reducing sulfur content to the lowest possible level. Ordinarily, selenium is not present in significant quantity as a residual and must be an additive if wanted in a steel.

Sulfur has an atomic weight of 32 and a melting point of 119°C (246°F). Selenium has an atomic weight of 79 and melts at 217°C (428°F). Both elements show appreciable liquid solubility in molten iron; but, because of their low boiling points, they cannot be held in molten iron or steel in large quantities. Nevertheless, the amounts that can be held are troublesome enough if unwanted. Upon solidification, both elements in iron will form iron compounds — namely, iron sulfide (FeS) and iron selenide (FeSe) — both of which have relatively low melting points (approximately 1200°C or 2200°F) as compared with the solidus of iron. Furthermore, these compounds are insoluble in solid iron and steel, and they form eutectic compositions with iron that have solidification points much lower than the compound. For example, the eutectic of iron and iron sulfide solidifies at 988°C (about 1810°F); and because of its low liquidus and its tendency to segregate, this particular eutectic is a potent cause of hot shortness or hot cracking. Because iron sulfide is virtually insoluble in solid iron or steel, this compound appears in the metal as nonmetallic inclusions. The inclusions are of globular shape in ingots and in castings; but they become highly elongated in wrought products because, although unmelted, they are plastic at hot working temperatures. In sheet or plate, as the inclusions are elongated, they also become flattened.

Quite early in dealing with the hot shortness problem caused by sulfur, steelmakers found that adding manganese would virtually eliminate the formation of low-melting iron sulfide. Manganese had a much stronger affinity for sulfur and formed manganous sulfide inclusions that had substantially higher melting temperatures. Manganese sulfide (MnS) inclusions ordinarily are gray in color, but because of the mass action exerted by the presence of so large a proportion of iron, a small amount of iron sulfide could form. However, the FeS was soluble in the MnS and although the color of the inclusions tended to change a little (by picking up a yellow or a brown tint), their melting temperature remained reasonably high. Therefore, the steel could be heated for hot working at temperatures normally used without incipient melting of the nonmetallic (sulfide) inclusions. A manganese content of four times the sulfur level was found sufficient to suppress most of the hot shortness in steel that could be caused by sulfur, but the amounts of manganese regularly included greatly exceeded this remedial level because of additional alloying benefits provided by the manganese. Consequently, for many years there was one ever-present reason to include manganese as an alloying element in steel — to control the composition of sulfide inclusions. Steels made with extremely low levels of manganese (e.g., Armco Ingot Iron) had to be made with very low residual sulfur, and even then they had to be hot worked at temperatures below the iron sulfide liquidus to avoid hot shortness.

For most carbon and low-alloy steels, a maximum of 0.04 percent is the most-often specified restriction for sulfur as a residual element. Manganese serves admirably as the antidote for the hot-shortness problem that surely would arise if this permissible residual sulfur were allowed to combine with iron and form low-melting FeS. The manganese sulfides that form present a substantial number of inclusions, and their size and shape in wrought sections will be dependent on steelmelting and casting practices as well as the amount of hot working.

Recent findings have suggested that excessive manganese sulfide may adversely affect the performance of steel under certain circumstances (see Technical Brief 26); but, despite these effects, there is a useful function of the element — at least as seen by those concerned with the machinability of steel. The most effective way of making steels that machine more easily is to increase the sulfur content. An entire group of *free-machining* or *sulfurized* carbon steels are regularly made to standard specifications that contain 0.10 to 0.30 percent sulfur. As cutting tools of almost every kind remove metal from the surface of this type of steel, the many manganese sulfide inclusions cause the chip to break into short pieces instead of continuing as lengthy, bulky turnings. Also, the inclusions appear to help lubricate the tip of the tool at the cutting edge, which minimizes galling or seizing. Consequently, the machined surface is inclined to be much smoother.

Selenium also imparts the same free-machining effect, but because of its higher cost, selenium ordinarily is used only in the more-expensive stainless steels. Because machining costs can be a major item in producing a component, the high-sulfur free-machining steels are widely used. In fact, the efficacy of sulfur in steel to improve machinability is so keenly regarded that manufacturers of mass-produced parts sometimes will issue a proprietary specification for a standard type of steel that is modified through the addition of a small amount of sulfur, perhaps in the range of 0.05 to 0.08 percent.

The welding engineer must be aware that although free-machining (sulfurized) steels may be selected to reduce machining costs, there does not appear to be any hesitancy on the part of the manufacturer to make use of welding to join components made of such steel. Welding can present problems with porosity and hot cracking when joining these steels, but with a knowledge of the factors involved and the metallurgical reactions that take place, one can usually develop a procedure to make welds that are suitable for service. The welding of free-machining steels (containing sulfur, phosphorus, lead, etc.) will be covered in more detail in Volume II, Chapter 13.

Silicon

Silicon is used mainly in steels because of its favorable characteristics as a deoxidizing agent. Not only is it sufficiently strong as an oxygen-getter to remove unwanted amounts of oxygen, but the oxide formed as inclusions in the metal ordinarily can be tolerated without significant adverse effects. The role of silicon as a deoxidizer will be described in some detail in Chapter 5, but a short review here can bring out additional benefits of silicon as an alloying element in certain types of steels.

Silicon has an atomic weight of 28 and a melting point of 1427°C (2600°F). The crystalline form of elemental silicon is diamond cubic. Silicon is not considered a metal, but rather a *metalloid*. Although the solubility of silicon in *molten* iron or steel is unlimited, *solid* iron will dissolve only up to about 15 percent. Above this level crystals of a compound, Fe_3Si_2, will appear in the matrix structure. The general effect of dissolved silicon is to lower the delta-to-gamma transformation range in iron, and to raise the gamma-to-alpha transformation range. However, at approximately 2.2 percent silicon, the gamma phase is entirely eliminated as an allotropic form, and the iron-silicon alloy retains the bcc crystalline form from room temperature to the melting point. Obviously, silicon is a strong ferrite former. If no silicon is added to steel during the melting operation, a very low residual amount (approximately 0.008 percent) ordinarily will appear in the chemical analyses. This trace amount is picked up from furnace and ladle refractories.

Silicon offers benefits as an alloying addition that have been utilized in certain steels, usually with silicon contents exceeding 0.3 percent. Silicon is a ferrite strengthener, and it is stronger in this respect than most other commonly used alloying elements (see Figure 4.14). Also, silicon is a strong promoter of hardenabilitiy, and proportions as high as about two percent have been added to heat-treatable alloy steels for this reason. As a ferrite-forming element, silicon can preserve the bcc crystalline structure of iron at all temperatures. This effect is used to great advantage in electrical steels, especially those types processed with controlled grain orientation. Finally, silicon tends to promote fluidity when steel is in the molten condition. This effect sometimes is helpful in pouring castings, and in certain fusion-welding processes. Of course, the extent to which any of these metallurgical effects can be employed will depend on maintaining a balance between the desired results, and the unwanted effects produced by the higher silicon content.

Copper

Copper, with an atomic weight of 64 and a melting point of 1083°C (1981°F), crystallizes in the fcc form. Copper and iron are completely soluble in the molten state; but, when as little as 0.1 percent carbon is present in the iron, complete solubility no longer exists. Upon solidification, about 10 percent copper will be retained in substitutional solid solution with delta and gamma iron. However, upon transformation to alpha iron, the solubility is markedly decreased (to possibly two or three percent); thus, upon slow cooling to room temperature under near-equilibrium conditions, much less than one percent copper will be retained in solution. The excess copper that is rejected from solution appears in the matrix as a copper-iron phase. Rapid cooling (particularly by quenching) will retain a much larger amount of copper in solid solution, and this is a feature that has been put to very good use in recent years when developing new types of high-strength low-alloy steels. Although copper is not included in Table 4.1, which lists elements as ferrite-formers and austenite-formers, copper is a weak austenite former.

TECHNICAL BRIEF 26:
Effects of Manganese Sulfide Inclusions in Steel

Recent findings regarding sulfide inclusions now show that, even though manganese sulfide is a relatively innocuous nonmetallic, the physical presence of too many of these inclusions, or of inclusions having unfavorable size and shape, will adversely affect the performance of steel being subjected to certain demanding operations. Briefly, elongated sulfide inclusions reduce the cold deformation capability of sheet steel in deep-drawing and in severe forming; they detract from fracture toughness; they make the steel anisotropic and thereby increase susceptibility to lamellar tearing; and they cause a variety of problems in welding. As evidence mounted on the undesirable effects of sulfides (and other kinds of nonmetallic inclusions) steelmakers moved to make steels for these demanding operations with reduced residual sulfur content and with sulfide inclusions of controlled size and shape, avoiding the highly elongated stringer inclusions as much as possible. By treating the molten metal with alloys and compounds containing calcium, magnesium, titanium, zirconium and rare-earth metals (cerium and lanthanum as contained in misch metal), the steelmaker is able to reduce residual sulfur to impressively small amounts; often levels in the range of 0.005 down to 0.001 percent are achieved. At this low level, very few sulfide inclusions appear in the steel. Also, these treatments often alter the morphology of the sulfide inclusions that do form in ways that are beneficial. The inclusions tend to have a higher melting point than the unadulterated manganese sulfide, and they are harder at hot working temperatures so that they do not elongate into stringers that induce anisotropy. Also, the rare earth metal sulfides are more stable under the heat-effect of welding.

Copper in steels has a long history that passes through several distinct stages. Prior to 1900, the element was looked upon as an injurious residual because of great difficulty with hot tearing or surface checking when hot rolling or forging steel that contained more than about one-half percent copper. This general experience gave copper a bad reputation as an alloying element in iron and steel, and every effort was made in selecting raw materials to reduce the residual level as much as possible. Because no steelmaking process could effectively remove copper from the melt, the residual level ordinarily was in the range of approximately 0.1 to 0.4 percent.

As metallurgical knowledge increased on the mechanism by which copper caused hot-working defects, it was realized that because of selective oxidation of iron at the high temperatures for rolling or forging, the copper tended to become concentrated on the surface beneath the scale. When the working temperature exceeded the melting point of copper (as was usually the case for most operations), liquid copper began intergranular penetration of the steel, whereupon the plastic strains of hot reduction tore these grain boundaries apart. Eventually it was learned that the addition of nickel, perhaps to the extent of one-half the level of residual copper, tended to nullify tearing by raising the temperature at which copper-nickel concentration on the surface would act on the grain boundaries. Thus, by having some residual nickel also present and by carefully regulating the hot-working temperature, the copper problem was essentially solved. Subsequently, steels with residual copper contents ranging up to perhaps 0.5 percent were put through hot-working operations without undue difficulty.

The second stage in the history of copper in steel started in 1913, when corrosion testing and service experience began to show that deterioration of steel by rusting was significantly delayed when approximately 0.2 percent or more of residual copper was present. This finding brought on the *copper-bearing steels*, which customarily were put before users as an option; thus, for steel that is to be used outdoors for structural purposes many purchaser's specifications included a requirement for 0.20 percent minimum copper. The anti-corrosion benefit of this small amount of copper was so consistent that even metallic coated steel sheet (e.g., galvanized) was produced from copper-bearing steel. Recognition of the corrosion resistance improvement made by copper continued until about 1930, when new findings were brought forth on the synergism that could be gained by teaming-up copper with small amounts of one or more additional elements. As mentioned when discussing phosphorus earlier, copper and phosphorus together have a synergisticly beneficial effect on atmospheric corrosion resistance; but, recent results from comprehensive test programs have demonstrated the effectiveness of using a variety of combinations of elements that include chromium, copper, molybdenum, nickel, phosphorus and vanadium. This corrosion research has been the basis for development of *weathering steels*, which are reported to have enhanced atmospheric corrosion resistance approximately two times that of copper-bearing steels, and four times that of steel containing 0.02 percent copper or less.

In addition to imparting increased atmospheric-corrosion resistance, copper as an alloying element in steel exerts additional specific effects, all of which are relatively mild. The ferrite strengthening effect of copper in steel is so small that the element did not deserve being included in Figure 4.14. When copper is in solution in a heat-treatable or hardenable steel, it does increase hardenability slightly by making the transformation of austenite more sluggish; however, copper is not used strictly as a transformation-hardenability promoter. A capability of copper in steel, now being actively utilized, is to serve in an alloy system for precipitation hardening (e.g., ASTM A 710). To perform this function, a copper content of about 1.25 percent is required, and this must be complemented with roughly the same amount of nickel. This new development in precipitation-hardening low-alloy steels should be of great interest to anyone concerned with high-strength weldments, because carbon is not employed as a hardening element; in fact, carbon is restricted to about 0.07 percent maximum. This kind of steel affords considerable opportunity to weld without preheating, and to avoid maintaining an elevated interpass temperature. This development might well be the third stage for the history of copper in steel — namely, significant usage as a strengthening element by precipitation hardening. The welding of this unique type of low-alloy steel will be reviewed in Volume II, Chapter 15.

Chromium

Chromium is a powerful alloying element in iron and steel, and it is added for three principal reasons: (1) it strongly increases the hardenability of steel, (2) it helps maintain the strength and hardness of steel when subjected to elevated temperatures during tempering treatment or in service, and (3) it markedly improves the corrosion resistance of iron and steel in oxidizing types of media. For the latter reason, chromium is the main alloying addition in stainless steels.

The atomic weight of chromium is 52, its melting point is 1875°C (3430°F), and its crystalline structure is the bcc form. The solubility of this element is unlimited in both molten and solid iron. There are three fundamental facts to be considered regarding the behavior of chromium as an alloying element in iron or steel. First, chromium has a strong affinity for carbon, and therefore the carbide particles formed are complex iron-chromium carbides instead of the simple iron or manganese carbide described earlier. Second, chromium combines readily with oxygen under most circumstances; and it can form a refractory nonmetallic compound, chromium oxide, at high temperatures, as well as a protective oxide surface film when present in sufficient amount in solution in the metal. Third, chromium is a strong ferrite former. Progressive additions of chromium gradually restrict the range of temperature over which the gamma phase exists; and, at about 13 percent chromium in iron, the gamma allotropic form no longer appears on heating or cooling. In steel, the amount of chromium needed to prevent the appearance of austenite will depend on the total content of austenite-forming elements also present; but, with low carbon content, the usual proportion of chromium for this effect is about 18 percent. At this level of chromium and above, the bcc crystalline structure is found in the alloy at all temperatures up to the melting point.

The amount of chromium as alloying additions in steel may range from a very small amount (possibly as little as 0.2 percent in the micro-alloyed carbon steels) to as much as 27 percent in the ferritic stainless steels. Chromium is so positive in its effects that the amount added is carefully calculated to accomplish specific objectives. In a great number of alloy steels, where chromium is added for increased hardenability, the primary consideration for welding that arises is the avoidance of excessive hardening or possibly cold cracking in heat-affected zones that are cooled rapidly. The welding procedure for these steels must be carefully planned as outlined in Volume II, Chapters 14 and 15.

Nickel

Nickel is also widely used as an alloying element in iron and steel, and often is added in combination with chromium. Nickel has an atomic weight of 59 and melts at 1453°C (2647°F). It has a fcc crystalline structure and is completely soluble in molten and in solid iron. In solid iron, the presence of nickel strongly promotes the appearance of the gamma (fcc) phase, and an addition of about 25 percent nickel will cause the alloy to retain this form at all temperatures. Nickel, therefore, is an austenite-former. Nickel dissolved in steel displays no inclination to form carbides or oxides because its affinity for carbon and oxygen is less than that of iron.

The hardenability of steel is increased by alloying with nickel, and for this reason nickel is present in many steels in amounts ranging from 0.25 to as much as five percent. It is a very satisfactory element for increasing hardenability because it often improves fracture toughness even while raising strength and hardness. Nickel is frequently used to obtain good toughness in steels at low temperatures and as much as nine percent may be employed for this purpose. In stainless steels, nickel may range from seven to 35 percent in steels of the austenitic type. The austenite-forming influence of nickel is also used in stainless steels to counterbalance the ferrite-forming influence of chromium when tailoring the microstructure of the martensitic stainless steels, and the duplex stainless steels. Nickel in the range of 30 to 40 percent is employed in iron to form the alloy called Invar, which has extremely low thermal expansion and contraction. Other than making a steel more hardenable, nickel as an alloying addition causes no difficulty in welding.

Molybdenum

Molybdenum has an atomic weight of 96 and melts at 2610°C (4730°F). The crystalline structure of this element is bcc, and when added to iron is a strong ferrite-former. As little as three percent molybdenum added to iron will retain the bcc form at all temperatures. As an alloying element in steel, molybdenum has a very strong tendency to form carbides, and it markedly increases hardenability; therefore, it is frequently used along with chromium and nickel to the extent of 0.25 to 0.50 percent. Molybdenum from 0.50 to 1.5 percent is often added to alloy steels to improve their strength and creep resistance when used in service at elevated temperatures. Stainless steels may contain molybdenum additions from 0.5 to 4 percent for a number of reasons. The austenitic type of stainless steels may contain this element for greater corrosion resistance to media that are likely to cause pitting.

Interest in molybdenum as an alloying element has been extended by its use in the microalloyed carbon steels, and in high-strength low-alloy steels where small additions, ranging from about 0.05 to 0.25 percent, increase strength and toughness through the influence of molybdenum on microstructure. Usually, the molybdenum addition is supplemented by an elevated manganese content, and possibly by a small addition of nickel. The objective of this microalloying is to suppress the formation of pearlite in the microstructure, or to reduce the size of the pearlite areas and produce finer carbide lamellae. With somewhat higher levels of molybdenum and other alloying elements, a more complex microstructure, which is called *acicular ferrite*, can form. This microstructure is illustrated in Chapters 9, 14 and 15, and the improved mechanical properties that it offers are discussed.

Niobium (Columbium)

Niobium (long-called columbium in the USA) has been used for many years as a carbide stabilizing element in stainless steels, and as a strengthening element in nickel-base and cobalt-base alloys for high-temperature service. Some use has been made of niobium-base alloys as refractory metals in special high-temperature applications, but, the addition of niobium in small amounts to carbon and low-alloy steels is a more recent development. Niobium has an atomic weight of 93, and melts at 2468°C (4474°F). Its crystalline structure is bcc, and when added to iron it is a ferrite-former. Niobium has a strong affinity for carbon, and a mild affinity for nitrogen and oxygen. Its propensity to form niobium carbides is quite strong, and for this reason, niobium can be used as a microalloying element to strengthen steel through the formation of particles of niobium intermetallic compounds. Strength is increased by precipitation of niobium carbides and carbonitride particles in the ferritic matrix microstructure, and the procedure for obtaining a controlled precipitate and satisfactory mechanical properties involves newly founded technology that continues to be studied and improved even at this time.

The amount of niobium required to generate an effective precipitate in carbon and low-alloy steels is relatively small; as little as 0.05 percent niobium can produce a significant increase in strength. Furthermore, when niobium and the resulting precipitate are properly controlled, the ferrite grain size is refined and this tends to improve toughness, especially at lower temperatures. Niobium is often incorporated in steels along with additions of vanadium and nitrogen. These alloying elements then interact with limited amounts of carbon present to form complex precipitates of niobium and vanadium carbonitrides. A number of high-strength low-alloy steels have been standardized by various agencies that contain differing ranges for niobium content, but all limit this element to relatively low levels; in fact, most have the maximum for niobium somewhat below 0.15 percent. As niobium exceeds the 0.1 percent level, some difficulty with cold cracking or with loss of weld zone toughness may be experienced in welding. Utilization of niobium in steel to gain strength, and perhaps to improve toughness, requires metallurgical knowledge of the phases that can appear in the microstructure with a given alloy system, and the temperatures that control their development.

The dissolution and precipitation behavior of niobium compounds differs from those of vanadium. On cooling austenite from a high temperature with the niobium in solution, niobium carbide (NbC) will begin to precipitate at about 1200°C (2200°F). Because of this relatively high precipitation temperature, the precipitated particles can coarsen appreciably and these would be ineffective for increasing strength. Optimum size and distribution of niobium carbide or carbonitride particles to increase strength and to refine the ferrite grain size is best achieved by hot working the steel under a closely controlled schedule for heating and finishing temperature, and possibly by accelerated cooling after completion of the hot working operation. The procedure employed in the rolling of plate is described in Chapter 5 under the subheading, Thermo-Mechanical Control Processing (T-CMP).

Vanadium

For many years, vanadium was added to alloy steels and to tool steels to increase hardenability, and to form complex carbides in the microstructure. The carbide particles were expected to add to the hardness to some extent, but were mainly counted on to improve the resistance of the steel to softening during heating for tempering, and during heating of tools in machining operations. The amount of vanadium added for this traditional purpose in alloy steels ranged from 0.10 to 0.15 percent, but could be as high as one percent in hi-speed steels for cutting tools.

More recently, good use has been made of vanadium in microalloyed carbon steels as well as high-strength low-alloy steels. Only a small amount of vanadium need be added (about 0.05 to 0.10 percent) for effective strengthening providing a substantial manganese content also is present. Vanadium has an atomic weight of 51, and a melting point of 1900°C (3450°F). It has a bcc form of crystalline structure, and its influence on the structure of iron is that of a ferrite-former. Vanadium has a mild affinity for oxygen, a somewhat stronger affinity for nitrogen, and a very strong tendency to form carbides. A benefit of vanadium deserving of mention is that it reduces austenite grain coarsening when steel is heated far above the Ac_3 temperature. Temperatures as high as 1350°C (2462°F) have been reported necessary for effective dissolution of vanadium nitride particles. Either controlled rolling, or heat treating can be used to strengthen steels alloyed with vanadium. Sometimes, a combination of controlled rolling and subsequent heat treating is employed.

Aluminum

Aluminum plays important roles in steel as a *deoxidizing agent*, as a *nitrogen fixation agent*, and as a *grain-refining additive*, but these functions require only very small additions of aluminum. Therefore, the manner in which aluminum serves these purposes is described in Chapter 5 when discussing killed steel, gas content of steel, and inherent austenite grain size. Although larger amounts of aluminum as an alloying addition presently are not used in carbon and low-alloy steels, metallurgists long have been intrigued by the chemical behavior of aluminum and the reasons for this high interest deserve explanation here. Aluminum is a remarkable metal because of the physical and chemical nature of the oxide that forms on its surface when exposed to air. Even though aluminum is thermodynamically reactive (as indicated by its position in the electromotive force series shown earlier in Table 3.18), the oxide film is solid, continuously bonded to the metal surface, and it provides highly effective resistance to further oxidation and corrosion. The film is only about five to 10 nm (0.2 to 0.4 µin.) thick, and if somehow damaged, it reforms immediately to prevent ingress by corrodants. This protective capability of aluminum has encouraged much experimentation over the years to see if aluminum could be added to iron in sufficient amount to impart this self-protective capability. The only alloying element found truly effective for making a "rustless" iron is chromium, and this is used in amounts of 10 percent or more in the family of alloys called stainless steels. Because of the cost of chromium, the search for a lower-cost alternative has been carried on in many laboratories. Unfortunately, when aluminum is added to iron in substantial quantity, there remains the matter of innate effects of aluminum that somehow must be overcome, or tolerated. First to be considered is the ferrite-forming effect, which is so strong that before any corrosion resistance is secured in iron by adding aluminum the alloy becomes completely and permanently "ferritic." Earlier, it was mentioned that only a little more than one percent aluminum in iron would produce this result. The solid solubility of aluminum in the bcc crystalline structure of alpha iron is approximately 35 percent, but from about 10 percent aluminum and beyond the structure of the solid solution becomes highly ordered (tending to form Fe_3Al) and this seriously detracts from the toughness of the alloy. Without the reversible alpha-gamma allotropic transformation mechanism, the ferritic Fe-Al alloy presents problems with grain coarsening, and there is no convenient means for refinement. Although iron-aluminum alloys show promise from the standpoints of oxidation (scaling) resistance, and, to a lesser extent, corrosion resistance, the limitations on toughness and ductility have stood in the way of commercial production and use. Nevertheless, research is continuing in this area.

Nitrogen

To consider nitrogen as an alloying element in steel, at first thought, may seem strange — particularly in view of frequent references regarding its influence as a contaminating residual in steel. Nitrogen is a gas that surrounds us in the atmosphere, and for many years steelmakers found it inescapable with the melting processes being used. Bessemer converters blew nitrogen through the molten charge with the air blast; open-hearth furnaces using preheated air passed it over the molten bath of metal; and electric-arc furnaces tended to dissociate nitrogen in the arc plasma into its nascent atomic form, which dissolved readily in the molten steel. Investigators do not agree on the amount of nitrogen that molten steel can take into solution, but it is very much larger than the amount that can be held in solution in solid metal, especially at room temperature. There is agreement, however, that after solidification the solubility of nitrogen in the solid steel undergoes sharp changes as the metal cools, particularly as the crystalline structure undergoes allotropic transformations. There is, of course, a marked decrease from liquid-solubility to solid-solubility when the bcc delta form of iron solidifies; but, when delta ferrite transforms to the fcc structure of gamma iron (i.e., austenite), the solid-solubility suddenly increases, albeit not to the liquid-solubility level. As the fcc form of solid cools, solubility gradually decreases; then, upon transformation to the bcc alpha or ferrite form, the solid-solubility is again sharply decreased. The bcc ferrite form of steel at a temperature just below the Ar_1 critical (approximately 727°C, or 1341°F) is able to hold only about 0.10 percent nitrogen in interstitial solution. Nitrogen in excess of this level is rejected from solution during transformation, whereupon it forms iron nitride and other complex nitride particles the compositions of which are determined by other alloying elements present, particularly those that also have an affinity for nitrogen.

The particles of iron nitride precipitated from solution just after the austenite-to-ferrite transformation in steel are believed to be Fe_4N, and because of the relatively high precipitation temperature, they are able to agglomerate and coarsen.

Therefore, these particular particles exert only minimal influence on mechanical properties. However, as the ferrite cools further, the interstitial solubility for nitrogen continues to decrease until, at room temperature, it reaches maximum solubility under equilibrium conditions, which is estimated to be only 0.001 percent or less. Of course, rapid cooling will retain more nitrogen in solution in the ferrite, and so the rate of cooling must be noted whenever residual nitrogen is under consideration. Also, the presence of other alloying elements will influence the amount of nitrogen that actually is retained in solid solution. Carbon, silicon and oxygen decrease nitrogen solubility in ferrite; whereas a number of alloying elements — including chromium, manganese, titanium, zirconium, and vanadium — increase solubility. Of course, carbon and oxygen are usually present to some extent, and these elements also have atom diameters close to that of nitrogen. They undoubtedly participate to a degree in the micromechanism that takes place, depending upon the amount of carbon and oxygen present and the presence of other alloying additives that have a strong affinity for these elements. Boron and hydrogen, because of their small atom diameter, also can participate in some manner if present. Technical Brief 27 presents a description of the way in which nitrogen is held in solution in the crystalline structure of ferrite, and the mechanism of changes that occur as temperature is lowered. This explanation can be helpful in understanding the two almost-opposing roles played by nitrogen in steel, which are: (1) nitrogen as an unwanted residual element, and (2) nitrogen as an alloying element utilized to increase strength and improve toughness.

Despite the seemingly small amount of nitrogen held in interstitial solution in ferrite, this residual can be a factor deserving of consideration in certain steels. This is particularly the case for low-carbon steels which are widely used in the form of sheet fabricated by deep-drawing, or as strip for difficult forming operations. These steels, unless treated with certain additives, are subject to aging, which manifests as an increase in hardness and a decrease in ductility and toughness. The extent of change in these properties can be sufficient to cause the steel to fail during deep drawing or severe forming. Because the steel is in a supersaturated state with respect to dissolved nitrogen (perhaps carbon and oxygen also), the aging mechanism can initiate upon standing at room temperature; or it can be accelerated. Aging is accelerated either by warming to temperatures up to about 260°C (500°F); or by modest straining, as will occur during uncoiling and flattening preparatory to fabrication. The mechanism producing the property changes is precipitation of nitrogen (and other solutes) as nitrides in a particularly effective state of very small size and fine dispersion. A low-carbon steel, after aging, may have its strength increased by 100 MPa (about 15 ksi), its hardness increased by approximately 25 units of HRB, and its ductility and toughness reduced by about one-half. Steels that are thoroughly deoxidized with aluminum, or with aluminum-and-titanium, are essentially non-aging. This is because both oxygen and nitrogen contents have been lowered during treatment of the molten metal, and residual quantities of these elements are effectively "tied up" as undissolved compounds. As steelmaking methods improved, particularly with the advent of vacuum degassing of large melts, residual nitrogen could be controlled in more straightforward eliminative fashion. Where further nitrogen reduction was wanted the level could be lowered into the range of about 0.005 to 0.001 percent, whereupon the effects of nitrogen were usually inconsequential.

At this point the second role played by nitrogen should be examined — that is, its utilization as an *alloying element*. This study must consider carefully all other alloying elements present, because nitrogen that is essentially alone in alloying steel would not be helpful or desirable. For this reason, a nitrogen addition always is accompanied by elements that have a stronger affinity for nitrogen than that of iron (e.g., boron, niobium, vanadium, titanium, zirconium). The nitrides and carbonitrides formed by these elements can be manipulated quite well, either through controlled hot working or by heat treatment, to appear in the microstructure in a distribution that exerts one or more of the following functions: grain refinement, microstructural modification, precipitation hardening and strengthening, and toughness improvement. No one alloying additive produces optimum results for all of these functions; therefore, research continues on establishing viable combinations or systems to serve specified requirements. When the decision is made to employ a measured addition of nitrogen in a steel as one of the alloying elements, this is easily accomplished by using either high-nitrogen ferromanganese, or nitrided electrolytic manganese as the raw material source. The latter raw

---

TECHNICAL BRIEF 27:
Nitrogen Retention in Steel

When the austenite-to-ferrite transformation occurs in steel, nitrogen that is in solid solution in the fcc crystalline structure may be entrapped in the newly formed bcc structure. The nitrogen encounters difficulty moving into new interstitial positions, because even though the bcc form is less closely packed than the fcc form, the coordination numbers mentioned in Chapter 2 for structures (8 and 12, respectively) suggest that the interstices of the bcc form offer smaller spaces in which foreign atoms can be lodged. Ferrite at a temperature just below the Ar_1 critical is able to accept some of the nitrogen atoms between two iron atoms situated on one of its axes; but this elongates the overall cubic structure along this axis, while the other two axes at 90° become a little shorter. The interstitial acceptance of the nitrogen atoms distorts unit cells of the bcc structure into cells of body-centered tetragonal (bct) structure. This bct phase often is identified on diagrams by the letter α'. This distortion will be illustrated later in Chapter 9, when martensitic microstructures are discussed: It suffices to say here that this interstitial solution of nitrogen (which probably includes some carbon and oxygen) strains the lattice commensurate with the percentages of these small-size atoms present and the amount of bct structure formed throughout the lattice. Furthermore, as the temperature falls, these solute atoms can be retained in supersaturated solid solution by rapid cooling. As will be explained shortly, they will attempt to precipitate out of solution during (1) slower cooling, (2) standing at room temperature or at slightly elevated temperatures, and (3) cold straining of the metal. When nitrogen combines with a transition metal element (e.g., iron, chromium, vanadium) an intermediate phase is formed that has metallic properties; whereas, with more electropositive metals (e.g., aluminum), nitrogen forms compounds of the nonmetallic type. Nitrogen values reported in the literature for steels must be carefully interpreted because they often do not differentiate between total contained nitrogen, combined nitrogen held as an intermetallic phase or a compound, and dissolved or free nitrogen held in solid solution. Bessemer steels undoubtedly contain the highest levels of nitrogen, and values ranging from 0.01 to 0.03 percent have been reported. Open-hearth steels are reported to range from 0.003 to 0.006, while BOF-produced steels may range from 0.003 to 0.008 percent.

material is an excellent choice because of its very high nitrogen content and virtual absence of other objectionable residual elements. Nitrogen in proportions up to about 0.04 percent are used in the microalloyed carbon and low-alloy steels, but a typical level is approximately 0.015 percent. Whatever level of nitrogen selected must be accompanied by commensurate amounts of strong nitride-forming elements. Certain ASTM standards for high-strength low-alloy steels advocate 0.03 percent maximum where nitrogen is used as an alloying addition.

Titanium

Titanium has acquired well deserved popularity as a corrosion-resisting metal, and as the base of titanium alloys with excellent strength-to-weight ratios. This usage overshadows the role of titanium as an alloying agent in steel; nevertheless, unique effects are obtained by adding very small percentages of titanium to selected steels in wrought form, in castings, and also as weld metal. To date, the benefits secured through titanium additions in the range of only 0.005 to 0.06 percent are impressive, and this alloying element is likely to figure even more prominently in future advances in the metallurgy of welding steels. The technology of its use is complex, and much research is continuing on metallurgical mechanisms involved. Awareness of current findings is important when employing titanium treated steels.

Titanium has an atomic weight of 48, and it melts at 1668 °C (3035 °F). Although the crystalline structure of titanium is hcp, when it is added to iron and steel, a very strong ferrite-forming effect (bcc) results. Approximately three percent titanium added to iron will completely suppress gamma phase (fcc) formation when the alpha phase (bcc) is heated above its normal α to γ transformation point of 921 °C (1674 °F). Equally important is the fact that iron and titanium readily combine to form brittle intermetallic phases. The compositions and temperatures over which these brittle phases form can be located on iron-titanium constitutional diagrams, where three phases are usually identified as Fe_2Ti, Fe_3Ti and $FeTi_2$. When a titanium addition to iron exceeds approximately three percent, the Fe_2Ti phase appears in the alpha iron's microstructural matrix providing that heating and cooling were conducted under equilibrium conditions.

When titanium is added to molten steel, a complex interplay takes place with the oxygen, nitrogen and carbon that is commonly present; because titanium has a strong affinity for all three of these elements. Therefore, it is difficult to confine the activity of titanium to only one function — that is, only deoxidizing, only nitride-forming, or only carbide-forming. The ultimate distribution of the titanium depends not only on circumstances that prevail at the time of its addition and handling of the melt up to the point of its solidification, but also on subsequent heating and cooling cycles to which the solid steel is subjected.

Because of titanium's strong affinity for oxygen, much of a small titanium addition to molten steel would be relegated to the formation of titanium oxide (TiO_2) unless the melt has a low oxygen content; for instance, approximately 0.01 percent. This low-oxygen state is often achieved by prior deoxidation that makes use of silicon for bulk removal followed by aluminum to reduce oxygen to a tolerable low level. Even with the melt in this low-oxygen state, some of the small titanium addition will be oxidized to TiO$_2$. Little attention had been paid to the titanium dioxide inclusions that remained entrapped in the solid steel, other than to strive for their elimination as unwanted nonmetallic inclusions. Recent findings, however, have shown that a suitable dispersion of very small titanium oxide particles in the melt can serve to nucleate finer grains during solidification and during subsequent microstructural transformations. Furthermore, this nucleation medium is permanent and effective during all thermal cycles imposed on the solid steel — even those cycles with peak temperatures approaching the melting point. Titanium treatment of molten steel to produce very small TiO$_2$ inclusions that remain dispersed in the solid steel is being pursued as a steelmaking technique for steels that are resistant to excessive HAZ grain coarsening by welds made with very high heat input.

The distribution of titanium that remains unoxidized following completion of solidification of the steel will be dependent on (1) the levels of nitrogen and carbon in the steel's composition, (2) whether any other strong nitride-forming elements are also present, and (3) the thermal cycles to which the steel is subjected before a final assessment is made of secondary phases and compounds in the microstructure. The tiny copper-orange colored particles that appear in titanium-treated steels are commonly labeled in the literature as "titanium carbonitrides," which suggests they are a compound of fixed composition; but, this is not the case. The copper-orange color of the particles identifies titanium nitride, but evidence also is obtained from selective metallographic etching and from microprobe analyses that varying amounts of titanium, carbon and nitrogen are incorporated in the particles. Titanium nitride and titanium carbide have a mutual solubility; therefore, the composition of "titanium carbonitrides" possibly should be designated as "TiC-TiN," or as "M$_x$(C,N)$_y$" where "M" represents titanium along with any other metal present that has similarly strong affinity for carbon and nitrogen, such as niobium, tantalum, vanadium and zirconium. The metallurgical behavior of these carbide-nitride particles during heating and cooling gives added reasons to believe that their composition is variable, and can be altered even when dispersed in solid steel.

Alloying steels with a small percentage of titanium has been practiced for many years. The titanium addition had been made in conjunction with small amounts of other alloying elements, such as copper, nickel and molybdenum, to produce one of the so-called *micro-alloyed steels*. When titanium is in solution in an austenitized steel, its hardenability is increased to a small extent commensurate with the level of titanium present. However, pre-existing titanium carbide particles do not dissolve as readily as those of manganese when the steel is heated above its Ac$_3$ critical temperature; therefore, the addition of titanium (albeit a hardenability promoter) will actually decrease hardenability if some carbon in the steel is tied up as undissolved titanium carbide particles in the microstructure.

Utilization of titanium principally as a former of nitride particles (TiN) is a more recently developed specialized alloying technique that requires close control of molten steel temperature, and the levels of oxygen, nitrogen and carbon in the melt prior to the addition of titanium. With a low level of oxygen to minimize TiO$_2$ formation, and a suitable level of nitrogen, the titanium will form extremely small disc-shaped titanium nitrides (TiN) which measure only about 200 Å, and are dispersed throughout the steel. During cooling through the Ar_3 to Ar_1 transformation, some titanium carbides will also tend to form. In general, during reheating, TiC particles can be redissolved in austenite more readily than TiN particles; so, here is the key to planning hot working and heat treating thermal cycles that favor the presence of TiN particles in the final microstructure of the steel. The presence of the titanium nitride platelets dispersed throughout the microstructure, in addition to increasing the strength of the steel, also enable it to resist grain coarsening when heated to a very high temperature for a short time; as will occur in the HAZ of a weld made with high heat input.

Certain alloying elements are not entirely compatible with titanium in steel because of opposing effects on metallurgical behavior during microstructural transformations. Niobium, for example, is not suitable to include with titanium in a steel's composition because niobium tends to lower the temperature range for transformation of austenite to a ferritic microstructure. Titanium, especially when present as precipitated titanium nitride particles, tends to raise the transformation range. Furthermore, niobium often has an adverse influence on the toughness of weld metal and the base metal HAZ of high heat-input welds.

Obviously, a very small percentage of titanium can have profound beneficial effects on the mechanical properties and the metallurgical behavior of steel providing the proper technology is applied in (1) making the titanium addition, (2) controlling the association of the titanium with other minor elements that may be present, and (3) manipulating by hot working and/or heat treatments the form of titaniferous constituents that appear in the final microstructure.

Boron

Boron is a nonmetallic element with an atomic weight of only 11 and a melting point of 2030°C (3686°F). As a very light element, its small atoms enter into interstitial solid solution in iron. Yet, the solubility of boron in iron and steel, whether the crystalline structure is in the bcc or the fcc form, is quite limited. The solid solubility is less than 0.1 percent, and any boron in excess of this soluble amount will appear in the microstructure as iron boride, an intermetallic compound having the formula Fe_2B. If other alloying elements, such as aluminum or chromium, also are present, these elements will not only affect the solubility level for boron, but will also enter into the composition of the intermetallic boride particles. Boron is added to alloy steels in very small amounts, perhaps as little as 0.003 percent, to increase hardenability. Boron is a very powerful promoter of hardenability on a weight percentage basis, but this effect is gained only with amounts of boron below approximately 0.010 percent. Any boron above this level is ineffective. Therefore, it has become a common practice to "needle" some alloy steels with a very small addition of boron to obtain increased hardenability. The benefit derived from this element often is measured and controlled by means of hardenability testing. Chemical analyses reveal such low levels of boron in the steel that the correlation between hardenability and boron level is not always clear.

Some use is made of boron in the high-alloy steels as an alloying element, and also as small additions to improve the hot-working properties of certain steels. Large proportions of boron, in the range of 0.5 to 2 percent, are added to a special kind of 18-8 austenitic stainless steel employed as control rods in nuclear reactors. The boron in the steel provides a number of unique advantages as a control over the neutron flux in the reactor. Earlier, in Chapter 3, the very high neutron-capture cross-section of boron was discussed. In carbon and low-alloy steels, boron is added in such small amounts that its effect is found principally in increased hardenability. Occasionally, some minute boron nitride particles can be detected among precipitates in the microstructure.

Cobalt

Cobalt has an atomic weight of 59 and a melting point of 1495°C (2723°F). Cobalt is completely soluble in iron and steel — in the molten state; and also in the solid state, where it forms a substitutional solution. The crystalline structure of cobalt is hcp, and when added to iron or steel it acts as a weak austenite-former. In fact, approximately 80 percent cobalt must be added to iron to form an alloy in which transformation to the ferrite (bcc) phase has been eliminated. Cobalt has less affinity for oxygen and less tendency to form carbides than does iron. It is an unusual element inasmuch as it decreases hardenability in steel. Nevertheless, it can be useful as an alloying element, since it produces a remarkable amount of solid-solution hardening when dissolved in either ferrite or austenite. Because of its high cost, cobalt is not ordinarily used in low-alloy steels. It is, however, included in the composition of the *maraging alloy steels* (ASTM A 538) to the extent of about eight percent because of its influence on the transformation of austenite as will be described in Volume II, Chapter 15.

Tungsten

The most frequent use of tungsten is made in tool steels, although recently it has been added with good results to a few of the highly alloyed heat-resisting steels. Tungsten is a very heavy element with an atomic weight of 184. Its melting point is 3410°C (6170°F), which is so unusually high that this property often must be given consideration when incorporating tungsten in melting or welding operations. The crystalline structure of tungsten is the bcc form. In steel, tungsten is a ferrite-former, has a strong tendency to form carbides (which are extremely hard), and increases hardenability. The reasons for adding tungsten to tool steel obviously are to increase hardenability and to form wear-resistant carbides, but tungsten is also effective in sustaining hardness as the metal rises in temperature to a red heat, a condition that is not uncommon at the tip of a cutting tool. Powdered tungsten carbide is used as an additive to steel weld metal being deposited as a wear-resisting surface layer. The process is controlled to minimize melting of the particles of powder; the principal objective being to secure a deposit with the extremely hard tungsten carbide particles uniformly dispersed in a matrix of steel weld metal. Of course, tungsten also can be added to the filler metal as an alloying element, and with a proper carbon content the weld deposit can form tungsten carbide as a constituent in the microstructure.

Lead

Earlier, in Technical Brief 19, lead served as an example of an element that was virtually insoluble in iron both in the liquid and the solid states. Yet, to overlook lead in reviewing alloying elements used in steel would be a mistake. Lead is added to some types of steel to improve machinability. Its presence can substantially increase cutting rates and extend tool life. Because the lead is insoluble in steel, both in liquid and solid form, the addition of this element must be made under conditions which will produce a fine, mechanical dispersion of lead in the steel matrix. For example, 0.15 to 0.35 percent by weight of very small lead shot may be gradually added to molten steel as it is cast into an ingot or some other solid form. With proper technique of addition, the lead is dispersed as minute globules which do not change the composition of the steel matrix, per se, and the mechanical properties of the steel remain essentially unchanged.

Lead is a low-melting, heavy element with an atomic weight of 207 and a melting point of 327°C (621°F). When making an addition of lead to steel, care must be exercised in avoiding not only the segregation or accumulation of lead under the force of gravity, but also the loss of lead by evaporation; because its boiling point is 1724°C (3137°F), and this is not far above the melting point of steel. Normally, however, fusion joining operations can be conducted on a leaded steel so that no difficulty is encountered with lead segregation, loss of lead by vaporization, or cracking from incipient melting. On the other hand, if unusual circumstances should produce a large weld melt composed mostly of base metal, under a hot arc and with very slow solidification rate, then some loss of the low-melting, volatile dispersoid must be anticipated.

Other Alloying Elements

From time to time, new steels will be publicized that contain additions of alloying elements not listed and described above. Sometimes the claims made are not fully supported, and the new product is short-lived. For example, a new free-machining steel containing tellurium was proposed, but it has not been produced in quantity. Uranium has been tried as an alloying addition to steel and to cast iron. Uranium is a strong sulfide-former that presumably will affect the nature of sulfide inclusions, but no particular advantages appear to be gained. While the possibility always exists that a useful element has been overlooked as an alloying additive, metallurgists at present are mainly devoting their efforts to exploring the synergism that can be secured through new combinations of familiar alloying elements. Following is an illustration of alloying technology for the future — technology that currently has not been commercialized.

It has been found possible to employ the inert gases as alloying additives despite what has been said earlier about their insolubility, etc. A technique now is being developed to make extremely fine-grained alloys through use of powder metallurgy (as the raw source), and rapid solidification (as the cooperative processing step). As the necessary metal powders are produced, argon or helium gas is entrained in the particles. Upon quickly melting these inert-gas-laden particles and quickly casting the melt into a rapid-solidification machine, the resulting alloy retains atoms of the inert gas, and because of their insolubility these atoms are forced into the most-primitive lattice dislocations and vacancy sites. The inert gas atoms stabilize sites and discourage coalescence. The sites then provide highly dispersed locations for subsequent precipitation of intermetallic compounds, such as carbides, in extremely small size. Also, the grains of this inert-gas "doped" alloy often are one-tenth the size normally found in the alloy when made without the inert gas addition. The properties of these unusual alloys have yet to be fully explored to document advantages. This example of a new steel under development is mentioned to point out the challenge that may lie ahead for welding, especially the fusion-welding processes. When joining these alloys, the inert gas atoms can have opportunity to accumulate and evolve as porosity. In addition to this unsoundness, the weld metal does not reproduce the extremely fine grain size and precipitation characteristics of the base metal.

BENEFIT OF REVERSING THE ALLOYING TREND

While significant changes can be made in the properties of irons and steels by adding relatively small amounts of alloying elements, the host element, iron, is a remarkable metal in its own right. In fact, for certain applications, iron could be more useful if left unadulterated; that is to say, *unalloyed*. This reasoning, in 1970, led to the invention of *interstitial-free* steel (commonly identified by the acronym "I-F"). This kind of steel (technically, an iron) has become widely used for metal articles and components manufactured by severe forming and by very deep drawing operations. The I-F steel's characteristics — high ductility, absence of yield-point elongation, and freedom from strain-aging — reduce breakage substantially during the severe plastic deformation that occurs in these operations. Sometimes, the I-F steel is welded during assembly of components or in preparing tailored blanks (as will be described on page 438), so there is reason to examine its weldability.

Iron can exhibit remarkable ductility if unaffected by other elements that are able to enter into solid solution — especially those elements having atom diameters small enough to enter interstitially (see Figures 4.7 and 4.8). However, this seemingly simple task of de-alloying iron must be weighed against the reality that certain foreign elements, although present interstitially to the extent of only a few parts per million, can cause measurable changes in the mechanical properties of iron. Commercial steelmaking practices which have evolved for making I-F steel entail the following essential steps:

(1) select raw materials that are devoid as possible of foreign elements, especially those that cannot be removed during the melting operation;
(2) minimize exposure of the molten iron to refractories from which foreign elements can be picked up;
(3) subject the melt to secondary refining operations (i.e., ladle metallurgy treatments) including vacuum degassing and decarburization; and
(4) add minimal amounts of titanium and/or niobium to scavenge what little carbon, oxygen and nitrogen might still be present in the ladle-treated melt before direct con-casting or pouring into ingots.

Typically, I-F steel contains an ultralow carbon content, and the total of residual carbon, phosphorus, oxygen and nitrogen probably will be less than 0.025 percent. Table 4.3 provides information on the composition and properties of I-F steel. Even though the residual elements that enter into interstitial solution have been restricted to extremely low levels, it is still necessary to somehow "tie up" or stabilize the very small amounts of C, O and N that remain. This is accomplished by microalloying — that is, adding a small amount of titanium, possibly some niobium, and perhaps aluminum — in order to form a minimal number of carbides, oxides and nitrides, which will be innocuous in the microstructure during plastic deformation.

Welding the I-F steel, whether as bare steel sheet or as a zinc-coated product, presents no particular problem; since it behaves essentially as a very-low-carbon "killed" steel. Resistance spot welding (RSW) is frequently used in the auto industry to assemble stamped and drawn body components made of I-F steel. Some use is made of tailored blanks, and autogenous butt welds prepared by the laser beam welding process (LBW) ordinarily will withstand the same amount of deformation in forming and drawing as the I-F steel comprising the blank.

The strength of I-F steel weld metal (autogenous) usually is a little higher than the base metal (presumably because of dislocation density in weld metal, since carbon is not present to contribute and hardening effect). Incidentally, for applications where somewhat higher strength is wanted in the I-F steel sheet, special heats of steel are produced with an elevated phosphorus content (~ 0.06 percent) rather than resort to carbon as a strengthening element. This modification obviously makes use of knowledge presented earlier and illustrated in Figure 4.14. Briefly, phosphorus appears to serve effectively as a ferrite-strengthening addition; and, if carbon content is extremely low, the metallurgical problems mentioned with steels of higher carbon content do not occur. To date, however, welding experience with elevated phosphorus type of I-F steel is limited. The popularity of this unique steel for auto and truck body components suggests that further reporting on the weldability of the higher-phosphorus content type I-F steel will be forthcoming.

Residual Elements

Steelmakers face a growing problem with pick-up of unwanted elements which may or may not significantly influence the qualities and properties of their products. These unwanted elements can be present in the ores and other raw materials procured for melting, but they are more likely to arrive with ferrous scrap metal that is being recycled. Light melting scrap from junked autos, switch gear, and other multi-component manufactured articles are especially troublesome because of the variety of nonferrous metals inconspicuously incorporated in small components. Great effort is made in sorting scrap and in eliminating contaminating types of metals and nonmetals, because some

Table 4.3 — Interstitial-Free I-F Steel[1]

Typical Chemical Composition, percent[2]

C	MN	P	S	Si	Ti	Nb[3]	Al	O	N	Other Elements[2]
0.004	0.12	0.010	0.010	0.01	0.07		0.04	0.003	0.004	Cu,Cr,Ni,Mo, etc.

Typical Mechanical properties of I-F Steel, and Sheet Formability

Ultimate Tensile Strength		0.2% Yield Strength		Elongation in 2 in.	Stretch Formability[4]	Drawing Formability[5]
MPa	ksi	MPa	ksi	Percent	n	r_m
345	50	170	25	40	0.235	2.10

(1) Not presently covered by a standard specification by any standardization body.
(2) Levels of residual elements commonly found in steel are quite low and not influential.
(3) When niobium (columbium) is included, it usually is added in conjunction with titanium, and the total of Ti plus Nb is approximately 0.08 percent.
(4) The *n* value is a strain-hardening coefficient determined by relationship between flow (yield) stress and strain. A high *n* value indicates good stretch-forming capability.
(5) The *r* value is indicative of resistance to thinning during deep drawing. Average normal anisotropy is designated r_m, while Δr is planar anisotropy or the propensity for earing to develop on the deep-drawn article.

of the elements picked up are not easily removed during melting operations. Those that cannot be removed are lowered in concentration only by dilution.

Residual elements that are carefully monitored by steelmakers because of possible adverse effects on steel qualities include antimony, arsenic, and tin. The latter element, tin, is a persistent problem because of the recycled tin plate scrap generated during the production of steel food and beverage cans — even though only a very small amount of the cans themselves are likely to return as recycled ferrous metal scrap. The tin plate scrap routinely is put through a de-tinning operation to improve its suitability for steelmaking, and to recover the valuable tin metal for further use.

The welding engineer must recognize that the residual-element threat is rekindled every time a welding process is applied, especially those processes employing fusion and the formation of weld metal. The miniaturized melting operation in welding always allows opportunity for other elements to enter the weld metal; sometimes they even invade the HAZ for the base metal by diffusion. These unforeseen, unwanted elements can be picked up during welding from flux, from shielding gases, or from other adjuvant materials. The overall problem can require much metallurgical insight in order to detect the causative element, determine its source, and devise a corrective measure. The problem may prove difficult because (1) the offending element may be one that is not routinely analyzed and monitored; (2) the concentration may be relatively low and seemingly innocuous, but still sufficient to exert adverse effects; and (3) the symptoms may occur erratically as changes are made that vary the source of the offending residual element. Fluxing materials deserve close attention because they often contain natural minerals, and the purity of these minerals is highly dependent on the locale from which they were obtained. As will be explained in Chapter 8, elements in flux frequently are transferred to weld metal during fusion welding operations.

SUGGESTED READING

Alloying, Edited by J. L. Walter, M. R. Jackson, and C. T. Sims, ASM International, Material Park, OH, 1988, ISBN-0-87170-326-2

The Effects of Residual, Impurity and Micro-Alloying Elements on Weldability and Weld Properties, Proceedings of London International Conference, November 1983, The Welding Institute, Abington, Cambridge, England, 1984, ISBN: 0-85300175-8

The Theory of Mild and Micro-Alloy Steels' Weldability, by Ivan Hrivnak, Welding Research Institute, Bratislava, Czechoslovakia, 1969, ISBN: 63-054-69-05

Phase Diagrams of Ternary Iron Alloys, Part 1, by V. Raghavan, 1986, ISBN: 0-87170-230-4, *Part 2*, by V. Raghavan, 1988, ISBN: 81-85307-00-8, *Part 3*, by V. Raghavan, 1988, ISBN: 81-85307-00-9; *Part 4*, (including *Phase Equilibria in Iron Ternary Alloys*), by V.G. Rivlin and G.V. Raynor, 1988, ISBN: 0-901462-34-9; *Part 5*, by V. Raghavan, 1989, ISBN: 81-85307-04-0; *Part 6*, by V. Raghavan, 1989, ISBN: 81-85307-12-1 and 81-85307-13-X, joint publications of ASM INTERNATIONAL, Material Park, OH and The Indian Institute of Metals

Chapter 5

In This Chapter...

GENERAL CATEGORIES OF IRON & STEEL 361
IRON PRODUCTION BY ORE REDUCTION 362
   Blast Furnace 362
   Direct Reduction Processes 365
CAST IRON 366
WROUGHT IRON 367
POWDER METALLURGY 367
STEELMAKING PROCESSES 368
   Significance of Acid and Basic Steelmaking 368
   Bessemer Converter 369
   Open Hearth Furnace 370
      Rimmed Steel 372
      Capped Steel 373
      Killed Steel 373
      Semikilled Steel 379
      Vacuum Deoxidized Steel 379
   Oxygen Steelmaking 380
   Basic Oxygen Steelmaking 380
      L-D Process 381
      Kaldo Process 383
      Off-Gas BOF 383
      Q-BOP Process 383
      Lance-Bubbling-Equilibrium 385
   Ladle Refining 385
      Slag Removal 385
      Mixing Capability 386
      Alloying Additions 386
      Vacuum Treatment 386
      Temperature Adjustment 387
      Desulfurization 388
   Electric-Arc Furnace 389
   Electric-Induction Furnace 389
   Electroslag Remelting 391
SPECIAL MELTING PROCESSES 392
   Vacuum Induction Melting 392
   Vacuum Consumable-Electrode Remelting 393
   Electron-Beam Melting 395
   Argon-Oxygen Decarburization (AOD) 396
FOUNDRY AND STEEL MILL OPERATIONS 397
   Ingot Steelmaking Practice 398
   Continuous Casting of Steel 400
HOT WORKING OPERATIONS 404
   Thermo-Mechanical Control Process (T-MCP) 406
COLD FINISHING 407
HEAT TREATMENT 408
CONT. COATING OF STRIP STEEL IN COILS 409
TYPES OF STEEL 410
   Carbon Steels 410
   Alloy Steels 411
      Construction 412
      Automotive, Aircraft, and Machinery 412
      Low-Temperature Service 413
      Elevated Temperature Service 413
   High-Alloy Steels 413
      Austenitic Manganese Steel 413
      Stainless Steels 414
      Heat-Resisting Steels 415
   Tool Steels 415
STANDARDS & SPECS. FOR STEELS 416
   Unified Numbering System 416
   AISI-SAE System of Std. Carbon & Alloy Steels 418
   ASTM Standards 420
   API Specifications 424
   Aerospace Material Specifications 426
   ASME Material Specifications 426
   AWS Specifications, Codes and Rules 429
CARBON AND ALLOY STEEL USED IN WELDED CONSTRUCTION 430
   Qualities of Steel Important to Welding 431
   Factors Affecting the Weldability of Steel 432
      Chemical Composition 432
      Mechanical Properties 434
      Metallurgical Structure 435
      Internal Soundness 435
      Cleanliness 436
FUTURE OF STEELS & THEIR WELDABILITY 437
   New Steels and Product Forms 437
   Dissimilar-Metal Welding 437
   Repair Welding — The Ultimate Challenge 439
SUGGESTED READING 439

Types of Steel and Their Manufacture

Knowledge of manufacturing methods used to produce iron and steel is very helpful in gaining a full understanding of the welding behavior of the various product forms employed to construct weldments. This learning step became increasingly important in the early 1990s, because the remarkable changes which were taking place in steelmaking also happened to control subtle aspects of chemical composition. Furthermore, innovations introduced in casting and in mill processing can produce specialized mechanical properties that are needed for particular service applications. The methods of producing these semi-finished product forms can greatly influence their behavior during subsequent weld fabrication, and the commercial methods must be closely examined to single out factors that deserve consideration when planning a welding procedure.

GENERAL CATEGORIES OF IRON AND STEEL

Iron is alloyed and processed to make a greater variety of construction materials than has been accomplished with any other metal. To facilitate a review of the basic processes from which iron and steel originate, the many product forms can be sorted into the following broad classes.

(1) Pig Iron
(2) Cast Iron
(3) Wrought Iron
(4) Metal Powder Sintered Iron
(5) Carbon Steel
(6) Alloy Steel
(7) High-Alloy Steel
(8) Tool Steel

Pig iron is the high carbon product of the blast furnace and is not cast directly into useful articles. It is used exclusively as a raw source of iron in subsequent melting and refining operations. *Cast iron* is the metal of the foundry, and it is highly regarded for ease of casting into molds. Although cast iron has a number of desirable mechanical properties — particularly, corrosion resistance — its very high carbon content puts certain limitations on joining by welding. *Wrought iron*, an early form which saw extensive application in pipe and structural bars, is no longer manufactured by its original method; but it continues to be produced in limited tonnage by a more efficient process which will be described later. *Powder metallurgy* is a process for making solid shapes from iron powder by sintering compacts. Fabrication into larger assemblies sometimes involves welding.

Carbon steel is a misnomer that survives through popular usage (since steel is an alloy of iron and carbon, *all* steel contains carbon). *Alloy steel* and *high-alloy steel* are common terms, but they have no standard definition among technical societies. This text will cover, as "alloy steel," a wide variety of low-alloy steels and steels that are alloyed mainly to obtain high strength and hardness. The popular stainless steels and heat-resisting steels are commonly accepted as constituting the "high-alloy steel" category; they are not covered in either Volume I or Volume II of *Welding Metallurgy*; hence the subtitle "Carbon and Alloy Steels."

Tool steel might be defined as any steel used in cutting or forming operations; however, there is a family of steels ranging from plain-carbon to highly alloyed compositions that are produced for use in manufacturing tools.

IRON PRODUCTION BY ORE REDUCTION

The starting point in producing iron and steel is the reduction of *iron ore*. Virtually all iron located in the earth's crust occurs principally in the form of oxides, sulfides, and carbonates of iron. Practically all of the iron and steel manufactured today is secured initially from iron ore of the oxide type. Although iron ore is so abundant that its supply is virtually limitless, the suitability of an ore for iron production depends largely on its iron content and the kind of impurities present. Ores vary greatly in composition depending on the region of the earth where they are found. For commercial production, the ore should have an iron content of at least 50 percent. The impurities in most ores are chiefly silicon oxide, aluminum oxide, and phosphorus pentoxide.

There are two main types of iron ore. *Hematite* (ferric oxide or Fe_2O_3), which is red in color and contains 70 percent iron, is found in northern Michigan and Minnesota. *Magnetite* (magnetic iron oxide or Fe_3O_4), which is black in color and contains 72 percent iron, was formerly mined in New York and Pennsylvania. The depletion of high-grade ore deposits in the United States has forced American steel producers to import substantial amounts of this raw material. One major move about 1950 was to develop the *taconite* types of ore located further north in Minnesota. These are rock-like minerals in which much of the iron is present as magnetite. Unfortunately, taconite requires extensive facilities for crushing, concentrating and pelletizing to produce a *beneficiated* ore pellet, or briquette, suitable for reduction in a furnace.

Metallic iron can be separated from iron ore in a variety of processes. The basic technique that accomplishes the "winning" of iron from ore with greatest economy is to use carbon as a reducing agent. The source of carbon can be coal, coke, oil, natural gas, or any other highly carbonaceous material. At a sufficiently high temperature, carbon combines with the oxygen held by the iron oxide to form carbon monoxide and carbon dioxide gases. In this process of reduction, the molten iron picks up a substantial carbon content, which must be decreased to much lower levels in subsequent iron- and steel-making operations. Hydrogen can also be used as the reducing gas, and this would avoid carbon pickup in the iron; however, the cost of hydrogen reduction is generally higher than operations using a carbonaceous agent.

Blast Furnace

Blast furnaces have been used worldwide for more than 500 years to extract metallic iron from its ores, and they continue to be used today to produce most of the iron tonnage for manufacturing cast iron and steel. Modern blast furnaces are large, complex facilities, but the metallurgical aspects of their operation can be described in simple steps. Even though iron produced by the blast furnace is not directly used as a construction material, there are features of its chemical composition that can transfer to finished iron and steel and cause problems in welding. The purity of ore, flux, and fuel must be carefully controlled as they are put through the blast furnace. Unwanted residual elements from the raw materials may be very difficult, if not impossible, to remove as the blast furnace iron is used in making finished iron and steel. Examples of these unwanted elements might include arsenic, copper, and tin.

A blast furnace consists essentially of a very large steel shell [Figure 5.1 (A)] perhaps 10 meters (30 ft) in diameter and 50 meters (150 ft) in height, lined with refractories. The raw materials are charged through a bell-and-hopper arrangement at the top of the furnace that allows continuous operation. Additions of iron ore, limestone, and coke are made periodically as required to keep the burden of the furnace at proper height. This charge of raw materials descends from the top of the furnace, as *tuyeres* near the bottom admit pressurized air that has been preheated to about 650°C (1200°F). The air burns a portion of the coke in a controlled manner to produce carbon monoxide gas which acts as a reducing agent to remove oxygen from the ore and other iron-bearing oxides present in the charge (e.g., mill scale). Heat generated by combustion of the coke supplies energy for the chemical reactions involved and for melting the liberated iron. A portion of the coke is absorbed in the molten iron and raises its carbon content to almost five percent. The limestone forms a fluid slag with coke ash, ore gangue, and other impurities. At this stage, the liquid iron and slag settle to the bottom of the furnace, where they accumulate on the hearth with the slag floating above the iron. The iron and slag are removed from the furnace through separate notches or tap holes at appropriate levels near the furnace bottom. Sometimes the air blast is enriched with purified oxygen to speed up the burning of the coke and to raise the furnace temperature. The addition of five percent oxygen to the air can increase iron output by approximately 20 percent.

Other operational innovations include the injection of fuel gas (coke-oven gas, or natural gas) into the furnace to substitute for some of the coke, and the injection of high-grade coal in pulverized form to increase iron output and reduce fuel costs. All of the carbon monoxide gas produced during the blast-furnace operation cannot be used in reducing oxides (because of equilibrium limitations). The mixture of carbon monoxide and carbon dioxide gases ascending to the top of the furnace is combustible; it is either burned as a flare or piped away for use as a fuel.

Chemical composition of iron from the blast furnace is typically 4.3% carbon, 1.0% manganese, 0.5% phosphorus, 0.05% sulfur, and 1.0% silicon. When the molten iron from the blast furnace is run into a casting machine, the iron fills a series of round-bottom molds with overlapping edges. The solidified sections when lying face-down look like a litter of pigs, and this accounts for the commonly used term *pig iron*. Because of its very high carbon content, pig iron is very hard and brittle.

To conserve thermal energy, the molten iron is often run into an insulated ladle or transfer car and is carried directly to a steelmaking facility. This form of blast furnace output is called hot metal. This practice allows treatment of the liquid iron to reduce its sulfur content. Ordinarily, sulfur is removed from the molten metal only by careful regulation of oxidizing conditions, the presence of a highly basic slag, and by allowing ample time for reactions to take place and transfer sulfur from the metal to the slag. However, injection of a desulfurizing agent through an insulated lance plunged deep into the molten metal produces vigorous action that can reduce sulfur in the hot metal from the 0.05 percent level to a level below 0.02 percent.

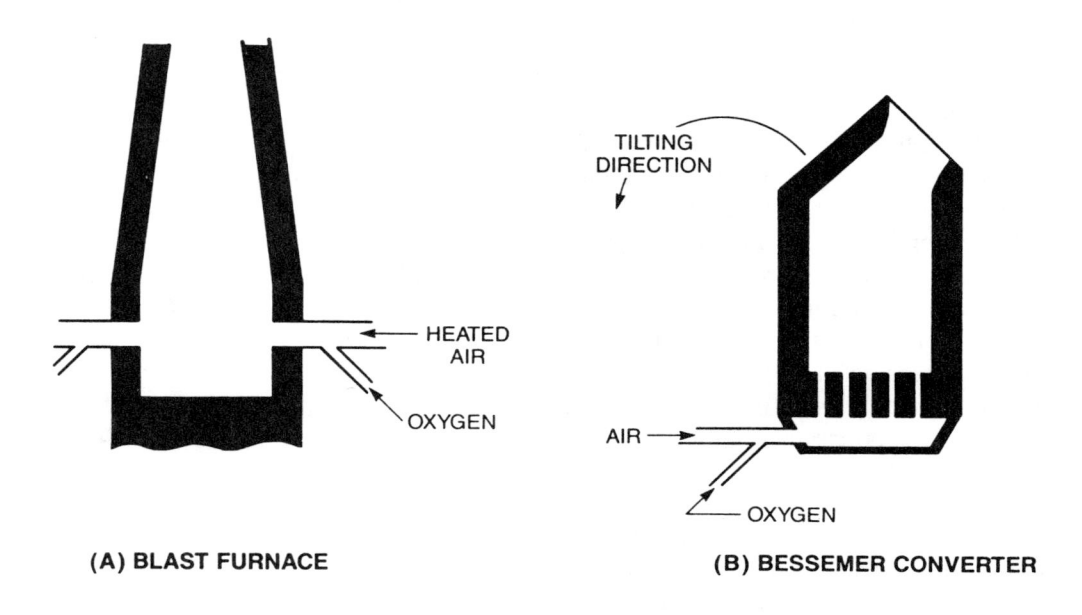

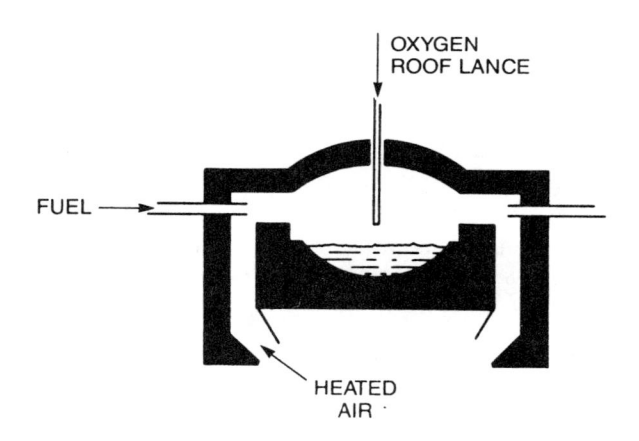

Figure 5.1 — Schematic cross-sections of (A) blast furnace for reduction of iron ore, (B) Bessemer converter for refining of pig iron into steel, and (C) open-hearth furnace for producing carbon and low-alloy steels

Also shown in each illustration is a means of introducing oxygen in each facility to react with carbon and produce heat, and to lower the carbon content of the metal

One of the first agents used in lance-desulfurization of hot metal was soda ash (sodium carbonate), but greatly improved results are now experienced with some form of calcium, or magnesium. Present practice often makes use of calcium carbide, lime, or magnesium-coke. Magnesium has a strong affinity for sulfur and it forms a high-melting-point stable sulfide. Because the boiling point of magnesium is 1107°C (2025°F), the addition of this desulfurizing agent to the molten iron is rather violent. Calcium has a higher boiling point (1440°C, or 2624°F) which tends to alleviate problems in adding the agent to the hot metal, and the sulfide compound formed by calcium is more stable than that formed by magnesium. The residual sulfur in the treated hot metal can then be decreased further in subsequent steelmelting. Practices for desulfurization have received much attention, because there is a clear need to hold the sulfur content of certain steels to very low levels for reasons related to mechanical properties and welding.

Direct Reduction Processes

Because of the great cost of a blast furnace which uses high-grade limestone and coke for its operation, steelmakers have attempted for many years to develop a process for the reduction of ore to iron that would require less extensive facilities and use less expensive fuel. A number of processes have been developed, and although generally referred to as *direct reduction* (DR), they differ considerably in equipment and process flow. They are usually identified by a trade designation.

Initially, a simple vertical shaft furnace was used to reduce pelletized ore to iron. Ore pellets were fed into the furnace at the top and natural gas was injected near the bottom to serve as fuel to heat the pellets to a high temperature, and to serve as the reducing agent. The reduced iron pellets (still solid, but porous) were discharged from the bottom of the furnace; hence the designation *DRI* (*direct reduced iron*). This concept worked, but many practical problems arose. The very hot pellets tended to form a "clinker" as they passed down the shaft, impeding their discharge. If the iron pellets were too hot emerging from the furnace and became exposed to air, they tended to oxidize extensively because of their porous structure.

Although the technical problems were largely overcome by equipment changes and improvements (such as making use of a rotating kiln as the furnace), successful use of direct reduction remains highly dependent on local costs for fuel gas, coal, and electricity. Perhaps the most critical factor in determining whether a DRI project is commercially viable is the form of carbon and heat energy to be used. Natural gas has been a popular source, but because of its rising cost, many developing processes have been turning to coal as the carbon/energy source. If steel scrap is plentiful nearby at a lower cost, this will also affect the DRI. During the early 1990s, circumstances favoring the erection of plants for producing DRI have been better outside the United States.

An early DRI process known as *R-N* reduces iron ore with an excess of carbonaceous material at approximately 1000°C (about 1850°F) in a brick-lined rotary kiln. Ore is crushed into lumps about 35 mm (1½ in.) in diameter, and a small amount of limestone is added to take up sulfur. The kiln is fired with oil or gas burners and the quantity of air entering the kiln is limited to ensure reducing conditions.

The reduced iron is separated magnetically from the discharge, then pressed into briquettes for steelmaking.

The *Strategic-Udy* process makes use of iron ore crushed into pebbles and fed into a rotary kiln, where it is mixed with carbonaceous material and limestone. Gas is used to heat this mixture to about 1100°C (2000°F). After partial reduction in the kiln, the charge is fed directly into an electric-arc furnace where the reduction process is completed.

The *H-Iron* and the *Nu-Iron* processes take finely divided high-grade ore and fluidize this powder in a stream of hot reducing gas in a fluidized-bed reactor. H-Iron employs hydrogen at a temperature of about 500°C (930°F) and at a pressure of 2.8 MPa (400 psi). Nu-Iron is reduced by carbon monoxide gas at a higher temperature, but at a lower pressure. The product of both processes is an iron powder that is compressed into chips or briquettes. All impurities in the ore are carried into the iron product.

Two DRI process that use natural gas are *Midrex* and *HYL*. The *Plasmared* process developed in Sweden feeds ore pellets down a shaft furnace and uses a plasma arc torch for generating and heating the reducing gas in a separate chamber. Either natural gas or coal can be used as the source of carbon.

In some of the newer DRI processes, the end product is molten iron rather than solid metal. The *KR* process has two chambers: an upper chamber for partial reduction of the ore, and a lower chamber where coal is gasified while the iron undergoes final reduction and melting.

A number of direct reduction facilities are operating in different parts of the world to supply iron for steelmaking; and in some cases the operation, although unfavorable from a cost standpoint, is continued because of the control it exercises over residual elements in the iron.

CAST IRON

Cast iron is the form of material favored by the foundry (over cast steel) for casting many kinds of articles. The reasons for this preference by the foundry are (1) the ease of producing cast iron in the *cupola*, (2) the lower melting temperature of cast iron, and (3) the fluidity of cast iron in the molten state. The cupola is constructed somewhat like a miniature blast furnace, but it is much smaller and less costly to operate. Three raw materials are charged into the cupola: coke as a source of heat, pig iron and steel scrap for metal, and limestone as a flux and slag covering. An unheated blast of air is forced through tuyeres in the lower part of the cupola to burn the coke and melt the charge.

The composition of cast iron may vary considerably depending upon the grade of pig iron and scrap used, but cast iron is generally high in carbon content (about 3.0 percent). By adjusting the remainder of the composition — that is, the Mn, P, S and Si — with additions of ferroalloys, etc., the mode and distribution of the carbon in the cast iron can be controlled to produce either *white* cast iron, *gray* cast iron, or *malleable* cast iron. These forms of cast iron differ greatly in mechanical properties and in welding behavior, and they will be discussed further in Chapter 14 of Volume II.

WROUGHT IRON

Although wrought iron may be considered as a material of bygone days, it is still being produced; and it can be given preference over more modern steels in certain applications, such as pipe, which may involve welding. Originally, wrought iron was produced in reverberatory hearth furnaces by melting pig iron and adding slag. The metal bath was decarburized with *roll scale* (iron oxide) and the slag and metal were puddled together into pasty balls. Next, the balls were compressed in a squeezing machine to remove excess slag, and to form a "bloom" suitable for rolling into bars or shapes. More modern methods of producing wrought iron involve mixing low-carbon steel with selected open-hearth-furnace slag.

Wrought iron consists of approximately one to four percent by weight of a slag component dispersed as elongated stringers in a matrix of low-carbon steel or iron. The slag consists mostly or iron oxide and some silica. The analysis of the iron matrix is about 0.03% carbon, 0.05% manganese, 0.10% phosphorus, 0.020% sulfur, and 0.01% silicon; the remainder is iron. The phosphorus content will be high if the steel comes from a Bessemer converter, and this high phosphorus level is a factor to be considered in welding. If the base steel is produced in almost any other steelmaking process, the phosphorus content is likely to be reasonably low. Wrought iron can be identified by macro-etch examination, which reveals numerous small stringers of non-metallic slag throughout the metal section.

POWDER METALLURGY

Whatever the source, powdered iron can be converted into useful articles or shapes by the technique known as *powder metallurgy*. This operation produces solid forms of iron and steel from metal powders without the need for melting furnaces, rolling mills, and other conventional steelmaking equipment. While powder metallurgy is a very promising way of producing metal articles, cost factors hamper its widespread use. First, premium powders must be used to secure the composition, particle size, shape and other properties needed to make satisfactory parts. Second, presses or rolling equipment are needed to compact the mixture of powders into a "green compact." At this stage, the compact is about 60 to 80 percent as dense as a solid mass of the same composition. If a press and dies are used for producing a particular shape, the die cost must be amortized over the number of parts produced.

The products of powder metallurgy are usually classified into three groups: ductile metal articles; brittle, refractory metal articles; and hard, cemented carbide articles for cutting tools and dies. Each product is made from specific mixtures of powders by compacting and *sintering*; the latter operation is executed in a controlled atmosphere or a vacuum furnace. If sintering is carried out with no liquid metal present in the compact, recrystallization of the metal or alloy powder is expected to "weld" the particles together to form a solid mass. A liquid metal is produced by the addition of a small proportion of lower-melting metal to the powders prior to forming the compact. The liquid metal "brazes" the solid metal particles together at the sintering temperature. An important aspect of the compact is *contiguity*, which is a quantitative measure of interphase contact. Contiguity can be defined as the fraction of internal surface area of a solid phase shared with particles of the same phase in a two-phase compact. Even sintering that includes a liquid phase depends on contiguity to provide rigidity to the compact and to minimize distortion.

Ordinarily, in the manufacture of iron and steel articles by powder metallurgy, no liquid metal is used in the compact. A sintering temperature well above 450°C (850°F) is used, because this is the approximate recrystallization temperature for iron. The efficiency of the sintering operation is aided by the use of finer powder, denser compacts, and higher sintering temperatures. The furnace atmosphere (or vacuum) must not permit any oxidation or allow any films of compound to form on the surfaces of the powder particles that would prevent bonding. It is difficult to obtain a sintered ductile metal article (made without liquid metal during sintering) which does not contain pores or microscopic internal voids. If the article produced by powder metallurgy is subsequently subjected to hot forging or rolling, this mechanical treatment should act to increase the density of the powder metallurgy product toward that of a solid metal. Density within 95 percent of solid iron can be secured by double-pressing and double-sintering. High velocity compaction (performed with explosives) can produce a density well above the 95 percent level.

STEELMAKING PROCESSES

Many processes are used for making weldable steel. Several of the processes described here, admittedly, are not in use today — certainly not in the United States. However, steel produced by these outdated processes may arrive at a fabrication site to be welded. This old, but still useful kind of material may be either (1) steel exported from a country with less-than-modern steelmaking facilities, (2) steel from an old and dismantled large structure from which sections have been salvaged, or (3) a steel structure of some vintage that is being rebuilt or repaired for further service. Many reports appear in the welding literature of unexpected fabrication behavior or service experience because the kind of steel involved was not fully recognized.

Significance of Acid and Basic Steelmaking

In specific steelmaking processes, the matter of acid versus basic steelmaking should be examined, because an understanding of these two terms is metallurgically important to both steelmaking and welding. The terms *acid* and *basic* are derived from the kind of refractory lining and the slag used in a process. Most of the nonmetallic compounds that are used in making refractory furnace linings or employed as a flux or slag can be classified as having either acid or basic (alkaline) characteristics when heated to the temperatures encountered in steelmaking. A material is classified by noting any tendency on its part to react with a strongly basic material like lime (CaO) or a decidedly acid material like silica (SiO_2). Unlike materials will react or attack each other, while like materials will not. A furnace operating with a basic-type slag will have a refractory lining made of basic materials, whereas a furnace using an acid-type slag will have a lining of acid materials. If an acid slag is used in a basic-lined furnace, the slag would quickly attack and damage the furnace lining. The common acid materials involved in steel-melting are silica (SiO_2) and phosphorus pentoxide (P_2O_5), while the basic materials are lime (CaO), burnt dolomite (MgO, CaO), iron oxide (FeO), and manganese oxide (MnO).

The important difference between acid and basic steelmaking processes is in their respective ability to rid the molten metal bath of residual phosphorus and sulfur. In the acid steelmaking furnace, there is no significant removal of phosphorus and sulfur because the acid slag cannot react chemically with these two elements. The charge of raw materials as a whole must meet the same maximum requirements specified for these two elements in the finished steel. This means high-grade ore and steel scrap must be used. For the most part, an acid-lined furnace functions mainly as a furnace to melt a charge, remove carbon, and hold the molten bath while nonmetallics rise from it and become part of the slag.

On the other hand, a basic steelmaking furnace can remove almost any reasonable amount of phosphorus and sulfur from the molten bath by means of its basic slag. The chemistry of this phase of steelmaking is quite complicated, but the fundamentals can be simply stated. Lime (CaO) is usually employed as the primary basic constituent in the slag. The removal of phosphorus is accomplished by oxidation. The phosphorus pentoxide that forms rises into the slag, where it combines with calcium in the lime to form calcium phosphate. Sulfur is removed in two ways: through the formation of manganese sulfide (MnS) that rises into the slag, and by direct absorption into the slag. In either case, the sulfur is held in the slag as calcium sulfide (CaS), with the calcium again being secured from the lime.

The formulation of slag and techniques for its control during steelmaking are as important as the handling of the molten metal. In steelmaking processes, close attention is given to the use of slag to protect the molten metal, and to control the metal composition. In some cases, slag must be efficiently skimmed from the molten bath to avoid possible *reversion* or transfer of phosphorus and/or sulfur back into the metal.

Bessemer Converter

The *converter process*, sometimes referred to as the pneumatic process for making steel, was invented by Sir Henry Bessemer in 1860. This process rapidly grew into worldwide use until it was eventually supplanted by the open hearth furnace in the early 1900s. The term Bessemer *converter* is used instead of furnace because it is not equipped with a source of heat to melt down a cold, solid charge. The converter must receive a charge of hot metal directly from the blast furnace. This was the main purpose of Bessemer's invention — to convert pig iron into steel and avoid the slow, laborious processes employing a crucible or a puddling furnace.

The converter can be lined with an acid or a basic type of refractory brick, and the type of lining is an important factor in determining how the converter operates to refine the pig iron in chemical composition. The acid-lined converter was more widely used in the United States, and this implies some guidelines for the composition of Bessemer steel, should any be encountered today in the course of welding fabrication. A small number of Bessemer converters are still in operation in other countries.

A typical Bessemer converter consists of a tilting vessel about three meters (9 ft) in diameter and four meters (12 ft) high that has a bottom perforated with many holes, each about 15 mm (.75 in.) in diameter, for passage of air into the vessel. (See Figure

5.1). While the converter is tilted on its side [i.e., Figure 5.1 (B) tilted leftward], molten pig iron is poured in. As the converter is righted, a blast of air (not heated) at 0.138 MPa (20 psi) is blown through the holes and into the molten metal. The oxygen in this air reacts first with the silicon and the manganese, removing them to the slag (iron-manganese silicate). The carbon is then oxidized to carbon monoxide, which in turn burns to carbon dioxide in a long blue flame from the mouth of the converter. The combustion of the silicon, manganese and carbon raises the temperature of the molten metal in the converter. When the carbon flame dies out — a point in the process known as the *carbon drop*, which occurs between 12 and 18 minutes after the start of the converting process — the vessel is ready to be emptied. Each batch of converted metal is called a blow, and is numbered consecutively for identification.

In an acid-lined converter, the silicon, manganese, and carbon contents can be reduced to extremely low percentages by the oxidation reaction — for example, 0.03 percent carbon, 0.05% manganese, and 0.01% silicon. The phosphorus and sulfur contained in the pig iron charge ordinarily would not be removed; this is because oxidation (the air blast) cannot be continued after the carbon drop, otherwise the temperature of the metal would fall too rapidly. The highly basic slag needed to hold the phosphorus and sulfur as calcium compounds is also absent. *Acid* Bessemer steel is characterized by high phosphorus content, up to 0.13 percent, and often by high sulfur content. Because of the tremendous quantity of air blown through the molten metal, the nitrogen content is much higher than in other kinds of steel.

The use of oxygen and steam as additions to the air blast, or oxygen-steam mixtures as an outright substitute for air, has brought about changes in the composition of Bessemer steel and has improved the rate of production of a blow. The most significant change is the lower nitrogen content of the oxygen-steam blown metal. Practically all Bessemer steel is plain carbon steel. The major tonnage is used as pipe, skelp, tin plate steel, steel wool wire, and free-machining types in bar form. In the latter material, sulfur in quantities ranging from 0.10 to 0.35 percent is added as the free-machining agent.

Although the basic-lined Bessemer converter is used very little in the United States, this type of converter performs in exactly the same way as the acid-lined converter, with one exception. The basic process uses a highly basic slag of lime added immediately after the silicon, manganese and carbon have been removed to aid in dephosphorizing the metal. The operational practice with the basic converter can have a strong influence on the chemical composition of the finished steel. Therefore, when steel to be welded is known to have been produced by the Bessemer process, it is imperative to know its full chemical composition, including residual elements like phosphorus and sulfur.

Open Hearth Furnace

From its status as the primary method of steel production in 1965, the open hearth furnace accounted for only about ten percent of the steel made in the United States as of 1993. This decrease was brought on mainly by the rapid introduction of *oxygen steelmaking*, but lessons learned with the open hearth furnace are an important prologue to methods being used today.

An open hearth furnace consists essentially of a large shallow pan or *hearth* about five meters (15 ft) wide, 15 meters (45 ft) long, and one meter (3 ft) deep as shown in Figure 5.1 (C). This hearth is usually lined with a basic refractory, such as dead-burned magnesite (MgO), to permit operation with a basic-type slag. The hearth is enclosed in a refractory-lined furnace with an arrangement of fuel burners and large internal ducts for the delivery of heated air and burning fuel over the bath of metal in the hearth. Figure 5.1 includes a schematic illustration of the open hearth furnace. An acid-type furnace lining and slag can also be used in an open hearth furnace, but the acid process has been employed only to a limited extent. This is because of the necessity for securing ore and other raw materials that are low in phosphorus and sulfur, since these residual elements cannot be eliminated during the acid steelmaking operation.

The open hearth is a batch furnace, and each batch or melt of steel tapped from the furnace is called a *heat*. The total time required to produce a heat of steel (weighing 100 to 300 tons) ranges from four to six hours. Heats are numbered consecutively, and their identification number is keyed with mill records containing all chemical analyses made on any material, as well as a complete processing history. By retaining the heat number of steel sections — whether in the form of plate, beams, bars, sheet, etc. — considerable information about the particular steel can be secured from the producing mill.

The fuel used in an open hearth furnace for melting a charge of raw materials can be powdered coal, gas, or petroleum oil. The air is preheated in dual *checker chambers* located below the hearth. These chambers operate on a regenerative cycle utilizing hot exhaust gases from the furnace. The making of a heat of open hearth steel can be broken down into three operational steps: charging and melting, refining, and finishing and deoxidation.

The raw materials charged onto the hearth for melting may consist of (1) cold materials entirely, such as pig iron, or iron and steel scrap; (2) hot molten metal, as supplied directly from a blast furnace or from a Bessemer converter; or (3) combinations of cold and molten metal. In addition to the metal charge, limestone is added as a flux and slag former. A suitable charge is melted down to form a bath about 75 cm (30 in.) deep, upon which floats a basic (lime) slag about 15 cm (6 in.) deep. The bath of molten metal and slag is open to the hot gases of combustion passing over it — hence the name *open hearth*.

In refining the molten bath of metal, liberal amounts of oxygen are made available by adding iron ore, mill scale, and limestone. Under these highly oxidizing conditions, the silicon, manganese, carbon and phosphorus in the bath are oxidized and their reaction products either rise into the slag or, in the case of carbon, escape as a gas. Sulfur is transferred to the slag by means previously described. The control of reactions in the metal bath and slag is a very complex operation and depends on the composition and temperature of the bath, as well as the condition of the slag. Lime is frequently added to counteract the acidic silica and phosphorus pentoxide entering the slag, thus maintaining sufficiently strong basicity to avoid reversion and other problems. Slag composition control is an important part of the refining period, just as it is in the shielded metal-arc welding process.

Finishing and deoxidation follow the refining period. The silicon, manganese, carbon, phosphorus and sulfur contents now have been reduced to the desired levels, but the molten metal bath is saturated with oxygen, probably as FeO. The amount of oxygen present depends to a large extent on the carbon content of the metal bath. The greatest amount of oxygen can be held when the carbon content is the lowest. At any level of carbon, however, the amount of oxygen present is far in excess of that which could be accommodated in the metal after solidification. Therefore, to avoid the formation of large gas pockets in the cast metal, a substantial portion of the oxygen must be removed. Five general practices have been used by steelmakers in handling a heat of molten steel to remove oxygen, and each practice produces a characteristically different kind of steel. The welding engineer must have an understanding of this phase of steelmaking in order to make adjustments in welding procedure to accommodate the particular kind of steel to be welded, or added as filler metal.

Rimmed Steel

Rimmed steel is the oldest method of dealing with oxygen in the molten steel, and it is still used in production today. To make rimmed steel, the oxygen-containing metal is poured into cast iron molds to produce ingots. The ingot commences to solidify by depositing a crust of almost pure iron on the walls and bottom of the mold. The remaining liquid steel is enriched slightly with carbon, which in turn reacts with some of the oxygen to form carbon monoxide gas. This gas is virtually insoluble in the molten metal and forms gas bubbles which attempt to rise and escape from the top of the ingot.

Chemical analysis of the slag at the end of the refining period lets the steelmaker know approximately how much oxygen is in the molten metal based on the balance that normally exists between the slag-and-metal oxygen contents. Much of the carbon monoxide gas does escape, and the effervescence produces a shower of sparks as the ingot cools. The remainder of the CO gas is trapped in the solidifying metal as deep-seated blowholes which will weld up and disappear during subsequent hot rolling. This cycle of (1) iron freezing on the mold walls, and (2) the segregated carbon reacting with the oxygen, continues until an outer *rim* of almost pure iron is formed, which on the ingot is about 75 mm (3 in.) thick. Finally, the *rimming* action stops with the loss of temperature and completion of solidification. Rimmed steel is characterized by a skin of nearly pure iron around the outer periphery of the cross section, and by a central core area into which is segregated practically all the carbon, phosphorus and sulfur. This pattern of segregation in the same ratio of rim-and-core dimensions persists through forging, rolling and forming operations of all kinds, hot or cold.

The chief advantage of rimmed steel is the excellent defect-free surface which can be produced with the aid of the pure-iron skin. This kind of steel is used mostly for making sheet and wire. Much of the wire is used in gas welding rods and as the core wire for covered arc-welding electrodes. Steel containing more than approximately 0.30 percent carbon cannot be furnished as rimmed steel, because the amount of oxygen held in the molten condition is not sufficient to support the rimming action during ingot solidification. Most rimmed steels contain less than 0.10 percent carbon, and they are characterized by a very low silicon content (less than 0.005 percent) and an absence of other deoxidizing elements, like aluminum. No attempts are made to add alloying elements which are oxidizable, since they would be lost to the oxygen in the metal. Low levels of manganese, perhaps up to 0.25 percent, are employed. The oxygen content of rimmed steel ranges from about 0.02 to 0.04 percent, while the nitrogen content is in the range of 0.002 to 0.005 percent.

Capped Steel

In making *capped steel*, the amount of oxygen is regulated in the molten metal so that the rimming action can be stopped at will. When poured into the ingot mold, the metal is allowed to rim weakly for a few minutes to build up a thick skin of sound metal. Then a heavy cap or chill plate (*mechanical capping*) can be locked atop the mold. The molten metal in the mold rises because of the formation of internal gas bubbles; and, when reaching the cap, the metal is chilled to form a solid crust which prevents further gas evolution. An addition of a deoxidizing element to the melt at the ingot top after it has been completely poured results in *chemical capping*. The composition and temperature of the molten steel must be closely controlled to avoid a high dissolved-oxygen content. The carbon content can be regulated within a range of 0.07 to 0.12 percent, and a small dose of deoxidizer may be added to the metal in the furnace or the ladle to adjust the quantity of oxygen held. Capped steels do not display the pronounced thick iron rim and segregate core found in rimmed steels. They are used in applications which require excellent surface, and where the heterogeneous composition and mechanical properties of rimmed steel should be reduced. Also, capped steels do not require the close control of deoxidation, as do the semikilled steels described below.

Killed Steel

As the term suggests, *killed steel* is made by removing or tying up the oxygen that saturates the molten metal prior to its solidification to prevent effervescence or rimming action during cooling. Removal of oxygen is accomplished in the furnace, in a ladle, or during the teeming of an ingot. After refining, the molten bath can be deoxidized by adding elements whose affinity for oxygen is greater than that of iron, and whose oxides can be eliminated from the bath. Some of the elements used for this purpose in carbon and low-alloy steels are silicon, aluminum, titanium, and zirconium. Silicon is the deoxidizer most often used, and it is usually handled as a ferrosilicon alloy. When added to the molten bath in the furnace, silicon reacts with the iron oxide (FeO) to form silica (SiO_2) which rises into the slag. Silicon deoxidation will remove the greater proportion of oxygen present in the steel.

More thorough deoxidation can be accomplished by adding stronger agents, such as aluminum, titanium, or calcium-silicon alloy. Many features of the steel are influenced by the particular deoxidation agent and practice followed, especially the quantity of the agent and the conditions of time, temperature, etc. which prevailed as the operation was performed. These features include: (1) the nature, number and size of nonmetallic inclusions retained in the solid steel (i.e., micro-cleanliness); (2) the propensity for grain-coarsening during heating to a high temperature (i.e., inherent austenitic grain size); and (3) the mechanical properties, especially toughness.

When killed steel is to be solidified as ingots, the deoxidized metal is poured into molds fitted with a refractory hot top. The deoxidized steel freezes to produce a sound ingot with a pronounced shrinkage cavity or pipe in the upper end of the ingot that is girdled by the hot top. This piped upper portion is sheared off during hot working operations, and is retained as scrap for remelting in a subsequent heat of steel.

The amount of oxygen dissolved in a molten bath of steel that has been exposed to air during steel melting depends on the level of carbon present. With 0.20 percent carbon or more in the melt, the oxygen content would be limited to about 0.02 percent, but the low-carbon melts (less than 0.2 percent carbon) generally absorb oxygen ranging from about 0.05 to 0.10 percent. When oxygen at levels approximately 0.02 percent or more is allowed to remain in the metal during solidification, gas pockets will usually form; this can adversely affect the mechanical properties of the steel, especially ductility and toughness. Large amounts of oxygen form copious amounts of oxide with iron, manganese, and other oxidizable elements in the steel, and these compounds will disperse throughout the solid steel as numerous nonmetallic inclusions.

Silicon as a Killing Agent

Silicon is often used as the deoxidizer in steelmaking. This agent is commonly handled as a ferroalloy containing 50 percent silicon. The amount of silicon to be added is calculated so that it will be slightly in excess of the quantity needed to combine with the oxygen that has been determined to be present in the molten metal. Too large an addition of silicon will elevate the final silicon content of the steel to an objectionable level. Silicon and oxygen react vigorously to liberate a large amount of heat, and the silicon dioxide (SiO_2) that first forms is virtually insoluble in the molten metal. Therefore, this compound must rise to escape in the slag atop the metal bath, or it will remain and become entrapped as inclusions in the solid steel. Because manganese is often present in the composition of the molten steel, the silicon dioxide inclusions will often occur as complex iron-manganese silicate compounds; but, if deoxidation is carried out with silicon alone, the final oxygen content of the steel will probably be in the vicinity of 0.01 percent. Steels that are deoxidized in this manner usually have a silicon content ranging from about 0.10 to 0.30 percent. If a lower oxygen content is desired, a larger amount of silicon must be added. A more effective deoxidation practice employs a stronger deoxidizing agent to supplement the silicon.

The choice of deoxidizing agents and their sequence of addition to the molten metal require careful planning in consideration of the possible metallurgical effects listed below:

(1) reversion of the oxides of certain elements back into the molten metal (e.g., phosphorus),

(2) the nature of nonmetallic inclusions that may be retained in the solid steel, and

(3) control of inherent grain size.

Deoxidation has broad ramifications, and much of the information in the literature on its many facets is related to steelmaking rather than fusion welding. The information that follows can also serve as a guide to the task of deoxidizing steel during fusion welding.

Manganese as a Killing Agent

Manganese is a modest deoxidizer, and a substantial addition is capable of reducing the 0.10 percent level of oxygen in low-carbon steel to about 0.05 percent by forming manganese oxide (MnO). Manganese and silicon added together act synergistically and can achieve an oxygen level of about 0.008 percent. The proportions of manganese and silicon in the silicate compound that forms can be controlled to achieve a nonmetallic composition that will remain liquid at the freezing temperature of the steel. This is advantageous in minimizing the nonmetallic inclusion content of the solid metal inasmuch as liquid nonmetallics coagulate and separate from the molten metal into a slag on the surface much better than solid nonmetallics.

Aluminum as a Killing Agent

Aluminum was used in steel as a deoxidizer before the year 1900. The strong affinity of aluminum for oxygen was well known at that time; but it was the Hall process, invented about 1890, that made aluminum plentiful and low enough in cost to be used as a deoxidizing agent in the high oxygen content steels being produced (e.g., Bessemer converter steel). As experience was gained using aluminum as a deoxidizer in steel in the early years, it was found that aluminum offered two additional advantages. First, aluminum has some capability to fix the nitrogen in the steel as aluminum nitride particles. Second and more importantly, when added in certain quantities, aluminum acted to restrict the growth (coarsening) of austenite grains when the steel was heated above its critical range (Ac_1 to Ac_3). Although this restriction of growth by austenite grains may seem to be a continuing advantage, there are instances where coarse-grain steel performs better than fine-grain steel (e.g., higher strength at elevated temperatures and increased hardenability). Therefore, steelmakers have developed practices for controlling both the size of austenite grains and the grain size of ferrite in the final product. Aluminum continues to serve as a highly effective agent in dealing with oxygen, nitrogen, and grain size, and its function in each is summarized below.

As a deoxidizer, aluminum is much stronger than silicon, and additions of aluminum to the molten steel can bring the oxygen level down to about 0.006 percent by forming aluminum oxide (Al_2O_3). Pure aluminum oxide has a melting point of 2072°C (3760°F), and when aluminum is used as the sole deoxidizer, it is likely to produce many solid, hard particles of corundum-like compound that have more difficulty escaping from the molten metal. Steelmakers often perform a preliminary deoxidation with silicon and some manganese, and this pre-treatment is followed by final deoxidation with aluminum. The resulting nonmetallic inclusions are complex aluminum-manganese silicate compounds that are fewer in number and less abrasive.

The choice of deoxidizing agents and the sequence of addition depends on many circumstances, starting with the product form, whether it is a casting, or a hot rolled or forged section, or weld metal. Consideration must be given to the required chemical composition, the inherent grain size, and mechanical properties. While aluminum offers many advantages as a deoxidizer, there are occasionally circumstances where its use would be objectionable. In such cases, the reduction in oxygen content must be accomplished with other agents (e.g., manganese, silicon, titanium, zirconium, calcium-silicon alloy).

Aluminum is not quite as strong as titanium or zirconium as a nitrogen fixation agent, but in the course of deoxidizing molten steel with aluminum, some of the aluminum will combine with the nitrogen present to form aluminum nitride. Ordinarily, molten steel can dissolve more nitrogen than can be accommodated in the solid steel without adversely affect soundness or mechanical properties. The presence of nitrogen in interstitial solution in the bcc lattice of ferrite will detract from toughness, and removal from the lattice by formation of aluminum nitride particles will relieve this effect. The aluminum nitride particles are usually submicroscopic, and they probably assist the Al_2O_3 particles in forming a dispersoid that acts to restrain austenite grain growth when heated above the Ac_1-Ac_3 critical range. Three distinct temperature ranges of aluminum nitride precipitation have been reported:

(1) precipitation of aluminum nitride starting at a temperature of about 900°C (1650°F) which produces plate-shaped particles, and which tend to coarsen because of the high temperature to reach sizes approaching 1000°C,
(2) intensive precipitation of smaller platelets or rod-like particles in a more dispersed distribution starting at about 750°C (1382°F), and
(3) very fine highly-dispersed AlN particles occurring over the temperature range of 650 down to 600°C (1202-1112°F).

These aluminum nitride particles can be dissolved by reheating to a temperature above approximately 1000°C (1832°F) providing their composition has not been altered by the presence of other strong nitride-forming elements, such as titanium, or vanadium.

Aluminum is the element most often used for making steels that have a fine *austenitic grain size* (sometimes referred to as "inherent grain size," but this terminology is not always applicable, as will be explained). Austenite grain size is not observed directly in carbon and low-alloy steels because the austenite phase transforms on cooling to ferrite and some form of carbide distribution (e.g., pearlite). There are several procedures for measuring austenite grain size, and these are described in Annex A3 of ASTM E 112. The *McQuaid-Ehn* test involves carburizing a specimen of the steel at a prescribed temperature and slow cooling to delineate the austenite grain boundaries in a hypereutectoid microstructure. This test is usually conducted using a carburizing temperature of 927°C (1700°F). Heat treatments applied to steel that involve solution treating (austenitizing) above the critical range seldom exceed this temperature. If the austenite grain size is fine (ASTM Nos. 5 to 8) at 927°C (1700°F), there is no question about having a fine-grain austenite if a lower temperature is employed. As explained in Chapter 2, the size of the austenite grains closely determines the size of the ferrite grains that form on transformation. Because ferrite grains are nucleated in the austenite, a fine-grained austenite will produce a finer grained, transformed structure than a coarse-grained austenite.

Many specifications for steel include specific requirements for fine austenite grain size, or coarse austenite grain size, and the responsibility for satisfying this requirement is put in the hands of the steelmaker. The terms for achieving a fine-grained steel cannot be included in the composition specification for the steel because a number of variables that control grain size are handled in accord with the plant's particular steelmelting process and procedure. The process for manufacturing fine-grained steel using an aluminum additive is described in Technical Brief 28.

Despite intensive research on austenite grain-size control, the mechanism involved is not yet completely understood. There is agreement that adding aluminum (or titanium, zirconium, etc.) forms a dispersoid of oxide and nitride particles that produces a fine austenite grain size on initial formation, and that this dispersoid continues to restrain grain growth while the austenite grains are heated to some particular high temperature. It is believed that the restraining particles in the dispersoid are oxides and nitrides that are submicroscopic in size, and are not those particles which are formed shortly after the aluminum (or other deoxidizer) is added to the steel and which later can be seen under the microscope. The effective submicroscopic particles are believed to form either just before or just after solidification, and they are distributed in the iron lattice virtually like a colloid. When an excess of deoxidizer is added, it is suspected that too little dissolved oxygen is being held in the steel at the proper time for the colloid-like dispersion of particles to be formed. Therefore, the restraining dispersoid of effective-size particles is absent, and the austenite grains of this steel

---

TECHNICAL BRIEF 28:
Manufacture of Fine-Grain Steel Using an Aluminum Additive

When a fine-grained steel is manufactured, the amount of oxygen and nitrogen remaining in the molten steel just before pouring is established, and a judicious amount of aluminum is added to tie up the oxygen as oxide particles, to fix much of the nitrogen as aluminum nitride, and to have a small excess of aluminum. In the solidified steel, the dispersion of submicroscopic oxides and nitrides acts as a permanent inhibitor of growth by the austenite grains and thus ensures a fine-grained steel to a temperature of 927°C, or 1700°F. Similar effect can be secured with specific additions of other elements, i.e., vanadium, titanium and zirconium, but aluminum remains the most popular additive for this task. However, when a fine-grained steel is heated to progressively higher temperatures above the critical range, as would be done for a hot-working operation, the inhibiting effect of the aluminum treatment suddenly diminishes at a temperature in the vicinity of 1075°C (1975°F) and rapid coarsening of austenite grains takes place as the temperature is increased, or as soaking time at temperature is extended. In fact, as illustrated in Figure 5.2, at temperatures exceeding approximately 1150°C (2100°F) the fine-grained steel will have produced larger austenite grains than a steel made with little or no aluminum (i.e., coarse-grained steelmaking practice). Austenite grains cannot be considered inherently fine, but they can be restrained in their growth to a given high temperature.

A substantial surplus of aluminum beyond the amount required to deoxidize and to fix the nitrogen present causes the steel to assume a coarse-grained character. To arrive at an approximation of the "judicious" amount of aluminum needed to produce a fine-grained steel, first consider that the percentage of aluminum (by weight) required to form Al_2O_3 is roughly equal to the percentage of oxygen present. Next, experience shows that approximately half the aluminum added to a melt of steel is dissipated by extraneous oxidation and other avenues of loss. Only 50 percent of the aluminum will reach the dissolved oxygen and nitrogen that is to be tied up. Consequently, a melt of low-carbon steel containing about 0.10 percent oxygen would be given an aluminum addition of four pounds (1.8 Kg) per ton (0.2 percent) to achieve a fine-grained austenite. If for some reason, twice this amount of aluminum were added (0.4 percent), the steel would probably produce a coarse-grained austenite.

can coarsen in normal fashion on heating. To make a fine-grained steel, the amount of aluminum (or Ti, V, Zr, etc.) added for a final deoxidation, and the amount of oxygen present must be within minimum and maximum limits that are determined mainly through experience with a particular melting process.

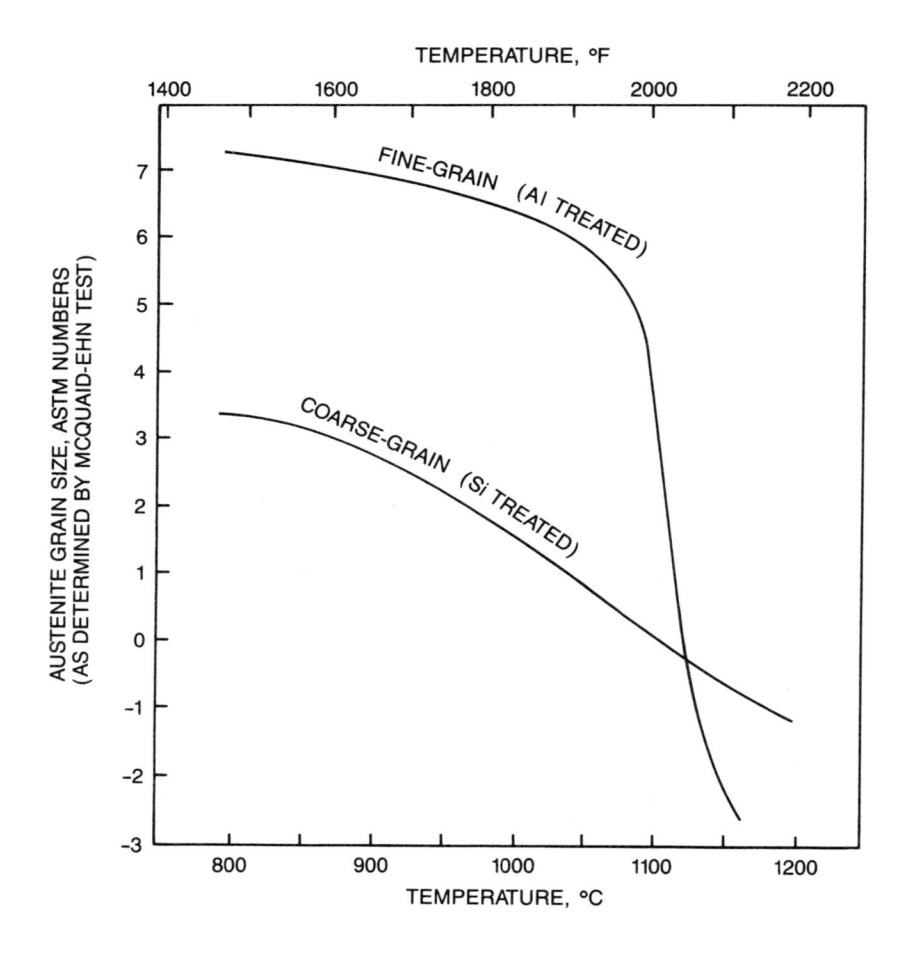

Figure 5.2 — Comparison of austenite grain coarsening tendencies of inherently fine-grained, and inherently coarse-grained carbon steels

When heated to progressively higher temperatures above the Ac_1-Ac_3 critical range, a fine-grained (aluminum treated) steel is shown to eventually produce larger austenite grains than a coarse-grained (silicon deoxidized) steel.

Semikilled Steel

As the name *semikilled* steel indicates, this method of dealing with dissolved oxygen represents a compromise between rimmed and killed steel processes. A small amount of a deoxidizing agent, generally aluminum or ferrosilicon, is added to the molten metal in the furnace, the ladle, or the ingot mold. The amount is sufficient to kill any rimming action, and leave some dissolved oxygen in the metal. The evolution of this oxygen, on solidification of the metal, tends to offset shrinkage and minimize the shrinkage cavity or pipe in the top of the ingot. Sometimes a heavy steel plate or cap is placed on the top of the ingot just after pouring to freeze a crust of solid metal over the top as is done in making capped steel. Any pipe that forms below this skin is shielded from the air and is kept free of oxide or slag. This cavity is called *secondary pipe*, and this internal cavity is welded up by subsequent hot working operations like the internal blowholes in capped steel. Semikilled steel has a number of desirable properties, including a high yield of usable material from the ingot. A comparison of the principal characteristics of rimmed, capped, semikilled and killed steels is shown in Table 5.1.

Vacuum Deoxidized Steel

Vacuum deoxidized steel requires special equipment which will be described in more detail under the heading *Ladle Refining*. The object of vacuum deoxidation is to extract dissolved oxygen from the molten steel without adding an element that forms nonmetallic inclusions. This can be done by adjusting the carbon content of the steel about 0.05 percent higher than is desired in the finished steel, and then subjecting the molten metal to vacuum pouring, or stream degassing conditions. Under these special conditions, the carbon reacts with the oxygen to form carbon monoxide. As the stream of molten metal enters the vacuum chamber, the CO gas quickly escapes from the steel, thus avoiding the formation of inclusions which would occur with the use of solid deoxidizers. As a result of the carbon-oxygen reaction, the carbon content of the metal drops approximately 0.05 percent to fall within the specified limits for this element. The vacuum environment has a pressure of about 200 microns at the start of the deoxidizing operation. Considerable pumping capacity is needed to draw off the outgassing of CO from the steel. Because deoxidizing elements or agents which form solid oxides are not used, steel finished by this technique is quite clean. As in vacuum degassing, the gas content (levels of hydrogen and nitrogen) is also lowered (See Table

Table 5.1 - Comparison of Rimmed, Capped, Semikilled and Killed Steels

	Rimmed	Capped	Semikilled	Killed
Carbon Content	Up to 0.3%	Up to 0.2%	Up to 1.0%	Up to 1.5%
Silicon Content	Less than 0.005%	Less than 0.01%	0.01/0.10%	Over 0.10%*
Primary Piping (shrinkage)	None	None	None	Severe
Porosity in Ingot	Considerable	Considerable	Some	None
Segration	Pronounced	Little	Little	Very Little

*Killed steel may have a very low silicon content if another deoxidizer, like aluminum, is employed.

5.2). After the vacuum-carbon deoxidation step, silicon and aluminum can be added to the molten steel to satisfy chemical specifications, and to control grain size and other properties.

Oxygen Steelmaking

In the earlier description of blast furnace and Bessemer converter operations, it was noted that purified oxygen along with the regular air supply is used to speed up those reactions which depend on oxidation. Because open hearth steelmaking was a rather time-consuming process, particularly when striving for low levels of silicon, manganese, and carbon in the bath, it was only natural for steelmakers to explore the oxygen injection technique. Figure 5.1 illustrates schematically the methods by which oxygen was introduced into the blast furnace, Bessemer converter, and open hearth furnace.

In work with the open hearth furnace a lance was inserted through the roof of the furnace and a jet of oxygen was injected directly into the molten metal. The roof-lance arrangement produced a vigorous reaction with oxidizable elements in the molten metal bath. The bath temperature was raised quickly by the heat of the reactions, and the products of reaction either passed off as gas (e.g., CO and CO_2), or were absorbed by the basic slag (e.g., silica, phosphorus pentoxide, etc.). Successful oxygen roof-lance practice required some modification of the furnace roof and better refractories to withstand the radiant heat generated by the intense reactions taking place in the bath. Oxygen-lancing could increase production from an open hearth furnace by as much as 40 percent. In addition to speeding up carbon removal and the melting of any added carbon steel scrap, oxygen-lancing could reduce the carbon content of the bath to unusually low levels, perhaps as low as 0.03 percent, when producing steels that required very low carbon content. This was less expensive than using the low-carbon ferroalloys for making additions to the steel. The results obtained with purified oxygen in steelmaking were so impressive, especially in speeding up open hearth furnace operation, that innovators turned full attention to developing a converter-type vessel designed to take maximum advantage of the oxygen technique.

Basic Oxygen Steelmaking

A major change occurred in high-volume steelmaking during the 1970s and '80s. Replacing the open hearth furnace, the *basic oxygen furnace* (BOF) now produces more than 60 percent of the steel consumed in the world. The BOF is not a furnace;

Table 5.2 - Gas Content of Alloy Steel Ingots

Gas	Alloy Steel, Regular Air-Melted Cast-In-Air	Alloy Steel, Vacuum-Stream Degassed Cast-In-Air	Alloy Steel, Vacuum-Carbon Deoxidized and Stream-Degassed
Hydrogen	10	1.5	0.8
Oxygen	45	35	20
Nitrogen	80	50	20

Gas Content of Steel - ppm

rather, it is a vessel with a refractory lining much like a Bessemer converter, and it is not ordinarily equipped with a source of heat for melting raw materials. The practice is to pour a partial charge of hot metal from the blast furnace into the BOF and then add cold pig iron and steel scrap. The additional heat needed to melt the cold, raw materials is provided by blowing a high-pressure stream of purified oxygen into the molten metal where oxidization of silicon, manganese, and carbon brings the entire contents to a temperature well above the melting point of iron. The oxygen is usually introduced through a lance inserted through the mouth of the BOF with the end positioned just above the surface of the molten bath.

Some BOF vessels have tuyeres in the bottom (similar to those in a Bessemer converter). The vessel design and designated operating procedure determine whether both oxygen-lance and bottom-tuyeres will be used for introducing the oxygen, with the latter supplied in very large quantities from an on-site purification and storage facility. The action of the oxygen jet is both chemical and physical. The oxygen immediately starts reactions leading to the formation of iron oxide that disperses rapidly throughout the bath. In this manner, the iron oxide carries the oxygen to the oxidizable alloying elements. Major advantages of the BOF process are flexibility in handling raw materials of many kinds and compositions, and the speed of removing oxidizable elements from the iron bath.

A basic-type refractory lining is used in the BOF even though removal of substantial amounts of silicon and phosphorus from the charge acts to produce acid-type slag. It is an accepted part of BOF steelmaking to counteract these acidic contributors by introducing strong basic fluxing materials, such as lime, during the operation to minimize attack on the vessel's lining. A number of different designs of basic oxygen vessels have been developed and put into use in various countries. These vessels have been identified with the names of the developer, or have been assigned a logo by the user. This review cannot include all of the different vessels now used in production, but examples of typical vessels are shown schematically in Figure 5.3 and outlines of vessel origin, mode of operation, metallurgical proficiency, and important supplementary facilities follow.

L-D Process

One of the first BOF developments originated about 1950 at Linz, Austria, and was named the *Linz-Donawitz oxygen injection process*. Its widespread use soon brought on the abbreviation L-D. The vessel used in the L-D process, Figure 5.3 (A), is basic lined, and is designed for top blowing. The oxygen lance is lowered to a predetermined position above a charge of hot metal and steel scrap, which might have a total weight of 150 tons, and the jet of high-pressure, high-velocity oxygen produces the vigorous reaction that can refine the charge to low-carbon steel in less than one hour. Sometimes powdered lime is metered through the lance and carried by the oxygen to the surface of the molten metal. This produces a foamy basic slag which is flushed off from time to time, taking away large amounts of silicon and phosphorus which have been removed from the bath by oxidation.

The use of purified oxygen and the vigorous agitation of the bath by formation of carbon monoxide during the blowing operation produces steel with a very low nitro-

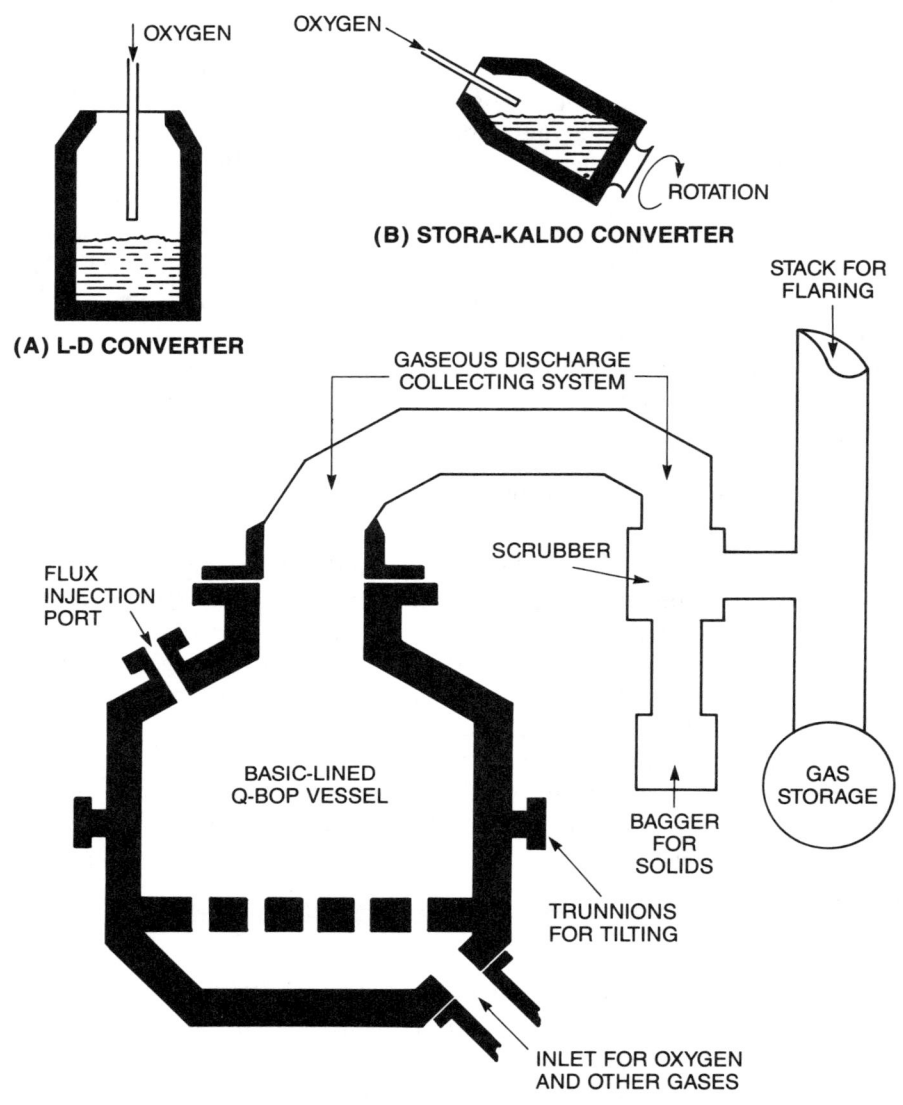

Figure 5.3 — Examples of steelmaking vessels that operate with the basic oxygen process

Steelmaking units of this kind are commonly called "BOF" converters, and they produce more than 60% of the carbon and low-alloy steels used worldwide.

gen content, as low as 0.002 percent, and a similarly low hydrogen content. The low level of these residual gases benefits the toughness of BOF-produced steels at low temperatures. The sulfur content of the steel depends on the amount of sulfur in the hot metal and scrap, and the measures taken during the BOF operation for removal. A sulfur content as low as 0.015 percent can be achieved if desired in the BOF output. A subsequent ladle desulfurization treatment can be applied to secure a lower level of residual sulfur. Approximately 45 cu m (1600 cu ft) of oxygen are used to produce each ton of steel by the L-D process.

Kaldo Process

The *Kaldo process*, another early BOF, had as its main feature the rapid circulation of slag against molten metal in a tilted, rotating vessel while a jet of oxygen from a lance impinged on the surface of the slag [Figure 5.3 (B)]. The Stora-Kaldo vessel is rotated at about 30 rpm to ensure uniform temperature distribution and to hasten steelmaking reactions. Approximately 55 cu m (2000 cu ft) of oxygen are consumed per ton of steel to reduce the silicon and carbon in a charge of hot metal to the low levels required in steel. The oxygen is transferred through the slag to the metal. Because carbon monoxide is burned to carbon dioxide inside the Kaldo vessel, this combustion produces enough additional heat to maintain the liquid state of the hot metal and to melt some cold scrap. The composition of the slag varies with the amounts of hot metal and scrap needed to accommodate the quantities of phosphorus and sulfur to be removed before the finished steel is cast.

Off-Gas BOF

OG-BOF is a widely-used designation in which the letters OG stand for off-gas. This steelmaking method does not differ from regular BOF; that is, the composition and properties of the steel are unaffected. The designation indicates a vessel of advanced design that is equipped to help protect the environment and to lower production costs. When steelmakers noted the large amount of carbon monoxide gas discharged from the mouth of a BOF vessel during the carbon-oxidation period of the blow and considered its high caloric value, they devised a hood system, as shown in the lower sketch of Figure 5.3, for controlled collection of smoke and hot gases during particular stages of the operation. Large ducts carry the smoke and gas from the collection hood over the mouth of the vessel to a fume scrubber. In some facilities, the fume hood includes coils of steel pipe carrying water which serve as a heat-exchanger to generate steam. The steam is piped to an insulated pressure vessel where it is stored for various uses. The clean gas (after scrubbing to remove particulate matter in the smoke) is piped to another pressure vessel for use as fuel. The amount of thermal energy discharged from the mouth of a BOF vessel is estimated at 738×10^6 J (700 000 Btu) for each ton of steel produced.

Q-BOP Process

Q-BOP is the acronym for *Quelle-Basic Oxygen Process*, one of the recently introduced BOF units for steelmaking as illustrated in Figure 5.3 (C). The equipment includes:

(1) basic-lined vessel capable of holding about 200 tons of metal,
(2) system for collecting smoke and waste gases discharged from the mouth of the vessel during oxygen blowing,
(3) combustion suppressor for maintaining carbonaceous gas in the form of carbon monoxide,
(4) flux injection device for maintaining a basic slag,
(5) slag skimmer,
(6) fume scrubber to remove particulate matter from discharge,
(7) stack for optional flaring of combustible carbon monoxide gas to carbon dioxide, and
(8) storage facility for compressing carbon monoxide and holding.

The Q-BOP vessel accepts hot metal from the blast furnace for its principal charge, and cold steel scrap can be included. As illustrated in Figure 5.3 (C), the Q-BOP vessel has tuyeres in the bottom for blowing oxygen. Natural gas is included with the blow if additional energy is needed. The Q-BOP unit is designed to operate with high efficiency. The steel produced by Q-BOP is typical of that from a BOF and one Q-BOP vessel is able to produce approximately as much steel as four open-hearth furnaces. The emission collection and control system minimizes environmental pollution.

The designation *SMP* stands for minimum slag refining process using the BOF, which is a practice advocated by Japanese steelmakers. This process shortens the time for producing a heat or blow of steel by using the minimum amount of slag needed to accommodate the acidic silica and phosphorus pentoxide generated from a molten bath. Heavy slag layers reduce metal yield by absorbing more iron. Slag also contributes to refractory wear in the BOF, and interferes with decarburization during the oxygen blow.

Preparatory steps for the low-slag process start with the hot metal from the blast furnace. To obtain effective dephosphorization, it is necessary to achieve a low silicon level in the hot metal by adding mill scale while the hot metal is in the transfer car from the blast furnace, and deslagging when the silicon is removed, but before appreciable drop in manganese or carbon contents.

Another effective method is to flow the hot metal through a trough-like furnace having two sections separated by a dam. In the first section, the metal is covered with a calcium oxide (highly basic) slag while iron oxide and oxygen are injected to remove silicon and keep the temperature at the desired level. The desiliconized hot metal (containing about 0.2 percent Si) then flows into the second section of the furnace while the dam holds back the calcium oxide (plus silica) slag. In the second section, a fresh slag made from briquetted soda ash (sodium carbonate) covers the metal while oxygen is blown through the metal to remove phosphorus, sulfur, silicon and nitrogen. Carbon oxidation is minimized by controlling oxygen flow and metal temperature. The hot metal, ready for the BOF, contains about 4% carbon, less than 0.02% phosphorus, and less than 0.01% sulfur. There is evidence that hot metal pretreated in this manner aids in obtaining clean steel from the BOF.

Lance-Bubbling-Equilibrium

Lance-bubbling-equilibrium (LBE) is a technique developed in France for use in a special type BOF that combines top-blowing with oxygen, and bottom-blowing with nitrogen. The nitrogen gas is admitted through a permeable refractory bottom, and this gas controls the blow reaction while a constant oxygen flow is maintained at the top of the bath. An improved rate of CO combustion to CO_2 is obtained, and this added heat is utilized in melting larger amounts of scrap. The oxygen blowing also reduces the amounts of phosphorus and sulfur in the bath. During the final stages of blowing, nitrogen (or argon) can be injected to form very small bubbles. These function as effective decarburization sites to reduce the carbon level to as low as 0.01% without excessive oxidation of the iron.

Ladle Refining

Sometimes called *secondary refining*, or *ladle metallurgy*, ladle refining can be carried out on molten steel from any kind of melting furnace, but greater use is made on steel from the BOF. The term *ladle metallurgy* is often emphasized because of the unique changes in composition and cleanliness of metal that can be accomplished with melting facilities already on hand in steel plants. There is no common design unit for ladle refining; each design is specially built to best fill the needs of a particular steel plant. Additional features are usually included as new installations are constructed. The following is a composite view of current ladle refining.

Ladle refining allows the primary steelmaking furnace (often the BOF) to be operated under conditions that favor high volume production at reduced operating costs. A properly planned and operated station allows greater productivity at the plant and produces better quality steel. Final adjustments in composition, homogeneity, cleanliness, temperature, etc., are left to be accomplished at a ladle treatment station. Steel can be adjusted at a *ladle station* to optimum temperature for whatever kind of casting is scheduled, and it can serve as a "cushion" between the steelmaking furnace and a continuous casting operation where temperature, degree of deoxidation, and other aspects of the molten metal are especially important.

A ladle station is more intricate than a melting furnace and the "station" may be the ladle itself, or a special chamber or enclosure which encapsulates the ladle while operations are performed on the melt. The ladles are specially designed for refining operations, and they often have a lining of high alumina refractory, rather than ordinary fireclay brick, to lengthen the time between relinings. These more complex ladles may be preheated unusually hot, as high as 1100°C (about 2000°F), prior to filling with molten metal and this avoids a marked temperature drop. The ladle has additional capacity over a given melt tonnage to assure freeboard sufficient for stirring and agitation without spilling over the sides, and to accept substantial additions for treatment and alloying. The ladle has become a sophisticated piece of steelmaking machinery.

Slag Removal

Slag removal is an important operation that must sometimes be accomplished quickly and thoroughly in conjunction with ladle refining. The slag on the metal as received from the steelmelting furnace can hold appreciable amounts of phosphorus and sulfur that under certain conditions may revert to the metal beneath. A variety of devices for skimming or deslagging the surface of molten steel include taphole valve systems, rod skimmers, and eccentric bottom pouring systems. All take the slag off the surface of the molten metal and divert it eventually to a slag pit for disposal.

Mixing Capability

Mixing capability at the ladle refining station ensures uniformity of composition when the steel is cast. Localized impingement of the oxygen jet in the BOF can cause some nonuniformity in the molten metal. This effect can persist despite vigorous agitation during the blow, particularly in a very large BOF. Various devices can be inserted in the ladle to mix the molten steel. One favored tool is a long perforated lance, covered with a permeable refractory sleeve, which is plunged deeply into the metal. A discharge of gas (often argon) causes immediate vigorous bubbling and mixing. Another technique, *capped argon bubbling*, admits argon through a porous plug in the bottom of the ladle for stirring. Sometimes a sealed lid covers the ladle and as the argon bubbles to the surface, the inert gas forms a protective atmosphere to prevent reoxidation. Another method of mixing consists of a large electrical coil in the "enclosure" or ladle station to accomplish induction stirring.

Alloying Additions

Addition of alloying elements is a frequent operation at the refining station because of the trend to micro-alloyed steels. Very small amounts of alloys are often added, and the final chemical composition must meet close limits. Because some of the alloying elements are oxidizable, the alloying operation is often carried out in unison with other treatments. For example, a lance may be inserted and argon gas bubbled through the metal to produce stirring, and at the same time the required amount of alloy is added. The raw alloy materials, in the form of small lumps, powder, or wire, are injected into the molten steel by feeding them down the lance along with the argon gas. Submerged injection is a good technique for consistent recovery of the alloying elements and predictable final chemical analysis. If the ladle is fitted with a sealed cover, or is held in a closed chamber, the argon accumulates as an inert atmosphere above the molten steel to minimize oxidation.

A technique for alloy additions introduced in the last decade involves a bell-shaped chamber lowered into the surface of the molten steel until it breaks through the slag layer. An inert gas is pumped into the bell, and alloy additions are made through an air-lock in the bell into clear molten metal. The molten steel is stirred at the time the alloys are added to ensure homogeneity.

Vacuum Treatment

Vacuum treatment of molten steel was mentioned as one of five methods of deoxidation. Benefits to be gained by subjecting the molten steel to a vacuum include lowering of the hydrogen content. The effectiveness of this treatment is dependent on the degree of vacuum and the manner in which the metal is exposed. The most efficient way of removing gases from the steel, particularly the dissolved hydrogen, is by *stream degassing*. For this operation, the full ladle of molten steel is clamped atop a second, empty ladle which is fitted with an air-tight cover that allows a vacuum to be drawn in the lower ladle. An interconnecting gate in the bottom of the upper ladle allows the metal to stream down into the lower vacuum-ladle and creates a large exposed surface for the mass of metal entering the vacuum-ladle. Continued vacuum pumping extracts the gases as they effervesce from the molten stream of metal.

Another system of vacuum treatment, *ladle degassing*, simply places an air-tight lid over a full ladle of molten steel, and the air is evacuated above the surface of the melt. Time must be allowed for dissolved gases to diffuse to the upper surface and effuse into the vacuum above. However, the holding time to degas steel which is resting quietly can be longer than most scheduling will permit. Ladle degassing is usually carried out by a three-step process: (1) establishing the vacuum above the molten metal, (2) stirring the melt to bring new metal to the surface and encourage effervescence, and (3) allowing a final quiet holding period to allow nonmetallic inclusions to rise from the molten steel and collect at the surface. Holding time will depend on many factors, including tonnage of metal in the ladle, depth of melt, temperature, stirring velocity and pattern, etc. Approximately 15 to 45 minutes is usually sufficient for complete vacuum-ladle treatment.

The benefit of vacuum treatment in lowering the levels of dissolved hydrogen, oxygen and nitrogen was illustrated earlier in Table 5.2. The reduction in hydrogen content by a factor of approximately ten avoids a number of difficulties, including flaking — a condition of internal fissures in the central areas of heavy sections, and cracking in weld heat-affected zones. The deleterious influence of hydrogen is discussed in detail in Volume II, Chapter 13.

Temperature Adjustment

Temperature adjustment of molten steel prior to casting has become a more demanding aspect of steelmaking, mainly because of the growing use of continuous casting. The reasons for this need of precise temperature control of molten steel forwarded to the pouring or casting stage will be given shortly. To lower the temperature of the molten metal, the ladle can be held to allow heat to dissipate if scheduling will permit the time required. Another quick, effective common practice is to add cold steel scrap of a favorable composition. Ordinarily, this closely-weighed addition is followed by stirring. To raise the temperature of metal in the ladle, an arc-heating unit can be operated at the surface of the molten metal to provide heat input. One type of station now in use puts the ladle of steel on a car to transfer it from the BOF to a casting unit. At an intermediate location, the car positions the ladle so that a non-magnetic (stainless steel) girth section of the ladle is surrounded by a closely fitted, large induction stirring coil. A cover is placed over the ladle, and three round graphite electrodes (each about 35 cm [14 in.] in diameter) are lowered through holes in the cover until within 25 mm (1 in.) of the molten metal surface. An arc between the electrodes and the metal surface then generates heat required to arrest the temperature drop, or to raise the melt temperature. Approximately 30 minutes are needed to increase the temperature each 40°C (70°F). This system enables the steelmaker to hold the molten metal at the proper temperature, and to maintain uniformity in composition and in temperature throughout the ladle's contents until the casting unit is ready to accept it for pouring.

Desulfurization

Desulfurization of steel is often performed during ladle refining. The need for this process depends on four factors: (1) the amount of sulfur in the hot metal from the blast furnace, (2) the amount contained in the steel scrap and other raw materials charged into the BOF, (3) the amount in the molten steel coming from the BOF, and (4) the amount permitted by the governing specification. Earlier in this chapter, when describing the transfer of hot metal from the blast furnace to a steelmaking facility via an insulated ladle or car, brief description was given of treatments with lanced injections of desulfurizing agents into the molten iron. Proper addition of one of these agents can bring the residual sulfur content of the hot metal to a level below 0.02 percent.

Some steels are improved by reducing the sulfur content as much as possible. An objective sometimes sought is to reduce sulfur below 0.010 percent, and this is not easily accomplished during the BOF operation for several reasons. The very high molten metal temperatures required to reduce sulfur are to be avoided because of vessel lining deterioration. Also, the relatively small amount of thin, basic slag on the surface of the bath is not adequate to hold the calcium sulfide that forms, and the speed of the BOF operation does not allow enough time for reaction between the molten metal and slag. Ladle refining is the practical stage to accomplish the required degree of desulfurization.

The sulfur problem can be handled during ladle refining in two different ways — (1) by adding an agent that aids in the reduction of sulfur content, and (2) by adding an agent that also exerts control over the composition, size, and shape of sulfide inclusions that may remain entrapped in the solid steel. The agent most frequently used to accomplish elimination of sulfur is calcium, usually in the form of calcium-silicon alloy. Since calcium is easily oxidized, the calcium-containing agent cannot be added until the molten steel has been given preliminary deoxidation with silicon or aluminum. Sufficient time must be allowed for calcium sulfide to form and remain in the slag atop the metal in the ladle. The use of calcium will exert some control over the composition and shape of sulfide inclusions as illustrated earlier in Figure 2.30.

Stronger agents for control of composition and shape of sulfide inclusions are the rare earth metals La, Ce, Pr and Nd. These elements form extremely stable sulfides and oxysulfides. To avoid loss of these elements before they can be utilized for sulfide formation, the molten steel must be thoroughly deoxidized prior to addition of a source, such as misch metal. The desirable feature of rare earth metal sulfides is a melting temperature of approximately 2200°C (4000°F) or higher. The rare earth oxysulfides have melting points no lower than about 1760°C (3200°F). In addition to assisting in the reduction of sulfur level, these sulfide-forming agents help form small, globular inclusions rather than elongated stringers in wrought, solid steel.

The ladle refining station is well suited to make additions of desulfurizing agents and to stir the molten metal to achieve uniform treatment. The importance of sulfur content in steels and the nature of sulfide inclusions in both base metal and weld metal will be discussed further in Volume II, Chapter 13.

Electric-Arc Furnace

Electric-arc furnaces have long been used to produce steel. Although very versatile, their cost of operation is generally higher than that of the open-hearth furnace. For this reason, electric-arc furnaces had been used mainly for melting alloy steels that were required to meet unusually high-quality standards and high-alloy steels that contained large amounts of oxidizable elements, such as the chromium in stainless steel. Usage of the electric-arc furnace in the production of carbon and low-alloy steels is growing steadily because of its capability for melting steel scrap and cold raw materials. The electric-arc furnace can be the choice for steelmaking without the large capital investment needed for a blast furnace or a BOF.

Electric-arc furnaces vary considerably in design and size. Foundries that produce steel castings operate electric-arc furnaces holding as little as 10 tons of metal. While most steel plants use furnaces of about 100 tons capacity, and furnaces have been constructed to melt as much as 400 tons. The giant-size furnaces use round graphite electrodes measuring about 71 cm (28 in.) in diameter, and the electrical power is rated in the order of 150 000 kva. The most common furnace has a circular, dished bottom for holding the metal bath. Three electrodes extend vertically through the roof of the furnace to provide arc heat to the surface of the molten metal. Furnace linings of burned magnetite (MgO-basic) or chrome brick (Cr_2O_3-weakly acid) are generally used, although foundries may employ acid linings of silica.

The materials charged to make a heat of steel may consist of scrap, pig iron, ore and alloys. The surface of the bath, except at the points where the arcs are burning, is covered with a blanket of slag. The electric-arc furnace offers excellent control of steelmaking reactions; because very high temperatures can be attained, close regulation of temperature can be exercised, and the bath cannot be contaminated by fuel.

Lancing with purified oxygen in arc furnaces speeds up to oxidation of the carbon or other oxidizable elements which must be removed from the bath. Large induction coils are sometimes installed beneath the hearth of the furnace to stir the molten metal and ensure uniformity of composition. Slag composition can be altered by addition of materials or by disposal of the "working slag" which removes unwanted oxides or residual elements and replacement with a finishing slag better suited for final operations. It is seldom necessary to transfer steel from the arc-melting furnace to a ladle refining station, unless vacuum degassing is wanted.

Electric-Induction Furnace

The high-frequency or coreless induction process is used in melting stainless steels, special tool steels, and other unusual alloys, such as cobalt-base or nickel-base heat resistant alloys. The furnace is usually of relatively small capacity and typically consists of a coil (for example, 40 turns, 12 in. [30 cm] in diameter, and 24 in. [60 cm] high) of water-cooled copper tubing, which acts as the primary of an air transformer. The secondary is the metal to be melted which is contained in a crucible within the copper coil. The crucible can be made of any good refractory, such as silica, alumina, magnesite, etc. A high-frequency current (1 kHz at 300 kW for a charge of one-half ton) passes through the coil which induces similar high-frequency currents in the outer portions of the charge. The raw materials charged consist of chunks of scrap and other melting stock whose size and shape fit the dimensions of the crucible. The heat generated in the outer parts of the charge is conducted to the central portion, and the entire charge usually melts within a couple of hours.

The strong stirring action created by the high-frequency current in the surrounding coil requires only a few minutes for refining and achieving homogeneity. The chemical composition of the metal poured is essentially that of the total materials charged, and no attempt is made to dephosphorize or desulfurize. Small losses of oxi-

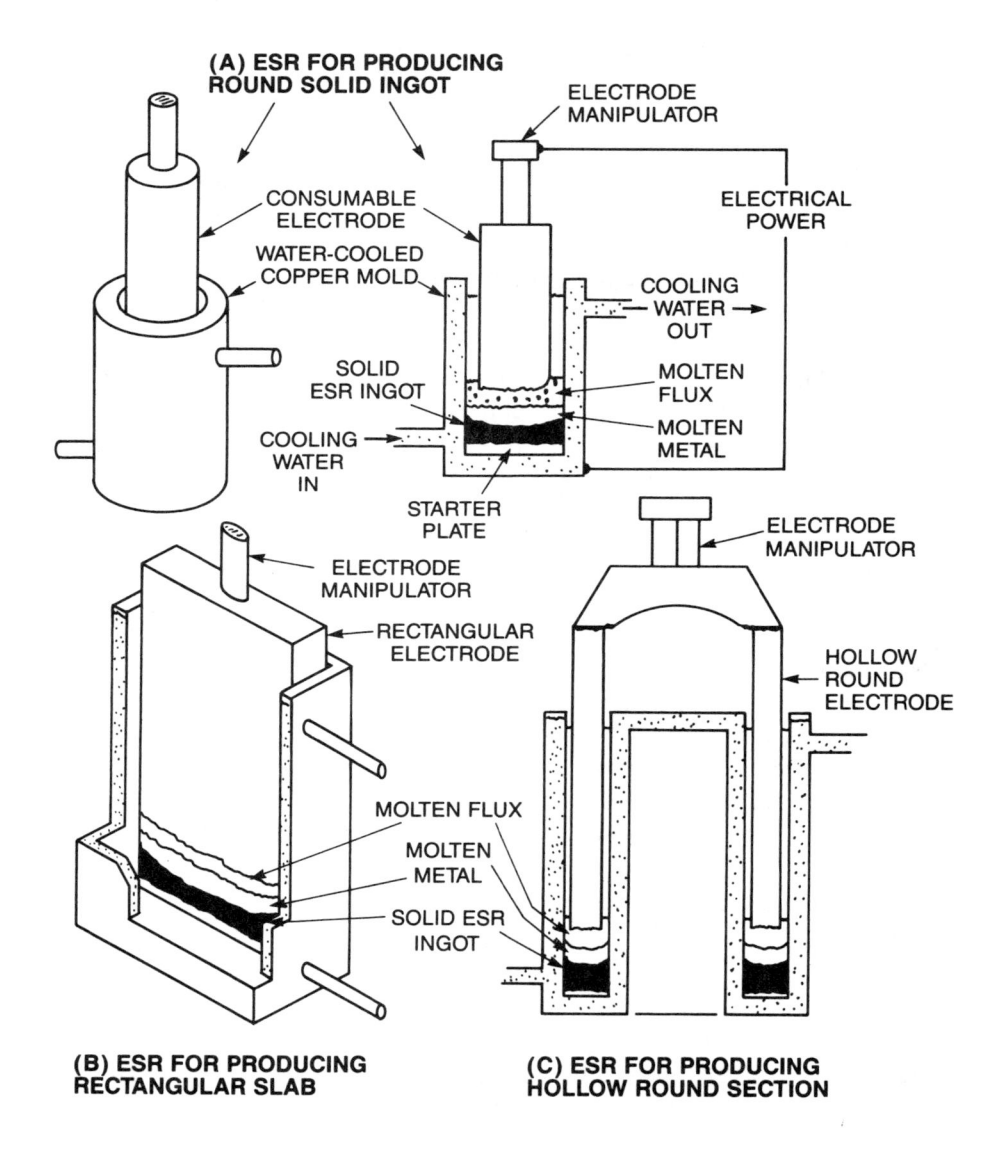

Figure 5.4 — Electroslag re-melting (ESR) furnaces for producing (A) round ingots, (B) rectangular slabs, and (C) hollow round sections

dizable elements like chromium, manganese or silicon occur. Melting costs confine the use of this furnace to the production of the more expensive alloyed materials.

Electroslag Remelting

In electroslag remelting (ESR) a large consumable bar or slab of solid metal is remelted by feeding it through a heavy slag layer atop a pool of molten metal, as schematically illustrated in Figure 5.4. The consumable component serves as an electrode, but it does not support an arc except for a brief starting period to convert the slag layer to the molten state. The melting operation is accomplished with heat generated by the flow of high-amperage electric current through the conductive molten slag. The molten metal from the electrode is then collected in a water-cooled copper mold, where the melt solidifies progressively (rather than accumulating as a bath that must be poured). The metal exists in the molten state a relatively short period of time. The contact between the melting droplets and the very hot refining slag can result in useful chemical reactions (such as desulfurization) and in removal of nonmetallic inclusions.

Although a somewhat costly process, ingots produced by ESR can have the following attributes:
(1) closely controlled chemical composition through selection of the electrode, and additions metered to the molten ingot during remelting;
(2) excellent cleanliness as evaluated by size and number of all kinds of nonmetallic inclusions;
(3) excellent ingot surface and internal soundness, which makes the material suitable for use as-cast, or after hot working to a desired shape; and
(4) isotropic mechanical properties, even after forging or rolling because of the relatively fined-grained, homogeneous ingot structure and favorable nonmetallic inclusion content.

Ordinarily, the electrode is a round solid bar, somewhat smaller in diameter than the round copper mold which is to be filled [Figure 5.4(A)]. However, various mold shapes can be employed. A rectangular mold, shown in Figure 5.4(B), forms a slab-shaped ingot suitable for rolling into plate. A mold with an open center as shown in Figure 5.4(C) forms a hollow round ingot suitable for processing into circular sections, hollow shafting, pipe, or tubing. The consumable electrode need not have the same general shape of the mold, but can consist of multiple electrodes of size and shape to provide a molten pool across the entire cross-section of the ingot in the mold. Additions are made by metering materials as wire or in granular form through the slag. The superheated slag readily melts the addition, and electromagnetic stirring in the molten pool produces a uniform distribution on the ingot's cross-section.

Some USA steelmakers produce carbon and low-alloy steels by the ESR process where one or more special attributes will justify higher melting costs. The isotropic mechanical properties are often the principal attributes sought by the fabricator and user, especially the improved ductility of rolled plate "through-its-thickness." The quality is often called the "z-direction" and is important in controlling the susceptibility of welded construction to develop *lamellar tearing*, a defect which will be described in detail in Volume II, Chapter 13. ESR furnaces range widely in size depending upon the kind of metal or alloy being produced. Costly high-alloy materials may be melted in units as small as one-hundred pounds capacity, whereas ingots for making large rolls, and for rolling into plate may be as heavy as 75 tons. Some of the ingot qualities are dependent on the nature of the electrode that is consumed, and may require use of an electrode made from specially produced steel to ensure that the final ESR ingot possesses all of the properties and qualities required.

SPECIAL MELTING PROCESSES

The special melting furnaces and processes that follow produce high-alloy steels, alloys of nickel, cobalt, titanium the refractory metals, etc. The higher operating costs of these processes limit their use for making carbon and low-alloy steels except when highly demanding requirements for composition, gas content, and cleanliness have to be met. General knowledge of their availability and capabilities is helpful when welding questions arise about the origin of materials.

An aim of these special processes is to minimize the residual amounts of oxygen, nitrogen and hydrogen in the finished material. Metal that is melted in air has ample opportunity to pick up significant quantities of these gases. Oxygen can be removed from the melt through the addition of strong deoxidizers; however, the oxides that form are not easily cleared from the melt, and the presence of nonmetallics in the solid material may be highly undesirable. Nitrogen is present in the air, and melting operations conducted in air allow the melt to pick up this gas. Nitrogen, unlike oxygen, is not easily removed from the melt by the addition of "getter" elements. Hydrogen is picked up from many sources in a melting operation; frequently from water vapor (humidity) in the air, moisture in (or on) raw materials, oil on scrap, and leakage of cooling water from furnace parts. Even though every practical precaution is taken to avoid sources of hydrogen when melting in an air atmosphere, an objectionable level of hydrogen may show up in the molten metal. Certain metals are avid absorbers of hydrogen and the properties of the solid metal may suffer badly from its presence. Titanium and its alloys are a prime example of this "hydrogen syndrome." A heavy slag layer, as used in the ESR process, is somewhat effective in shielding the molten metal from the atmosphere, but does not reduce the levels of oxygen, nitrogen, and hydrogen in the metal coming from the consumable electrode. An electrode made of vacuum degassed steel is used when very low levels of gas are required in the ESR ingot. The following special melting processes make use of a vacuum as an integral part of their melting scheme, and provisions are made for pouring the molten metal into the required mold without exposure to air. A vacuum is maintained over the metal continually during melting and pouring until solidification has occurred in a mold.

Vacuum Induction Melting

Vacuum induction melting, (VIM), was the earliest commercial process for melting under a vacuum. The electric-induction furnace is encased in a chamber, and the entire operation after charging is conducted "in vacuo." Sampling of molten metal, and necessary additions, is made via air-locks, and pouring into an ingot mold is performed without exposure to air. Control of melt composition during VIM is excellent since no significant losses can occur from oxidation, and the process does not depend on complex steelmaking reactions. Virgin raw materials of known analysis are melted in a crucible set within the induction coil, and a minimum of slag develops on the surface of the molten bath. Electromagnetic stirring continually brings fresh metal to the surface, so that gaseous impurities are effectively removed. The stirring or circulation of the metal is gentle, and no significant amount of slag is churned into the bath. The bath tends to float out nonmetallic inclusions during the melting period, and this action is improved when the molten metal is held for a short time after the electrical power is cut off. A recent innovation in VIM is to reduce the nonmetallic inclusion content by pouring the metal from the crucible through a filter. The filtering systems are proprietary devices which are not yet freely described by users.

The relatively high cost of VIM operation confines this melting process to high-alloy materials, but the availability of small furnaces and the qualities of the product make VIM very attractive for special filler metals used in welding. For example, when welding high-strength alloy steels used in nuclear or aerospace constructions, there can be a need for a small quantity of best-possible filler metal in the form of bare solid wire. This need can be filled through VIM production.

Vacuum Consumable-Electrode Remelting

This process is often designated VAR (*vacuum arc remelt*) and is widely used for producing high-alloy materials. Some high-quality special filler metal for joining the carbon and low-alloy steels is also produced by this process. The VAR process is similar to ESR describe earlier. Instead of using a heavy blanket of slag to protect the molten metal, as illustrated for ESR in Figure 5.4, the VAR process is performed in a vacuum. Figure 5.5 schematically illustrates a typical VAR unit and shows strong similarity to metal-arc welding, wherein a consumable electrode is melted into a pool on

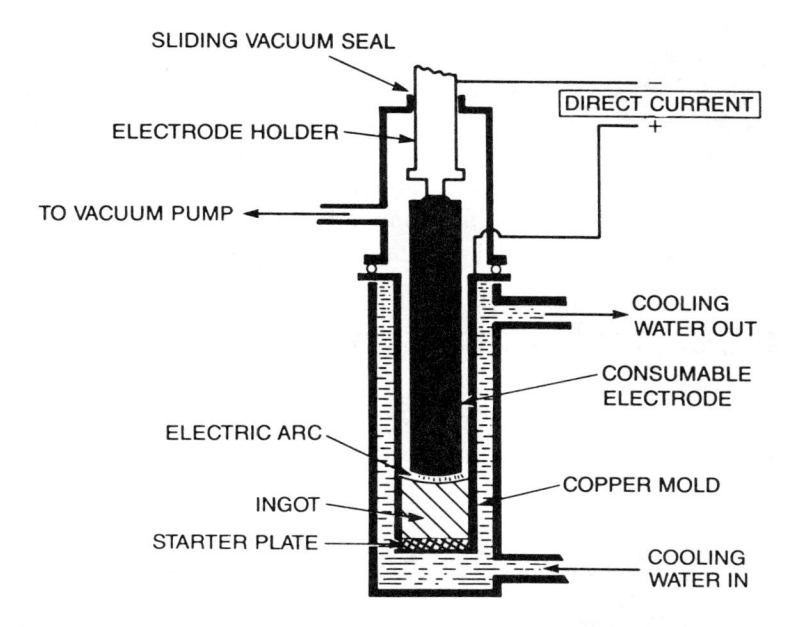

Figure 5.5 — Vacuum consumable-electrode remelting (VAR) furnace

a base that chills and solidifies the deposit. A VAR furnace has four principal components: (1) upper electrode feeding holder, (2) lower water-cooled metal crucible or mold, (3) pumping system for maintaining a vacuum in the chamber that encloses the melting-end of the electrode and the mold, and (4) electrical power source.

The electrode is connected to the negative terminal of a dc power source. The electrode may be quite substantial in size, a diameter of 40 cm (16 in.) is commonplace. Furnaces capable of melting electrodes as large as 1.25 m (50 in.) and weighing five tons or more have been built, and this is not necessarily the practical limit. The power source delivers many hundreds of amperes of current at a suitably low voltage to melt these electrodes. The water-cooled metal crucible is commonly made of copper and forms the mold for the ingot. The mold is connected to the positive terminal of the power source. The electrode is lowered as it melts without breaking the vacuum.

When the VAR furnace is in operation, the metal is transferred from the electrode to the bottom of the mold as a fine spray of superheated droplets in a uniform flow. The melting rate can be very rapid. A metal plate of a similar composition to the ingot being made is sometimes placed on the bottom of the mold to receive the initial deposit of melted metal. The molten pool forms under the electrode and solidification takes place progressively and in an upward direction as additional metal is added to the pool. Little difficulty is experienced with center porosity and segregation, which are troublesome conditions in ingots cast in the conventional manner. During melting of the electrode and transfer of metal to the pool, degassing, dissociation of objectionable compounds, and deoxidation takes place. All gaseous products given off are extracted from the chamber by the pumping system which runs constantly to maintain a vacuum in the order of one to 50 microns.

Small losses occur in certain alloying elements in steel during the VAR operation; for example, manganese. These losses are anticipated and compensated for by a higher content of the volatile elements in the electrode.

VAR melted steel is recognized for its improved quality. The gas content is a very low level. Table 5.3 shows typical gas contents for an alloy steel melted by four different electric furnace melting processes. The vacuum-melted steels are markedly cleaner than air-melted steels when rated both on size and frequency of nonmetallic inclusions. While air-melted steels are customarily evaluated by average field ratings, the vacuum-melted steels are often rated on the basis of the worst field in any specimen. Vacuum-melted steels display better fatigue properties, improved impact tough-

Table 5.3 - Typical Gas Contents of Alloy Steel Ingots Melted by Four Electric Furnace Processes

Melting Process	Oxygen	Nitrogen	Hydrogen
Regular air-arc furnace	45	80	10
Vacuum induction furnace (VIM)	4	3	< 1
Vacuum consumable-electrode remelting furnace (VAR)	4	50	1
Vacuum consumable-electrode remelting furnace using vacuum induction melted electrode (VIM + VAR)	3	3	< 1

Gas Content - ppm

ness, greater tensile ductility, and higher stress-rupture and creep properties. Large sections of VAR steels are homogeneous in chemical composition.

Studies have been made of the loss of metals and metalloids under vacuum melting conditions. At a vacuum of 10 microns, practically no phosphorus or sulfur is removed because these elements are strongly associated with iron in the molten condition and must be removed through chemical reaction. Zinc, magnesium, calcium and bismuth are easily evaporated, even at pressures as high as 100 mm. Some lead, tin, copper, and manganese will be lost at this pressure, but removal will be more rapid at the vacuum level where the VAR normally operates. Aluminum and silicon react with many crucible materials to form oxides that have markedly lower volatilization points, and large amounts can be lost into the vacuum. Pure magnesia crucibles are resistant to attack by aluminum or silicon in molten iron. No loss of nickel, cobalt, molybdenum, vanadium, or titanium occurs. A small loss of chromium occurs under very hard vacuum conditions.

The electrode used in VAR may be vacuum-melted steel with the lowest possible residual amount of undesirable elements, gas, and nonmetallic inclusions. Preparing the consumable electrode by VIM or VAR, and then again melting as a final VAR operation into a mold of desired size and shape, is not unusual, and these added steps in electrode preparation are reflected in the final cost of the product.

The majority of VAR units consume a single electrode as shown in Figure 5.5. A new type of furnace now undergoing trial employs two horizonal oppose electrodes. This kind of unit has been designated *vacuum-arc double-electrode remelting*, (VADER). Instead of feeding a single consumable electrode down into the mold, VADER has two electrodes arcing against each other. This arrangement is reported to offer a higher melting rate and allows closer control over the metal being deposited in the molten pool. The pool is shallow, and a finer microstructure results from more-rapid solidification. The improvement in microstructure can be helpful when limited hot working is applied to the ingot.

Electron-Beam Melting

Electron beam bombardment was put into use in a commercial "furnace" in the late 1950s for remelting and refining special metals and alloys. Much of the EB melting technology sprang from the electron-beam welding process (EBW), which is described and illustrated in Chapter 6. The melting unit is much larger and incorporates features required for metal production and casting. This melting operation is now called *electron-beam cold-hearth refining*. The molten metal flows along a shallow trough made of water-cooled copper that forms a series of gravity-fed pools. Each dammed pool is heated and kept molten by an individual electron-beam gun. The entire system is encased in a vacuum chamber. After the molten metal trickles out of the last pool through a notch in the final dam, it enters a water-cooled copper mold to solidify as a round or a rectangular ingot. The vacuum chamber is maintained at a pressure in the vicinity of 0.1 micron. No ceramic refractories are used in constructing the system, and there are at least two ways to remove any undesirable elements in the melting stock: (1) gaseous impurities effuse from the shallow stream of molten metal as it flows from one pool to the next, and (2) nonmetallics rise to the surface and are diverted by slag-skimming devices built into the course of troughs and pools. The designs of electron-beam melting units are proprietary, so details of construction and operation are not available.

Several electron-beam melting facilities are presently operated in the USA, and they are used mainly for remelting titanium scrap. The EB remelting process is highly efficient for ridding the material of contaminating gases, solid oxides, and even fragments of tungsten and other dense tool bit material. Ingots as heavy as five tons are being formed by progressive solidification in round, or slab-like water-cooled copper molds. This process is not regularly used for producing carbon and low-alloy steels as operating costs for the process are relatively high. However, the procurement of specially produced steel and its price can be justified by the very high quality of the metal produced.

Argon-Oxygen Decarburization (AOD)

The *AOD process* was developed for producing irons, steels and other alloys with high chromium content, perhaps containing about 10 percent chromium or more. Prior to introduction of the AOD vessel, high-chromium alloys were melted in the electric-arc furnace when large tonnage was required. As the raw materials were melted, it became necessary to remove carbon from the bath, and this was done essentially by oxidation. The oxidizing techniques included the charging of various solid oxides and blowing with oxygen, but this resulted in oxidation of the chromium present. After the carbon content was reduced to the desired level (and lost to the atmosphere as CO and CO_2), reducing agents had to be added to the furnace to regain the chromium as metal, or it would be lost as oxide in the slag. A favored reducing agent was low-carbon ferrosilicon. This melting-and-refining procedure required materials, time, and close attention to secure properly melted steel from the electric arc furnace. Several hours or more were needed for producing a typical production heat of stainless steel.

A major breakthrough in steelmaking technology started about 1970 with understanding of the equilibrium relationship between carbon and chromium levels in the molten bath, and the bath temperature. In addition to controlling temperature and oxidizing conditions in accordance with the known equilibria, reduction of the partial pressure of the carbon monoxide bubbles in the bath is achieved by blowing an argon-oxygen mixture through the melt. This results in an efficient method of refining molten chromium-containing alloys. The AOD operation is like the BOF steelmaking process, with additional technology because of the chromium involved.

Today, more than half of the high-chromium steels and alloys produced are made in AOD vessels. These are pear-shaped refractory-lined converters, somewhat like the Q-BOP illustrated earlier in Figure 5.3. Unrefined molten metal from an arc-furnace is poured into the AOD vessel while it is tilted on its side to receive the charge. As the vessel returns to an upright position, tuyeres in the bottom allow gas mixtures of argon, oxygen, nitrogen, air, and sometimes steam, to flow upward through the metal. Precise control of the gas mixture, as determined by the temperature level of the bath, can accomplish decarbrization with minimal oxidation of the chromium present. Air is blown through the tuyeres while the vessel is horizontal to keep these passages cool and open.

As the vessel is righted to a vertical position, argon is substituted for the air as the tuyeres become submerged. Oxygen is quickly added to the argon to achieve an argon-oxygen ratio of about 1:3. The argon and oxygen percentages and their flow rate must be carefully programmed to reduce the carbon content to the required level and at the same time to maintain bath temperature within the narrow range dictated by equilibrium conditions. When the bath is within the required composition limits, argon alone is put through the tuyeres to lower the temperature to the desired level for pouring. Nitrogen can be added to the gas mixture if an elevated level of this element is required by the composition specification. The continued blowing of argon, in addition to reducing bath temperature, also improves metal quality by decreasing dissolved gas and oxide contents. The entire AOD refining operation can be completed in less than an hour, a substantial improvement over the time required to accomplish the same results in an electric-arc furnace. AOD converters range in size from small units of five or 10 tons capacity suitable for use in a foundry making castings, to very large vessels holding 50 tons or more for steel plant operations.

FOUNDRY AND STEEL MILL OPERATIONS

Refined molten steel from a melting process is ordinarily utilized in two different ways: (1) it can be cast directly into an article of the desired shape which is the function of the foundry, or (2) it can be poured into a form which requires further hot or cold deformation to produce the shape desired by the user. This has largely been the provenance of the steel plant because of the substantial capital investment required in rolling mills, extrusion presses, draw benches, and other heavy equipment. Interestingly, in today's technology, foundries sometimes perform hot work on their castings, and steel plants sometimes pour steel into shapes and apply no further deformation or work.

Foundries use many kinds of furnaces for melting iron and steel to be cast into a shaped-mold. The selection of a furnace is guided by several considerations: (1) the composition of the alloy to be cast, (2) the amount of melt required, and (3) the quality standards to be met. Foundries may be equipped with the cupola, open hearth, electric-arc furnace, AOD vessel, electric-induction furnace, etc. The shaped mold into which the metal is poured is usually made of molded sand, but ceramic and metal molds can be used. The solid metal castings are often subjected to heat treatment to improve mechanical properties. The practice of applying a small degree of hot work to the castings by *hot isostatic pressing* (HIP) has been instituted where freedom from internal microshrinkage must be ensured.

In addition to treatment of castings after they have been made, new technology is being utilized in the foundry in pouring the castings themselves. The innovations include new materials for making molds, and entirely new casting techniques for achieving directional solidification and crystal growth control.

Because of the frequent application of welding to cast components to assemble them into welded construction, it is important to understand the methods used to produce different types of castings and their inherent properties. In general, steel castings have welding characteristics comparable to wrought steel of the same composition; however, subtle differences between cast and the wrought forms require special considerations in planning welding procedures. These differences are explained from a metallurgical viewpoint in chapters that follow.

Steel plants, with integrated facilities ranging from the blast furnace to rolling mills, originally evolved to melt large quantities of steel, and to produce semi-finished wrought sections for further manufacturing and construction. Some of the wrought forms being produced by steel plants are virtually ready-for-use, as examples: grinding balls, railroad rails, I-beams, and wire used for filler metal in welding. Steel plant operations have expanded over the years in a number of ways, and several new ventures are associated with welding. A growing number of smaller steel plants have facilities which are designed to provide steel to a limited sector of the market. One example is a plant that produces only the deformed-section bars for concrete reinforcement. Some plants have been built to feed steel to a particular manufacturing operation that consumes a large amount of steel in a particular product form. A fabricator of pressure vessels and heat exchangers may have its own steel plant within the company to supply most of its plate and tubing needs.

The most profound change that has occurred in producing wrought steel over the past 20 years is the use of *continuous casting* instead of *ingot practice* as the first step in converting molten steel to solid form. Prior to about 1960, virtually all molten steel at the steel plant was poured into large molds made of thick cast iron to form an ingot. The cross-sectional shapes of ingots were square, rectangular, octagonal, etc., to facilitate hot working by forging or rolling to the next required shape.

Mill processing, when started with an ingot, is a complex and costly processing route. Many factors must be carefully controlled to obtain an ingot that is free of injurious surface defects and that does not have unacceptable internal unsoundness or composition heterogeneity. Furthermore, much thermal energy is expended to heat the steel at various stages for hot working operations. It is more advantageous to cast the molten metal directly to near-finish size; or at least to a billet or bar much smaller in size than the usual large ingot.

Research and development began about 1950 on a machine that would accept molten steel at the top and discharge an endless length of solid bar or slab at the bottom, which could be cut "on-the-move" to required lengths. Today about half of all wrought steel produced starts in continuous-cast form. Some producers utilize continuous casting for almost 85 percent of their output. Certain steel shapes or sections cannot be continuously cast with available equipment or technology.

Approximately half of the wrought steel handled today originates in ingot form, and discussion of mill operations follows, starting with ingot pouring and mill practice commencing with an ingot.

Ingot Steelmaking Practice

The molds into which molten steel is teemed to form ingots are made of cast iron with walls a thick as 15 to 25 cm (6 to 10 in.). Most molds are open at both ends and are placed flush on a thick cast iron "stool" that serves as the bottom while the mold is being filled from the top. Some splashing occurs as the stream of molten metal fills the mold, particularly when the stream initially descends upon the bare stool. This splatter tends to adhere to the mold wall and if not remelted by the rising molten metal, usually transfers as a "scab" to the ingot surface.

One practice to avoid scabs is to "bottom-pour" by using a special arrangement of fountain head and runner to fill the mold quietly from the bottom. The fountain head absorbs the shock of the metal stream coming from the ladle, and this arrangement assures a good ingot surface. However, the presence of refractory lining in the fountain head and the runner provides a source of nonmetallic inclusions to the flowing metal. The nonmetallic inclusions have little opportunity to float out of the steel once it has entered the mold.

Because of the high cost of bottom-pouring, most ingots are made in top-poured molds. An intermediate pouring box or *tundish* is sometimes held above the mold to better control the stream of metal entering the mold. The inside surface of the mold is frequently coated with tar, aluminum paint, or some proprietary coating, to help form a smooth ingot surface free of defects.

Ingot size and shape is determined by the kind of wrought section to be produced from the ingot. Rectangular ingots, often weighing as much as fifteen tons, are used for making slabs and are rolled into plate and other flat-rolled forms. Square ingots are used for billets, structural shapes, bars, and rods. The walls of molds often are corrugated or fluted, with corners rounded, to prevent surface cracking on the ingot. Study of solidification patterns prompted these features for particular mold shapes. In order that the ingot may be easily stripped (removed) from the mold after solidification, the ingot section is tapered over its length. The molds used in making rimmed or semi-killed steels may be either big-end-up, or big-end-down.

Killed steel is always poured into big-end-up molds, and a refractory-lined "hot-top" is fitted on the top of the mold before teeming with molten steel. This hot-top acts as a reservoir for a small volume of molten steel. As the molten steel in the central area of the ingot below shrinks as it solidifies, the molten steel in the hot-top feeds or replaces the shrinkage which would otherwise form a hollow or "pipe" in the top-center of the ingot.

When a very large ingot of killed steel is completely teemed, additional heat may be supplied by shoveling a quantity of slow-acting exothermic mixture atop the metal. This keeps the metal in the hot-top molten. Special hot-topping devices have been developed that make use of a small electric arc which dwells on the top surface. A layer of cinders or slag placed atop the metal in the hot-top as an insulting blanket is effective in minimizing the depth of pipe cavity that forms in the ingot beneath. This confines the pipe to the hot-topped portion of the ingot. The technique employed is important because it determines the amount of steel lost as scrap when the hot-reduced wrought section from the ingot is cropped (trimmed) to remove all pipe cavity from this upper end. In Volume 2, Chapter 13, an explanation will be given for *laminations* in wrought steel that originate from the pipe or shrinkage cavity and have remained in the product because the steelmaker failed to crop sufficiently on the product from the top of the ingot.

Ingot steelmaking, in addition to being costly because of the facilities involved and the energy consumed, presents other challenges of a metallurgical nature that can be of concern in welding. A certain amount of chemical composition heterogeneity is usually found in products from the ingot because of the basic mechanisms involved in the relatively slow solidification of large masses of metal. Sometimes the result of elemental partitioning or segregation in the ingot is favorable, as in the case of "rimmed steel" describe earlier. Normal composition heterogeneity must be kept in mind because of the influence of actual composition of the finished product on its properties and behavior in welding.

Steelmakers regularly follow a long-standing practice in sampling, analyzing, and reporting the chemical composition of the steel produced. The analysis commonly reported for steel is the *ladle* or *heat analysis*. Samples for these determinations are taken by momentarily holding a metal spoon in the stream of molten metal from the ladle during teeming. This spooned metal is poured into a small mold, and the solidified block is called a *ladle sample*. These samples are taken during the pouring of the first ingot, the middle ingot of the mold string, and from the last few ingots poured. The analytical values reported as the ladle analyses for the heat are those selected by the steelmaker after examining results from all ladle samples. The "fading" of oxidizable elements from the final molten metal coming from the ladle may require diverting or downgrading one or more of the final ingots poured.

The ladle or heat analyses reported by the steelmaker represents an average composition. Because of the tendency for certain elements to segregate during solidification, or to undergo a small decrease in amount from the molten metal in the ladle, the finished steel products eventually produced from the ingots can display slightly different values for alloying and residual elements. The extent of differences between ladle (heat) analyses and check analyses on the products depends upon many factors, including: (1) ingot size, (2) metal pouring temperature and total time of metal in the ladle, (3) location of sampling for check analyses, and (4) the particular element being studied.

The AISI has prepared tables of check analysis tolerances for most of the alloying and residual elements, and these tolerances are widely used by both steelmakers and specification writers. Steelmakers ordinarily analyze their ladle samples for a large number of elements for a complete understanding of each particular heat of material, and for preparation in reporting whatever elements are requested by the purchaser. Any check analyses made at the steel mill on products become part of the analytical record of each heat.

Continuous Casting of Steel

Continuous casting was put into commercial steelmaking use almost 30 years ago, but its use is looked upon as a new, important event even today because of continued significant advances being made in this operation. Continuous casting is referred to as "C-casting," and the machines as "C-casters." These machines require a large capital investment by the steelmaker, but they take the place of a costly operating segment of a steel mill where ingot practice would be used. In addition to the significant economic advantage of C-casting for the initial solid steel, the wrought products produced by this method display metallurgical improvements that will be pointed out later.

The simplest form into which steel can be C-cast is the small rectangular bar. The steelmaker also produces large-diameter rounds, square blooms, and large rectangular slabs. These are the intermediate sections needed for production of shapes and forms ordered by steel fabricators and manufacturers. Early techniques for securing a continuous solid strand of billet or slab involved pouring the molten steel from the ladle downward into a vertical mold with an open bottom. The mold was made of copper, and passages were included behind the inner wall of the mold to provide generous water cooling.

To achieve solidification of the section being C-cast, the vertical mold was rather long (perhaps as long as 90 cm or 36 in.). After emerging from this mold, the C-cast strand consisted of a solid shell with a liquid center. Upon emerging from the bottom of the mold, the bare, red-hot strand was sprayed with water as it was guided downward by rolls. When the strand reached a point where the entire cross-section had solidified, the steelmaker arranged to cut the strand into desired lengths. The distance from the top of the mold to the point where the strand becomes fully solidified is regarded as the *metallurgical length*, and this could be as much as 40 meters (130 ft). Somewhere along the latter part of the metallurgical length, the strand had to be gently curved to assume a convenient, horizontal run-out direction for cutting and handling. The early C-casting machines were of considerable height and required a building several stories high. More modern machines avoid the awkward, very tall facility for housing the operation by either (1) accelerating cooling of the strand and curving it to a horizontal position at an earlier point, or (2) making use of a curved mold that discharges the strand almost horizontally.

As illustrated in Figure 5.6, a typical C-casting operation with a vertical mold uses some of the equipment that is used in steelmaking, but for the C-casting operation, the equipment components have been interrelated following considerable research and development efforts. There must be close communication between the C-casting shop and the melting shop, which supplies the ladles of finished, ready-to-cast molten steel. It is imperative to maintain a balance between melting-and-refining furnace output and consumption at the C-caster. If the caster is shut down for lack of molten steel, start-up requires use of a "dummy bar" to lead the newly poured metal slowly through the mold. If the melt shop runs too far ahead of the C-caster, a group of prepared molds must be ready to permit pouring the ladle of steel into ingots. However, continuity of operation is important, and many variables are constantly monitored to avoid even momentary stoppage. As many as several hundred sensors feed input to a computer, which in turn calls for adjustments in vital functions to keep the system synchronized. C-casting is maintained at a constant rate, and peripheral conditions are varied to hold the prescribed strand speed. These conditions include the temperature of the steel in the ladle, the amount of cooling accomplished in the mold, and the amount of cooling achieved by water sprayed below the mold.

When a crane delivers a full ladle of molten steel from the melt shop to the C-caster, the ladle is set on a turret, which rotates and positions the ladle above a tundish. The turret allows another full ladle to be brought at the proper time to stand ready in the next turret position; this setup keeps the tundish properly filled. The tundish is used to maintain a uniform supply of molten steel above the mold, and this supply can reach 50 tons. The level of liquid steel in the tundish, usually about one meter (40 in.), is monitored by a radiation-type detector, which ensures a level within a range of about 25 mm (1 in.). This close level control maintains a uniform ferrostatic head on the mold. The liquid steel level in the mold is held within a narrower range, perhaps only five mm (0.2 in.).

402   Chapter 5

Because the ladle of steel has come from a treatment station and is completely refined, deoxidized, and homogenized, every effort is made to transfer the steel to the mold with minimal exposure to air. The ladle is often covered to minimize temperature drop, and the interior below the cover is flooded with an inert gas. A ceramic shroud with a flowing shield of inert gas surrounds the stream from the ladle nozzle to the tundish. The tundish is also usually covered, the interior is flooded with inert gas, and a ceramic shroud protects the stream pouring from the tundish into the mold. A system of dams and skimmers may be used on the tundish or at the mold entrance to remove any slag that has formed on the molten steel and divert it into a slag pot.

The mold itself oscillates in a rapid up-and-down motion, moving about one cm (0.4 in.) in a vertical plane at a frequency of 20 to 200 cycles per minute. This motion prevents the solidifying steel from freezing tightly on the mold wall. A proprietary powder is metered into the mold entrance along with the molten steel to coat the mold walls and to serve as a lubricant as the steel solidifies and the molded steel section moves toward the lower exit. A long train of rollers beneath the mold comprises a

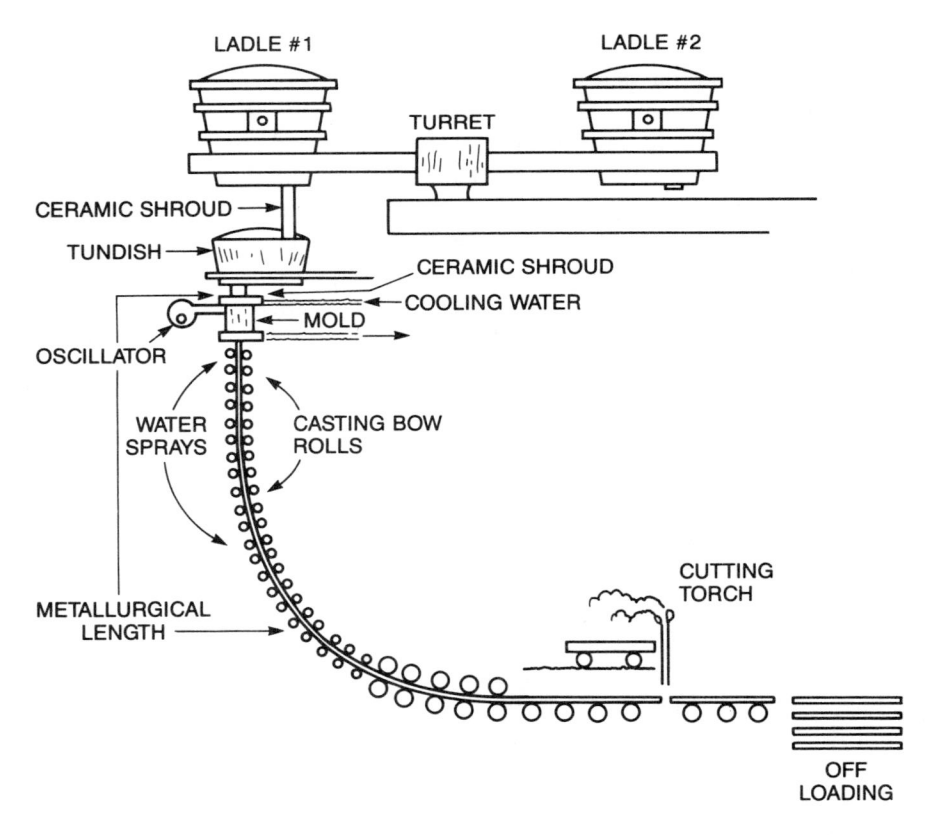

Figure 5.6 — Schematic illustration of a single-strand machine for the continuous casting of steel

carefully engineered system that facilitates further cooling using a mist of water and compressed air. The rollers gradually curve the strand in a horizontal plane for convenience in cutting and loading for transport. At the upper portion of this metallurgical length, gentle handling is very important to avoid breakout of liquid metal from the center of the strand, and to prevent internal crack formation.

The product from a C-caster requires very little cropping of scrap material. An overall yield of almost 95 percent usable solid steel from the melt is common. This figure compares with about 80 percent yield using ingot practice because of the removal of hot-topped sections and extreme bottom portions, and because of the surface conditioning required to remove scabs and other defects from the ingots and from the hot-reduced products. Furthermore, the C-cast product has good homogeneity because of the relatively small mass of molten metal that has solidified. The surface of the cast section is usually so good that conditioning is not needed. In fact, the C-cast blooms or slabs, if kept hot enough, are ready for rolling on the hot mill. When the cast steel is transferred directly to the hot rolling mill, almost half of the thermal energy ordinarily needed to achieve the high temperature required for hot rolling or forging is saved. However, if the C-cast material must be surface conditioned or must be soaked at a prescribed temperature prior to forging or rolling; facilities are available for these operations.

Commercial C-casting was begun with a machine that produced only one strand of a small billet section, but modern equipment is capable of producing multiple strands of very large sections. Machines can C-cast round bars as large as 25 cm (10 in.) in diameter, blooms and billets as large as 50 cm (20 in.) square, and rectangular slabs up to 30 cm (12 in.) thick and 200 cm (80 in.) wide. A remarkable new development in mold design for the rectangular sections allows the width of the moving slab being cast to be changed over a very short length from perhaps 100 to 200 cm (40 to 80 in.) in graduated steps of about 50 mm (2 in.).

The speed of C-casting machines can range from about 0.5 to 2 meters per minute (20 to 80 in. per minute) depending on the section being cast and the design of the machine. Although the aim is to cast as rapidly as possible to maximize tonnage from the unit, the need to obtain good surface, to achieve freedom from internal defects, and to minimize danger from molten metal breakout has limited casting speed to about two meters per minute (80 in. per minute) with today's technology.

Some C-casting machines, particularly those designed to cast smaller billet and bar sections, fill a number of molds simultaneously under a multi-stream tundish. C-casting as many as eight strands of 10 cm (4 in.) square billets is not unusual, and the production of at least two strands of slab simultaneously is becoming quite common.

Close attention is being paid to the metallurgical qualities of C-cast steel, and many recently developed technological innovations are being used. For example, when especially large rectangular slab sections are being C-cast, electromagnetic stirring may be applied to the metal in the ladle, the metal in the tundish, or the molten metal in the upper portion of the mold. This stirring ensures homogeneity of composition and helps minimize mechanical property variations in the final wrought product. Another example of technological innovation applied to C-casting is in the production of sheet steels. This subject is discussed in Technical Brief 29.

HOT WORKING OPERATIONS

Ingots, after being stripped from their molds, may be allowed to cool to ambient temperature if further processing is not carried out at the time. However, cooling must be done carefully with certain types of steel to prevent developing internal cracks in the large mass with its coarse microstructure. Common practice is to bury crack-susceptible ingots in a bed of powdered insulating material. The reheating of a cold ingot must also be done slowly and uniformly. When scheduling permits, to avoid cracking and to conserve heat in a newly produced ingot, the hot ingot [perhaps still at a temperature of about 600°C (1100°F)] is often delivered immediately after stripping to the soaking pits. These pit-like furnaces allow the ingots to stand vertically with good circulation of heating gases surrounding them. After heating to the proper temperature for hot working, the ingot is subjected to forging, pressing, or rolling to produce a smaller shape and size.

Forging or pressing of ingots after heating to about 1200°C (2200°F) can be either an operation to rough-form an article, such as a turbine shaft, or can be the production of an intermediate or semifinished form, such as a bloom, billet, or slab. Forging and pressing are generally used as preparatory operations when handling highly alloyed steels that are difficult to hot work without tearing or cracking. In addition to steam hammers (for forging) and hydraulic presses, a *rotary forging machine* has been intro-

---

TECHNICAL BRIEF 29:
Continuous Casting of Thin Steel Strip

There is firm belief among steelmakers that C-casting can be extended into a new product area where even greater economic gains can be realized. The targeted area is thin strip and sheet steel. Considerable savings would be found in countries with large manufacturing capability for automobiles, appliances, and other sheet stock hardware. To produce steel sheet using ingot practice, most of the major steel plant facilities are required: soaking pits and rolling mills for blooming, slabbing, and hot rolling strip. The C-casting of slabs currently eliminates the need for the blooming and slabbing mill. If thin strip could be continuously cast directly, virtually all hot mill facilities could be eliminated from the procedure for making cold rolled sheet steel.

Much development work is underway to cast molten steel directly into thin strip. Methods proposed to date involve C-casting the molten steel on a spinning chilled metal roller, or casting the steel between two closely spaced endless metal belts that are both moving in the same direction, or pouring molten steel between two large drums, closely spaced, at their entry or "pinch point." These two basic concepts have been evaluated with a number of modifications. In one case, the spinning roller is partly immersed in the melt so that a thin film is dragged out of the melt. This solid film is then continuously stripped off the roller farther along its circumference. Generous water cooling is applied to remove the heat and to obtain rapid solidification.

Most investigators feel that C-cast thin steel will display improved properties compared to similar steel produced by present-day methods. This belief stems from the benefits of rapid solidification, which include (1) a more complete solid solution of alloying elements, (2) less segregation of residual elements, (3) reduced size and number of nonmetallic inclusions, and (4) firmer control of grain size. Work with an experimental C-caster has produced strip as thin as 1.5 mm (0.06 in.) thick and up to 800 mm (32 in.) in width. This strip in coils is nearing the size and quality needed for cold rolling into thinner strip and sheet.

duced for mill usage. The rotary forge can accomplish reduction so vigorously that the workpiece experiences little drop in temperature during the operation. The machine has four high-powered hammers that strike the ingot surface at about 200 strokes per minute simultaneously on all sides to exert equal pressure. The resulting hot work reduces the cross section and increases the length. The greater tonnage of ingot steel of the carbon and low alloy types, however, is hot worked by first rolling on a blooming mill.

Hot rolling mills can be generally classified into *roughing mills* and *finishing mills*, and the resultant products have acquired the commonly used names outlined in Table 5.4. Blooms, billets, slabs, and sheet bar are semifinished products of roughing mills, and because they are usually intended for further hot working, they are not made to close tolerances. These *semifinished* products are often subjected to surface conditioning operations, such as oxygen scarfing, mechanical chipping, or grinding to remove seams, laps, and other surface imperfections.

Hot rolled bars, plates, structural shapes, rods, and strip are considered finished rolled products because they are expected to meet stated dimensional tolerances. The surface of these products is expected to be reasonably smooth and free from defects; however, it will be covered with a black oxide scale in the hot rolled condition. Special shapes may be produced from round billets by hot extrusion in a special press. This may be done because the section is too difficult to hot roll or because the shape must meet close dimensional tolerances.

Although hot rolling mills may be called blooming mills, bar mills, and rod mills, they vary greatly in size, method of operation, and the kinds of rolls used. If a mill is

Table 5.4 - Common Terms for Hot-Rolled Mill Products

From Roughing Mills

Blooms	Large square or rectangular sections with rounded-corners at least 225 cm^2 (about 36 $in.^2$) and larger for further hot working
Billets	Square sections with rounded corners ranging from 10 cm^2 (about 1½ $in.^2$) to approximately 225 cm^2 (about 26 $in.^2$) for further hot working
Slabs	Rectangular sections at least four cm in thickness (about 1½ in.) with width at least twice the thickness. Commonly about 25 × 125 cm (approximately 10 × 50 inches) for further hot rolling into plate or strip

From Finishing Mills

Bars	Smooth rounds, squares, hexagons, flats, etc., rolled to specified dimensional tolerances
Plates	Smooth flat sections with minimum thickness of about five mm (approximately 0.2 in.) and of substantial width
Structural Shapes	I-beams, channels, angles, etc.
Rods	Smooth rounds ranging from a minimum diameter of about five mm (approximately 0.2 in.) to a maximum of about 15 mm (approximately ⅝ in.). The hot rolled product usually is coiled
Strip	Smooth flat-rolled ranging in thickness from about one to 30 mm (approximately 0.04 to 1¼ in.) and ranging in width from about 60 to 250 cm (about 24 to 96 in.) and hot coiled as a continuous length. Frequently used as semi-finished material for cold rolling to strip and sheet

termed an 18-inch bar mill, for example, this indicates the distance between the driving pinions on the roughing stands. This dimension gives an approximation of the mill's size and capability and does not indicate the maximum size of bar produced.

A *reversing mill* consists of only one frame or stand of rolls. Only two rolls, one above the other, are normally used. These rolls can be quickly adjusted to produce material of varying thickness, and their direction of rotation can be quickly reversed. Many roughing mills are the reversing type. An ingot is reduced to a smaller cross section by passing it back and forth through the stand while the rolls move closer together in each successive pass. A *continuous mill*, on the other hand, consists of a number of roll stands. The distance between each pair of rolls is fixed, and this pass dimension decreases progressively from the first stand to the last. The stands may be placed in line for straight-through production of the rolled product or may be arranged side by side so that the product shuttles back and forth. These mill stands may contain three rolls aligned vertically. The product first passes through the lower pair of rolls and is then returned through the upper pair. Vertical rolls are sometimes used to control the width of the product as it emerges from a pair of horizontal rolls.

The hot rolling operation affects the material in a number of ways. It joins deep-seated clean blowholes and any secondary pipe, reduces the grain size of the steel, improves toughness and ductility, makes the material more homogeneous in chemical composition (but does not eliminate marked segregation), and imparts directionality (anisotropy) in the steel (like fiber in timber). The degree of directionality that develops in the wrought product depends on many features of a given steel and the nature of the hot working operation that has been applied.

Cooling the steel products after they have been hot rolled to finished dimensions is carried out according to the steel's metallurgical characteristics. Low-carbon steel types are simply cooled in air on racks to maintain straightness. Highly hardenable steels are slowly cooled by burying the bars or shapes in an insulating material or by placing them under an insulating cover.

Thermo-Mechanical Control Process (T-MCP)

T-MCP is a method of hot rolling or forging steel whereby mechanical properties produced in the product are virtually equivalent to those obtained by heat treating conventionally rolled or forged steel. T-MCP is a refined metallurgical concept featuring controlled hot working, accelerated cooling, and microalloyed steel compositions. Coordination of these three factors enables the steelmaker to produce a fine-grained steel product in which the microstructure has a favorable distribution of constituents to obtain good mechanical properties.

When a steel product, such as plate, is manufactured with a T-MCP, the rough section of steel (i.e., the slab) is heated to a temperature regularly used for hot working operations. This temperature [about 1200°C (2200°F)] is sufficiently high to dissolve carbides and carbonitrides in the austenitized microstructure. The early part of hot working is performed in a normal manner, but just prior to final hot work reduction (commonly referred to as the finishing pass) the temperature of the steel is controlled by the rate of working and by cooling to reach a relatively low level for a hot working operation [perhaps about 900°C (1650°F)]. Plastic deformation of the steel at this lower temperature markedly refines the austenite grains and retards precipitation. Also, when hot work is completed and accelerated cooling carries the steel to the critical transformation range for the particular composition, austenite-to-ferrite transformation produces fine ferrite grains and a precipitate of fine, highly dispersed carbonitrides.

For maximum effectiveness, final hot working may be continued as the steel drops below the Ar_3 critical temperature of the steel being processed. The rolling mill or forging equipment must be sufficiently rugged and capable of deforming the steel at the abnormally low hot working temperature.

Deformation at temperatures within the critical range, where ferrite forms, produces strain hardening, and the ferrite lattice is riddled with dislocations to the extent that the microstructure appears textured. Extremely small, highly dispersed niobium carbonitrides can be generated in a very fine-grained ferrite by T-MCP, and the steel section will have comparatively high strength and good toughness. Optimum particle size and dispersal of carbonitrides is obtained at a working finishing temperature in the vicinity of 775°C (1425°F). T-MCP provides a way of gaining desired strength and toughness in wrought steel products at a lower cost compared to applying a heat treatment. Additional unique property improvements often appear in the T-MCP material.

Microalloyed steels with higher strength and improved toughness achieved by T-MCP have been available from a limited number of steelmakers for some time. The success of these proprietary products encouraged the ASTM to develop a standard specification for killed steel plates of this kind for pressure vessels. The efforts of this work can be seen in ASTM A 841/A 841M, *Steel Plates for Pressure Vessels Produced by the Thermo-Mechanical Control Process (T-MCP)*. This document gives only broad limits for chemical composition and mechanical property requirements. Nevertheless, it represents a firm move toward agreement on requirements for a steel processed by the T-MCP method, and it provides the terminology to identify key steps in a planned system of steel alloying and processing. The welding of T-MCP steels will be covered in Volume II, Chapter 14.

COLD FINISHING

The term cold finishing can apply to a number of final operations in a steel mill such as polishing, straightening, and grinding. However, there are two methods commonly used to reduce the size or cross-sectional shape of a piece of steel while it is cold. The first method is *cold rolling*, or the rolling of cold material between heavy, hardened rolls in the same manner as hot rolling. The amount of reduction that can be made in a cold-rolling pass is much less than can be made by hot rolling. The second method of cold finishing, *cold drawing*, is the drawing of the material through an opening in a die. Cold-drawing dies consist of a block of extremely hard, wear resistant material containing a tapered hole. The smaller orifice conforms to the desired final size and shape of the product. The bar or rod to be cold drawn is coated with a lubricant; it is then pulled or pushed into the larger orifice and drawn completely through the die.

These cold finishing operations are used for a number of reasons: (1) better accuracy and uniformity of cross-sectional size and shape, (2) smoother surface finish, (3) increased strength at a small sacrifice in ductility, and (4) easier production of complex shapes.

It is customary to cold roll sheet and strip. Bars can be cold rolled or drawn, although drawing is the most widely used method. The smallest diameter rod that can be commercially hot rolled is about five mm (0.2 in.). To produce smaller material, the hot rolled rods are reduced by cold drawing; this results in obtaining *wire*.

HEAT TREATMENT

Heating furnaces are used extensively in producing steel. *Annealing* treatments are customarily applied to remove work-hardening if a soft material is desired. *Hardening* and *tempering* operations are used to improve strength and toughness properties prior to shipping the material to a fabricator. These heat treatments affect the mechanical properties of the steel by altering its microstructure. Descriptions of these treatments and their effects on microstructure and properties will be presented in Volume II, Chapter 12.

Furnaces for heat treating steel differ greatly in their temperature capability, kind of interior atmosphere, and method of operation. They range from simple open-fired furnaces heated with gas or oil to elaborate electrically heated furnaces making use of protective (nonscaling) atmospheres. Some even provide a chamber for cooling the heated steel to room temperature while protecting it with a special atmosphere. *Bright annealed steel* is a product that is heat treated in a controlled-atmosphere furnace and requires no scale removal operation. The annealed steel has a surface that is as bright as (or brighter than) the material appeared when it entered the furnace.

Heat treating furnaces are also used in steel production to change the chemical composition of solid steel, particularly its carbon or sulfur content. The nitrogen content can also be altered by similar technique. To achieve a uniform change in carbon, sulfur, or nitrogen content, the steel section must be quite thin, and all of the surface area must be exposed to the furnace atmosphere. Sheet or strip stock or fine wire is best suited for these special treatments.

Flat-rolled steel in coils must be rewound with a separating element between the wraps or layers to secure an open coil. With the surface of the steel so exposed, the carbon content may be decreased or increased to any desired level; from less than 0.005 percent to 1.0 percent. To produce a very low carbon steel, the material would be annealed at a temperature of 700°C (1300°F) in an atmosphere of hydrogen, carbon dioxide, or water vapor for about ten hours. These gases remove carbon from the steel by the following chemical reactions at the high temperature:

	CARBON		GAS		PRODUCT
(1)	C	+	$2H_2$	$\Rightarrow$	CH_4 (methane)
(2)	C	+	CO_2	$\Rightarrow$	2CO (carbon monoxide)
(3)	C	+	H_2O	$\Rightarrow$	H_2 + CO (hydrogen and carbon monoxide)

In practice, several gases are passed over the coil as it is heated. The operation may be started with a 96% nitrogen/4% hydrogen mixture; the hydrogen is then increased to

18%. Finally, steam is injected to accomplish decarburization. Nitrogen can be removed from the steel by a similar annealing treatment using an atmosphere of moist hydrogen. The moisture is believed to act as a catalyst for the following reaction:

$$\underline{\text{NITROGEN}} \quad + \quad \underline{\text{HYDROGEN}} \quad \Rightarrow \quad \underline{\text{AMMONIA}}$$
$$2N \quad + \quad 3H_2 \quad \Rightarrow \quad 2NH_3$$

The removal of nitrogen, however, proceeds at a much slower rate than does decarburization.

Furnace atmospheres bearing a carbonaceous gas are used to add carbon to the steel. Composition control through heat treatment of the solid steel is a great advantage in steelmaking because it permits steel to be melted, hot worked, and cold finished while at a favorable level of carbon and nitrogen for these production operations. Finally, the steel can be altered in carbon content to the level best suited for fabrication or for service. Also, the open-coil, controlled-atmosphere annealing treatment is capable of decarburizing the steel to a low level that is not economically obtainable through melting processes.

CONTINUOUS COATING OF STRIP STEEL IN COILS

Almost half of the steel produced in the world is used in the form of flat-rolled (sheet or strip) stock, much of it ranging in thickness from 0.5 to 2.5 mm (0.02 to 0.10 in.). Because plain steel can easily corrode, a surface coating must often be applied. Painting was the earliest protection for steel articles. However, when greater protection is required, a metallic coating of aluminum, cadmium, chromium, copper, lead, nickel, tin, or zinc is applied by hot dipping, thermal spraying, or electroplating.

Prior to 1900, it was realized that coils of steel strip could be coated continuously, sheared into sheet sections, and finally fabricated into an article with the corrosion protection in place on the surface. The first commercial product to be made by this technique was galvanized steel sheet, made by passing steel strip continuously through a bath of molten zinc. This product was followed by others that used a hot-dipping technique to apply other metallic coatings.

The continuous coating of steel strip has benefitted from greatly improved technology. Available products now include a wide array of coatings of metals, alloys, and multi-layer combinations. Coating thickness may be controlled by using air knives, and different thicknesses may be obtained on the opposite sides of a given strip. One-side-coated strip is also regularly produced.

The surface of metallic-coated strip is made with various finishes obtained by alloying the coating, by controlling its solidification, and by subsequent cold rolling. Because metallic coatings made by the hot dip method are often thicker than needed for certain applications, electroplating of steel strip is receiving increased attention. This process can deposit very thin coatings with precise control of thickness and appearance. The savings achieved by using metallic-coated steel to fabricate articles (rather than applying protection to the finished article) have proved so substantial that demand arose for *prepainted* steel strip and sheet. Consequently, a large portion of the cold rolled, flat-rolled material produced today is either metallic coated or prepainted continuously at a mill.

The presence of a metallic coating or a paint film on the surface of strip or sheet can present a number of welding problems which vary in nature depending on the kind of coating, its thickness, and the welding process being applied. Difficulties include weld cracking, blowholes in the weld metal, pollution of the work place, damage to the surface appearance, excessive deterioration of welding electrodes, and the inability to join parts using any of the welding processes.

TYPES OF STEEL

A list of irons and steels was presented at the start of this chapter. The description of the manufacturing processes used in the production of these materials makes it possible to review in more detail four broad categories of commercial steels: carbon, alloy, high alloy, and tool steels. It is entirely possible to encounter all four classes of steels in a single welded article.

For many years, steelmakers and users referred to *grades* of steel when identifying the various carbon and alloy steels. Because grades were sometimes perceived as differing quality levels, use of the term *type* to identify steels has been encouraged. Although many recently issued specifications designate steels by type numbers, some long-established specifications, standards, and codes may still be denoted by grade.

The terms *specification* and *standard* are also used interchangeably. A specification is a statement of the attributes that a material must possess to be acceptable. The required attributes, in the case of steel, are often elements in the chemical composition that must fall within specified ranges or that must meet prescribed maxima or minima. In lieu of composition requirements, mechanical properties, such as tensile strength, ductility, or hardness, may be specified, and the chemical composition of the material is determined by the supplier. In addition to composition and mechanical properties, a specification may state requirements for hardenability, weldability, grain size, cleanliness, soundness, magnetic properties, and corrosion resistance. If the specification covers shapes or forms of semifinished steel, the requirements may include dimensional tolerances and surface finish features. The more comprehensive specifications contain detailed information on the test procedure to be followed in evaluating the required attributes.

The most widely used specifications for steel are found in the *ASTM Standards*. The ASTM promotes consistent usage of terms; it regularly includes definitions for terms such as standard, specification, proposal, classification, and test method.

Carbon Steels

Many of the names used to describe various kinds of steel were adopted long ago through common usage, and their selection shows little logic based on present technical knowledge. For example, cast iron is a high-carbon-content product containing more carbon as an alloying element than cast steel. Some modern steels contain virtually no carbon (less than 0.005 percent) and are essentially iron. Despite terminology variations, the carbon steel classification is generally accepted for all commercial irons and plain steels.

Carbon primarily controls the response of steel to hardening heat treatments. However, even carbon steels frequently receive modest additions of other alloying elements like manganese, silicon, sulfur, and copper. These elements may impart properties to the steel that are equally as important as its ability to harden. Less frequently, small amounts of titanium, aluminum, or zirconium may also be present in carbon steel. Their purpose may be to deoxidize the steel or to gain other beneficial effects. Unless the addition of supplementary elements is made to a specified minimum or range, the carbon steel classification is applied.

Carbon steel can be manufactured in a number of different furnaces, and it can be furnished as rimmed, capped, semikilled, or killed steel. Both the furnace or the process used and the deoxidation practice affect the characteristics and properties of the steel. The greatest change in properties, however, is achieved by having different amounts of carbon present in the composition. Increasing the carbon content leads to marked increases in hardness and strength. Table 5.5 provides a brief classification of carbon steels according to carbon content and indicates how usage varies with the changes in composition and properties.

Free-machining steels containing additions of sulfur, or lead must be given close attention when considering welding. Machinability is an important characteristic when mass producing metal components and can be a deciding factor in selecting a steel. Although a free-machining steel can be welded, satisfactory results are ensured only when its metallurgical behavior in a given application is completely understood. Refer to Chapters 13 and 14 of Volume II for more information.

Alloy Steels

Alloy steels were a natural development from plain carbon steels. Relatively small additions of alloying elements produce remarkable improvements in properties, but no single element is appropriate in all applications. Alloying elements are added to steel

Table 5.5 - Classification and Usage of Carbon Steels

Percent Carbon Content	Common Name	Typical Hardness, Rockwell	Examples of Usage
0.005 max	Open-coil decarburized steel	35-B	One-coat enameling sheet, non-aging drawing sheet
0.03 max	Armco ingot iron	45-B	Enameling, galvanizing, and deep-drawing sheets
0.15 max	Low-carbon steel	60-B	General purpose strip and sheet for auto bodies, frames, wheels, etc., and welding electrodes
0.15/0.35	Mild steel	90-B	Structural shapes, plates bars, etc.
0.35/0.60	Medium-carbon st	25-C*	Machine parts, etc.
0.60/1.0	High-carbon steel	40-C*	Tools, dies, railroad rails, and springs

*As heat treated

to (1) improve mechanical properties (strength and toughness), (2) increase or decrease the propensity to hardening during heat treatment, (3) alter magnetic properties, and (4) retard corrosion. The more commonly used elements in alloy steels are manganese and silicon (in larger amounts than in carbon steels) and nickel, chromium, molybdenum, copper, and vanadium. The function of these and other alloying elements is described in Chapter 4, and welding behavior will be covered in Volume II, Chapter 15.

Alloy steels are defined as (1) those in which the maximum specified for alloying elements exceeds one or more of the following: manganese 1.65%, silicon 0.60%, copper 0.60% or (2) those in which a definite range or minimum quantity of the following alloying elements is required: aluminum, molybdenum, nickel, titanium, tungsten, vanadium, or zirconium. In marketing steel, two different bases have been used to set prices, a carbon steel and an alloy steel base. The standards and tolerances for alloy steel products are more demanding than those for carbon steel, and therefore the alloy steel products are more expensive. There is, however, a group of steels that contain alloying elements and yet may be sold as carbon steels. This group is designated *high strength low alloy* (HSLA) steels. Steels in this group are usually produced to specified mechanical properties, and the compositions are determined by their producers. From the standpoint of welding metallurgy, however, these are considered to be alloy steels, and their behavior during welding will be discussed in Volume II, Chapter 15.

A large number of alloy steels are on the market, and new types are being introduced regularly. Much alloy steel is used in the following four areas of application, where welding is the principal joining process.

Construction

Construction steels are applied to reduce weight in railroad cars, truck frames, ships, crane booms, and other articles, and comprise a group of alloy steels commonly referred to as *high tensile steels*. These steels have higher tensile and yield strengths than those found in mild steel or carbon structural steel. They permit the use of smaller cross sections to meet given load requirements. To minimize cost, a minimum amount of alloying is used to gain the desired strength. Heat treatment may be used in conjunction with alloying to gain high strength. The carbon content is usually held low to avoid difficulty in welding. Many high tensile steels have a proprietary chemical composition and are often introduced under trade names. After a particular alloy steel has proved useful, it is normally included in one or more published standards.

Automotive, Aircraft, and Machinery

Steels for automotive, aircraft, and machinery applications have been standardized by the AISI and the SAE. These steels are presented by composition requirements, and they feature a number of alloying elements and combinations of these elements. Although the AISI-SAE steels were originally standardized for producers of automobiles and tractors, these steels have become widely recognized and are used extensively in many kinds of machinery, vehicles, aircraft, and hardware. Detailed chemical composition requirements are given in Volume II, Chapters 14 and 15.

Low-Temperature Service

Steels for low-temperature service have been developed by alloying to provide the improved toughness required for many applications. Examples include road grading equipment operating at -30°C (-20°F) during the winter months, aircraft flying at high altitudes where the temperature is regularly -50°C (-60°F), and pressure vessels holding liquefied gases at -196°C (-320°F). Very low-temperature applications are referred to as *cryogenic service*. Ordinary carbon steels exhibit poor ductility and especially poor impact toughness at very low temperatures, and they are generally regarded as unreliable for critical applications. Nickel is often added to steel to improve toughness at low temperatures. While an addition of one percent nickel might accomplish the desired improvement in steels used at moderately low temperatures, those expected to display good toughness in cryogenic service can require additions of nickel from five to nine percent.

Elevated Temperature Service

Steels for elevated temperature service are uniquely alloyed to provide good strength at elevated temperatures. Typical applications that make extensive use of welding include steam superheaters, chemical processing retorts, and oil refining towers. Carbon steels rapidly lose their strength as the temperature increases above about 500°C (950°F). A variety of alloy steels that possess good strength at temperatures up to 600°C (1100°F) are available. Some of these steels contain only small additions of chromium or molybdenum. The more complex steels contain significant amounts of chromium, molybdenum, nickel, and vanadium. In steels intended to serve near 600°C (1100°F), the chromium addition may be as high as nine percent.

High-Alloy Steels

When the level of chromium, nickel, or manganese in steel is 10 percent or higher, the material is considered a *high-alloy steel*. This is a new term that is gradually finding wider acceptance. Smaller groups of steels under the high alloy classification have also acquired common names through commercial usage.

Austenitic Manganese Steel

Austenitic manganese steel was invented by an English metallurgist named Hadfield in 1882, when he found that a manganese addition of 10 percent or more along with a very high carbon content made an unusual steel; it offered good toughness along with an exceptional ability to harden while undergoing cold working. This steel proved to be of considerable value in services where abrasion resistance was needed or where impact loading occurred and deformation of the steel part had to be minimized. Examples include railroad track crossovers and switch points, and equipment used for digging or handling heavy rocky deposits. The term austenitic was obtained from the crystalline structure that this steel was able to maintain because of its high manganese content. Research on Hadfield's simple composition of 1.0 to 1.4 percent carbon and 10 to 14 percent manganese has shown that small additions of other alloying elements, such as nickel, molybdenum, and vanadium, can improve impact toughness considerably. The manner in which these alloy additions alter the microstructure and accomplish this toughness improvement will be discussed in Volume II, Chapter 15.

Stainless Steels

Stainless Steels are highly alloyed iron base metals with an usual ability to resist attack by corrosive media at atmospheric or elevated temperatures. Although they are particularly useful in cooking utensils and auto body trim, their unique properties serve many needs in industry, frequently in welded construction. Their unusual resistance to corrosive attack is primarily due to chromium content, which is 10 percent or higher in most types of stainless steel. Nickel is also used as an alloying element, ranging in content from two percent to 35 percent. Consequently, stainless steels are grouped into two main categories: plain chromium (or nearly so) and chromium-nickel (with the remainder essentially iron).

Approximately 40 types of wrought stainless steel have been given standard type numbers and are covered by composition specifications organized by the AISI. Although the AISI type numbers are widely used by other organizations, they sometimes have slightly different composition limits. Many stainless steels are produced which do not have an AISI standard type number. These compositions may represent slight modifications of AISI standard compositions or they may be proprietary stainless steels with substantial additions of various alloying elements to secure specific properties or behavior in fabricating. The AISI issues a type number and composition specification for a steel only when it is made by more than one producer and has achieved substantial usage.

Stainless steels in the form of castings have been given designations and composition limits by the ACI Division of SFS. The compositions differ slightly from comparable wrought types because of the control that must be exercised over pouring and flowing characteristics of the molten metal, and to avoid cracking or fissuring during solidification.

Because of the high cost of chromium and the substantial amount that must be added to iron to achieve stainless-like behavior, much research has been directed toward finding a lower cost substitute. Aluminum and silicon, when added to iron, induce some corrosion and scaling (oxidation) resistance, but to date no real alternative to the chromium has been found. However, stainless steels with just enough chromium to confer a needed degree of corrosion resistance have been developed. This minimum amount appears to be slightly above 10 percent. These types provide cost effective material for many high volume applications (e.g., the catalytic converter chambers for automobiles). Welding a steel containing 10 percent chromium or higher (i.e., a stainless steel) calls for certain welding processes and techniques that differ from those used on the carbon and low-alloy steels.

In addition to the recognized corrosion resistance that stainless steels possess, a number of other useful properties or characteristics are also offered by certain types. Stainless steels are often selected for their toughness at sub-zero temperatures, their good strength at elevated temperatures, or their ability to remain nonmagnetic under a variety of conditions. Some alloys are hardened by simple, low-temperature precipitation heat treatments and thus avoid many problems associated with the hardening of carbon and alloy steels by quenching. The reasons for selecting a stainless steel must justify its higher cost, which may be approximately ten times that of carbon steel. However, when all aspects of fabricating, treating, and service performance are considered, it is common to find stainless steel components properly incorporated in many kinds of welded construction.

Heat-Resisting Steels

Heat-resisting steels consist of a number of groups of high-alloy steels developed for high temperature usage. At least four different classes, which offer various combinations of oxidation resistance and high-temperature strength, are available. Coal gasification plants and petrochemical reactors are two examples of applications making increasing use of heat-resisting steel. The gas turbine has been particularly demanding in its need for stronger metals at its operating temperatures. Turbines are now working in many different forms: the turbo-supercharger, the aircraft jet propulsion unit, the turbo-propeller unit, and as a turbine-generator for driving ships, locomotives, trucks, and electrical power units. The heart of each turbine, regardless of design, is the bladed rotor, which is driven by the hot gases. The working efficiency of the gas turbine increases with operating temperature. To achieve acceptable efficiency, the turbine must operate with a gas temperature of at least 550°C (1000°F). Present gas turbines regularly operate in the range of 700 to 900°C (1300 to 1650°F). Even higher operating temperatures are being explored.

Another important application for high strength, heat-resisting steels is the main steam piping and superheater tubing in modern high pressure, high-temperature steam electrical power generating plants. As the planned operating temperature of these steam power plants exceeds about 600°C (1100°F), it is necessary in most cases to depart from use of the alloy steels to the austenitic Cr-Ni high-alloy steels.

The simpler high strength, heat-resisting steels are almost identical to the standard types of stainless steels, or they are distinguished by small modifications in chemical composition. However, some of the newer materials are so highly alloyed with cobalt, molybdenum, tungsten, niobium, and titanium that they cannot be hot forged or rolled, and therefore must be poured as castings. In some cases, the iron content of the earlier materials has been replaced with nickel or cobalt. Because the materials cannot be called steels, they are termed *super alloys*. Nevertheless, the designer often considers welding to be the best means of joining these components. Although it is necessary to determine how to weld these complex alloys, this subject is not within the scope of this text.

Tool Steels

Any steel that is used in a cutting or forming operation might be considered a *tool steel*. As a result, a family of steels is manufactured specifically for making tools. These steels range from plain carbon to highly alloyed steels, and they are produced with special mill practices to secure the high quality required. Tool steels must be sound and must have excellent cleanliness to avoid imperfections on highly polished die surfaces.

The term *tools* covers cutting tools, punches, shear blades, and dies for forming, drawing, extruding, forging, rolling, and molding. These tools may be used on either hot or cold material; the material may be soft or hard and may be a metal or nonmetal. Virtually all tool steels are sold in the annealed condition to facilitate shaping into the required form and are then hardened by heat treatment. There is also a group of soft, low-carbon tool steels that can be cold formed into tools and dies, then carburized to produce a hard, wear- resistant surface.

The AISI and SAE have jointly devised a system of classifying tool steels into categories based upon their alloy composition, method of hardening, and principal application. More than a dozen principal groups ranging from carbon steels to complex alloy steels have been categorized. The carbon steels contain about 0.60 to 1.5 percent carbon for securing hardness in the heat treated condition. The alloy steels, in addition to containing a substantial carbon content, include moderate amounts of alloying elements such as chromium, molybdenum, and tungsten. Some of the terms for these AISI-SAE types of tool steels are molybdenum high speed, tungsten high speed, chromium hot work, tungsten hot work, air-hardening medium alloy cold work, oil-hardening cold work, shock-resisting, low-carbon mold, and water-hardening tool steels.

Tool steels are sometimes welded for assembly or repair. In general, the utmost care is required in selecting a welding process and in planning the welding procedure to avoid difficulty. Nevertheless, the welding of certain tool steels has been so successful that the application of these steels has been extended to some important high-strength welded structures (e.g., portions of military aircraft frames). As will be explained in Chapter 15 of Volume II, a knowledge of the metallurgy of these steels is essential before beginning a welding operation.

STANDARDS AND SPECIFICATIONS FOR STEELS

The various kinds of steel are most often identified by a type designation or a specification number. However, specifications are issued by many organizations or agencies, including technical groups, consumers, governmental agencies, and standardization institutes. There are more than 150 major standards organizations in the U.S. alone, and these have issued and maintain more than 25 000 standards. Only a portion of these are standards for steels, but it is increasingly important to use a recognized designation and to specify complete, correct chemical composition limits.

Unified Numbering System

The SAE and ASTM realized the growing magnitude of the identification problem, and in 1967 they jointly proposed development of a unified system for designating metals and alloys. With the help of representatives from industry and technical organizations, and funding from AMMRC, the first edition of an SAE-ASTM handbook was produced in 1975. It contained an index of chemically-similar specifications for commercial metals and alloys along with a description of the chemical composition of each. The handbook, entitled *Metals and Alloys in the Unified Numbering System*, followed a practice for numbering metals and alloys originally set forth in a 1974 document issued as SAE J 1086 and ASTM E 527. The sixth edition of the handbook, issued in 1993, is identified by SAE as HS-1086 FEB93 and by ASTM as DS-56E.

The unified numbering system uses alphanumeric designations consisting of a single-letter prefix followed by five digits. Seventeen series of metals and alloys are established in the handbook; each bears a letter prefix which, in most cases, identifies the family of materials. For example, A is for aluminum and its alloys, S identifies stainless steels, and Z is for zinc alloys. The five digits, whenever feasible, incorporate identification numbers from the material's originating or existing system. For example, the widely recognized carbon steel presently identified as AISI 1020 has been assigned the designation UNS G10200. However, each series is independent with regard to selection of digits, and a UNS number may bear no relationship to the original identification of a material.

The UNS handbook contains more than four thousand unified designations for commercial metals and alloys, including more than five hundred welding and brazing filler metals. Table 5.6 lists the UNS system organized in the sixth edition. The handbook also includes a cross reference of commonly known documents and trade designations which parallel those covered by the UNS numbers. The current sixth edition of the handbook is also available as computer software developed by SAE and released as UNSearch Version 3.0. Its format allows rapid electronic searching by all criteria of the entire metal and alloy data bank, and it permits in-house information gathered by the software user to be added to the file.

Table 5.6 - SAE-ASTM Unified Numbering System

Metals and alloys listed in the UNS system of SAE-ASTM assigned to date and included in the Sixth Edition, 1993, of SAE HS-1086, and of ASTM DS-56

UNS Number*	Metals and/or Alloys in each Series
AXXXXX	Aluminum and Aluminum Alloys
CXXXXX	Copper and Copper Alloys
EXXXXX	Rare Earth and Similar Metals and Alloys
FXXXXX	Cast Irons
GXXXXX	AISI and SAE Carbon and Alloy Steels
HXXXXX	AISI and SAE H-Steels
JXXXXX	Cast Steels (except Tool Steels)
KXXXXX	Miscellaneous Steels and Ferrous Alloys
LXXXXX	Low Melting Metals and Alloys
MXXXXX	Miscellaneous Nonferrous Metals and Alloys
NXXXXX	Nickel and Nickel Alloys
PXXXXX	Precious Metals and Alloys
RXXXXX	Reactive and Refractory Metals and Alloys
SXXXXX	Heat and Corrosion Resistant Steels (including Stainless, Valve Steels, and Iron-Base "Superalloys")
TXXXXX	Tool Steels, Wrought and Cast
WXXXXX	Welding Filler Metals
ZXXXXX	Zinc and Zinc Alloys

*The five "X" marks represent UNS Numbers and are replaced by five digits when they serve to identify a specific material in the series.

Welding filler metals in the W series are those whose compositions are determined by their weld deposit analysis (e.g., covered, flux-cored, and other composite electrodes). Those welding filler metals whose compositions are determined by their analysis (e.g., solid bare wire or rods) are located in the appropriate series according to their base composition. For example, a solid bronze brazing alloy would be found in the Cxxxxx series, *Copper* and *Copper Alloys*. Weld filler metals promulgated by the AWS are included in the UNS handbook, and each listing is cross referenced wherever it corresponds to a specification of ASME, AMS, or other organization.

The Unified Numbering System has been influential in promoting cooperation between bodies that assign designations to metals and alloys and in correlating any identification that has been given to commercial materials. However, a UNS designation is not a specification; it is only a unified identifier of a metal or alloy. The chemical compositions given in the UNS handbook are for identification purposes only, and are not to be used in lieu of the originator's composition limits.

AISI-SAE System of Standard Carbon and Alloy Steels

The most widely used specifications for designating carbon and alloy steels, and for providing complete composition requirements, are the standards of the AISI and SAE. The designations developed under the AISI-SAE system have been incorporated in the documents of many other organizations. First issued in 1912, these widely recognized steels have increased in number through the years, and their composition requirements have been kept in excellent order by periodic revision under the joint efforts of AISI and SAE. These two organizations each publish a form of the AISI-SAE steels system, because they have different ways of determining the eligibility of a steel for listing. Otherwise the designations and composition requirements are essentially identical to one another.

The AISI and SAE provide steel designations and composition limits; this is only a portion of the information needed to define a steel product. For this reason, when steel products are purchased, a more comprehensive specification is often imposed.

Table 5.7 outlines the AISI-SAE system for categorizing and designating carbon and alloy steels. A four-digit number (five or six in a few cases) is assigned to each steel. The first two digits indicate the general nature of the alloy combination or composition. The last two digits indicate the approximate mean of the range of carbon content. A spread of carbon ranges is provided for most alloy combinations. Care must be taken when searching the AISI and SAE systems to ascertain the composition limits for a particular steel, because the proper limits may vary with the product form. The ranges or limits cited are generally those set for bars, billets, blooms, and slabs.

Plates are covered by a separate set of composition limits; the ranges for some of the elements are slightly broader. This adjustment has been necessary because the large rectangular ingots from which plates are produced are more likely to have a greater degree of segregation of certain elements by normal solidification mechanics than smaller ingots. Steels that exhibit specific hardenability when heat treated (identified by an H suffix) also have a separate set of composition requirements. Here the permissible ranges for certain elements are slightly greater to permit adjusting the steel's composition to ensure the required response in hardenability testing.

Steels in the AISI-SAE system may be withdrawn from the listing of either organization because of insufficient usage or other ineligibility. Deletions occur more often in the SAE listing than in that of the AISI. Because the modification or repair of existing structures often involves welding steels produced years ago, the outline in Table 5.7 includes all known categories of AISI-SAE carbon and alloy steels, even if not on current lists. Records on existing structures may disclose the specifications under which steel was purchased; this is the first clue in ascertaining the composition of steel to be welded. To further interpret AISI-SAE designations, particularly when they are gleaned from older records, Table 5.8 lists prefix and suffix letters used in the past. Only a few of these letters continue to be used today. A listing of AISI-SAE carbon steels and their composition limits will be included in Volume II, Chapter 14, while a similar table for alloy steels will appear in Chapter 15.

Because a large portion of steel tonnage is sold in the form of sheet, and because the chemical composition and steelmaking practice for this product form must be specially tailored to secure required fabricating characteristics, the AISI-SAE systems

Table 5.7 - AISI-SAE System for Designation of Carbon and Alloy Steels*

Description	AISI-SAE Designation	UNS Identifier Number**
Low-Carbon Steels for Wire and Rods	100X	G100XO
Carbon Steels	10XX	G10XXX
Carbon Steels, Resulfurized (Free Machining)	11XX	G11XXX
Carbon Steels, Resulfurized & Rephosphorized	12XX	G121XX
Manganese Alloy Steels with Mn 1.60 to 1.90%	13XX	G13XXX
Manganese Steels with Mn Maximum over 1.0%	15XX	G15XXX
Nickel Alloy Steels	2XXX	
Nickel-Chromium Alloy Steels	31XX	G31XXX
High Nickel-Chromium Alloy Steels	33XX	G33XXX
Carbon-Molybdenum Alloy Steels	40XX	G40XXX
Chromium-Molybdenum Alloy Steels	41XX	G41XXX
Chromium-Nickel-Molybdenum Alloy Steels	43XX	G43XXX
Nickel-Molybdenum Alloy Steels	46XX	G46XXX
Nickel-Chromium-Molybdenum Alloy Steels	47XX	G47XXX
High Nickel-Molybdenum Alloy Steels	48XX	G48XXX
Low-Chromium Alloy Steels	50XX	G50XXX
Chromium Alloy Steels	51XX	G51XXX
High-Carbon Chromium Alloy Steels	51X00	G51986
High-Carbon Chromium Alloy Steels	52100	G52986
Chromium-Vanadium Alloy Steels	61XX	G61XXX
Chromium-Vanadium-Aluminum Alloy Steels	E71400	G71406
Nickel-Chromium-Molybdenum-Boron Alloy Steel	81B45	G81451
Low Nickel-Chromium-Molybdenum Alloy Steels	86XX	G86XXX
Low Nickel-Chromium-Molybdenum Alloy Steels	87XX	G87XXX
Nickel-Chromium-Molybdenum Alloy Steels	8822	G88220
Silicon-Manganese Spring Steels	92XX	G92XXX
Silicon-Manganese-Chromium Spring Steels	92XX	G92XXX
Nickel-Chromium-Molybdenum Alloy Steels	93XX	G93XXX
Nickel-Chromium-Molybdenum Alloy Steels	98XX	G98XXX
Boron Containing Steels	XXBXX	
Boron-Vanadium Containing Steels	XXBVXX	
Lead Containing Steels (Free Machining)	XXLXX	

*Categories of Composition, Ranges and Limits for Elements
**All "X" marks shown in the UNS Identifier Number following the alphabetical identification of each series are replaced by a digit when a specific steel in the series is singled out.

include a number of standard carbon and alloy sheet steels in separate tables. In fact, separate specifications have been issued. SAE first issued J410.c, then in 1982 superseded it with J 1392, *Steel: High Strength Hot Rolled Sheet and Strip, Cold Rolled Sheet, and Coated Sheet*. More than a decade ago, AISI proposed a generic system for grouping and designating sheet steels, which is explained in Technical Brief 30.

ASTM Standards

Comprehensive specifications for steels, covering many features and qualities in addition to chemical composition, are issued by the ASTM. This organization is the world's largest source of voluntary consensus standards. The ASTM has an extensive staff that annually publishes a book of standards; it encompasses a broad array of materials (including nonmetals and metals), test methods, analytical procedures, and recommended practices. The current edition is a 69-volume set, which is organized into 16 sections, one of which is a separate index. Table 5.10 lists only a few of the sections and volumes that cover areas of interest. ASTM standards are used so widely that they are certain to be encountered in welding operations.

An ASTM standard represents a common viewpoint of those parties concerned with it provisions, and its use is purely voluntary. Each standard is given a fixed alphanumeric identification consisting of a prefix letter (indicating the general kind of material) followed by an arbitrary identifier number. This specification designation is followed by a dash and two digits indicating the year of adoption or last revision. For example, A 285-90 is the *ASTM Standard Specification for Pressure Vessel Plates, Carbon Steel, Low and Intermediate Tensile Strength*. Steel supplied to this specifica-

Table 5.8 - AISI-SAE System for Characterization of Carbon and Alloy Steels

Prefix Letter*	Meaning
A	Basic Open-Hearth Alloy Steel
B	Acid Bessemer Carbon Steel
C	Basic Open-Hearth Carbon Steel
D	Acid Open-Hearth Carbon Steel
E	Basic Electric Furnace Steel
M	Merchant Quality Steel
TS	Tentative Standard Steel
Q	Forging Quality, or Special Requirements
R	Re-rolling Quality Billets

Suffix Letter*	Meaning
A	Restricted Chemical Composition
B	Bearing Quality Steel
C	Guaranteed Segregation Limits
D	Specified Discard
E	Macro-Etch Tests
F	Rifle Barrel Quality
G	Limited Austenitic Grain Size
H	Guaranteed Hardenability
I	Non-Metallic Inclusions Requirements
J	Fracture Test
T	Extensometer Test
V	Aircraft Quality or Magnetic Particle Test

*Prefix and Suffix Letters for Indicating Manufacturing Practice or Qualities Required

tion may be made by killed, semikilled, capped, or rimmed steel practices. The plates are used for making fusion-welded pressure vessels.

Like many ASTM standards, A 285 covers more than one grade, and the grades are differentiated by tensile strength and chemical requirements. In fact, the three grades available under A 285 have been given individual UNS identifier numbers: Grade A (UNS K01700), Grade B (UNS K02200), and Grade C (UNS K02801). Only a small number of the ASTM standards for carbon and alloy steels make use of AISI-SAE designations and composition limits. The three grades of A 285 must conform to chemical requirements prescribed in the ASTM document. Although the limits are less demanding than those for a comparable AISI-SAE type, ASTM A 285 has proved to be a satisfactory specification.

An important feature of ASTM standards is the availability of generic specifications that provide requirements for many items commonly required when procuring materials. These include ordering information, testing and retesting methods, dimensional tolerances, quality requirements, repair of defects in the semifinished product, marking guidelines, and loading requirements. For example, ASTM Specification A 20-92a covers *General Requirements for Steel Plates for Pressure Vessels*, and the provisions of A 20 are included in A 285 as additional requirements. However, if a requirement of A 285 is in conflict with the provisions of A 20, then the requirements of A 285 prevail. Therefore, as a general rule, the principal or parent specification requirements apply over those of any conflicting generic specification that may be involved.

Table 5.9 - AISI[1] and SAE[2] Systems for Classification of High-Strength Carbon and Alloy Steel Sheet

(A comparison of the two systems of classification)

Yield Strength in ksi[3]		Chemical Composition[4]		Carbon Level[5]		Deoxidation Practice[6]	
AISI	SAE	AISI	SAE	AISI	SAE	AISI	SAE
035	035		A	H		O	O
040	040		B	L		K	K
045	045	S	C			F	F
050	050		S				
055		W	W				
060	060						
065			X				
070	070	X	Y				
080	080		Z				
100		D					
120							
140							
160							
190							

Notes:
1. AISI AU-603D, High-Strength Sheet Steel Source Guide (0290-4M-RI), February 1990.
2. SAE J1392, Steel, High-Strength, Hot Rolled Sheet and Strip, Cold Rolled Sheet, and Coated Sheet, May 1982.
3. When yield strength is indicated in S.I. units, four digits are necessary.
4. A, B, C, and S of SAE system represent various combinations of C, Mn, and N and P contents. These are grouped under "S" in AISI System. Letter designations X, Y, and Z in SAE system represent microalloyed steels with 10 ksi, 15 ksi, and 20 ksi respective spreads between yield and tensile strength. These are grouped under "X" in AISI system. The weathering types (W) are equivalent in the two systems. There is no SAE equivalent for the AISI dual-phase (D) chemical composition.
5. Value of carbon levels for H and L designations vary with the grade of steel in SAE.
6. Deoxidation practices are indicated by the same letters in the AISI and SAE systems; namely, O = nonkilled, K = killed, and F = killed plus inclusion control.

ASTM A 20 is a useful standard for augmenting a material specification because it includes a number of standardized supplementary requirements. These requirements are identified by the prefix S. They cover a variety of topics including requirements for testing tensile properties, evaluating impact and fracture toughness, and examining for internal soundness. Other requirements associated with steelmaking include vacuum treatment (S 1), product check analysis (S 2), vacuum carbon-deoxidation (S 17), limits on unspecified elements (S 18), and restricted chemical requirements (S 19).

ASTM A 20 indicates that steels covered by this specification are intended to be suitable for fusion welding; however, no welding tests or weldability requirements are

Table 5.10 - ASTM Annual Book of Standards - 1993 Edition

Section*	Volume*	Title	Number of Standards Contained	ISBN No.
1		**Iron and Steel Products**		
	01.01	Steel–Piping, Tubing, Fittings	131	0-8031-1901-1
	01.02	Ferrous Castings; Ferroalloys	116	0-8031-1902-X
	01.03	Steel–Plate, Sheet, Strip, Wire	88	0-8031-1903-8
	01.04	Steel–Structural, Reinforcing, Pressure Vessel, Railway	139	0-8031-1904-6
	01.05	Steel–Bars, Forgings, Bearings, Chain, Springs	101	0-8031-1905-4
	01.06	Coated Steel Products	120	0-8031-1906-2
	01.07	Shipbuilding	134	0-8031-1907-0
3		**Metals Test Methods and Analytical Procedures**		
	03.01	Metals–Mechanical Testing; Elevated and Low-Temperature Tests; Metallography	117	0-8031-1915-1
	03.02	Wear and Erosion; Metal Corrosion	79	0-8031-1916-X
	03.03	Nondestructive Testing	123	0-8031-1917-8
	03.04	Magnetic Properties; Metallic Materials for Thermostats, Electrical Resistance, Heating, Contacts	112	0-8031-1918-6
	03.05	Analytical Chemistry for Metals, Ores, and Related Materials (I): A 751 - E 354	73	0-8031-1919-4
	03.06	Analytical Chemistry for Metals, Ores, and Related Materials (II): E 356; Surface Analysis	123	0-8031-1920-8
12		**Nuclear, Solar, and Geothermal Energy**		
	12.01	Nuclear Energy (I)	110	0-8031-1965-8
	12.02	Nuclear (II), Solar, and Geothermal Energy	128	0-8031-1966-6
13		**Medical Devices and Services**		
	13.01	Medical Devices; Emergency Medical Svs.	197	0-8031-1967-4
14		**General Methods and Instrumentation**		
	14.01	Analytical Methods–Spectroscopy; Chromatography; Computerized Systems	73	0-8031-1969-0
	14.02	General Test Methods, Nonmetal; Laboratory Apparatus; Statistical Methods; Forensic Sciences	203	0-8031-1970-4
	14.03	Temperature Measurement	37	0-8031-1971-2
15		**General Products, Chemical Specialties, and End Use Products**		
	15.01	Refractories; Carbon and Graphite Products; Activated Carbon	132	0-8031-1973-9
		Index		
	00.01	Index, Subject Index; Alphanumeric List	1060 pp	0-8031-1982-8
		Complete Set of 69 ASTM Volumes	1500	0-8031-1983-6

*Partial list of Sections and Volumes selected for their pertinency to the study of carbon and alloy steels

TECHNICAL BRIEF 30:
The AISI System for Generic Designation of Sheet Steels

More than a decade ago the American Iron and Steel Institute (AISI) proposed a generic system for grouping and designating sheet steels; it is outlined in Table 5.9. Complete details can be obtained from AISI AU-603D, *High Strength Sheet Steel Source Guide (0290-4M-RI)*. In this document, the AISI does not provide chemical composition specifications but only proposes designators that allow steels to be grouped. It facilitates cross referencing to other specifications, such as those of the ASTM and SAE.

AISI AU-603D develops alphanumeric designators for each steel by examining three basic components in the following order:
  (1) the minimum yield strength in ksi using three digits (when SI units are used, as many as four digits may be necessary),
  (2) a chemical composition classification using one letter, and
  (3) a classification for the deoxidation practice followed in making the steel.

Therefore, a designator has a total of five characters, unless the material is a dual-phase steel. In this case, a six-character designator is used in the following order:
  (1) a chemical composition classification using one letter,
  (2) a classification for the deoxidation practice using one letter,
  (3) the minimum tensile strength in ksi using three digits (when SI units are used, four digits are necessary), and
  (4) the suffix T, signifying that tensile strength is indicated.

The AISI defines high-strength steel sheet as having a minimum yield strength of 240 MPa (35 ksi). The information provided by an AISI or SAE designator for steel sheet can be quite important from a welding standpoint. In the AISI system covering all except the dual-phase steels, the fourth character in the designation (the first letter) has the following meaning:

CATEGORIES OF SHEET STEEL
S = structural quality steel
X = low alloy steel
W = weathering steel
D = dual-phase steel

Structural quality steel contains C, Mn, P, and N in various combinations. Low-alloy steels have compositions that primarily contain Nb, Ti, and V, either singly or in combinations. To achieve specific properties, other elemental additions such as Si and Zr are possible. Weathering steel compositions contain various combinations of Si, P, Cu, Ni, and Cr. The W-steels have an atmospheric corrosion resistance that is greater than that of conventional low-carbon steel. The dual-phase steels are high strength, low-alloy types that have a microstructure consisting of martensite or other transformation product dispersed in a ferrite matrix. These microstructural aspects are explained in subsequent chapters, and the welding of dual-phase steels will be discussed in Volume II, Chapter 15.

Because fusion welding of steel usually requires that the weld metal be deoxidized (to avoid porosity), it is important to know the steelmaking practice followed in producing the sheet steel. The significance of the fifth character (second letter) in the AISI designator is as follows:

DEOXIDATION PRACTICE
O = nonkilled
K = killed
F = killed plus inclusion control

The *AISI High Strength Sheet Steel Source Guide* helps identify various kinds of high-strength sheet steel using the system described above. It also cites trade names and indicates sources of high-strength sheet steel in North America.

set forth. Weldability concerns are addressed in some ASTM standards. For example, ASTM A 735, *Standard Specification for Pressure Vessel Plates, Low Carbon Manganese-Molybdenum-Columbium Alloy Steel, for Moderate and Lower Temperature Service*, indicates "S73 - Weldability, S73.1, Weldability tests shall be conducted." However, the weldability requirements must be established by agreement between steel supplier and purchaser.

Another generic ASTM standard, A 6-92a, provides *General Requirements for Rolled Steel Plates, Shapes, Sheet Piling, and Bars for Structural Use*. This document is referenced in a number of ASTM specifications covering various steel products used in structural applications. The general requirements of ASTM A 6 for structural applications are less demanding than those of A 20 for pressure vessel usage in areas such as check analyses for chemical composition and test methods for tension properties. ASTM A 6 makes no mention of the suitability of the steel for fusion welding (as does ASTM A 20), even though explicit requirements are contained in A 6 concerning arc welding repair of imperfections. However, ASTM A 6 includes some of the supplementary requirements mentioned earlier for A 20 at the option of the steel purchaser.

Each ASTM standard must be reviewed every five years but is subject to revision at any time. A two-digit suffix number immediately following the designation indicates the year of adoption. If the standard has been reapproved without change, the year of reapproval is placed within parentheses. A lower case letter following the year of adoption or reapproval indicates more than one issue during that year. A superscript epsilon (ε) indicates an editorial change since the last revision or reapproval. When using ASTM standards, it is important to refer to the current issue. However, if an earlier version played a key role in a situation, that issue must also be reviewed.

The ASTM has an ongoing effort to support those who rely on the metric system of units. A standing committee (E-43) issued its first *Metric Practice Guide* in 1964, and this eventually became ASTM E 380, *Standard for Metric Practice*, in 1968. The current ASTM document is E 380-92, *Practice for Use of the International System of Units* (SI) The *Moderized Metric System*. Initially, ASTM standards were supplemented with equivalent metric units wherever the English units appeared. As particular standards were reviewed in detail, they were issued with a new kind of identification. The original designation was retained, but following a slash mark, the designation was repeated and the suffix M was added. Examples include ASTM A 6 / A 6M and ASTM A 285 / A 285M. Where a separate completely metric standard is provided, the original designation is retained and the suffix M is added (e.g., ASTM A 568M).

API Specifications

Because the petroleum industry extensively uses welded steel tanks, vessels, and tubular goods, the API has prepared and issued material specifications and standards for welding. Table 5.11 contains some API specifications covering various product forms and kinds of steels. Specification 5L is particularly useful because of the large tonnage of line pipe used in petroleum transport. The importance of welding when laying cross-country pipeline and when fabricating related equipment has prompted the widely recognized API Standard 1104, which will be reviewed in Volume II, Chapter 14.

Table 5.11 - API Specifications Governing Materials*

API Numerical Designation	API Document Stock No.	Title	Coverage
Spec 2B	811-00300	Specification for Fabricated Structural Steel Pipe, Fourth Edition; July 1990	Structural steel pipe fabricated from plate for use in the construction of welded offshore fixed platforms
Spec 2H	811-00540	Specification for Carbon Manganese Steel Plate for Offshore Platform Tubular Joints, Sixth Edition; July 1, 1990	Intermediate strength steel plates up to 75 mm (3 in.) thick for use in welded tubular construction of offshore platforms, in selected critical portions which must resist impact loading, and lamellar tearing
Spec 5L	811-02000	Specification for Line Pipe, 40th Edition; November 1, 1992	Seamless and welded line pipe. Processes of manufacture, chemical and physical requirements, and methods test included
Spec 5L3	811-02800	Recommended Practice for Conducting Drop-Weight Tear Tests on Line Pipe, Second Edition; March 1, 1978	Recommended method for conducting drop-weight tear tests on line (20 in.) OD and larger with wall thickness $\leq$ 19 mm (0.750 in.)
Spec 12D	811-06200	Specification for Field Welded Tanks for Storage of Production Liquids, 10th Edition; January 1, 1982; Supplement 2, May 1985	Covers material, design, fabrication, and erection requirements for vertical, cylindrical, above-ground, welded steel tanks for production service
Spec 12F	811-06400	Specification for Shop Welded Tanks for Storage of Production Liquids, 10th Edition; June 1, 1988; Errata, July 1, 1990	Covers material, design, construction requirements for vertical, cylindrical, above-ground, shop-welded steel tanks for production service
API 510	822-51000	Pressure Vessel Inspection Code – Maintenance, Inspection, Rating, Repair, and Alteration, 7th Edition; March 1992	Covers maintenance, inspection, repair, alteration, and rerating procedures for pressure vessels used by petroleum, chemical, and related process industries
Std 620	822-62000	Design and Construction of Large, Welded, Low-Pressure Storage Tanks, 8th Edition; June 1990	Covers design and construction of large, welded, field-assembled storage tanks operated at gas pressure of 103 kPa (15 psig) and less, down to an internal gas pressure close to atmospheric pressure, with metal temperatures not greater than 121°C (250°F)
Std 650	822-65000	Welded Steel Tanks for Oil Storage, 8th Edition, November 1988; Addendum 1, May 1992	Covers material, design fabrication, erection, and testing requirements of welded steel storage tanks with internal pressures approximating atmospheric pressure, including various sizes and designs
Publ 920	822-92000	Prevention of Brittle Fracture of Pressure Vessels, 1st Edition; March 1990	Provides guidance on the selection of steels for new pressure vessels and in the inspection and operation of existing pressure vessels to minimize the probability of brittle fracture caused by low toughness at temperatures below 49°C (120°F)
Publ 941	822-94100	Steels for Hydrogen Service at Temperatures and Pressures in Petroleum Refineries and Petrochemical Plants, 4th Edition, April 1990	Provides suggested operating limits for steels used in equipment at petroleum refineries and petrochemical plants in which hydrogen or hydrogen-containing fluids are processed at elevated temperatures and pressures

*List of selected specifications issued by the American Petroleum Institute covering steel tanks, vessels, and tubular goods

Specification 5L covers both seamless and welded steel pipe in 11 grades, with minimum yield strengths ranging from 172 MPa (25 ksi) to 551 MPa (80 ksi). The steel compositions needed to achieve the required yield strengths represent both carbon steels and the low-alloy steels. The current Spec. 5L, the fortieth edition (1992), is highly detailed in its requirements for pipe strength, size tolerances, weld quality, and methods of test and inspection. Chemical composition requirements allow the steelmaker freedom in determining the optimum arrangement of elements added for alloying and those added for deoxidation purposes.

Aerospace Material Specifications

SAE also publishes the Aerospace Material Specifications (AMS). These are complete specifications designed for procurement purposes. There are more than two thousand specifications in force, and they cover many products, materials, processes, and practices involved in aerospace construction and maintenance, including accessories, chemicals, fabricated parts, nonmetals, and metals. AMS numbers usually consist of four digits, and an attempt has been made to reserve certain number series for particular kinds of materials. As examples, copper alloys are found in AMS 4500 to 4740, while miscellaneous nonferrous alloys are numbered AMS 4750 to 4893. Whenever a revised AMS is issued, a suffix letter is added to the four-digit number starting with A and proceeding alphabetically with each subsequent revision.

ASME Material Specifications

The ASME maintains a *Boiler and Pressure Vessel Code (B&PV Code)*, and a *Code for Pressure Piping*, both of which contain material standards and specifications that cover carbon and alloy steels. These codes are recognized by ANSI, and are highly regarded for their conservatism and solid technical basis. The ASME codes govern design, construction, maintenance, inspection, and care of power boilers, heating boilers, nuclear power plant components, pressure piping systems, and pressure vessels operating at 103 kPa (15 psi) and higher. The codes are prepared and updated by volunteer committees, and they have achieved wide acceptance and usage.

The ASME *Boiler and Pressure Vessel Code* has been frequently referenced in the safety regulations of most states and major U.S. cities. It is also included by various federal agencies as part of their regulations, and is often used by authorities of other countries. The ASME code is unique in that it requires third party inspection independent of the fabricator and user; it is conducted by an inspector commissioned by the National Board of Boiler and Pressure Vessel Inspectors (NBBPVI).

The ASME code is published in 11 sections as outlined in Table 5.12. A new edition of the code is issued by the ASME every three years. Users may submit inquiries to the ASME at any time for clarification of requirements or to pose other questions. Inquiries dealing with significant topics are given a case number, and both the question and ASME's response are regularly published in *ASME Mechanical Engineering* and in separate booklet form as a supplement to the code. Subsequently, these case interpretations are carefully reviewed to determine revisions to the next editions of the code. As a result, it is important to work with the current issue of the code and to review any case interpretations that may have been published.

Section II contains selected standards and specifications for materials and is divided into three parts. Part A contains ferrous material specifications, Part B covers non-ferrous materials, and Part C covers welding rods, electrodes and filler metals. The code-adopted material specifications of the ASME are highly regarded because they are, for the most part, selected from the ASTM standards; the ASME version in Section II retains the ASTM designation but is given the prefix S. The date of adoption or last revision of the standard by ASTM is not included in the ASME designation, but it is understood that the latest edition is in force. Therefore, Standard

Table 5.12 - ASME Boiler and Pressure Vessel Code

Organization of Sections and Their Contents

Section No.	Title	Coverage
I	Power Boilers	Construction of power, electric and miniature boilers, and high-temperature boilers used in stationary service. Also, power boilers used in locomotive, portable and traction service.
II	Material Specifications	Code-adopted standards and specifications for ferrous and non-materials, welding rods, electrodes, and filler metals
III	Nuclear Power Plant Components	Seven subsections covering (1) general requirements, (2) Class 1 components, (3) Class 2 components, (4) Class 3 components, (5) Class MC components, (6) component supports, and (7) core support structures.
IV	Heating Boilers	A construction code covering design, fabrication, installation and inspection of steam heating and hot water supply boilers directly fired by oil, gas, electricity or coal.
V	Nondestructive Examination	NDT methods accepted for use under the Code.
VI	Recommended Rules for Care and Operation of Heating Boilers	Guide to owners of steel and cast iron heating boilers regarding maintenance and repair.
VII	Recommended Rules for Care of Power Boilers	A guide similar to that in Section VI covering stationary, portable, and traction-type power boilers.
VIII	Pressure Vessels, Division 1	Basic rules for construction, design, fabrication, inspection and certification of pressure vessels. Rules formulated on basis of design principles and construction practices applicable to vessels for pressures up to 20.7 MPa (3000 psi).
VIII	Pressure Vessels, Division 2	Division 2 provides an alternative to the minimum construction requirements of Division 1. Division 2 rules are more restrictive in the choice of materials, but they permit higher design stress intensity values in the range of temperatures over which the design stress intensity value is controlled by the ultimate or yield strength. Division 2 rules cover vessels installed at stationary locations.
IX	Welding and Brazing Qualifications	Relates to the qualification of welders and welding operators and the procedures to be followed to comply with the Code.
X	Fiberglass-Reinforced Plastic Pressure Vessels	A recent construction Code established general specifications for the glass and resin used in fabrication, and qualification procedures. Limits are given for permissible service conditions.
XI	Rules for Inservice Inspection of Nuclear Power Plant Components	Requirements for maintaining a nuclear power plant in a safe and expeditious manner, and for returning a plant to service following an outage.

A 285-85, mentioned earlier, has been adopted by the ASME and is identified in Section II as simply SA 285.

Material specifications of the *Boiler and Pressure Vessel Code* are not confined to those of the ASTM. Whenever ASTM has not issued a corresponding standard for a material that has come into pressure vessel or boiler usage, a suitable document may be adopted from another organization. For example, in the case of filler metals for welding, selection is made from the standards of the AWS, and the AWS designation is usually given the prefix SF in identifying the ASME specification in Section II.

The inclusion of a particular type of steel in Section II of the ASME code does not automatically sanction its use in a boiler or pressure vessel, nor does the absence of a steel prohibit its use. Each lot of steel used in code-governed construction must be tested prior to use, but if the steel type is included in an ASME material specification, the required testing may be greatly minimized.

Section IX of the code provides a general outline of the testing procedure used to qualify a new steel. However, considerable testing as deemed necessary by a guiding committee of ASME may be required before the steel is approved for use in a boiler or vessel.

Table 5.13 - ASME B31; Code for Pressure Piping

Organization of Sections and Their Contents

Section	Title	Coverage
B 31.1	Power Piping	Power and auxiliary service systems for electric generation stations; industrial and institutional plants; central and district heating plants; and district heating systems.
B 31.2	Fuel Gas Piping	Systems for fuel gases such as natural gas, manufactured gas, liquefied petroleum gas (LPG) – air mixtures above the upper combustible limits.
B 31.3	Chemical Plant and Petroleum Refinery Piping	All piping within the property limits of facilities engaged in processing or handling of chemical, petroleum, or related products. Also applies to piping systems that handle all fluids, including fluidized solids, and all types of service including raw, intermediate, and finished chemicals; oil and other petroleum product;s gas; steam; air; water; and refrigerants, except as specifically excluded
B 31.4	Liquid Petroleum Transportation Piping Systems	Piping for transporting liquid petroleum between producers' lease facilities, tank farms, natural gas processing plants, refineries, stations, terminals and other delivery and receiving points.
B 31.5	Refrigeration Piping	Piping systems for refrigerant and brine at temperatures as low as -196 °C (-320 °F), whether erected on the premises or factory assembled. Does not include (1) self-contained or unit refrigeration systems subject to requirements of Underwriters' Laboratories or any other nationally recognized tesing laboratory, (2) water piping, or (3) piping designed for external or internal pressure not exceeding 103 kPa (15 psig) regardless of size.
B 31.8	Gas Transmission & Distribution Piping Systems	Gas compressor stations, gas metering and regulation stations, gas mains, and service lines up to the outlet of the customer's meter set assembly. Gas storage lines and gas storage equipment of the close-pipe type that is either fabricated or forged from pipe, or fabricated from pipe and fittings.

The *ASME Code for Pressure Piping*, B 31, is published in six sections as outlined in Table 5.13. Each section prescribes requirements for design, materials, fabrication, erection, testing, and inspection of a designated piping system. Third party inspection is not required under the Code for Pressure Piping. However, pressure piping external to a boiler is covered by the ASME B&PV code and therefore third party inspection is required. Some sections of the Code for Pressure Piping require qualifications to be performed in accordance with other documents. Examples of documents that add to the total requirement are Section IX of the ASME B&PV Code, and API Standard 1104.

AWS Specifications, Codes and Rules

The AWS has issued a number of specifications, standards, codes, recommended practices, and guides dealing with welded construction, materials for welding, and

Table 5.14 – AWS Codes, Specifications and Standards*

AWS Documents Dealing with Basic Activities Involved in Welding

A2.4	Standard Symbols for Welding, Brazing, and Nondestructive Examination
A3.0	Standard Welding Terms and Definitions
B4.0	Mechanical Testing of Welds, Guide for
B1.11	Visual Inspection of Welds, Guide for

AWS Documents Covering the Welding of Iron and Steel in Applications

D1.1	Structural Welding Code – Steel
D1.3	Structural Welding Code – Sheet Steel
D1.4	Structural Welding Code – Reinforcing Steel
D1.5	Bridge Welding Code
D3.5	Steel Hull Welding, Guide for
D3.6	Underwater Welding, Specification for
D10.12	Welding Low Carbon Steel Pipe, Recommended Practices and Procedures for
D11.2	Welding Iron Castings, Guide for
D14.3	Earthmoving and Construction Equipment, Specifications for Welding
D15.1	Railroad Welding Specification – Cars and Locomotives

AWS Documents Covering Filler Metals and Fluxes

A5.01	Filler Metal Procurement Guidelines
A5.1	Carbon Steel Electrodes for Shielded Metal Arc Welding, Specifications for
A5.2	Carbon and Low Alloy Steel Rods for Oxyfuel Gas Welding, Specification for
A5.5	Low Alloy Steel Covered Arc Welding Electrodes, Specification for
A5.8	Filler Metals for Brazing, and Braze Welding, Specification for
A5.13	Solid Surfacing Welding Rods and Electrodes, Specification for
A5.15	Welding Electrodes and Rods for Cast Iron, Specification for
A5.17	Carbon Steel Electrodes and Fluxes for Submerged Arc Welding, Specification for
A5.18	Carbon Steel Filler Metals for Gas Shielded Arc Welding, Specification for
A5.20	Carbon Steel Electrodes for Flux Cored Arc Welding, Specification for
A5.23	Low Alloy Steel Electrodes and Fluxes for Submerged Arc Welding, Specification for
A5.25	Carbon and Low Alloy Steel Electrodes and Fluxes for Electroslag Welding, Specification for
A5.26	Carbon and Low Alloy Steel Electrodes for Electrogas Welding, Specification for
A5.28	Low Alloy Steel Filler Metals for Gas Shielded Arc Welding, Specification for
A5.29	Low Alloy Steel Electrodes for Flux Cored Arc Welding, Specification for
A5.30	Consumable Inserts, Specification for

*Selected AWS documents pertaining to the welding of carbon and alloy steels. For a complete list of the many publications, audio/visual training aids, computer software, etc., available, see current issue of AWS Publications Catalog.

welding procedures. With certain exceptions, the AWS does not promulgate types of steels but concentrates on proper welding and testing procedures for articles constructed from steels standardized by other organizations. Notable exceptions, however, are the steel types presented in the AWS specifications for filler metals. These specifications will be discussed in Volume II, Chapter 10. A number of frequently used AWS documents that deal with steels, their fabrication by welding, and inspection methods for weldments are listed in Table 5.14. Many others are listed in the AWS Publications Catalog.

CARBON AND ALLOY STEEL USED IN WELDED CONSTRUCTION

Although weldability is an important characteristic that must be often considered when selecting a steel, most specifications do not address this characteristic in their scope of properties and requirements. This absence of detail must not be construed as oversight or lack-of-need. Instead, weldability is not regularly included because of the difficulty in clearly defining the required behavior in welding, permissible procedures, tests to be performed, and required weld properties. Furthermore, the usable welding processes are so far-ranging that complete coverage of weldability is likely to produce a cumbersome document. The great breadth and depth needed for weldability coverage can be illustrated by singling out low-carbon, plain steels, which are regarded as quite weldable. Yet, when welded by certain processes, and particularly when welded at high speed, even these steels can exhibit welding shortcomings (as will be discussed in Volume II, Chapter 14) and some fabricators may elect to ensure satisfactory welds through a narrowly-written specification.

A limited number of steel standards convey some information regarding suitability for welding. This may be found in the form of a brief commentary, or perhaps as a symbol that is keyed to a weldability scale. However, this meager information is seldom augmented by description of permissible welding procedures or tests for evaluation of suitability for welding. Only a few of the widely used codes or specifications issued by national organizations include guidance on the welding of specific types of steel for certain applications. One example is the *ASME Boiler and Pressure Code* which in Section VIII requires that materials used in the construction of welded pressure vessels must have good weldability. The fabricator is given wide latitude in selecting a welding process when constructing a vessel from the designated steel. Nevertheless, the procedure finally established must produce welds that pass the requirements of Section IX of the Code, Welding and Brazing Qualifications. This freedom in determining procedure for welding or brazing is necessary to accommodate the diversity of fabrication conditions. It also encourages progress in devising new and improved welding and brazing procedures.

Another example of mandated guidance, somewhat more detailed, can be found in the *ASME Code for Pressure Piping*. In Section B 31.1, Chemical Plant and Petroleum Piping, a wide variety of steels are recognized in Chapter III (Materials) for refinery plant usage. To control the fabrication of these steels by welding, eleven classes of steels have been grouped according to practices deemed necessary for preheating or postweld heat treatment (see P-steels, Appendix D). However, this arrangement provides only basic guidance regarding proper welding, and many additional details must be reviewed to plan a procedure that will ensure satisfactory welds. Indeed, the details to be considered are so numerous that a code or material specification containing all of them would be a cumbersome document.

Qualities of Steel Important to Welding

Quality is a term frequently used to cover features of a steel to establish its suitability. A steel's quality can be assessed by a number of lineaments and properties, but appraisals are based more often on one or more of the following:

(1) homogeneity of chemical composition,
(2) uniformity of mechanical properties,
(3) level of fracture toughness,
(4) degree of freedom from surface defects,
(5) maximum size and number of nonmetallic inclusions,
(6) internal soundness, and
(7) nature and uniformity of microstructure.

These features have been found from long experience to be important for particular steel applications because of the manner in which the steel will be fabricated, and the performance expected in service. Attention given to these qualities has proved so effective that a practice in ordering steel developed in which descriptors are applied to bundle the requirements into a convenient, general term; hence the marketing of steels as *aircraft quality*, *axle-shaft quality*, etc. This ordering practice is often used for steel plate, where the six commonly recognized terms are: regular quality, structural quality, cold pressing quality, cold flanging quality, forging quality, and pressure-vessel quality.

Although the steel producer states the areas of application for which each quality classification is believed to be suited, the purchaser must make an educated guess on what should be ordered (and paid for) to ensure steel that fabricates satisfactorily, and is fit-to-serve in the intended application. Even when a particular quality classification is selected, the information routinely offered by steel supplier regarding weldability is likely to be quite general. Consequently, personnel responsible for welding during fabrication operations will be pressed to search available resources for information, and to thoroughly examine any governing codes or specifications for guidance.

An example of a current steel specification in which qualities are addressed by required standards and by directives is ASTM A 737 / A 737M, *Standard Specification for Pressure Vessel Plates, High Strength Low Alloy Steel*. This specification covers steel plates for service in welded pressure vessels and piping components where high strength and improved toughness are required. The maximum thickness of plates produced under ASTM A 737 / A 737M is limited only by the ability of the steelmaker to meet the mechanical property requirements. ASTM A 20 / A 20M (with its fusion welding requirement mentioned earlier) is an applicable document, and A 737 / A 737M assumes suitable welding procedures.

The steel supplied under A 737 / A 737M may be made by any of the following processes: open hearth, basic oxygen, or electric furnace. However, the steel must be killed and made by a practice that ensures a fine austenitic grain size. Two low-alloy steel compositions are offered under A 737 / A 737M: Grade B, which is niobium-containing, and Grade C, which contains vanadium and nitrogen and may have niobium present to a maximum of 0.05 percent. Both grades of plate must be heat treated by normalizing (with the usual air cooling), but the purchaser may approve faster cooling providing a tempering treatment at 595 to 705°C (1100 to 1300°F) is subsequently applied. These combinations of composition and heat treatment are designed to provide plate with two levels of strength and with good toughness. However, the purchaser must indicate whether impact tests are required to evaluate toughness.

If desired, Charpy V-notch impact tests can be ordered for the A 737 / A 737M steel by applying supplementary requirement S5 from ASTM A 20 / A 20M. The purchaser can also specify supplementary requirements for vacuum treatment (S1), ultrasonic examination (S8, S11 or S12), vacuum carbon deoxidation (S17), and special product analysis (S2). The most recent issue of A 737 / A 737M provides two additional optional supplementary requirements: S74, reporting residual elements (copper, chromium, molybdenum, and nickel), and S76, a carbon equivalent equation governing product analysis. Inclusion of the equation indicates concern about weldability, and the specified CE maximum of 0.52 percent will be explained in Volume II, Chapter 14.

Although ASTM A 737 / A 737M applied in its entirety is a complete steel specification, it does not provide complete assurance regarding weldability. Even with (1) the product form confined to plate, (2) the end use clearly designated as pressure vessels, and (3) the steel composition limited to two relatively simple types of alloying element arrangement, weldability cannot be put aside as a capability already under full control. Records on the actual steel received for fabrication should be scrutinized in terms of planned welding procedures to avoid problems during fabrication and in subsequent service.

Factors Affecting the Weldability of Steel

A number of general qualities of steel can have a significant effect on weldability and are described below, along with a brief explanation of how each quality might intermesh with weldability. More detailed information on interactions between a steel's qualities and its behavior in welding can be found in Chapters 13, 14 and 15.

Chemical Composition

Chemical composition should first be viewed as a general quality of steel because it provides great insight into welding properties. A large number of steels are proprietary and are often represented by trade names. However, it is helpful to understand each steel based on its chemical composition. An understanding based on chemical composition shapes a keener appreciation of the influence of alloying and residual elements on welding properties, and experience from welding existing steels serves as a valuable reference when a new composition is encountered.

Assessment of weldability for a given steel and decisions regarding proper welding procedure can be assisted by knowledge of the steel's production method starting with melting and pouring, because these are the practices that control many aspects of chemical composition. Mention was made earlier that some elements can fade in the molten steel in the ladle, and some elements can segregate in an ingot during solidification. Consequently, if a given steel has certain influential elements reported in its ladle or heat analyses as being close to specified limits, the user must consider their distribution in the product form to be welded.

As a practical illustration, consider steel plate 75 mm (3 in.) thick purchased to ASTM A 36 / A 36M specifications. This material (UNS K02600) should meet tensile strength requirements of 400 to 550 MPa (58 to 80 ksi) and have a minimum yield point of 250 MPa (36 ksi). To ensure meeting these tensile requirements in this relatively thick plate, the carbon content of the steel might run close to the permissible maximum of 0.27 percent. If the heat analysis for this steel reported carbon to be 0.27 percent, a check analysis of the plate product for carbon content would permit 0.04 percent over this limit (per ASTM guidelines). Therefore, an assessment of welding behavior should be based on handling steel with a carbon content as high as 0.31 percent. Although a seemingly small increase, carbon is a highly influential alloying element, and this factor may require careful consideration.

On the other hand, if the 75 mm (3 in.) thick steel plate were produced by direct slab-casting instead of ingot steelmaking, then the likelihood of carbon check analyses exceeding the ladle determination by as much as 0.04 percent would be diminished. Furthermore, if the plate were purchased to the requirements of ASTM A 285 / A 285M, this specification has the carbon maximum set at 0.28 percent, and this limit applies to both *heat* and *product* analyses. Therefore, the welding of steel with a carbon content near or above the 0.30 percent level would not have been considered. The significance of the 0.30 percent carbon level in shaping the weldability of plain steels will be explained in Volume II, Chapter 14.

A quality aspect of steel composition that is not easily assessed by evaluating routine chemical element reports is the condition of *oxidation*, or as commonly stated *deoxidation*. Certain elements with a high affinity for oxygen at steelmaking temperature are added to accomplish deoxidation; the common elements are silicon, aluminum, and titanium. When a steel is fully killed by adding one or more of these agents to the molten metal, subsequent chemical analyses usually report a residual amount of the deoxidizing element(s) present in the steel.

Table 5.1 shows residual silicon values for various kinds of carbon steel deoxidized with silicon additions. Although silicon content is not always included in the mill chemical analysis report for a carbon steel, this content is one indication of whether a steel is rimmed, semikilled, or fully killed. If the steel has been deoxidized by the vacuum technique, no residual amounts of deoxidation elements will be found by chemical analyses. Determination of oxygen content is not a simple analytical procedure, and the value secured requires careful consideration when interpreting the degree of deoxidation. For this reason, specifications seeking a fully killed steel should either (1) require silicon, aluminum, or other deoxidizing elements to be present in a range that ensures a fully killed steel, or (2) state that the steel must show no reaction between carbon and oxygen upon solidification by virtue of sufficient deoxidizing agents having been added, or by having been subjected to vacuum treatment.

For a number of reasons, killed steel is considered the highest quality, and it generally gives fewer welding problems. The rimmed and semikilled steels, when fusion welded, can display porosity in the weld metal unless sufficient deoxidizing agent is available to handle the contained oxygen. However, because the rimmed and semikilled steels offer certain advantages in aspects other than welding, they often are encountered in fabricating and manufacturing operations (even where welding is planned). Consequently, fusion welding operations on steel may require consideration of the state of oxidation of the base metals and filler metals involved.

Many other aspects of chemical composition affect steel quality. Even residual elements call for control that is not necessarily clear in the purchasing specification. For example, many steel standards require that the sulfur content be held to a maximum of 0.05 percent. However, this residual element appears in steel in the form of nonmetallic sulfide inclusions which, if present in large number or size, can adversely affect cold forming capacity, accentuate anisotropy in mechanical properties, and lower the level of fracture toughness. These unwelcome effects have led steelmakers to minimize sulfur levels. Consequently, the sulfur content seldom exceeds half the permissible maximum; it is often 0.02 percent or less. Some steels, especially those for applications where cold forming capacity and impact toughness are paramount, may contain less than 0.01 percent sulfur.

Despite the number of specifications aimed at controlling steelmaking, the final composition is shaped by many factors. These include the raw materials (especially when steel scrap is included) and the melting process used. Many additional elements beyond those specified can be present at low levels, and these are termed *residual*, *trace*, *minor*, or *unspecified* elements. Chapter 13 of Volume II will contain information on minor elements in steel that can affect the metallurgy of welding, sometimes with ill effects, and sometimes with beneficial influence.

Mechanical Properties

Mechanical properties are used to delineate the kind of steel ordered from the mill. Many material specifications have tensile requirements, and where the quality of the steel deserves special consideration, the number of tension tests and their location and orientation in the product may also be detailed.

For example, to determine compliance and uniformity of tensile properties in high strength steel plate, the specification may require tension test coupons to be taken from (1) each plate rolled, (2) each end of the plate, and (3) the longitudinal axis of the tension-test specimen transverse to the final rolling direction of the plate. If the hot rolled plate is to be heat treated, the procedure for securing the tension-test coupons may be especially detailed to ensure that the specimens are truly representative of the plate material. This quality assessment by tension testing or other mechanical tests can have a bearing on weldability.

An illustration of possible relationship between tension test results and weldability can be found when differences are observed in test values of specimens taken from rolled steel plate with their longitudinal axes (1) parallel to the principal rolling direction ("X" axis), (2) transverse to the principal rolling direction ("Y" axis), and (3) through the thickness of the plate ("Z" axis). Values for both strength and ductility can differ appreciably between these three axes, and ductility (particularly as judged by reduction-in-area values) can be a sensitive criterion of quality. If ductility determined through the thickness (Z-axis direction) is low, a weldment fabricated from the plate can be quite susceptible to the problem termed lamellar tearing. To minimize susceptibility to lamellar tearing through material quality, the mechanical properties (chiefly ductility) should be controlled by chemical composition, steelmaking practice, and hot working procedures.

The level of fracture toughness exhibited by a steel is a mechanical property or quality aspect experiencing ongoing test development. For many years, impact testing was favored for evaluation of notch toughness, and the Charpy V-notch coupon became the standard specimen for this purpose. However, recent work indicates that impact tests with small notched specimens are only useful for comparing different steels; the results cannot directly appraise the serviceability of a steel in a given structure. Research on fracture toughness of steel has produced several new test specimens and methods, and some of these have been standardized (e.g., ASTM E 208, E 338, E 436, E 561, E 604, E 740, and E 813). Despite varying test methods, there is strong agreement that fracture toughness is an important quality, especially in steel used for welded structures.

Metallurgical Structure

Metallurgical structure or microstructure is another steel feature that mirrors steel quality. The microstructure of steel is primarily shaped by chemical composition, but many conditions of steelmaking, casting, and hot working exert profound influences over details such as inherent grain size, grain shape, homogeneity, and phase distribution. Control of these details requires care during steelmaking, and sometimes added processing steps, such as heat treatment, must be taken. Although this treatment increases the cost of steel, the improved performance resulting from a controlled microstructure often justifies the higher cost. The weldability of steel, even with the fusion welding processes, can be influenced by the microstructure of the base metals being joined. This effect occurs primarily by the way the prior microstructure influences the newly formed microstructure of the base metal heat-affected zone adjacent to the weld.

Internal Soundness

Internal soundness, and degree of *freedom from surface defects* are two qualities frequently addressed by specifications. Attention was centered many years ago on these aspects of steel products because they were difficult to evaluate with existing nondestructive inspection methods. In fact, steel plate for boiler or pressure vessel construction was specified to be either *firebox quality* or *flange quality*. These terms were used for many years to differentiate steels that could pass a nick-break test to establish freedom from internal defects beyond certain dimensions; those passing the test were labeled firebox quality.

More definitive requirements now apply in conducting nondestructive examination of steel products, and limits for size and number of internal and surface imperfections are specified. Many specifications now include restrictions on surface conditioning by grinding; they may provide the procedures for repairing deeper surface imperfections by metal removal, welding, and surface finishing. Per ASTM A 6 / A 6M, the steelmaker is given limitations on metal removal from the various products and has directives on the process and procedure to be used when repairing by welding. Repairs are subject to final inspection to ensure the following:

(1) imperfections have been completely removed,

(2) specified metal removal limits have not been exceeded,

(3) established welding procedures have been followed, and

(4) weld deposits are of acceptable quality.

Cleanliness

Cleanliness of the steel (i.e., the number and size of nonmetallic inclusions distributed throughout) is another internal quality that can be defined by specification. Manufacturers of ball and roller bearings were the first to recognize the importance of steel cleanliness, and they purchase steel to demanding cleanliness standards. A variety of methods for evaluating cleanliness were developed as steelmakers and bearing manufacturers collaborated in achieving bearing-quality steel. The improved steelmaking methods that grew out of this collaboration has been put to use making cleaner steels for the many articles that require a highly polished surface which must be free of imperfections caused by nonmetallic inclusions.

ASTM E 45 describes a number of methods for determining the size, shape, quantity and type of nonmetallic inclusions in steel, and for establishing a quantitative rating for the cleanliness of the material being examined. Standard practices are given both for macroscopic and for microscopical methods. The macroscopic methods include *macroetching*, *fracturing*, *step-down machining*, and *magnetic particle examination*, but these methods are not capable ordinarily of establishing the types of nonmetallics revealed. Microscopical methods are based on metallographic examination, and although more labor-intensive they offer a number of advantages. Extremely small inclusions can be seen, and the character or type can be assessed. When type identification is wanted, selective-etching, use of polarized illumination, or microprobe analysis can be employed to establish composition, as for example, see Figure 2.30. Also, quantitative assessment of the number of inclusions within specified ranges of size can be made by metallographic image analysis (see Figure 2.25).

Nonmetallic inclusions can be classified into two groups: *exogenous* and *indigenous*. Exogenous inclusions are those of external origination, such as entrapped slag (bits of refractory material from furnace and ladle linings), and from other devices contacted by the molten metal. The number and size of exogenous inclusions depend on the skill of the steelmaker in handling the molten steel in the melting and pouring operations (prior to solidification). Indigenous inclusions are those resulting from chemical reactions that occur in the molten metal during the steelmaking process. These inclusions are usually particles of complex sulfide compounds as residual sulfur combines with various elements present in the steel, and complex oxides as deoxidizing elements combine with the dissolved oxygen present in the molten steel. If the nitrogen content of the molten steel is also relatively high and elements with a strong affinity for nitrogen, such as titanium and zirconium, are present, then nitride particles can also appear in the steel as indigenous inclusions. Much has been accomplished in steelmaking toward producing cleaner steels, but nonmetallic inclusions continue to be a quality factor deserving of close attention. Inclusions of certain kinds will be shown in later chapters to be a root cause of a number of metallurgical problems in steel, including anisotropy in mechanical properties, low ductility, poor fracture toughness, fissuring in weld heat-affected zones, and lamellar tearing during welding.

THE FUTURE OF STEELS AND THEIR WELDABILITY

Knowledge of types of steels — their composition, properties, surface character, metallurgical behavior, etc. — is becoming increasingly important in making welds that are fit for the intended application of a weldment. This need for greater depth-of-knowledge concerning the materials being fabricated by welding is spurred by several distinct trends and activities that deserve comment here.

New Steels and Product Forms

"New" steels customarily have been pictured as those containing further additions of alloying elements that provide new or improved properties. Steels that are new in this contemporary sense will be introduced in Chapters 14 and 15 where influence of their alloy contents on weldability will be discussed. However, a survey of new steels being utilized today will find a growing number that embody newness or novel properties which have little or no dependency on alloying; yet, their innovative features can have a significant influence on behavior in welding; therefore, preparation of welding procedures likely will require knowledge of the steel beyond the recognized effects of alloying elements present in the composition. Any one of the microalloyed types of steel carrying the designation, "T-MCP," could serve as an example of a new steel which offers useful mechanical properties that are not obtained straightway through alloying. T-MCP signifies that thermo-mechanical controlled processing was applied in producing the particular product form in hand, as explained earlier in this chapter. Fabricators of weldments that include a T-MCP steel should appreciate that in-depth knowledge of its microstructure provides the rationale to guide preparation of welding procedures that preserve the enhanced mechanical properties in the weld HAZ.

Another new kind of steel deserving scrutiny utilizes titanium as a deoxidizing agent, as explained in Chapter 4. This earlier information will be supplemented in later chapters to fully explain how both titanium oxides, and a small surplus of titanium, can benefit microstructure in the HAZ of a weld by restriction of grain growth and thereby improve toughness. This exploitation of compounds of titanium (sometimes aided by a very small amount of boron) also has been carried into the control of weld metal microstructure to secure improved toughness. Consequently, familiarization with the microstructure of steel and its variation with composition, deoxidation treatment, hot working and cooling treatments, etc., can be very helpful. Indeed, understanding of its genesis in steelmelting will lead to greater freedom in selecting welding parameters instead of imposing of limitations.

Dissimilar-Metal Welding

Welds between different types of steels, or between steel and an entirely different metal or alloy, have been a small but important part of weldment production. These joining operations have called for very particular examination of all materials and factors involved in order to establish compatibility of the dissimilar materials, as well as the applicability of the chosen process and procedure. Furthermore, the longevity of the weld under given service conditions often deserves study. Experience in making dissimilar-metal welds, using steel as one of the workpieces, will be reviewed in appropriate chapters of Volume II. Attention will be directed here to a novel metalworking activity which is rapidly growing in industry, and which will intensify the challenge of dissimilar-metal welding. This growing activity involves the "tailored blank," which is a flat sheet of steel cut to the pattern best suited for deep drawing or forming into a particular article or component. Auto and appliance manufacturers have eagerly turned to the use of welded tailored blanks; because any manner of sheet sections can be joined to the sides of the main blank, thereby avoiding the need to cut the required blank from a larger piece of sheet. The cost of welding butt joints between the flat sheet sections is more than repaid by not using wide widths of sheet and by reducing scrap loss.

Tailored blanks, altered in shape by welded appendages, have taken an interesting turn that brings up the matter of dissimilar-metal welding. Designers have been quick to realize that further savings can be achieved by reducing the thickness of sections in the blank wherever the finished component will allow a thinner sheet portion. This move immediately brings up the matter of welding together dissimilar thicknesses of sheet in a given blank. Three particular welding processes are being favored for making the smooth, full-section butt welds required in the blank — namely, the mash-seam variation of resistance seam welding (RSEW, see page 531); the autogenous form of laser beam welding (LBW, see page 567); and the high-frequency induction method of pressure welding [a combination of IW (page 531) and HPW (page 589)]. A debate currently is underway on the relative merits of the RSEW mash-seam process and the LBW autogenous fusion process. Developments in methods for welding blanks are fast-moving because of high interest within the automotive industry.

Both the RSEW mash seam process and the LBW autogenous fusion process have met the present challenge of dissimilar thickness, but the challenge alluded to earlier has intensified, because designers are already calling for sections of higher-strength steels, and metallic coated steels for those sections of the blank where the component would be improved by their presence. Since the more-expensive steel is limited to those logical portions of the component, a minimal increase is incurred in material costs. Welding personnel may have to adjust welding parameters to accommodate the dissimilar metals involved; but this usually is accomplished more easily when welding flat sheet in a repetitive, mechanized operation to produce blanks, rather than accomplishing the welding by joining odd-shaped, formed workpieces in the article or component.

To date, welding has met the dissimilar metal challenge whether the dissimilarity involves thickness of sheet, type of steel, or kind of steel surface finish. Naturally, adjustments must be made in welding procedure to accommodate whatever arrangement is incorporated in a blank as tailored, and this requires good familiarity with the process employed, whether it is RSEW, LBW or some other process suitable for making smooth, full-section welds.

The successful use of tailored blanks suggests that designers soon will call for sections of markedly different materials in tailored blanks. For example, a small section of stainless steel possibly could solve a problem with life expectancy of a component, and the inclusion of only a small section of the corrosion-resisting material exactly where needed would incur less cost than all-stainless-steel construction. Herein lies a need for a knowledge of welding metallurgy. Welding personnel must be aware that the introduction of stainless steel into a blank of carbon or low-alloy steel puts RSEW mash seam, and LBW autogenous welding on an unequal footing. The content of Chapter 6 should provide fundamental assistance in this regard.

Repair Welding —The Ultimate Challenge

Repair of an article or a structure by welding can present challenges in every aspect of the operation. Often the task must be completed with little time to locate information on the steel originally employed and the welding procedure followed, if records indeed are still available. Tests to identify the steel and evaluate its weldability also may be too time-consuming. It is not unusual to have a repair welding procedure that is much more complex than the one originally followed because circumstances may not allow preheating, or postweld heat treatment. Also, the only welding position possible may be unfavorable (e.g., overhead) and this may adversely affect weld quality and properties. Obstacles to weld repair are best overcome or circumvented through knowledge of the steel to be welded and the capabilities of the process to be employed.

SUGGESTED READING

The Making, Shaping and Treating of Steel, Tenth Edition, by W. T. Lankford, Jr., N. L. Samways, R. F. Craven, and H. E. McGannon; Association of Iron and Steel Engineers, Pittsburgh, PA, 1985, ISBN: 0930767-00-4.

Engineering Properties of Steel, by Philip Harvey; ASM International, Materials Park, OH, 1982, ISBN: 0-87170-144-8.

Properties and Selection: Iron and Steels, and High-Performance Alloys, Metals Handbook, Volume I, 10th Edition; ibid, 1990, ISBN: 0-87170-377-7.

Chapter 6

In This Chapter...

- **SOLID-STATE WELDING (SSW)** **444**
- **FUSION WELDING** **445**
- **BRAZING AND SOLDERING** **445**
- **HEAT SOURCES FOR WELDING & CUTTING**......**448**
  - Electrical Heat Generation 448
    - Electric Arc 448
    - Electron Beam...459
  - Electric Resistance..461
  - Electromagnetic Radiation 461
    - Laser Beams 463
  - Chemical Heat Generation....................................465
  - Mechanical Heat Generation.................................466
- **THE WELDING AND CUTTING PROCESSES** **467**
  - Arc Welding Process (AW).....................................467
    - Power Sources for Arc Welding.........................467
    - Auxiliary Equipment for Arc Welding 470
    - Basic Forms of Arc Welding...............................471
    - Shielded Metal Arc Welding (SMAW).................472
    - Stud Arc Welding (SW) 477
    - Gas Tungsten Arc Welding (GTAW)....................478
    - Gas Metal Arc Welding (GMAW)........................489
    - Flux Cored Arc Welding (FCAW)........................501
    - Submerged Arc Welding (SAW) 505
    - Plasma Arc Welding (PAW) 511
    - Percussion Welding (PEW) 514
    - Magnetically Impelled Arc Welding.....................515
    - Welding Arc Technology 517
  - Resistance Welding Processes..............................520
    - Resistance Spot Welding (RSW)........................522
    - Resistance Seam Welding (RSEW) 531
    - Projection Welding (PW)....................................532
    - Upset Welding (UW)...533
    - Flash Welding (FW) 537
    - Electrical Metal-Explosion Welding Process......541
    - Induction Welding (IW) 542
    - Electroslag Welding (ESW)................................542
  - Electron Beam Welding (EBW).............................548
    - Welding in a High Vacuum (EBW-HV) 550
    - Welding in a Medium Vacuum (EBW-MV)..........565
    - Nonvacuum Electron Beam Welding (EBW-NV).565
    - Tracking Joints During Electron Beam Welding..565
  - Laser Beam Welding (LBW)..................................568
    - Nature of Laser Beams & Plasma Generation 569
    - Basic Techniques in Laser Welding....................571
    - Attributes of Laser Welding................................571
    - Laser Welding Difficulties and Defects 572
    - Shielding Gas Effects in Laser Welding...............575
    - Filler Wire Feeding in Laser Welding..................576
  - Oxyfuel Welding (OFW) 582
    - Oxyacetylene Welding (OAW) 582
  - Thermite Welding (TW) 585
- **SOLID-STATE WELDING PROCESSES**................**587**
  - Hot Pressure Welding (HPW) 589
  - Induction Welding (IW)..592
  - Friction Welding (FRW)..593
    - Inertia-Drive Friction Welding............................594
    - Direct-Drive Friction Welding 595
    - Materials Suited for Friction Welding 595
    - Mechanical Properties of Friction Welds 598
    - Other Forms of Friction Welding.........................599
  - Explosion Welding (EXW) 607
  - Diffusion Welding (DFW).......................................612
  - Ultrasonic Welding (USW) 613
  - Cold Welding (CW)...616
  - Electrostatic Bonding..617
  - Electrodeposition Welding....................................619
- **BRAZING AND SOLDERING PROCESSES**...........**620**
  - Brazing Processes (B)...620
  - Soldering (S)...624
- **SURFACING BY WELDING & THERM. SPRAY** **628**
  - Buildup..628
  - Buttering 628
  - Hardfacing 629
  - Overlaying and Cladding......................................632
- **THERMAL CUTTING PROCESSES (TC)**...............**633**
  - Metallurgical Effects of Thermal Cutting 633
  - Oxygen Cutting Processes (OC)............................634
    - Oxyfuel Gas Cutting (OFC)................................635
    - Chemical Flux Cutting (FOC) 636
    - Metal Powder Cutting (POC) 636
  - Electric Arc Cutting Processes (AC) 637
    - Air Carbon Arc Cutting (CAC-C).........................639
    - Plasma Arc Cutting (PAC) 640
  - Electron Beam Cutting (EBC) 643
  - Laser Beam Cutting (LBC) 647
- **SUGGESTED READING** **651**

Welding Methods and Processes

New welding processes are constantly being sought to accomplish a number of general objectives, including: (1) improved properties approximating those of the base metal, (2) better joint efficiency and integrity, (3) improved weld appearance and (4) reduced production time and cost. Welding technology has improved greatly, although problems still exist with heat-affected zones, surface discoloration, distortion, residual stresses, and disruption of geometric shape. Yet, despite these and other shortcomings, welding continues to advance as the favored joining method because of its many advantages over adhesive bonding and mechanical attachment.

Approximately 100 different welding, thermal cutting, and allied processes have been developed. Much of the equipment employed with these processes is marketed under trade names applied by each manufacturer. While trade jargon may be easier to recall, precise welding terminology is necessary for engineering specifications and directions. Considerable agreement exists among nations regarding terms, but a truly universal terminology for welding is not yet formalized. Therefore, *Standard Welding Terms and Definitions* (ANSI/AWS A3.0) the current AWS standard, will be followed as closely as possible throughout this text.

The task of gaining familiarity with a hundred different welding and related processes calls for a well-planned approach. Most authors agree that bona fide welding processes fall into two basic classifications — one is *solid-state welding* (although some prefer the term "solid phase" welding), and the second is *fusion welding*. Although brazing and soldering produce weld-like connections in metals, all agree that these processes are sufficiently different, from the standpoint of the filler metal used and in the nature of the finished connection, that they should be classified and treated separately. The *Master Chart of Welding and Allied Processes*, published by AWS and shown in Figure 6.1, serves to group the recognized processes and to assign standard definitions and designations. However, the satellites of basis process groups in the AWS Chart are based on long-held common views rather than strict interpretation of the joining method involved. AWS has recognized both approaches in the Master Chart and has provided more definitive classification of the processes by publishing the three diagrams contained in Figure 6.2. Even these contain irregularities; for instance, the resistance welding processes *RSW* and *RSEW* are classified both as solid-state welding in Figure 6.2-A and as fusion welding in Figure 6.2-B. Projection welding (PW) is classified only as fusion welding despite its similarity to resistance spot welding. Despite the ambiguities, these AWS diagrams are useful in bringing the broad array of established processes into view, and in providing standard designations and alphabetical symbols for their identification.

A *weld* is defined by AWS as "a localized coalescence of metals or nonmetals produced by heating the materials to a welding temperature, with or without the applica-

Chapter 6

MASTER CHART OF WELDING AND ALLIED PROCESSES

gas metal arc welding	GMAW
–pulsed arc	GMAW-P
–short circuiting arc	GMAW-S
gas tungsten arc welding	GTAW
–pulsed arc	GTAW-P
plasma arc welding	PAW
shielded metal arc welding	SMAW
stud arc welding	SW
submerged arc welding	SAW
–series	SAW-S

block brazing	BB
diffusion brazing	CAB
dip brazing	DB
exothermic brazing	EXB
flow brazing	FLOW
furnace brazing	FB
induction brazing	IB
infrared brazing	IRB
resistance brazing	RB
torch brazing	TB
twin carbon arc brazing	TCAB

electron beam welding	EBW
–high vacuum	EBW-HV
–medium vacuum	EBW-MV
–nonvacuum	EBW-NV
electroslag welding	ESW
flow welding	FLOW
induction welding	IW
laser beam welding	LBW
percussion welding	PEW
thermite welding	TW

air acetylene welding	AAW
oxyacetylene welding	OAW
oxyhydrogen welding	OHW
pressure gas welding	PGW

Central diagram: WELDING PROCESSES connected to ARC WELDING (AW), BRAZING (B), OTHER WELDING, OXYFUEL GAS WELDING (OFW), ALLIED PROCESSES (continued next page), RESISTANCE WELDING (RW), SOLDERING (S), SOLID-STATE WELDING (SSW).

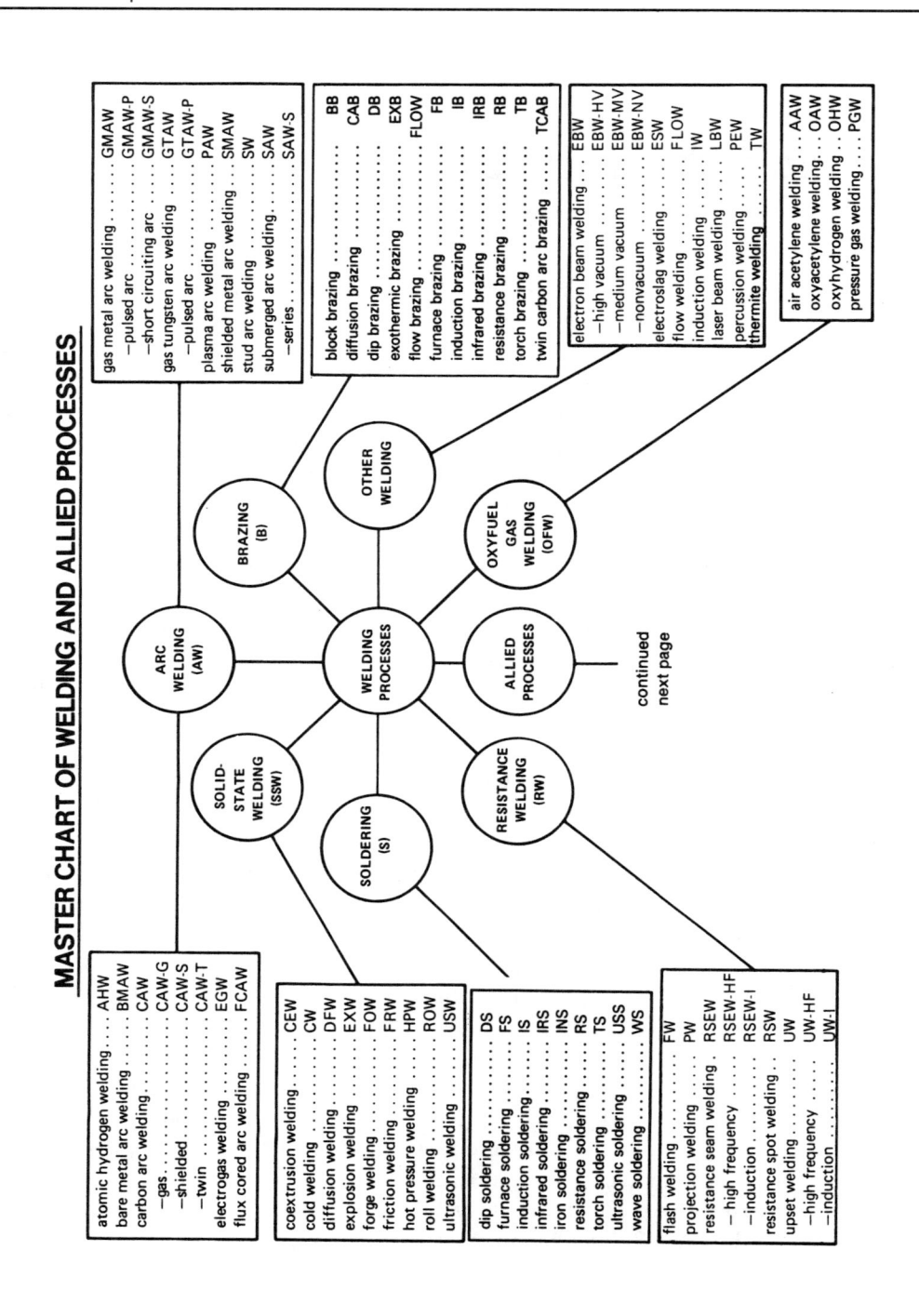

atomic hydrogen welding	AHW
bare metal arc welding	BMAW
carbon arc welding	CAW
–gas	CAW-G
–shielded	CAW-S
–twin	CAW-T
electrogas welding	EGW
flux cored arc welding	FCAW

coextrusion welding	CEW
cold welding	CW
diffusion welding	DFW
explosion welding	EXW
forge welding	FOW
friction welding	FRW
hot pressure welding	HPW
roll welding	ROW
ultrasonic welding	USW

dip soldering	DS
furnace soldering	FS
induction soldering	IS
infrared soldering	IRS
iron soldering	INS
resistance soldering	RS
torch soldering	TS
ultrasonic soldering	USS
wave soldering	WS

flash welding	FW
projection welding	PW
resistance seam welding	RSEW
– high frequency	RSEW-HF
–induction	RSEW-I
resistance spot welding	RSW
upset welding	UW
–high frequency	UW-HF
–induction	UW-I

Figure 6.1 — Master Chart of Welding and Allied Processes
(Reprinted from ANSI/AWS A3.0-89, *Standard Welding Terms and Definitions*)

Welding Methods and Processes   **443**

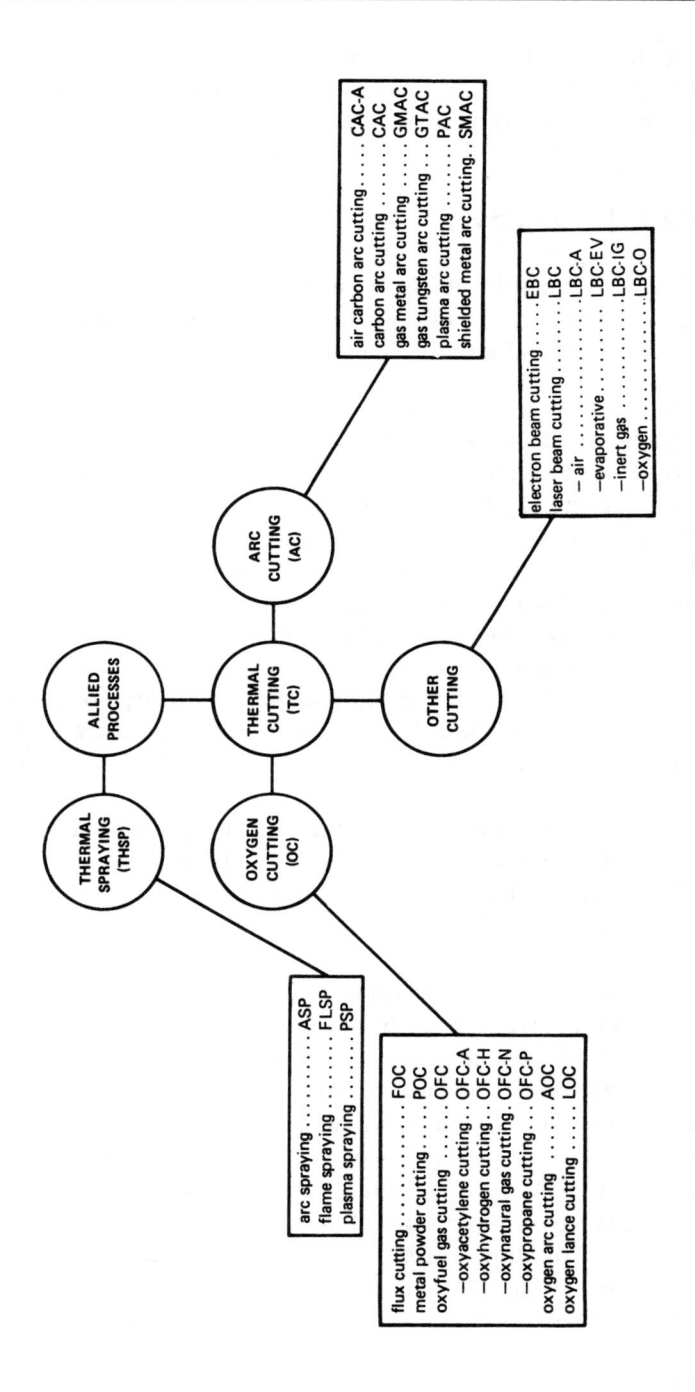

```
                              ┌─────────────────────────────────────┐
                              │ air carbon arc cutting ....... CAC-A │
                              │ carbon arc cutting ........... CAC   │
                              │ gas metal arc cutting ........ GMAC  │
                              │ gas tungsten arc cutting ..... GTAC  │
                              │ plasma arc cutting ........... PAC   │
                              │ shielded metal arc cutting ... SMAC  │
                              └─────────────────────────────────────┘
                                           │
                                           │         ┌─────────────────────────────┐
                                           │         │ electron beam cutting . EBC │
                                           │         │ laser beam cutting .... LBC │
                                           │         │  - air ............... LBC-A │
                                           │         │  - evaporative ...... LBC-EV │
                                           │         │  - inert gas ........ LBC-IG │
                                           │         │  - oxygen ............ LBC-O │
                                           │         └─────────────────────────────┘
                                         ( ARC )
                                         (CUTTING)
                                         ( (AC) )
                                            │
                ( ALLIED )               ( THERMAL )             ( OTHER  )
                (PROCESSES)──────────────( CUTTING )─────────────(CUTTING )
                                         (  (TC)   )
                     │                       │
                ( THERMAL )              ( OXYGEN )
                (SPRAYING )              (CUTTING )
                ( (THSP)  )              (  (OC)  )
                     │                       │
         ┌───────────────────────┐           │
         │ arc spraying ..... ASP │           │
         │ flame spraying .. FLSP │           │
         │ plasma spraying . PSP  │           │
         └───────────────────────┘           │
                                  ┌──────────────────────────────────┐
                                  │ flux cutting ............... FOC │
                                  │ metal powder cutting ....... POC │
                                  │ oxyfuel gas cutting ........ OFC │
                                  │  - oxyacetylene cutting .. OFC-A │
                                  │  - oxyhydrogen cutting ... OFC-H │
                                  │  - oxynatural gas cutting. OFC-N │
                                  │  - oxypropane cutting .... OFC-P │
                                  │ oxygen arc cutting ......... AOC │
                                  │ oxygen lance cutting ....... LOC │
                                  └──────────────────────────────────┘
```

Figure 6.1 — Master Chart of Welding and Allied Processes (continued)

tion of pressure, or by the application of pressure alone and with or without the use of filler metal." This broad definition encompasses both solid state and fusion welding processes, as well as brazing and soldering.

SOLID-STATE WELDING (SSW)

Solid-state welding is defined by AWS as "a group of welding processes which produce coalescence by the application of pressure at a temperature below the melting temperatures of the base metal and the filler metal."

Solid-state welding in its simplest form requires only that the atoms of one workpiece be brought sufficiently close to the atoms of another workpiece to permit the interatomic forces to bridge the interface. (Atoms and their interatomic forces are described in Chapter 2.) There are two reasons why so simple a weld is difficult to attain. First, metal surfaces are rough on an atomic scale. Even a finely ground, flat surface presents hills and valleys that are thousands of atoms deep. Under light pressure, only a few opposing peaks of metal surfaces actually touch. The number of contacting surfaces can be increased by applying more pressure; but a second barrier is encountered, a layer of oxides and occluded gas which is present on all metal surfaces under normal atmospheric conditions. Depending on the composition of the metal and its previous manufacturing history, these layers may be several hundred atom layers thick. For welding to occur, these layers must be mechanically disrupted, dissolved, chemically reduced, or floated away. Early blacksmiths dealt with this problem by heating iron and steel to the "sweating" temperature. This incipient melting at the surface initiated a break-up of the oxide film. The severe hot plastic deformation applied during forging dispersed the oxides and forced intimate contact between the metal workpieces, accomplishing an acceptable weld for primitive objects. Although forge welding was practiced for several thousand years, it was not until sometime in the nineteenth century that the importance of surface oxides was recognized. Solid-state processes succeeded with the development of the furnace operations with hot reducing atmospheres or vacuums to remove the surface oxides and occluded gases. Techniques were also devised for localized plastic deformation of the interface between the workpieces in order to rupture the oxide films without disrupting the overall sectional shapes. Heating was found to lower the amount of deformation required. However, with some ductile metals (e.g., aluminum and copper) it is possible to achieve cold welding (CW) by deformation alone. For most metals, a temperature halfway to the melting point or above (on an absolute temperature scale, such as °K) was found to greatly assist the formation of a solid-state weld. The earliest commercial processes introduced beyond simple forge welding (FOW) were upset welding (UW), roll welding (ROW), and hot pressure welding (HPW). These were followed by the resistance welding processes (RSW, RSEW and PW), which employed some fusion between the workpieces, at least at the start of the welding operation. Friction welding (FRW) research was conducted, ultrasonic welding (USW) emerged and became a viable commercial process (at least with certain nonferrous metals), and explosion welding became a practicable, useful process.

FUSION WELDING

Fusion welding is defined by AWS as "any welding process that uses fusion of the base metal to make the weld." Within this terse AWS definition are three fundamental considerations which should be kept in mind when planning the application of a fusion welding process:

(1) Localized fusion of the base metal and allowing metal to coalesce is a most expedient way of breaking up oxide barriers on the surface.
(2) The high temperature required to reach the molten state introduces a host of metallurgical conditions that must be dealt with to ensure a satisfactory weld; these include solidification of the molten metal without defects, control of absorbed gases, and avoidance of unfavorable microstructural transformations during heating and cooling.
(3) Localized fusion, and possibly the deposition of filler metal, produces a weld that must be controlled to solidify in an acceptable shape; unless pressure is applied during the process to eliminate the molten phase.

Fusion welding processes recognized by AWS are shown in Figure 6.2-B. The concept of joining two piece parts by fusing their edges began in 1881 with the handheld carbon arc. About seven years later a consumable bare steel rod replaced the carbon electrode. Fusion welding with gas torches began in 1901 following commercial production of acetylene (circa 1892), and oxygen (circa 1895). Although Figure 6.2-B shows a proliferation of fusion welding processes, including the use of an electron or laser beam for localized fusion, the electric arc remains the most adaptable energy source for welding.

BRAZING AND SOLDERING

Brazing (B) is defined by AWS as "a group of welding processes that produces coalescence of materials by heating them to the brazing temperature in the presence of a filler metal having a liquidus above 450°C (840°F) and below the solidus of the base metal. The filler metal is distributed between the closely fitted faying surfaces of the joint by capillary action." AWS also provides a definition for braze welding, which is similar to that for brazing except, in this variation, "the filler metal is not distributed in the joint by capillary action."

Soldering (S) is defined by AWS as "a group of welding processes that produces coalescence of materials by heating them to the soldering temperature and by using filler metal having a liquidus not exceeding 450°C (840°F) and below the solidus of the base metals. The filler metal is distributed between the closely fitted faying surfaces of the joint by capillary action."

Although both brazing and soldering involve heating the base metal and using a dissimilar filler metal, they are viewed as different processes because the temperatures required call for distinctly different filler alloys and because the joints produced have distinctively different mechanical properties. The filler metal may be deposited in an open joint between the piece parts of heated (but still solid) base metal, or it may be distributed between closely abutting faces by capillary attraction between the molten filler metal and the heated base metal. To assure adhesion of the base metal, molten

SOLID-STATE WELDING CLASSIFICATION CHART

*Pressure normal to faying surfaces

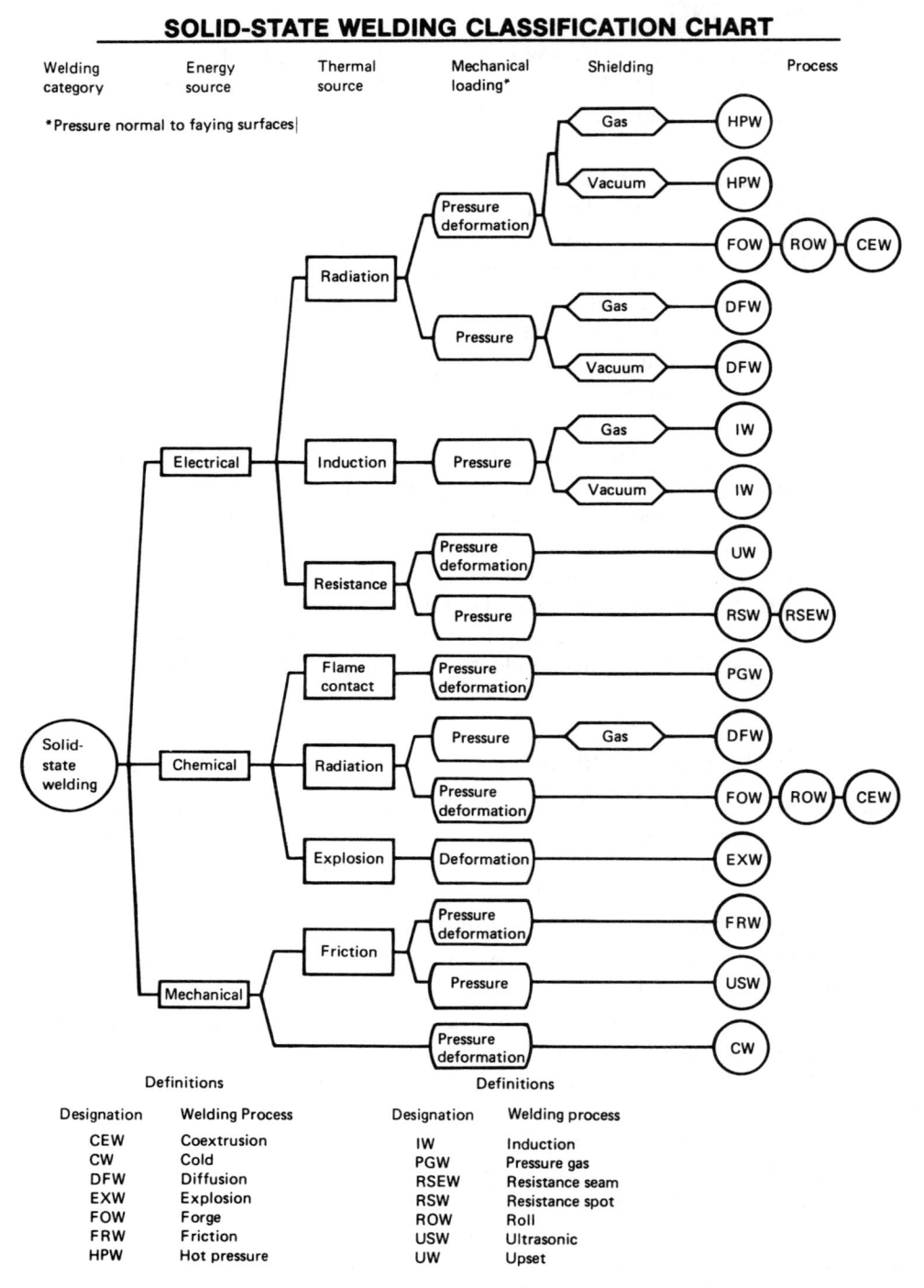

Designation	Welding Process	Designation	Welding process
CEW	Coextrusion	IW	Induction
CW	Cold	PGW	Pressure gas
DFW	Diffusion	RSEW	Resistance seam
EXW	Explosion	RSW	Resistance spot
FOW	Forge	ROW	Roll
FRW	Friction	USW	Ultrasonic
HPW	Hot pressure	UW	Upset

Figure 6.2-A — Classification diagram for solid-state welding processes
(Reprinted from ANSI/AWS A3.0-94 – *Standard Welding Terms and Definitions*)

FUSION WELDING CLASSIFICATION CHART

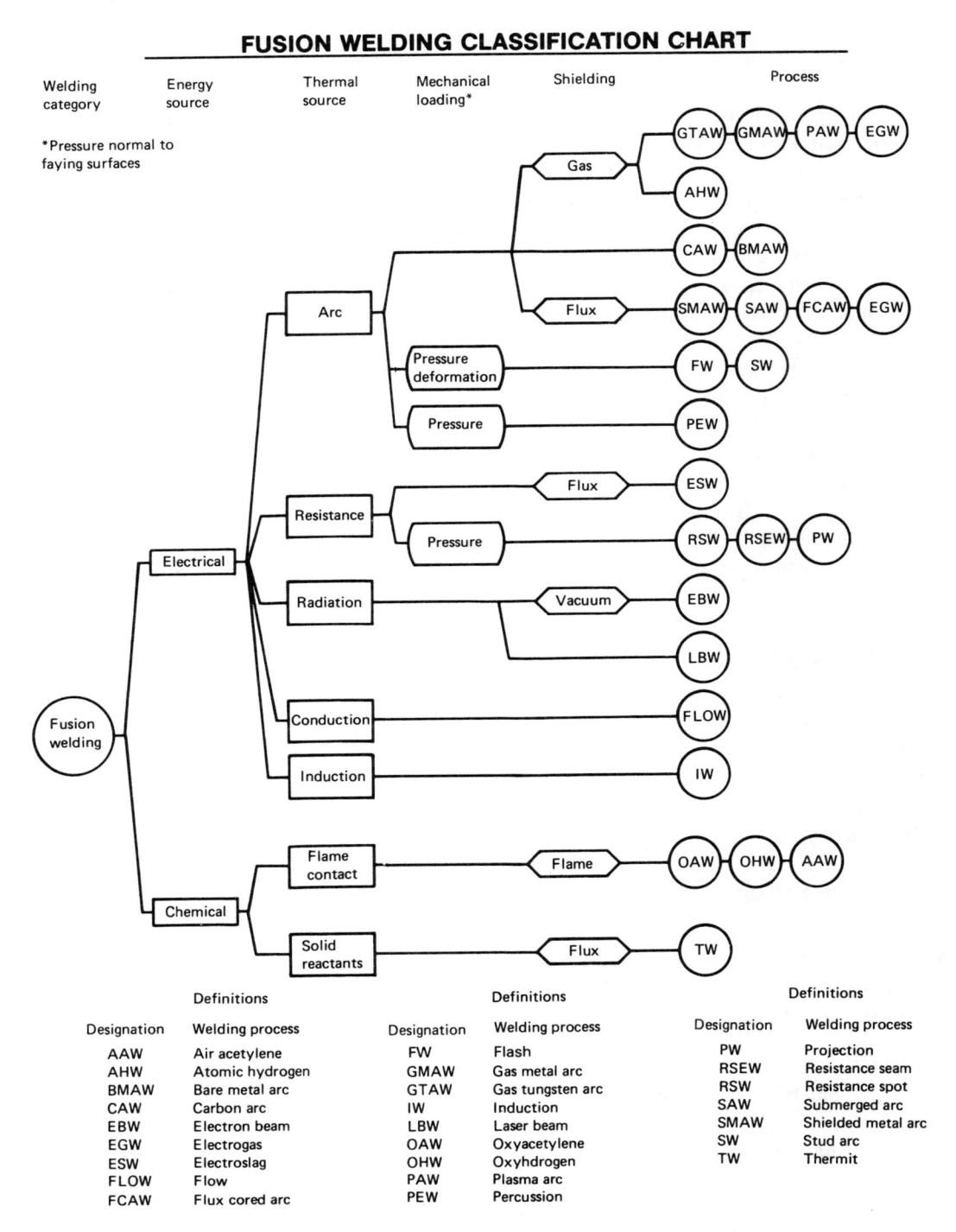

Designation	Welding process	Designation	Welding process	Designation	Welding process
AAW	Air acetylene	FW	Flash	PW	Projection
AHW	Atomic hydrogen	GMAW	Gas metal arc	RSEW	Resistance seam
BMAW	Bare metal arc	GTAW	Gas tungsten arc	RSW	Resistance spot
CAW	Carbon arc	IW	Induction	SAW	Submerged arc
EBW	Electron beam	LBW	Laser beam	SMAW	Shielded metal arc
EGW	Electrogas	OAW	Oxyacetylene	SW	Stud arc
ESW	Electroslag	OHW	Oxyhydrogen	TW	Thermit
FLOW	Flow	PAW	Plasma arc		
FCAW	Flux cored arc	PEW	Percussion		

Figure 6.2-B — Classification diagram for fusion welding processes
(Reprinted from ANSI/AWS A3.0-94 – *Standard Welding Terms and Definitions*)

filler metal must *wet* its surface. Also, "dewetting" must not occur. This phenomena sometimes is encountered in soldering when the filler metal initially wets the base metal surfaces, then retracts into globules that just rest on the surfaces. Dewetting can result from the development of an oxide film on the surface of the base metal caused by oxygen dissolved in the molten filler metal.

The higher temperatures used in brazing encourage diffusion of atoms into the solid metal, and this is frequently observable as an alloy change in the microstructure. Under proper conditions virtually all of the molten filler metal can diffuse into the base metal(s) leaving little or no solidified filler metal in the completed joint. This is termed *diffusion brazing* — a technique preferred for joining articles made of high temperature alloys, because it provides strength at temperatures exceeding the melting point of the initial filler alloy.

The diffusion that can occur in the brazing temperature regime also may be used to produce a liquid eutectic composition from solid base metal and a solid filler metal to bond the joint. Called *eutectic brazing*, this technique causes liquid metal to form in situ by diffusion at the interface between the base metal and a pre-placed film of selected filler metal. Temperature and time must be regulated to control the amount of liquid eutectic alloy that forms to wet the joint and to complete the joint on cooling. There is seldom reason to employ eutectic brazing with the carbon and alloy steels.

Classification of the brazing and soldering processes by AWS according to thermal energy, shielding, etc., is diagrammed in Figure 6.2-C.

HEAT SOURCES FOR WELDING AND CUTTING

Heating is an essential step with almost all welding processes. For this reason, a review of heat generation should be useful before examining the individual processes. Three basic sources are used to obtain the thermal energy needed for welding: (1) electrical, (2) chemical, (3) mechanical. The electrical and chemical methods are the most commonly used. Mechanically produced heat, as developed by friction, impact, and deformation has limited applications.

Electrical Heat Generation

Four electrical methods are used to produce heat for welding: (1) the electric arc, (2) the electron beam, (3) electrical resistance, and (4) electromagnetic radiation. Sometimes "induction heating" is listed as a distinct method, but the actual heating results from the *resistance* of the metal to the flow of an induced electric current.

Electric Arc

The electric arc preceded all of the others as a heat source for fusion welding. An arc is a low-voltage, high-current sustained electrical discharge across a gap between poles (usually the electrode and the base metal). In addition to providing a highly concentrated source of heat, the arc also can be used to remove refractory oxide films from the surfaces to be welded, to assist the transfer of the molten electrode as drops, and to help support and shape the weld pool.

The characteristics of an arc will vary significantly depending on the electric current, pole materials, gases in the gap between the poles, etc. These features profound-

Welding Methods and Processes

BRAZING AND SOLDERING CLASSIFICATION CHART

*Pressure normal to faying surfaces

Definitions

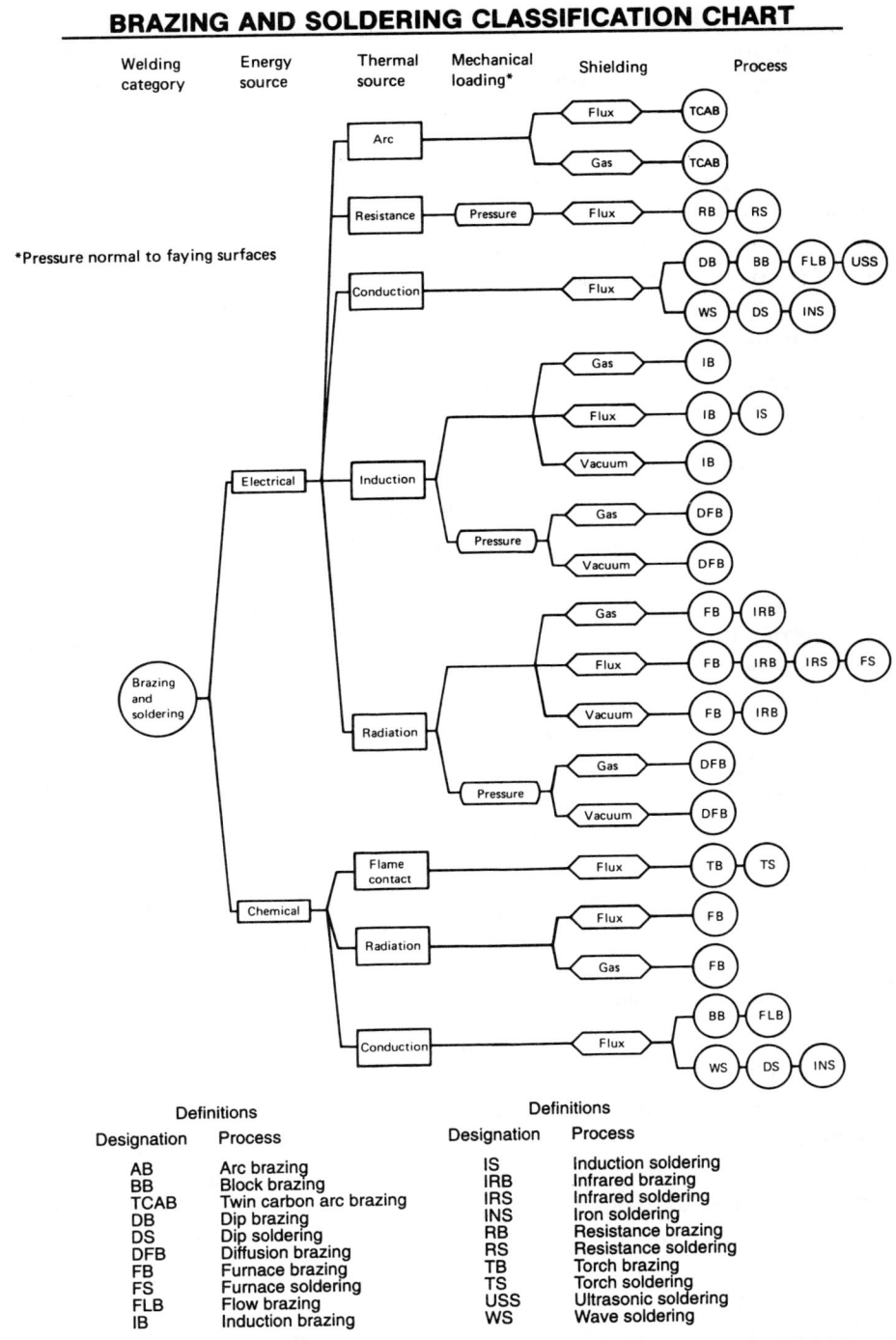

Designation	Process
AB	Arc brazing
BB	Block brazing
TCAB	Twin carbon arc brazing
DB	Dip brazing
DS	Dip soldering
DFB	Diffusion brazing
FB	Furnace brazing
FS	Furnace soldering
FLB	Flow brazing
IB	Induction brazing
IS	Induction soldering
IRB	Infrared brazing
IRS	Infrared soldering
INS	Iron soldering
RB	Resistance brazing
RS	Resistance soldering
TB	Torch brazing
TS	Torch soldering
USS	Ultrasonic soldering
WS	Wave soldering

Figure 6.2-C — Classification diagram for brazing and soldering processes

(Reprinted from ANSI/AWS A3.0-94 - *Standard Welding Terms and Definitions*)

ly affect arc initiation, stability, energy density, temperature, spectral emission, and other characteristics. In welding, the most common arc configuration is *point-to-plane*; the point being the arcing tip of the electrode, and the plane being the base metal. Welding arcs can be further classified in two basic groups — the first with one pole as a thermionic *nonconsumable* electrode, such as tungsten, and the second with one pole as a *consumable* electrode which also serves as a source of filler metal. Welding arcs have a number of common features, and a review of them should provide a helpful background not only for comparing the arc welding processes, but also for explaining how they affect the metallurgical characteristics of the welds.

Arc with a Nonconsumable Electrode

A welding arc in the usual point-to-plane configuration consists of a round column of flame between the poles. The arc column is somewhat constricted at the electrode and it spreads outward when approaching the base metal. The column, as shown schematically in Figure 6.3, has two concentric axial zones which have only hazy demarcation between them; an inner zone called the *plasma*, and an outer zone called the *mantle*. The principal path of current conduction is the plasma because of the highly ionized state of various gases and vaporized solids in this zone. (See Technical Brief 31 for more details on ionization.) The supply of electrons from the welding power source flowing from the tip of the cathodic electrode is attracted to the anodic base metal where they impact and release their kinetic energy as heat. Positively charged (ionized) atoms in the plasma are drawn to the electrode where they impact and their energy is manifested as heat. The high temperature of the plasma is a consequence of the extremely rapid molten imparted by electrical potential to charged particles (i.e., electrons and ions) because whenever each particle is slowed or stopped by collision its kinetic energy is liberated as heat. The temperature in the mantle is somewhat lower, and here is where the previously disassociated molecules and ionized atoms return to their normal state and liberate the energy acquired during their activation. Metals that were vaporized in the plasma often are able to return to a condensed state in the mantle. Consequently, very light deposits of metals as their oxides may be found on the base metal adjacent to the weld after the arc has traveled onward.

Initiation of an arc for welding is an important first step, because a careless start can result in difficulties with weld quality (see Technical Brief 32). Once the welding arc is initiated, as current flows through the arc, the ionized plasma does not behave like a metallic conductor, for which a fixed relationship exists between amperage and voltage in accordance with Ohm's law. Instead, the arc presents a non-linear form; at a low level of current, the volt-amperage relationship has a negative slope. Then, as current is increased toward the high levels commonly used in the arc welding processes, the volt-amperage curve becomes flat, and later begins to rise because of increasing resistance in the plasma. Although most of the electrical energy supplied to the arc is initially utilized as kinetic energy, a small, but significant portion of energy is changed to other forms; including electromagnetic radiation, electrostatic effects, and magnetic forces. These various forms are utilized to gain certain operating refinements which will be detailed in descriptions of the welding processes.

Welding Methods and Processes 451

The temperature of an arc mainly depends on the level of current being conducted, but temperature also is influenced by the kind of matter that enters into formation of the plasma. A very small welding arc can be operated with current as low as a few amperes, but the arc is a versatile heat source and welding current can be as high as several thousand amperes when required. The plasma temperature as surveyed by special emission can range from 5000 to 30 000 °K. Higher arc temperature, perhaps as

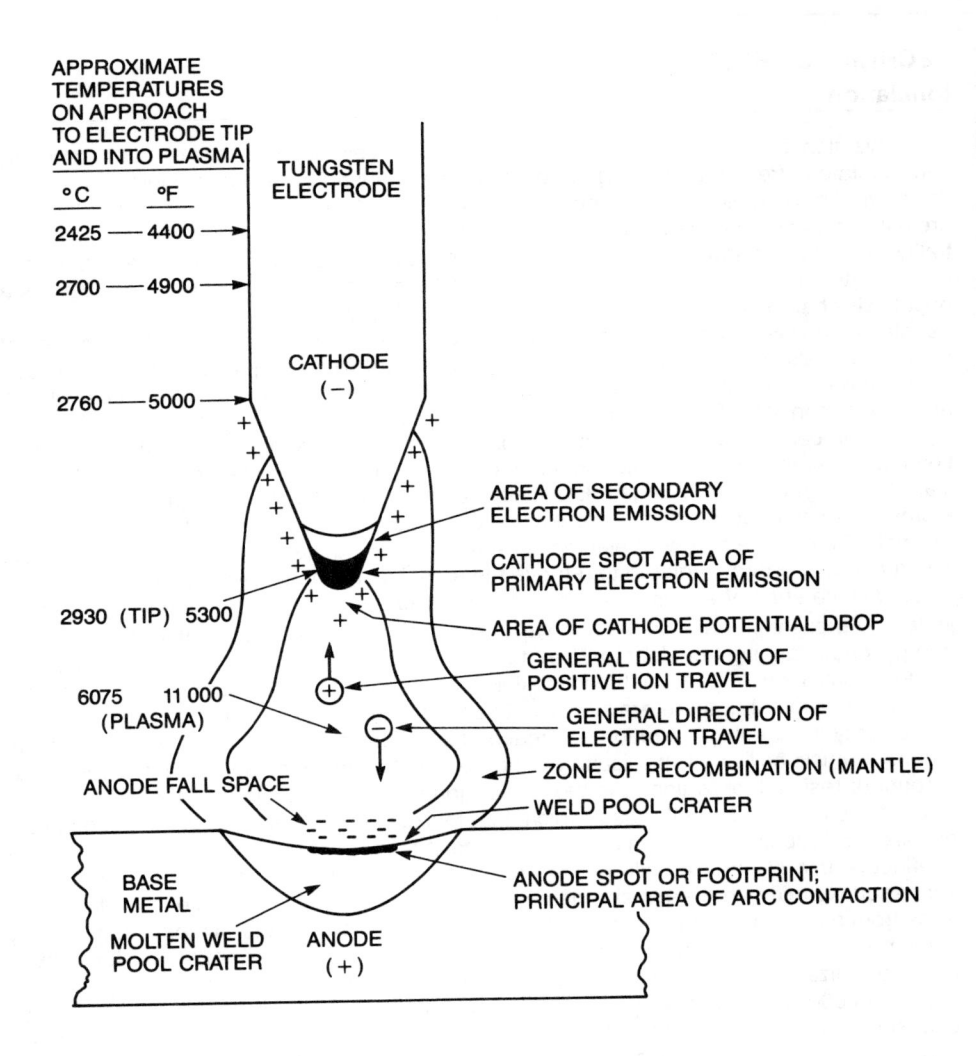

Figure 6.3 — Welding arc and component functions

Schematic illustration of non-consumable electrode operating with direct current and electrode negative (DCEN; sometimes referred to as "straight polarity") showing electrical current flow and typical temperatures at electrode tip and in arc column.

high as 50 000 °K, can be obtained by special power loading, or by constricting the diameter of the plasma using special techniques which will be outlined when describing the *plasma arc welding* process.

Besides the temperature difference between the very hot plasma and the somewhat cooler mantle, there are important variations in heat generation along the axial length of the arc between the poles. Three contiguous areas of heat generation exist along the

---

TECHNICAL BRIEF 31:
Ionization

Ionization, outlined earlier in Chapter 2, is a special elementary state of matter, and this special state is frequently developed to some extent during a number of welding processes. Atoms, molecules, and compounds become ionized whenever one or more electrons are either removed, or are added to the outer shell of atoms to change their electrical neutrality to a charged state. Removal of electron(s) results in a positively charged atom, and in this state it is an ion (called a cation). The addition of electron(s) to an atom produces a negatively charged atom, in which case the ion is called an anion. Ionization of metals, fluxes, slags, shielding gases, and contaminants can play important roles in the behavior of these materials during a particular welding operation. This subject deserves continued attention because of frequent involvement in the processes, and the variety of metallurgical events that can stem from having a material present in the electrically charged state.

The process by which atoms or groups of atoms become ionized has many ramifications, but a simplified explanation that serves welding interest can be centered on the ionization of a single atom of a gas. Electricity cannot be conducted through a gas when its atoms under normal conditions are electrically neutral, but any atom of the gas can, for example, be made a carrier of electric current by removing electron(s) from its outer shell. The amount of energy required to remove the most loosely bound electron from the outermost valence shell of atoms is measured in electron volts (one eV is equal to 1.6×10^{-19} joules). Various elemental gases and vaporized solids were shown in Table 2.2 to require energy inputs ranging from four to 25 eV to remove an electron in a given element and thus produce ionization. In general, the ionization potential of elements decreases with increasing atomic weight, but there are significant variations in this general trend.

Energy for causing ionization can reach atoms from different sources and by different means. Particle collision is particularly effective. Most gases and vapors have a very small amount of residual ionization, and the application of voltage accelerates the charged particles that chance to be in the gas, which then ionizes nearby neutral atoms by the impact of colliding. Propelling a stream of electrons from a thermionic cathode into the gas is highly effective use of particle collision for producing ionization. Electromagnetic radiation of very short wavelength (x-rays and gamma rays) causes ionization by ejecting photoelectrons from atoms, which in turn carry sufficient energy to ionize neutral atoms that may lay in their path. Simply heating atoms, even those of a gas or vapor, is not an efficient way of causing ionization because one eV equates to a temperature of approximately 7700 °K (nearly 14 000 °F). Yet, in welding, intense heating of materials by an electric arc, or by exposure to a concentrated laser beam not only produces a certain amount of volatilization, but the degree of thermal ionization which occurs in the vapors causes them to behave as an electrically conductive plasma.

A noticeable trend in welding is the employment of energy sources of greater intensity to speed up heating and melting so as to shorten operational time. Consequently, an increasing degree of ionization probably will be encountered in the course of welding with certain processes in the future. Various phenomena related to ionization are described in review of the processes.

arc — at the anode, in the plasma, and at the cathode. The energy liberated in each of these three areas is the product of the current flowing and the potential drop across the particular area. The plasma represents about 90 percent of the arc's length in the central portion, and no significant temperature variation is ordinarily found along this portion. Two important areas exist at the extremities of the arc immediately adjacent to the arc's two poles, and these areas commonly display significant differences in heat generation. These areas are known respectively as the *anode fall space* and the *cathode fall space*, and their relative output of heat is dependent on a number of conditions under which the arc operates; especially the kind of electrode being used, and the polarity of the welding current. Energy is transferred from the arc to the metal being welded by six principal mechanisms:

(1) conversion of kinetic energy in electrons and ions,
(2) work function at the surface,
(3) radiation from the arc,
(4) thermal conduction,
(5) thermal convection, and
(6) resistance heating in the electrode.

The first two mechanisms listed, conversion of kinetic energy and work function, are the main conveyors of energy into the workpieces. Resistance heating in the electrode is a significant contributor of heat to the workpieces only when a consumable electrode is used and there is appreciable electrode extension beyond the current contact.

The practical importance of differential energy generation at the anode and the cathode of an arc can be seen by further examination of Figure 6.3 where a nonconsumable tungsten electrode is shown operating on DCEN current. Under these welding conditions, almost twice as much heat is generated at the positive pole (anodic base metal) than at the negative pole (cathodic electrode). This fortunate situation delivers maximum heat where it is most needed, that is, at the relatively massive base metal where melting is to be accomplished. The electrode is heated both by resistance to passage of the welding current, and by bombardment of positive ions that move through the plasma from the anode. However, the relatively heavy positive ions move more slowly than electrons, and the ions dissipate some of their kinetic energy by forming a sheath around the electrode tip. Nevertheless, the heat contributed to the electrode promotes thermionic electron emission from the tip. Electrons are emitted mainly from a cathode spot on the electrode tip and they travel as a concentrated stream at high velocity to the anodic base metal where their kinetic energy is converted to heat.

As electrons leave the surface of the cathodic electrode, they absorb heat much in the manner of boiling or evaporative action, and this mechanism acts to reduce the temperature of the electrode. Consequently, on DCEN polarity, the arcing end of the electrode has a somewhat lower temperature than is found at mid-length between the tip and the externally cooled collet. This favorable cooling effect at the tip extends the operating life of the electrode. Therefore, DCEN (electrode negative) is the favored polarity for welding processes that use a tungsten electrode with dc current.

Further refinement of arc welding with a tungsten electrode operating on DCEN is commonly accomplished by tapering the arcing end of the electrode to localize the cathode spot. Also, the thermionic emissivity of the electrode can be improved by utilizing a tungsten alloy which includes an addition of metal oxides that have a low work function. Oxides of thorium, cerium and lanthanum are a few of the additives that provide this improvement. The fundamentals of the arc with a nonconsumable electrode as outlined here are the bases of the technology that has evolved for the highly successful gas tungsten arc process (GTAW). The energy put into the electrodes and workpieces is dispersed by six modes of heat transfer or consumption, mentioned here because of their close association with the arc:

(1) conduction through the workpieces,
(2) convection,
(3) radiation,
(4) vaporization of superheated metal,
(5) electron or ion emission, and
(6) melting.

The foremost purpose of the energy from the arc is, of course, to melt the consumable electrode and portions of base metal to form a weld. In the case of the base metal, this requires heating the metal above the liquidus and supplying the latent heat of fusion. Relatively little of the energy put into the weld pool and heated base metal is dissipated by ion emission, radiation, or convection. Most of the energy is dispersed by conduction to cooler areas of base metal (and to any fixture in intimate contact with the base metal near the weld that can act as a heat sink). The rise in temperature during welding and the conduction of heat away from the weld is very important, because of the influence on properties of the weld and its heat-affected zone (see Chapter 7).

When polarity with direct current is reversed from that illustrated in Figure 6.3, that is, when the arrangement is made DCEP (electrode positive), a greater amount of heat is liberated at the electrode than at the workpieces. This change in heat distribution presents problems in welding with a nonconsumable electrode. The electrode is heated by positive ions and its own resistance to the flow of current, and no cooling effect takes place on or near the tip from electron evaporation. To prevent melting, larger diameter electrodes are needed to provide more mass for absorbing the greater amount of heat, reduce resistance heating, and to improve conduction of the excessive heat away from the tip. Lower currents are necessary as well. Nevertheless, a very useful phenomenon accompanies operation of the arc on DCEP; a cleaning effect, sometimes called "cathodic etching" or "arc cleaning," is exerted on the base metal over an area beneath the DCEP arc. This is believed to take place when relatively heavy positive gas ions strike the cathodic base metal and selectively break up any oxides on the surface by kinetic energy exchange, whereupon the emission of electrons from the base metal acts to disperse the fragments. This cleaning effect is effectively utilized when welding materials that have a refractory oxide film on their surface, such as aluminum, magnesium, and aluminum-coated steel. Cathodic etching has the unusual capability of traveling over the edges of base metal sections and also entering tightly abutting joints or cracks to accomplish oxide removal. The cleaning action is most effective when argon serves as shielding gas. Helium is ineffective because of low particle mass and sputtering yield, but a small addition of argon to a helium shield can remedy this shortcoming.

When alternating current (ac) is used for welding (e.g. 60 Hz current used in the USA), the electro-mechanics just reviewed for the direct current (dc) arcs still apply in a general way, but the arc is extinguished at each half cycle when the current falls to zero. With modern square-wave power supplies, re-ignition is straightforward, since the plasma (ionized gas) is sustained during reversal. However, with conventional "sine wave" machines, the plasma may be lost as the current falls to zero, and methods must be used to re-establish electron flow from the cathode. This is simple when using a thermionic electrode, an excellent source of electrons. Other methods are necessary to initiate flow when the base metal becomes the cathode. The present solution is to incorporate capacitors or other energy sources in the electrical power system to ensure arc current with a "balanced" (sine) wave form.

Although the ac arc is somewhat more troublesome to operate as a heat source for welding, it does offer distinct advantages. The elimination of fixed magnetic forces that cause "arc blow" and "arc deflection" often gives good reason to use the ac arc. Also, the cleaning effect described earlier holds true for ac arc operation as well, although it occurs only during electrode-positive polarity in the cycle.

Arc with Consumable Electrode

Arc phenomena when using a consumable metal electrode are somewhat different than outlined above for the nonconsumable electrode. When a bare steel rod first was tried in 1881 instead of a carbon electrode, the arc, although difficult to start, operat-

---

TECHNICAL BRIEF 32:
Initiating a Welding Arc

A fundamental problem that must be overcome to start the welding arc is reluctance of un-ionized gases to conduct electrical current. Of course, immediately after constituents in the gap become ionized, conductivity will increase to a level where welding current will flow even with a potential as low as ten volts; but spontaneous ignition of an arc in an air gap by sheer current would require a hazardous level of electrical potential — perhaps several thousand volts.

Two techniques of welding arc initiation are commonly used; one is to produce momentary short-circuiting of the electrode and base metal, and the other is to impose high-voltage for a very short period across the electrode/base metal gap either as high-frequently ac, or as discharge from a capacitor. When "striking-the-arc" is employed, the tip of the electrode touches the base metal, and metal at the tip is heated very quickly to a high temperature by contact resistance to the passage of current. A "fuse action" then occurs to produce a very short gap in which the process of ionization begins. This allows the electrode to be withdrawn slightly to maintain an arc by which welding can proceed. With touch-starting, an electrical potential of about 60 volts minimum is needed for proper flow of current after striking the arc. However, current flow through the arc quickly produces further ionization, and the voltage required for continued current flow usually decreases to somewhere in the range of 15 to 40 volts depending on the materials involved and the arc length.

The arc voltage under these closed-circuit conditions is proportional to the length of the arc gap, with the shorter arc operating at a lower voltage. The proportionality ratio is dependent on the kind of gases and vaporized constituents in the plasma, but this feature of welding current flow provides a basis for automatic arc-length adjustment through voltage control. The touch-start technique with a tungsten electrode is not favored, because contamination at the tip occurs very easily with resulting adverse effects.

ed reasonably well and the electrode quickly melted and was deposited in the weld pool. Of course, without shielding from the air, the weld metal was heavily oxidized, porous, and brittle. Although the bare consumable electrode could be operated with direct current on either polarity, DCEN appeared to be the better arrangement because higher deposited rates were achieved than with DCEP. Alternating current could not be used with the bare electrode because the arc was too unstable.

Flux-covered consumable electrodes were introduced in 1907, and they immediately brought needed improvements in arc stability and weld quality. As experience with coated and covered steel electrodes grew, it was found possible to make a number of helpful changes in welding behavior through flux formulation. The distribution of heat at the positive and negative poles could be altered, and as compared with the patterns found with the nonconsumable electrode or the bare steel electrode, the heat distribution could be reversed. This explains the widespread usage of certain classifications of covered steel electrodes to accomplish good fusion and deep joint penetration if needed by operating on DCEP. Full scientific explanation for the mechanism by which covering ingredients affect heat generation at the positive and negative poles of the arc is still lacking.

As flux-covered consumable electrodes developed, their applications expanded. Covering formulations were devised to control both the mode of melting at the electrode tip and the form of metal transfer from electrode to weld pool. While metal melted at the tip is often mentioned as being "droplets," additional modes have been characterized as globular (massive drops), spray (many tiny droplets), and short-circuiting (drops that momentarily bridge the arc gap). The mode of melting and form of transfer are very important aspects on a consumable electrode welding operation because of influence on many features; including, amount of spatter, extent of air aspiration into the arc, the ease of depositing filler metal into a joint (especially when out-of-position), the shape of the completed weld, the composition of the weld metal, and weld soundness. Studies of electrode melting and transfer have made use of cine photography and required frame speed in the range of 4000 to 8000 frames per second to clearly observe the different modes and forms.

The various forces that propel molten droplets from the electrode tip to the weld pool have been studied to improve understanding of the mechanics of deposition; particularly when the electrode is being used in the overhead position. The forces vary in strength with different operating conditions, but the principal ones include (1) a plasma jet, (2) pinch effect, (3) evolving gas pressure, (4) explosive filament evaporation, and (5) electromagnetic action. The plasma jet flows axially in the center of the arc column as ionization produces a high density of ions and electrons, and these particles stream at sonic speed toward one of the poles, thus serving as a transporter. Pinch effect arises from the circular magnetic field generated around the arc column, which acts to squeeze the axial conductor (i.e., the plasma). As illustrated in Figure 6.4, when a droplet forms at the tip of a consumable electrode the compressive magnetic force squeezes this molten portion and a local constriction develops adjacent to the electrode tip. The magnetic forces acting perpendicular to the walls of the constriction act to pinch off the liquid connection and the droplet is propelled away from the electrode toward the base metal. During this rapid heating at the tip, small amounts of gases in the metal greatly increase in pressure and this can be a pneumatic-assist to the launching of droplets. Also, as the molten connection between electrode tip and droplet necks down, the very high density of current passing through the remaining filament can cause explosive evaporation and this adds to the propelling pressure behind the droplet. After this initial launch or push, the droplet can be accelerated by the plasma

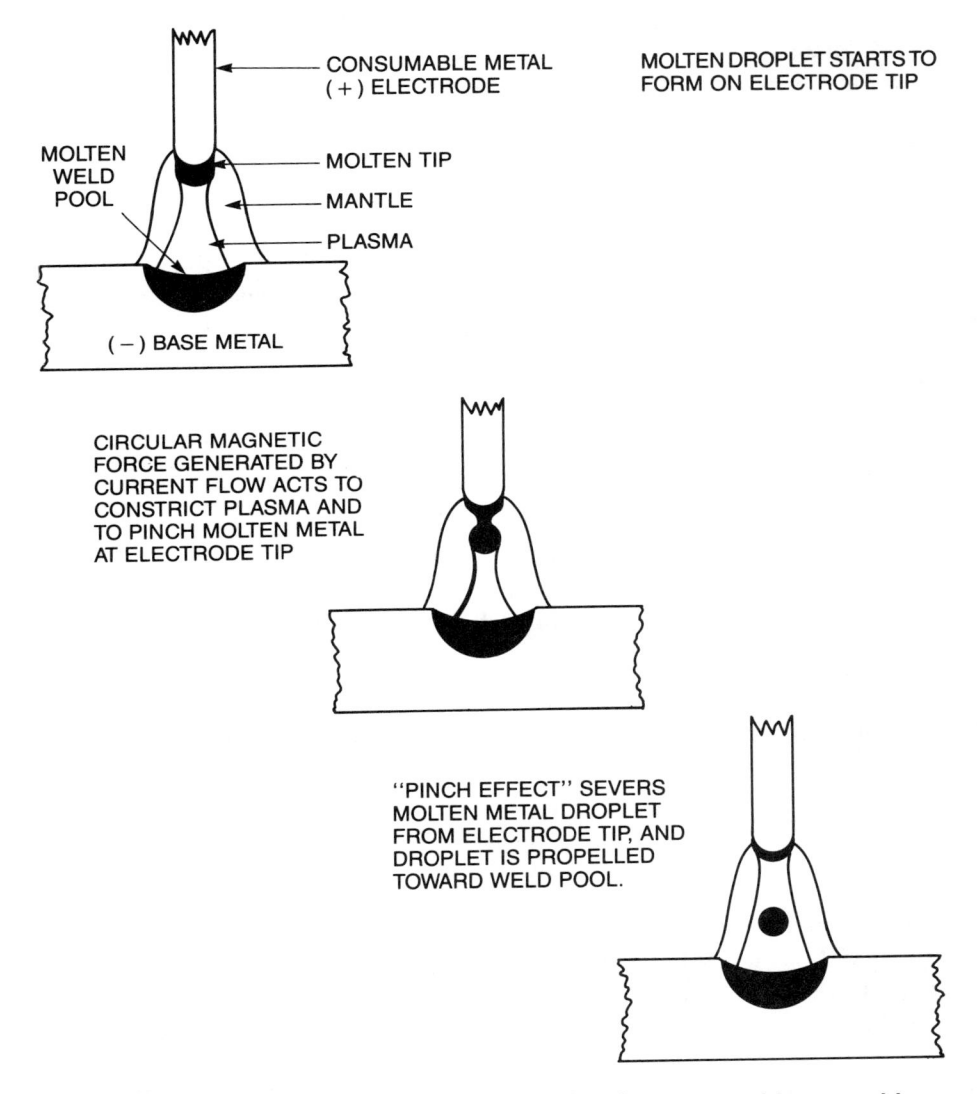

Figure 6.4 — Pinch effect during deposition of a consumable arc welding electrode

Schematic illustrations of molten metal droplet deposition from consumable electrode when arc welding with DCEP.

jet, and by electromagnetic action as the current diverges around the freely moving droplet. As each droplet comes very close to the base metal, a slight gravitational effect probably provides some additional attractive force. Finally, as the droplet enters the weld pool, the usual surface tension forces that exist in a liquid will hold the droplet as added metal. Of course, when the electrode is deposited out-of-position the metal transferred to the weld pool must solidify quickly before gravitational force causes the molten metal to sag excessively, or to flow (drip) out of the weld pool.

The high temperature of the arc and ease of bringing it in contact with (or in close proximity to) the metal to be melted make it a favored heat source, particularly for fusion welding. Power in the arc must be controlled at a sufficiently high level to insure localized fusion of the base metal, and to control the weld microstructure and heat-affected zone. Power in the arc is computed in watts using the following simple equation:

$$W = IE \qquad \text{(Eq. 6.1)}$$

where

W = Watts (Power)
I = Current in amperes conducted by the arc
E = Potential drop in volts across the arc

However, power used to do useful work and to regulate the heating and cooling rate is calculated as energy input to the weld, and is expressed as energy per unit length of weld. The following equation is widely used to calculate the heat (energy) developed by an arc during a welding operation.

$$J = \frac{IE}{S} \qquad \text{(Eq. 6.2)}$$

where

J = Joules per millimeter
I = Current in amperes conducted by the arc
E = Potential in volts (arc voltage)
S = Velocity of weld travel in mm/second

For example, arc welding with a current of 300 amperes and a voltage of 20 volts will liberate 6000 joules (per second). If the weld heat travel speed is 2.5 mm/second (about six ipm) the arc would make available to the weld in the quantity of 2.4 kJ/mm (61 kJ/inch). Prior to wider usage of the metric system, the energy available from the arc was calculated in terms of *joules per inch* as follows:

$$\text{J/in.} = \frac{I \text{ (amperes)} \times E \text{ (volts)} \times 6}{S \text{ (weld travel speed in ipm)}} \qquad \text{(Eq. 6.3)}$$

The equation above computes the total power of the arc made available per unit length of weld. It is *not* the heat input absorbed by the metal, as often implied by published material. Considerable heat can be lost by radiation, convection and conduction, and by spatter. Fortunately, the arc welding processes are able to capture the arc's energy as heat to the extent of 20 to 85 percent. Efficiency of arc energy utilization is high with the submerged arc process (SAW) because the arc is blanketed by molten flux. Efficiency is somewhat lower with shielded metal arc welding (SMAW), where the arc is exposed; and it is lowest with gas tungsten arc welding (GTAW), where loss by radiation alone can exceed 20 percent of the arc energy due to little or no self-absorption in the arc column and transparent shielding gas. In addition to differences in energy utilization among the various arc welding processes, the joint configuration, the material being welded, and the welding technique can affect melting efficiency and other heat effects. However, studies of heat flow in welding have found that with a given welding process, and for travel speeds within the usual range for ordinary welding procedures, the arc energy per unit length of weld is a useful parameter to correlate with weldability results. Consequently, further use will be made of the heat input formulae given above, particularly when discussing the welding of carbon and alloy steels in Volume II, Chapters 14 and 15.

Electron Beam

A high velocity stream of electrons as an energy source offers distinct advantages — namely, very deep penetration, precise control over the shape and size of the workpiece area to be heated, a high rate of heat input produced by the power density of the beam, and a narrow heat-affected zone. Consequently, the electron beam is now widely used for welding both thick and thin sections; and in addition has proved very useful for piercing very small diameter holes, for cutting, and for surface treatment of metals.

A disadvantage of the electron beam process is the need for a vacuum, at least in the beam-generating portion of the equipment; this adds to capital and operating costs. The means of generating a beam is similar to that of a cathode-ray or television tube (CRT); using an electron "gun" that must operate in a hard vacuum (perhaps better than 0.0133 Pa, or 10^{-4} torr). The gun incorporates a heated cathode to provide the electron source and a high voltage to accelerate the electrons toward an anode (this can be the base metal) at a velocity which easily can be about two-thirds the speed of light. By sending this stream of electrons through an electromagnetic lens, a beam can be focused to a spot of desired size, even as small as 0.1 mm (approximately 0.004 in.). The beam can also be deflected electromagnetically, allowing it to be programmed for following complex joint configurations. With a "self-accelerating" gun, it is not necessary for the workpiece to be the anode. Instead, an intermediate anode with a hole in the center allows electrons to pass through, and the emerging beam can be directed to impinge on workpieces some distance away.

An electron beam is more efficient in a vacuum because it minimizes collisions with molecules of gas, which can decelerate and defocus it. The vacuum also provides a sparse atmosphere that avoids significant oxidation and contamination of heated metal and molten metal, although some minor compositional losses can occur by evaporation and out-gassing. In operation, the electrons strike the surface of metal, releasing the bulk of their kinetic energy as heat. Quickly, a small diameter "keyhole" can be produced that rapidly melts completely through the workpiece. This penetration by the beam is so rapid that only a small amount of heat is conducted into the sidewall of the keyhole. For example, steel plate 50 mm in thickness (about two inches) can be penetrated in about one second to form the keyhole. Given a tightly abutting joint, the beam can be moved along the joint centerline so that the keyhole melts the faying interface as it advances, forcing the molten metal to the rear of the hole where it solidifies to form a weld.

The power of an electron beam also is measured in watts; it is the product of the accelerating voltage and the beam current. However the voltage is much higher than that of an arc — somewhere in the range of 30 to 200 kV; and the current is much lower — from only a few milliamperes when welding very thin material to perhaps 1.5 amperes for thick sections. When calculating rate of heat input during welding, the *available* energy in an electron beam can be calculated with the following simple equation:

$$J/mm = \frac{IE}{S} = \frac{P}{S} \qquad (Eq.\ 6.4)$$

where

J/mm	=	Joules per millimeter
I	=	Beam current in amperes
E	=	Accelerating potential in volts
P	=	Beam power in watts or joules per second
S	=	Travel speed in millimeters per second

The power supplied by the electron gun may not seem high; but bear in mind that the beam can be focused to a very small spot, producing extremely high power densities. For example, a spot 0.8 mm (about 0.03 in.) in diameter may receive 30 kW of power, which represents a power density of about 60 kW/mm^2 (approximately 38 MW/in^2). In fact, the power density of the electron beam is so high that the beam often is oscillated, or moved in a circular pattern, as it advances in order to obtain a wider fusion zone and to stir the molten weld pool so that unwanted gases can escape. Also, welding speeds can be very high.

Although a hard vacuum is required in the high-voltage gun, a soft vacuum (perhaps only 6.7 Pa, or 0.05 torr) is quite satisfactory for welding in the work chamber. For this reason many EBW machines are designed with two-chambers with the gun in a hard vacuum chamber and the electron beam being delivered through a special interconnecting orifice into a soft vacuum work chamber. Although some beam power may be dissipated into a soft vacuum, the shorter pumping time for the work chamber is an attractive incentive. Even with a soft vacuum there is no significant oxidation or contamination of the fused metal, because the amount of unwanted gases present is less than found in commercially pure inert gas. Furthermore, the electron beam can be emitted through a special orifice directly into the atmosphere if a high energy loss is acceptable and if arrangement can be made to shield the heated and melted areas on the workpieces from oxidation. Because beam energy would be dissipated by passing through air or shielding gases, the shortest possible electron travel distance is necessary between the gun's orifice and the work.

Another disadvantage of the electron beam when used as a heat source is the X-radiation generated at the work, which requires shielding for protection of personnel in the vicinity. Shielding is usually incorporated in the design and construction materials of vacuum chamber machines, but it is a more demanding task when dealing with nonvacuum application.

Electric Resistance

Metal is heated by the passage of an electric current because of its resistance just as described earlier for the electric arc. Because iron and steel have a modest resistance to current flow at room temperature (unlike aluminum or copper) resistance heating is a viable method for a number of welding processes. When an electric current flows between two contacting pieces of metal, heat is developed at particular locations, in the body of each piece, and especially at the contacting surfaces. The amount of heating depends on the inherent electrical resistivity of the metal and the contact resistance. Contact resistance between two pieces of metal depends upon a number of factors, among them the inherent or characteristic resistivity of the metals, the pressure forcing the pieces together, and the presence on the surfaces of oxides or compounds of high electrical resistivity. The quantity of heat produced is determined by the amount of current passed through the pieces, the overall resistance which the assembly of pieces offers to the passage of current, and the length of time the current flows, as shown in the following equation:

$$W = I^2 R t \qquad \text{(Eq. 6.5)}$$

where

$W =$ Heat in watt seconds or joules
$I \ =$ Current in amperes
$R =$ Resistance of metal system in ohms
$t \ =$ Time in seconds

Resistance heating is the basic method of heating in the family of processes identified earlier in Figure 6.1 as *Resistance Welding* (RW). For those processes where electric current is *induced* into the metal (induction heating), heat generation follows the above equation. However, the current density will probably vary within the cross section of the metal member because the depth to which the current is induced and its distribution will depend on the frequency of the current and the resistive and magnetic properties of the metal. In general, higher frequency will induce current that is more concentrated close to the surface of the metal section.

Resistance heating also can be generated in materials other than metals, such as molten slags. In electroslag welding (ESW), for example, resistance heating is sufficiently high so that heat can be generated not only to keep the slag molten, but also melt an adjacent body of metal.

Electromagnetic Radiation

Electromagnetic radiation is a general term incorporating a broad array of phenomena that range from cosmic rays to radio waves, including all-important visible light. Radiation has a dual nature; it behaves in many ways like a particle, but it also has wave properties. Because of its particle-like nature, radiation is envisioned as extremely small finite bundles of energy called quanta or photons. A number of forms of radiation are used by the welding industry. The complete spectrum of electromagnetic radiation is scaled on Figure 6.5. Gamma radiation and x-rays are used in nondestructive examination of welds; infrared rays from quartz lamps are used for heating workpieces when brazing or diffusion welding (DFW); and, of course, ultraviolet rays are copiously generated by arc welding. A small portion of the spectrum near its middle, where visible light and neighboring ultraviolet and infrared radiations are scaled in a narrow band, is where lasers produce radiation with characteristics so unique that they are used in the fusion welding process called *laser beam welding* (LBW).

All electromagnetic radiation travels with exactly the same velocity in a vacuum: 186 280 miles per second (3×10^{10} cm per second), commonly referred to as "the speed of light." Unlike the electron beam, which quickly dissipates its energy when passing through air, a beam of visible light will travel through air at almost the same speed as in a vacuum. If light is projected through water, the speed is reduced by almost 25 percent. Common glass reduces the speed by about 35 percent. Light can be directed to opaque or solid matter where, on collision, its photons will liberate their kinetic energy as heat; for example, as in focussed sunlight.

An important characteristic of light emitted from ordinary sources is that it spreads during travel — in fact, the intensity of ordinary light is inversely proportional to the

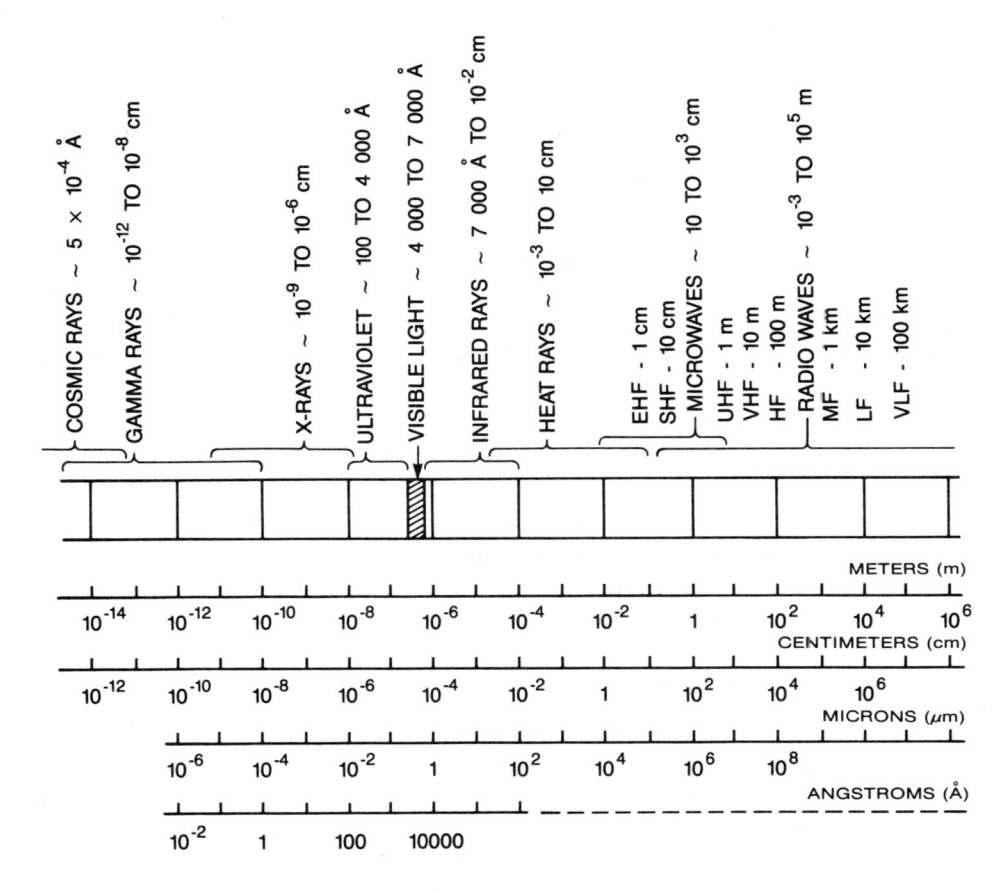

Figure 6.5 — Electromagnetic radiation spectrum showing approximate ranges of wavelengths for regions of interest

Several metric measuring scales are provided because various SI units are used in the literature on welding.

square of the distance traveled. Consequently, it is difficult to condense ordinary light and collect the energy for heating materials to a high temperature.

Laser Beams

Invention of the laser in 1960 provided a device with which energy from a primary source (electricity) could be converted into a high-energy beam of radiation which when focused would generate sufficient power density to melt metals. The output beam of the laser is uniquely suited for concentrating, either by transmitting or reflective optics because it is virtually monochromatic (single wavelength), coherent (all waves in phase), and closely collimated (minimal divergence or spreading as it travels). Under ideal conditions, an optical system can focus laser beams to diffraction dimensions, a spot diameter as small as the wavelength of the radiation. The power density could exceed 1.5 GW/m^2 (about one MW/in.^{2}). In some cases, the beam can be transmitted by a fiber optic cable.

A laser has three basic components: (1) a laser material or medium, (2) an excitation system, and (3) a resonant cavity. The *lasing medium* can be either a rod-like crystal, or a tube of gases. The first lasers were of the solid state type; and, initially, single crystals of ruby served as the lasing medium. In this decade, glass rods doped with *neodymium,* or *neodymium in yttrium-aluminum-garnet* (Nd-YAG) were developed to improve operating efficiency. Excitation in solid-state lasers is usually accomplished with high-energy flash lamps that closely surround the rod-like member and provide intense white light to start the laser in operation. One mirrored end is only partially reflective, and a proportion of the amplified radiation escapes through this exit end of the laser rod. Hence the acronym, LASER, arose from the following description of the mechanics of the operation; Light Amplification by Stimulated Emission of Radiation. Early solid state lasers produced a pulsing beam because most of the electrical energy put into the device had to removed as heat. This inefficiency, along with other reasons, limited solid state lasers to low levels of beam power, usually about one kW maximum, and continuous beam output was not easily maintained. Nevertheless, these "pulsed wave" lasers were very useful for precision welding of thin sections and small diameter wires.

The latest solid state lasers have been designed to avoid the heat problem, and can be operated in a "continuous wave" mode. YAG (CW) lasers with beam power as high as 2 kW have been built for welding and cutting.

There are reasons to note the wavelength of radiation produced by lasers. The beam from a ruby laser has a wavelength of 0.69 µm (the near-infrared region) and the angle of divergence can be less than 0.75 degree for a particular laser. The neodymium glass and the Nd-YAG lasers produce a beam with 1.06 µm wavelength, and with little divergence.

The *excimer laser* is pulsing laser introduced in the 1990s that holds strong promise for welding and cutting. It employs a special mixture of gases as the lasing medium instead of a solid rod. The mixture may consist of a noble gas (krypton or xenon) and a halogen (either fluorine or chlorine). Excimers are short-lived dimeric molecules that exist when the gases are energized to an excited state, and on return to the ground state pulses of ultraviolet radiation are emitted. The beam from a KrF mixture has a wavelength of 0.248 µm, while the XeCl mixture produces a beam of 0.308 µm wavelength. The excimer laser beams are not visible to the human eye, and the path of the beam must be detected by a special technique.

Continuous wave lasers are mostly of the gas type, and are built in either of two basic designs — *axial flow* and *transverse flow*. Both are suitable for continuous operation, and units have been built in power exceeding 25kW. The lasing medium is usually carbon dioxide mixed with gases such as helium and nitrogen. The gas is contained in a tube, and is excited with a high-voltage dc glow discharge. Nitrogen readily accepts this energy and achieves vibrational levels that are close to those of carbon dioxide; consequently, the CO_2 molecules are excited by resonant energy exchange. The CO_2 molecules then undergo a downward transition in energy level, and emit photons that form a laser beam with a wavelength of 10.6 µm (infrared radiation). Helium in the gas mixture significantly expedites the lasing transition. Carbon monoxide is being investigated as a gas for the laser, because it offers the high power capability of CO_2, and its radiation wavelength appears advantageous for fiber optic delivery. Cooling of the lasing gas is important because the electrical glow discharge produces heat, and the lasing action becomes inefficient if the gas temperature rises above approximately 200°C (about 400°F).

The 10.6 µm laser beam is focused with lenses made of zinc selenide or potassium chloride. The beam can be reflected by gold-plated copper mirrors and directed through a system of lenses to the target area on the workpiece. The beam can be passed through a wide variety of transparent materials and atmospheres with little loss in power before it impinges on the workpiece to release its energy.

Although the laser beam could be focused to a very small spot in order to produce high power density, metal sections under many circumstances could not be penetrated by melting to form a keyhole as described earlier for electron-beam welding. Two factors significantly affect the absorption of energy by the workpiece. First, a shiny metal surface reflects light at the wavelengths of laser beams. This problem can often be overcome by blackening the surface of the workpiece to be heated. Second, when sufficient energy is absorbed to melt a cavity preparatory to keyhole development, metal vapor is produced. This vapor can absorb enough energy to become a plasma with sufficient reflectivity to buffer further absorption and halt penetration. The plasma that develops in the penetration cavity of aluminum and copper is especially reflective. One remedial measure to overcome the buffering effect of the plasma is to increase the power density beyond a critical threshold whereupon keyholing occurs. Another practical solution to the penetration problem is to use a jet of supplementary gas to sweep the obstructive plasma from the cavity or keyhole and achieve complete penetration without unusually high power density.

An entirely new approach to welding and cutting using electromagnetic energy is presently under development as *microwave plasma*. This process capitalizes on the efficacy of electromagnetic radiation for heating combined with the large amount of energy that can be concentrated in a gas through ionization. Microwaves in the spectrum band from about 10 to 10_3 cm wavelength are a powerful heat source, but they cannot be applied directly to metal because of sporadic breakdown of their electromagnetic energy (as can be witnessed when a metallic object is unwittingly used in a microwave oven). Nevertheless, a unique system has been devised for transferring microwave energy to a plasma stream. Microwaves from a magnetron generator are transmitted via coaxial cable or a waveguide into a tuned cavity where a stream of argon gas is ionized and emitted as a columnar plasma approximately 2 mm (0.08 in.) in diameter. The energy of the plasma is easily regulated, and temperatures in the order of 4000 °C (7200 °F) appear to serve well for welding. Higher temperature can be arranged for cutting, and the velocity of the plasma flow is readily controlled. The significant advantage of a system that will generate a plasma stream using microwaves (as compared with power beams from a laser or from an electron gun) would be the very efficient conversion of microwave energy into heat to form the high-temperature plasma.

Chemical Heat Generation

Only a limited number of exothermic chemical reactions can be utilized as a heatsource for welding. To be suitable for fusion welding, a reaction between substances must progress quickly, reach a temperature well above the melting point of the metal workpiece, and offer a high heat content. Other considerations involve availability, safety in handling, cost, diverse metallurgical effects, etc.

The flame formed by a combustible gas was one of the first successful forms of chemical heat generation used for welding. Although acetylene, hydrogen and propane each held promise as a fuel gas, it was quickly found that, when burning these gases, purified oxygen was needed to achieve a flame temperature high enough to serve as a localized, moving heat source. Of these, only the oxyacetylene flame at about 3200°C (approximately 5800°F) proved completely adequate for welding ferrous metals. Even though propane offers a higher caloric value than acetylene, propane requires a larger volume of oxygen for combustion. This dilutes the flame and reduces its temperature to about 2900°C (approximately 5200°F). This small difference in the intensity of the heat source has a significant effect on welding proficiency. Consequently, the oxypropane flame is used for heating iron and steel rather than fusion welding. Hydrogen also is limited in its welding application, because its flame temperature is in the order of 2800°C (about 5000°F).

About 15 percent of the heat from the flame is transferred by radiation. Most is by contact between the hot gases flowing from the flame and the cooler surface of the workpiece where conduction of heat takes place. Naturally, the quantity of hot gas being issued is another factor that enters into the question of heating power. The term *combustion intensity* has been used to characterize a flame and takes into account both flame temperature and flame speed (burning velocity). Combustion of the fuel gas may occur in one or more steps, giving rise to distinct zones in the flame. Each zone has a different temperature, a different set of gaseous combustion products, and different amounts of entrained air. By controlling the proportions of fuel and oxygen, the general conditions within specific zones of a hot flame can be adjusted to be *oxidizing, reducing,* or *neutral*. The oxyacetylene flame is a prime example of this multizone combustion behavior, and the manner in which this can be used to advantage will be discussed when the OAW process is reviewed.

Chemical heat generation by solid substances is put to use in the thermite welding (TW) process where powdered aluminum and iron oxide are pre-mixed together, and after ignition by an external heat source to initiate a vigorous, self-supporting exothermic reduction reaction the mixture produces superheated metallic iron, which is utilized as a heat source to melt base metal surfaces and to provide filler metal to form a weld between these surfaces in the workpieces. Solid-substance heat sources can also be used for localized heat treatment of metal sections; for example, to apply stress-relief heat treatment to the area containing a weld. For this purpose, silicon often replaces the aluminum to obtain an exothermic reaction of slower rate and lower temperature.

Mechanical Heat Generation

Heat for welding can be generated mechanically by utilizing friction, impact, or deformation of metal. Frictional heat is developed by rubbing two surfaces together under proper combinations of speed, pressure and repetitive motion until a temperature conducive to welding is attained. The impact of two colliding sections of metal will develop heat, but tremendous collision velocity and minimum recoil is necessary to release sufficient kinetic energy as heat to play a real role in a welding process. When deformation is utilized as a means of heating, the extent of deformation in the metal workpiece can be *elastic*, in which case many cycles of load application must be rapidly applied to obtain even a modest temperature rise; or it can be *plastic,* in which case the temperature rise will be commensurate with the degree of deformation and the rate at which it occurs. Naturally, increasing deformation at a more rapid rate results in faster temperature rise in the deformed volume of metal.

Frictional heat for welding can be obtained by rapidly rotating the surface of one workpiece under pressure against a stationary surface of another workpiece until a high temperature is attained at the faying surfaces. The frictional forces appear to develop in several successive stages and can be described as (1) dry sliding friction, (2) localized seizure-and-rupture, and (3) plastic deformation and viscous flow. At the start of dry sliding motion, asperities that act to interlock the contacting surfaces are rubbed flat, and any contaminants on the surfaces such as oxides are scuffed away. During the second stage, tiny areas on the surfaces are brought into more intimate contact where atomic and molecular attractive forces can promote seizure, but these minute areas are almost immediately ruptured as the sliding motion continues. The amount of heat that develops during these first two stages depends on the kind of metal in the workpieces, the surface finish on each piece, the amount of pressure keeping them in contact, and the velocity of relative motion. However, most of the heat required to accomplish welding is generated in the third stage. This is when plastic deformation occurs in the heated metal immediately adjacent to the faying surfaces, and a viscous flow of surface debris and metal is exuded from the joint as the rising temperature lowers the strength of the metal. At the conclusion of this third stage in the friction welding process, the pressure can be increased to upset a greater amount of metal from the joint to ensure its quality, or to control overall dimensions of the welded workpiece.

Heat generated by elastic and plastic deformation of metal contributes to the welding mechanism in several processes. For example, in the ultrasonic welding process

(USW), a high-frequency (about 20 kHz) vibration is supplied to the workpieces which causes localized, reciprocating lateral slip at the faying surfaces. Like friction welding on a microscopic scale, the onset of welding is marked by breakdown of asperities, dispersion of oxide film, and localized plastic flow resulting in intimate contact between the workpiece surfaces.

A more clamorous example of heating of metal by plastic deformation is found with the explosion welding process (EXW), an operation which involves bringing together the surfaces of two workpieces at extremely high velocity. Naturally, the kinetic energy released at impact will cause deformation and heat. The rise in metal temperature assists in forming a weld, however, the EXW process deforms the surfaces to be welded at a very high rate, producing a phenomenon termed *jetting*. The jetting mechanism plays a major role in the process.

THE WELDING AND CUTTING PROCESSES

Virtually all known welding and thermal cutting processes are applicable to carbon and low-alloy steels, and they are summarized here. Currently used welding processes have become highly developed, and much detail would be required to completely cover their many variations and refinements. Space is not available in this volume for a complete description of each process, the equipment employed, or the procedures involved. This information can be readily obtained from one of the handbooks or other publications listed in *Suggested Reading* at the end of this chapter. Instead of repeating the information in those publications, this volume will concentrate on the metallurgy of each process; the ways in which specific operating aspects influence the chemical, physical, and mechanical properties of welds produced; and the effects of these features on the performance of the weld in service.

Arc Welding Process

This review starts with the *arc welding* (AW) process because it has been the most adaptable to the joining requirements of a great variety of fabricating and manufacturing operations. Dozens of AW processes have been developed, including manual, semi-automatic, machine, and automatic; and innovations in utilizing the arc in new welding and cutting processes have not been exhausted. The arc can be tailored with respect to size, temperature, and other forces that it generates, and this versatility accounts for much of its widespread usage. The extremely high temperature that the arc can produce permits it to supply a large amount of heat to a small area and to localize welding effects. The surrounding atmosphere can vary greatly depending on the process, as will potential troublesome contaminants. It is necessary, therefore, to review processes on an individual basis inasmuch as variations in materials, equipment, and operational procedures often result in significantly different metallurgical events during the making of a weld; and these variations will be reflected in the final features and properties of welds.

Power Sources for Arc Welding

Electric current for welding arcs is supplied by *arc welding machines*, often called *power sources*. They are designed specifically for arc welding processes because arcs operate at relatively low voltages and high currents. The power source could be a large storage battery; but for the convenience of continuous, steady-state electrical power from generating station lines, a unit that steps down the primary line voltage to the 15-35 volt range needed to properly sustain a welding arc is preferred. The open circuit (no current) voltage may be as high as 80 V. Higher voltages generally are regarded as hazardous because for the possibility of severe electrical shock. Welding power sources may deliver three distinctly different kinds of electrical power for welding: (1) alternating current at the same frequency as the primary current (60Hz in the USA), (2) direct current, and (3) pulsed current with various wave forms and frequencies. Moreover, power sources have two general classes of operating characteristics that affect their performance — *static characteristics*, which are measured by output voltage and output current; and *dynamic characteristics*, which are determined by response to transient conditions caused by momentary changes in the welding arc. The first characteristic is assessed by the relationship maintained in the output between the voltage and the amperage, both before and after the arc is established. This relationship can be categorized by a plotted curve, and in this way power sources can be described as offering either *drooping arc voltage* or *constant arc voltage*.

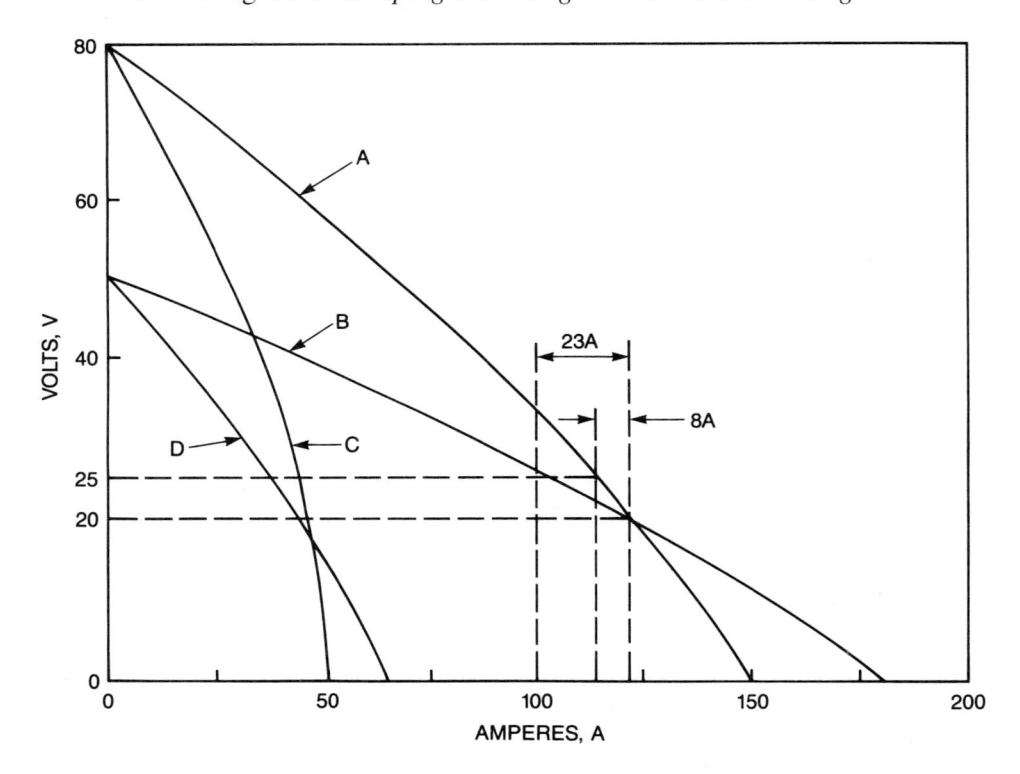

Figure 6.6 — Typical volt-ampere curves characteristic of "drooping type" power source

This particular power source has adjustable open-circuit voltage that can range from 50 to 80 V, and also has adjustable current output. (Reprinted from *AWS Welding Handbook, Vol 2, Welding Processes*, Eighth Edition, Chapter 1, Arc Welding Power Sources, Figure 1.13)

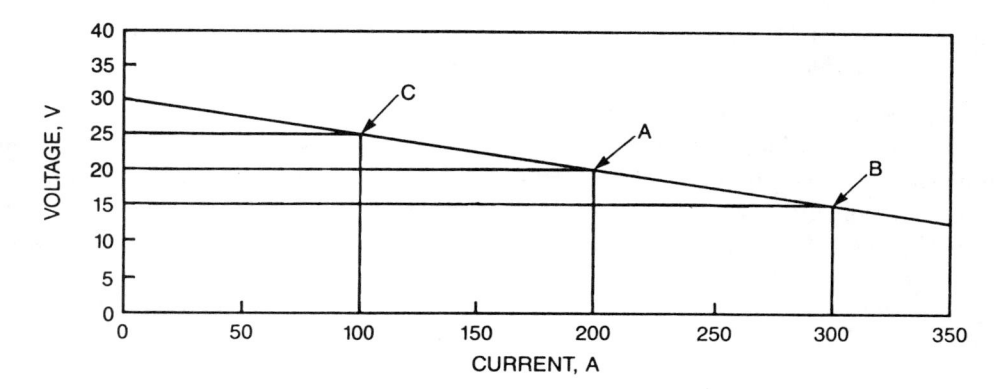

Figure 6.7 — Volt-ampere relationship for a typical "constant voltage" type of power source

(Reprinted from *AWS Welding Handbook, Vol 2, Welding Processes*, Eighth Edition, Chapter 1, Arc Welding Power Sources, Figure 1.14)

The first, the "drooper," has a pronounced negative slope in its volt-ampere output curves. Figure 6.6 illustrates curves typical of a drooping-voltage power source. Note that the percent changes in amperage are much smaller than changes in voltage. The machine illustrated has an adjustable open-circuit voltage (stand-by voltage when no welding current is flowing) that can range from 50 to 80 V. It is also equipped with controls that allow the slope of the V-A curve to be altered. For example, with the voltage set at the highest level, this power source can be operated per curve "A", where a change in voltage from 20 to 25 closed-circuit (arc) volts (25 percent) would result in a decrease in current from 123 to 115 A (6.5 percent). This change in current is relatively small, and the amount of heat being liberated by the arc to perform welding would not be changed appreciably. If even a smaller change in heat from the arc were desirable, the power source could be set to produce a steeper V-A slope (curve "C") whereupon a variation in arc length that changed the voltage would vary the current over a narrower range. This kind of power source is sometimes referred to as a "constant current" machine. A welding machine with drooping characteristics is preferred for manual arc welding because the small changes in arc length which occur as the electrode is manipulated will not significantly change the welding current. If some change in current is desirable, (as for out-of-position welding) it can be accomplished with controlled changes in arc length by selecting a more shallow slope in the machine output.

The second welding machine, the "constant voltage" power source, is characterized by the V-A curve shown in Figure 6.7. Although the voltage output is not constant, the curve has much less negative slope than the drooper power source just examined. With the constant voltage power source, a small change in voltage causes a large change in current. This characteristic is useful with consumable-electrode arc welding processes that use constant wire feed (i.e. constant melting rate), because the arc automatically tends to maintain a constant length. Changes in the gun-to-work distance cause slight but proportional changes in voltage and greater inverse changes in current. Fortunately, with a fixed deposition rate, the changes in current compensate for changes in electrode extension, keeping the arc length fixed.

Alternating Current Power Sources

Welding machines that supply ac are mostly of the transformer type, operating on single-phase line current and stepping down the primary voltage to levels suitable for welding. Typically, open-circuit voltages will be in the range of 50 to 80 V, but may be as high as 100 in some countries. Since voltage levels above 80 pose an electrical hazard they should be equipped with an automatic "hot start" control that provides the higher voltage only for a fraction of a second to initiate the arc and commence the welding. The ac welding power sources make use of various mechanical and electrical devices for controlling output, such as moving cores, tapped secondary coils, saturable reactors, thyristors, etc. Some ac machines, such as those using *thyristors*, can be used to change the conventional sine wave to a "square" wave to help in starting and stabilizing the arc. Also, the ac wave form can be imbalanced to change the proportion of positive and negative polarity.

Direct Current Power Sources

Originally, welding machines that supplied dc were rotating generators driven by electric motors or by internal combustion engines. Now, most are transformer-rectifier units, many of which have a convenient switch to by-pass the rectifier and provide ac for welding if needed. The V-A characteristic of a dc power source can be drooping (constant current), or it can be constant voltage. Transformer-rectifier power sources are popular because they have no moving parts, are easy to maintain and are relatively inexpensive. They also are very adaptable, providing (1) dynamic inductance to control arc stability and mode of metal transfer from consumable solid metal electrodes, (2) pulsed direct current for achieving a required mode of metal deposition, and (3) when equipped with solid-state transistors and inverters, different frequencies and wave forms in pulsing direct current.

Auxiliary Equipment for Arc Welding

To minimize the occurrence of defects at the beginning and ends of welds, a number of pieces of equipment are available to help initiate and to stop the welding operation. The starting point is vulnerable because of potential difficulties in initiating the arc and quickly arriving at steady-state conditions. However, when the arc is suddenly extinguished, the molten weld pool at the terminus, called the *crater*, may not solidify properly. A *starting weld tab* and a *runoff weld tab* extending from the ends of the joint will help avoid incorporating these two points of potential difficulty in the finished weld; however, the use of these tabs is often impractical.

One effective means of arc initiation is to superimpose high-frequency, high-voltage, low current on the power supply output. Most commonly used is the spark-gap oscillator. The frequency may range from 50 to 3000 kHz, while the potential can be in the order of several thousand volts. It provides extremely high electrical potential at the electrode tip to discharge a spark and create the ionized gas path needed to initiate a welding arc. High-frequency superimposed current is very helpful with tungsten electrodes to eliminate the touch-starting that easily causes electrode contamination. For arc initiation only, the high-frequency current is applied for a fraction of a second at the start of welding. Although easy to use, high-frequency stabilization can produce troublesome radiation effects and radio interference if not handled in accordance with strict guidelines on electrical connections, transmission, and shielding.

Controls can also be included in the welding equipment to taper down the current at the end of a weld in order to achieve uniform profile, and to fill the crater to avoid defects. A foot pedal provides current control for some manual welding, and controls for programmed automatic welding include components for current adjustment.

Basic Forms of Arc Welding

There are three basic ways of using an arc to provide heat for welding. One is to establish an arc between two poles or electrodes and bring the arc into close proximity of a joint to melt the metal and produce a weld. This arrangement is used in the *atomic hydrogen* welding process (AHW) with a pair of tungsten electrodes supporting an arc shielded with hydrogen gas. A second way of arc welding is to have a single electrode as one pole for the arc and the base metal the second, opposite pole. In this way the arc is always in intimate contact with the base metal to be heated and fused. This method can be very efficient and, with consumable electrodes, can provide an external source of filler metal as well. The third arc welding method is to position the two pieces of base metal to be joined with a very small gap between them, each piece serving as an electrical pole of an arc initiated between them. This relatively simple method is used in arc *stud welding* (SW), and for making a butt joint between the ends of light-wall tubing. The arc must be sustained long enough to superficially melt the opposing surfaces, but a mechanical action usually must follow to close the joint and allow solidification of a weld.

The second of these methods, in which the arc is established between an electrode and the workpiece, is clearly the most important. Electrodes for arc welding were mentioned earlier as being either *nonconsumable*, or *consumable*. The term, nonconsumable, implies more or less permanent material that does not enter into the welding operation in any way other than acting as a pole to support the arc. More importantly, the nonconsumable electrode materials, if transferred to the workpiece, would be major contaminants. The most commonly used materials are carbon and tungsten, both of which have high melting temperatures and very high boiling points.

Although carbon electrodes are seldom used today for arc welding, they are extensively used in arc cutting and gouging operations. The disintegration of carbon electrodes by electrical forces deserves examination because of potentially harmful metallurgical effects. Whenever a carbon electrode serves as the positive pole for an arc (DCEP, or the positive half of an ac cycle), some carbon atoms are freed from the surface as ions and they are propelled to the cathodic base metal. This loss of carbon atoms represents electrode erosion. Carbon ions that reach a molten pool of some metals, especially iron, are absorbed very efficiently. The carbon content of iron or steel melted in this manner easily can rise above one percent. This mechanism requires consideration when making use of air carbon arc cutting and gouging (CAC-A).

During the early years of welding, consumable electrodes were merely sheared strips of sheet, but steel rod and straightened-and-cut wire soon followed. It did not take long for users of arc welding to realize that bare steel electrodes were unsatisfactory for arc initiation and stability, and that in general the welds produced were overly porous, poorly shaped, and quite brittle. A modest improvement in arc stability was forthcoming as light films or coatings of various chemical compounds were applied to the rod or wire surface. One continuous method that was widely used in those early years involved a rod dipped in hydrated lime, Ca(OH)$_2$, prior to cold drawing it into wire for electrodes. The presence of calcium on the electrode surface helped stabilize the arc and improve the appearance of the deposited weld, but it did not remedy the brittleness problem.

Bare or lightly coated consumable electrodes are no longer in use. When welding with a bare, unprotected electrode, sufficient oxygen and nitrogen from the surrounding air infiltrate the arc and exert a number of harmful effects, including loss of certain alloying elements, poor deposit shape, porosity, and embrittlement of the weld metal. Embrittlement is mainly a result of quantities of oxygen and nitrogen absorbed by the metal it passes as droplets through the arc, and when it is molten in the pool under the arc. The quality of the weld can be significantly improved by providing some shielding from the oxygen and nitrogen in the air. Therefore, virtually all modern arc welding processes include some form of protection for (1) the heated end of the electrode, (2) the plasma through which the melted electrode metal is being transferred, (3) the molten weld pool, and (4) the newly solidified, but still very hot, weld bead.

As base metal is melted under a consumable electrode to form a weld, the amount of base metal melted depends on many factors, such as polarity of current, travel speed, arc voltage, type of shielding supplied to the arc and mass of base metal. The molten volume on the base metal opposite the electrode tip and arc is called the *weld pool*. The pool usually is slightly dished because of the pressure or force exerted by the arc. This dished shape is called the weld pool *crater*, and its depth is determined by many factors associated with the electrode, welding conditions, and the base metal. The molten metal from the electrode that enters the weld pool quickly commingles with that melted from the base metal. A relatively homogeneous mixture of metal from these two sources is achieved quickly through the electromagnetic stirring that occurs in the weld pool. Therefore, a single weld deposit is uniform in composition from side to side and from top to bottom. Most arc welds, however, have a very narrow zone of metal between the weld metal and the base metal that is called the *superheat meltback zone*. This zone may be only about 0.05 mm (0.002 in.) in width, and it represents base metal that was melted in situ by the superheated weld deposit. Its composition will remain that of the base metal, because the stirred weld pool moves away along the weld pass before melting takes place. The width of this zone increases as the heat input of the arc welding operation becomes greater. The composition, microstructure, and mechanical properties of this meltback zone will be examined in more detail in Volume II, Chapters 9 and 15.

Shielded Metal Arc Welding (SMAW)

For many years, *shielded metal arc welding* (SMAW), performed manually, was the most widely used fusion welding process for steel fabrication, but over the past decade other more efficient consumable electrode processes have been rapidly overtaking SMAW. The SMAW process, as defined by AWS, uses a consumable electrode consisting of a core that produces filler metal, and a flux covering that provides shielding during arc deposition. In shoptalk, the manual process is commonly referred to as "stick electrode" welding. The electrode's core can be solid metal wire, cast rod, or metal tube filled with powdered metals and alloys. The flux covering on the core can be applied by dipping, but more often a relatively heavy covering is extruded on the core. Flux coverings consist of carefully prepared mixtures of many powdered minerals and metals that perform a number of functions during welding. These will be summarized here, but the formulation of coverings and the specific functions of ingredients are discussed in detail in Chapter 8.

AWS specifications for covered arc welding electrodes can be found in AWS A5.1 for carbon steel classifications, and in AWS A5.5 for low alloy steel classifications. These standardized electrodes will be reviewed in detail in Volume II, Chapter 10.

Ingredients in the electrode covering can be selected to provide gaseous shielding, arc stabilization, fluxing action, and slag formation. Much alloying of the weld deposit is accomplished using powdered metals in the covering, which transfer into the weld deposit during deposition. In fact, most SMAW electrodes used for carbon and low-alloy steels consist of a low-carbon plain steel core wire with virtually all alloying done via additions in the covering. Therefore, care must be taken that a full measure of flux covering, which may comprise as much as one-quarter of the electrode by weight, is always intact on each increment of the core as the deposition proceeds.

In Figure 6.8, the SMAW process is diagrammed to illustrate the consumable covered electrode depositing filler metal as a bead-on-plate weld in the flat position; the essential functions are noted. Covered electrodes for manual welding are made in standard sizes that range from about two to 10 mm ($3/32$ to $3/8$ in. in the USA); defining the diameter of the metal core. The thickness of the flux covering is determined by the requirements of the specification to which the electrode is marketed and its handling characteristics. This range of electrode sizes can be used for welding base metal ranging in thickness from thin sheet to heavy plate. Electrodes as large as 19 mm ($3/4$ in.) are made for special applications, such as filling large cavities in castings, or modifying the shape of metalworking dies. They are cut to standard lengths depending on their size and proposed usage. The most common length in the USA is 350 mm (14 inches). Very small sizes may be shorter to avoid excessive heating from resistance to the current being passed along their length, and larger sizes may be longer. One end of the electrode is stripped of covering for about 25 mm (1 inch) to permit gripping in a holder to provide electrical contact. The opposite end usually has the covering chamfered, with the exposed end of the core cleaned of flux to permit touch starting the welding arc. When the arc is struck with a SMAW electrode, the first small increment of filler metal deposited will not receive normal shielding, nor will it contain intended additions of deoxidizing and alloying elements. This can cause porosity and inadequate alloying at the very start of a weld pass. The importance of this small deficiency depends on many aspects of the weldment being produced and is discussed further in Volume II, Chapter 13. This problem can be aggravated on restarting a partially used electrode if covering fragmentation results in a completely bare core for a short length of the electrode.

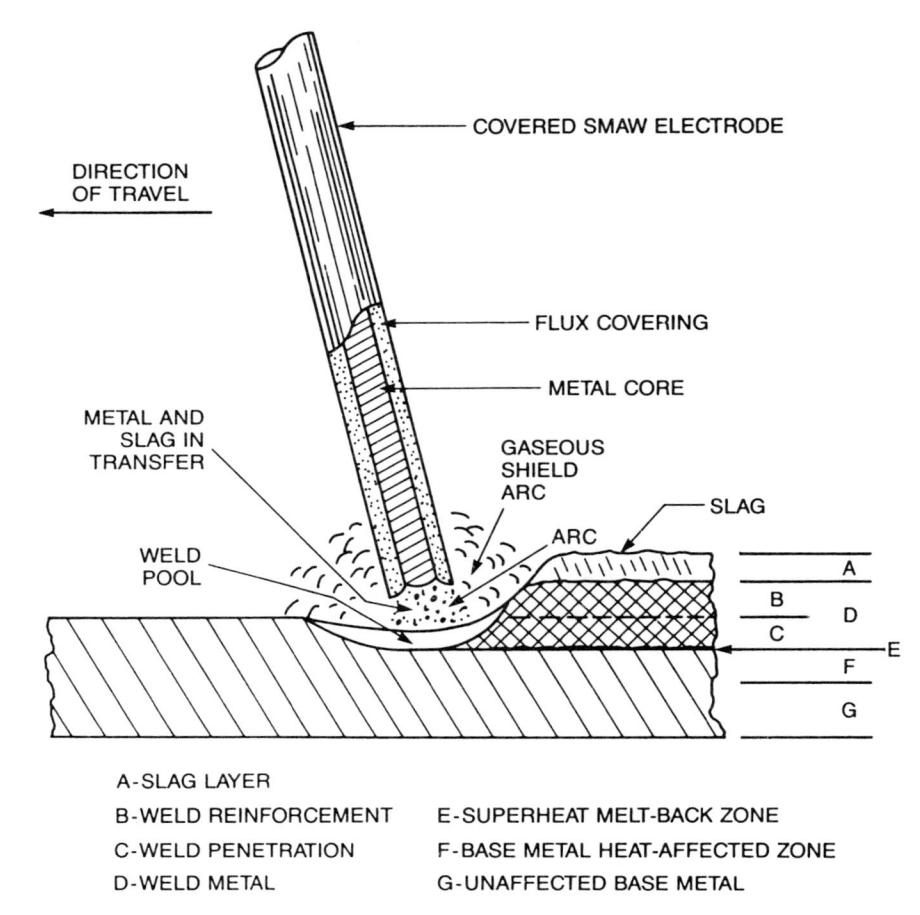

Figure 6.8 — Diagrammatic sketch of the shielded metal arc welding process (SMAW) using a consumable flux covered electrode

The coverings on most SMAW electrodes are formulated to resist melting just enough to form a short conical projection around and beyond the arcing core, thus accomplishing two major objectives — providing greater protection of the melting end of the metal core, which helps direct the flight of molten droplets toward the weld pool; and preventing short-circuiting of the electrode should it happen to touch the base metal surface. The mode of metal transfer from SMAW electrodes to the weld pool has been studied intensively because of its role in establishing many of the deposition characteristics of a given type of electrode. A number of specialized techniques were developed for measurement of droplet size, arc pressure, molten droplet viscosity, etc. These features have been correlated with welding current, arc voltage, and other more abstruse operating conditions, such as the sound of the arc. The electrode covering and its particular formulation of ingredients are highly influential in establishing the mode of transfer and this subject is discussed in Chapter 8.

The temperature of molten droplets as they leave the tip of the electrode is an important facet of arc welding metallurgy because of unusual chemical reactions that can take place between the metal droplet and a covering of slag before the droplet is deposited in the weld pool. Determination of droplet temperature has been difficult because of the intensely hot plasma which surrounds the relatively small droplets. Also, the slag that covers the droplets has an emissivity quite different from that of the metal, and the droplet is in existence for a very short time. Calorimetric measurements and special thermocouples inserted along the centerline of the core have established that the temperature of droplets from a steel SMAW electrode is usually in the range of approximately 1900 to 2400°C (3452 to 4352°F), which is much higher than the bath temperatures during steelmaking. This high droplet temperature helps explain the very rapid changes in deposited metal composition that can occur as covered electrodes are deposited. The very short duration of slag-metal contact may not allow chemical equilibria to be reached, but intimacy of contact and high temperature strongly favor rapid reactions. The temperature of molten droplets is dependent to some extent on the welding parameters employed, but temperature differences are relatively small and the high temperature range mentioned above is not likely to be breached. Significant slag-metal reactions that are known to occur during the SMAW process are reviewed in Chapter 8 when the functions of electrode covering ingredients are discussed.

"Drag rod" is shop jargon for a SMAW electrode made with thicker-than-ordinary covering to simplify manual welding in the flat and horizontal positions. It permits the welder to drag the end of the electrode as it is melted by the arc along the weld pass line, thereby reducing the degree of skill required. The electrode is usually made longer (about 800 mm or 28 inches) to achieve a weld pass of reasonable length, to minimize resetting and restarting, and to minimize loss from unusable-stubs. Welds made with the drag technique should follow a carefully planned procedure, and qualities of the weld should be monitored to ensure fitness-for-purpose.

"Gravity welding" is a method for depositing SMAW electrode in the flat and horizontal positions that is virtually mechanized by use of simple, portable equipment. Less operator training is needed for gravity welding as compared with the regular manual SMAW process. The method is commonly used in shipbuilding to make short horizontal fillet welds that are required in great numbers on secondary structural elements of a ship's hull and decks. A tripod supports the grip-end of the electrode while the arcing end is held against the workpieces and is guided along the weld pass line either by a holder that slides downward under the force of gravity, or by a spring-loaded holder. The electrodes are usually specially made in long lengths to produce welds of desired lengths. Because the angle between the electrode and the base metal surface changes as the electrode is deposited, there will be a small, gradual variation in weld pass size and in joint penetration along the weld length. These quality aspects ordinarily will require that a prescribed procedure be followed and that weld inspection be conducted to ensure compliance with construction specifications.

Much of the published data for mechanical properties of weld metal are obtained from test welds made in the flat position. This position is most amenable for achieving optimum shielding, freedom from imperfections, and uniform microstructure. When welding out of position, however, the welder must exercise considerable skill in the weaving and whipping manipulations of the electrode to produce a weld pass of satisfactory shape and size. These motions of the electrode and the arc, in general, expose more of the metal at the surface of a shallow weld pool, and cause some disturbance of the gaseous shielding over the pool. Consequently, there is a tendency for weld metal deposited out of position to pick up more oxygen and nitrogen from the atmosphere, and these residual gases have adverse influence on weld metal ductility and toughness. Weld metal deposited out of position also is prone to greater loss of easily oxidized elements (e.g., Si, Cr, Al) intended to be present for some purpose.

Hydrogen is a troublesome element in SMAW covered electrodes. However, certain coverings on SMAW electrodes clearly benefit from the presence of hydrogen-containing materials in their formulation. For example, the cellulosic type of covering widely used on carbon steel electrodes (e.g., E 6010) incorporates as much as 30 percent organic material; and it must be carefully baked during electrode manufacturing to retain the small but important percentage of combined moisture necessary to achieve the deposition characteristics preferred by welders. When this kind of covering becomes "dried out" in storage, it is not unusual for welders to immerse the electrodes in water briefly to regain the desired operating characteristics. On the other hand, with alloy steels and stainless steels, hydrogen can be the bane of SMAW welds. For this reason, low-hydrogen electrodes were developed using covering ingredients with lowest possible water-of-constitution, baked at high temperatures to reduce water-of-crystallization as much as possible, and shipped in ways that would minimize moisture pick-up from the atmosphere. The moisture content was expected in some cases to be below 0.20 percent to classify as a low-hydrogen electrode. Because electrode coverings tend to pick up moisture (they are hygroscopic), users of SMAW must be keenly aware of the practices to be followed when hydrogen must be controlled in a given welding operation. The hydrogen problem receives considerable attention in Chapters 10, 12, 13, 14 and 15. The problem of hydrogen (moisture) in SMAW has been minimized somewhat with the development of coverings having improved resistance to moisture pickup, and these moisture-resistant electrodes are given additional labels by their respective manufacturers.

Another general problem of metallurgical nature that can arise in welding with SMAW covered electrodes is the presence of a slag layer during the operation. Electrode coverings are carefully formulated to generate a slag that quickly separates from molten metal and floats atop the molten weld. Of course, avoiding entrapment of this slag in manually made welds is highly dependent on the manipulative skill of the welders, but success in making a weld that is free of injurious slag also depends on the characteristics of the slag. As explained in Chapter 8, the slag must not be so fluid at welding temperature that it fails to adequately cover the molten metal. Yet, the slag must not be so viscous that it fails to escape from the overlaps of molten metal that occur as the electrode is deposited with weaving or whipping motions. Finally, the solidified slag should be easily and cleanly removed from the solid weld pass, and should not leave tenacious particles in crevices. Slag particles that are somehow tightly held on the surface of a weld, instead of being remelted and floated away when a subsequent additional weld pass is deposited, may remain entrapped and become a weld defect.

Stud Arc Welding (SW)

The need to join various kinds of studs or other small articles to surfaces of larger members has led to the development of specialized semiautomatic and automatic operations which make use of various welding processes. These include resistance and friction welding, but interest here is directed to arc welding; both shielded and unshielded. Example of an important application of stud arc welding is the attachment of shear connector studs on the flanges of steel beams when making composite steel-concrete decks for bridges.

In the shielded stud arc welding process, a welding gun holds the stud with the base end positioned on the desired surface location of the workpiece. Initially, the stud is held in light contact with the workpiece, but when welding current is supplied the stud is lifted slightly to allow initiation of an arc. The arcing period is very short, perhaps ranging from 0.1 to 1.0 second — only enough time to melt the end of the stud and to form a shallow weld pool beneath the stud. When current flow is stopped, the gun plunges the stud end into the pool and holds the stud in position while the molten metal solidifies. Shielding of the arc has been given much innovative attention. Often a ceramic ferrule is pre-placed to restrict infiltration of air to the weld area and to confine molten metal when the stud is plunged into the pool. The lower edge of the ferrule may have a few small notches to act as vents for hot gases that must escape. The arc end of the stud is often provided with a small quantity of flux and deoxidizer to suppress porosity in the weld. Aluminum in some convenient form is commonly used as the deoxidizer, and the quantity can be very small because of the short arcing time.

The stud arc welding process develops sound, fully fused welds in carbon and low-alloy steels even when the surface is galvanized. The very short arcing time of stud welding produces a narrow HAZ that cools rapidly, producing microstructures in hardenable steels that have low fracture toughness. For this reason, some steels may require preheating for proper stud application.

Stud arc welding without shielding can be accomplished with smaller studs by using arcing times so short (1-12 milliseconds) that only minimal oxidation takes place. For these studs, the required electrical current is supplied by a capacitor power source; which discharges the needed current at a potential of approximately 100 to 200 volts. Because the arcing time is so very short, no ferrule, flux or deoxidizer is employed. Several techniques are used to initiate the arc in capacitor discharge stud welding, and the technique employed determines both the mechanism of gun operation and the stud end configuration at its welding base. However in all three methods, the arcing time is very short before the stud base is plunged onto the workpiece.

An attribute of capacitor discharge welding is its capability for welding a stud to a workpiece surface with very little weld penetration and HAZ in the workpiece. Consequently, studs can be welded to one side of steel sheet with little or no distortion or temper coloration developing on the opposite (show) side. As previously mentioned for stud arc welding, very satisfactory welds with adequate toughness can be made on many steels, but the very rapid cooling of the weld and the HAZ can lead to a toughness problem with hardenable steels.

Gas Tungsten Arc Welding (GTAW)

This process uses a small-diameter tungsten electrode, and a stream of argon or helium gas to surround and protect both the electrode and the weld pool. The process acquired the early designation, "TIG" welding, signifying "tungsten inert gas." However small amounts of other gases, such as hydrogen or nitrogen, can also assist in gaining certain desirable effects. Consequently, GTAW is the designation standardized by AWS.

Figure 6.9 illustrates a typical GTAW torch, and the rudiments of a manual welding operation. The process is readily mechanized, and can be operated automatically, but because a nonconsumable electrode is employed, special techniques are needed to add filler metal to the weld pool. One technique is the addition of cold, bare welding rod into the arc as shown in Figure 6.9. In mechanized and automatic welding, filler metal can be fed from a spool. The filler wire can also be preheated to a high temperature by electrical resistance heating just before entering the arc. This technique is called the gas tungsten arc *hot wire* system. It increases the rate of filler metal melting to permit faster welding, and it can be used effectively for surface overlay operation. Often, when welding a butt joint in thin material in which the faying surfaces are precisely fitted with virtually no joint clearance, the GTAW process can be applied with complete penetration to make a full-section weld without adding any filler metal. When the work pieces are firmly held in a square-edge butt joint, the spread of heat ahead of the weld pool will close any small joint clearance and will cause the base metal edges to upset and increase in thickness. After the edges are fused and solidified, the weld thickness will be very close to that of the original base metal sections at the joint. GTAW is admirably suited for making autogenous square butt welds, corner welds, and edge welds. For many joint configurations where a preset amount of filler metal must be added to achieve the required weld shape and size, a protruding edge or lip of base metal can be included in the joint penetration to build up the desired weld section or reinforcement. The additional filler metal required can also be preplaced as a consumable insert in the joint, and this technique is often used in making butt joints for girth welds in pipe. These techniques are focused, of course, on making a weld of desired size and shape, but it must be kept in mind that the composition of the weld metal will depend on the component materials from which it is derived. In general, base metals are manufactured to different chemical specifications than are welding rods and wire. In some cases, base metal composition can be metallurgically unsuited for making weld metal. The problems and inadequacies that can arise from contributing too much base metal to the final weld composition will be discussed in Volume II, Chapters 13, 14 and 15. The ways in which the chemical composition of welding rod and consumable inserts are modified to assist in arriving at a suitable GTAW weld composition will be covered in Chapter 10 of Volume II. The composition of consumable inserts toward final weld composition is sufficiently important to warrant the publication of ANSI/AWS A5.30, *Specification for Consumable Inserts*, which includes composition requirements for inserts of mild steel (Group A) and chromium-molybdenum steels (Group B). Reasons for offering certain insert compositions with elevated silicon content and with additions of strong deoxidizers (aluminum, titanium, and zirconium) will become clear as the metallurgy of GTAW is discussed.

Copious details are published on equipment for manual and automatic GTAW, and on procedures for a wide variety of base metal configurations and joints. In general, the welds produced are quite satisfactory, and they have some metallurgical characteristics that are unique. However, the process is so clean and free of adjuvant materials (e.g., fluxes) that its successful operation depends on normal occurrence of a number of chemical and physical actions. Also its operation is more likely to be troubled by small metallurgical conditions that would be inconsequential with other fusion welding processes. Recommended practices for GTAW can be found in AWS C5.5.

The principal weld qualities which GTAW can provide are weld metal composition control, and weld metal soundness and cleanliness. With proper gas shielding, it is possible to fuse base metal and to introduce filler metal without significant alteration in chemical composition. There need not be any significant loss of oxidizable elements, or any meaningful pick-up of oxygen or nitrogen by the weld metal. There can be small composition changes, but this would occur only because the process produces weld metal that is almost devoid of nonmetallic inclusions. Metal melted during GTAW operations ordinarily allows its nonmetallic inclusions to escape to the surface, where they may distribute as a film or accumulate as small patches of slag. The weld metal is usually so well protected that virtually no new nonmetallic inclusions are formed. Of course, if the nonmetallic inclusions present include elements reported in the total chemical composition for the base metal, analysis of the weld metal could show a small decrease in the levels of these elements in the weld metal. For example, a total residual titanium content in a steel base metal likely has some of this titanium

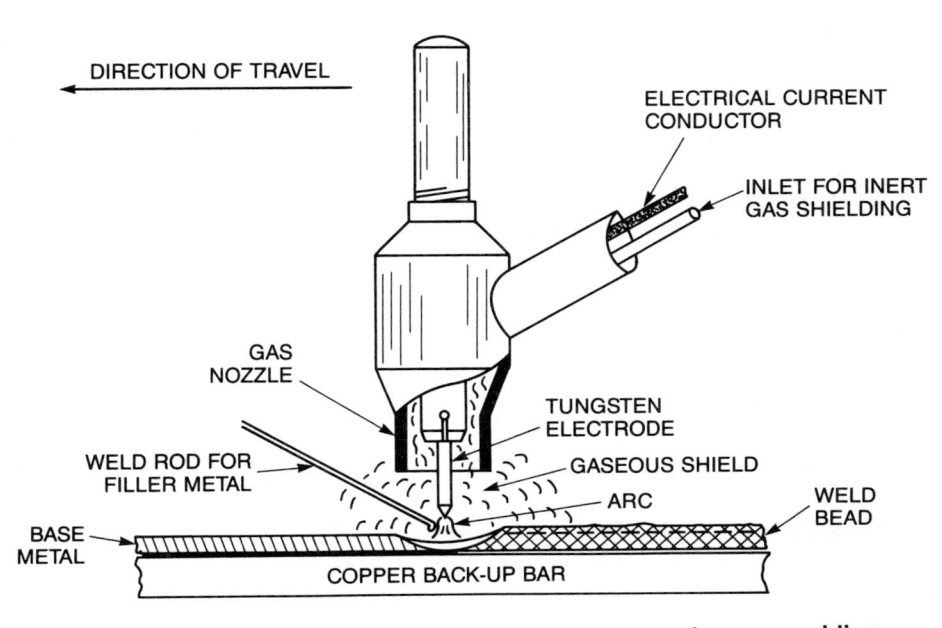

Figure 6.9 — Diagrammatic sketch of the gas tungsten arc welding process (GTAW) with a welding rod for filler metal

tied up as complex oxide-nitride-carbide inclusions. These are floated to the surface during fusion by GTAW. The outstanding cleanliness of the weld metal is a highly desirable attribute, and the small loss of elements through inclusion flotation is easily compensated when necessary. Some base metals can contain elements that will behave in an unfavorable manner when fused under the sterile conditions of GTAW. For example, steel that is not completely deoxidized (e.g., rimmed steel) is likely to have porosity in an autogenous weld because of the continuance of the carbon-oxygen reaction mentioned in Chapter 5. As another example, a steel may have been treated during steelmaking with special additives to improve a particular physical or mechanical property (other than weldability). These additives sometimes include elements that form high-melting-point oxides, such as calcium, magnesium and zirconium. When these refractory-type oxides are present in the steel, they may contribute to conditions on the surface of the weld pool that interfere with the consistency of the GTAW operation. A problem of this nature will be discussed before leaving the subject of GTAW. Success with GTAW in making a weld that is suitable for its intended service depends on both proper handling of the many procedural details that are covered by the literature on this process, and on assessment of the particular application for conditions that will adversely affect the normalcy of chemical and physical actions during the process.

Power Sources for Gas Tungsten Arc Welding

Most GTAW operations make use of welding machines of the drooping voltage (constant current) type, and either dc or ac may be used. Welding operations on clean, bare steel usually employ steady direct current on straight polarity (DCEN). Surface oxides melt and flow readily on DCEN, and the dc arc is quite stable. Pulsed dc has been found to be advantageous for girth welds in pipe and tubing that is fixed in a horizontal position. Pulsing at a relatively slow rate, perhaps several pulses per second, allows the arc to travel the circumference throughout the out-of-position fusion operation without the need to vary parameters. A low rate of pulsing allows brief cooling periods for the weld pool, thus avoiding melt-through at small gaps in joint clearance. In addition to changing the pulse rate, modern power sources allow independent control over all aspects of pulsed dc, including rate of current rise, duration of high current, rate of current decrease, and duration of low current. Very rapidly pulsed current imposes vigorous electromagnetic forces that act to stir the weld pool. This agitation of the pool appears to reduce the formation of porosity in the weld metal. Also, GTAW-P results in greater joint penetration without increasing the width of the weld at the top. In the 1990s researchers have studied the effect of pulsing current on weld joint penetration, and they have found the weld depth to increase steadily until the pulse frequency reached about 3000 Hz, after which no further increase was derived from pulsing rate alone.

Many power sources for GTAW are equipped to supply high-frequency current for arc initiation without the need to touch or scratch the base metal surface with the tungsten electrode. Touch-starting the GTAW arc on steel invites pick-up of contamination on the electrode tip, and an acquired contaminant is not exuviated because the electrode is more or less nonconsumable. The contaminant usually causes increasing arc disturbance as the welding proceeds. When a touch-start must be performed, the likelihood of electrode contamination can be minimized by starting the arc on a small piece of graphite or copper that is grounded to the steel base metal to be welded. New techniques for starting the arc are possible with recently introduced rectifier-type power sources, such as the thyristor controlled dc type. One technique permits the welder to place the tungsten electrode directly on the workpiece. Next, a starting switch allows a flow of low-level current to preheat the electrode tip and promote electron emission. When the electrode is lifted from the surface, an increase in voltage triggers an increase in current to the required welding level whereupon an arc is established. Therefore, this technique does not entail the high-current short-circuit condition that ordinarily causes electrode contamination.

Ordinarily, when welding bare steel, there is no need to operate the dc on DCEP, which is the polarity that produces greater heat at the tungsten electrode than at the base metal. Also, alternating current serves no special purpose for GTAW on bare steel, but many steel products that have metallic coatings can use ac to advantage. Aluminum-coated steel is a good example of a product form that requires some degree of compromise between the kind of welding current usually used for bare steel, and that which serves best for aluminum. Often, some experimentation is necessary to select a power source that is acceptable for welding metallic coated steels, because many of the newer coatings are actually alloys that have been chemically treated to alter their surface composition for the sake of appearance or finishing (e.g., painting). Because the tungsten electrode becomes very hot in the GTAW operation, and is therefore an efficient electron emitter, no particular difficulty is encountered in maintaining an arc with alternating current. There can be a problem with partial rectification of ac when welding some of the metallic coated steels; however, this can be overcome by using a power source designed to equalize the positive and negative portions of the cyclic current wave.

Electrodes (Nonconsumable) for GTAW

Several types of tungsten-alloy electrodes are available for GTAW, and these differ significantly in cost, life, and operating characteristics. "Pure" tungsten may cause problems with alternating current because the arcing end tends to melt, and this molten metal will often accumulate until it forms a globular droplet that clings to the still-solid tip. This formation is accentuated by higher welding current. The molten droplet often allows the arc to spread over the electrode end (somewhat erratically) and fails to concentrate the plasma on the weld pool in the desired manner. The molten end can even detach droplets of tungsten, which then enter the weld pool and remain entrapped in the weld as brittle, metallic inclusions, easily detectible by radiographic examination. Tungsten electrodes for GTAW should be ground smooth to reduce heating by contact resistance between the gripping collet in the torch and the electrode surface, and to improve heat conduction out of the electrode through the collet.

When the tungsten electrode is alloyed with electron-emissive agents a number vital functions involved in supporting an arc are greatly improved. ANSI/AWS A5.12 is a specification for seven classifications of tungsten and tungsten alloy electrodes both for welding and for cutting. The EWP classification covers plain tungsten of at least 99.5 percent purity. Five classifications are provided in which a specific emissive oxide is required in a prescribed amount; including the oxides of cerium, lanthanum, thorium and zirconium. A catch-all classification is added for electrodes that contain some (unspecified, but identified) oxide or elemental additive to the tungsten base that improves its suitability as an electrode; but, the tungsten level must be 94.5 percent minimum.

The electrode designations in A5.12 have elemental symbols and suffix digits that, in some instances, indicate the nominal amount of a particular oxide contained in each; namely, EWCe-2 (ceria), EWLa-1 (lanthana), EWTh-1 and EWTh-2 (thoria). An earlier issue of the specification listed "EWTh-3," which had an internal lateral segment of thoria along the electrode length, but this classification is not included in the current specification. The EWZr-1 alloy is required to contain only 0.15 to 0.40 percent zirconia. The presence of electron-emissive oxides and agents improves the electrode's starting ability, arc stability, current carrying capacity, usable life, resistance to contamination, and ability to maintain a sharpened or conical tip for concentrating the arc. The tapers on pointed or conical electrode tips usually have included angles ranging from 30° to 120° depending on the degree of arc concentration desired. The tip shape of the tungsten electrode has a strong influence on the character of the arc and the overall geometry of the weld produced, particularly the depth and shape of joint penetration. Rather than leaving the shape of the tip to chance, common practice is to machine grind the end of the electrode to produce a specific pointed or conical shape.

Shielding Gases for GTAW

Much information appears in the literature on shielding gases for GTAW; because, not only is the gas is a major cost item, it also is a significant factor in controlling the nature of the arc produced and the weld obtained. Furthermore, selection of a gas for a particular GTAW operation must often be made from the kinds of gases and mixtures commercially available, and on-site mixing of a preferred combination may not be easily accomplished. Shielding gases deserve study because the influence the character of the arc, and this in turn determines much regarding weld size, shape, soundness, and properties.

There are a number of differences in the character of the arc when using argon versus helium as the shielding gas. These differences can be used to advantage when selecting a gas to be employed in a specific application of GTAW. A comparison of the characteristics found when using argon, or helium as the shielding gas is provided in Table 6.1. In procedure planning, it is not unusual to use a particular argon-helium mixture to obtain certain helpful effects from each gas to the degree that the mixture provides. For this reason, a number of custom (pre-mixed) combinations are marketed by gas suppliers.

Experimentation has been carried out with small additions of other gases to the shields on inert gases to improve some particular characteristic of the arc, or to obtain a certain result in welding. Findings from work of this nature are not well documented in the literature. Hydrogen is an example of an additive that can be useful in either argon or helium shielding, but it must be used with care. Less than five percent hydrogen as an addition generally results in higher arc voltage, increased joint penetration, and faster welding travel speed without the occurrence of undercut on some types of steel, but under some circumstances the hydrogen addition can cause weld porosity, and it may produce hydrogen-induced cracking in welds made in hardenable steels. Shielding gas technology is given a more searching review in Chapter 8.

GTAW Process Control

Recent research on the GTAW process is aimed at developing improved systems of control for machine and automatic welding. A common requirement in these operations, particularly for autogenous welds in sheet or thin plate, is to gain uniform root penetration along the weld length, and to have satisfactory weld face and root bead formation. When complete joint penetration must be accomplished without the aid of backing strip, sometimes difficulty is encountered with variations in the weld along its length; either incomplete joint penetration, or excessive penetration which results in a root bead that is too large and nonuniform, if not actually melted through and leaving a perforation.

Early studies of GTAW process control dealt mainly with the arc and its welding performance as affected by shielding gas selection and arc voltage regulation. Some work on weld penetration control experimented with the use of infrared sensors positioned below abutting joints to signal the extent of root penetration. Current research, which may provide new approaches to GTAW control in real-time, is concentrating on several features of the molten weld pool, including:

(1) localized surface depression in the weld pool beneath the arc,
(2) surface humping of molten metal on the periphery of the pool, and
(3) oscillation of the molten metal in the pool.

Table 6.1 — Characteristics of Argon and Helium Gases When Shielding the GTAW Process

General Characteristics When Shielding with Argon (Ar)	Performance Item	General Characteristics When Shielding with Helium (He)
Easier than with He	Arc Initiation	Somewhat more difficult than with Ar
Lower than with He, which can be helpful when welding thin sections. Less change in arc voltage with variation in arc gap	Arc Voltage	Higher than with Ar. Arc is definitely hotter with He, which is helpful in welding thick sections
Best stability of all gases	Arc Stability	Good stability, but not equal to Ar
Ar, heavier than air, requires less gas than He to shield in flat and the horizontal positions	Protection of Weld	He, being lighter than air, may require more gas to properly shield the weld; except perhaps when welding overhead
Ar preferred for manual welding because of less voltage change with variation in arc length	Manual Welding	He can be used, but variation in arc length causes significant arc voltage change that affects heat input to weld
Ar can be used, but He is better suited to avoid porosity and undercut	Automatic Welding	Much automatic welding is based upon arc-voltage control and makes use of He because of arc length sensitivity
Ar more effective than He when utilizing positive ions from electrode on reverse polarity to remove surface films of oxide on base metal	Cleaning Action	He alone is not satisfactory to accomplish positive ion removal of surface oxide from base metal but small addition of Ar makes mixed Ar-He shielding gas quite effective

Occurrence of the above features and interactions between them is believed to determine in large part the shape of the weld produced in a given application. Techniques for quantitative evaluation of these features have been developed, and measurements obtained appear usable as operating parameters for closed-loop control systems.

Surface depression in the molten weld pool beneath the arcing electrode has been measured in the laboratory during welding by digital radiography and correlated with arc force determinations. With welding currents below about 200 A, virtually no pool depression exists, but above this current level arc force has a quadratic dependence on current and in this higher current regime pool depression and weld penetration are closely related. Published research speculates on the feasibility of a real-time system which would monitor pool depression and in turn would control weld penetration continuously.

Surface humping around the outer edge of the molten pool can also be measured, but the hump consists of molten metal in a dynamic state of displacement from the depression beneath the arc. Although the extent of humping could be used as an indicator of weld penetration, the surface depression appears better suited for the purpose.

Oscillation of the molten weld pool during welding has been studied using various techniques to stimulate its molten and to explore several different means for measuring this feature. The oscillation frequency of a weld pool on a partially penetrated joint might lie somewhere in the range of 100 to 400 Hz depending upon its diameter, but a substantial decrease in frequency usually occurs as the weld pool completely penetrates the joint and the diameter of the penetration on the root side increases to about one-half that on the face (somewhat like a stretched membrane). This abrupt change in oscillation behavior is being considered as a basis for penetration signalling in a control system. However, a number of practical problems must be overcome before such a system would be applicable when welding at the travel speeds commonly used in industry.

Even when all precautions are taken in following a pre-planned procedure for a GTAW operation to ensure regularity in results, difficulty with inconsistent weld penetration still may arise. The cause and the mechanics of the problem in this case can be essentially metallurgical in nature. When the difficulty is of this origin, complete rationalization is not yet at hand. While this form of the penetration problem is more often encountered in welding the stainless steels, the carbon and alloy steels have also shown some vulnerability.

Briefly, melting depth into base metal can vary between different heats (melt lots) of the same type of steel. Great care can be taken to closely calculate (or establish by trials) the depth of penetration desired in a weld, and often the intent is to obtain the same result as the procedure is repeated on many workpieces. Yet, as successive lots of steel representing different heats (and sometimes from different sources) are put through the same welding operation, an occasional lot may contain welds having unacceptable penetration. There may be a spasmodic lack of penetration along the length of the weld, but more often the entire length of weld will display inadequate penetration simply because melting did not extend to depth normally obtained with the established welding procedure. Sometimes, a particular lot of steel will have a proclivity for abnormally deep penetration, forming too-large a weld zone or unwanted melt-through of the joint.

These differences in weld penetration with fixed GTAW operating conditions are caused by the presence of certain minor elements in the steel, or by a change in the level of minor elements usually present. This form of penetration problem is complex because differences in weld penetration are caused by either the presence of minor elements or a change in the level of a minor element. The problem is complex because several different mechanisms, each capable of influencing weld penetration, apparently can operate concurrently. These mechanisms exert their effect through changes in one or more of the following process features:

(1) arc configuration, including plasma jet,
(2) arc voltage,
(3) arc stability,
(4) work function at the anode and cathode surfaces,
(5) surface tension of molten metal in the weld pool,
(6) surface tension gradient in the weld pool,
(7) weld pool depression depth,
(8) circulation of molten metal in the weld pool, and
(9) gas shear coupling between plasma gases and weld pool.

The features listed above affect the magnitude and distribution of heat in the base metal. The configuration of the GTAW arc is not uniform along its axis, but flares symmetrically to a greater diameter above the base metal because of the larger anodic contraction region. If a minor element causes the contraction region to grow, the energy of the arc will be distributed over a greater area. This in turn increases the weld width and decreases depth of penetration. If a minor element causes the arc voltage to decrease, the arc wattage will be reduced and a smaller weld with less penetration will result. Arc instability will reduce the arc energy and interfere with heat transfer to the base metal. If a minor element enters into the arc by vaporization and causes contraction of the arc anode footprint, this will concentrate the space charge and voltage distribution, resulting in deeper weld penetration. Some minor elements increase surface tension, while others cause a decrease. When the effect is to decrease surface tension, the weld pool crater grows deeper, and more arc energy is able to enter the base metal to accomplish deeper penetration. Conversely, an increase in surface tension results in a shallower weld pool crater; and as the depth the pool decreases, so does the joint penetration, even though the volume of metal melted may remain unchanged. As surface tension is altered, circulation of molten metal in the weld pool can undergo a surprising change that affects penetration depth. Weld-pool motion is produced by electromotive or Lorentz forces, buoyancy force, plasma jet impingement, and by surface tension gradient. The phenomenon weld-pool motion is discussed further in Technical Brief 33. Incidentally, ordinary natural thermal convection in the weld pool is believed to be negligible; however, circulation produced by surface tension gradient apparently is a significant transporter of heat via the weld pool and accomplishes even further melting of the base metal.

Identification of the minor elements that influence weld penetration behavior during GTAW operations is not complete. They may be contained in the base metal, filler

TECHNICAL BRIEF 33:
Effects of Weld Pool Circulation on Penetration Depth in GTAW

Circulation of metal in the weld pool usually follows the pattern called Marangoni convection. This pattern, illustrated in Figure 6.10, develops because the temperature of metal at the surface of the weld pool is hotter in the center, under the arc; and its temperature decreases as the liquid-solid interface is approached. Normally, the surface tension of the molten metal decreases as its temperature is raised. Therefore, the cooler surface metal around the weld pool's edges (with its higher surface tension) will draw metal from the hotter center area and set up the customary Marangoni convection pattern shown in Figure 6.10(A) which helps produce a wider, less-penetrating weld.

The pattern shown in Figure 6.10(B) occurs when a minor element causes surface tension to increase in the center of the weld pool. Molten metal is drawn toward the center by the changed gradient, and the hotter metal under the arc is thrust downward. This increases the depth of melting as indicated schematically. Exactly how minor elements increase surface tension as molten metal temperature is raised is not fully understood, but their influence in this direction can be quite pronounced. Sulfur, for example, produces a positive surface tension coefficient; that is, it causes decreased surface tension at the liquidus of iron. However, as the temperature of the molten metal is raised, the surface tension increases. Positive coefficients are believed to result from a surface-active element becoming less of a surface segregate and surfactant as the temperature rises, thus reversing its initial influence at the melting point.

The cause of abnormally shallow weld penetration, or of irregular penetration in GTAW operations, is sometimes readily seen on the surface of the molten weld pool as small beads, islands, or patches of slag-like material. This material clearly comes from within the base metal (or other consumables that are used), where it existed either as nonmetallic inclusions that floated to the surface of the weld pool or as an element that found sufficient oxygen to form an oxide. This slag-like material interferes with heat input to the weld pool, either by its physical presence on the surface as it moves under the arc, or by a thermionic effect that changes the energy distribution in the arc and reduces temperature at the anodic metal. The material usually remains on the surface of the solidified weld, sometimes taking the form of a tenacious, glassy bead which is quite hard but friable. More often the material collects as a small island or patch, as illustrated in Figure 6.11, and is dispersed intermittently along the length of the weld. Ordinarily, the weld is of normal microstructure and soundness beneath each slag-like bead or patch, although the surface configuration may be slightly different. The principal problem caused by this material is the nature of the weld root surface when complete penetration through the base metal is intended to form a uniform face reinforcement.

Proof that this slag-like material originates from the base metal or other consumable can be gained whenever an initial weld exhibiting the telltale beads or patches is surface-cleaned and remelted by the same GTAW operation. The second fusion of just the weld metal produces none of the slag-like material. Microanalysis of beads and patches removed from weld surfaces usually reveals the material to be either one of the refractory-type metal oxides or a complex compound containing a substantial percentage of such a metal along with other oxidizable alloying elements. When shallow or irregular penetration occurs accompanied by these symptoms, a solution to the problem may lie in dispersing this foreign obstruction to allow normal, consistent heat-input by the arc. Sometimes this can be accomplished by positioning the GTAW torch at a small drag angle so that the flow of shielding gas sweeps the accumulation to the trailing edge of the pool, where it can solidify with minimal disturbance to the arc and heat-input. When a stronger measure is needed to disperse the slag-like material, the use of pulsed welding current, or magnetic arc oscillation will assist in this direction. Usually, trials must be conducted to determine what modifications to the GTAW procedure are needed to achieve improved and acceptable results for a given material.

metal, consumable inserts, etc. Also, some minor elements are not simply residuals; they can be additives remaining from treatments applied during melting to improve a particular property. Special additives are used to improve such properties as hot workability, fracture toughness, resistance to lamellar tearing, retention of fine grain, and others. Sometimes the urgency of overcoming an obstacle in manufacturing may encourage use of a special additive despite limited knowledge of its composition and side effects. Although features of the GTAW process that take part in determining weld penetration have been outlined above, the mechanism by which specific minor elements exert their influence in particular metals and alloys has not been clearly

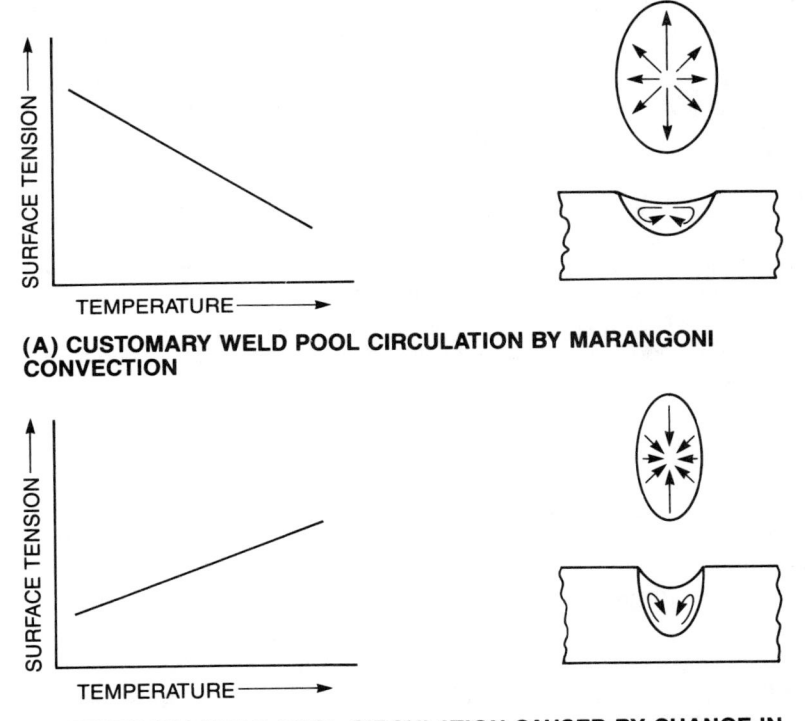

(A) CUSTOMARY WELD POOL CIRCULATION BY MARANGONI CONVECTION

(B) REVERSED WELD POOL CIRCULATION CAUSED BY CHANGE IN SURFACE TENSION GRADIENT FROM MINOR ELEMENT PRESENCE IN BASE METAL

Figure 6.10 — Two patterns of weld pool circulation which occur during GTAW fusion of base metal

These sketches are based on observations made through high-speed cinephotography. Upper illustration (A) shows customary Marangoni convection circulation where hotter metal with lower surface tension is transported radially to promote the melting of a wider, less-penetrating weld. Lower illustration (B) shows pattern which occurs when a minor element causes surface tension to increase in the pool's center and molten metal is drawn inward and thrust downward. The hotter metal transported in this manner produces deeper weld penetration.

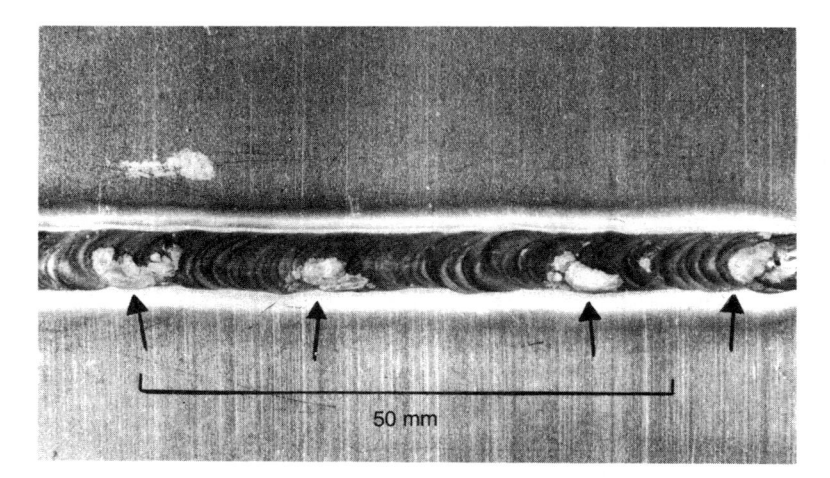

Figure 6.11 — Slag-like material on the face of GTA weld that accumulated on weld pool during autogenous fusion and was distributed intermittently on the weld

The presence of particles or patches of slag-like material in sufficient quantity can strongly interfere with weld penetration. The materials contributing to the slag have been found in the base metal, or in the filler metal added to the weld.

established. In general, elements that readily form oxides with relatively high melting points, such as aluminum, calcium, cerium, lanthanum, magnesium and zirconium, have been found to impede weld penetration. Elements having the opposite influence — that is, a tendency to increase the depth of weld penetration, or at least to exhibit a remedial effect — include bromine, chlorine, fluorine, selenium and sulfur. As might be expected, there can be competition between elements from these opposing groups. Therefore, when a material is encountered that offers very shallow or irregular weld penetration, sometimes an element can be interjected to remedy the problem. For example, a small amount of hydrogen might be added to the shielding gas to increase arc voltage and form a hotter arc which possibly will overpower the impeding effect. Halide salts are generally an effective remedy when applied as a very light film on the base metal or other consumable, but this fluxing technique is seldom practicable. Small additions of special gas to the inert gas shield are easy to arrange, and investigative work has been conducted wherein small amounts of chlorine, sulfur dioxide, sulfur hexafluoride, etc., were employed. These additive gases have proven extremely effective in increasing weld penetration; unfortunately, they are toxic, and the safety of personnel in the welding operations area prevents their use. Sulfur is so effective as an alleviative element that some steels intended for extensive automatic GTAW operations are specified to have a minimum level of sulfur as a safeguard for consistent weld penetration.

Gas Metal Arc Welding (GMAW)

The GMAW process allows welds to be made with continuous deposition of filler metal from a consumable electrode, and it avoids the chore of cleaning a solidified slag from the surface of each weld pass. The basic arrangement of components and their function in a GMAW torch is illustrated in Figure 6.12; this basic arrangement serves both for manual and for automatic welding. Upon its initial commercial introduction, the process was dubbed "MIG" welding to signify *metal inert gas* and to set it apart from "TIG" welding, reviewed earlier. As reactive gases were added to the gaseous shielding, AWS adopted the acronym GMAW as the standard designation. The wire used as an electrode for GMAW is standardized by AWS in two specifications: AWS A5.18 for carbon-steel filler metals, and AWS A5.28 for low-alloy steel filler metals. Incidentally, the filler metals in each of these specifications also can be used in the GTAW process, either spooled or in straightened-and-cut lengths. The low-alloy steel filler metals can be either solid wire or metal cored. A more detailed discussion of electrodes for GMAW will appear in Volume II, Chapter 10.

When the GMAW process was introduced, argon was favored as the shielding gas for welding steel, because it permitted the development of very stable, spatter-free metal transfer. However, this type of transfer was possible only when the electrical current supplied to the electrode was DCEP and above a rather high amperage threshold, called the *transition current*. For example, with a steel electrode of 1.6 mm (1/16 in.) size, the minimum welding current value is about 275 A. Below this current level, the

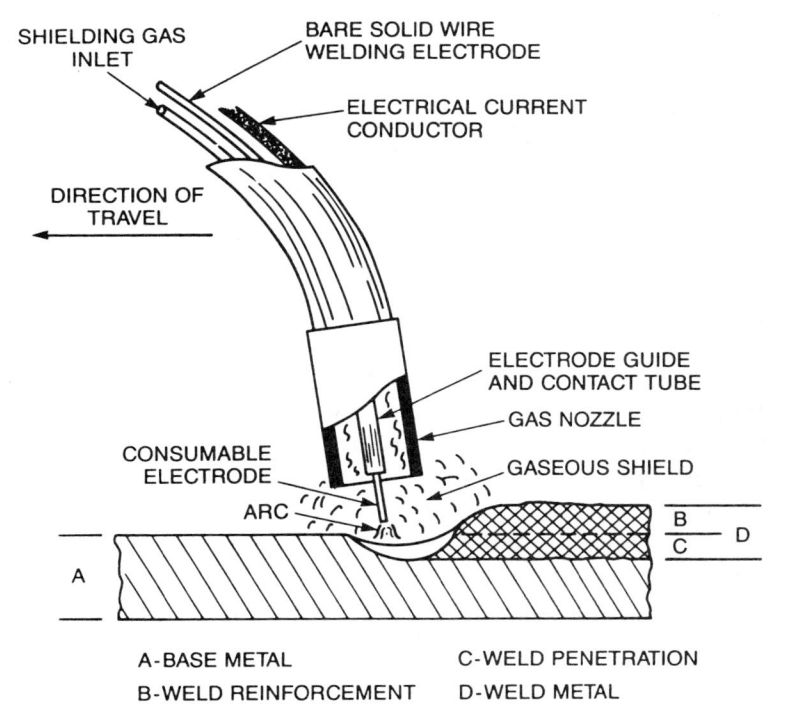

Figure 6.12 — **Diagrammatic sketch of the gas metal arc welding process (GMAW)**

electrode melts by forming sizable molten globules at the rate of about 12 per second, and much spatter and erratic arc behavior occurs as the globules of metal are transferred to the weld pool. When the welding current is above the critical threshold, the globular melting and transferring changes over a relatively narrow range to an *axial spray* of fine droplets at a release rate approaching several hundred droplets per second, allowing an arc with more stability and less spatter.

Early trials with GMAW using helium for shielding were discouraging because higher voltage was required to start the arc and to maintain a spray-like metal transfer, but even with the best adjustment of current and voltage the arc was not satisfactorily stable, and the amount of spatter was excessive. Yet, as explained in Chapter 8, additions of helium to shielding mixtures in which argon comprises the major component have proved very effective. From the beginning, GMAW operation on DCEP has generally provided better results than DCEN polarity.

DCEP not only delivers greater heat input to the cathodic base metal, for good penetration and a fluid weld pool, it also provides a positive-ion cleaning effect. However, when pure argon is used to weld steels, this effect is very random. The arc tends to wander and produce a nonuniform weld bead that often contains some porosity. Also, the surface cleaning action varies and leaves irregularly wetted areas along the toes of the weld bead. The introduction of small amounts of oxygen, or of carbon dioxide gas into the argon shield accomplishes a dramatic improvement. As little as 0.5 percent oxygen in the argon produces a uniformly oxidized surface on the base metal, and this ensures confinement of the arc to a clear weld pool. The same improved results are secured by adding carbon dioxide to the argon in amounts not exceeding about 10 percent. With these gas mixtures, the arc is steady and produces a uniform, sound weld deposit. Pre-mixed bottled gases are available containing one, two, three, and five percent oxygen in argon. Larger percentages of oxygen are not advocated because no further improvement takes place in the GMAW operation, and oxidizable elements from steel weld metal are lost. With oxygen content at five percent or less, adjustments are easily made by enriching the electrode composition with such elements as manganese, silicon, and aluminum. The low cost of carbon dioxide encourages maximum use of this gas for shielding, but as the amount of CO_2 in argon exceeds about 10 percent, the globular form of metal transfer reappears even at high current density. The high current density on the electrode needed for axial spray transfer produces rather deep penetration. Therefore, for welding steels, the process is usable only on heavy sheet (perhaps three millimeters or about one-half inch minimum thickness) and on plate; and it is more or less limited to the flat and horizontal positions.

Alternating current is not used for GMAW. Because the arc is extinguished and reestablished between each half-cycle of ac, it is seriously handicapped without the ionizing agents regularly used in the coverings on SMAW electrodes, or the emissive tungsten electrode for GTAW. The surfaces of bare wires for the GMAW electrode can be treated with various chemical compounds to obtain stable arcs with ac, but circumstances do not encourage marketing an electrode of this kind.

Axial Spray Transfer

The GMAW process operating on DCEP at high current density commonly makes use of the constant-voltage type power source described earlier (see Figure 6.7). By selecting an electrode feed rate and setting the arc voltage, the welding current is regulated by the power source to maintain a constant arc length. In this manner, a semi-automatic arc welding process is also provided for manual welding. For fully automatic GMAW operations, special power sources have been introduced which can detect precursory electrical events in arc striking and in short circuiting, and can control the output of power in a manner (via current waveform manipulation) that avoids defects and improves spatter suppression. Thusly, the self-regulating arc makes the GMAW process well suited for machine and automatic welding operations.

The mode of metal transfer must be kept in mind as the procedure details are set. Three functions are involved in the GMAW electrode deposition process; namely, electrical current flow, heat flow, and mass transfer. Of the three, the flow of electrical current dominates the behavior of the process. The current generates heat for melting the electrode and generates the forces that control the mode of mass transfer. The manner in which the weld pool receives the molten mass from the electrode is very important. Although several distinctive modes of transfer, which are widely recognized, will be described here, there is ordinarily a globular transfer, or vice versa, as operating conditions are changed from a set of parameters that favors one or the other. To obtain consistent results when making repetitive welds in the same kind of joint, it is necessary to select the most suitable form of electrode deposition and to operate within parameters that ensure maintaining the desired transfer mode. Fortunately, behavior of the welding arc and the mode of transfer taking place can be interpreted from instantaneous current and voltage fluctuations, and these provide practicable means of closed-loop control to maintain operational consistency.

When axial spray transfer is performed at very high levels of current, melting of the electrode assumes a characteristic form which is called *streaming spray transfer*. When undergoing this somewhat modified mode of axial spray, the electrode produces a slender column of molten metal that extends a short distance from its pointed tip toward the weld pool. At its free terminus, the streaming liquid column separates into a dense stream of many fine droplet that continue flight to the weld pool. The length of this liquid column is dependent on current density and may extend, for example, to about six mm (¼ in.). This stream of liquid metal must be given consideration when welding with high current and with relatively low arc voltage (which is tantamount to a short arc gap). Under these circumstances, the liquid column may reach the surface of the weld pool, whereupon a momentary short circuit occurs with the usual unwanted results.

Globular Transfer

GMAW arcs shielded with CO_2 characteristically transfer the melting electrode in globular form. Despite this mode of transfer, CO_2 shielding can be used in machine welding by keeping the tip of the melting electrode below the general level of the molten weld pool, a position not easily held in manual welding. This precise tip position must be maintained constantly to have molten globules that leave the electrode be captured by the weld pool; otherwise, many of the globules will escape as heavy spatter. Although no threshold level for current is involved in this machine welding technique, a relatively high current must be used to create a suitably deep crater in the weld pool.

When CO_2 shielding is used, even at high current density, the globules leaving the electrode are projected nonaxially. As each globule forms, it is supported to some extent by electromagnetic repulsive force, and the pinch effect which regularly occurs with a substantially inert gas shield is absent. Therefore, the globule in a CO_2 shield grows until gravity causes detachment, or until the molten globule touches the weld pool and is pulled off by surface tension force. During this short-circuit interval there is the strong possibility of the globule being superheated and dispersed as spatter. These operating circumstances call for close control of welding parameters, and the CO_2 shielded machine welding operation is limited to the flat and horizontal positions. Nevertheless, CO_2 shielded globular transfer provides a new economical process for machine and automatic welding, and fabricators of steel structural components have recognized its advantages over processes that use a solid flux or slag. The weld bead deposited by GMAW shielded CO_2 gas has a rather rough surface appearance. Also, it regularly develops a broad penetration pattern, and this facilitates the melting of abutted root faces in a joint.

Short Circuiting Transfer

This variation of the GMAW process was developed to allow for thinner sections of base metal, and to make it easier for welders to work on joints in all positions. Although globular melting does occur at the electrode tip in this short-circuit mode of metal transfer, power sources are designed to overcome the problems ordinarily associated with globular transfer. When originally introduced, the process was given a variety of trade names by equipment suppliers, but AWS adopted "short circuiting transfer" as standard terminology. Labels that persist to some extent include "short arc," "shorting arc," "dip transfer," and "micro-wire" welding — all of which have some accordance with this particular variation, which is now termed *short circuit gas metal arc welding* and is identified by the letter designation GMAW-S.

Globular transfer became a more useful mode in GMAW following the development of a power source with a unique form of dynamic current regulation. Figure 6.13 illustrates schematically four critical steps in the short-circuiting transfer mechanism. In the first step, the arc melts a globule on the end of the electrode. Upon reaching a large enough size, the globule (as shown in the second step) bridges the very short arc gap between the electrode tip and the weld pool. Ordinarily, under this short-circuit condition, a surge of current would flow and superheat the molten column between electrode and pool, causing it to explode into spatter. However, the new power source immediately limits the current surge through the molten, short-circuiting column by virtue of a reactance or inductance built into the power source circuitry. The flow of current provided by the power source at this stage is regulated to be sufficient for resistance heating of the molten column of metal, and to generate enough pinch effect (as illustrated in the third step) to sever the column and allow the molten metal to be drawn by capillary forces into the weld pool. This action reopens the gap between the electrode tip and weld pool and allows the re-establishment of an arc (fourth step). This sequence of four steps continues as a repetitive mechanism from arc stage to short-circuit stage at a rate that may exceed 100 cycles per second. This rate is too rapid for the eye to detect, but the short circuiting mechanism pro-

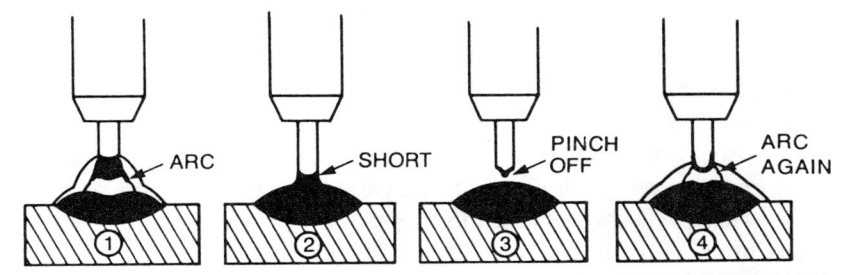

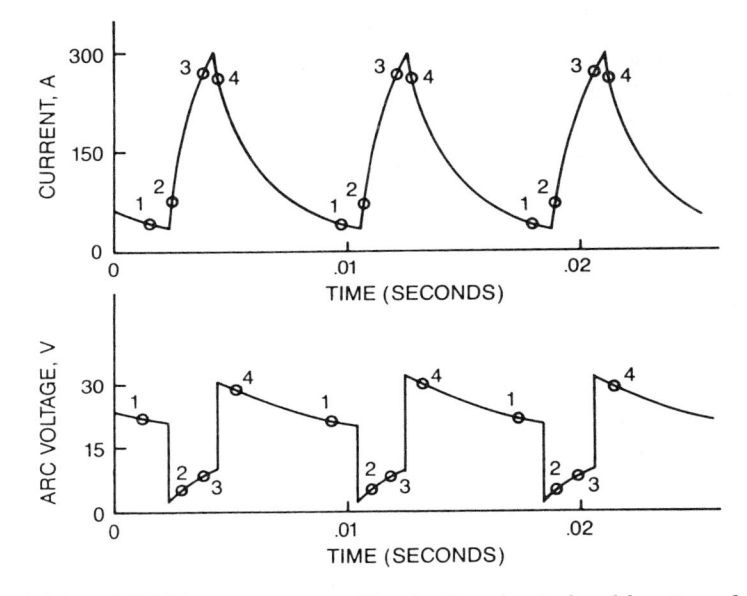

Figure 6.13 — GMAW process operating in the short circuiting transfer mode (now designated GMAW-S)

duces a buzzing sound that is plainly heard. The welding current and polarity employed with GMAW-S is DCEP.

The short circuiting transfer mode is very efficient for welding thin sections of base metal. Smaller sizes of electrode wire and new electrode feed units are available to assist in this direction. The transfer mechanism enables welders to (a) control penetration into the joint to avoid excessive melt-through, (b) bridge sizeable joint root openings where fit-up was poor, and (c) weld in all positions with almost equal ease. By using a small size electrode (perhaps as small as 0.5 mm, or about 0.02 in.) and fine-tuning the power source to provide a low level of dynamically regulated current, GMAW-S with short circuiting transfer enables welders to join sections as thin as 0.7 mm (about 24 gage). While the less expensive CO_2 serves very well for shielding the carbon and low-alloy steels, argon-carbon dioxide mixtures improve conditions for welding thinner sections of base metal.

The capabilities of GMAW-S with short circuiting transfer, as outlined, are so advantageous that this variation of the process is applied to many kinds of structures and articles, especially those requiring deposition of weld metal in all positions (e.g., horizontal fixed-position girth joints in pipelines). Weld metal deposited by this variation of the process displays normal strength and good toughness, and no particular difficulty is encountered in conforming to various codes and standards for welding construction. However, short circuiting transfer can leave small areas of incomplete fusion at the interface between base metal and weld metal. These areas occur when the GMAW operation is performed with relatively low heat input for the joint being welded, as likely would be the case when penetration into the base metal is to be minimized, or when welding steels thicker than six mm (¼ in.). The unfused areas are difficult to detect by NDE because of their crack-like formation along the interface (often in an angular orientation), as illustrated in Figure 6.14. Yet, during most kinds of mechanical testing, they clearly open up to reveal the incomplete fusion. The threat of this particular defect sharply dampened enthusiasm for short circuiting GMAW-S. Many code-writing bodies either forbade its use for welding heavy sections of steel, or called for stringent testing of closely-specified and controlled procedures before permitting its use.

Pulsed Spray Transfer

Pulsed gas metal arc welding (GMAW-P) is a unique form of the GMAW process that allows axial spray transfer at much lower *average* currents by limiting the hot,

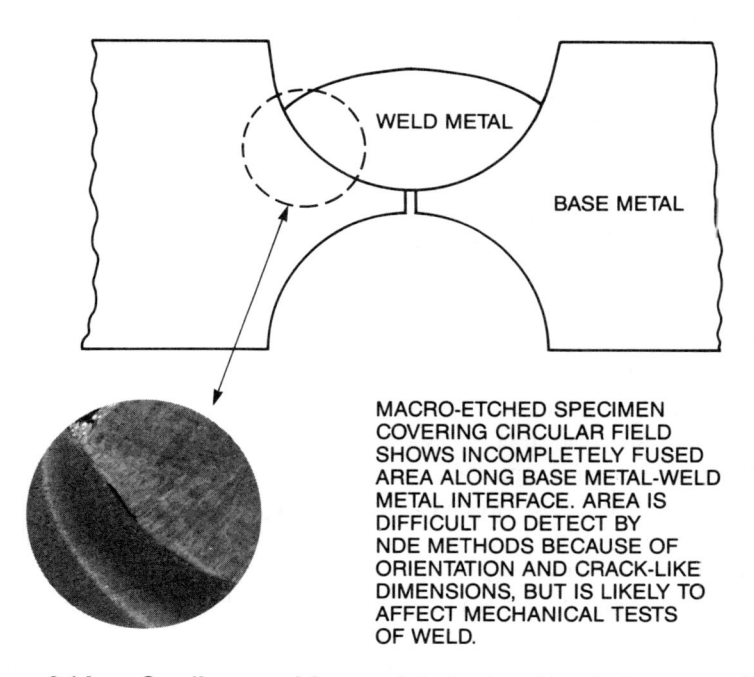

Figure 6.14 — Small area of incomplete fusion found along interface between carbon steel base metal and weld metal deposited by GMAW-S process operating in short circuiting transfer mode

forceful mode of transfer to brief intervals. These intervals are separated by periods of arcing with only modest current — too low to melt a full globule on the electrode tip. The amperage of the peak current exceeds the transition current threshold for axial spray transfer. Consequently, molten metal is transferred in free flight from the electrode to the weld pool only during peak-current pulses. The rate of electrode deposition is controlled by the frequency of peak-current pulsing; and this, along with electrode size and levels of amperage, determines weld penetration and other features of the weld.

The simplest power sources for pulsed spray welding use transformer-rectifiers equipped to apply pulses of dc in either the familiar sinusoidal form or in chopped forms, and always DCEP. The levels of amperage for background current and for peak current pulses are adjusted independently. Transformer-rectifier pulsed dc power sources have frequency capabilities of 60 or 120 Hz, or of submultiples of line current frequency. Figure 6.15 is a schematic illustration of the pulsed spray welding mode of

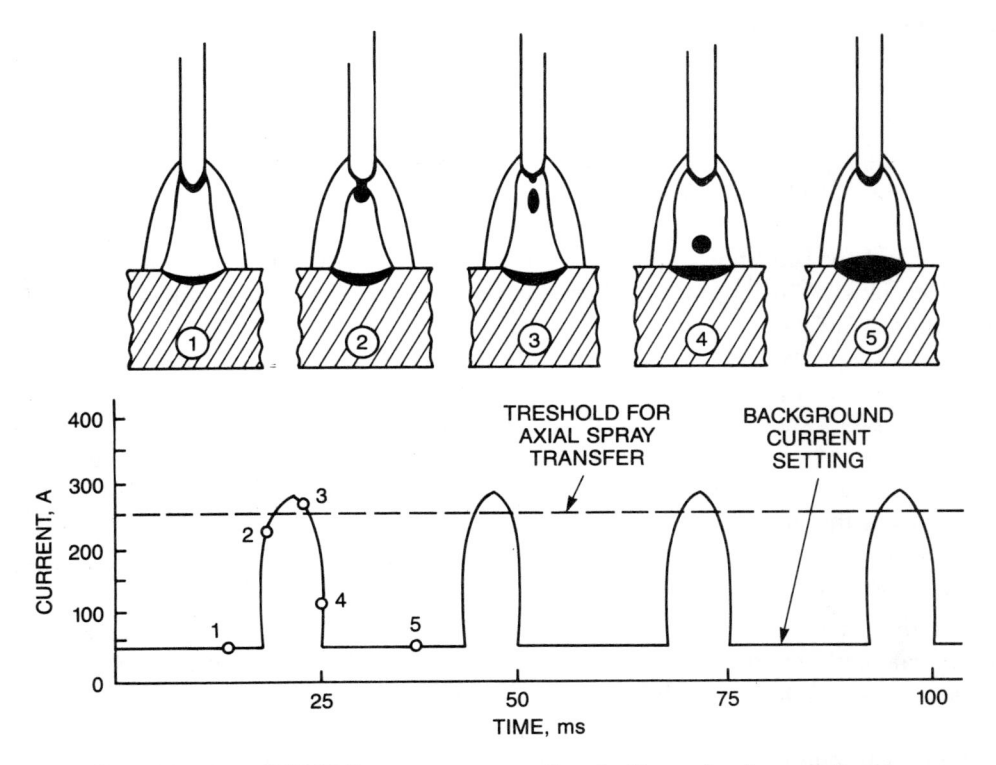

Figure 6.15 — GMAW-P process operating in the pulsed spray transfer mode

These schematic sketches illustrate the deposition of a 1.2 mm (0.045 in.) size carbon steel electrode using DCEP welding current and argon + CO_2 gas mixture for shielding. The transformer-rectifier power source pulses at 30 Hz with 8.3 ms pulse duration and a peak current amplitude of 275 A with background current set at 40 A and approximately 25 ms duration. Five cyclic steps of electrode melting are shown in the upper sketches, and a droplet of molten metal is separated from the electrode during the peak current flow of each pulse.

metal transfer which shows the synchronization between current pulsing and metal droplet transfer. This is a highly advantageous form of GMAW-P that facilitates welding of thin sections and heavy sections alike, in all positions. The background current for pulsed spray welding is usually set at a level commensurate with the electrode size to maintain an arc at all times, to preserve gas ionization, to generate some melting on the electrode tip preparatory to droplet formation and projection by the next pulse of peak current, and to maintain a molten weld pool of desired size. The amplitude of the threshold or transition current for the peak pulses depends on the size of electrode being used, as well as the amount of energy required to complete the formation of a tiny molten droplet on the electrode tip, to pinch off the droplet, and to project it axially toward the weld pool. Each pulse of current above the transition supplies enough power to free one droplet of metal from the electrode.

A practical problem confronts users of pulsed spray welding, that of ascertaining the best parameters for achieving a desired welding result. Although a printed guide is usually provided with power sources to assist in selecting the parametric values, the number of items that must be coordinated to gain an acceptable operation requires considerable estimating. Sometimes trial welding is necessary to fix workable settings for all of the variables.

The likelihood of having localized lack-of-fusion areas, as illustrated in Figure 6.14 for the GMAW-S short circuiting transfer mode, is practically eliminated with pulsed spray GMAW-P welding. Melting and penetration of the base metal is ordinarily regulated through the amperage levels and durations set for background current periods and for peak power pulses. Where unusual circumstances might cause some localized lack-of-fusion, improvements have been made in control of the process that appear to avoid the problem. For example, when a groove in thick steel, out of position, requires weave beads, power to the arc is likely to be restricted to hold deposited metal in proper placement regardless of position. Deposition of beads in this groove will encounter a shorter arc gap as the electrode oscillates toward the groove faces, and adequate fusion possibly will not take place at the weld-base metal interface. Power sources for pulsed arc welding have been developed that incorporate closed-loop control systems that use arc voltage signals to sense small changes in arc length, and to compensate with appropriate response in either pulse frequency, or in mean current to ensure proper fusion.

Study of heat input during GMAW-P with pulsed spray transfer shows that methods of calculation with the usual arithmetic mean arc voltage and current readings give values that are significantly lower than the actual heat input. This fact can be of practical metallurgical importance when attempting to closely predict HAZ cooling rates and microstructural changes using a data base originally secured with nonpulse welding; albeit the GMAW process. When instantaneous power values during pulsing are employed in calculations, the difference between pulsed and nonpulsed GMAW is minimized. The thermal efficiency of the various forms of GMAW have been compared over a range of electrode melting rates. The short circuiting mode of transfer has the highest thermal efficiency; about 95 percent, whereas the nonpulsed spray and the pulsed spray transfer modes both had a thermal efficiency of about 85 percent at low electrode melting rates, and about 80 percent at high rates.

Synergic pulsed welding is a more systematized form of pulsed spray GMAW-P. Hard-wired analog electronic circuitry is coupled to the pulsed power source to relate four principal current parameters directly to the electrode feed rate: the pulse amplitude and duration, and the background amplitude and duration. A special electrode feed unit is available that pulse-feeds wire in precise synchronization with intermittent transfer of filler metal into the weld pool. This unit makes non-slip delivery of wire to the welding gun by using a rubber treaded capstan that also serves as the armature of a low-inertia motor. A number of "synergic" programs can be pre-set to accommodate commonly used welding procedures, or special programs can be tailored and memorized for procedures frequently used in a particular shop. Synergic pulsed welding makes use of a sophisticated all-transistor pulsed dc power source capable of a wide variety of pulsed-current wave shapes, frequencies, amplitudes and durations. A square-wave current effectively provides peak pulsing, and the frequency of pulsing can be varied at will. Peak pulses can be limited to less than 10 ms, allowing hundreds of peak-pulses and droplet transfers per second. Such power sources supply required parameters for any viable electrode size, gas combination, heat input, etc. Other types of pulsed power source use inverters instead of the conventional transformers. Primary ac line power is rectified into dc, then inverted back into higher-frequency ac, and finally rectified into dc. In addition to offering an attractive high power factor (about 0.8 to 1.0, as compared with 0.5 to 0.8 for the transformer-rectifier types), the inverter power sources are smaller, lighter, and more capable of controlling many welding parameters.

The expanded capabilities of the transistor-type and inverter-type pulsed dc power sources allow optimum performance in a given welding operation. They incorporate microcomputers and various pieces of peripheral equipment to store preprogrammed standard procedures, software for special procedures, and digital read-outs to completely take charge of and automatically organize the welding to be performed. Thus, a "one-knob" control is all that is required to vary the power source output and the wire feed unit in concert.

Narrow Groove Welding with GMAW

Narrow groove welding (originally known as "narrow gap welding") is an automated GMAW operation sometimes performed on joints in thick sections, perhaps, 25-300 mm, or about one to 12 inches. It requires base-metal joint preparation with a rather narrow, machined opening between the opposing faces of V-grooves or U-grooves. The purpose of the narrow opening is simply to minimize the volume of metal that must be deposited in order to fill the groove weld. This decreases welding cost, reduces stresses and distortion, and usually results in improved mechanical properties. Experience shows that although these advantages are worthwhile, the weld passes are more susceptible to lack-of-fusion at the steep groove faces; consequently, a number of innovations described in the literature can be incorporated in the GMAW process to ensure proper fusion between the groove faces and the weld metal. The innovations can be summarized as:

(1) mechanical oscillation of the electrode from side-to-side,

(2) circular movement of the electrode by locating it off-center,

(3) simultaneous feeding of small twin electrodes side-by-side,

(4) feeding two intertwined electrode wires to cause the arc to rotate as the electrodes melt,

(5) mechanical deformation of the electrode in a manner that causes the arcing tip to oscillate or to spiral as it is melted, and

(6) application of pulsing (DCEP) current.

Narrow gap welding by GMAW can be performed on either single or double groove joints in all positions, providing high travel speeds and low heat input are employed to produce a small weld pool that solidifies quickly. The flat position is preferred because it allows adjustment of parameters to obtain the best possible weld quality. The mechanical properties of narrow gap welds made by GMAW are generally better than those made by other arc processes (e.g., SMAW, SAW) using the traditional, wider groove joints. The multiple, small, rapidly cooled passes made by GMAW assist in achieving good weld metal toughness, but sometimes these welds have a higher hardness in their heat-affected zones than is acceptable. This problem can be dealt with either by preheating or by post-weld heat treatment if the steel cannot be changed to one with a lower carbon equivalent.

Difficulties and Defects with GMAW

The growing usage of GMAW, especially in machine and automatic welding operations, emphasizes the proficiency of the process and the advantages of its several modes of metal transfer in making welds. However, a GMAW procedure requires that a larger number of process parameters be given attention as compared with other arc welding processes. If a GMAW operation is undertaken without proper coordination of power source, auxiliary equipment, electrode, shielding gas, controls for welding variables, weld examination, etc., difficulty can easily arise. Also, vigilance must be maintained to avoid base metals that are not suited for welding by GMAW, to guard against abnormal conditions, and to not have oversights on any of the vital steps. Several difficulties are mentioned here to illustrate problems of a metallurgical nature that might be encountered.

Hydrogen embrittlement and *hydrogen-induced porosity* have been known of occur with GMAW even though the process is touted as being "low hydrogen." Indeed, manufacturing practices for wire used in making solid electrodes, and strip for tubular composite electrodes follow low-hydrogen guidelines. The gases for shielding are produced to low-hydrogen standards.

Although AWS specifications dealing with electrodes for GMAW require their surface to be free of foreign matter that would adversely affect the welding properties or the operation of the welding equipment, residues that somehow contaminate the electrode surface are known to have provided harmful amounts of hydrogen. Because the electrodes employed are small in size, a relatively large surface area is fused when forming a given volume of weld metal. Producers of wire for bare solid electrodes are well aware of the need for very clean wire, but surface lubricants are necessary for cold drawing the wire, and these lubricants often represent some form of hydrocarbon. Removal of the drawing compound sometimes leaves a hydrocarbon residue on the wire surface, and while this may be undesirable because of its hydrogen potential, the lubricity provided by a thin film may be wanted for easier feeding of the electrode to the torch. Considerable care is taken by electrode producers in packaging their product to keep it clean, but mishaps involving hydrocarbons sometimes occur as the spools of electrode are carried to the feeder and put to use.

Spooled solid electrodes are often plated with a very thin copper coating to assure good electrical contact in the GMAW torch. Although a copper-coated surface may appear bright and clean to the eye, whether this finish is actually clean and free of hydrogenous contaminant is a question that deserves a close look. Copper has good resistance of atmospheric corrosion, but in contact with iron, copper serves as the protected (cathodic) side of galvanic coupling. Under adverse circumstances accelerated rusting can develop on a steel electrode beneath a copper coating. The rust is not easily seen, and hydrogen is contained in the hydrated oxide. Wire sometimes is cold drawn after copper plating using soap or other die-lubricants, and their films are removed by a cleaning operation prior to spooling. Cases are on record of surface residues on the GMAW electrode that contain elements like sulfur, chlorine and potassium in sufficient amount to cause erratic arc behavior.

Shielding gases supplied in cylinders are usually of sufficient purity (low dewpoint) to avoid difficulty with hydrogen, but there is a tendency for the moisture content of a gas to rise as the cylinder is emptied. Rubber and polytene hose should not be used for gas lines because moisture can be absorbed by the gas through the walls of these materials. Gas lines preferably should be metal, polyvinyl chloride (PVC) or polytetrafluoroethene (PTFE).

Leakage of water vapor into the shielding gas through poorly connected joints in the body of water-cooled GMAW torches is another possible source of hydrogen. The persistency of hydrogen to enter the arc, the weld metal, and even the heat-affected base metal, will be discussed in detail in Chapter 13 of Volume II.

Porosity in welds made by GMAW can also be caused by oxygen and/or nitrogen when proper attention is not given to the amounts of deoxidizers and nitrogen-getters needed to deal with residual quantities of these gases in the filler and base metals. While the GMAW process will ordinarily avoid harmful pick up of oxygen and nitrogen from the air atmosphere, lack of attention to regular operational details can result in a porosity problem. Spatter on the *inside* of the gas nozzle can cause porosity when the buildup of these outcast attachments produces a turbulent flow of shielding gas that entrains air from the ambient atmosphere. The amounts of oxygen and nitrogen picked up through this disturbance can easily exceed the deoxidation and nitrogen-fixation capabilities of the weld composition. The shape of the weld bead also can play a part in porosity retention; a relatively narrow, deeply penetrating weld has a longer "escape path" for any gas that is not tied up by chemical actions. In the event that porosity is unexpectedly found in a weld made by GMAW, investigation should first be directed to hydrogen because experience in welding steel has found hydrogen to be the more frequently encountered cause.

Difficulties related to weld tracking and weld composition homogeneity may result from unusual profile which can occur at the base of a bead deposited by GMAW. With axial spray transfer in a shield of argon gas, or a gas mixture containing a high proportion of argon, joint penetration is likely to have a narrow, deep finger-like projection of weld metal at its base, and a wider elliptical-shaped area above as shown in Figure 6.16. The relative ease of ionizing argon results in a high energy core of plas-

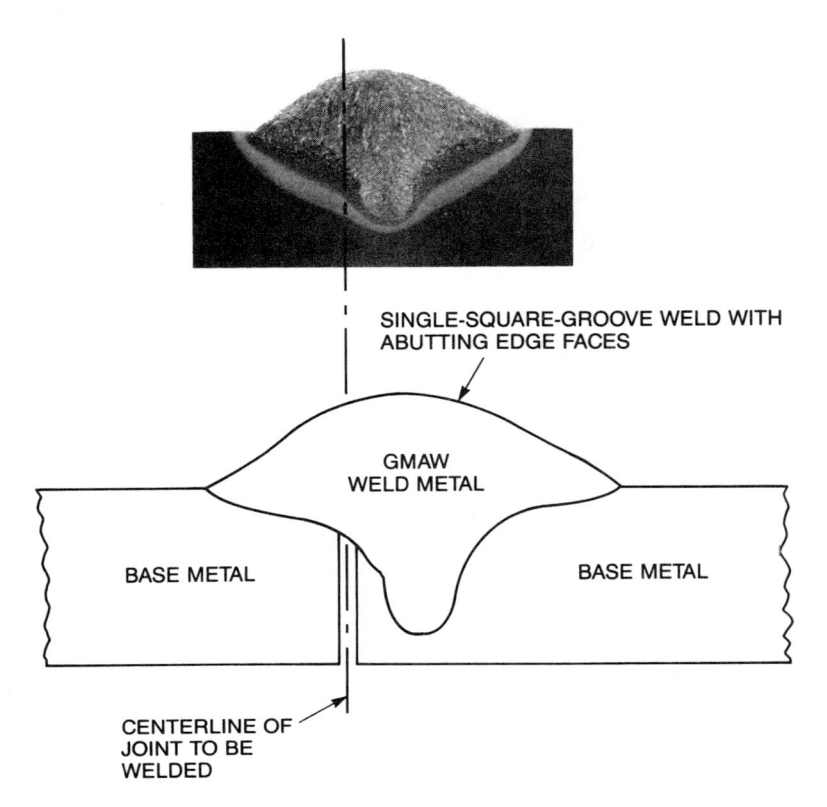

Figure 6.16 — Penetration pattern typical of a weld bead deposited by GMAW operating with axial spray transfer using argon shielding gas

This penetration profile requires precise tracking of joint centerline to ensure that the deep, finger-like projection at the root of the bead fuses the abutting edge faces and does not leave much of the joint unwelded as illustrated above. Remedial measures are described in Chapter 8.

ma in the arc which dwells on the approximate center of the weld pool and forms the finger-like projection. In a square-groove weld with little or no root opening where the abutting faces are expected to be fused during penetration, a GMAW weld deposit of this particular shape may leave too much unfused root if the weld pass is made slightly off center. The addition of helium or carbon dioxide to the argon shielding will change the penetration a broader, almost parabolic penetration profile.

The homogeneity of chemical composition in a GMAW weld can be influenced by its penetration profile. Deposited beads that have a deep finger-like projection from the base of an elliptical area tend to be richer in base metal in the lower portion, and richer in filler metal in the upper portion. In fact, the lower finger-like projection of weld metal sometimes will display tiny islands of unmixed base metal composition dispersed in the weld metal matrix. This heterogeneity is more easily observed metallographically if the filler metal and the base metal are quite dissimilar in composition;

e.g., a carbon steel base metal welded with a stainless steel electrode. Since some base metal compositions are not favorable in the form of weld metal, for example, a steel of medium carbon content, the composition, microstructure, and mechanical properties of a finger-like projection at the base of a GMAW deposit that extends into the base metal may deserve special attention. An elevated carbon content in the projection of weld metal could conceivably induce cracking susceptibility, or significantly reduce the toughness of this portion.

Flux Cored Arc Welding (FCAW)

Fluxing materials for arc welding can be contained inside of an electrode fabricated in a tubular configuration. This arrangement of flux within the electrode, which leaves a bare outer metal surface for electrical contact, has proven quite satisfactory; in fact, two kinds of flux cored electrodes have evolved. One type requires that gas shielding be supplied through a nozzle surrounding the electrode, much in the same manner as in the GMAW process. The second type dispenses with the use of shielding gas supplied externally. Instead, gas-generating materials are included in the flux formulation. Most such wires are made from thin narrow steel strip that is roll formed into a U-shaped configuration and filled with the powdered fluxing materials. This filling must be without gaps, or "holidays," because of the vital functions performed by the core materials — namely fluxing, deoxidation, alloying, gas generation, and protective slag formation. Once filled, the strip is tightly closed into a round tubular outer shape. Various methods are used to fold different amounts of steel strip within the wall of the round tube along with the powdered core material. The core material may comprise about 20 percent of the electrode on a weight basis, which is approximately the same as found on covered SMAW electrodes. To produce the small sizes of flux-cored electrodes required for FCAW, the tubular electrode must be cold drawn like wire to reduce its diameter. This operation compacts the powdered core material so that it does not escape whenever the electrode is mechanically cut off. Although cold drawing tightens the longitudinal closure seam, it does not hermetically seal the seam. The electrode manufacturer must take precautions to prevent excessive moisture or drawing lubricants from entering the core through this mechanical seam, because they will adversely affect the use of the welding electrode. Care also must be exercised to ensure a clean surface. The electrode ordinarily is not copper coated because this operation would require an aqueous plating bath, and the bath solution would be likely to enter the mechanical seam in the electrode and cause great disturbance during welding. A limited amount of FCAW electrodes are produced using steel tube in which the longitudinal mechanical seam is replaced by a welded seam. This is accomplished by filling a relatively large diameter pre-welded tube from an open end with the powdered core mixture, and then reducing the diameter of the filled tube to the smaller sizes required for FCAW electrodes. This is a more costly method of FCAW electrode production, but it does avoid any possibility of lubricant and moisture pickup by the core through a mechanical seam.

Flux systems have been developed which provide all functions necessary for FCAW electrodes to exhibit good behavior in arc welding, and to produce welds of acceptable quality and mechanical properties for many classifications and applications. Although AWS issued separate specifications for carbon steel FCAW electrodes (AWS A5.20), and for alloy steel FCAW electrodes (AWS A5.29), the gas shielded and the self-shielded types are intermixed in these documents. An electrode supplier may offer several different electrodes under his proprietary designations for a single AWS classification, each electrode providing slightly different operating characteristics to satisfy specific applications. Twelve different classifications of carbon steel electrodes presently are prescribed in AWS A5.20, and a suffix on the electrode designation indicates a general grouping of shielding and usability characteristics. Five of these groupings are employed to classify low alloy steel electrodes in AWS A5.29. Despite the multiplexity of electrode types in AWS A5.20 and A5.29, these documents probably are the most widely used specifications, world-wide, for FCAW electrodes. A review of the FCAW electrodes covered by these AWS specifications will be found

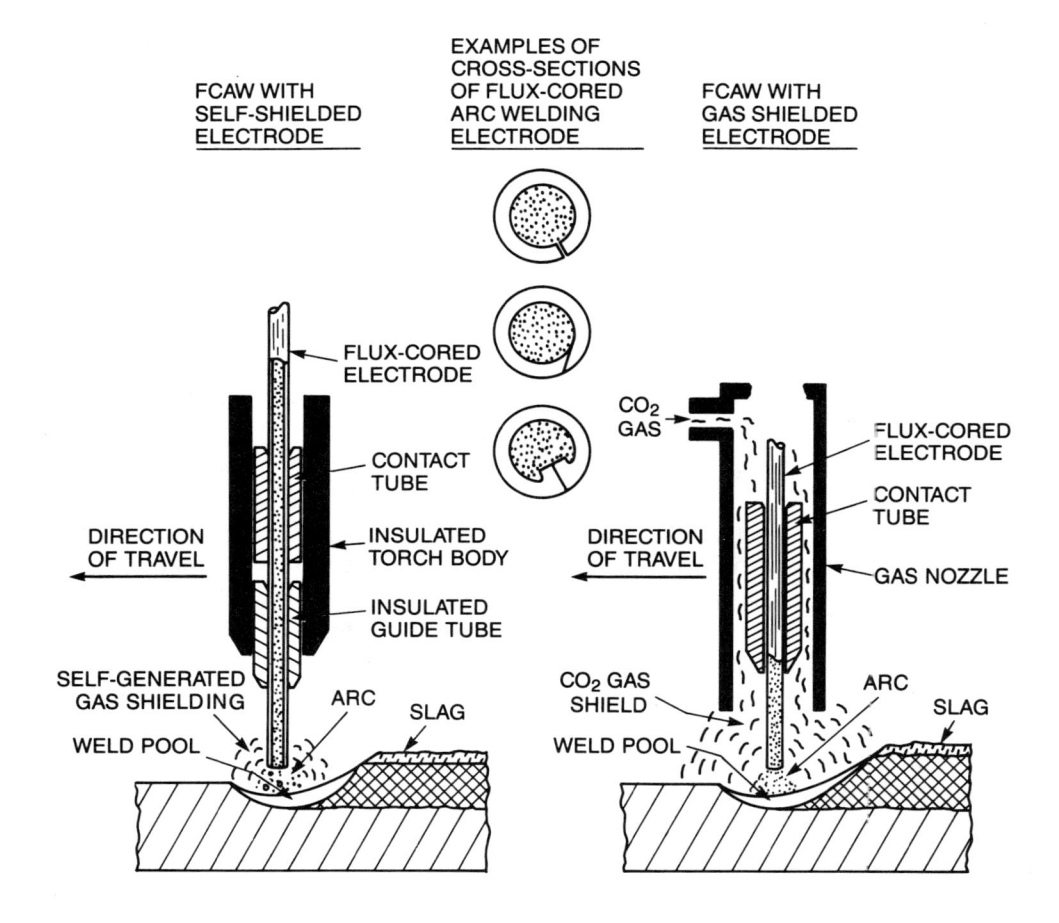

Figure 6.17 — Diagrammatic sketches of the flux cored arc welding process (FCAW)

Sketch on left illustrates a self-shielded electrode (FCAW-S) during deposition. Sketch on right shows a gas shielded type electrode (FCAW-G) being deposited with CO_2 shielding gas.

in Volume II, Chapter 10 on filler metal. Figure 6.17 illustrates FCAW electrodes of both the gas-shielded and the self-shielded types, and the two different kinds of torches employed with each for the arc welding operation.

Gas shielded FCAW-G electrodes differ from SMAW electrode coverings principally in (1) the elimination of the binder constituents, and (2) the elimination of gas-generating constituents for the gas-shielded type. During deposition of a gas-shielded FCAW-G electrode, the solid steel strip in the electrode is melted by the arc and transferred as spray or small globules. When there is a tendency to form large globules, problems with short-circuiting and excessive spatter usually arise. Alloy powders in the core material melt in the arc and transfer directly to the weld pool where deoxidation and alloying takes place. Gas shielded type electrodes *must not be deposited* without the designated externally-supplied gas because the core formulation has neither the gas-generating constituents, nor the amount of deoxidizing elements needed to cope with exposure to the air atmosphere. In fact, the amounts of manganese and silicon are so carefully apportioned in the electrodes of the gas shielded type that changing from CO_2 shielding to a more-protective argon-CO_2 mixture will be reflected in the strength and toughness of the weld metal because of higher recovery of these elements in the weld metal. When low alloy steel electrodes are deposited under the required shielding gas, the recovery of oxidizable elements (e.g., chromium) is approximately like the recovery percentages found with SMAW covered electrodes.

Self-shielded FCAW-S electrodes are highly desirable because of the cost and practicable problems associated with supplying external gas for shielding. Self-shielded electrodes have attained a high level of success in both the carbon steel and the low alloy steel varieties. This type electrode not only avoids the need to provide external shielding gas, but strong air currents are less likely to have an adverse effect on welding behavior or on weld properties. Although some self-shielded electrodes contain gas-generating constituents in their core, primary protection against the ill-effects of oxygen and nitrogen pick-up from the surrounding air atmosphere is provided by a substantial addition of aluminum (and possibly other deoxidizers) in the core material. Self-shielded electrodes usually rely heavily on an aluminum addition in the core material; therefore, the AWS specifications for weld metal deposited by this type of electrode either impose no restriction on aluminum content, or they set a maximum of 1.8 percent for some classifications. Evidence of the effectiveness of aluminum (and some titanium) for controlling residual oxygen and nitrogen can be seen in analyses of weld metal deposited by gas shielded and by self-shielded electrodes as presented in Table 6.2. It is not unusual for weld metal from self-shielded electrodes to contain low residual oxygen and nitrogen contents, but this does not automatically provide better impact toughness. Many aspects of welding procedure with FCAW electrodes influence the toughness of the weld metal, and in general self-shielded electrodes require greater control in welding to ensure meeting impact test requirements.

In view of the self-shielded electrode's dependence upon its strong deoxidation capacity, it should be obvious that electrodes of this type *should not be deposited* with externally-supplied gas shielding. If this added gaseous protection was given to the self-shielded electrode, the aluminum content (plus other deoxidizing elements) would be inordinately high in the weld deposit; perhaps high enough to suppress austenite formation an thus produce abnormally coarse grains in the microstructure.

Procedures used with FCAW electrodes, both the gas shielded and the self-shielded types, must be carefully tailored for this process and the particular joint to be welded. For example, certain FCAW electrodes are designed specifically for single pass welds on steel that is covered by mill scale, rust, or that have other strong demands for deoxidation (e.g., rimmed, or capped steel). These electrodes contain higher levels of manganese, silicon, and other deoxidizing elements to ensure meeting radiographic quality requirements. They should not be used for multiple-pass welds because beyond the first pass the levels of manganese, silicon, etc., may rise too high to obtain normal ductility and toughness. Conversely, a FCAW electrode of an AWS classification designed for single-and-multiple pass welding may not contain sufficient deoxidizers to overcome the oxygen pick-up from surface scale or rust, and its use may encounter some difficulty with porosity in the initial pass. Hence, the reason developing and standardizing the several classes of single pass electrodes.

Another procedural feature of FCAW that receives attention when using self-shielded electrodes is the length of electrode extension (unmelted) beyond the electrical contact tube to the electrode tip at the arc. This electrode extension (commonly called "stick out") can be set at a specific length that can range from as little as 16 mm (⅝ in.) to as much as 100 mm (about four in.). The longer extensions are quite feasible because the torch or electrode holder does not have a gas nozzle that requires the torch to be held close to the workpiece to ensure proper gas coverage. Longer electrode extension increases the temperature rise of the solid electrode by resistance heating, and as a result the deposition rate is increased. This has strong appeal from a production standpoint, but several additional effects of longer electrode extension must be watched carefully. Increasing extension decreases weld penetration, so if a minimum depth is required an adjustment in joint preparation or in welding parameters must be made. Increasing extension usually produces larger droplets of molten metal as the electrode melts, and this change in transfer mode allows more nitrogen to be absorbed from the ambient air atmosphere. Increased nitrogen content produces a larger number of alu-

Table 6.2 — Typical Chemical Compositions of Weld Metal Deposited by Flux Cored Arc Welding Electrodes

AWS Classification Type of Electrode	E70T-1 Gas Shielded*	E70T-4 Self-Shielded**
Carbon	0.12	0.20
Manganese	1.50	0.75
Phosphorus	0.015	0.015
Sulfur	0.015	0.005
Silicon	0.40	0.25
Aluminum	0.005	1.00
Titanium	0.02	0.10
Oxygen	0.070	0.025
Nitrogen	0.015	0.025
Shielding gas used externally	CO_2	None

* *process variation designated FCAW-G*
***process variation designated FCAW-S*

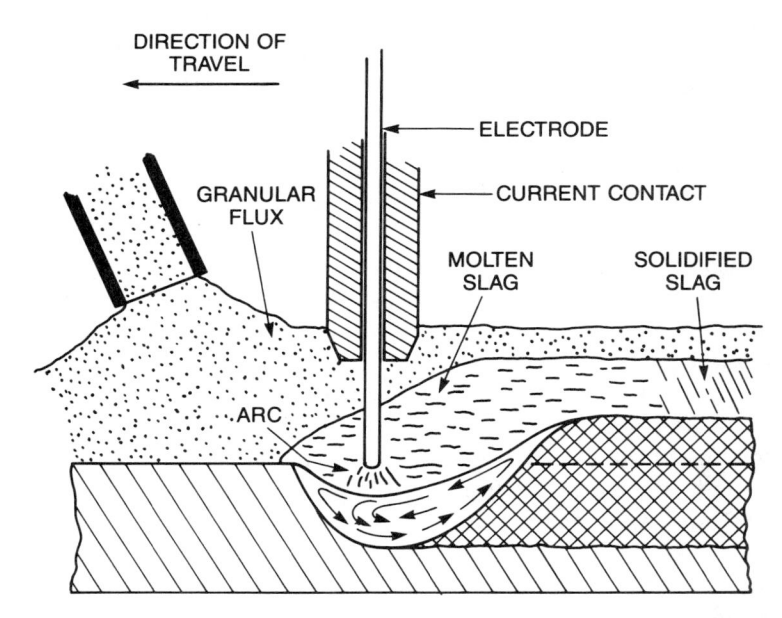

Figure 6.18 — **Diagrammatic sketch of the submerged arc welding process (SAW)**

Sketch illustrates electrode deposition on a thick plate. Arrows drawn on weld pool show usual hydrodynamic motion of molten metal.

minum nitride inclusions, along with a greater number of other nonmetallic inclusions, and higher-nitrogen weld metal deposited with longer electrode extension has exhibited lower fracture toughness. Proper electrode manipulation during deposition can be troublesome with longer extension because the electrode tends to squirm as its free length is heated. Sometimes an insulated section of guide tube is provided beyond the contact exit point (as shown in Figure 6.17) to steady the position of the electrode.

Submerged Arc Welding (SAW)

This process was introduced as a machine welding process - particularly one that could weld thick sections. Instead of permitting an air gap to exist between the electrode and the base metal, the tip of the electrode is submerged in a mound of finely granulated flux that is heaped along the line which eventually will become the weld pass. The process can be applied semi-automatically with a manual electrode feeder, but more often is a machine or an automatic welding operation. The SAW process is widely used today, particularly for welding thick sections, yet it can be applied to plate as thin as approximately five mm (about 0.200 in.). Very impressive capabilities can be demonstrated for high rates of travel, for high rates of metal deposition, and for surfacing.

Figure 6.18 illustrates the SAW process in rudimentary form, but after four decades of development and use there are a number of variations in arrangement of electrodes and sources of filler metal. These variations can involve the use of multiple electrodes, often arcing above the common pool. Also, powdered metals can be metered above the joint, or included with the flux to be melted and uniformly alloyed in the weld deposit. The user, and his exactness in carrying out the details of SAW procedures probably exerts more influence on the final outcome of welding than is the case with any other arc welding process.

The SAW process is unique because very high welding currents can be employed without experiencing a violent arc. The current applied is often four or five times as high as that used with SMAW or GMAW, but the submerged arc behaves nicely beneath the blanket of molten flux without spatter or entrainment of air as would occur with gaseous shielding. The high currents generate considerable heat which leads to deep penetration of base metal, fast deposition of the electrode wire, and permits fast travel speed. Remarkable control can be exercised over the contour of the deposit and the extent of penetration into the base metal. Both of these factors have important effects on the chemical composition and metallurgical properties of the weld metal. Through regulation of travel speed, voltage, current, and other conditions, the electrode may be deposited as a wide, nearly-flat bead with shallow penetration into the base metal, or on the other extreme may consist of a narrow, high-crown reinforcement with very deep penetration into the base metal. The deeply penetrating welds may contain as much as 70 percent melted base metal. The flat, shallow penetrating welds mentioned first may contain as little as 10 to 20 percent base metal. Obviously, the analysis of the weld can be varied widely through selection of base metal, electrode, flux composition, and the proportions of each that enter into the weld with a given welding procedure.

Current for SAW can be either dc or ac. Direct current generally is operated on DCEP to secure deeper joint penetration. With DCEN, the melting rate of the electrode will be increased, but penetration will be decreased. Welding is most conveniently performed in the flat position, although some SAW application is done in the horizontal position. Vertical and overhead deposition is impracticable because of the high fluidity of the molten flux and weld metal. Some form of backing is required to prevent fall-through of the weld metal from a completely penetrated joint. Multiple electrodes can be used in various arrangements to deposit weld metal more rapidly or to surface an area. The electrodes may be positioned side-by-side, or in tandem, and their arcs can be grounded through the base metal, or between electrodes via the weld pool. Some use has been made of flat strip as the consumable electrode when employing SAW for surfacing. Sometimes a supplementary filler metal — either solid or tubular powder-filled — is fed into the arc to increase the rate of metal deposition.

Fluxes for SAW

Fluxes that are used to form the protective molten shield for the SAW arc and to cover the weld metal are acid, neutral, or basic in nature. Their composition and functions actually are more complex and demanding than those used in steelmelting because they must accomplish proper venting, shielding, alloying, removal of solidified flux, etc., while welding is in progress. Of course, no change can be made in the flux applied during the SAW operation (as is often the practice in steelmaking). Even the size of granular flux particles and the height of the unmelted flux burden above the molten flux and weld metal is important in allowing gases to escape and to assist in contouring the top surface of the weld bead. The metallurgy involved in SAW flux preparation and use

> **TECHNICAL BRIEF 34:**
> **Hot Cracking Susceptibility in Submerged Arc Welds**
>
> Submerged Arc Welding is characterized by the relatively slow solidification of a rather large mass of molten metal. As a result, problems can arise with hot cracking, coarse grain microstructure, and segregation. This is because the molten metal at times may not have a favorable cross-section for proper, progressive solidification. Deeply penetrating weld beads can be susceptible to cracking in the center, as illustrated in Figure 6.19; and the cracking does not necessarily extend to the weld surface. Weld bead shape is a highly influential factor, and the depth-to-width ratio of a bead often can be used as an index related to cracking propensity. The depth-to-width ratio of the bead illustrated in Figure 6.19 is approximately 1.5:1, which is a high ratio indicative of conditions that foster cracking susceptibility. Another area in a weld bead that exhibits sensitivity to hot cracking is found when the bead shape produced by SAW has a "flare angle" as indicated in Figure 6.20. This bead shape develops when secondary melting takes place above the deeper penetration, and segregation tends to occur along the front where two solidification patterns meet. Susceptibility to crack initiation in the center of the weld bead or at the pivot of a flare angle depends on the degree of segregation that develops in a given weld metal composition, as well as the solidification rate as determined by welding conditions. Hot cracking susceptibility of steel weld metal under adverse solidification conditions can be correlated with weld composition using the following equation developed by TWI with the trans-varestraint test to calculate units of crack susceptibility (UCS):
>
> $$UCS = 230^*C + 190\,S + 75\,P + 45\,Nb - 12.3\,Si - 5.4\,Mn - 1 \quad \text{(Eq. 6.6)}$$
>
> *When carbon content is less than 0.08 percent; use a value of 0.08.*
>
> UCS = < 10 – High resistance to cracking
> UCS = > 30 – Strong susceptibility to cracking
>
> The above equation portrays carbon and sulfur as the principal elements that promote hot cracking. Although phosphorus and niobium are less deleterious, these elements are sometimes present in significant amounts in certain steel base metals (e.g.; weathering steels, and HSLA steels), and a weld made with deep penetration can pick up an appreciable quantity in its composition from the base metal. Manganese and silicon reduce susceptibility to hot cracking; although high silicon contents above about 0.6 percent may act adversely, particularly if carbon is unusually high, or manganese is unusually low.
>
> Further examination of SAW weld beads of different shapes shows that the pattern of grain solidification is a contributing factor in hot cracking susceptibility. When columnar grains grow inward from the fusion line toward the center, as illustrated earlier in Figure 6.19, and as can be seen again in Figure 6.21, interdendritic segregation of low-melting phases and compounds is encouraged, and these reach highest concentration along the final central plane where the opposing solidification fronts meet. In addition to the influence of interdendritic segregation on cracking susceptibility, the profile of the upper surface of the weld can introduce stress concentration which increases cracking propensity. The marked concavity of the weld in Figure 6.21(A) undoubtedly added to the conditions which promoted centerline cracking.

is sufficiently important to treat this subject in detail in Chapter 8. Therefore, only a brief outline is given here to highlight key aspects of SAW flux technology.

SAW fluxes are categorized as acid, neutral, or basic. The chemical nature or basicity of the flux in the molten state has a profound bearing on its functions during welding, and on the properties of the weld metal, especially its toughness. A *Basicity Index* can be calculated for a given SAW flux composition that quantifies its basicity, acidity, or near-neutrality. These formulae will be discussed in Chapter 8. Rather than just classifying the fluxes, AWS uses a system under which fluxes are matched with specific electrodes and are given designations indicative of the weld metal composition, mechanical properties, and other weld metal features obtained under specific test conditions. AWS has issued separate specifications for electrodes and fluxes used for SAW of carbon steel (AWS A5.17), and for low alloy steel (AWS A5.23), but these documents cover only a portion of the electrode-flux combinations being used in industry today. Many of the fluxes are proprietary formulations that are designed to be used with a specific electrode composition (which also may be a proprietary type of steel) to meet the requirements of a particular kind of service or an important application.

SAW fluxes can be manufactured in a number of ways; and they can be categorized as either *mechanically mixed*, *bonded*, *agglomerated*, or *fused*. Mechanical mixtures are not generally used because the heavier particles can settle and segregate as the flux is shipped and handled. Those containing powdered metals as constituents ordinarily must be bonded or agglomerated, because the metallic constituents would be oxidized at temperatures used in manufacturing fused fluxes. The recovery of oxidizable metallic elements in the weld metal (as might be contained in either the elec-

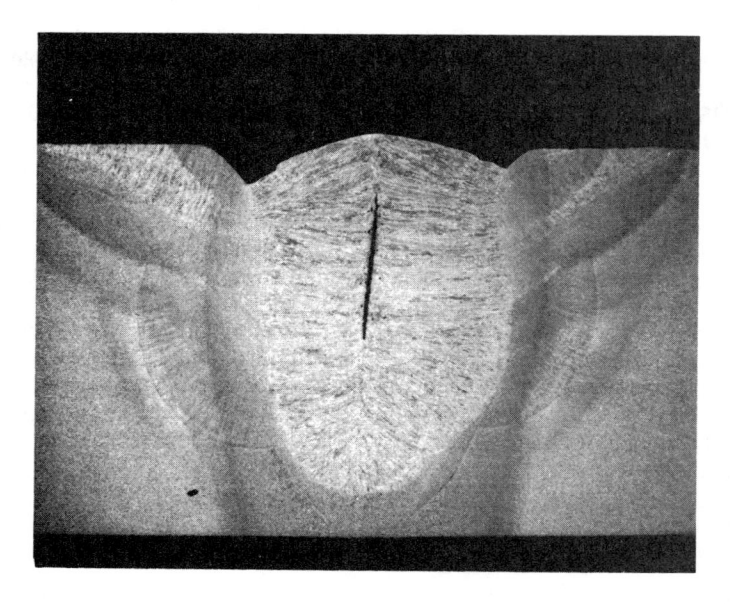

Figure 6.19 — Deeply penetrating weld made by SAW process with hot cracking located in center of weld

Welds with a high depth/width ratio may have unfavorable bulbous cross-sectional shape which is susceptible to cracking at center from micro-shrinkage and segregation of low-melting constituents. Note that the cracking does not extend to surface of weld where its presence could be easily detected. (Courtesy of The Welding Institute, England)

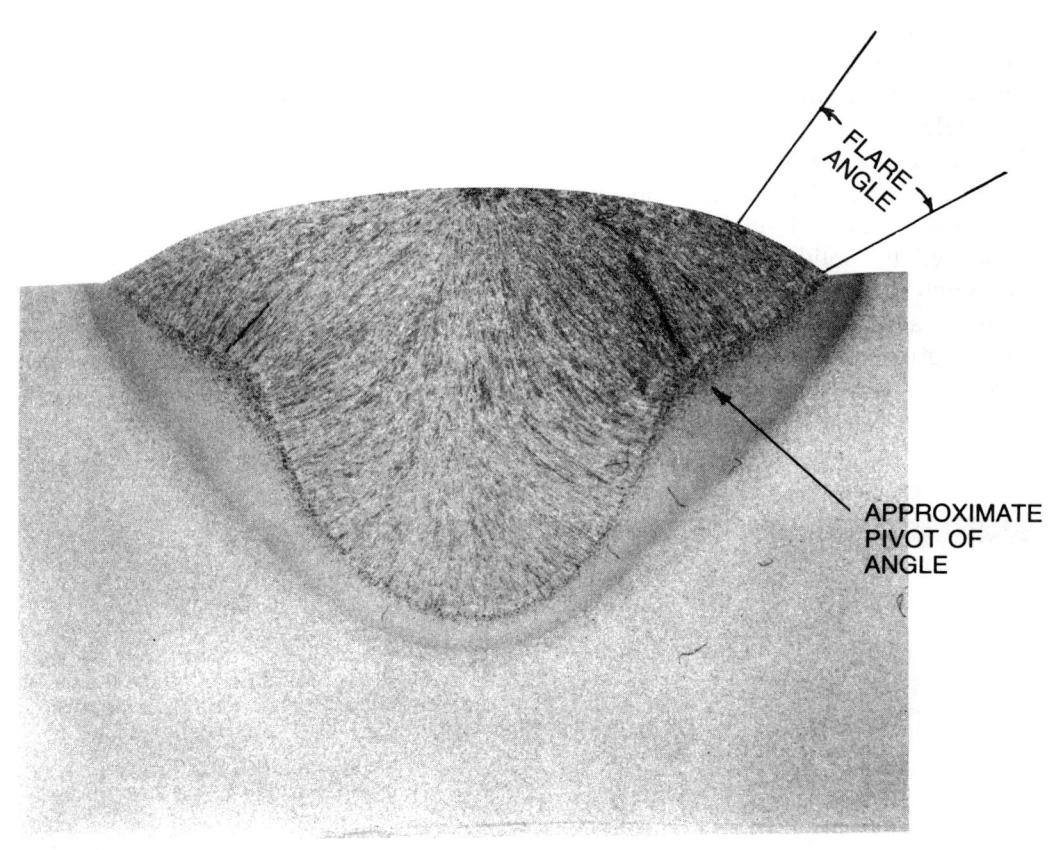

Figure 6.20 — Weld made by SAW process with flare angle in bead shape, and hot cracking along the juncture of two solidification fronts
(Courtesy of The Welding Institute, England)

trode or the flux) in general is similar to that experienced in covered electrode metal-arc-welding.

As a rough rule, the amount of flux melted during the SAW operation will be equal to the weight of the electrode deposited. However, this relationship depends on the composition of the flux and the welding conditions. For example, the welding voltage affects the amount of flux melted. Longer arc lengths which accompany higher welding voltages can impart more heat to the flux than shorter arcs. Therefore, if an alloy-containing flux is being used to contribute to the composition of the weld metal, the welding conditions (particularly the voltage or arc length) must be carefully controlled to maintain a uniform flux-to-metal melting ratio and thus ensure consistent chemical composition along the length of the weld.

Electrodes and Filler Metal for SAW

Electrodes and filler metals will be examined in detail in Volume II, Chapter 10; but here an overview will cover the diverse product forms used in SAW operations, and some features that influence weld quality and properties. Most carbon steel SAW electrodes are solid round wire because this is the lowest-cost form, but certain steels are available as coils of tubular composite electrodes. The composite form permits manufacture of small quantities, and the composition (especially for low-alloy steel SAW electrodes) can be set precisely through the metal powders included in the core. This kind of electrode can be made on the same production facilities used for flux cored electrodes. Chemical composition requirements both in AWS A5.17 and AWS A5.23 are separated into (1) limits for solid electrodes, and (2) limits for *weld metal* deposited by solid electrode-flux and by composite electrode-flux combinations. Although solid wire electrodes and composite electrodes may be interchangeable from a chemical-composition standpoint, they require different welding parameters to produce a given size and shape of weld. Under the same welding conditions, composite electrodes produce a sound, homogenous weld metal, but penetration usually is less than achieved with a solid wire electrode.

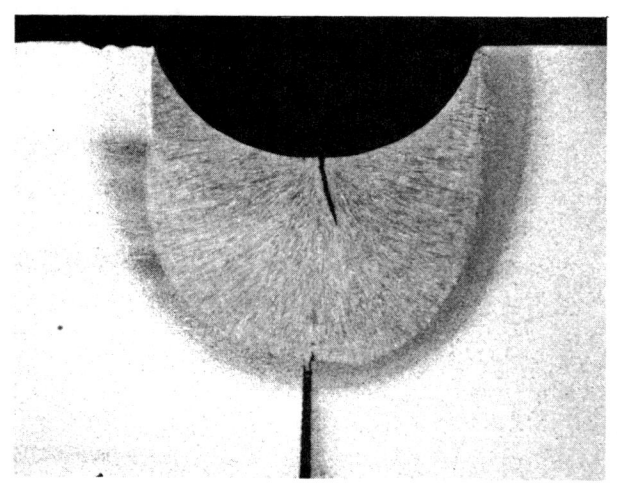

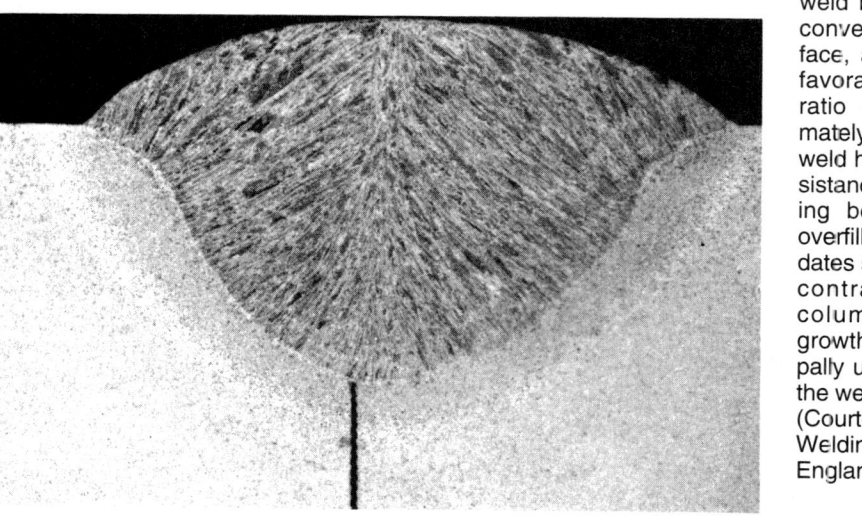

Figure 6.21 — Importance of weld bead shape in SAW process to control hot cracking susceptibility

Two double-square-groove butt joints in heavy steel plate are shown after deposition of initial SAW welds. Specimen A (above) has a weld depth-to-width ratio of approximately 0.67, and the concave top surface of the bead developed a hot crack along the centerline between opposing formations of columnar grains. Specimen B (below) has a weld bead with a convex top surface, and a more favorable D/W ratio of approximately 0.55. This weld has good resistance to cracking because the overfill accommodates solidification contraction and columnar grain growth is principally upward from the weld interface. (Courtesy of The Welding Institute, England)

Weld metal produced by SAW is clean and sound because of the excellent protection afforded by the blanket of molten slag, the deoxidation that can be effectively arranged via electrode or flux composition, and the relatively hot, fluid weld pool that allows gases and impurities to escape from the molten metal. SAW also can function as a low-hydrogen process, unless the flux has been contaminated with moisture or the wire has been contaminated with hydrogenous compounds (e.g., soap, oil, grease). However, problems can arise in SAW with hot cracking, coarse grain microstructure, and segregation as a result of the relatively slow solidification of a large mass of molten metal (See Technical Brief 34).

The penetration capability of SAW is highly advantageous because it frequently permits "one-side-welding," which can be a major saving when making butt welds in large sections of steel plate. *Weld backing* is a specific consideration in the use of the SAW process whenever joint penetration is complete. A relatively large volume of very fluid, molten metal usually is generated, and it remains molten for an appreciable period of time. Therefore, the weld must be supported somehow until solidification at the root of the weld bead is complete. However, difficulty may be encountered in obtaining a root face with acceptable reinforcement and contour. Several methods have been developed to cradle the molten weld metal as it completely penetrates the root of the joint. They include (1) steel backing strips or bars, (2) copper backing bars, (3) flux backing, and (4) backing tapes. With the first method, a strip or a bar of base metal is provided to accept some weld penetration and thus become a part of the welded assembly. Because this solid backing member is welded in place, it creates two unfused notches (located at each side of the weld between the backing member and the underside of the base metal). These notches, under certain service circumstances, have been known to propagate upward through the weld and deserve the kind of consideration given lack-of-fusion at a weld root. The backing member and its inherent notches sometimes is removed to leave a notch-free weld root. The copper backing bar, flux, or backing tape provide only temporary support while the weld is molten, and after solidification this support is removed. The efficacy of each method in creating a satisfactory weld root face is highly dependent on how it is employed. Excessive penetration onto a copper backing bar is liable to result in cracking at the weld root as molten copper penetrates the grain boundaries of the hot steel. Flux backing and backing tapes are viable methods of supporting SAW weld penetration, but they also require judicious application.

Plasma Arc Welding (PAW)

Experience with the highly successful GTAW process spawned ideas on how a stream of gas might be used to concentrate the plasma of an arc instead of letting it flare openly, and to greatly increase the energy level of the plasma in a controlled manner. These objectives have been achieved in the PAW torch by providing a special gas nozzle around a tungsten electrode operating on DCEN. The constricted plasma that forms at the tip of the electrode becomes highly ionized as electrical power is supplied at much higher levels than used for conventional welding arcs.

As illustrated in Figure 6.22, two basically different arc modes are embodied in plasma torches, namely: (1) the *transferred* arc, and (2) the *nontransferred* arc. With

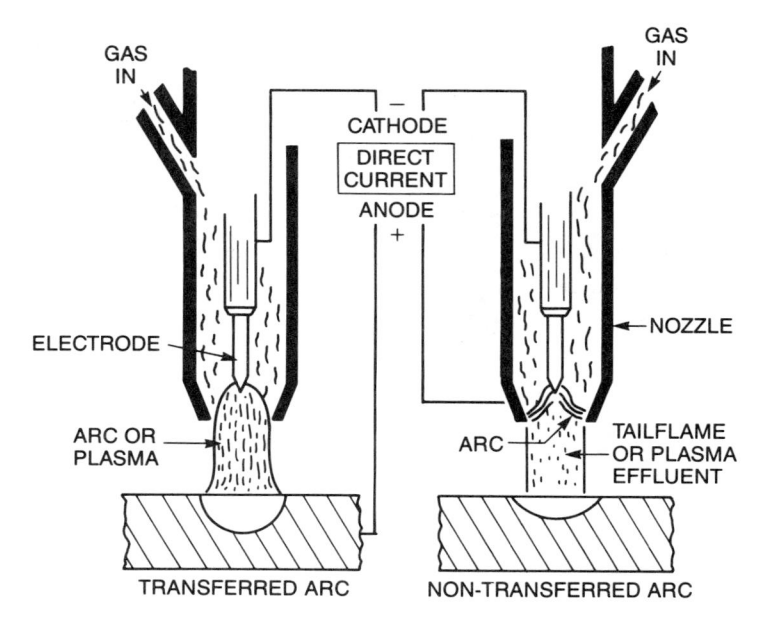

Figure 6.22 — Two kinds of plasma arc welding torches for the PAW process: transferred arc type (Left), and non-transferred arc type (Right)

the transferred arc torch, the plasma is electrically directed from the cathodic tungsten electrode to a workpiece which has been grounded as the anode. The plasma emitted from this torch is somewhat like a stiff, stable jet, which can be firmly directed to a specific small area on the workpiece to accomplish melting. With the nontransferred arc mode, the arc is contained within the body of he torch by positioning it between the tip of the DCEN tungsten electrode and an anodic component, which often is the metal nozzle that directs and constricts the flow of gas. In this case, the gas passing through the plasma undergoes intense thermal ionization, thermal expansion, and is emitted from the torch as a tailflame with substantial velocity. This tailflame is not as constricted or as stiff as the transferred plasma, nor does it have as much thermal capacity. The transferred arc can have a plasma temperature as high as 35 000°C (about 63 000°F) as compared with the free plasma temperature of 6075°C (11 000°F) indicated in Figure 6.3. Conventional welding arcs are relatively diffuse and the plasma can be easily distorted or expanded by moving gases or other small forces such as a magnetic field. This diffuseness limits the temperature of the ordinary free arc to about 10 000°C (about 18 000°F). Theoretically, with a constricted or confined plasma, almost unlimited temperature should be possible, but an upper limit of about 110 000°C (roughly 200 000°F) probably separates the thermal plasma from thermonuclear plasma.

Any gas, or mixture of gases that will not adversely affect the tungsten electrode can be used in a plasma arc torch. The particular gas employed will depend on the nature of the operation. In addition to the inert gases, argon and helium, good use is made of mixtures containing nitrogen and hydrogen. The diatomic (molecular) gases like nitrogen (N_2) and hydrogen (H_2) have particularly good heat transfer characteristics in the plasma torch because they absorb large quantities of energy on dissociation and liberate this energy when recombination occurs.

The energy or heat extracted from an impinging gas stream is governed by the energy it absorbed earlier in the plasma column. However, at the extreme temperatures found in the plasma, additional energy supplied to a gas may not be surrendered readily when striking the work surface. Some gases may transfer heat at a rate equivalent to that anticipated at a much lower temperature. Although gases exhibiting this behavior cool less rapidly, they are capable of sustaining high heat transfer rates for a longer period of time. In addition to the orifice gas used to form the constricted plasma or tailflame, most PAW torches provide an auxiliary shielding gas through an outer nozzle to protect the weld pool from the atmosphere.

Gases used for the plasma and for shielding, and their flow rates, strongly affect weld penetration, the shape of the melted zone, and the suitability of the weld bead after its solidification with respect of adhering slag, undercutting, surface roughness, etc. Much of the technology that evolved when applying the GTAW process concerning the effects of various gases, circulation of metal in the weld pool (Marangoni convection), etc., also holds true in the PAW process. Mixtures of two or more gases for the plasma stream are often needed to secure optimum results. Gas used for shielding exerts some effects in conjunction with those of the plasma, particularly when welding the root of a groove weld. Hydrogen is often considered as an additive in small amount to argon both in the plasma gas, and in shielding because it assists in achieving higher weld travel speeds than with argon alone. Yet, it must be noted that hydrogen in the plasma gas contributes to the total hydrogen content of the weld metal.

An advantageous characteristic of the plasma arc process frequently useful in welding is that only small changes in heat input to the work occur with variations in torch-to-work distance. This is unlike conventional arcs in which the electrode and the work serve as the two electrical poles. The immutable nature of the constricted plasma can be helpful in maintaining weld uniformity when the torch is not able to precisely track an irregular joint. The plasma arc process also can form a "keyhole" when making a single-square groove weld and thus accomplish joining in a single pass. The resulting welds tend to have a wine-glass shape cross-section because the keyhole technique forms a comparatively narrow penetration to the root of the joint while surface heating produces some bowl-shaped melting. To minimize the size of the bowl at the surface, the plasma should be no larger in diameter than deemed necessary. The great energy available in the plasma jet, or in the plasma tailflame allows welding at a fast rate of travel, and this reduces total heat input and decreases distortion.

The power that can be handled by plasma arc torches ranges from only a few watts, as might be used for welding very thin sheet, up to several hundred kilowatts, as would be used for cutting operations. Much of the energy is the result of very high plasma voltages, on the order of 75 to 250 volts. Therefore, specially designed power sources are necessary for plasma arc operations. The PAW process often is favored for joining very thin material where a welding current less than about five amperes would be sufficient for fusing the metal. Electrical energy is converted to heat in the plasma arc torch at an efficiency of about 90 percent.

From a metallurgical viewpoint, welds made with the plasma arc process are much like those made with the gas tungsten arc welding process. It does not provide fluxing agents or deoxidizers that might be required to properly weld certain steels, so the remedial measures employed with GTAW should be considered for PAW. The high temperature plasma also makes it possible to spray high melting-point metals. Particles which can be melted without decomposition (including nonmetallics) can be metered into the plasma stream and accelerated to high velocity. When the molten particles strike the work surface, they form a fused coating or overlay. The material may be a high melting-point metal like tungsten, or even a refractory nonmetallic compound like zirconium oxide. An important feature of the plasma arc process when applying a coating or an overlay is its ability to minimize melting of the workpiece or substrate. Low dilution of the deposit is especially important when highly alloyed consumables are being applied, and pick-up of base metal tends to degrade the deposit.

Percussion Welding (PEW)

The percussion welding process makes use of an *unshielded* arc to weld very small members to each other, or to attach a small member to the surface of a larger workpiece with minimal distortion or discoloration from heating. Electrical power for the arc is supplied by discharge of a capacitor, and the workpieces are brought together so rapidly that the motion amounts to impact, hence *percussion* in the process designation. In addition to the unique metallurgical aspects of PEW, innovative systems are used in initiate the arc and to provide the required amount of thermal energy in only one millisecond or less to melt the opposing surfaces to be welded together.

The PEW process is conducted as an automatic operation with a hand-held gun, or with a machine, and is widely used for making electrical connections with very small size wire of many metals and alloys, and for applying small studs to the back side of finished decorative metal trim without defacing the show side.

Three different methods are used for precise initiation of the arc to accurately control its extremely short duration. In one method, the capacitor has a sufficiently high potential (perhaps from one to six kV) to overcome resistance of the air as the workpieces rapidly close. A second method imposes auxiliary high-frequency high-voltage ac on the welding current from the capacitor. The third method involves a nib or tiny projection on one of the workpieces to serve as a disintegrating fuse as the workpieces make contact at the very start of impact. Arc duration is very important because of the very small amount of metal to be melted. Figure 6.23 provides an example of the exacting coordination of welding current and percussion required to take place during the extremely short period of time for execution of the entire process operation.

The percussive motion for PEW can be supplied by a spring, a cam, or an electromagnet. Means must be provided at the completion of the process to prevent rebound after impact to avoid putting tensile stress on the weld as it solidifies and cools.

The times involved in melting and solidification are so short that PEW is ordinarily performed in air, but protective gas shielding can be provided for metals that are prone to form refractory oxides. Despite the extremely short arcing period, microchanges in chemical composition of the weld metal can occur because of the very high temperature to which the opposing surfaces are heated. These changes, along with the very rapid solidification and cooling have been found to produce unusual microstructures in the weld. Nevertheless, the mechanical properties of welds made by PEW are usually satisfactory.

Magnetically Impelled Arc Welding

In about 1890, discovery was made that applied magnetic fields can manipulate an electric arc in many ways, but no real use was made of this knowledge in welding until about 1940, when equipment was devised to rapidly rotate an arc between the ends of

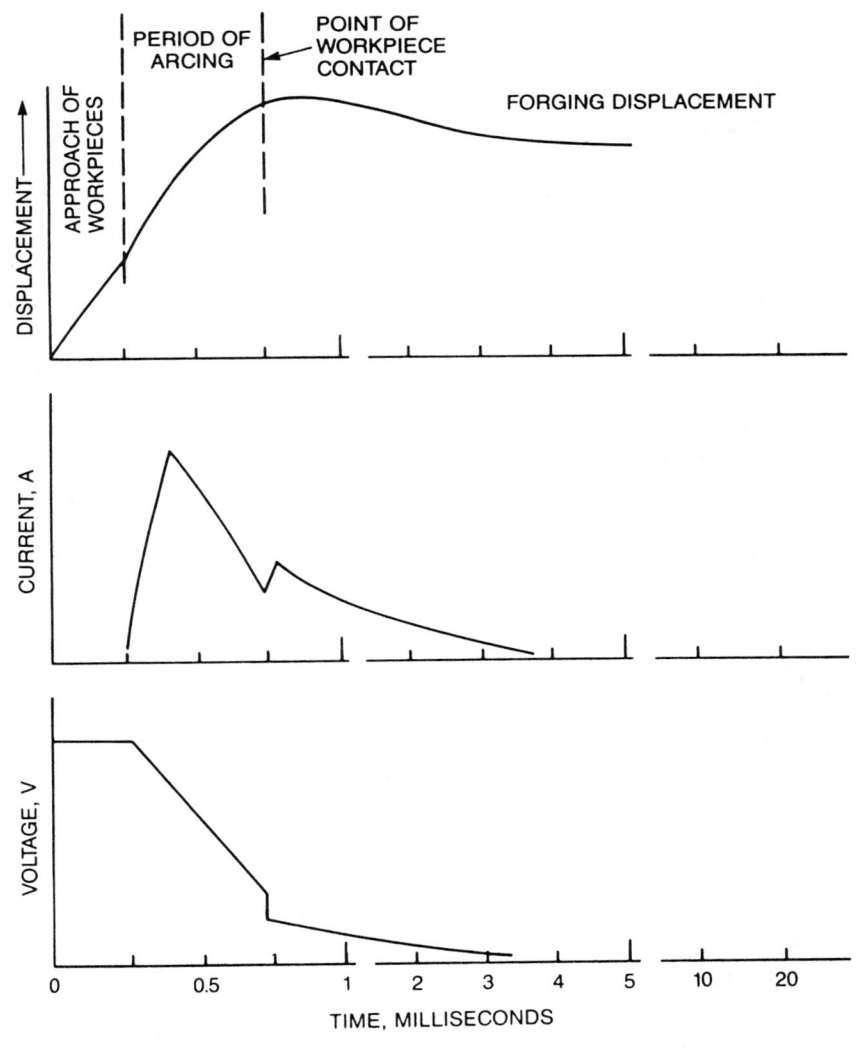

Figure 6.23 — Percussion welding process (PEW) diagrammed to illustrate coordination between voltage, current, and displacement

Of the total welding time, contact of the workpieces occurred at about 0.75 ms, after an arcing period of only 0.5ms. An oscillograph is employed to record the very rapid sequence of events in this process.

two circular workpieces held in abutting position with a short gap between them. This initial industrial experimentation matured as a process for making arc butt welds in tubing. Additional worthwhile applications of electromagnetic arc manipulation followed to accomplish arc deflection, constriction, and shaping in various welding and cutting processes.

Magnetically Impelled Arc Butt Welding

The acronym MIAB commonly identifies this machine-welding process, which is used to make girth butt welds in relatively thin-wall tubing. To illustrate the MIAB welding process and typical conditions for making a butt weld in tubular workpieces, Figure 6.24 presents schematics of the principle of arc rotation, a weld cycle record showing the main stages, and photographs of a MIAB weld in carbon steel tubing. The advantages of this process are rapid automated welding, low energy consumption, low spatter, modest upset or flash, and low material consumption. MIAB welding is capable of making high quality welds, and it competes with flash welding (FW) and friction welding (FRW); although MIAB is not fully competitive in handling relatively thick sections.

Shielding gases are sometimes used in the MIAB process, but their benefits are not as consistent as might be anticipated. The use of argon, helium, carbon dioxide, and mixtures of gases including some hydrogen often cause unexpected difficulties with arc mobility. Only nitrogen has been found to have a consistent effect in promoting arc rotation.

Although MIAB machines are not yet being used in large number in industry, the process offers attributes that can be utilized in metallurgical control of welding. For example, capability to impell the localized arc repeatedly around the joint before welding conditions are reached allows considerable freedom in preheating steel workpieces. Machines have been built to weld high-pressure gas transmission steel piping in the field under extreme weather conditions at temperature as low as -40°C (-40°F) where preheating is a vital step in the welding procedure.

Magnetically Impelled Arc Fusion Welding

This process, identified by the acronym, MIAF, is similar to MIAB except no force is employed to close a gap between opposing melted surfaces. Instead, the rapidly moving arc is supported between an encircling nonconsumable electrode (made of copper and graphite) and an edge joint which is melted through a number of passes by the arc. The stationary assembly of workpieces-and-electrode is in turn encircled by a dc electrical coil that provides the magnetic force to cause a dc arc to rotate between the electrode and the edge joint. Figure 6.25 illustrates use of the MIAF process to make an edge-flange weld.

A number of passes or revolutions of the arc are usually made to melt workpieces and form a MIAF weld, but the overall welding time is short because the arc travel speed is usually in the order of 40 m/min. (about 130 ft/min.). High welding current must be supplied to the arc to accomplish the required melting in a short time (perhaps about 2000 A). Current is limited by the tendency of spatter to be expelled from the joint by magnetic forces. Consequently, the size of the weld that can be produced also is limited. Both circular and non-circular assemblies of workpieces can be welded by MIAF.

Welding Arc Technology

Brief mention is made here of arc welding processes that have almost disappeared from industrial usage. However, their technology includes metallurgical knowledge that is helpful in developing a well rounded understanding of phenomena in processes now in regular use.

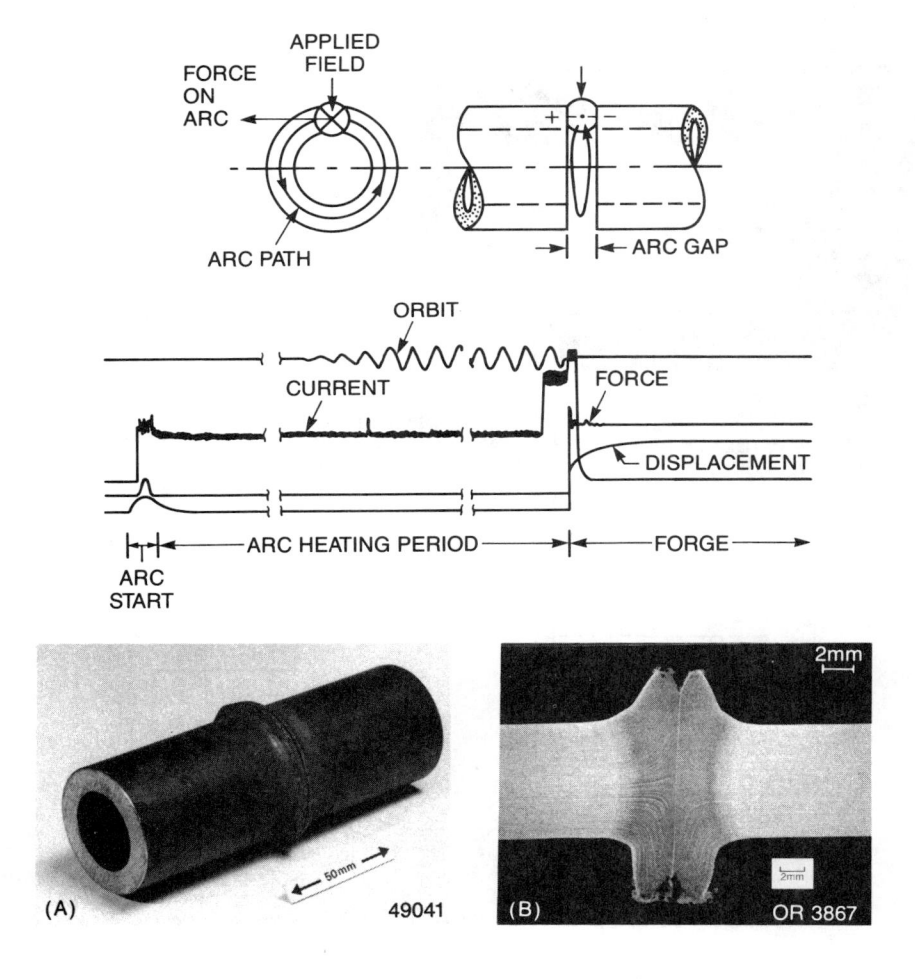

Figure 6.24 — Magnetically impelled arc butt welding process; commonly designated MIAB

In photograph (A), a MIAB weld is shown in round steel tubing 51 mm (2 in.) OD with 9,5 mm (0.375 in.) WT. To start the welding procedure, as diagrammed above, abutting ends of tubes are separated by a small gap. An arc between the end faces rotates around the circumference until the faces become molten. When the arc is extinguished, upset force is shown in photograph (B) of a longitudinal etched cross-section. (Courtesy of The Welding Institute, England)

518   Chapter 6

Atomic Hydrogen Arc Welding (AHW)

This process was the forerunner of gas shielded arc welding, but it has been largely displaced by other gas shielded processes that require less skill to apply, and that offer greater productivity. In the AHW process, an arc between two tungsten electrodes in a manual torch is used to dissociate a stream of hydrogen gas before it impinges on the metal to be melted. Dissociation of molecular hydrogen (H_2) into atomic or nascent form involves the absorption of substantial thermal energy. As the atoms leave the arc and reach the workpieces, the energy absorbed during dissociation is released as the atoms recombine and return to molecular form. Metal upon which the hydrogen stream impinges is readily melted and forms a weld pool. The weld area

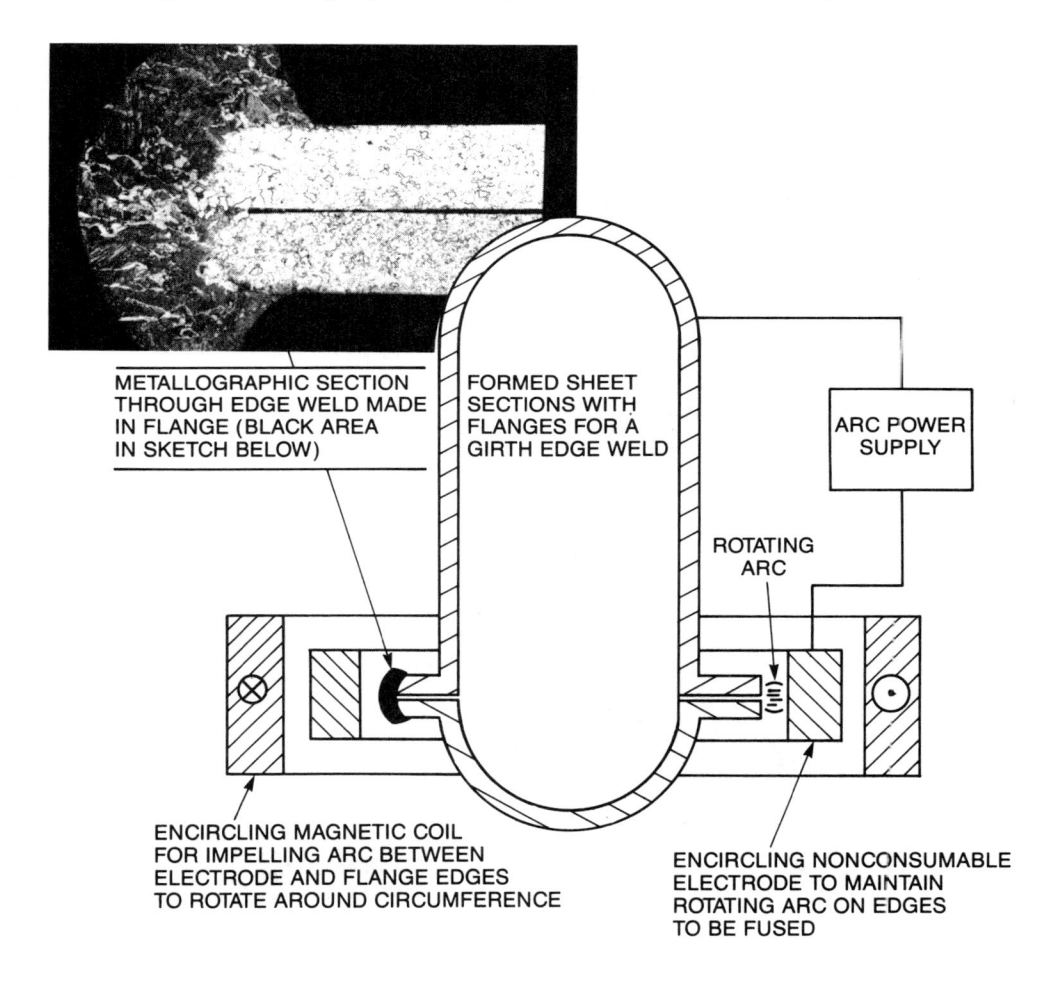

Figure 6.25 — **Magnetically impelled arc fusion welding process; commonly designated MIAF**

The rapidly moving arc between the encircling non-consumable electrode and the flanges of the workpieces fuses a small volume of metal to form an autogenous edge weld. (Courtesy of The Welding Institute, England)

is surrounded by hot hydrogen, and therefore is effectively shielded from the air. At the outer fringes of the shield, the hydrogen vigorously combines with oxygen in the air to form water vapor. Small amounts of iron oxide on the surface of the steel may be reduced to iron by the hot hydrogen.

Care must be taken to avoid applying the AHW process where the hydrogen gas can exert unfavorable metallurgical effects on the weld metal, or on the base metal heat-affected zones. For example, porosity in the weld metal can be encountered when welding steels that contain significant levels of sulfur and selenium. The hydrogen gas reacts with these elements in the molten weld pool to form hydrogen sulfide or hydrogen selenide gases, both of which are almost insoluble in the molten metal. Also, hardenable steels that have a sensitivity to hydrogen embrittlement and cracking must be preheated and slowly cooled after welding, and possibly given a postweld heat treatment to alleviate the effects of hydrogen picked up during the AHW operation.

Carbon Arc Technology

Although *carbon arc welding* (CAW) is hardly ever used today for welding, carbon electrodes (graphitic) are used extensively for arc cutting and gouging operations (CAC-A). Electrodes are often copper coated to improve electrical contact. Most electrodes are round rods in standardized diameters, but flat and half-round shapes are available to gouge rectangular grooves.

An arc between a carbon electrode and steel involves electrical phenomena that can impart a drastic composition change to steel which can easily lead to serious consequences. The basic description of carbon arc technology given here is pertinent to any operation in which a single carbon electrode serves as one pole for an arc against base metal. Furthermore, with carbon electrodes so readily available, welders with the best-of-intentions will occasionally make carbon arc welds in steel. Although the weld may have very satisfactory appearance, it is likely that the carbon content of the weld metal has been increased to a level that seriously impairs fracture toughness.

A dc arc between a carbon electrode and steel base metal operates in a more stable manner on DCEP (reverse polarity). For this reason, carbon arc cutting and gouging operations (usually CAC-A) are performed with dc as DCEP polarity. In this case, the current flow carries ionized carbon atoms from the anodic electrode to the cathodic base metal where the carbon is rapidly absorbed by the molten pool. This mechanism can easily raise the carbon content of the molten steel to more than one percent.

On the other hand, if the dc arc held with the carbon electrode could be properly maintained with DCEN, the carbon pick up mechanism would be largely avoided. Even so, there is an ever-present danger of contaminating the weld in steel with carbon, for example, by accidently touching the tip of the carbon electrode on the base metal or in the weld pool.

Alternating current can be used to produce a stable arc with a carbon electrode, but only if the carbon (graphite) contains an addition of arc stabilizing materials. The ac operation is not usually as efficient as when using dc, but carbon from the electrode is picked up less rapidly because the transport of carbon ions to the base metal occurs only during the reverse polarity half of the ac cycle.

Resistance Welding Processes

A group of welding processes is designated as *resistance welding (RW)* because heat is generated by the passage of current through and between the solid workpieces. Additionally, pressure is applied at some stage in the operation. These two aspects — (1) conduction through the solid metal workpieces, and (2) the application of force — make the RW processes a closely allied group.

Several basic facts about resistance heating should be kept in mind when examining the RW processes. When electrical resistance was reviewed earlier as one of the basic heat sources for welding, two particular locations of heat generation were detailed. The passage of current through the solid body of the workpieces produces heat as determined by the electrical resistivity of the metal at any given condition of temperature and time. Yet, much greater heat generation takes place at surfaces in contact. Current does not flow as efficiently across the interface of two surfaces in contact as it does through solid metal. As the interface or contacting surfaces become more imperfect, that is, are roughened or are coated with oxide, the contact resistance will increase and a larger amount of heat will be generated by a given current. Usually, current will tend to flow through the path of lowest resistance. Therefore, it is possible to localize current flow across an interface by applying a highly localized force to reduce its contact resistance enough to engender conduction. The applied force, in addition to localizing current flow and heating also holds the workpieces together, and may be instrumental in determining how the metal that is melted will be distributed at the weld.

When resistance welding makes use of alternating current, the frequency of the current plays a role in determining the path of current flow in the workpieces. With high frequency ac, perhaps above about 10 kHz, alternating current will follow the path of lowest impedance or inductive reactance rather than the path of lowest resistance. The pathway of lowered inductive reactance is that closest to the return conductor, and this phenomenon concentrates current flow along surfaces that are in close approach (but not in contact). This tendency of high frequency alternating current flow is termed *proximity effect* (commonly called "skin effect") and is especially pronounced at ac frequencies above 100 kHz. Proximity effect is utilized in several important RW processes to be described shortly that are widely used in industry.

All resistance welding processes involve a coordinated application of current (usually ac of high amperage and low voltage) of the proper magnitude for the correct length of time. The current passes through a closed circuit made up of the secondary circuit of the welding machine, the electrodes, and the workpieces. Force is employed to hold the workpieces together, localize the flow of current, help confine molten metal in the localized area, and in some cases to finally expel much of the molten metal in completing the weld. A number of the RW processes are illustrated schematically in Figure 6.26 to show the various ways in which current is introduced to the workpieces and how force is applied. Contact resistance is the greatest contributor of heat for the welding operation, but as many as a half-dozen separate electrical resistivities also affect the utilization of the current.

The schematic in Figure 6.26(A) for spot welding two pieces of steel sheet includes seven contributors to total resistance. If examination of spot welding is started from the top electrode, there is (1) a low resistance in the body of the copper alloy electrode that nevertheless generates a small amount of heat, (2) next a modest contact resistance is found between the tip of the top electrode and the surface of the top sheet of steel, (3) conduction through the solid top sheet of steel encounters some resistance and produces a small amount of heat, (4) highest contact resistance occurs

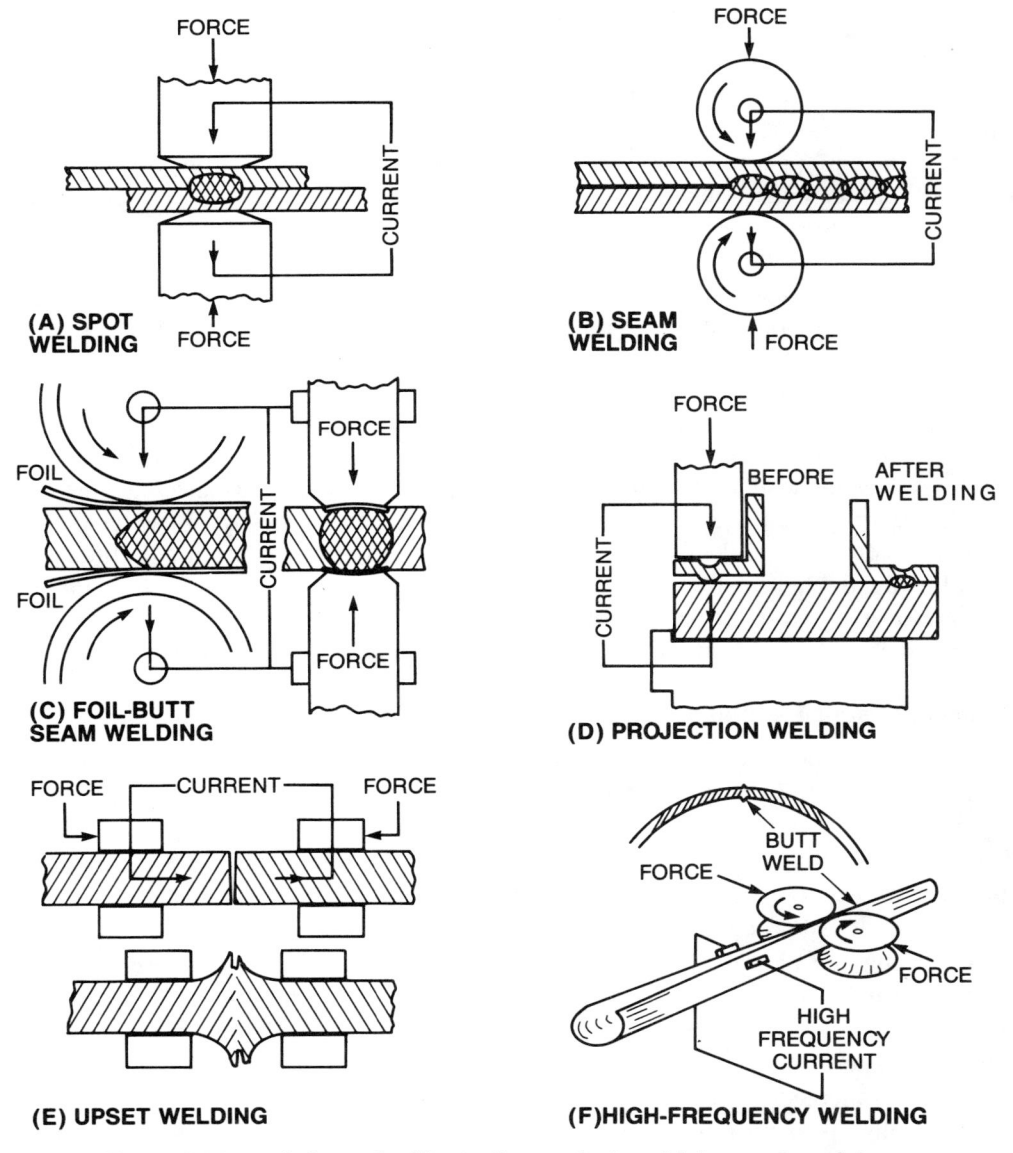

Figure 6.26 — Schematic illustrations of six widely used resistance welding processes

The sketches show the various methods employed to introduce electrical current into the workpieces to accomplish localized resistance heating, and to apply force to control the final disposition of heated metal, some of which can be molten.

at the interface between the two pieces of sheet and the local heat generation is sufficient to melt a nugget of metal, (5) conduction through the bottom sheet also produces a small amount of heat, (6) again contact resistance between the bottom sheet and the tip of the bottom electrode produces some heat, and (7) a very small amount of resistance heating occurs in the body of the bottom electrode. Body resistances in the electrodes and the steel sheets are governed by their resistivity and temperature, which is why spot welding electrodes usually are water-cooled. Contact resistance between electrode tip and work surface should be minimized, and this explains not only the practice of electrode cooling, but also the use of high-conductivity alloys for the electrodes. The contact resistance presented by a surface on the workpiece often is measured with an apparatus that determines *surface resistance*. This is important information that sometimes must be monitored on a production-run of material to be certain that wide deviations are not occurring from lot-to-lot that will cause the spot weld nuggets to vary excessively in size. Yet, even this feature is not quite so simple because actual contact resistance at the interface of the workpieces changes during the welding operation. Initially, contact resistance is determined by kind of metal, surface finish, condition, electrode force, etc., but during welding the "spreading resistance" changes markedly as affected by the rupturing of surface oxide films, and the development of current flow across the interface. The presence of a low-melting metallic coating (e.g., zinc) on the surfaces of the steel sheet can change both the initial contact resistance, as well as substantially lowering the spreading resistance, at least during the early part of the welding operation. Most RW processes must have their parameters pre-set, and the operation is performed so rapidly that seemingly little time would be available to make corrective adjustments once the welding operation is initiated. However, over the last decade, devices and systems have been developed to automatically adjust parameters during the operation to ensure an acceptable weld.

Resistance Spot Welding (RSW)

Resistance spot welding is one of the earliest RW processes, and presently is the most widely used. RSW exemplifies the attractiveness of the RW processes inasmuch as no mechanical preparation of the workpieces is usually necessary, no filler metal is required, the equipment can be portable or can be fixed-base, the welding operation itself is completed in only a few seconds, and the entire operation can easily be mechanized. The RSW process is limited to the joining of faying surfaces or overlapping edges by making one or more spot welds between them. This kind of joint is often favored in the assembly of articles made from sheet. Also, more than two layers of base metal can be assembled in a single application of the RSW process and a weld nugget will form at each workpiece-to-workpiece interface. Multiple welds can be made simultaneously on a single machine by parallel and series spot welding. Non-critical welds in many articles are made with rudimentary machines, and assurance that a spot weld nugget actually has been made at the hidden interface between two workpieces sometimes is assessed simply by observing expulsion or splashing during the operation, which is a small shower of sparks emitted from the interface of the workpieces indicating that more-than-enough welding current has been passed to fuse a weld nugget. As a more-discerning check on weld quality, sample welds can be subjected to a destructive "peel test" in which a chisel is used to start separation of the lap joint, and then one workpiece is peeled away from the other. A test that pulls the nugget completely out of one workpiece as a full-thickness button is regarded as satisfactory. Although this test method is widely used for assessing spot welds in low-carbon steels, it is not as applicable to the higher-strength sheet steels, especially the low-alloy types. Spot welds in these steels, even when sound and of proper size, often tend to fracture through the weld nugget thereby failing to pull a button out of either workpiece. Metallographic examination of the weld nugget is very effective for evaluating the quality of a resistance spot weld. This method provides reliable information on weld size, shape, extent of penetration into the workpieces, microstructure, and whether any defects threaten the structural integrity. Because of cost, metallographic examination is used mainly as part of overall RSW quality control. For making spot welds in demanding applications, such as aircraft, resistance welding machines are more sophisticated and instrumented. Many more mundane applications have acquired a high degree of refinement by virtue of extensive usage and experience. The resistance spot welding of auto bodies perhaps is the foremost example. A good grasp of RSW requires not only a study of available equipment and the capability of each machine, but also a familiarity with the technology of the process.

In a typical RSW operation, as illustrated in Figure 6.26 (A), the pieces to be joined are clamped between electrodes which apply force and welding current. The selection of an electrode is of considerable importance to minimize change in the working face by deformation. As the electrode becomes alloyed with the material in the workpieces (or with a metallic coating on the workpiece surfaces), deterioration takes place from repeated heating and cooling and pick-up of metal particles. For example, the two electrodes in Figure 6.27-A represent typical truncated cone-faced electrodes before use (on left), and after substantial use (on right) in a resistance spot welding operation. Usage has not only roughened the face of the electrode, but the contact area of the face now is 2½ times larger than the original face. To continue making acceptable spot welds, the welding current and/or time would have to be increased above the levels originally employed. The alternative is to replace the used electrode, or to redress its face.

Force is applied by the electrodes with different systems involving levers, cams, springs, hydraulic cylinders, and pneumatic cylinders. One requirement in selecting the force system and planning the spot welding schedule is the rate of electrode "follow-up" during the flow of welding current. The physical volume and strength of heated metal between the electrodes quickly changes, and this requires a movable electrode first to allow slight separation for a brief time to accommodate thermal expansion. When the heated metal undergoes softening, the electrodes must follow-up to maintain enough force on the outer surfaces of the workpieces to avoid substantial increase in contact resistance between the electrodes and the workpieces. The escape of fused metal also must be prevented.

The qualities sought in a resistance spot weld depend upon the required appearance of the welded article and its intended service. Two features of a spot weld that are predominant to quality are its diameter and penetration. Generally, the fused nugget diameter should be 3½ to 5 times the thickness of the thinnest sheet member; while its *penetration* should be at least 20 percent of that thickness, but not more than about 80 percent. Figure 6.27-B is a macrograph of a resistance spot weld between two pieces of bare steel sheet. Surface indentation is minimal, the nugget diameter is approximately 4.6 times the sheet thickness, and the nugget has penetrated about 60 percent into the thickness of each sheet. Also, there is virtually no sheet separation and no evidence of expulsion. Finally, the solidified nugget has no defects or discontinuities.

Phases of the RSW Operation

The complete RSW operation consists of three basic steps — namely:
(1) squeeze time, the interval for the initial application of force before welding currents initiated;
(2) weld time, the period of current flow; and
(3) hold time, the period after the flow of current to allow the weld nugget to solidify.

Figure 6.27-A — Typical resistance spot welding electrodes, with truncated cone flat face, before use in RSW operation (left) and after substantial RSW use (right)

After making many resistance spot welds, the contact area of the electrode face has increased by about 2½ times. If continued use is made of the original welding current and time, a point is reached where decreased heat generation is certain to jeopardize the formation of proper size nuggets. (Courtesy of The Welding Institute, England)

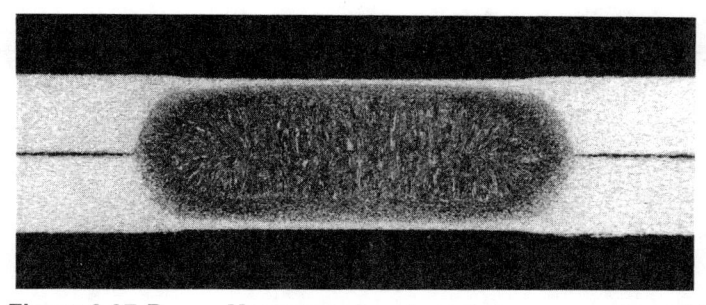

Figure 6.27-B — Macrograph of resistance spot weld between two sheets of low-carbon bare steel showing fused nugget development from standpoints of size (diameter) and penetration, and with minimal indentation of outer surfaces of the sheets

(Courtesy of The Welding Institute, England)

Figure 6.27-C illustrates continuous records of significant electrical and physical changes that transpire in making a resistance spot weld. The data are recorded on a time-scale graduated in 1/60 s units because 60 Hz ac welding current was used.

Changes of interest in Figure 6.27-C occur mainly during the first two steps. During squeeze time (1), the *force* record (A) over its programmed ten-cycle period shows a smoothly rising load through the electrodes with a slightly diminishing rate of application. When the weld time (2) starts and a couple of cycles of current are passed, the force rises at a somewhat faster rate because of thermal expansion in the workpieces. After a total of about five cycles of current have been delivered, the force curve shows a perceptible decrease or dip because softening of the workpieces is taking place with temperature rise and the electrode force must follow-up to regain and to attain its programmed level, which usually is accomplished after more time. Force is maintained for a short starting period in the hold time (3) while the nugget solidifies and the weld area is cooled by the electrodes; then force drops to zero when the electrodes are retracted.

Continued examination of Figure 6.27-C will find the record for *current* (B) with 16 cycles of 60-Hz ac. Some feedback controls are capable of terminating the weld time at a point where adequate heating has been accomplished, and this prevents expulsion or splashing. Record of the *electrode movement* (C) shows a rapid head approach and then a steady mechanical squeeze (1), which continues for a cycle or two into the weld time (2); whereupon thermal expansion of the workpieces causes the electrodes to move apart slightly for the early portion of the weld time, but when nugget melting and metal softening are nearing completion, closing resumes. *Voltage* (D) measures the total resistance of all materials and their surfaces between the electrodes. It is a dependent variable which has been used very successfully in devices that automatically monitor-and-correct energy input to ensure a spot weld of required size. Unfortunately automatic voltage-drop sensing systems do not work in a reliable manner when spot welding steel sheet with a metallic coating (e.g., galvanized) because of erratic voltage signals during the operation.

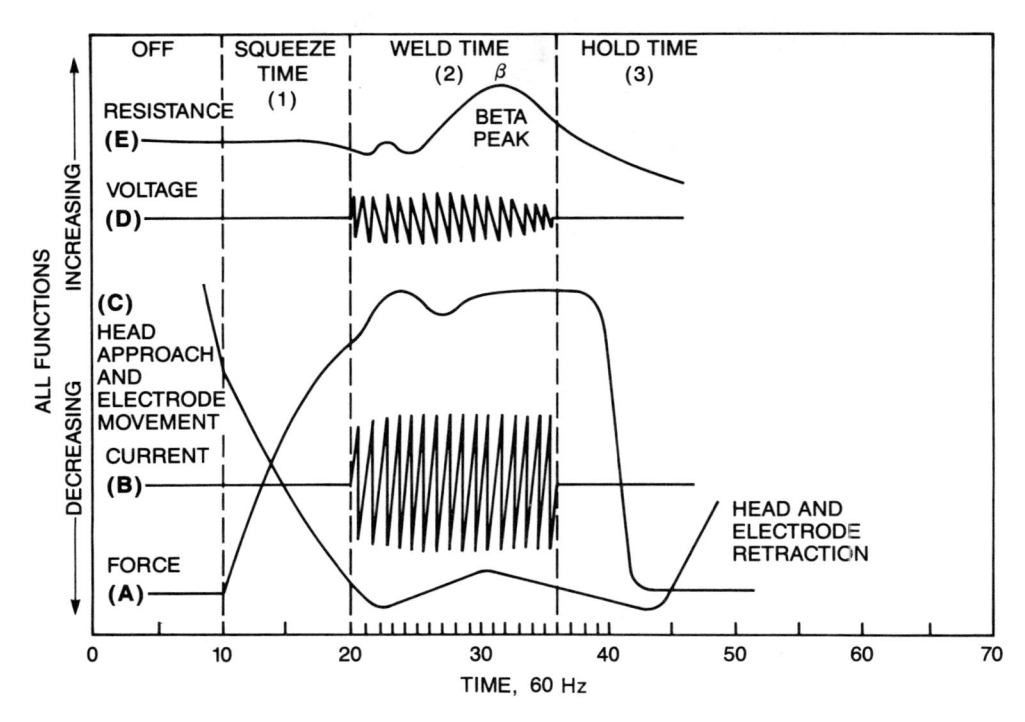

Figure 6.27-C — Continuous monitoring of welding current, voltage, force, electrode-to-electrode resistance, and electrode movement during resistance spot welding

These data would be typical of that recorded during the resistance spot welding of two overlapping sheets of 1.6 mm (0.0625 in.) thick low-carbon bare steel.

Resistance measurements made from electrode-to-electrode (E) are charted in Figure 6.27-C, and are often referred to as *dynamic resistance* because of the rapid changes that take place. Starting with squeeze time (1), dynamic resistance decreases slightly, the result of surface asperities being flattened. Their collapse is more pronounced when current is initiated (2) and the dynamic resistance decreases more rapidly, following which resistance rises because of increasing resistivity in the heated metal. The *rate* of resistivity increase at this stage is an excellent indicator of whether a good spot weld of acceptable size will be made. An increase of approximately two to ten percent per cycle is believed to be necessary. Less than two percent per cycle usually results in underwelding. More than 10 percent per cycle is likely to produce expulsion (splashing). The maximum rise has been designated the *beta peak* (ß) which indicates the start of molten nugget formation. With continued flow of current, the nugget enlarges and causes decreasing dynamic resistance over the last portion of the weld time (2). As the weld and surrounding metal continues to cool during the hold time (3), the dynamic resistance steadily decreases. Studies of dynamic resistance have aided the development of in-process spot weld controls that perform very well on bare steel. Unfortunately, when steel sheet has a metallic coating of a lower-melting metal (e.g., zinc), the orderly progression of events just outlined that determine the shape and change of the resistance between the electrodes no longer hold.

During production operations many variables require attention. For example, a spot weld being made close to one-or-more welds already in position will have some of its intended welding current shunted through the existing solid, cold spot welds. Consequently, a greater amount of energy is applied to allow for the diversion of some current through the nearby welds.

RSW with Coated Steels

Metallic coated steels is a subject with broad scope because of the great diversity of coatings now available. Modern coatings not only have various nonferrous alloy compositions, but the metallurgical structure of the coatings and the nature of their surfaces are sufficiently different to characterize individually the manner in which they react during spot welding. Coatings of tin, lead, aluminum, and alloys of these metals are used to some extent, but the workhorse metal for protecting steel against corrosion continues to be zinc. However, procedures for welding galvanized steel sheet are only of limited assistance because of the many varieties of zinc-coated materials now being employed.

The composition of zinc-type coatings on steel sheet can range from essentially plain zinc to carefully formulated zinc alloys containing aluminum, silicon, magnesium, tin, lead, and other minor elements that have been added to improve adherence, corrosion resistance, surface appearance, cold forming capability, etc. The use of aluminum as an addition has been carried to the extent of providing an alloy coating consisting of approximately 55% Al, 43% Zn, and 2% Si. This metallic coating has the elevated temperature resistance of aluminum coatings, and the galvanic protection characteristics of zinc coatings. Silicon is present to control growth of the intermetallic compounds of aluminum, zinc and iron that form as a layer along the interface between the alloy coating and the steel. From time-to-time, new varieties of zinc coating are introduced by producers, often under a trade-name, and with emphasis placed on their attributes and properties more than details of coating composition, weldability, etc. Sometimes the new product will contain small additions of special elements to the zinc, as examples, high-purity aluminum, and mischmetal (lanthanum and cerium). Other new products may make a greater departure from plain zinc and the coating possibly will represent alloys of zinc-and-nickel, or zinc-and-manganese. Nevertheless, the majority of metallic coated steel using zinc as the principal protective element is essentially coated with plain zinc, although minor elements may be present for control or finish purposes.

Zinc coatings, applied by dipping steel in molten zinc, will develop zinc-iron alloys and intermetallic compounds as an intermediate layer at the coating-steel interface. The thickness of this layer depends on the time of exposure at the elevated temperature, and whether any further heating was applied to the product (e.g., galvannealed steel sheet). These zinc-iron alloys and intermetallic compounds have electrical and mechanical properties that are quite different from those of essentially unalloyed zinc. Most zinc-coated steel sheet is made on continuous, high-speed production lines to specific coating weight limits. The material ordinarily is coated on both sides, although one-side-coated material also is available. Chemical baths condition the surface to make it ready for painting. In most cases, these surface treatments detract from weldability. In addition to straightforward chemical treatments, such as chromate passivation and phosphate treatment, the zinc coating may be anodized, which is an electrolytic process that further adversely affects weldability.

The overall thickness of zinc coatings is important, particularly in extending useful life under corrosive conditions, and in its effect on weldability (problems associated with resistance spot welding of zinc-coated steels are discussed in Technical Brief 35). Zinc coating thickness is most often stipulated as weight of zinc per unit area of surface, but care must be taken to establish whether a given weight of zinc is the *total* amount present on two sides of a coated sheet. Table 6.3 contains information on several varieties of zinc coatings and various ways of expressing their thickness. The spot welding characteristics of electrogalvanized sheet steel are similar to those of hot-dip product, and are commensurate with the thickness of zinc coatings on the surfaces.

Besides research directed to the RSW process and its application to zinc-coated steels, equal effort has been given to ensuring consistent weldability in coated steel products. In the course of this work, the need arose for a test method to evaluate product weldability. Several different testing procedures were proposed by large users of zinc-coated steel, which then generated much activity on the part of steel producers and users alike. A pertinent finding in this work concerned the permissible range of welding current found in particular lots of product for making acceptable spot welds. Most bare steel sheet has a spread of several thousand amperes at a fixed weld time between the making of a minimum acceptable size nugget and the first nugget size at which expulsion is produced, whereas zinc-coated steels usually have a weld current range of only a couple thousand amperes, or less. Those lots of zinc-coated steel that

Table 6.3 — Thickness and Weight of Zinc Coatings on Steel[a]

	Approximate Thickness as Measured by Microscopy		Approximate Weight of Zinc on One Surface		Coating Designation Per ASTM A 525[b]
	μm	Inch	g/m^2	oz/ft^2	
Heavy Hot-Dip Zinc Coating	200	0.0083	1525	5.0	
Heavy Zinc Coating on Sheet by Hot-Dip Process	53	0.0021	381	1.25	G235
Medium Zinc Coating on Sheet by Hot-Dip Process	38	0.0015	275	0.90	G165
Light Zinc Coating on Sheet by Hot-Dip Process	13	0.0005	100	0.33	G60
Very Light Zinc Coating by Electrogalvanizing Process	3	0.0001	25	0.08	
Zinc-Iron Alloyed Coating	8	0.0003	61	0.20	A40

a. As an approximation of the thickness of zinc coating to be dealt with in resistance spot welding, each 100 g of zinc per m^2 forms a layer about 13 μm thick (i.e., each 0.6 oz of zinc per ft^2 forms about one-thousanth of an inch of coating thickness).
b. Coating designations in ASTM A 525 are based on two-side equally coated steel sheet produced by the hot-dip process, and coating weights given in this standard are the total weight of zinc on both sides. Therefore, one-sided coated, and differentially coated steel sheet require modified handling to determine coating weight on a specific surface. The prefix "G" indicates regular hot-dip zinc coating, while prefix "A" indicates zinc-iron alloyed type coating.

TECHNICAL BRIEF 35:
Problems in Spot Welding Zinc-Coated Steel

Zinc-coated steel sheet presents two general problems in resistance spot welding. First, the working faces of the copper alloy electrodes become contaminated with zinc causing softening, enlargement and deterioration of the tip. This, in turn, reduces the density of current being delivered through the area where a weld nugget is wanted. Successive welds will suffer reduction in nugget size and eventually will be unacceptable. Second, the soft, low-melting-temperature zinc coating reduces both the static contact resistance of the workpiece surface, and its dynamic resistance as welding current is delivered. Therefore, substantially higher current must be used to form weld nuggets of proper size. To give an indication of the magnitude of the problem in high-production operations (e.g., auto bodies), a set of spot welding electrodes used on bare steel can make more than twenty thousand welds before nugget size becomes undersize and unacceptable. When welding zinc-coated steel of the same thickness under the best of conditions, only two or three thousand acceptable welds might be made. Here it must be appreciated that the energy for forming a spot weld nugget in zinc-coated steel ordinarily requires both welding current and time to be 20 to 50 percent higher. Also, the electrode force must be 10 to 25 percent greater than for bare steel. Consequently, electrode deterioration is accelerated by the welding parameters that must be held in addition to zinc contamination. Generally, thinner zinc coatings markedly lessen difficulties. For metallurgical reasons, coatings that are primarily zinc-iron alloys or other intermetallic compounds are not as detrimental to resistance spot welding as is free zinc.

Much effort has been directed to resolution of the problem in spot welding zinc-coated steel, and reported research includes the following approaches for improving electrode tip life:

(1) new electrode alloys,
(2) modifications of electrode tip shape,
(3) plating the electrode tip to resist zinc pickup,
(4) refrigeration of electrode coolant,
(5) upslope of welding current to improve initial seating of electrode, and
(6) use of feedback or adaptive control.

Although these remedial measures were of some help in alleviating the zinc problem, none gave the extent of improvement sought for high-production operations. The last item listed, adaptive control, covers attempts to employ feedback signals during each weld to increase welding current or time as needed to maintain the required nugget size as the electrode tips deteriorate. However, the adaptive control systems developed for RSW of bare steel have yet to demonstrate consistent capability of zinc-coated steel to warrant extensive installation for high-production operations.

A practical measure which has proved effective and economical for RSW operations on zinc-coated steel is the use of small, replaceable "caps" to serve as the tip on an electrode shank. These are often manufactured of RWMA Class II copper alloy with a RWMA type "E" tip shape that has a flat face on a 45° truncated cone. The cap allows the internal cooling water to be brought consistently close to the electrode's working face. After the maximum number of acceptable spot welds have been made, the cap permits a quick change to a new tip.

---

had the narrowest permissible weld current ranges at commonly used weld times appeared to have the poorest weldability under production conditions.

Lobe curves were brought into use by this weldability research to illustrate the range of welding current between acceptable size welds and larger welds giving the first indication of expulsion. Figure 6.28 shows the current range established at different fixed weld times for a typical bare steel sheet and a zinc-coated steel sheet manufactured by the continuous hot-dip process. A marked difference exists in the width of the current ranges for the bare and the coated materials. The width of current range is considered indicative of weldability, with a wider lobe preferred because greater current-decrease can be tolerated before welds of unacceptable small size are made.

Aluminum-coated steel sheet is produced in large tonnage on continuous hot-dip coating lines similar to those used for galvanizing. Aluminum provides a protective coating on steel principally by its inherent corrosion resistance, and does not serve as a sacrificial metal to offer galvanic protection for iron in most environments. However, aluminum-coated steel has markedly better resistance to oxidation at elevated temperatures, hence its widespread application in heat exchangers, automotive exhaust systems, and ovens. Aluminum-coated steel presents the same problem with electrode deterioration as outlined for zinc-coated steel, and tip face enlargement and

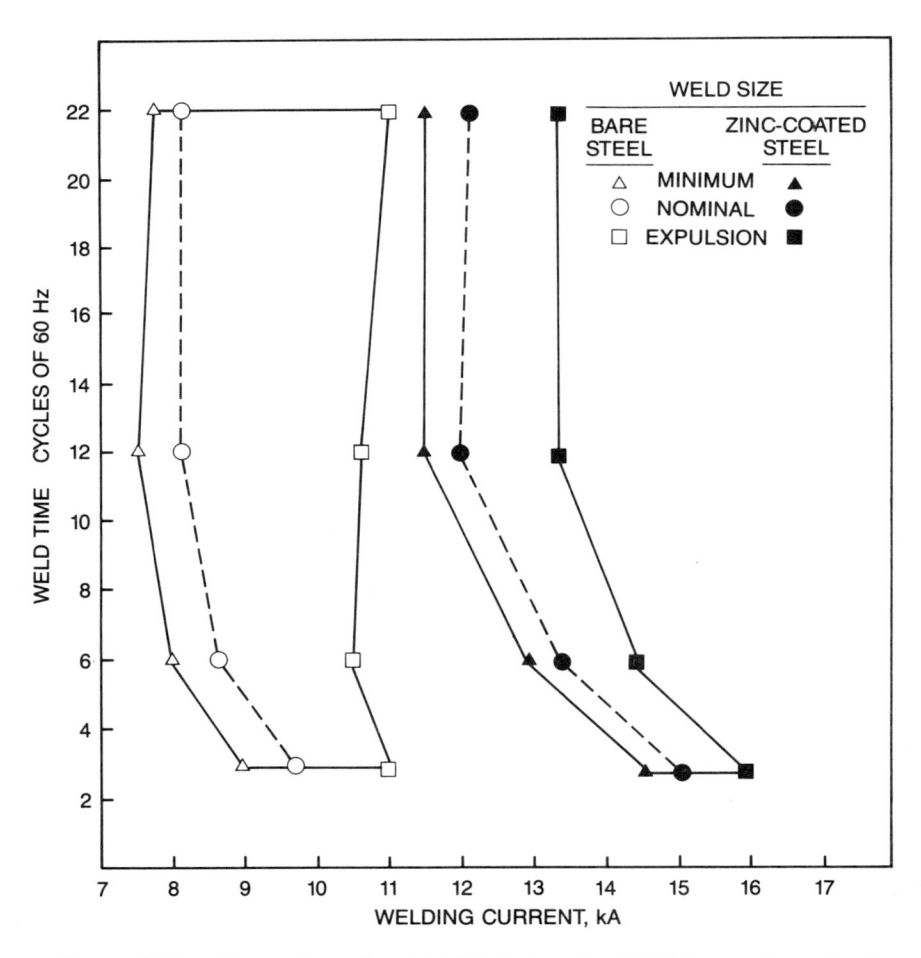

Figure 6.28 — Examples of weldability lobes for RSW low-carbon steel sheet in bare form, and when zinc-coated in conformance with ASTM A 525, G 60

deterioration occurs at a more rapid rate. Consequently, higher welding current and/or longer weld time is needed with a set of electrodes at an earlier point in a production run with aluminum-coated steel. Feedback and adaptive controls presently available do not work in a reliable manner on aluminum-coated steel, so RSW operations must be carried on with the same intensive electrode maintenance practices as described for galvanized steels.

Resistance Seam Welding (RSEW)

Resistance seam welding uses the principles of spot welding with a modification in equipment. Seam welds can consist of overlapping spot welds made progressively along an overlapping joint as shown in Figure 6.26 (B), but there are several additional variations. Electrodes on a seam welding machine can consist of two narrow wheels of copper alloy between which the lapped edges of the workpieces are passed. The circular electrodes usually rotate at constant speed and hold the edges together under constant force. The welding current is usually provided in carefully timed pulses to form distinct overlapping nuggets of weld metal. When the current interruptions are sufficiently long to form a series of separated nuggets, the operation is known as *roll-spot* welding. Some seam welding machines designed to weld at high speeds may employ a continuous flow of current to the workpieces. Consequently, a continuous fused weld area is formed between the overlapping edges. With this process variation, there can be a tendency for arcing to occur at a minute point between an electrode wheel and the outer surface of the assembly on the exit side. The arcing can be of sufficient intensity to cause superficial melting of the sheet surface, and possibly of the electrode surface. The RSEW process sometimes makes use of a single wheel electrode running over the lapped edges of workpieces with a copper alloy mandrel as the backing electrode. This arrangement serves for making a seam in relatively small diameter tubing.

Mash seam welding is a variation of the RSEW process that is used when the double-thickness joint of overlapping edges is not wanted. This is done by precisely feeding a small overlap of only about one to one-and-a-half times the sheet thickness between the electrode wheels. The wheels must be wide enough to completely cover the overlap. High electrode force and continuous welding current are applied, and the joint is mashed flat to the required thickness, which commonly will range from about 120 to 150 percent of the sheet thickness. Continuous seams with good appearance can be produced by this process variation, but force, current, speed, overlap, etc., must be accurately controlled to achieve consistent fusion and uniform appearance along the length of the weld. However, the faces of the weld may display irregularities that would require grinding or polishing if the seam is to be undetectable beneath paint.

Metal finish seam welding is a process variation that employs an overlapping joint, but the weld produced has a flat "show-side" with good surface finish and a back side with a slightly projecting continuous edge. A wide electrode wheel makes the flat mash weld on the show side, while a wheel with a slight chamfer on its working face applies the necessary force from the back side. This process variation also makes use of high electrode force and continuous welding current. Precise control must be maintained of all operating conditions, particularly alignment between the edge of the sheet that is mashed into the show side and the apex of the angle formed by the chamfer on the face of the wheel applying force to the back side.

Foil butt seam welding is a modification of the RSEW process as illustrated in Figure 6.26 (C). The edges of the sheets to be joined are butted together and a very thin, narrow strip of foil is introduced between one or both of the wheel electrodes and the workpieces. The foil acts as a bridge to distribute welding current to both edges of the joint; it acts to increase electrical resistance and contribute to welding heat; and it helps to contain the molten weld nugget as it grows. The foil also serves as filler metal to produce a flush or slightly reinforced weld. Machines for foil butt seam welding are specially built for this process. The foil usually is bare steel, and this provides one way of avoiding zinc contamination and the rapid deterioration of electrode wheels when welding galvanized steels. The surface of the weld itself will be devoid of zinc where the foil is applied, but this can be remedied by zinc spraying or painting.

Projection Welding (PW)

This resistance welding process is illustrated in Figure 6.26 (D) in simplified form. The process, where applicable, offers distinct advantages over spot welding from a number of standpoints, but preparatory to welding the workpieces require a projection or small embossment on one or both contacting surfaces of the workpieces. This may add to costs or may be impracticable for some reason. The projection or embossment localizes the force applied to the workpieces and concentrates the current flow through a particular area. The projection also reduces the amount of current and force required to produce a fused weld nugget and allows thicker sections to be welded. Other advantages include (1) less required overlap because current is concentrated by the projection (2) current shunting is not a problem regardless of close spacing of projections, (3) marking on the outer surface after welding can be greatly minimized by using a large flat-face electrode at the "show-side", and having the projection for welding on the other workpiece, (4) multiple welds can be made simultaneously during one welding cycle and (5) less difficulty is encountered with electrode deterioration when projection welding the metallic coated steels (e.g., galvanized) because the large flat-face electrodes are not subjected to localized heating.

The necessary projection can be formed on a sheet of metal by stamping or embossing, or on a solid section by forging or machining. Projections on sheet metal can range in form from small spherical domes to special oval shapes, but they must be carefully designed for the specific application. The desired technique is to apply just enough current to create fusion just before the projection collapses from loss of strength during heating. Excessive force will cause premature collapse of the projection, and, although some melting may occur around the periphery of the weld area, the central area will likely be incompletely fused. Many factors ranging from kind of material to the design of the projection enter into determination of an optimum welding schedule for a given application. The final parameters employed usually are based on trial welds. Metallographic examination is highly desirable for assessment of projection welds to be certain that proper fusion was accomplished. Strength and peel tests may be somewhat misleading if only peripheral fusion has been achieved inasmuch as strength values may appear adequate, and a button may pull out of a surface when peeling the weld apart. Nevertheless, ring-shaped weld nuggets indicate inadequate melting and as such would be quite susceptible to further small decrease in energy input.

Resistance welding of crossed round wires is a form of projection welding that is widely used in fabricating wire mesh for many applications. Crossed wires require no projection preparation since they make pin-point contact at their intersection. The welding schedule for current, time and electrode force is dependent on the kind of steel wire, but these parameters are determined to a large extent by the amount that the wires are to be compressed together by welding. The extent of compression has been called "mutual indentation" and "set-down," and is calculated as the ratio of the decrease in joint height as a result of welding to the diameter of the smaller wire, but is usually expressed as a percentage. The strength of a crossed wire weld requires a special jig for application of tensile load, but strength generally increases with greater percent set-down. A larger flash is exuded from the joint with increasing set-down. Therefore, the amount of set-down desired is likely to be a comprise between the weld strength required and appearance (amount of flash and its roughness). A set-down of 15 percent is usually regarded as minimal; 30 percent is often the aim; and 50 percent is occasionally used where appearance is of little concern, but strength is of importance. Multiple welds are often made in an assembly of crossed wires in high-production operations. To ensure proper welding, the same electrode force must be applied to each joint. Sometimes flat platens are used as the electrodes for these assemblies, but problems can arise with surface marking and variations in weld size when the electrodes do not apply uniform force on each joint. Improved results are obtained with individual shaped-electrodes that are spring loaded to apply uniform force and fast follow-up as the weld is made.

Upset Welding (UW)

A butt joint welded by the UW process, as illustrated in Figure 6.26 (E), commonly involves two sections of the same cross section joined end-to-end. Each workpiece is clamped in a pair of jaws made of copper alloy and held in required position with their end surfaces in contact. Welding current and axial force are applied simultaneously. Sufficient heat is generated by contact resistance at the junction between the workpieces to the flow of high-amperage, low-voltage current (usually supplied from a transformer) to bring the metal in the immediate vicinity of the junction to the melting temperature. Although some fusion may occur, this process accomplishes welding essentially in the solid state. The metal at the joint is heated at least to a temperature where recrystallization can take place rapidly across the faying surfaces. Force applied during the heating period causes upsetting, and this action can produce several benefits: (1) any molten metal and oxides which may have formed are easily forced out of the juncture, and (2) recrystallization across the interface is speeded up by the hot work, and possibly some grain refinement also takes place. The upset can remain as reinforcement, or can be trimmed away to regain the original cross section.

Upset butt welds in carbon and low-alloy steels are used for joining coils of steel rod for continuous wire drawing as well as for fabricating a wide variety of products made from bars, strip, and tubing. The upset butt process involves relatively slow heating and no measures are taken to protect the joint from air. Consequently, a generous upset is required to exude oxidized metal. For this principal reason, other butt welding processes (e.g., flash, percussion, and friction welding) are often preferred.

Continuous Upset Seam Welding

Continuous welding of a longitudinal joint in pipe and tubing has undergone revolutionary developments since first use more than a century ago. When originally introduced, furnace-heated strip called "skelp" was passed through a bell-shaped die to force the abutting edges together. The "weld" that resulted was not expected to be pressure-tight, and therefore this product was labeled "mechanical tubing." A limited amount of pipe and tubing is produced today by this furnace-buttweld process (see ASTM A 512).

Most tubular products now use electric resistance heating in a continuous production operation to bring the abutting edges of a longitudinal joint to proper welding temperature. The product made with this heating method is widely known as "ERW" tubing (see ASTM A 513). The weld quality of ERW tubular products permits use in pressure service.

When manufacturing ERW pipe and tubing, the flow of electric current and the metallurgy of welding are inexorably linked. Ordinarily, current for resistance heating is high-amperage, low-voltage 60 Hz ac, which is introduced by wheel electrodes, one on each side of the joint and in contact with the outer surface as close to the edge of the joint as possible. While much of the current follows the path of solid steel around the circumference of the tube from one electrode to the other, the close proximity of the wheel electrodes to each other encourages enough current to short-circuit across the abutting joint to heat the edges by contact resistance. As the metal at the joint reaches welding temperature, squeeze rolls on each side of the tube apply force to form an upset butt weld. Upsetting occurs as a highly localized projection (called "flash") on both the outside and the inside of the tube. The flash is trimmed as required with cutting tools on the outside surface and/or the inside surface.

The ERW process using 60 Hz ac has an innate limitation on the speed at which the tube can move through the welding unit. Alternating current generates heat in accord with I^2R (Equation 6.5), but in this case I varies from nil to peak during each half-cycle of current. Consequently, resistance heating is most intense at current peaks, and this cyclic energy input produces "stitch effect" because the stitch is the hottest portion of the weld. If too few cycles (stitches) and too little energy are delivered to a given area because the metal is moving rapidly, the molten stitches will not overlap and the weld will have minute, periodic interruptions along its length. With 60 Hz ac, the tube travel speed must not exceed about 27 m/min. (about 90 ft/min.) to avoid problems arising from stitch effect. One recourse that will permit higher travel speed is to employ an ac power source with a frequency of 180, or 360 Hz. Direct current has been used to some extent as a means of avoiding stitch effect, but problems with other conditions can arise with dc power.

The mechanics of closing the heated edges of the joint as the tube passes through the squeeze rolls requires close control. It is desirable to have the outside corners of the edges meet and start to undergo upsetting first, and to have the remainder of the joint follow to completion. An inverted vee of about five degrees reduces the tendency for electromotive forces to expel molten metal from the joint. Furthermore, the majority of flash is upset to the outside of the seam where it is more easily trimmed off.

A noteworthy step in welding ERW piping and tubing was taken when high-frequency, alternating current (HF ac) was successfully employed for resistance heating of the rapidly moving workpiece. Greater welding speed without stitching became feasible by increasing the frequency of ac to one kHz and higher. In fact, additional benefits were secured when frequency was increased almost into the radio-frequency range, that is, 10 kHz to 500 kHz. The first advantage gained with use of HF ac is found in the mode of current flow through the workpiece to accomplish resistance heating. HF ac flows along the path of lowest inductive reactance rather than lowest resistance. This effect can be manipulated with *proximity conductors*, and with *impeders* to cause the major flow of current to follow a designated path and to concentrate along a surface (skin effect) where resistance heating is desired. The proximity effect often is obtained by positioning the workpieces and electrodes, or by placing an extra conductor along (but insulated from) the path where resistance heating to high temperature is required. An impeder is a device that controls impedance in nearby conductors through its permeability. Impeders for HF ac resistance welding are made of magnetic ceramics or "ferrites".

Figure 6.29 illustrates the essentials of making a longitudinal upset butt weld in tubing by the resistance seam welding process using HF ac power (RSEW-HF). The two small electrodes that straddle the joint introduce HF ac of about 400 kHz at approximately 100 V. Sliding electrodes are usually preferred over wheel electrodes because only light pressure on the electrodes is needed to introduce the high amperage current. Despite the light pressure, which greatly minimizes electrode wear, the current transfer is very efficient, even when scale or rust is on the outside surface of the formed strip. The current flows along a highly localized path down one side of the vee and back along the other side as a result of proximity effect. The edges of the strip between the electrodes and the closure point downstream are heated superficially by the localized current flow, and at the point of closure the depth of heating is usually

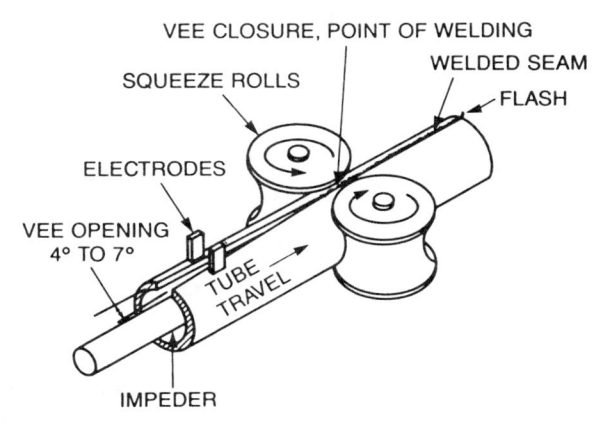

Figure 6.29 — Resistance seam welding a longitudinal butt joint in tubing by electrode contact introduction of high-frequency, alternating current (RSEW-HF)

only about 0.75 mm (0.03 in.). The molten layer that forms on the surfaces of the opposing edges, along with any oxide on the molten metal, is exuded as flash on both the top and bottom of the weld. The overall area of weld, upset section and heat-affected zone ordinarily is quite narrow.

Figure 6.30 illustrates possible paths of HF ac flow when upset butt welding the longitudinal seam in a tube. The path of least inductive reactance is around the inside of the tube (ABC), but flow via this path is minimized by positioning an impeder within the tube as seen in Figure 6.29, and also in Figure 6.31. When located just beneath the closure point between the squeeze rolls, the impeder can improve welding efficiency by as much as 25 percent.

Another means of applying HF ac power to continuously weld a longitudinal butt seam in tubing is to induce the current into the workpiece with an induction coil as illustrated in Figure 6.31. Because the current is magnetically induced into the tube, no electrodes contact the tube. This avoids the risk of surface marking by contacting electrodes. Induced current flows around the outside surface of the tube and on its opposing edges along the open vee. The induction coil method is not as efficient as contact electrodes for introducing welding power, but the use of an impeder as shown in

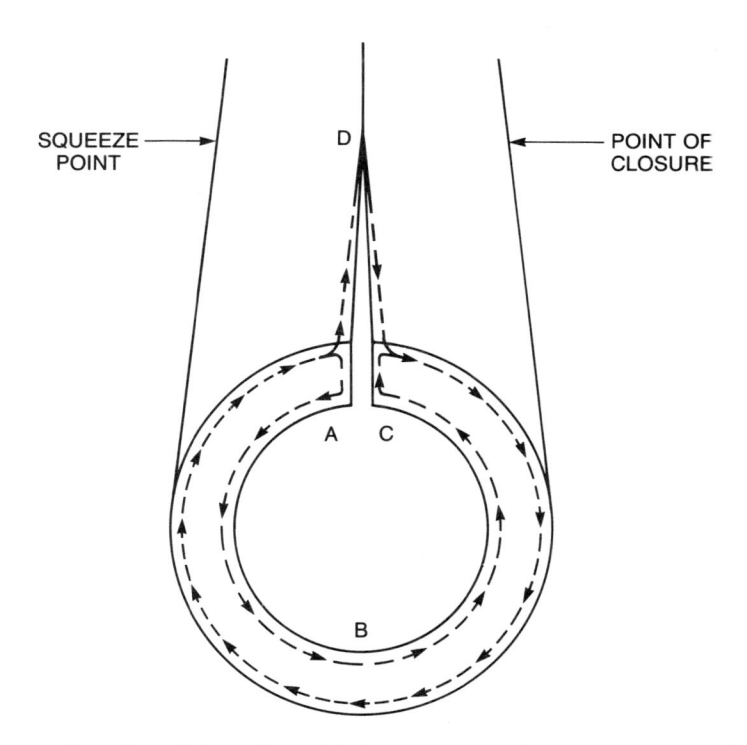

Figure 6.30 — Possible paths of high-frequency, alternating current flow when welding a steel tube with a longitudinal butt seam (RSEW-HF)

Two principal paths of current flow are shown on the above drawing as ABC, and ADC. An impeder can be used to favor flow along the edges of the vee (ADC).

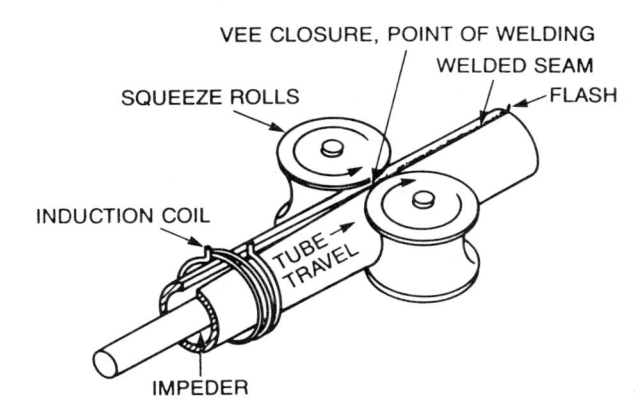

Figure 6.31 — Resistance seam welding a longitudinal butt joint in tubing using an induction coil to introduce high-frequency, alternating current (RSEW-I)

Figure 6.31 can help improve efficiency. The induction method is favored when welding the seam in small-diameter or in thin-wall tubing, and for welding metallic-coated steel tubing.

As experience was gained with the unique characteristics of HF ac, a number of process variations were introduced in resistance welding. These variations include both continuous operations for welding different kinds of longitudinal joints, as well as welding joints of finite length, such as the joining of abutting ends of two steel strips. Figure 6.32 illustrates a few examples where HF ac is employed for its heating efficiency, ease of automation, and consistent weld formation.

Flash Welding (FW)

Flash welding has a unique means of generating heat that is instrumental in making high quality welds in a wide variety of joints and workpiece shapes. The heating mechanism has been variously defined as an "arc," and as "resistance heating;" with the latter now more widely accepted as proper interpretation of phenomena observed during the flash welding process.

Although seemingly similar to upset welding, the distinct heating mechanism of flash welding, prior to application of upset force during the process provides more consistent weld quality. The workpieces are firmly held in copper jaws for alignment and manipulation. Initially, the workpieces are brought into very light contact. The abutting surfaces need not be carefully machined to make intimate contact over their entire cross section; in fact, slight misfit can be helpful. High amperage at low voltage ac is passed across the juncture of the workpieces producing a spectacular shower of sparks that emanates from the joint. Current flows at high density through small areas wherever the irregular abutting surfaces make contact. The points of contact are usually referred to as "asperities." Resistance heating causes most of the rapid temperature rise as short circuiting of current takes place through the asperities. The heat is so intense at these very small contacts that molten metal is violently ejected from the joint; hence the shower of sparks. The departing particles of metal leave small cavities in the faying surfaces in which minute transient arcs are formed. They are short-lived because the voltage is ordinarily too low to sustain an arc, and the cavity in which the arc forms soon closes as new asperities are short circuited. Flashing, therefore, is the rapid, superficial melting that takes place at small points coming into contact with most of the temperature rise the result of resistance heating. Although fusion develops over the abutting faces of the workpieces, the molten metal is substantially eliminated from the interface during the upset period. The final bonding should probably be considered solid-state welding.

Much research has been reported on optimum technique for moving the two workpieces toward each other to secure best possible flash weld quality. As the workpieces

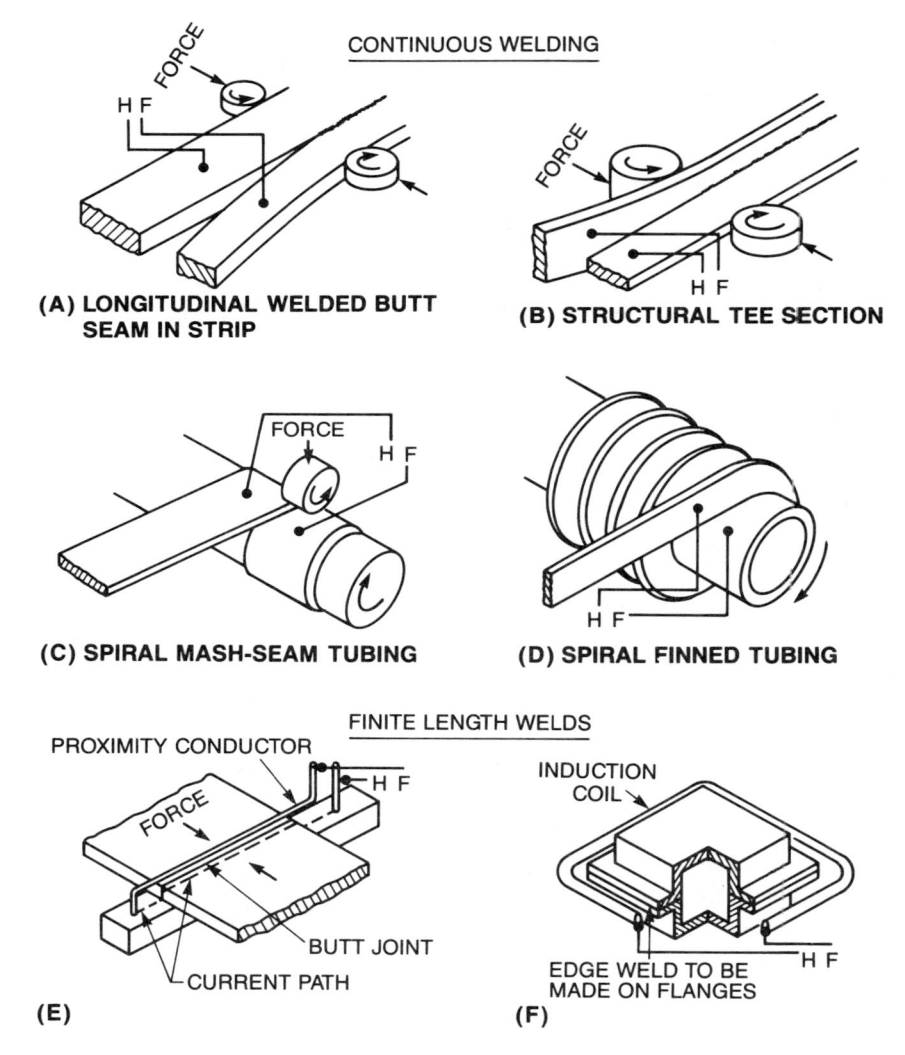

CONTINUOUS WELDING

(A) LONGITUDINAL WELDED BUTT SEAM IN STRIP

(B) STRUCTURAL TEE SECTION

(C) SPIRAL MASH-SEAM TUBING

(D) SPIRAL FINNED TUBING

FINITE LENGTH WELDS

(E)

(F)

Figure 6.32 — Examples of resistance welding processes using high-frequency, alternating current

make contact and flashing initiates, movement of one workpiece to maintain proper flashing and to generate the required welding temperature starts at a constant rate, but then accelerates in the final stage. This linear motion to close the joint usually follows a parabolic curve. As soon as the faying surfaces are completely molten, upsetting force is applied to produce a flash weld as illustrated in Figure 6.33. Among the many refinements developed in operation of the flash welding process is the finding that continued flow of current for a very short interval following the initiation of upsetting is very helpful because the I^2R heating aids the mechanics of upsetting and assists in making a weld free of defects.

Figure 6.33 illustrates two external features indicative of a properly made flash weld. First, the flash-and-upset peaks at the weld include a thin, central fin of metal (marked "A") which represents metal exuded during upset from the superficially melted faying surfaces. This fin ordinarily would contain most of the oxides and contaminants that formed during heating and melting. Second, the substantial peaks of metal (marked "B") that were upset during the application of forging pressure have been forced upward at a sharp angle of slope in relation to the longitudinal axis of the workpieces (perhaps at a slope of 50 to 80°). This weld profile suggests that an adequate amount of metal and oxides and contaminants have been forced out of the weld zone. The metal in the upset can be removed by trimming it flush to the cross section of the workpiece.

Flash welding efficiently produces welds of good quality, and the process is widely used. Some applications subject the finished weld to rigorous examination and testing, such as the welds that join components in gas turbine shafts. Other applications require flash welds to withstand substantial cold deformation; for example, girth-welds in bands of steel strip that are roll-formed into auto wheel rims. These demand-

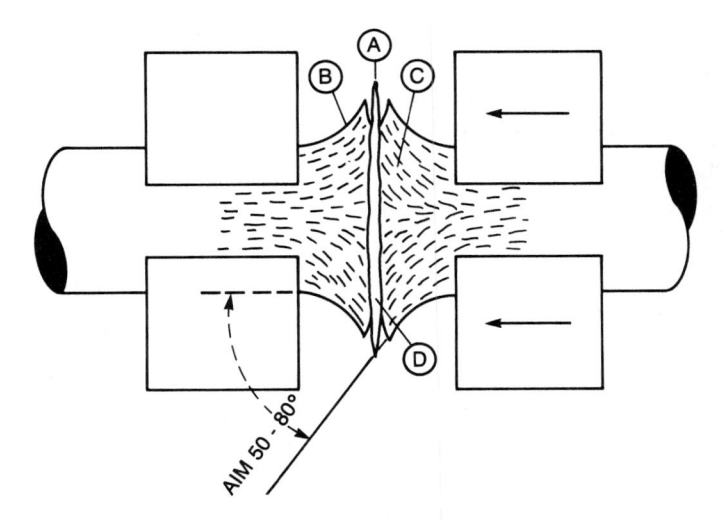

Figure 6.33 — Typical flash weld configuration in steel schematically showing (A) central fin of metal which was molten when exuded from joint, (B) metal displaced by upsetting force, (C) flow lines that can be seen metallographically, and (D) weld zone

ing applications have highlighted three metallurgical conditions that can exist in flash welds which might adversely affect performance; they are: (1) upturned fiber, (2) clusters of microscopic oxide inclusions, and (3) oxygen pick-up.

To illustrate the "upturned fiber" condition, the weld in Figure 6.33 has been decorated schematically with flow lines as might be observed by macro etching a longitudinal specimen. Welds with pronounced flow lines (anisotropy or directionality as described in earlier chapters and indicated in the area marked "C" in Figure 6.33) when stressed in the short-transverse or Z-direction; may exhibit fracture toughness significantly lower than expected.

Clusters of microscopic oxide inclusions in flash welds can be a problem; particularly in applications where the weldment is subjected to severe cold deformation and weld soundness must not be jeopardized in a manner that would permit leakage of gas under pressure through the weld. The circumstances in flash welding that cause inclusion clusters have been studied, and the auto wheel rim application serves to illustrate the way in which a weld leakage problem can arise. Figure 6.34 outlines six steps in the making and examination of a flash-welded, cold-formed wheel rim which proved unsuitable because of leakage pathways through the weld. Briefly, voltage control is important during the heating (flashing) period because it controls the degree of violence with which molten particles are ejected from between the faying surfaces. More importantly, voltage also determines the size of the cavities left in the molten surfaces as the particles are ejected. Although the amount of oxygen in the atmosphere in contact with these surfaces is reduced by the burning of the metal particles being ejected, enough oxygen remains to cause some oxidation of the molten metal, including the surfaces within the cavities. When flashing conditions cause deeper cavities (from high energy input, and especially from high voltage), the oxides formed within these deeper pockets are not easily eliminated during upsetting. Consequently, clusters of oxide can remain along the weld center-plane, and these areas are prone to tearing during cold deformation. The tears will appear as distinctive flat facets in a specimen that is broken completely through the weld center-plane. Those concerned with the soundness and pressure-containing integrity of flash welds call these torn areas either "flat fractures" or "penetrators."

Oxygen pick-up will occur to some degree during the flashing period, and the oxygen of the weld zone (marked "D" in Figure 6.33) will be elevated. With most steels, particularly the low-carbon varieties, oxygen-related effects are likely to be inconsequential. However, the higher carbon, heat treatable types of steel can undergo some decarburization in the weld zone. In some cases, small white areas of ferrite (almost devoid of carbides) may appear in the microstructure of the weld zone because of the increase in oxygen content of the metal. The increased oxygen level could adversely affect the fracture toughness of flash welds in higher strength steels. Little use is made of gaseous protection in flash welding because of practical problems in applying a controlled atmosphere to the joint during the tumultuous flashing period.

Adaptive controls are now available for automatic flash welding machines that sense vital process parameters, and as they are compared with programmed reference signals compensating adjustments are made as the FW operation proceeds.

Electrical Metal-Explosion Welding Process

The explosion of metal by "fuse action" is a method of heating that embodies both resistance heating and the arc. Although not widely used, this process offers unique advantages for metal-to-nonmetallic assemblies that are very difficult to join by other processes. Metal explosion occurs when a high-energy electrical storage unit (a capacitor bank) is discharged through a fine wire or a thin foil of metal and vaporizes the metal with explosive force in a fraction of a millisecond, thus producing a plasma under high pressure. As the vapor pressure decreases, an arc is eventually established

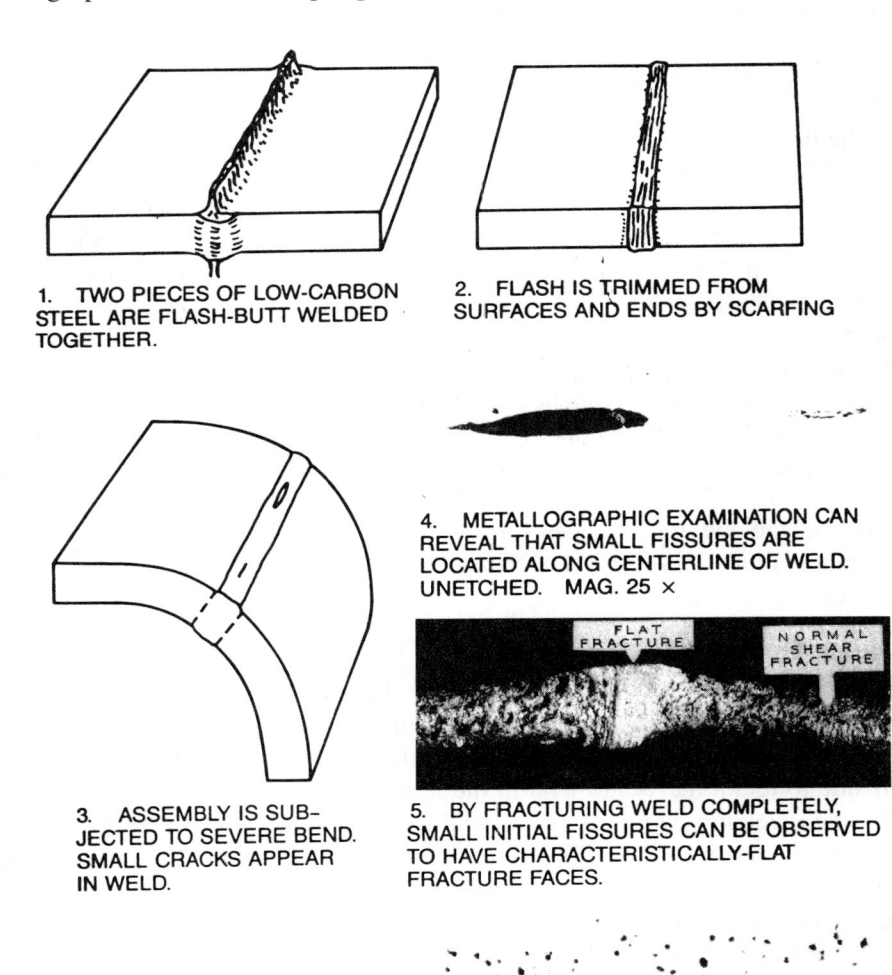

1. TWO PIECES OF LOW-CARBON STEEL ARE FLASH-BUTT WELDED TOGETHER.

2. FLASH IS TRIMMED FROM SURFACES AND ENDS BY SCARFING

3. ASSEMBLY IS SUBJECTED TO SEVERE BEND. SMALL CRACKS APPEAR IN WELD.

4. METALLOGRAPHIC EXAMINATION CAN REVEAL THAT SMALL FISSURES ARE LOCATED ALONG CENTERLINE OF WELD. UNETCHED. MAG. 25 ×

5. BY FRACTURING WELD COMPLETELY, SMALL INITIAL FISSURES CAN BE OBSERVED TO HAVE CHARACTERISTICALLY-FLAT FRACTURE FACES.

6. HIGH MAGNIFICATION REVEALS LOCALIZED ENTRAPMENT OF OXIDES ALONG WELD CENTERLINE. FLAT FRACTURES INITIATE AND PROPAGATE READILY IN THESE AREAS.

Figure 6.34 — Flat-fractures or penetrators in a flash weld in low-carbon steel caused by clusters of entrapped oxide

and a final flow of current occurs. Temperatures in the order of a million degrees have been obtained by exploding fine wire. With thin foil positioned between abutting ends of rod or wire sections to be joined, electrical explosion of the metal foil can provide superficial melting that will serve to establish a bond. Of course, force must be exerted on the workpieces to withstand the explosion, and to bring the molten interfaces quickly together. Since the electric current passes through the inter-positioned foil or wire only, the process can be applied to joints in metals, nonmetals, and composite assemblies involving ceramics, glass, or graphite. The heat generated during the explosion is extremely localized with minimal oxidation and heat effects.

Induction Welding (IW)

Induction welding describes any process that produces *coalescence* of metals through the heat obtained from the *resistance* of the work to *induced* electric current, with or without the application of force. Induction welding can involve either solid-state, or fusion welding. Common to all IW operations is an induction coil, but coil design can vary greatly depending upon the nature of the workpieces and the kind of joint to be welded. Coils can encircle workpieces with one or more loops as seen in Figures 6.31 and 6.32, or a "pancake coil" can be supported just above the surface into which current is to be induced. The depth to which current penetrates, and, consequently, the depth of the heated zone, can be regulated by the frequency of the alternating current used. Lower frequency ac generates a greater depth and uniformity of heating. The means by which induction heating can be utilized to accomplish welding are so varied that a given operation usually is tailored as an individual case to achieve the kind of weld required. In addition to judicious selection of equipment and careful engineering of the operation, successful implementation of induction welding regularly relies on metallurgical appraisal at critical steps to be certain that no influential variable has been overlooked.

Electroslag Welding (ESW)

Electroslag welding is designed to reduce the cost of welding an open square-groove butt joint in thick plate held in essentially the vertical position. ESW is more cost efficient because it allows the gap to be filled in one pass. A dam or shoe is held on each side to retain the weld metal in the joint. Figure 6.35 illustrates the essentials of ESW when welding two thick plates held in the vertical position. However, use of ESW can be extended to the welding of various types of joints as required for sections of different shapes; as examples, a corner joint, and a circumferential joint in a cylinder.

The ESW process presently has a disquieting reputation because of reported defects encountered during the making and testing of welds, and because of a weld fracture found in a main girder of an important highway bridge which was thoroughly investigated and which drew much attention. Some of the problems that have beset ESW are the result of improper procedure when dealing with interruptions in the continuity of the ESW operation on a given joint. Points where ESW is initiated, or where the operation is interrupted and re-started require repair welds that must be carefully planned and executed. A few problems encountered are peculiar to the ESW process, and will be pointed out in this review along with means of avoidance. Also, problems arise when commonplace metallurgical features present in ESW welds are not given

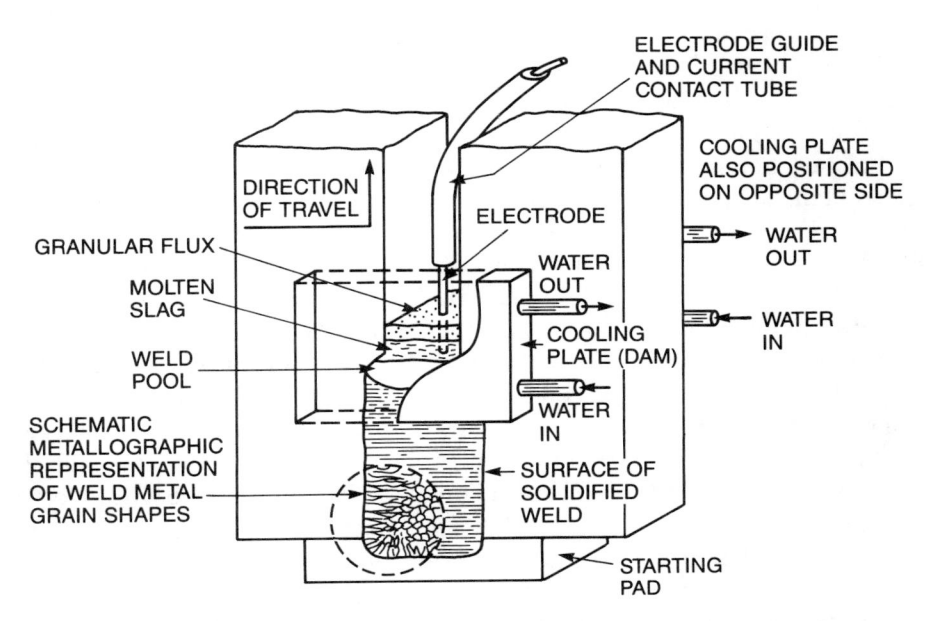

Figure 6.35 — Electroslag welding process (ESW) shown schematically as applied to a butt joint in thick plate

appropriate consideration. These features and the problems they engender will be mentioned, but more detailed information will be presented in Volume II, Chapter 13.

The ESW operation illustrated in Figure 6.35 is started on a pad at the foot of the joint where a layer of granulated flux or slag has been placed. The electrode tip draws an arc for a short interval to melt a layer of slag, whereupon the molten slag quenches the arc and the operation proceeds by resistance heating with the current (either dc or ac) flowing from the electrode tip, through the molten slag, to the weld pool immediately underneath. The electrode entering the slag is continuously melted by heating through its own resistivity and by the more intense heat generated by electrical resistance in the conductive slag. Droplets melted from the electrode fall through the slag by gravitational force into the weld pool beneath. The molten slag also serves very effectively to shield the melting electrode, droplets, and weld pool from air, and the flux can be formulated to carry out favorable (steelmaking) reactions during the welding operation because the time of contact between molten slag and metal is relatively long for a welding operation. The molten slag depth is maintained at about 50 mm (2 in.) by additions of granulated material to replace that lost by leakage between the copper shoes and the weldment surfaces. The slag, which may reach a temperature as high as 2000° C (3632° F), is given a turbulent circulatory motion by the electrical current flow and this assists penetration as the opposing workpiece faces are melted.

The ESW process has a number of attributes, among them, economy in welding thick sections in a single pass, a minimum of angular distortion after welding, and good weld soundness and cleanliness insofar as porosity and nonmetallic inclusion content are concerned. However, it is necessary to (1) remove nonmetallic residue in the joint at the start of the weld and reweld the area, (2) maintain a continuous operation (without interruption of melting and solidification) because of almost-certain unsoundness in the weld at any intermediate re-start point, (3) anticipate a weld metal composition containing approximately 50 percent base metal because of extensive base metal penetration, and (4) accept relatively soft, course microstructures in as-welded weld metal and base metal heat-affected zones because high thermal input per mass of metal results in much slower cooling than ordinarily experienced in welding.

ESW generally uses one or more consumable bare electrodes that are directed into the molten slag by nonconsumable contact guides. Different electrode forms are available. One variation involves a *consumable* guide and contact tube to contribute metal to the weld pool during the welding operation. For this "consumable guide welding", a length of heavy-wall steel pipe is placed in the joint gap extending from the starting pad to a point above the workpieces where the tube is held and electrical contact established. This guide tube is insulated from the groove faces either by nonmetallic shims, or by a heavy coating of flux on the outer surface of the tube. A bare wire electrode is fed continuously down the tube, and both the electrode and the tube entering the molten slag are melted into the weld pool. Sometimes, the wire electrode may be the metal cored variety. AWS A5.25 is a specification for solid and composite metal cored electrodes and for fluxes used in electroslag welding of carbon and high-strength, low-alloy steels. Consumable guide welding has been a much-used method because it simplified the mechanics of applying ESW. Another variation of ESW uses a long, narrow plate of appropriate size for an electrode and feeds it downward into the molten slag to contribute generously to the weld pool, but this procedure has not seen wide usage.

A problem arises with nondestructive examination of ESW welds by ultrasonic testing (UT) and by radiography. The use of these methods is made difficult by the very coarse microstructures that are typically present in the as-welded condition. The coarse grains cause strong attenuation of applied ultrasonic energy during UT and small defects are often obscured in the scattered transmission. Radiography encounters the usual problems associated with resolving defects in thick sections, and the coarse grains can degrade the sharpness of images registered on film. Because porosity, slag pipes, and cracking may be present, NDT can require extraordinary steps to determine whether quality requirements for a given weld are being met.

Cracking Susceptibility in Electroslag Welds

Several kinds of cracking can occur in ESW welds. To avoid this form of defect, a proven welding procedure must be followed closely. Also, diligence must be exercised to ensure that small details concerning welding equipment and materials are being properly controlled and monitored.

Hot cracking in the center of ESW welds is the form of cracking most often encountered. Its occurrence is attributable to the microstructural conditions much like those illustrated earlier for SAW welds where the mechanics of grain formation and the segregation of low-melting minor elements and compounds result in an increase in hot cracking susceptibility. However, circumstances during the ESW process differ enough from SAW to warrant closer examination of this cracking problem. In the case of the larger ESW weld pool, solidification proceeds from surfaces of solid base metal and the cooled retaining shoes. Columnar grains grow slowly from these surfaces with the long axis of the grains approximately perpendicular to the liquid-solid interface through which heat is flowing to cooler regions. When the weld pool is relatively deep as compared to its width, the columnar grains will meet at the weld center almost head-on, that is, at an obtuse angle (see example from SAW in Figures 6.19 and 6.21 (A)]. Under these solidification conditions, low-melting elements and compounds in the weld composition will reach a high level of intergranular concentration by segregation in the relatively small amount of remaining grain boundary area on final solidification. Consequently, hot cracking susceptibility at the weld center will be markedly increased.

On the other hand, if the ESW weld pool is shallow the columnar grains grow upward, almost side-by-side, and the acute angle of approach to each other distributes low-melting segregates on a greater grain boundary area in lesser concentrations; consequently, cracking susceptibility is not likely to be increased [see example from SAW in Figures 6.20 and 6.21 (B)].

The angle at which columnar grains meet at the center of an ESW weld can be related to a "form factor" which is calculated from weld pool dimensions. This factor is the ratio of the pool's width and its maximum depth. Welds that are relatively shallow with columnar grains meeting each other at an acute angle will have a high form factor (typically greater than about 2.0, and the weld zone will have a low susceptibility to hot cracking at its center. Conversely, welds with a low form factor, perhaps about 1.0 or lower, will solidify with columnar grains growing from two opposing sides of the relatively deep pool. When the grains meet at the weld center at an obtuse angle, the cracking propensity of the weld is likely to be high.

Although a form factor of at least 1.5 is to be desired for ESW welds, this criterion does not completely control hot cracking because susceptibility is also dependent on the levels of certain minor and residual elements in the weld composition that form low-melting phases during solidification. The carbon content of the weld is especially important since hot cracking susceptibility increases with higher carbon levels. Weld metal formed during the ESW process can include an appreciable percentage of the base metal composition because of deep penetration into the edges of the workpieces. Many types of steels provided as thick plate have carbon contents near the permissible maxima; therefore the selection of ESW filler metal may require careful consideration of its carbon content to avoid arriving at an unsuitable level of carbon in the weld composition.

The effects of current and voltage during the resistance heating and melting in an ESW operation are not the same as experienced with an electric arc. As the amperage of welding current is increased, the molten weld pool becomes deeper. Consequently, higher amperage acts to reduce the form factor and this increases sensitivity to hot cracking in the weld. Voltage of the welding current is a very important variable in the ESW process, and is a primary means of controlling the depth of the weld pool. Increasing voltage will increase both the depth of the weld pool and the width of the weld zone. The wider weld can equate to a higher form factor, and this would increase cracking resistance. Welding voltage is usually in the order of 30 to 55 volts. Lower voltage may permit an arc to form between the electrode and the weld pool, and this is not the desired mode for operating the process. The conductivity of the molten slag is another variable deserving of consideration because higher conductivity decreases the resistance heating and results in a shallower weld pool and less base metal penetration.

Cracking from hydrogen also can be a threat to ESW steel weldments because of the nature of the microstructure sometimes formed in the weld metal, and potential sources of hydrogen in an ESW operation. The microstructure that lends itself to hydrogen cracking is one that contains a coarse network of proeutectoid ferrite outlining prior-austenite grains. Otherwise, the matrix structure consists of ferrite and carbides distributed in a Widmanstätten pattern. The ferrite veining has been found to be highly susceptible to hydrogen cracking, which is surprising in view of the relatively low hardness of this phase. The microhardness of the ferrite has been established at HV 115, as compared to HV 200 for the matrix. In a number of ESW steel welds that contained ferrite vein cracking, this defect could not be detected by UT or by radiography because of the tightness of the fissures and their irregular path. The cracks ranged in size from 0.05 mm (0.001 in.) to as large as 18 mm (¾ in.). They usually initiated near the fusion line and ran in discontinuous fashion toward the weld center. A quantitative hydrogen value for cracking threshold could not be established because the level of restraint in the weldment was a major factor in the initiation and the extent of the ferrite vein cracking. However, the threshold temperature at which cracking initiated in an ESW (hydrogen-containing) steel weld during uninterrupted cooling under ordinary conditions was determined to be approximately 125°C (257°F). Potential sources of hydrogen in an ESW operation include damp flux, humid ambient atmosphere that condenses as moisture on the water-cooled retaining shoes, hydrogenous contaminants on the electrode or any other filler metal, and wet asbestos that is sometimes used to caulk the crevice between the copper retaining shoes and the surface of the workpieces. Even though no arc is present in the ESW process to act as the highly efficient dissociator of molecular hydrogen, apparently enough atomic hydrogen can be formed by the high temperature of the operation to enter the metal in the weld pool and eventually cause cracking in ferrite veining of a solidification microstructure.

Another form of cracking sometimes encountered in ESW steel weldments stems from copper which is accidentally melted on the surface of inadequately-cooled retaining shoes. The copper shoe, because of the relatively low melting point of copper, can attain a molten surface at the time it is in contact with newly-solidified weld metal, or with heat-affected base metal; whereupon the liquid copper penetrates the grain boundaries of the solid steel via the liquid-metal embrittlement mechanism and causes the boundaries to separate as shrinkage stress develops.

In considerable research conducted to find ways of improving the toughness of ESW steel welds, no easily-applied techniques have evolved for obtaining substantially improved microstructures in the weld metal or in the base metal HAZ. Even in the plain low-carbon steels, low fracture toughness continues to be a matter of concern. When ESW is used for welding higher-strength heat treated steels, a complete post-weld heat treatment is ordinarily required to obtain the desired properties in the weld areas. Some progress has been made toward weld metal toughness improvement by use of alumina-rich fluxes. Perhaps the most effective means developed to date for improving weld metal toughness is to alloy the weld composition with small amounts of nickel and molybdenum in combinations.

Electrogas Welding (EGW)

The electrogas welding process, although actually an electric arc process, is described here because it evolved as an offshoot of the electroslag welding process and much of the technology and metallurgy of ESW can be applied to the EGW process. EGW was designed to machine weld square-groove butt joints held in the vertical position in relatively thin steel plate (perhaps 12 to 40 mm; about ½ to 1½ in. thick), below those thicknesses which can be welded by ESW. A large amount of the thinner plate fabrication is involved in hulls of ships and barges, framework for buildings, storage tanks, etc.

Electrogas welds are made by feeding one or more solid wire, metal cored, or flux cored electrodes through a nonconsumable guide and contact tube to an arc above the weld pool. AWS Specification A5.26 covers consumables used for electrogas welding of carbon and low-alloy steels. Externally supplied gas may, or may not be provided, depending on the kind of electrode employed. The shielding, therefore, can range from only gas protection (a mixture of argon and carbon dioxide flooded into the joint gap just above the weld pool) to a more complex form of shielding that may make use of a thin layer of slag from an electrode similar to that used for FCAW. The flux cored electrode also may provide some gaseous shielding. Water-cooled copper shoes are held on each side of the joint to confine the molten weld metal in the gap.

Although EGW closely resembles ESW, the arc provides a higher temperature heat source that accomplishes faster welding, and the weld pool regularly has a more favorable form factor. Electrode oscillation can be used to minimize the weld pool depth and thereby secure a grain solidification pattern that is not susceptible to weld metal hot cracking in the center area. Also, oscillation of the electrode assists in securing fusion of the groove faces of the workpieces. When the EGW operation is initiated or stopped prematurely, restarting is not likely to produce unsound weld metal, a major problem with ESW. Because the molten weld pool is smaller, the grain sizes of weld metal and base metal heat-affected zones of electrogas welds tend to be smaller than those of electroslag welds. However, because the EGW weld usually is made in a single pass, the heat input is greater than would be involved in making a multi-pass weld by any other process. Therefore, EGW requires that attention be given to the strength and toughness of welds in steels that are likely to be adversely affected by the relatively slow cooling rate.

Welds properly made by the electrogas process ordinarily are of high quality and free of harmful discontinuities. However, the faces of the square groove joint (with appropriate amount of gap) are parallel to the axis of the electrode, and care is needed to ensure that complete fusion occurs across both opposing faces of the joint to avoid lack of fusion, entrapped slag, or porosity. There have been no reports by those using EGW on steels of encountering unusual weld metal microstructures (such as ferrite veining), or of unusual susceptibility to hydrogen cracking as reported for the ESW process.

Electron Beam Welding (EBW)

In earlier discussion of heat sources for welding, a high-velocity stream of electrons was described as an effective means of heating metals; because the kinetic energy of the elementary particles is converted into thermal energy as they impact the metal. Since about 1915, this feature had to be reckoned with when operating the x-ray tube. However, commercial utilization of electron beam welding (EBW) did not occur until about 1960, when it was recognized that the cleanliness of a vacuum and the very deep penetration of a beam of electrons justified costly equipment for welding heavy sections of high-alloy steels and reactive metals. Later use of EBW for fabricating steel parts of cars and trucks during the 1970s clearly demonstrated the mass-production capabilities of this process.

The electron "gun" is the heart of any EBW machine, and the majority of machines make use of the self-accelerating type of gun illustrated schematically in Figure 6.36. Work-accelerated guns (anode workpiece) seldom are used because variations in workpiece surface condition, and in gun-to-work distance, etc., cause changes in welding conditions and produce irregularities in the weld. Guns of the self-accelerating type (cathode filament for electron emission) are commonly discussed in terms of operating voltage. Low voltage guns operate below 60 kV, while high voltage guns operate above this level. Because beam power is the product of the accelerating voltage and the beam current, the high-voltage guns operate at lower beam currents for equivalent power. Selection of gun voltage is dependent upon a number of factors in the welding operation. A high accelerating voltage is used with lower beam current to obtain maximum focal range, increase joint penetration and aid in production of narrow, parallel-sided welds. Although the depth of penetration and the amount of metal melted to form the weld can be changed by varying the weld travel speed most control in the EBW process is obtained through manipulation of the beam focal point, the beam spot size, oscillation of the beam, and pulsing the beam. The operating system of the latest EBW machines use computerized numerical control for precise application and repeatability of complex programmed machine operation. Three variations of EBW have evolved that differ mainly with respect to vacuum arrangement and have been designated as:

- EBW-HV – Hard vacuum in the work chamber (10^{-6} to 10^{-3} torr); often the same vacuum as drawn in the gun column.
- EBW-MV – Soft, moderate, or partial vacuum in a separate work chamber (10^{-3} to 25 torr), but hard vacuum in gun column.
- EBW-NV – No vacuum provided for workpieces, but hard vacuum in continuously pumped gun column.

The EBW-HV machine (illustrated in Figure 6.36) offers all the advantages of the electron beam process; however, depending on the size of the work chamber, a prolonged pumping time may be required to produce the needed hard vacuum. To decrease pumping time and thereby increase welding production rate, the dual-vacuum EBW-MV machine was developed as illustrated in Figure 6.37. The hard vacuum in the gun column, once drawn, can be preserved by means of the isolation valve that closes whenever the work chamber is opened to the atmosphere to load workpieces. The medium vacuum drawn in the work chamber reduces pumping time, and is quite

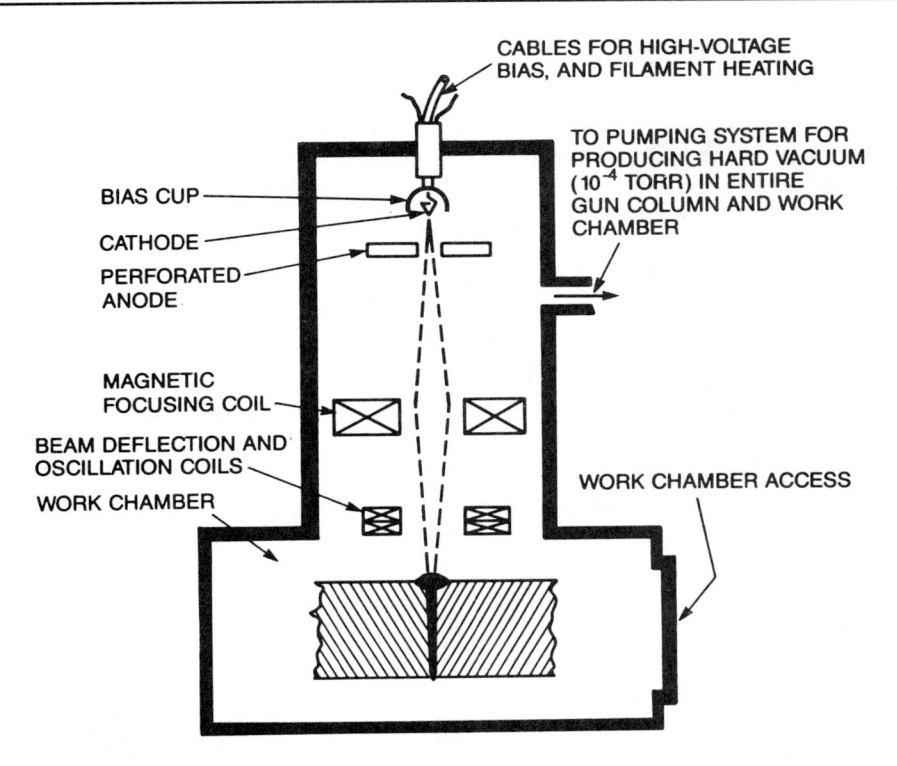

Figure 6.36 — Electron beam welding machine shown schematically for the original high-vacuum form of the process (EBW-HV)

adequate for protecting both the electron beam and the weld. Maintaining different degrees of vacuum in the gun column (hard) and the work chamber (medium or soft) is not difficult because the opening between the column and the chamber for passage of the electron beam is very small, and furthermore, gaseous atoms and molecules in the work chamber must have a favorable trajectory to pass through this opening since there is virtually no driving force from the pressure difference. Highly automated dual-vacuum EBW-MV machines use auxiliary loading chambers that interlock with the work chamber. This arrangement permits a partial vacuum to be drawn well in advance of the welding stage. Most guns for EBW in vacuo are mounted atop the work chamber, as shown in Figures 6.36 and 6.37, and the welding is performed in the flat position. When welding sections thicker than about 25 mm (1 in.) in a single pass, drop-through can occur. Positioning the gun and joint horizontally will assist in overcoming this drop-through tendency.

The nonvacuum variation of electron beam welding, EBW-NV, as illustrated in Figure 6.38, does not provide a high-purity environment — that is, unless some form of protective gas shielding is provided to displace the ambient air atmosphere. The distinct advantage of EBW-NV is the absence of a work chamber that requires vacuum pumping, consequently, production rate is greatly increased. However, certain necessary precautions must be taken to shield personnel from the x-rays emitted from the work during the welding operation. Although EBW-NV requires considerable engineering for a given application, several complete units have been installed for mass-production of specific articles, such as certain automotive components. EBW-NV machines with beam power up to 60 kW are being used and newer machines offer power up to 150 kW. The higher power is needed to weld heavier sections, and to compensate for some dissipation of the beam's energy as it passes through air or other gases at atmospheric pressure.

Welding in a High Vacuum (EBW-HV)

Electron beam welds made in high vacuum can be very narrow with deep penetration and with parallel sides. The penetration-to-width ratio can easily be as high as 25:1 when welding thick material. This is perhaps three or four times greater than the penetration-width ratio secured with any arc process. This unique weld zone shape has several advantages, but also can cause certain weld defects. To control weld shape, first to be considered is the extent of penetration in the usual square-groove butt joint with tightly abutting faces. The joint can be either partially penetrated by the weld, or completely penetrated, as desired. The focal point of the electron beam with respect to the thickness of the joint will be influential when making both kinds of welds, but

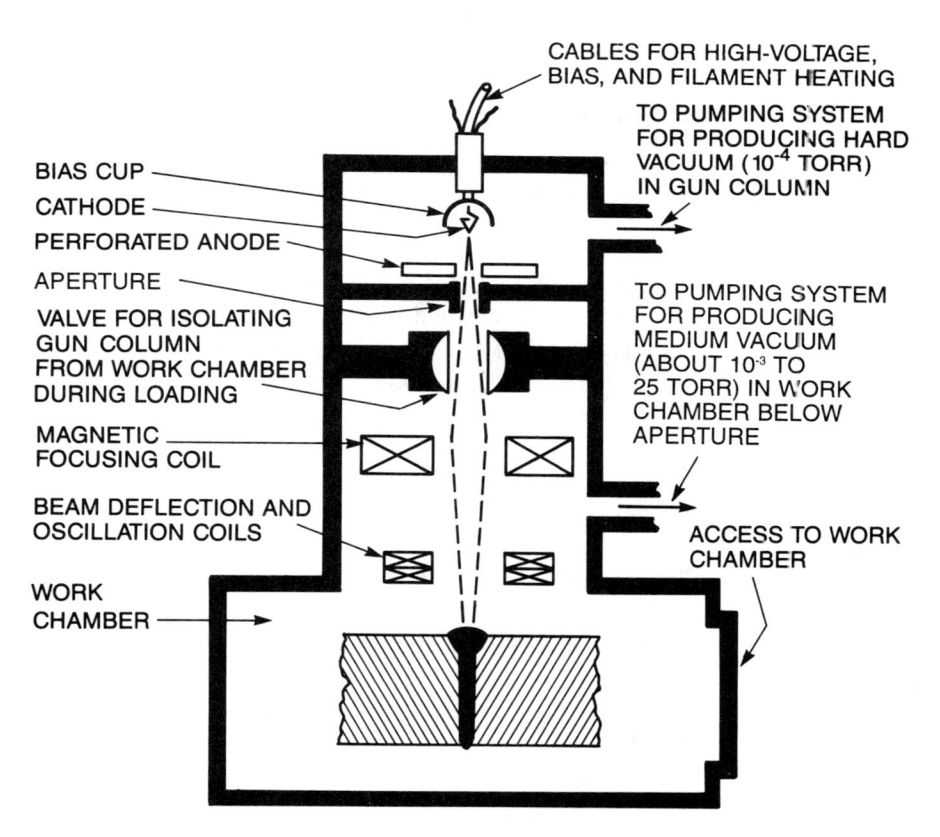

Figure 6.37 — Electron beam welding machine of the dual-vacuum type shown schematically with the gun column under a hard vacuum, and the work chamber at a lesser vacuum (EBW-MV)

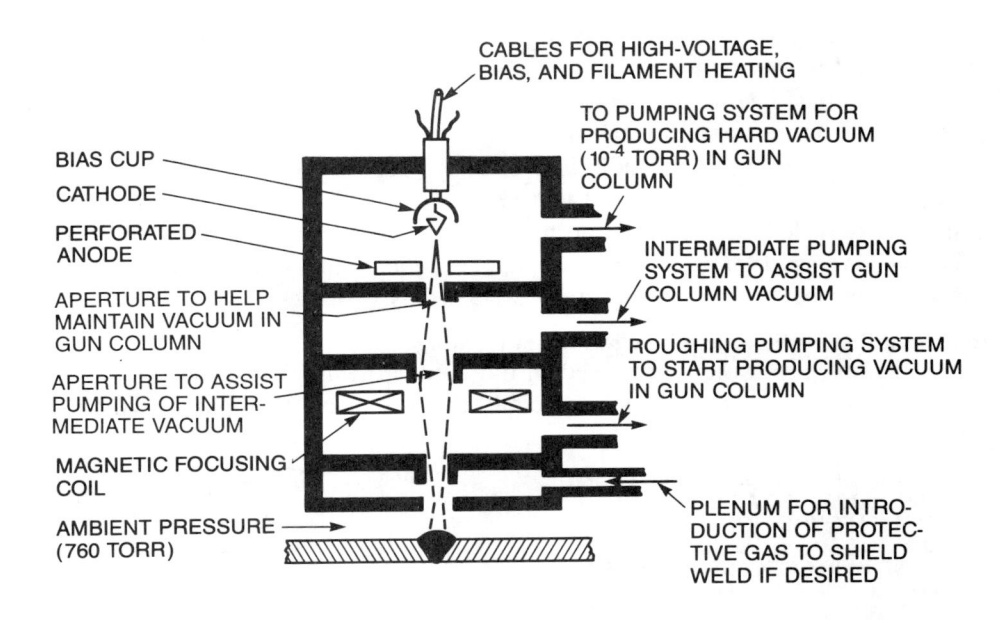

Figure 6.38 — Electron beam gun (shown schematically) for welding workpieces at atmospheric pressure; the EBW process termed non-vacuum (EBW-NV)

results will differ in each with a given focus pattern. With a partial penetration weld, sharp focus of the beam to mid-thickness of the joint will produce optimum heat input, maximizing the efficiency of the power applied. Overfocusing or underfocusing the beam in relation to the section being melted increases the beam diameter, which reduces power density; and this usually lowers the rate of heat input. However, with a complete-penetration weld, a sharp mid-thickness focus may actually *reduce* heat-input for a given power level, because the amount of nontransferred power carried completely through the joint can be as much as 50 percent. Nevertheless, the beam power can easily be adjusted to provide the proper level, and there are additional conditions that must be considered when deciding on the focal point of the beam for welding a particular joint.

Keyhole Technique in Electron Beam Welding

The keyhole technique is employed in the majority of applications of the EBW process to accomplish complete joint penetration in a single weld pass. Consequently, much research has been directed to the mechanics of keyhole formation and to related phenomena that affect the efficacy of its operation. An electron beam focused to the usual spot size, a little less than one mm (0.04 in.) in diameter, can provide enough energy to not only melt the metal beneath the point of impingement, but also cause it to vaporize. The reaction force of the vapor depresses the molten area beneath, and this can quickly lead to the making of a fully penetrating keyhole, as illustrated in Figure 6.39, in about one second. As this liquid-walled keyhole is moved along the centerline of the butt joint, liquid metal flowing from the front to the back of the advancing keyhole solidifies to produce a fusion weld that is narrow, parallel sided, and completely penetrating. The liquid metal builds up on the trailing rim of the keyhole both at its top and bottom opening, leaving small formations on the weld face and root. Often this eliminates the need to add filler metal in order to avoid underfilled weld sections. This presumes, of course, that the welds will be terminated on a runoff tab in which the underfilled portion of the joint will not be objectionable. Also, the keyhole technique is preferred for avoiding internal porosity and other defects that are prone to occur near the root of partial penetration EBW welds.

Partial Joint Penetration when Electron Beam Welding

Partial joint penetration employed in the course of an EBW procedure requires close attention to weld zone shape, especially the root of all weld passes whether the

SQUARE GROOVE BUTT JOINT(1) IN PLATE IS WELDED AS ELECTRON BEAM (2) FORMS LIQUID-WALLED HOLE THROUGH THE JOINT (3) AND WORK PIECES ARE MOVED IN DIRECTION OF ARROW (4). FULL PENETRATION WELD (5) SOLIDIFIES AT SHORT DISTANCE BEHIND MOVING HOLE (6), WHILE A PORTION OF ELECTRON BEAM PASSES COMPLETELY THROUGH THE KEYHOLE (7).

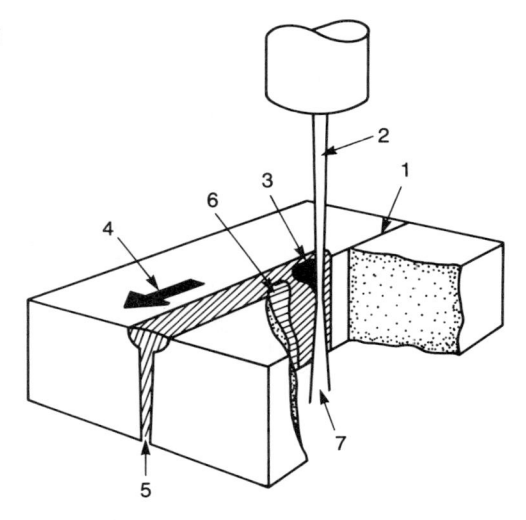

TRANSVERSE SECTION THROUGH WELD IN PLATE ETCHED TO REVEAL FUSION ZONE AND HEAT-AFFECTED ZONES.

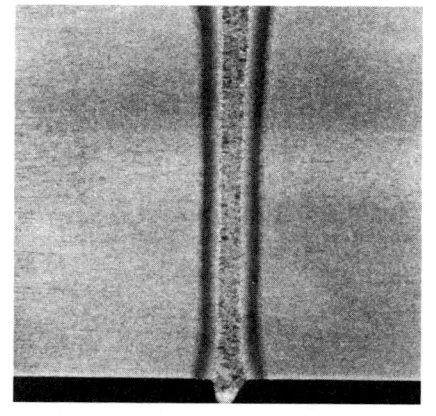

Figure 6.39 — Schematic illustration of the "keyhole" formed during single-pass, full-penetration welding of a square groove butt joint in thick steel plate by the electron beam process (EBW-HV)

(Courtesy of The Welding Institute, England)

initial incomplete joint penetration or subsequent passes as required to complete the weld. A sharp penetration tip at the root, as illustrated in Figure 6.40, is likely to result in "spiking," which is a series of voids and cold shuts at the root along the length of the weld. The unsoundness may not be detectable in a transverse sectioned specimen, as in Figure 6.40 (upper right), but a longitudinal specimen (lower left) sectioned vertically through its center will reveal the repetitious character of defects along the root. Spiking is believed to arise from dynamic imbalance between the surface tension and vapor pressure forces acting in the penetration cavity during welding. Abrupt changes in molten metal distribution interferes with the electron beam and causes fluctuations in penetration depth, which can be seen along the fusion boundary of Figure

Figure 6.40 — Spiking problem which may occur when making partial joint penetration welds by EBW

In the photo to the right, the transverse specimen has been etched to show partial section penetration weld made in C-Mn steel plate 70 mm (2¾ in.) thick by EBW-HV. The fusion zone has a dagger-sharp root. A longitudinal section through the root has been etched and shown in the photo below. Spiking during welding by the electron beam has formed voids and cold shuts along the sharp root of the fusion zone. Selection of proper welding parameters can avoid the spiking problem.

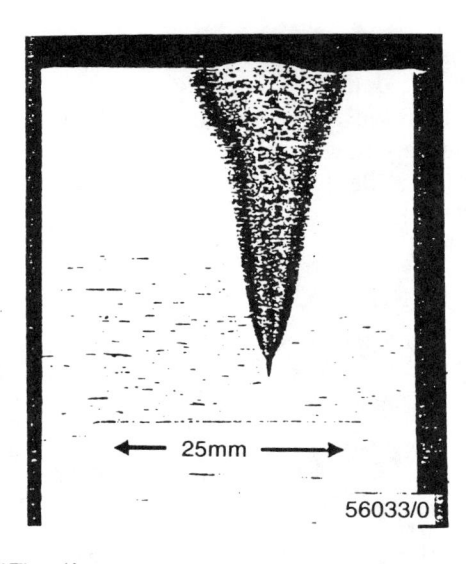

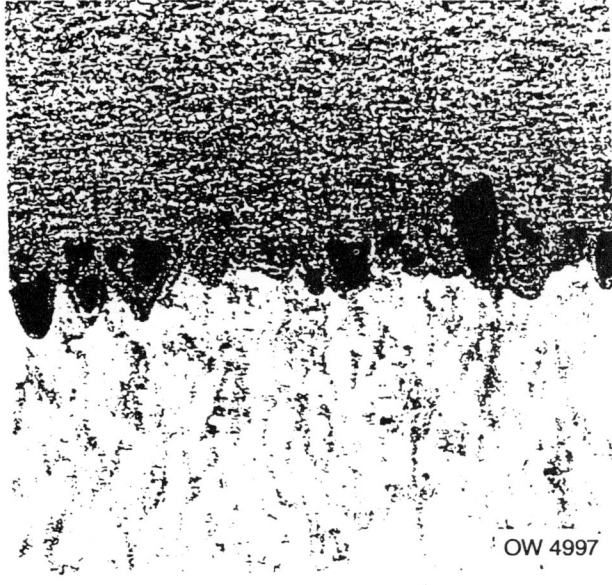

6.40 (lower left). Because vapor generated in the penetration cavity cannot vent through the bottom of the weld as it does when a keyhole is formed during welding, the gaseous material is forced to pass entirely through the electron beam outward into the vacuum chamber. The varying amounts of material making this passage cause the beam focal position to fluctuate and the result is unsoundness at the weld root along the fusion boundary. By adjusting welding parameters to form a well rounded root, as shown in Figure 6.41 (A), spiking and its related defects can be avoided [see Figure 6.41 (B)]. However, adjustment of parameters to accomplish the rounded root must be done with care because excessively round penetration can represent a bulge in the lower portion of the weld zone as illustrated in Figure 6.42. This weld shape encourages solidification cracking, and the formation of a shrinkage void. The latter type of defect can be seen at mid-depth of the weld.

Detection of EBW Defects

The detection of defects in welds made by the EBW process, and the diagnosis of their causes has required groundbreaking work because of the uniqueness of this relatively new process. A particular region of an EBW weld that has shown susceptibility to defects is the end of the weld where the electron beam is extinguished without a runoff tab. Instantaneous cut-off of beam power generally results in an underfilled, poorly formed, unsound terminal area. Instead, power can be gradually

(A) TRANSVERSE SECTION

(B) LONGITUDINAL SECTION THROUGH CENTER OF WELD; ENLARGED 5X

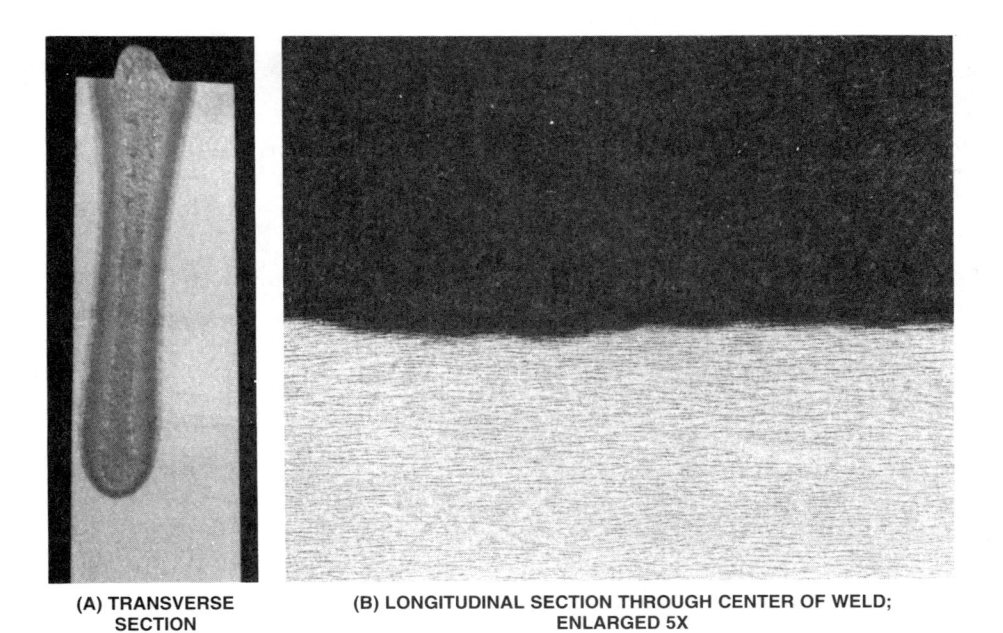

Figure 6.41 — Avoidance of spiking during partial joint penetration welding in steel plate by EBW-HV through control of weld zone root formation

Welded plate is C-Mn steel 70 mm (2¾ in.) thick. Transverse etched specimen (A) displays fusion zone with properly rounded root and parallel sides (without bulging) produced by selected welding parameters. Longitudinal etched specimen (B) shows root to be free of voids or cracking. (Courtesy of The Welding Institute, England)

reduced while travelling to a stopping point. The end of a weld terminated in this manner is called the "fade-out" region. This is the usual completion method in circumferential welds on round sections or in cylinders where the start is overlapped by the weld end. This region is susceptible to the formation of porosity, voids and cracking unless judicious adjustment is made of welding parameters. Considered in developing optimum programming for the fade-out region are (1) beam focus, (2) beam oscillation frequency and pattern, (3) beam current reduction rate, (4) travel speed change, and (5) fade-out region distance.

Figure 6.43 illustrates defects in the fade-out region, where a weld start is overlapped and unfavorable welding parameters are used to complete the weld. In this typical example, as welding conditions were altered to gradually reduce the penetration and to terminate the weld, large voids formed in the center of the weld overlap just above the root (see section A–A′). Further changes produced a sharp tip at the root during the overlap (see section B–B′), and beam spiking occurred which caused defects along a portion of the weld root. Defects in the terminal region of welds are avoidable, but a fade-out technique must be tailored for each application. EBW-HV welding, although offering complete avoidance of adverse effects from atmospheric nitrogen and oxygen, can be beset with problems that emanate from the metal being welded. The very high energy concentration in the electron beam tends to boil off some metal during melting to produce the required weld zone. As vapors leave the weld pool and disperse in the work chamber, the high velocity electrons of the beam collide with some of the evaporated atoms and remove valence electrons; thus leaving these atom in a cationic state. Naturally, these ionized atoms are drawn toward the cathodic filament of the electron gun. Should a certain amount of the ionized metal vapor reach the cathodic area of the gun, then arcing (also known as "flashover" or "gun discharge") will occur. This malfunction may last for only a few microseconds and then self-quench, producing an insignificant blip on the current trace, and probably accompanied

Figure 6.42 — Micro-shrinkage void caused by unfavorable bulging in width profile of EBW-HV weld

C-Mn steel plate 70 mm (2¾ in.) thick with partial section penetration fusion zone that has properly rounded root. However, bulged shape in lower portion has caused large void to form by micro-shrinkage. The narrower width in fusion zone just above the bulge has solidified first and blocked downward feeding of molten metal during solidification. (Courtesy of The Welding Institute, England)

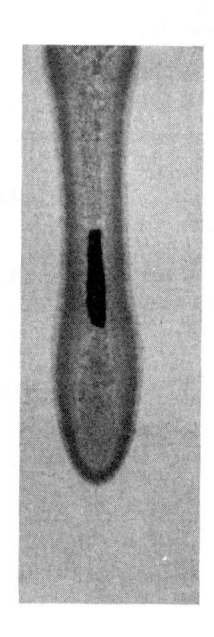

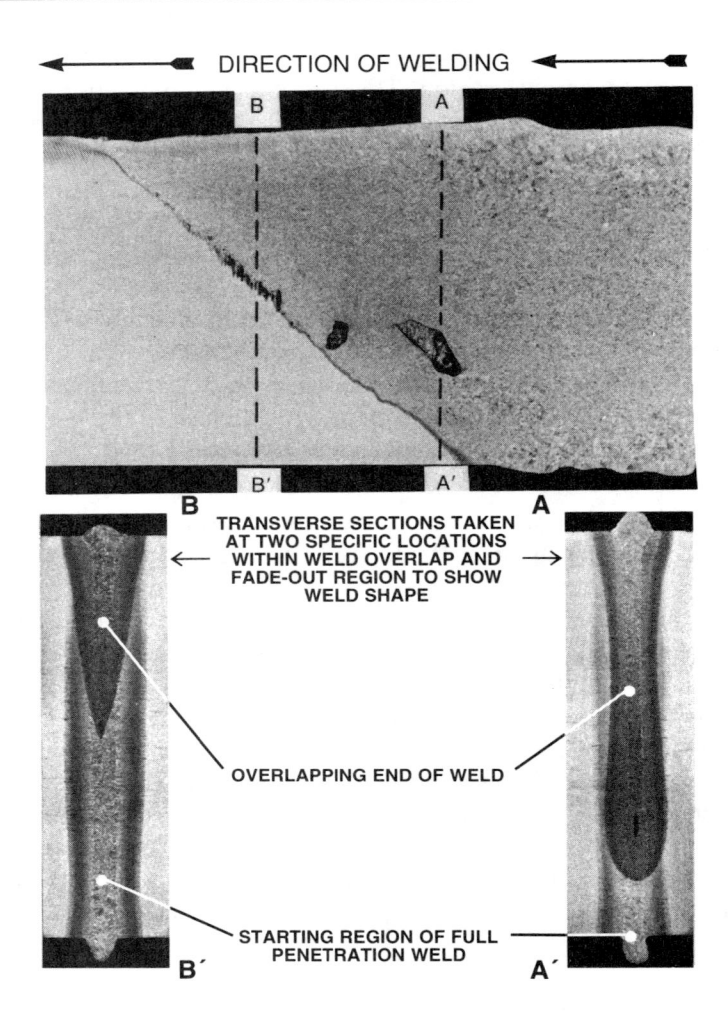

Figure 6.43 — **Voids in fade-out or termination region of EBW-HV single-square-groove weld made with complete joint penetration**

C-Mn steel plate 70 mm (2¾ in.) thick has been sectioned longitudinally (upper photo) at overlap of weld start and stop and etched to show sloping interface produced by power fade-out. Transverse sections A-A' and B-B' in lower photos from a replicate welded specimen show voids from beam spiking and microshrinkage which occurred during unfavorable power reduction procedure when stopping the weld. (Courtesy of The Welding Institute, England)

by only a small splash in the weld pool. If the discharge period is prolonged, defects are likely to be formed in the weld. Also, the power supply may be tripped from arcing and this can cause defects that extend through the thickness of the weld. Inadequate weld penetration frequently results from short-duration gun discharge.

As an increasing variety of materials are involved in workpieces, arcing has become more troublesome (See Technical Brief 36); but even when arcing does not occur, EBW welds sometimes contain porosity. Both welding procedure and base

> **TECHNICAL BRIEF 36:**
> **Preventing Arcing in Electron Beam Welding**
>
> A number of measures have been taken in new EBW machines and in rebuilding older models to avoid or to suppress the problem of arcing (a.k.a. flashover or gun discharge). Guns have been constructed with a tilted cathode in a bent column, and with magnetic traps incorporated in regular gun columns. New configurations of ribbon cathodes have been devised that are not as sensitive to ionic particle impingement. Also, new power sources are equipped with electronic regulators between the source and the gun to control the beam current by discriminating between minor and major discharge events. Magnetic traps, when installed in a gun's column as an auxiliary device, as illustrated in Figure 6.44, have proved effective in preventing arcing.
>
> Cleaning and servicing of EBW machines are very important in reducing the likelihood of electron gun arcing during welding. The surfaces of everything exposed in the vacuum chamber should be kept free of dust and extraneous matter. Acetone is the solvent preferred for removal of foreign substances. Chlorinated hydrocarbon solvents should not be used. Sometimes a broad out-of-focus beam can be used to sweep the surface of the area to be welded to vaporize oxides and other trace impurities thus leaving a clean surface.
>
> In general, the avoidance of defects calls for consideration of the metallurgical characteristics of the metal to be welded in light of the unusual electromechanical aspects of the EBW process; all of which can be compounded by execution of the welding operation in a vacuum. The first objective in making a weld by EBW is to penetrate a given joint with a melted zone of proper cross-sectional shape. This is achieved only when welding parameters are selected that overcome unfavorable effects from the high energy density of the welding beam, stray magnetic fields, ionization of vapors, etc.
>
> During melting of the weld, watch must be maintained for unwanted reactions that might arise in the weld pool. From a materials viewpoint, if a steel other than one that is fully killed (deoxidized) is welded, there is a possibility of the spontaneous chemical reaction between carbon and available oxygen in the steel to form carbon monoxide gas. Generation of CO gas will agitate the molten weld pool to cause spattering, and this insoluble gas likely will form porosity in the solidified weld. The addition of filler metal containing strong deoxidizing agents usually combats the porosity problem. The CO gas entering the work chamber also can cause arcing. The level of contained oxygen in steel above which arcing is likely to occur is about 0.02 percent; rimmed steel exceeds this level, while semi-killed and capped steels range from about 0.01 to 0.03 percent. In the low-alloy steels, only chromium present as an alloying element appears to mildly accentuate the arcing problem. Sulfur and selenium in free-machining steels have been potent causes of arcing, as well as porosity in the weld metal; which is not surprising in view of the relatively low boiling temperature of these elements.
>
> Solidification of the weld zone deserves forethought to avoid defects; especially if the zone will have the typical narrow, deep cross-sectional shape. Although the usual mechanisms will be operative during solidification of an EBW weld, the mechanics of solidification are carried out on a much faster time-scale as compared with that of most other welding processes. Consequently, a new array of microstructures, distributions of low-melting segregates, discontinuities and defects are found in EBW welds.

metal composition affect porosity occurrence. Randomly distributed minor porosity in welds of killed steels appears to be related to oxygen content regardless of the deoxidizing agents used to treat the steel. Other residual gases, such hydrogen and nitrogen, do not seem to enter into porosity formation unless their level is abnormally high. The unacceptable level of oxygen with regard to possible porosity in welds of carbon and

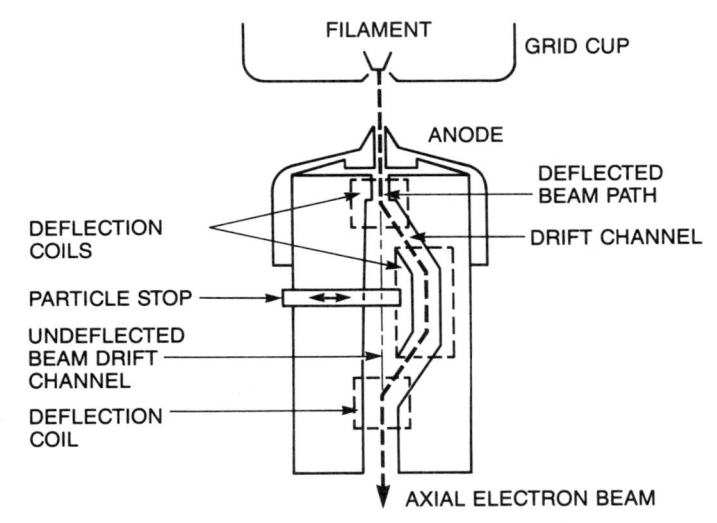

(A) SCHEMATIC DIAGRAM FOR INSTALLATION OF MAGNETIC TRAP IN GUN COLUMN AT POSITION JUST BELOW PERFORATED ANODE.

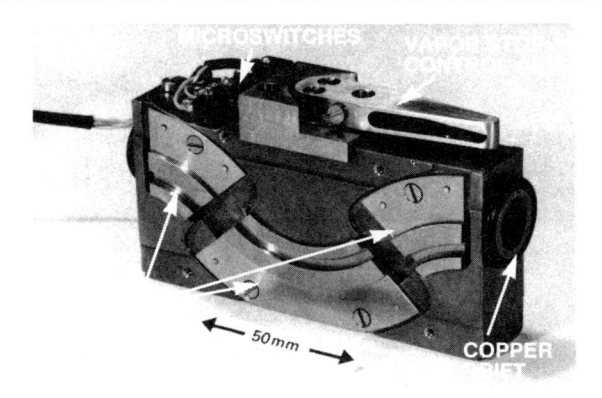

(B) MAGNETIC TRAP ASSEMBLY SHOWING POLE PIECES, DRIFT-CHANNEL APERTURE, AND CONTROL ARM FOR PARTICLE STOP.

Figure 6.44 — Magnetic trap developed for EBW machines. Installation in EB gun column can prevent sporadic flashover or gun discharge

This three-bend magnetic trap blocks line-of-fLight of foreign particles that are attracted to the cathode and cause gun discharge. (Courtesy of The Welding Institute, England)

low-alloy steels depends on the EBW procedure. Thicker base metal sections and faster travel speeds increase the likelihood of porosity formation. For example, a 75 mm (3 in.) thick carbon steel plate welded at a relatively slow speed (e.g., 100 to 150 mm/min or four to six ipm), can tolerate as much as 0.01 percent oxygen without producing a porous weld; but, at 250 mm/min (10 ipm), the oxygen should be limited to about 0.003 percent. This is a low level of oxygen in a steel, and it is likely that vacuum degassing during steelmaking would be required to meet this oxygen restriction. Consequently, if a steel to be welded contains an oxygen content that appears to be too high, porosity in the EBW weld can probably be avoided by welding at a low travel speed, which would also favor a broader cross-sectional weld shape. Data correlating thickness-speed-oxygen appear in EBW literature.

The rapid heating and cooling of the typical weld made by EBW is also reflected in the narrow heat-affected zones formed in the base metal. Despite their narrowness, the microstructure, mechanical properties, and the soundness of these zones deserve scrutiny. Welds in hardenable steels may call for preheating to avoid cracking; however, the threat of cold cracking is lessened by the virtual absence of hydrogen. The microstructure of EBW weld metal differs in several respects from that found in other fusion welds, and metallographic studies have assisted in rationalizing the variations that occur with welding procedures. As EBW welds solidify, as would be expected, grains form in the weld zone by epitaxial growth on the grains of solid base metal at the fusion boundaries, and then grow toward the weld center. The mode of solidification structure is strongly dependent on weld travel speed. At low travel speeds (e.g., below 150 mm/min, or six ipm), "raft" microstructure develops as illustrated in Figure 6.45-A (A). This structure, designated Type I by investigators, consists essentially of columnar grains that develop parallel to the welding direction. At the weld center, a band of highly directional columnar grains were formed as final solidification took place. When weld travel speed was raised to 250 mm/min (10 ipm), the columniation of grains became noticeably stronger as shown in Figure 6.45-A (B), and was designated Type II, "elongated." When weld travel speed was increased further to 300 mm/min (12 ipm), the weld microstructure seemingly became equiaxed, as shown in Figure 6.45-A (C), and was designated Type III, "equiaxed." However, longitudinal specimens prepared later disclosed that the elongated grains became tilted at an angle to the top surface of the weld as travel speed was increased beyond about 200 mm/min (8 ipm). Figures 6.45-B (E) and (F) illustrates this tilt. Therefore, the grains in the planar specimens appeared equiaxed because elongated grains were cut through their cross-section. A fourth solidification mode, which developed at a very high travel speed, displayed a marked change in both plan-section and longitudinal metallographic specimens. Figure 6.45-A (D) illustrates Type IV "interlocking" microstructure observed in the center of a plan-section specimen from a weld made at a speed of 500 mm/min (20 ipm). The columnar grain growth and orientation has persisted to the centerline where a direct interlocking of grains occurred. A longitudinal section from this weld through its centerline is shown in Figure 6.45-B (G). As in most fusion welding processes, the shape of the EBW weld pool is the key to understanding the solidification mode that develops, and pool shape is mainly dependent on welding speed.

The raft micro-structure (Type I) in EBW welds has been associated with generally poor mechanical properties, particularly when certain interdendritic or intergranular segregates are present in abundance. The fully interlocking structure (Type IV) increases susceptibility to hot cracking during solidification because this mode favors centerline segregation. Earlier, when discussing hot cracking in submerged arc welds a testing system for evaluating cracking susceptibility was described, and an equation derived from transvarestraint test data for calculating "units of cracking susceptibility" (UCS) from weld metal composition was given as Equation 6.6. A testing system devised by TWI for EBW uses a simple plate specimen 20 mm (0.320 in.) thick, which is prestrained mechanically with a lever-acting jig. The following equation is suggested for assessing the probable extent of cracking in the test specimen from chemical composition:

$$\text{Cracking Index}^* = 52.2\ C + 1972\ S + P\ (4268\ C - 285) - Mn\ (1135\ S - 9) - 21.2$$

(Eq. 6.7)

* *Crack length in mm at 5080 kgf load.*

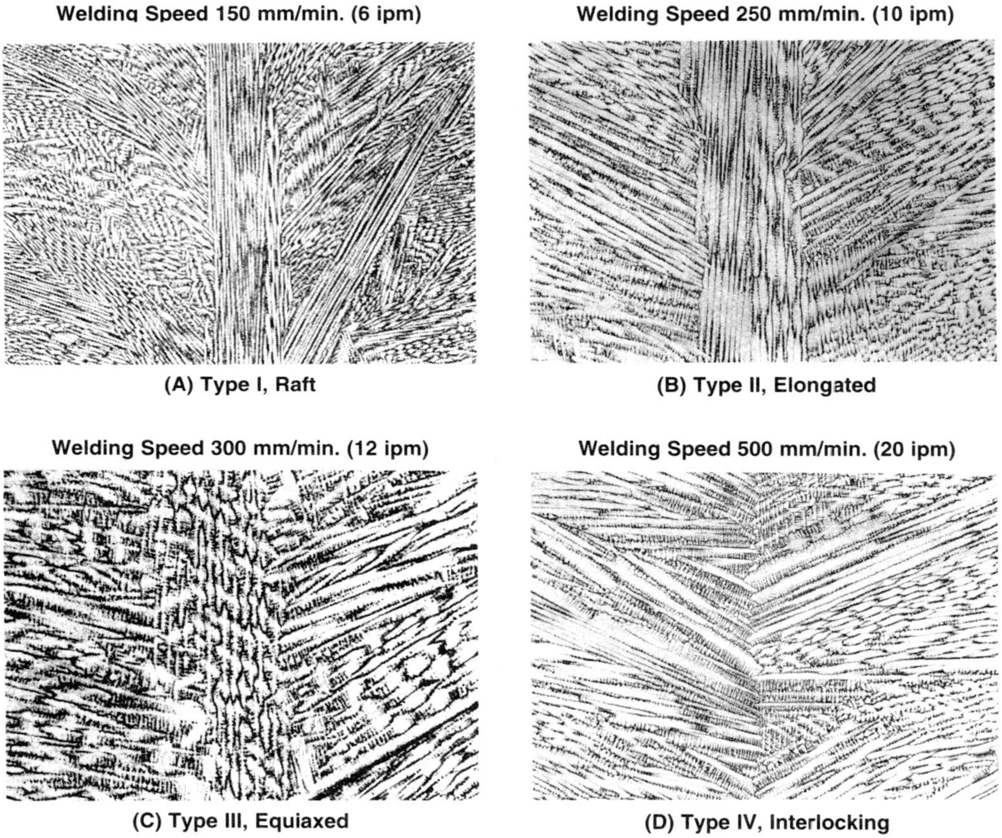

Welding Speed 150 mm/min. (6 ipm)

(A) Type I, Raft

Welding Speed 250 mm/min. (10 ipm)

(B) Type II, Elongated

Welding Speed 300 mm/min. (12 ipm)

(C) Type III, Equiaxed

Welding Speed 500 mm/min. (20 ipm)

(D) Type IV, Interlocking

Figure 6.45-A — Four types of EBW solidification mode observed metallographically in plan-section specimens

C-Mn steel plate 25 mm (1 in.) thick welded at four different travel speeds to determine influence of rate. Complete joint penetration in single-square-groove welds. (Nital etchant; original magnification: 100X; courtesy of The Welding Institute, England)

This testing procedure for EBW and the index derived for a given steel composition give only a preliminary indication of its susceptibility to solidification cracking when welding a real joint. It is noteworthy, however, that here again, carbon and sulfur induce the strongest adverse influence, but, of course, many other factors will affect results including weld travel speed, weld width, and weld contour, as well as the solidification microstructure.

The very narrow, parallel-sided EBW welds shrink much less than welds made with an arc process, and very little angular distortion occurs. For these reasons, EBW is favored for the assembly of near-finished, machined components. However, the narrow EBW weld requires accurate, unwavering tracking of the joint by the electron beam to accomplish complete penetration and form a weld zone of desired size and cross-sectional shape. The joint made ready for EBW welding should be precisely prepared with close-fitting abutting faces. A study has been made of the amount of gap that can be permitted in a square-groove butt join to be welded autogenously (i.e., without filler metal) without adversely affecting weld quality. Approximately one percent of the joint thickness is the maximum gap that can be tolerated.

The welding beam must follow square-groove butt joints very accurately, and must not be deflected after their entry of a keyhole at the top surfaces of the workpieces. Beam deflection is not a likely problem when welding thin sections of steel

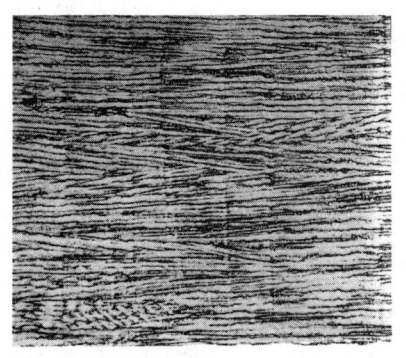

(E) Welding Speed 150 mm/min. (6 ipm)

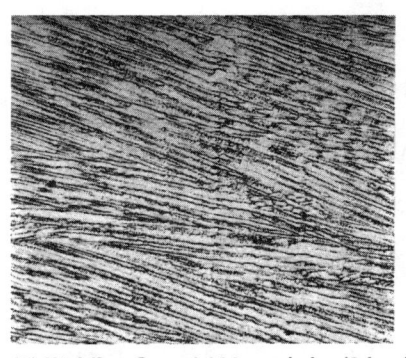

(F) Welding Speed 200 mm/min. (8 ipm)

Figure 6.45-B — Three welded plates selected from those prepared for metallographic examination in Figure 6.45-A, but now prepared as longitudinal specimens through the weld center to illustrate the role of sectioning in proper appraisal of solidification microstructure.

(Courtesy of The Welding Institute, England)

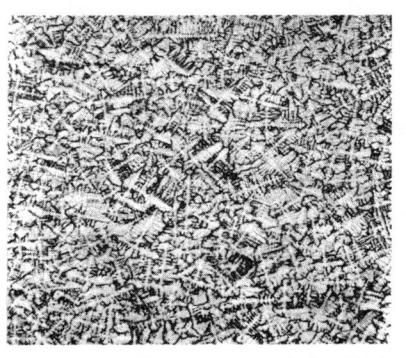

(G) Welding Speed 500 mm/min. (20 ipm)

(less than about three mm, or ⅛ inch in thickness), but in thicker sections a "missed joint" below the surface of the work is possible and not easily detected by NDT because of tight closure and its perpendicular position in the welded section. Examination by UT appears to provide the best assurance of finding a "missed joint" that may be present. Figure 6.46 presents two examples of welds between thick steel plates where beam deflection resulted in partly missed joints. The consequences of this condition, if undetected, can be severe.

Because the charged particles of a beam are easily deflected by a magnetic field, beam deflection during the EBW operation can be caused by magnetism in the metal workpiece or in metal fixturing. Two different magnetic fields can be encountered, namely: (1) residual magnetism in a ferromagnetic material, and (2) thermoelectric

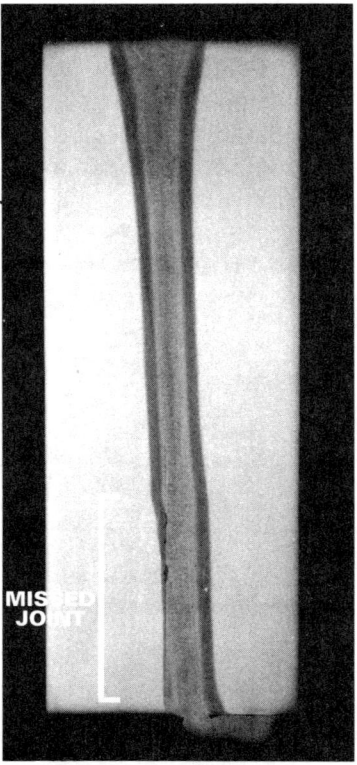

(A) Weld in alloy steel, 2¼Cr-1Mo, where a transverse flux density of 3.5 gauss parallel to joint plane caused small beam deflection toward right side.

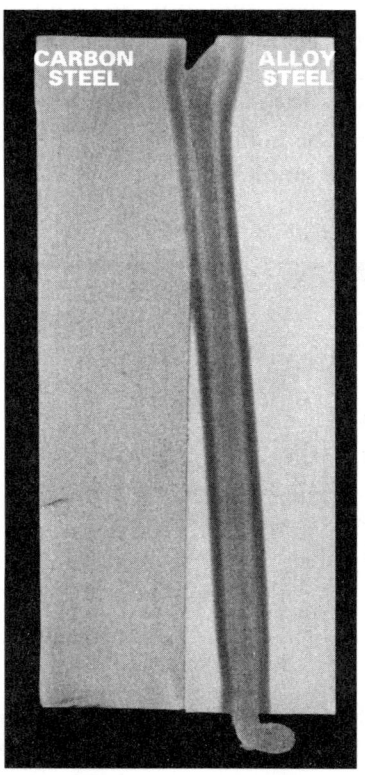

(B) Weld between dissimilar steel types; carbon steel, and 2¼Cr-1Mo alloy steel, where magnetism generated by thermoelectric current flow caused marked beam deflection toward right side.

Figure 6.46 — Incomplete joint penetration caused by unanticipated beam deflection in single-square-groove welds made by EBW in 150 mm (6 in.) thick steel plate.

(Courtesy of The Welding Institute, England)

currents between the hot junction (molten wall of the keyhole) and a cold junction (cooler metal elsewhere in the workpiece) when the joint is between dissimilar metals.

Residual magnetism might exist in any ferromagnetic material. It can originate from lifting electromagnets on a crane, holding workpieces on magnetic tables, mechanical shock in the presence of a magnetic field (even the Earth's field), or prior welding using direct current. This has encouraged the practice of routinely scanning workpieces with a search coil or a Hall probe to measure the strength and direction of any residual magnetic fields in ferritic steels. A search coil or a Hall probe usually is employed because residual magnetism can be patchy and a good degree of precision is needed to assess flux density over the range of about 0.5 to 10 gauss that can be significant in this problem. Generally, a strategy can be easily developed for dealing with residual magnetism. The workpieces might be carefully degaussed, or sometimes magnetic shielding can be arranged to reduce the effect of a persistent magnetic field. The shield might be a tube of soft ferromagnetic material (e.g., soft iron) fitted between the end of the EBW gun column and the workpieces because most of the deflecting field is usually above the work's surface.

Magnetism originating from thermoelectrically-generated current is a more complex problem because the thermoelectric current circulates within coupled dissimilar workpieces during the welding operation and is independent of the electron beam current. Magnetic flux density close to the weld pool during welding of dissimilar steels can measure as high as 10 gauss, suggesting that a current in the order of a hundred amperes was flowing in a closed path between the point of welding (the hot junction) and elsewhere in the workpiece (the cooler junction). After welding is stopped, any thermoelectric current in the weldment will start to decay, but several minutes may elapse before the current falls to zero.

The strength of the magnetic field that accompanies thermoelectric current is controlled by the following three factors:

(1) thermoelectric potential of the dissimilar coupled materials,

(2) electrical resistance of the current path, and

(3) magnetic permeability of the materials (whether ferromagnetic or not).

Even if a combination of materials has a potential difference, it is not inevitable that beam deflection will result in EBW welding. Although measurement of potential difference is a worthwhile preliminary step, the best procedure is to make a prototype weld and determine by examination the extent and direction of any beam deviation.

Chromium is frequently used as an alloying element in steels, and only a few percent difference in chromium content at a joint can produce emf strong enough to cause appreciable beam deflection [see Figure 6.46 (B)]. However, the complexity of beam behavior will be appreciated by examining the weld in this Figure over its complete penetration depth. The weld in Figure 6.46 (B) over its full depth is, in fact, slightly S-shaped. This subtlety suggests that a magnetic field above the top surfaces of the workpieces deflected the electron beam away from the vertical plane of the butt joint toward the right side in the photograph. Once inside the steel and melting its keyhole, the beam penetrated in a straight path for about 25 mm (1 in.), but at a small angle from the vertical. Further penetration by the beam then followed a path that curved slightly to the left in the photograph until the beam exited through the bottom surface of the alloy steel workpiece. Analysis of this beam behavior suggested that the beam was initially deflected to the right by a flux density of +8 gauss above the workpieces, and by -8 gauss inside the alloy steel section. These equal, but opposite magnetic fields are probably generated by the circulatory nature of thermoelectric current. When EBW is applied to dissimilar steel square-groove butt joints in very thick sections, it is not unusual to find beam deviation following a curved path to some extent, which is added reason to make a prototype weld to ascertain the pattern of penetration that will be produced by the beam.

Beam deviation is a vexing EBW problem; especially when welding dissimilar materials, thick sections, narrow weld zones, varying gaps in square-groove butt joints, etc. Many suggested remedies can be found in the literature, and new ones undoubtedly will continue to be developed. A tube of soft ferromagnetic metal, mentioned earlier, to shield the beam during its passage from gun-to-joint surface can be effective, but it will not nullify the effect of an internal magnetic field located within the base metal sections to be welded. The electrical resistance offered by materials in the pathway through which thermoelectric current will flow offers some control over the extent of beam deflection. The strength of the magnetic field is commensurate with the level of current flow; therefore, an increase in the electrical resistivity of the materials in the pathway by elevating their temperature will reduce current flow and decrease the strength of the magnetic field. This preheating remedy is limited in overall effect, but may be sufficient to alleviate a minor beam deviation.

The third means of controlling the strength of magnetic fields that cause beam deviation is to alter the magnetic permeability of the base metals. This can be done with carbon and low-alloy steel, since they are ferromagnetic, by controlled hot working and/or heat treatment. However, these treatments seldom are practicable for solving a beam deviation problem.

Performing EBW at a higher voltage will act to reduce beam deflection, that is, providing the kind of EBW machine will allow a change in operating voltage. A study of this possible remedy found that the bend radius of the beam will decrease with increasing accelerating voltage in the following manner:

$$r = \frac{33.721\sqrt{V_r}}{B} \quad \text{(Eq. 6.8)}$$

where

r = Radius of beam curvature in mm
B = Flux density in gauss
V_r = Relative corrected accelerating voltage [computed as $V_r = 1 + (9.78515 \times 10^{-7} \times V)$]

Another technique for dealing with beam deviation is to accept the deflection and to tilt-position the joint into the pathway of the beam. First, the pattern of the weld zone produced by the deflected beam must be established with a prototype weld that is sectioned into transverse slices to determine departure of the fused area from the abutting joint. Among the precautions to be taken in preparing a trial weld is to ensure that the anticipated thermoelectric current is flowing by having the correct temperature difference between the weld pool and cooler metal elsewhere.

Welding in a Medium Vacuum (EBW-MV)

The dual-vacuum type EBW-MV type machine, illustrated in Figure 6.37, maintains a hard vacuum by necessity in the electron gun column, but operates with only a medium vacuum in the work chamber. Machines of this kind have increased production capability because lengthy hard-vacuum diffusion pumping is avoided with each opening of the work chamber to re-load with workpieces for welding. Faster-acting mechanical pumping can produce the medium vacuum in the work chamber. Various systems involving stop-valves, sliding seals, multiple work chambers, etc., are used to isolate the gun column and preserve its hard vacuum during the parts-loading step, but the basic feature that connects the gun column and the work chamber is a very small aperture or orifice that allows a tightly columned beam of electrons to enter the work chamber and perform the welding operation. In general, the technology described above for welding in a high vacuum also applies to welding in a medium vacuum: with one notable exception; namely, much less difficulty is experienced with arcing or gun discharge when welding in the EBW-MV type machine. The small connecting aperture, while open to permit passage of the electron beam for the welding, also greatly reduces the number of cationic particles created in the work chamber that reach the cathodic filament of the gun.

Nonvacuum Electron Beam Welding (EBW-NV)

The nonvacuum EBW type machine was schematically illustrated in Figure 6.38. The EBW-NV process is usually carried out in a specially built facility that includes (1) and EB gun of proper power to weld specific joints in given workpieces, (2) an arrangement for providing shielding gas when necessary to ensure required weld quality, (3) protection of all nearby personnel from emitted radiation, and (4) mechanism for the beam to travel along the joint to be welded at required speeds, and to track the centerline of the joint. The need for precise tracking is eased to some degree by the fact that welds made by EBW-NV are ordinarily slightly wider than those made in vacuo. At atmospheric pressure, scattering of electrons in the welding beam by collision with gas molecules causes a significant increase in beam diameter; therefore, beam energy is distributed over a broader area. The beam power reaching the workpieces to form a weld is highly dependent on the distance that the beam must travel, and the nature of the intervening gas. With a 60 kV gun, the maximum standoff distance is usually less than 50 mm (2 in.). The weld penetration depth required is an important factor in establishing the maximum gun-to-work distance.

Tracking Joints During Electron Beam Welding

Tracking joints when welding with the EBW process usually requires extra attention. Most joints are the square-groove butt type with tightly abutting faces and the narrowness of the typical weld zone does not permit much deviation from the joint centerline. A wide variety of techniques and devices are available for precisely tracking joints. Lines can be scribed on the top surfaces of workpieces when ready for welding to outline proper lateral location of the finished weld toe-to-toe. This simple technique does not adjust the electron beam position during welding to take care of changes in fit-up of the workpieces, or their movement. Therefore, increasing attention is being given to devices that keep the beam centered on the joint from start to finish during welding.

Mechanical probes have been used to lead the welding beam along the joint. Optical tracking with light-sensitive detectors can sense the joint, but problems with metal vapor deposition on the optical elements limit its use. In a similar tracking system, the joint is scanned with an electron beam and evaluated by x-ray emission.

Recent developments in real-time tracking systems make use of back-scattered electrons and the principle that a reduction in the quantity of electrons reflected will occur as an auxiliary electron search-beam passes over the joint. During welding, the search-beam periodically is swept across the joint ahead of the weld pool or keyhole as a precisely-timed pulse. Back-scattered electrons emitted during the momentary excursion of the beam are collected and analyzed to fix the exact position of the joint for the oncoming welding operation. These systems include a detector plate for collecting back-scattered electrons from the top surface of the workpieces, equipment for analyzing the collection, and auxiliary devices for altering the welding beam position and its operating parameters to maintain the desired weld penetration and pattern.

Much has been learned about back-scattered electrons, techniques for reflecting them from suitable metallic surfaces, and methods for collecting them without erroneous signals or interference. As an example, the joint tracking head in Figure 6.47-A has a hidden collector which is reached by reflected back-scattered electrons. This system is able to precisely track a curvilinear edge-flange joint in two pieces of 9.5 mm (⅜ in.) thick mild steel that has an inserted shim of 0.8 mm (⁵⁄₁₆ in.) thick steel as filler metal at a travel speed of 125 mm/min (5 ipm) without having the weld encroach on the outer edges [see photograph (B) in Figure 6.47-A].

Back-scattered electrons also have been used to generate moving images of the EBW operation, and a picture of 512×512 pixel quality can be put before the welding operator on a CRT. A special detector mounted inside the work chamber scans the workpiece surfaces at the joint with a 4kW search-beam of 0.2 mm (0.008 in.) diameter at the tremendous speed of 400 m/sec (almost 1000 mph) and provides a picture of the weld pool and surrounding surfaces.

Joint tracking in real-time has obvious advantages for automatic correction of welding parameters to obtain properly fused welds consistently. In addition to the unique cross-sectional shape of electron beam welds, the process has remarkable capabilities for welding different types of joints. Figure 6.47-B illustrates a few examples of unusual joint configurations that can be welded by the EBW process. The remarkable penetration of EBW can be used to advantage when a "blind joint" must be welded, that is, one where the abutting edges to be fused are hidden behind a solid member. The electron beam often provides a solution to this problem through its penetrating power because the beam can keyhole through the top solid member (as shown in the two upper illustrations of Figure 6.47-B) and accomplish the necessary welding in the member(s) beneath. In certain joints, the electron beam can be deliberately refracted to penetrate along a curved path and in this way produce a curved fusion zone. For example, the T-joint shown in lower-left of Figure 6.47-B can be welded from one side by directing the beam at a low angle with respect to the head of the "T" configuration. The beam will pass under the leg of the "T" and produce a similar weld metal fillet on the opposite side. In essence, the T-joint has been welded in a single pass using a "bent" keyhole. Two (or more) separated joints can be welded simultaneously by a single beam if all joints are properly placed in the path of the beam. For example, the illustration in lower-right of Figure 6.47-B shows an assembly that requires three square groove butt joints to be welded, and the middle joint actually is hidden within the article. By proper adjustment of welding parameters, a single electron beam can fuse all three tiered butt joints in a single pass.

The repair of small cracks in metal articles has become a major EBW activity because of unique traits of the process. Metal components subjected to mechanical service that have developed small fatigue cracks can be refused and finished by adding weld metal. If filler metal is needed, a cap strip can be placed atop the crack and melted into the weld. Sometimes the electron beam is broadened, or oscillated from side-to-side to ensure complete fusion of the cracked area. EBW offers a precise, welding capability that involves minimal heat input, and this avoids many of the problems associated with weld repair by an arc process (shrinkage, distortion, heat-affected zone degradation, etc.).

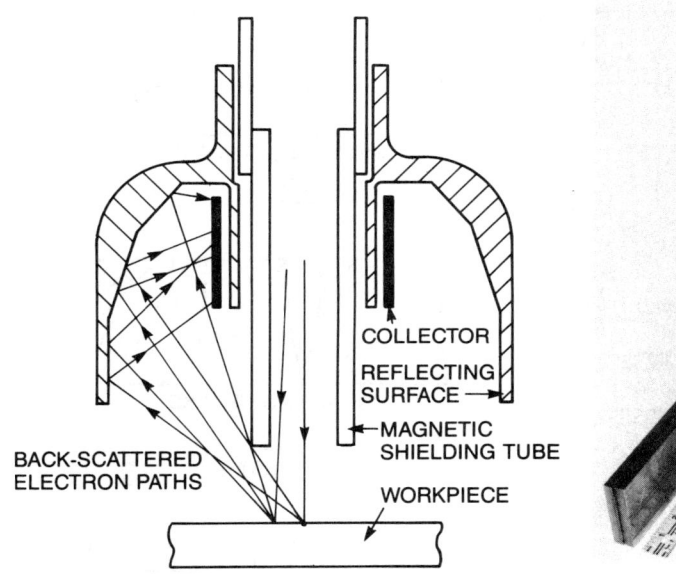

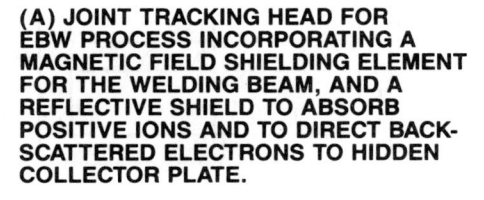

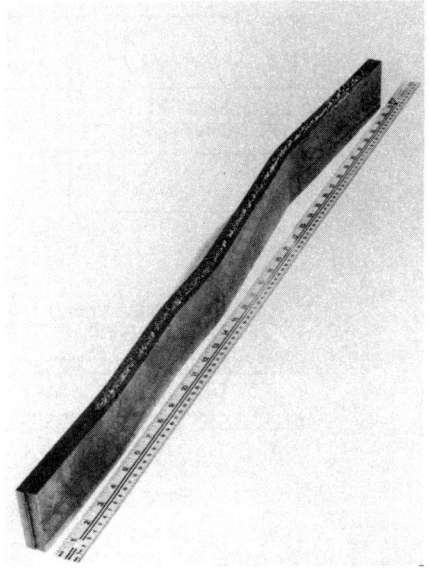

(A) JOINT TRACKING HEAD FOR EBW PROCESS INCORPORATING A MAGNETIC FIELD SHIELDING ELEMENT FOR THE WELDING BEAM, AND A REFLECTIVE SHIELD TO ABSORB POSITIVE IONS AND TO DIRECT BACK-SCATTERED ELECTRONS TO HIDDEN COLLECTOR PLATE.

(B) MILD STEEL TEST SPECIMEN TO SIMULATE A CURVILINEAR EDGE-FLANGE JOINT WITH INSERTED SHIM WELDED BY EBW PROCESS USING REAL-TIME JOINT TRACKING SYSTEM TO FOLLOW THE CURVING JOINT.

Figure 6.47-A — Joint Tracking system for EBW which operates on real time

(Courtesy of The Welding Institute, England)

Laser Beam Welding (LBW)

The use of early lasers for welding was confined to very thin metal and fine wires because they were low-powered, solid-state devices and often could be operated only in a pulsing mode. Usage started in the electronics industry for welding hermetically sealed devices, and for making circuit connections. The introduction of gas lasers about 1970 provided substantially higher power and continuous-wave mode, allowing laser beam welding to be extended to thicker sections.

LBW and EBW are sometimes called the "power beam" technologies because of similarities in certain capabilities, and these two processes are often given coequal consideration when selecting a welding process for a particular application. While both LBW and EBW are most effective when producing welds by the deep-penetration (keyhole) technique, their capabilities for accomplishing full penetration in thick steel sections are not equal given the kinds of welding units presently available. Gas lasers with power ratings up to 10 kW are now fairly common in industry, and the

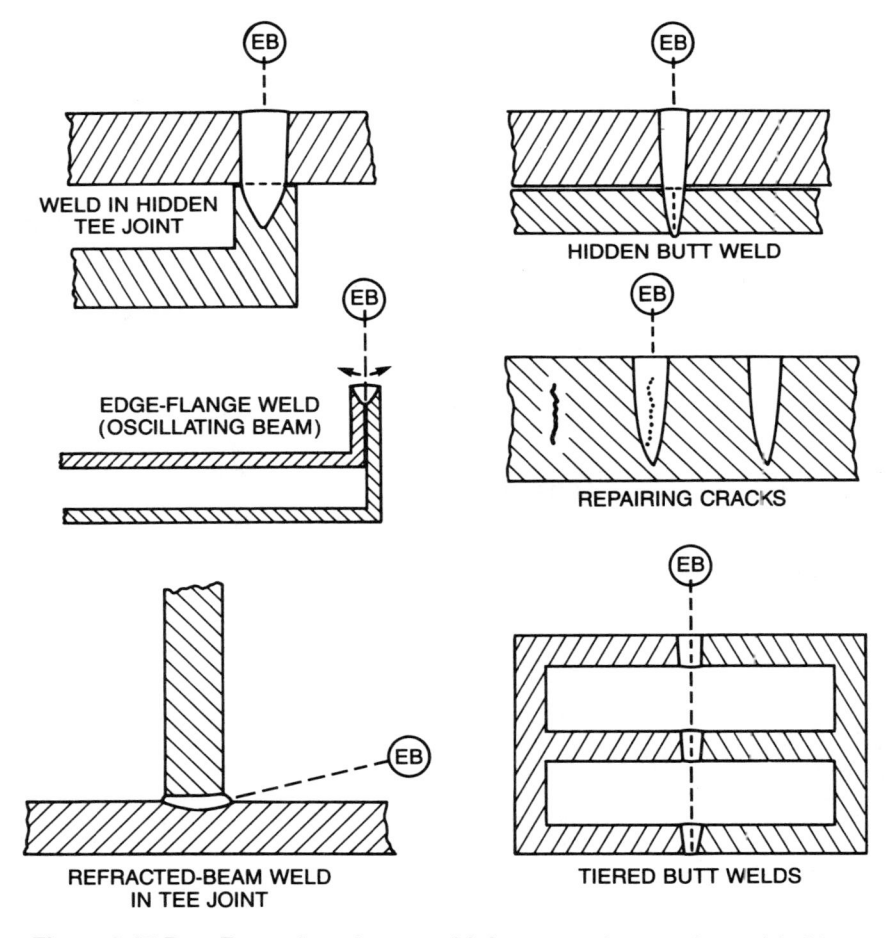

Figure 6.47-B — Examples of unusual joint types that can be welded by EBW processes

maximum thickness square-groove butt joint found feasible to weld by LBW in a single pass is about 25 mm (1 in.). A small number of gas lasers with power ratings up to 45 kW presently are sited in industry, but their capability with respect to maximum depth of penetration has yet to be clearly established. By comparison, EBW has been shown capable of fully penetrating a butt joint in steel up to about 250 mm (10 in.) thick. The modest penetration capability of the laser has not seriously hampered its exploitation. However, the comparatively high cost of large LBW units, and the uncharted areas in which they might be applied, have made fabricators cautious in extending usage of LBW in this direction.

The LBW process is attractive because of the uncomplicated nature of the laser beam for heating and melting metals, as well as its overall suitability for highly automated and repetitive joining operations. Laser welding does not require electrodes, flux, or other ancillary items. The laser beam can be projected through air; through a controlled atmosphere of selected gases surrounding the workpieces; or even through the transparent wall of an enclosure, in which the workpieces are hermetically sealed to maintain a vacuum or some other protective atmosphere. The laser permits great latitude in source-to-workpiece distance, and a single laser beam source can power a number of work stations. Also, if necessary, the outputs from more than one laser can be superimposed to provide a beam with greater total energy-density.

The Nature of Laser Beams and Plasma Generation

Laser beams have unique characteristics because of their monochromatic, coherent, and minimally divergent radiation. These qualities allow laser beams to be projected great distances without significant loss of intensity and to be focused to extremely small spots. The energy-density at the focal point is a function of the laser's power and the spot size; although not equalling that of an electron beam, the energy-density of a laser beam can be sufficient to melt metal or even to cause its vaporization. As will be discussed momentarily, this metal vapor can be either an asset or a hindrance for welding, depending on its position in the path of the beam.

When a laser beam strikes a metal workpiece initially, much of its energy can be reflected away unless the workpiece surfaces are blackened. However, there is an energy threshold at which a breakdown occurs in the reflectivity of the metal's surface. Beyond this threshold, a much larger proportion of the beam's energy is transferred to the metal, quickly melting and vaporizing the metal and easily piercing a keyhole. The surface temperature of the molten metal has been estimated to exceed 3000°C (5432°F), which is above the boiling point of iron and most of the alloying elements commonly used in carbon and low-alloy steels. The energy-density threshold at which the breakdown appears to occur spontaneously (about 500 microseconds) is approximately 10^9 W/cm^2.

As metal vaporizes, the metal vapor rises from the weld as a cloud, or plume, and tends to become a thermally-ionized plasma; as such, it can rapidly absorb and then release energy from the beam. The state of this plume — that is, its nature and disposition — determines whether welding will be assisted or hindered as the operation progresses. Under some circumstances, the plume of plasma virtually "couples" the beam to the molten metal of the workpiece. Unfortunately, as plasma generation continues, the plume that rises from the weld can absorb an increasing amount of the beam's energy and then radiate it in all direction above the workpiece. This amounts to partial blockage of the beam, which decreases melting and also causes a widening of the fusion zone at the upper surface (i.e., a "nailhead" weld profile).

As an energy-input barrier, the plume problem varies in severity depending on several factors — the metal being welded, the energy-density of the laser beam, the travel speed of welding, and whether the plume is allowed to develop in uncontrolled fashion. Aluminum has presented more of a problem with plasma plume than has steel. For this reason, the use of LBW on aluminum and on aluminum-coated steel has been limited in the past. However, research has shown that the problem with aluminum can be alleviated by use of a higher-power laser and, particularly, by focusing the beam to a smaller spot to increase its energy-density. In this way, power transmission to the weld is improved, and reasonable penetration is achieved. Systems presently are being developed that utilize paraboloidal mirrors which can reduce the focal spot to about one-half its usual size, thus increasing the beam's power density threefold. Another remedy applied recently to aluminum to improve performance of the LBW process involves pretreatment of the surface with a beam from a XeCl excimer laser. This preliminary treatment roughens the oxide film that is commonly present on the surface of aluminum, and heat input through the roughened surface is substantially improved. Of course, as soon as the surface becomes molten, the effectiveness of this remedy diminishes.

Uncoated plain steel is not nearly as troublesome as is aluminum, or aluminum-coated steel products, from the standpoint of plasma generation during LBW. Nevertheless, the presence of certain elements in the plume of vapor above the weld — elements that yield easily to thermal ionization — can cause a significant change in beam energy input. In fact, if zinc from a galvanized coating were to enter the vapor, the number of electrons formed could reach a population sufficient to totally absorb the laser beam and decouple it from the workpiece.

A number of strategies may be taken in order to exercise enough control over plasma development to allow the LBW operation to proceed. Under some circumstances, for example, an increase in beam power will permit higher weld travel speed and enable the keyhole to outrun the plasma plume, thus decreasing its interference. Also, very rapid pulsing of the laser beam can serve to limit the duration of energy peaks above the plasma-generating level, and make welding possible. Yet another strategy is to use helium as a shielding gas above the keyhole; but a more effective use of inert gas would be to direct a gas jet across the point of welding, in a manner that will sweep the plasma plume off the top of the weld and improve access of the beam to the workpieces.

Certain features of a laser beam that affect welding performance can differ when comparing the outputs from one design laser to another. For example, the beam delivered by any laser will not have uniform energy over its cross-section. Laser beams can have as many as four different transverse electric and magnetic distributions. The simplest mode is TEM_{00} in which the energy density over the circular cross-section is approximately Gaussian. Other beam modes can be multimodal (designated as TEM_{01}, TEM_{11}, and TEM_{10}), and these are not as focusable as the TEM_{00} mode. The beam from a laser can also be polarized, and this will cause the radiation to interact asymmetrically with the metal surface. Precisely aligned parallel polarization can almost double the heating efficiency of a beam as it is guided along a joint. Experience has shown that focusing the beam to a point slightly below the upper surface of the workpiece is most effective for transferring energy to the metal and accomplishing penetration. LBW can produce a fusion zone by keyholing that has a penetration-to-width ratio as high as 6:1 — somewhat higher than can be achieved by an arc welding process, but not as narrow as obtained by electron beam welding in vacuo.

Another feature of a laser beam that determines the efficacy of its energy transfer to the workpieces is the manner of operation — whether pulsed or continuous. Either manner of operation can be secured from both the solid-state lasers and the gas lasers. The wavelength of radiation emitted by each type of laser is mentioned in their respective descriptions, but this feature has relatively little influence on welding performance. In contrast, the manner of operation (pulsed- versus continuous-beam) can have significant effects. In addition to determining the amount of energy delivered per unit of time, the selection of a pulsed beam versus a continuous beam can influence the extent of plasma formation, which in turn becomes a major factor in controlling heat input to the workpieces. Beam manipulation through optical systems is presently under intense study to develop lasers with improved capabilities and to meet special requirements. New lasers will be capable of beam spinning, oscillation, movement over small-dimensional rasters, cyclic defocusing, etc.

Basic Techniques in Laser Welding

In general, two basically different techniques have evolved for joining metal workpieces by the LBW process. The most frequently used already has been briefly described as *penetration* welding. The second technique commonly is called *conduction limited* welding, so named because the laser merely directs its beam to a surface where the photon energy is absorbed (albeit rather inefficiently). The metal beneath the surface is heated by thermal conduction from the hot surface. The penetration technique is favored because of its many advantages; however, the conduction technique is more likely to be employed when a pulsed laser beam is being applied to thin material to form a seam weld of fused overlapping spots. Conduction limited welding also is the more likely choice when fusing overlays on a substrate; as it affords precise control over melting of the filler wire or metal-powder additives that comprise the new surfacing, with minimal melting of the substrate. The laser can be applied with the conductive technique to weld thick sections as well, by fusing filler metal into beveled groove joints; however, the thermal efficiency would be so poor that a rather large and costly laser would be needed, and production by this technique would be less efficient than could be obtained with one of the arc welding processes. Therefore, more attention will be directed to the penetration (keyhole) technique in the discussion that follows.

Attributes of Laser Welding

Three basic features of a laser beam promote its use as a welding heat source: its high energy density, its small focal spot, and its keyhole penetration ability. These features are utilized in the LBW process, and the generalized capabilities achieved are summarized below:

(1) Process may be automated by pre-programmed procedures with good reproducibility.
(2) Welding may be performed autogenously or with a filler-metal addition.
(3) Unimpeded beam delivery is achieved through any kind of gaseous atmosphere at any pressure; certainly a vacuum is not required.
(4) Travel speed for welding can be fast.
(5) Fused zones can be confined to narrow width.
(6) Heat-affected zones in base metal can be narrow.
(7) Cooling rates during welding are ordinarily fast, but can be decreased by preheating.
(8) Shrinkage and distortion from welding are greatly minimized.
(9) Dissimilarity in electrical and magnetic properties between the two sections being joined are not influential.

The above capabilities portend a bright future for the LBW process, but careful consideration must be given to the ways in which these capabilities are integrated in the total welding procedure for a given application. Under unfavorable circumstances, any one of these capabilities can prove instead to be a troublemaker. For example, the fact that LBW travel speed can be fast and thereby minimize the width of the fusion zone to alleviate shrinkage and distortion would ordinarily be looked upon as advantageous. Yet, narrow welds made at fast travel speed without filler metal can invite cracking, lack-of-fusion, porosity, and unacceptable weld shape. Such pitfalls are illustrated by a series of transverse macrographs that follow made from LBW welds in steel.

Laser Welding Difficulties and Defects

Figure 6.48 (A) is a weld with an acceptable face and root profile formed by penetration (keyhole) laser welding; however, Figure 6.48 (B) illustrates a similarly formed weld that is not as satisfactory. This is because the square-groove butt joint had a gap of 0.2 mm (0.008 in.) prior to autogenous welding, and this left the face and root of the weld with a total underfill of roughly 10 percent. A more serious condition, which resulted from misalignment between the tracking of the laser beam and the centerline of the butt joint, is illustrated by the weld cross section in Figure 6.48 (C). In this weld, an offset of only 0.75 mm (0.03 in.) between the centerlines of the beam and the joint has caused lack-of-fusion through three-quarters of the base metal thickness. When autogenous LBW is employed, the attention that must be given to joint fit-up and to joint tracking will depend to a large extent on aspects of the procedure that control the width of the fusion zone (e.g., weld travel speed, beam manipulation, etc.). Joint tracking in LBW operations often must be more precise than ordinarily required with the arc welding processes. The usual tolerance for misalignment for autogenous LBW of a tightly fitted butt joint is somewhat less than one-half of the focal spot size.

Laser welds are not especially prone to cracking and porosity, but circumstances can arise that cause these forms of unsoundness to occur. Figure 6.49 shows sections from two laser welds in 12.5 mm (½ in.) thick steel that illustrate the effect of increasing weld travel speed to the point of producing a fusion zone too narrow for proper weld solidification. The weld in Figure 6.49 (left) made with travel speed of 0.75 m/min. (30 ipm) is free of internal defects, although its very narrow width certainly calls for accurate following of the joint centerline by the keyholing laser beam. The weld in Figure 6.49 (right), made at a 50 percent higher weld speed, contains a small centerline solidification crack and occasional porosity in the very narrow root area. Higher weld travel speed had the expected effect on the depth-to-width ratio of the fusion zone and on the width of the HAZ in the base metal. Metallographic and SEM examination of the centerline cracking revealed this defect to be solidification cracking in the region of segregation between opposing columnar grains and dendrites, which is a finding that parallels earlier reporting of cracking in electron beam welds. As helpful guidance, formulae developed to appraise steel compositions for solidification cracking susceptibility in EBW can also be used to predict cracking susceptibility in welds made by LBW.

Knowledge about porosity in laser welds comes mostly from laboratory test work, because relatively little production laser welding done in industry has required methodical porosity evaluation. Laboratory work indicates that although the LBW process does not appear to have any inherent feature that would regularly cause porosity to appear in the weld fusion zone, a surprising number of LBW test specimens have displayed porosity to some extent. Progress has been made in determining the cause of certain forms of porosity found in laser welds, but a full understanding is not yet in hand. Two types of porosity have been recognized with LBW. One consists of relatively small scattered globular pores [as shown in Figure 6.49 (right)] that appear to occur more frequently in the lower portion of the fusion zone, and occasional welds have contained these scattered pores in sizes large enough to be considered objectionable. This form of porosity appears to be caused by the oxygen content of the steel. Rimmed and semikilled steels have shown the expected propensity to develop this kind of porosity in

(A) Acceptable Weld

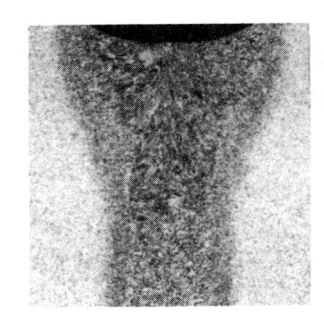

(B) Weld Underfilled
Gap of 0.2 mm (0.008 in.) in fit-up of square abutting edges caused about 10% underfill.

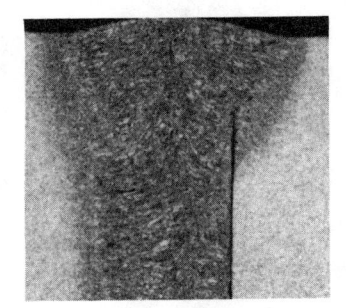

(C) Lack of Fusion
Approximately three-quarters of section unwelded because of 0.75 mm (0.03 in.) misalignment between tracking centerline of beam and joint centerline.

Figure 6.48 — Laser welds (autogenous LBW) of differing quality made in thin steel plate

Single-square-groove butt joints in 4 mm (5/32 in.) thick steel illustrating importance of fit-up, and of accuracy in joint tracking for achieving complete joint penetration. Welds made with 5 kW axial flow CO_2 laser with 1.5 m/min. (60 ipm) travel speed with helium shielding flowing at 30 l/min. coaxial to laser beam. (Courtesy of The Welding Institute, England)

their autogenous welds, but steps taken to deoxidize the fused metal are usually effective in preventing porosity formation.

One particular operating condition when using the keyhole technique in LBW has been found to leave scattered porosity (more often in the lower portion of the weld) regardless of the type of steel. At high keyhole travel speeds, the root portion of the keyhole has been found to curve slightly backward, away from direct action by the laser beam. When this curved shape develops, the keyhole is held open only by radiation reflected from its front wall, and the rear wall may become unstable and entrap some porosity during abnormal progression of solidification.

The conduction-limited technique for LBW offers favorable conditions for avoiding porosity in partial-penetration welds. A weld pass made in this manner by LBW is not troubled by "spiking," a phenomenon described earlier for EBW and illustrated in Figure 6.43. Apparently, as a laser beam generates metal vapor and penetrates to produce a cavity, the beam is attenuated to some extent as it nears the bottom of the cavity. This attenuation produces a rounded root which favors the escape of gases and decreases the likelihood of defects, as can occur with a very sharp terminus at the root of a partial-penentration electron beam weld.

The second type of porosity found in laser welds in steel occurs as very small pores lying in the weld zone immediately adjacent to the fusion boundary. This porosity is believed to result from the decomposition of nonmetallic inclusions in the base metal, as melting occurs under the high intensity radiation on the wall of the keyhole. At slow travel speed, the gas produced combines into larger bubbles which escape through the molten metal. Surprisingly, however, a weld made in the same steel at high speed may

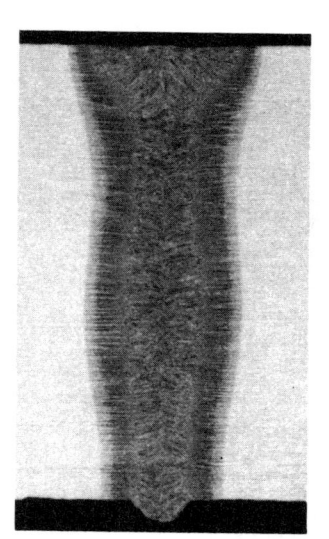

(Left) Acceptable Weld: No internal defects; weld travel speed = 0.75 m/min. (30 ipm); depth-to-width ratio ≈ 3.2.

(Right) Defective Weld: Internal centerline cracking, and occasional porosity in the root area of weld. Weld travel speed = 1.10 m/min. (43 ipm); depth-to-width ratio ≈ 4.8.

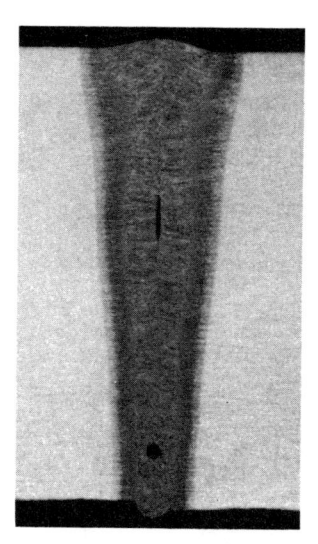

Figure 6.49 — Laser welds (autogenous LBW) in single-square-groove joints illustrating influence of fusion zone profile on susceptibility to cracking and porosity

Complete joint penetration welds in 12.5 mm (½ in.) thick steel plate made with 10 kW continuous wave CO_2 transverse flow laser using helium shielding on face and root, and with plasma control jet device. Laser beam focused on workpiece surface. (Courtesy of The Welding Institute, England)

display no fusion boundary porosity, because there may be insufficient time for decomposition of the inclusions at the weld-metal/base-metal interface. The nature of nonmetallic inclusions in the base metal is also a determining factor in the occurence of fusion-boundary porosity. For LBW, steels that are fully killed with silicon are less apt to have welds containing porosity than steels fully killed with aluminum. This trend is contrary to what might be anticipated from knowledge of the oxides formed by these two deoxidizers, so more research is needed to resolve this anomalous porosity occurrence.

Shielding Gas Effects in Laser Welding

The laser beam can pass through any gas or mixture of gases at any pressure (including a vacuum); however, the nature of the vapor that forms in a plume above the weld affects not only the depth of weld penetration, but also the occurrence of porosity in the weld. A variety of benefits are gained by excluding air during LBW operations on steel, which has prompted studies comparing the virtues of argon, helium, and carbon dioxide for shielding. Helium shielding is favored for LBW because it allows deeper weld penetration than does argon. While carbon dioxide gas is attractive because of its lower cost, welds made under CO_2 shielding have only about 70 percent of the penetration obtained with helium, and they have a somewhat greater propensity for porosity formation when the weld metal is not fully deoxidized. Nevertheless, CO_2 with appropriate procedural adjustments is regarded as a viable

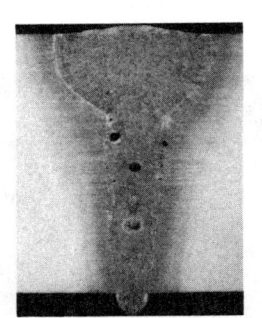

(A) 5.2 kW @ 0.3 m/min. (12 ipm)　　(B) 7.2 kW @ 0.48 m/min. (19 ipm)　　(C) 9.1 kW @ 0.9 m/min. (35 ipm)

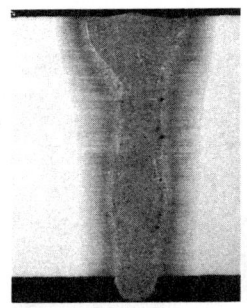

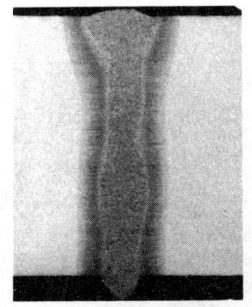

Figure 6.50 — Laser welds (autogenous LBW) illustrating influence of heat input and travel speed on occurrence of porosity

Melt-pass specimens made in 12.5 mm (½ in.) thick HSLA fully-killed steel having the following chemical analysis:

C	0.17	V	0.005
Mn	1.37	Nb	0.022
P	0.014	Al	0.023
S	0.033	O	0.002
Si	0.46	N	0.0076

Beam power reaching the workpiece was measured by absorption bolometer and was adjusted for each travel speed to achieve keyhole penetration. Welding beam provided by 10 kW continuous wave CO_2 transverse flow laser under helium shielding and helium plasma control jet with beam focus at workpiece surface. (Courtesy of The Welding Institute, England)

alternative for inert gas in providing shielding for LBW welds in steel. Figure 6.51 shows a gas box positioned to provide helium shielding to protect the molten weld pool, the keyhole, and the wire being fed as filler metal.

Figure 6.50 illustrates the two forms of porosity described above in transverse macrographs from 12.5 mm (½ in.) thick laser welds in HSLA plate steel. Specimen Figure 6.50 (A), welded at a relatively slow speed, displayed several moderate-size pores scattered in the lower portion of the fusion zone. Specimen Figure 6.50 (B), melted with the travel speed increased by about 50 percent, displayed about half a dozen very small pores typical of fusion boundary porosity. When the travel speed was increased by another 50 percent [Figure 6.50 (C)], the weld was free of porosity. Changes in travel speed also had the expected influence on fusion zone profile; the slowest speed caused the "nailhead" shape, and the fastest travel speed produced a very narrow, almost straight-sided profile.

Filler Wire Feeding in Laser Welding

In metal sections cut to size and shape by shearing or slitting, irregularities can detract from joint fit-up, and autogenous laser welds on production workpieces sometimes are not acceptable. Generally, if the gap between the workpieces exceeds about two mm (0.08 in.), the usual laser beam will pass between the edges without producing enough melting to bridge the gap with weld metal. Smaller gaps in joints ordinarily can be welded if the beam is manipulated in some manner to broaden its path. Experience suggests that gaps up to about 0.12 mm (0.005 in.) are tolerable with the

Figure 6.51 — Laser beam welding (LBW) using filler wire
Wire nozzle is shown at left presenting wire at about a 45° angle. A gas shielding device obscures the view of the weld pool and laser beam.

normal beam in a straight path, but the amount of weld underfill resulting from a gap will be a direct function of its dimensions.

The use of filler metal is a logical way to compensate for gaps in an abutting joint. Furthermore, the use of filler metal offers an opportunity to modify the chemical composition of the fusion zone when desired. There will be occasions when a given base metal requires deoxidation, and this can be accomplished through filler metal selection. Also, early mechanical-property tests on laser welds have indicated some inadequacy in fusion-zone toughness, especially with those plain steels which have a carbon content above approximately 0.20 percent. Work to date suggests that the most pragmatic way to alleviate this weld toughness problem is to modify the weld composition, using filler metal that contains additions of alloying elements known to improve weld metal toughness. Although small-diameter bare wire fed from a spool is the logical choice of filler metal for LBW, early experience showed that the wire had to be kept away from close encounter with the laser beam because the wire would quickly melt and vaporize, and this would generate excessive plasma which could divert much of the beam's power as scattered radiation.

Trailing-Edge Feeding

Study of techniques for introducing filler wire into the fusion zone of a LBW weld disclosed important metallurgical details. Because a keyhole offers a trailing weld pool that is longer than its frontal weld pool, the technique examined first was to feed wire from a regular feeding unit through a nozzle into the back end of the molten weld pool, as illustrated schematically in Figure 6.52. In the course of this work, three factors were found to be important: (1) the angle, Θ, between the vertical centerline of the beam and the longitudinal axis of the wire nozzle; (2) the distance, A, between the two points where the laser beam and the filler wire would intersect with the top surface of the solid workpiece; and (3) the wire feed rate, R. For the wire angle, Θ, 30° to 60° was found to be permissible. To avoid premature melting and vaporization of the wire, the distance A had to be at least two mm (0.08 in.). The lengthy trailing weld pool appeared to allow introduction of the wire at distances up to six mm (¼ in.); beyond this separation distance, difficulty was experienced with unmelted wire caught in the solidifying weld. The optimum separation distance, A, appeared to be about four mm (5/32 in.); however, when welds made with filler wire fed in this manner were sectioned and examined metallographically, there was considerable evidence (as shown in Figure 6.53) that commingling of the melted filler wire with the weld metal in the pool was incomplete. If the only reason for adding filler metal is to compensate for gap in the joint, this technique possibly would be considered satisfactory since the weld was sound. Yet, there often is good reason to strive for complete assimilation of the filler wire into the weld metal in order to achieve a homogeneous composition.

Leading-Edge Feeding

Study of techniques for adding filler metal in LBW welds next explored the introduction of wire into the leading edge of the weld pool. This work established that two slightly different techniques were feasible. In the first, the wire could be directed into the molten pool just within its leading edge. Of course, the very short distance between the edge of the molten pool and the laser beam requires precise positioning to avoid premature melting and possibly some vapor formation. In developing a second technique, it was found that a steep angle of entry by the wire into the pool (i.e., small Θ) greatly reduced the likelihood of the wire intersecting the beam. Figure 6.54 illustrates the effectiveness of this second filler wire technique by comparing transverse metallographic sections from two laser welds — one made without filler wire, and the other with wire. The square-groove butt joint had a deliberately set gap of 0.75 mm (0.03 in.) causing the weld in Figure 6.54 (A) made without filler wire, not unexpectedly, to have a severely underfilled weld face. When filler wire was fed to the same joint at a feed rate that provided a volume of metal slightly larger than the gap space, the weld profile was like that shown in Figure 6.54 (B). Although this etched specimen displays an upper area bounded by a solidification pattern related to the filler metal addition, there does not appear to be significant difference in chemical composition between the upper and the lower areas of this weld.

An improved modification of the technique for introducing wire into the front of the weld pool evolved when the 1.2 mm (0.045 in.) diameter wire inadvertently entered the gap between the abutting faces of the square-groove joint at a location

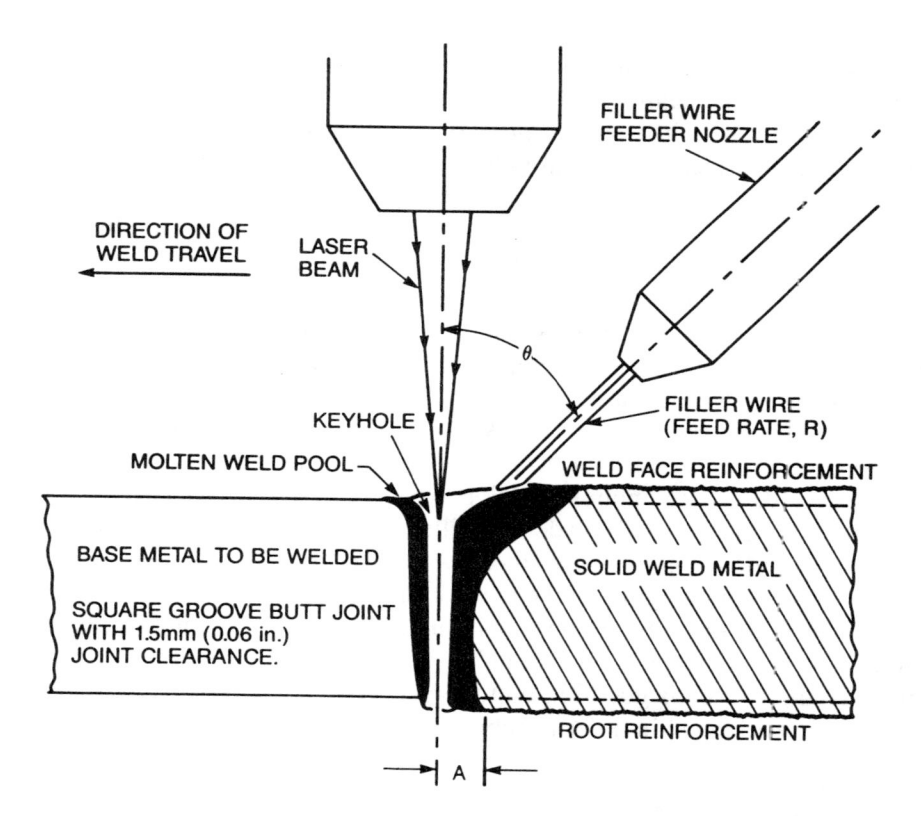

Figure 6.52 — Laser beam welding (LBW) with filler wire introduced into trailing end of weld pool

This schematic illustration includes several important details that are discussed in the text, but the addition of filler metal to the trailing end of the weld pool surrounding the keyhole is not favored because of weld composition heterogeneity.

below the top of the molten weld pool and the surfaces of the workpieces. By feeding wire "within-the-gap" to the front of the advancing keyhole, as illustrated schematically in Figure 6.55, the gap served as a mechanical guide to deliver the wire to the center of the weld; also, a direct encounter between the wire and the laser beam was avoided by feeding the wire below the top of the keyhole. This arrangement allows greater freedom in setting the nozzle angle (Θ) and permits the nozzle to be positioned a small, but helpful distance ahead of the hot laser beam and weld pool. The location at which the wire passes through the joint gap and penetrates the front wall of the molten keyhole can be anywhere below the top surface of the workpieces and above their mid-thickness. If the wire is directed to a point that is too low, whiskers of unmelted wire are likely to be found protruding from the weld root surface. An incentive to consider this within-the-gap technique for feeding filler wire is the greater tolerance permitted for joint fit-up. Of course, no portion of the gap during the welding operation can be smaller than the wire diameter, because feeding rate might be fric-

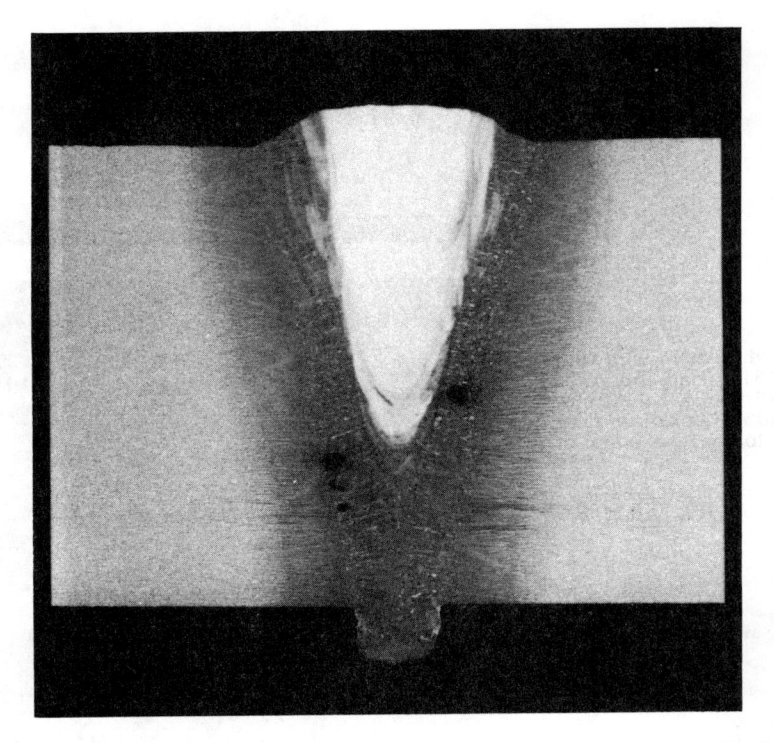

Figure 6.53 — Weld composition heterogeneity in weld made by LBW with filler wire added to trailing end of weld pool

Transverse etched specimen prepared from weld made in 8 mm (5/16 in.) thick steel using a 5 kW fast axial flow CO_2 laser to keyhole a single-square-groove joint with 1.2 mm (0.045 in.) size filler wire of double-deoxidized (Sl + Ti) variety. Travel speed (R) 0.35 m/min. (14 ipm). Comprehensive helium gas shielding of weld coaxial flow around the beam plus helium flow to box covering weld zone fore and aft point of welding; including the entry of filler wire into the trailing end of weld pool. Marked variations in etching on the weld fusion zone indicate significant changes in weld composition. (Courtesy of The Welding Institute, England)

tion-reduced or even jammed. The volume of wire fed into the gapped joint needs to be pre-set according to gap width measurement. The gap probably can vary in width by ± 0.25 mm (0.010 in.) without causing difficulty.

Although single-pass deep-penetration (i.e., keyhole) welding is the most efficient way to laser weld, circumstances can arise where the available laser has insufficient beam power to accomplish keyholing in the given thickness of base metal. In these circumstances, multi-pass welding with a weld pool is a viable alternative. Figure 6.56

(A) Laser weld made without filler wire addition. Gap of 0.75 mm (0.03 in.) in joint caused severely underfilled weld face.

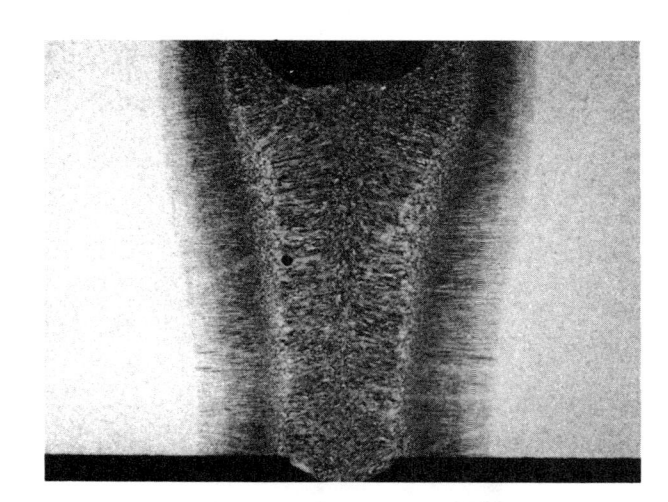

(B) Laser weld made in joint with 0.75 mm (0.03 in.) gap, but with 1.9 m/min. (75 in./min.) of 1.2 mm (0.045 in.) double-deoxidized filler wire added to leading edge of weld pool.

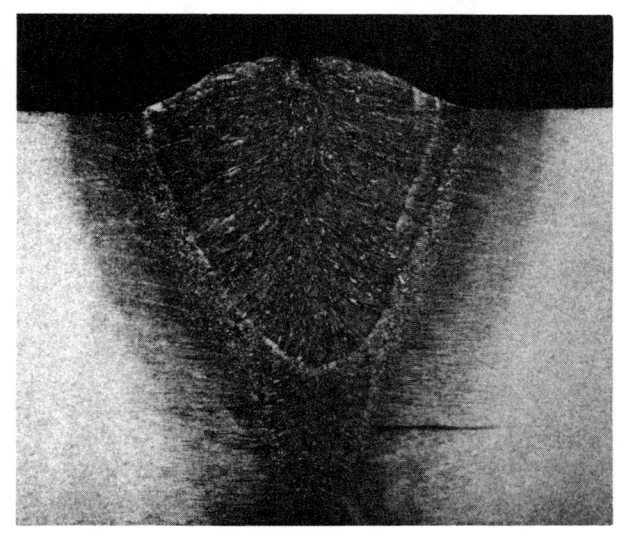

Figure 6.54 — Comparison of welds made by LBW; (A) autogenously, and (B) with filler wire added to the leading edge of the weld pool

The 5 kW axial flow CO_2 laser described for Figure 6.53, and the same welding conditions were used in making the specimens illustrated here; except for a deliberate gap between the abutting faces of the single-square-groove joint and the change in filler metal addition. Text explains why the leading-edge technique for adding filler wire is preferred. (Courtesy of The Welding Institute, England)

illustrates a two-pass welding procedure with a 5 kW CO_2, fast-axial-flow constant-wave laser on 12.5 mm (½ in.) thick steel using an autogenous initial pass through approximately half the thickness. A second pass with filler wire (at slightly slower weld travel speed) filled the open preparation and completed the weld. Figure 6.57 illustrates a similar procedure incorporating an autogenous weld and subsequent filler-pass welds. This specimen, a laser weld in 25 mm (1 in.) thick steel, required a total of four passes.

Hot wire welding, as mentioned earlier when describing the use of filler wire in the GTAW process, has been used to some extent with LBW for increasing the deposition rate. This system will also allow the weld travel speed to be increased by as much as 50 percent as compared with the same operation using cold wire feed. Incorporation of hot wire filler into the LBW operation requires much greater care in coordinating process parameters in order to avoid defects such as lack of fusion. The highest useful temperature for resistance-preheating steel filler wire appears to be approximately 650°C (about 1200°F). Higher temperature in the filler wire causes a variety of disturbances that often result in weld defects.

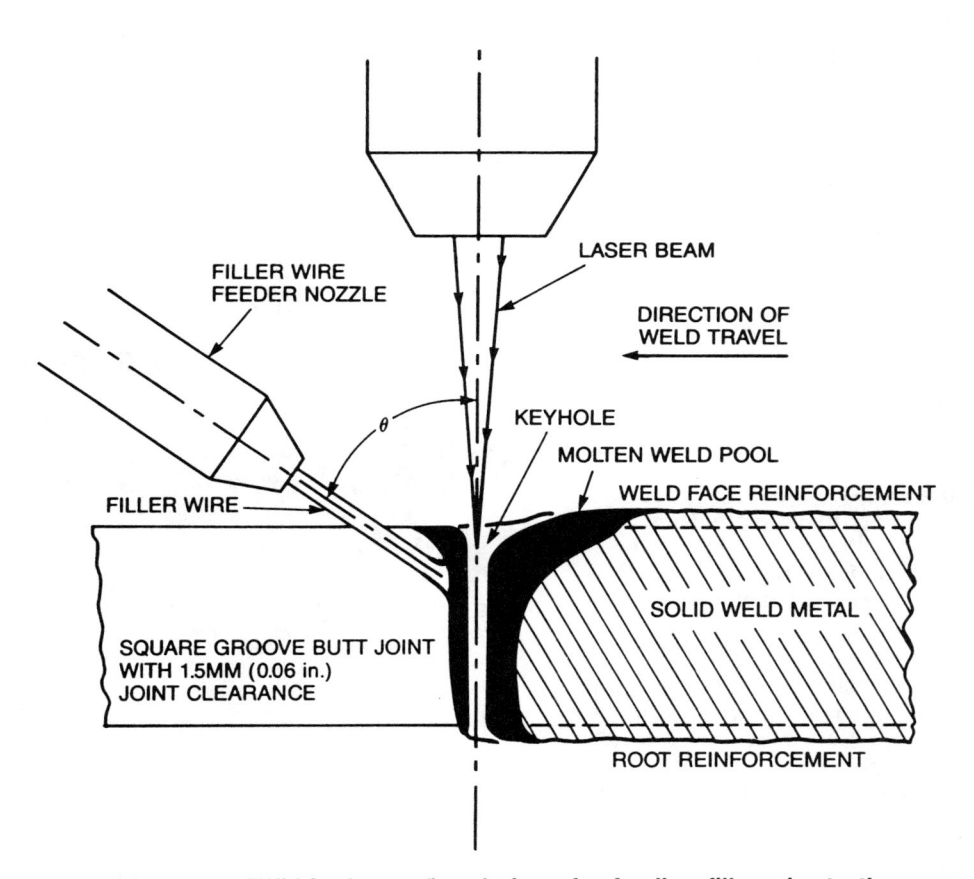

Figure 6.55 — "Within-the-gap" technique for feeding filler wire to the front edge of weld pool during LBW

Oxyfuel Welding (OFW)

Combustion of gases from a torch provided one of the first successful processes for welding metals. However, commercial realization came about only after purified oxygen was made available in quantity at affordable cost, hence the term *oxyfuel gas welding*. Fuel gases burned with air do not produce the combustion intensity needed for localized fusion of metals at a reasonable rate to form a weld. In fact, for fusion welding ferrous alloys, only *acetylene* burned with oxygen produces a temperature sufficiently high (3200°C, or 5800°F) to provide a viable fusion welding process.

Oxyacetylene Welding

The oxyacetylene flame does not approach the temperatures obtainable with the electric arc, and therefore the OAW operation is characterized by a large total heat input to the workpieces because of slow weld travel speed. Nevertheless, oxyacetylene welding, mainly because of its versatility, continues to serve as a useful welding process.

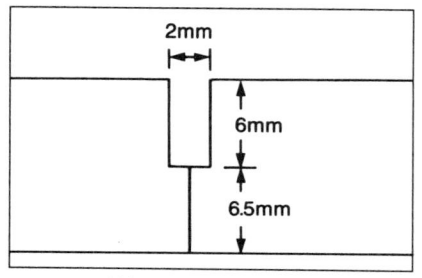

(A) Joint Preparation

(B) Autogenous Root Pass

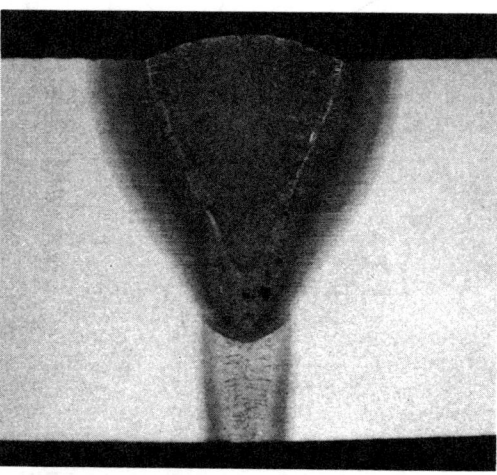

(C) Final Pass with Filler Wire in Open Groove

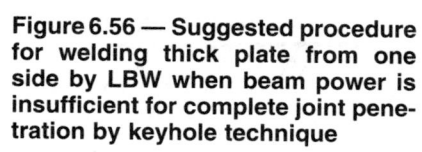

Figure 6.56 — Suggested procedure for welding thick plate from one side by LBW when beam power is insufficient for complete joint penetration by keyhole technique

(A) is drawing of single step-edge shape butt joint. (B) autogenous root pass made with weld travel speed of 0.5 m/min. (20 ipm) to weld lower-half of 12.5 mm (½ in.) thick steel plate. (C) shows final pass to complete weld made with travel speed of 0.35 m/min. (14 ipm), and with 1.2 mm (0.045 in.) size filler wire feeding at rate of 3.75 m/min. (148 ipm). (Courtesy of The Welding Institute, England)

Despite the simplicity of oxyacetylene gas welding, if the rudiments of this process are not understood and properly controlled, welds produced in steels can acquire profound shortcomings without the welder being aware of their occurrence during the making of the weld. The character of the oxyacetylene flame can be varied by changing the relative proportions of acetylene and oxygen issued at the torch tip. As illustrated schematically in Figure 6.58, when equal volumes of acetylene and oxygen are burned, the flame is called a *neutral flame*, and it consists of two distinct parts; a small, bright inner cone where the two gases combine to form carbon monoxide (CO) and hydrogen (H_2), and a larger outer flame where an additional 1½ volumes of oxygen picked up from the air are used to burn the CO and H_2 to carbon dioxide (CO_2) and water vapor (H_2O). When the amount of acetylene is slightly increased over the one-to-one mixture, a *carburizing* or *reducing* flame is issued from the torch. This adds a third zone to the flame, an irregular feather or fringe surrounding the inner core

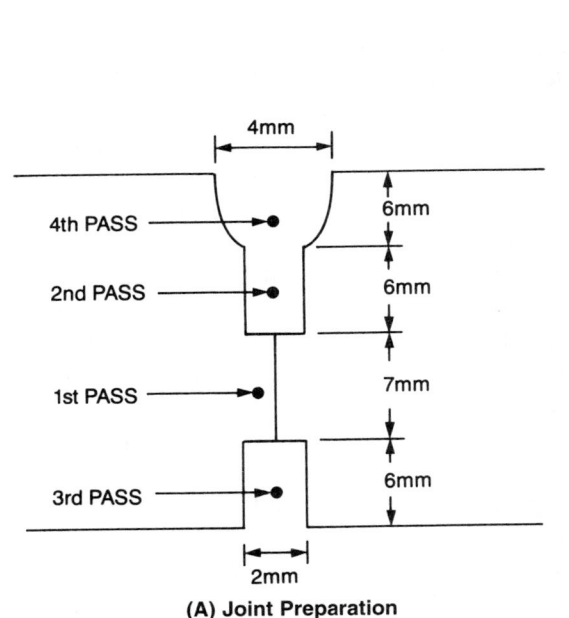

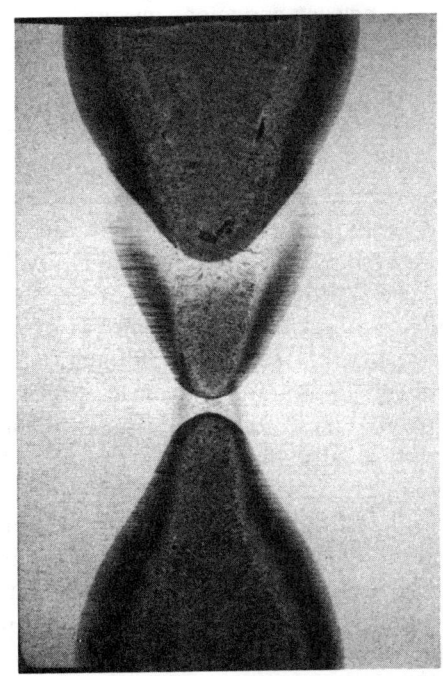

(A) Joint Preparation   (B) Completed Weld Made in Four Passes

Figure 6.57 — Suggested LBW procedure for laser weld in thick plate when limited beam power requires multiple passes

(A) illustrates joint preparation for 25 mm (1 in.) thick steel plate which requires an initial autogenous pass made with keyhole technique to provide a base for three subsequent passes with filler metal to complete the weld. Root pass was made at travel speed of 0.5 m/min (20 ipm). Second and third passes were made at travel speed of 0.35 m/min (14 ipm), and a filler wire of 1.2 mm (0.045 in.) size was added to the weld pool at a rate of 3.75 m/min (148 ipm). Fourth and final pass required a slightly slower travel speed (0.30 m/min, or 12 ipm), and a somewhat higher filler wire feeding rate (4.74 m/min, or 187 ipm). (B) is a transverse full-thickness specimen etched to show the pattern of four passes in the completed weld. (Courtesy of The Welding Institute, England)

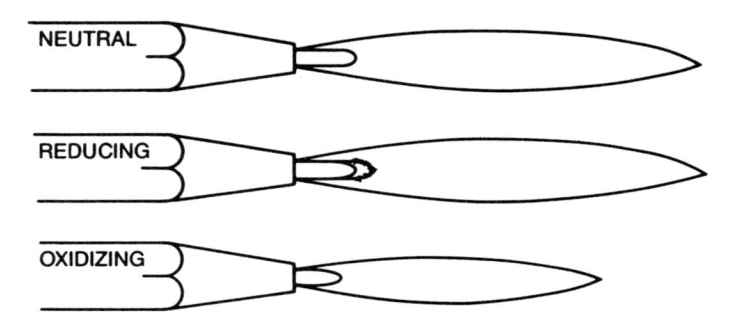

Figure 6.58 — Three types of oxyacetylene flames obtained by controlling the proportions of gases

Familiarization with the differences in character of the flames (by their colors, shapes, and audible sound) is gained by observing the flame being issued from the torch as oxygen and acetylene proportions are varied above, and below a one-to-one mixture.

caused by free carbon combining with oxygen. On the other hand, when the amount of oxygen is increased in excess of the one-to-one mixture, the flame becomes oxidizing. In this case, only an inner cone and outer flame are again present, but the shape and color of the inner cone can be easily distinguished from the cone of a neutral flame.

Oxyacetylene welding requires welder training to make sound, acceptable welds because skillful coordination of torch, weld pool and welding rod is required. As this skill is being exercised, a number of metallurgical reactions may be taking place. In fact, the carburization action mentioned above actually will facilitate the making of a weld by suppressing the formation of oxide on the surface of the molten metal and improving its fluidity, but, in reality, the weld that forms can be unsuitable for reason of abnormally high carbon content. The so-called neutral flame, even though it envelopes the molten weld pool, still allows some oxidation to take place because of the amount of air aspirated into the flame. Many low-carbon steel welding rods are rimmed steel, and although the familiar carbon-oxygen reaction takes place in the molten weld pool, much of the gas that is produced escapes from the weld instead of becoming entrapped as porosity because of the relatively slow solidification rate. When porosity is to be minimized as much as possible, either the base metal, or the welding rod, or both should contain adequate deoxidants. Silicon and manganese are favored for deoxidation purposes because their oxides form a fluid slag that is unobtrusive. When significant amount of elements are present in the fused weld metal that form viscous, high-melting-temperature slags (e.g., aluminum, chromium, and titanium), a flux may have to be applied during welding to avoid oxide entrapment. A limited number of standardized compositions for carbon and low alloy steel rods for oxyfuel gas welding can be found in ANSI/AWS A 5.2.

Oxyacetylene torches have proved so versatile as a means of providing heat that they have been carried into other joining processes. Brazing is a prime example. The temperature requirements for brazing are not as demanding as for fusion welding steel, so for this reason hydrogen, propane and natural gas are sometimes substituted for acetylene.

Pressure Gas Welding (PGW)

This is another process that uses the oxyacetylene flame for heating. Two variations of the PGW process have been employed in industry: (1) the closed-joint method, and (2) the open-joint method. In the closed-joint method, two workpieces are butted together under moderate force and heated until a predetermined amount of upsetting has occurred at the joint. In the open-joint method, the faces to be joined are individually heated until they become molten. At this time they are brought into contact and force is applied to produce upsetting. In the closed-joint method, since the metal does not reach its melting point, solid-state welding occurs. In the open-joint method, the molten metal that formed on the faces is squeezed from the joint as force is applied and produces a flash during solidification somewhat like the central fin on a flash weld as illustrated earlier in Figure 6.33. Some of the PGW operations in industry have changed from oxyacetylene flame heating to induction heating because this electrical method is faster, and can be applied with precision on each pair of workpieces. Induction welding (IW) is described in more detail later as a solid-state welding process.

Two important points in the oxyacetylene gas welding procedure applied to steel are: (1) adjustment of the proper type flame, that is neutral, reducing, or oxidizing, and (2) selection of the proper kind of flux, if any is to be used. Both of these points are determined by the type of steel to be welded. The character of the flame can exert significant chemical action on the metal being fused. An oxidizing flame ordinarily is not used when welding steel because it vigorously produces a slag of any elements present that are easily oxidized. The presence of the slag on the surface of the molten weld pool usually is a hindrance, and loss of the elements from the composition of the weld metal may be detrimental to the properties of the weld. On the other hand, a carburizing or reducing flame can increase the carbon content of steel weld metal. Some steels containing a high carbon content should be welded with a carburizing flame to maintain the desired level of carbon in the weld metal. A rough method of estimating the carburizing potential of the reducing flame is to note the length of its feather in relation to the inner cone. A 2X excess-acetylene flame has a feather twice the length of the inner cone, and will have considerable potential to add carbon the molten weld pool of steel. Much of the oxyacetylene gas welding on structural steel is performed with a neutral flame because the carbon content is not particularly high, and raising the carbon content of the weld metal is likely to be detrimental to mechanical properties, particularly fracture toughness. The character of the flame's inner cone is influential because heat output or combustion intensity is highest at its tip, and therefore the tip of the inner cone regularly is brought into close proximity to the metal to be heated. Secondary combustion, which occurs in the large outer envelope of flame, does not produce as high a temperature as found at the tip of the inner cone.

Thermite Welding (TW)

Thermite welding makes use of a chemical reaction between solids as a heat source to bring workpieces to a welding temperature, as well as utilizing the reaction to produce molten metal that serves as filler metal. The process owes its name to the *thermit* mixture which was developed by Goldschmidt in 1895. Presently, the process is not often used in production welding because other processes are more efficient, but thermite welding has seen much use through the years for making butt welds between lengths of railroad rails, for joining very thick sections of cast iron and steel castings, and for joining very large size steel reinforcing bars embedded in concrete structures. The unique technique involved in thermite welding and the nature of the weld metal produced by the process hold good reasons to understand how the process operates, and to be familiar with the chemical composition and properties of a thermite weld. Circumstances can arise today where thermite welding is the best process to fill special needs. Repair of massive sections that have cracked in large machines is an application where thermite welding offers advantages.

The basic thermite mixture consists of powdered aluminum and iron oxide in proportions of about one part aluminum to three parts iron oxide by weight, respectively. When this mixture is heated locally above about 1100°C (2000°F), usually using a special ignition powder, a vigorous exothermic reduction reaction takes place which proceeds rapidly through the mass of mixed powders. The aluminum, having a much greater affinity for oxygen than does iron, reduces the iron oxide and liberates a considerable amount of heat. The entire mass becomes molten, with the iron settling to the bottom and the aluminum oxide floating to the top. Prior to a thermite welding operation, a selected mixture is placed in a refractory crucible above the pieces to be welded and the joint enclosed in clay. In practice, additional metals and compounds often are included in the thermite mixture to alloy the iron and to improve its properties. Four different mixtures are commonly used: (1) plain thermit, (2) steel thermit, (3) cast-iron thermit, and (4) wear-resistant thermit. Variations of these aluminothermic mixtures are tailored for specific applications. For example, wear-resistant thermite is available as a number of mixtures that produce alloyed iron with increasing levels of hardness to match that of the base metal being joined. The raw materials incorporated in the thermite mixture must be carefully selected for purity inasmuch as the aluminum will reduce most of the oxides present, including those of undesirable elements such as phosphorus and sulfur. The reduced elements become part of the iron alloy composition.

The temperature created by the thermite reaction is about 3150°C (approximately 5650°F). Actually, the molten metal reaches only about 2400°C (4350°F) because of heat loss through the crucible, and the energy required to melt the pieces of metal included in the mixture as an additional source of iron, or of alloying elements and deoxidizers. The molten metal from the thermite reaction in the crucible is guided to the joint to be welded and contained in the joint by a refractory clay or a sand mold which is fastened around the workpieces. The joint has a gap to facilitate entry of the molten metal; and the thermite metal, by virtue of its superheat, melts a portion of the base metal with which it comes into contact. Upon solidification of this melt, the weld is completed. The microstructure of a thermite weld and the adjacent heat-affected zone depends on the chemical composition of the base metal and the weld metal, and also on the cooling rate of the welded section following solidification. Consequently, the procedure for thermite welding hardenable steels and certain compositions of cast iron may include preheating, controlled cooling following the thermite reaction, and possibly post-weld heat treatment.

SOLID-STATE WELDING PROCESSES

The *solid-state* welding processes are not often used on carbon and low-alloy steels because they are somewhat time-consuming, and usually require a specialized facility in which to perform the operation. Furthermore, considerable care is required in the preparation of the workpieces, the execution of the operation, and the application of NDT to ensure that the resulting weld is fit-for-purpose. Ordinarily, some prototype work is required to be certain of the applicability of the selected process, and the integrity of welds made by the procedure adopted. Nevertheless, unique advantages are offered by solid-state welding, among them: joints inaccessible to ordinary fusion processes can be solid-state welded, the time required for welding is independent of the joint area, a multiplicity of joints can be welded in a single operation, opportunities for making dissimilar-metal joints are increased, and problems with fusion welding processes, are virtually eliminated.

Designations adopted by AWS for the individual processes classified under the solid-state welding method were presented in Figure 6.2-A, but additional process names are used by various segments of industry. These include "diffusion bonding," isostatic bonding," and "electromagnetic bonding." Diffusion bonding of titanium has been used extensively in aircraft and aerospace vehicle construction. This usage has been supported by substantial technological development that is now well documented in the literature. Much of the guidance and the findings reported deal with alloys other than steels, but sufficient attention has been given to commercially pure iron and to commonly-used types of steel to provide guidance on the selection of a process and planning of a procedure.

The AWS definition for solid-state welding implies that no melting of the base metal occurs, and states that no brazing metal is used. A number of processes listed under the solid-state classification are in accord with this definition and diffusion welding, outlined later, is an example. Some processes do not adhere strictly to the definition's guidelines and the principal kind of noncompliance is the occurrence of *some melting* during execution of a given process. Some of these processes are highly dependent on a small amount of melting taking place as a key step in the success of the welding operation. Proper control is not likely in these processes without an understanding of this key step, even though it may involve only a trifling amount of molten metal. The deliberate employment of molten metal to aid the making of a solid-state weld in its early stages has fostered a distinctive technique which is called "liquid phase diffusion bonding," "eutectic bonding," or "activated diffusion bonding." Before examining this specialized technique, a review will be made of the rudiments of straightforward solid-state welding to explain why this seemingly-simple method has fostered a number of different approaches.

Basically, solid-state welding of metals proceeds when two surfaces are brought into highly intimate contact, and interatomic bonding across the interface between the two workpieces is not impeded by nonmetallic surface films or chemisorbed foreign elements. These two clear-cut requirements can be fulfilled, but the required steps may not be convenient or economical. Consequently, industry has not made widespread use of solid-state welding.

The first requirement, contact *intimacy*, can be assisted by preparation of very smooth surfaces, flat or congruent, on the faces of the workpieces that will be faying during the welding operation. To count on prepared surfaces to immediately provide the required degree of intimacy simply by bringing them into contact is quite impractical. An effective alternative is to apply force to attain the required intimacy. Force can be applied in a number of ways and at different levels to accomplish one or more of the following: (1) simply bring the faying surfaces into close contact and cause a certain amount of flattening of asperities, (2) cause localized or superficial deformation of the interfacial surfaces in a manner that accomplishes congruency of the two surfaces, and (3) produce bulk deformation of the workpieces in a manner that ensures intimacy at their junction. The extent to which interfacial intimacy is achieved will have a profound influence on how rapidly and completely the solid-state weld is formed. The level of force applied and the degree of deformation often is a carefully preplanned feature of a particular process.

The second requirement of the solid-state welding method, avoidance of interference to the development of interatomic bonds by surface films of oxides, nitrides, or other compounds (even by absorbed gases), also can be accomplished in several ways. The surfaces to be welded can be stripped of nonmetallic films, etc., prior to welding, and this "sterile" surface state preserved as much as possible by quickly putting the surfaces into contact under pressure to exclude air. Sometimes an inert atmosphere is provided to retard restoration of surface films. The most effective technique is to place the workpieces in a hard vacuum and raise the temperature to remove adsorbed gases and eventually break down the surface films. The vacuum technique has the added advantage of keeping the surfaces in a clean state while solid-state welding progresses. With some metals and alloys, a practical technique for dealing with the surface film is to count on deformation at the interface to rupture and fragment the barrier film and thus facilitate interatomic bonding as much as possible. The integrity of the weld that develops will depend on the thickness and character of the surface film, and the efficacy of the deformation in fragmenting the film to allow bonding forces to develop. The additional measure which must be applied to some degree is to increase *temperature*. This operating parameter assists solid-state welding in several ways because almost all of the mechanisms that take place are temperature-sensitive. Raising the temperature can rid the surface film of moisture and other chemisorbed contaminants that add to barrier effect. It also can reduce the strength of the metal and thereby increase the extent of plastic deformation. This will aid in fragmenting surface films. Even at low levels of pressure, as asperities are flattened, creep at elevated temperature can increase the area of intimate contact. Of course, if the surfaces are not protected from air as the temperature is raised, the oxide surface film on iron and steel will increase in thickness and become a more formidable barrier. As temperature is raised even higher, fragmented oxide particles along the interface can be dispersed by diffusion. Generally, this requires a temperature exceeding about one-half of the metal's melting point. Iron and carbon steel start to dissolve oxide at a reasonable rate under protective conditions when the temperature exceeds about 800°C (1472°F). Therefore, this temperature will serve as a useful reference point when devising operating parameters for the solid-state welding of iron or plain steel. As the temperature is raised above the recrystallization temperature for the metal being welded, a new configuration of grains will be formed to aid in eliminating vestiges of the junction between the workpieces.

Temperature, pressure, and time to achieve solid-state welding are adjusted to suit the materials involved, the nature of the workpieces, and the facility and technique being employed. In general, an effective temperature for solid-state welding is at least 0.7 of the melting point of the metal in degrees absolute. For iron and plain steel, this suggests a temperature somewhere above 950°C (1742°F), which is sufficiently high to provide a full-annealing or a normalizing heat treatment, a fact to be taken into consideration when contemplating the solid-state welding of steels that have already been heat treated to gain particular mechanical properties. The level of pressure applied will depend on the nature of the workpieces, the procedure for dealing with surface conditions, and the overall deformation in the workpiece that can be tolerated. When minimal deformation is an important objective, success in forming an acceptable weld becomes more dependent on diffusion phenomena and the operation is considered *diffusion welding* (DFW). A bonding time of 10 to 30 minutes commonly is allowed at operating temperature and pressure to ensure achieving the required intimacy at the faying surfaces, the development of interatomic bonding, and the disappearance of interfacial oxide particles and voids.

When *liquid phase diffusion bonding* technique is employed, a dissimilar filler metal is placed between the faying surfaces as an intermediate layer. This metal is not a lower-melting brazing filler metal, but is a metal or an alloy that will actively diffuse into the surfaces of the workpieces, and upon doing so, will form an alloy that for a short period will have a melting point *below* the bonding temperature being applied. As a consequence of this alloy formation by diffusion, molten metal is created at the interface that flows under little pressure to fill voids and *temporarily* provides intimacy of contact by virtue of its wetting of both surfaces and serving as molten linkage. From this point onward in the process at the bonding temperature a unique further action takes place. The pre-positioned filler metal or alloy continues its diffusion into the base metal of the workpieces until its concentration decreases and the melting point of the linkage alloy rises *above* the operating temperature. Even after solidification, diffusion continues to be promoted by temperature and time, and often the joint properties approach those of the base metal because only a trace of the filler metal or alloy remains at the interface, having functioned as a diffusion welding aid, rather than a bona fide brazing metal.

Hot Pressure Welding (HPW)

Hot pressure welding is so broad in its nature that a number of processes could carry this designation. For explicitness, more definitive terms usually are employed. *Forge welding (FOW)* has been practiced throughout recorded history as a way of joining pieces of metal, but the development of other processes over the past century gradually has relegated forge welding to the assembly of articles where the relatively crude appearance of a hammered joint is aesthetically pleasing, or gives the appearance of authentic antiquity. Nevertheless, forge welding deserves description because a weld made by this process might be encountered in an article made at a much earlier date, and because forging knowledge gathered through the years has been put to use in several hot pressure welding processes presently in use.

Forge welding consists of heating the metal workpieces to be joined, but mainly aiming at heating their surfaces to a welding temperature at which they can coalesce. For iron and steel, this temperature is a "yellow" heat (i.e., above approximately 850°C, or 1600°F). When these heated surfaces are in contact and the workpieces hammered together with an appreciable amount of hot plastic deformation, a weld is formed at their junction. The cleanliness and integrity of the weld can vary greatly. The scarves, as the surfaces to be welded are called, often are heated in an open fire of coke or gas, or in a furnace with an air atmosphere. Consequently, the heated surfaces can become rather heavily oxidized. It is true that a carburizing forge fire will create a liquid layer of high-carbon, low-oxygen metal on the surface which may facilitate welding. However, the carburization usually leads to a hard microstructure along the weld interface, and its brittleness is difficult to control. An oxidizing fire or furnace atmosphere creates a film of oxide which may become liquid if heated sufficiently high, but this surface condition should be recognized as a barrier to welding. During an early period of forge welding, sand was sprinkled on the heated surfaces to serve as a flux because it diluted iron oxide by forming silicates. As a result, less oxygen diffused into the material than without a flux. When solid surfaces were hammered together, fluid or viscous slag was squeezed from the joint more-or-less completely. Wrought iron is quite adaptable to forge welding because the silicate slag inclusions contained in this material act as a self-contained flux to protect the heated surfaces and to facilitate the welding during hot plastic deformation. The principal problem in forge welding steel is the uncertainty of dispersing oxides from the interface between the workpieces to allow interatomic bonding to develop. Hammering, either manually, or by a power-driven hammer, can accomplish the fragmentation and dispersal necessary, but close control must be exercised over the amount of metal movement at the faying surfaces during the hammering operation - and this is not a simple task. The common course of action has been to produce as much deformation as permissible in effort to gain as much metal-to-metal intimacy as possible and thus facilitate weld formation.

Die forging welding is a process with improved control over conditions that develop at the interface between the workpieces, but the process has limited usefulness and considerable development work is required for each particular assembly where applicable. Briefly, heated workpieces that are partially-formed or rough-shaped are placed above a die cavity having a desired shape. Forging pressure is applied to rapidly force the workpieces to fill the die cavity; but, in doing so the faying surfaces of the hot workpieces must undergo deformation that accomplishes fragmentation of any oxide film and receive sufficient pressure to become congruent and develop the interatomic bonding that constitutes welding. When die forge welding is properly planned and executed, production rate can be quite high. The process also offers opportunity to use dissimilar metals in the welded assembly.

Roll welding (ROW) is a process that uses pressure and deformation exerted by a rolling operation, such as carried out between a pair of rolls in a rolling mill. Much steel-clad plate is produced by hot rolling a pack or sandwich, which consists of a thick steel slab with thinner plates of stainless steel, or of nickel on one or both slab surfaces (see ASTM Standards A 263, A 264 and A 265 for various kinds of steel-clad plate, sheet and strip). The peripheral edges of a pack for rolling usually are sealed by a fusion welding process prior to heating to minimize the formation of oxide on the faying surfaces inside the pack. Sometimes the surfaces of the steel slab are plated with metal (e.g., nickel) for further protection against oxide formation and to aid welding by diffusion. The pack is heated to approximately 1200°C (about 2200°F), which is well below the melting point of any of the materials involved, and hot reduced to the required gauge by rolling. Under these conditions of hot working, solid-state welding occurs completely over the faying surfaces within the pack. Demarcation in microstructure and chemical composition at the interface between the materials depends on the metals and alloys involved, and on any prolonged heat treatment applied after rolling.

Coextrusion welding (CEW), as the name infers, is a hot pressure welding process carried out as an extrusion operation. Welding occurs readily during this process principally because of the great increase in surface area in mutual contact. The chief utilization of coextrusion is to produce bi-metal shapes in long lengths, and metal-clad tubing. Dissimilar metal slugs or tube rounds must be preconstructed of the required materials and made ready for heating and extrusion. The mechanics of the operation that foster welding are the compressive force or pressure exerted as the composite piece is forced through the extrusion die, and the amount of surface deformation that takes place, which fragments any surface films and diminishes the concentration of nonmetallic compounds along the interfacial surfaces. The process is put to use in specialized operations, and considerable development work is devoted to optimizing each procedure to be followed in material preparation, heating, extrusion, and NDT of the product. Ordinarily, no melting is needed at the interface of a dissimilar metal coextruded product because the intimacy achieved by heat, pressure, and plastic deformation usually is sufficient to produce solid-state welding. If a liquid phase is advantageous because of the nature of the metals being joined, a plating or a layer of dissimilar metal can be included at the interface as a planned step in the operation.

Diffusion welding (DFW), sometimes called "diffusion bonding," and "solid-phase bonding," is listed here under hot pressure welding because essentially it utilizes the solid-state welding method. It is distinguished mainly by minimal deformation of the workpieces as they undergo diffusion welding. This diminution of deformation is achieved by use of relatively low pressure, but, this, in turn, requires clean, smooth, close-fitting faying surfaces, restricted temperature to avoid gross plastic flow, and longer hold-time at temperature to permit creep and diffusion mechanisms to operate and thus bond the surfaces together to a degree that amounts to a weld. Generally, the temperature selected for diffusion welding is set in relation to the melting point of the materials involved. The minimum temperature usually is 0.5 of the lowest melting point represented, and most metals and alloys are best diffusion welded in the temperature range between 0.6 and 0.8 of the melting point in degrees absolute. Time at the diffusion welding temperature is not as critical because although the rates of important reactions occur exponentially with temperature, they progress with the square root of time, and therefore its relevance is less-and-less as the operation progresses. With the wide variety of metals and alloys joined by this process, times at diffusion welding temperature may range from less than a minute to several hours or more. The carbon and low-alloy steels usually require at least ten minutes, but time will depend on intimacy at the interface and surface conditions. Pressure can be quite low when fit-up is good and surface conditions are favorable for welding, and as little as seven MPa (about one ksi) could be adequate under ideal circumstances. The surfaces of the workpieces to be welded need not be mirror-smooth or polished; in fact, some roughness appears to promote welding through the interfacial shear displacement that occurs as the heated surfaces are forced together under pressure. Of course, these surfaces must be very clean prior to assembly. Thin oxide films are not as formidable a barrier to welding as are mobile compounds such as oil left by fingerprints. During heating and time at diffusion temperature, a protective environment of some kind normally is provided. This can range from (1) tightly abutting fit-up that excludes as much air as possible, (2) good fit-up with peripheral edges sealed by an oxidation-resistant compound, (3) envelopment in an air-tight metal packaging that permits application of pressure, (4) a protective gas atmosphere, and (5) a vacuum in the order of 10^{-3} torr. Use of a vacuum, although costly, greatly enhances the viability of the process. Diffusion welding can be used to assemble articles that would be quite difficult to fabricate by other welding processes, and it is not possible to generalize on where the process is likely to be found in use. Development work is vital to the procedure for each application to ensure acceptable welds.

Induction Welding (IW)

This process could have been included among the hot pressure welding processes because, for the most part, its only distinctive feature is the use of electrical resistance for heating the workpieces by low-frequency induced current. Mention was made of this method of heating when discussing pressure gas welding (PGW), a process most often conducted as a fusion welding operation to ensure a weld reasonably free of entrapped oxide. Induction heating is highly regarded for welding because heat can be generated rapidly at the localized joint, and can be precisely controlled, but must allow some soaking time to equalize temperature distribution. The frequency of current used in IW is an important factor in determining the maximum heating rate. Protective or inert atmosphere for shielding the heated areas can be easily arranged. However, the heating rate can be so rapid that shielding from the air may be unnecessary. Inductor coils for heating each pair of parts of a given design are readily fashioned from copper tubing, and a wide variety of machines have been built for holding the parts and applying force to accomplish welding. A good example of the usefulness of IW is the making of butt joints between lengths of small-diameter, thick-wall steel tubing as used in boilers and superheaters.

The time required to make an induction weld is mainly dependent on (1) material, (2) joint design, (3) fit-up and surface condition of workpieces at the joint, (4) amount of pressure at the interface through force applied to the parts, and (5) heating rate and temperature level attained. Low-frequency ac was mentioned as the kind of power employed because the objective is to avoid the skin heating effect induced by high-frequency current. Uniform heating of steel is encouraged by the deeper penetration of low-frequency induced current, and by the fact that heating regularly is carried by temperatures above the Curie point for welding. When the magnetic coupling is lost above the Curie temperature, the early rapid heating of surfaces in close proximity to the inductor will be decreased, whereupon the inner material will have opportunity to catch up and achieve a more uniform temperature in the cross-section of the workpieces. Most induction welding is performed with equipment designed and perfected for a particular repetitive hot pressure welding operation.

Electromagnetic impact welding is a specialized variation of induction welding that has been used to some extent in industry for the welding of lap joints in small components, and for socket joints in tubing. The joint in the workpieces is encircled by a heavier-than-usual inductor coil (because of very high current to be handled in the final stage). High-frequency ac is passed through the inductor to rapidly heat the joint area to a temperature somewhat below the melting point of the material. When this stage is reached in the operation, a bank of capacitors discharges current through the inductor coil to produce a strong electromagnetic inductor pulse for only a few microseconds. The imposed pressure may be in the order of 350 MPa (about 50 ksi), and this produces a highly localized forging action to accomplish welding. This process is employed on relatively small components that are usually circumferentially continuous.

Friction Welding (FRW)

Friction welding is categorized as a solid-state process, but here again is a process where a film of molten metal may be produced between the surfaces to be joined at an intermediate stage of the operation. Although any melted metal will ordinarily be completely displaced from the completed weld, the molten metal during its short existence can perform a helpful function in the course of accomplishing the weld. For many years, the only kind of friction welding practiced in industry utilized the heat generated by *rotating* one workpiece while it was held forcibly in contact with another that was usually stationary, although it could be rotated in the opposite direction. Because of the rotary motion, at least one of the two workpieces has to be circular in cross-section at the joint. Two different forms of this rotational method are being widely used today; one is termed *inertia-drive friction welding*, and the other *direct-drive friction welding*, which is sometimes called "continuous-drive friction welding." Both forms are well suited for repetitive manufacturing operations because the welding can be fully mechanized and operating parameters pre-programmed. When the workpieces are heated to proper temperature by friction at the joint, typical welds are characterized by moderate upset of plastically deformed metal on each side of the joint, and a localized flash that is upset or exuded from the joint interface. Heat-affected zones in the metal on each side of the joint are usually quite narrow. FRW is an attractive process for components of size and shape that can be accommodated by a friction welding machine because no filler metal is required, shielding with a protective gas is not required in most applications involving carbon and low-alloy steels, and welding operators need only limited instruction. Generally, welds made by the rotational forms of friction welding are of good quality, and replication in large-scale production is excellent.

Research in recent years has produced a number of additional friction welding process forms that involve motions other than rotation. These have broadened the application of friction welding to accommodate workpieces of various shapes, and have also made surfacing a viable, unique operation. These new innovations in friction welding will be reviewed after a close examination is made here of the well-established rotational form of the process.

Although friction welding parameters must be tailored for each application to particular workpieces, operating parameters can be selected and coordinated from a fairly wide range of force, rotating speed, and time to make an acceptable weld. The force must be high enough to keep the faying surfaces in contact and to produce frictional heating at the desired rate, which also must be fast enough to minimize oxidation of the faying surfaces. The pressure of compressive force applied to hold the workpieces in contact is determined by the spindle speed, the desired heating time, and whether a specified amount of axial shortening of the two workpieces must have occurred after the weldment is completed. Inadequate force limits heating, and little axial shortening takes place. Excessive force produces rapid heating to a high temperature, tends to form melted metal, and causes rapid axial shortening. Typical basic conditions for welding steel include force in the range of 30 to 60 MPa (about 4350 to 8700 psi), and a peripheral velocity of at least 90 m/min. (about 300 ft/min.). If tangential velocity is below approximately 75 m/min. (250 ft/min.), the torque soon rises to a very high level and difficulties can arise with insecure clamping of the workpieces, nonuniform upsetting, and metal tearing on the faying surfaces.

The quality of a friction weld in steel and its mechanical properties are usually improved by application of a higher forging force at the conclusion of the heating period. For this final forging stage, the force might be increased to about 75 to 150 MPa (roughly 11 to 22 ksi). Typical heating time for relatively small workpiece ranges from five to ten seconds. The time must be sufficient to permit thermal conduction to any central portion of the end faces where there is less motion to generate frictional heating. However, the heating time must not be so long as to allow excessive upsetting and axial shortening, or the development of an abnormal HAZ along the length of the weldment on each side of the weld. Rotating speed is usually the least critical variable, but if speed is excessive, overheating can occur at faying surfaces and this can have a detrimental effect on the mechanical properties of the weld.

Inertia-Drive Friction Welding

The inertia-drive form of rotational friction welding has one workpiece clamped in a chuck connected to a heavy flywheel. The flywheel is accelerated to a rotating speed that has been determined by trial welds, or can be calculated from available formulae. The rotating flywheel, therefore, is a source of stored mechanical energy that is available for conversion to heat. After reaching the required speed. the flywheel is disengaged from the driving motor, and the rotating workpiece is brought to bear against the stationary workpiece under compressive force. As faying surfaces rub together, the flywheel energy is transposed to heat by friction. As the flywheel speed decreases, a stage is reached where less heat is being generated at the interface than is being dissipated by conduction into the workpieces. As the heated ends of the workpieces at the joint begin to upset under the applied force, the torque starts to increase rapidly. If frictional heating has brought the interfacial area to a welding temperature, the rising torque causes rotation of the flywheel to suddenly stop and a solid-state weld is formed. Additional force is often applied during this final stage to increase the amount the upset and flash, which usually improves the quality and the mechanical properties of the weld. The energy made available for a particular welding operation will depend on the size, shape, weight and rotating speed of the flywheel. The flywheel itself and/or its operating speed can be changed to exert control over the size and shape of the weld and its heat-affected zones. Figure 6.59 illustrates the interactive effect of flywheel total energy, force, and flywheel speed on the pattern of heat that develops at the joint interface and the shape of the upset flash at the surface.

Direct-Drive Friction Welding

In this form of rotational friction welding, one workpiece is clamped in a motor-driven chuck that rotates continuously at a constant selected speed during the heating period. Ordinarily, a controlled compressive force is maintained on the joint between two workpieces to cause frictional heating until either a predetermined time has elapsed, or until a preset axial shortening occurs. At this point, the rotating workpiece is made to stop (either by disconnect of a drive clutch, or by application of braking), and a higher force is often applied to increase the upset of the heated metal at the joint.

Direct-drive friction welding has been broadened in its application both to very large size, and to extremely small size sections. As machines were developed for use in these outer reaches, remarkably different operating parameters had to be satisfied to produce satisfactory welds. For example, a machine to weld large diameter shafts and tubes that represented sections as large as 1480 cm^2 (229 in.^{2}) was required to have a force capacity of 20 MN (4.5×10^6 lb f). Conversely, when friction welding was scaled downward to the microfriction welding range, the parameters were shifted to markedly lower levels. Tangential velocity, for example, had to be several times as high as used for average size workpieces (often in the vicinity of 300 m/min., or about 1000 ft/min.). When very small diameter wires were to be butt welded (e.g., two mm; about 0.08 in. diameter round), the rotational speed had to be in the order of 47 000 rpm. Wires smaller than one mm (0.04 in.) diameter round are successfully friction welded, but rotational speed must be increased above 125 000 rpm. To obtain the rotating speeds required for microfriction welding, machines make use of an air turbine to drive the chuck holding the moving workpiece. Compressive force on the micro-size joint often is applied at a selected rising rate and the final forging force might be only about 600 Nf (approximately 135 lb f). The total heating and welding time for accomplishing a microfriction weld is usually less than one second.

Materials Suited for Friction Welding

In general, friction welding is a very satisfactory process for joining steels, and it offers several metallurgical advantages among its attributes. Virtually any steel that can be hot forged can be friction welded, and many dissimilar metals and alloys can be welded to steel. The process allows considerable latitude in steel composition for making welds that are sound, but the steel's analysis has a strong relationship with the mechanical properties of the weld and its HAZ, and actual analysis of the workpieces deserves this attention. If the steel exceeds certain levels of alloying or residual elements, unsuitable hardness, bend ductility, or fracture toughness can develop in the as-welded condition. No formulae are yet well established for evaluating the suitability

ENERGY

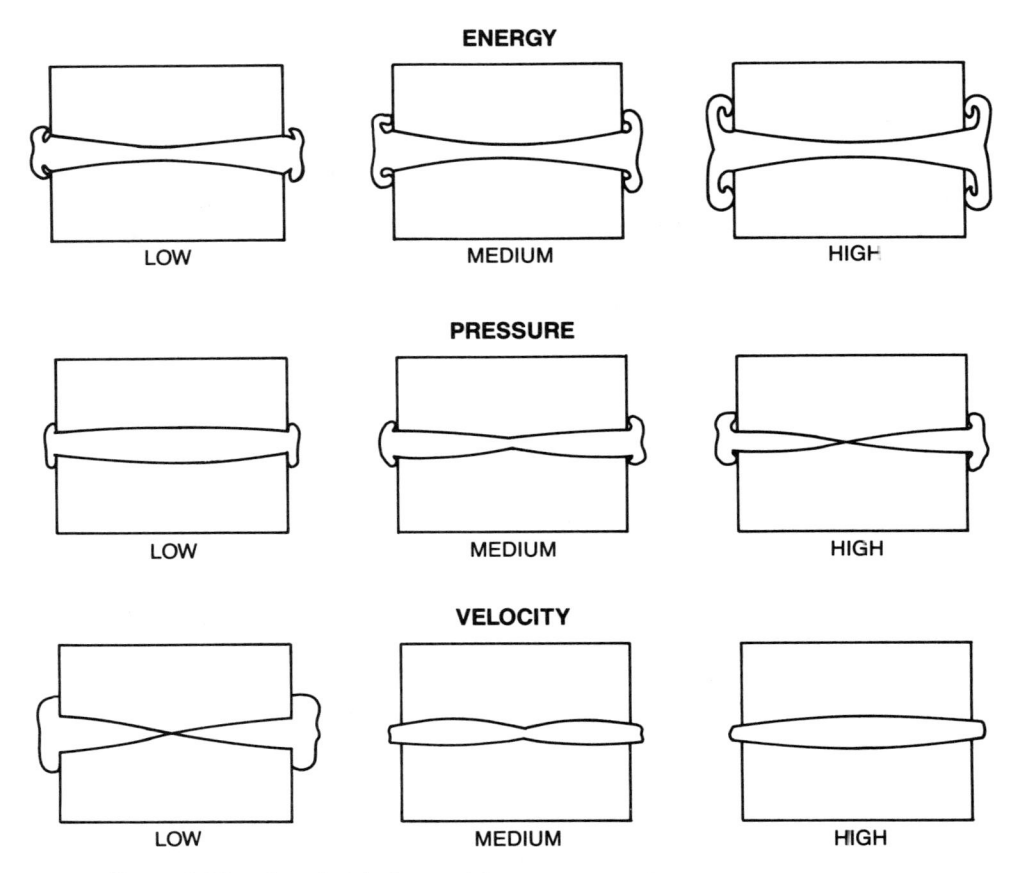

| LOW | MEDIUM | HIGH |

PRESSURE

| LOW | MEDIUM | HIGH |

VELOCITY

| LOW | MEDIUM | HIGH |

Figure 6.59 — Inertia friction welding patterns of heat and upset or flash at the weld produced by variations in energy, pressure and velocity

(Reprinted from *AWS Welding Handbook, Vol. 2, Welding Processes* Eighth Edition, Chapter 23, Friction Welding, Figure 23.10)

of steel compositions for friction welding in terms of their potential weld strength, hardness, ductility, or toughness.

Carbon Content and FRW Suitability

Carbon content is the first element that requires consideration because of its profound influence on hardness and toughness. Plain steel and low-alloy steels containing as much as 0.6 percent carbon can be friction welded without great danger of cold cracking, but in the as-welded condition steels containing more than about 0.2 percent carbon will tend to form welds and heat-affected zones that have high hardness, poor bend ductility, and low fracture toughness. Significant amounts of alloying elements in the composition that promote hardenability would be expected to aggravate poor mechanical properties.

Nonmetallic inclusions can be a determining factor in a genuine appraisal of the suitability of a steel for FRW. The free-machining steels, and wrought iron, because of the many inclusions regularly present in these materials should be welded with great caution. Redistribution of the inclusions during plastic deformation at the weld creates planes of weakness like the "upturned fiber" condition described earlier for flash welds and illustrated in Figure 6.33. When the residual sulfur content of most other types of steel exceeds approximately 0.025 percent, there is a strong likelihood that toughness will be adversely affected by an inclusion-dominant mechanism during fracture propagation.

Research conducted with carbon and low-alloy steels showed that steelmaking practice had a strong influence on the ductility and toughness of friction welds made by both inertia-drive and the direct-drive forms of the FRW process. Inclusion content and morphology as controlled by steelmelting process, deoxidation practice, and special finishing treatments were clearly established as major influential factors. As would be expected, a reduction in number and size of nonmetallic inclusions in the steel workpieces gave better ductility in bend tests, and greater energy absorption values in Charpy-V impact tests. This work found that steels produced by electroslag remelting, and those treated with calcium for inclusion control produced friction welds that had much better ductility and toughness.

FRW Suitability for Dissimilar-Metal Welds

The capability of the friction welding process for making welds between dissimilar metals is being put to wide use in industry. Some of the metals and alloys being welded to steel by FRW represent combinations that are metallurgically incompatible when attempting to use processes that rely on fusion between the dissimilar workpieces. The FRW process is outstanding in its success on dissimilar metal welds because melting of the faying surfaces is not necessary, and even if some molten metal does form it is likely to be eliminated from the joint during application of forging force. Of course, attention should be given to the FRW parameters employed to ensure minimal mechanical mixing and/or diffusion at the interface between incompatible materials. Steel friction welded to aluminum gives a good demonstration of the capability of FRW to make acceptable dissimilar metal welds. Other welding processes that allow these two metals to commingle during fusion, or by diffusion encounter difficulty with the Al-Fe intermetallic compound that forms because of its extreme brittleness.

The capability of the FRW process to weld incompatible dissimilar materials also has been extended by the development of modified procedures for workpiece preparation. In the main, these methods require one of the workpieces to be fitted with an intermediate layer of a third material that is compatible with both of the materials in the members to be joined. The intermediate layer serves as a buffer or interlayer. After a "three element" weld is completed, the intermediate buffer remains as a thin metallic bonding between the incompatible metals in the weldment. The buffer or barrier layers can be applied by electroplating, or fitted as a solid wafer to the end face of one of the workpieces. Buffers of chromium, nickel, and silver have been used successfully in dissimilar metal welds with steel. The buffer metal selected is determined by the requirement that it must be compatible with both the steel and the other metal incorporated in the weldment.

When hardenable steels are friction welded, preheating the workpieces can be an effective measure for reducing hardness in the as-welded condition, but preheating is seldom a viable remedy for low ductility, or poor toughness. Postweld heat treatment is a more reliable and effective means of achieving acceptable weld mechanical properties. Since most friction welded assemblies are relatively small and can be conveniently processed with widely available facilities, a postweld heat treatment that is tailored for a given steel type can be performed to regain the kind of microstructure throughout the weldment that provides the required strength, hardness, and toughness. In some cases, a postweld heat treatment applied locally by induction heating makes the weld and HAZ acceptable at a lower cost.

Welds made by FRW, even when in steels that are quite hardenable, show little susceptibility to cold cracking. This infrequency of cracking occurrence can be credited mainly to two situations that will ordinarily prevail with the FRW process: (1) hydrogen is unlikely to be increased in the weld area by pick-up from a hydrogenous source during welding; and (2) solidified weld metal, which often has a microstructure that is sensitive to cold cracking, will not be present because if formed during welding it will be expelled along with the normal flash. Yet, it has been reported that some steels, both wrought and cast, may contain sufficient residual hydrogen to cause hydrogen-induced cold cracking in heavy section, high strength friction welded components. In one case of unexpected difficulty, hydrogen cracks initiated in the center of the weld and propagated radially outward. Cracks also were found in the HAZ away from the weld bond line. This hydrogen cold cracking appeared to developed after an incubation period of about six weeks, and the cracks grew under the driving force of tensile residual stresses (i.e., no loads had been applied). The nature and position of the cracks required careful application of NDT techniques for detection.

Friction welds usually have much narrower weld zones compared with electrical resistance flash or upset butt welds because the conversion of mechanical energy to heat takes place at the interfacial rubbing surfaces, and heating of adjacent metal takes place entirely by conduction. Shorter welding time, of course, produces narrower weld areas. A typical narrow weld in a C-Mn steel is illustrated in Figure 6.60. The absence of a central fin between the upset flash from the two workpieces indicates an absence of melting during this particular FRW operation (as contrasted with the configuration sketched for the flash weld in Figure 6.33).

Mechanical Properties of Friction Welds

The mechanical properties of friction welds in steels have not been studied or reported as extensively as for other long-established processes. In general, the strength of good quality rotational friction welds is dependent on the steel's composition and the microstructure that forms as dictated by the cooling rate after welding. Ordinarily, strength gives no surprise; however, weld toughness (as evaluated by Charpy-V notch impact testing) sometimes gives cause for concern. Instrumentation for charting the principal variables when making rotational friction welds is being put to increasing use to study interplay between functions to determine the influence on mechanical properties. Figure 6.61 charts principal operating parameters for both the inertia-drive and the direct-drive forms of rotational friction welding to show approximately how these parameters are varied from the start of a weld to its completion. More sophisticated instrumentation is now available to acquire detailed data in real-time during the welding operation, including the amount of torque, energy consumption rate, and

> **TECHNICAL BRIEF 37:**
> **Friction Welding in Undersea Applications:**
>
> A note will be included here regarding an unusual application of friction welding in which hydrogen and hardening could be troublesome, but are handled adeptly in carrying out the rotational form of the process. A portable direct-drive type machine is being used to attach steel studs when needed to the subsea members of existing offshore marine structures. While friction welding avoids any problem with hydrogen pick-up in this underwater operation, accelerated cooling of the heated weld area by the surrounding seawater could result in unacceptable hardening in the weld and HAZ. The remedial technique successfully used is to fit a shielding collar of plastic foam on the extreme end of the stud before welding it to the steel structural member. Thermal insulation provided by the thick collar to the weld and adjacent heated areas is sufficient to block quenching action by the seawater.

deceleration rate. Computerized acquisition of data in real-time allows immediate analysis and microprocessor control of the welding operation to maintain the critical functions within limits established for the particular workpieces; in this way, each weld's configuration and its mechanical properties are optimized.

Data gathered by instrumentation of the direct-drive form of rotational friction welding has established some rudimentary guidelines for making steel welds that have improved ductility and toughness. Impact energy absorption values can be improved by welding with lower rotating speed and reduced deceleration from time of clutch release to complete stoppage. This refinement in operating technique appears to cause a slight hot working effect that, in turn, causes any austenitized microstructure to transform on cooling to a slightly finer microstructure; consequently, toughness is improved.

Another innovation in FRW technique that has been successful in improving weld toughness in selected steels is called "post rotational twist." Work to date shows that twisting a freshly made weld exerts a small amount of plastic work in the areas of the weldment that have been austenitized by the heat of welding. This micro-deformation of the austenite encourages transformation to a finer acicular ferrite structure on cooling. The extent of toughness improvement with this kind of microstructure can be enough to bring Charpy-V notch values above the minimum specified for a given type of hot rolled steel, but the steel being welded-and-twisted must have a composition that is favorable for friction welding and is amenable to toughness improvement via grain refinement. The amount of twisting need to make an improvement in toughness is a minimum of about 15 degrees, and this must be precisely executed immediately after arrest of the rotating workpiece as the weld is completed and while an austenitic microstructure still exists in the hottest area of the weldment. A delay of more than a couple of seconds prior to twisting will make the action ineffective, and in some steels might cause interfacial microfissures to form in the weld. Twisting as much as 85 degrees has been carried out without difficulty in studies of post rotational twist.

Other Forms of Friction Welding

As the virtues of solid-state welds made by friction welding became widely recognized, laboratories that developed friction welding machines started to experiment with various forms of sliding motion to generate frictional heating for welding. This effort was directed toward applications where the physical shape or length of workpieces could not be accommodated by conventional inertia-drive or direct-drive forms of friction welding. This work has already probed the feasibility of using a half-dozen different motions, which for brevity can be termed as (1) radial, (2) orbital, (3) sur-

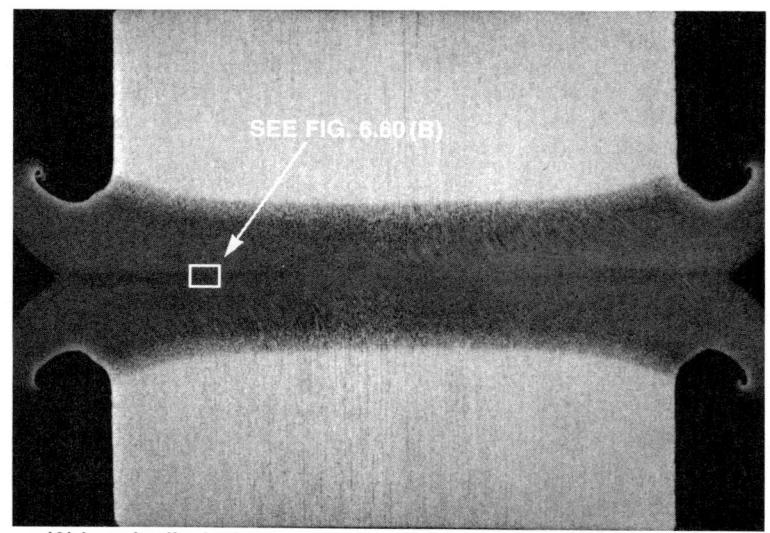

(A) Longitudinal macrosection through bar with weld area at center

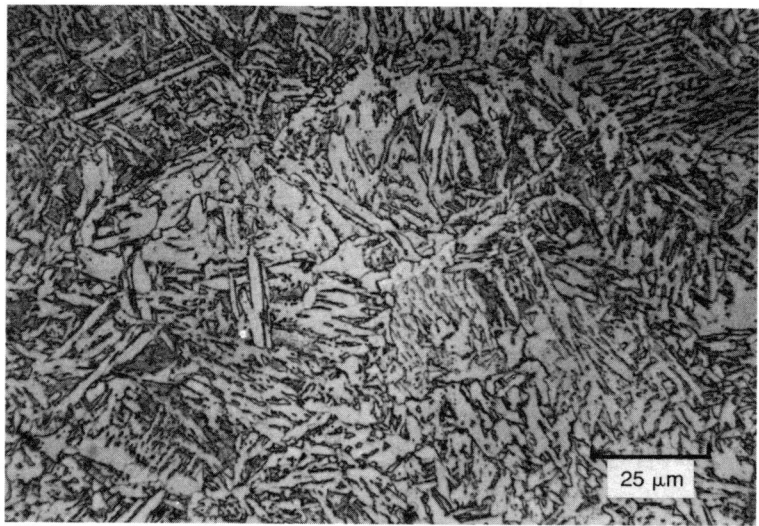

(B) Photomicrograph of weld microstructure in area indicated above on macrosection specimen. Original magnification 800X

Figure 6.60 — Friction weld made by continuous drive technique in butt joint of 25 mm (1 in.) diameter round bar of steel containing 0.14% carbon

(Courtesy of The Welding Institute)

facing, (4) linear reciprocating, (5) co-extrusion, and (6) hydropillar. Although process development with these motions has been slow because of practical problems with machine design and operation, one new process has been put to commercial use; namely, "radial friction welding". Several others are close to completion of operational

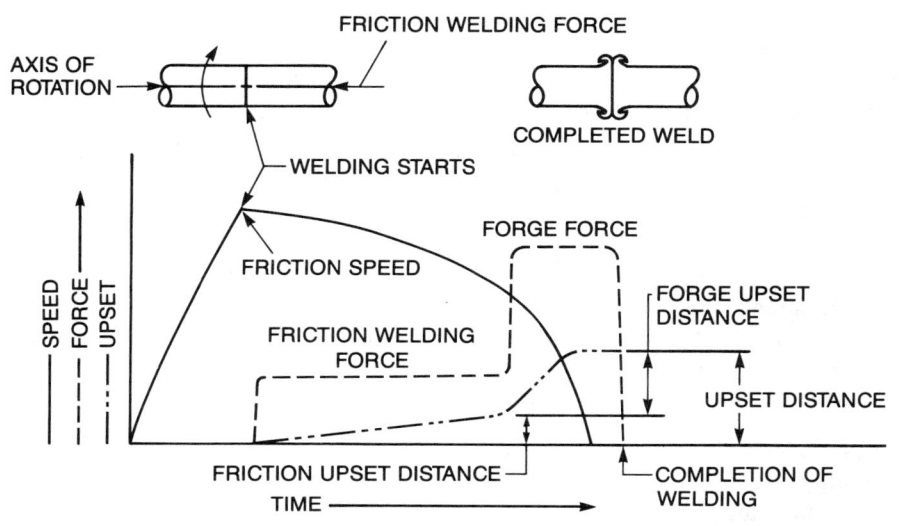

(A) GENERALIZED DIAGRAM OF INERTIA FRICTION WELDING.

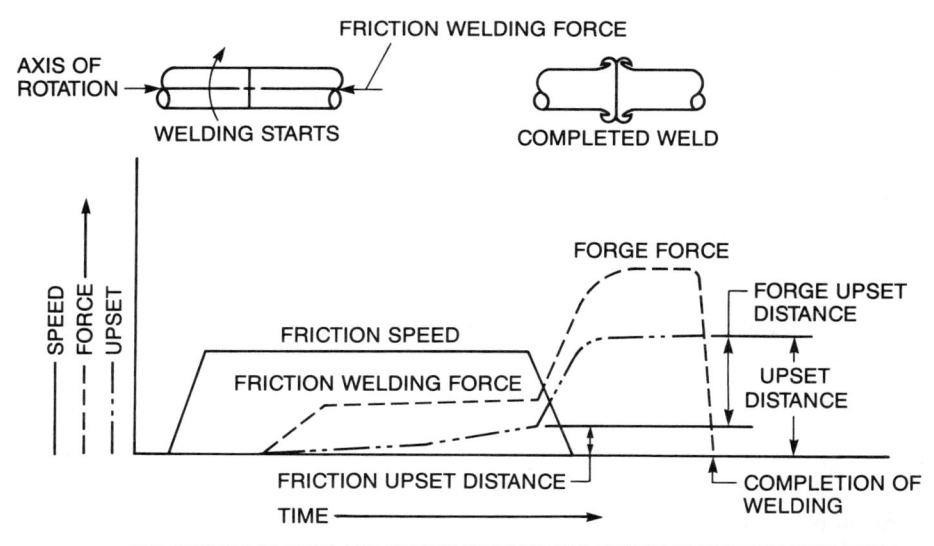

(B) GENERALIZED DIAGRAM OF DIRECT DRIVE FRICTION WELDING.

Figure 6.61 — Principal operating parameters in rotating friction welding charted to show variations during successive stages

(Reprinted from ANSI/AWS A3.0-89, *Standard Welding Terms and Definitions*, Figures 35 and 36)

Radial Friction Welding

For butt joints in steel tubes or bars, this method uses consumable steel rings to serve as the joining member as shown in Figure 6.62. By rotating only the ring, it is not necessary to rotate either of the workpieces to joined. After the ring has been rotated to bring itself and the prepared surfaces of the workpieces with which it is in contact to welding temperature, high compressive force is applied radially to the ring to cause this metal to friction weld to both workpieces. The ring can be forged into the grooved butt joint from the outside as shown in Figure 6.33, or the ring can be positioned and rotated on the inside of a butt joint in pipe or tubing and thus require radial expansion. This FRW technique can be advantageous when joining long lengths of stock since neither of the lengths need to be rotated at the high speed required for friction welding. If the lengths of workpieces are not straight, or have been pre-formed to a specified shape along their length, it still may be possible to insert and position an internal expandable plug in tubular workpieces to resist the radial compressive force during the operation. Among the many possible applications of radial friction welding would be the attachment of a projecting collar to the surface of a round shaft, as illustrated by the longitudinal sectioned sample weld in Figure 6.62 (C).

Surfacing by Friction Welding

Surfacing by friction welding is now under active study because it expands the realm of metals, alloys, and even metal-matrix composites that can be applied as an unadulterated layer of overlay on a substrate. This is a distinct metallurgical advantage over other surfacing processes that involve fusion where some dilution of the deposited material with melted base metal cannot be completely prevented. Friction surfacing avoids problems that can arise from melting a consumable in air, with the complex mechanics of solidification of deposits, with hydrogen pick-up and cold cracking, and with unfavorable surface conditions associated with beads and layers deposited by a melting operation. Friction surfacing is performed as a mechanized operation using a purpose-built machine, or by modifying a machine tool, such as a milling machine.

Friction surfacing employs a rotating consumable, which can be a solid round metal bar or a metal tube filled with selected powdered material. The free end of the consumable is brought to bear under pressure on the surface of a substrate (workpiece). Because the rotating consumable is smaller in cross-section, its temperature at the working end rises faster from frictional heating than does the surface of the workpiece; thus the workpiece serves as a heat sink and minimizes the size of a HAZ. As the heated end of the rotating consumable becomes plasticized a bulbous upset tends to form, and as the consumable traverses the substrate under pressure it deposits a layer of metal that is soundly welded to the workpiece's surface. There is virtually no penetration of the overlay into the surface of the workpiece, and the solid-state weld at the interface entails virtually no interalloying.

Figure 6.63 illustrates the essentials of the friction surfacing technique, and shows a machine in operation. Operating guidelines and parameters for this process are being investigated and catalogued. The consumable in Figure 6.63 is a 25 mm (1 in.) diameter round bar of mild steel which is depositing a layer about two mm (0.08 in.) thick at a rate of approximately 4.5 kg/hr (roughly ten pounds per hour) by rotating at 975 rpm under an axial compressive force of 28 kN (about 6300 lb f) at a travel rate of 4.9 mm/s (about 12 ipm). Higher rotating speed tends to give a thinner deposit, but acts also to improve the quality of the weld along the interface with the workpiece. A lower axial force increases the deposit thickness, but tends to reduce the effective bond width.

A problem can be encountered with the friction-deposited layer as illustrated in Figure 6.64-A (A) and (B). A lack of bonding can occur along each edge of the

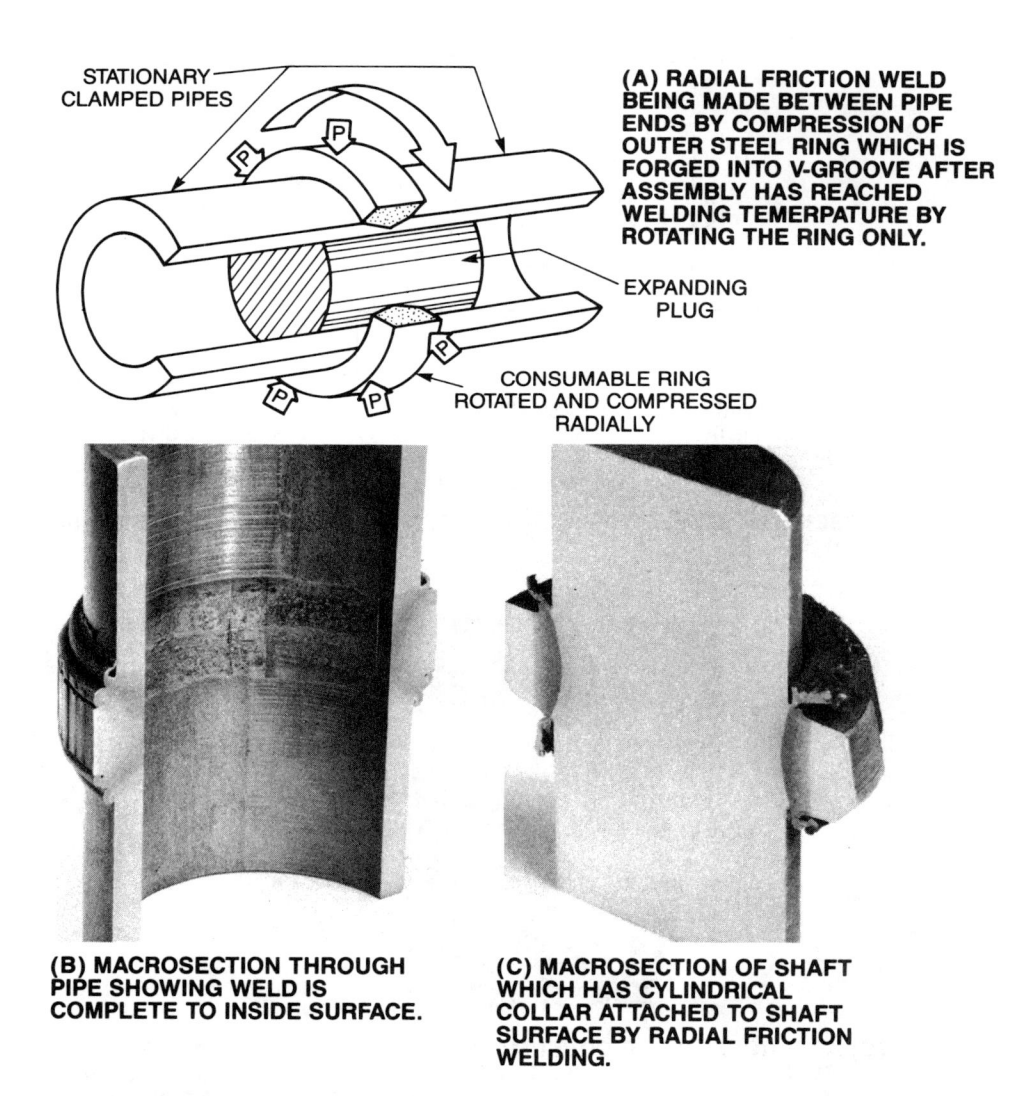

(A) RADIAL FRICTION WELD BEING MADE BETWEEN PIPE ENDS BY COMPRESSION OF OUTER STEEL RING WHICH IS FORGED INTO V-GROOVE AFTER ASSEMBLY HAS REACHED WELDING TEMERPATURE BY ROTATING THE RING ONLY.

(B) MACROSECTION THROUGH PIPE SHOWING WELD IS COMPLETE TO INSIDE SURFACE.

(C) MACROSECTION OF SHAFT WHICH HAS CYLINDRICAL COLLAR ATTACHED TO SHAFT SURFACE BY RADIAL FRICTION WELDING.

Figure 6.62 — Radial friction welding

(A) illustrates essentials of this technique in one possible arrangement for radial friction butt weld in steel pipe. (B) is longitudinal specimen etched to show cross-section of weld. (C) illustrates use of radial friction welding to join a collar to the surface of a solid shaft. (Courtesy of The Welding Institute, England)

deposited layer a short distance inward from the toes. This condition reduces the effective width of the bonded layer to less than the diameter of the rotating consumable. This problem arises because there is a reduction in bonding force at the free edges of the layer inasmuch as the rotating consumable cannot transmit the needed pressure through the plasticized metal that is extruded from beneath its peripheral edge. Along the middle of the interface between surface layer and workpiece, the weld is complete and the bond appears free of discontinuities as can be seen in Figure 6.64-A (C).

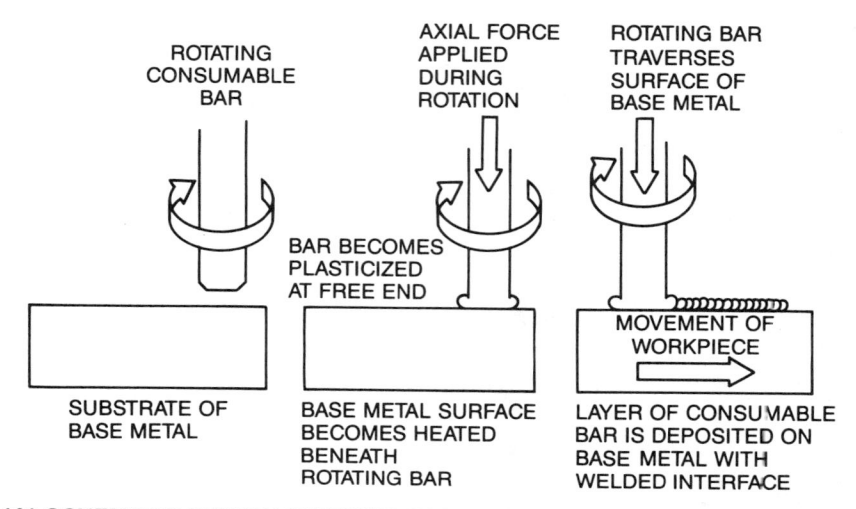

(A) SCHEMATIC SKETCH SHOWING ESSENTIALS OF FRICTION SURFACING.

(B) ARRANGEMENT OF CONSUMABLE (VERTICAL BAR) AND SUBSTRATE (PLATE BOLTED IN HORIZONTAL POSITION BENEATH BAR) SHOWING HOT PLASTICIZED BAR END AND DEPOSIT LEFT ON BASE METAL SURFACE WITH TRAVEL OF OPERATION.

Figure 6.63 — Surfacing by friction welding

(Courtesy of The Welding Institute, England)

Metallographic examination at higher magnification disclosed slight differences in microstructure from place-to-place in the surface layer which apparently are caused by variations in actual temperature, force and deformation as deposit is kneaded by the rotating consumable onto the workpiece's surface. Figure 6.64-B (D) shows a somewhat finer microstructure along the surface of the deposited layer to a depth of about 65 μm (about 0.0026 in.). Examination along the interface at the high magnification confirmed that the middle portion is free of any discontinuity, as shown in the photomicrograph of Figure 6.64-B (E). A number of remedies are presently being studied to avoid the small unbonded intrusions at the toes of the deposited layer.

Orbital Friction Welding

During orbital friction welding, workpieces are brought together in abutting position under compressive force and one is repeatedly moved in a prescribed orbital path that produces uniform frictional heating across the faying surfaces that eventually will become welded. The workpieces may be non-circular and the orbital motion will be one that equalizes sliding speed over the surface areas to be welded together. A machine can be used that rotates both workpieces in the same direction and at the same speed, but by having the rotational axis of one piece initially offset with respect to the other the relative motion between faying surfaces will generate frictional heating. During deceleration to the instant of welding, the offset workpiece can be returned to a common axis for the two workpieces to satisfy alignment requirements.

Linear Reciprocating Friction Welding

Friction welding with a reciprocating motion of small amplitude and with suitable frequency to develop sufficient frictional heat for welding has been studied, and the operating conditions needed for making welds in steel have been identified. This kind of motion has been pursued as basis for a FRW process because of its potential to accommodate non-circular workpieces. A machine has demonstrated feasibility by making welds in mild steel non-round members with cross-sections as large as 1000 mm^2 (about 1½ sq in.) in approximately five seconds.

In development work with linear reciprocating motion, a sinusoidal movement was employed (i.e., zero to maximum velocity in one direction, which is then repeated in the opposite direction). Examination of welds indicated that the sliding speed between the faying surfaces of workpieces should be as high, or even higher than used in rotational friction welding to ensure uniformity of heating and stable metal flow. The minimum sliding speed appeared to be approximately 1 m/sec (about 40 in./sec). The motion frequency and amplitude depended on the nature of the workpieces, but typical linear reciprocating motion might include an amplitude of ± 2 mm (0.08 in.) and a repeating cycle of 40 Hz. Compressive force applied to the area to be welded would be dependent on the time desired for the complete operation. The procedure for setting operating parameters usually involved first selecting a welding time, and then determining the necessary force for the particular reciprocating sliding motion to meet the time criterion. Completed welds have protruding flash around their perimeter. In addition to displaying excellent weld quality, the two workpieces could be stopped in precise angular alignment at the moment of solid-state welding.

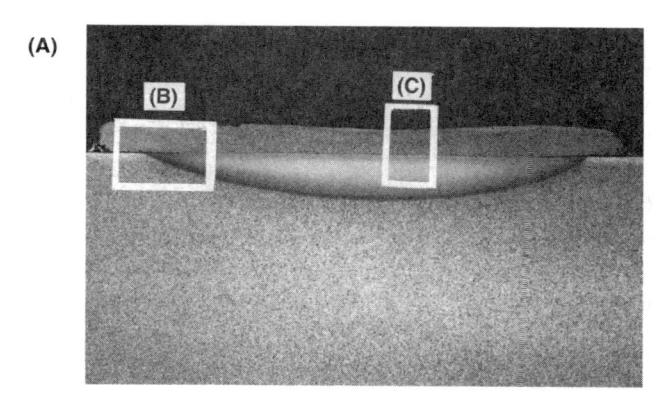

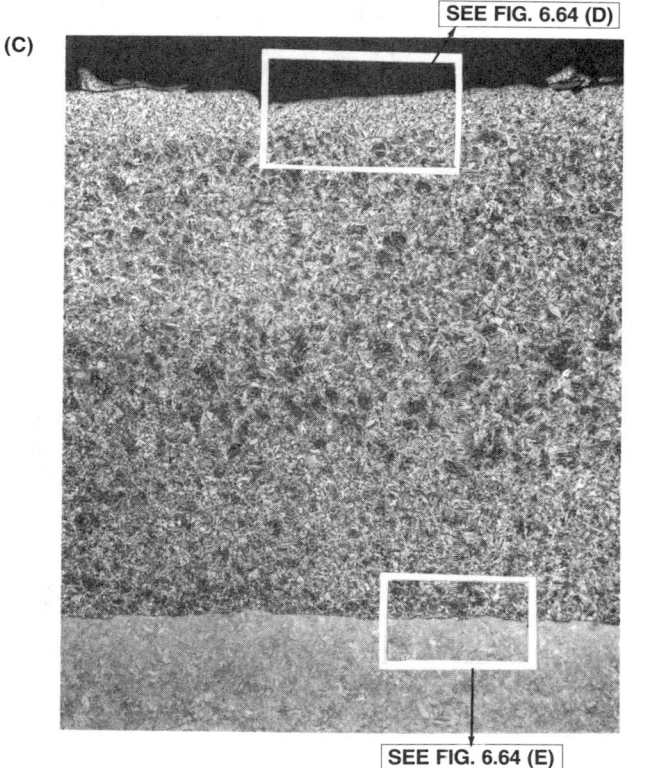

Figure 6.64-A — Friction surfaced deposit sectioned and examined metallographically

(A) is photomicrograph showing deposit 1.6 mm (1/16 in.) thick and 19 mm (3/4 in.) wide of mild steel on a Ni-Cr-Mo alloy steel containing 0.4% carbon. Two boxed areas have been selected for examination at higher magnification. (B) shows the terminus of an unbonded edge at a magnification of 70X. (C) shows the microstructure at 70X of the entire thickness of deposit, and two areas (one at the deposit surface, and the other at the deposit-base metal interface) boxed for examination at a magnification of 500X in Figure 6.64-B. (Courtesy of The Welding Institute, England)

Co-Extrusion Friction Welding

Extrusion of a cold section through a die that causes a very high shear strain by reducing the cross-section will generate considerable heat by friction between the metal section and the die. The temperature of the die can be regulated as desired by a cooling system. Heat in the metal section can be regulated by the design of the die, the amount of reduction imposed on the metal section, and the rate at which the section passes through the die. The application of a layer of dissimilar metal on the workpiece section, or even multiple layers, and performing the reduction operation through the die can find a solid-state weld formed between the layer(s) and the workpiece. In addition to producing the elevated temperature needed to produce the co-extrusion friction weld, the reduction operation through the die tends to rupture oxide films that may be on the faying surfaces during the severe plastic deformation. The break-up of oxide films also assists in accomplishing a solid-state weld, and this action sometimes can be accentuated by an axial differential in feeding rates when clad layer(s) are supplied as a separate sheath of metal to the workpieces as the assembly enters the die.

Hydropillar Friction Welding

This newly proposed friction welding technique is presently under development for welding a variety of joints, including those of some length between abutting plates and circumferential joints between cylindrical sections. The hydropillar process involves forcing a rapidly rotating round bar vertically into a narrow gap in a square-groove butt joint. The bar continues to rotate against the opposing edges of the workpieces, and the surfaces of both the round bar and the workpieces are quickly heated to the molten state. This generation of molten metal produces hydrostatic pressure against all surfaces including on top of the pillar of molten metal (hence the name,"hydropillar"). The rotating bar can remain fixed at a location to make a plug weld by feeding the bar into a cavity, or the bar can be moved along a joint and fed to make a continuous weld.

The hydropillar process deserves investigation and development because of advantages already proved by rotational friction welding, and the potential for making friction welds in thick plate at low cost. The process has demonstrated capability to enter holes in steel plate 100 mm (4 in.) deep and fill them with molten metal in 20 seconds. Also, it appears feasible to weld plate as thick as 300 mm (12 in.) with dissimilar filler metal and have little interalloying at the weld workpiece interface. Hydropillar welds in plate are likely to cause minimal distortion in a weldment because of absence of angularity in the final weld zone, and little or no need for multilayer filler metal deposition.

Explosion Welding (EXW)

Engineers engaged in metalworking long have been intrigued with explosives as a concentrated form of energy that could be utilized through mechanical action to accomplish objectives. Explosives are relatively low in cost for the energy provided, and they can supply almost any quantity of energy as required for a given operation. Probably the first practical application of explosive energy for welding was the "explosive stud," which is a metal member designed to be driven by a cartridge through the barrel of a special tool into a metal surface. The imbedded stud has high holding power as if welded in place, although no detailed examinations have been reported of interfacial bonding or welding observed between the driven stud and the base metal in which it is imbedded. Further examples of welding that can occur between an imbedded projectile and a base metal are found when examining armor

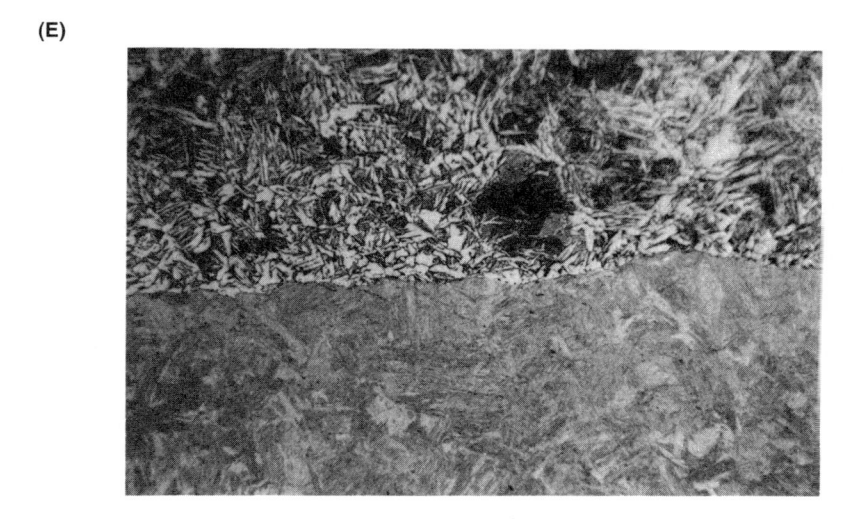

(D)

(E)

Figure 6.64-B — Friction surfaced deposit examined metallographically in boxed areas of Figure 6.64-A for variations in microstructure

(D) is boxed area at surface of deposit where the microstructure is somewhat finer. (E) is boxed area straddling the deposit-base metal interface. Operating parameters for this friction surfaced deposit using 25 mm (1 in.) diameter round bar were 975 rpm, 28 kN axial force, and 4.9 mm/s (12 in./min) travel speed. (Nital etchant; original magnification: 100X; courtesy of The Welding Institute, England)

piercing projectiles (non-explosive type) that have partially penetrated a steel plate, and of bomb fragments that have struck and penetrated a metal surface to some depth.

The explosion welding process has been used to produce unique varieties of steel plate, sheet and strip that are clad with dissimilar metals and alloys. The uniqueness of the explosion-welded clad steel products is in the character of the weld that forms at the interface between metal cladding and the steel substrate, and the minimal mechanical mixing and diffusion of dissimilar compositions that occur at the weld. While explosion welding might be viewed simply as cold pressure welding, unusual mechanics are brought into play during its operation.

Figure 6.65 schematically illustrates the basics of explosion welding; here a high-alloy steel plate is being welded onto the top surface of a heavier plain steel plate. The prime plate or flyer plate, which will serve as a corrosion resistant cladding, can be held parallel to the base plate or target plate with a small standoff as shown, or the prime plate can be arranged at a small angle (possibly five to 15°) with the vertex at the starting end where the explosive layer is detonated. The angular arrangement sel-

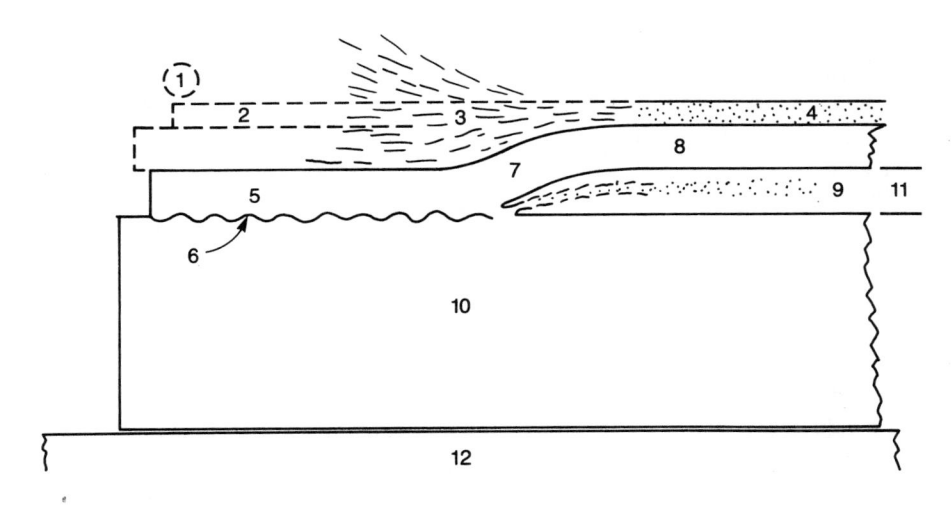

LEGEND:

1. DETONATOR
2. STARTING POSITION OF LAYER OF EXPLOSIVE
3. AREA OF HIGH PRESSURE FROM EXPLOSION
4. LAYER OF STILL-SOLID EXPLOSIVE
5. PRIME OR FLYER PLATE WELDED TO BASE PLATE
6. WAVY INTERFACE OF WELD BETWEEN CLADDING AND BASE PLATE
7. PRIME PLATE UNDERGOING COLLISION WITH BASE PLATE UNDER EXPLOSIVE VELOCITY AND FORCE
8. PRIME PLATE IN STANDOFF POSITION; USUALLY ABOUT 1.5mm (0.06 in.)
9. HYDRODYNAMICALLY JETTED STREAM OF METAL PARTICLES
10. BASE OR TARGET PLATE OF STEEL
11. STANDOFF DISTANCE
12. BACKING OR BED TO SUPPORT BASE PLATE DURING EXPLOSION

Figure 6.65 — Explosion welding process (EXW) as employed for applying a thin plate of dissimilar metal to the surface of a thicker steel plate

Essential details of EXW are shown schematically, and most dimensions are out-of-proportion for easier visualization.

dom is used because long plates elevate the finishing end to a standoff that is too high, and, as a consequence, the collision of prime plate with the base plate may be so violent that cracking and spalling can occur. An essential feature of explosive welding is that the detonation of the explosive must not occur at once over it entirety, but instead it must initiate across the width under the detonator and progress uniformly and steadily along the length of the assembly, albeit at an extremely rapid rate. The detonation velocity will depend on the kind of explosive employed and the amount of inert material included in the mixture. Ammonium nitrate is considered a slow speed explosive, but its detonation velocity can range from 1500 to 5000 m/s (about 5000 to 16 000 ft/s). Trinitrotoluene (TNT) is a high velocity explosive and can detonate as fast as 7500 m/s (about 25 000 ft/s). The advancing implosive force on the prime plate progressively deflects this member downward at a velocity that can be in the order of 150 to 300 m/s (about 500 to 1000 ft/s), and the pressure that occurs momentarily at the interface has been estimated to reach as high as 7×10^5 MPa (approximately a million psi). The pressure generated is directly proportional to the density of the explosive, and the square of the detonation velocity.

When detonation velocity and standoff distance are properly controlled, a highly unusual phenomenon, *jetting*, occurs at the apex of the collision angle between the two plates. The surfaces of each plate are deformed so quickly that they are heated and forced to flow hydrodynamically, and in doing so they issue a spray of metal from the apex (as shown schematically in Figure 6.65 as detail No. 9). The welding capability of EXW is largely related to this jet formation. There is a minimum detonation velocity and a minimum collision angle below which jetting does not occur, and in the absence of jetting proper welding along the interface becomes less certain, although not unachievable. Several reasons can be given for the efficacy of welding under jetting conditions. During the mechanism of deformation-plasticizing-jetting, the colliding plates are swept free of surface oxides because their surfaces are the first metal to be expelled as jetted material. As plasticized metal commences to flow at the interface, it finds the collision region somewhat of an obstacle because compressive force on this region is higher than the dynamic yield strength of the metal. This localized impedance to hydrodynamic flow causes the movement of plasticized metal to become turbulent. However, the metal flow is rhythmic, and it registers in the weld as a repeated wave-like pattern along the weld interface. The size of the waves and their wavelength varies, of course, with operating parameters, but the wave patterns along the interface usually have amplitudes in the range of 0.1 to four mm (0.004 to 0.16 in.) and wavelengths of 0.25 to 5 mm (0.01 to 0.2 in.). The microstructure of this wavy weld zone ordinarily has no features indicating that any melting has occurred prior to the faying surfaces coming together and welding. Yet, the frozen evidence of the wavy interface suggests that the surface temperatures reached during deformation and collision is fairly close to the melting point. Analyses of jetted particles from the joint show that both the prime plate and the base plate contributed metal to their composition.

No real evidence of melting of the workpieces ordinarily is observed in the weld zone. Because the mass of cold plates rapidly quenches the weld zone, no significant diffusion appears at the interface between dissimilar metals. The virtual absence of inter-alloying by melting or diffusion at the weld interface allows metals to be applied to steel that are considered incompatible by fusion welding. Although microstructures differ greatly in the weld zone because of the wide variety of dissimilar metals and alloys joined by EXW, two particular features are commonly observed. Figure 6.66 has been sketched to first show the wavy pattern that commonly exists along the interface. When the explosive energy applied to the assembly for welding is substantially greater than required, or the detonation proceeds at an excessive rate, there is a tendency for the waves along the interface to form a curl on their crest. This feature is also included in Figure 6.66. When the waves are curled over at the crest, it is not unusual to find that some jetted material has been entrapped in small pockets beneath the crests. If the pockets are small and not numerous, they usually do not significantly degrade mechanical properties of the weld. Of course, in addition to any unsoundness created by the pockets, any jetted particles imbedded in the microstructure can detract from weld toughness if they represent an especially brittle inter-alloyed composition. When welding aluminum to steel, or titanium to steel, jetted particles are likely to represent extremely brittle intermetallic compounds, and every effort should be made to minimize their entrapment. Fortunately, these brittle compounds ordinarily do not form by diffusion at the interface of EXW welds, and the sound weld existing between any pockets normally has good toughness. In fact, experience with the EXW process has been so favorable that steel plate clad with aluminum has been cut into long slices of appropriate thickness to serve as transition sections when a component of aluminum must be welded to one of steel by a fusion welding process.

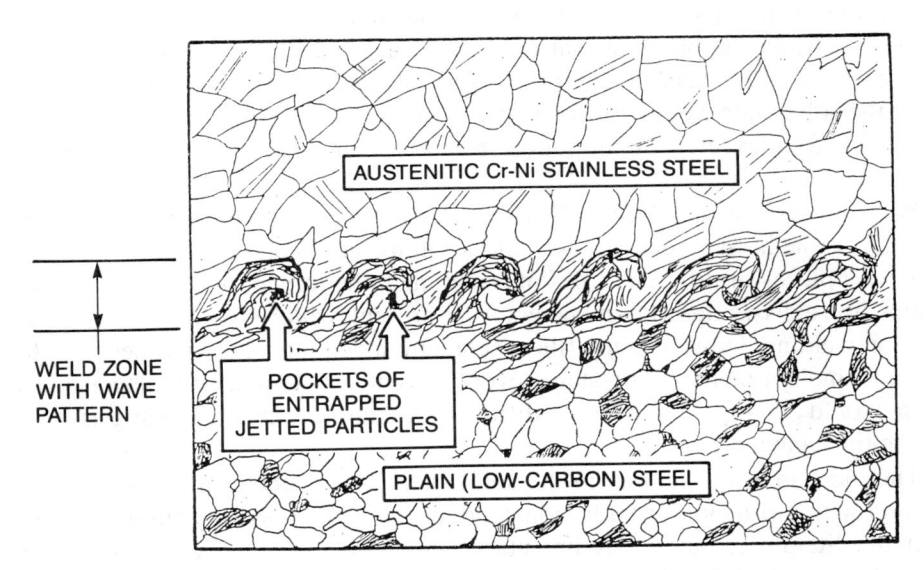

Figure 6.66 — Schematic of an explosion weld (EXW) between an austenitic Cr-Ni stainless steel cladding and a plain steel base plate

Microstructural details of the EXW bond have been sketched as would be observed at a magnification of about 50X to show typical wave pattern along the cladding-base metal interface. The curled crests along the wavy interface are shown to contain occasional small entrapped particles.

Testing of welds made by the EXW process often requires special forms of mechanical test specimens because a principal aim is to establish the quality of bonding along the interface. ASTM Standard A 263 gives guidance on tension and shear specimens that can be employed, but a number of specimens of novel design are proposed in the literature that can be very informative for given applications. The "chisel test" commonly is used for qualitative evaluation when establishing EXW operating parameters. Mechanical testing often is supplemented by metallogaphic examination. The metallographic specimen should present a plane parallel to the direction of detonation and perpendicular to the surfaces of the weldment. On this plane, the wave pattern at the interface can be observed. A regular, well-formed wave pattern is indicative of a good weld.

Diffusion Welding (DFW)

A number of processes used in industry have acquired names such as solid-state bonding, hot press bonding, and pressure welding, but these are shop-language synonyms for the term diffusion welding adopted as standard by AWS. This process, as defined by AWS, produces a weld "by the application of pressure at elevated temperature with no macroscopic deformation or relative motion of the workpieces. A filler metal may be inserted between the faying surfaces."

The AWS definition could have included the stipulation that no melting should occur during the process; i.e., the weld must be accomplished by solid-state phenomena. To have "diffusion" take place in the usual concept of this metallurgical term, a significant dissimilarity should exist in the composition of materials at the joint. Often the workpieces are of similar materials, and the weld formed at their interface is simply dependent on the bonding that can take place between intimate faying surfaces under pressure at elevated temperature. Recrystallization may sometimes take place during the solid-state welding mechanics, and there may be dispersal of minor foreign compounds (oxides, etc.) by migration from the interfacial surfaces into the body of the weldment.

When dissimilar metals and alloys are present at the joint, then diffusion, per se, does play a role in the process. Dissimilarity in material composition may be presented by workpieces of different types of materials to be joined, or if they are of similar composition a dissimilar filler metal could be present, which would legitimize the term diffusion welding. However, any filler metal employed must not melt because the formation of molten metal at any stage, even if eventually completely dispersed by diffusion, would classify the process as diffusion brazing, or eutectic brazing as described early in this chapter.

The mechanics of welding in the solid-state, both without diffusion of composition, and with diffusion as an additional aid, have been intensively studied and reported in the literature, but the rudimentary stages that take place under pressure at elevated temperature can be summarized as:

(1) deformation of asperities on contacting surfaces and break-up of surface films by the applied pressure to form an interfacial boundary;
(2) dissolution of oxides, etc., at the interfacial boundary, and grain boundary migration; and

(3) continued time-dependent plastic deformation at the interfacial boundary which normally will eliminate voids, promote atomic bond development, produce diffusion between regions of dissimilar composition, cause grain boundary migration and recrystallization.

Pressure is important during the first stage of diffusion welding because pressure establishes the extent of intimate contact to facilitate the mechanics of the two following stages. Many different systems are used to apply pressure to assemblies of workpieces for welding and the selection will depend on structural aspects of the weldment, and whether a special atmosphere of gas or a vacuum must be used to control surface films. The compressive force at the joint interface of steel workpieces may be as high as 20 MPa (2.9 ksi) to ensure a particular level of weld quality. However, the most important variable overall is temperature, since it controls the rates at which the various pertinent functions progress to states befitting the desired weld quality. Although higher temperatures increase the rate of reactions that contribute to making a diffusion weld, there are usually good reasons to use the lowest usable temperature. The minimum temperature is generally approximately one-half the melting point of the lowest melting material involved (often identified as "T_s") expressed in degrees Kelvin. Most diffusion welding is conducted at a temperature somewhere in the range of 0.5 to 0.7 of T_S °K.

Diffusion welding is not often employed as the principal joining process for articles of carbon and low-alloy steels because the operation usually involves high equipment costs, and has a relatively slow output. Nevertheless, DFW advantages that sometimes make it first-choice for complex components to which the usual fusion welding or resistance welding processes are not readily applied. When general considerations single out DFW as applicable, a number of metallurgical benefits are included among its advantages; namely:

(1) Localized zones of heat-affected microstructure are not produced in weldments made by DFW, and the elevated temperature selected for the process can often provide a beneficial heat treatment to many steels.
(2) Uniform heating of an assembly of workpieces for DFW and subsequent cooling incurs little distortion in the weldment, and minimal short-range residual stresses in the weld area.
(3) The solid-state mechanics of DFW permits considerable freedom in using dissimilar metals and alloys in a weldment, and depending on the temperature and time employed, without development of intolerable development of brittle intermetallic compounds.
(4) Defects that are prone to occur in fusion welds will not be normally encountered in weldments made by DFW, and voids left at the interfacial boundary of DFW welds constitute a problem that is usually easily remedied.

Ultrasonic Welding (USW)

Ultrasonic welding began with an observation made during research on electrical resistance spot welding in 1950. Experiments were being conducted to determine possible benefits of applying ultrasonic energy to the electrodes as the RSW process was performed on sheet metal. Surprisingly, the workpieces were found welded together after one of the tests where current had not been delivered by the spot welding machine. Study of the phenomenon quickly found that welding occurred because of reciprocating micro-sliding at ultrasonic frequency under pressure at the faying surfaces. Temperature rise in workpieces appeared insignificant, but the mechanical action produced the necessary break down of asperities, and the dispersion of surface films to bring about intimate contact between nascent surfaces and to weld them together.

Several different kinds of ultrasonic welding machines are now available to accommodate various forms of material to be joined. The material is generally in the form of very thin sheet, foil, or fine wire. While ferrous alloys can be welded by the ultrasonic process, nonferrous metals and alloys are much easier to weld. Much of the thin sheet, foil and fine wire that call for assembly by welding are alloys of aluminum, copper and gold, and consequently USW is a natural contender for the welding task.

Figure 6.67 is an elementary illustration of ultrasonic welding used for making a spot weld between two thin pieces of sheet metal. The workpieces are clamped together under compressive force between an anvil and a vibrating probe, often called a "sonotrode." The probe is coupled to a magnetostrictive transducer which drives the probe at a selected frequency somewhere in the range of 10 to 80 kHz, with 20 kHz a commonly used frequency. Transfer of energy from the probe to the faying surfaces of the workpieces is an important part of the technology of the process. Sliding motion and superficial deformation is not wanted between the probe tip and the outer surface of the workpiece. Instead, effective delivery of ultrasonic energy must be made to the faying surfaces to be welded. The superficial micro-deformation on these faying surfaces to form a solid-state weld will be dependent on a number of factors, including the kind of transducer in the ultrasonic welding machine, the choice of metal for the probe tip, its shape and its surface finish, the clamping force applied, and the amplitude of lateral vibratory motion. Proper arrangement of equipment and operating parameters will deliver 70 to 90 percent of the USW machine's power to the surfaces being welded. The time required to make an ultrasonic weld will depend on the thickness of the workpieces, as well as the clamping force and the power being delivered; but normally the time can be one second, or less. In fact, one-half second is a typical weld time. Determination of optimum welding procedure for a given application is heavily dependent on making test welds using progressive changes in force, power and time, rather than applying formulae or having tables for predicting parameters.

In addition to making spot welds in thin sheet, as shown in Figure 6.67, the USW process can be used to make continuous roller seam welds, line welds, ring welds, cross-wire welds, wedge wire welds, and ball-spot welds. The need for these different types arises as a process is employed in splicing foil, encapsulating electronic packages, fabricating light structural members, connecting electrical circuitry, closing tube ends, and a multitude of other applications.

Ultrasonic welding offers obvious metallurgical advantages, including: the absence of marked heat effects, the ability to join dissimilar metals and alloys without the formation of a substantial intermixed or diffusion zone having unfavorable properties, the absence of fluxes or shielding gases, and the avoidance of spitting, out-

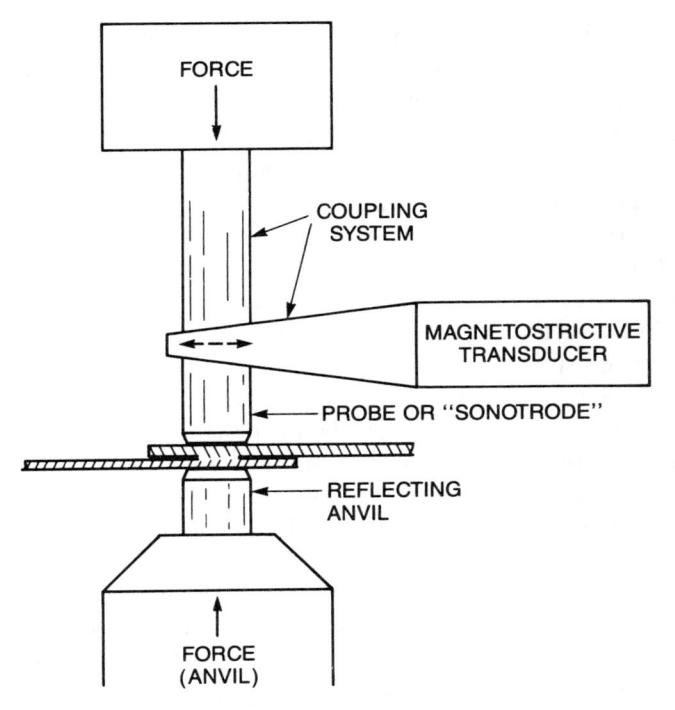

Figure 6.67 — **Ultrasonic welding process (USW) showing essential equipment and mechanics of operation for making a spot weld between two thin overlapping sheets**

gassing, or other emissions that might be undesirable when welding under unusual conditions, for example, when encapsulating an assembly *in vacuo*.

Steels are not the only alloys that are somewhat difficult to weld by the USW process. Very soft metals, like lead and tin, absorb and dissipate much of the ultrasonic energy before it reaches the interfacial surfaces where the weld is to be formed. Hard metals and alloys of almost any composition also give much difficulty.

Study of the mechanism by which welding takes place in the USW process reveals that the induced ultrasonic energy first causes any films on the faying surfaces of the workpieces to be disrupted or destroyed. The lateral motion appears particularly effective in this action. When the nascent metal surfaces come together under pressure, welding takes place, but it is believed that at this instant of initial contact the continuation of reciprocating sliding for a very short period facilitates the solid-state bonding. However, it is suspected that under some circumstances lengthy continuation of vibratory motion can cause the newly-made weld to fail by fatigue.

Metallographic examination of welds made by USW has found internal plastic flow in a superficial zone at the interface of the joined workpieces. The metal in this extremely shallow zone appears to have remained solid during the welding operation. Although some of the ultrasonic energy is dissipated in the form of heat, the temperature developed in the workpieces along the faying surfaces apparently does not cause melting on any ordinary microscopic scale. Many joints display plastic flow in metal at the interface, and it is not uncommon to find some evidence of temperature rise. Welds in metals having a relatively low recrystallization temperature may display recrystallization and growth of the grains at the weld interface. Sometimes evidence of diffusion between dissimilar metals appears, but the extent usually is very limited.

Ultrasonic welds often have a number of common features even though internal microstructural details will vary with the particular metals or alloys in the assembly. The outer surface of the workpiece against which the vibrating probe was impressed generally will display a slightly roughened appearance because of the combined compressive and shear forces imposed upon it. Spot welds ordinarily register as an ellipse because of the linear reciprocal motion of the vibrating probe. Indentation left in the surface by the probe usually is very slight, perhaps, less than five percent of the section thickness. The actual weld at the interface of the workpieces cannot be judged for size or shape from outer surface markings. Nor can the integrity of ultrasonic welds be easily judged by metallographic examination. Some welds may have an interface that is unremarkable, and the faying surfaces appear to have come together without the occurrence of any mechanical or metallurgical actions. If positive signs of recrystallization, and/or diffusion across the interface are present, these features ordinarily imply that a true weld exists. Experience suggests that some form of mechanical testing, such as tension, or tension-shear testing, can give a firmer indication of weld integrity. Steady production of acceptable welds depends on the mechanical precision of the equipment being used and its consistent application to the workpieces. Normal variations in the material being welded seldom affect weld quality.

Cold Welding (CW)

Metals have a remarkable ability to form a weld-like bond when absolutely clean surfaces are brought into very intimate contact even when near room temperature. This bonding phenomenon has always been an incentive to see how it could be utilized in simple, low-cost joining operations. A welding process operable at room temperature obviously would not degrade the properties of metals being joined by the effects of heating; hence the impetus to strive for *cold welding* when possible. One economical way of preparing a metal surface so that oxide or other films will not constitute a barrier to cold welding is to carry out plastic deformation in a manner that disrupts any surface film and brings fresh, uncontaminated metal together under sufficient pressure to ensure the required intimacy. Experience has shown that considerable deformation has to be engendered in most applications to achieve the conditions conducive to cold welding. Also, cold welding has been found more successful with soft metals that have a low hardening rate. In fact, nonferrous metals that have a face-centered cubic crystallographic structure appear to be the best candidates for cold welding. Aluminum and copper lend themselves readily to the process. Soft commercially pure iron can be cold welded by taking extra precautions, but only very limited success has been achieved to date in cold welding steels, even the softer types.

Cold welding might seem a relatively simple solid-state process, but knowledge of its fundamental metallurgy is helpful in guiding key steps that must be carried out with precision to obtain a good quality weld. The CW process not only can be performed at room temperature on amenable metals, but success in making an acceptable weld is not dependent on a temperature rise during the deformation stage. Therefore, no evidence is ordinarily found when examining cold welds metallographically for:

(1) jetting or other features that are promoted by increasing temperature,
(2) microstructure changes from heat-effect, and
(3) diffusion between dissimilar metal workpieces.

The bond line or interface between the workpieces appears as devoid of interaction as does a simple, fine-line grain boundary. Of course, the microstructure in the weld area will display evidence of plastic deformation, such as, elongated grains, slip lines, and twinning. These signs of cold work will vary with the nature of the metal, and the kind and amount of work executed during welding of a given joint.

Figure 6.68 illustrates a number of typical lap and butt welds that can be made by the CW process. The configurations after welding are severely indented or deformed, and this appearance may not be acceptable in a particular article. Butt joints usually produce sufficient weld flash to permit removal and restoration of a flush surface at the weld. An important fact to note about butt welds is that flow lines in wrought materials are likely to be upturned immediately adjacent to the weld (a condition illustrated earlier in Figure 6.33 for a butt weld made by the flash welding process). Thus, the degree of anisotropy in the material and the severity of "upturned fiber" at the weld will determine whether ductility and toughness at the weld will be adversely affected.

A number of manual tools for cold welding small sections are available. One tool resembles a pair of sturdy pliers which can grip the ends of small rods or wires and make a butt weld simply through upsetting the ends against each other. Hydraulic presses also are available for the assembly of somewhat larger parts. For example, the capsules for silicon rectifiers can be hermetically sealed by cold welding, and if the operation is performed in a vacuum or in an inert atmosphere, the sealed capsule retains the protective atmosphere for service. Cold welding, although an attractive process from a cost standpoint, often is passed over when selecting a production joining process because the completed weld does not exhibit easily-interpreted external signs that the expected bonding indeed has taken place at the internal interface.

Electrostatic Bonding

Electrostatic bonding is an unusual solid-state joining process for joining metals to *glass*, *ceramics*, and *semiconductors*. None of the fusion or the solid-state welding processes described to this point are able to make a weld with acceptable properties between a metal and a nonmetal. The welding engineer should be aware of this process, which at first was known as "field assisted bonding," because of increasing usage of composites and nonmetals as materials in modern manufacturing. Many present-day articles require hermeticity, and while metal-to-glass joints can be sealed with adhesive, these glued joints sometimes are short-lived under rigorous service conditions. Electrostatic bonding, although still undergoing intensive development in the laboratory, has proved a viable process for making a weld-like bond between various metals and glass, ceramics, and semiconductors as required in the production of solar packages. Electrostatic bonding makes use of binding forces that can be arranged between unlike atoms described earlier in Chapter 2 as ionic and covalent crystal bonding.

618 Chapter 6

The electrostatic bonding process is able to join metals to glass at temperatures well below the softening point of the glass. The components to be joined must have very smooth polished surfaces. The workpieces are placed together in an electrostatic bonding machine where they are automatically put through the following steps:

(1) preheat to a selected temperature in the range of about 300 to 600°C (approximately 575 to 1100°F),

(A) LAP WELD, BOTH SIDES INDENTED

(B) LAP WELD, ONE SIDE INDENTED

(C) EDGE WELD, BOTH SIDES INDENTED

MANDREL

(D) BUTT JOINT IN TUBING BEFORE AND AFTER WELDING

(E) DRAW WELD

(F) LAPPED WIRE, BEFORE AND AFTER WELDING

(G) MASH CAP JOINT

FLASH

(H BUTT JOINT IN SOLID STOCK, BEFORE AND AFTER WELDING

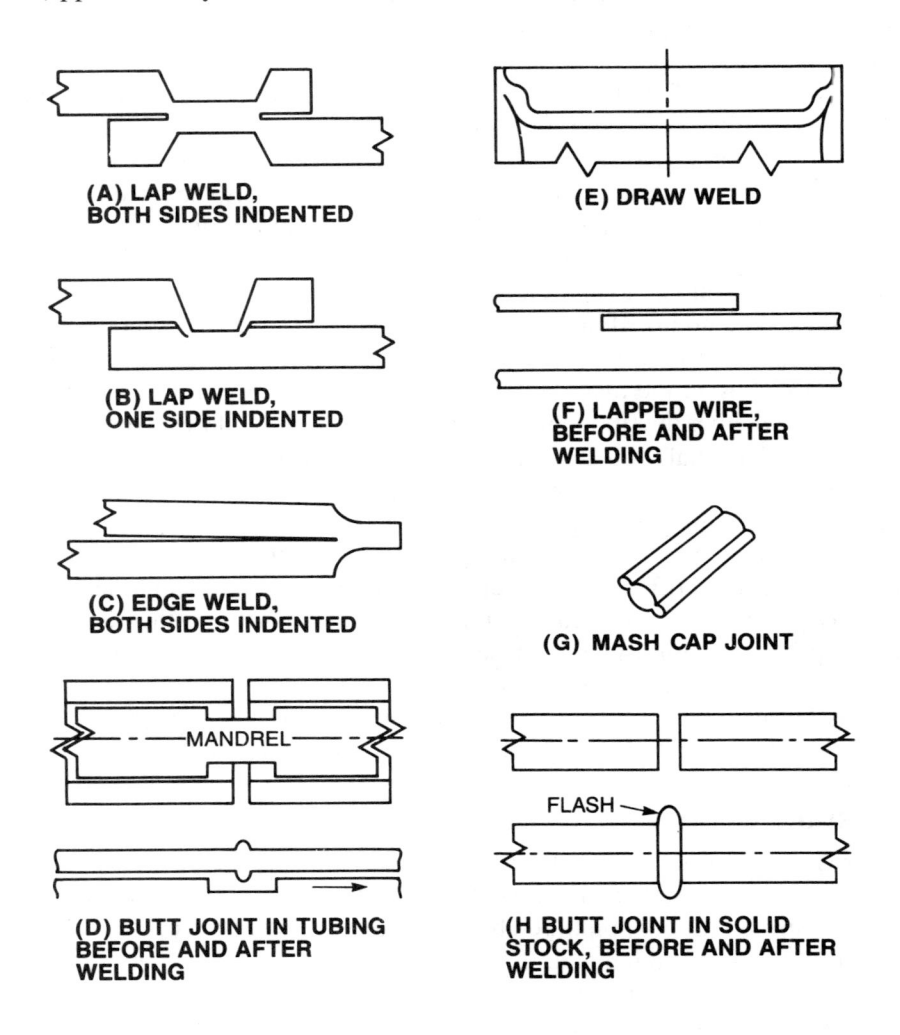

Figure 6.68 — **Application of the cold welding process (CW) to a variety of joints in metal workpieces**

(A) lap weld, both sides indented; (B) lap weld, one side indented; (C) edge weld, both sides indented; (D) butt joint in tubing, before and after welding; (E) draw weld; (F) overlapped ends of wire, before and after welding; (G) mash cap joint; (H) butt joint in solid stock, before and after welding. (*AWS Welding Handbook, Vol. 2 Welding Processes*, Eighth Edition, Chapter 29, cold Welding, Figure 29.11)

(2) apply pressure that is dependent on the kind of glass or nonmetal involved (the amount of pressure is usually lower than that used in diffusion bonding),

(3) apply high dc voltage across the workpieces (perhaps in the range of 200 to 3000 V) with the metal at positive potential and glass at negative potential, and

(4) hold under the selected conditions of temperature, pressure, and voltage (ordinarily, milliamperage current will be flowing) for a time period that may range from 10 seconds to five minutes as required for bonding to take place.

The exact mechanism by which a direct chemical bond develops between the metal and the glass or other kind of nonmetal has yet to be established, but research to date suggests that under the applied high voltage bonding takes place by complex migration of positive and negative ions between the metal and nonmetal surfaces. As a net effect, the polarized dc field introduces additional metal ions into the nonmetal to produce an anodic bond. Also, there is some evidence that during the operating steps listed above, a process of surface oxidation and electrochemical etching is taking place which closes micro-gaps between the faying surfaces, thus bringing the surfaces into more intimate contact.

In general, bond strength improves with higher preheat temperature, higher pressure, and longer time, but optimum parameters must be developed by trials for each dissimilar metal-nonmetal application. Lowest possible bonding temperature often is preferred to avoid cracking in glass or ceramic as the assembly cools and stress arises from difference in thermal contraction between the metal and nonmetal. Benefits of conducting the entire electrostatic bonding program in a vacuum (10^{-4} torr), are being explored, and it appears that the vacuum makes the process more tolerant of surface contaminants and of variations in component fit-up. Also, wider ranges of operating parameters appear possible in vacuo. Electrostatic bonding is not operable for metal-to-metal bonding because the necessary electric fields that must form in the interfacial micro-gaps to generate ionic migration cannot develop in a metal-to-metal assembly.

Electrodeposition Welding

Electrodeposition welding is an electroplating operation which, under some circumstances, can serve as a welding process. Electrodeposition welding has been employed only to a limited extent, but is capable of joining a wide variety of metals and should not be overlooked as a means of solving a difficult joining problem. Also, it can join a metal to a nonmetal workpiece, providing the latter is an electrical conductor, or has a metal surface film previously deposited by some suitable technique. Electrodeposition welding involves the deposition of a metal film by plating on the surfaces of two or more workpieces as they are securely held in close proximity. The gap between the pieces should be as small as possible, perhaps as little as 12 microns (about 0.0005 in.) apart. The plated metal will bridge this gap and then build up to the desired thickness to provide joint continuity. The required thickness of the plated metal will depend on the nature of the components, the joint design, and the strength needed at the plated joint. However, plated coatings for joining typically are in the order of 75 microns (about 0.003 in.) thick.

Electrodeposition welding offers the advantages to be expected of a process that does not involve heating. Distortion is eliminated, the base metal adjacent to the joint is not heat-affected, and dissimilar metal joints present no particular difficulty. Components of steel having a UTS of 700 MPa (100 ksi) are reported to have been joined by electrodeposition of a nickel alloy with the resultant weld displaying a strength in excess of 620 MPa (about 90 ksi). Further opportunity to secure stronger plating for joining is afforded by heat treatment applied after electrodeposition. During electrodeposition, layers of a number of selected metals can be deposited. Subsequent application of a heat treatment of proper temperature and programmed time will eliminate any threat of hydrogen embrittlement, and can produce diffusion between the metals in the lamellar plating and with the workpieces. Diffusion can form a new homogeneous alloy in the weld area. Under these circumstances, the electrodeposition process actually becomes diffusion welding (DFW), as described earlier. The composition of the new alloy formed "in situ" can be controlled by the kind and amount of metals and alloys deposited in the multilayered plating, and it may offer properties superior to those offered by any single layer metal or alloy.

Use of electrodeposition welding is restricted by a number of limitations and problems. The size and shape of the parts to be joined are important considerations since the joint must be immersed in the electrolyte and the anode must be positioned so as to properly "throw" the deposit over the joint area. The plating conditions must be properly controlled to deposit a sound coating. The possibility of hydrogen embrittlement of the coating and/or the parts during the plating operation should not be overlooked. The electrodeposition process has been used mostly for joining small assemblies which do not require high weld joint efficiency, such as wave guides for radar equipment.

BRAZING AND SOLDERING PROCESSES

Brazing and soldering are classified in Figure 6.2-C as categories distinct from fusion welding, and from solid-state welding. This separation is made because brazing and soldering are based on a liquid-solid joining mechanism, that is, a method whereby the base metal workpieces are heated, but *not melted*, and a dissimilar *molten* filler metal is introduced or somehow generated in the joint to coalesce with the base metal during cooling. Brazing differs from soldering mainly in the filler metals used, the temperatures employed, and the physical and mechanical properties of completed joints. These differences set the processes clearly apart in industrial utilization. Also, a metallurgical perspective finds reasons to discuss brazing and soldering separately.

Brazing Processes

Brazing encompasses a group of almost a dozen processes that produce coalescence in a close-fitting joint by heating the base metal to a suitable temperature that is below its melting point, and applying molten dissimilar filler metal that is drawn into the joint by capillary attraction to fill it. A stipulation that defines brazing is that the filler metal shall have a liquidus above 450°C (840°F), but below the solidus of the base metal. Silver alloy filler metals often are used, and this operation sometimes has been called "silver soldering," which is inappropriate inasmuch as the silver alloy filler metals have liquidus temperatures above 450°C (840°F): Therefore the operation is *brazing*.

Another closely related process, *braze welding*, makes use of a filler metal having a liquidus above 450°C (840°F) and below the solidus of the base metal, but differs from brazing in that the filler metal does *not* flow into the joint by capillary attraction. Instead, the filler metal is deposited in an open joint from a welding rod or a consumable electrode. Even though the filler metal is melted in making in addition, the base metals are not melted. The bonding that takes place between the filler metal and the heated, still-solid base metal is much the same as occurs in brazing. There are circumstances in braze welding, particularly when an arc is used as the heat source, where some melting of the base metal takes place. This small amount of melted metal is assimilated by the weld metal. Ordinarily, the oxyacetylene torch is used as the heat source and this allows closer control of base metal surface temperature so that melting does not occur. Even then, there is likely to be more interfacial diffusion and penetration of the base metal grain boundaries during braze welding than in brazing.

Brazing is economically attractive for the assembly of parts, and most types of iron and steel can be readily joined by one of its processes. Special procedures may be required for handling cast iron, free-machining (high sulfur content) steels, high-carbon steels, and certain alloy steels. While some of the metallurgical problems that arise with fusion welding can be avoided by using a brazing process, there are critical steps in the brazing operation where deviations can adversely affect joint quality. The unacceptable conditions that can develop range from inadequate bonding because of failure-to-wet, to lack of gas or liquid tightness because of failure-to-fill the joint. The defects encountered often are peculiar to a brazed joint and bear little or no resemblance to defects encountered in welding. AWS has published its *Brazing Handbook (BRH), Standard for Brazing Procedure and Performance Qualification* (AWS B2.2), and *Recommended Practices for Design, Manufacture, and Inspection of Critical Brazed Components* (AWS C3.3) and other brazing-related standards.

Joints that are brazed usually involve lapped or scarfed surfaces of relatively large area with small thickness. Because brazing filler metal is lower in strength than steel base metal, an attempt is made to hold the intermediate layer of filler metal to a minimum in order to obtain maximum joint strengths. Joint clearance is spaced to permit capillary flow. This spacing may be in the range of 0.05 to 0.25 mm (0.002 to 0.010 in.) depending on the brazing alloy being used and the brazing technique. In a typical brazing operation, the surfaces to be joined are cleaned to remove contaminants and loose oxide, and are coated with flux. The flux, when molten, is intended to dissolve surface oxides. The joint area is heated until the flux melts whereupon the cleaned base metal is protected against further oxidation by a layer of liquid flux. Brazing alloy is melted at some point on the surface of the joint area. The capillary attraction between the base metal and the filler alloy is several times higher than that between the base metal and the flux. The liquid flux in the joint is displaced by the liquid filler alloy. The joint after cooling to room temperature will be found filled with solid brazing alloy, and the flux, now also solidified, will be found on the outside of the joint.

Brazed joints usually are made with relatively small clearances. The viscosity of the filler alloy, therefore, is almost as important a factor as surface tension and wetting ability. Low viscosity is a desirable characteristic of brazing alloys since a viscous filler metal might not have sufficient capillary attraction to penetrate close-fitting joints. A great variety of techniques can be employed in brazing, but each technique includes three principal steps: (1) surface preparation, (2) heat application, and (3) brazing filler metal introduction.

Surface preparation is important because the surface must be clean and free of oxide films that would hamper wetting by the molten brazing alloy. The presence of other stable compounds like carbides, sulfides, and nitrides on the surface also can hinder wetting, although to a lesser degree. The necessary degree of surface cleanliness usually is achieved by one of four common practices: (1) application of a flux prior to or during heating, (2) use of a gaseous atmosphere that actively reduces surface oxides, (3) use of a protective atmosphere or a vacuum that prevents further oxidation and depends on a mild flux to remove any oxide already present, and (4) employment of a brazing filler metal containing alloy additions that impart a self-fluxing capability (e.g., boron, lithium, and phosphorus).

Heating of steel workpieces for brazing can be accomplished with a gas torch, in a furnace, by electrical resistance methods (including current passed through the work, induction coils, and resistance blankets), by an electric arc, radiant quartz lamp, laser beam, or in a molten dip bath. The selection of a heating method is determined by many factors, including the nature of the workpieces, the brazing alloy being applied, the number of assemblies to be brazed, and the costs involved. The gas torch is commonly used on a small number of assemblies where flux can be applied, or where flux can be injected into the gas flame, or the filler metal is self-fluxing. Furnace heating is more often used for high production operations. Induction heating is excellent for repetitive operations and is favored when very rapid heating is desired. Arc, resistance, laser, and dip brazing are less widely used. Furnace heating is favored when (1) the parts to be assembled can be jigged to hold them in proper position, (2) the brazing filler metal can be preplaced, or (3) when exacting atmosphere control during heating and brazing is required.

Brazing in a special furnace atmosphere or in a vacuum chamber is widely used. Active atmospheres, like dry hydrogen which does not require a supplementary flux, are particularly popular for mass-production of high quality brazed assemblies, or *brazements* as they are called. This technique maintains the workpieces in a clean, unoxidized condition. In fact, the surface may be brightened considerably by exposure to the hot hydrogen atmosphere and may make further finishing of the surface on the brazement unnecessary. Argon is a popular gas for use as a furnace brazing atmosphere. It can be dried to a low dew point and will provide excellent protection against oxidation. It cannot, however, reduce oxide that is already present on the surfaces of the workpieces, so a self-fluxing filler metal or a supplementary flux must be employed. Other gases that can be used under certain circumstances in furnace brazing atmospheres include carbon monoxide, carbon dioxide, nitrogen, and methane.

Their reactive and protective qualities insofar as brazing is concerned will be discussed in Chapter 8. Vacuum furnace brazing is performed at pressure ranging from 0.5 down to 10^{-4} torr depending on the amount of protection needed against further oxidation. Vacuum conditions will not completely eliminate oxide already on the surface of steel, but will greatly facilitate brazing by a self-fluxing alloy filler metal.

Filler metal for brazing can be introduced in a variety of forms and by a number of different techniques. AWS A5.8 is a *Specification for Filler Metals for Brazing* that classifies a large number of popular compositions into six groups that represent alloys of aluminum, silver, gold, copper, nickel, cobalt, and magnesium as filler metals. A number of alloys (mostly of the precious metals) also are available for "vacuum service," and are used mostly in brazing electronic devices. These brazements require that the filler metal will undergo the melting and flowing operation without spattering, and that residual elements with a high vapor pressure (e.g., Cd, Zn, Pb, C, Mg, Sb, K, etc.) be restricted to very low levels in the alloy to minimize "emitter" activity during operation of the device. Requirements are provided for chemical composition, forms, sizes, testing and certification. A guide is appended to assist in the selection of best filler metal classification for particular applications. Among the guidance included in this AWS document is a table of solidus, liquidus, and brazing temperature ranges. The brazing temperature range deserves consideration because of liquation during the solidification process. When brazing with preplaced filler metal of alloys having a wide temperature range between solidus and liquidus temperatures, the several constituents of the filler metals tend to separate during melting. Under these circumstances, the lower-melting constituent will flow, leaving behind a "skull" of the high-melting constituent in the joint gap. This mechanism is undesirable since the unmelted skull does not contribute in proper proportion to the filler metal entering the actual brazed joint. However, where fit-up is poor, a filler metal with a wide brazing temperature range usually will fill a joint more easily.

The most popular brazing alloys for steel are either copper-base, or silver-base. Copper and silver, when alloyed with elements like zinc, tin, cadmium, silicon, phosphorus, and manganese, produce a variety of brazing alloys with brazing temperature ranges having a minimum temperature as low as about 600°C (1100°F) and as high as 1150°C (2100°F). At the lower end of this temperature spectrum are the silver-base alloys. In the mid-range of 875 to 1050°C (about 1600 to 1900°F) are various kinds of bronze. The temperature employed in a brazing operation generally is substantially above the liquidus of the filler metal, often at a good flow temperature established for the alloy. Brazing with pure copper, for example, is carried out at approximately 1120°C (about 2050°F). Sometimes a brazing alloy can be selected that has a usable temperature which corresponds with a favorable annealing or normalizing treatment temperature for the steel in the workpieces. Thus two operations, brazing and heat treatment, can be carried out together.

Metallographic examination of the bond between brazing filler metal and the base metal usually will show that a certain amount of mutual diffusion has taken place to produce zones of intermediate alloy compositions. Intimate wetting of the base metal surface is facilitated by having this surface diffusion occur quickly. The rate of diffusion is determined by the composition of the brazing filler metal and the brazing temperature. By holding the brazed assembly at elevated temperature for a long period of time, it is possible to promote complete disappearance of the brazing filler metal from the joint by diffusion and thus almost defy detection. However, the properties of joints in carbon and alloy steels treated in this manner, although somewhat stronger, are not sufficiently improved to make the treatment commercially attractive. The presence of tin, zinc, cadmium, and boron in the brazing filler metal will accelerate wetting. Elements of small atomic size, like boron, can diffuse rather quickly into the base metal. Furthermore, iron has a strong affinity for these elements and quickly forms intermetallic compounds with them. Therefore, it is important to know the manner in which the brazing filler metal migrates or diffuses into the base metal. Some elements, like copper and zinc, diffuse quite rapidly into the *grain boundaries* of steel, but their migration into the crystal lattice of the grains themselves is much slower. This tendency for intergranular penetration by a brazing filler metal is dependent on the level of stress in a workpiece as the filler metal wets the solid surfaces. Residual stresses in the workpiece from cold forming operations can be a significant driving force for the intergranular penetration phenomenon. In the absence of stress, brazing with pure copper, and with brazing filler alloys containing zinc to join workpieces of steel is feasible and is widely practiced. Excessive *intergranular* penetration can result in a loss of toughness in the base metal zone adjacent to the brazed joint. Some elements, such as boron, which form an intermetallic compound with iron when they diffuse from the filler metal into the base metal also can reduce toughness.

Two noteworthy kinds of special brazing filler metals are *self-fluxing alloys*, and *reactive brazing alloys*. The self-fluxing types contain small additions of elements that make the alloy very active and provide ability to dissolve or flux the oxides on the base metal surfaces. This capability can be secured with elements like lithium and boron. Lithium is particularly effective in silver-base brazing alloys. An addition of just half a percent is sufficient to make the alloy self-fluxing in an inert gas furnace atmosphere. Boron is more often used in nickel-base alloys that are designed for operations above 1050°C (about 1950°F). Reactive brazing alloys contain additions of elements intended to depress the melting point and which are somehow eliminated or greatly reduced in amount in the later stages of the brazing operation to raise the remelting temperature of the filler metal in the joint. The depressant elements can be eliminated by diffusion into the base metal, by volatilization, or by chemical reaction with the other constituents in the filler metal. In general, raising the remelting temperature of the filler metal in the joint will improve its capability to withstand elevated temperature service.

Soldering

Soldering is a very old process — used by the Romans starting about 50 B.C. to make longitudinal and circumferential lap joints in water pipes made of lead, using solder composed of approximately 50 percent tin and 50 percent lead to bond and seal the joints. Soldering is much like brazing except that the operation is conducted at a lower order of temperatures with filler metals having a liquidus *not exceeding* 450°C (840°F) and below the solidus temperature of the base metals involved.

Soldering is not favored as a process to be used in making joints in steel where *strength* is an important property. In general, solder filler metals have strengths that are substantially below most base metals, and for this reason the metals and alloys used as filler metals sometimes are referred to as "soft solders" (in contrast to the term "hard solders" inappropriately applied to silver brazing alloys). In addition to the relatively low strength of the filler metal, per se, the bond between solder and the base metal usually is lower in strength than is achieved by brazing, or by welding. Therefore, most soldered joints are lap joints, with the overlap at least 3X thickness to compensate for bond strength. Butt joints seldom are used because of their limited strength. In fact, many soldered joints in sheet material are designed with interlocking edges so that the base metal carries the load, and the solder is applied for reasons other than establishing strength. The virtues of soldering are found in its capability for accomplishing a bond at modest temperatures, for easily providing a liquid-tight seal in a lap joint, for filling the gaps in interlocking joints and making them rigid, for providing a flush, smooth surface, and for improving the electrical and thermal conductivity of mechanical joints. Any one of these attributes can be a deciding factor when selecting a method of joining. Consequently, soldering remains a useful process both in repair operations, and in manufacturing.

More than a half-dozen different processes are available for soldering and most of them have been classified and given AWS designations in Figure 6.2-C. All processes have the following essential steps in their operation (but not necessarily in the same sequence). First, the base metal surface must be cleaned and stripped of heavy oxides and any other compounds that can interfere with wetting by the molten solder. This vital cleaning step cannot be accomplished with a flux alone. Second, the base metal is heated to a temperature where the flux and molten solder will "wet" its surface and will remain bonded after the solder has solidified. Special equipment and systems have been devised to introduce ultrasonic energy into the surface of workpieces via the molten solder when dip soldering (DS), or when using a soldering iron (INS) to break up and disperse oxide films on some base metals. In this manner, the workpiece surface is wetted by solder without the aid of a flux. In general, the temperature at which solder will bond to base metal need not be as high as the liquidus of the solder. Third, the solder, when melted by some means, is introduced to the joint either by preplacement, or by utilizing capillary attraction. If the base metal surface has been properly cleaned and heated to accept wetting by the solder, the joint after solidification will have developed a metallic bond at interfaces between the solder and the base metal. Metallographic examination of this bond usually finds evidence of very shallow diffusion along the interface. During wetting, the solder dissolves (not melts) the base metal superficially and this mechanism forms a very narrow interalloyed zone. The diffusion that takes place does not necessarily strengthen the bond between solder and base metal, but it does signal effective wetting, which is a requisite initial action toward proper bonding. Procedures to be followed with various soldering processes are described in most welding handbooks, and the AWS *Welding Handbook, Volume 2*, Chapter 13, is quite detailed. Automated soldering is a highly effective route toward reducing the cost of assembly by soldering, and improving the consistency and quality of joints, and a number of industries have developed highly specialized methods to suit their manufacturing needs.

Solder compositions used as filler metal are, for the most part, alloys of tin and lead, and they serve for the majority of applications. These two metals play markedly different roles in establishing the properties of a particular solder composition. Tin is the active metal in the compositional system, and it is tin that promotes wetting by the solder, and bonding upon solidification. Lead contributes nothing toward wetting and bonding, but is used in specific percentages in Sn-Pb alloy compositions to provide solders with different solidification ranges that are best suited for particular soldering operations. These variations in lead content also alter the mechanical properties of the solders. The requirements applied to equipment assembled with the aid of solder seldom are directed to strength, but deal more often with corrosion resistance, and with ease of wetting the base metal to which it will be applied. Some applications call for "lead free" soldered joints and best possible corrosion resistance. Commercially pure tin (99.8 percent Sn), called "straits tin," or "block tin," is often used for these applications because pure tin solder has excellent wetting capability on iron and steel, good corrosion resistance, and other favorable properties. However, the cost of tin is significantly higher than for lead, and those applications that consume large quantities of solder are more likely to employ solder with the highest lead content that still provides the needed wetting capability and other required properties or characteristics.

Figure 6.69 illustrates the tin-lead phase diagram on which a number of Sn-Pb solder compositions have been indicated. Lead is shown to have a melting temperature of 327°C (620°F), while tin melts at 232°C (450°F); so obviously every tin-lead alloy will comply with the AWS temperature limitation of 450°C (840°F) for any filler metal used in soldering. In addition to showing the relatively small differences in liquidus temperatures among the solders, the phase diagram shows their respective solidification ranges. This can be an important physical characteristic in certain soldering operations. When using solder on auto bodies to fill dents by wiping the solder flush, and when wiping joints in the bell-and-socket connections of cast iron drain pipe, it is very helpful to use solder with wide solidification range. This allows the solder to be shaped while still in a pasty state when cooling through its solidification range. As should be expected because of this extended liquation period, solders with wide solidification ranges are more likely to develop hot cracking (through the still-molten areas) if the joints are moved during cooling before solidification is completed. Other soldering operations may call for a filler metal that freezes quickly during cooling, and this behavior is obtained by using the eutectic composition (63 percent Sn - 37 percent Pb), or perhaps the 70-30 Alloy favored for dip soldering. The microstructure of eutectic solder is lamellar in appearance, and it possesses the best mechanical properties of the tin-lead alloys. Accelerating the cooling of soldered joints results in a small increase in strength. Soldered joints may lose a small amount of strength as they age. Preheating also reduces strength. This is believed due a coarser grain size in the solder metal.

Other alloying elements are used in solders to secure different melting and freezing temperatures, to improve mechanical properties, to increase toughness at low temperatures, to aid corrosion resistance, etc. Briefly, a small amount of antimony is often added for greater strength. Silver has been used in low percentages to increase strength. Cadmium sometimes is used to replace part of the tin. Certain residual elements such as aluminum and zinc are avoided because they can adversely affect the

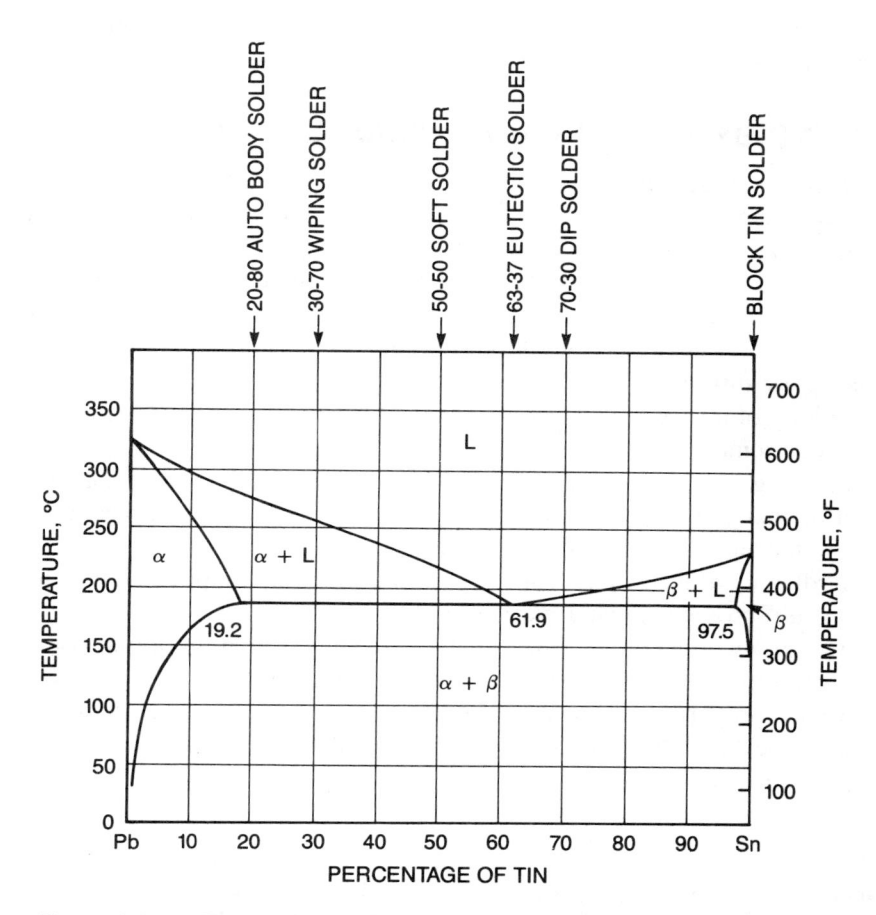

Figure 6.69 — Phase diagram for the tin-lead alloy system showing temperatures of liquidus and solidus for commonly used solders

wetting ability and flowing characteristics of molten solders by forming tenacious oxides on the surface of the molten solder. Iron inhibits the flow of molten solder by forming high-melting intermetallic compounds with tin. Consequently, specifications for solder often include maxima for certain residual elements. A standard long used to provide composition specifications for solders is ASTM B 32 for *Solder Metals*. ASTM also has issued ASTM B 284 for *Rosin Flux-Cored Solder*, and ASTM B 486 for *Paste Solder*.

The fluxes used in soldering are often proprietary formulae, but sometimes simple chemical compounds or acids can be employed. Fluxes can be classified into three general categories: (1) inorganic fluxes (acids and salts comprise the most active fluxes), (2) organic fluxes (organic acids, bases, and hydrohalides are moderately active fluxes), and (3) rosin fluxes (these depend mostly upon abietic acid and are the least active fluxes). Fluxes are selected to perform their vital function on a particular base metal to be soldered, but further consideration must be given to the nature of the fluxed residue that may remain on/or near the joint. Some of these residues can be electrically conductive, or quite corrosive to any materials in which they remain in contact. Removal of flux residue may be vital to the serviceability of the soldered assembly.

SURFACING BY WELDING AND THERMAL SPRAYING

Surfacing is the standard AWS general term for many different welding, brazing and spraying operations that heretofore have been called "buildup," "buttering," "cladding," "overlaying," "hardfacing," "metal spraying," etc. Any application of one or more layers of filler metal to the surface of base metal by welding, brazing, or spraying to obtain desired dimensions, or properties (as opposed to joining base metal sections together) is now considered to fall in the broad category of *surfacing*.

Reasons for surfacing can be summarized as:
(1) Buildup of a base metal section to repair shape or to fulfill dimensional requirements.
(2) Buttering of a base metal surface with a layer(s) of dissimilar composition that serves as a transition layer when a welded joint subsequently is made against the buttered surface.
(3) Hardfacing of base metal with a layer of dissimilar composition which, in itself, provides greater wear resistance.
(4) Overlaying or cladding of base metal with a layer(s) of dissimilar composition that provides new or improved chemical, physical, or metallurgical properties. These might involve corrosion resistance, magnetic properties, electrical properties, micro-cleanliness, and thermal fatigue performance.

Buildup

This term denotes the application of weld metal to a base metal surface to comply with dimensional requirements, or possibly to conveniently alter the shape of the section. As an example, a boss is added where a threaded hole is to be made and the section surrounding the hole must be thickened. Buildup commonly is performed to restore a corroded or worn surface to proper dimensions, or to allow restoration of a required surface finish. Generally, the filler metal selected for deposition is one that is compatible with the base metal, often being the same as would be employed for making a welded or a braze welded joint. The process selected for buildup might be any of those regularly used for making a joint, but because a larger quantity of filler metal often is applied in a surfacing operation the procedure frequently is tailored to provide a higher rate of deposition. Also, parameters are adjusted and other conditions altered to promote a flat, smooth surface on the metal as-deposited to minimize grinding or other finishing operations, if required.

Buttering

Buttering is similar to buildup, except that buttering usually is confined to a scarf or edge that subsequently is to become assembled in a welded or a brazed joint and the buttered layer of dissimilar composition is intended to satisfy a *metallurgical* purpose rather than merely add to the base metal section. Buttering virtually doubles the cost of joining; yet it can be a realistic solution to difficult joining problems. As an example, consider a weldment of highly hardenable steel base metal that when arc welded would require preheating to avoid underbead cracking and a postweld heat treatment to restore toughness to the heat-affected zones. Circumstances may be such that neither preheating or postweld heat treatment will be feasible in final operations on the weldment. In this case, edge preparations on this hardenable base metal in advance of final operations could be given a thick buttered layer using the required procedure. However, the metal deposited would be a suitable composition that allowed further arc welding without preheat and not require postweld treatment. Further arc welding against these buttered edges in making joints would be done with the care needed to confine significant heat effects *within* the buttered layer. Of course, the strength of joints made by this procedure must be considered beforehand. Many different circumstances can arise where a buttered layer serves as a transitional member in the joining procedure and remains as a structural element in the completed joint. The role played by the buttered layer may involve heat-affect on the microstructure, pick-up of alloying elements, buffering of magnetic properties, etc.

Hardfacing

Surfacing with metals that provide greater resistance to abrasion, deformation, erosion, fretting, galling, and other ravages of wear in service has become a major industrial activity because of the following advantages:

(1) The metal can be applied to area exactly where needed.
(2) A judicious selection can be made from many filler metals that are available to provide compositions that develop mechanical properties in hardfacing layers that are better suited to resist service conditions.
(3) Application of the hardfacing can be performed in the shop, or in the field, and can be repeated as often as needed if wear continues.
(4) Hardfacing can be multi-layers with different filler metals being used for each stratum to provide the most suitable composition and properties for a bonding layer, bulk buildup, backing layer, and wear resistant surfacing.
(5) Costly alloys are candidates for use as filler metal because relatively small quantities can serve to surface a large area.
(6) The compositions, microstructures, and properties of hardfacing are limitless because of the great variety of metals and alloys available, and the many processes at hand to apply them.

Surfacing metals are available as bare welding rods, covered electrodes, coiled wire, paste, and metal powders. AWS Specifications A5.13 and A5.21 have been issued for surfacing rods and electrodes (both bare, and covered) in solid form, and in composite form, respectively; but many more surfacing filler metals of proprietary composition and form are available from suppliers, and their catalogs can be consulted for information on these products. The selection of a hardfacing composition requires study of many details concerning (1) the nature and function of the article or component on which the hardfacing is to be applied, (2) service conditions to which the hardfacing will be exposed, and (3) the projected life of the hardfacing and the cost-effectiveness of this means of improving performance, or extending useful service life.

A tremendous variety of products presently are hardfaced because of cost and the advantage of having a tough, cheap substrate which supports an expensive, often brittle surface that is subjected to wear. Furthermore, when excessive wear is experienced unexpectedly, surfacing with a hardfacing filler metal usually is a practicable recourse. The literature is replete with information on selection of hardfacing materials, and on procedures to apply them. Process selection is an important part of procedure — not only because of the control which process exercises over the composition and properties of hardfacing, but also because of deposition costs. For many years, oxyacetylene gas welding and the arc processes (e.g., SMAW, FCAW, GMAW) served to carry out most hardfacing operations. Today, however, almost any of the welding, brazing and spraying processes are likely to be utilized in some manner. In general, applications that call for a relatively thick layer of hardfacing are done with a fusion welding process. Brazing is used where a thin, solid section of hard material is to be joined to the surface of a workpiece over a localized area where wear is anticipated, or braze welding is used to deposit a layer of hard bronze.

Processes for *thermal spraying* (THSP) were not examined in earlier reviews of welding processes in this chapter, but several were classified and given designations in Figure 6.1. These included flame spraying (FLSP), electric arc spraying (EASP), and plasma spraying (PSP). Spraying usually is employed for depositing relatively thin layers of hardfacing, but spraying today has become a much more proficient process with advent of the plasma spray torch. Spraying is no longer confined to applying metals with a relatively low melting temperature, as had been the case with aluminum and zinc where "metallizing" torches for the FLSP and EASP processes were used to produce coatings on steel for corrosion protection, or to produce buildup layers. Both the transferred and the non-transferred plasma torches are variants from the PAW welding process, and are capable of providing high temperature and heat input required to melt any kind of metal or alloy in any form, and also to melt a controlled amount of the substrate being hardfaced. Metal powder is the favored form of filler metal for plasma spraying, and the process is often conducted as a mechanized operation. Plasma spraying has very useful capabilities for producing both thin and thick hardfacing that is homogeneous and is properly fused to the base metal. Hypersonic thermal spraying employs a new kind of oxyfuel torch that generates gas velocities in the order of 1830 m/s (6000 ft/s) which is more than five times the speed of sound. This torch utilizes almost any kind of coating material in powdered form, and exposes the particles to a flame temperature higher than 5000°C (2760°F) as they are propelled at very high velocity onto a substrate to form a layer of the desired material. The high-velocity oxyfuel torch can use acetylene, hydrogen, propane, etc., for fuel, and the selected gas is burned with purified oxygen. "HVOF" sprayed coatings benefit from the flattening of heated particles as they impinge on the substrate to form a dense coating with excellent bond strength. A metallurgical aspect of this new process that requires consideration for a given application is the relatively high heat input to the coated workpiece.

Laser surfacing is now emerging as a process which can use powder, wire, or strip as filler metal to produce homogeneous fused surface layers. The laser can also be used to remelt and consolidate a surfacing layer previously applied by another process. The laser offers a precisely controlled heat source that achieves consistent melting of the filler metal and controlled fusion of the base metal surface. Availability of equipment and operating costs will sometimes limit application of the laser for surfacing operations. Procedures and techniques for laser surfacing are still in the development stage at this time.

Procedures for hardfacing presently are moving in promising directions, not only to new processes (e.g., friction surfacing, as described earlier), but also by devising *systems* consisting of combined, or articulated processes. For example, a system has come into use that employs gas metal arc welding to deposit a layer of weld metal that serves only as a matrix for very hard particles that are distributed as uniformly as possible throughout the surfacing layer. The GMAW deposit can be a medium-carbon low-alloy steel weld metal, while tungsten carbide can serve as the hard particles to impart abrasion resistance. Powered tungsten carbide (approximately 325 mesh) is injected into the molten weld pool during the GMAW deposition of the surfacing matrix. Care must be taken in arranging injection of the powder to minimize the melting of particles by the arc, and yet to secure fairly uniform distribution of solid particles in the matrix.

Dilution of a surfacing deposit by melted base metal is a basic aspect of hardfacing operations that calls for consideration as the two principal components, the filler metal and the melted portion of base metal, commingle to form the surfacing layer. Dilution can be quite significant in determining the composition of the hardface for two reasons: first, a relatively small volume of filler metal ordinarily is deposited; therefore, penetration by melting into the base metal can contribute an appreciable proportion of base metal to the hardfacing layer. With arc welding processes using consumable electrodes, dilution can exceed about 15 percent even with operating parameters at optimum settings to minimize penetration. Hot wire surfacing expertly applied usually can hold the dilution level to somewhat less than 15 percent. Second, the steel being hardfaced ordinarily is much leaner in alloy content than the filler metal; consequently, the iron base of steels is a potent diluent in reducing the alloy content of the hardfacing layer. This downward shift of alloy content can detract significantly from the service capability of the hardfacing, and this is reason why interest always has remained high for surfacing processes that can apply filler metal with virtually no dilution.

Ion implantation is a process for surfacing or "surface modification" that does not involve welding or spraying, but deserves mention for its unique capabilities, and because of consequences that might ensue if ion implanted surfaces became involved in subsequent welding operations. Ion implantation is able to introduce alloying elements into an unheated, solid metal surface unfettered by the familiar metallurgical constraints outlined for alloying in Chapter 4. The usual guidelines for solid solubility, diffusion, etc., simply do not hold for the ion implantation process. Elements for alloying the surface of a workpiece can be selected regardless of their atom size, physical properties, chemical characteristics, etc., and forcibly implanted into the surface of a metal or alloy. There are, however, operational limitations that restrict the elements that can be used in the process. Nevertheless, freedom from the solid solubility limits allows highly unusual alloyed surfaces to be created that offer a wide variety of properties and characteristics.

When ion implantation is performed on a workpiece, a source of ions must be created by use of an accelerating voltage that may range from several keV to MeV. A high vacuum must also be provided, and the surface to receive the implantation must usually be positioned for direct exposure to the ionic beam; although various techniques are being developed to circumvent this line-of-sight requirement. Ions of gaseous elements are generated most readily, and nitrogen is a very cooperative alloying element. Carbon and boron are also willing participants in ion implantation. Metal ions are more difficult to generate, and presently the process has made use of those elements that form volatile compounds (e.g. halides of aluminum, chromium, and copper). One feature of an ion implanted surface to be kept in mind is that the treatment extends only to a very shallow depth; typically from 0.1 to 0.5 micron (about 0.04 to 0.2×10^{-4} in.). The surface can exhibit highly unusual properties that may markedly affect welding behavior during subsequent application of a conventional welding process that is surface-sensitive (e.g., B, FRW, RSW, SSW), but whether the relatively small amount of implanted alloying element will influence the application of a fusion welding process is problematical.

The implantation of surfaces that makes unusual alterations in composition, and the deposition of thin films by new processes is moving forward very rapidly. Personnel concerned with welding must be alert to recognize surfaces on workpieces that have been treated by these new methods, and to have an understanding of their physical make-up, properties, and chemical composition.

Overlaying and Cladding

Overlaying is a term used sometimes in place of *surfacing* when a dissimilar filler metal provides a layer having improved properties (other than wear resistance) on a base metal. The term "cladding" is more descriptive of a thin solid section bonded to the surface of a heavier section. Earlier in this chapter mention was made of cladding steel plate by explosion welding, or by roll welding two (or more) solid sections together.

Overlaying steel workpieces is much like hardfacing except that the filler metal being deposited by welding, brazing, or spraying is not necessarily a composition that attains a high level of hardness. In fact, the overlay deposited on carbon and low-alloy steel workpieces is often some type of stainless steel, and its purpose is to provide corrosion resisting surfacing on a section that otherwise would be attacked by the service environment. While corrosion protection is the most frequent reason for overlaying, other special purposes can be fulfilled. For example, where a section must be made of a particular steel to utilize its strength, but areas on the surface of the section must have reduced friction characteristics to sustain sliding loads, an overlay of bronze might be applied by any of the processes used for braze welding. As another example of special purpose overlay, the gas tungsten-arc process and the plasma arc process are notable for their ability to produce deposits of weld metal with a low content of very small nonmetallic inclusions. This capability to produce very clean metal has been utilized in overlaying steel rolls that will have their surfaces ground-and-polished to a mirror finish, and must not have surface imperfections caused by stringers of nonmetallic inclusion.

Effective surfacing is sometimes provided by a very thin film or coating of metallic or nonmetallic material deposited by vapor deposition. Heat sources much like those employed in the welding processes are used to produce the required vapor for the particular deposition process. Again, welding personnel should have some knowledge of the nature of these vapor deposited surface films on workpieces in order to anticipate their possible involvement and effects in subsequent welding operations. The pulsed excimer laser is being used to "flash evaporate" complex materials (e.g., alloys, composites, ceramics) and then allow the vapor to condense on a workpiece's surface where a very thin, undiluted film is wanted. The plasma torch can also be employed to vaporize some of these materials where the higher temperature of the excimer laser is not required. Ion beam steering of the vaporized material to the surface of the workpiece is a promising new development. Perhaps the zenith of surface treatment is the *diamond* film which can be deposited by thermal decomposition of a hydrocarbon gas, and inducing the carbon atoms to firmly bond to a substrate where they crystallize in the familiar tetragonal structure of the diamond. These new avenues of surfacing technology eliminate two problems long endured with many of the welding processes; namely, dilution of the surfacing deposit with base metal, and development of a heat-affected zone in the base metal beneath the surfacing layer.

THERMAL CUTTING PROCESSES

Apart from joining processes, high temperature heat sources can be used to cut, and to perform shaping operations like beveling, gouging, scarfing, piercing, and washing. These operations may be performed on steel plates, sheet, bars, castings, and forgings to prepare them for subsequent welding, or they may be incorporated at almost any stage of the welding procedure to aid in producing a finished weldment. As a simple commonly used example, a single-bevel-groove butt joint in plate with an abutting root face may be fusion welded first from the open or beveled face side. Next, any unfused abutting root joint can be thermally gouged out from the back side to ensure complete penetration by subsequent deposition of filler metal on the back side. Most of the heat sources developed for welding have been adapted for cutting, and some are so effective that they now compete with machine tool operations regularly used for blanking, small hole drilling, and trimming. The metallurgical effects of the thermal cutting processes are similar to those of welding, but unusual phenomena also can occur that can influence the character of welds made in thermally-cut surfaces.

Metallurgical Effects of Thermal Cutting

Whether severing, gouging a shallow groove, or shaping the metal by some other action, all thermal cutting processes accomplish their objectives by *melting* metal in the workpiece. Processes differ, however, in the amount of metal that is melted and how it is displaced, and, more importantly, they differ in the amount of molten metal left to solidify on the cut surface of the workpiece. In fact, such differences occur even as the *procedure* is varied with a given process. Consequently, it is important to understand exactly how specific processes are intended to operate, and investigate how a given process and procedure actually perform in a particular operation. Two aspects to be watched closely are: (1) the amount of metal melted and whether its chemical composition becomes altered during melting, and (2) the amount of melted metal retained on the cut surface of the workpiece and allowed to resolidify as part of the section. The importance of these aspects will become apparent as the various thermal cutting processes are reviewed.

Oxygen Cutting Processes (OC)

Shop jargon describes work being performed with the oxygen cutting torch or oxygen lance as "burning," but this is a nonstandard term for *oxygen cutting*. Nevertheless, burning or oxidation is the primary reaction put to use in oxygen cutting. Iron, when heated to about 870°C (1600°F) or higher, combines readily with purified oxygen to form iron oxide. This chemical reaction takes place very rapidly, it generates a high temperature, and it liberates a large amount of heat. Several different reactions are possible as affected by the amount of oxygen being supplied to the iron during the oxidation reaction. The reactions that may take place expressed as balanced chemical equations are as follows:

$$Fe + 0.5\, O_2 \rightarrow FeO \qquad (Eq.\ 6.9)$$

The product of this first reaction is ferrous oxide which contains 77 percent iron and 23 percent oxygen. Heat liberated by the reaction is 65 k cal/mole, or 272 k J/mole (258 Btu/mole). This particular reaction is likely to be transitory, if it occurs at all, because when ample oxygen is available the reaction is strongly inclined to proceed in accord with the second equation below.

$$3\,Fe + 2\,O_2 \rightarrow Fe_3O_4 \qquad (Eq.\ 6.10)$$

This reaction produces magnetite, a black iron oxide containing 72 percent iron and 28 percent oxygen. The heat released is 267 k cal/mole, or 1117 k J/mole (1059 Btu/mole), which predominates the first reaction shown above.

$$2\,Fe + 1.5\,O_2 \rightarrow Fe_2O_3 \qquad (Eq.\ 6.11)$$

This third reaction occurs when the quantity of oxygen available for the reaction is substantially in excess of that to produce the black oxide, magnetite. Occurrence of the third reaction will be evident from a red iron oxide (hematite) being ejected from the kerf along with the otherwise black oxide. Hematite contains 70 percent iron and 30 percent oxygen, and its formation releases a slightly smaller amount of heat, 197 k cal/mole (781 Btu/mole).

Differences in the quantity of heat liberated by each of the three reactions described above serve to emphasize the importance of adjusting the flow of oxygen commensurate with cutting conditions to obtain proper efficiency in the operation. The feature of the iron-oxygen reaction which makes oxygen cutting so successful is the very high rate of reaction. When iron is heated to 870°C (1600°F) or higher, even in just a small spot under the preheat flames of a torch, oxidation of the iron can begin almost instantly and proceed with great speed. The rate of energy liberation is so fast that heat from the reaction produces an intense local temperature, which in turn acts to melt a considerable amount of iron adjacent to the focal point of the reaction. Material discharged from the kerf ordinarily consists of about two-thirds iron oxide, and one-third unoxidized iron that was melted and carried out of the kerf by the stream of unused oxygen. Purity of the oxygen used in cutting is an important factor in sustaining an efficient operation. Purity of 99.5 percent or better is the norm for oxygen supplied commercially, and this ensures the cutting efficiency regularly expected. A decrease of only one percent in oxygen purity will reduce the maximum obtainable cutting speed by about 15 percent, and will raise oxygen consumption by as much as 25 percent. Oxygen with purity below 95 percent will not produce acceptable ignition and properly sustain cutting action. The purity of oxygen in the cutting stream can be decreased by abnormal operation of a torch that allows air to aspirate into the stream.

Alloying elements in iron, as found in steels, affect the oxygen cutting operation in several different ways (See Technical Brief 38). Also cast iron, despite its rather simple composition, is resistant to regular oxygen cutting procedures because of the high carbon content. Compared to iron the slower oxidation rate of carbon as graphite and iron carbide hinders the oxidation of the iron matrix. In addition to this difficulty in the cutting operation, per se, the heat-affected zones adjacent to the kerf can develop cracking.

Steels cut by an oxygen process can display unusual changes in alloy composition or distribution on a microscopic scale at the cut surface. These changes can be detected by their effect on microstructure, and can be analyzed quantitatively by localized sampling, or by the use of the electron microprobe analyzer. Significant increases in carbon and nickel contents can occur at an oxygen-cut surface because of selective oxidation of elements in the steel during cutting. Carbon and nickel oxidize at a slightly slower rate than iron; consequently, these elements tend to concentrate in molten steel on the cut surface. Increased carbon, because of the potency of this element in alloying steel, usually can be detected by a larger amount of iron carbide phase appearing in the microstructure at the cut surface. The carbon may migrate inward a short distance from the cut surface because its small atomic diameter permits a high rate of diffusion. Nickel, if present as an alloying element, tends to remain concentrated immediately adjacent to the cut surface. A small superficial decrease in chromium sometimes will occur because of preferential oxidation of this alloying element over the iron. These changes in alloy composition at the cut surface generally are regarded as minor effects. Yet, circumstances occasionally will make the increase in carbon content a problem.

Oxyfuel Gas Cutting (OFC)

A number of fuel gases may be used in combination with oxygen for flames to preheat the steel in a localized spot the ignition temperature. Acetylene (OFC-A) is most widely used for this purpose, but good use also is made of natural gas (OFC-N), propane (OFC-P), and hydrogen (OFC-H). Special hydrocarbon gases also are available that offer advantages in handling, flame temperature, energy release rate, cost, etc. Gasoline also is being used with a torch designed for this fuel.

The conventional oxygen cutting torch is equipped with a tip that contains a number of orifices from which small oxygas flames burn continuously for heating, and a central orifice from which a stream of high-pressure oxygen can be released. When operating the torch, a small spot on the workpiece surface is first preheated to a bright red heat. When the steel is at this ignition temperature, the flow of high-pressure oxygen is started and an immediate cutting action initiates. The kerf cut through the section will continue to advance as long as the oxygen stream is supplied. Heat from the oxidation reaction is so intense that often the oxygas heating flames are unnecessary once the operation gets underway. However, the outer circle of heating flames helps prevent aspiration of air into the oxygen cutting stream.

Iron when heated to the ignition temperature will quickly combine vigorously with oxygen to produce considerable heat. It is not necessary, therefore, that a substantial area of the surface of the steel be heated to a red heat (approximately 870°C, 1600°F) to start cutting. Indeed, a tiny droplet of red hot metal will serve nicely. If the workpiece is already at or above the ignition temperature, as would be the case immediately following a hot rolling or a forging operation, preheating with an oxygas flame is not necessary. The oxygen lance (LOC) is a simple blowpipe that provides only the high-pressure oxygen stream for severing large, red-hot steel sections. When scarfing the surface of a workpiece, a small molten drop of steel can be introduced for quick starting, and the torch tip is designed to deliver an oxygen stream at a low angle to the surface. This produces a shallow groove, and all oxide and molten metal except for a thin film on the scarfed surface is swept away by the excess oxygen. Consequently, the heat-affected zones left by *scarfing* ordinarily are quite shallow, and they cool very rapidly. Lance cutting, on the other hand, ordinarily produces copious amounts of molten metal, much of which is trapped on the relatively rough cut surface where it solidifies and produces a substantial heat-affected zone.

Chemical Flux Cutting (FOC)

This process can be used for cutting steels that form a refractory slag by using a conventional oxyfuel cutting torch and one additional piece of equipment. A flux feeding unit is employed to inject a powdered flux into the line carrying the high-pressure oxygen stream. The flux need not be complex. For example, sodium bicarbonate serves effectively in disposing of the viscous refractory slag that forms in the kerf of high-alloy steels (such as the stainless steels). While this process was developed for oxygen cutting the stainless steels, which contain a large amount of chromium, it does see occasional use on certain alloy steels that contain aluminum, and on cast iron. Chemical flux cutting is not applicable to metals like copper, nickel, Monel, Inconel, and so forth, because these do not contain enough oxidizable iron, manganese, or chromium to generate the heat required.

Metal Powder Cutting (POC)

Use is made of powdered iron in oxygen cutting, scarfing, and gouging operations when severing alloy steels containing elements that form an obstructive refractory slag, or where instantaneous starting is desired. The metal powder process use oxyfuel (oxygen) cutting equipment supplemented by a unit that feeds iron-rich powder through a tube to the torch tip where it is directed into the oxygen cutting stream. The

> **TECHNICAL BRIEF 38:**
> **Effects of Alloying Elements in Steel on Oxygen Cutting**
>
> Certain commonly used alloying elements are oxidized as readily as iron, and generally they cause no problem in cutting the carbon and low-alloy steels. Manganese, silicon, and vanadium fall into this category. Some alloying elements, however, are not as readily oxidized as iron, and when present in steel in substantial quantities, they limit the amount of iron available for reaction with the oxygen stream and thereby reduce heat liberation. The elements nickel, copper, and cobalt are of this variety.
>
> Another more frequently encountered difficulty arises when alloying elements are present that are readily oxidized, but their oxides are highly refractory. Chromium and aluminum typify elements of this kind. When appreciable amounts of elements are present that form refractory oxides, a protective slag forms over the molten surface being cut. This slag shield can be so effective in hampering further oxidation that the alloy steel cannot be effectively cut with regular oxygen cutting procedure. Chromium levels about 10 percent and higher produce this obstructive film of refractory oxide. Therefore, oxygen cutting of the stainless steels is an unwieldy operation. Corrective measures have been developed to overcome this problem, and they can be grouped as either stopgap techniques that can be applied immediately still using the regular cutting torch, or as special processes that involve different cutting equipment.
>
> Several techniques that can be immediately applied are based on the introduction of more iron to the reaction point at the kerf to increase heating and to dilute the obstructive refractory slag. Additional preheating and torch oscillation can provide greater heat in an attempt to encourage the refractory slag to disperse, but this technique is usually of limited assistance. One effective technique is to use a "waster plate" of plain steel that is placed atop the alloy steel workpiece to feed more iron into the oxygen stream. Also, a plain steel bare rod can be fed into the kerf as the cutting operation is initiated and progresses, but this technique requires considerable operator skill. Although these techniques allow oxygen cutting to sever alloy steels, they produce a cut edge that is quite rough.
>
> Special oxygen cutting systems devised to overcome the refractory slag problem will be described in the text; these systems include *chemical flux cutting* (FOC), and *metal powder cutting* (POC). Furthermore, it can be noted here that these processes offer additional advantages, such as instantaneous starting of a kerf or a scarfed area (without preheating). Also, the advent of arc cutting processes, particularly, *plasma arc cutting* (PAC), have afforded many more measures for handling the thermal cutting of alloy steels and other alloys.

metal powder must be carried through the tube by an atmosphere of compressed air or nitrogen for if compressed oxygen were used, or if the powder was injected into the oxygen hose line, spontaneous combustion of the metal powder likely would occur. As the metal powder enters the oxygen stream at the torch tip, it burns and produces considerable heat. The viscous refractory slag that forms on some high-alloy steels is removed by a combination of melting and dilution with iron-rich oxide. Instantaneous starting on a cold workpiece is possible providing a burst of metal powder is metered into the oxygen stream just in advance of the torch being applied to the workpiece.

Electric Arc Cutting Processes (AC)

Arc cutting can be divided into two categories: (1) processes that employ the arc alone to provide sufficient heat to melt through the workpiece, and (2) processes that supplement the arc and its melting action with a compressed gas delivered as a stream. The latter processes are more efficient since the gas stream plays an important role in removing the molten melted metal from the kerf. The gases used in the arc cutting processes may range from a highly chemically active gas like oxygen, to an inert gas, for example, argon. Even hydrogen, nitrogen, and compressed air sometimes are employed in certain processes. A variety of arc torches are described in the literature for gouging, piercing, etc., in addition to cutting a kerf.

Simple cutting processes are carried out with the same equipment as used for welding. *Carbon arc cutting* (CAC) is the earliest form of arc cutting, and is performed with a graphite electrode operating on DCEN, and at a level of current much higher than would be used for a welding operation. Cutting of the workpiece actually is a matter of melting a kerf through the section. Removal of molten metal from the kerf occurs primarily by the force of gravity, although the removal is supplemented to some extent by the pressure and other forces exerted by the arc itself. Electrodes designed for effective arc cutting are made for *shielded metal arc cutting (SMAC)*. This process ordinarily is performed with DCEN, but some electrodes can be operated on alternating current. When thick plate is cut by SMAC, the flux covering serves to insulate the electrode from the sides of the kerf as it is inserted deeper into the opening to accomplish melting through the thickness. The covering also permits higher current to be supplied to the arc and serves to concentrate the heat in a small area on the workpiece for effective melting.

Gas metal arc cutting (GMAC) can be performed with an electrode holder or "gun" as illustrated earlier in Figure 6.12 for welding. The small size bare electrode wire used to deposit filler metal will also serve to support a cutting arc that operates with reasonable efficiency providing the proper combination of electrode feed, voltage, current, travel speed, and shielding gas is employed. While a number of gases can be used for shielding, argon is perhaps the best suited for the cutting operation. The advantages of argon lie in the operating characteristics of the arc rather than any need to shield the surfaces of the kerf. During GMAC operations, the end of the electrode actually extends into the kerf through the entire workpiece thickness. The arc is established between the leading side of the electrode extension and the edge of the workpiece through the entire depth of the kerf. The plasma oscillates in a vertical direction within the kerf to melt the workpiece from top to bottom. The molten metal leaves the kerf at the bottom under the force of gravity and the arc forces. The process is capable of cutting any kind of material, but cut edges of the workpieces are quite rough and the operating cost does not permit extensive usage of carbon and low-alloy steels.

Gas tungsten arc cutting (GTAC) stems from the welding process, GTAW, illustrated earlier in Figure 6.9. Generally, the electrode holder used for cutting must be suitable for carrying current well above the range normally used for welding. A mixture of argon and hydrogen is often used for shielding to secure increased heat from the arc, but some success has been achieved with nitrogen alone, and this will reduce operating cost. The sides of a kerf cut through workpieces are inclined to be rough, particularly with manual cutting, but scaling of the edges will be minimized by the shielding and cooling afforded by the gas stream. Gravity, arc forces, and the modest flow of shielding gas are the factors that discharge molten metal from the bottom of the kerf. Therefore, it is not practical to employ this process on a workpiece thicker than about 12 mm (½ in.). In addition to the roughness of the kerf, the cost of the operation discourages regular use of GTAC on the carbon and low-alloy steels.

Unique cutting processes have been devised by combining electric arc and oxygen cutting into a single operational system. Oxygen arc cutting (AOC) uses a tubular steel electrode bearing a flux covering. The electrode is gripped in a special holder which passes current to the electrode for striking the arc, but immediately after the arc is established a stream of compressed oxygen is fed through the tubular electrode to the workpiece. The intense heat of the arc is supplemented by an oxidation reaction when applied to ferrous alloys. The operator must feed the consumable electrode into the kerf to sustain the cutting operation. The kinetic energy of the oxygen stream expels the oxide and molten metal from the bottom of the kerf. Any metal or alloy can be severed by this process, even those that do not contain oxidizable elements. A modification of oxygen arc cutting has been developed in which the tubular steel electrode bears a proprietary exothermic covering. The end of the electrode that makes contact with the workpiece can be ignited with the aid of a 12-volt battery. As the exothermic reaction takes place in the covering to preheat an area of the workpiece, a stream of oxygen performs the usual cutting of a kerf through the section. The principal advantage of this process is its portability and minimum of equipment, which does not include the usual power source.

The surface of a kerf made by manually cutting with an electric arc, or with an oxygen-arc process is ordinarily quite rough, but sometimes these processes are the most expedient ones available.

Air Carbon Arc Cutting (CAC-A)

This widely used, effective process evolved with the discovery that molten metal beneath a carbon arc on the surface of a workpiece could be quickly swept away as rapidly as it formed by a jet of compressed air without extinguishing the arc. This function allows CAC-A to sever sections of metal by repeated gouging passes, but other processes that have penetration capability to cut the entire thickness of a section in a single pass are more efficient for severing bars, shaping plates, etc.

The tool for the CAC-A process is a relatively simple electrode holder for a single carbon electrode with orifices in the holder to direct streams of compressed air at a specific angle toward the point where a weld pool would normally form. The compressed air is released as required for removal of molten metal beneath the arc. The carbon electrode is often copper plated to improve its current carrying characteristics and to make it more durable. Electrodes may range in size from four mm (5/32 in.) to as large as 18 mm (about ¾ in.), which allows current from 200 to as high as 1500 A to be employed. Higher arc current accomplishes more rapid removal of metal from the workpiece because all of the melting is produced by heating from the arc. Removal of metal by the air stream (supplied at a pressure of about 560 to 700 kPa, or 80 to 100 psi) is purely a pneumatic or mechanical action. The operation can be performed manually, or it can be mechanized.

Operation of the CAC-A process is normally carried out with an arc welding power source supplying current on DCEP because the arc would be quite unstable on

DCEN. This matter of polarity introduces an important phenomenon which deserves close attention. As the carbon arc is operated on DCEP, ionized carbon atoms leave the electrode and are attracted to the cathodic workpiece where these atoms rapidly carburize molten metal beneath the arc. This carbon-enriching mechanism cannot be avoided, and it is important that all molten (carburized) metal be removed from the gouge or kerf. Some use is made of alternating current with this process, but the same carburizing mechanism occurs during the reverse-polarity half-cycle of current flow. Several operating conditions must be held in reasonable control during the CAC-A process to be certain that all molten metal is swept from the surface of the workpiece. Air pressure and flow must be adequate. Arc voltage must be sufficiently high to permit a relatively long arc and thus allow the air stream to sweep below the tip of the electrode. When the operation is performed with proper conditions, the surface of the gouge normally will show no signs of carburized metal. However, if operation conditions are not correct, if the forward travel of the arc is irregular, or if any condition permits molten metal to remain, this metal (assuming it to be ferrous) invariably will be highly carburized. Carburized metal left behind on the workpiece can be recognized by its dull gray-black color. This appearance is in contrast with the bright blue color of a properly made groove. Although carburized metal on the surface can be removed by grinding, it is more efficient to carry out the CAC-A operation within prescribed conditions and avoid the retention of the undesirable metal. It cannot be assumed that carburized metal on the surface of a gouge or kerf will be assimilated or alloyed with filler metal subsequently deposited by a welding process. Although the electromagnetic stirring of the weld pool by an arc will promote this action, instances have been found (as illustrated in Volume II, Chapter 13) where the carburized metal from the gouging operation has been dispersed throughout the weld metal as tiny islands of higher-carbon metal. The importance of a microstructural condition like this will depend, of course, on the kind of weld metal, the size and location of the islands, and the performance expected of the weld in service.

Plasma Arc Cutting (PAC)

The plasma arc cutting process is a spin-off from plasma arc welding. Although either the nontransferred arc type of torch, or the transferred arc type (see Figure 6.22) can be adapted for cutting, the transferred arc is used most often because it is more efficient in carrying energy to the workpiece where heat is utilized in melting, and because a more forceful tailflame can be generated for driving the molten metal free of the kerf.

Arc energy is the principal factor that determines the thickness of steel that can be cut by PAC, and while most use is made of the process in cutting sections in the range of about five to 75 mm (approximately $\frac{3}{16}$ to 3 in.), PAC torches are available to cut sections at least as thick as 150 mm (6 in.). This range of sections can require current from about 70 to 1000 A, and open-circuit voltage of at least 120 V. Very thick sections may require open-circuit voltage as high as 400 V; consequently, strict electrical safety measures must be followed when using the process. The process can be carried out manually, or can be mechanized, and the latter operation is capable of producing a good quality, square, smooth edge with a minimum of adhering dross. Because cutting is accomplished by melting, and does not ordinarily make use of oxidation as a principal source of heat, PAC machine cutting facilities are favored in many shops that must cut a variety of metals and alloys, some of which contain only a small content of oxidizable elements.

Selection of gases to form the PAC plasma, and for secondary shielding requires consideration of many factors. Argon-hydrogen mixtures have been found more effective for the cutting (plasma) gas than argon alone. Nitrogen has proved suitable for cutting steel, Much experimentation was carried on with additions of oxygen to the cutting gas to include oxidation as part of the heating and metal removal mechanism. When this appeared to achieve some success, air was substituted for the oxygen-containing gas mixtures. While the quality of cutting performed with air suffered a little, the use of air became popular for cutting steel because of substantially reduced operating cost. Gases used as a secondary flow for nozzle protection and for workpiece shielding include argon, argon-hydrogen, and carbon dioxide as well as air. Introduction of the cutting gas into the torch with a tangential or swirling motion has an effect on the top edges of the kerf. As the cutting gas flows from torch to workpiece with clockwise rotation (as viewed from above) a more efficient transfer of arc energy takes place on the right side of the kerf (as viewed when facing in the direction of travel). This produces a square top edge on the right side of the kerf, and a slightly rounded edge at the left side. This difference in the top edges requires consideration when selecting a particular torch that embodies a swirling injector for gas. Whether, the swirling motion is clockwise, or counter-clockwise, and the direction of travel will determine whether the square top edge is on the trimmed workpiece, or on scrap to be discarded. There is also a tendency for PAC kerfs to be slightly wider at their top, and this may produce a very slight bevel on each side of severed workpieces.

Success on the PAC process is largely attributable to developments in torch design and the mode of operation with particular gases for plasma formation and for secondary shielding. Important design changes were made in the internal shape of the nozzle chamber or plenum to markedly constrict the plasma. The smaller diameter tailflame that issued from the torch had a higher temperature, it produced a narrower kerf, and it permitted faster travel speed. In some torches, plasma constriction was achieved by introducing the cutting gas into the plenum with a stronger swirling movement around the electrode. This mode of gas motion not only constricted the plasma, but also forced some cool un-ionized gas to move outward and form a boundary layer against the inner wall of the nozzle to reduce its temperature and extend operating life. Redesign also allowed secondary shielding gas to be added to the protective boundary layer.

A surprisingly large improvement in the cutting capabilities of the PAC process took place when "damp" gases were noted to improve the quality of kerf edges, and this observation was carried further by actually injecting water as a secondary medium in the torch as illustrated in Figure 6.70. The water is also given a swirling motion, and only about 10 percent of the amount injected is flashed into vapor. As the vapor encounters the plasma, some of it may be dissociated into hydrogen and oxygen. The energy absorbed to accomplish this dissociation subsequently is released at the workpiece when recombination occurs. Presence of the water has a profound effect in constricting the tailflame. Also, water injected with a swirling motion tends to be moved

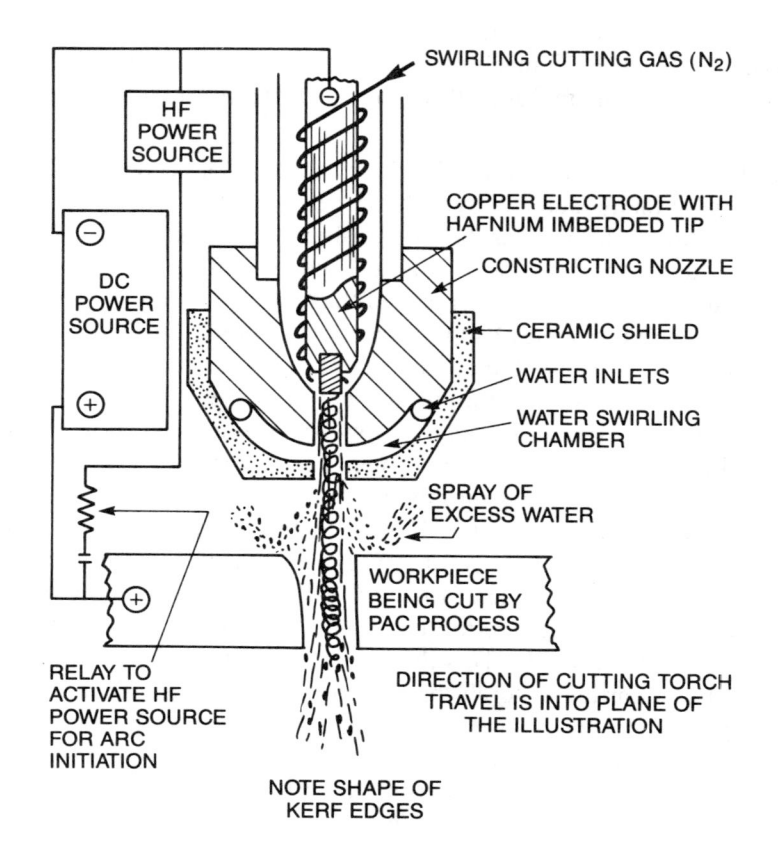

Figure 6.70 — Schematic of the plasma arc cutting process (PAC) showing components of a torch incorporating water injection for tail flame constriction and for fume suppression

radially outward and contributes to a boundary layer at the inner wall of the nozzle. When water is metered into the torch in proper amount, a portion will exit the torch nozzle as a conical spray surrounding the tailflame. The droplets of water impinge on the top surface of the workpiece to assist in reducing oxidation of the kerf edges, and in suppressing smoke, fumes and noise.

Many shops use a water table on which the workpiece can be supported and submerged with its top surface just below the water level. The PAC operation is then performed with the torch just above the water. The tailflame will cut a kerf through the submerged workpiece with a minimum of surface oxidation, smoke, fumes noise, and workpiece distortion.

Tungsten has served as the electrode in PAC torches, but hafnium and zirconium give longer operating life as an electrode, especially when air or an oxygen-containing gas is used as the cutting gas. To reduce electrode cost, a "pill" of costly metal (e.g., hafnium) can be embedded in the tip of a copper holder to serve as an electrode as shown in Figure 6.70.

Metallurgical effects of plasma arc cutting on steel are those to be expected when rapidly melting a narrow kerf, and all melted material is ejected by the jet-like tailflame leaving only a very thin molten film on the kerf walls. Naturally, this process will produce heat-affected zones in the solid steel adjacent to the kerf. The depth of these zones, their contour, and the microstructural and mechanical property changes within them vary greatly with the PAC procedure, and the kind of steel being cut. Briefly, the kinds of gases employed for the cutting plasma, and the use of water injection, does not appear to have noteworthy chemical effects on the cut edges, such as the carbon enrichment that occurs with oxyfuel gas cutting, or the carburization that can occur with air carbon arc gouging. Because of the high temperature provided by the PAC tailflame, cutting travel speed can be relatively fast, and the heat-effects observed are likely to be similar to those from high speed arc welding processes, but of course, no molten filler metal is deposited on the kerf wall during cutting that would serve as a heat source and thus retard cooling. Conversely, if water injection, and/or a water table are employed, cooling of the heat-affected zone can be very rapid. Since a narrow zone in the solid steel has been heated above the critical range, when dealing with carbon and low-alloy steels the commonly encountered consequences of rapid cooling can be (1) hardening, (2) loss of toughness, and (3) cold cracking. Whether these troublesome conditions manifest to a significant degree depends greatly on the composition of the steel being cut. One fundamental point to be kept in mind concerning any thermal process that cuts full thickness in a single pass is that the HAZ will extend from top to bottom surface in an unbroken pattern, that is, the zone does not ordinarily have a series of smaller, overlapping zones as found in a multipass weld. Figure 6.71 illustrates typical heat-affected zones on the cut edges of a carbon steel, and allows the depth and contour of the zone produced by plasma arc cutting to be compared with the zone from oxyfuel gas cutting. Metallographic examination of the PAC edge, as shown in Figure 6.72, shows the change in microstructure that occurred from the heat-effect of cutting. Although a marked change occurred in the microstructure which undoubtedly raised the hardness level at the edge, no cracking is seen. When the composition of the steel has the potential for producing a very hard, crack-susceptible microstructure, the usual remedial measures are preheating, and/or post-cut heating.

Electron Beam Cutting (EBC)

The unique ability of the electron beam to completely pierce or penetrate metal has been described earlier when discussing electron beam welding. Although this penetration capability can be used for cutting, other thermal cutting processes are much cheaper and, in most cases, more satisfactory. Consequently, the electron beam seldom is used strictly for metal cutting. Nevertheless, the unique penetration capability of the electron beam when focused on a very small spot makes EBC an outstanding process for making very small, clean holes in any kind of metal or alloy in workpiece thicknesses ranging from about 0.05 to eight mm (0.002 to 0.3 in.). Electron beam drilling is used because it offers a number of advantages over conventional mechanical drilling when many small holes must be drilled, when the material has poor machinability because of high hardness, or when the hole geometry is not a simple circle and instead is an oval or some special shape. EB drilling includes special features in its process technology that should be noted by welding personnel.

The machines for EB hole drilling usually are the dual-vacuum type (illustrated earlier in Figure 6.37), and additional devices often are installed to overcome unwanted emissions that increase the risk of arcing or gun discharge. One such device is an arrangement of counter-rotating, synchronized, overlapping disks with windows for beam passage that are positioned in the upper part of the work chamber to collect emitted debris. Scrapers sometimes are installed to remove continuously the deposited material. This arrangement, along with the vacuum pumping applied directly to the work chamber, avoids undue occurrence of gun discharge. The vacuum maintained in the work chamber for the drilling operation is not as hard as for welding, and is usually in the order of 10^{-2} torr. The EB hole drilling machines have a relatively high accelerating potential (e.g., 120 kV), and are designed to deliver a penetrating beam as a precisely controlled very short burst, perhaps with a duration of only 0.05 to 100

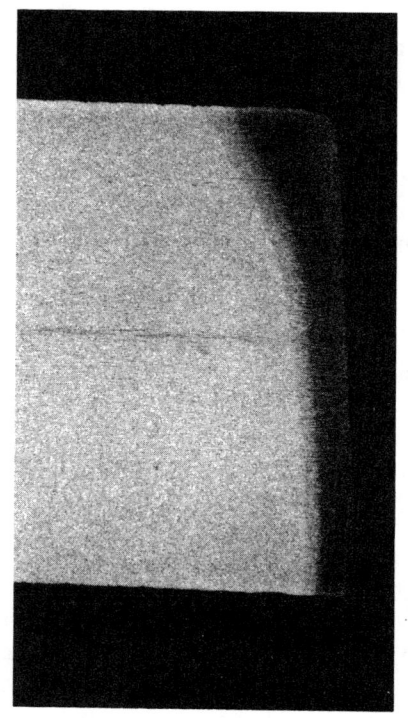

Figure 6.71 — **Macrographs comparing the torch cut edges of carbon steel plate 12.5 mm (½ in.) thick.**

Cuts made by (left) oxyfuel gas cutting (OFC), and by (right) plasma arc cutting (PAC) with dry argon–15% hydrogen cutting gas. Steel contains 0.15 carbon, 1.41 manganese and 0.37 silicon. (Magnification 10X; Courtesy of The Welding Institute, England)

Figure 6.72 — Microstructural changes in 0.15% carbon steel at kerf edge cut by PAC process

Original banded microstructure of steel plate in Figure 6.71 at left side of photomicrograph has gradually been changed to a homogenized acicular microstructure at right side by heat input from PAC, and little evidence remains on kerf face of metal that was melted. (Nital etchant; original magnification: 100X; courtesy of The Welding Institute, England)

milliseconds. The hole is drilled by one burst, the beam power having been established by trial holes. The fine focus of the beam may produce a power density of perhaps 10^6 W/mm^2 (6×10^8 W/in.^{2}), and the total power may reach 12 kW for a single pulse.

Drilling holes in metal with an electron beam became a viable operation with the discovery of an augmentive effect provided by a special backing material. As illustrated in Figure 6.73, a sharply focused beam in the first schematic (A) heats and melts a spot on the upper surface of the workpiece. Continued application of the beam (B) starts to pierce a hole in the workpiece and the pressure of escaping metal vapor maintains an open hole with a film of molten metal clinging to the wall of the hole. In a very short time, the beam penetrates the workpiece (C) in the fashion of a keyhole. If the beam is extinguished at this point, the molten metal on the sidewall of the hole would run toward the bottom pulled by gravity, would likely bridge the small opening, and solidify as a plug of weld metal. Instead, as can be seen in (D) of Figure 6.73, the special backing material receives the emerging electron beam from the hole and immediately starts to vaporize. This develops sufficient vapor pressure to expel virtually all molten metal from the hole in the workpiece leaving only a thin film on the wall which solidifies after the beam completes its pulse. The backing material must have some substance to avoid being completely penetrated by the beam, and yet is must quickly

646 Chapter 6

yield gaseous material to clear out the hole. Backing materials are proprietary composites. One example is composed of a mixture of three parts powdered metals (for substance) and one part hardened silicone rubber (for quickly producing a high vapor pressure). The EB pulse must be timed to just penetrate the workpiece and to gasify a small cavity in the backing material so that it serves as a pressure chamber.

The holes drilled by the electron beam can range in size from 0.05 to 1.0 mm (0.002 to 0.04 in.) in diameter. The smallest hole that can be drilled in a workpiece is more-or-less governed by its thickness, and this minimum hole size is about one-fifteenth of the workpiece thickness. The holes are drilled quickly with one burst of the

(A) PULSE OF ELECTRONS COMMENCES TO HEAT WORKPIECE AND MELT SPOT ON SURFACE.

(B) ELECTRON BEAM STARTS TO PIERCE HOLE IN WORKPIECE, WHICH IS HELD OPEN BY PRESSURE OF METAL VAPORS PRODUCED BY HIGH POWER DENSITY OF BEAM.

(C) ELECTRON BEAM HAS JUST PENETRATED WORKPIECE AND HOLE STILL IS HELD OPEN BY PRESSURE OF VAPORIZED METAL, BUT APPRECIABLE MOLTEN METAL REMAINS ON WALL OF HOLE.

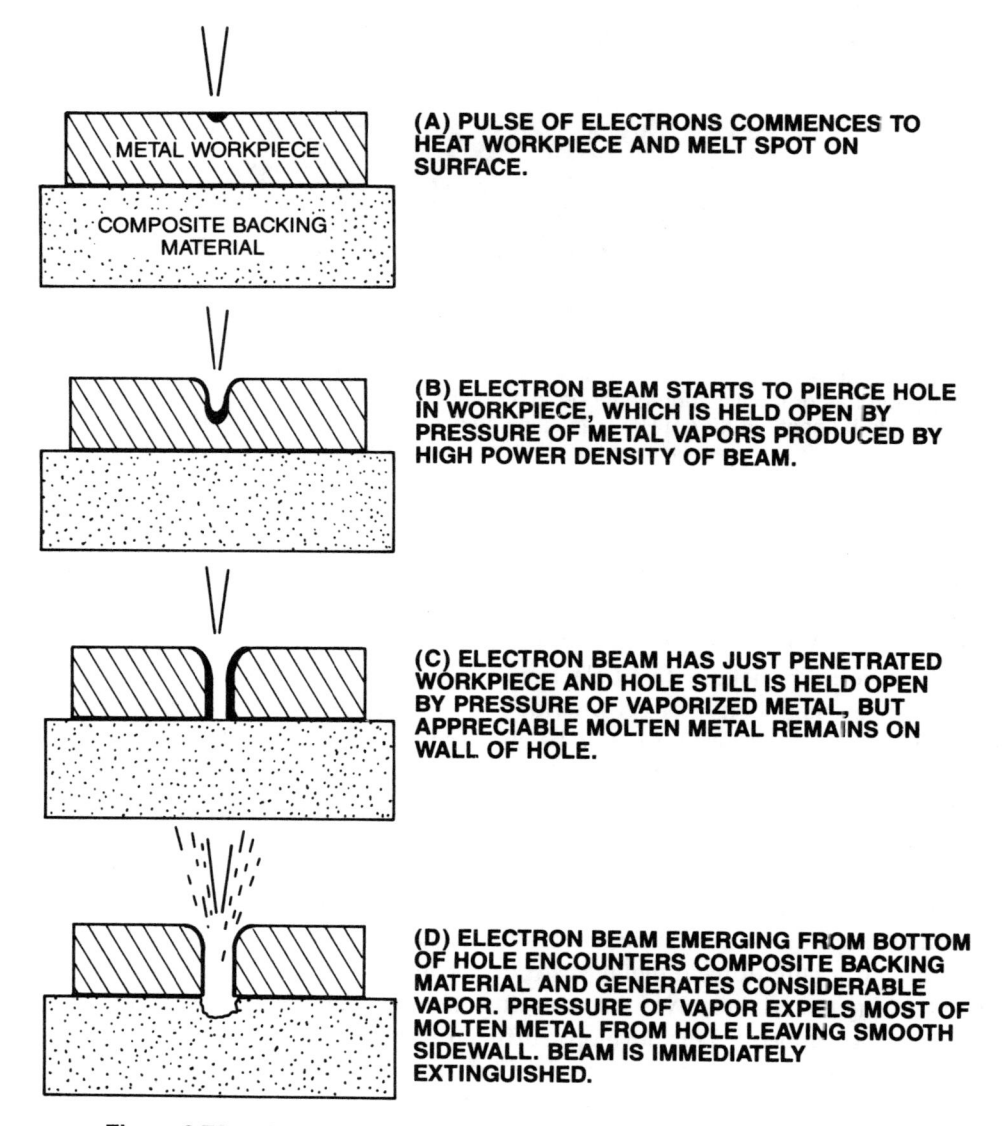

(D) ELECTRON BEAM EMERGING FROM BOTTOM OF HOLE ENCOUNTERS COMPOSITE BACKING MATERIAL AND GENERATES CONSIDERABLE VAPOR. PRESSURE OF VAPOR EXPELS MOST OF MOLTEN METAL FROM HOLE LEAVING SMOOTH SIDEWALL. BEAM IS IMMEDIATELY EXTINGUISHED.

Figure 6.73 — Hole drilling by the electron beam process

electron beam, and the beam can be rapidly and precisely repositioned by deflection coils for multiple hole drilling in a prescribed pattern if desired, rather than by moving the workpiece. Highly sophisticated computer-controlled workpiece manipulators have been devised for EB hole drilling on-the-fly, and as many as four thousand holes per second have been made in metal foil to serve as a perforated screen. The closest approach of multiple holes is about two diameters from center-to-center. Slight rounding of the hole edge on the upper surface is a common occurrence, but the hole shape (including taper, straightness, or otherwise) can be controlled by the beam shape and its focal point with respect to the workpiece thickness. An unexpected advantage found in EB hole drilling is ability to drill holes at an acute angle to the workpiece surface, with angles as small as 20 degrees having been demonstrated.

Since the holes drilled by an electron beam involve melting, and not all of the molten metal is eliminated by the pressured discharge of vaporized backing material through the hole, there will be a very thin layer of metal remaining on the wall that will solidify. Also, a very narrow heat-affected zone will surround each hole. The total depth of fusion plus heat-affected zone from the hole wall will depend on parameters employed in the drilling operation, but this depth can be as shallow as 0.025 mm (0.001 in.). The soundness and microstructural character of these zones has yet to be described in detail in the literature.

Laser Beam Cutting (LBC)

Laser beam cutting has a broad range of applications from very thin metal sections to substantially thicker material (e.g., heavy sheet and plate). This far-ranging usage has virtually separated the process into two different kinds of equipment and procedures. Thin metal members can be severed by melting and vaporization with a finely focused beam from a low-power solid-state laser. They are very effective for precision cutting of very thin material, such as the trimming of resistor wires for microelectronic devices, and for profiling metal foils. The thick sections call for the higher-power gas laser and a gas-jet at the cutting head to assist in producing a clean kerf as illustrated in Figure 6.74. Best results for cutting steel are obtained by using oxygen for the gas jet. In addition to utilizing the high temperature provided by the focused laser beam, the oxygen-assisted LBC-O operation is enhanced by the extremely rapid oxidation reaction that occurs between the heated iron and oxygen. A large amount of additional heat is produced in the kerf by this reaction, and the gas-jet also effectively discharges molten material and vapors through the bottom of the kerf. As a result, much thicker sections of steel can be cut, and the travel speed can be increased substantially. The kerf made by LBC-O is quite narrow, the edges are fairly smooth, and little or no beads of dross or metal solidify along the bottom edges of the kerf. The fast travel speed is desirable from a productivity standpoint, and the relatively low heat input minimizes warping of flat plate during cutting.

No firm guideline can be given for selecting either a lower-power solid-state laser, or a higher-power gas laser for cutting sections of intermediate thickness. Some interaction exists. For example, the solid-state laser can make use of the gas-jet under certain circumstances. Also, the gas laser can be operated in a pulsing mode (similar to the solid-state laser) when peaks of power are needed to gain complete penetration through a thick section. Much of the literature on laser beam cutting follows this loose separation of thin section and thick section operations.

Commercial LBC systems consist of a number of principal parts including (1) the laser beam generator, (2) power source, (3) beam delivery system, (4) focusing and cutting head, (5) mechanism for providing relative movement between beam and workpiece, (6) special control systems for automation, and (7) sensing devices to monitor cutting action during the LBC operation. Capital investment in a laser cutting facility generally is higher than for other more familiar thermal cutting processes. However the process can be used effectively for cutting the carbon and low-alloy steels.

Thin section cuts in steel, from foil up to about 1.5 mm (0.06 in.) in thickness, usually is performed with the Nd:YAG laser. The ouput power from this solid-state, pulsing type laser may range from about 300 W up to about 1 kW. Because the beam can be

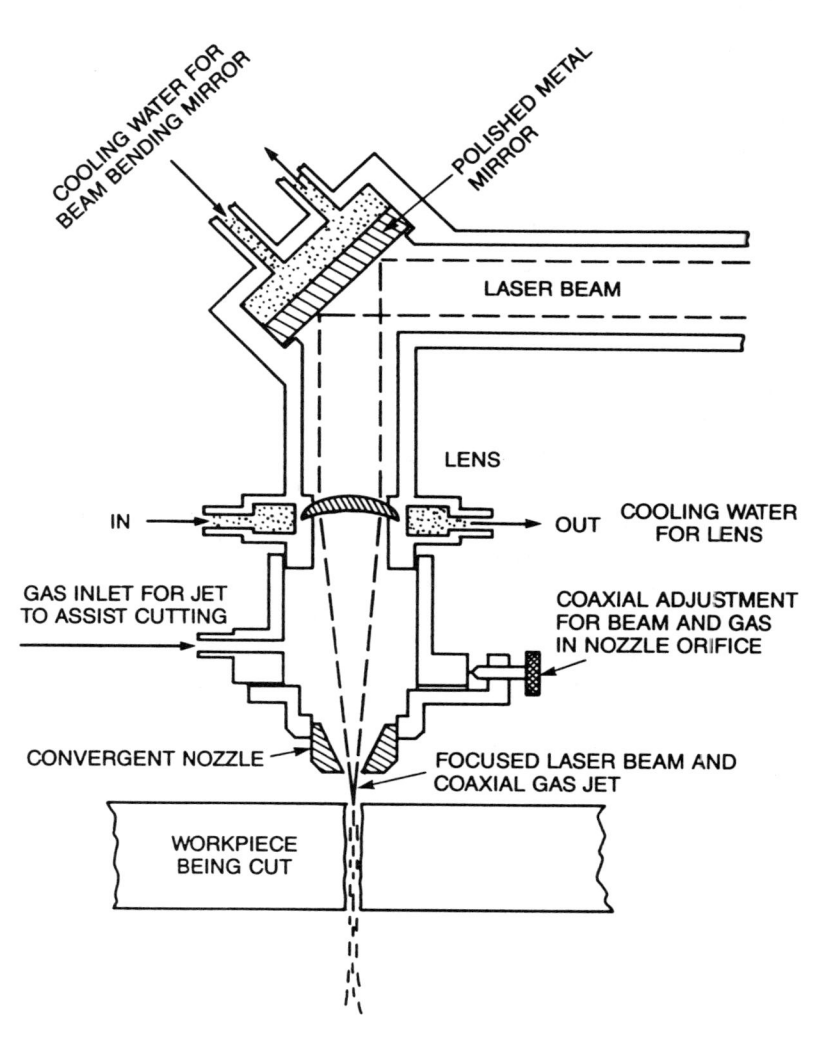

Figure 6.74 — Laser beam cutting (LBC) with coaxial gas jet to assist cutting action

focused to a spot as small as 0.1 mm (0.004 in.), cutting ordinarily is accomplished by the combined actions of melting and vaporization. Sometimes a pulsing beam can be given an enhanced wave form with spikes of peak power to achieve complete penetration through the section. Special controls are available to alter the wave shape of the pulsing beam and to obtain circular polarization. Travel speed in cutting depends on laser power and workpiece thickness, but the travel rates when using the melting-and-vaporization method of cutting are not as fast as can be obtained using an oxygen-jet assist. Nevertheless, when the steel is thin, the travel rate even without gas-jet assistance can be quite acceptable. For example, steel sheet one mm (0.04 in.) thick can be cut at a travel speed of approximately one m/min. (40 ipm). Multiple layers of sheet cannot be "stack-cut" with this method of operation because too much molten metal is retained on the walls of the kerf, and the edges tend to weld together during cooling.

With a gas-jet supplementing the laser beam, especially one using oxygen, cutting speed can be increased substantially, the kerf will be clean, and stack-cutting is possible. The kerf left by LBC in thin material is quite narrow, and usually is not much wider than the diameter of the spot melted by the focused beam on the workpiece surface. However, kerf width is governed by a number of factors, including beam diameter, pulse form, cutting speed, material thickness, and edge smoothness requirements. The LBC process is so proficient on thin material that light-wall tubing often has cutouts that require manipulation in three dimensions. LBC with the solid-state laser also can be used for small hole drilling in thin material since the beam can be focused to a very small diameter. Its penetration capability is increased by pulsing with an enhanced (spiked) wave form. Although LBC is not as proficient as the electron beam process for small hole drilling, the fact that a vacuum chamber is not needed for LBC is strong incentive to consider this process. A YAG laser of 250-W power capacity has demonstrated the feasibility of drilling holes as small as 0.025 mm (0.001 in.) in diameter in ferrous material using fiber-optic beam-delivery to a section 2.7 mm (0.105 in.) thick. Hole drilling with this system also permitted drilling these holes at angles up to 20 degrees off the perpendicular.

Thick sections of steel are cut with the beam from a CO_2 gas laser, which may range in power from about one kW to as high as 15 kW, but most common is a power rating near three kW, mostly for reason of equipment cost. The gas laser can be either axial flow, or transverse flow, and although they are continuous wave type generators, the beam often is pulsed by electronic control to obtain complete penetration at higher cutting speed, and to minimize the kerf width. Almost always, a gas-jet is included to secure a clear kerf and, as indicated in Figure 6.74, the favored kind of gas jet uses a coaxial flow of oxygen around the beam. The assistance provided by this kind of gas jet increases the permissible cutting travel rate by a factor of ten over the rate normally obtainable with the melting-and-vaporization (LBC-EV) technique. LBC-O cutting with the gas laser is applicable up to 50 mm (2 in.) in thickness of steel, but most of the LBC-O cutting in industry is performed on steel thinner than 12.5 mm (½ in.). When steel thickness exceeds 12.5 mm (½ in.), the beam must be tailored in a number of ways to secure best possible effectiveness. Energy distribution in the beam is arranged in the Gaussian mode (TEM_{00}) to facilitate focusing to a small spot on the workpiece. The focal point of the beam must be positioned just below the top surface of the workpiece. Pulsing with an enhanced wave form also may be necessary. Even then, a laser with greater power may be needed, and this can substantially increase capital investment in equipment.

In the LBC of thin material, the cutting head can be conveniently moved over a stationary workpiece (if desired) because the Nd:YAG laser can deliver its 1.06 μm wavelength beam through waveguides or through flexible optical fiber to the movable cutting heat with very little loss of power. However, the beam from a CO_2 gas laser for thick section cutting has a longer wavelength of 10.6 μm which means an unacceptable loss of power if transmission is attempted through optical fiber. When the beam must move in more than one direction, or is to be transmitted to serve more than one workstation, a sophisticated system of light pipes with water-cooled, gold plated mirrors made of copper must be arranged to bend the beam by reflection. Consequently, it is more convenient to position a stationary beam and to move the workpiece in precise fashion on a mechanized table beneath the beam. The mass of workpiece and table together presents an inertia problem with the high cutting speeds now possible with LBC-O.

Suitability of Laser-Cut Edges for Welding

Edges on steel workpieces cut by LBC can be satisfactory to fit into a square-butt joint for subsequent welding by any appropriate fusion welding process. Commercial usage of LBC has been fostered by larger companies with facilities for computer-controlled, tailored cutting of shapes from sheet or plate, and for production runs of closely replicated workpieces. When a weldment fabricator "jobs out" the cutting operation, he or she avoids the relatively high capital investment of a laser powerful enough to cut thick plate. However, subtle metallurgical problems can arise if close technical communication is not maintained during this outsourcing. A gas-type laser of about 3 kW capacity is needed to cut steel plate as thick as approximately 19 mm (¾ in.); yet, the evaporative technique (LBC-EV) is not efficient on plate thicker than approximately 10 mm (⅜ in.), so the operation usually is assisted by a supplementary gas stream or jet to rapidly and continuously remove melted metal and dross from the kerf. The kind of gas employed will affect the quality of the cut edge. Therefore, it behooves the weldment fabricator to advise the LBC operator whether the cut edges of the steel will receive any additional grinding or machining before being welded.

When the LBC operation is gas-assisted, the low pressure stream or the high pressure jet is blown coaxially with the laser beam as illustrated earlier in Figure 6.74. Gases used for this purpose include air, argon, carbon dioxide, helium and oxygen. The selection of a particular gas should be guided by a number of considerations, including (1) achieving the optimum cutting speed on the plate thickness to satisfy requirements for edge squareness and plate warpage, (2) the type of steel being cut and its hardenability, and (3) the permissible amount of oxide on the faces of the kerf. In addition to being a significant factor in costing the LBC operation, the gas chosen will influence the quality of the kerf, which will constitute the edge of the workpiece unless further metal is removed. Studies on the role of gases in assisting the LBC operation show how cutting speed can be increased, edge quality improved, warpage minimized, and sections cut to closer dimensional tolerances through proper gas selection.

Oxygen as the assist-gas (LBC-O) allows faster cutting speed, but it is likely to increase kerf roughness and thicken the oxide film on the faces of the kerf. Early findings indicate that LBC-O performed on high-chromium steels can leave the kerf face oxidized to the extent that post-cut cleaning must be carried out to ensure freedom from difficulty in subsequent welding. Use of an inert gas jet (LBC-IG) produces cut edges of excellent quality; but, in addition to the high gas cost, the cutting speed is slower than with LBC-O. Air as the assisting jet (LBC-A) is attractive from a cost standpoint; but the cutting speed is not likely to be as high as with LBC-O, and the kerf quality is not as good as with LBC-IG. A carbon dioxide jet produces results somewhat like those obtained with a jet of air. Early results with nitrogen as a high pressure jet produced very encouraging results from the standpoints of cut smoothness, freedom from dross, and absence of bottom edge burrs; although cutting speed was lower than obtainable with LBC-O.

Some cutting operators make a practice of accelerating cooling by some means immediately after the cut is made to minimize warpage and to improve the dimensional precision of the cut sections. It behooves the weldment fabricator to advise all concerned regarding the type of steel being provided for LBC, so that forethought can be given to the extent and the degree of hardening that will take place in the HAZ adjacent to the cut faces. Increased hardness along the edges from accelerated cooling can lead to a variety of difficulties in subsequent cold forming, or in welding operations.

SUGGESTED READING

Welding Processes, Volume II, Welding Handbook, Eighth Edition; American Welding Society, Miami, FL; 1991, Library of Congress No. 90-085465; ISBN: 0-87171-354-3

Welding Process Technology, by P.T. Houldcroft; Cambridge University Press, Cambridge, England; 1977, Library of Congress No. 76-47408; ISBN: 0-521-21530-7

Source Book on Innovative Welding Processes, edited by M.M. Schwartz; American Society for Metals, Metals Park, OH; 1981, Library of Congress No. 81-3535; ISBN: 0-87170-105-7

Welding Processes and Power Sources, by E.R. Pierre; Power Publications Company, Appleton, WI; 1974, Library of Congress No. 73-86974

The Theory and Technique of Soldering and Brazing of Piping Systems, by H.A. Sosnin; Nibco Incorporated, Elkhart, IN, 1971, Library of Congress No. 72-181079

Solidification Cracking of Ferritic Steels During Submerged-Arc Welding, by N. Bailey and S.B. Jones; The Welding Institute, Abington, Cambridge, England, 27 pages; 1977, ISBN: 0-85300113-8, AWS Code SCFS

Chapter 7

In This Chapter...

TERMS AND DEFINITIONS **653**
   Heat 653
   Temperature 654
   Thermal Flow 654
   Conduction, Convection, and Radiation 655
   Enthalpy and Entropy 656
TEMPERATURE AND TIME IN WELDING **656**
   Heat Flow Equations 657
   Heat Source Characterization 658
   Rate of Heating 659
   Heating Potential of Energy Sources 661
      OAW Heating Potential 662
      AW Heating Potential 662
      FRW Heating Potential 662
      EBW Heating Potential 663
      LBW Heating Potential 663
      Electrical Resistance Heating Potential ... 663
   Peak Temperatures 664
      Defining the Weld Zone 665
      Numerical Modeling of Temperatures 666

   Temperature Distributions 668
      Effects of Temp. Distribution on Cooling Rate 670
      Special Considerations Regarding Temp. Dist. .. 672
   Time at Temperature 674
      Furnace Heating for Welding Simulation 675
      Temper Color as an Indicant 675
      Resistance Heating for Welding Simulation 676
   Cooling Rate of Heated Zones 677
      Correllation with Heat Input 677
      Workpiece Pre-Weld Temperature 681
      Instantaneous Cooling Rate 683
   Influence of Travel Speed on Weld Zone Size 686
   Hardness/Microstructure/Cool. Rate Relat'nship ... 693
      Cooling End Points 696
CONTROL OF TEMP. IN FUSION WELDING **698**
   Predictive Diagrams for Fus. Weld'g Parameters ... 699
   Mathematical Modeling of Fusion Welding 699
   Sensory Sys. for Adaptive Cont'l of Fus. Weld'g 701
   Adaptive Contr'l of Solid-State Weld'g Processes ... 703
SUGGESTED READING **705**

Temperature Changes in Welding

Heat plays a fundamental role in most of the welding processes, as well as in thermal cutting. Therefore, an understanding of the processes is greatly aided by good comprehension of exactly what is meant by terms such as *heat*, *temperature*, *heat source*, *thermal flow*, *heating rate*, *cooling cycle*, etc.

In addition, thermodynamical terms such as *enthalpy* and *entropy* will be encountered in perusal of the welding literature. Originally, the name *thermodynamics* was given to a branch of science dealing with transformation of heat into mechanical work, or vice versa. The thermodynamical perspective ignores the precise mechanism of the various transformations, and makes use of principles which are independent of the particular mechanism. Laws conceived through studies of thermodynamics currently have very broad application.

TERMS AND DEFINITIONS

The terms which follow, along with others, have been used earlier — particularly when discussing the processes in Chapter 6. Also, they appear frequently in the literature suggested at the conclusion of each chapter to provide more detailed coverage of subjects. There are good reasons to make clear the meaning of these key terms before energy transformations and temperature changes are methodically examined, so that one may see how they relate to the metallurgical aspects of making a weld.

Heat

Heat is a general term for the energy that resides in the to-and-fro (kinetic) motion of atoms in any substance. This thermal energy can be transferred from one body to another simply through contact (via conduction), through transmission of radiation (transference of photons involving emission-and-capture), or by convection (fluid motion of hotter masses of matter). Also, thermal energy can be transformed into other forms of energy (e.g., mechanical, electrical), and other forms of energy can be transformed into heat. Joule, in 1840, explained the relationships between heat and electrical, mechanical, and chemical energies, and he contributed greatly to the physical science of thermodynamics. From laws set forth by thermodynamics, it is recognized that when heat is transformed into any other kind of energy, the total quantity of energy remains invariable; that is, the quantity of heat which disappears is equivalent to the quantity of the other kind of energy produced, and vice versa (i.e., first law of thermodynamics). Simply stated, energy can be converted from one form to another, but it cannot be destroyed or created.

Temperature

Temperature is the level of intensity of thermal energy that exists in a substance. Unfortunately, different scales of temperature measurement (each graduated in "degrees" of different unit size and indexed by different numbers) have evolved through the years (e.g.,°C,°F,°K) and several continue to be favored by various segments of the scientific community and in industry, much to the inconvenience of many. An important fact about temperature to be kept in mind, regardless of the measurement scale used, is that the level of kinetic energy or intensity of temperature gives no indication of the total amount of energy held within a given volume of a substance. For example, a minute quantity of a metal can exist at an extremely high temperature, but its limited volume stores only a small quantity of heat. Conversely, a large amount of a substance can contain a large quantity of heat, but the quantity is calculable only when the intensity as measured by degrees is known and the volume of the substance is considered. Furthermore, this large quantity of energy is available directly as heat only for use at a temperature below that of the substance. There is no means of extracting this kinetic energy directly in "concentrated" form and utilizing it at a higher temperature.

Thermal Flow

Thermal flow will take place between substances in contact, or in close proximity, whenever their temperature levels differ. The transference of thermal energy always will be from the substance at the higher temperature level to that at the lower temperature level (regardless of the quantities of thermal energy held by each, respectively); that is, the flow of thermal energy will continue in this direction, from the warmer or hotter substance to the cooler one, until a temperature difference no longer exists. The rate of thermal flow will be determined by the extent of difference between the levels of temperature in the two substances (or by the extent of temperature difference between two locations in a single body of a substance). *Temperature gradient* is the difference in temperature between two points a stated unit distance apart. For example, if welding has been performed on a steel plate and the weld is at a temperature of 1500°C (2732°F), whereas a point in the plate 25 mm (1 in.) away from the weld is 100°C (212°F), then the temperature gradient is simply calculated as 1500 – 100 = 1400°C per 25 mm, which is a gradient of 56°C/mm (or 2732 – 212 = 2520°F per inch, which is a gradient of 2.52°F/mil).

Heat energy is much like water; that is, it always flows downhill and the declivity of the slope controls the velocity of run-off. While the rate at which heat flows depends primarily on the steepness of the temperature hill, it also is controlled secondarily by the capability of the materials encountered enroute to perform this transfer of energy. The transfer rate of thermal energy by a material was explained in Chapter 3 as a characteristic physical property for each material, and was termed *thermal conductivity*. In welding, very steep temperature gradients often exist — not only between the heat source and the workpieces, but also within the workpieces themselves. Thermal conductivity is significant because it can differ between the materials employed, and thermal conductivity in a given material changes as temperature is raised or lowered. Also, heat capacity can be a factor that requires consideration when thermal flow occurs under non-steady-state conditions.

Conduction, Convection, and Radiation

Conduction, convection, and radiation are the three general processes by which heat energy can be transported from a source or location to a cooler area, and brief descriptions of these processes were given in Chapter 3. Each process is markedly different, and the complete mechanics of each calls for much more description than permitted by space here. Heat flow problems that arise in welding sometimes require analyses that depend on very subtle aspects of transport mechanics. Conduction is the process whereby kinetic energy is directly communicated from one atom to a neighboring one of lower energy level until equilibrium is reached. In solids, the particular thermal conductivity of a given material governs the heat flow rate under the temperature gradient existing at each moment. The rate of transmission will be proportional to the temperature gradient, and the quantity of heat transferred in unit time will be determined by the area of cross section through which the conduction process operates.

Heat flow in welding can present complex situations that almost defy analysis by calculation. This is the case when heat flow is periodic, such as the flow of heat coming from a welding arc which is weaved from side to side as the electrode travels forward. To avoid complex and speculative computations, controlled welding experiments that make use of actual temperature measurements or of calorimetric determinations often are performed to solve real problems. When compared with other physical processes, heat conduction in a solid is a relatively slow process.

In liquids or gases, heat transport is somewhat faster, because additional mechanisms are likely to enter into the transference process. In liquids, the convection process is self-starting; this is because increasing temperature usually causes expansion of a substance, and this reduces density. Consequently, heated liquid acts to rise. This results in convection currents which can quickly transport heated material upward, where it can commingle with material of a lower temperature. Furthermore, cooler material, because of its greater density, acts to descend thus creating a circulatory system. Liquids subjected to localized heating also exhibit convection currents that are driven by surface tension gradient. This mechanism, *Marangoni effect*, can play a significant role in creating circulating currents in molten weld metal, and currents from this effect were described in the GTAW process in Chapter 6.

In gases, conduction is augmented by diffusion of atoms (or molecules) having a high temperature into areas of lower temperature. There is a continual interchange of hotter gas A into cooler gas B, because the atoms or molecules are not held by bonding forces (as in a crystalline solid). Also, the gas atoms or molecules are relatively far apart, and there is relatively little viscosity existing between them. The rate of interchange between hotter and cooler gas will depend on the mean-free-path, which is the distance each atom or molecule can travel before collision, and on the *average velocity of translation*, which varies as the square root of the temperature. The thermal conductivity of a gas, strictly speaking, is very difficult to determine experimentally. Values (corrected for diffusion and radiation effects) were reported for five gases of interest in Table 3.11.

Although radiation was reported in Chapter 3 to play a limited role in most of the welding processes, there are circumstances where knowledge of the mechanics of this energy transport process can be used to advantage in controlling temperature changes.

The rate of heat transference by radiation is not easily predicted from calculations. The surface of the hotter body can vary significantly in emissivity, and absorption of photons of energy by the cooler body depends on its surface as well. A metal surface coated with lampblack can absorb about 95 percent of radiant energy falling on it, whereas the absorptivity of a brightly polished metal surface can be as low as five percent. The emissivity of a surface tends to vary directly with its absorptivity. The dependence of radiation heat transfer on emissivity and absorptivity must be taken into consideration in certain circumstances in welding, such as the selection of transparent materials for the focusing lens of lasers, or the use of a radiometer type of instrument for measuring radiation being emitted from an object to measure its temperature, and perhaps to scan its surface locally to determine the approximate distribution of heat in the object.

Enthalpy and Entropy

Enthalpy is a thermodynamic measure of the energy content of a system per unit mass; while *entropy* is a measure of the degree of disorder in a system undergoing change which represents the energy unavailable for useful work. Entropy, quite logically, is higher with greater disorder. The second law of thermodynamics states that the entropy always increases in any spontaneous change: in the limit, entropy remains constant if the change takes place reversibly; and it never decreases spontaneously. Entropy is lowered by the application of a magnetic field when it induces ordering, and entropy is lowered by reduction in temperature. When a liquid freezes to form a solid (e.g., the solidification of weld metal), this transformation, expressed in strictly fundamental terms, is the change in free energy, that is, the enthalpy and entropy of the system.

TEMPERATURE AND TIME IN WELDING

Unusual combinations of temperature and time must be dealt with in welding. The temperature changes during welding are wider and more drastic than in any other metallurgical process for several reasons:
(1) welding heat sources are hotter than most that are commonly used in industry for heating,
(2) welding operations are carried out so rapidly that very steep temperature gradients are established between the weld area and surrounding base metal, and
(3) both base metal and any fixturing in intimate contact act as highly efficient heat sinks which promote cooling rates as fast as permitted by the thermal conductivity of the metals involved.

The temperature changes in welding, for the most part, are not a favorable treatment for steels. There are exceptions, however, and it can be helpful to know when and where the heat effects of welding can be beneficial.

Fusion and resistance welding methods, in particular, produce temperature changes that frequently give the metallurgist cause for concern. Nevertheless, progress is being made — both through better control of temperature changes in welding, and through modifications of the composition of steels so that their metallurgical behavior is more tolerant of the thermal cycles encountered with modern welding processes.

Undesirable effects from the thermal cycle of welding may occur in the zone of solidified metal (i.e., the *weld metal*), or in heat-affected base metal (i.e., the *heat-affected zone*). In fact, with some steels the heat-affected zone may be more prone to develop an unfavorable condition than the weld metal. The following aspects of welding thermal cycles will be outlined in a general way:

(1) rate of heating,
(2) heating potential of energy sources,
(3) peak temperatures,
(4) temperature distributions,
(5) time at temperatures,
(6) cooling rates of heated zones,
(7) cooling end points, and
(8) repeated heating by multiple passes and multiple-impulse welding.

When the above aspects are known in detail, much can be done to predict or to explain the effects of welding on a given steel. Although specific effects will be discussed in appropriate detail later, several are mentioned here to emphasize the importance of understanding temperature changes during welding.

Localized melting, or the amount of fusion penetration into a joint being welded is controlled mainly by heat input; but other subtle features of the heat source also can exert an effect. The penetration into the base metal, in turn, can have a profound influence on the amount of alloying or mixing that occurs between the base metal and added filler metal. The peak temperature reached by the weld area can determine the extent of grain growth in heat-affected zones, as well as the amount of softening that may take place in the heat-affected zones of work-hardened base metal. The expansion and possible warpage of the workpieces is dependent on the extent of heating, particularly the temperature uniformity.

Perhaps the most important effect of temperature changes in welding carbon and low-alloy steels is the degree of hardening which may take place in the base metal heat-affected zones, and the influence of this property change on fracture toughness, and on susceptibility to hydrogen cold cracking. Hardening that takes place in steel from welding is highly dependent on composition and microstructure, and both the rate of heating and, especially, the rate of cooling will affect the outcome. Steel compositions that undergo transformation hardening (outlined in Chapter 4 as one of five hardening mechanisms) are most sensitive to cooling rate. Repeated heating during welding from multiple passes may cause some compositions to respond with precipation hardening, while others may be softened by this thermal cycle.

Heat Flow Equations

The flow of heat during welding has been studied by many investigators to provide mathematical relationships between the nature of the heat source and the resulting cycle of heating and cooling in specific locations in the workpieces. Rosenthal, as early as 1935, offered solutions to heat flow equations for a moving heat source. Although scientifically sound, they did not provide a pragmatic system that could be readily utilized in industry. Adams, starting in 1958, made significant contributions to methods for calculating peak temperatures and cooling rates for welds — both in thick and in thin plates. The literature contains many analytical equations that deal mostly with arc welding because of widespread usage of the SMAW and SAW processes in the past. The equations defining thermal cycles range from simple expressions that give only rough approximations to complex statements that include a variety of variables that affect heating and heat flow in specific processes and applications.

Many of the proposed equations cannot handle all of the variables associated with modern welding processes. For example, heat sources are now more complex than assumed, and more consideration must be given to the efficiency of heat transfer to the workpieces under the operating conditions of a particular procedure.

Heat Source Characterization

Investigators often envisioned the heat source as a "moving point." However, there are at least three common geometrical forms, and each requires a different assessment of heating efficiency and the resulting thermal cycle at the weld. Lower-current welding arcs are regarded as a surface heat source and can be treated mathematically as a *point* heat source. However, heat sources that produce a keyhole during welding (e.g., EBW, LBW, PAW), obviously should be treated as a *line* heat source. Some arcs produce deeply cupped weld pools and have unusually high heat transfer efficiency. These must be treated as a compromise between a point source and a line source. Broad heat sources that envelope an area on a surface (e.g., a large oxyfuel flame) are considered to be *plane* heat sources.

In addition to considering the geometric form of the heat source, differentiation must be made between a *continuous* heat source, an *instantaneous* heat source, and a *periodic* heat source. The electric arc and the oxyfuel flame are two examples of a continuous heat source which are kept more-or-less in constant motion relative to the workpieces. Electric resistance spot welding, electrical metal-explosion welding, arc stud welding, and rotational friction welding are examples of an instantaneous heat source since the heat for welding is liberated at a fixed location in the workpieces over a relatively short period of time. Periodic or cyclic heat sources are like those used in resistance (roller) seam welding with on-and-off current, and in pulsed arc welding. Literature on thermal flow in welding ranges from entirely theoretical studies based on heat flow theory to experimental methods entailing the welding of test pieces and temperature versus time measurements. To simplify the mathematics, much of the work on heat flow in welding has been directed to arc welding with a single electrode in point-to-plane arrangement — i.e., a continuous, moving, point heat source. Unfortunately, a complete system for quantitative prediction of the total effect of welding parameters on the size and shape of the weld zone, and on the thermal cycles in and nearby the weld, has yet to be achieved.

The geometry of the workpieces and the type of joint to which the heat source delivers its energy poses a major question, even in straight-line, single-square-groove welds between flat sections. For purpose of modeling calculations, it is necessary to first establish whether heat delivered to the weld will flow away into the base metal in three general directions (i.e., three-dimensional heat flow), or will flow only in two general directions (two-dimensional heat flow). Most welding, particularly when performed on thick plate, involves heat flow in three general directions: (1) through the thickness, (2) transverse to the travel direction of the heat source, and (3) parallel to the direction of weld travel.

Three-dimensional heat flow necessitates the use of complex equations and extensive calculations for modeling. The welding of a relatively thin plate — that is, one which is completely penetrated by the weld pass — can be modeled in accordance with two-dimensional heat flow equations, since there is no base metal remaining through the thickness to act as a heat sink. Consequently, the mathematical treatment is simplified. Sometimes, investigators will depreciate probable heat flow through the thickness to avoid the complexity of three-dimensional analysis, and they usually refer to their calculations as "pseudo two-dimensional treatment." A method for estimating the error between 3-D and 2-D heat transfer analyses of a temperature field has been proposed by investigators to improve the actuality of results obtained by a 2-D heat transfer analysis. Their method of reducing error in a 2-D approximation is based on a modification of the true power density distribution function in the temperature field.

Physical differences between laboratory specimens and calculations, and conditions in actual weldments also must be considered. Thermal patterns often are calculated on the basis of a uniform, straight-line joint with steady heat input along its length and without end-effects. Since the time-dependent distribution of temperature at a weld presents problems in diagrammatic portrayal, presentations are often simplified with diagrams depicting *quasi-stationary* conditions. To be of value, a mathematical system must deal with the nature of the heat source, geometry of the joint, geometry of the weld zone, contact with fixturing, variations in mass at the extremities of joints, etc. Attention here will be centered on the fundamentals involved to provide general guidance on whatever course is followed when regulating heating and cooling to control metallurgical changes and final conditions.

Rate of Heating

The rate of heating with a given welding process depends on two basic features — the intensity or temperature level of the heat source, and the efficiency of heat transfer from the source into the workpieces. Of course, the source must be able to supply a sufficient volume of energy, so that flow to the workpieces can occur without the source itself suffering a decrease in temperature. In the text which follows, several commonly used welding processes employing entirely different methods of heat generation will be compared from the standpoint of the rate of heating that is likely to occur in the workpieces. The operating temperature or energy intensity of the heat source in each process will establish the steepness of the temperature gradient between the heat source and the work. A higher temperature at the source means a steeper temperature gradient between the source and the work (which is presumed to be at or near room temperature); therefore, a faster heating rate should prevail, providing that nothing unusual interferes with the process. Welding processes vary in their efficiency of heat transfer; and, as pointed out in Chapter 6, significant interference can arise in the operation of certain processes (e.g., laser welding).

A modest temperature level is provided by the oxyfuel welding process (OFW) using an oxyacetylene flame. This flame has a temperature at the tip of its inner blue cone of approximately 3200°C (5800°F). If the highest obtainable heating rate is desired, then the welder must be careful to hold the tip of the inner blue cone immediately adjacent to the surface of the workpieces, since this is the position that provides the steepest temperature gradient. The heat transferred from the burning gases in the flame is distributed over an area on the surface of the work. Heating the surface with an oxyacetylene flame is not particularly efficient; therefore, the rate of heating is comparatively slow and fusion penetration is sluggish. In fusion welding with an oxyacetylene flame, a small root opening is pre-set in the joint when complete penetration of the workpieces is required. This allows the hot gases of the flame to sweep along the edge surfaces to be heated and melted, thus minimizing the time required for heating. Unfortunately, much of the heat available in the oxyacetylene flame is lost to the surrounding environment by a combination of rapid air flow, radiation, and convection. Also, the comparatively modest temperature of the flame and slow heating rate of the weld area allows much time for heat to be conducted into the workpieces beyond the edges to be fused, and this wayward heat represents wasted energy that is not utilized in melting a weld zone.

Compared to welding with an oxyfuel flame, arc welding with a single electrode in point-to-plane arrangement will provide a significantly faster rate of heating. Two important differences between the two processes explain arc welding's faster heating rate. First, the temperature of the arc is considerably higher than that of the oxyacetylene flame (6075°C or 11 000°F for the ordinary arc, which is about twice that of the oxyacetylene flame). Second, the arc is in constant contact with the base metal throughout the welding operation, and no unusual manipulation is required by a welder to keep the arc in this position for efficient transfer of arc energy. Furthermore, most of the useful heat of the arc is generated at the anode and cathode drops, and one of these always will be in intimate contact with the base metal. Of course, not all of the heat generated by the arc is absorbed by the workpieces. About one-fourth of the heat developed by the arc is lost to the surrounding atmosphere, and this includes heat consumed in vaporizing some metal from the weld pool as well as from the electrode.

The efficacy of a welding arc in putting its energy into the workpieces can be measured. This *arc efficiency* is calculated as the percent energy transferred to the workpieces compared with the total energy supplied by the power source. Arc efficiency varies considerably among the arc welding processes. Differences between arc *output energy* versus *input energy* were mentioned briefly in Chapter 6 in an early review of the arc as a welding heat source.

When commonly used arc welding processes are compared, the GTAW process is found to range in efficiency from about 20 to 50 percent depending on operating conditions. SMAW, GMAW and FCAW processes are roughly equal in arc efficiency, ranging from about 65 to 85 percent. The SAW process ordinarily has excellent arc efficiency because the complete, thick cover of flux prevents heat loss by radiation. Of course, some energy is consumed in melting the granulated flux to provide the molten slag blanket; even so, the amount of arc energy that enters the workpieces can exceed 90 percent.

High arc efficiency means power economy; and if the efficiency for a particular process and procedure is known, a better appraisal can be made of heating and cooling rates at the weld. This is the basic information needed to anticipate a number of important metallurgical changes that occur during welding. Many aspects of the arc in a welding operation can affect its input efficiency. Arc length is an essential variable. Longer arc columns allow greater heat loss to the surroundings. Arcs that operate partly within a depressed weld pool are able to transfer more of their energy into the cup-like crater of molten metal. This is especially true if the arc has penetrated to form a keyhole, as is often the case with the PAW process.

When electrical resistance spot welding is compared with the oxyfuel and arc welding processes, it is found that in RSW the metal itself acts as the heat source by transforming electrical energy delivered through the electrodes into heat. Heat that is primary to resistance spot welding (RSW) develops at two principal locations — in the body of each workpiece, where the metal's electrical resistance determines the amount of heat generated by a given flow of current; and at contacting surfaces between overlapping workpieces, where the electrical resistivity is affected by the many factors described earlier in Chapter 6 for the RSW process. Heat is also developed at the contacting interface between the electrodes and the surfaces of the workpieces; but this is unwanted heating, and it is minimized by circulating a coolant (often water) to the electrodes. More heat is generated at the area of contact between work surfaces, which is the location where melting and coalescence of the metal is desired. The transformation of electrical energy into heat as the current flows is virtually instantaneous; and by providing an adequate measured flow of current, the metal can be heated from room temperature to the molten state within a small fraction of a second, if desired. The heating rate easily can be so fast as to find the assembly of workpieces not yet positioned to confine the molten metal generated, and much of it could be lost by expulsion.

In actual RSW practice, the heating rate is deliberately limited through control of current level and time to satisfy a number of practical considerations. From a metallurgical perspective, heating rate by resistance to current flow can be so rapid that abnormal microstructural conditions can be created. For example, a steel consisting of grains of ferrite and scattered globular carbides when very rapidly heated can transform to austenite in a narrow heat-affected zone adjacent to the weld nugget. However, the heating (and sequential cooling) may be so rapid as to allow only partial dissolution of the carbide particles when the austenitized structure exists. The resulting heterogeneous microstructure likely would have undesirable mechanical properties (as will be explained in Chapter 13 of Volume II). Nevertheless, the very rapid rate of heating and coalescence are principal reasons for the popularity of the resistance welding processes, since it fosters high productivity.

Heating Potential of Energy Sources

Study of heating capability calls for examination beyond the temperature level of the source and the amount of power being made available. In general, the factors that determine the effectiveness of a source are *energy intensity*, *energy density*, and *transfer efficiency*. The most effective heat sources for welding are those that can:

(1) transform a given form of power into thermal energy having a high temperature intensity,
(2) focus or concentrate the energy being delivered as a relatively small containment, point, or front on the metal,

(3) dispense power without any particular limitation on amount and convert this energy by a direct mechanism into heat in the metal, and
(4) avoid phenomena that interfere with the delivery of power, the conversion to thermal energy, and transference to the metal.

An examination of energy utilization in a number of familiar welding processes will illustrate the importance of these factors.

OAW Heating Potential

In the oxyacetylene welding process (OAW) heat is generated *outside* the workpieces, and hot gases of the flame transfer their thermal energy mostly by conduction as they impinge upon the metal surface. A limiting factor which may require consideration is the *combustion velocity* of the gas mixture. Burning gases are subject to a maximum flame propagation rate, and this restricts both the temperature of the flame and the quantity of heat that can be secured under a given set of combustion conditions. Of the total heat provided by the oxyacetylene flame, much is liberated in the outer zone where carbon monoxide and hydrogen are burned with oxygen that is picked up from the atmosphere. This outer flame serves for preheating the workpieces, but does not contribute directly to the energy density of the primary or *working* flame (i.e., the inner reaction zone where $C_2 H_2$ plus O_2 yields CO and H_2). The heat input to workpieces by the overall combustion of gases is limited by a transfer efficiency of only about 35 percent.

AW Heating Potential

In arc welding, (AW) the central core of the plasma column ranges from about 5000 to 30 000°K. The arc decreases in temperature toward the outer boundary of its mantle. Temperature distribution also varies along the arc column length and radially toward its periphery depending on the level of current, kind of electrode, whether a flux is present, nature of the surrounding atmosphere, and distribution of the cathodic and anodic poles. Under ordinary atmospheric conditions, as the current is increased the arc column grows in diameter and energy losses begin to increase significantly. The aforementioned temperature of 30 000°K is a practical limit for most welding arcs, but the temperature can be elevated to at least 50 000°K by constricting the diameter of the arc column through use of plasma torch technology.

FRW Heating Potential

Friction welding (FRW) has a high heating potential initially when small surface irregularities touch each other at their summits. These points in contact heat very rapidly at first and quickly commence to flow plastically. As more of the surfaces are brought into actual contact, power must be increased to continue the movement between the two surfaces; whereupon frictional heat generation rate increases and the temperature of the metal at the surfaces soon rises above its melting point. However, liquid metal acts as a lubricant and its formation causes heating rate to decrease. Consequently, the upper limit of temperature is virtually fixed by the melting point of the metals involved, and the surface temperature will not rise above this level because of the lubricity of the molten metal. Care must be taken with heating rate to ensure that sufficient time has been allowed for the conduction of adequate heat to melt all surface areas. In some cases, heat is allowed to diffuse into the adjacent solid metal to obtain a slower cooling rate at the weld after completion of the operation.

EBW Heating Potential

The *electron beam* (in vacuo) is unique as a welding and cutting heat source because of the extremely high power density that can be effected. In fact, this power density is higher than any other continuously operating heat source. This capability became possible with the development of heavy-duty electron guns and magnetic lens systems that can focus the beam to a very small spot. Power at the EB gun is established by the accelerating voltage and the beam current, and beam powers up to 200 kW are common. When the beam is focused to a spot as small as 0.1 mm (0.004 in.) in diameter, its power density can exceed a billion watts per square millimeter (10^9 W / mm^2). The efficiency of energy transfer can be very high because the kinetic energy of electrons impacting either solid or liquid metal is converted directly into heat. EBW machines using a vacuum can produce a keyhole in steel as thick as 300 mm (12 in.) very rapidly, because the evaporation of the superheated metal is aided by the very low pressure in the vacuum chamber.

LBW Heating Potential

The *laser beam* as a heat source is analogous to the electron beam inasmuch as its power density is determined by the power capacity of the laser and the size of the spot to which the beam is focused. Industrial lasers have power ratings as high as 25 kW. When the beam is focused to a very small spot, there does not appear to be any restriction on the amount of photon energy that can be directed to the target area. The highest power density is delivered by pulses of a laser beam, and a spot diameter as small as the wavelength of the radiation being projected is theoretically possible. In actual laser welding practice, beam spot diameters are limited to the range of about one to 10 mm (0.04 to 0.4 in.). Although extremely high power densities (e.g., 10^9 W / mm^2 for a pulsed beam) can be produced, LBW is not likely to be conducted at a level this high. This is because the energy would be imposed at a rate much faster than could be conducted to heat the required areas of metal at the joint. At power densities two or three orders of magnitude lower, the photon energy still can quickly vaporize metal to form a keyhole and to serve as a heat source for fusing an abutting joint. Very high power density is more likely to be utilized when the beam is used for drilling holes or cutting.

Electrical Resistance Heating Potential

Electrical resistance heating deserves brief comment even when not used as the principal source of heat for welding, per se. Resistance heating can be applied to assist other welding processes — i.e., use of a long electrode extension in the SAW, GMAW and FCAW processes to increase electrode deposition rate. Regardless of the technique employed for introducing current into metal, the power density can range from a level that simply warms the metal, up to a point where explosive volatilization of the metal occurs. The amount of heat generated by resistance can be calculated because the controlling factors, current, voltage, and time, can be regulated. For example, during resistance spot welding, by passing 12 000 amperes of current at five volts through a workpiece for 0.1 second, 6000 watt-seconds or joules of energy are supplied to provide energy equal to 1434 calories, which would be sufficient to melt 9.3 grams (0.02 lb) of steel if utilized without loss by conduction. In addition to controlling the total amount of energy and the time over which it is released, electrical current can be directed to flow through particular areas of the workpieces where resistance heating is wanted. This directional control can be exercised through a number of means; including the location of contacting electrodes, the use of high-frequency alternating current to produce a low-inductance path, the use of proximity conductors, the use of impeders, and the choice of ac frequency for induced current to regulate the depth of current penetration into the workpieces. Resistance heating is versatile and adaptable, and its power density is virtually unrestricted.

Peak Temperatures

The maximum or peak temperature reached in workpieces subjected to a heat source is determined by the rate of heat input and the rate of heat loss. As long as the rate of heat input exceeds the rate of heat loss, the temperature will continue to rise. When heat input equals heat loss, the peak temperature is achieved. As the temperature is raised at this location in a workpiece, the temperature also rises at all regions in the vicinity following the physical laws of heat conduction. Therefore, the heat supplied for welding must be sufficient not only to melt the volume of metal required, but also to match the amount of heat removed by conduction. This means that a slow rate of heat input requires more heat to melt a required volume of metal. This drawback of oxyfuel welding was pointed out earlier.

With a moving heat source, peak temperature attainment is a quasi-stationary situation, and the steadily decreasing effect of the retreating heat source must be taken into consideration. Under these conditions, when the peak temperature is reached at a given location, cooling commences at the location immediately thereafter. The temperature of greatest interest may not necessarily be the maximum reached at any location during the welding operation. Of course, the maximum temperature ordinarily is expected to occur in the molten weld zone. Frequently, a location in the HAZ of the workpieces and the peak temperature reached therein is of greater interest. A particular HAZ location at a specified distance from the weld zone may deserve consideration because of metallurgical events that can take place as certain temperature levels are exceeded.

Figures 7.1 and 7.2 show typical temperature distribution during the making of two arc fusion welds using markedly different weld travel speeds. The temperatures range from ambient in the base metal, to a very high peak temperature in the fusion zone. The steepness of temperature gradients vary depending on their direction from the heat source, the difference in temperature levels at their extremities, and the span between. Surface temperatures of the molten weld pool in steel have been reported to range from 1650 to 2400°C (about 3000 to 4350°F). In the case of power beam processes (EBW and LBW), surface temperature of the metal upon which the beam impinges can, of course, reach the boiling point, which for steel would be about 3000°C (5432°F).

The influence of welding travel speed should be noted when comparing Figures 7.1 and 7.2. As the moving heat source increases its travel speed, the isotherms are closer to the heat source in all directions, particularly those ahead of the source, meaning steeper temperature gradients. Often as weld travel speed is increased, the power input of the heat source also must be increased to provide the required amount of melting, and this tends to raise the temperature intensity of the (arc) heat source. Consequently, the ultimate peak temperature (in the weld zone) usually is raised by some small amount as can be noted by comparing Figures 7.1 and 7.2.

Defining the Weld Zone

The term *weld zone* has been favored when referring to that portion of a weld that is melted during its making, and then solidifies on cooling. AWS terms any melted portion the *weld metal*, and differentiates between (1) *mixed zone*, which is that portion of the weld metal consisting of a mixture of base metal and filler metal, and (2) *unmixed zone*, which is a thin boundary layer of melted base metal adjacent to the

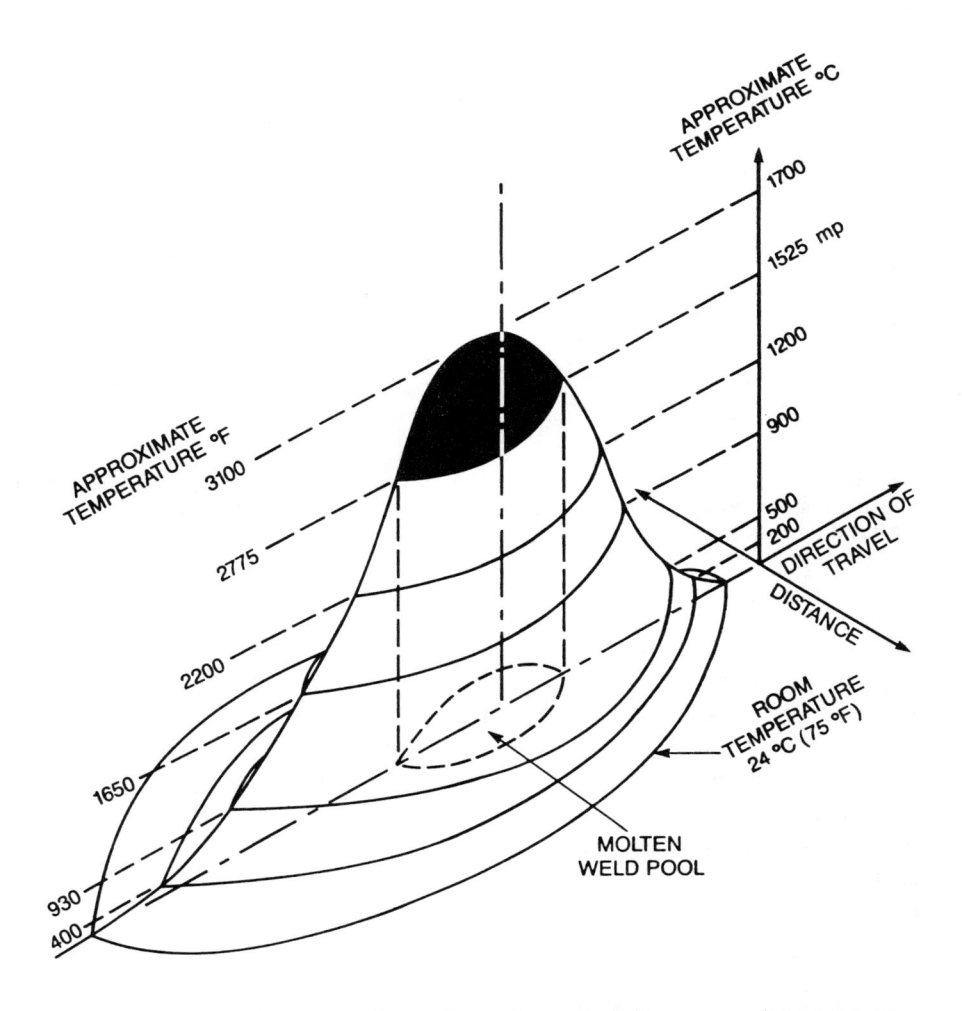

Figure 7.1 — Schematic illustration of quasi-stationary surface temperature profile produced by a welding heat source moving at medium travel speed across a steel plate

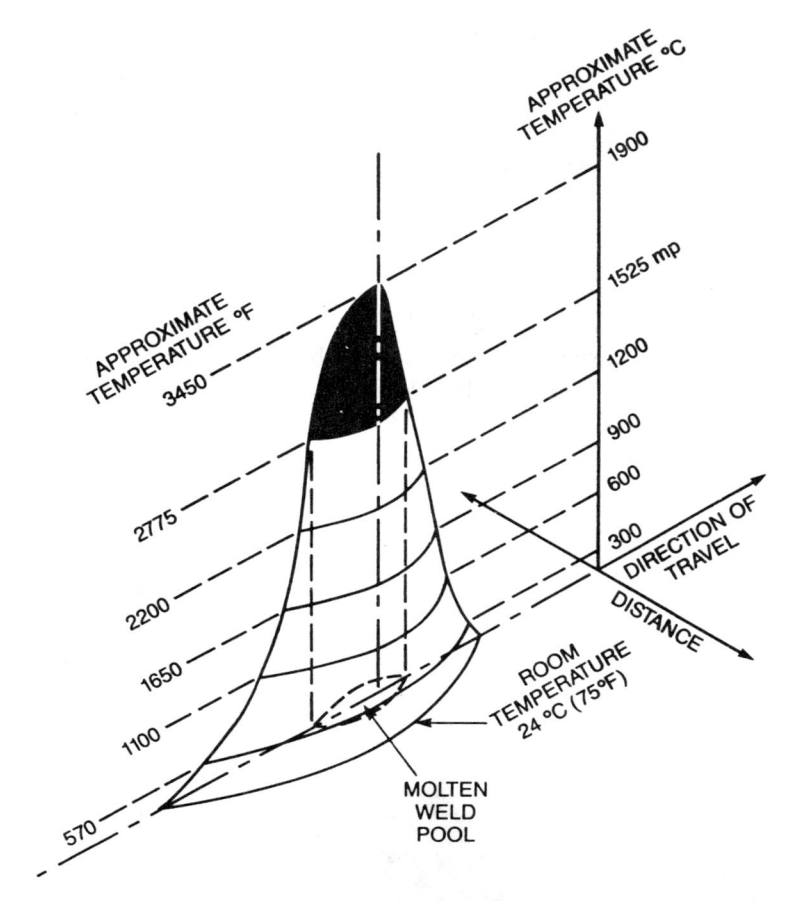

Figure 7.2 — Schematic illustration of quasi-stationary surface temperature profile produced by a welding heat source moving at high travel speed across a steel plate

weld interface that solidified without mixing with the remaining weld metal. Of course, with some welding processes, there is no "melted and re-solidified" metal, per se. This is either because none is formed during the procedure, or because any metal melted during the procedure is exuded from the interface between the workpieces. Also, with some welding processes, the weld metal or weld zone is a nugget of re-solidified metal. The quantity of heat held by a volume of weld metal will be singled out as an important factor when attention turns to the cooling rate of heated zones. Whether the melted metal is mixed or unmixed does not significantly affect cooling rate; so, to simplify discussion, the entire melted portion will be considered the "weld zone" as its role in temperature changes in welding is explored.

Numerical Modeling of Temperatures

Peak temperatures reached at specific locations in a weld can be important starting points in predicting the consequences of continued thermal flow in a weldment.

Difficulty in measuring peak temperatures at specific locations in various kinds of welds led many investigators to pursue numerical modeling and thereby develop complete distribution models that would depict temperature at finite points of interest at any time during heating and cooling. The literature now contains many analytical equations and solutions proposed for welding thermal models; including the quasi-stationary state. As an example of effort in this direction, the equation below emerged from the work of Adams to predict the distribution of peak temperature in the base metal of single pass, full penetration butt fusion welds in sheet or plate:

$$\frac{1}{T_p - T_o} = \frac{4.13 \, _pCtY}{H_{net}} + \frac{1}{T_m - T_o} \qquad \text{(Eq 7.1)}$$

Where

- T_p = Peak or maximum temperature (°C) at a distance, Y (mm), from the weld fusion boundary in the base metal heat-affected areas
- T_o = Initial uniform temperature of sheet or plate
- T_m = Melting temperature (°C) specifically, the liquidus of the metal being welded
- H_{net} = Net energy input from a welding arc to the base metal determined as;

$$\frac{f_1 \, E \, I}{V}$$

Where
- f_1 is heat transfer efficiency
- E is closed circuit (arc) voltage
- I is welding current amperage
- V is travel speed of arc in mm/s

(Note: often, f_1 is assumed to be approximately 0.9 as a convenient shortcut in performing calculations.)

- p = Density of metal being welded, g/mm³
- C = Specific heat of solid metal, J/g (°C)
- $_pC$ = Volumetric specific heat, J/mm³ (°C)
- t = Thickness of sheet or plate, mm
- Y = Perpendicular distance from fusion line to point for which peak temperature is being calculated

When the above equation is solved, the peak temperature at the weld fusion boundary is equal to the melting temperature of the metal, which is essentially correct; however, the equation is intended to be solved for peak temperatures reached at specific locations in the adjacent base metal. As a demonstration of this capability, the equation has been applied to an arc weld made in five mm (0.200 in.) thick steel using the following parameters:

- T_o = 25°C
- T_m = 1510°C
- $_pC$ = 0.0044 J/mm³ (°C)
- f_1 = 0.9
- E = 20 arc volts
- I = 200 amperes
- V = 5 mm/s
- H_{net} = 720 J/mm

As would be expected, the peak temperatures reached in the base metal decrease with greater distance from the fusion boundary. The T_p calculated for Y of 1.5 mm (0.06 in.) is 1184°C (2163°F), while Y of 3.0 mm (0.12 in.) is 976°C (1789°F). These calculated temperatures will be found in general agreement with measured values cited in later discussion of heat-affected zones. This kind of information on peak temperatures allows an estimate to be made of the width of any heat-affected zone by considering the minimum temperature needed to cause a specified microstructural change in the given metal. Also, the quantitative influence of preheat to widen any heat-affected zone can be ascertained. While equations of this kind are helpful in gaining a fundamental grasp of the roles played by basic variables, other factors encountered in real weldments can quickly invalidate their applicability. Also, the physical constants of metals change with temperature, and allowance for such changes (if known) requires lengthy computations.

The precise calculation of temperatures in steel is more difficult than with other metals, like aluminum and copper, because of the significant difference in thermal conductivity of the two major forms of crystallographic structure that can exist in the heated areas (i.e., bcc below the Ae_1 and fcc above the Ae_3). Consideration of this reversible phase transformation, which actually takes place over a range of temperature in steel, puts so much guess work into computations that its influence on thermal diffusivity in a steel weldment often is ignored and nominal values for thermal conductivity are used.

Temperature Distributions

The temperature distribution during most welding operations is in almost continual change as the workpieces are heated and cooled, until the entire weldment cools to ambient temperature. It is this dynamic behavior which makes graphic presentation of welding temperature distribution so difficult. A *quasi-stationary* portrayal was employed for Figures 7.1 and 7.2, and this is commonly practiced. Diagrams of this kind show an *instantaneous* distribution of temperatures at a specified time in the course of welding — usually at the moment peak temperature is reached in the heat-affected base metal immediately adjacent to the weld. As the heat source continues to travel, and as heat is conducted to cooler areas, the temperature distribution changes as dictated by a number of conditions, some of which have been included as factors in Equation 7.1. The extent of superheating in the weld zone and the thermal conductivity of the solid base metal play a part in establishing the distribution of temperature at each point in time.

Figure 7.3-A is an elementary presentation that originated in the work of Carslaw. It illustrates the rapid changes in temperature distribution during the making and cooling of a hypothetical weld in plate using (for simplification) an instantaneous source of heat. The straight lines marked "A" in the diagram depict temperature distribution that would be found if no thermal conductivity took place. The temperature of the weld zone over its entire cross-section is shown to be above the melting point of the metal, and the base metal temperature is shown unchanged. This is unrealistic, of course, so examination must be made of the distribution pattern that actually prevails in a real fusion weld in specific locations at particular times. After cessation of instantaneous heat input, Curve "B" indicates that heat flowing into the cooler, still-solid base metal causes the actual peak temperature of the weld zone to be lower by a small amount. Also, Curve "B" shows a small decreasing gradient in weld zone temperature in the direction of the fusion boundaries because of heat flow into the base metal.

The first location of interest in the still-solid base metal of Figure 7.3-A is that point immediately adjacent to the boundary of the weld zone, which has been designated No. 1. Here, heat flowing from the weld zone has raised the base metal temperature to a level almost equal to that of the boundary metal in the weld zone. A short distance farther away from the weld, in the location designated No. 2, Curve "B" shows the base metal temperature has risen only one-half as high as found at location No. 1. Even farther away from the weld zone, at location No. 3, the rise is only one-third that found at location No. 2. At location No. 4, the temperature rise is almost imperceptible. Yet, with delay of another second, the temperature distribution, as indicated by Curve "C", starts to shift in a manner that is commonly found in temperature changes in a weldment. The base metal immediately adjacent to the weld zone is continuing to drop in temperature while the plate a little farther away (location No. 2) is actually rising. This manner of change can be seen in Curves "D" and "E" as they pass location Nos. 3 and 4, although the temperature changes are not as great. The curves in Figure 7.3-A are informative not only regarding heating as a weld is made, but also with respect to cooling.

Temperature gradient has a profound influence on thermal flow. Figure 7.3-A deserves re-examination for this feature, because the gradient is by no means constant at any moment during welding. In Figure 7.3-A, Points 1, 2, 3, and 4 are 25 mm (1 in.) apart. At the time corresponding to curve "B," the temperature gradient is T_1 degrees per unit distance between Points 1 and 2. Further, at the time corresponding to curve

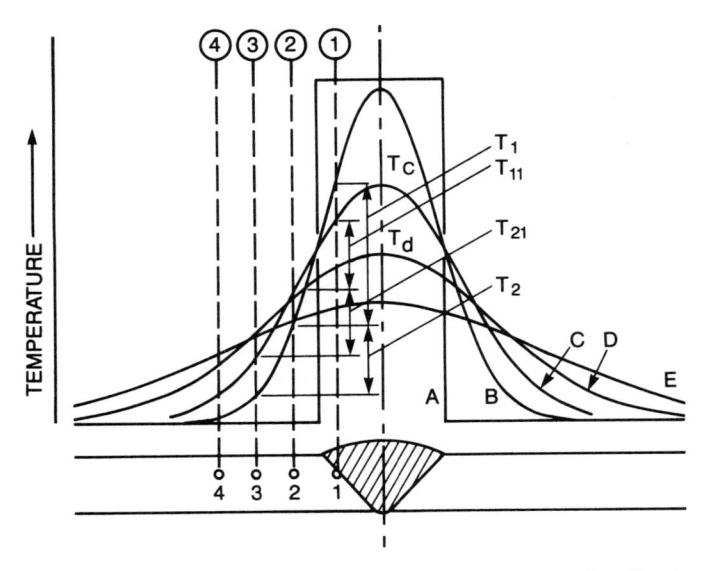

Figure 7.3-A — illustration of changing temperature distribution in a hypothetical plate weldment made with an instantaneous heat source after four one-second intervals of cooling time (curves B, C, D and E)

(Data according to Carslaw)

670 Chapter 7

"C," the temperature gradient between Points 1 and 2 has decreased to T_{11} degrees per unit distance. Also, between Points 2 and 3 the gradient has decreased to T_{21} degrees per unit distance. The steeper the temperature distribution curve, the higher is the gradient — which is merely a statement of part of the law of heat conduction.

Effects of Temperature Distribution on Cooling Rate

A resultant of temperature gradient often of concern is the cooling rate that follows. Referring again to Figure 7.3-A, the temperature in the center of the weld zone will be seen to fall from T_c to T_d in the second between the recording of curves "C" and "D." The weld zone therefore cooled at the rate of $T_c - T_d$ degrees per second. During the next second, the temperature of the weld zone did not fall as much — that is, the rate of cooling decreased.

To help explain basic rules concerning cooling rate, Figure 7.3-B has been prepared using the temperature-time calculations employed earlier in Figure 7.3-A. Curve No. 1 in Figure 7.3-B shows the temperature at the Point 1 location as the hypothetical instantaneous heat source brought the metal to the molten state and then allowed it to cool freely. Because the temperature gradient is very steep between the melt and the cold base metal, the metal at Point 1 starts to cool very rapidly during the first second after the heat source is extinguished. During the initial second, the temperature of the span of metal between Locations 1 and 2 (on Figure 7.3-A) rises rapidly by conduction from the molten weld zone, and although cooling commences before the

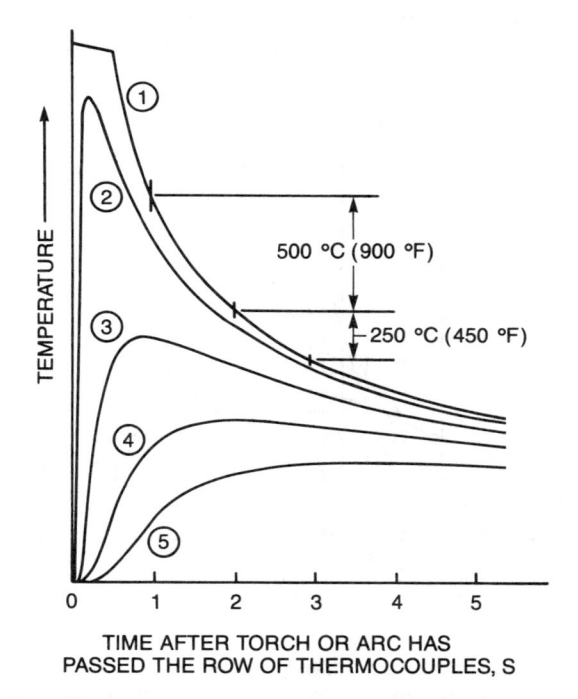

Figure 7.3-B — The temperatures in Figure 7.3A are re-plotted as a function of time for four different distances in the base metal moving away from the weld

first second has elapsed, Curve No. 2 shows that the temperature at Point 2 location is just below that of Point 1. The metal in the next span, moving from Point 2 to Point 3, is heated during the first second by conduction from the hot zone immediately adjacent to the weld, but temperature rise at Point 3 after one second is only about half that of Point 2. Metal in the next span to Point 4 is heated to a lesser extent (as shown by Curve 4), and the adjacent span just beyond Point 4 receives very little heat during the first second. The temperatures that exist in the heat-affected metal spanning between the points are, of course, subject to either positive, or negative gradients depending on distance from the weld and the amount of time that elapsed after start of welding.

Incidentally, the curves in both Figure 7.3-A and 7.3-B are based on calculations made long before temperatures were being measured by high-speed pyrometry during actual welding. Yet, the general theoretical bases of these diagrams have been found to yield results when applied to some kinds of fusion welds that agree within approximately five percent of actual temperature measurements.

The temperature-time curves of Figure 7.3-B portray the rise and fall of temperature at unitized locations in a plate that is subjected to the heat of welding. Some general observations can be drawn from the calculated curves. In a welded plate, with other things being equal, the higher the peak temperature reached at a point during welding, the faster the cooling rate will be after cooling has begun. Also, the heating and cooling rates become markedly slower the greater the distance from the weld zone. Temperature rise is most rapid in the heat-affected base metal at the start of welding, when the flame or arc dwells against cold metal; nevertheless, it is rare for difficulties to occur because of too rapid heating. Sometimes, during the welding of very brittle metals such as white cast iron, the initial application of a heat source to cold workpieces will cause superficial cracking; but as continued heat input brings the metal above its melting point, the surface crazing is eradicated and the slower heating rate inward from this weld zone is less likely to cause further cracking.

The cooling of heat-affected zones is of great importance in welding, and many reasons arise that call for a prediction of cooling rate at a specific location; usually a narrow zone immediately adjacent to the weld is of greatest interest. The curves of Figure 7.3-B allow various quantitative assessments to be made of cooling rate. For example, the three intercepts marked on Curve No. 1 (which correspond to the first three seconds of time that elapsed after start of welding) show that in the interval between one and two seconds the temperature at Point 1 on the welded plate was calculated to drop by 500 Celsius degrees (900°F). Therefore, the cooling rate can be stated simply as 500°C (900°F) per second. Between the second and third seconds of elapsed time, the drop in temperature was somewhat smaller; only 250°C (450°F), and therefore the cooling rate during this period of time was 250°C (450°F) per second.

The need often arises to assess quantitatively the rate of cooling to correlate with a particular metallurgical property change, and several different means of quantitative assessment are used by investigators in their study of various phenomena. These different forms of expressing cooling rate will be encountered in the welding literature. The curves in Figure 7.3-B will serve in following the several expressions listed below that are commonly used in quantitative evaluation of cooling rate:

(1) the number of degrees that the temperature drops over a stated period of time,

(2) the number of seconds that elapse as temperature falls through a stated range of temperature, and

(3) instantaneous cooling rate expressed as degrees per second as the temperature falls past a stated temperature level.

While the temperature at Location 1 in Figure 7.3B is shown by its cooling curve to have dropped by 500°C (900°F) during the one-second interval between one and two seconds on the abscissa, there might be reason to consider that the temperature fell by 750 Celsius degrees (1350°F) over the two seconds that elapsed between one and three seconds on the abscissa. Consequently, the average cooling rate over this two-second period would be expressed as 375°C (675°F) per second. For some purposes, an average cooling rate over a time period of lengthy duration might serve very well. When elapsed time is used for assessment, obviously, two levels of temperature must be selected to mark a range for measuring the cooling time. Increasing use is being made of instantaneous cooling rate, which usually expresses cooling in degrees per (one) second of time. This method of quantitative evaluation requires selection of a temperature level that is pertinent to the objective of the assessment; and, as will be noted shortly, investigators sometimes choose different temperature levels to best serve a given objective. Also, a time shorter than one second may be used to establish the instantaneous rate; and, perhaps for many purposes in the study of welding, the smaller the time interval the better the assessment. For example, metal in the molten weld zone may drop one degree in one-thousandth of a second (0.001 s) very shortly after the start of cooling, which would be expressed as an instantaneous cooling rate of 1000 degrees per second; that is at a temperature level of 1700°C (3100°F). An instantaneous cooling rate always must be accompanied by the temperature level at which it is calculated. Instantaneous cooling rate is frequently used as the indicant of temperature change because averaged cooling rates calculated for a range of temperature can veil or conceal the significance of cooling rate as it is correlated with other kinds of determination.

Special Considerations Regarding Temperature Distribution

A number of general observations should be kept in mind when calculating temperature distributions. The molten weld pool requires somewhat different analytical treatment than does solid metal, because the molten weld is subject to anisotropic flow; that is, circulation can be promoted either by magnetic stirring or by thermal convection. Consequently, motion in the molten pool will affect temperature distribution. Also, with a weld pool that moves continuously along a pass, energy absorption at the leading edge is affected by the latent heat of fusion, which is released during solidification. The latent energy is given up mostly at the trailing edge of the weld pool, which helps account for the more elongated shape toward this end of the pool. This is only one of several minor factors affecting temperature in a weld pool that sometimes is not included in mathematical appraisal. Temperature distribution in the weld metal assumes greater importance when unusual cross-sectional weld-zone shapes are encountered. For example, a finger-like penetration can extend from the root of some weld zones made by the arc processes. Other arc weld zones may have a pattern, called *secondary penetration*, extending outward from the upper portion; or possibly a bulbous configuration may develop at the mid-portion. The asymmetric weld zone shape sketched for the weld in Figure 7.4 made by SAW is not particularly unusual for this process (also see Figure 6.20). When weld zones become misshapen by the temperature distribution that prevails during their formation, the pattern of the base metal heat-affected zone may also be altered from the norm. The SAW weld sketched in Figure 7.4 was made experimentally with imbedded thermocouples in the base metal at three different locations adjacent to the (anticipated) fusion boundary. Temperatures recorded at locations identified as "S," "B," and "R" have been plotted from virtually the start of welding until almost a minute has elapsed. The plotted curves show that the HAZ beneath the secondary penetration (location "S") had undergone a slightly slower rate of heating during welding, and it cooled at a noticeably slower rate than at locations "B" and "R." Peak temperatures for the three locations were in the range of 1425 to 1480°C (approximately 2600 to 2700°F) before each location began precipitous cooling. Whether differences of the magnitude shown in Figure 7.4 for thermal cycles will cause significant change in microstructure is very much dependent upon the kind of steel being welded. The presence of abnormally affected microstructure in the base metal, albeit a very narrow area at the weld zone

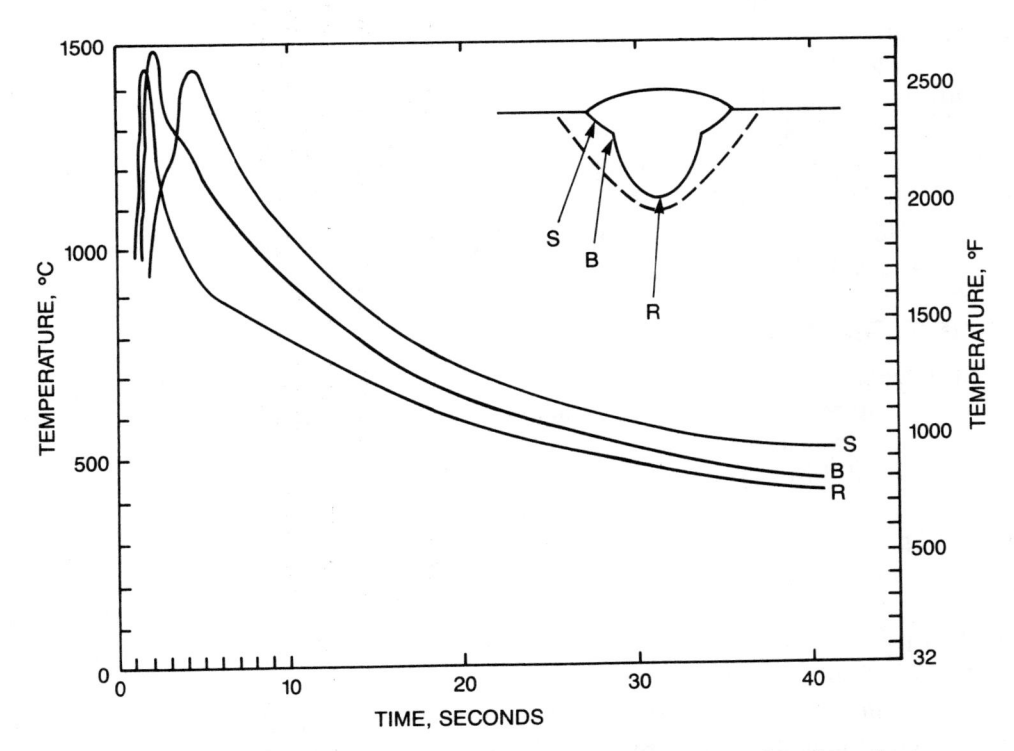

Figure 7.4 — Variations found in thermal cycles at three locations in a weld made by SAW as determined with thermocouples imbedded in the base metal immediately adjacent to the weld zone boundary

(Courtesy of The Welding Institute, England)

boundary, can influence the mechanical behavior of welds in high-strength alloy steels in demanding applications.

Temperature distribution in the base metal heat-affected zone can be calculated with stronger confidence than for the weld zone, because solid metal behaves almost like an isotropic medium with respect to heat flow. Of course, consideration must be given to the many conditions that exert influence on heat flow — including (1) multi-directional flow as dictated by base metal thickness and the depth of weld penetration, (2) form of heat source and its efficiency, (3) presence of heat sinks, and (4) detailed welding parameters (travel speed, etc.). Peak temperatures are not always the item of principal interest. Many metallurgical phenomena that produce property changes are initiated and continue at temperatures below the peak. *Recrystallization* with abnormal grain growth, and *strain-aging embrittlement* are just two examples of metallurgical mechanisms that can occur during exposure to lower temperatures (and exposure for longer time) and deserve study because of influence on mechanical properties.

Although the welding thermal cycle often is reviewed as a relatively simple array of dynamic temperature excursions (as shown in Figure 7.3, for example), most welding operations produce far more complex temperature distributions. The aggregated thermal effects of rapid multi-pass fusion welding is a common example. Temperature distribution also is altered when a welding process is closely followed by a postweld heating torch or by induction heating in order to control cooling or to apply tempering. Tandem arrangement of two or more electrodes is another example not uncommon in multi-electrode submerged arc welding. More advanced procedures in modern manufacturing can include two or more different welding processes applied in tandem. For instance, the deposition of filler metal by the GMAW process to fill a groove might be followed immediately by the GTAW process to remelt the weld face and improve its reinforcement profile. Temperature distribution in welds made with a combination of processes depends not only on the operating parameters employed, but also on sequencing and timing in their application.

Time At Temperature

The length of time during which a heated area of a weldment remains at a particular temperature depends simply on maintaining an even balance between heat input and heat loss in that area. Usually, in most welding operations, temperature at specific areas quickly rises to a maximum and then begins to fall rather rapidly because the heat source either moves or is extinguished and the base metal and fixturing constitute plentiful heat sink. Prolonged exposure to a maximum temperature would be found in processes employing furnace heating of the entire assembly of workpieces (e.g., diffusion welding).

The influence of time at temperature on metallurgical phenomena in welding steels should not be evaluated by itself, but must be considered in combination with rate-of-heating and peak temperature. These three aspects of the welding thermal cycle deserve close attention when welding steels using advanced production procedures because of the inclination to increase the speed of welding as much as possible to reduce costs. Although this trend toward shorter overall weld time is not likely to change peak temperature significantly, it can markedly increase the rate of heating and reduce time at temperature in heat-affected zones. As a consequence of this rapid temperature rise and short dwell time at peak temperature, some steels may develop a microstructure in the HAZ that is prone to become semi-hardened. This microstructural condition in steel develops when a constituent containing cementite undergoes only partial dissolution during heating because too little "soaking time" is spent above the Ac_3 critical temperature. As a consequence, localized areas in the microstructure surrounding particles of undissolved cementite become abnormally rich in carbon content, and these high-carbon areas easily may form martensite in a matrix of unhardened microstructure on cooling; hence the expression *semi-hardened*. This microstructural condition can detract significantly from ductility and toughness.

Much welding research has been directed over a long period of time to the development of test methods to simulate the microstructure found in weld heat-affected zones of steels to expedite study of mechanical properties. The first pieces of information sought for establishing a synthetic weld-effect test were (1) peak temperature reached in a HAZ, and (2) time at temperature for the first increment of still-solid base metal adjacent to the fusion boundary. Early calculations and pyrometric measurements suggested that the peak temperature was probably in the range of about 1375 to 1480°C (2500 to 2700°F). The time at temperature was recognized to vary with the process employed and its operating parameters.

Furnace Heating for Welding Simulation

Very early test development work made use of short-time furnace heating. When small specimens were heated to approximately 1425°C (2600°F) considerably greater grain coarsening in the microstructure was observed as compared with a counterpart weld HAZ. Apparently the rapidity of heating during welding, the minimal time spent at peak temperature, and of course the rapid cooling that followed did not allow the grain growth and other coarsening effects to materialize in the microstructure at the peak temperatures estimated to occur in the weld HAZ. One early test system using furnace-heated specimens established that a temperature of 1150°C (2100°F) appeared to cause the same extent of grain coarsening observed in the HAZ of steels welded by the SMAW process (see Jominy Hardness-Bend Ductility Weldability Method). While this arbitrary adjustment in furnace temperature for heating specimens appeared to be a step in the right direction to obtain the expected peak temperature effect, there remained a wide disparity in time at temperature for the synthetic specimen as compared exposure of the HAZ of a real weld. The longer furnace time, albeit at 1150°C (2100°F), produced other changes in the microstructures of some steels.

Temper Color as an Indicant

Temper colors on bare, clean, bright steel provide a crude but useful method for estimating time and temperature of exposure of heat-affected areas in weldments — at least, as judged from *surface* appearance. When a weld is made by localized heating, temper colors ranging from black, through blue, brown, and straw coloration will run in bands parallel to the long axis of the weld after the weld has cooled. These variations in colors arise from the different thicknesses of oxide films that form on the surface of iron and steel when heated in air. Sand blasting or pickling are effective for preparing a surface on which temper colors from welding are to be observed. The colors formed on iron and carbon steel by progressively higher temperatures are listed in Table 7.1.

Temper colors, therefore, can give a rough indication of the maximum temperature imposed on the base metal at varying distances from the weld. If two different welds are compared for temper colors, and the brown-purple transition is found closer to the edge of the weld in the first plate, it can be concluded that the weld in the first plate was heated more rapidly and cooled faster than that in the second plate. Another helpful piece of information that can be obtained by temper color examination concerns uniformity of heating and cooling, particularly when tooling is being utilized as heat sink. Lack-of-contact between workpiece and tooling over localized areas often will register as darker temper coloration on the surface where tooling did not extract heat as expected from the weldment.

Resistance Heating for Welding Simulation

Today, better results are obtained with rapid electrical resistance heating of specimens to more closely simulate thermal cycles that occur in the base metal HAZ. A commercial unit called the "Gleeble" employs the electrical resistance heating technique, and is capable of producing (1) very rapid heating, (2) short holding time at peak temperature, and (3) controlled cooling to closely simulate a defined thermal cycle as might be anticipated in a specific base metal section with a given welding process. Application of multiple cycles to a specimen to simulate multi-pass welding is also feasible with the "weld thermal simulator" units now available.

An important fact that remains concerning heat-affected microstructure in the peak temperature increment of real welds is that the grain coarsening is seldom as great as might be judged from temperatures actually measured in this region of the HAZ. Among possible explanations advanced for this difference in general behavior, the most perceptive deals with the various minute particles of secondary phases and precipitates in the matrix microstructure that are not dissolved as early as suggested by constitution or phase diagrams. The rate of heating is too fast, and the time above the dissolution temperature level simply is too short. Consequently, these particles exist in the matrix longer than normally expected, and they serve to inhibit grain growth until dissolution occurs. At this point in time the heat-affected area already is rapidly cooling to a temperature level that is too low to cause grain coarsening. Earlier, in Chapter 4, brief mention was made of alloying steel to form difficult-to-dissolve minute particles (i.e., titanium carbonitrides) in the microstructure to maximize this mechanism for restricting grain growth. This steelmaking practice will be examined in more detail in Chapters 14 and 15 when modern low-alloy steels are reviewed.

Table 7.1 - Temper Colors Formed on Iron and Carbon Steel

Color Formed on Surface	Approximate Temperature At Which Color Forms	
	°C	°F
Light Straw	200	400
Tan	230	450
Brown	275	525
Purple	300	575
Dark Blue	315	600
Black	425 and higher	800 and higher

Cooling Rate of Heated Zones

Understanding *cooling* rates in the weld zone and heat-affected areas is more important than understanding *heating* rates. Cooling is more complex, however, because many more features of a weldment and welding procedure exert an influence, and because the rate of cooling has a dominating effect on the final microstructure and properties. Three general features have a profound influence on the cooling rates at various locations in a weldment:

(1) the weld zone, or nugget, which serves as a heat source during the course of the welding process;

(2) the mass of base metal, along with any intimately contacting metal fixtures that can provide a heat sink; and

(3) the initial temperature of the base metal and fixtures.

Although these three features require individual review, attention also must be given to their interrelationships as factors controlling cooling rate.

Correllation with Heat Input

As a heat source, weld zones and nuggets are not as quantifiable as the originating sources used in welding. When heat-originating sources were described in Chapter 6 (gas flame, electric arc, laser beam, etc.) and formulae were given for calculating their energy output, attention was directed to the efficiency of energy transfer with the various sources. This is because the amount of energy that actually enters the workpieces in the form of heat will differ among the sources, and any change in actual heat input to the workpieces will affect cooling rates in their heat-affected areas. Chapter 6 also introduced a correlation between calculated energy of the originating heat source and the HAZ cooling rate. With the electric arc, for example, the available energy was calculated as the product of current and arc voltage and expressed as watts of power. To indicate how much power from the arc was being made available to the weld, the apportioning of joules per unit length of weld was calculated according to the simple formulae given in Equation 6.2 (J/mm) and in Equation 6.3 (J/in.). Such calculations have fostered attempts to control cooling rates by monitoring "arc energy input" (See Technical Brief 39). This technique met with limited success, however, and researchers eventually abandoned it in search of a better method for predicting and controlling cooling rates.

Portrayal of HAZ cooling rate is difficult because of the need for a unified means of representation that can be readily correlated with the energy made available during welding. Figure 7.5 illustrates the heating and cooling cycles found at five HAZ points located at increasing distances from the weld centerline. The uppermost curve represents the thermal cycle with a peak heating temperature of 1365°C (2490°F) measured at a distance of 10 mm (0.39 in.) from the weld centerline. This temperature is a little lower than those found in a SAW as illustrated in Figure 7.4, a difference probably caused by small variations in the distance from thermocouple tip to the fusion boundary. The remaining four curves in Figure 7.5 show the thermal cycles found at distances of 11, 14, 18 and 25 mm (0.45, 0.55, 0.70 and 1.00 inch) from the weld centerline, and, as would be expected, these curves all show decreasing rates for both heating and cooling in the HAZ as compared with the 10 mm (0.39 in.) location. Three

> **TECHNICAL BRIEF 39:**
> **Controlling HAZ Cooling Rate via Arc Energy Output**
>
> About 1940, a practice began of stipulating limits for heat source energy when welding certain steels. The assumption was made that energy *input* to the workpieces, although not likely to be 100 percent of the source *output*, would be a linear function of weld time on each unit length. Therefore, a straightforward correlation was anticipated between calculated source energy per unit weld length and the cooling rate in the base metal heat-affected zone. Because the SMAW and SAW processes were most popular at the time, an outpouring of tables, nomographs, slide rules, and diagrams appeared in the literature purportedly showing the relationship between "arc energy input" (actually, arc energy *output* was being calculated) and HAZ cooling rate. In general, higher welding current and slower travel speed produced greater heat input per unit length of weld; and this, in turn, produced slower cooling rates in the HAZ. The practice in welding certain highly hardenable steels was to set a minimum "heat input" for a particular joint to be welded and thus ensure slow cooling to reduce hardening, whereas certain other steels were assigned a maximum limit to avoid degradation of mechanical properties in their HAZ.
>
> Relationships between "heat input" and HAZ cooling rates in published material were confirmed by several different investigative methods — by actual temperature measurements with thermocouples imbedded in the HAZ; by calculations based on electrical analogy with heat flow; and by comparing hardness measurements in heat-affected zones with hardness curves for Jominy bars, in which the cooling rates simulated the thermal cycles of welding. Many early guides to HAZ cooling rate control became entrenched in the welding literature, and some continue to appear in current documents covering the welding of certain steels. However, it must be recognized that much of the available data were developed for weldments made by the SMAW and SAW processes using rather limited ranges of current, arc voltage, and travel speed. The assumption of general linear relationship between arc energy and HAZ cooling rate eventually proved to be too all-embracing and inexact for the purpose of predicting and controlling HAZ cooling rate.

general observations which can be made regarding the information presented in Figure 7.5 are:

(1) that peak temperatures decrease rapidly with increasing distance from the weld centerline,
(2) that the length of time required to reach peak temperature increases with increasing distance from the weld centerline, and
(3) that the rate of heating and cooling both decrease with increasing distance from the weld centerline.

The HAZ *immediately* adjacent to the weld zone is of great importance in carbon and alloy steels, because this is the increment where microstructure and mechanical properties undergo significant change as a result of the welding thermal cycle. The width over which this important HAZ microstructure extends will be influenced by several of the welding parameters. The peak temperature of the still-solid HAZ immediately adjacent to the fusion boundary appears to approach the incipient melting temperature of the base metal. This can be observed metallographically by noting the condition of grain boundaries and sulfide inclusions in the base metal that lie in very close proximity to the fusion boundary. While the true peak temperature on the base metal side of the interface approaches the incipient melting temperature, the duration of this peak temperature is very short and it falls very rapidly.

Many proposals have been made by investigators on how the rate of cooling in the first increment of the HAZ should be characterized to permit meaningful correlation with welding energy. The rate of temperature decrease is in constant change during the entire cooling period, and an average rate for the entire cooling curve is unusable for correlating metallurgical properties with energy. The most significant period in the descent from peak temperatures in the HAZ in terms of microstructures formed in various carbon and low-alloy steels is within the range of approximately 800 to 500°C (about 1475 to 930°F). This led some investigators to concentrate their attention on this intermediate portion of the HAZ cooling curve, and they used the precise length of time that elapsed during cooling through a specific temperature range as the cooling rate indicant. For example, the nomograph in Figure 7.6 was developed through thermal flow calculations to portray the relationship between cooling time in seconds and heat input when using different plate thickness and joint geometry (i.e., bead-on-plate conditions that entail thermal flow in only two directions, as compared with a fillet weld in a T-joint that would result in three-dimensional thermal flow). The nomograph indication that a fillet-welded T-joint cools faster than a bead-on-plate

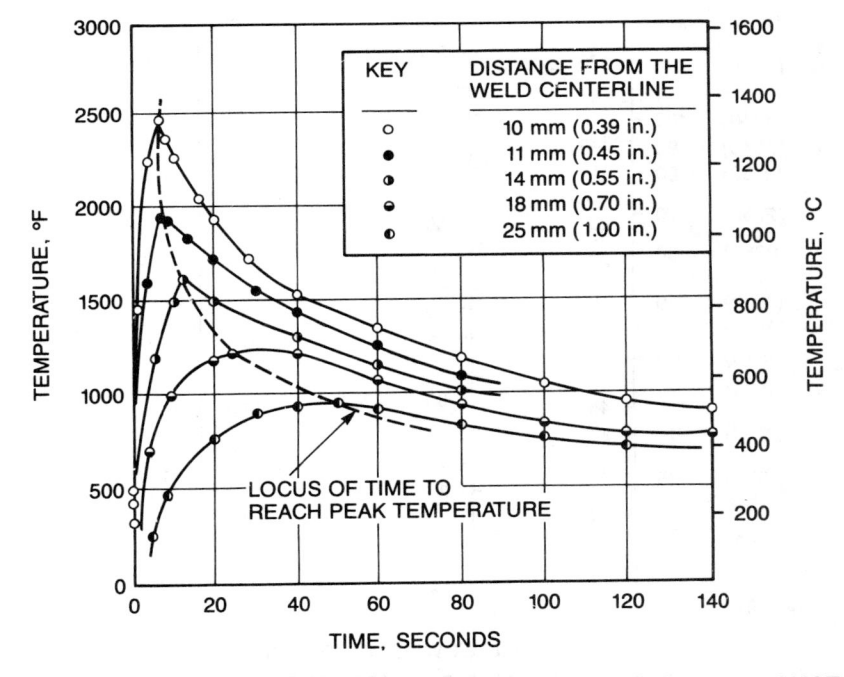

Figure 7.5 — Curves illustrating thermal cycles found in base metal HAZ at five specific points located at increasing distances from the centerline of a weld made by SMAW

Steel plate 13 mm (½ in.) thick, at an initial temperature of 27°C (80°F), welded with arc energy of 3940 J/mm, or 100 kJ/in. (Reprinted from *AWS Welding Handbook*, Vol. 1, Fundamentals of Welding, Seventh Edition, Chapter 3, Heat Flow in Welding, Figure 3.1)

weld is in agreement with earlier discussions about the influence of base metal mass and heat flow directions.

Re-examination of Figure 7.5 and its cooling curves will find that 90 seconds elapsed as the weld HAZ at the 10 mm location cooled from 800 to 500°C (1472 to 932°F). The nomograph in Figure 7.6 indicates a calculated cooling time of 60 seconds. The column of cooling times in Figure 7.6 also has the label "Δt_{8-5}" in its heading, which investigators world-wide use to signify cooling time in seconds for the HAZ immediately adjacent to the weld zone to pass through the 800 to 500°C temperature range. At first thought, this difference in time for the HAZ to cool through

HEAT INPUT IN JOULES	COOLING TIME IN RANGE FROM 800 TO 500 °C (1472 TO 932 °F)	PLATE THICKNESS FOR CARBON AND ALLOY STEELS
(J/mm) kJ/in. kJ/in.	SECONDS (Δt_{8-5})	mm in.

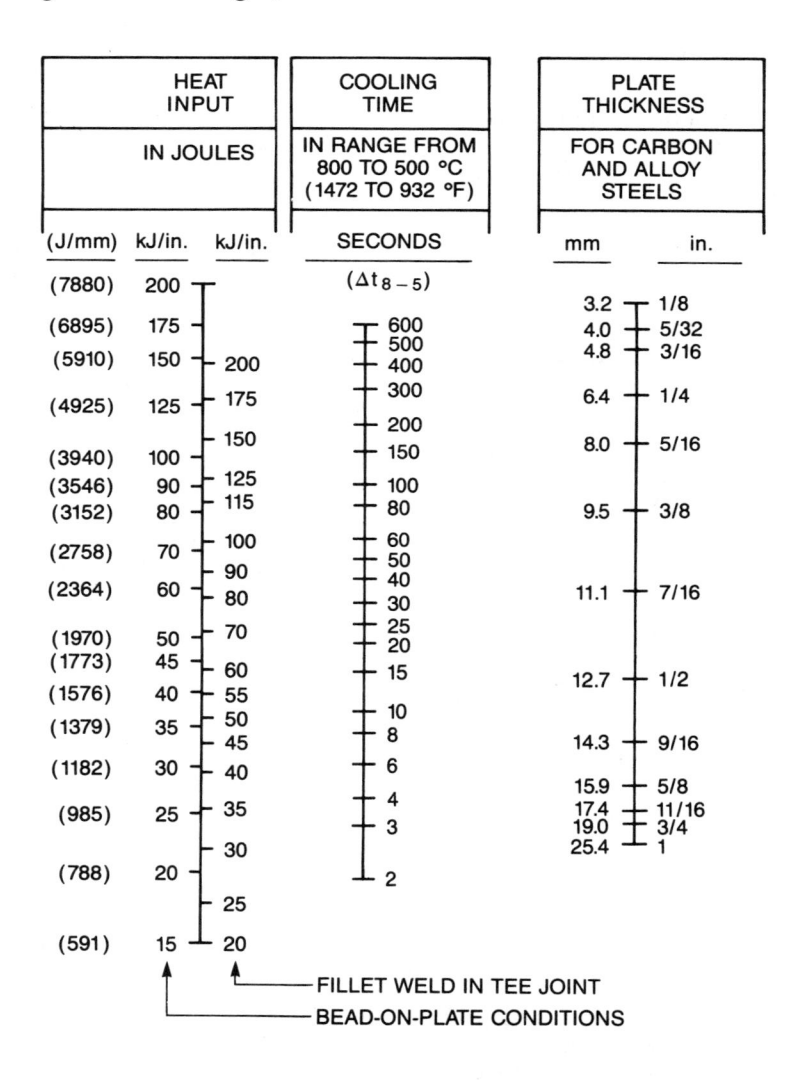

Figure 7.6 — Nomograph for calculating time that elapses between 800 and 500°C (1472 and 932°F) when base metal HAZ immediately adjacent to weld cools after an arc welding operation

(Data according to Inagaki and Sekiguchi)

this important temperature range might be regarded as demonstration of the difficulty in closely calculating thermal flow in real weldments. However, certain aspects of welding procedures have been found to alter heat input and heat distribution in the workpieces, and differences in the procedures for welds in Figures 7.5 and 7.6 easily could account for the difference in cooling time. These influential aspects of welding procedure will bear further discussion.

Virtually all tables of rates and diagrams in the literature for weld cooling cycles consider mass and geometry factors. To give a perspective of the quantitative effect of base metal mass, Figure 7.7 provides temperature information on three thicknesses of steel plate welded by the SMAW process. The cooling curves in each plate were determined using imbedded thermocouples. In each plate thickness, the initial temperature was close to room temperature (RT), and the same heat input, 1850 J/mm (47 000 J/in.), was employed in depositing a weld; a level of arc energy typical for the SMAW process. The cooling curves in Figure 7.7 show faster rates as plate thickness increases, and the Δt_{8-5} measured value for the 6 mm (¼ in.) plate was 51 seconds, while the 13 mm (½ in.) plate was 24 seconds, and the 25 mm (1 in.) plate was 10 seconds. Calculated Δt_{8-5} values using the nomograph in Figure 7.6 for these SMAW welds were 52, 17, and nine seconds, respectively. Thus, measured and calculated values here are in closer agreement than found earlier for the weld in Figure 7.5 at its 10 mm (0.39 in.) location. Some difference in HAZ cooling times for the three thicknesses of plate can be seen in the upper range of temperature [i.e., approximately 900°C (1652°F and above)], but the important temperature range of 800 down to 500°C (1472 to 932°F) is obviously markedly affected, and the thicker plate cools much more rapidly in its HAZ. The practical value of this knowledge about cooling rate in the HAZ will become clear when various steels are examined to note the influence of cooling rate on microstructural behavior.

Workpiece Pre-Weld Temperature

Another factor that received attention during early work on HAZ cooling rate is the initial temperature of the workpieces, especially to secure a quantitative perspective on the influence of preheating to slightly elevated temperatures. Figure 7.8 shows the results of two welding tests conducted with 13 mm (½ in.) thick steel plate. One plate, initially at room temperature before welding, cooled as shown by the solid-line curve. The second plate, preheated to 260°C (500°F), produced the slower rate (dashed-line) cooling curve. The measured Δt_{8-5} value for the plate welded when at RT is 23 seconds (nomograph in Figure 7.6 predicted 18 s), whereas the plate preheated to 260°C (500°F) has a 58 second lapse of time. This difference in cooling time between the RT and the preheated plates is quite significant in altering the HAZ microstructure of many steels. The application of preheat to temperatures intermediate between RT and 260°C (500°F) would result in cooling curves falling somewhere within the area bordered by the two curves in Figure 7.8. Preheating to a temperature higher than 260°C (500°F) would result in cooling rates slower than the upper dashed-line curve of Figure 7.8. The need to know the HAZ cooling rate with best possible exactness has been a never-ending challenge in welding steels because of the variety of problems that can arise. In certain steels, HAZ cooling must be slow enough to avoid excessively hardened microstructure that would be lacking in toughness, and would be susceptible to cold cracking. Whereas, other steels might require sufficiently-fast cooling in the HAZ to re-establish a kind of microstructure with favorable toughness.

Preheating, and the use of higher heat input to regulate HAZ cooling rate and thus favor softer, tougher microstructures in many welded steels quickly entered into practice when planning welding procedures. Nevertheless, the challenge remained to devise a simple, reliable system to quantify the factors involved, establish their inter-relationships, and to facilitate calculation of welding parameters that would ensure a controlled outcome with respect to HAZ cooling rate and the resulting microstructure and properties. The elapsed time for passing through a particular temperature range (e.g., Δt_{8-5}) proved unwidely as a cooling rate indicant, and reporting on its correlation with microstructural behavior dwindled before it became part of a welding control system used by industry.

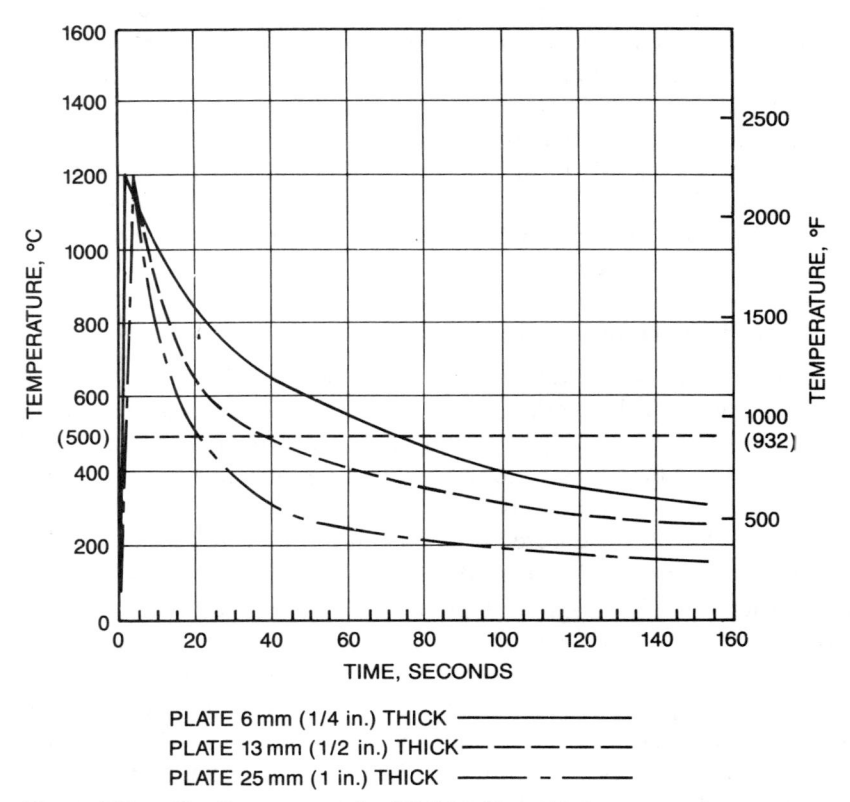

PLATE 6 mm (1/4 in.) THICK ————————
PLATE 13 mm (1/2 in.) THICK — — — — —
PLATE 25 mm (1 in.) THICK ——— - ———

Figure 7.7 — Cooling curves for HAZ in three thicknesses of steel plate, each welded by SMAW, showing influence of base metal mass

Plates at initial temperature of 27°C (80°F) welded with arc energy of 1850 J/mm (47 kJ/in.)

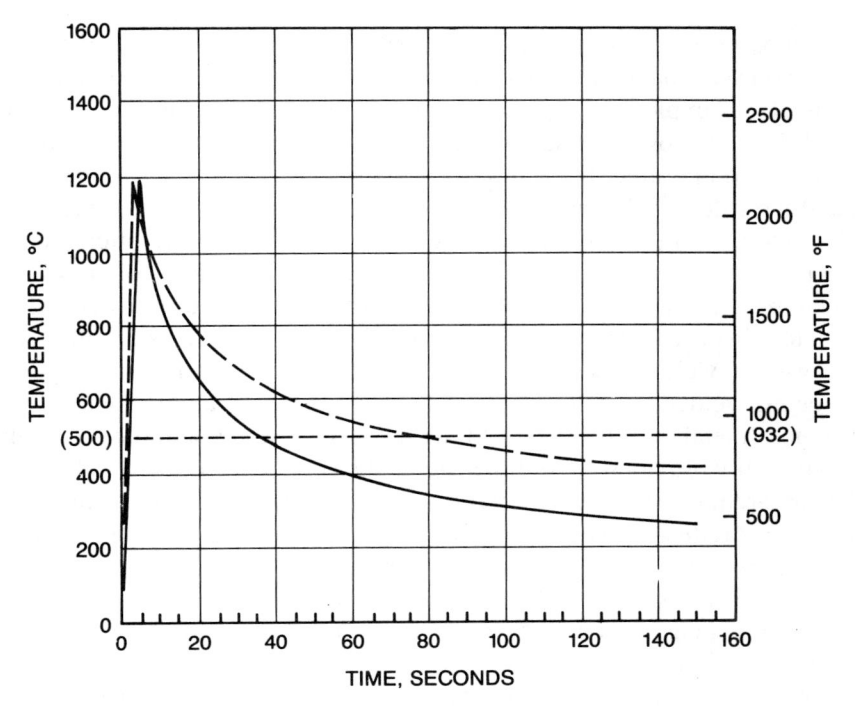

SOLID CURVE ——————— PLATE WELDED WHEN AT INITIAL TEMPERATURE OF 27 °C (80 °F)

DASHED CURVE — — — PLATE PREHEATED TO 260 °C (500 °F) PRIOR TO WELDING

Figure 7.8 — Influence of preheating as found in cooling curves for SMAW process on 13 mm(½ in.) thick steel plate using arc energy of 1970 J/mm (50 kJ/in.)

Instantaneous Cooling Rate

Instantaneous cooling rate — the rate of temperature change occurring at a particular instant — was the next choice of investigators as an index for HAZ cooling rate, and the first work to appear in the literature used 704°C (1300°F) as the designated temperature level for this kind of assessment. In other work, the favored temperature level was 538°C (1000°F). Figure 7.9 provides a diagram with a scale of instantaneous cooling rates on the vertical coordinate that can be encountered in the HAZ of 13 mm (½ in.) thick steel plate when applying the SMAW process over a wide range of arc energies. Curves are provided for the two temperature levels of interest, and it will be noted that the instantaneous cooling rates at the 704°C (1300°F) level are faster because of the steeper temperature gradients between the HAZ and the unaffected base metal. As heat input is increased to arc energies that might be considered excessive for a 13 mm (½ in.) thick plate, the instantaneous cooling rate of the HAZ for both temperature levels selected for measurement are decreased only to a minimal extent. This is the case here because of the somewhat limited heat-sink capacity of a 13 mm (½ in.) thick plate.

Despite disagreement among investigators on the best temperature level at which to assess instantaneous HAZ cooling rate, continued studies began to include more of the conditions encountered in real weldments in order to ascertain their influence. Figure 7.10 presents data on welds made by the SMAW process in 13 mm (½ in.) thick steel plate. Two simple joint geometries are represented, the butt joint, and the fillet weld, and the test program also covered the influence of weld metal volume on the instantaneous cooling rate. In each of the tests for Figure 7.10, when making the butt welds and the fillet welds, the workpieces were at room temperature prior to deposition of a weld pass. In actual practice, the workpiece temperature often rises as welding progresses because the weldment is not given time to cool to RT before a subsequent pass is deposited; therefore, the interpass temperature may constitute a preheated state. This is a frequently encountered situation that makes calculation of cooling rates for real weldments so uncertain. However, if a plate in which a grooved butt joint is being welded is at RT, the last pass cools more rapidly than the first pass because more metal is in proximity to the point of welding, and the metal already in the weld possibly provides three-dimensional thermal flow from the weld area. At this stage of research, the observation was made that the volume or mass of metal within a 75 mm (3 in.) radius of the point of welding can influence cooling rate in the HAZ over the temperature range in which important microstructural transformation can occur. However, mass beyond this radius is too remote to influence cooling through the temperature range that affects transformation mechanisms. In Figure 7.10, the slower HAZ cooling rate observed in butt welds, as compared with fillet welds, is attributable to three-dimensional thermal flow from the fillet weld. The comparison of

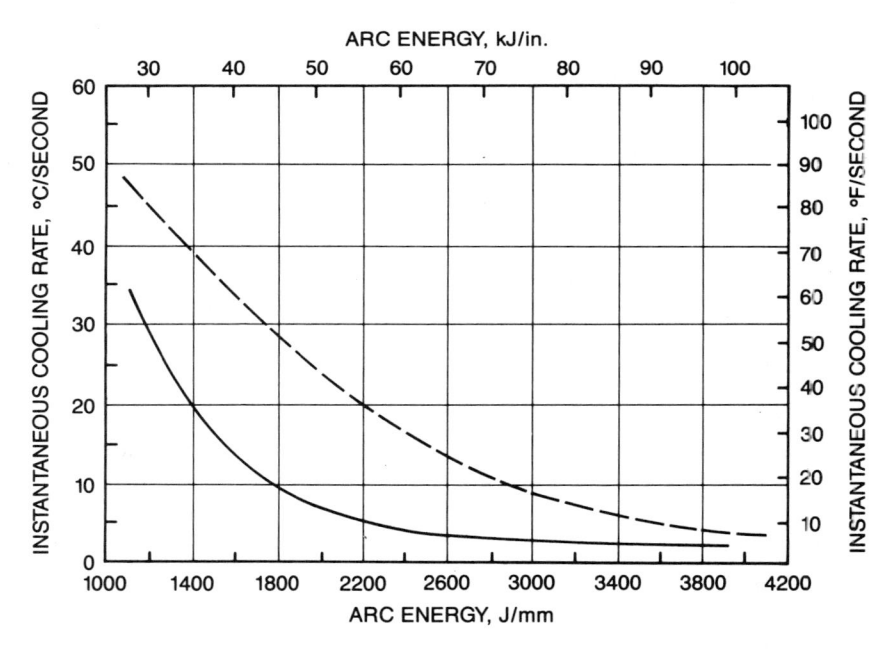

SOLID CURVE ——————— INSTANTANEOUS COOLING RATE MEASURED AT 538 °C (1000 °F)
DASHED CURVE — — — INSTANTANEOUS COOLING RATE MEASURED AT 704 °C (1300 °F)

Figure 7.9 — Instantaneous cooling rates found in HAZ of 13 mm (½ in.) thick steel plate welded by SMAW using a wide range of arc energies

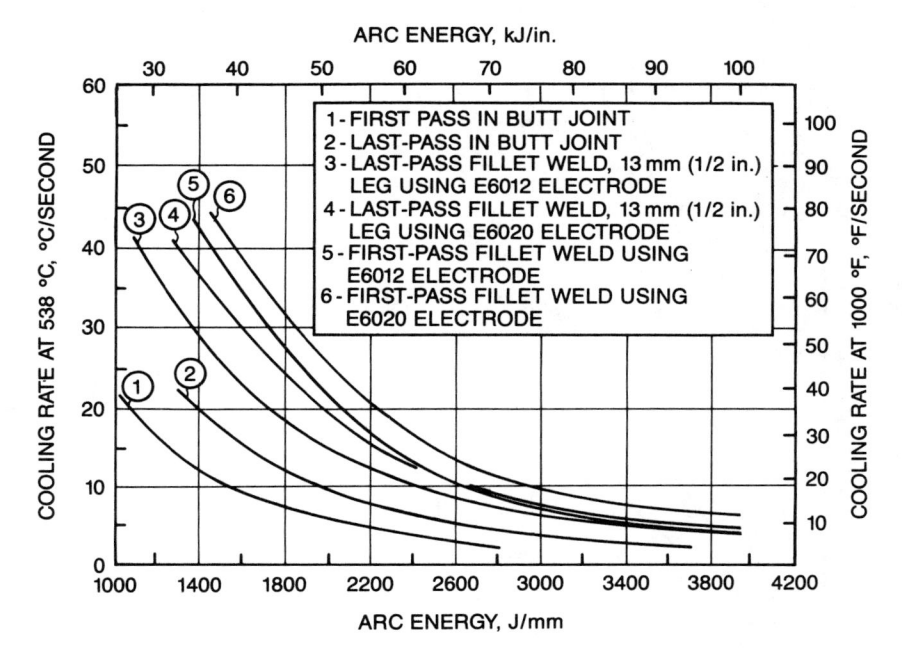

Figures 7.10 — **Instantaneous cooling rates found in HAZ of 13 mm (½ in) thick steel plate welded by SMAW showing influence of joint geometry and mass of metal in proximity of joint with changes in level of arc energy**

(Data according to Hess, et al, at RPI)

E 6020 and E 6012 SMAW electrodes used for fillet welds shows difference in HAZ cooling rate that is not significant, and demonstrates that the electrode classification (and even the size) has little effect on the relationship between arc energy and HAZ cooling rate.

Mention was made earlier that higher welding current and slower travel speed produce greater heat input per unit length of weld and therefore cooling rates in the HAZ are lowered. However, surprising findings came to light when travel speed was studied separately as a parameter in arc welding procedures, especially when the travel speed was increased far beyond the moderate speeds employed with the SMAW process. Much of the research described to this point has been carried out with the SMAW process with travel speeds in the range of approximately two to five mm/s (about four to 12 ipm), and under these conditions it did appear that energy input from the arc was a linear function of arc energy per unit length, and that the resulting heat input directly controlled HAZ cooling rate. Thus it appear that the only considerations remaining were (1) metal mass within the 75 mm (3 in.) radius of the weld, and (2) the pre-weld temperature of this metal acting as a heat sink. However, when travel was increased to speeds that ranged as much as five times faster, the first effect noted in welds was a substantial increase in the depth of the weld zone or joint penetration as travel speed increased. Figure 7.11 illustrates this propensity for deeper penetration with faster travel speed with fixed levels of arc energy. This relationship between arc

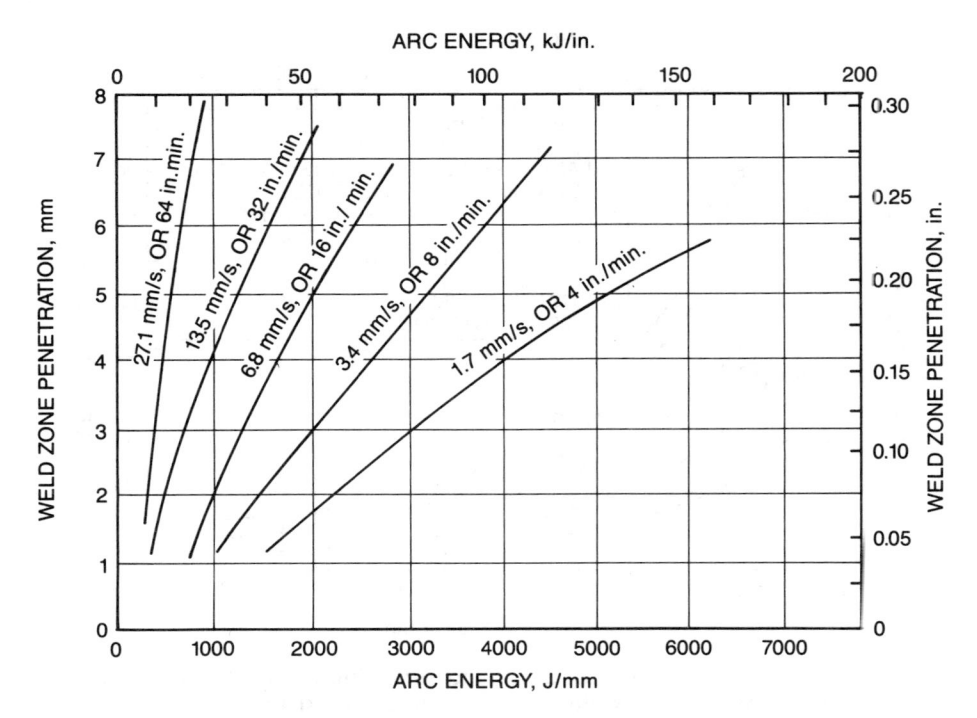

Figure 7.11 — Effect of weld travel speed in consumable electrode arc welding processes on depth of melting or weld zone penetration as found over a wide range of arc energies

(Data according to Jackson and Goodwin)

energy, travel speed, and penetration by the weld zone was reported to hold true for the SMAW, GMAW and SAW processes. Of course, travel speeds employed with the SMAW and GMAW processes overlapped those for SAW only to a small extent.

Influence of Travel Speed on Weld Zone Size

As travel speed was studied further by measuring weld zone cross-sections in specimens welded by various arc processes, it was found that the area of the weld zone was enlarged by faster travel speed at any given level of arc energy. Figure 7.12 illustrates schematically the marked difference in both depth of weld zone penetration and in weld zone area in weld specimen A made a faster travel speed than weld specimen B but with both welds having the same arc energy available (approximately 1780 J/mm, or about 45 kJ/in.). Weld A had almost five times as much area of melted weld zone as compared to weld B. Apparently, faster travel speed allows a greater portion of the arc energy per unit length to be utilized in melting metal rather than just extending the magnitude of heating into the base metal beyond the boundary of the fused zone. Yet, it should be noted that the overall or total energy input to welds A and B of Figure 7.12 appears somewhat the same because even though weld A has greater weld zone area, weld B has a substantially larger heat-affected (but not melted) zone.

Because the melted zone serves as the primary heat source for that important first increment of HAZ adjacent to the fusion boundary, investigators began to examine more carefully the relationship between weld travel speed and the HAZ cooling rate. Figure 7.13 provides additional weld zone area data secured from tests using the SMAW process at its usual travel speeds up to about 6 mm/s (approximately 14 ipm), and then employing the SAW process over the same range, but extending the SAW tests to roughly twice the travel speed. Results for these two processes in the replicated ranges of travel speed and arc energy were virtually identical. When the two welds illustrated in Figure 7.12 are re-examined in light of the relationship diagrammed in Figure 7.13, the measured weld zone areas differ somewhat from predictions. Weld A of Figure 7.12 had a measured weld zone area of 119.4 mm² (0.185 in.²) as compared with a predicted area of about 87 mm² (0.135 in.²); whereas weld B had a measured area of 28.4 mm² (0.044 in.²) and a predicted area of 42 mm² (0.065 in.²). These examples illustrate the degree of approximation that possibly must be indulged as currently available diagrams are put to use. Nevertheless, the trend of results obtained by actual testing and by prediction are consistent in this case.

Work by many investigators shows that voltage in consumable electrode arc welding processes has no significant effect on the total fused area of the weld zone, and that current and travel speed are the controlling factors. Analysis of data for the SAW process found the fused area directly related to the 1.716 power of the current and inversely related to the first power of the travel speed as shown by Figure 7.14, which includes an equation for the data.

The efficiencies of the various welding processes to transfer energy developed by their heat source into the workpieces has been mentioned in a general way during earlier discussion. It will be helpful to have more definitive expressions of efficiency at

WELD ZONE AREA	WELDING CURRENT	ARC VOLTAGE	TRAVEL SPEED	ARC ENERGY
	A	V		
A — 119.4 mm², OR 0.185 in.²	800	26	11.6 mm/s OR 27.4 in./min.	1793 J/mm OR 45.5 kJ/in.
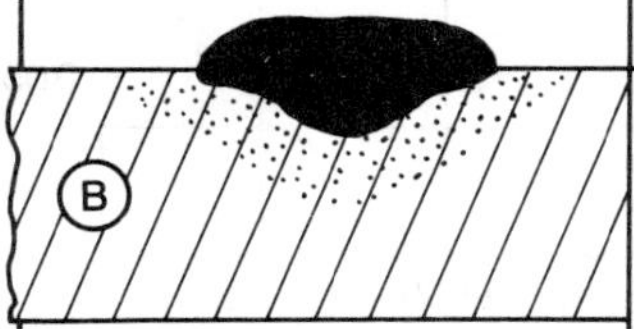 B — 28.4 mm², OR 0.044 in.²	135	26	1.98 mm/s OR 4.7 in./min.	1772 J/mm OR 44.8 kJ/in.

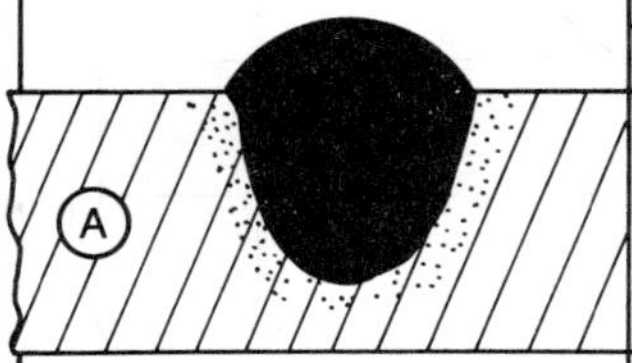

Figure 7.12 — **Difference in area of weld zone in welds made by SAW process using same level of arc energy, but at different travel speeds**

(Data according to Jackson and Goodwin)

hand as predictive systems for HAZ cooling rate are examined. Figure 7.15 illustrates the effort of one team of investigators to quantify arc efficiency for the popular arc welding processes. A chart with log-log plotting is used because of the broad range of arc energies covered (about 300 to 12 000 cal/s, or one to 50 kVA). The factor for efficiency in transferring available arc energy into the workpieces was given the symbol "η" by the investigators. Test data for welding mild steel with the three processes studied fall into bands, which have a constant slope of 45 degrees. As reference, dashed-lines have been drawn along the same slope to mark 100 percent, and 50 percent arc efficiency. The submerged arc process (SAW) has highest efficiency with all measured results falling in a narrow band extending from 90 to 99 percent efficiency. The shielded metal arc process (SMAW) and gas metal arc process (GMAW) are not as efficient in energy transfer, and results fall within a band extending from 66 to 85 percent efficiency. Gas tungsten arc welding (GTAW) has the lowest arc efficiency of the several processes studied, and results ranged from 21 to 48 percent depending on the operating conditions selected (e.g., arc length, polarity, electrode size and tip shape, shielding, etc.) to achieve specific results. Consequently, energy transfer efficiency will require close attention as a variable factor when modeling a given welding operation.

At this point, it would be well to examine a few systems proposed by investigators to calculate thermal flow in weldments to allow prediction of the volumes of weld zone and HAZ produced at the joint, and ultimately to predict the cooling rates of

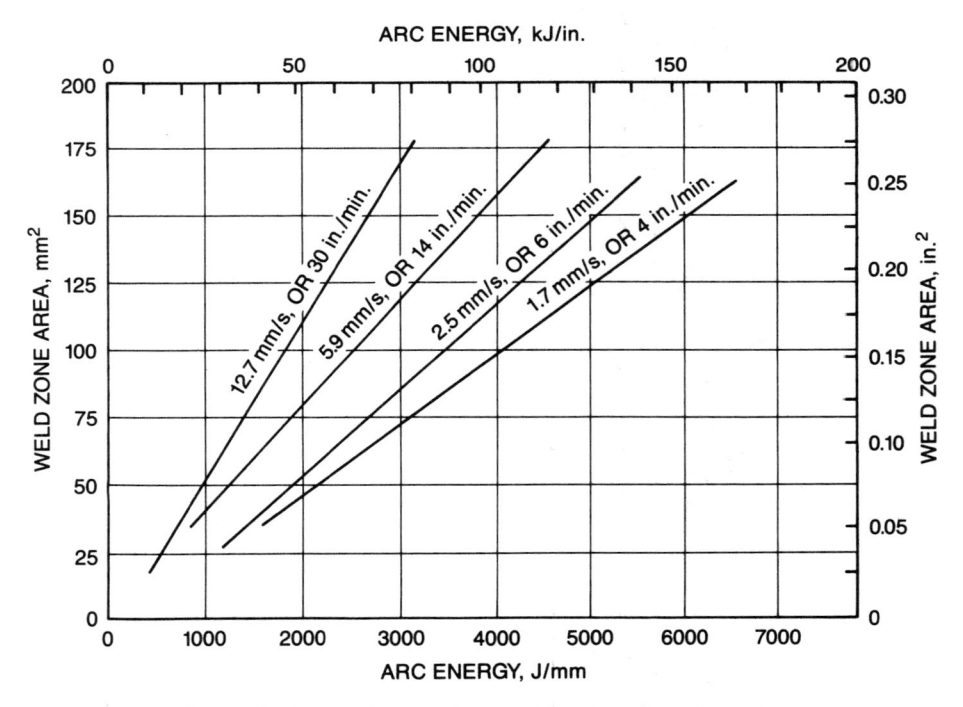

Figure 7.13 — Effect of weld travel speed in SMAW and SAW processes on area of fused metal formed in weld zone as found over a range of arc energies
(Data according to Jackson and Goodwin)

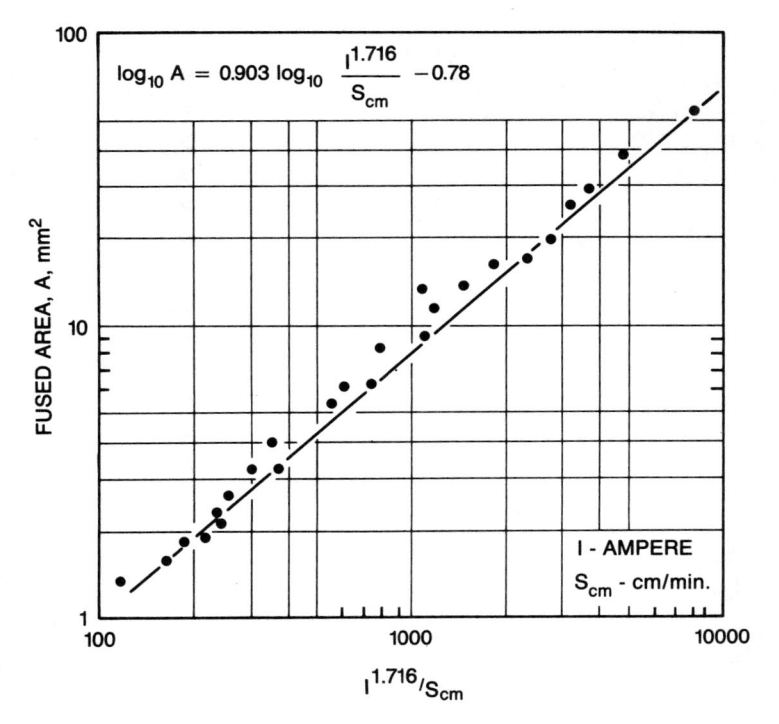

Figure 7.14 — Effect of welding current and speed of travel on cross-sectional area of weld zone when using SAW process
(Courtesy of The Welding Institute, England)

these zones. One early equation developed by Rosenthal in 1941, shown below, has often served as basic mathematical modeling:

$$R_t = \frac{2\pi k v (t-t_o)^2}{Q} \quad \text{(Eq 7.2)}$$

where

R_t = Cooling rate at t of a weld bead on a semi-infinite plate (°C/s)
k = Thermal conductivity of the steel (cal/cm/s/°C)
v = Speed of welding (cm/s)
t = Instantaneous temperature at heat source point (°C)
t_o = Initial plate temperature (°C)
Q = Heat transferred to plate (cal/s)

Dorshu in 1968 confirmed the validity of this Rosenthal heat transfer equation for weld-centerline cooling in thick plate by making experimental welds with the GMAW process and injecting the tip of a thermocouple into the surface of the trailing end of the weld pool. Temperature descent was recorded from the molten state to nearly room temperature.

Both the Rosenthal equation, given above, and the Adams equation for peak temperature, presented earlier in Equation 7.1, take weld travel speed into consideration; however, the weighing of a factor for travel speed deserves further refinement.

A number of investigators striving to calculate temperature distribution in welds made with a moving point (arc) heat source opted for dimensionless equations and diagrams. A diagram derived from Rosenthal's equations and prepared by Christensen, et al, is shown in Figure 7.16. From this dimensionless diagram, the cross-sectional areas of weld zone, and the "recrystallized" portion of the base metal heat-affected zone are predicted by an operating parameter assigned the symbol "n". Shinoda and Doherty wrote the following equation for this factor:

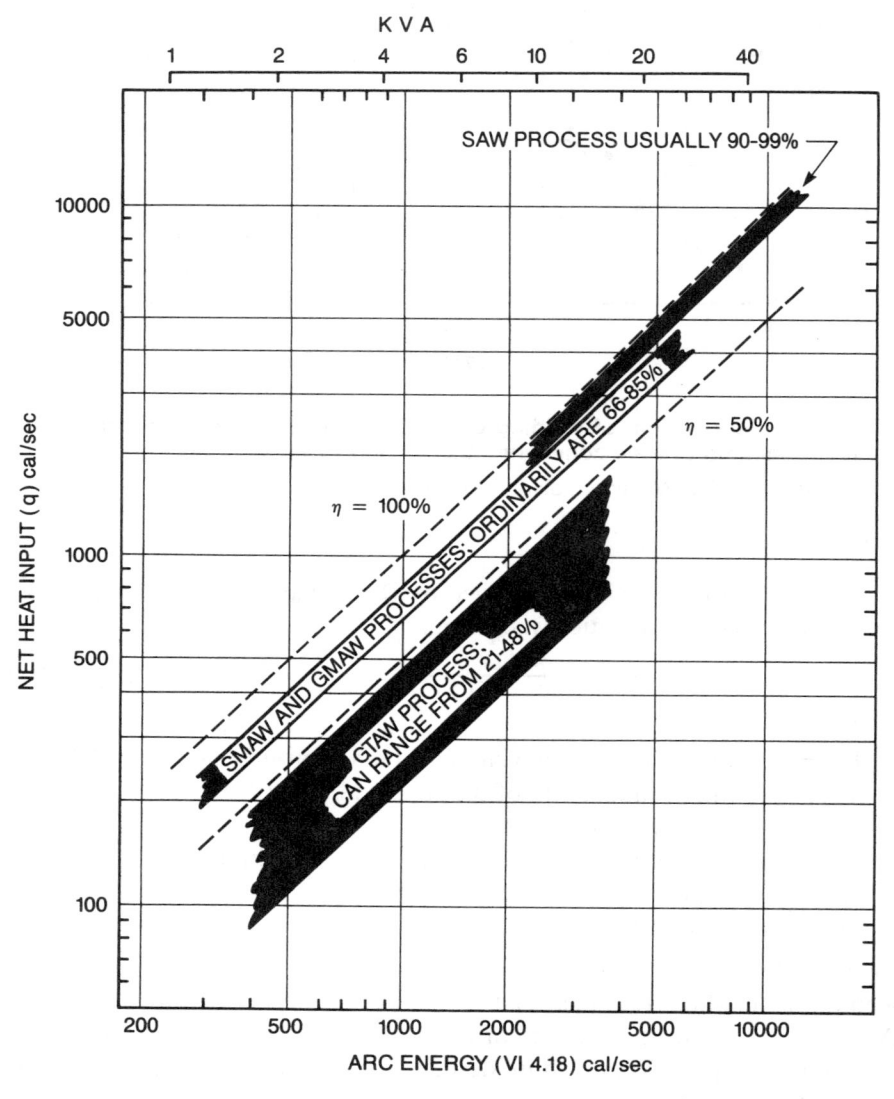

Figure 7.15 — Ranges of arc efficiency ($\eta = {4.18q}/{VI} \times 100\%$)
Ranges found with SMAW, GMAW, GTAW and SAW processes when measured under most operating conditions (Data according to Christensen, Davies and Gjermundsen)

$$n = \frac{qS_{cm}}{4 \pi k \alpha (t_f - t_o)} \qquad \text{(Eq 7.3)}$$

where

- n = operating parameter
- q = net heat input (cal/s)
- k = thermal conductivity of plate (cal/cm²/cm/°C/s; steel ~ 0.10)
- α = thermal diffusivity = $k/p\ C_p$ (α cm² °C/s; steel ~ 0.075)
- t_f = melting point of plate, °C (steel ~ 1520°C)
- t_o = initial temperature of plate, °C
- p = density of plate
- C_p = specific heat of plate (cal/g °C; steel ~ 0.17)
- S_{cm} = weld travel speed, cm/s

Measured gross and net cross-sectional areas are plotted in Figure 7.16 for several arc welding processes. In the case of steel, each test is represented by two points in the two-part diagram, an open symbol giving the weld area at the corresponding value of n computed at t_f = 1520°C (2768°F), and a filled symbol giving the recrystallized metal area at a different value of n computed at 910°C (1670°F). These plotted points show good agreement with the isothermal contours (solid-line curves) based on theoretical values.

Christensen also measured welded specimens for width and depth of weld beads and the recrystallized zone, and calculated the hemi-circular contours required by theory. At low values of operating parameter, the widths proved greater than predicted, but at high values of n, the trend was reversed. Although the general trends were the same, it was concluded that the width of weld zones and recrystallized zones could not be as reliably predicted as their cross-sectional areas. Penetration proved difficult to predict and appeared to be as dependent on the welding process as on the operating parameter. Temperature in the molten pool was estimated to range as high as 2100°C (3812°F), but the distribution of peak temperature was shown to be dependent on the presence of direct currents in the weld pool, which also influenced weld zone shape.

Christensen found cooling rate prediction based on the finite area of the heat source, that is, the shape and dimensions of the weld pool to be quite difficult. Forces exerted by the arc on the molten weld pool became increasingly important as power input was raised and preferential thermal currents flowing in the pool were reflected not only in weld zone shape, but also in temperature pattern in the adjacent base metal. Of course, temperature aberrations were less pronounced at some distance from the fusion boundary, but usually greatest interest in temperature and subsequent cooling rate is directed to the first increment of the HAZ immediately adjacent to the weld zone.

Another published diagram for predicting weld zone area made use of a parameter representing welding current and travel speed, and is shown in Figure 7.17. Here, the ratio between the square of welding current to travel speed (expressed as time per unit length of weld) has been correlated with weld zone area using log-log coordinates. The plotted data indicates no significant difference between the performance of SMAW as compared to SAW. A relationship similar to that shown for SMAW and SAW was stated to have been reported for the GMAW process. When published data for welds made in another laboratory with the SAW process were plotted on Figure 7.17, these additional results and the predictions of the diagram were in general agreement.

Since the size of the heat-affected (recrystallized) zone and the weld (fused) zone are related by similar mathematical equations it appears that the weld zone area could

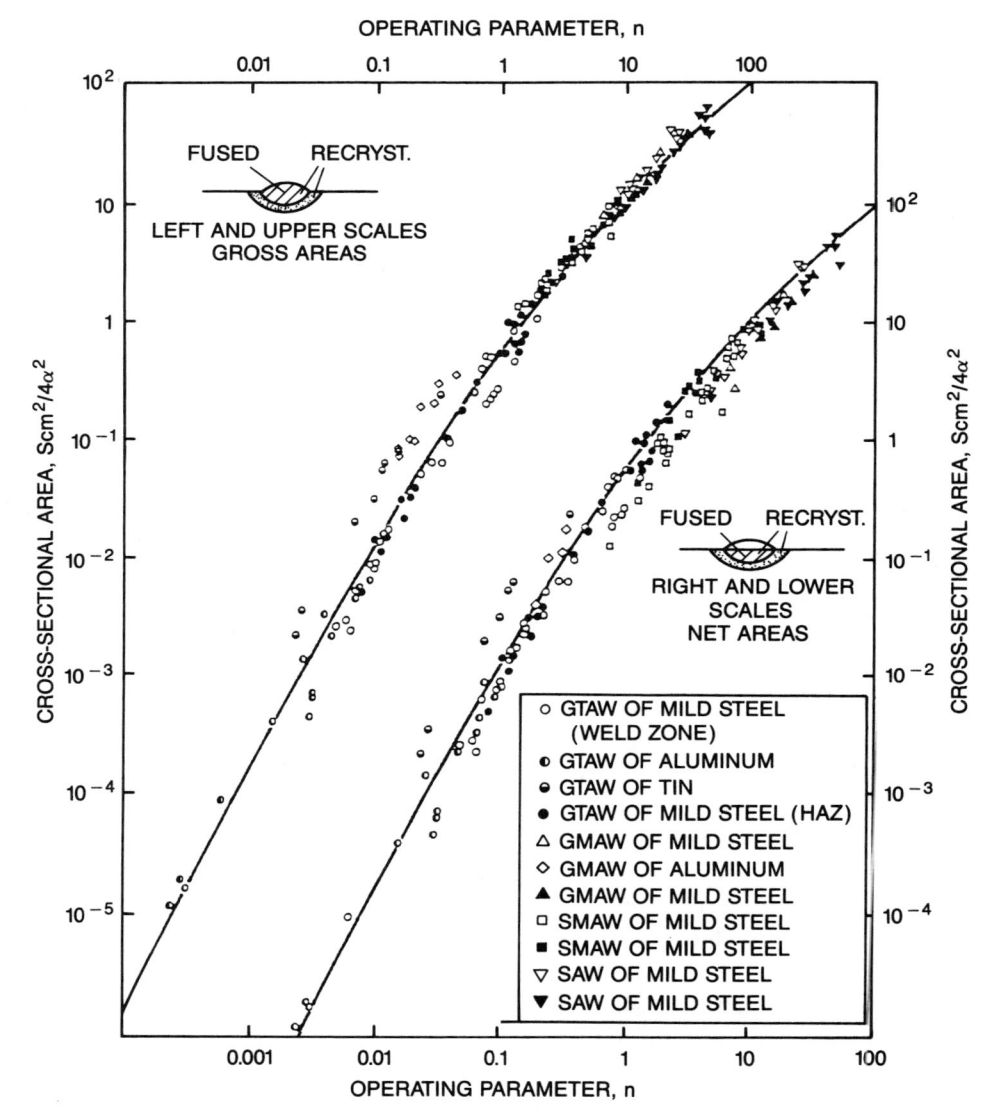

Figure 7.16 — **Fused zone and recrystallized HAZ cross-sectional areas predicated by mathematical equation (solid lines) as compared with plotted points representing measurements from tests**

For weld tests on steels, open symbols give weld zone area at corresponding value of n computed $tf=1520°C$ (2768°F) while filled symbols give the recrystallized HAZ area at different value of n computed at $tf=910°C$, or 1670°F. (Data according to Christensen, Davies and Gjermundsen)

be considered the thermal energy source for the heat-affected zone. The reasonableness of this assumption is demonstrated in Figure 7.18 which shows the relationship between weld zone cross-sectional area and instantaneous cooling rates at two temperature levels, specifically, 704°C (1300°F), and at 538°C (1000°F) which were taken from temperatures measured with thermocouples imbedded in the HAZ. The cross-hatched bands of the diagram enclose more than a dozen data points within the boundaries of the bands scattered along their lengths. The relatively narrow widths of the bands indicate reasonable consistency of determinations, and the log-log plotting produced straight bands with a slope of 45°.

Hardness/Microstructure/Cooling Rate Relationship

Hardness testing has been proposed as a practical indicator for HAZ cooling rate because a given steel when continuously cooled at various rates from above its critical range will transform to particular microstructures each having intrinsic levels of hardness; that is, both the microstructure and its hardness are firmly determined by the rate of continuous cooling that prevails as the austenitized steel undergoes transformation. In general, faster cooling rates produce harder microstructure (up to the maximum obtainable hardness for each steel). This hardness-cooling rate relationship is reliable in its repeatability, and is dependent on the given steel's *hardenability*, which is a characteric of carbon and low-alloy steels that is governed primarily by their composition and the austenite grain size.

A number of tests have been developed for quantifying the hardenability of a steel. The Jominy end-quench specimen is used in a standardized test method described in ASTM A 255, and considerable Jominy test data are available in handbooks since extensive use is made of these data in selecting steels for their hardening capability. The

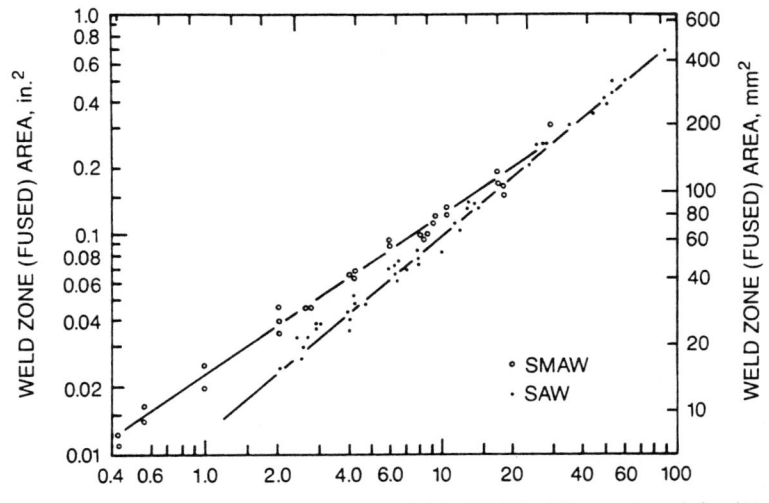

Figure 7.17 — Relationship between weld zone area and parameter I^2/S, the ratio of welding current squared to weld travel speed

(Data according to Stout, et al)

cooling rates which occur at specific distances along the length of the Jominy specimen during end-quenching have been measured, and Figure 7.19 plots the instantaneous cooling rates at the 704°C (1300°F) temperature level. Note that the end-face of the specimen (at left on the diagram) undergoes very rapid cooling when quenched by the stream of water flowing against it, and a precisely reproducible decline in temperature occurs along the length of the specimen, at least to a distance of approximately 65 mm (about 2½ inches) where the rate has decreased to only 2°C (3.6°F) per second.

Ordinarily, the furnace temperature for austenitizing the Jominy specimen to conduct hardenability evaluation is 83°C (150°F) above the Ac_3 critical point for the steel

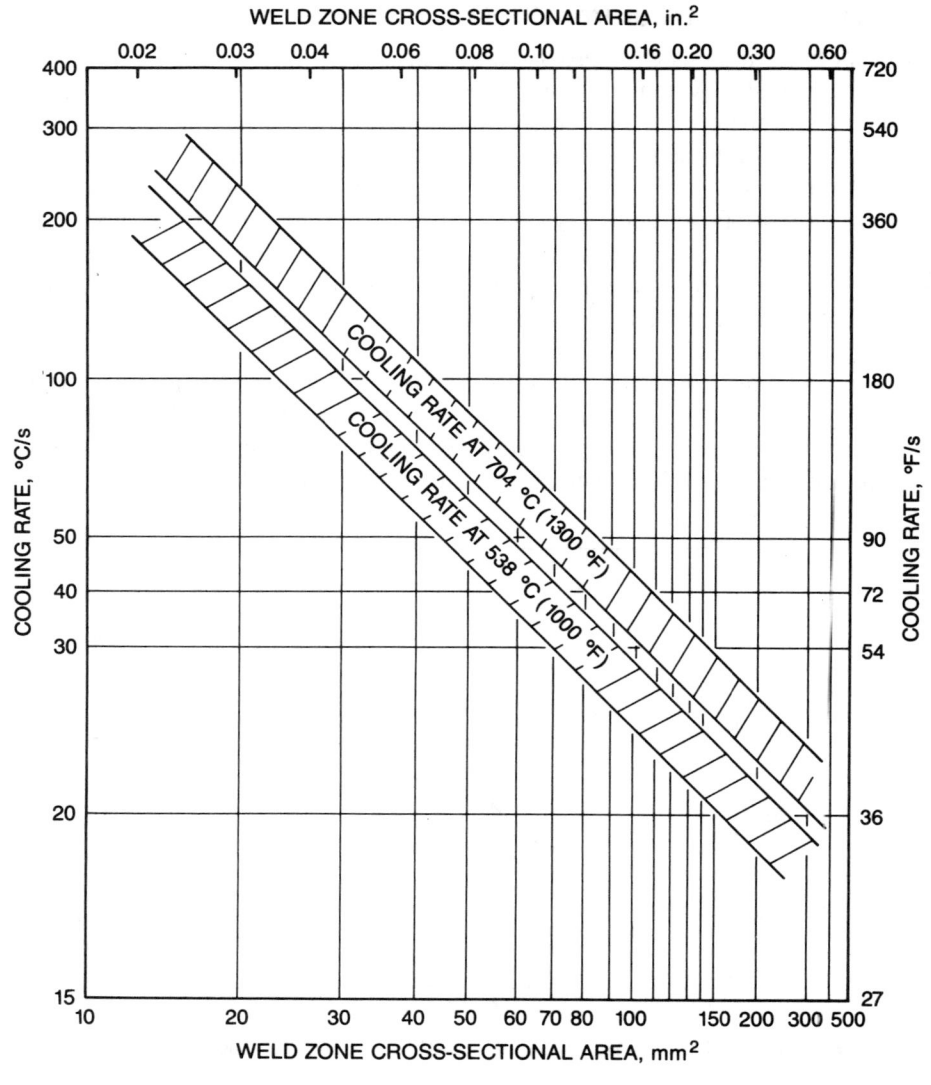

Figure 7.18 — **Relationship between weld zone cross-sectional area and instantaneous cooling rate in HAZ at two temperature levels where microstructural transformations frequently occur in steels**

being tested. Therefore, when following this practice, Jominy austenitizing temperatures range from about 845 to 925°C (1550 to 1700°F), the chosen temperature being dependent on the type of steel. However, as explained earlier when discussing peak temperature actually reached in the HAZ of a weld, the Jominy specimen used in a welding evaluation should be austenitized at a temperature of 1150°C (2100°F) in order to match more closely the effects of welding thermal cycles on the microstructure in the HAZ of a weldment. This somewhat higher starting temperature in the Jominy specimen does not cause significant change in the position of the cooling curve in Figure 7.19 depicting instantaneous rate at 704°C (1300°F), but this higher austenitizing temperature is likely to produce a large austenite grain size which will tend to increase the hardness of microstructures formed at each Jominy position (i.e., harden-

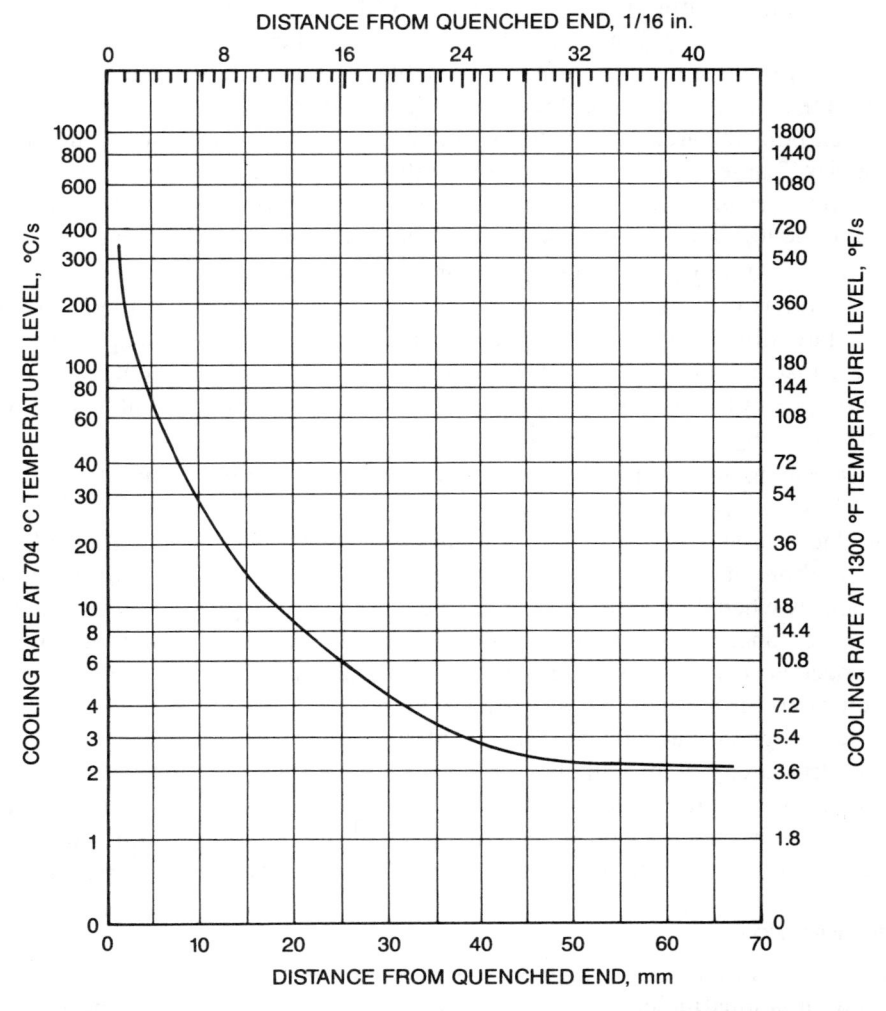

Figure 7.19 — Cooling rates (instantaneous) in the Jominy end-quench test specimen measured just beyond the quenched end-face to distances up to 65 mm (2½ in.) along the length of the specimen

ability is increased). Consequently, published Jominy hardenability data (based on the usual lower austenitiizing temperature) might be misleading if a hardness value located on the Jominy curve is expected to provide an indication of the HAZ cooling rate. The actual HAZ cooling rate likely would be somewhat slower than indicated by the conventional, published Jominy hardenability curve for the given steel.

A problem of a metallurgical nature can sometimes be encountered in securing an accurate hardness value for the weldment's HAZ. There is, of course, the matter of making a hardness determination in the very first increment of heat-affected base metal just outside the fusion boundary where the fastest cooling rate occurs. Although Rockwell hardness, or Vickers hardness determination can be made with standard indenter and a test load of 10 kgf, when studying welds with narrow HAZs a microhardness test method using a smaller test load should be considered in order to confine the indentation within the first increment of the HAZ. Of course, microhardness determinations in the HAZ of welds made at high speed can encounter microstructural heterogeneity when too little time is available for complete dissolution of carbide particles in the austenite and for diffusion of the carbon to achieve homogeneity. Then the small carbon-rich areas may tend to produce abnormally-hard microstructures in a softer matrix. Large hardness indentations will tend to give average values, whereas microhardness readings can vary considerably depending on the nature of the microstructure in the localized area where the microhardness test indentation is located.

The use of Jominy specimen positions (expressed either in millimeters or in $\frac{1}{16}$ inch units) has proved to be so convenient that many tabulations simply provide a "Jominy Hardening Factor" instead of a stated cooling rate. This is an appropriate tactic because instead of citing cooling rate at some arbitrarily-selected temperature level, the Jominy position represents the entire continuous cooling cycle and any particular indicant of cooling rate can be obtained if needed.

Another metallurgical problem may arise when the Jominy system of ascertaining HAZ cooling rate is applied to welds made at very fast travel speed. Faster cooling rates produce harder microstructures only to a point. The *maximum obtainable hardness* in carbon and alloy steels is governed strictly by their carbon content. Consequently, when the maximum obtainable hardness for a given steel is found in its heat-affected zone, the Jominy hardenability curve indicates only the minimum cooling rate needed to form that microstructure. The actual rate could have been faster inasmuch as no further increase of hardness in the given steel is possible. Also, when cooling rate is greatly accelerated, some untransformed austenite may be retained in the microstructure, which is likely to reduce the hardness to a small extent. Despite these metallurgical quirks, the Jominy concept is valid and useful. In fact, many of the heat-input requirements incorporated in presently used specifications for welding hardenable steels stem from testing that included HAZ hardness-Jominy correlation.

Cooling End Points

The temperature level at which a weld ceases to cool may be the result of any one of several situations including: (1) application of a subsequent weld pass, (2) preheating and/or maintenance of an elevated interpass temperature, and (3) immediate application of postweld heat treatment. If the fall of temperature in a cooling weldment is arrested, the delay in cooling can have important effects on weld cracking susceptibility and on weld properties. Usually, the effects of delayed cooling are desirable, but sometimes the effects should be avoided. The type of steel being welded plays a vital role. When the transformation of a newly formed HAZ is arrested at an elevated temperature level because of preheating or interpass temperature control, consideration must be given to the specific effects likely to be produced in the base metal. For example, this arrest at an elevated temperature can reduce residual stresses, or it can reduce the likelihood of cold cracking in the HAZ. However, cooling arrest when welding some steels, especially at a temperature above approximately 260°C (500°F), can adversely affect properties such as toughness or weld strength.

Arrest of temperature on cooling by making another weld pass is a very common occurrence which can introduce many variations in the welding thermal cycle. When the cycle must be controlled to avoid either having the weld zone or the HAZ drop too low in temperature, or remain at elevated temperatures too long, a temperature range should be set both for preheating and for the interpass temperature. For critical situations, sometimes a postweld holding period also is required to be within a specified temperature range. This precaution of specifying a temperature range minimizes the possibility of preheating to an unusually high temperature which might cause abnormally slow cooling, and it prevents having multiple weld passes made in such rapid sequence that heat accumulates in the weldment and produces effects similar to those of an excessive preheat temperature. By specifying a reasonably wide preheat and interpass temperature range appropriate for a particular weldment — for example, setting the range at 200 to 315°C (about 400 to 600°F) — there is assurance of having the weld zone and HAZ cool continuously until they are no hotter than 315°C (600°F). Knowing this maximum temperature, along with the heat-sink capacity of the weldment and its intimately contacting tooling, will assist in estimating probable cooling rates and the time spent in the interpass temperature range.

Preheating and elevated interpass temperature reduces the cooling rate in a weld by lowering the temperature gradient between the weld zone and the unheated base metal. Instead of the total drop in temperature of 1676°C (3025°F) that a weld made in base metal at room temperature must undergo (i.e., from 1700 to 24°C, or from 3100 to 75°F), the temperature in a weldment preheated to 316°C (600°F) drops only 1384°C (2500°F). Consequently, cooling rate is reduced, particularly at intermediate temperature levels deemed to be important for reasons of microstructural transformation, such as occurs in the temperature range of 704 to 538°C (1300 to 1000°F). In multipass welding, succeeding passes often are made on base metal that has been preheated by preceding passes because the time between passes was short enough so that the weldment did not cool to ambient temperature. Of course, if preheating has not been applied, the first pass is made on base metal at ambient temperature and this might put the weld in a precarious condition right from the start. If cold cracking developed during the rapid cooling of this first pass, further multipass welding with carefully regulated interpass temperature to obtain a sound weld would be done in vain, hence the importance of preheating.

When welds are made outdoors in the winter and the base metal is cold, the cooling rate can be excessive because of the steep temperature gradient, and the weldment might cool to temperatures where threatening metallurgical phenomena are possible. Harder microstructures, increased residual stresses, and an increased risk of hydrogen-induced cold cracking are a few of the potential consequences. While preheating and interpass temperature control coordinated together is a remedy for low ambient temperature, circumstances sometimes necessitate welding on very cold steel. When constraints are imposed on preheating, satisfactory welds can be ensured only by selecting the proper type and kind of steel, the welding process, and the welding parameters.

Measurements of the effectiveness of preheating in reducing cooling rate have been made for most processes. Since preheating simply acts to reduce the temperature gradient between the weld zone and the unaffected base metal, the cooling rate will comply with established equations for heat flow. Although the mathematics involved are complex, it can be stated the cooling rates accompanying a definite preheat temperature will bear a relationship to those occurring in a plate initially at room temperature. For example, if the rates for three-dimensional cooling in a plate at room temperature were assigned a value of 1.00, the ratios for other plate temperatures would be as shown below:

PLATE TEMPERATURE		RATIO
°C	°F	
0	32	1.10
24	75	1.00
93	200	0.85
204	400	0.67

Use of the above ratios can be of practical assistance in judging the effectiveness of various ambient temperatures, or preheating levels in controlling cooling rates.

CONTROL OF TEMPERATURE IN FUSION WELDING

The many fusion welding processes that use a moving point heat source applied with mechanized travel present two challenges; namely, (1) to predict the operating parameters needed to produce a required weld, and (2) to maintain consistency over the entire course of the mechanized operation. The latter is required despite the occurrence of unanticipated conditions that can disturb the balanced relationship between heat input and the weld being produced.

Prediction of operating parameters to produce the required weld has been a continual challenge because available guidelines ordinarily provide only approximations that must be fine-tuned. Consequently, trial welds often must be made with "cut-and-try" adjustments to parameters to achieve acceptable results.

Consistency in the mechanized fusion welding operation presents the greater challenge because the moving point heat source traverses a joint of some length and is expected to produce a weld zone of required geometry (and possibly must control the cooling rate in the base metal's HAZ at a prescribed rate). A variety of disturbing effects and conditions can arise in production welding operations if preventive measures are inadequate. These troublesome conditions range from localized variations in joint fit-up to deviations of the point heat source from the required travel path. Also, transient metallurgical phenomena may begin somewhere along the weld pass and interfere with proper heat input, such as experienced with localized slag-patch development illustrated eariler in Figure 6.11.

Discussion of the welding thermal cycle preceeding this concluding portion of the Chapter has mainly emphasized its relationship to weld zone geometry and weld cooling rate because they determine the microstructure and properties of both the weld zone and the base metal HAZ. Since other features of a weld are also influenced by its thermal history; including distortion and residual stress development, there is good reason to review ways and means of predicting the heat input needed under defined operating conditions to produce a specified weld, and to ensure uniformity along its entire length.

Predictive Diagrams For Fusion Welding Parameters

A limited number of diagrams have been published for arc welding processes that outline principal operating settings for making welds in particular kinds of joints. Parametric envelopes are developed with the aid of equations to coordinate the operating settings required for making welds of acceptable quality, and then their validity is usually confirmed by examination of trial weld specimens. A parametric tolerance envelope can be helpful and save time to predict operating settings when planning a welding procedure, but each diagram is limited to the type of joint, welding process, base metal thickness, etc., for which computations and trial welds were made. Figure 7.20 is an example of a parametric envelope diagram covering the SAW process when making double-square-groove welds (without preheating) in 13 mm (½ in.) thick steel plate. Of course, changes in the nature of the joint to be welded would likely require alteration in the predicted settings for current, arc voltage, and travel speed. Another advantage of the diagram is the ready visualization of trends in the qualities of welds produced as changes are made in any of the three principal operating parameters of the process. In Figure 7.20, the influence of arc voltage changes can easily be judged from the relative areas of the three envelopes for usable parameters.

Figure 7.20 indicates that when voltage is raised to 30 V, the face reinforcement is reduced to acceptable heights and the bead face width increases to give a much larger envelope of permissible conditions. Further increase in arc voltage to 35 V leads to the formation of undercut at the higher travel speeds and causes excessively wide bead faces at low travel speeds. Consequently, the tolerance envelope at 35 V is quite small. After parametric tolerances have been established for making welds of acceptable size, shape, penetration, etc., steps can be taken to determine what further restriction of the envelope is necessary to ensure achieving the proper cooling rate and for welds with required mechanical properties. Weld toughness can be a property that calls for testing and correlation of results with cooling rate and microstructure. When a predictive diagram like that in Figure 7.20 is available, the procedural conditions for obtaining welds with optimum mechanical properties can be delineated within the envelope selected for use.

Mathematical Modeling Of Fusion Welding

To avoid limitations of a diagrammatic presentation of usable welding parameters, investigators turned to mathematical modeling in attempt to provide comprehensive equations that would allow computation of temperature at important locations in a weld at pertinent times. These studies often take a fundamental approach to thermal energy and its flow in a weldment, and apply either a *finite-element method* of analysis, or the *finite-difference method* as is regularly used in fluid-flow analysis. The complexities in welding and changes in the thermal properties of metal during heating and cooling call for many assumptions to be made in mathematical modeling. Nevertheless, formulations on heat-flow modeling can be found in the welding literature that deal with the entire spectrum of events taking place in a welding operation involving the transient heat state; including:

(1) heat source and its input efficiency,
(2) peak temperature reached in the weld zone,
(3) size of weld zone formed and its general shape,
(4) solidification rate and cooling of weld zone,
(5) temperature reached in the base metal HAZ at specific distances from the fusion boundary and at any selected time, and
(6) cooling rates at all heated areas in the base metal HAZ.

Most of the proposed equations give consideration to the factors known to influence heat input in welding and thermal flow in metals, including the following:

(1) type of metal or alloy used as base metal,
(2) thickness of base metal,
(3) type of joint and size and shape of weld,
(4) welding process and characterization of its heat source,
(5) welding current level,
(6) welding voltage when arc is used, and
(7) weld travel speed.

All told, the equations proposed in the literature are intended to provide information like that listed below since these are often wanted beforehand for assessment purposes.

(1) weld zone microstructure,
(2) base metal microstructure at particular locations in its HAZ,
(3) hot cracking probability in weld zone,
(4) thermal strains in the general area of the weld, and
(5) residual stress levels associated with the weld.

In general, the equations in current literature are not claimed to produce values that serve as precise operating settings for making a particular weld. Mostly, the equations are promoted as mathematical tools to carry out analyses of thermal input and heat flow during application of a particular welding process. The analytical results than can be used to conceptualize the weld which will be formed, and to portray its complete thrermal cycle. This conceptualization is then put to further use by probing for unfavorable aspects of weldment design, fixturing of workpieces, welding process selection, procedure planning, and accelerated cooling by heat sinks. While this kind of scrutiny can bring about overall improvement in the production of weldments, the mathematical approach does not provide the real-time control of temperature for which a need arises in many welding operations.

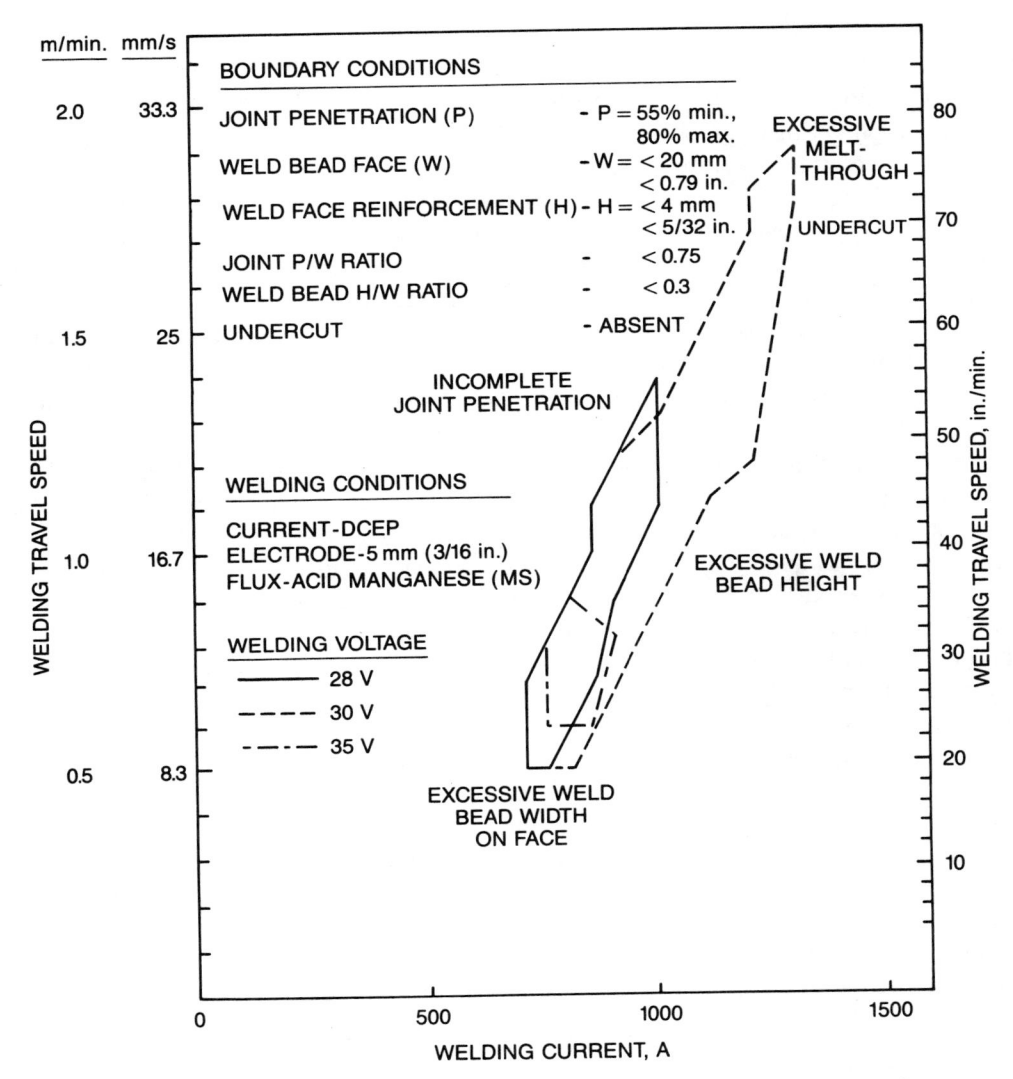

Figure 7.20 — **Parametric tolerance envelopes for SAW process to make double-square-groove welds in 13 mm(½ in.) thick steel plate**
(Courtesy of The Welding Institute, England)

Sensory Systems for Adaptive Control of Fusion Welding

Users of fusion welding processes with mechanized travel along the weld pass have long awaited the arrival of affordable, reliable adaptive control systems that can deal with the unanticipated and imponderable disturbances that arise all too frequently in production operations. Examples of troublesome conditions that are virtually unavoidable when fabricating commercial weldments include (1) change in thickness when traveling from one plate to another, (2) change in temperature of the base metal because of nonuniform preheating, or because of differing heat-sink capacity, and

(3) changing physical properties as the kind of steel in the workpieces is varied (as required by design) along the length of the joint. These and many other variable conditions can require that welding parameters be shifted to maintain weld uniformity.

Development of in-process, real-time regulation of welding is a challenging task, and perusal of the literature covering this area will find that many difficulties are encountered in selecting a sensing device to "marry" to a microprocessor and thus provide a fully integrated system. Much attention has been given to key influential welding output signals that can be measured by a sensor and then employed to quickly alter operating settings and continue making the required weld without significant variation. The weld zone cross-sectional shape and area, described earlier as having a relationship with HAZ cooling rate, would serve nicely as an output to be translated into a regulated input, but presently there is no practical way of continuously gauging the weld zone's cross-sectional shape and area during the welding operation.

A number of adaptive control systems using a variety of sensing devices will be reviewed in a concluding chapter on automation of welding, but it is appropriate here to mention that infrared radiation from a weld immediately after it has been made has been proposed as quantitative thermal output that can be utilized as basis for adaptive control. Temperature at the centerline of a weld, and in the base metal HAZ at a selected distance from the weld toe are being used to monitor the temperature field associated with a weld, and to adjust inputs as required to obtain the necessary outputs. Both fiber optic thermographic sensing, and infrared cameras have been applied to scan the weldment surface along a line trailing the end of the weld pool, and along a line in the HAZ parallel to travel of the heat source. Integration of system components for IR sensing of surface temperature, computing internal heating and cooling rates in workpieces, and controlling output-input relationships requires extensive mathematical modeling. A number of assumptions and discretionary decisions must ordinarily be made to simplify the modeling task; such as basing calculations on infinite plate sections, homogeneous and isotropic material in the workpieces, invariant thermal properties, entirely conductive heat flow and no surface losses, etc. A certain amount of experimentation is often required to fine-tune a control system for a given weldment.

During operation of an adaptive control system, a number of practical difficulties can be encountered. These include nonlinearities, modeling shortcomings, and system response time that is not fast enough to accomplish the needed degree of corrective action when a troublesome condition is encountered. Naturally, the latter problem becomes more acute as weld travel speed increases. Although IR sensing can provide satisfactory resolution of temperature fields, this method of thermal monitoring has a longer-than-desirable sampling period. Consequently, full compensation for an unavoidable joint condition may require a couple of seconds time before complete alteration of operating settings is accomplished. To shorten response time for achieving acceptable adjustment of conditions, the IR sensing point has been set as close as possible to the welding heat source. However, this arrangement tends to increase the amount of unimportant physical and electronic "noise" from the welding operation, and this noise must be filtered out before accepting the feed-back as significant signaling.

An innovation being considered to improve efficiency in making output corrections from input secured through IR measurement is to employ two heat sources in tandem in fairly close proximity. The first source operates under steady-state conditions regardless of any disturbances encountered along the joint being welded while the second following source (which is operating at a lower level of energy generation) will execute the necessary adjustment, upward or downward as required, to effect corrections as rapidly as possible. Another method of measuring surface temperature on a molten weld pool is called *optical spectral radiometric/laser reflectance*. This is a high-resolution technique that is able to map the thermal contours surrounding the arc on the pool surface. So far, studies with this method have been applied only to a stationary weld pool made by GTAW. Although not ready for use as part of an adaptive control system, the OSR/LR method has demonstrated a capability to provide a remarkably detailed surface temperature profile of a weld pool, and to distinguish the effects of minor elements in steels that alter molten metal convective flow in the pool. This is the Marangoni convection described in Chapter 6 (see Technical Brief 33) that is now recognized to affect weld zone penetration and shape.

Research on adaptive control systems has been greatly aided by new developments in microelectronic devices, but the crux of challenge lies in the need for sensors that can detect a key feature which delineates weld quality and to immediately feed output regarding this feature from every increment along the weld length. Techniques for noninvasive, in-process measuring of weld features are only in early stages of development and application. Innovations in this area should be followed closely because the introduction of successful systems into commercial welding operations is likely to take place quickly. Incentives can be found for adoption of adaptive control both in weld quality improvement, and in reducing costs for weld repair.

Adaptive Control of Solid-State Welding Processes

For more than 30 years, users of solid-state welding processes have sought feedback or adaptive controls. This has been especially true for the resistance spot welding process (RSW). Nothing can be more disturbing to a manufacturer than to have a structure assembled by RSW fail prematurely for lack of fused-nugget formation or true bonding at the hidden interface between metal sections. Incidents of this kind dramatize the chief NDE problem with most welds made by a solid-state process — the preliminary visual examination that is so widely and frequently used to inspect fusion welds usually is of very little value in assessing a solid-state weld; because the weld, per se, is concealed between the workpieces. Of course, certain bits of helpful evidence regarding a solid-state weld can be gleaned from actions during the operation, or from telltale features that can be observed on the weldment. Expulsion or spitting from between the faying surfaces during spot welding has been used for a long time as a signal that melting has been accomplished. However, the extent of melting, and the amount of fused metal retained to form a weld nugget is always a moot question when expulsion is the only assessment employed. Visual indicators of some bonding having taken place during flash welding (FW) and friction welding (FRW) were mentioned in Chapter 6 during description of these processes.

Temperature measurement has been somewhat overlooked as a means of monitoring and controlling the making of solid-state welds; inasmuch as the systems that have been adopted by industry for in-process regulation of RSW, FW and FRW generally measure changes in either mechanical displacements, electrical phenomena, or acoustic emissions. Nevertheless, temperature changes during RSW have been extensively studied. Literature covering this effort is very helpful in gaining an in-depth understanding of resistance welding processes, but the work directed to temperature change has not been fruitful in producing in-process, real-time control systems that can be put to use in industry.

The avenues of approach used by investigators to follow the transient temperature changes during the RSW process can be summarized as follows:

(1) mathematical modeling to equate RSW operating conditions for a specific weld to heat generation at the intended weld location,
(2) infrared emission from the electrode surface by thermo-vision apparatus during the welding period,
(3) indications from thermosensitive paint on the workpiece surfaces as recorded by high-speed cinematography, and
(4) arrangements of thermocouples at the electrode face-workpiece surface for measuring temperature in real-time.

These innovative schemes have shown some promise for adaptive control of spot welds between bare steel sheets, but to date steels with some form of metallic coating (i.e., aluminum or zinc) continue to baffle the best of the temperature-based systems.

The challenge continues for devising a simple, rugged, and reliable adaptive-control system for the RSW process when welding metallic coated steel — particularly in the automotive industry. Cars and trucks today use increasing amounts of coated sheet and strip steels (as described on page 409), but the electrode deterioration problem in welding these steels (as outlined on pages 527-531) continues without a real breakthrough in sight.

SUGGESTED READING

Rosenthal, D. D., "Mathematical Theory of Heat Distribution During Welding and Cutting," *Welding Journal*, 20 (5): 220s-234s; May 1941.

Adams, Jr., C. M., "Cooling Rates and Peak Temperatures in Fusion Welding," *Welding Journal*, 37 (5): 210s-215s; May 1958.

Christensen, N., Davies, V. del., and Gjermundsen, K., "Distribution of Temperatures in Arc Welding," *British Welding Journal*, 12 (2): 54-75; February 1965.

Myers, P. S., Uyehara, O. A., and Borman, G. L., "Fundamentals of Heat Flow in Welding," *Welding Research Council Bulletin*, No. 123: 46 pp; July 1967.

Dorschu, K. E., "Control of Cooling Rates in Steel Weld Metal," *Welding Journal*, 47 (2): 49s-62s; February 1968.

Bradstreet. B. J., "Effect of Welding Conditions on Cooling Rate and Hardness in the Heat-Affected Zone," *Welding Journal*, 48 (11): 499s-504s; November 1969.

Boillot, J. P., Cielo, P., Bégin, G., Michel, C., Lessard, M., Fafard, P., and Villemure, D., "Adaptive Welding by Fiber Optic Thermographic Sensing: an analysis of thermal and instrumental considerations," *Welding Journal*, 64 (7): 209s-217s; July 1985.

Doumanidis, C. C., and Hardt, D. E., "Simultaneous in-process Control of Heat-Affected Zone and Cooling Rate During Arc Welding," *Welding Journal*, 69 (5): 186-196s; May 1990.

Heat Effects of Welding, Dieter Radaj, Springer-Verlog, Berlin, Germany, 1992, ISBN 3-540-54820-3.

Jones, B. K., Emery, A. F., and Marburger, S. J., "An Analytical and Experimental Study of the Effects of Welding Parameters of Fusion Welds," *Welding Journal*, 72 (2): 51s-59s; February 1993.

Chapter 8

In This Chapter...

OXIDATION OF IRON	708
OXIDATION OF STEEL	710
Carbon/Oxygen Reaction in Molten Steel	710
Oxidation of Solid Steel	713
PREVENTING OXIDATION DURING WELDING	714
Shielding Slags	714
Fluxes	716
Controlled Atmospheres	717
Vacuum	718
Technique	720
Deoxidizers	721
Protective Surface Alloys	721
Liquid Blankets	722
SHIELDING THE JOINING PROCESSES FROM AIR	722
Carbon Arc Welding (CAW)	722
Metal Arc Welding	723
Covered Electrodes	724
Generic Electrode Cover'gs & Typical Formulas	731
Flux Cored Electrodes	738
Submerged Arc Welding (SAW)	744
Types of SAW Fluxes & Their Classification	744
Methods of Manufacturing SAW Fluxes	748
Physical Chemistry of Fluxes in SAW Process	750
Transfer of Elements Between SAW Flux/Slag and Weld Metal	758
Electroslag Welding (ESW)	765
Gas Shielded Arc Welding	766
Argon	768
Helium	768
Carbon Dioxide	769
Propane	770
Nitrogen	770
Hydrogen	770
Miscellaneous Gas Additives	771
Gas Tungsten Arc Welding (GTAW)	771
Gas Metal Arc Welding (GMAW)	772
Globular Transfer in GMAW	775
Repelled Transfer in GMAW	775
Projected Transfer in GMAW	776
Streaming or Axial-Spray Transfer in GMAW	776
Pulsed Spray Transfer in GMAW-P	777
Rotating Droplet (Kinking) Transfer in GMAW	778
Explosive Drop Transfer in GMAW	778
Short Circuiting Transfer in GMAW-S	778
Flux Cored Arc Welding (FCAW)	779
Plasma Arc Welding (PAW)	780
Electrogas Welding (EGW)	780
Other Welding Processes Using Gas Shielding	781
Laser Beam Welding (LBW)	781
Non-Vacuum Electr'n Beam Weld'g (EBW-NV)	781
Protecting Brazing Processes From Air	782
Soldering — Ways to Achieve Bonding	784
Fluxes for Soldering	784
Mech. Means to Accomplish Solder Bonding	784
SUGGESTED READING	785

Fluxes, Slags and Gases for Shielding

The earth's atmosphere is composed of about four-fifths nitrogen and one-fifth oxygen. When exposed to air, most metals will combine with oxygen, and to a lesser extent with nitrogen — especially when these metals are heated, and particularly when they are in the molten state. The rate of oxide formation will vary widely with the different metals, but even a thin film of oxide on the surface of metals can cause difficulty. In fact, a layer of oxide can actually prevent the joining of two workpieces by welding, brazing, or soldering. Other problems may occur as well. Quantities of oxide, if entrapped in a weld, may lower its strength significantly. Molten droplets from a consumable electrode may become coated with oxide during transfer to the weld zone and may be prevented from adhering or commingling. In some instances, a portion of the oxides formed may dissolve in the molten metal and cause embrittlement of the solid weld. When nitrogen is absorbed in molten iron and steel, its presence also can adversely affect toughness unless it combines with a strong nitride-forming element and is held in the metal as solid, innocuous, nonmetallic particles.

A few metals exhibit little or no tendency to form oxides when heated for welding. Consequently, these metals can be welded without having to cope with the oxidation problem. Lead and platinum are two examples. In the case of lead, the oxide forms at a very slow rate; therefore, an excessive quantity seldom accumulates during a melting operation. Platinum, however, shows no tendency to form oxide. This is because the dissociation pressure of platinum oxide in air at welding temperatures is sufficient to cause immediate dissociation into metal and oxygen.

The ill effects of nitrogen are not as widely recognized because they often are overshadowed by the oxidation problem. Nevertheless, many metals will react with nitrogen to form a nitride. This compound may exist as a surface film, or it may become entrapped in the metal like an oxide. Nitrogen gas also may dissolve in the molten metal, as does oxygen.

Established welding processes, and the procedures advocated for their application, generally have been planned to operate in a manner that provides adequate protection from oxidation and nitrification. Since metals differ in their reactivity with oxygen and nitrogen, application of specific processes is limited to those metals that can be safely welded under the degree of protection afforded by the process. Yet, if proper control is not exercised, oxygen and nitrogen can interfere with the making of a quality weld. Sometimes the interference is subtle and not detected immediately; in such cases, the weld may appear to be sound, but its toughness may have been seriously degraded.

The effects of oxygen and nitrogen are far-reaching, and because iron is one of the more easily oxidized metals, the oxidation (and nitrification) problem will be reviewed starting with basic aspects and extending to practicable means of control. The subject in connection with iron, carbon steel, and alloy steel will be pursued in three logical steps; (1) the mechanism of oxidation, (2) means of prevention, and (3) incorporation of preventive measures in welding processes.

OXIDATION OF IRON

The mechanism of oxidation and its effects can be illustrated more easily by first examining iron, then extending this study to steel. Picture molten iron in a crucible exposed to air [Figure 8.1(A)]. Oxygen molecules (O_2) of the air are impinging on the surface of the iron melt; and, upon making contact, the molecules are dissociated into two atoms of oxygen. A portion of the nascent oxygen atoms then dissolve in the surface of the molten iron [Figure 8.1(B)]. Because there is an abundant supply of oxygen available to the surface of the melt, oxygen atoms continue to be absorbed by the iron and to diffuse deeper into the melt [Figure 8.1(C)]. This diffusion process is quite rapid.

The melt of iron cannot hold more than approximately 0.2 percent by weight of oxygen in solution (at the melting temperature of iron). Oxygen in solution lowers the melting point of iron and eventually produces a melting range. Although molten iron that contains approximately 0.2 percent oxygen represents a saturated solution, oxygen atoms will continue to be attracted by the iron and will enter into the surface layer of the melt. Now the surface layer is no longer a single liquid composed of iron and a small amount of oxygen. Instead, a new substance is formed — namely, a liquid containing 22 percent oxygen which is lower in density than the saturated solution of oxygen in iron. The new substance accumulates and forms a layer on the surface [Figure 8.1(D)]. Air no longer is in contact with metallic iron. The oxygen molecules of the air now impinge on the new surface layer, which is the compound, iron oxide (FeO). Actually, this new compound may contain 22 to 28 percent oxygen or more, and therefore is not necessarily of the precise composition implied by the chemical formula, FeO. Some of the impinging air molecules split and enter the FeO surface layer just as occurred earlier on the iron surface. The oxygen atoms in the surface layer of FeO diffuse deeper, and when they reach free atoms of iron, they form more of the iron oxide compound. Again the process is rapid, and a crucible of molten iron exposed to air may be converted to liquid iron oxide in a relatively short time if conditions are favorable.

If a lid is placed on the crucible as shown in Figure 8.1(E) and no oxygen reaches the iron, there can be no oxidation. A lid is not easily placed over welding operations, but the weld can be covered with a protective slag equivalent to a lid. Also, instead of a solid lid or a layer of protective slag, air can be kept out of the crucible by a flow of gas that does not react with iron — for example, helium [Figure 8.1(F)]. A gas shield is provided in some welding processes — such as PAW, GTAW and GMAW — whereas a protective atmosphere is provided in the SMAW and FCAW processes by combustible or volatile materials in the electrode covering or core. Because these covering and core materials often contain carbonaceous ingredients, the gaseous shield generated is likely to consist of carbon monoxide (CO) and carbon dioxide (CO_2).

Three possible circumstances should be considered in studying Figure 8.1. First, if the iron melt is at or near its boiling point, iron atoms may be rising to the surface faster than the oxygen atoms arrive. Under these conditions, a fume of iron and oxygen molecules will be observed rising from the top of the crucible. The surface of the molten iron will be clean, which, incidentally, may be the condition of molten globules passing through a welding arc. Second, not all gases react with iron in the same manner as oxygen. Nitrogen and hydrogen ordinarily are present in their molecular form (N_2 and H_2) above the iron melt and must undergo dissociation into atomic form

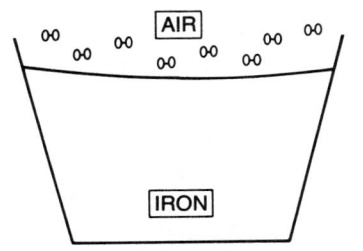

A. MOLECULAR OXYGEN (O_2) IN THE AIR ATMOSPHERE COMES IN CONTACT WITH THE MOLTEN IRON SURFACE.

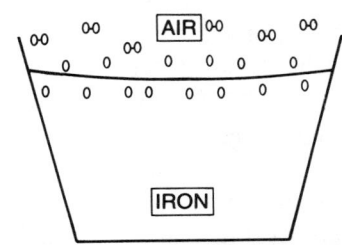

B. OXYGEN MOLECULES IMPINGING ON THE IRON DISSOCIATE INTO ATOMS AND QUICKLY DISSOLVE INTO THE IRON THROUGH THE SURFACE.

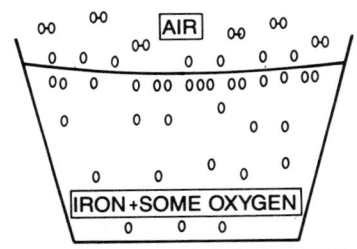

C. OXYGEN ATOMS DIFFUSE FROM A CONCENTRATION AT THE SURFACE TOWARD THE BOTTOM OF THE MELT.

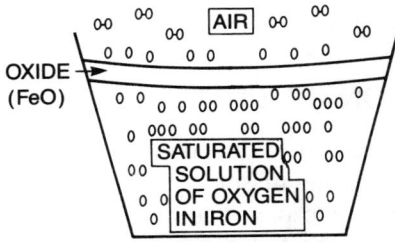

D. A LAYER OF IRON OXIDE WITH VARIABLE OXYGEN CONTENT FORMS ON THE SURFACE OF THE MOLTEN IRON, WHILE OXYGEN CONTINUES TO ENTER THE OXIDE LAYER AND DIFFUSE THROUGHOUT THE MELT.

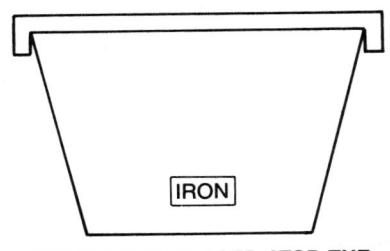

E. A SOLID LID IS PLACED ATOP THE FULL CRUCIBLE TO EXCLUDE AIR AND PREVENT OXIDATION.

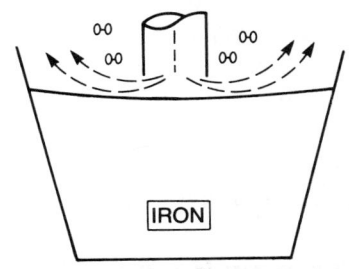

F. A FLOW OF INERT GAS PREVENTS AIR FROM REACHING THE SURFACE OF THE MOLTEN IRON AND THUS OXIDATION IS PREVENTED.

Figure 8.1 — Four stages in the oxidation of molten iron in a crucible (A to D, inclusive), and two means of oxidation prevention (E and F)

to dissolve. Initially, nitrogen and hydrogen atoms are more likely to go into solution in the molten iron rather than to form a compound as does oxygen. Of course, if a sufficient quantity of nitrogen is present, iron nitride compound can form. Other gases, such as argon and helium, do not dissolve to a significant extent in iron. Carbon monoxide (CO) has virtually no solubility in molten iron but, when in contact with the molten iron, it will dissociate and contribute carbon to the melt, thus forming carbon dioxide (CO_2). Third, the solubility of oxygen in iron, and any other gases that have some solubility, increases as the temperature is raised. Dissolution and entry into the molten iron becomes more rapid as temperature is raised, and diffusion of the gas atoms throughout the melt will take place at a faster rate.

Increasing the pressure of a gas above the molten iron also increases the amount of gas entering the metal. Much of the information published on gas solubility is based on one atmosphere pressure. When oxygen enters molten iron to the full extent of its limited solubility, the manner in which the oxygen atoms exist in the iron is uncertain. There is evidence that so-called dissolved oxygen in iron is in fact only loosely held in chemical combination with iron. Even when the solubility limit of 0.16 percent at a temperature level of 1540°C (2800°F) is exceeded, the liquid oxide phase that forms will have an oxygen content of 22.6 percent initially, but the percent oxygen will vary with temperature (and with pressure).

OXIDATION OF STEEL

If steel (an alloy of carbon in iron) replaces the pure iron in Figure 8.1, oxygen will dissociate and dissolve just as occurred with iron. Now, however, a new circumstance comes into play that requires more detailed schematic illustration, and this is provided in Figure 8.2. As this figure is examined for details, it must be kept in mind that the amount of oxygen that can be contained in solution in molten steel will be markedly influenced by its carbon content. As shown by the graph in Figure 8.3, adding carbon to the extent of about 0.1 percent sharply decreases the oxygen solubility limit from 0.16 percent to about 0.03 percent. Nevertheless, even this smaller amount of oxygen can be objectionable in steels under a number of circumstances. When the steel is in the molten state, there is a propensity for the carbon to react with the oxygen present to form carbon monoxide (CO), a gas that is virtually insoluble. Earlier in Chapter 5 on steelmaking, the necessity for dealing with this small amount of oxygen to avoid gas pockets and blowholes was described in connection with the production of rimmed, capped and killed steels. The rudiments involved are illustrated schematically in Figure 8.2.

Carbon/Oxygen Reaction in Molten Steel

Even though the amount of oxygen entering molten steel is relatively small, the exothermic reaction between the oxygen and carbon is vigorous and liberates a substantial amount of heat (about 28 kcal/mol). In Figure 8.2, the air atmosphere has supplied oxygen molecules in Location 1; and at the layer of slag (FeO), shown to already exist on the surface of the molten steel, dissociation into atoms of oxygen has occurred. Upon entering the slag layer (Locations 2 and 3), the oxygen diffuses

Fluxes, Slags, and Gases for Shielding   711

① AIR ATMOSPHERE
② HIGH OXYGEN CONTENT SLAG
③ LOW OXYGEN CONTENT SLAG
④ HIGH OXYGEN, LOW CARBON METAL
⑤ CARBON MONOXIDE GAS BUBBLES
⑥ LOW OXYGEN METAL

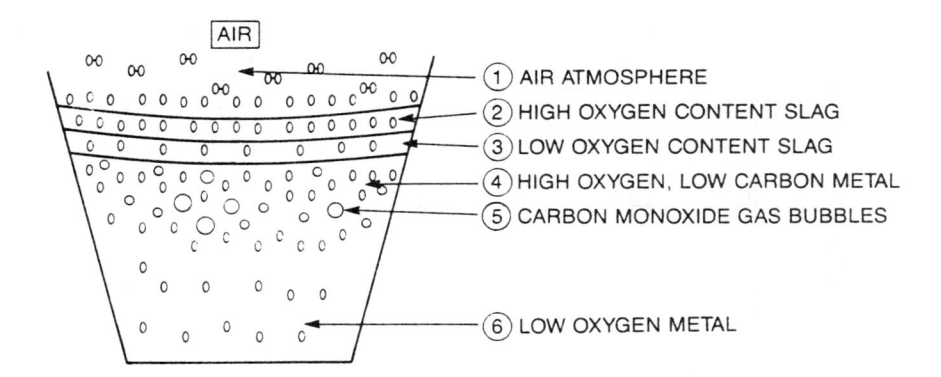

① AIR ATMOSPHERE CONTAINING MOLECULAR OXYGEN WHICH BECOMES DISSOCIATED UPON CONTACT WITH THE MELT.
② REGION OF HIGH OXYGEN CONTENT IN SLAG (POSSIBLY 30% OXYGEN).
③ REGION OF LOWER OXYGEN CONTENT IN SLAG (POSSIBLY 20% OXYGEN).
④ REGION IN STEEL MELT CONTAINING HIGH OXYGEN CONTENT AND LOWERED CARBON CONTENT (POSSIBLY 0.15% OXYGEN).
⑤ BUBBLES OF CARBON MONOXIDE (CO) GAS FORMED IN STEEL MELT BY OXYGEN REACTING WITH CARBON PRESENT.
⑥ REGION OF RELATIVELY LOW OXYGEN CONTENT IN STEEL MELT; PERHAPS ABOUT 0.1% OXYGEN PRESENT.

Figure 8.2 — Conditions encountered when cooling a melt of steel that has been exposed to air and that contains dissolved oxygen

through and enters the melt of steel. This is the point where steel differs from the pure iron illustrated in Figure 8.1; because oxygen, after entering the steel at Location 4, tends to react with the carbon present to form carbon monoxide gas (Location 5). Because of its virtual insolubility in the melt, the carbon monoxide attempts to escape by bubbling upward out of the melt. The extent of the C + O reaction and the dispersal of the CO gas will vary greatly depending on the temperature and stirring action of the melt, the levels of carbon and oxygen present, and the pressure of the gaseous atmosphere above the ladle. The reaction can be so mild that only a few small gas pockets will form in the steel during solidification. On the other hand, the reaction can be so profuse that bubbles of CO gas will effervesce from the surface of the molten slag (or metal) with a frothing action. A ladle of molten steel has been used in Figure 8.2 to simplify the illustration of oxygen finding carbon in molten steel, but a weld zone in the molten state is subject to the same reaction.

Predicting the approximate amount of CO gas likely to be generated in molten weld metal, and the probable form of its dispersal, is difficult because of the complex conditions peculiar to the welding processes. Weld metal temperature can be significantly higher than steel in a ladle, but the time available for the C + O reaction to occur during welding is very short. In a weld, the depth of molten metal through which bubbles of CO gas must rise to escape is relatively shallow. Also, molten weld metal is

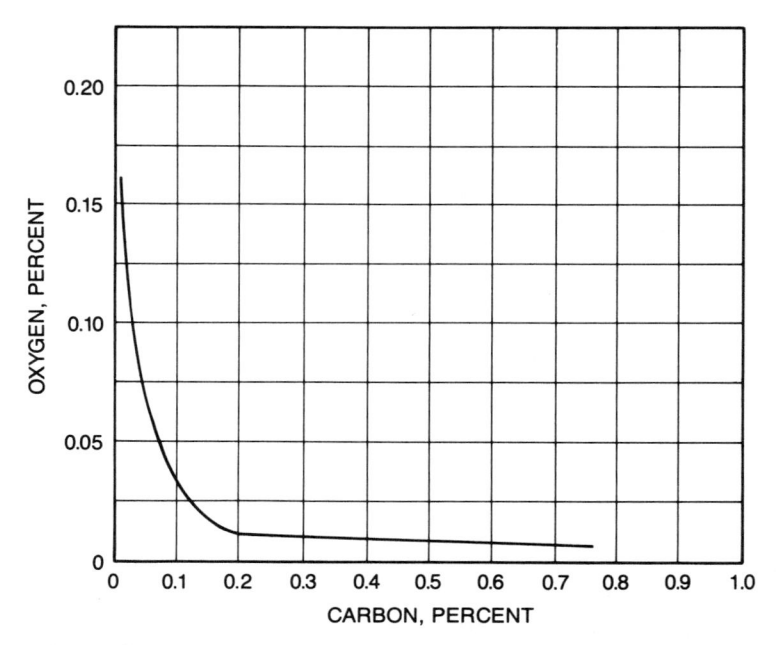

Figure 8.3 — Solubility of oxygen in molten steel under one atmosphere pressure as affected by carbon content (equilibrium oxygen values)

likely to be quite fluid and to have low surface tension because of high temperature and the frequent use of flux on the surface, or use of a protective gas shield over the weld pool. Nevertheless, solidification of a weld is quite rapid, and if the C + O reaction occurs there is a strong possibility that some CO gas will be entrapped to form porosity in the solid weld, or will form blowholes in the surface, rather than completely effervesce and leave the solid weld metal sound. This threat to weld soundness is always present when *steel* weld metal contains dissolved oxygen, and this is the reason for frequent use of deoxidizers that are introduced during steel welding operations.

Despite extensive studies of weld metal deoxidation, there remains a real need for more precise quantitative deoxidation control with many of the welding processes when applied to certain steels. The general practice has been to *thoroughly* deoxidize weld metal by small additions of elements like aluminum, silicon, titanium and zirconium to avoid porosity or blowholes, and ensure freedom from these defects by having a small excess of deoxidizing elements present in the weld metal when molten, and as it solidifies. Early examples of this deoxidation practice can be seen in certain welding rods for the oxyfuel gas welding process, and in the generous additions of ferrosilicon to the formulae for coverings on SMAW electrodes. This leaning toward ample deoxidizer additions to ensure thorough deoxidation has led to the making of welding consumables that contain higher-than-ordinary levels of deoxidizing elements. These consumables will be discussed in Volume II, Chapter 10; but it should be pointed out here that strongly deoxidized filler metal must be selected with care when welding certain types of high-strength, low-alloy steels. When matching strength weld metal for these steels contain high levels of deoxidizing elements, particularly silicon, weld metal toughness may be adversely affected.

Oxidation of Solid Steel

The oxidation of heated iron or steel proceeds much more slowly in the solid state than in the molten condition, partly because of the lower temperature, and because rates of surface absorption of oxygen and its diffusion through the solid metal are lower. Although oxygen atoms are small enough in diameter to enter into interstitial solution in the crystallographic structure of solid iron, their size does not allow easy diffusion through the lattice (as in the case of hydrogen). A highly effective retardant to oxidation in solid steel by an oxygen-containing atmosphere is the formation of solid oxides on the surface. It is called *scale* when it forms at an elevated temperature and is black or dark red in color, and it is called *rust* when it forms as a relatively loose and flaky red or orange colored oxide at lower temperatures.

The chemical composition of the oxide layer that forms on the surface of steel, whether it is scale or rust, can contain metal oxides other than those of iron. This is particularly true of alloy steels. Aluminum, chromium, and silicon, when present as alloying elements, will alter the composition of the surface oxide; and often a more protective type of scale or rust will be formed. Molybdenum, on the other hand, when present in very large amounts, will enter into the surface oxide formation and sometimes will cause accelerated scaling. This effect arises because molybdenum oxides have unusually low evaporation temperatures.

Freshly cleaned iron and steel are coated with a thin, invisible layer of oxide. When the metal is heated in air, the thickness of the oxide layer increases and the surface displays temper colors as described in Chapter 7. The physical and chemical characteristics of the oxide can vary considerably depending on the composition of the base metal, the cleanliness of the metal surface prior to exposure, the nature of the atmosphere, and the temperature-time cycle to which the metal is exposed. Many of the low-alloy steels have been concocted to display greater resistance to scaling and rusting, and therefore may require special consideration when removing their surface oxides prior to welding. Carbon and low-alloy steels do not present as severe a problem with surface oxides as are encountered with metals and alloys that form a highly protective oxide surface film. Aluminum, magnesium, titanium and stainless steels are examples of such materials.

The atmosphere to which heated iron or steel is exposed determines the characteristics of the surface film that forms. In addition to the oxide film which can form on the surface when oxygen is available, other more complex surface films — hydroxides, sulfides, and carbonates — can result from the presence of compounds containing hydrogen, sulfur, and carbon dioxide. Surface films may also contain a layer of adsorbed gases extracted from the atmosphere. These additional elements and compounds, when incorporated in the surface film, may either increase or decrease the difficulty of dealing with the surface film during welding. Examples are included in the discussion that follows.

PREVENTING OXIDATION DURING WELDING

This Chapter covers fluxes, slags, and gases for shielding molten or heated metal from the effects of exposure to air during welding. There is good reason to study shielding — both from a general viewpoint, and in some detail as incorporated in the welding processes. Even though iron and steel do not combine with oxygen and nitrogen in the air as vigorously as the more reactive metals, relatively small amounts of these gases entering the weld zone can influence weld soundness and mechanical properties. Furthermore, the various methods used to minimize the effects of oxygen, nitrogen, and other gases picked up during welding differ in efficiency and can impart distinctive influences. As welding has been carried into unusual environments, innovative shielding measures have been developed to protect the operations adequately. A few examples of unusual environments include (1) welding underwater, where the shielding must not only offer protection from water and steam, but sometimes must avoid quenching of the weld, (2) sub-sea hyperbaric (i.e., dry) welding, where high pressure and humidity in the chamber can influence arc welding processes in a number of ways, and (3) welding outdoors during hostile weather where strong winds, low ambient temperatures, and rain or snow would threaten the integrity of welds made by a commonly used process operated with regular shielding. The material being welded also can call for innovation in shielding. For example, prepainted or metallic coated steel sheet may require fume extraction from the immediate vicinity of the weld at the same time that shielding is provided to ensure a sound weld.

The following measures for avoiding exposure to air are presently used in welding processes: shielding slags, fluxes, controlled atmospheres, vacuum, technique, deoxidizers, protective surface alloys, and liquid blankets.

Shielding Slags

Welding slags, in addition to offering low permeability to gases, have other features tailored for the specific welding process being used. In general, consideration must be given to melting point, density, viscosity, surface tension, and ease of removal of the slag from the weld after cooling. Briefly, the slag must not freeze at a higher temperature than the weld metal. If slag freezes on the groove faces ahead of the weld metal, the slag layer will prevent coalescence of the weld metal with the underlying base metal. The slag must be less dense than the weld metal so that it floats and separates from the metal. This density requirement is easy to meet when welding steel; however, when welding the light metals, greater care in selection of slag constituents is required.

The viscosity of molten slag for welding is not easy to measure, only in recent years has a laboratory apparatus become available for making quantitative viscosity determinations. Until the advent of this viscosity measuring apparatus, slags were judged by various experimental techniques and described by qualitative terms ranging from *creamy* (watery), *thin*, and *thick*. Viscosity in a liquid was described in Chapter 3 as the force required to move one of two parallel plates at unit velocity, the plates being separated by a unit thickness of the liquid under consideration. The information in Chapter 3 regarding measurement of viscosity in molten metals also applies to molten slags. *Kinematic viscosity* is the ratio of viscosity to density, and its unit of measurement is the *stoke*.

If a slag has high viscosity during the welding operation, it will flow slowly. Atoms and molecules cannot move rapidly in a viscous slag; consequently, the slag will tend to be quite impermeable to gases from the atmosphere above it, and the molten metal will be well protected if it is completely covered by the slag. Of course, welding operations are carried out so quickly that little time is available for oxygen from the air to diffuse through the slag to the weld metal. Ordinarily, larger quantities of oxygen will enter the weld metal from other sources, such as the oxides in the slag-forming ingredients that surrender oxygen by dissociation at high temperatures, or by reduction when coming into contact with iron or other elements in the steel melt.

The viscosity of molten slag has considerable practical importance in the metal-arc welding processes. Heavy layers of slag, whether of very high or of very low viscosity, are difficult to manipulate in the vertical and overhead positions. Too low viscosity is undesirable for these positions because the slag will tend to run off of the weld metal and drip. A slag of higher viscosity, therefore, can be used to advantage in vertical and overhead welding. The viscosity of slag can be controlled by varying its composition or the temperature. Maintaining a "cold" melt (by using a relatively low energy input) usually causes a noticeable increase in viscosity, but the principal means of viscosity control is through precise formulation of the slag composition. Small additions of certain ingredients, such as calcium fluoride, can markedly lower the slag viscosity. For this reason, discussion of formulae for slags will require the welding process and any other special operating conditions to be specified.

Slag, when cooled to room temperature, may be wholly or partly uncrystallized. During cooling, the atoms distributed at random in the liquid slag become more and more difficult to move, and the molten material becomes more viscous. Yet, no crystals may appear in the seemingly solid slag; or possibly crystals of one or two substances will appear in minor proportions. Atoms in such slag, therefore, are distributed wholly or partly at random, have characteristic arrangements, but are still not of the rigid sort manifested by crystalline solids. Slag that cools to room temperature without crystallizing is considered a supercooled liquid or a glass. Crystalline structures in nonmetallic substances have characteristic coefficients of contraction during cooling which usually differ from those of steel. Consequently, slag with some crystallinity may be inclined to break free of the steel weldment during cooling, thus aiding slag removal and cleaning.

Surface tension is another property of the welding slag which, like viscosity, often had been discussed in general terms because of the lack of quantitative values. As outlined in Chapter 3, surface tension is the unit force required to break the surface film on a liquid by stretching; or, when two liquids are in contact, surface tension is the unit force required to effect a separation of the two liquids. Surface tension usually is expressed as dynes per centimeter or as ergs per cm^2. To illustrate the role of surface tension in liquids of any kind, two very familiar liquids can be examined; mercury and water. Mercury at room temperature has low viscosity, but it has high surface tension. Because of high surface tension, globules of mercury coalesce readily; but they will not penetrate crevices, and mercury will not wet any materials. Water at room temperature has both low viscosity and low surface tension. Water wets many materials, and will penetrate tight crevices. Mercury can be held in a cloth handkerchief because of its high surface tension, but water will pass through holes in the mesh of the cloth. Molten metals have relatively high surface tension, and this property is not indicative of viscosity. Molten iron has about twenty times the surface tension of water at room temperature. In general, surface tension decreases as the temperature of a liquid increases. Most of the remarks made concerning the viscosity and surface tension of molten metals also apply to molten slags. The wetting capability of molten slag improves as its surface tension is lowered.

The surface tension of molten weld metal can be influenced by the surface tension of slag floating upon its surface. For example, when an arc seam weld is to be made by a slag-shielded process, and the weld profile is to have a concave weld face with full penetration of any crevice at the root, the welding parameters ordinarily would include a relatively high energy input; however, these profile objectives would be accomplished more efficiently if the covering slag has low surface tension by virtue of its composition. A low surface tension slag will increase the wetting capability of the weld metal. Slags with a substantial content of titanium dioxide (TiO_2) or flourspar (CaF_2) have extra-ordinarily low surface tension. Little will be said about the composition of shielding slags at this point other than they are made up of compounds that are stable at the welding temperature. These compounds usually are oxides of elements like Ca, Si, Mn, and Al that do not dissociate easily at high temperatures and are not reduced by molten iron. A shielding slag is made by proportioning a number of oxides so that a material with the desired melting point, viscosity, surface tension, and so forth is obtained at the welding temperature. Formulae for typical slags used in welding processes will be reviewed shortly.

Fluxes

While it often is difficult to make a sharp distinction between shielding slags and fluxes, there is enough difference between the two types of materials to warrant separate review. In the preceding discussion of slags, it was emphasized that a shielding slag was inactive or inert, and the molten metal upon which it floated was protected from oxidation simply by the barrier that it provided. In most cases, a *fluxing* type of slag is more likely to be employed. By fluxing, a material is implied which is physically and/or chemically active both in shielding the metal from oxidation and in disposing of any oxide that forms as a result of inefficient shielding or through poor manipulation.

A flux deals with oxides by commingling with the oxide to form a slag of more favorable melting point and viscosity, i.e., multiphase mixing. A pool of molten metal exposed to the air atmosphere may have a refractory oxide formed over its surface that obstructs the welding operation. A second material added to this oxide covering (which also might be an oxide in some cases), can create a slag that will be less of a hindrance to the welding operation. The bulk of the new slag covering may be greater than the original oxide, but because it has a lower melting point and lower viscosity, manipulation during welding is made easier and the weld can be completed satisfactorily. Molten aluminum when exposed to air quickly forms a very undesirable oxide film. The composition of the oxide is Al_2O_3 and its melting point is about 2000°C (roughly 3700°F). If silica (SiO_2) is added to this oxide film, the Al_2O_3 and the SiO_2 combine to form an aluminum silicate compound that has a much lower melting point and viscosity.

A second distinct method by which a flux can deal with an oxide on a molten metal is to dissolve the oxide. The properties of the molten flux covering — meaning its melting point and viscosity — may not be altered significantly as it dissolves oxide. It was mentioned above that molten aluminum when exposed to air formed an oxide with a relatively high melting point, and that a lower melting slag could be obtained by adding silica. A more efficient way of dealing with Al_2O_3 is to use a flux containing chloride or fluoride salts because these materials have a very high solubility for aluminum oxide.

A third method for disposing of oxides on the surface of a metal is to reduce them — that is, to chemically dissociate the oxide into oxygen and metal, which returns the metal to the weld pool. Naturally, this requires that the flux contain an element or a compound which gives a greater heat of oxide formation than the metal in the molten pool; or in other words, an element or a compound which has a greater affinity for the oxygen. Usually, it is not difficult to find a reducing agent to serve as a flux. However, it is necessary to consider the nature and properties of the newly created oxide. Obviously, a less-workable oxide covering on the metal is not desirable. For example, iron oxide (FeO) on the surface of a pool of molten iron can be reduced by sprinkling it with powdered aluminum, but the resulting aluminum oxide (Al_2O_3) would be less desirable than the iron oxide. Actually, several schemes can be used to deal with the oxide problem. The first is to select a reducing agent that forms a satisfactory, low-melting oxide. If this is not feasible, a reducing agent might be available that forms a gaseous oxide which would escape into the atmosphere. Carbon has been used as a reducing agent, and powdered graphite might serve in some cases. If the only materials capable of the reduction reaction formed oxides of high melting points, then a composite flux probably could be formulated. A composite flux would contain an agent capable of reducing the metal oxide and also a fluxing material to commingle with the newly created oxide to form a slag with acceptable properties.

Controlled Atmospheres

Molten metal can be protected from oxidation simply by replacing the atmosphere of air with a nonoxidizing gas, but this method is sometimes difficult to achieve in practice. Controlled atmospheres are not often used in melting operations in furnaces because of the difficulty of keeping the melting chamber gas-tight; however, controlled atmospheres are widely used in heat-treating operations to prevent metal parts from scaling. Welding processes make use of both partial and complete gas shielding.

A gaseous atmosphere that is nonoxidizing to a molten weld pool may be one of several varieties. It can be an inert or nonreactive gas like helium or argon, which forms no known compounds. It can be a gas that is the product of a chemical reaction and has no tendency to further react with the molten metal. Carbon dioxide is a good example of a reacted gas that is used as an artificial atmosphere in welding. In strong contrast to these two kinds of gases, a third might be a *reducing gas*. This is a gas that will not only protect the heated metal from oxidation, but also will reduce any oxide that may already exist on the metal surface — in the same manner as a solid reducing type of flux. Hydrogen, for example, is a reducing gas.

Before a gas can be selected for use as a controlled atmosphere to protect a molten weld pool from oxidation, its properties and its effect on the weld metal to be shielded must be determined. Naturally, factors such as cost, ease of handling, physiological effect on personnel, stability at elevated temperatures, and so forth, must receive consideration. A most important factor which must not be overlooked is the solubility of the gas in the molten metal being protected; because, if a substantial quantity of the gas enters the molten metal, it is apt to cause outgassing during solidification. The inert gases helium and argon have practically no solubility in most metals and are commonly used in the inert-gas shielded-arc welding processes. Carbon dioxide also is virtually insoluble in molten steel, and is widely used as the gaseous shield in a number of welding processes. However, carbon dioxide is not completely protective to iron at high temperatures and will allow some oxidation of the surface. Also, some loss of oxidizable elements from alloy steel will occur when welded under a carbon dioxide atmosphere. Carbon monoxide would offer greater protection from oxidation; however, besides being toxic, it would act to carburize iron and steel under many circumstances. Nitrogen often is indicated to be a useful protective gas in welding. This is true only in a limited area of operations. Although nitrogen does not attack molten iron with the vigor of air (because air contains oxygen), nitrogen does form nitrides on the surface, and it does dissolve in the molten metal to some extent. With certain iron alloys, stainless steel, for example, this limited action by nitrogen may be an acceptable improvement over air when a gas backing is needed for a welded joint. However, nitrogen is not used for shielding on the top side of an arc weld. This is because the arc, with its high temperature, would dissociate molecular nitrogen into nascent atomic nitrogen, and "nitrification" of the molten weld pool would be greatly increased. Hydrogen forms an excellent protective atmosphere because of its reducing capability, and for this reason it is widely used as an atmosphere in furnaces for heat treating and brazing operations. The atmosphere may be 100 percent hydrogen, or the hydrogen may be only one gas in a mixture of other gases. The use of hydrogen over *molten* metal is undesirable because this gas is soluble to some extent in virtually all metals, and a sufficient amount may dissolve to cause problems. The role of hydrogen in atomic hydrogen arc welding (AHW) was described in Chapter 6. The subject of hydrogen will be discussed further in Volume II, Chapter 13, when the absorption of gases and the formation of gas pockets and cracks in welded steel is reviewed.

Vacuum

An effective means of avoiding oxidation when processing heated metal is to eliminate the surrounding atmosphere. Even a mediocre vacuum might cause less oxide formation than an artificial atmosphere which is 99.9 percent inert gas. Furthermore, it may be cheaper to pump down a chamber to a usable vacuum. A "vacuum" is described in terms that indicate its reduced pressure as compared with atmospheric pressure. The reduced pressure of a given vacuum is indicative of the residual gas still present. The SI unit for expressing the pressure of a chamber is the pascal (Pa), and the internationally established reference for pressure by the earth's atmosphere is 101 325 Pa. Because one pascal represents a very small pressure (i.e., only one newton per square meter), multiples of the pascal (kPa and MPa) are more convenient. However, industry for many years expressed vacuums in terms of the height of a mercury column in an absolute mercury manometer. A unit of pressure called the *torr* represented the pressure exerted to raise the mercury column by one mm in height. Yet, vacuums often are described in the literature by other metric units, such as *microns* (μm), in the mercury manometer. Table 8.1 shows popular ways of expressing gas pressure ranging from atmospheric to absolute vacuum, and indicates the levels where steelmelting, brazing, electron beam welding, and other operations requiring oxidation protection are performed.

A perspective of oxidation potential when relying on a vacuum for protection can be gained by first visualizing air at standard atmospheric pressure (101 325 Pa, or 760 torr, or 14.7 psi) surrounding a piece of metal. Approximately 2.7×10^{19} molecules of gas (mostly nitrogen and oxygen) are present in each cubic centimeter of this atmosphere. The molecules are in constant motion, but consider the mean free path between collisions of the molecules and with the metal. At standard atmospheric pressure, the population of molecules is so dense that they move only 1×10^{-4} mm before impacting another molecule or the surface of the metal. Consequently, an ample supply of nitrogen and oxygen reaches the surface of the metal for chemical reactions to take place.

When the standard atmosphere is pumped down to a vacuum of 10^{-5} torr, the concentration of gas molecules remaining will be only 0.000 001 percent, and the mean free path of molecule movement will be increased to about 25 mm (one inch). Therefore, the occurrence of reactions involving contact of the gas molecules with the metal's surface will be greatly reduced. If the vacuum is pumped to a lower pressure, perhaps 10^{-7} torr, the population of molecules would be reduced by a factor of 10^{10} and would number about 2.7×10^{9} per cm³. At this reduced level, the mean free path would be about 14 m (approximately 46 ft) and reactions at the metal surface become less

Table 8.1 - Degree of Vacuum Used to Protect Metal During Various Processes

SI Units, Pa	Height of Mercury Column mm, or Torr	Microns (μm)	Percent of Gas Molecules Remaining*	Comments on Occurrence or Usage
101 325.0	760	760×10^3		Standard atmospheric pressure on earth (14.7 psia)
3361.2	25	25×10^3		Upper limit of partial vacuums
133.3	1	1000		Beginning of "fine" vacuums
66.7	0.5	500	0.05	Vacuum brazing at this level
13.3	0.1	100	1.3×10^{-2}	and below
6.7	0.05	50		Vacuum melting furnaces (50 to 1 μm)
0.133	10^{-3}	1	1.0×10^{-4}	Beginning of high vacuums
0.0133	10^{-4}	0.1		Least degree of vacuum required for operation of EBW gun
0.0013	10^{-5}	0.01	1.0×10^{-6}	Approximate lower level for EBW
0.13 mPa	10^{-6}	0.001		Beginning of "ultra-high" vacuums 1 nm (millimicron)
	10^{-7}	10^{-4}	1.0×10^{-10}	Ultra-high vacuum
	10^{-8}	10^{-5}		Ultra-high vacuum
	10^{-9}	10^{-6}		Approximate level of best vacuum achieved in laboratory apparatus
	$<10^{-14}$	$<10^{-11}$		Outer space in our universe
	0.000....	10^{-00}		Absolute vacuum

*Total concentration of gas molecules remaining at the indicated pressure

frequent. Table 8.1 includes the approximate concentrations of gas in vacuums of varying degree. These calculations show that air remaining as a contaminant in the vacuums employed to protect certain operations is often lower than in inert gas atmospheres provided for shielding. The biggest deterrents to the use of a vacuum for protecting a joining operation are the cost and the inconvenience of getting the workpieces into a vacuum chamber and then accomplishing the joining as an automated operation. Nevertheless, vacuum brazing — and electron beam welding, in particular — have flourished in recent years. This growing usage is largely due to the design and construction of efficient equipment with innovative loading arrangements and rapid pumping to the required level of vacuum. For example, the dual-vacuum EBW-MV system, illustrated earlier in Figure 6.37, has figured prominently in the utilization of the electron beam welding process for mass production.

A question regarding the feasibility of employing a vacuum is whether the joining process will operate satisfactorily in that environment. A welding arc is not readily operated in a vacuum because the plasma requires ions, usually supplied as a gas at the start of the operation. Problems with corona discharge and other electrical phenomena can arise when the pressure of the atmosphere surrounding the arc is reduced to a very low level. Heating for welding *in vacuo* cannot make use of a gas flame; however, the electron beam, electrical resistance, and the laser beam can be used without difficulty.

A unique feature of any joining process conducted in vacuo is the degassing or purification mechanism which occurs. When the process involves fusion, dissolved gases and volatile impurities evaporate and are drawn off by the pumping system. This mechanism assists in obtaining porosity-free metal after solidification. The vacuum not only prevents oxidation of heated surfaces, but may even bring about the dissociation of oxides present before the operation began. Under normal conditions at room temperature, metal surfaces are covered with films of oxides, water vapor, and possibly other contaminants. These are adsorbed on the surface and held tightly by the unsatisfied bonding forces of the metal atoms at the surface. When two metal surfaces are brought into contact, their adsorbed surface layers effectively prevent cohesion or "cold welding." However, if the adsorbed surface layers are removed by exposure in a high vacuum — particularly if the surfaces are gently heated while in the vacuum to increase the degree of film evaporation — then the two metal surfaces will display a marked tendency to "cold weld" during contact. Proven joining processes operated in vacuo regularly rely on additional process features such as heat, pressure or filler metal addition to ensure consistent joint properties.

Technique

There are a number of welding processes that combat the oxidation problem through a particular feature of joint design, heating cycle, or operating conditions. One example is the resistance spot welding process applied to sheet where the fused nugget forms within the confines of the two workpieces. Because the molten nugget makes only minimal contact with air, ordinarily no real oxidation problem exists. The outer surfaces of the sheets are in contact with cool electrodes of high thermal conductivity, and these provide the heated (but not molten) sheet surface with some protection from oxidation.

Another technique disposes of the products of oxidation that form during welding by upsetting or by forging the weld to expel the undesirable compounds. In forge welding and in resistance upset welding, as examples, no particular precautions are taken to prevent oxidation of the metal surfaces during the initial stages of the operation. However, an important part of the final stage is to bring the molten or nearly-molten ends together under pressure to extrude the oxide-bearing metal out of the weld area.

In the closing portion of this chapter, a number of techniques regularly employed in commonly used processes will be discussed to show the efficacy of joint design, upsetting, minimal heating time, oxide diffusion, and other means of overcoming the oxidation problem.

Deoxidizers

The functions of a deoxidizer have been described in earlier chapters. It can be stated here that the addition of elements to steel that are capable of oxidizing preferentially, or of removing oxygen from existing iron oxide, provides a practical way of dealing with the oxidation problem, but only in the (molten) weld zone. The addition of a deoxidizer to the steel being welded does not protect it from developing a thicker surface oxide film. In fact, the presence of a strong oxygen-getter element in the steel may encourage surface oxide formation, which can be a greater hindrance to welding. Of course, other steps can be taken in conjunction with the use of a deoxidizer to prevent ready access of air to the heated metal at the weld and thus assist in minimizing oxidation of both solid and molten metal. A deoxidizer in the steel can play a vital role in the portion that is melted during welding and eventually becomes the weld zone.

Appraisal of an element to serve as a deoxidizer in steel can start with examination of the *free energy of formation* (ΔF) for the oxide system of a given element for comparison with that for iron oxide. For example, silicon generates approximately twice as much thermal energy as compared with iron oxide when silica (SiO_2) forms as compared with iron oxide. The silica that forms is quite stabile, and it either tends to rise to the surface, or remains to become entrapped in the solidifying metal. A rough ranking of commonly used deoxidizers for steel — based on their free energy of formation for oxides at 2000°K expressed as ΔF in cal/g-mol O_2, for the system, and starting with the most powerful — is calcium (CaO, −193 720), zirconium (ZrO_2, −170 737), aluminum (Al_2O_3, −165 000), titanium (TiO_2, −134 908), silicon (SiO_2, −130 300) and manganese (MnO, −112 000).

Choosing the element to serve as a deoxidizer and the amount added requires review of many factors. Furthermore, the consideration given to each factor will depend on whether the deoxidizer addition is to be made to base metal to filler metal, or directly to the weld pool. Selection of deoxidizing elements will bear further discussion in Chapters 14 and 15, during review of welding procedures for different steels.

Protective Surface Alloys

Oxidation of the base metal can be prevented in some joining processes by coating or plating the surface with a metal or alloy that has a lesser affinity for oxygen. Iron or steel can be plated with nickel, for example, to provide a surface that will melt with less evidence of oxide formation, at least in the early stages of heating. Nickel plating is especially helpful when elements like aluminum or chromium are present as alloying elements in the steel because they assist in forming oxides on the surface during heating that are quite refractory. As soon as substantial melting occurs at the surface, a nickel plating would be assimilated or dissolved and the protective effect would no longer exist. Consequently, protective surface alloys are more often used in brazing or braze-welding operations. Here, base metals like steel may be coated or plated with copper, nickel, silver, and gold to facilitate bonding. Surfacing by thermal spraying on steel sometimes makes use of an initial sprayed film of molybdenum metal to combat oxidation of the iron in the surface and thus improve the bonding of the next layer of surfacing metal.

Liquid Blankets

Sometimes a welding operation which is just borderline with respect to oxidation difficulty can be improved by modest protective means. Consequently, liquid media that have no real fluxing capability, and that do not provide complete protection from oxidation may be used. For example, resistance spot welding of certain metals and alloys may be performed while the workpieces and the electrode tips are submerged in a bath of kerosene, silicone oil, water, or carbon tetrachloride. The liquid blanket employed will depend, of course, on the metal being welded and the reaction that can be permitted between the heated metal and the liquid. Final selection of a liquid ordinarily is made only after welded specimens have been carefully examined to be certain that a harmful degree of surface deterioration or alteration has not been produced in lieu of surface oxidation.

SHIELDING THE JOINING PROCESSES FROM AIR

Now that the principal functions of slags, fluxes, gases, etc., have been described, attention can be centered upon the methods of oxidation protection incorporated in the joining processes. Description of the processes in Chapter 6 included some information on the shielding afforded by each process, but this chapter will focus entirely on shielding. A good grasp of the principles will be useful both in effectively carrying out the commonly-used processes, and in modifying a process to meet the demands of special situations.

Carbon Arc Welding

Carbon arc welding (CAW) presently is used only to a very limited extent, but much was learned about shielding during the early development of arc welding (circa 1925) when CAW was popular. When the carbon electrode is employed, besides ions of carbon and metal in the arc, there are substances derived from the air, such as the unstable compound CN, and the oxides of carbon. In Chapter 6 when describing welding with nonconsumable electrodes, and when describing "air carbon arc cutting," it was emphasized that distribution of the ionized particles in the arc are highly dependent on the kind of welding current being used, and, if direct current, the polarity involved. When the carbon arc is operated on DCEP, carbon ions tend to leave the electrode and flow toward the weld pool (negative pole or cathode) where they impinge upon the molten metal. Consequently, with this polarity (DCEP), the weld metal is subjected to a strong carburizing mechanism. A very long arc can have an atmosphere that is oxidizing because of the larger amount of air aspirated into the arc mantle, whereas a carbon arc with a normal or shorter arc gap will have an atmosphere with a preponderance of carbonaceous gases (mostly CO and CO_2).

Welders using the carbon arc to fusion-weld iron and steel learned to control the nature of the welding atmosphere to some degree by resorting to simple methods. The oxidizing effect of air aspirated into the arc was reduced by inserting a string of combustible material into its mantle to combine with at least some of the oxygen in the air. If the string consisted of tightly rolled-up paper, it burned to form water vapor and mainly carbon dioxide, both of which are more protective of the molten steel than oxygen. The string was fed into the upper part of the arc, which was the narrowest part, and where the largest amount of air was present. By removing a large portion of the uncombined oxygen from the arc, the combustible material sometimes permitted welding to be performed without a flux. When more effective protection was needed in carbon arc welding, the string of combustible material was impregnated with slag-forming and fluxing ingredients. As the string burned, these ingredients melted and performed their functions right at the point where they were most needed. The nature of the slag and flux ingredients depended, of course, on the metal being welded. For steels, minerals such as clay and asbestos were used for forming the slag, and fluorspar was favored as the flux. From this simple beginning, shielding the arc with gases and protecting the molten metal with slag and flux developed into a highly refined and complex technology.

Metal Arc Welding

Consumable electrode arc welding with bare steel (stick) electrodes preceded the use of covered or coated electrodes. While fluxes sometimes were applied to the groove faces of the joint when welding with bare electrodes, it became obvious that covering the electrode with slag-forming and fluxing materials was the most practical way of applying these agents to the lengthy joints that often had to be welded. Yet, there are special metallurgical problems where arc welding with a bare electrode, or one that has only a light dipped coating, can provide a solution. As one example, steel weld metal with the lowest possible silicon content has the best resistance against dissolution by a bath of molten zinc (as would be used for applying hot dip zinc coatings). Consequently, when the low silicon iron pots used to hold molten zinc are to be arc welded, electrodes bearing typical flux coverings usually are avoided because they deposit weld metal deoxidized mainly by silicon. Bare metal electrodes of low-carbon content steel will deposit weld metal with a very low silicon content. This undeoxidized weld metal likely will contain some porosity and blowholes, but the weld metal will not be as rapidly dissolved by the molten zinc. This remedy for the weld metal problem in zinc pots (i.e., using bare, rimmed steel electrodes) has been used for many years despite the difficulty of maintaining a stable arc. More recently, the GMAW process and a low-carbon, low-silicon steel electrode has been used to obtain a weld with less porosity and blowholes, and it also provides smoother arc operation and weld bead deposition.

Covered Electrodes

Covered electrodes for the shielded metal arc process (SMAW) were described in a general way in Chapter 6 when examining the process. The electrodes consist of a center rod or wire covered with suitable solid flux applied by wrapping, dipping, or extrusion. The extrusion process is used for the majority of electrodes manufactured today, although any one of the three methods will make a satisfactory covering. Formulation of a welding electrode covering has gradually advanced from an art controlled by rule-of-thumb to a science that draws heavily from knowledge of the physical chemistry of steelmaking. Ingredients incorporated in a typical covering applied by the extrusion method would include bulk materials like cellulose, limestone, rutile, fluorspar, and titania. Fibrous materials like asbestos are used for binding, along with various kinds of clay and talc for plasticizing and aiding slipping during extrusion. Powdered metals are added for deoxidation and alloying the weld deposit, and the formulation is completed by adding liquid sodium or potassium silicate to act as a cementing agent or binder. The dry materials are granulated or powdered to controlled particle sizes to ensure ease of extrusion, and to avoid cracking of the covering during the baking operation that follows extrusion to remove the moisture from the silicate binder. The amount of covering on an electrode will vary depending on its intended use, particularly the welding position for which the electrode is designed. The weight of the covering on an electrode may range from five to 70 percent of the weight of the core wire, and possibly higher in some unusual cases. Heat absorbed from the arc to melt the covering, therefore, will vary from an insignificant amount to a substantial amount that is commensurate with the covering weight and composition.

The covering on an electrode improves its operation in a number of ways. When an electrode is melting, heat is concentrated on the core wire and spreads through the covering by conduction. Consequently, although the melting temperature of the covering mixture is usually lower than that of the core wire, the metal core melts ahead of the covering and this forms a cylindrical shell at the melting tip that helps protect the upper part of the arc. Because the major portion of air that enters the arc is aspirated into the upper mantel, the cone-shaped cavity formed by the covering shell around the melting tip is instrumental in reducing the amount of oxygen and nitrogen in the plasma. With electrodes of the "contact" or "drag" type, this covering shell may extend beyond the core wire by about 12 mm (½ in.) and this extension shields the arc almost completely from the air. Flux-covered carbon electrodes for the CAW process were not successful, because the graphite electrode is consumed too slowly and the covering disappeared prematurely.

Figure 8.4 is a schematic of the arcing end of a SMAW electrode. As the core wire is melted, the covering also is melted by heat from the arc and the hot core wire. Chemical reactions between the molten metal and the covering material take place initially as they mix at their peripheral juncture uppermost in the protective conical shell. If metal particles have been included in the covering formulation, they also become molten and commingle with the molten metal coming from the core wire. The molten droplets that detach and travel to the weld pool are surprisingly homogeneous in composition, and the composition from drop-to-drop is remarkably consistent, providing the covering composition is consistent along the length of the electrode. The molten

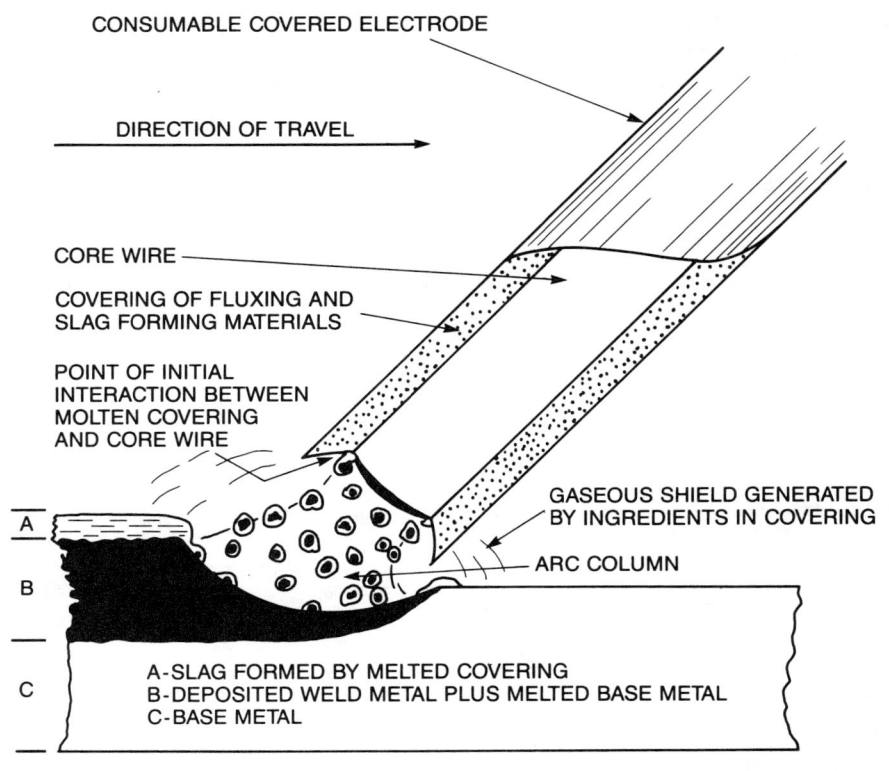

Figure 8.4 — Typical deposition mechanics of covered electrode in SMAW process

Schematic illustration shows how droplets of molten metal from melting core wire are protected by film of molten flux-and-slag from covering during transfer from electrode to weld pool.

droplets probably are in a state of rapid circulation because of electromagnetic forces generated by the flow of current, and this likely accounts for their homogeneity. Of course, as the molten droplets enter the weld pool, this metal is subjected to further vigorous stirring in the pool. Only in unusual cases is there compositional heterogeneity after the weld pool has solidified. Consequently, the addition of powdered metals to the electrode covering for purposes of alloying and deoxidation has become common practice in manufacturing electrodes for the SMAW process.

Functions of Covering Ingredients

The respective functions of covering ingredients deserve review here. Table 8.2 lists minerals, non-minerals, and metallic materials often included in coverings for SMAW steel electrodes. These ingredients have been grouped into seven categories according to the principal reason for their incorporation in a covering. It should be recognized that their roles often are multi-functional; for example, titania is listed as an arc stabilizer, but it also is a slag former and serves as a slipping agent in the extrusion process. A limited number of arc stabilizers are listed, because many of the ingredients in the other six categories also assist in stabilizing the arc. Many kinds of powdered metals are used in electrode coverings to accomplish deoxidation of the weld deposit and to tailor its alloy content; however, only a small number are listed in Table 8.2. The brief remarks in the table will be supplemented by several paragraphs that follow.

The composition of a SMAW electrode covering almost wholly determines the operating characteristics of a steel electrode. In comparison, the type of steel in the

Table 8.2 - Materials Used in Coverings on Steel Electrodes for SMAW Arc Welding Process

Common Name	Technical Name and/or Approximate Chemical Formula	Remarks
Arc Stabilizers		
Titania	TiO_2	Frequently used form of purified titanium oxide
Potassium oxalate	$K_2C_2O_4$	Infrequently used
Lithium carbonate	Li_2CO_3	Infrequently used
Gas-Forming Materials		
Cellulose	Purified wood pulp $C_6H_{10}O_5$	Principal ingredient in "cellulosic" electrodes
Wood flour	Raw wood pulp $C_nH_nO_n$	
Limestone	$CaCO_3$	Produces CO and CO_2 during welding and forms basic slag
Fluxing Agents		
Cryolite	Na_3AlF_6	Strong fluxing agent
Barium fluoride	BaF_2	
Lithium fluoride	LiF	Very effective flux
Lithium chloride	$LiCl$	Infrequently used
Witherite	$BaCO_3$	Generates CO and CO_2 gases, but then becomes strong flux
Fluorspar	Fluorite CaF_2	Strong fluxing agent
Slag-Forming Materials		
Bauxite	Alumina Al_2O_3	Raises melting temperature and increases viscosity of slag
Feldspar	Alkali type $K_nNa_nAlSi_3O_8$ Plagioclases $CaAl_2Si_2O_8$	
Fluorspar	Fluorite CaF_2	Markedly decreases viscosity of molten slag
Ilmenite	$FeTiO_3$	Impure form of titanium oxide
Rutile	TiO_2 (10% Fe)	Unrefined form of titanium oxide. Mainstay of "rutile" electrodes
Silica flour	Cristobalite SiO_2	Strong acid slag former
Wollastonite	Calcium silicate $CaSiO_3$	
Dolomite	Magnesite $CaMg(CO_3)_2$	Often used for forming slag when melting steel in furnace, but seldom included in electrode coverings
Zirconia	Zirconium oxide ZrO_2	Infrequently used
Slag-Forming Materials (continued)		
Magnetite	Iron oxide Fe_3O_4	Magnetic iron oxide
Periclase	Magnesium oxide MgO	Raises melting temperature and increases viscosity of molten slag
Pyrolusite	Manganese dioxide MnO_2	
Slipping Agents to Aid Extrusion		
Bentonite clay	Montmorillonite $Al_2Si_4O_{10}(OH)_2$	Used where water-of-constitution can be tolerated
Kaolin clay	Kaolinite $Al_2Si_2O_5(OH)$	
Mica	Muscovite $KAl_2(Si_3Al)O_{10}(OH)_2$	
Talc	Soapstone $Mg_3Si_4O_{10}(OH)_2$	
Glycerin	Glycerol $C_3H_5(OH)_3$	Trihydric alcohol
Binding Agents		
Sodium silicate	Water glass $Na_2O_nSiO_2(OH)_n$	Agent most often used
Potassium silicate	$K_2O_nSiO_2(OH)_n$	
Asbestos	Crysotile $Mg_3Si_2O_5(OH)_4$	Improves durability baked covering, and mixes w/ slag when melted during welding
Dextrin	Starch $C_6H_{10}O_5$	
Gum arabic	Acacia $C_nO_nH_n$	
Sugar	$C_n(OH)_n$	
Alloying and Deoxidizing Elements		
Ferrosilicon	Usually 50% Si + Fe	Silicon is deoxidizer and alloying element
Ferroaluminum	Usually 85% Al + Fe	Strong deoxidizer
Ferrotitanium	Usually 40% Ti + Fe	Strong deoxidizer and grain-refining agent
Zirconium alloy	40% Zr + 40% Si + Fe	Deoxidizer
Electro-manganese	Mn = 100%	Most common alloying element
Electro-nickel	Ni = 100%	
Chromium metal	Cr = 100%	
Ferromanganese	Std. type Mn = 80% + Fe	

core wire, that is, its alloy composition and its level of deoxidation (i.e., rimmed, as compared with killed steel) has very little influence on electrode behavior during deposition. Covered electrodes gave tremendous impetus to metal arc welding, because the readily ionizable compounds in the covering of SMAW electrodes made the arc more stable and easier to maintain. The rate of metal deposition — sometimes called the "burnoff rate," but properly referred to as the *melting rate* — and the depth of penetration of the weld bead into the base metal can be controlled to some extent through the covering composition. Even the shape of the deposited bead — that is, its concavity or convexity, and its surface smoothness — can be influenced to an appreciable degree. Technical Brief 40 presents short history of the development of the SMAW electrode.

Limestone is a mineral often included in electrode coverings to generate protective gases. It is an essential ingredient in the coverings for low-hydrogen and stainless steel electrodes. Cellulose or other gas-forming organic materials cannot be used in the low-hydrogen electrode covering because too much hydrogen is generated. Also, the mixture of CO and CO_2 gases from organic material allows too much carbon the be picked up by the high chromium content (stainless) steels. The chemical formula for limestone is $CaCO_3$. At temperatures above approximately 500°C (about 950°F) the limestone becomes calcined, which means that it gives up about 44 percent of its weight as carbon dioxide gas. This gas is surprisingly good protection against oxidation, even with electrodes containing a large amount of chromium or other oxidizable

---

TECHNICAL BRIEF 40:
Development of the SMAW Electrode

When covered electrode development first started, the initial step beyond the bare metal electrode was simply an extra-heavy dip of lime applied to a sul-coated wire just before the final wire drawing draft. This crude method was soon rendered obsolete by the development of light coverings, applied by dipping or spraying.

The need for a means of protecting the arc and the molten weld metal from the surrounding atmosphere was met by adding ingredients to the covering which generated gases at the electrode tip. The gases had to be practically nonoxidizing to the molten metal droplets coming from the electrode and the molten weld pool. The first gases put to this use came from the combustion, volatilization, or dissociation of compounds. Cellulose, as ground wood flour, or purified alpha flock, was one of the earliest ingredients that provided gases giving reasonable protection to the carbon and low-alloy steels. Like most organic compounds, cellulose boils at temperatures below about 650°C (1200°F); and, upon entering the welding arc, it instantly becomes a gas that combines with oxygen at the outer edges of the mantel. The expansion of gases from the covering exerts a jet action to drive air from the vicinity of the arc.

The gaseous products from the volatilization and combustion of the cellulose in an electrode covering are carbon monoxide, carbon dioxide, hydrogen and water vapor. Note that organic compounds generate hydrogenous gases, and these allow hydrogen to be picked up by the molten metal. The harmful influence of hydrogen in the weld was not recognized for a long time, but it will be discussed in Chapter 13 of Volume II.

Some electrodes rely primarily on combustible materials in their covering to provide oxidation protection. A very thin blanket of slag, or none at all, is formed, which makes it easy to manipulate these electrodes in overhead welding.

alloying elements. The lime (CaO) that remains after releasing the carbonaceous gas is an excellent arc stabilizer, flux, and slag-former; therefore general esteem is held for limestone or other compounds that supply calcium oxide to the welding slag. The value of employing a basic or "lime" slag has been recognized for many years in steelmelting because of its ability to absorb and hold unwanted residual phosphorus and sulfur. Although arc welding operations hardly provide enough time for reactions between the molten metal and the slag to proceed toward substantial phosphorus and sulfur removal, having calcium oxide present in the slag maintains the potential for removing these residual elements. Also, a *basic* welding slag has additional virtues related to microstructure of the weld metal, its oxygen content, and nonmetallic inclusion content. These benefits have been seen quite clearly in weld metal deposited under a basic flux in the submerged arc process, and will be pointed out as fluxes for the SAW process are discussed.

Fluxing agents and slag-forming materials are listed in Table 8.2 as two separate groups, but difficulty in making a clear-cut distinction was pointed out earlier. The principal function of a flux is to remove any undesirable oxides that may have formed despite the protection afforded by a gaseous shield and/or by a slag blanket. The chemical compounds listed in Table 8.2 as fluxing agents are particularly effective for this function, but in reality many of the materials in the other listed groups also have fluxing capability. Consequently, there is no great need to apportion a substantial percentage of the covering mixture to ingredients that function solely as a flux. Calcium oxide, which is produced when limestone becomes calcined, has been mentioned as having fluxing capability, particularly for the oxides that form on iron and steel. As another example, sodium oxide, which becomes available from the frequently-used sodium silicate binder, acts as a flux.

When the various covering ingredients that do not gasify become molten and flow over the heated workpiece surfaces and weld metal as a blanket, they perform both the fluxing and shielding functions. The molten flux-slag blanket is pushed ahead on the groove faces of the joint by the advancing arc. The molten weld pool opposite the electrode will be more or less free of the molten blanket because of the positive pressure exerted by the plasma. On the heated groove faces just beyond the weld pool, the flux-slag blanket dissolves iron oxide (including rust and mill scale if not excessively thick). As the iron oxide enters the blanket, it combines with silica (SiO_2) to form iron silicate (Fe_2SiO_4). This iron compound has a lower melting temperature than steel and, unlike iron oxide, is insoluble in molten steel. The thicker the slag blanket the less opportunity there will be for diffusion of oxygen from outer parts of the blanket to the metal beneath. Since FeO in the blanket is mostly combined as iron silicate or as iron manganese silicate ($2MnO \cdot FeO \cdot SiO_2$) which melt in the range of about 1200 to 1300°C (approximately 2200 to 2400°F), the iron oxide can no longer act as a transporter of oxygen to the metal and cause its oxidation. When appreciable amounts of compounds of calcium, silicon and manganese are present, the slag blanket becomes almost impermeable to oxygen.

Slag-forming ingredients in the electrode covering usually make up the bulk of the molten blanket. If too low in viscosity, the blanket may not remain on the hot surfaces, but will run off and drip away. When the slag blanket is too viscous, it is likely to impede the flow of molten metal and some slag can become entrapped in the weld zone. Coverings usually have such a multiplicity of ingredients, and their interactions are so complex and synergistic, that the effects of covering composition changes on viscosity, etc., cannot be closely predicted. To achieve optimum covering formulation for a particular kind of electrode, the manufacturer must make numerous trial coverings and then conduct tests for electrode operating performance. Welding test procedure is governed by the intended classification of the electrode, and many features require observation and measurement to be certain the electrode will meet the requirements of applicable specifications. Additional qualities usually must be scrutinized to ensure that handling characteristics have been properly tailored to please experienced, discriminating welders. For instance, proper formulation of the electrode covering also must take into consideration the friability of the slag blanket after cooling. Easy removal, or even spontaneous self-break-away, is highly advantageous. Decisions made in the course of electrode-covering development and testing are largely judgmental; there are no established means of quantitative measurement of most operating characteristics deemed to be important.

Powdered metals and alloys used for deoxidation and for weld metal alloying are listed last in Table 8.2. It is practicable to incorporate both deoxidizers and alloying additions in the covering; because, as the electrode is deposited, these materials are transferred to the weld metal. When the elements added to the covering are oxidizable, their transfer may not be 100 percent efficient, but when reasonably good protection from oxidation is provided, recoveries will approximate the values given in Table 8.3. Note that a number of the elements (i.e., beryllium, titanium, zirconium, and aluminum) are transferred with very poor recovery. These elements are such avid oxygen "getters" that they will expend themselves by reducing some of the oxides present in the fluxing and slag forming ingredients of the blanket. For example, the addition of a substantial amount of aluminum not only will introduce some aluminum into the weld deposit, but a portion of the aluminum will reduce oxides of manganese and silicon

Table 8.3 - Recovery of Elements From Coverings of Electrodes Deposited by SMAW Process

Principal Element To Be Added To Weld Deposit	Form of Material In Electrode Covering	Approximate Recovery of Element, %
Carbon	Graphite	75
Manganese	Ferromanganese	75
Phosphorus	Ferrophosphorus	100
Sulfur	Iron sulfide	15
Silicon	Ferrosilicon	50
Chromium	Ferrochromium	95
Nickel	Electro-nickel	100
Copper	Copper metal	100
Niobium	Ferroniobium	70
Titanium	Ferrotitanium	5
Molybdenum	Ferromolybdenum	80
Vanadium	Ferrovanadium	75
Beryllium	Copper-beryllium alloy	0
Boron	Ferroboron	2
Nitrogen	Nitrided manganese metal	50
Tungsten	Ferrotungsten	80
Aluminum	Ferroaluminum	20
Zirconium	Nickel-zirconium alloy	5

present. Consequently, the weld composition beneath the slag formed by the aluminum-containing covering may have higher levels of manganese and silicon than is desired. When highly oxidizable ingredients are included in the covering formulation, a careful study must be made of their total effects.

Powdered metals sometimes impose unexpected constraints in manufacturing covered electrodes because of the intense reactivity that certain metals or alloys may exhibit when exposed to aqueous chemicals. For example, powdered silicon metal, and even ferrosilicon containing more than 50 percent silicon, will react with aqueous sodium and potassium silicates and generate considerable heat. This reaction will cause the moist, plasticized covering mixture to harden prematurely, and this could impede the extrusion covering operation if it occurred to an appreciable degree. Consequently, as indicated in Table 8.2, powdered ferrosilicon for deoxidation purposes usually is the 50 percent silicon variety. Manufacturing techniques for using powdered metals and alloys must take into consideration such items as (1) reactivity with air and water, (2) particle size influence, (3) composition effects, and (4) the stabilization of powder particles during the mixing and extrusion operations.

In general, the recovery of an element from the electrode covering will be similar to the percent recovery of the same element when present in the electrode core wire. However, the product form in which a given element is incorporated in the covering can affect its recovery in the weld deposit. Particle size of the ingredient will also affect efficiency of transfer. For example, carbon added to the covering as lampblack, which has an extremely fine particle size, will be recovered at a much lower percentage in the weld deposit than if present in the form of graphite. Also, aluminum present as aluminum metal will not be transferred as efficiently as when added as ferroaluminum. Very finely powdered metals and alloys of the oxidizable kind usually will not be recovered in as high a percentage as when coarser particles are employed. Most additions of alloying elements are powdered to pass through a 100-mesh sieve.

Adding deoxidizers and alloying elements through the electrode covering has become common practice because it usually is more economical to purchase a less-expensive type of core wire, and to subsequently enrich the weld deposit with required alloying elements by incorporating them in the covering as powdered materials. As a typical example, an electrode producer can purchase plain-carbon steel wire containing 0.95 percent carbon at a comparatively low cost. By adding high-carbon ferrochromium to the electrode covering, an excellent high-carbon chromium alloy steel weld deposit will be obtained for surfacing applications. The difficulties involved in producing high-carbon chromium alloy steel wire demand that the steel producer set a fairly high selling price for the product. Another advantage for electrode producers by making alloying element additions through the covering is the opportunity to closely regulate the alloy composition of deposited weld metal using a given batch of core wire from a heat of known chemical composition. Procurement of core wire containing the necessary allying elements at requisite levels ordinarily would consume an inordinate amount of time.

Generic Electrode Coverings and Typical Formulas

The composition of coverings on arc welding electrodes gradually has become quite complex. At least five separate aspects of manufacturing and electrode application influence the development of covering formulas for each classification, namely: (1) suitability for production, (2) handling durability and storage stability, (3) arc welding characteristics, (4) slag removal, and (5) metallurgical effects on the deposited weld metal. Although the formulation of electrode coverings is handled by electrode producers, a general knowledge of the subject should be helpful to anyone concerned with arc welding.

Specifications for a number of covered arc welding electrodes of steel that are frequently used to weld carbon and low-alloy steels, as issued by the AWS, were listed in Table 5.14 of Chapter 5. Complete requirements for carbon steel electrodes can be found in ANSI/AWS A5.1, while those for low alloy steel electrodes are contained in AWS A5.5. The requirements in these specifications concern the operating characteristics of the electrode, the mechanical properties of the deposited weld metal, and (only when necessary) chemical analysis limitations for the weld metal. The composition of the electrode covering is left to the creativity of the electrode producer. The covering on each electrode classification usually falls into a general type to provide the expected operating characteristics. The various electrode classes will be discussed in detail in Volume II, Chapter 10; but at this point formulas for electrode coverings will be examined to see how materials are apportioned to satisfy all of the electrode manufacturing and usability requirements described above. At least a dozen generic varieties of coverings are found on the commonly used electrodes. Table 8.4 lists five selected types that will serve to illustrate the ingredients employed in the coverings' mixtures, covering formulas, and the chemical composition of each covering as a whole. This information is illustrative only; no commercial covering is likely to have a composition exactly as shown in Table 8.4.

High Cellulose, Sodium Containing, GasShielded Coverings

Electrodes of the E 6010 classification typify this generic kind of electrode covering. They have been very widely used because a reasonable amount of oxidation protection is provided without a thick slag blanket being formed on the weld deposit. The substantial portion of cellulose in this covering generates a large amount of carbon monoxide and carbon dioxide, and also some hydrogen. The hydrogen can be objectionable in the gaseous shield when welding certain kinds of steels, particularly those that harden in the heat-affected zone. Nevertheless, this cellulosic type of covering is favored by welders because of the deeply penetrating, forceful arc and the spray deposition of weld metal. These operating characteristics are helpful in making many of the welds commonly required in fabricated structures, such as buildings, bridges, and pipelines. The forceful arc is generated by the carbonaceous gases and the hydrogen in the arc atmosphere. These gases have high ionization potentials (see Table 2.2), and they foster relatively high arc voltage, which, in turn, produces high arc energy. Titania in the cellulosic type of covering first acts as an arc stabilizer, but the TiO_2 is retained to assist in forming a thin slag. The cooled slag is quite friable, and is readily cleaned off the weld surface. Ferrosilicon is included in the covering as a deoxidizer, and a small amount of ferromanganese often is present to compensate for the small loss of manganese by oxidation. With sodium silicate as the covering binder, the electrode is usable only with direct current, and deposition normally is made with DCEP. A counterpart covering is made that also can be operated on alternating current by using potassium silicate as the binder (e.g., E 6011). The potassium ions are more efficient arc stabilizers than sodium ions, and this enables the electrode to operate smoothly on ac. This distinction between coverings usable only on dc as compared with the same general type of covering for either ac or dc operation holds true for many of the electrode classifications using the cellulosic type of covering.

High Titania, Sodium Containing, Gas-Slag Shielded Coverings

Electrodes of the E 6012 classification provide an example of this kind of electrode covering. A typical covering formula includes only a small proportion of cellulose, but a large amount of a titaniferous ingredient, usually the natural mineral, rutile. This covering provides a noticeably heavier slag blanket on the weld, but the protection against oxidation is improved only a small amount because of the nature of the slag-forming materials. Siliceous slag formers like clay, feldspar, and asbestos are included to combine with the rutile and form a welding slag of proper viscosity.

Table 8.4 – Typical Covering Formulas for Steel SMAW Electrodes

	E6010 High cellulose, sodium, gas shielded	E6012 High titania, sodium, gas-slag shielded	E6020 High iron oxide, slag shielded	E7015 Low hydrogen, sodium, gas-slag shielded	E7018 Low hydrogen potassium, gas-slag shielded, iron powder
Part I – Material Formulas (Parts by Weight)					
Cellulose	25	5			
Limestone				40	30
Fluorspar				15	10
Rutile		55	20		
Titania	12				
Asbestos	15	10	15	10	8
Iron oxide		1	30		
Clay		10	5	5	2
Iron powder					35
Ferrosilicon	2	2	2	5	5
Ferromanganes	4	4	6	4	4
Sodium silicate	60	40	70	25	
Potassium silicate					25
Part II – Chemical Composition (Percent After Baking)					
CaO				25.5	14.4
TiO_2	10.1	46.0	15.4		
CaF_2				15.2	11 0
SiO_2	47.0	23.6	40.0	20.0	20 5
Al_2O_3		5.0	2.3	2.8	2.0
MgO	3.2	2.0	2.8	1.2	2.0
Na_3AlF_3				5.7	5.0
FeO	1.3	7.0	30.7		
Na_2O	5.1	2.4	3.3	1.7	
K_2O					1.2
Si	1.5	1.5	1.0	2.8	2.5
Mn	2.8	2.5	4.0	2.0	1.8
Fe				2.8	28.5
CO and CO_2				20.2	12.0
Volatile Matter	25.0	5.0			
Moisture	4.0	2.0	0.5	0.1	0.1

Because the slag is slightly viscous, it supports molten metal to assist deposition in joints with a root opening, and in all positions. The arc is not forceful, but is very stable because of the large amount of effective ion-generating material present in the covering. Electrodes with this type of covering can be operated either on DCEN, or on ac. The counterpart electrode covering, which makes use of potassium silicate for binding, has even greater arc stability (E 6013), and can be operated on ac, or on dc with either polarity.

High Iron Oxide, Slag Shielded Coverings

Electrodes of the E 6020 classification have a covering that is typical of this generic kind. The operating functions during welding and the protection from oxidation are obtained with all-mineral ingredients. The materials commonly included are oxides of iron, manganese, silicon and titanium. Iron oxide usually is present in largest proportion. Ferrosilicon is present for deoxidation, and some form of manganese often is present to replace the amount expected to be lost by oxidation. The mineral ingredients form a noticeably heavier slag layer than most SMAW electrodes, so absence of a protective gas shield is not a problem. The thick slag layer allows high welding currents (either ac or DCEN) to be employed to the extent that welders refer to these electrodes as "hot rods." The slag and metal are quite fluid because of the high temperature generated by the high current, so these electrodes ordinarily are limited to the flat and horizontal welding positions, and on relatively thick base metal. The electrode deposits weld metal as a spray transfer, and the deeply penetrating weld zone usually has excellent soundness, good mechanical properties, and is characterized by very fine ripple marks on the weld metal surface. The slag after solidification will be found to have a honeycomb of very small holes on its underside, and the thick layer is easily removed from the weld surface.

Low-Hydrogen, Sodium Containing, Gas-Slag Shielded Coverings

Electrodes of the E 7015 classification evolved shortly after discovery of the cold cracking problem in the heat-affected zones of hardenable steels when they were arc welded under an atmosphere containing a significant amount of hydrogen. This type of electrode covering was the first produced to avoid the threat of hydrogen-induced cracking in the HAZ, and the earliest coverings employed were the all-mineral formulas that were regularly being used on stainless steel SMAW electrodes. As can be seen from the E 7015 covering in Table 8.4, the most important change accomplished by applying this kind of covering was replacement of the frequently used cellulose with powdered limestone to perform the gas-generating function. Also, the all-mineral make-up allowed the proportion of sodium silicate binding agent to be reduced. Abandonment of cellulose in the covering allowed higher electrode baking temperature and, naturally, this favored a reduction in the level of hydrogen retained in the form of moisture in the covering.

As the role of hydrogen in the HAZ cracking problem became clear, researchers started to explore ways of quantifying the amount of hydrogen in electrode coverings in an effort to establish the maxima that would be required to avoid cold cracking risk imparted by the coverings on various classifications of electrodes. Early measuring work became centered on a simple thermal extraction test. The procedure called for a small sample of the covering to be heated to 982°C (1800°F) in a tube furnace. A flow of oxygen carried gases evolved from the covering to a collection system where water vapor (and any hydrogenous gases converted to H_2O) would be weighed. All-mineral coverings being used in that early period (about 1943) were found to range from approximately 0.4 to 1.0 percent by weight in moisture content. This measuring procedure eventually matured into the *Moisture Test* presently included in AWS specifications for carbon and low alloy steel covered electrodes (ANSI/AWS A5.1 and AWS A5.5). In these specifications, the maximum moisture content of low-hydrogen coverings on carbon steel electrodes is 0.6 percent, while low alloy steel electrodes have requirements ranging from 0.4 percent maximum for the lowest strength classifications (e.g., E 7015) to a maximum of 0.10 percent for one of the highest strength classifications (i.e., E 12018-M1).

Experience with the covering moisture test over many years showed worthwhile correlation between covering moisture values and HAZ cold cracking propensity. Therefore, the AWS test method and the adopted moisture content requirements were continued as one of the criteria for the quality assessment of low-hydrogen coverings on various classifications of steel SMAW electrodes. However, anomalous results were encountered from time to time which suggested that the relationship between covering moisture content and hydrogen-induced weld cracking was being influenced by circumstances or variables that were not yet identified. The occasional inconsonance was sufficiently complex to instigate many studies over the past several decades which were aimed at hydrogen, its sources, and its role in causing cold cracking. A number of these studies examined SMAW electrode coverings and the ingredients used in their compounding to assess more comprehensively the matter of moisture as a source of hydrogen.

As individual ingredients of coverings were examined, pertinent findings were gained on the various ways in which moisture could be held in the electrode covering, and on the likelihood that a particular form of moisture could contribute to hydrogen in the weld. The moisture in coverings was found to be held in the following general forms:

(1) Mechanically entrapped moisture, which is retained after baking the moist covering on the electrode, or is absorbed during exposure to ambient atmosphere that contains some level of humidity.

(2) Water of crystallization, which may have been part of the usual crystalline character of an ingredient, or which may have formed a hydrated compound as the moist covering is mixed and then baked.

(3) Water of constitution, which is a very stable component of some hydroxides and minerals used as ingredients, and which may resist dissociation until heated to a very high temperature; often a red heat, and sometimes near the volatilization point.

In order for covering moisture to contribute hydrogen to the weld, the water had to be carried into the arc, where it would dissociate and provide atomic hydrogen to the molten metal. Of course, the moisture test reported the total of hydrogenous gases (including moisture) released from the covering after heating to 982°C (1800°F). However, the question remained — what would happen to the moisture when the covering on an electrode during the welding operation became heated as it approached the arc at the electrode tip? Would the moisture be be effectively driven off and never reach the arc? Or would the water be so intimately bonded in a compound that it would resist dispersal and easily enter the arc? Obviously, the dispensation of moisture in the covering would be highly dependent on its form in the covering at the time of weld deposition.

A principal repository for moisture in the covering on low-hydrogen electrodes is the silicate binding agent, which can hold water of crystallization. Baking the covered electrode at temperatures as high as 480°C (about 900°F) will not remove all of the retained moisture, but temperatures of about 260°C (500°F) and above are usually satisfactory to eliminate most of the moisture; although in some cases a temperature as high as 427°C (800°F) may be needed to meet the maximum moisture requirement. Clay is often used as a plasticizer in the wet covering mix and will assist coating of the core wire by extrusion. The amount of clay and its kind must be given consideration because this mineral ingredient regularly includes a substantial amount of water of constitution in the structure of its solid particles. Furthermore, different kinds of clay lose their tightly bound water at different temperatures during heating, and the rate of release also varies. Koalin clay, for example, will retain much more water of constitution in a baked covering than a bentonite clay. Asbestos is useful ingredient to improve the durability of the baked covering, and it aids in forming a protective welding slag. All varieties of asbestos hold water of constitution, and complete elimination of water from this kind of mineral requires firing at temperatures well above those suitable for electrode baking operations. For this reason, some minerals, like asbestos, are pre-fired before incorporating them in a low-hydrogen covering. As the technology for low-hydrogen coverings progressed, materials were selected with an eye to the tightly bound moisture content in the as-manufactured (baked) covering. Highest possible baking temperatures were employed to rid the covering of as much moisture as possible without cracking the covering or causing it to become overly fragile. It did not appear helpful to retain a small amount of moisture in the all-mineral type of covering in order to achieve smooth operating performance, as is sometimes the case with cellulosic-type electrode coverings.

Low-hydrogen SMAW electrodes are commonly packed for shipping and storage in hermetically sealed containers to maintain their moisture level, but it is recognized that coverings can pick up moisture after their container is opened and the electrodes are exposed to the ambient atmosphere. The amount of moisture picked up is dependent on several factors; namely:

(1) hygroscopicity of the particular covering as governed by its ingredients, manufacturing procedure, and final structural state in the as-manufactured condition;

(2) conditions of exposure, which include such factors as humidity in the ambient atmosphere, temperature, air circulation, etc.; and

(3) duration of exposure.

This recognition of potential moisture pickup by low-hydrogen coverings prompted many safeguards to be established by those concerned with the manufacture of these electrodes and their use. Several of these safeguards are described in Technical Brief 41.

A notably different approach to the question of hydrogen was taken by striving to measure the amount of hydrogen that actually entered the weld metal deposited by a

SMAW electrode — more specifically, the amount of *diffusible hydrogen* accumulated in the weld deposit from all sources. Technical organizations that advocate diffusible hydrogen as the more meaningful determinant employ various ways of collecting the hydrogen from a test specimen, and these differences have interfered to some degree with reaching agreement on the conclusions that can be drawn from certain studies. Nevertheless, the unit of measurement used most often in conducting a diffusible hydrogen test is the number of milliliters of hydrogen evolved per 100 grams of deposited weld metal; usually written as mL/100 g (incidentally, one mL/100 g = 0.89 ppm). ANSI/AWS A4.3 covers *Standard Methods for Determination of the Diffusible Hydrogen Content of Martensitic, Bainitic, and Ferritic Steel Weld Metal Produced by Arc Welding.*

Studies have been conducted to ascertain whether a correlation exists between (1) the moisture content of the covering on selected low-hydrogen electrodes (determined

---

TECHNICAL BRIEF 41:
Safeguards for Handling Low-Hydrogen SMAW Electrodes

The tendency for moisture pickup in low-hydrogen electrode coverings has prompted manufacturers to establish various safeguards for the handling of these electrodes. In addition to hermetically sealed containers, various maxima were set for the length of time electrodes could be exposed to ambient conditions before deposition as weld metal. Arrangements were often made to put electrodes from opened containers into ovens or heated cabinets until needed for welding. Heated portable holders to protect electrodes at the site of welding were sometimes used. Conditions are stipulated in the aforementioned AWS electrode specifications for typical storage conditions for covered arc welding electrodes, and drying conditions were set for re-baking electrodes when deemed necessary to restore the required low moisture level in their coverings. Additionally, there was a suspicion that high humidity in the air atmosphere during the time of electrode deposition could also contribute to hydrogen pickup in the weld.

Many investigations were conducted to assess the degree of importance for the above factors, but so many variables influenced findings that the potentials of specific moisture-contributing conditions could not be firmly quantified. However, test data supported the general observation that moisture picked up during exposure (sometimes called "re-hydrated" moisture) was not likely to be as potent toward inducing weld cracking as the moisture contained in the covering as manufactured. This suggested that moisture absorbed during exposure was being held mechanically, or tended to form hydrates that were weakly bonded. The AWS specification for low-alloy steel-covered electrodes (AWS A5.5) provides recommended moisture limits for electrodes after atmospheric exposure that are slightly higher than the maxima for electrodes as manufactured, or as reconditioned by baking.

The attention given to safeguarding the performance of low-hydrogen electrodes encouraged the development of "low-moisture-absorbing" coverings, and an electrode bearing a covering of this kind is now identified in ANSI/AWS A5.1 with the suffix, R, added to its regular classification (e.g., E7015R). Supplemental requirements are provided for an electrode designated with an R (low-moisture-absorbing). The electrodes to be tested must be exposed to a controlled test environment of 80 percent relative humidity at a temperature of 26.7°C (80°F) for not less than nine hours. Although the moisture content of the "R covering" can be determined by any suitable method, a referee procedure is provided in ANSI/AWS A5.1 under the caption, Absorbed Moisture Test. Details are given for the environment in the test chamber, the exposure procedure, and moisture content limits after exposure (referred to as the "as-exposed" condition).

by the AWS Moisture Test), and (2) the diffusible hydrogen value obtained on weld metal deposited by these electrodes (using the AWS Diffusible Hydrogen Test). Investigators have found a general relationship which has been both charted and expressed in equation form in the literature. At low values from the two kinds of test — that is, below about 0.3 percent moisture in the covering, which corresponded to about 8 mL/100 g of diffusible hydrogen — the relationship showed promising consistency. Above these levels, however, the ratio of diffusible hydrogen to covering moisture increases, and the data display considerable scatter. Despite deviations observed thus far in correlation between the values obtained from covering moisture test and the weld diffusible hydrogen test, the data being secured are serving good purpose. Variability is likely to diminish as test procedures undergo continued improvement and standardization.

Further discussion of moisture test values obtained for coverings will be undertaken in Volume II, Chapter 10, when specific classifications of SMAW electrodes are examined to determine the collaborative roles played by core wire and covering when the electrode is deposited as filler metal. Because hydrogen is the culprit that can accentuate the problem of cold cracking, diffusible hydrogen and data from tests will figure prominently in discussion of this complex metallurgical problem in Chapter 13 of Volume II.

Additional advantages are offered by the low-hydrogen type of covering on SMAW electrodes that are distinct from the problem of avoiding hydrogen-induced cold cracking. Benefits are realized when welding (1) steels to be vitreous enameled, (2) free-machining steels of the high sulfur-content variety, and (3) steels to serve in weldments at low temperature where weld toughness is important. Details of these and other low-hydrogen SMAW electrode applications will be given in appropriate subsequent chapters, mostly in Chapters 13, 14 and 15 of Volume II.

The formula for an all-mineral, low-hydrogen type covering given in Table 8.4 (i.e., E 7015) is a typical so-called "lime" type, meaning that it contains limestone. This ingredient, upon being calcined by heat from the arc generates the protective gaseous shield. The "lime" that remains joins other solid ingredients to make up the slag blanket over the weld. This particular covering is suitable for welding deposition only on direct current (DCEP). For ac operation, a "titania" type low-hydrogen covering has been developed that employs substantial amounts of materials containing titaniferous compounds and potassium compounds. This type of covering is applied to the E7016 electrode.

Iron Powder Containing Coverings

Iron powder is an ingredient frequently used in SMAW electrode coverings, and the amount incorporated in covering formulas may range from as little as five percent to as much as constituting one-half of the covering weight. Iron powder imparts a number of desirable operating characteristics to the electrode. It adds metal ions to the arc and improves stability of the plasma. Iron powder that melts and remains unoxidized will join the weld pool and increase the amount of metal deposited. At least a half-dozen different types of covering have been modified with an addition of iron powder to gain these improvements in operating characteristics. Even the popular E6010 electrode with its high-cellulose, gas-shielded covering may contain a small amount of iron powder to impart smoother arc characteristics. The low-hydrogen, iron powder type covering listed in Table 8.4 (i.e., E 7018) has demonstrated good welding characteristics under a wide variety of conditions and is applied to a number of classes of steel electrodes.

Flux Cored Electrodes

The development of flux cored electrodes to permit continuous feeding of spooled, bare-surface filler metal through a hand-held gun, or through a mechanized welding torch, was reviewed in Chapter 6 when describing the flux cored arc welding (FCAW) process. Initially, it was believed that the materials and formulations used for external coverings on SMAW electrodes also would serve for the new FCAW electrode. Of course, there would be no need for binding agents, particularly the aqueous forms of sodium and potassium silicates, inasmuch as dry, powdered forms easily could be substituted for those compounds considered indispensable. The core-filling operation when manufacturing the FCAW electrode demanded a freely flowing, uniformly mixed, dry, powdery material to ensure a properly filled core. When the flux was sealed in the core cavity, there were no requirements for bonding, but the mechanical seam in the electrode called for a flux mixture that was not highly hygroscopic. Most of the ingredients used in formulating flux mixtures for the core can be found among those listed in Table 8.2.

The new arrangement of filler metal and flux in the FCAW electrode might seem a simple, reversed replacement for the covered filler metal in SMAW electrodes, but experience soon revealed that the mechanics of melting, and the flux-metal interactions during deposition of a FCAW electrode differed from that of SMAW electrodes. With most FCAW electrodes, the tubular metal sheath is melted by the arc slightly in advance of the core material, apparently because the sheath serves as an electrical pole, whereas the core material must be melted by radiated and conducted thermal energy from the arc. With metal droplets melting off the tubular metal sheath being projected past the unmelted core material, most of the metal-flux reactions must take place in the weld pool. Nevertheless, the FCAW electrode allowed copious introduction of ionizing, fluxing, slag-forming, and gas-forming ingredients into the arc, and this provided (1) a smooth, stable plasma, (2) control over the mode of metal transfer, (3) operation with alternating current, if desired, (4) a slag layer of whatever thickness was required to aid in controlling the contour of the weld deposit to ease all-position welding, and (5) simplification of the task of formulating a mixture for the core that was low in moisture (and other hydrogenous compounds). Regarding the low-hydrogen aspect, it must be kept in mind that, although the core material consisted of ingredients that initially were very dry, and likely had low hygroscopicity, the mechanical seam in the tubular sheath could be a point of entry for moisture over a period of time.

After more experience with the FCAW process, formulation of flux mixtures broadened into the two different kinds commonly designated as *gas shielded* and *self-shielded*, and which now are identified by the abbreviations FCAW-G, and FCAW-S, respectively. The latter designation signifies that no external shielding gas is provided to the electrode. Both types of FCAW electrodes offered the five advantages outlined above, and together they made FCAW electrodes a unique form of filler metal for open-arc fusion welding. Both kinds of FCAW electrodes were warmly received in industry, and a number of specific types were developed to provide the operating characteristics preferred by welders and welding operators, and to provide the weld properties sought by engineers. Standardized types of carbon steel, and of low alloy steel FCAW electrodes are covered by ANSI/AWS A5.20, and by AWS A5.29, respectively, and the AWS specification in force for carbon steel electrodes, issued in 1979, classifies twelve different types of electrodes on the basis of the following:

(1) whether CO_2 is used as a separate shielding gas,

(2) suitability for either single- or multiple-pass applications,

(3) type of current,

(4) position of welding, and

(5) as-welded mechanical properties of the weld metal.

Any electrode classified under one classification shall not be classified under any other classification of this specification.

AWS plans to issue an updated specification for carbon steel FCAW electrodes (ANSI/AWS A5.20-94) which will add two more types of electrodes, and also will allow three of the electrodes to be each classified under two classification; specifically, (1) the original classification that made use of 100 percent CO_2 shielding gas, and (2) a new classification that is to be identified by an "M" suffix on the AWS designation, which means that the electrode has been classified with a 75 to 80 percent argon/balance CO_2 mixture of shielding gas.

The core material formulation for a FCAW electrode is fully as complex as outlined earlier for the covering on SMAW electrodes, so it should be helpful to examine the metallurgical behavior of various types of FCAW electrodes both with externally supplied shielding by particular gases, and without external shielding (i.e., self-shielded). Examples of typical core material formulations for four types of FCAW electrodes are given in Table 8.5. Research on core material has been very actively pursued by the electrode manufacturers as evidenced by the proliferation of proprietary electrodes that are tailored to provide closely defined operating characteristics. Although designed to fill a particular application need, most of these electrodes are qualified to be designated under an AWS classification(s).

Gas-Shielded FCAW-G Electrodes

The AWS T-1 and T-5 type core formulations in Table 8.5 exemplify gas-shielded FCAW electrodes, and are designed to be deposited with externally-supplied shielding gas. Carbon dioxide is most often used for shielding because of its low cost. Plain CO_2 tends to promote globular transfer of metal from many of the gas shielded electrodes; however, for the spray transfer mode a gas mixture of argon and carbon dioxide may be necessary. The argon-containing mixture most commonly employed is 75 percent Ar and 25 percent CO_2. The higher the percentage of argon in the shielding gas, the higher will be the transfer efficiencies of alloying and deoxidizing elements from the electrode. This effect of shielding gas make-up on weld metal composition must be considered because unusually good recovery of alloying elements might result in unwanted high strength and hardenability in the weld metal, or excessively high levels of deoxidizing elements might seriously reduce weld toughness.

Gas selection for a gas-shielded FCAW electrode depends on the carbon content required in the weld metal. With 100 percent CO_2 shielding gas, the atmosphere of CO and CO_2 which normally forms in-and-around the arc can act as *carburizing*, or *decarburizing* depending on the level of carbon in the electrode and the weld deposit. If the carbon is very low, perhaps below 0.05 percent, the molten weld pool will tend to pick up a small amount of carbon from the shielding atmosphere and this carburization reaction may raise the weld metal carbon level to about 0.08 to 0.10 percent. On the other hand, if carbon is greater than about 0.10 percent, the molten weld pool is likely to undergo some loss of carbon. This can be of concern, particularly if the electrode is of the hardfacing variety and success depends on maintaining a high carbon content in the weld metal. Substitution of argon for a portion or all of the CO_2, decreases this tendency to alter carbon content, and the degree of mitigation depends on the percentage of inert gas in the gaseous shield.

Another aspect of gas shielding for FCAW electrodes is the uniformity of gas flow around the electrode and over the weld pool. The usual precautions to prevent accumulation of metal spatter or slag fouling the torch nozzle must be taken. Of greater importance, particularly when welding outdoors or in a drafty shop, is that wind flowing across the workpiece surface not deflect or distort the shielding and allow air to reach the melting tip of the electrode. This condition can develop without being noticed by the welder or an observer. Tests have shown that a side wind of only one m/sec (a little more than two mph) can have a significant adverse influence on the toughness of steel weld metal. Localized sampling of a CO_2 gas shield flowing at 15 liters/min (32 ft³/h) showed that a cross wind of 0.73 m/sec (1.5 mph) could dilute the

Table 8.5 – Typical Flux Core Formulas for Steel FCAW Electrodes

	AWS Designation			
	EXXT-1	**EXXT-3**	**EXXT-5**	**EXXT-8**
Type of Electrode	Gas Shielded	Self-Shielded	Gas Shielded	Self-Shielded
Essential operating characteristics and application	Spray transfer DCEP, suitable for multiple pass welding, medium slag	Spray transfer, DCEP, suitable for single-pass welding of thin base metal	Globular transfer, single- and multiple-pass welding, thin slag blanket	All-position deposition, DCEN, good low-temperature toughness

Core Ingredients	Percentage of Ingredients in Core by Weight			
Rutile	30	5	15	10
Fluorspar	10	25	25	25
Limestone		5		10
Feldspar	7	10	15	5
Cryolite	5	5	15	5
Dolomite	15			
Magnetite	3			
Sodium silicate	5		5	
Silica			5	
Ferrosilicon	5	10	5	10
Ferromanganes	10	15	10	15
Ferrotitanium		5		5
Ferroaluminum		20		15
Iron powder	10		5	

CO_2 shield in the vicinity of the electrode tip (extended 10 mm, or about ⅜ in.) with as much as 25 percent of air. While this dilution of the shield results in increased oxidation, the *nitrogen* pick up that occurs is even more damaging to weld metal mechanical properties. The amount of nitrogen in the weld metal will depend on many aspects of a given welding operation, and there is no driving force to eliminate nitrogen. The final weld metal nitrogen content is increased by (1) the amount available from melted base metal, (2) the amount of air that enters the arc and surrounding atmosphere, (3) greater stick-out of electrode beyond the gas nozzle, (4) higher arc voltage (i.e., longer arc), (5) higher welding current, and (6) a welding position or technique that entails wider electrode weaving. This sensitivity of the gas shielded FCAW process calls for assured protection against strong drafts or wind and consideration of procedural parameters because alloying elements with a strong affinity for nitrogen ordinarily are not present in the electrode in sufficient quantity to render it innocuous.

Self-Shielded FCAW-S Electrodes

Electrodes of the FCAW-S type are attractive for open-arc welding because external gas is eliminated, and the gun or torch is simpler than the kind used for gas-shielded welding electrodes. Each of the self-shielded electrode types must be selected with care, however, because deposited metal under some circumstances can be deficient in toughness. Although oxidation has been repeatedly mentioned as a problem that must be controlled when welding steel, *nitrogen pick up* and its effects actually present a greater challenge in the self-shielded process. Typical levels of nitrogen illustrate changes that can occur when FCAW-S electrodes are deposited as weld metal. Many steel base metals contain only about 0.005 percent nitrogen. Steel weld metal deposited by the SMAW-S process regularly picks up a small amount of nitrogen to reach approximately 0.015 percent. Gas shielded FCAW-G electrodes, even when deposited under recommended conditions, often deposit weld metal that is slightly higher, perhaps about 0.02 percent nitrogen. If the nitrogen content of steel weld metal exceeds about 0.03 percent, unless a strong nitride-forming alloying element is present to tie up the nitrogen, soundness and toughness may be adversely affected. Weld metal deposited by self-shielded FCAW-S electrodes that have a core not properly designed and formulated to deal with nitrogen can reach levels as high as 0.08 percent, which results in porosity and seriously impaired toughness. A maximum safe level of nitrogen cannot be suggested for deposits from FCAW-S electrodes because many features of the weld metal determine its sensitivity to nitrogen content. Present state-of-the-art FCAW-S electrodes deposit weld metal that has a nitrogen content of about 0.025 percent. Because this is a portentous level of nitrogen, the electrode maker often includes an element in the core mixture that has a strong affinity for nitrogen. Nitrogen-getters that have performed best with FCAW-S steel electrodes are aluminum, titanium, and zirconium. Unfortunately, when these elements are incorporated in the core mixture at the levels needed for adequate nitrogen-gettering, they have not been as innocuous as when employed for deoxidation. While most nitrogen getters are quite effective in preventing porosity, the weld metal toughness, remains inconsistent.

Self-shielded electrode deposition research reveals how the mechanics of electrode melting, metal transfer, and shielding control nitrogen pick up in the weld metal. Three particular areas in the FCAW-S process are involved in the absorption of available nitrogen. These areas have been identified as (1) the heated, but still solid exterior of the metal sheath just above the arcing tip, (2) the molten droplets of metal being transferred from the melting tip to the weld pool, and (3) the weld pool and still-hot trailing weld bead. The pick up of nitrogen at each area is influenced by a surprising number of electrode features and operating conditions related to shielding as summarized in the following paragraphs.

Self-generated gaseous shielding provided by the FCAW-S electrode must be used in moderation because an excess of gas-forming ingredients in the core mixture causes disruption of the arc and creates too much spatter. Organic materials for gas generation usually cause unstable arcing, and the hydrogen regularly present in such ingredients should be avoided. Limestone and other mineral carbonates are effective gas generators, but the total amount present in the core mixture usually must be limited to no more than approximately 15 percent. Despite the relatively small amount of self-generated gas shielding, drafts or wind are much less likely to affect welding performance than with gas-shielded FCAW-G electrodes. This advantage of the self-shielded type electrode has been closely studied, quantified, and documented because of its importance in the erection of steel framework for buildings, bridges, and other weldments fabricated outdoor.

Fluidity of melted flux and slag formed by the core mixture of FCAW electrodes is important to ensure good coverage over the heated arcing end of the electrode and the droplets being melted from the metal sheath. If the molten flux-slag material is too viscous, or is too fluid, proper coverage will not take place and both oxidation and nitrification will increase. The electrode manufacturer formulates the core mixture to obtain proper fluidity. This is not an easy task because of the range of operating parameters over which the electrode is required to perform satisfactorily. The core ingredients must melt and form a low viscosity fluid at a temperature of approximately 1000°C (about 1800°F) while inside the electrode sheath, but they must resist wholesale vaporization in the arc where the temperature exceeds 2800°C (about 5000°F). In addition to providing proper coverage during deposition, the core mixture must add ionizing elements to the plasma for smooth, stable operation of the arc, and eventually the solid slag that is formed on the weld bead must be easily removed. Formulations for the core of self-shielded FCAW-S electrodes rely heavily on fluorspar (calcium fluoride) for strong fluxing action, and for adjusting the fluidity of the resulting slag to secure the desired coverage.

Mode of droplet transfer from the FCAW-S type electrode is important in terms of nitrogen pickup. There does not appear to be unanimity of opinion among investigators regarding size of droplets to minimize nitrogen absorption during transfer from electrode to weld pool. Some state that a spray transfer of fine droplets will absorb less nitrogen, while others prefer larger droplets that almost border on globular transfer. Nevertheless, all agree that, if the molten drops grow so large that they exceed the diameter of the electrode and in so doing exceed the reach of good protection from self-generated gas shielding, then nitrogen pickup will be increased substantially. Investigators agree that coverage of the molten droplets (whether fine spray, or larger globules) by a slag of proper fluidity is critical in minimizing nitrogen pick up. Fine spray droplets are described as being more easily protected. Longer electrode extension often is employed to increase metal deposition rate in FCAW operations, but a consequence is to produce larger metal droplets. In one particular closely controlled experiment with FCAW-S electrodes, an extension of 15 mm (0.59 in.) beyond the contact tube produced steel weld metal containing 0.022 percent nitrogen. When the extension was doubled to 30 mm (1.18 in.) with the other parameters essentially constant, the weld metal nitrogen content was doubled (i.e., 0.042 percent). Aluminum present in the core mixture was transferred in the amount of 0.57 percent in the weld metal deposited with both the 15 mm and the 30 mm extensions. This level of aluminum is far in excess of the stoichiometric amount needed to tie up the nitrogen as particles of aluminum nitride (AlN), and as a result the amount of dissolved or free nitrogen in each case was less than 0.0002 percent. Further findings from this experimental work will be given shortly when discussing the microstructure of weld deposits and their nonmetallic inclusions.

Techniques for deposition of self-shielded FCAW-S electrodes influence nitrogen pickup. Lower nitrogen levels in weld metal are achieved by using the shortest practical electrode extension and the shortest possible arc. The latter minimizes, of course, the time-of-flight of molten particles from electrode to weld pool thus reducing exposure to nitrogen-bearing plasma and atmosphere. Position of welding affects weld metal nitrogen content with vertical welding inviting greater pick up. Interestingly, overhead welding produces weld metal with least nitrogen pick up; apparently this occurs because the protective gases generated are hot and rise upward to hug the workpiece surfaces, and because the weld pool must be kept very small. Overhead welding reduces the exposure of molten metal to available nitrogen. Deeper weld penetration reduces the nitrogen content of the weld metal by dilution; this assumes that the base metal contains very little nitrogen. Incidentally, FCAW-S electrodes do not produce deep penetration welds.

Composition of weld metal from FCAW-S electrodes, as mentioned earlier, ordinarily will include one or more strong "nitrogen-getter" alloying elements (e.g., aluminum, titanium, zirconium). The amount contained in the core mixture and transferred into the weld deposit usually exceeds the stoichiometric amount needed to tie up the residual nitrogen because of uncertainty over the final total nitrogen likely to be encountered. It is not unusual when aluminum is used to find it in the weld metal in the range of ½ to one percent. With aluminum present to this extent, the amount of free nitrogen in the weld metal is likely to be only one or two parts per million. Even with the nitrogen tied up virtually completely as a nonmetallic compound, the toughness problem can be of concern.

Microstructure and cleanliness of weld metal from self-shielded FCAW-S electrodes have been examined metallographically to seek an answer to the toughness problem. No significant difference in microstructure was found when weld metal with low residual nitrogen (<0.02 percent) was compared with weld metal containing a relatively high nitrogen content (about 0.04 percent). Although weld metals containing the low and high nitrogen contents both contained numerous small (approximately 0.10 to 0.98 μm) aluminum nitride inclusions, the high-nitrogen weld metal exhibited more than half of its volume of nitride particles as larger inclusions (with a diameter greater than one micron). Because nonmetallic inclusions are known to play a role in fracture mechanisms, the conclusion was reached that microvoid coalescence occurred more readily at the larger nitride particles and this provided a greater number of nucleation sites for fracture initiation. Small nitride particles, less than about 0.75 μm in diameter, even when present in larger number, were not as damaging.

Submerged Arc Welding (SAW)

The shielding provided in submerged arc welding (SAW) may appear to be relatively simple, straightforward slag blanketing, but this part of the SAW process has developed into a refined, complex technology. As indicated earlier in Chapter 6 when describing the SAW process, the user regularly deals with at least three independent components that unite or commingle to make up the composition of the weld metal, namely; (1) base metal, (2) electrode, and (3) flux. Furthermore, the proportion of metal contributed by each of these components will depend on many details of the procedure applied. There is greater freedom-of-choice in planning a procedure for SAW than any other arc process. The user must understand the likely effects of decisions made regarding welding consumables and operating parameters. To cover the SAW process, electrodes will be reviewed in Volume II, Chapter 10; and base metal and weld compositions in Chapters 14 and 15. Here, in Chapter 8, the granulated material commonly called "flux" for shielding will be examined. Flux complexity has increased over past years, and research has provided insight on the actions and reactions that take place as this component performs its functions during welding.

Types of SAW Fluxes and Their Classification

A number of technical organizations around the world, including the AWS and the IIW, have proposed systems for the classification of SAW fluxes and abbreviations for their designations. The designations tend to be unwieldy because of (1) the very wide variety of flux compositions, (2) the unavailability of information on most of the fluxes regarding their composition or the ingredients, (3) the need to consider a particular electrode along with the given flux to satisfy a set of requirements for which the flux-electrode combination is designed.

The AWS Committee on Filler Metal found classification of SAW fluxes by chemical composition to be unworkable for specification purposes, and instead, devised a system under which a flux can be given designations that are indicative of weld deposit composition, mechanical properties, and other weld metal features obtained when used in combination with different AWS standard classifications of electrodes under specific test conditions. With this AWS system, a flux may carry a number of AWS classifications; each indicating conformance with prescribed requirements when teamed up with a specific standard electrode classification. AWS has issued separate specifications for electrodes and fluxes used for SAW of carbon steel (ANSI/AWS A5.17), and for low alloy steel (ANSI/AWS A5.23), but these documents cover only a portion of the electrode-flux combinations being used in industry. Many of the fluxes are proprietary formulations that are designed to be used with a specific electrode composition (which also may be a proprietary type of steel) to meet the requirements of a particular kind of service or an important application. The AWS system will be outlined in Volume II, Chapter 10, when SAW electrodes are reviewed.

Fluxes for SAW are made up mostly of oxides which when melted together form a substantial layer of slag that completely blankets the arc and weld pool, and the trailing weld bead. The earliest SAW fluxes were relatively simple mixtures of granulated metal oxides, and often included limited amounts of powdered metals and alloys for deoxidation, and for modest alloying of the weld deposit. These mixtures were intended to be free of gas-forming ingredients because the arc between the electrode and weld pool was completely submerged in the molten blanket, and a protective gas shield was not needed. Today, the composition of a flux for SAW can be equally as complex as a covering on a SMAW electrode or the core mixture in a FCAW electrode. There is added reason, however, for gaining a deeper understanding of ingredients commonly used in fluxes for SAW and their functions and effects because of the more frequent, direct involvement of the user in flux selection and careful planning of welding parameters. The oxides presently used in SAW fluxes are mostly those of metals, such as calcium, potassium, sodium, silicon, aluminum, titanium, magnesium, manganese, and zirconium. The sources of these oxides can be naturally occurring minerals, beneficiated materials, technical-grade compounds, and chemically pure reagents. Many useful SAW flux ingredients already appear in Table 8.2. Others being used by flux manufacturers are not made public knowledge because this business is highly competitive and flux formulae are trade secrets. Patent literature gives only a vague description of commercial flux compositions.

As a flux is formulated by selecting mineral ingredients, metallic materials, and other additives (which will be mentioned later), the finished product usually falls into one of the following broad categories which are either standard AWS terms or commonly-used expressions intended to give preliminary insight into the general nature of the flux.

Neutral SAW Fluxes

Fluxes that do not contain additions of metals, alloys, or compounds that increase the levels of silicon, manganese, or other alloying elements in the weld deposit to any appreciable extent are considered to be neutral. Whether small changes occur in certain alloying elements, such as carbon, will depend on the oxides and other ingredients that constitute the flux and how the given formulation reacts at different levels of temperature during the welding operation. When a neutral flux is employed, deoxidation and any required alloying of the weld deposit is usually accomplished through the electrode composition. Of course, if the steel being welded has a large percentage of an element that is wanted in the weld deposit (e.g., manganese), the base metal could be the source for the increased amount providing weld penetration and resulting dilution are properly planned and accomplished.

Although neutral fluxes are intended to allow the deposition of carbon steel weld metal without significant composition changes, especially in the manganese and silicon contents of the weld metal, there can be exceptions to this expected performance. Some neutral fluxes decompose to some extent after being melted by the arc and they release oxygen, which can result in lower recoveries of carbon, manganese and silicon. On the other hand, some neutral fluxes that contain manganese silicate can undergo enough decomposition of this ingredient to increase the manganese and silicon contents even though no metallic form of these elements was present in the flux formulation. These changes may be consistent over a wide range of welding parameters; although the arc voltage during welding can often be an influential factor. Any changes in the carbon, manganese and silicon levels in carbon steel weld must be carefully evaluated because of their relationship to the mechanical properties of the weld. A useful index for flux neutrality that can foretell changes in manganese and silicon levels is known as the *Wall Neutrality Number*, which is described in Technical Brief 42.

Neutral fluxes offer the advantage of maintaining more-or-less unchanged weld metal composition even over a range of SAW operating conditions; however, their selection must be governed by conditions to be accommodated in the joint to be welded and the welding procedure. Since neutral fluxes contain little or no deoxidizer, the deoxidation capability must be provided by the electrode and/or the base metal composition. Therefore, close watch must be maintained for potential problems with porosity and weld cracking. For example, if single pass welds are made on steel plate that has a heavy oxide scale on its surface, the likelihood of having some porosity and blowholes in the weld will be quite strong.

Another advantage of a neutral SAW flux is that the composition of the metal being deposited is not subject to marked variation by changes in the quantity of flux melted per unit volume of melted electrode. Ordinarily, for each unit weight of electrode melted, an equal weight of flux is melted, but this one-to-one relationship between flux and electrode is often altered by the particular welding parameters applied. An increase in arc voltage, or a decrease in current during welding will result in a greater amount of flux melted and the flux-to-electrode ratio will exceed 1.0. On the contrary, if the arc voltage is lowered (i.e., arc length becomes shorter), or the welding current is increased, the flux-to-electrode consumption ratio will decrease, possibly to as low as 0.7. In any given SAW application, knowledge of the flux-to-electrode ratio is a first step toward close prediction of the composition of deposited weld metal. This ratio assumes even greater importance as more active fluxes, and especially the alloy fluxes are employed.

Active SAW Fluxes

Active fluxes contain significant amounts of silicon, and possibly manganese, usually as ferroalloys. The purpose of these additions is to reduce propensity to porosity in the weld zone by ensuring adequate deoxidation of metal in the weld zone. A manganese addition can be helpful in avoiding hot cracking susceptibility; particularly when welding steel that has an elevated residual sulfur content. Active fluxes are used often with electrodes that have low manganese and silicon contents, which can be the case for electrodes made from wire of the rimmed or semiskilled types of steel. There can be a worthwhile cost incentive to provide the necessary percentages of manganese and silicon via an active flux rather than obtain them by using a killed steel wire. Of course, when an active flux is employed for the aforementioned purposes, or for increasing the strength of the weld, consideration must be given to a number of conditions that will arise during the welding operation; including (1) the flux-to-electrode melting ratio, (2) transfer efficiency of the needed elements from the ingredients included in the flux, and (3) the amount of base metal that will constitute a part of the weld zone (i.e., percent dilution). From earlier remarks regarding the use of neutral fluxes, it should be understandable that active fluxes are favored for welding heavily rusted or scaled steel. The quantities of deoxidizing agents added to the flux usually can be adjusted to take care of the amount of iron oxide picked up from the surfaces of the steel which are actually melted during welding.

Alloy SAW Fluxes

So-called "alloy fluxes" contain substantial amounts of metallic ingredients to add alloying elements to the composition of the weld zone. The alloy additions may be relatively small as might be required to form a low-alloy steel weld, or the additions may be quite large. Alloy fluxes are widely used. They also offer cost incentive. Some fabricators regularly employ a single, standardized type of carbon steel electrode for many different weldments that require alloy steel weld metal. They have an alloy flux specially mixed for each application to provide the specified alloy steel weld composition, and to accomplish proper deoxidation of the weld metal. This practice demands good knowledge of transfer efficiencies; and close control of joint preparation and welding parameters to ensure correct content and homogeneity of alloying elements throughout the length of all weld beads. Despite the vigorous electromagnetic stirring that takes place in the weld pool, large amounts of metals incorporated in the flux may not be distributed uniformly in weld deposits if penetration is deep and if travel speed is high. Under these welding conditions, a slightly higher concentration of elements being added through the flux may occur near the weld surface as compared to the fusion line. To avoid this heterogeneity, weld metal that secures its alloy content through the flux should be deposited as relatively shallow beads and very high travel speed should be avoided. Arc length (i.e., arc voltage) should be held as constant as possible because this operating parameter determines the amount of flux melted in

---

TECHNICAL BRIEF 42:
The Wall Neutrality Number

The Wall Neutrality Number (N) is an index for flux neutrality that can be used for guidance in anticipating changes in the manganese and silicon contents of carbon steel weld metal when the flux is employed with commonly used SAW operating parameters. The "N" index is determined by a procedure that is detailed in ANSI/AWS A5.17 whereby the neutrality of the flux is assessed by making two weld pads through stipulated procedures which differ only by having the second pad made with an arc voltage that is 8 V higher than the first. These pads are chemically analyzed for manganese and silicon contents to ascertain differences between them; regardless of whether increases or decreases are found. The value on N is calculated by the following equation:

$$N = 100 \left(|\Delta\% \text{ Si}| + |\Delta\% \text{ Mn}| \right) \quad \text{(Eq. 8.1)}$$

where
   $\Delta\%$ Si is the change in weight percent of silicon, and
   $\Delta\%$ Mn is the change in weight percent of manganese.

The Wall Neutrality Number is the absolute value (ignoring positive or negative signs). A flux-electrode combination that produces weld metal composition changes that equate to a N value 40 or lower is considered to signify a virtually neutral flux, whereas N greater than 40 indicates an active flux. The Wall Neutrality Number does not address alloy fluxes.

proportion to the quantity of electrode deposited, and unregulated changes in flux-to-electrode melting ratio can cause appreciable variations in weld composition.

Methods of Manufacturing SAW Fluxes

Manufacturing methods play a part in shaping the overall character of SAW fluxes. The method used to put the raw ingredients together, and the nature and size distribution of the granular particles, will be influential during welding. Very fine particles tend to reduce the volume of flux melted, and will restrict the escape of gases from the melting flux and the molten metal. As the ingredients of the flux melt and commingle, the molten slag produced will have a number of important properties. These properties include freezing temperature, viscosity, basicity, and chemical activity with the molten metal from the electrode and the weld pool. These properties become factors in controlling important aspects of the weld, such as general shape, penetration, surface contour, undercutting, etc. Manufacturing methods for making flux have acquired five frequently applied labels that indicate how the ingredients were put together, namely: (1) fused flux, (2) agglomerated flux, (3) bonded flux, (4) mechanically mixed flux, and (5) recycled flux. The terms "agglomerated" and "bonded" do not always carry the same meaning in technical literature and flux advertisements. The manufacturing method builds in certain features that often require consideration during the planning of a SAW operation. The following brief descriptions provide general guidance.

Fused Fluxes

Fused fluxes are made by melting the entire mixture of ingredients in a furnace at 1500°C (about 2700°F) or higher. The molten material is stirred for homogenization, and then solidified by quenching in water or by chilling on a metal plate. The cake-like fragments are crushed to powder and screened through standard sieves to secure granular particles within a prescribed size range (e.g., 12 to 200 mesh). Although this is a costly manufacturing method, the flux has the attribute that each particle is homogeneous and of the same composition. Because the particles are glass-like solids, they are not especially hygroscopic. The flux is easier to keep dry which helps to avoid the cold cracking problem when welds are made in hardenable steels. However, fused fluxes do not necessarily produce welds with the lowest hydrogen content. Research has shown that the fused silicate lattice may contain trapped hydroxyl (OH) ions. Although only limited amounts of these ions are present, nevertheless they contribute to the hydrogen that enters the weld metal under the submerged arc.

Fused fluxes usually fall into the "neutral" flux category because melting the mixture during manufacturing would oxidize most metallic ingredients if they were included. Consequently, metallic ingredients are not included in the mixture before fusing since they would not be free later to perform as a deoxidizer or an alloying element. Of course, powdered metals and alloys can be added to the prepared fused flux to change it to the "active" category, but this practice must include thorough mixing of the metallic addition into the bulk, and stringent precautions that no settling of heavier metallic particles occurs during transport or handling of the flux.

Fused fluxes are manufactured sometimes with substantial additions of manganese silicate, and/or calcium silicate plus silica, as an alternative to making use of the oxidizable ferroalloys of manganese and silicon. When a flux which includes these compounds is fused, the fusion operation does not hinder their action to cause modest increase in the weld metal content of manganese and/or silicon. Consequently, fused fluxes of this nature can be suitable for joining some steels with the SAW process using an electrode made of rimmed or semikilled steel.

"Foamed flux" is a variety of fused flux that has been used to some extent for SAW. Foamed flux is made by including ingredients during manufacture that cause foaming during solidification from the molten state. The crushed and screened product has relatively low density, and lower thermal conductivity. Because of these properties, a lesser proportion of foamed flux is melted during the SAW operation.

Agglomerated Fluxes

Agglomerated fluxes are not pre-fused; instead, the mixture of ingredients receives a ceramic or a mineral material addition as a cementing agent. To accomplish cementation, the mixture with the agent present may require heating to about 600°C (about 1100°F). This temperature allows inclusion of metallic ingredients without significant deterioration by oxidation. Consequently, agglomerated fluxes may be of the neutral, active, or the alloy variety as desired. After cementation, the cooled cakes of flux are crushed and screened to produce the specified range of mesh sizes.

A number of reasons can be given for agglomerating the ingredients in a flux. Regardless of the particle sizes of the ingredients (some might be extremely fine powder), the agglomerated flux has all constituents, both metallic and nonmetallic, present virtually uniformly in particles of usable sizes. There will be no tendency for heavier ingredients to settle and cause the flux to become non-uniform, and there will be no particular problems with including very fine powder materials. Agglomerated fluxes perform well at high welding current, but not as well as the fused fluxes.

Bonded fluxes also make use of binding agents to hold all ingredients together until solid cakes of flux can be crushed and screened. The bonding materials used most often are the aqueous silicates of sodium and potassium. The dry ingredients, including metallic materials, are thoroughly mixed with the liquid silicate and then dried by baking at approximately 300°C (about 600°F) or higher. The dried, bonded cake is crushed and screened just like the agglomerated flux. Usually, crushing is a little easier because of the nature of the bonding agent. Bonded fluxes sometimes have a greater tendency than agglomerated flux to pick up moisture; therefore, bonded flux may require better protection during storage and handling.

Mechanically Mixed Fluxes

Mechanically mixed fluxes can be prepared simply by stirring dry ingredients together until the mixture is uniform. The mixing operation normally does not change the particle sizes of the ingredients. Often, mechanically mixed flux particles have a wide range of mesh sizes. Ingredients must be selected carefully so that very coarse and very fine particle materials do not constitute too much of the total flux because this could interfere with proper SAW operation. Also, mechanically mixed fluxes must be handled and stored without exposure to sustained vibration that could cause settling of heavier ingredients (usually the metallic ingredients). This can cause serious heterogeneity in the flux and resulting troublesome composition variations in the weld metal. Mishandled flux can become so non-uniform because of settling that even the SAW operation will appear abnormal. Adverse operating conditions — such as erratic changes in arc stability, spasmodic undercutting, irregular bead appearance, etc. — indicate that a non-uniform flux is being used, and that some variation in weld metal composition is likely to be occurring. Obviously, these are some of the reasons for the development of fused, agglomerated and bonded fluxes.

Recycled Fluxes

Recycled fluxes represent an unusual category. This label is not applied to the *unmelted* flux that can be retrieved from a completed SAW weld pass as is frequently done in many fabricating operations. Yet this action to conserve a consumable may lead to difficulty if the character of the flux being used is not given proper consideration. Many fabricators have systems for removing the excess flux from a weld just completed, and for transporting it to a hopper from which further distribution can be made to points for subsequent recovery. This recovery operation normally includes screening the flux being collected to remove any pieces of fused slag which broke loose from atop the weld bead. Precautions to be taken in collecting and transporting the unmelted granular flux to the hopper will depend on the kind of flux being retrieved; this refers to the general category (i.e., neutral, active, or alloy) and the method of manufacture (i.e., fused, mechanically mixed, etc.). Settling or separation of heavier particles can occur in certain fluxes as they are transported to the hopper by vacuum, compressed air, or vibratory means incorporated in the recovery system. Remixing of the collected flux is one corrective step that might avoid the problem.

The label *recycled* ordinarily is applied to a flux that is reconstituted from a fused slag that is taken from the surface of SAW welds, and which is recrushed and screened to the desired particle sizes. This reprocessing usually is performed by a company that specializes in the operation because, in addition to the crushing and controlled screening required, the weld-fused slag ordinarily will require corrective additions. Most of the oxide contents in the slag will be reasonably close to those of the original slag, and this provides a base upon which to rebuild a usable flux. Large-scale users of SAW are attracted to recycled flux because of savings through return of their fused slags to a recrusher company.

Admixture of recrushed slag to reconstitute a usable flux requires knowledge of the composition of the slag, and clear objectives for the final composition and nature of flux required for further SAW operations. Recycled fluxes should be identified with the designation of the original flux, and a full description of the recycled product. Some codes prohibit the use of recycled flux, while others limit the proportion that can be mixed with original unused product.

Physical Chemistry of Fluxes in the SAW Process

The above descriptions of flux categories and of manufacturing methods provide a general perspective which now can be sharpened by examining physical chemistry and metallurgical aspects of different fluxes. The chemical reactions that occur between flux or slag and the molten metal in the SAW process are as important as those in steelmaking. While some of the reactions that occur during steelmaking also take place to a degree in SAW, the welding process involves (1) shorter time span for the reactions to take place, (2) greatly increased exposure of metal to flux or slag because of droplet transfer, (3) higher peak temperatures during contact between flux or slag and the metal and (4) intimate mixing because of weld pool stirring. The mechanics of deposition during SAW entails droplets of molten metal from the tip of the electrode passing through or around plasma that is submerged within the molten slag just above the arc cavity that shapes the weld pool crater. When the SAW process was introduced almost 60 years ago, users were pleased to find a solid metal weld with a protective slag as a layer over the entire weld face. Yet, with continued application of the process and increasing demands concerning weld composition and properties, strong need arose for studies of the deposition mechanism in minute detail, and in-depth rationalization of the fluxes that formed the protective slag. Significant findings to date will be summarized in this chapter.

The early fluxes employed for SAW were "acid" in composition — or, at least, they were judged to be so from knowledge acquired in steelmaking (see section titled, "Significance of Acid and Basic Steelmaking," in Chapter 5). Fluxes of this kind appeared quite satisfactory in view of their ability to perform the shielding function rather well even with high welding currents. This capability facilitated good weld penetration in thick sections, and assisted in gaining higher metal deposition rates. The solidified acid-type slag could be easily lifted from the weld face leaving the surface quite clean albeit temper colored with very thin oxide film. As the use of SAW was extended from welding steels with only modest toughness requirements to more demanding applications requiring high toughness, it became evident that a better understanding of electrode deposition mechanics and the transfer of elements between flux or slag and the weld metal was sorely needed. This was required, not only for closer control of alloying elements, but also because minor elements transferred from the flux appeared to be involved in establishing the toughness of weld metals.

Deposition of steel electrodes during SAW have been studied by cineradiology techniques to determine the sizes and shapes of droplets leaving the electrode tip and passing through the arc to the weld pool. The time-of-flight of the droplets was also measured. Lower welding currents form larger, fluctuating-shape droplets that tend to slide down the walls of the arc cavity or crater into the weld pool. Higher welding currents tend to form much finer droplets that stream from a pointed electrode tip and transfer directly into the molten weld pool. Although the time required for a droplet to accomplish this transfer is extremely short, perhaps less than a tenth of a second, there is a time difference between large and small droplets (because of employing lower, or higher welding current, respectively).

Elements and compounds from the flux and the electrode that enter the plasma become highly ionized, and reactions between these ions and molten metal droplets occur at an extremely high rate. Nevertheless, the reactions are not ordinarily given sufficient time to proceed to completion or equilibrium as is the case in steelmaking operations. During the very short welding exposure of molten droplets to the intensely high temperature of the plasma, the thermochemical reactions between flux (or slag) and the metal droplets are supplemented by electrochemical effects generated by the high density of flowing electric current, and also by a polarity influence when welding is being performed with dc power. This initial period of extremely rapid and complex reactions is followed by a period of possibly several seconds during which the molten slag and metal are in contact, but now the intimacy of contact is greatly lessened through normal density-driven or gravitational separation and coalescence of these immiscible liquids into a layer of molten slag atop a layer of molten metal. During this period, the temperature has dropped into the steelmelting realm, and this has encouraged some investigators to apply steelmelting chemistry in their studies of the SAW process.

As the temperature of a newly formed SAW weld pass continues to decrease, the metal solidifies first, and is followed by hardening of the slag as an undercooled liquid (although crystalline solid can form in some kinds of slags). This is the point in the process where bonding starts to develop at the interface between the metal and the slag, and the nature of this bond determines the eventual ease (or difficulty) of detaching the slag and cleaning the weld surface, which can be a practical problem of considerable importance.

Some investigators view the SAW process as having a range of "effective reaction temperature" of about 1700 to 2000°C (approximately 3100 to 3600°F) as deduced from studies of mass action and equilibrium constants. This probably is a realistic approach to slag-metal chemical activity at the lower temperatures where kinetic interactions are likely to predominate, but the wide range calculated for effective reaction temperature suggests that many of the reactions analyzed in this manner had not achieved a state of equilibrium. Consequently, continued attention is being given to the thermodynamics envisioned at plasma (arc) temperatures, and to observed electrochemical effects when rationalizing the transfer of elements between slag and metal. Of course, these uncertainties regarding actual conditions of temperature, time, and electrodynamics that prevail during the initial or formative stage of weld composition are compounded by variations induced through the different welding parameters employed in SAW procedures, and through the wide variety of materials involved (i.e., fluxes, electrodes, and base metals).

During early usage of SAW, there were disquieting reports of low toughness in welds made by the process, even in the carbon steels. In addition to studies initiated on weld microstructure to determine possible reasons for this property shortcoming, manufacturers of SAW fluxes began to explore a wider variety of compounds for flux formulation. This effort produced fluxes that formed a slag that was either less "acidic," or was classed as "basic" (at least, according to steelmelting parlance). Weld metal produced with SAW fluxes of the latter character were found to have improved toughness, and this finding encouraged continued effort to solve the toughness problem through flux formulation. Understandably, attention soon was centered on the *basicity* of fluxes — a characteristic which is discussed further in Technical Brief 43.

Compositional Classification of SAW Fluxes

Analysis of a SAW flux is laborious because of the necessity to differentiate between elements present in metallic form, or as a nonmetallic compound. Adding to this analytical difficulty is the matter of the source of various compounds found by chemical analyses. The particular raw materials selected by the flux manufacturer to provide the needed compounds may also include other elements and compounds. While these may not be requisites, they may contribute to the qualities of the flux. On the other hand, these subordinate elements and compounds may only be "hitch-hikers" that simply add to the complexity of the analytical results. Analyzing and characterizing commercial SAW fluxes has been so onerous a task that the U.S. Navy recently

TECHNICAL BRIEF 43:
Assessing Basicity of SAW Fluxes

Early attempts at basicity assessment for SAW flux was largely a judgmental appraisal based on a few established facts concerning the behavior of compounds when heated to very high temperatures, as surely would be the case in a welding arc. Compounds known to form an acid slag included the oxides of silicon, titanium, and zirconium. Basic slag-forming oxides included those of calcium, magnesium, manganese, iron, potassium and sodium. Alumina (Al_2O_3) was known to be amphoteric inasmuch as it acts as an acid oxide when incorporated in a basic slag, but will exert basicity in an acid-type slag.

From a fundamental physical chemistry viewpoint, basicity is determined by the ease with which oxides dissociate into a metallic cation and an oxygen anion. Oxides that dissociate easily form a basic slag. Silicon dioxide (SiO_2, silica sand) is the classic example of an acid ingredient because of its reluctance to dissociate when melted. Conversely, calcium oxide (CaO_2, lime) is a prime basic flux ingredient. Calcium fluoride (CaF_2, fluorspar) had been viewed as a basic compound in fluxes, but this potent slag-thinner is now regarded as virtually neutral. Whether a flux-slag is acid, is basic, or is virtually neutral can be approximated by calculating the ratio of the total amount of basic components to the acid components. A ratio less than 1.0 might indicate an acidic flux or slag, whereas a basic flux or slag would have a ratio that exceeds parity. However, this simple assessment does not include adjustment on a molar basis; nor does it recognize the relative powers of the oxides included in the computation. Discernment of basicity potentials for compounds has been given attention by a number of investigators and will be reviewed shortly.

Basic fluxes were generally found at the onset of development to have operating characteristics not as smooth and desirable as the long-used acid type fluxes. The basic fluxes were not as amenable to handling high welding currents. However, this operating performance did not deter their use when good toughness had to be secured in the weld metal. Often the welding procedure called for relatively small, multiple passes to be made with low welding current as further assurance that toughness would be favored. A practical drawback to the use of a basic flux was the significantly greater effort required to clean the solidified slag from the weld, especially when the face of a weld pass was within a grooved joint. Nevertheless, the improvement in weld toughness gained by the basic type of flux provided ample incentive to continue research and development in the direction of basic flux formulations.

When the user of SAW with concerns regarding weld toughness turned to the task of finding a "basic" type of flux that would satisfy requirements, the search often proved to be frustrating. Suppliers of flux did not regularly reveal the chemical makeup of their products nor did they give an indication of flux basicity. The suppliers preferred to describe their flux in terms of operating qualities for particular kinds of SAW applications. This marketing policy prevails in much of the current sales practice. Yet, information on flux basicity is sometimes vital to the success of SAW in fabricating a steel weldment, and it may have to be obtained by inquiry to the manufacturer, or by chemical analysis.

sponsored a project at National Institute of Standards and Technology (NIST) to compare the suitability of x-ray fluorescence, optical emission spectroscopy, and x-ray diffraction for analyzing fluxes. This work, reported in the open literature, also included sieve-analysis procedures. A search was made for particle segregation during bag filling and from shipping transport, but none was found.

As the variety of SAW fluxes grew to fill the broadening usage of the SAW process, technical organizations were pressured to come forth with an orderly classification system that would guide users in selecting the type of flux best suited for depositing weld metal that would satisfy all requirements. The diversity in combinations of compounds that constitute the better part of each flux made classification difficult. In 1974, the IIW proposed a classification system based on five categories of generic-like types of SAW flux for welding steel as outlined in Table 8.6. Although this IIW system is limited both in scope and in detail, it continues to this day to be useful in a rudimentary way. The principal compounds in the five kinds of IIW classified fluxes range from a typical acid calcium silicate formulation (CS), through less acidic formulations that also include oxides of calcium, manganese, and magnesium, etc., and finally to a strongly basic formulation (BF). In addition to the efforts toward orderly classification of flux types, much research was being conducted on representative fluxes of the various types, especially with regard to their control of weld composition, and the toughness of weld metal produced. This activity brought forth the general finding that the basicity of a SAW flux exercised significant influence over the transfer or exchange of certain elements between flux-slag and the weld metal. This influence apparently was an important factor in the amount of oxygen gained and/or retained in the weld metal, and high oxygen levels were associated with poor weld toughness. Furthermore, higher oxygen levels were found in weld metals deposited with the more acidic fluxes.

At this point in studies of SAW fluxes, there was agreement that their basicity exerted important effects on (1) operating characteristics during the process (2) composition of the weld metal, including the levels of residual or minor elements (e.g., oxygen), (3) mechanical properties of the weld metal, especially its toughness, and (4) ease of slag removal from the weld surface. Because adverse results in any one of these aspects of a SAW application can be very troublesome, it was obvious that some means had to be devised for quantitative assessment of flux basicity, and to determine the degree of correlation between this property and the aforementioned conditions or properties on which improvement was sorely needed.

Computed Basicity Assessment of SAW Fluxes

A number of empirical formulas for quantifying the basicity of welding flux appear in the literature, but perhaps the one most frequently employed to evaluate a SAW flux is the following as proposed by Tuliani, Boniszewski and Eaton:

$$BI = \frac{CaO + CaF_2 + MgO + K_2O + Na_2O + Li_2O + BaO + SrO + \frac{1}{2}(MnO + FeO)}{SiO_2 + \frac{1}{2}(Al_2O_3 + TiO_2 + ZrO_2)}$$

(Eq. 8.2)

Table 8.6 – IIW Classification of Fluxes for Submerged Arc Welding of Steel*

Symbol	Principal Compounds	Common Name to Designate Kind
CS	CaO + MgO + SiO$_2$ > 50%	Calcium silicate
MS	MnO + SiO$_2$ > 50%	Manganese silicate
AR	Al$_2$O$_3$ + TiO$_2$ > 45%	Aluminate rutile
AB	Al$_2$O$_3$ + CaO + MgO >45% (Al$_2$O$_3$ 2 ≥ 0%)	Aluminate basic
BF	CaO + MgO + MnO + CaF$_2$ > 50% (SiO$_2$ ≤ 20%)	Basic fluorides

*Classifications as proposed in IIW Document X11A-88-74

where

BI is a "basicity index," and

values for chemical compound components are mass percent.

When BI determined with Equation 8.2 is less than one, the flux is acid. When BI is 1.0 to 1.5, the flux is essentially "neutral" meaning that it is neutral with respect to basicity (not the neutral category described earlier for fluxes devoid of metallic alloying and deoxidizing ingredients). As BI exceeds 1.5, the term "semibasic" is sometimes applied until an index of about 2.5 is reached. Fluxes with a BI above the 2.5 level are commonly called "basic" or "fully basic."

To illustrate the range of basicity found among the types of commonly used SAW fluxes, as indexed by a BI value, Table 8.7 presents information on typical fluxes patterned after analytical determinations made on flux from several manufacturers. These examples illustrate the diversity of compositions employed as a base for obtaining certain operating characteristics, and the differences in basicity (as indicated by their BI values calculated with equation 8.2). Flux "CS" has the strongest acid formulation with its BI calculated to be 0.45, is typical when calcium silicate is the base ingredient. The most basic flux, "BF", has a BI of 3.35. SAW flux formulations with a base mixture of CaO, MnO and CaF_2, and with less than half of the total ingredients consisting of SiO_2, can have a BI value of 4.0 or higher if wanted. However, while very high flux/slag basicity may favor certain compositional aspects of the final weld and

Table 8.7 – Typical Flux Compositions for Submerged Arc Welding of Steel

Examples of Slag-Forming Portion Only

IIW Designation	CS	MS	AB	CS	BF
Description	Strongly acid, calcium silicate	Weakly acid, manganese silicate	Neutral, aluminate basic	Neutral, calcium silicate	Basic fluoride
Manufacturing Method	Bonded	Bonded	Bonded	Fused	Agglomerated
Attributes	Tolerates high current. Easy slag removal	Use for fast weld travel speed	Can use on steel with rust or mill scale if formulated "active"	Helps gain good weld metal toughness	Helps gain good weld metal toughnes
Disadvantages	High oxygen pick-up	Works best on relatively clean steel	Use only with dc power	Slag removal more difficult	Slag removal more difficult

Composition of Nonmetal Portion of Flux (wt %)

CaO	15		3	30	18
CaF_2	5	4	12	10	28
MgO	3	15	16	5	22
Fe_2O_3	1	3	2	2	2
MnO	2	28	10	4	4
Na_2O	1	1	1		
SiO_2	40	45	16	40	14
TiO_2	15			4	2
ZrO_2				2	2
Al_2O_3	18	4	40	3	8
Basicity Index (BI)	0.45	0.75	1.06	1.09	3.55

enhance its toughness, the flux may not be satisfactory from standpoint of operating characteristics, weld porosity, and slag detachability.

Optical Basicity Assessment of SAW Fluxes

Optical basicity is another method of evaluating the basicity of a molten slag. Reports in the literature deal mostly with studies of slags used in metal melting processes. Determination of optical basicity for welding slags to secure an "OB index" is a more recent application. Proponents of the OB method claim that a non-zero index is obtained for a flux or slag that correlates more closely than the BI from Equation 8.2 with a number of aspects of weld metal composition, and with weld metal impact toughness. The optical basicity method uses spectroscopic analysis of compounds derived from a fused flux to establish the identity of the compounds present, and then proceeds with ionic theory to assign finite basicity values to each. Equations are next employed to weigh the finite values for the oxides and arrive at a basicity index for the lot. Because spectroscopic determinations are part of the procedure, this basicity determination is commonly called the "optical basicity" index.

Briefly, the OB index is based on the interactions that occur after dissociation of compounds in a molten flux or slag between the electronegatively charged oxygen anions and the electropositively charged metal cations. The electron donor power of the oxygen anion is considered in relation to a probe ion that is added to allow a spectroscope to observe the "red" shift in the UV bands of the probe ion. This shift results from electron donation by the oxygen anion, and is caused by expansion of the s-orbital, which in physical chemistry is termed the *nephelauxetic* effect. Measurements of this spectral shift have been translated into a scale for the OB index. If a comparison is made of indexes obtained with the empirical BI equation in Equation 8.2 and the OB index on the same fluxes, BI of 0.03 (strongly acid) would correspond to an OB index of approximately 525×10^3, while a BI of 7.5 (fully basic) would correspond to an OB index of approximately 725×10^3.

Basicity evaluation by this optical method can be carried out on a flux or slag employed with any of the welding processes. A recent study was, in fact, performed with flux cored arc welding with a series of specially prepared fluxes incorporated in FCAW-G electrodes of the T-1 and T-5 classifications. These fluxes represented variations of acidic formulations for the T-1 electrodes, and basic formulations for the T-5 electrodes. Both BI and OB indexes were determined for the fluxes, and these indexes were plotted against data obtained on weld metal deposited in steel test plates welded in accordance with the requirements of AWS A5.20, but with three different sets of operating parameters. As would be expected, weld metal oxygen contents varied from a relatively high level (e.g., 0.08 percent) with the most acidic T-1 flux formulation to low oxygen levels with the basic T-5 formulations. Correlation between weld metal oxygen content and BI for the T-1 acid type fluxes was not good. Considerable variability appeared in the plotted data, and a regression coefficient of only 0.34 was calculated. On the other hand, correlation between the oxygen contents of these selfsame acid fluxes and their OB index was excellent — with a regression coefficient of 0.82.

Correlation between basicity and sulfur content of weld metal in this study with flux cored arc welding also confirmed the expected general trend of finding reduced sulfur contents with increasing basicity. Again, in this study, the OB index for both acid and basic type fluxes correlated much better with weld metal sulfur content than did BI. As the last testing step in this study, basicity was correlated with toughness of weld metal deposited by the T-5 classification GTAW-G electrodes. Charpy V-notch impact tests conducted at -29°C (-20°F) produced data that followed the expected trend of finding higher energy absorption values with increasing flux basicity, and again the OB index correlated somewhat more closely than did the BI (with regression coefficients of 0.87 and 0.71, respectively).

In concluding this study of flux-slag basicity in the FCAW process, it was pointed out by the investigators that the OB index probably correlated more closely with weld metal features of interest because it differs from the BI equation (Eq. 8.2) in quantitative assessment of the flux compounds for their contribution to basicity. As an example, calcium fluoride (CaF_2) is neutral in optical basicity computations, whereas the BI equation regards CaF_2 as a basic factor with no limitation. Also, manganese oxide (MnO) and iron oxide (FeO) are factored in the BI equation as basic contributors, although of lesser power; but in the OB index determination these oxides, because of their thermodynamic instability detract from the total content of basic oxides.

The optical basicity method for evaluating a flux or slag is a somewhat complex procedure that requires perusal of references in the literature to assemble needed equations for arriving at the OB index for a given flux or slag. Nevertheless, this index appears to correlate remarkably well with weld metal features that have shown some degree of relationship with flux-slag basicity. The OB method should be applicable with any of the welding processes that operate with a flux or slag, and it could be very helpful when employed on the SAW process under discussion here. Hopefully, further studies will confirm its usefulness, and perhaps lead to simplifying its application.

Physical Characteristics of SAW Fluxes and Slags

Many of the fluxes currently used with the SAW process were created by making trial welds in the laboratory, and then refining the flux formulation as judged from welding experience reported by users. Operating characteristics and slag detachability were very important, and only those fluxes that produced welds with acceptable appearance were marketable. Of course, many welds from test plates, and from actual weldments fabricated in the field were analyzed for composition, examined metallographically, and tested for mechanical properties.

More recently, laboratory work has included the measurement of certain properties and characteristics of flux in an effort to reduce the amount of testing required to develop new and improved fluxes. The properties of the granulated flux at room temperature prior to use in welding has been assessed by determining bulk density, sizes of particles and their distribution, flowability of the granular mass through hoppers and distribution tubes, segregation tendency through settling, and composition (both by chemical and mineral analyses). Special laboratory methods were devised to observe and to measure melting behavior during very rapid heating to simulate SAW conditions. Properties studied included melting range, viscosity at various temperatures when molten, surface tension, molten density, electrical conductivity, and thermal contraction upon cooling. Findings from this work has shown that the various kinds of flux do not differ a great deal in softening temperature (usually about 1325°C, or 2420°F), in melting temperature (commonly about 1500°C, or 2732 °F), and in surface tension and viscosity just after becoming molten. Surface tension often measures about 275 dynes/cm, which is only about one-sixth the surface tension of molten steel. The viscosity of molten slag is very close to that of steel, measuring perhaps only a few centipoise, which is not quite as fluid as water. However, when molten slag viscosity is examined closely, the MS (manganese silicate) type slags are somewhat more fluid than the CS (calcium silicate) types, which explains why MS fluxes are favored for high speed SAW operations.

Thermal contraction determinations on solidified slag during cooling helps explain the greater difficulty in removing basic flux slag from the weld compared with the acid types. Basic slags have a coefficient of thermal contraction which more closely approaches that of steel when cooling through the temperature range of about 500 down to 50°C (932 to 122 °F). Consequently, the solidified basic slag does not free itself from the steel weld surface as easily as acid slags where the coefficient of thermal contraction is about 10 percent lower than that of steel. Zirconium oxide is sometimes included in a small amount to basic flux formulations to improve slag detachability. Two of the flux compositions in Table 8.7 — IIW designations *CS* for "neutral calcium silicate," and *BF* for "basic fluoride" — contain a two-percent addition of ZrO_2.

Transfer of Elements Between SAW Flux/Slag and Weld Metal

In addition to the endeavor described above to quantify the various physical, thermal, and electrical properties of fluxes and thus establish a more scientific relationship among their operating characteristics, even greater effort has been directed toward rationalizing the influence of flux composition on the weld metal chemical composition, microstructure, nonmetallic inclusion content, and mechanical properties. Considerable research has been directed to the transfer of elements to and from the weld metal, and voluminous information and data are in the literature on the SAW process. Much of this information is of interest only to the flux manufacturers because of their obligation to ensure that a supplied flux formulated for use with particular electrodes will deposit weld metal that meets composition requirements, etc. However, users of SAW can benefit from a general knowledge of element transfer to guide their selection of a SAW flux and electrode for welding a given base metal, and to improve their control of weld composition and properties. Also, the user must occasionally deal with problems that arise from unexpected pickup of elements in welds, and from losses of vital alloying elements during abnormal operating circumstances. Four possible sources of elements will require attention — namely (1) flux, (2) electrode and any additional filler metal, (3) base metal and any permanent backing metal that is penetrated, and (4) environment, which is normally the ambient atmosphere that can contribute oxygen, nitrogen, and hydrogen from the air that permeates the granulated flux overburden above the joint to be welded.

Almost any element can be added to the weld metal via the flux. The percentage desired in the weld composition may present a problem. The transfer of oxidizable elements from the flux (and also from the electrode) during the SAW process is much like the recovery percentages given earlier in Table 8.3 for the SMAW process. However, the importance of SAW flux type and composition will arise when the aim is to limit the amount of a specific element entering the weld even when using a neutral flux that contains no deliberate addition of the element. Silicon, oxygen and nitrogen are virtually impossible to avoid handling in the SAW process. Although silicon is easily avoided in ferroalloy form, it is present as a nonmetallic compound in a number of the ingredients considered indispensable in SAW fluxes. Oxygen is ever-present because of the less-than-complete shielding from the atmosphere by the flux, and because many flux ingredients include oxides in their make-up. Nitrogen, unless present in a nitrided ingredient, comes mainly from the atmosphere and presents little or no problem with pickup by the weld metal.

The trio of solid materials in the SAW process that shape up the weld composition — flux, electrode, and base metal — has been studied in many different combinations, and the literature contains information on inter-related reactions that affect weld composition, and on the profound effects of operating parameters. Attention in this chapter will continue to be centered on the influence of flux, per se, on element transfer. The metals and compounds added to flux to accomplish deoxidation, alloying, and control of weld metal properties will be discussed in Volume II, Chapter 10 when filler metals are reviewed.

The following brief paragraphs summarize the more significant findings from studies of the transfer of elements between flux and weld metal when applying the SAW process to carbon and low-alloy steels. In addition to noting the principal mineral constituents that comprise a particular type of flux and the manner in which each "system" affects element transfer, close attention is given to flux basicity inasmuch as this feature has been shown to influence the transfer of certain elements. Several elements that deserve close attention are normally present only at very low levels in the weld metal composition (e.g., oxygen, sulfur, hydrogen), but they can exert profound adverse effects on the mechanical properties and cracking susceptibility if tolerable levels are exceeded in a particular weld metal.

Carbon

Carbon content of weld metal deposited by the SAW process will be close to the total contributed by flux, electrode, and base metal. However, very close prediction of weld metal carbon content is quite difficult. Fluxes for hardfacing operations sometimes contain a substantial amount of carbon-containing ingredients to produce a high-carbon weld deposit. Transfer efficiency of this additional carbon will be dependent on the form of ingredient incorporated in the flux — just as described for the SMAW process. Transfer of carbon from the flux is more predictable when it is present as a powdered high-carbon ferroalloy, and when the particles are not extremely fine. A flux with high oxygen content, such as found with acid fluxes that include easily dissociated oxides, is likely to transfer carbon with decreased efficiency.

Manganese

Manganese is looked upon as an alloying element in steel weld metal that is generally beneficial. There is seldom need to restrict the manganese content to a very low level; consequently, a small pickup of this element from the SAW flux is not likely to be objectionable. Also, about 25 percent of the total manganese available will be lost by oxidation, by reaction with oxides in the flux, and by vaporization. Because of regular loss to this extent, and the frequent use of a low-manganese (inexpensive) steel as the SAW electrode, there is often need to include some form of manganese in the flux.

When weld composition specifications require the manganese content within prescribed limits, careful preplanning of manganese sources is required. Recovery of manganese in the weld metal may vary from the expected amount despite taking all recognized influences into consideration, because SAW operating parameters exert some effect. Low welding current and slow travel speed have been observed to cause a greater loss of manganese, while higher currents and faster travel speed result in better recovery. The amounts of calcium, manganese and iron contained in compounds of the flux ingredients appear to have weak influence, but the trends are not consistent and the small variations that occur have yet to be rationalized. Increasing basicity in a neutral flux (i.e., flux without a deliberate addition of metallic manganese) usually lowers the loss of manganese by a small amount. Flux compositions that include large proportions of manganese oxides and other manganous compounds, such as the manganese silicate type flux, usually raise the weld metal manganese content a little above the expected level. The presence of aluminum oxide (Al_2O_3) as an ingredient in the IIW "AB" type of flux is a powerful aid to the transfer of manganese into the weld metal from any manganese silicate that may be present.

The addition of strong deoxidizing agents to a SAW flux (e.g., calcium disilicide), thus making it an "active" type, tends to reduce the loss of manganese by a small percentage. These changes in manganese transfer efficiency or recovery are relatively minor, but they may require consideration when the weld metal manganese content must be held within a relatively narrow composition range.

Phosphorus

Phosphorus pickup in SAW weld metal is generally undesirable, and the level of this residual element can usually be predicted quite closely from the amounts present in the flux, electrode and base metal. Computation is assisted by 100 percent recovery of all phosphorus in the three sources, but it is necessary to know the flux/electrode melting ratio and the percent dilution of the weld by the base metal. When a low level of phosphorus is required in the weld metal, the only assured method of achieving a stated maximum is to have the total phosphorus contributions below this amount. Phosphorus contents of the electrode and the base metal are usually readily available, but the flux may require careful analytical work to be certain that no ingredient is carrying an unusually high phosphorus content. For example, chemically purified titanium oxide (TiO_2), which sometimes is included in a formulation, may contain an inordinate amount of phosphorus. All phosphorus, regardless of ingredient form, will enter the weld metal during the SAW process. Even the fully basic fluxes, which are touted to have the ability to remove phosphorus during steelmaking, appear unable to lessen phosphorus recovery in SAW weld metal. Apparently the oxidation period during the SAW operation is not of sufficient duration to form phosphorus pentoxide. This compound, when formed, and with sufficient time, could enter the slag where calcium phosphate would form and hold the phosphorus out of the weld metal.

Sulfur

Sulfur in weld metal is generally looked upon as an undesirable residual element; and much attention is given not only to conform to some specified maximum, but also, in certain instances, to obtain the lowest possible sulfur content. The sulfur content of SAW weld metal often can be estimated from its total in the electrode and the base metal. The ingredients used in fluxes seldom contain significant amounts of sulfur. Ordinarily, all of the sulfur available to the weld metal will be retained, but fully basic SAW fluxes can accomplish some reduction in sulfur content; perhaps just enough to make the selection of a basic type flux worthwhile. This has been the general experience even though removal of sulfur by the reactions that occur in steelmaking with a basic flux does not appear thermodynamically possible. Research on sulfur removal through SAW flux formulation is continuing because of anomalous findings. One recent study of the desulfurization capabilities of various kinds of SAW flux found that a manganese silicate acid-type flux accomplished somewhat greater desulfurization than a calcium silicate flux of relatively high basicity. This finding suggested that a manganous compound in flux may act in a more basic manner than expected from the BI value calculated from Equation 8.2.

Silicon

Silicon is commonly used as the primary deoxidizer in steels, including weld metal, and silicon serves very well in a wide variety of situations. Yet, there is a growing awareness that the silicon content should be limited to lower levels for good toughness in the higher-strength weld metals. Adequate deoxidation can be ensured with a relatively low silicon content by augmenting with other deoxidizing elements such as titanium and aluminum. The very small particles of non-metallic inclusions formed by titanium and aluminum can exert added benefit by nucleating a microstructure of fine, acicular ferrite, which has good toughness. The inclusion volume of aluminum oxide particles must be limited, because they can adversely affect weld metal toughness if present in large number.

It is easy to obtain the desired level of silicon in the weld metal. The transfer efficiency of silicon during a SAW operation is approximately 50 percent, and this is good enough to achieve a high level of silicon, if wanted. When the silicon content of the weld metal must be held within a relatively narrow range, the basicity of the flux must be considered. A flux that is "acid" in character by virtue of containing a high proportion of silica (SiO_2) will contribute silicon to the weld metal. Consequently, an acid-type flux that is formulated with metallic silicon-containing ingredients, such as ferrosilicon and silico manganese, in amounts calculated to satisfy normal transfer efficiency is likely to overshoot the maximum silicon level set for the weld metal. If the proportion of silica in the flux is reduced to change the character of the flux to a fully basic type, the weld metal silicon content could be lowered to about one-half of the level obtained with the acid flux formulation. Incidentally, if the flux also contains substantial amounts of other oxidizable metals, such as chromium (present as ferrochromium, or as electrolytic chromium metal), then pickup of the silicon in the weld metal will be further increased; especially if the flux is the acid type.

Oxygen

Oxygen pickup in the weld metal during the SAW operation starts in the arc cavity where free anionic oxygen is formed from some of the flux compounds as they dissociate when heated to the extremely high temperature. Silica (SiO_2) plays a leading role in this thermal dissociation and ionization mechanism to provide oxygen anions. Air entrapped in the granulated flux contributes little oxygen to the weld metal. Dissociation of oxygen-bearing compounds in the flux while in the arc column is the major source of oxygen. As the arc travels forward and temperature decreases in the trailing weld pool, the available oxygen recombines with compounds and with any deoxidizers present in the flux, the electrode or the base metal. This recombination favors the element with the highest free energy of formation, and then would proceed with others as dictated by free energy and mass action considerations. The strongest deoxidizer likely to be present in the flux would be calcium (easily added as calcium silicide), probably followed by magnesium, aluminum, zirconium, titanium, silicon, and manganese. SAW weld metal can have an oxygen content as high as 0.2 percent, which likely would cause the carbon-oxygen reaction to occur and produce porosity in the weld metal. Oxygen exceeding about 0.05 percent would be sufficient to reduce toughness in higher strength weld metals. Two important facts have been established regarding oxygen content and the role of deoxidizers. First, acid fluxes foster higher oxygen contents (greater than 0.05 percent) in SAW weld metals. Second, despite best efforts with deoxidizers to tie up the oxygen, the toughness of weld metal often is adversely affected. This degradation can be attributed to (1) high residual oxygen content, (2) nonmetallic deoxidation products that are too numerous and relatively coarse in size, and (3) an excess of deoxidizing element that does not combine with oxygen and remains as an alloying element with unfavorable influence on toughness. This latter effect (3) often is exerted by excess aluminum. Conversely, basic fluxes produce weld metal with substantially lower oxygen content, usually in the range of about 0.015 to 0.05 percent. When a basic flux is employed to obtain a low level of oxygen, perhaps below about 0.025 percent, care must be taken to reduce the additions of metallic deoxidizing ingredients to avoid the unfavorable alloying effect mentioned above as Item 3 for acid fluxes. With judicious additions of deoxidizers that produce very fine oxide particles, especially the oxides formed by titanium, not only is the oxygen content reduced, but nonmetallics remaining in the weld metal act as nucleating sites to encourage the formation of a very fine grained, acicular ferrite microstructure. There is some evidence that a very low oxygen content in steel weld metal may not be desirable if the acicular microstructure is wanted. When oxygen is below about 0.01 percent, it appears that introduction of deoxidizers like aluminum and titanium will not form enough oxide particles to nucleate the acicular microstructure.

Nitrogen

Nitrogen pickup by weld metal during SAW is not a problem; because only a minuscule amount of air infiltrates the granulated flux burden atop the submerged arc operation, and this air is effectively kept away from the arc by the layer of molten flux or slag that forms. Ordinarily the pickup of nitrogen by SAW weld metal is less than occurs with the open arc welding processes. It was mentioned that weld metal made by the SMAW process has about 0.015 percent nitrogen, also the FCAW process with gas shielding has about 0.02 percent, and the FCAW process with self-shielding electrodes has about 0.025 percent nitrogen. A survey of composition data for many samples of SAW weld metal made with a wide variety of fluxes and operating parameters found the nitrogen was consistently in the range of about 0.008 to 0.014 percent, indicating that the pick-up during SAW over the amounts regularly present in electrodes and base metals is trivial. Of course, if additional nitrogen is wanted in SAW weld metal, it can be obtained by including a nitrogen-rich ingredient in the flux, for example, nitrided manganese metal in powdered form.

Hydrogen

Hydrogen should not be a problem in the SAW process, because no significant amounts of moisture or other hydrogenous compounds should be incorporated in the flux or involved elsewhere in the consumables or the operating environment; hence the frequent mention of SAW as a "low-hydrogen" process. Yet, problems with hydrogen arise during SAW from time to time because (1) the welder fails to guard against involvement of moisture, oils, etc.; and (2) the SAW process normally operates in a manner that is twice as effective as SMAW in capturing and dissolving available hydrogen. These two dispositions of SAW deserve examination here, at least from the standpoint of fluxes.

Literature on SAW includes ample warnings about keeping flux clean and dry, but only meager information is provided on moisture content of the fluxes as supplied by the manufacturer, and their hygroscopicity. It should be appreciated that flux manufactured by bonding with aqueous silicates have the highest propensity to pick up moisture if exposed to a humid atmosphere. When put in the hands of the user, SAW fluxes appear to have satisfactorily low moisture contents. Two types of *fused* flux analyzed by the AWS "Coating Moisture Test" (see AWS A5.1 or A5.5) were found to contain only 0.006 and 0.008 percent moisture, respectively. Fused type flux should display least propensity to pick up moisture. Two fluxes of the *bonded* type were analyzed and found to contain only 0.018 and 0.039 percent moisture, respectively. Bonded fluxes probably represent the most widely used kind of product. In addition to carefully storing and handling the flux to avoid moisture pick-up and contamination by hydrogenous compounds, the welder should avoid using SAW flux in a very humid atmosphere which can introduce some moisture to the welding operation. AWS A5.17 and A5.23A do not specify a limit for moisture content, because many conditions and circumstances beyond the province of these specifications for electrodes and fluxes determine whether hydrogen-induced problems will occur. Examples of such conditions include:

(1) the amount of moisture actually in the flux at the time of use,
(2) entrapped grease and other hydrogenous compounds on the surface of a solid electrode, or contained within the core of a composite electrode,
(3) the prevailing humidity in the atmosphere at the welding site,
(4) the amount of moisture as condensate on the surfaces of the joint to be welded,
(5) the composition and properties of the weld metal and base metal involved and their susceptibility to porosity or cold cracking, and
(6) the kind of joint and amount of restraint imposed during welding.

In reality, hydrogen can present a problem in SAW operations, and close attention must be maintained to guard against pickup of an amount that will cause porosity, or cold cracking, or detract from fracture toughness. This attention is vital when welding high-strength steels. Post-weld investigation of a problem suspected to be caused by hydrogen pickup is not easily conducted; because, unless extraordinary precautionary steps are taken, the damaging hydrogen can diffuse and escape from a weld before direct measurement can be accomplished. Combinations of electrode and flux can be subjected to one of the diffusible hydrogen test methods given in ANSI/AWS A4.3. This standardized test was mentioned earlier when examination was made of flux coverings on SMAW electrodes. An in-depth discussion of diffusible hydrogen, including its significance in SAW operations, will be undertaken in Volume II, Chapter 13.

The reason hydrogen can enter weld metal during the SAW process (as compared with SMAW and other open-arc processes) stems from having the submerged-arc heat and melt the moisture-containing flux while an excess of flux remains unmelted as an overburden. This covering rather effectively prevents the moisture or steam from being driven away. The amount of hydrogen that enters the arc and eventually dissolves in the molten weld pool depends on many factors, most of which have yet to be studied methodically to establish quantitative data. ANSI/AWS A4.3 dealing with diffusible hydrogen determination contains information on the amount of hydrogen likely to be found in weld metal from an E 7018 SMAW electrode when the moisture content of the covering is known. AWS has not published similar information for SAW, but there is experimental evidence that weld metal in the SAW process will contain approximately twice as much hydrogen for a given flux moisture content. There is reasonably good agreement among investigators that a properly baked and handled E 7018 electrode with a covering moisture content of about 0.1 percent will produce weld metal having a diffusible hydrogen content of about four mL/100 gm; whereas SAW weld metal made under flux bearing approximately the same moisture content of 0.1 percent will have diffusible hydrogen somewhere in the range of about eight to 12 mL/100 gm. Study of fully basic SAW fluxes of the bonded type manufactured with a high temperature baking step, and then used with proper welding practice found deposited weld metal to contain no more than five mL/100 gm of diffusible hydrogen.

Flux composition has some influence on the amount of hydrogen entering the weld metal under a given set of operating conditions. Increased basicity of the flux from "neutral" to strongly basic (i.e., about three BI) can reduce the total hydrogen by about 30 percent, and in doing so a greater proportion of the hydrogen appears to be transferred to the slag thus reducing the amount of diffusible hydrogen in the weld metal. If carbonates are incorporated in the flux as an ingredient, they generate carbon dioxide in the arc cavity which acts to reduce the partial pressure of hydrogen in the plasma, reducing the amount of hydrogen that dissolves in the molten weld metal. Welding conditions that produced 14 mL/100 gm of hydrogen in SAW weld metal when repeated with about 7.5 percent CO_2 in the flux lowered the diffusible hydrogen content to less than two mL/100 gm. Flux manufacturers are continuing to study the use of various carbonates as possible helpful ingredients in commercial SAW fluxes.

Titanium

Titanium, added as a deoxidizer to steel, has been mentioned frequently in earlier chapters. This element has been studied extensively in research on fluxes as an agent for obtaining finer grain size to enhance weld metal toughness. In addition to the very high free energy of titanium to tie up oxygen as titanium dioxide (TiO_2, $\Delta F = -160$ k), titanium is a strong nitrogen getter and this property holds strong promise for improving the mechanical properties of weld metal. The technology for adding the small, but critical amount of titanium to weld via a SAW flux to carry out the grain refinement mechanism is quite complex. There continues to be need for close control of the level of titanium and the nature of particles formed as described in some detail in Chapter 4 under the heading of *Titanium* as an alloying element. Furthermore, in SAW weld metal, there is apparently a need to know the microstructure that will constitute the primary phase during solidification. While it is feasible to accomplish this desirable grain refining effect by an addition of titanium in appropriate form to the SAW flux, consideration must be given to the compositions of the electrode that will be employed and the base metal to be welded. A coordinated approach with flux and electrode to grain refinement in weld metal through titanium inoculation will be included in Volume II, Chapter 10, when SAW electrodes are reviewed as filler metal.

Electroslag Welding

Early use in the USA of the electroslag welding process (ESW) employed acid-type fluxes that have been formulated for the SAW process, but about five to 20 percent of fluorspar (CaF_2) was usually added to increase fluidity of the molten slag above the surface of the weld pool. AWS now has a specification for electrodes and fluxes which can be used in ESW of carbon and low-alloy steels (ANSI/AWS A5.25); however, this document does not cover the composition of the flux. Instead, fluxes are classified on the basis of mechanical properties of a weld deposit made with a particular electrode. Consequently, manufacturers of ESW fluxes regularly qualify their products without revealing proprietary information about composition.

Flux for ESW, in addition to having a suitable melting point and viscosity, must be electrically conductive in order to generate the resistance heating needed to produce penetration into the base metal, and to melt the electrode and any other consumables involved. The blanket of molten slag formed by the flux is in contact with molten metal for a much longer period of time than in the SAW process, and this ordinarily would allow more extensive slag-metal reactions to take place. Yet, there are good reasons to avoid interactions between slag and metal. No attempt is made to add alloying elements to the weld metal via the flux because only a relatively small amount of slag is consumed compared with the amount of weld metal formed. Flux consumption is very roughly five to 10 percent of the weight of metal deposited. Furthermore, when fresh flux is added, it is not necessarily metered uniformly with melting of the electrode and the formation of the weld. Needed alloying elements for the weld are contained in alloy steel electrodes of solid wire, bar, or plate, or in metal cored electrodes. A supplementary alloyed consumable in some practical form also can be used by uniformly feeding it into the weld pool as welding progresses.

The flux used in ESW is an important part of the process even though only a small amount is consumed. Although a starting pad regularly is used to accept the abnormal initial increment of ESW weld deposit, it is very important that the process assume a stable operating mode before reaching the actual joint to be welded. To ensure quickly achieving the proper state of operation, a "starting flux" often is employed by placing a small amount of this special flux on the starting pad. Starting fluxes contain ingredients that provide stabilizing agents for the initial arc. These agents generate the molten flux layer, reduce flux melting temperature, and increase slag viscosity — all of which assists in gaining a satisfactory start. As soon as stable operating conditions are achieved, a "running flux" is introduced to protect the principal weld. Briefly, the running flux must melt and then quench the starting arc to (1) generate heat by resistance to the flow of electrical current, (2) protect the molten weld pool from the atmosphere, (3) circulate freely to distribute heat uniformly over the entire joint, (4) remain chemically stable and inert so that composition of the weld does not vary along its length, and (5) encourage flow of molten metal to all corners and crevices of the dammed joint without excessive molten slag leakage. A typical ESW flux for welding steel could be formulated from minerals and compounds (as listed earlier in Table 8.2) to have the chemical composition shown in Table 8.8. This nonmetallic flux may have to be augmented by metallic deoxidizing agents in the electrode or in a supplementary filler metal to prevent the carbon-oxygen reaction from occurring in the weld metal.

Running flux for the ESW process continues to receive attention because its formulation can alleviate problems stemming from the coarse grain microstructure that develops as a consequence of high heat input and slow cooling of the weld. Development efforts have been centered on the oxygen content of the weld metal because ordinarily the oxygen level is surprisingly low, perhaps only 0.003 to 0.015 percent. The small number of oxide particles that form at this oxygen level to act as nucleation sites for austenite grain formation is insufficient to produce a finer grain size. This observation is bolstered by the finding that an electrogas weld (EGW) made at equivalent energy input, but with a weld-metal oxygen content of about 0.075 percent has a much finer grain structure. When a consumable guide tube is used for ESW, the weld metal oxygen content can increase to about 0.03 percent, and this has been accompanied by a finer grain microstructure. Work has been conducted on modification of fluxes to change their composition and to alter their basicity to capitalize on the oxide-nucleation concept. There is some evidence that alumina-rich flux compositions give better toughness values, but the generally low level of toughness continues to be reason for a cautious approach to making use of the ESW process.

Gas Shielded Arc Welding

Almost a dozen arc welding processes use a gas that is piped to surround the arc and shield it from the air. Atomic hydrogen welding (AHW) was the first arc process of this kind, even though the primary purpose of the hydrogen was to carry the arc energy to the workpieces. Gas carbon arc welding (CAW-G) came next, but the advent of gas tungsten arc welding (GTAW) was the process that clearly demonstrated the advantages of gas shielding for arc-and-weld protection from air. The success of GTAW encouraged development of additional processes using gas for shielding.

Table 8.8 – Typical Flux for Electroslag Welding of Steel

Ingredient	Chemical Formula	Approximate Percent By Weight
Lime	CaO	5
Fluorspar	CaF_2	15
Magnesia	MgO	5
Pyrolusite	MnO_2	20
Alumina	Al_2O_3	20
Silica	SiO_2	35

Present technology of gas shielding has grown to be almost as complex as that for flux shielding, but the relatively small number of gases suitable for safe use tends to restrict freedom in this area. The externally supplied gaseous shield around the arc now is more likely to be a mixture of gases because of increased knowledge about the proficiency of each in acting as a shielding medium, arc stabilizer, alloying addition, and fluxing agent. Also, new findings have shown that a mixture of gases can exert very helpful control over metal transfer mechanics in a consumable electrode process.

Gases employed for shielding arc welding processes are listed in Table 8.9 along with pertinent physical properties. Below air, the next three gases listed — argon, helium, and carbon dioxide — are the most frequently used, either individually or in mixtures. The remaining gases in the Table are additives to one of the aforementioned three, or to a mixture thereof. Note that oxygen and nitrogen, the very gases that are usually intended to be excluded from the arc and weld area, are listed as useful additives. To be acceptable for welding operations, a gas must be (1) regularly produced in a purity that is suitable for protecting particular metals and alloys, (2) safe for personnel to transport in bottled gaseous form, or in liquefied form, and safe to discharge at a specified flow rate through a welding torch into the atmosphere of the workplace, and (3) available at a cost per unit consumed that is acceptable for the benefits it provides. The selection of a gas, or a mixture of gases, is guided by several general considerations; first, the physical and chemical properties of the gas; second, the operating characteristics that each gas imparts to a particular process; and third, the kind of metal or alloy that the gas is expected to protect. The information that follows on shielding gases has been tailored to apply mainly to the carbon and low-alloy steels simply to condense the substantial technology now on hand. After individual gases have been examined for chemical and physical properties pertinent to their use for shielding, a number of popular welding processes selected from Chapter 6 will be re-examined to

Table 8.9 - Gases Used in Shielding Welding Arcs

Name of Gas	Chemical Symbol	Atomic Weight	Molecular Weight	Specific Gravity	Thermal Conductivity*	Remarks
Air	$N_2 + O_2$	–	–	1.00	0.6	
Argon	Ar	40	40	1.38	0.4	Inert, heavy
Helium	He	4	4	0.138	3.3	Inert, light
Carbon dioxide	CO_2	–	44	1.53	0.4	Mildly oxidizing
Propane	C_3H_8	–	44	1.52	0.4	Hydrocarbon fuel
Hydrogen	H_2	1	2	0.069	4.1	Flammable, light
Oxygen	O_2	16	32	1.108	0.6	Strongly oxidizing
Nitrogen	N_2	14	28	0.969	0.6	Strongly nitriding
Nitric oxide	NO	–	30	1.34	0.6	Suppresses ozone
Carbon monoxide	CO	–	28	0.969	0.6	Carburizing, toxic
Chlorine	Cl_2	35	71	2.49	0.2	Active, toxic
Hydrogen sulfide	H_2S	–	34	1.193	0.3	Fluxing, toxic
Sulfur dioxide	SO_2	–	64	2.269		Fluxing, toxic
Sulfur hexafluoride	SF_6	–	146	5.118		Fluxing, toxic

*Thermal conductivity at RT (20°C) in cal/sq cm/cm/sec/°C × 10^{-4}

gain a broader perspective on the ways in which shielding gas selection and introduction influences process operating mode.

Argon

The earth's atmosphere consists mainly of a mixture of about 21 percent oxygen and 78 percent nitrogen and includes about one percent argon and other trace gases. Argon is easily separated as a sub-product during large-scale production of purified oxygen and nitrogen by the fractionation process. With air as the source, argon is plentiful and is available at modest cost. It is a colorless, odorless, tasteless, monatomic gas, and is one of the inert gases in the first column of the periodic table (Table 2.4). Argon forms no compounds (i.e., is chemically inert) because its outer electron shell (M, or no. 3) of each atom has a complete octet of electrons that satisfy sub-shell orbitals s and p (see Table of electron configurations in Appendix X). Therefore, argon atoms have no valence.

During the air separation process, argon can retain a small amount of nitrogen (approximately 0.09 percent) and traces of oxygen and hydrogen (about 0.001 percent each). If commercial-grade argon with these impurities is used for shielding an arc, various problems can arise, including weld metal porosity, cold cracking, and difficulty in achieving normal weld travel speed. Therefore, *welding-grade argon* is regularly produced to 99.995 percent purity. The dew point of gas for shielding the arc should be −60°C (about −75 °F) or lower, which amounts to 11.4 ppm water vapor by volume.

Argon is a relatively heavy gas, heavier than air by about one-third, and ten times heavier than helium. The specific gravity of argon favors somewhat lower flow rates for protective shielding of the arc processes. The proper flow rate is determined by a number of factors related to torch design, the Reynolds number of its nozzle, welding procedure and parameters, ambient conditions, etc. Generally, only enough gas should be provided to exclude air from the weld area because excessive gas not only is costly, but also may adversely affect welding conditions and resulting weld quality. Schleiren shadowgraphs show that the specific gravity of shielding gas is only a minor factor in protecting the arc and weld pool because the arc heat causes a marked expansion and decrease in density so that all gases behave in about the same manner in the immediate vicinity of the arc. After the gas leaves the arc area, specific gravity can have more influence on how the gas dissipates as it provides protection to the trailing weld and to the joint ahead.

Argon requires about the same amount of electrical energy as do most other gases for ionization (see Table 2.2), but it is important to note that argon at 15.7 eV is less than helium at 24.9 eV. This lower ionization energy for argon facilitates starting an arc and, with argon available to its plasma, the arc will operate at a relatively low closed circuit voltage for any given current value. Furthermore, the change in voltage with changing arc length will be comparatively small.

Helium

Although there is only a trace of helium in the atmosphere (one part in 186 000), and mere traces in certain minerals, a natural gas produced by certain wells located in Canada, Russia, and the USA is the major source of helium. The Texas Panhandle has wells that supply helium-bearing natural gas that ranges from about one to seven percent of this valuable inert gas. The Federal Government through its Bureau of Mines handles all helium production in the USA because of its strategic importance. The availability of helium for welding purposes can change with national circumstances. Separation of helium from natural gas is accomplished by liquefaction and fractionation, and welding-grade helium has a purity better than 99.99 percent. Helium for the GTAW process must be very dry (−60°C, or −75°F dew point or lower) for the same reasons mentioned earlier for welding-grade argon.

Helium is the lightest inert gas. It has only twice the specific gravity of hydrogen, and one-tenth that of argon. Helium is monatomic, and is inert because its single electron shell (K, or No. 1) holds a pair of electrons that completely satisfies the s orbital. Helium is colorless, odorless, and tasteless. Although breathable in gas mixtures that include oxygen for therapeutic reasons, helium is not life-sustaining. The very low specific gravity of helium makes it rise quickly in air. Also, its high thermal conductivity causes a stream of helium to heat rapidly throughout, and this adds to the tendency of the gas to rise in air. For these reasons, flow rates recommended for a helium shield are often higher than for argon, or for a mixture of helium and other heavier gases. On the other hand, the low specific gravity of helium and its tendency to rise sometimes can be helpful when welding overhead.

The relatively high electrical energy required to ionize helium (24.9 eV) makes arc starting in a plain helium shield somewhat difficult unless either a high open-circuit voltage is provided, or a high-frequency ac current is superimposed on the welding current at the start. The operating voltage of an arc in helium will be higher than an arc in argon. Therefore, a greater amount of heat will be generated by the arc in helium at any given level of current.

Carbon Dioxide

Lower gas cost is the principal incentive for using carbon dioxide (CO_2) to shield GMAW and FCAW. CO_2 is manufactured mostly by controlled burning of oil, coke or other carbonaceous fuels. The collected CO_2 gas is cleaned and purified to provide a grade suitable for weld shielding. The principal contaminant in CO_2 is moisture. Maximum moisture content for welding-grade CO_2 is a dew point no higher than −40°C (−40°F). Much of the CO_2 sold as welding grade has a dew point in the vicinity of −57°C (−70°F). Even when virtually pure and devoid of moisture, CO_2 does not completely prevent oxidation of molten steel because this gas tends to dissociate into carbon monoxide and oxygen. Dissociation occurs at about 650°C (1200°F), so the arc temperature is more than adequate to cause dissociation. The amount of oxygen that is liberated in the immediate vicinity of an unusually long arc may run as high as 25 percent. In the nearby cooler areas surrounding the arc where a temperature of approximately 350°C (about 650°F) exists, the carbon monoxide and oxygen recombine in an exothermic reaction to again form CO_2. The CO_2 gas also can react directly with heated iron to form iron oxide and liberate carbon monoxide. CO_2 offers better protection from oxidation when the arc length is very short, but even then, it is not sufficiently protective to use with the GTAW process because of accelerated deterioration of its nonconsumable tungsten electrode.

Propane

This gas, and other hydrocarbon fuel gases, had been used to a limited extent as shielding for metal arc welding where the operation was mechanized, and where the weldments were non-critical. They effectively minimized oxidation and nitrification, but depending on conditions of flow and introduction at the arc, some carbon pick-up easily could occur in steel weld metal. A significant increase in carbon content could cause a serious toughness decrease. The improved availability of CO_2 has supplanted the use of hydrocarbon gases for shielding. In non-arc welding processes, fuel gases like propane, acetylene, etc., are often used to produce a heating flame, and the gaseous products of combustion are sometimes controlled to provide shielding from air.

Nitrogen

Because purified nitrogen is readily available in large quantities at relatively low cost as a by-product from the production of purified oxygen by the fractionation process, it is tempting to consider its potential as a shielding gas since it would provide non-oxidizing protection. In fact, nitrogen is used extensively for this reason in back-filling pipelines and tanks to avoid explosions and other incidents that could occur if oxygen was present within the container. Although nitrogen can be used to shield certain nonferrous metals, iron and steel have a solubility for nitrogen, and whether it will be acceptable as a shield in welding depends on the operating conditions. If nitrogen coming into contact with the molten steel is the usual molecular nitrogen (N_2), the amount of nitrogen absorbed is likely to be relatively low, perhaps only about 0.05 percent. However, if welding is being carried out with an arc, the amount of nitrogen absorbed could be increased more than tenfold. The ordinary molecular nitrogen (N_2) is readily dissociated by an arc, and this nascent form of nitrogen (single atoms) is much more easily absorbed by the steel. This explains why nitrogen sometimes can be used for shielding the underside of a joint being welded where the gas is isolated from the arc, but cannot be used in shielding the arc and molten weld pool on the face side of the weldment.

Hydrogen

This gas offers its own set of unique properties as shielding for arc welding. It is quite flammable, of course, and burns quickly as it meets with oxygen in the air. Hydrogen served very well as shielding in the atomic hydrogen welding process, but it is hardly suitable for use in any of the commonly used arc welding processes because of its flammability. Hydrogen is used as a minor additive to the inert gas shield in the GTAW process when applied to certain high-alloy materials. Small amounts of hydrogen, perhaps from two to five percent, when added to an inert gas will raise the closed-circuit voltage. For example, the GTAW process operating with plain argon shielding may have an arc voltage of 10 V. An addition of five percent hydrogen to the gas shield could raise the arc voltage to approximately 12 V. As a result, the arc will generate more heat and weld penetration is increased. Although these benefits are very helpful when welding high-alloy materials, hydrogen is seldom added to the shielding employed for carbon and low alloy steels. In fact, if an addition of hydrogen is made to the shielding gas when welding a hardenable steel, hydrogen-induced cracking might be encountered. Hydrogen added beyond the five-percent level is also likely to cause porosity in the weld metal, so the small benefits of the addition may not outweigh the risk of weld quality degradation.

Miscellaneous Gas Additives

Table 8.9 lists gases that have been used as small additions to some shielding gases. Oxygen frequently is added to argon shielding for the GMAW process to improve the contour of the weld bead and the shape of penetration. The problem of undercutting also can be alleviated by using oxygen, or an oxygen-bearing gas like CO_2 as an additive. The "oxygen potential" of gas mixtures will be discussed as shielding for the GMAW process is examined. Nitrogen is added in special cases where increased nitrogen in the weld metal can be utilized as an alloying addition. Also, during gas shielded arc welding, ultraviolet radiation from the arc forms small amounts of ozone (O_3). The concentration of ozone is usually very low, and it is ordinarily detectable only close to the arc. However, ozone is very irritating when inhaled by humans. A small addition of hydrogen or of nitric oxide gas (perhaps only 0.03 percent) to the shielding being used for the GMAW process has been found to reduce the generation of ozone. Hydrogen reacts with oxygen to produce water vapor, which in turn accelerates photodecay of the ozone. Chlorine is mentioned in the literature as a powerful fluxing additive to inert gas; this is particularly the case when welding aluminum. However, the toxicity of chlorine has discouraged experimentation when welding steel. The sulfur-bearing gases, even as very small additions, impart a remarkable increase in fluxing ability to a gaseous shield for welding steel. These gases also are toxic and must be used with caution.

Gas Tungsten Arc Welding (GTAW)

The GTAW process usually is shielded either by argon or by helium, or by mixtures of these two gases; but, small amounts of other gases (hydrogen, nitrogen, sulfurous gases) may be added to obtain certain benefits. Shielding with gases containing substantial amounts of carbon dioxide or nitrogen is not workable for GTAW because of accelerated tungsten electrode deterioration. The significant difference in welding voltage for an arc operating in argon as compared with one in helium was mentioned earlier. This matter of arc voltage, along with the difference in thermal conductivity between argon and helium exerts important effects in the GTAW process. Argon quickly ionizes to produce a plasma, but the main thrust of this plasma undergoes less expansion because of the relatively low thermal conductivity of argon. Therefore, the arc usually consists of a very hot inner plasma or core and an outer plasma or mantle at a lesser temperature level. Typically, the argon-arc produces a localized area of deeper penetration in the base metal below the arc centerline as illustrated in Figure 8.5. For a given arc length, the argon arc has less energy than the helium arc; therefore, the total area of penetration is less. The localized high temperature along the centerline of the argon arc helps loosen oxides on the surface, and this plays a part in the better surface cleaning action of the argon arc. Unfortunately, this localized temperature effect is mainly responsible for a pronounced tendency to undercut the edges of the weld. The heat from the arc is not uniformly distributed over a sufficiently broad area to allow molten metal to flow into the undercut before freezing takes place.

As also illustrated in Figure 8.5, the helium arc produces an almost parabolic-shaped weld penetration. The relatively high thermal conductivity of helium greatly improves the uniformity of thermal ionization throughout the plasma cross-section. This results in more uniform current density across the arc, and avoids localized heating and deeper weld penetration along the weld centerline. The voltage drop across the arc is increased by the more rapid discharge of electrical energy, and, as a result, the arc produces a high energy input over a broad area. Figure 8.5 illustrates the changes in character of the arc and in the shape of weld penetration produced when adding one percent hydrogen to the argon shield, or when using a mixture of 50 percent helium plus 50 percent argon. The arc and penetration changes are quite in line with the gas mixtures employed. Discretion must be exercised in making use of hydrogen in an argon-based mixture for welding steel. While hydrogen in the range of about three to eight percent sometimes is employed when welding austenitic high-alloy steels, this amount of hydrogen in a shield when welding hardenable steel could result in cold cracking.

Gas Metal Arc Welding (GMAW)

Success of the GMAW process must be credited in large measure to progress made in shielding technology. The GMAW process now is being employed in many applications where SMAW formerly served because gas mixtures for GMAW, along with new power sources, permit the process to be applied with equal facility to metal sections ranging from thin to thick. Furthermore, use of GMAW with properly selected welding parameters, particularly optimum shielding gas mixture, enables the welder or the welding operator to accomplish the following:

(1) make welds with minimum spatter, controlled penetration, and good contour;

(2) make welds at a high duty cycle, at fast travel speed, in all positions, with high electrode deposition rate, and with high electrode efficiency;

(3) make welds that have excellent internal soundness, and with acceptable mechanical properties; and

(4) make welds with a minimum of slag on the surface, and with satisfactory general appearance.

The proficiency with which all of the above can be performed will vary in a given application, particularly with the gas shielding employed. When the GMAW process was described in Chapter 6, the role of shielding gas was reviewed in a preliminary way, but is deserving of further coverage. In addition to a more-searching examination of gas mixtures successfully employed, it is important to point out unfavorable kinds of gas shielding to forestall difficulties. In Chapter 6, three different basic modes of metal transfer that have been developed to adapt the process to particular welding situations were described; namely, (1) axial-spray transfer, (2) globular transfer, and (3) short-circuit transfer. Two additional modes also were described that represented modifications of spray transfer; namely, pulsed spray mode, and synergic pulsed welding. These five are useful modes of transfer, and one is routinely selected to fit the circumstances of a given welding operation. However, several other modes of metal

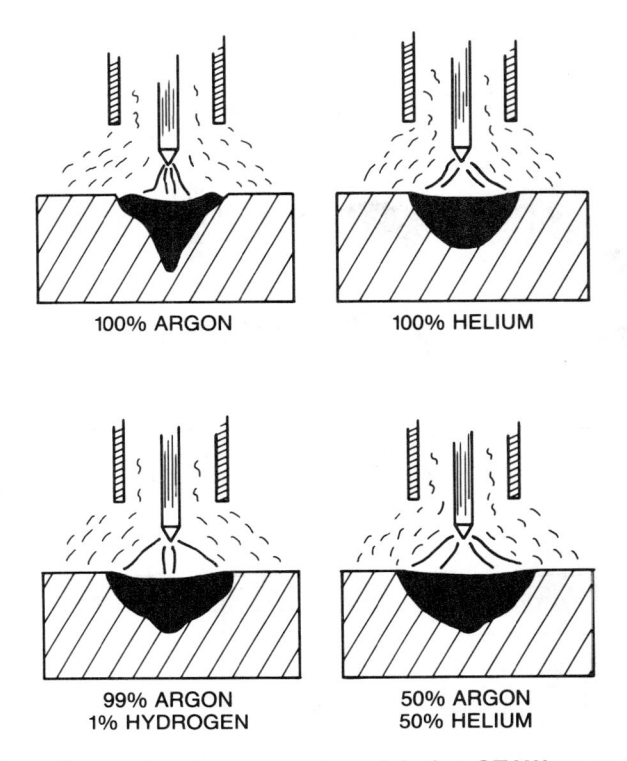

Figure 8.5 — Shape of weld zone produced during GTAW process using four different shielding gases

transfer that are quite unfavorable have been identified, and information on why they occur and how to supplant them with a more suitable mode will be explained shortly. The IIW, to encourage a universal terminology and understanding of the different modes of metal transfer during GMAW, proposed seven specific types as illustrated in Figure 8.6. Some investigators feel that a simpler classification could be made by dividing the modes of transfer into just two groups called "free-flight transfer" and "short-circuit transfer." The first four IIW types shown in the upper half of Figure 8.6 would be classed as clear-cut, free-flight transfer modes, while (E) and (F) in the lower half of the illustration are considered special types of the spray-transfer mode. Only (G), which now is designated as short-circuit transfer (GMAW-S) in the USA, does not involve free flight of molten metal in the form of droplets from electrode to weld pool. The mode of metal transfer that actually takes place during GMAW-S is determined by complex interactions arising from composition of the electrode, electrical parameters, and the kind of shielding gas. These interactions involve a variety of forces, including gravity, plasma pressure, electromagnetic effects, and Lorenz forces; but, both shielding gas flow and the nature of gas, also exert effects. The transfer of metal by all these modes takes place so rapidly that high-speed cine photography must be used to observe the mechanics that operate in each case.

A recent investigative approach taken at NIST for sensing the mode of metal transfer from a steel electrode in the GMAW process makes use of a microcomputer

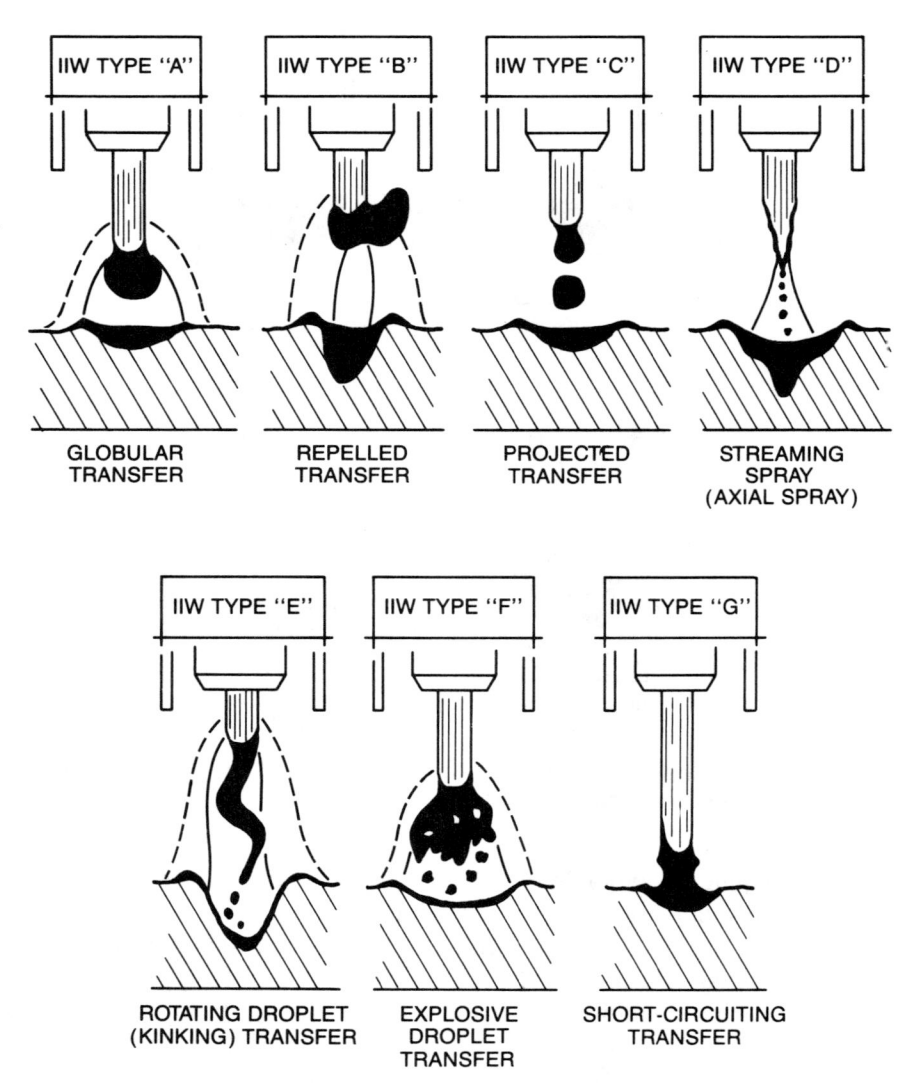

Figure 8.6 — Seven different modes of metal transfer from consumable electrode to weld pool during GMAW process with labeling provided by IIW

with capability for fast capture of current, voltage, and other data during welding. Statistical analysis, Fourier transforms, amplitude-histograms, and peak-searching algorithms brought out several parameters that have dynamic characteristic outputs that could be correlated with metal transfer modes, and with other important aspects of welding, such as the amount of spatter. No single parameter appeared suitable for monitoring all of the metal transfer modes. Fourier transforms had high sensitivity to all modes except spray transfer. Spray transfer was monitored best through integrated current amplitude histograms. If metal transfer modes can be firmly characterized by particular welding parameters, this capability, despite the complexity of the dynamic output, could be used for real-time automatic control of the desired kind of transfer. Also, the suitability of shielding gas mixtures could be evaluated with greater sureness.

Globular Transfer in GMAW

Transfer of metal from the electrode to the weld pool during GMAW as molten globules, for the most part, is not a favorable mode because weld penetration is shallow, much spatter occurs, and the deposited weld bead tends to be irregular. This mode of transfer is prone to occur when welding current is relatively low, but arc voltage is high enough to avoid short circuiting transfer: The dominant forces that shape this mode of transfer are gravity and surface tension. Gravitational force pulls the molten metal mass toward the arcing end of the electrode, but surface tension of the molten metal resists detachment of a droplet. When the size and weight of the bulbous molten end overpowers surface tension and Nernst's pinch force constricts and necks the molten metal at the electrode tip, a globular drop (often larger in diameter that the electrode itself) is separated and falls into the weld pool. The transfer frequency of molten metal drops will vary with levels of current and arc voltage, but the transfer of drops is not particularly rapid. The size of the drops, however, produces splashing when falling into the weld pool, which causes spatter. Steps usually are taken to shift to another transfer mode that is better suited for the joint to be welded. Forces generated by electrical parameters can be manipulated to improve globular transfer; for example, as the current is increased the size of drops become smaller because surface tension decreases as the temperature of the molten metal at the electrode tip is raised. Also, the frequency of drop formation-and-detachment per unit of time is increased. Globular transfer is likely to be the mode of transfer when using CO_2 shielding. Changes in the electrical parameters as can now be obtained with new power supplies will improve transfer conditions. Substantial use of argon in the otherwise CO_2 gas shield is an effective recourse. With a 50-50 percent argon-CO_2 mixture, the pinch force increases and the size of drops being transferred is decreased. Of course, if gas cost is not an important consideration, the shielding can be changed to argon with a small addition of CO_2 or of oxygen. The gas mixtures will shift the transfer mode toward streaming (axial) spray, but whether this IIW type "D" mode actually evolves also will be dependent upon having a relatively high welding current. With an argon-based shield containing additions of CO_2, O_2, or He, changes occur that are related to the properties of these additive gases. The argon, however, imparts pinch-effect, and promotes spray-type transfer. Ionization of the argon plasma occurs readily, less spatter is produced, and the transfer of metal from the electrode to the weld pool is much smoother.

Repelled Transfer in GMAW

IIW type "B" transfer is a transitory mode between the short-circuit and spray-transfer modes. The sketch of type "B" indicates why repelled transfer can be very troublesome when attempting to control weld penetration and weld bead appearance. This mode occurs when electromagnetic force generated by welding current is directed upward and almost equals gravitational force. This unfavorable resultant of forces develops when welding current concentrates over a localized area on the lower portion of the molten drop. A change in shielding gas to a mixture rich in argon will assist in moving away from repelled transfer; but, a change in electrical parameters (e.g., using higher current) usually is more effective in avoiding the repelled-transfer mode.

Projected Transfer In GMAW

IIW type "C" projected transfer is a tolerable mode providing the interaction between the various forces is favorable and can be maintained as a stable operating condition. Projected transfer can be viewed as an improved form of globular transfer. The drops of molten metal are smaller in diameter than that of the electrode, and the drops are accelerated rapidly toward the weld pool by electromagnetic force rather than the lower velocity produced by gravity. Projected transfer is achieved by ensuring a sufficiently high current to avoid globular transfer. Projected transfer is possible with CO_2 gas shielding, but a gas mixture rich in argon broadens the range of permissible current and voltage.

Streaming or Axial-Spray Transfer in GMAW

IIW type "D" transfer is the very useful mode that operates when current has been increased above the operating threshold outlined in terms of electrical conditions and shielding gas in Chapter 6. The transition to this mode is dramatic because of the manner in which the melting electrode tip becomes pointed and a steady stream of droplets of metal are driven at high velocity toward the weld pool. This manner of melting by the electrode is caused primarily by strong electromagnetic forces which are proportional to the square of the welding current being passed through the arc. Under the influence of these forces, the tip of the electrode over a short distance back from the tip becomes an anodic zone and the surface melts rapidly to produce the sharply pointed molten end. The flight of droplets leaving the tip and moving toward the cathodic weld pool are concentrated in a small central area of the plasma, and this constitutes a stiff, stable arc with exceptionally good penetration capability directly beneath the area of droplet impingement.

Mention was made in Chapter 6 of the improvement in axial-spray transfer by adding small amounts of either carbon dioxide or oxygen to an argon shield. This aspect of shielding gas formulation now has advanced to a quantitative assessment of "oxygen potential," and ways of evaluating the influence of oxygen or of oxygen-containing gases in an argon-based shield. A number of formulas appear in the literature for calculating an oxygen potential (OP) for argon-based shielding gas mixtures, and the latest of these also take into consideration the oxygen contents of the filler metal and the base metal. Briefly, to employ the axial-spray transfer mode properly, an argon-based shield should contain at least two percent oxygen, or at least five percent CO_2. Studies have been conducted with oxygen additions ranging up to five percent, and CO_2 up to 20 percent to observe effects, including degree of arc stability, wetting of the base metal (particularly along the toes of the weld bead), amount of spatter, and the amount of surface slag that forms on the weld. Data also have been gathered on the increased loss of manganese and silicon that accompanies rising oxygen potential. In general, the losses of these two primary alloying and deoxidizing elements in steel increase with rising OP. With oxygen as the additive, the loss increases linearly, but with CO_2 as the additive, the loss does not follow a linear relationship. Analyses of

GMAW weld metal for oxygen content found that CO_2 as the additive gave lower residual oxygen levels than when oxygen was used as the additive. This research found that oxygen potential deserved attention not only for its influence on operating characteristics, but also because of its effects on weld metal mechanical properties. With very low OP and with resulting low residual oxygen in the weld metal, strength tends to increase because of greater recovery of manganese and silicon. However, weld metal toughness was adversely affected because very low residual oxygen did not form minute oxide inclusions in sufficient number to nucleate the formation of acicular ferrite microstructure. As oxygen potential of shielding gas increases, weld deposits assume a more convex contour, the amount of surface slag increases, but the wetting of the base metals is improved. With more than about six percent oxygen, or more than about 15 percent CO_2, welding conditions start to deteriorate; and with more than about 30 percent CO_2, a true axial-spray transfer does not develop.

A steel electrode deposited by GMAW with a helium gas shield does not undergo strong pinch effect. The molten metal attempts to transfer in the globular mode. The deposit under helium will be noticeably wider than with argon, and the area of penetration is larger, although not as deep as the center-penetration of a deposit under the argon shield. GMAW is not practiced with straight helium gas for the shield because the arc is very hot, and the operation is disturbed by poor transfer mechanics, spatter, and other undesirable features.

Research on shielding gas mixtures for GMAW shows much promise of being able to improve the axial-spray transfer mode to permit faster travel speed, increased rate of metal deposition, and deeper penetration if desired. Recent work found a mixture of 65 Ar + 26.5 He + 8 CO_2 + 0.5 O_2 in volume percentages broadened the ranges for travel speed, metal deposition rate, and penetration depth to almost double those for simpler mixtures of an argon-based shield with additions of O_2 and CO_2. Ordinarily, more than eight percent CO_2 in an argon-based shield leads to the deterioration of operating characteristics. Yet, when helium is included, the mixture improves the characteristics to a surprising degree, hence the reason for continued research on shielding gas mixtures. Normally, the cost of shielding gas (Ar + O_2 or CO_2) is only about four percent of the total for labor, energy and consumables for welding plain steel by GMAW. If gas with a substantial He content is employed, the gas mixture cost might rise to about 15 percent. However, if a new mixture can increase welding speed significantly, a saving can be made because labor represents about 75 percent of the total welding costs.

Pulsed Spray Transfer in GMAW-P

In addition to efforts to extend the capabilities of the axial-spray transfer mode as described above, attention also is being given to optimizing the pulsed spray arc transfer mode through use of argon-based mixtures of shielding gas. The most commonly used gas mixture at present for the pulsed-spray transfer mode (GMAW-P) is 95 percent Ar and 5 percent O_2, but users should be alert for technical reports of progress on new mixtures. Producers of gases for shielding arc welding presently market mixtures which they advocate for certain GMAW operations (actually the mode of transfer). However, the composition of the gas mixture sometimes is not disclosed.

As progress is made with new mixtures, it may be necessary to make increased use of mixing-type-flowmeters on individual tanks of gases to capitalize on new-found shielding gas technology.

Rotating Droplet (Kinking) Transfer in GMAW

This is an unusual (and generally unfavorable) form of axial-spray transfer illustrated in Figure 8.6 as IIW type "E". Rotating droplet transfer develops when arc energy (current and arc voltage) is increased substantially above the normal axial-spray threshold. Increased electrode stick-out beyond the electrical contact also contributes to the occurrence of type "E" transfer. With the greater arc energy and resistance heating from stick-out, the electrode quickly melts over an abnormal length and this molten column rotates in a helical spiral because of dynamic electromagnetic forces surrounding the arc. At the lower tip of this rotating molten column, a stream of droplets are projected toward the cathodic base metal. The welding current needed to generate this rotating droplet mode is about 30 percent higher than that representing the axial-spray threshold for a given set of conditions. A high rate of electrode melting is experienced, and the arc energy and metal deposition is distributed over a broader area than with ordinary spray-arc mode. Although this form of transfer is reported to be useful for weld metal overlays and for depositing large fillet welds in heavy sections, its application obviously must be carried out with thorough pre-testing of welds, and with monitoring to control the quality of welds produced. The shielding gas must be one that is argon-rich, but helpful additives are yet to be clearly identified and their optimum percentage established.

Explosive Drop Transfer in GMAW

Explosive droplet transfer, IIW type "F", obviously is a transfer mode to be avoided if a sound weld is to be made. A number of improperly selected welding parameters or procedural items can cause this abnormal mode of electrode melting. For example, extremely high current, or incorrect shielding gas for the operating conditions (e.g., CO_2 shields being used with very high current and high arc voltage). Ordinarily, the welder or welding operator will be made aware of this form-of transfer by highly disruptive behavior that produces more spatter than deposited weld metal.

Short Circuiting Transfer in GMAW-S

Short circuiting transfer, now identified as GMAW-S, illustrated in Figure 8.6 as IIW type "G," also was discussed in Chapter 6 and illustrated in more detail in Figure 6.13. This earlier review pointed out the role played by CO_2 shielding gas, and by mixtures of CO_2 and argon. Attention has shifted away from the short circuiting transfer mode of GMAW-S to the pulsed spray mode of GMAW-P because of concerns over the occurrence of lack-of-fusion defects along the weld metal-base metal interface of welds made with short circuiting transfer. Nevertheless, this form of transfer can serve extremely well in certain applications if the welds are fit-for-purpose.

Figure 8.7 summarizes the information just offered on the influence of shielding gases in the GMAW process, and it includes additional information regarding the kind of welding current, and polarity when dc.

Flux Cored Arc Welding (FCAW)

Flux cored arc welding using gas-shielded type electrodes (FCAW-G) was discussed earlier in this chapter from the standpoint of fluxes employed in the core. Shielding during welding either with CO_2, or with a mixture of argon and CO_2 was given some attention in Chapter 6 when the FCAW process was covered. Research on optimizing the deposition of FCAW-G electrodes through the shielding gas has not been a strong activity simply because flux ingredients in the core exert more pronounced and

SHIELDING GAS	KIND OF WELDING CURRENT		
	DIRECT CURRENT		ALTERNATING CURRENT
	ELECTRODE POSITIVE (DCEP)	ELECTRODE NEGATIVE (DCEN)	
100% ARGON	ARC WANDERS, FINE SPRAY TRANSFER	WILD ARC, POOR GLOBULAR TRANSFER	ALTERNATING GLOBULAR AND SPRAY TRANSFER
98% ARGON 2% OXYGEN	STIFF ARC, GOOD SPRAY TRANSFER	FAIR SPRAY TRANSFER	GOOD SPRAY TRANSFER IF ELECTRODE EMISSIVE COATED
100% HELIUM	"HOT" ARC, GLOBULAR TRANSFER	POOR GLOBULAR TRANSFER	
50% ARGON 50% HELIUM	DEEP PENETRATION FAIR SPRAY TRANSFER	GLOBULAR TRANSFER	
100% CARBON DIOXIDE	GLOBULAR TRANSFER MUCH SPATTER		SPRAY TRANSFER IF ELECTRODE EMISSIVE COATED
90% ARGON 10% CARBON DIOXIDE	FINE GLOBULAR TRANSFER	(SHORT-CIRCUITING TRANSFER CAN USE 75% Ar + 25% CO_2)	
100% CARBON DIOXIDE SHORT-CIRCUITING TRANSFER	SEE FIG. 6.13		

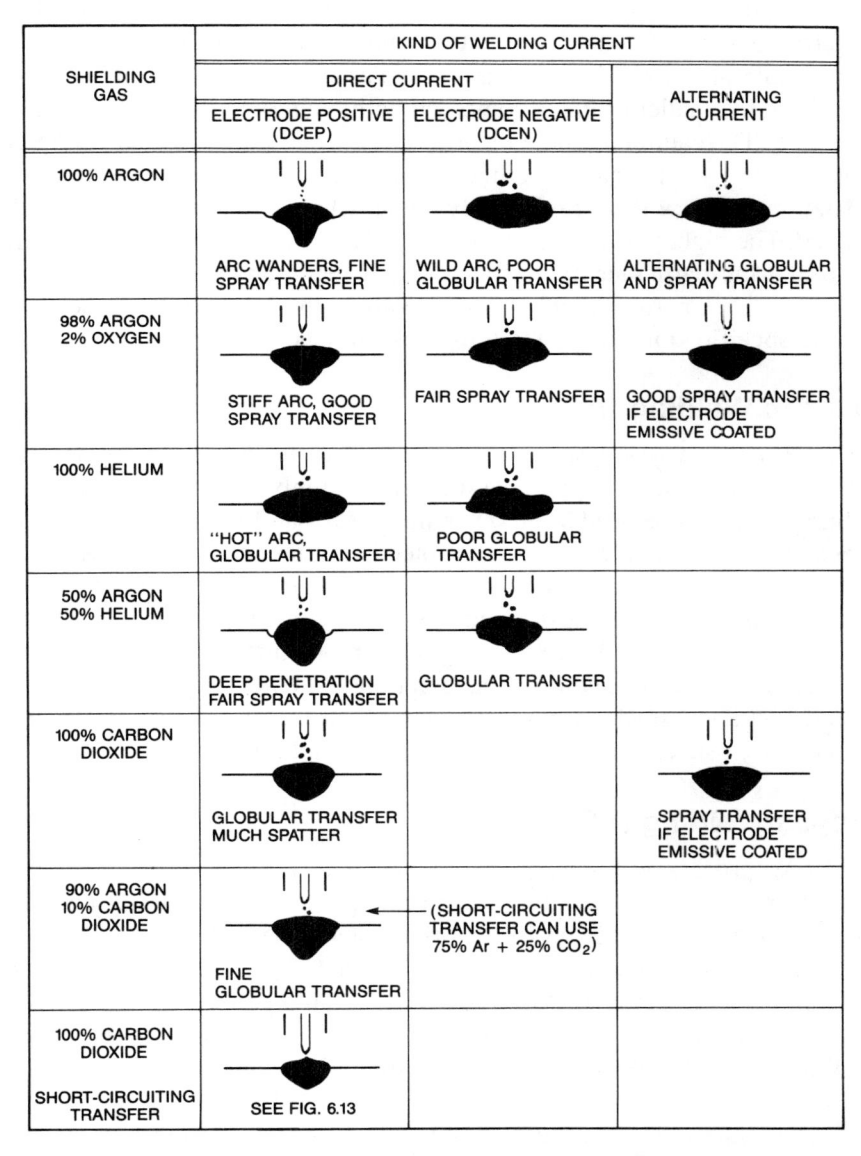

Figure 8.7 — Typical weld bead deposit shapes in GMAW process when using different shielding gases and a variety of welding parameters

controllable effects than from shielding gases. Also, CO_2 offers lowest gas cost, therefore much effort has been directed toward developing core formulations that work well with CO_2 shielding. This strategy puts a product in the hands of the users that carries clear instructions regarding the shielding gas to employ for protection during deposition. If other gas-shielding mixtures were employed, the results could be abnormal.

Literature on FCAW-G electrodes, and even ANSI/AWS A5.20 and A5.29, cautions users regarding the employment of argon-CO_2 shielding. AWS qualification procedures are mainly based on welding tests with CO_2 shielding; yet, it is recognized that welding behavior of FCAW-G electrodes usually improves under an argon-CO_2 shield. The preferred gas mixture is 75 percent Ar and 25 percent CO_2, and this mixture is widely available pre-mixed in cylinders. This gas mixture assists in obtaining spray-like deposition of smaller droplets compared with the more globular deposition under CO_2 shielding. The cautionary note regarding the use of argon-rich shielding gas concerns the recovery of manganese, silicon, and other oxidizable elements in the weld metal. Their recovery will be significantly higher when argon-containing shielding gas is employed. The higher percentage recovery depends on the proportion of inert gas in the shield and the particular oxidizable element being analyzed. If recovery of elements like Mn, Si, Cr, Al, Ti, and V is unusually high, the strength of the weld metal may exceed specification limits, and toughness may be adversely affected.

Plasma Arc Welding (PAW)

The PAW process was described in Chapter 6 with some detail regarding orifice gases that transfer the arc energy to the workpieces. Only brief mention was made that an auxiliary gas could be used as an outer shield surrounding the plasma and that this auxiliary shield could provide increased protection to the metal being fused. Since the auxiliary gas does not contact the tungsten electrode, the limitations on gas selection as outlined for the GTAW process do not apply to the PAW process. Gas selection for auxiliary shielding is guided primarily by the kind of metal being welded, but other factors, such as cost, often require consideration. When employing PAW for joining carbon and low-alloy steels, argon, argon-and-hydrogen mixtures, and even carbon dioxide have been successfully used for auxiliary shielding.

Electrogas Welding (EGW)

The EGW process was described in Chapter 6, but without including detailed information on the role of shielding gas in this process. AWS A5.26 covering consumables for electrogas welding of steel briefly states that externally-supplied shielding-gas for welding qualification specimens shall be CO_2 for all classifications of weld metal except for a low-alloy Ni-Mo-V steel where an argon + CO_2 mixture is permitted. Other gases may be used for qualification testing when agreed upon between supplier and purchaser. Stringent demands regarding shielding gas are not made for the EGW process because characteristics tailored for depositing *beads* of weld metal are more-or-less unimportant. With EGW, the weld metal is continuously collected in the weld pool between the faces of the square groove in the workpieces and the confining copper shoes or dams. In fact, a "gas box" sometimes is built to surround the weld joint and to ride upwards with the weld pool level as the operation progresses. The box maintains an atmosphere of shielding gas under slight positive pressure to exclude air. Whether this auxiliary atmosphere is CO_2, or an argon-rich mixture makes little difference in weld soundness or appearance, but weld metal composition can be affected by oxidizable element recovery just as explained for gas-shielded type FCAW electrodes.

Other Welding Processes Using Gas Shielding

Utilization of gases as an artificial atmosphere during welding to protect against the adverse effects of air always has been foremost in thinking of ways to improve the welding processes. Long before shielding was realized in atomic hydrogen welding (AHW) as mentioned earlier, oxyfuel gas welding (OFW) showed in a simple way that even an atmosphere of spent combustible gases offered some protection. In oxyacetylene welding (OAW), the importance of knowing whether the gas flame has been adjusted to be *neutral*, *oxidizing*, or *reducing* was described and illustrated in Chapter 6; see Figure 6.58. The protection afforded by gases of combustion has been put to use in other processes, for example, pressure gas welding (PGW). Surveying the practices in industry finds that special techniques with gases have been instituted in a number of processes when the results justify the means. Upset welding (UW), flash welding (FW), and friction welding (FRW) are just a few of the familiar processes that for certain applications have been improved by the user with some kind of shielding gas. However, it remained for the arc welding processes to clearly demonstrate the dramatic improvements that could be accomplished with auxiliary gases, both in negating the effects of air, and in installing new, helpful reactions during the welding operation. This experience has strongly encouraged development work with non-arc processes to examine possible benefits by using an auxiliary gas shield. Examples of effort in this direction are briefly described below:

Laser Beam Welding (LBW)

Laser beam welding is a good example of a welding process that has great potential, and yet the phenomenon of plasma formation can seriously interfere with its application under certain circumstances and render it almost unusable. The development of plasma jet sweeping with auxiliary gas was mentioned in Chapter 6 for the LBW process, and it has been established that helium is the best when welding with continuous wave CO_2 lasers. Helium provides excellent shielding of the weld pool from oxidation, and more importantly, it is less prone than other gases to plasma formation by reaction with the focused laser beam. Weld penetration of laser welds shielded with argon appears to be adversely affected. Argon apparently introduces stronger attenuation effects on the incident laser beam.

Non-Vacuum Electron Beam Welding (EBW-NV)

Nonvacuum electron beam welding is an interesting example of a process that can be greatly improved by an auxiliary shielding gas. Research has shown that shielding gases able to provide improvement can range from *very dry air* to an inert gas. Naturally, inert gas provides the greatest improvement, but its cost- effectiveness must carefully weighed against that of cheaper gases. Helium has been used in much of the research conducted thus far, but only further research will determine whether acceptable improvement can be gained with less-expensive gases. Perhaps improvement can be attained by an entirely different EB gun-to-workpiece arrangement. Research is presently underway and the welding literature must be monitored for progress.

Protecting Brazing Processes From Air

Brazing operations make use of virtually every known means of protecting the metal being joined from oxidation. Fluxes are most commonly used, but shielding with a controlled atmosphere often is the technique employed for large-scale production. The popular fluxes are borate compounds of sodium, potassium, and lithium, and these serve best when the operation is performed at temperatures about 760°C (1400°F) and higher. These compounds have good solubility for oxides formed on steel, and they provide a protecting molten blanket on the metal workpieces during lengthy brazing cycles. Compounds of fluorine have better oxide removal capacity. Sodium, potassium, and lithium fluorides form very active fluxes and they react readily with most metal oxides at elevated temperatures. The fluorine compounds often are incorporated in fluxes intended to react with chromium oxide and other refractory oxides formed from alloying elements in the base metal. Fluorides also are effective in increasing the capillary flow of brazing alloys. Chlorides have fluxing characteristics similar to those of fluorides, but the chlorides are not quite as effective. Boric acid in the calcined form is a commonly used base for brazing fluxes. It has the ability to improve the viscosity of the melted flux at brazing temperature to avoid excessive runoff, and it facilitates removal of flux residue from the brazement surface after cooling. AWS in its *Brazing Handbook* has classified fluxes into 15 categories according to their chemical type, application, and use on various base metals; but no attempt is made to set composition limits.

When brazing of steel was reviewed as a joining process earlier, various ways of avoiding the need for flux by employing a controlled atmosphere in a heating furnace was outlined. The remedial measures ranged from no atmosphere whatsoever (a vacuum of at least 0.5 torr), the use of inert gases, and finally to the oxide-reducing gas,

Table 8.10 - Controlled Atmospheres Used in Furnace Brazing Carbon and Low-Alloy Steels				
AWS Brazing Atmosphere Type No.	2	4	5	7
AGA Class No.	102	301	601	
Description	Rich Exothermic Based	Lean Endothermic Based	Dissociate Ammonia	Hydrogen (Purified)
Approximate Composition in Volume Percent				
H_2	15	37	75	100
N_2	70	45	25	
CO	10	17.3		
CO_2	4.5	0.4		
CH_4	0.5	0.3		
Moisture Requirement, Maximum (Dew Point Temperatures)				
°C	–	0	-51	-60
°F	–	32	-60	-75

hydrogen. The generation and use of prepared gases for a furnace chamber at brazing temperatures is a highly developed and complex technology. Important safety precautions must be taken with the large volumes of gas, many of which are combustible and explosive. The literature covering this subject should be very thoroughly perused before attempting usage. AGA and AWS have simplified the array of gas mixtures used in industry for protective atmospheres by organizing basic classes and assigning generic names and identification numbers. Several kinds of atmospheres often used in the furnace brazing of carbon and low-alloy steels are given in Table 8.10.

An *exothermic* gas like AWS 2 or AGA 102 sometimes is selected for copper brazing of low-carbon steel because it is (1) low in cost, (2) adequately reducing to maintain bright surfaces and to promote wetting, (3) relatively low in sooting potential, and (4) easily prepared in a gas generator. Sometimes this gas requires refrigeration to reduce its dew-point to ensure bright steel surfaces. The potential of this gas for carburizing or for decarburizing the steel usually is tolerable when brazing most steels. If the gas mixture has a high dew point, then its decarburization potential increases substantially. If a particular carbon level must be maintained in the steel, particularly a moderately high carbon content (e.g., about 0.40 percent C), this can be done by decreasing the dew point and also removing the carbon dioxide. Generator prepared gases that are *endothermic* (e.g., AWS 4, or AGA 301), sometimes are used when decarburization must be avoided, and an atmosphere with greater oxide-reducing capability is wanted.

Dissociated ammonia (AWS 5 or AGA 601) is a medium-cost gas that is prepared by passing ammonia (NH_3) through a dissociating unit consisting essentially of a pressure vessel filled with nickel shot that serves as a catalyst. The vessel and its contents are heated to about 925°C (1700°F) where the ammonia dissociates into a mixture of 75 percent H_2 and 25 percent N_2. Anhydrous ammonia is used, but if the dew point of the dissociated gas is not sufficiently low, the gas can be passed through a molecular-sieve absorber to remove water vapor, carbon dioxide, and any undissociated ammonia. The dew point of the dissociated gas typically is -40°C (-40°F) or lower, the CO_2 content is below about 0.15 percent, and residual ammonia is less than 0.03 percent. The high hydrogen content and low dew point of the gas gives the heated atmosphere a strong oxide-reducing potential, hence its popularity for furnace brazing steel.

Attention frequently is directed to the amount of water vapor contained in gases, especially when dealing with the controlled atmosphere in a brazing furnace, and the *dew point* is quoted as a quantitative measure of its moisture content. This term signifies the exact temperature (at a given pressure and a constant water vapor content) at which the saturation water pressure of a given parcel of gas is equal to the actual vapor pressure of the contained water vapor. In other words, the slightest cooling from this temperature immediately causes moisture to precipitate (as dew). Measurement of moisture content in a gas can be accurately determined with comparatively simple instruments, and is favored for the convenience of this method over gas analyses with other more elaborate apparatus. It should be appreciated that a dew point of -40 °C (-40°F) indicates that the gas contains only 66 ppm of moisture. A dew point of -60 °C (-75°F) corresponds to a moisture content of only six ppm.

Soldering — Ways to Achieve Bonding

From the variety of soldering process variations shown in Figure 6.2-C, it might seem that a number of methods likely are put to use to remove the surface film of oxide that ordinarily is present on iron and steel. Yet, this is not the case with soldering. Most of the many methods for dealing with surface oxide described for the welding processes are not applicable. Also, some of the protective or oxide-removal methods employed for brazing cannot be used for soldering. The low temperature regime of the soldering process is the principal reason that ways of dealing with surface oxide are sharply limited. Nevertheless, the removal of oxide from the surface of iron or steel is essential to achieving a satisfactory bond between solder and the workpiece as described on page 625.

Application of an active flux, whether in solid or liquid form, is the usual means incorporated in soldering processes to overcome the surface oxide obstacle. Atmospheres of protective reducing gases, as employed for brazing, are not effective at the lower temperatures of soldering operations. The only other means that has been used to some extent for dealing with oxide on ferrous materials during soldering is localized mechanical treatment.

Fluxes for Soldering

Fluxes for soldering can be divided into three broad categories — inorganic fluxes, organic fluxes, and noncorrosive protectors. Removal of surface oxide from iron and steel preparatory to wetting by solder is easier with the inorganic compounds, which usually are an active chloride compound, or an acid. The acid can be either hydrochloric or phosphoric — but not nitric. This is because nitric acid, under some circumstances, behaves as an "oxidizing" reagent; and it will actually thicken rather than remove the oxide film (see page 277 for further explanation). The remaining two flux categories, organic compounds and noncorrosive protective substances, are not particularly effective as flux for soldering iron and steel. Many proprietary fluxes marketed for soldering iron and steel are mixtures of active inorganic compounds along with some of the organic types. It is important to have some knowledge of flux composition and its corrosion potential, because this information must be considered in planning flux removal and cleaning of the finished workpiece after soldering.

Mechanical Means to Accomplish Solder Bonding

Mechanical treatment, applied locally to the workpiece surface concurrently with a particular surface area immersed or covered with molten solder, was discovered long ago to be an alternative for fluxing. Initially, the mechanical treatment simply amounted to vigorous wire brushing with a stainless steel wire brush applied to the area through the molten solder. The brushing treatment was intended to disrupt the oxide surface film and allow the molten solder to wet the area stripped by abrasion from the wire brush. The brushing treatment was a more-or-less crude technique that did not see wide usage, but it encouraged further investigation of mechanical means of oxide removal from a workpiece surface beneath molten solder.

Soldering technology now employs mechanical treatment by utilizing ultrasonic energy to break up the surface oxide and permit wetting by the molten solder. The ultrasonic technique has been adapted to "soldering irons" for manual operations, as well as molten solder baths for dip-soldering and wave-soldering operations. Sometimes ultrasonic energy is applied to supplement the effectiveness of a less-active type of flux; in this way, the post-soldering cleanup step can be carried out more easily, or possibly eliminated.

SUGGESTED READING

Fluxes and Slags in Welding, by C. E. Jackson, WRC Bulletin No. 190, December 1973

An Introduction to Welding Fluxes for Mild and Low Alloy Steels, by M. L. E. Davis, The Welding Institute, Abington, Cambridge, England, 16 pp, 1981, ISBN 0 853 001 456, AWS Code IWFAS

"Shielding Gases for Gas Metal Arc Welding," by N. Stenbacka and K. A. Persson, AWS *Welding Journal*, v 68, no. 11, pp 41-47, November 1989

Brazing Handbook, American Welding Society; 493 pp, AWS Code BRH; 1991; ISBN: 0-87171-359-4.

Chapter 9

In This Chapter...

FUSION WELDS 787
  Solidification of Weld Metal 788
    Modes of Primary Solidification Structure 794
  The Weld Zone 798
  The Unmixed Zone 799
  The Partially Melted Zone 800
  The Heat-Affected Zone 801
  Unaffected Base Metal 802

SOLID-STATE WELDS 802

MICROSTRUCTURAL TRANSFORMATIONS IN SOLID IRON AND STEEL 802
  Phase Changes in Steel 803
    Ferrite 804
    Austenite 804
    Cementite 806
    Pearlite 807
    Widmanstätten Pattern 809
  Microstructural Changes in Steel During Heating 810
  Microstructures Formed in Steel During Cooling 816
    Martensitic Microstructures 818
  Isothermal Transformation of Austenite 824
    Pearlite Formation Isothermally 826
    Bainite Formation Isothermally 828
    Martensite Formation 828
  Reappraisal of Microstructures Formed in Steel 831
    Upper Bainite 832
    Lower Bainite 833
  Importance of Critical Cooling Rate 834
  Importance of Delay-Time Before Austenite Transformation 836

  Martensite: Implications in Welding 836
    Temperature Range for Martensite Formation.....837
    Quantitative Prediction of Martensite Format'n ...839
    Martensite Hardness Rationale 841
    Martensite Format'n Monitoring by Acoustic Emission Signals 842
  IT Diagrams: Summation of Usefulness 843
  Transformat'n of Austenite During Cont. Cooling....848

PREDICTION OF MICROSTRUCTURES IN THE HEAT-AFFECTED ZONES OF WELDS 851
  Jominy Method of Predicting HAZ Microstructure ...852
  Mathematical Approach to Prejudging HAZ Suitability 856

TRANSFORMATIONS IN WELD METAL 857
  Continuous Cooling Transformation Diagrams for Weld Metal 867
    Importance of Weld Metal Composition 870
    Role of Grain Size in Weld Metal 872
    Influence of Nonmetallic Inclusions in Weld Metal 872

STUDY OF A TYPICAL FUSION WELD IN STEEL 876
  Making Welds with Good Toughness 883
  The Challenge of Optimizing Weld'g Procedures....888
    Base Metal 889
    Weld Metal 890
    Welding Process and Procedure 890
    Weldment Property Testing 890
    Nondestructive Examination 890

SUGGESTED READING 891

Simple Welds in Iron and Steel

The preceding eight chapters of this first volume provided a background extracted from many sciences and technologies known to be involved in welding — structures of materials, alloying of metals, properties of metals and alloys, heat generation and flow, etc. Now, this basic information can be put to specific use. As with previous editions, Chapter 9 of this text carries the title, *Simple* Welds in Iron and Steel. But, in truth, no weld is "simple;" a multiplicity of events must be relied on in its making. When these events are properly controlled and manifolded, a weld of definable nature is produced. Of course, there are many variations in welds — primarily because of the different processes employed, but also because of the particular procedure followed. This chapter is intended to provide a meaningful perspective of *typical* welds in iron and steel. Details will be studied to see how workpiece conditions and operating parameters integrate to establish the properties of a weld. The prototypical welds examined will serve as bases for comparison when, later, the multitude of different steels and welding procedures are taken up in Volume II.

Certain mechanisms and phenomena that operate during the making of a weld are difficult to make plain by photomicrographs. This is because of variations and continuing microstructural transformations that occur for various reasons. Therefore, schematics will be employed in this chapter, at least at the start, for illustrating the basic metallurgical action to be conveyed. This approach should make photomicrographs of actual in-situ conditions easier to interpret later in studies of particular steels.

FUSION WELDS

The majority of processes in use today involve melting of the workpieces. Most welds require melting a substantial amount of the base metal, and possibly adding molten filler metal. Eventually, there is need to anticipate effects stemming from the amount of melting that takes place in a given operation; but, for discussion here, weld metal will be treated in general terms, and its role in welding will be examined first. Later, attention will be given to base metal, which is subjected to a gradation of temperatures ranging from just below the liquidus downward to a temperature that has no significant effect on the base metal. However, there is good reason to look very closely at the so-called interface between the weld metal and base metal, because this area is not entirely without features. Rather, the base-metal/weld-metal interface often has discrete, very narrow zones in which the heating and cooling cycles and the maximum temperature have resulted in microstructures that differ from the norm for weld metal or for heat-affected base metal. These interfacial zones often present a mingling of phases and constituents in unique arrays. Although quite innocuous in most cases, such mingling can be influential in establishing cracking susceptibility and mechanical properties in welds made in high-strength steels. Consequently, the unmixed zone and the partially-melted zone will be included in this examination of the typical fusion weld.

Solidification of Weld Metal

If the unique feature in the solidification of a fusion weld, *epitaxial growth*, were ignored, fusion welding could be likened to making a casting. When a casting solidifies in a metal mold, the first crystals of the casting that form at its outer surface (i.e., against the mold wall) nucleate heterogeneously. Grains then grow inward on these randomly oriented nuclei in a direction perpendicular to the mold wall. Unlike solidification in a casting, however, molten weld metal need not undergo initial nucleation during solidification, because the wetted face of the base metal provides the crystal lattice on which atoms of the cooling weld metal can site themselves and form grains. The degree of coherency between the grains sited in the base metal and those in the weld metal depends on similarities in their chemical composition, as well as their crystallographic structure. Also, allotropic transformation occurring in either the base metal or the weld metal can mask prior epitaxy in microstructures at the base-metal/weld-metal interface. Indeed, allotropic transformation will occur with most of the carbon and low-alloy steel compositions. Welding travel speed is one of the factors that affects the manner in which epitaxial growth proceeds at the interface.

Figure 9.1 consists of four schematic drawings which show epitaxial growth of grains from base metal into solidified weld metal. Simplified circumstances are assumed in these drawings to assist in illustrating epitaxy and the variations in its mechanics as found in weld metal. The first, Figure 9.1-A, shows two sections of base metal with abutting edges that are to be welded autogeneously by an arc. The grain size in these sections has been enlarged to help demonstrate the phenomena that occur at the base-metal/weld-metal interface during solidification. The effects of three different rates of travel are demonstrated. To simplify illustration and discussion, the metal being welded is assumed to have a ferritic (bcc) microstructure which does not

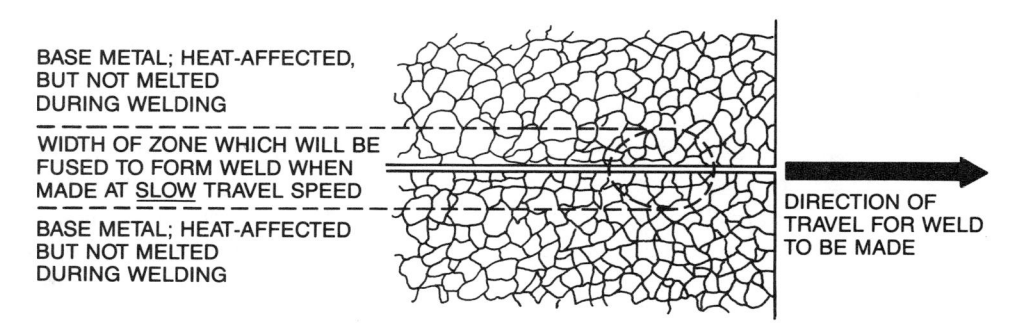

Figure 9.1-A — Schematic illustration of two plates forming a single-square-groove joint to be welded autogenously by an arc process.

Grains are sketched in the base metal to permit later illustration of epitaxial grain growth into the weld zone. Shape of weld pool and width of weld are indicated by broken lines as will be formed during *slow* welding travel speed.

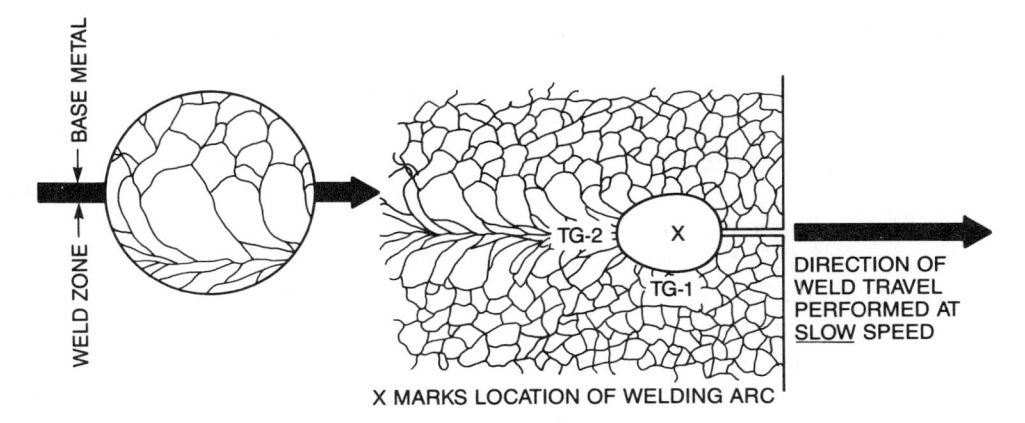

Figure 9.1-B — Plates from Figure 9.1-A after welding at *slow* welding travel speed

Epitaxial grain growth from base metal into weld zone (planar mode) undergoes curving directionality induced by rotation in steepness of thermal gradients (TG-1 to TG-2) as weld progressed slowly along the joint.

undergo crystallographic transformation during heating or cooling. Localized grain growth, which is to be expected in the base metal heat-affected zone, is not shown at this time to facilitate a comparison of weld zone details.

Figure 9.1-B shows the weld made at *slow* travel speed. The weld pool is almost circular in shape, and it forms a trailing fused weld zone which is relatively wide. Grains sited in the base metal with an exposed face on the base-metal/weld-metal interface are wet by the molten weld metal, and they act as a substrate on which growth of solid phase can take place. In general, grain growth will proceed in a direction perpendicular to isotherms and parallel to thermal gradients. As the weld zone cools and solidification initiates epitaxially, grains grow toward the weld centerline in columnar form, and in their later stage of growth they gradually curve in the direction of the traveling weld pool. Initially, solidification is a continuation of the entire face of each exposed base metal grain along the interface with no change in crystallographic structure. The epitaxial grain formation shown in Figure 9.1-B is called the *planar* growth mode, because each grain extending into the weld zone is a continuum of the parent grain in the base metal. In this particular steel, grain growth is most rapid in the <100> crystallographic direction. However, growth is competitive, and many points of blockage and nucleation of new grains can be seen. The curve in the columnar form developed because thermal gradients surrounding the weld pool gradually changed in slope as the pool traveled along its path. When heat from the arc first formed the weld pool, the steepest thermal gradient existed from the pool to the transverse location marked *TG-1* on the sketch. However, as the pool moved farther away, the thermal gradients changed in rotational fashion; and the thermal gradient from the longitudinal location *TG-2*, in the weld zone behind the trailing end of the weld pool, finally became operative. This explains the development of elongated grains lying almost parallel to the weld centerline in the last metal to solidify. The planar mode of epitaxial grain growth is observed infrequently in welds made in the carbon and low-alloy steels, because a number of circumstances usually are encountered that alter the solidification mechanism. Hindrance to the planar mode of grain growth comes mainly from the relatively fast rate of cooling that is typical of welds, and from the multiplicity of elements commonly present in steels. The ways in which cooling rate, composition, and other circumstances affect the solidification structure deserve study because of their frequent involvement in shaping the structure of real welds.

Figure 9.1-C illustrates the weld made at *moderate* travel speed, which is more typical of welds made in commercial practice. Although grain growth into the weld metal is epitaxial, the finer grain-structure of the weld zone has developed via *cellular* mode. The cells are microscopic, pencil-shaped protrusions of solid metal that freeze epitaxially on the exposed faces of base-metal grains along the base-metal/weld-metal interface. The growth of these cells into narrow, columnar grains is controlled by the combined influence of a number of factors which complicate the solidification mechanism. The shape of grains in the weld zone is determined by the geometric shape of the weld pool, a feature that is also established by the collective effects of a number of factors. The weld pool in Figure 9.1-C made at a moderate travel speed has a teardrop shape with its angular sector trailing behind the arc. This difference in shape from the oval pool produced at slow travel speed has occurred because the capability to conduct heat to cooler surroundings has been lowered at the weld centerline. More specifically, from the welding arc to the trailing end of the pool, thermal gradients are not increased as much in the longitudinal directions as they are in the transverse directions. The result is a longer, narrower molten pool and an increased cooling rate.

Increasing the cooling rate of the molten weld metal often leads to supercooling. Supercooling favors nearly simultaneous development of many cells on the faces of base metal grains along the interface. Also, supercooling increases *competitive growth* among the grains. This is the process by which grains with less-favorably oriented

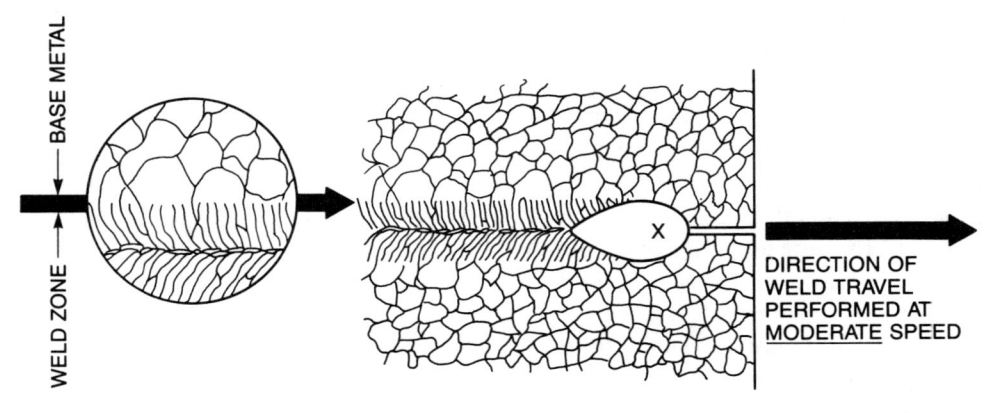

Figure 9.1-C — Similar plates and joint as shown in Figure 9.1-A, but welded at *moderate* welding travel speed

Epitaxial growth of grains into weld zone is the cellular mode because of constitutional supercooling promoted by faster cooling rate.

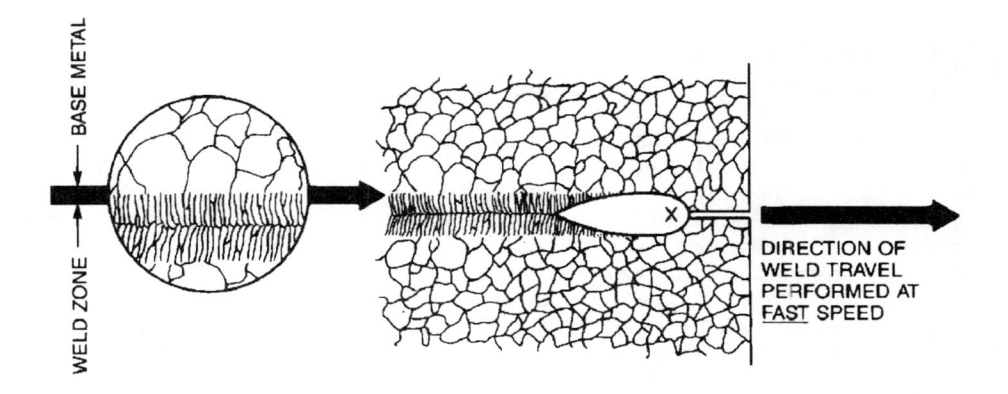

Figure 9.1-D — Similar plates and joint as shown in Figure 9.1-A, but welded at *fast* travel speed

With fast welding travel speed, weld pool has pronounced teardrop shape, weld zone is much narrower, and grain growth into weld zone is epitaxial, but with fine cellular solidification substructure. Grain growth is from base metal toward weld center with virtually no curve in the shape of grains.

growth directions are blocked by grains better oriented for continued growth. The most influential factor that promotes the cellular mode of grain growth is termed *constitutional supercooling*, because it involves compositional differences between the cores and the peripheries of individual cells. Constitutional supercooling is more likely with compositions that have a large temperature difference between their solidus and liquidus. With such compositions, microsegregation occurs during solidification, which steadily increases the solute concentration in liquid remaining ahead of and alongside each cell. The liquidus temperature of the enriched liquid is lowered continuously, increasing the difference between solidus and liquidus temperatures. Steels, in general, have appreciable temperature difference between solidus and liquidus, and certain alloying and residual elements can increase this difference significantly. Thus, conditions in most carbon and low-alloy steels favor the cellular-grain-growth mode of weld solidification.

Figure 9.1-D presents the weld made at *fast* travel speed. The weld pool has a pronounced teardrop shape; the width of the fused zone is much narrower; and the high rate of cooling that accompanies the fast welding travel speed has produced an even finer cellular mode of epitaxial grain growth, with very little curvature of the grains toward the direction of the receding weld pool. Welds having this microstructural configuration can be susceptible to centerline hot cracking if, during solidification, the amount of low-melting solutes is increased significantly in the last metal to solidify along the centerline.

Study of weld-zone primary grain growth has been carried far beyond the three simple exercises illustrated in Figure 9.1, and much effort has been directed toward identifying and quantifying the factors that exert control over weld structure. Two additional forms of primary weld structure that occur, although less frequently, have

792 Chapter 9

> **TECHNICAL BRIEF 44:**
> **Solidification Structures in Steel Fusion Welds**
>
> Fusion welding calls for specialized knowledge of how molten metals solidify, and the nature of microstructures formed as affected by composition and freezing conditions. Unlike base metals, most of which are hot worked and/or heat treated in the course of their production, weld metal is most often used in the as-deposited condition; although it often receives localized reheating during multipass welding. A further unlikeness that might not be easily discerned in the final microstructure lies in the solidification of the weld fusion zone which initiates by epitaxial growth on the crystalline structure of the base metal. This is a unique mechanism peculiar to welding for nucleating the first grains of solid metal to form. While epitaxial solidification soon gives way to competitive growth, nevertheless, basic metallurgical conditions that are developed through epitaxy at the weld boundary can be of considerable importance in establishing the qualities of the weld. Some of the conditions which must be considered, despite not being readily discernable because of subsequent transformations during cooling, include: (1) the size of grains in the initial epitaxially induced layer of solid metal, (2) the degree of columniation produced by privileged or selective directional growth, (3) the extent of micro-segregation of low-melting elements and compounds occurring either interdendritically, or intergranularly, and (4) disruptive distribution of various alloying elements by diffusion, and of impurities concentrating by reason of insolubility.
>
> Four kinds of solidification structures that start in the epitaxially-induced initial layer of solid weld metal have been identified, and their modes of growth have been designated as; (1) planar, (2) cellular, (3) cellular dendritic, and (4) columnar dendritic (Figure 9.2). The nature of early solidification in the weld zone is highly dependent on two factors; the amount of solute in the weld composition, and a parameter which represents the combined influence of temperature gradient (G) in the direction of solidification as divided by the rate of advance (R) of the solidification front. This parameter has been designated the *G/R ratio* (See Figure 9.3).
>
> The manner in which these factors influence the mode of weld metal solidification can be seen in Figure 9.3, where for any given solute content the weld microstructure is shown to become more dendritic as the G/R ratio decreases.
>
> Many of the problems that arise in fusion welding with cracking, porosity, localized embrittlement, and abnormal weld appearance have their origin in the solidification process. Pursuit of explanation of a troublesome weld condition is expedited by starting from the root-cause, which can lie in the initial or primary structures that form on solidification.

been termed *cellular dendritic* and *columnar dendritic*; these will be illustrated shortly. If attention is confined to carbon and low-alloy steel weld zones, a number of general observations can be outlined with regard to weld zone solidification microstructures. (See Technical Brief 44.)

Although they may vary greatly in size and shape, all primary grains that solidify at the base-metal/weld-metal interface exhibit epitaxial growth. Sometimes, the grains growing into the weld zone will have a distinctive substructure. The size of grains on which the epitaxial growth initiates determines the maximum cross-section size of the primary grains in the weld zone. Columnar grains in the weld zone can be especially long if conditions favor growth in a particular direction. Primary weld zone grains can be smaller than the base metal grains on which they nucleate (e.g., cellular mode of grain formation). Also, it is possible to nucleate fresh grains on minute particles in the fused weld metal; titanium oxide particles, for example, can form small primary grains. Of course, the grains in carbon and low-alloy steel welds are subject to refinement later from crystallographic transformations that occur during cooling in the solid state, inasmuch as the great majority of steel compositions are characterized by allotropy. Nevertheless, the nature of primary microstructures, as determined by their mode of solidification, will have strong influence upon the microstructures found after all subsequent allotropic transformations have taken place. Therefore, determining the mode of primary grain solidification and formation is an important first step in establishing the final microstructure in the weld zone.

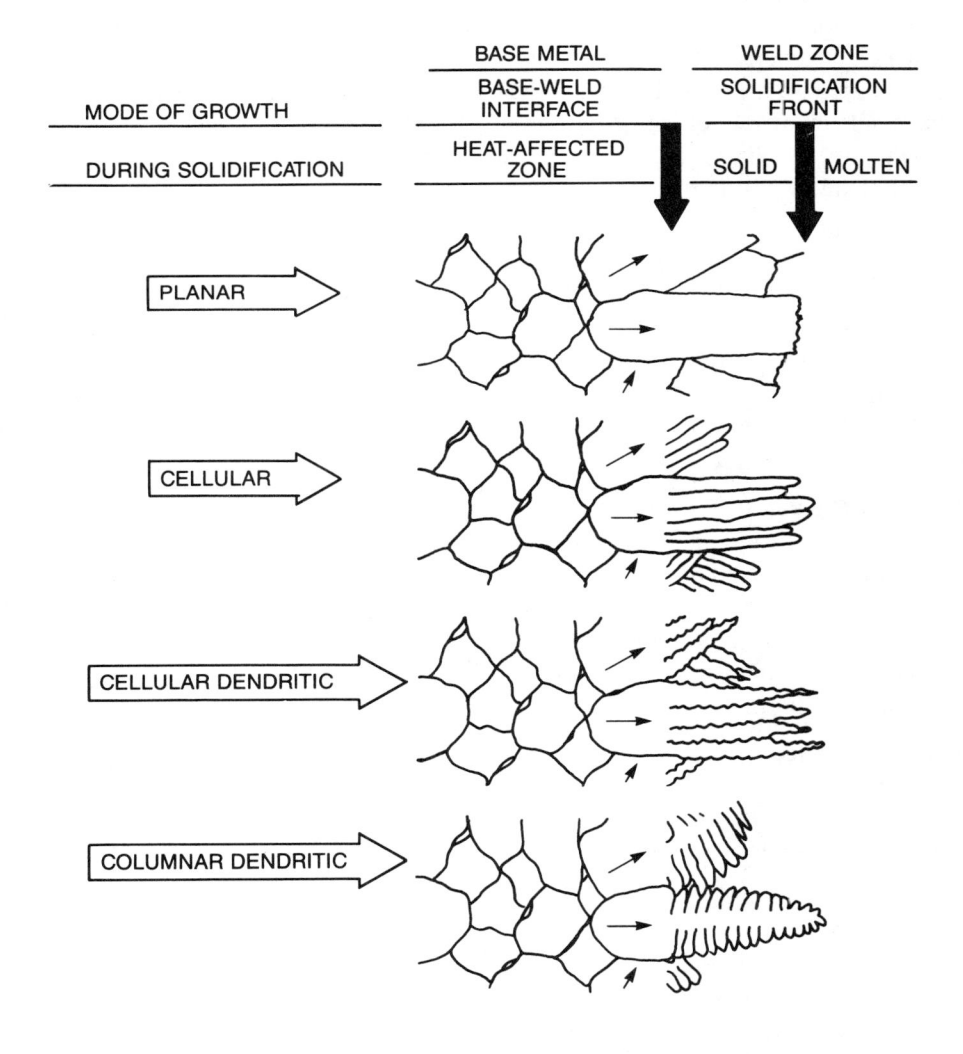

Figure 9.2 — Modes of solidification structure found in weld metal and their common designations

Arrows in base metal grains at interface with weld zone indicate preferred growth direction

Modes of Primary Solidification Structure

Composition of the weld melt — specifically, its content of solute elements — has been noted for exerting strong influence on weld zone microstructure through constitutional supercooling. In fusion welding the carbon and low-alloy steels, typical weld compositions contain alloying and residual elements that favor microsegregation and supercooling during solidification. Only in pure metals, and in a limited number of special steels, will microsegregation *not* develop during solidification of the weld zone. Figure 9.2 gives a perspective of four distinctive modes of primary epitaxial grain growth that can occur during the solidification process. The schematic sketches of the four modes are arranged in order of their occurrence starting with the planar mode, which is generally found when the weld melt has little or no propensity to undergo microsegregation. The cellular and the dendritic modes occur in weld compositions that have sufficient solute atoms to promote the microsegregation mechanism and constitutional supercooling, but whether these mechanisms are permitted to operate to a detectable extent will depend on welding conditions.

Welding Conditions that Influence Solidification Mode

The particular solidification mode that develops in the weld zone during freezing is governed essentially by three interrelated aspects of the weld which are dictated by the nature of the joint and welding conditions employed in the operation — namely: (1) welding travel speed; (2) shape of the weld pool, that is, its ovalness or the acuity of the trailing portion as a teardrop shape takes form at high travel speeds; and (3) temperature gradient in the direction of solidification. The manner in which welding conditions affect these three aspects deserves more detailed discussion, as follows.

Temperature Gradients

Temperature gradients are those extending over distances from welding arc to the pool boundary where solidification is taking place. The symbol, G, has been used by most investigators to identify temperature gradient. However, the steepness of the temperature gradient alone does not correlate well with solidification mode. Travel speed and weld pool shape also must be considered when assessing the influence of temperature gradient.

Welding Travel Speed

Welding travel speed commonly is identified by the letter, v, although some investigators prefer the term "travel velocity." This must not be confused with "solidification velocity," which is a fundamental aspect of the freezing of the weld zone.

Weld Pool Shape

Shape of the weld pool was discussed earlier as a welding condition that influences the weld-zone primary microstructure. The shape of a weld pool in open-arc welding is determined mainly by welding travel speed, as illustrated in Figure 9.1; but a key parameter affecting solidification mode is the rate of advance of the solidification front in the weld zone — the "solidification velocity" mentioned above. A parameter, R, is used to represent movement of this front; however, different approaches have been used by investigators in arriving at a value for R. Generally welding travel speed, v, is multiplied by a value representative of the shape of the weld pool to provide a meaningful R value. The shape of the weld pool is quantified principally by considering the

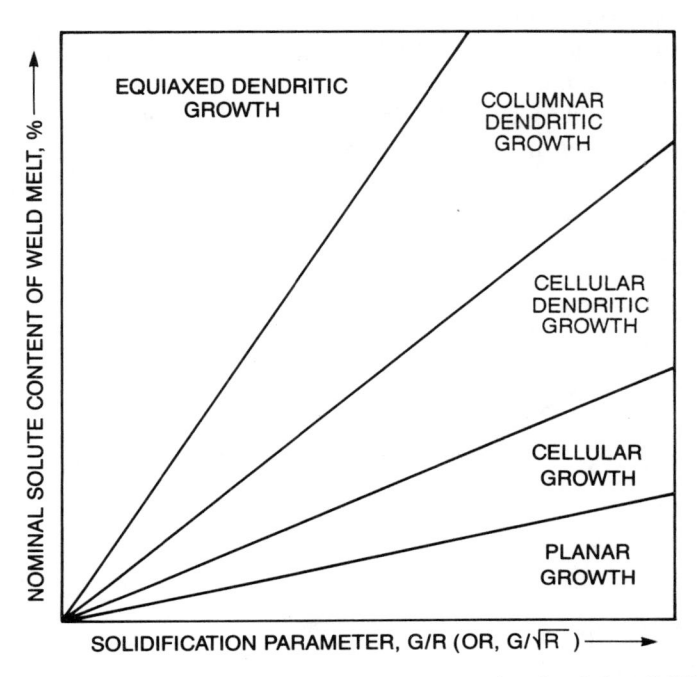

Figure 9.3 — General predictions on modes of epitaxial solidification structure based on relationship between solute content of weld metal and the combined solidification parameter

angle (usually identified as ø) formed by a tangent drawn from a parallel in the welding direction toward a given point on the boundary of the weld pool. Regardless of the method employed for analyzing ø, fast welding travel speed produces a steep G, and when G is divided by R, the ratio serves as a combined parameter that correlates with the nominal solute content of the weld metal to define the primary solidification microstructure that is likely to form. Difference of opinion exists on whether R or $\sqrt{R}$ should be used; but this does not alter the general prediction of growth mode, which is diagrammed in Figure 9.3. It is difficult in devising a welding procedure to control G and R independently; but, in general, a fast welding speed is tantamount to setting a steep G. As roughly indicated by Figure 9.3, for a given solute element content in the weld melt, the solidification microstructure becomes more dendritic as the value for G/R ratio decreases. The weld cooling rate then determines the size and spacing of cells and dendrites.

Figure 9.4 is a photomicrograph centered on the fusion boundary between the weld zone and base metal of a weld to show how the zones that have been discussed and schematically illustrated to this point can be observed metallographically. In this example, the material is a wrought high-silicon ferritic electrical steel which has a very low residual element content. Allotropic transformations do not mask the primary microstructures that result from autogenous fusion welding. The photomicrograph has been marked off to indicate four zones that can be deduced from examination of microstructure *and* nonmetallic inclusions. The base metal, Zone 4, has a microstructure of small grains of ferrite. The heat of welding has caused grain growth in this base metal within Zone 3 immediately adjacent to the weld zone. Confirmation that this zone has remained solid is gained from observing the presence of elongated nonmetallic inclusions within this zone, which are also seen in the wrought base metal (Zone 4). The agglomeration of nonmetallics into tiny nodules in Zone 2 indicates fusion. Solidification has taken place epitaxially, as should be expected, and because

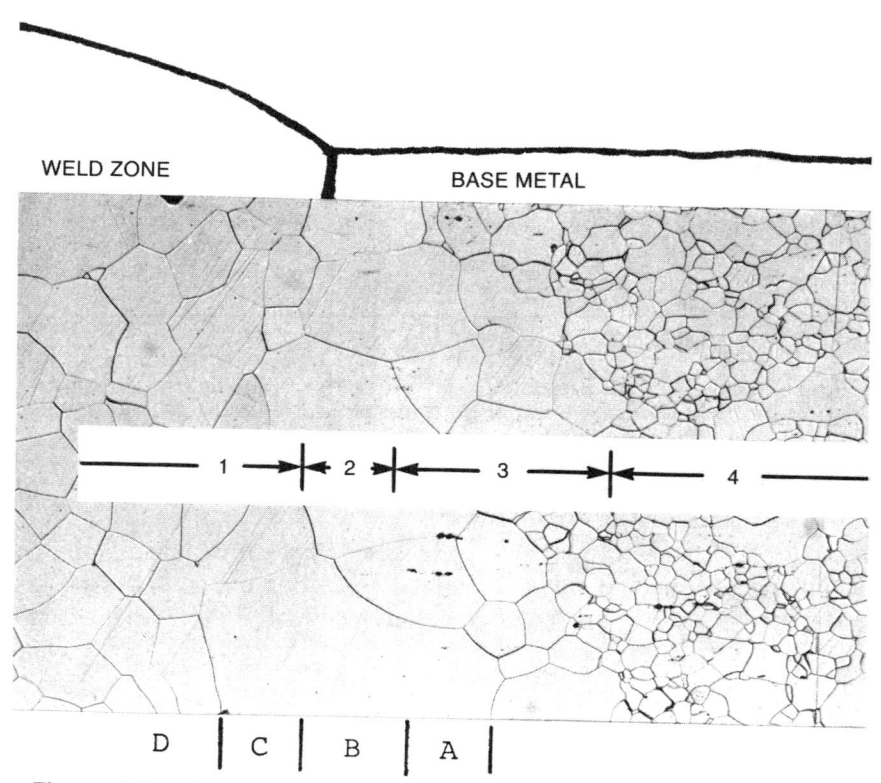

Figure 9.4 — Photomicrograph of transverse (vertical) metallographic specimen prepared through weld made by GTAW process with filler metal addition in high-silicon "ferritic" electrical steel illustrating epitaxial grain growth at onset of weld solidification

In addition to weld zone epitaxy, specific observations can be drawn from the above photomicrograph as guided by demarcations in grain structure and nonmetallic inclusions. Within the numbered zone, the following is found: (1) weld zone of mixed filler and base metals, (2) weld zone of unmixed base metal composition, (3) base metal HAZ with coarsened grains, and (4) base metal with unchanged microstructure.

The single grain located above each letter represents the following: (A) a coarsened grain in base metal HAZ that remained solid, (B) epitaxial resolidified grain of base metal composition, (C) continued epitaxial extension of grain by solidifying weld metal composition, and (D) epitaxial extension of grain stopped by formation of competitive grains.

this steel is not prone to constitutional supercooling (i.e., microsegregation) the mode of coarse grain growth into the weld zone is planar. In the final stage of weld zone solidification, Zone 1, the epitaxial planar growth is blocked by nucleation and competitive growth of smaller ferrite grains.

One other phenomenon in fusion welding that is explained in part by solidification mechanics is the formation of ripples on the surface of the solid weld pass, or weld bead. This phenomenon is addressed in greater detail within Technical Brief 45.

Discussion to this point dealing with the "weld zone," and the mechanics of its freezing has dealt with a melt that either was produced autogenously, or was introduced as filler metal, or both; in which case the zone is termed the *mixed zone*. The weld zone is stirred by various forces to attain a more-or-less homogeneous state of composition regardless of its sources. A methodical examination of a fusion weld differentiates several additional "zones." Figure 9.5 has been prepared to show the general location, originating sources of composition, commonly-applied terminology, and relationship to temperature distribution before a detailed description is provided of these additional zones. This figure has been drawn to identify five distinct zones, and allow remarks regarding the general character of microstructure found in each zone. This illustration is based on a weld in low-alloy steel base metal containing 1¼ percent chromium, using a SMAW electrode containing only residual chromium (i.e., 0.05 percent), but with high manganese (i.e., 1¼ percent). These composition differences between base metal and weld metal assist in the identification of zones by microprobe analyses.

TECHNICAL BRIEF 45:
The Solidification Mechanics of Weld Surface Patterns

The formation of ripples on the surface of a weld bead is a fairly common phenomenon, and these surface patterns are usually accepted as an innocuous feature of a fusion weld. However, efforts must be made occasionally to minimize these ripples, or to make their pattern finer and more regular. Sometimes the finer ripple is wanted simply for the sake of appearance. Other times, under specific circumstances, resistance to fatigue crack initiation can be improved by avoiding the notch-effect of coarse ripples.

Ripples are caused by the small-time-scale discontinuous nature of solidification, and each ripple can outline the weld pool shape at a given moment as the mechanics of solidification operate. The solidification front velocity, R, fluctuates cyclically above and below a mean value of growth rate which has been set mainly by welding travel speed. Weld melt composition also enters into the picture because the solute content in the solidifying metal varies slightly from solute-rich to solute-lean with fluctuation in R. This variation in solute content can be detected as band-lines in the weld microstructure by etching a metallographic specimen quite heavily. The band-lines will be concordant with ripples on the weld surface.

With less solute in the composition of the weld melt, and with faster travel speed, fainter and finer ripples are formed on the weld surface. Also, less solute-veining or band-lines will be seen in the microstructure. Ripples can also be caused by perturbations of the weld pool as cyclic variations occur in the arc forces. When pulsing welding current is employed, a repeated pattern of secondary ripples may be superimposed on the regular continuing pattern.

The Weld Zone

The *weld zone*, identified as Zone 1, the *mixed zone,* in Figure 9.5, already has received much attention, and has been described with regard to sources of the metal, homogeneity, solidification mechanism and primary microstructure, etc. An aspect of

①- WELD ZONE (MIXED ZONE) (CONTAINS Mn 1.25%, Cr 0.05%)
②- UNMIXED ZONE (SUPERHEAT MELT-BACK)
③- PARTIALLY-MELTED ZONE
④- HEAT-AFFECTED ZONE (OF BASE METAL)
⑤- UNAFFECTED BASE METAL

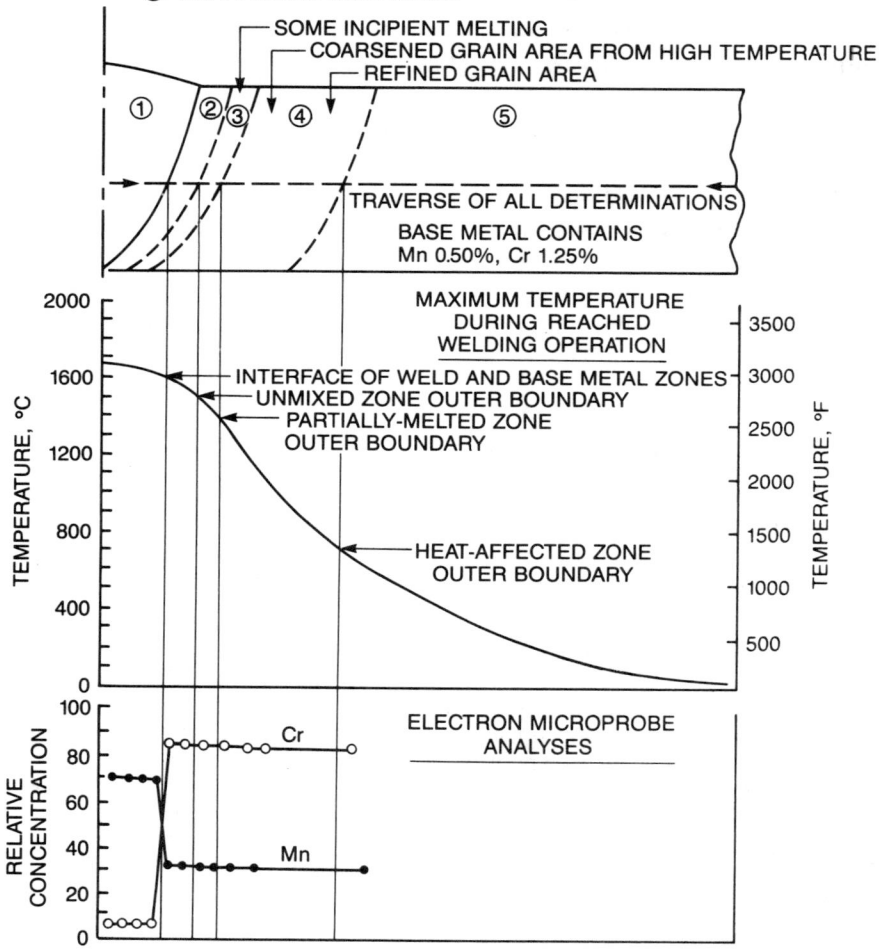

Figure 9.5 — Schematic panorama of distinct zones found in fusion welds

Also shown are maximum temperatures of exposure by the base metal and the weld during the fusion operation, and composition determinations in the weld zone and base metal HAZ made by microprobe.

the mixed zone to be emphasized here is the importance of the microstructure eventually produced after all temperature excursions have been completed. After initial solidification of the weld zone, the rate of cooling of the primary microstructure is important because of allotropic transformation that is likely to occur, and the influence of cooling rate on the nature of the microstructure that results from a transformation. Sometimes, the cool-down of the weld zone after solidification is the only temperature excursion to be considered, but often the effects of a more complex thermal history must be weighed. After initial solidification of the primary microstructure, the weld zone may be reheated to some temperature level as a result of multipass welding, or by application of a post-weld heat treatment. The importance of studying microstructural changes in the weld metal, per se, has been emphasized repeatedly because of the different primary structures that serve as a starting point. Furthermore, even though the weld metal and the base metal in a weldment may be subjected to the same post-weld thermal cycling, these two components of a weld commonly display differences in microstructure that persist because of their dissimilar origins and compositions. The microstructure of actual weld metals of steel will be examined in detail shortly.

The Unmixed Zone

The very narrow zone immediately adjacent to the weld zone, identified as Zone 2 in Figure 9.5, consists of a boundary layer of melted *base metal* that was not stirred or mixed with the weld zone described above. This boundary layer is treated as a distinct zone even though it could be appropriately considered as part of the "weld zone" because of being melted, and initiating epitaxial growth of grains which grow into the weld zone as solidification took place. Nevertheless, the fact that the *unmixed zone* has the *base metal* composition must be kept in mind.

The unmixed zone also has been called the "superheat melt-back zone" since it is produced by thermal energy contained in the weld zone in excess of that required to form an adequate molten weld. As the welding heat source moves along the joint with its concomitant melting, penetration, and stirring, there follows for a fleeting period of time a flow of excess heat from the weld zone into the base metal that raises the temperature of the first increment above its melting point. Now, however, the penetration and stirring from the heat source have moved onward and the weld zone is dropping in temperature. The maximum temperature reached in the weld metal and the base metal of the weldment in Figure 9.5 is shown by the temperature distribution curve. The temperature of the molten metal in the mixed zone is indicated to be somewhere in the range of 1600-1700°C (2912-3092°F), which is well above the liquidus of approximately 1525°C (2775°F) for the steel composition. As this excess thermal energy is conducted across the penetration-interface, the nearest layer of base metal rises in temperature, and as approximately 1525°C (2775°F) is exceeded, this boundary layer becomes molten. The volume of the weld melt and its temperature determine the depth to which melting (unstirred) will progress in the base metal beyond the initial interface. Unmixed zones are wider with processes that produce a large volume of weld melt and which is effectively heated, such as the SAW process. With commonly used arc fusion processes, the depth or thickness of the unmixed zone can range from about 0.05 to 2.5 mm (approximately 0.002 to 0.10 in.). As should be expected with other fusion welding processes, unmixed zones that are narrower will be found when examining welds made by the EBW and LBW processes, whereas the ESW process produces an unmixed zone that is quite wide.

Detection of unmixed zones requires very close observation and special etching because their composition is the same as the base metal. Even with similar base and weld compositions, however, there can be subtle changes in microstructure, reaction to etching reagents, and differences in nonmetallic inclusions; and any one of these subtle changes may indicate the limits of the unmixed zone. Reliable location of the boundary between the unmixed and the mixed zones in a steel weldment often can be made with the microprobe analyzer described in Chapter 2. Indeed, the weld portrayed in Figure 9.5 has sufficient dissimilarity in composition between weld zone and unmixed zone to allow microprobe analyses to indicate their interface. The abrupt decrease in manganese level and increase in chromium level clearly locates the mixed-unmixed boundary. Precision in locating this boundary by chemical analysis is dependent, of course, on the center-to-center spacing of the microanalyzed spots and their sampling diameter.

The unmixed zones in most steel weldments seldom cause problems, and they ordinarily can be overlooked or ignored. Yet, there are circumstances with steel weldments where this zone can detract from performance, for example, as has occurred with welds in certain low-alloy steels that were expected to exhibit toughness during explosion bulge testing at cryogenic temperatures. In such cases, the unmixed zone proved more susceptible to brittle fracture than was acceptable. In general, a composition optimized for wrought base metal is not likely to be optimum for weld metal, even if the latter exists only as a very thin boundary layer (unmixed weld zone) in the weldment. As greater demands for toughness under service conditions at low temperatures or at high loading rate are placed upon fusion welds, the unmixed zone requires more attention and consideration.

The Partially Melted Zone

The *partially melted zone*, identified as Zone 3 in Figure 9.5, is sited immediately adjacent to the unmixed zone, and does not always develop in every fusion weld made in steel. Its occurrence depends on a base metal chemical composition and/or microstructure in which small areas are melted by exposure to a temperature that approaches the usual liquidus of approximately 1525°C (2775°F). Exposure time during welding is brief, of course, and since the liquidus temperature just mentioned is based on relatively slow heating, the rapid heating during welding possibly will not convert any of the metal to the liquid state. The lowest temperature that will produce some melting, albeit partial melting in highly localized microstructural features or areas, will be dependent on the composition of the steel and its particular microstructure at the time of welding.

Carbon steels ordinarily show little or no evidence of a partially melted zone in fusion welds, but alloy steels sometimes contain microstructural features that undergo localized melting when heated. Elements added to alloy steels such as manganese and nickel can depress and separate the liquidus and solidus of the steel enough to increase the possibility of a partially melted zone at a fusion weld. Minor elements such as silicon, sulfur and niobium are more likely to participate in the formation of a partially melted zone. Localized melting can be damaging, either by adversely affecting the final microstructure and metal properties in this zone, or by causing hot tearing or cracking in the localized areas.

The incipient melting that occurs in the partially melted zone sometimes is called "liquation." It can be initiated by low-melting segregates in the microstructure, particularly in grain boundaries, or by nonmetallic inclusions. For example, complex sulfide inclusions melt at temperatures below the liquidus of the bulk steel, and can form fissures which propagate as cracks when the weld is cooled, or stressed during service. As should be expected, the occurrence of a partially melted zone, its severity in terms of depth or thickness, and the prevalence of partially melted areas depend on both the welding heat input and the composition of the steel; particularly the presence of residual elements that form low-melting microstructural constituents, segregates, or nonmetallic inclusions. Although partially melted zones do not cause widespread difficulty their formation in fusion welds must be recognized so that remedial measures taken when deemed necessary.

The Heat-Affected Zone

Beyond the zones in the base metal that involve melting is a much broader area called the *heat-affected zone* (HAZ). The HAZ is identified as Zone 4 in Figure 9.5. The partially-melted zone (Zone 3) might be considered part of the HAZ because the zone has remained essentially solid (except for incipient melting), and it certainly is "heat-affected." AWS defines the "heat-affected zone" as the "portion of base metal whose mechanical properties or microstructure have been altered by the heat of welding, brazing, soldering, or thermal cutting." Yet, some feel that the HAZ only encompasses that portion of base metal which undergoes solid-state microstructural changes. This view ignores that fact that the heat-effect of welding can alter the mechanical properties of steel without causing a detectable change in microstructure (The effects of stress relief heat treatment, and the mechanisms of recovery and relaxation will be addressed in Volume II, Chapter 12). To avoid any confusion, the HAZ in the present text will be the entire zone in which *properties* and/or *microstructure* are altered by the heat of welding, as defined by AWS. When a partially-melted zone develops, it will be treated as part of the HAZ and significant abnormalities will be given specific attention.

The curve in Figure 9.5, indicating the maximum temperatures expected in the fusion weld made by SMAW along a traverse extending from the center of the weld zone to unaffected metal, ranged from a high of about 1675°C (approximately 3050°F) in the weld zone, down to lower critical (i.e., the Ac_1, indicated to be a temperature of 727°C, or about 1341°F). These temperatures and their probable effects are speculative because of the strong influence of rapid heating rate and short time at temperature as discussed in Chapter 7. The composition of a given steel determines its exact liquidus and solidus temperatures, and the critical temperatures and ranges at which allotropic transformation in microstructure will take place. It also establishes properties that control response to recovery and relaxation.

Most of the carbon and low-alloy steels undergo the reversible allotropic transformations that are innate to their iron base, and this behavior controls the majority of microstructural alterations from the heat-effect of welding. Although the Ac_1 temperature is cited as the lower limit of the critical range on heating where microstructural transformation starts, even lower temperatures can, under certain conditions, cause a microstructural change in some types of iron and steel, especially when in the cold-worked condition. Some metallurgical actions that are possible at temperatures from 727°C (1341°F) down to about 400°C) (750°F) include precipation hardening, strain-aging embrittlement, and critical grain growth. Prediction of the exact temperature at which a particular change in microstructure or properties will take place during the heating and cooling cycle of welding is quite difficult, so most temperatures and ranges cited should be accepted as closest possible approximations unless suitable tests can be conducted to establish precise temperatures.

Unaffected Base Metal

Although not changed in microstructure or properties, the base material is significant to the HAZ because its microstructure can influence its transformation during the *heating* portion of the thermal cycle. Important are its grain size, banding of any kind, center segregation, and kind and number of nonmetallic inclusions. For example, if the base metal carbides have agglomerated as large, spheroidized particles, the start of austenite transformation and their dissolution will be delayed until a temperature somewhat higher than 727°C (1341°F) is reached (see explanatory information in Chapter 4 and Figure 4.12 regarding the role of carbon in steel). Furthermore, there are likely to be localized concentrations of carbon in areas where the temperature has risen only to a maximum that is within the Ac_1 - Ac_3 critical range. These high carbon areas can transform into a very hard microstructures when the HAZ cools to room temperature. The nature of nonmetallic inclusions present also are important because, as mentioned earlier, certain kinds of sulfide inclusions can undergo incipient melting and help create a partially-melted zone.

SOLID-STATE WELDS

Much of the metallurgical description concerning the microstructural changes in steels during fusion welding also can be applied to solid-state welds; even without a weld zone (solidified weld melt). Steels respond microstructurally to heat-effect in the same manner as described for fusion welding, but the different thermal excursions and rates of temperature change for each solid-state process (as outlined in Chapter 6) must be surveyed step-by-step to rationalize the microstructure ultimately produced at the weld. Solid-state welding processes of importance will be given in-depth examination in later chapters.

MICROSTRUCTURAL TRANSFORMATIONS IN SOLID IRON AND STEEL

The microstructure of metals has been given only preliminary attention thus far, even though the underlying role of crystallographic structure and the allotropy of iron was described in Chapter 2, and phase changes as dictated in iron and steel were covered in Chapter 4. In these earlier discussions, cooling to the temperature levels where phases transformed was assumed to be slow, that is, the rate approached equilibrium conditions. Slow cooling was emphasized when using the phase diagram in Figure 4.12. Even though such extremely slow cooling seldom, if ever, occurs in welding, the phases that appear under these near-equilibrium conditions can serve as bases on which to build a comprehensive description of the microstructures that develop from transformations during the unique thermal cycles of welding. Also, the microstructures of weld zones and of the heat-affected base metals often are so unlike in appearance and properties, and in their behavior during subsequent thermal cycling, that these two components of a weldment must be discussed separately.

As described in earlier chapters, iron, unalloyed, can exist in two crystallographic forms, body-centered cubic (bcc) and face-centered cubic (fcc): but, because allotropic transformation from one form to the other occurs reversibly at *two* widely separated temperature levels below the melting point of iron, *three* designations have been assigned for identification of the crystallographic forms to indicate their range of temperature existence. At room temperature and below, iron is bcc and this form has been designated *alpha* iron until it transforms on heating at 912°C (1674°F) to the fcc form; whereupon the designation *gamma* iron applies. When the fcc form is heated to 1394°C (2541°F), transformation returns the structure to bcc and the designation *delta* iron applies to the solid phase until it melts on heating to 1538°C (2800°F).

Commercially pure iron at room temperature has modest strength, although its plasticity allows strength to be increased by cold working. Also, the allotropic alpha-gamma transformation allows grain refinement in the microstructure by controlled heating and cooling, although even a highly refined grain size increases strength only to a limited extent. Another means of strengthening pure iron is to markedly increase the number of dislocations in its crystallographic structure by some means such as dislocation-riddling. Although this has been demonstrated with iron whiskers, it is not a feasible way of strengthening or hardening the more-substantial sections ordinarily used in construction and manufacturing. Luckily, small amounts of carbon (i.e., less than one percent) can be added to iron in order to produce remarkable changes in mechanical properties. The lower atomic weight of carbon provides a relatively large number of atoms for any given weight-percentage, and therefore the carbon content must be controlled with reasonable precision to obtain desired results. Ordinarily, close attention is given to each hundredth-of-a-percentile carbon introduced.

Phase Changes in Steel

In steels, iron exists in two crystallographic forms: (1) the bcc structure which when existing at a temperature of 912°C (1674°F) and below is called "ferrite," but specifically is *alpha ferrite* (α Fe), but when existing at a high temperature just below the solidus is designated *delta ferrite* (δ Fe), and (2) the fcc structure which is called *austenite* (γ Fe) instead of gamma phase when carbon is present. *Cementite* is the name given to the hard iron carbide compound (Fe_3C) that forms in steel because of the chemical affinity that iron holds for carbon. These forms of iron deserve illustration as viewed metallographically because they appear frequently in various configurations in the microstructure of steels. However, initial presentation of microstructural appearance will be based on steel that has been slowly cooled from the molten state to room temperature.

Ferrite

Ferrite, the bcc phase, has been shown in earlier illustrations (Figures 2.23, 2.24 and 9.4) to explain such phenomena as recrystallization, grain size, and epitaxial grain formation. Figure 9.6 is presented here to show the general character of a completely ferritic microstructure in a commercial low-carbon steel that contains so little carbon that no cementite or carbide particles have formed as a secondary phase. The very small amount of carbon present is in solid solution. The ferrite grains in the photomicrograph vary in size, have random shapes and have no preferred orientation. They are without any particular feature in their flat, polished-and-etched faces. The ferrite phase will be appearing in a wide variety of configurations as study of steel microstructures continues.

Austenite

Austenite, the fcc phase, appears in carbon and low-alloy steels only above the Ac_3 temperature of the particular steel composition; then it exists up to either the austenitic-to-delta phase transformation range, or to the melting range in steels containing more than about 0.60 percent carbon. Preparation of a photomicrograph of this austenitic microstructure requires hot-stage metallography, which is costly and presents a variety of problems. An alternative to hot-stage metallography is to use a high-alloy Cr-Ni stainless steel that retains its austenitic microstructure to room temperature and lower. This substitution has been made in preparing Figure 9.7, to show the gen-

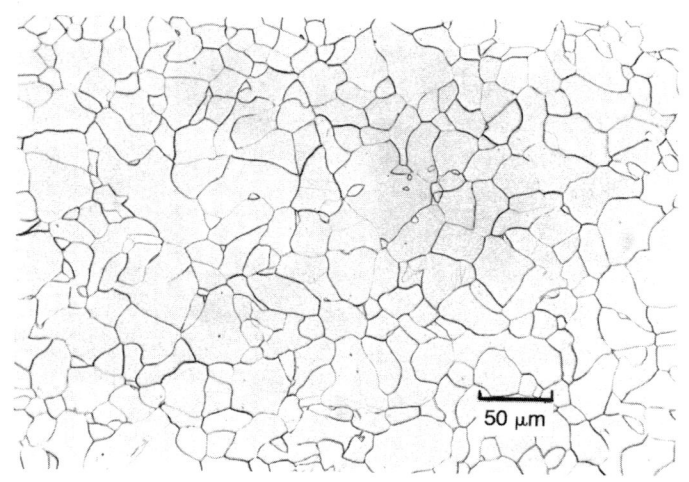

Figure 9.6 — Microstructure of ferrite as found in slowly cooled steel containing a very low carbon content

(Nital etchant; original magnification: 320X; courtesy of The Welding Institute, England)

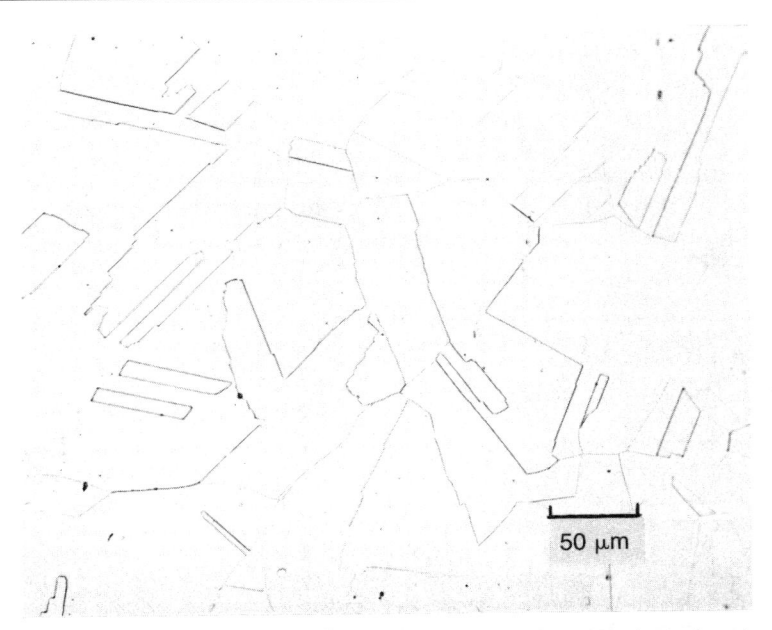

Figure 9.7 — **Microstructure of austenite as might be viewed metallographically in carbon and low-alloy steels when heated to a temperature level above their Ac$_3$ point and below any point of melting or of austenite-to-delta phase transformation**

Specimen on which photomicrograph was prepared actually is austenitic Cr-Ni stainless steel at room temperature where austenitic microstructure is retained. See text for reason to substitute. (mixed-acids etchant; original magnification: 250X)

eral appearance of grains of austenite. They are much like that of ferrite, except that the austenite grains tend to have a more angular shape. This tendency arises because of different directions of preferred nucleation-and-growth in the fcc structure. The austenite grains in Figure 9.7 also display some twinning, but this crystallographic condition is not likely to be present in austenite above the Ac$_3$ temperature. A very important dissimilarity between ferrite and austenite, which is not easily discernable through metallographic observation is their great difference in solid solubility for carbon. Whereas ferrite at room temperature can hold no more than about 0.008 percent carbon in solution, the favorable arrangement of atoms in the fcc crystalline structure provides spaces in the lattice for generous solubility of carbon in interstitial solution [as shown by the limit line (SE) in Figure 4.12]. The importance of this carbon solubility in austenite in relation to the mechanism of transformation hardening will become clear when the cooling of an austenitized microstructure is reviewed. (See Technical Brief 46.) Brief mention will be made here that austenite in small amounts can be retained in the microstructure of some steels, either by reason of alloy composition, of extremely fast cooling from the austenitized state, or possibly both. Retained austenite and its effects will be discussed in later chapters as specific steels are reviewed for microstructural behavior, and as particular problems with dimensional stability, toughness, cracking susceptibility, etc., are examined.

Cementite

Cementite or *iron carbide* appears in many configurations depending on the initial distribution of carbon in the austenite, the cooling rate of the austenite during transformation to a body-centered type crystalline structure, and the temperature level at

---

TECHNICAL BRIEF 46:
Austenite: The Mother of Microstructures

Whether molten weld metal of the carbon and low-alloy steels solidifies as delta ferrite, or as gamma austenite, as dictated mainly by carbon content is not particularly important since the microstructure will inevitably become austenite at a high temperature during cooling. The occurrence of austenite, so-named to honor William Roberts-Austen, is surely in place below the ranges for solidification and for the disappearance of the delta phase (see iron-iron carbide phase diagram presented earlier in Figure 4.12, extreme upper-left portion).

Austenite is a single-phase constituent consisting of the gamma (fcc) form of iron with carbon held in interstitial solid solution. Austenite, during cooling to room temperature must somehow transform to the alpha (bcc) form of ferrite to achieve phase stability, and in so doing must expel virtually all of the carbon atoms inasmuch as the newly formed ferrite phase has almost no solubility for carbon. The ejected carbon atoms have a proclivity to quickly form cementite, an iron carbide (Fe_3C) compound, but other elements present that also have an affinity for carbon, such as manganese and chromium, can replace some of the iron in the cementite to form more complex metal-carbide particles.

The splendid metallurgical merit of iron lies in the many ways the transformation of austenite can be manipulated to create microstructures in steel that are essentially ferritic, but in which the dispersal of cementite can be controlled to obtain a great variety of mechanical properties. Very small amounts of austenite can sometimes be retained in the otherwise ferritic microstructure at or near room temperature, but this austenite is now in a metastabile temperature realm. Retained austenite, whenever present, should be considered for its possible effects on mechanical properties and other behavioral characteristics because despite the small volume fraction present, it can be highly influential. Sometimes retained austenite can be a potentially disturbing phase (e.g., increased susceptibility to cold cracking, impaired dimensional stability). On the other hand, retained austenite can be beneficial in properly controlled circumstances, such as improving the toughness in certain low-alloy steels at low service temperatures.

The transformation of austenite during slow continuous cooling is a well-understood mechanism involving nucleation, diffusion and growth; all of which require adequate time to transpire. The resultant microstructure is ferrite, and the cementitic phase created by carbon in the steel is distributed in the eutectoid configuration with the ferrite phase in colonies called *pearlite*. This is regarded as an equilibrium microstructural constituent.

Departure from slow cooling of the austenite to other cycles will still cause the austenite to transform to ferrite, but another mechanism can be involved and markedly different forms of carbide dispersal will be produced. Accelerated cooling is the most common circumstance that causes the austenite to transform under non-equilibrium conditions, and rapid cooling is certainly the usual case in fusion welding. Other microstructures that can evolve from austenite are called *bainite, martensite, acicular ferrite*, etc. These must be recognized when they appear in toto, and also when two or more of these phases and constituents exist in a complex mixture. For this reason, austenite is looked upon as the "mother" of microstructures.

which cooling is stopped. The configurations of cementite can range from a dispersion of sub-microscopic size particles to large globules and several unusual configurations in between. The globular or spheriodal carbides are not as large as the grains of ferrite that make up the matrix of the microstructure. Cementite was described in Chapter 4 to be a metastable phase, and the fact was pointed out that the final stable form of carbon in the iron-carbon alloy system should be free carbon. Although free or uncombined carbon (usually in the form of graphite) often appears in the microstructure of many cast irons because of their much higher carbon contents, in steels the carbon ordinarily remains firmly bonded to iron (or to other carbide-forming alloying elements that may be present, and that often have a stronger affinity for carbon than does iron). Yet, as also described, somewhat unusual circumstances involving the composition of a given steel, and long-time service at high temperature can lead to a breakdown of cementite and allow free carbon to appear as a final stable phase in the microstructure. This movement toward true equilibrium— which will be discussed in Volume II, Chapter 13, under the heading "Graphitization" — has very undesirable effects on a *steel* weldment.

Pearlite

Pearlite is the microstructural configuration in which cementite is most often found in steels. This is a two-phase eutectoid structure which consists of alternating platelets or lamellae of cementite and ferrite as illustrated in Figure 9.8. Pearlite forms when austenite is cooled through the Ar_3 - Ar_1 transformation range at a rate that is not too

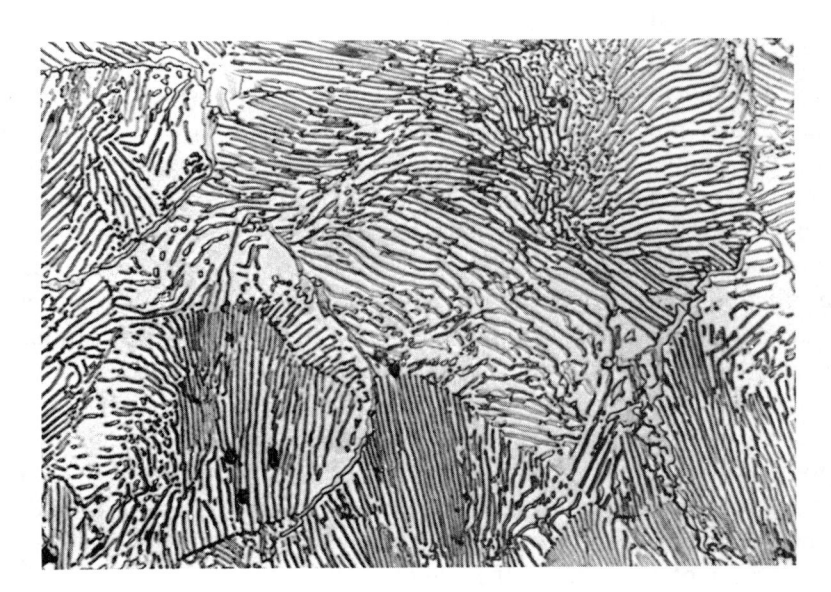

Figure 9.8 — Microstructure of pearlite as formed in steel containing 0.85% carbon after being fully austenitized and then cooled slowly through the eutectoid transformation at the Ar_1 point

A small amount of hypereutectoid cementite (Fe_3C) is present in the grain boundaries of the austenite from which the pearlite transformed. (Nital etchant; original magnification: 500X)

fast to allow the nucleation-and-growth process by which pearlite is formed. Examination of the phase diagram in Figure 4.12, and application of the lever law, indicates that as proeutectoid ferrite, or cementite forms by transformation of the austenite, the carbon content of the remaining austenite is either increased, or decreased by diffusion toward the eutectoid amount. For discussion at the moment, the iron-iron carbide alloy will be regarded as having a eutectoid composition of 0.77 percent carbon, and that austenite of this carbon content transforms during slow cooling when the Ar_1 temperature of 727°C (1341°F) is reached. However, a number of factors are operative that shift the eutectoid composition to slightly higher, or lower carbon contents.

The mechanics of austenite-to-pearlite transformation during cooling has been studied intensively because of the influence of this constituent on mechanical properties. Briefly, pearlite formation starts with the precipitation of nuclei of cementite on austenite grain boundaries or other suitable microstructural sites. The entire sequence of the eutectoid transformation to pearlite is accompanied by evolution of heat. This subtle exothermic action occurs as hypoeutectoid ferrite is formed during cooling through the Ar_3 to Ar_1 critical range, and as hypereutectoid cementite is formed during cooling through the Ar_{cm} to $Ar_{1,3}$ critical range. Consequently, under conditions of very slow cooling, temperature measurements will display a noticeable arrest in cooling rate during transformation in the same fashion as observed during the solidification of other eutectics. As precipitated cementite nuclei increase in size to form platelets, carbon diffuses inward from the surrounding austenite (which contains 0.77 percent carbon) toward the flat faces of the cementite platelet to provide the 6.67 percent carbon needed to form the compound Fe_3C. This diffusion reduces the carbon in the adjacent austenite and causes a platelet of ferrite to form which in turn rejects carbon because of this phase's virtual lack of carbon solubility (i.e., ~0.02 percent). The rejected carbon is believed to encourage nucleation of another parallel carbide platelet. As the mechanism repeats, the alternate lamellae of carbide and ferrite platelets increase in number to form a "colony" having lamellae orientation that has a relationship with the austenite grain in which it develops. The colony also grows edgewise by diffusion of carbon in the vicinity of its laminated interface with the austenite. This radial enlargement of colonies continues until multiple colonies meet at mutual interfaces forming "nodules" of pearlite, each of which may be positioned over portions of several prior austenite grains. Faults and imperfections are common in the lamellar structure of the colonies and nodules, and this can be seen in Figure 9.8. Because pearlite formation is diffusion-controlled, austenite transformation to this microstructure is limited to relatively low rates of cooling. The actual rate of cooling determines the amount of hypoeutectoid ferrite, or of hypereutectoid cementite that is able to precede the formation of pearlite, and the rate of cooling also determines the thickness of the lamellae in the pearlite. When austenitized steel is cooled at a moderate rate as for example in still air, the pearlite may be so fine that high magnification must be used to resolve the ferrite and cementite lamellae. Consequently, when photomicrographs are made at lower magnification to reveal other microstructural features (e.g., grain size), often the lamellar character of the pearlite is not discernible.

Widmanstätten Pattern

Widmanstätten pattern was first noticed in the microstructure of nickel-iron meteorites that had been sectioned for metallographic examination. Once this microstructure became familiar, and an appreciation was gained of the mechanism by which it forms, the structure was found to occur in many kinds of alloys, including steel, after certain heat treatments. Although seen infrequently in steel, there is good reason to recognize its occurrence in steel, and to be aware that toughness may be very poor when this structure forms. Widmanstätten microstructure in steel is apt to develop with coarse-grained austenite and slow cooling, for example, when austenite is cooled from an unusually high temperature. Cast steel, and overheated wrought steel are known to sometimes form this microstructure, but the weld HAZ in steel is also a potential location when very high welding heat input is employed and slow cooling rate prevails.

The characteristic appearance of Widmanstätten structure is due to the peculiar shape and distribution of proeutectoid ferrite. The ferrite has a strong inclination to precipitate from coarse austenite grains as long grains of ferrite having a very low aspect ratio. These are distributed both as parallel laths, and as a crisscrossed or basket-weaved pattern. These patterns arise because the ferrite is nucleating and growing preferentially along certain crystallographic planes in the austenite (mostly the octahedral planes). The lattice of Widmanstätten ferrite is crystallographically oriented with that of the parent austenite phase. The mechanics of this kind of ferrite formation has been studied and found to involve the nucleation of very broad plates of ferrite which are able to thicken by a "ledge-growth" mechanism. In this manner, the broad face of a ferrite plate can advance perpendicularly to the interface with austenite and achieve greater thickness dimension than found in the lamellae of pearlite. A photomicrograph of Widmanstätten pattern in the microstructure of steel (observed in the HAZ of a weld made with high heat input) is presented later in Figure 9.44.

Martensite

Martensite is the hardened microstructure that forms in an austenitized steel when cooled at a rate too fast to permit the nucleation-and-growth mechanism by which pearlite is formed. Instead, transformation of austenite to martensite takes place by a very rapid diffusionless mechanism after undercooling to a markedly lower temperature. As the austenite transforms to martensite, carbon is trapped as solute atoms in an interstitial solid solution. The mechanics of austenite-to-martensite transformation is unique, and is a very important behavioral aspect of steel. Many features of martensite and its transformation mechanics will be examined when the morphology of this microstructure is reviewed later.

While martensite can be briefly described as a supersaturated solid solution of *carbon* in a very fine grained matrix, tempering martensite by heating to a temperature below the eutectoid temperature results in the formation of free cementite. At tempering temperatures below about 260°C (500°F), the cementite usually consists of submicroscopic particles distributed throughout the matrix, but above this temperature the particles increase in size and eventually become small spheroids that are plainly seen during metallographic examination.

Microstructural Changes in Steel During Heating

Microstructural changes during *heating* are important because conditions created in various areas or zones at the end-point of heating often set the stage for the nature of transformation events that follow during cooling. While heat-treating operations applied to steel are carried out at controlled temperatures to secure particular microstructures and properties, welding introduces the broadest possible range of temperatures, a gradient of end-points of heating, from which cooling starts. However, certain end-point temperatures that lie somewhere between (1) the high peak that produces the liquation found in the partially-melted zone, and (2) much lower temperature further away from the weld that produces no perceptible change in microstructure, will be found to have significant effect on the microstructures finally produced at given points in the HAZ of the base metal. These effects can be studied methodically by heating small steel specimens to selected temperature levels found within the HAZ and cooling them at a rate similar to that occurring in the HAZ. Some of these intermediate temperatures produce effects that are beneficial to the microstructure and its properties, whereas others are quite detrimental.

The phase diagram in Figure 4.12 does not serve well as a guide to the microstructure existing in steel at rising levels of temperature during heating. It must be recalled that phase diagrams are plotted from results obtained on specimens *cooled very slowly* from the *molten state* to room temperature. Heating by most welding processes is quite fast, and attention must often be centered on a zone that is not heated to the molten state. Consequently, microstructural changes during the heating stage are not likely to be as predicted by an equilibrium phase diagram and the applied lever law.

To illustrate the manner in which microstructural changes occur in steel during heating, a plain steel of AISI 1020 (UNS G10200) has been selected to demonstrate what happens at notable temperature levels within the range to which the HAZ of a weldment is exposed. Figure 9.9 is a photomicrograph of the pearlite-and-ferrite microstructure typical of a plain steel containing 0.20 percent carbon. As a general rule, when a plain steel is slowly cooled to produce a microstructure containing pearlite, the percentage of this phase likely to be present can be approximated by multiplying the carbon content by 125. In this case, the steel with 0.20 percent carbon does display about 25 percent pearlite as predicted, while the major portion of the microstructure (75 percent) consists of grains of hypoeutectoid ferrite.

If the steel in Figure 9.9 containing 0.20 percent carbon with its microstructure of pearlite and ferrite is heated rather quickly, as usually occurs during welding, the Ac_1-Ac_3 temperature range over which the microstructure transforms to austenite will tend to be somewhat higher than the Ar_3-Ar_1 range found during cooling of an austenitized microstructure. The extent of the elevation on heating is dependent not only on the heating rate, but also on the size and composition of the cementite platelets in the pearlite or other shapes of carbide particles that must be dissolved. With higher rates of heating, the temperature of the steel will rise well above the eutectoid level without all of the pearlite having been transformed to austenite. Cementite that is present as thicker platelets in the pearlite, or as large spheriodal particles requires longer time or higher temperature before it becomes uniformly dissolved in the austenite. Also, if carbide-forming alloying elements are present in the steel's composition, such as

Figure 9.9 — Microstructure of pearlite and ferrite typical of a plain steel containing 0.20% carbon

This AISI 1020 (UNS G10200) steel had been cooled slowly in air and transformation on passing through its critical range produced a microstructure consisting of about 25% pearlite (dark patches) and 75% ferrite (light-colored grains). Lamellae in the pearlite are too fine to be plainly resolved in each nodule at the magnification employed for the photomicrograph. (Nital etchant; original magnification: 500X)

chromium, vanadium, niobium, etc., the complex carbides containing these elements will be more resistant to dissolution. Although manganese enters into the composition of cementite, this commonly used alloying element does not significantly affect its dissolution in austenite. It must be recognized that when pearlite undergoes transformation on heating, the austenite initially formed has a carbon content of 0.77 percent (i.e., the same as carbon in the eutectoid microstructure), but the carbon content drops to 0.20 percent as proeutectoid ferrite transforms to austenite, and as increasing temperature and prolonged time allow the carbon to diffuse throughout the austenitized microstructure. If the microstructure of a steel being heated consists of globules of cementite in a ferritic matrix, these more massive carbide particles represent a concentration of 6.67percent carbon that must be uniformly dissolved in the austenite. These forewarnings about the influence of heating rate, and the mode of cementite in the microstructure of the steel being heated are important when striving for an austenitized microstructure with homogeneous carbon distribution. They help explain why instructions for properly heat treating a steel will regularly call for a heating temperature that is about 50°C (100°F) *above* the reported Ac_3 critical temperature for the particular steel being treated. Also, a generous soaking time is usually stipulated to be certain that homogeneity is achieved in the austenite before cooling commences. In welding, unfortunately, much of the metallurgical treatment from heating is just the opposite; meaning that the heating rate is very rapid, peak temperature varies greatly with the location in the HAZ, and time at temperature is usually very short.

Figures 9.10-A and 9.10-B consist of a schematic series to show how the pearlitic microstructure of the 0.20 percent carbon steel of Figure 9.9 changes at particular tem-

MICROSTRUCTURE OBSERVED AT TEMPERATURE

WHEN RAPIDLY HEATED TO THE TEMPERATURES INDICATED BELOW, THE MICROSTRUCTURES SCHEMATICALLY SHOWN AT LEFT WILL EXIST IN THE 0.20% CARBON STEEL ILLUSTRATED EARLIER IN FIG. 9.9

(A) AT 727 °C (1341 °F)
ALTHOUGH THE STEEL IS NOW HEATED TO ITS REPORTED EUTECTOID TEMPERATURE, NO PERCEPTIBLE CHANGE HAS OCCURRED IN THE MICROSTRUCTURE OF PEARLITE AND FERRITE BECAUSE OF RAPID HEATING.

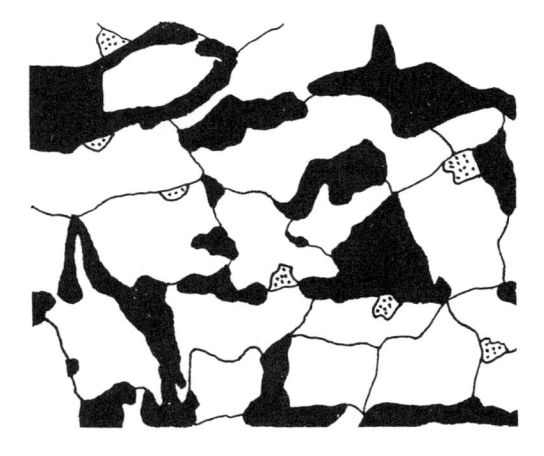

(B) AT 800 °C (1472 °F)
THIS TEMPERATURE IS WELL WITHIN THE TRANSFORMATION RANGE, YET, ONLY THE PEARLITE HAS TRANSFORMED TO AUSTENITE. EARLY SIGNS OF AUSTENITE NUCLEATION IN FERRITE (AT GRAIN BOUNDARIES) CAN BE SEEN. THE AUSTENITE GRAINS FORMED FROM PEARLITE STILL CONTAIN 0.77% CARBON, WHICH IS THE EUTECTOID COMPOSITION. THESE AREAS HAVE BEEN COLORED SOLID-BLACK, BUT ACTUALLY THERE ARE MANY SMALL GRAINS OF HIGH-CARBON AUSTENITE PRESENT. THE VERY SMALL GRAINS OF AUSTENITE JUST INITIATING IN VARIOUS BOUNDARIES ARE COLORED-SPOTTED BECAUSE THEY ARE OF EXTREMELY LOW CARBON CONTENT.

Figure 9.10-A — Effect of temperature level on the microstructure of plain steel containing 0.20% carbon when heated as described for a series of four specimens

(A) is photomicrograph at 500X of steel illustrated earlier in Figure 9.9. The microstructure of pearlite and ferrite is typical of this steel in the hot rolled and air cooled condition. Reheating to 727°C (1341°F) does not produce a perceptible microstructural change. (B) is the microstructure of the same steel now schematically illustrated to show microstructural changes on heating within the steel's transformation range (Ac_1 to Ac_3).

perature levels during heating. While this an elementary illustration of atypical microstructures that will be produced by the abortive austenitizing conditions of welding, nevertheless, these are the microstructures from which a portion of the final HAZ will develop through transformation during cooling.

Specimen A of Figure 9.10-A has been heated to the equilibrium eutectoid temperature, but due to the rapid rate of heating the microstructure at 727°C (1341°F) is unchanged from that shown in Figure 9.9, and upon cooling again to room temperature will remain unchanged. However, if the steel was held for a long period at the eutectoid temperature, austenite would begin to form in the pearlite, and some of the cementite in the pearlite would start to agglomerate into spheroidal carbides.

Specimen B of Figure 9.10-A, which was heated to 800°C (1472°F), now represents a microstructural condition at that temperature which can give cause for concern over mechanical properties. The localized (blackened) areas of high-carbon austenite

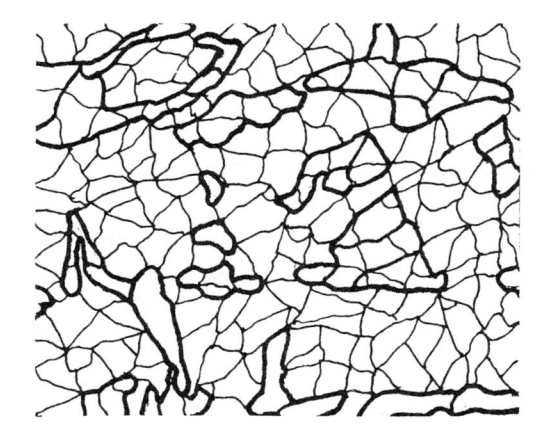

(C) AT 871 °C (1600 °F)

THE STEEL HAS AN Ac₃ TEMPERATURE OF ABOUT 850 °C (1560 °F), SO UPON BEING HEATED JUST ABOVE THE TRANSFORMATION RANGE, AUSTENITIZATION IS COMPLETE AND CARBON FROM DISSOLUTION OF THE PEARLITE NOW IS UNIFORMLY DISTRIBUTED THROUGHOUT THE AUSTENITE GRAINS TO THE EXTENT OF 0.20%. NOTE THAT THE AUSTENITE GRAIN SIZE IS QUITE SMALL. GRAIN BOUNDARIES THAT ORIGINALLY DELINEATED COLONIES AND NODULES OF PEARLITE HAVE BEEN SCHEMATICALLY INDICATED AS THICKENED LINES, BUT ORDINARILY THESE BOUNDARIES CANNOT BE DISTINGUISHED IN HOT-STAGE METALLOGRAPHY FROM THOSE OF NEWLY FORMED AUSTENITE GRAINS.

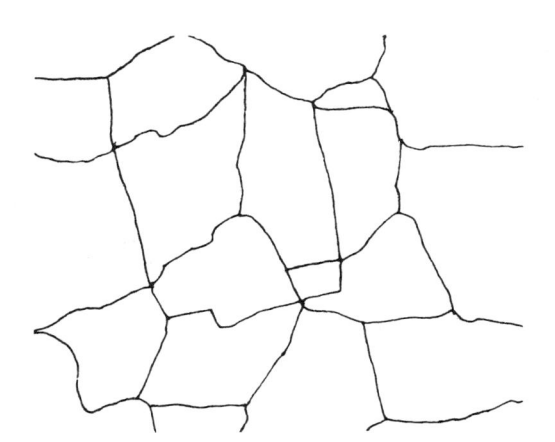

(D) AT 1200 °C (2200 °F)

AS INDICATED EARLIER IN FIG. 9.5, THIS RELATIVELY HIGH TEMPERATURE NOT ONLY CAUSES COMPLETE AUSTENITIZATION, BUT SIGNIFICANT GRAIN COARSENING HAS OCCURRED WHICH IS LIKELY TO PERSIST IN SOME DEGREE IN THE MICROSTRUCTURE FORMED UPON COOLING TO ROOM TEMPERATURE.

Figure 9.10-B — Remaining two illustrations, (C) and (D) shown schematically, from series of four, depicting noticeable changes in microstructure when 0.20% carbon steel is heated to temperatures above its transformation range

formed from pearlite have not been given time for carbon to diffuse into grains of low-carbon austenite that are just starting to form. The high-carbon austenite, upon cooling to room temperature, will be inclined to produce a harder microstructure that also may have poor toughness. These harder areas will exist in the softer ferrite that either did not undergo transformation during heating to within the Ac_1 - Ac_3 range, or that received no significant increase in carbon content during the relatively short exposure to the peak temperature of 800°C (1472°F).

The microstructure illustrated in Specimen C of Figure 9.10-B relates to the "refined grain area" of the HAZ indicated on the sketch presented earlier in Figure 9.5. Because of grain refinement the toughness of microstructures that form from the very fine grain austenite upon cooling can be better than that of the original base metal.

The austenite in Specimen D is coarsened by exposure to high temperature, and this is typical of the innermost portion of the HAZ (indicated in Figure 9.5) where temperatures may range from 1100 to 1400°C (about 2000 to 2550°F). The extent of grain coarsening depends on both thermal cycle and the metallurgical behavior of the steel. Larger grain size in the austenite increases the propensity of this portion of the HAZ to produce microstructures having high hardness and lower toughness. In Chapters 14 and 15 of Volume II, when welding procedures for different types of steel are discussed, attention will be given to means by which grain coarsening can be minimized in the higher temperature portion of the HAZ.

The previous illustrations are patterned after microstructures found in specimens in which the entire bulk has been heated to a specific temperature. While some of the zones of a weld within which unusual microstructural changes take place are extremely narrow, the presence of an unfavorable microstructure can be as threatening as a single weak link in an otherwise strong chain.

In discussion thus far on transformations that take place in solid iron and steel during heating, the Fe-Fe$_3$C phase diagram of Figure 4.12 has been used for approximations of temperatures at which transformations are expected to occur. This diagram is based on temperature measurements made during extremely slow cooling and detection of heat evolutions at critical points over the transformation ranges. Metallographic examination of microstructures then played a confirming role in phase identification. Several other methods can be used to secure temperature data needed to construct a phase diagram for the Fe-Fe$_3$C system. Abrupt changes in electrical resistance as the temperature varied has been used to a limited extent in plotting phase diagrams. Another method is to determine the hardness found in steel specimens after heating to selected temperatures and quenching. The hardness values are indirect indications of the nature of the microstructure at the elevated temperature prior to quenching. The volumetric changes that occur during transformation are particularly informative, and dilatometers have been utilized for dimensional measurements while specimens are heated to high temperatures, and during cooling. This method is quite effective because, as explained in Chapter 2 under *Thermal Expansion and Contraction,* metals undergo significant changes in volume as their crystalline structure is changed from one lattice arrangement to another. An ASTM Standard Test Method for linear thermal measurement can be found in E 228, and in E 289. The sudden changes in specimen dimension over-and-above the normal thermal expansion and contraction not only signal the temperature at which transformations start, and are completed, but they also can give a semi-quantitative indication of additional aspects of microstructural change. Figure 9.11 illustrates the rudiments of the dilatometer method. Linear curves are presented for expansion and contraction in specimen length of four familiar materials, namely, aluminum, copper, iron and AISI 1020 steel, during relatively slow heating to high temperatures. As would be predicted by the coefficients for thermal expansion given earlier in Table 3.12, of the four materials examined here, aluminum has the highest coefficient as indicated by the steepest curve in Figure 9.11. As also mentioned earlier, metals tend to exhibit slightly higher coefficients as their temperature rises, and for this reason the expansion plottings in Figure 9.11 curve slightly upward instead of being strictly linear. More to the point, the changing length of the aluminum and the copper specimens during heating and cooling follows a single curve, because these metals do not undergo a transformation in their fcc crystallographic structure during heating and cooling. On the other hand, iron, and the AISI 1020 steel display significant abrupt changes in length (and volume) during heating and cooling. Iron (as shown by the dashed-line curve) is seen to expand at a rate that has a slightly increasing coefficient, but when the critical point at a temperature of 912°C (1674°F) is reached, the transformation from bcc to fcc crystalline structure (alpha to gamma form of iron) causes a sudden decrease in length. This crystallographic alteration brings overall thermal expansion to a halt until all of the transformation has taken place at this temperature level. Following this transformation, the

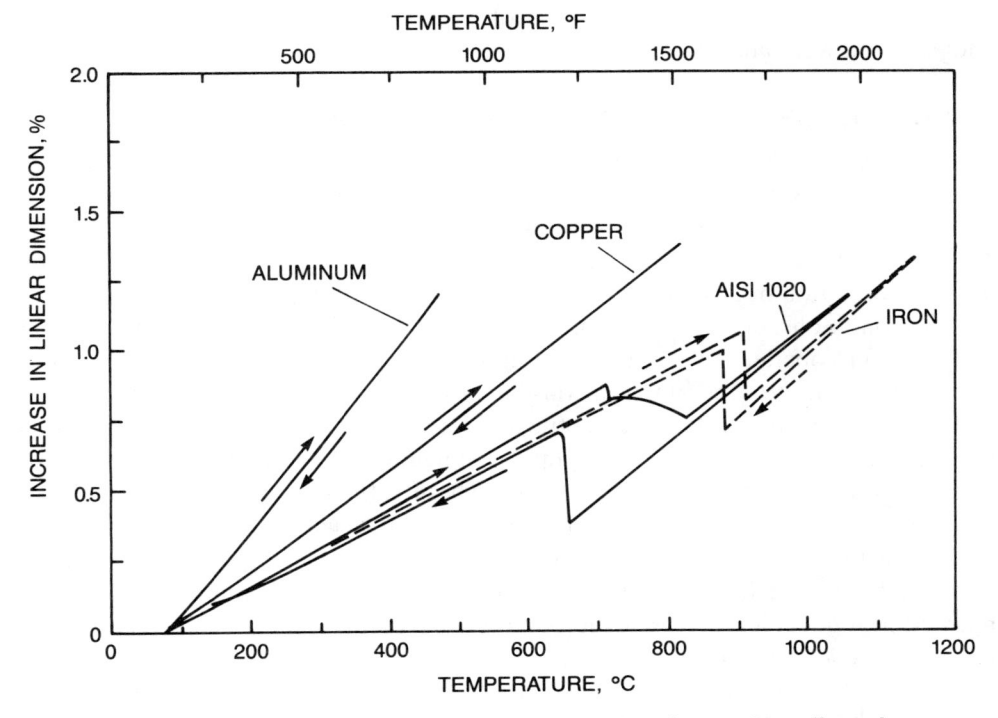

Figure 9.11 — Dilatometer linear curves for heating and cooling aluminum, copper, iron and AISI 1020 steel

816 Chapter 9

gamma phase iron on further heating displays the higher coefficient of the fcc crystalline structure. During cooling, because the transformation is allotropic, contraction of iron almost re-traces the expansion plotting — except for a slight undercooling, which is most evident when passing through the fcc to bcc transformation point.

The dilatometric behavior of AISI 1020 steel in Figure 9.11 is entirely in accord with the Fe-Fe$_3$C phase diagram. The increase in length as heating is started from room temperature has the coefficient of its bcc structure until the eutectoid temperature (Ac$_1$) of about 727°C (1341°F) is reached, whereupon a small abrupt decrease is registered as pearlite in the microstructure undergoes transformation to austenite. As temperature rises through the Ac$_1$ to Ac$_3$ transformation range, the specimen gradually contracts as the ferrite in the microstructure transforms to austenite. When completely austenitized as the temperature exceeds the Ac$_3$ level at about 850°C (1560°F), linear expansion of the specimen is resumed, but at the higher coefficient of the fcc crystalline structure. During cooling, the austenite in the AISI 1020 steel tends to undercool more than does the gamma phase of iron, and the austenite does not transform to pearlite and ferrite until the temperature drops to almost 650°C (approximately 1200°F). Because of increased metastability of the undercooled austenite, the transformation to a microstructure of pearlite and ferrite takes place over the next drop of about 25°C (about 45°F). Herein lies the challenge of understanding and controlling the behavior of steel. The extent of undercooling of the austenite that occurs under a given set of conditions will have profound effect on the transformation mechanism, and this, in turn, will determine the nature of the microstructure formed, mechanical properties, and volumetric changes; all of which play key roles in establishing the integrity of a weld and its fitness-for-service.

Microstructures Formed in Steel During Cooling

When steel is heated to make the microstructure austenitic, and then cooled, the *rate of cooling* exerts strong influence over the mechanism by which the austenite can transform from a fcc crystalline structure to the bcc structure which exists below the Ar$_1$ temperature. The "pearlitic" distribution of cementite described earlier requires the transformation of austenite to occur by *nucleation-and-growth*. Under equilibrium cooling conditions, or nearly so, ample time is available for the nucleation-and-growth mechanism to operate. Even when cooling is moderately accelerated, nucleation-and-growth is able to proceed but, as a consequence of undercooling, the pearlite colonies and nodules are smaller, their cementite and ferrite lamellae are finer, and the proportion of hypoeutectoid ferrite making up the remainder of the microstructure decreases and the grains of ferrite become smaller. These refinements in pearlite and ferrite cause modest increases in strength and hardness of the steel. (See Technical Brief 47.)

As austenitized steel is cooled at progressively increasing rates, the aforementioned refinements in the pearlite microstructure continue until a cooling rate is reached beyond which the nucleation-and-growth mechanism is unable to operate. At the *critical cooling rate* for the steel in hand, the austenite transforms by an entirely different mechanism, and the resultant microstructure has an entirely different morphology. The mechanism which operates at and beyond the critical cooling rate is *transformation hardening*; and the microstructure produced is *martensite*, the hardened structure mentioned earlier. Steels of a given carbon content with a martensite microstructure have much greater strength and hardness as compared with any mixture of ferrite and pearlite. However, the ductility and toughness of martensite can be quite low. The transformation hardening mechanism was illustrated earlier in Chapter 4 (Figure 4.11), and martensite was briefly discussed when reviewing carbon as the principal alloying element in steel. Nevertheless, the martensitic microstructure is a highly effective first-step toward obtaining optimum combinations of strength, hardness and toughness by a subsequent tempering treatment. Thus, two essential steps for heat treating steel are (1) heating to austenitize and then cooling rapidly to obtain a martensitic microstructure, and (2) reheating to a lower temperature to temper the martensite and improve both ductility and toughness; even though the temper-

---

TECHNICAL BRIEF 47
Pearlitic Microstructures

Pearlite is an easily identified microstructural constituent in steel, and was originally given the name *pearlyte* by H.M. Howe because its lamellar appearance when examined metallographically resembles mother-of-pearl. Pearlite is the eutectoid reaction product that forms by the time-dependent nucleation-diffusion-growth mechanism as austenitized steel cools at a relatively slow rate through the Ar_1 critical temperature (approximately 725 °C, or 1340 °F).

During transformation of the austenite, carbon atoms diffuse away from regions that have become hypoeutectoid ferrite, and they migrate to regions that eventually become pearlite colonies. Each grain-like pearlite colony consists of very thin alternating platelets of ferrite and cementite arranged as roughly parallel lamellae (see Figures 9.8 and 9.9). The crystallographic orientation of the ferrite and cementite phases are largely the same, and therefore pearlite really is the manifestation of two interpenetrating single crystals.

During the formation of pearlite, in addition to diffusion of the carbon atoms, iron atoms also make short-range transfer across the interface between the austenite and pearlite to accomplish necessary micro-changes in composition and in crystalline structure.

With increasing cooling rate, the temperature at which the austenite transforms is lowered somewhat, and the interlamellar platelet spacing becomes finer. Finally, with substantially faster cooling, a point is reached where atom-by-atom short-range diffusion is not possible and pearlite is not formed. Instead, the austenite with its fcc crystalline structure is forced to undergo transformation to ferrite (bcc) through a mechanism of cooperative displacement or by shearing. This fundamental change in transformation mechanism produces markedly different microstructural constituents; such as bainite, martensite and acicular ferrite.

Pearlite is seldom found in the microstructure of weld metal, but it is commonplace in the wrought and cast products of carbon and low-alloy steels. In fact, there is a general group called "pearlitic steels" because the carbon in these steels usually is distributed in the cementitic lamellae of pearlite as a result of regular cooling in the steel mill or foundry. This nomenclature is applied simply to give the user a rough idea of the steel's mechanical properties. Of course, the pearlitic microstructure can be changed by heat treatment if so desired.

Steel with a pearlitic microstructure will have strength and hardness that are commensurate with its carbon content, but will not be the highest levels obtainable. Toughness probably will be adequate for most purposes, but it is important to recognize those pearlitic steels that undergo a significant degradation in toughness at lower temperatures. A basic trait of pearlite is to cause an abrupt transition in toughness at a level not far below room temperature. If the microstructure consists of coarse pearlite colonies and ferrite grains, the toughness transition may take place somewhat above room temperature.

ing treatment will sacrifice some strength and hardness. Numerous alloy steels have been developed to fill *hardenability* requirements and improve consistency of response to heat treatment. The critical cooling rates for securing the all-important martensitic microstructure range from approximately that achieved by water quenching to the much slower rate of cooling in air. Many of these steels have type numbers and are listed among current standard types. Some types with high hardenability are described as "air hardening," while those of lesser hardenability are called "oil hardening," or "water hardening" steels. Not all steels being welded are in a hardened-and-tempered condition, and many steels have only modest hardenability, that is, their critical cooling rate required to cause martensite to form will be a relatively fast rate. Yet, circumstances in welding often contrive to form hardened zones in steel weldments, and this presents a general problem for reasons outlined below:

(1) Cooling rates in weld heat-affected zones can be surprisingly fast and easily can exceed the critical cooling rate required for martensite formation.
(2) Certain steels have chemical compositions that induce high hardenability, or in other words, a relatively slow critical cooling rate.
(3) Microstructural conditions created in steel by prior processing treatment, or caused by the heating cycle in welding can produce heterogeneity of composition in austenite and may result in localized martensite formation.
(4) Welding procedures do not necessarily include reheating that will temper any martensite that forms.
(5) The mechanism of martensite formation and the properties of this microstructure generate circumstances that may foster spontaneous cold cracking.

Much of the research described in preceding chapters has dealt with the challenge of welding steels that may develop hardened zones during welding and in which hardenability is virtually the antithesis of weldability. In fact, one accepted definition of critical cooling rate is "the minimum rate of continuous cooling for preventing undesirable transformations." The word *undesirable* indicates that cooling at a slower-than-critical rate will allow nucleation and growth of pearlite, and this is not in keeping with the objectives of heat treatment for steels. Yet, in welding technology, softer microstructures in the weld zone and HAZ after cooling to room temperature is ordinarily the objective when planning the welding procedure.

Martensitic Microstructures

Martensite and *martensitic transformation* have become generic terms because steel is not the only alloy able to undergo this unique mechanism of crystallographic change. Martensite transformations can occur in several nonferrous systems (e.g., copper-aluminum, and gold-cadmium), and also in some nonmetal systems, such as silica, and in zirconia. Attention here, however, is centered strictly on martensite in iron-base alloys where this microstructure is formed by the diffusionless, displacive, shear mechanism in which metal atoms in undercooled fcc austenite execute a highly cooperative shift in lattice position in their attempt to change quickly to the bcc structure of ferrite. This change from fcc toward bcc does not require thermal energy for activation, and, once initiated, the austenite-to-martensite transformation propagates almost at the velocity of an elastic wave in the iron lattice (about 10^5 cm/s, or 4×10^4 in./s). This extremely rapid crystallographic change traps carbon atoms that were previously dissolved as an interstitial solid solution in the fcc austenite, and the carbon is then held in a state of supersaturation in the martensite.

Figures 9.12-A, 9.12-B, and 9.12-C present three schematic sketches to make clear the following:

(1) why the bcc lattice of ferrite has virtually no solid solubility (either substitutionally, or interstitially) for carbon;

(2) how carbon is rather easily held in interstitial solid solution in the fcc lattice of austenite; and

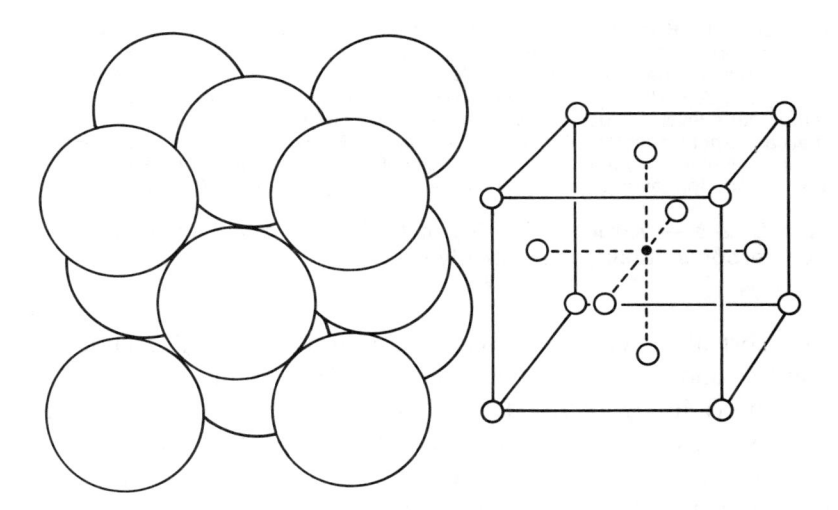

(A) AUSTENITE - Face-centered cubic crystalline form of iron which can dissolve approximately 2.1% carbon at temperature of 1154°C (2109°F), which is point "E" on Fe-Fe$_3$C phase diagram in Fig. 4.12. Although atoms in fcc austenite are more closely packed than in bcc ferrite, the octahedral interstices of the fcc lattice offers holes approximately 1.04Å in diameter in which carbon atoms are able to lodge as illustrated despite their estimated diameter of 1.54Å when in pure elemental aggregate form.

Figure 9.12-A — Arrangement of atoms in crystallographic structures of steel; (A) is austenite, the fcc form with a dissolved carbon atom in the preferred site

Schematic figures at left show atoms as spheres, frequently in close proximity as they are believed to exist, but this representation conceals dissolved carbon atoms at the octahedral interstice of the fcc lattice. Therefore, figures at right clearly indicate favored interstices where (because of lattice hole size) carbon atoms lodge most often. Open circles are iron atoms. Carbon atoms are solid black.

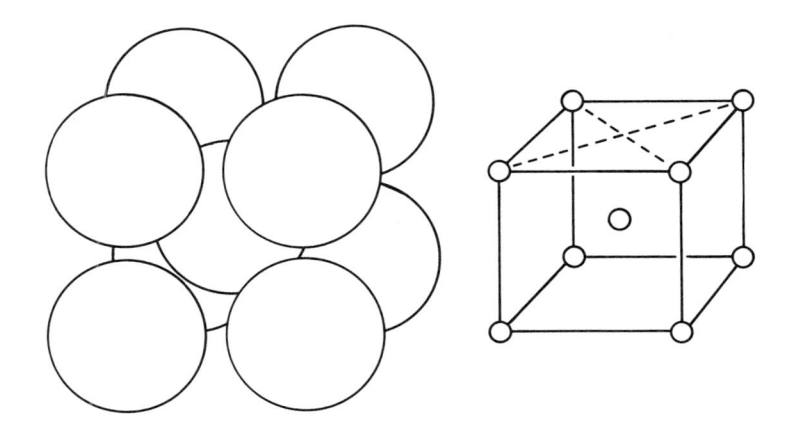

(B) FERRITE - Body-centered cubic crystalline form. Lattice locations almost large enough to hold carbon atom interstitially is octahedral site, and one is indicated by dashed-line intersect on top face of cube where hole is approximately 0.72 Å in diameter. Ferrite can dissolve 0.0218% carbon at the eutectoid temperature level, but upon cooling to room temperature the amount of carbon that can be held in interstitial solution is very limited. Many investigators report a RT solubility limit of only about 0.008%, yet calculations based upon free energy of iron carbide formation suggest the limit actually may be below 0.007 ppm.

Figure 9.12-B — Arrangement (B) of ferrite, the bcc form of iron, in which carbon solubility is extremely low

(3) where most of the carbon atoms are sited in the crystalline lattice of martensite, and some of the consequences of this malformation.

Ferrite, as shown in Figure 9.12-B, does not have a "close-packed" lattice, but its interstitial "holes" are a little too small to accept a carbon atom. Raising the temperature increases the size of these holes and accounts for the slightly increased carbon solubility that is found up to the eutectoid temperature. Application of a tensile stress also influences to a small extent the solubility of carbon, and the mobility of carbon atoms as the bcc lattice attempts to accommodate these solute atoms. Austenite, shown in Figure 9.12-A, also has a number of interstitial holes at the octahedral and tetrahedral sites in its fcc lattice, and both sites offer larger interstices than any in the bcc lattice. The favored location in austenite to hold a carbon atom in solid solution is believed to be the center of the cubic crystal as shown in the schematic of the lattice on the right side of Figure 9.12-A, and thusly the fcc lattice is able to accept a goodly amount of carbon interstitially. Martensite has a body-centered *tetragonal* lattice as shown by the illustration of a single crystal in Figure 9.12-C, but all of the crystalline structure of martensite does not necessarily consist of bct crystals - only those that have entrapped carbon atoms during the decomposition of the austenite. The remainder of the lattice constituting the martensite is essentially bcc. The proportion of bct in the martensitic structure is directly dependent, therefore, on its carbon content.

When austenite transforms to martensite the bct crystals that entrap carbon undergo an increase in their vertical lattice parameter because the carbon atom forces greater separation of the iron atoms above and below this interstitial location. A very slight decrease occurs in the horizontal lattice parameter providing no carbon atom intrudes on this axis. The bct crystal form is one of the Bravais lattices illustrated earlier in Figure 2.5, but its occurrence in martensite is an aberration attributable to the virtual inability of bcc ferrite to accept carbon atoms interstitially. Furthermore, the relatively low temperature range over which martensite transformation takes place prevents carbon from exercising enough atom mobility to escape the confinement in the bct lattice by precipitating and allowing the distorted crystals to assume their normal bcc form. Yet, this metallurgical aspect of martensite can be manipulated by alloying to decrease the extent of undercooling before the start of transformation to martensite — as will be explained in Volume II, Chapter 15. A number of important, practical facts about martensite have been explained by intensive study of its total crystallography and morphology using x-ray diffraction techniques and electron microscopy at very high magnification.

Martensitic microstructures are strikingly different from those of nucleation-and-growth transformation to pearlite. A typical optical photomicrograph of martensite in a medium carbon steel is presented in Figure 9.13, and, as can be seen, the general morphology is *acicular*. This needle-like appearance is dependent to some degree upon the composition of the steel, the etching technique applied to the metallographic specimen, and the extent to which the martensite has remained untempered (i.e., free of incipient precipitation of carbon or cementite on internal crystallographic planes). The needle-like configuration can actually appear to some degree without the benefit of etching on

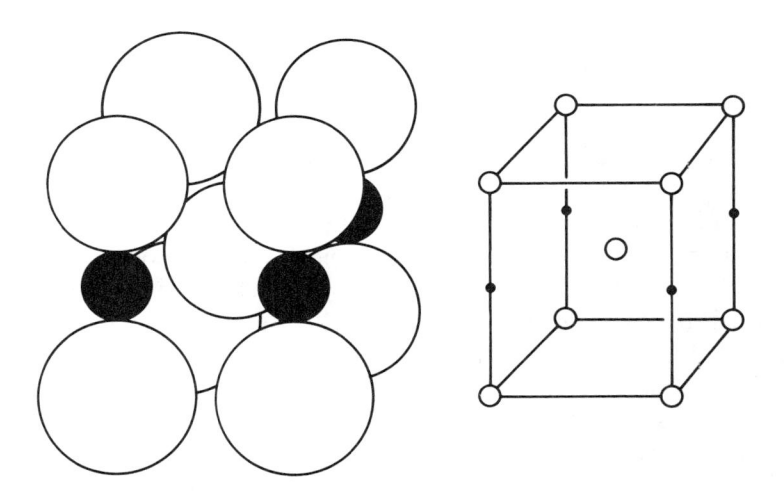

(C) MARTENSITE - Body-centered tetragonal crystalline structure formed when austenite is drastically undercooled and transforms by diffusionless, shear mechanism. Carbon atoms are trapped as illustrated and they force lattice to become tetragonal rather than cubic. Not all crystals hold four trapped carbon atoms as illustrated because the number of sites occupied is directly dependent upon the carbon content of the steel. Besides volume increase accompanying the fcc to bcc change in crystalline structure, additional expansion occurs commensurate with the number of crystals distorted into the tetragonal form.

Figure 9.12-C — Arrangement (C) of martensite, the hardened structure in which carbon atoms distort the body-centered cubic arrangement into the body-centered tetragonal (bct) form

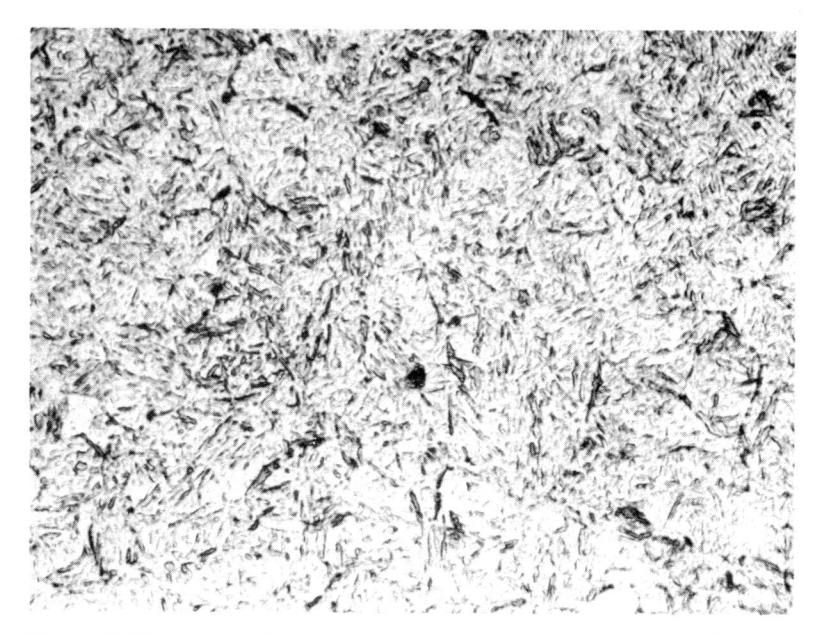

Figure 9.13 — Typical acicular appearance of martensite as observed metallographically in the microstructure of a hardened steel

(Picral etchant; original magnification: 500X)

a polished (but protected) surface of a specimen that subsequently has been subjected to martensitic transformation. Explanation for this surface phenomenon can be found in the mechanism by which martensite forms and disrupts a polished surface.

Figure 9.14 shows schematically the typical surface morphology of martensite (upper sketch), and the shear mechanism (lower drawing) which produces plate-like crystals of bct structure. There is a close crystallographic relationship between the parent austenite and the resulting martensite because of the instantaneous, diffusionless change in positions of atoms as the fcc form shifts by shear to the bct form. The martensite crystals are flat and not very thick, but they often reach considerable length; this accounts for the needlelike appearance on an intersecting surface. This common crystal shape is called *lath* martensite, and groups of laths are referred to as sheaves, or packets. When the basic unit crystals are very thin, the microstructure is called *sheet* martensite, and when thicker, lenticular-shape units form, the microstructure is called *plate* martensite. Lath martensite has a large population of dislocations in its crystalline structure, which comes as no surprise in view of its development under conditions of undercooling and very rapid transformation. Plate martensite may occur sparsely and randomly in a matrix of lath martensite. There is a tendency for plates to develop internal twins that have a spacing of only 50 to 100 Å. Plate martensite is more prevalent in steels with higher carbon contents, and the occurrence of twinning in the plates increases noticeably when the carbon content exceeds about 0.30 percent.

Figure 9.15 includes a photomicrograph illustrating martensite in steel containing 0.34 percent carbon where some plate martensite (indicated by arrows from P) can be seen in a matrix of lath martensite (marked L). Examination of this microstructure at very high magnification finds that some of the plates contain internal twins as shown by the TEM micrograph. This detail in the character of martensite is quite significant because susceptibility to cold cracking from hydrogen embrittlement appears to correlate closely with the extent of twinning as carbon content increases. This aspect of martensite will be discussed further in Volume II, Chapter 13, when cold cracking and the influence of hydrogen is reviewed.

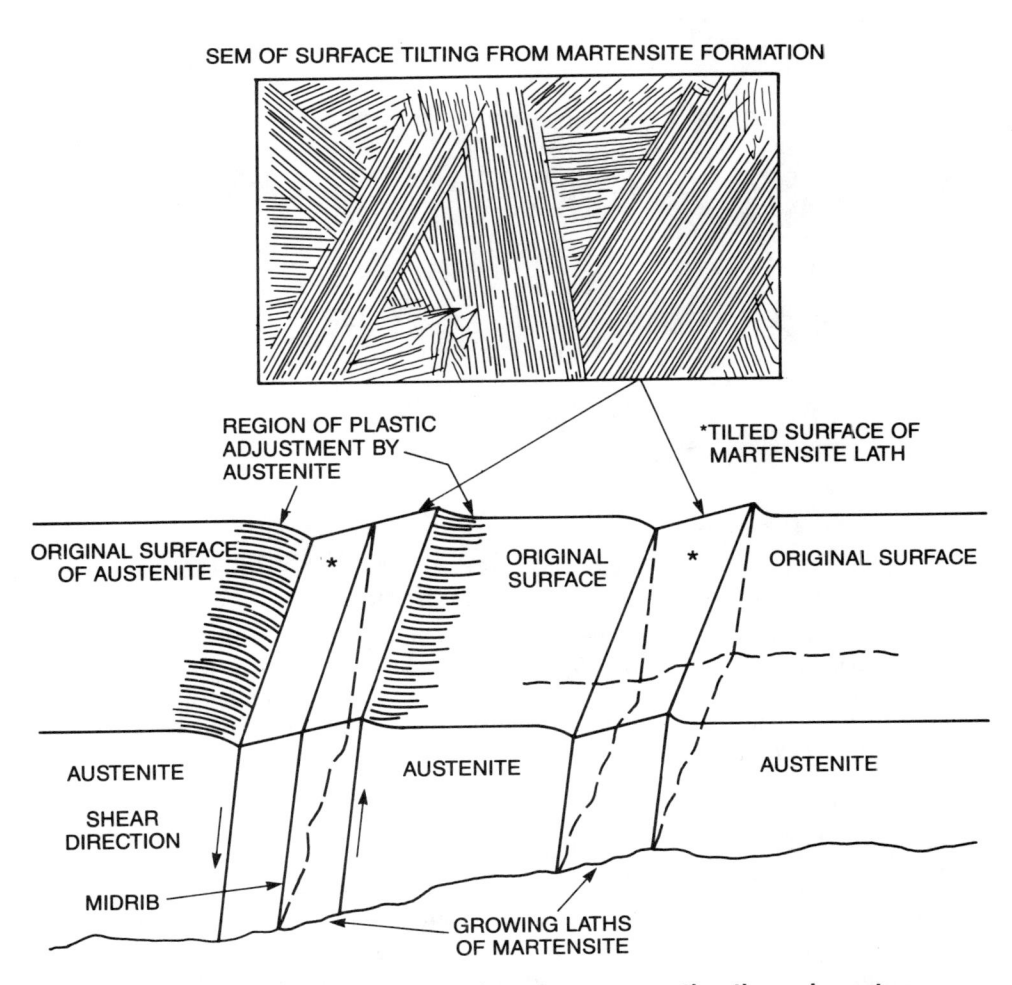

Figure 9.14 — Schematic illustration of a cross-section through martensite and the shear mechanism which forms long, thin sheets or platelets of body-centered tetragonal crystalline structure

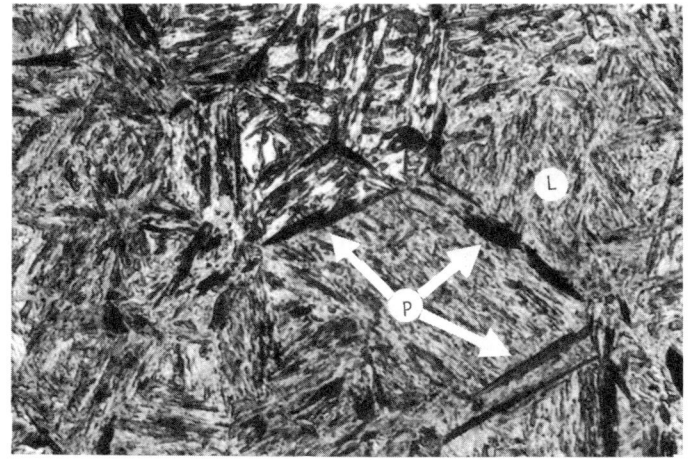

Figure 9.15 — Martensite in steel containing 0.34% carbon
(Upper) optical photomicrograph shows some plate martensite (P) in a matrix of lath martensite (L). (Lower) transmission electron micrograph (TEM) shows twinning in a plate of martensite as indicated by arrow.

Isothermal Transformation of Austenite

All previous discussions of the transformation of austenite to ferrite, pearlite, and martensite have been based on cooling the austenitized steel *continuously*. Although a noticeable change in rate of cooling can occur during transformation because of the exothermic nature of phase changes, cooling was not deliberately halted at any intermediate level of temperature. The transformation of austenite also can be studied by deliberate use of step-cooling; under conditions of *isothermal transformation*.

Although not always directly applicable to the circumstances under which steel is heat treated, or welded, diagrams that portray the results of isothermal transformation provide very helpful basic information. One result was the recognition of a microstructure called *bainite*, which consists of ferrite and finely dispersed cementite that forms under conditions that lie between the nucleation-and-growth mechanics of pearlite formation and the shear-type mechanism by which martensite is formed. The bainitic microstructure is produced commercially in heat-treated steel articles by an operation known as *austempering*. Also, there will be circumstances in welding operations where high preheating or interpass temperatures will cause similar microstructural effects.

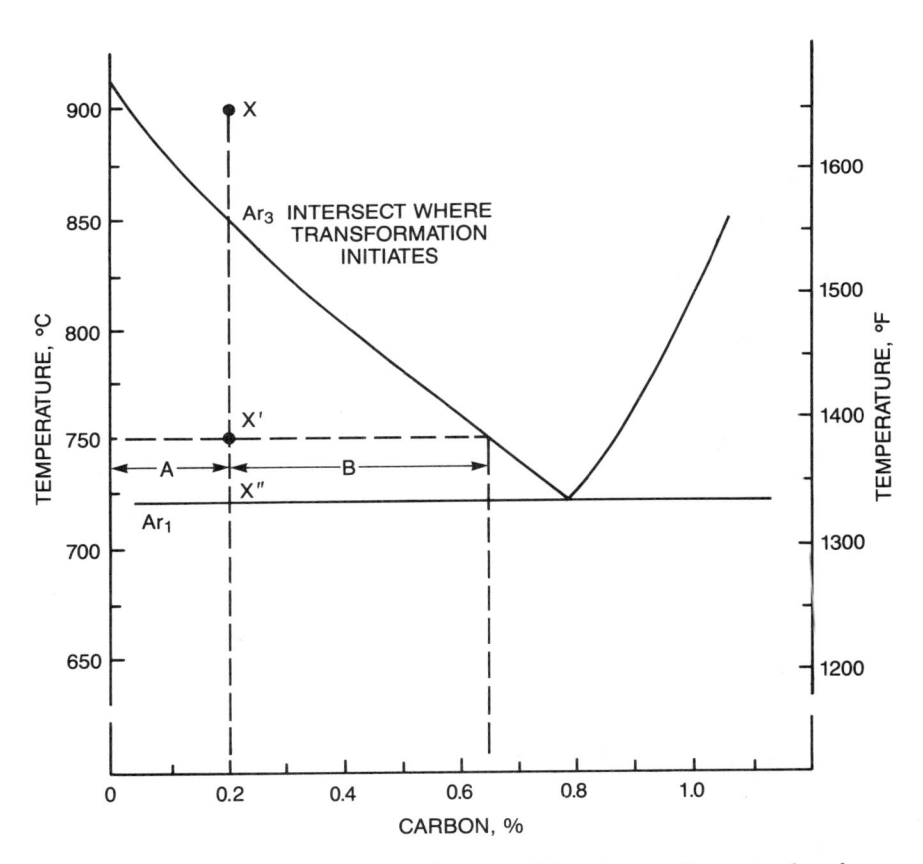

Figure 9.16 — Portion of iron-iron carbide phase diagram showing events in 0.20% carbon steel during cooling from the austenitic state

Very slow cooling of austenite from point X (900°C, or 1650°F) allows ferrite formation to start on reaching a temperature of 850°C (1560°F), which is the Ar_3 point of the steel's transformation range. At point marked X' (750°C, or 1380°F), approximately two-thirds of the microstructure has transformed to ferrite, and the remaining austenite now contains approximately 0.65% carbon as indicated by application of the lever law.

To compare the results of conventional continuous cooling with isothermal treatment, the AISI 1020 steel containing 0.20 percent carbon will serve nicely. Steels of other carbon contents will behave basically in the same manner as the AISI 1020 steel except that (1) the upper critical (A_3) temperature will vary with the carbon content, (2) the rates of cooling required to cause undercooling prior to austenite transformation can be slower as the carbon content increases, and (3) cementite instead of ferrite is the first phase to nucleate in the austenite when the latter contains more carbon than the eutectoid composition during slow cooling.

Pearlite Formation Isothermally

On the portion of the Fe-Fe$_3$C phase diagram in Figure 9.16, the position marked X depicts the 0.20 percent carbon AISI 1020 steel at a temperature of 900°C (1650°F)

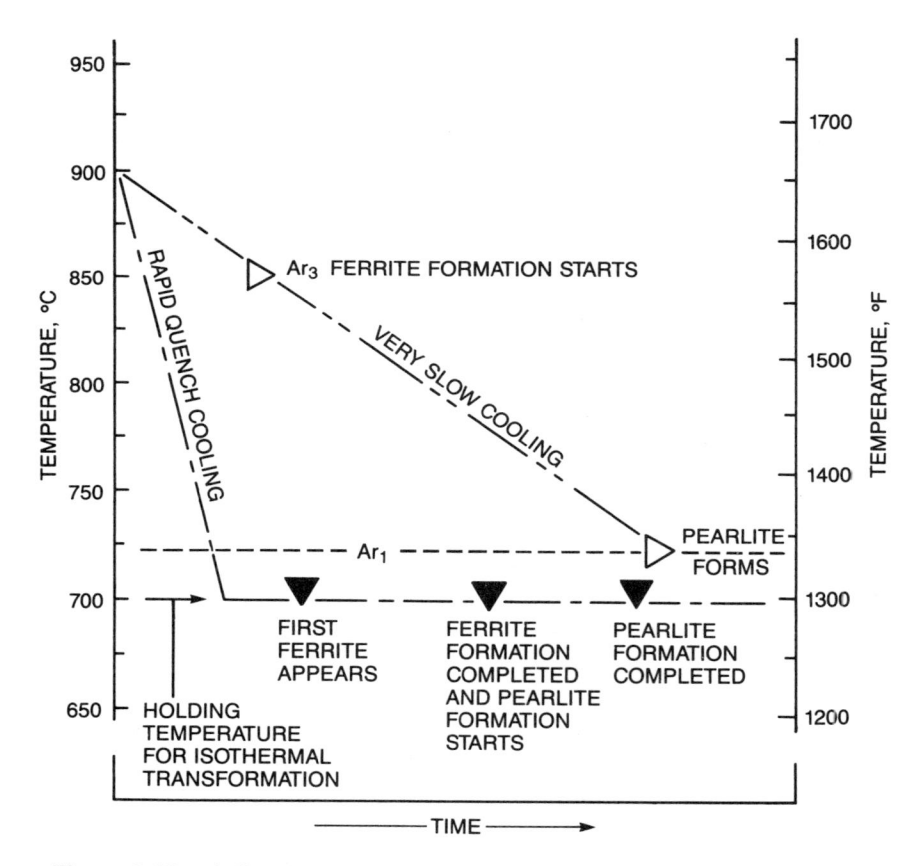

Figure 9.17 — Influence of cooling rate on transformation of austenite to ferrite in steel

This schematic diagram shows qualitatively that during very rapid cooling of austenite from 900°C (1650°F) no ferrite is formed at the holding temperature of 700°C (1292°F) until a couple of seconds have elapsed; then ferrite formation proceeds. When formation of hypoeutectoid ferrite is complete, transformation of remaining austenite to pearlite commences and is completed after a period of time under the isothermal conditions.

where its microstructure would consist of grains of austenite (each grain containing 0.20 percent carbon). Slowly cooled, perhaps at a rate of one degree Celsius per minute (about 2°F/min.), ferrite starts to form as the temperature reaches the Ar_3 level at about 850°C (1560°F). Ferrite grains continue to form and grow by nucleation-and-growth as the temperature drops. When the temperature level of 750°C (1380°F) is reached at X′, the lever law will indicate that the amount of austenite still untransformed in the microstructure is equal to [(A/A + B)×100] percent, or about 30 percent. Also, the austenite has been enriched with carbon because of the limited solubility of carbon in ferrite. The lever law indicates the carbon content of the austenite at 750°C (1380°F) to be approximately 0.65 percent. Further slow cooling to Ar_1 (marked X′′), causes the amount of untransformed austenite to decrease to about 25 percent, and the balance of the microstructure (about 75 percent) to have transformed to ferrite; the carbon content of the austenite has increased to approximately 0.8 percent, the eutectoid composition. As the heat of transformation is released at the Ar_1 level, all of the austenite transforms to pearlite. No further changes in microstructure occur as the steel cools to room temperature.

Figure 9.17 is a rudimentary diagram of events that occur during *isothermal transformation*. When the austenitized steel containing 0.20 percent carbon is cooled very rapidly from 900°C (1650°F) to a temperature just below the Ar_1 level, perhaps by quenching a very thin specimen into a bath of molten metal or salt being held at the lower temperature, the steel reaches the holding temperature remaining *100 percent austenite*! The diagram also bears a reminder that during very slow continuous cooling some of the austenite began to transform to ferrite when the steel reached a temperature of 850°C (1560°F) because this ordinarily is the Ar_3 point for the 0.20 percent carbon steel. The austenite, when very rapidly cooled to 700°C (1292°F), does not start to transform to ferrite until a short period of time has passed, perhaps, a couple of seconds. Held at this temperature, ferrite grains continue to form and after approximately five minutes all of the hypoeutectoid ferrite will have formed. During this transformation, the austenite is enriched with carbon by diffusion from the ferrite and approaches the eutectoid carbon content. After all of the hypoeutectoid ferrite has formed, transformation of the austenite to pearlite commences. At this relatively high transformation temperature, the pearlite formation requires substantial time to proceed to completion, perhaps as much as three hours.

When isothermal transformation of austenitized steel is carried out at still lower constant temperatures a number of phenomena are changed in a systematic way; including (1) incubation time prior to start of transformation, (2) rate of transformation, and (3) the spacing of lamellae in any pearlite that is formed. Briefly, study of isothermal transformation established that austenite will transform at most subcritical temperatures if held long enough, and the time required to complete the transformation and the nature of the microstructural product will vary in a systematic way with the isothermal temperature. Studies of this kind laid the foundation for a new kind of diagram (circa 1930) that remains popular and useful today. Originally, the diagram was referred to as the *isothermal transformation diagram*, or the *time-temperature-transformation diagram*. It portrayed microstructures produced at temperatures below the critical range, and the time required for transformation of the austenite to be started and completed. Soon, however, abbreviated references came into use; including *T-T-T diagram*, *S-curve*, and *C-curve*. These designations were proposed as the diagram underwent reshaping by continuing research, and they actually mark important metallurgical milestones in the course of this work. For instance, "S-curve" was proposed when early diagrams displayed curving lines that resembled a rough letter "S." Eventually, the lower portion of the S-curve was found incorrect. Investigators had been concentrating on the "nucleation-and-growth" of pearlite formation and had overlooked the real mechanism by which martensite is formed. Steps taken later to correct the diagram explain why the designation "C-curve" evolved. Some of these early designations persist in the literature, but the proper terminology today is *isothermal transformation diagram*, or for brevity, the "IT diagram."

To illustrate events in diagram development, Figure 9.18 has been prepared to show the upper portion of the IT diagram for transformation of austenite in a plain carbon eutectoid steel (0.77 percent carbon content). This composition is chosen because it has fewer boundary curves since no hypoeutectoid ferrite or hypereutectoid cementite is produced prior to pearlite formation. Figure 9.18 shows general trends in microstructural alterations as isothermal transformation temperatures are lowered, and indicates an abrupt change in transformation below approximately 200°C (400°F) when martensite is produced. The diagram shows that temperature reductions below 700°C (1300°F) cause a steady decline in the time required to start and complete pearlite formation. Pearlite is produced in the shortest time at about 550°C (1022°F), at which point the start of pearlite formation begins in about one second and ends in a little less than ten seconds. This forms a "nose" on the start-curve of the IT diagram. At the same time as IT temperatures are lowered toward nose of the transformation curve, the lamellae in pearlite become finer.

Bainite Formation Isothermally

As the IT temperature is lowered below the nose of the transformation start-curve, pearlite cannot form because the rate of carbon diffusion is too slow for the nucleation-and-growth process to produce pearlite. Nor are the circumstances conducive to forming martensite. Therefore, at temperatures from about 500 down to 250°C (932 to 482°F) the decomposition of austenite takes place by transitional mechanisms between that for pearlite and for martensite, and these can be generally described as massive progressive transformations rather than the nucleation-growth-diffusion mechanism by which pearlite forms, or the shear mechanism of martensite formation. The transformation products that form at temperatures within this transitional IT range are called *bainite*. The start-curve for transformation of austenite to bainite often forms a "bay" beneath the nose of the IT curve and above the temperature for the start of martensite formation. This bay portrays the increasing difficulty of the austenite to undergo transformation at lower temperatures, and in Figure 9.18 the greatest delay is shown to occur at about 250°C (482°F). The extent or depth of the bay will be shown to have importance in welding steels of high hardenability.

Martensite Formation

When early investigators of isothermal transformation in steels extended their studies to still lower temperatures; that is, isothermal levels below 250°C (482°F),

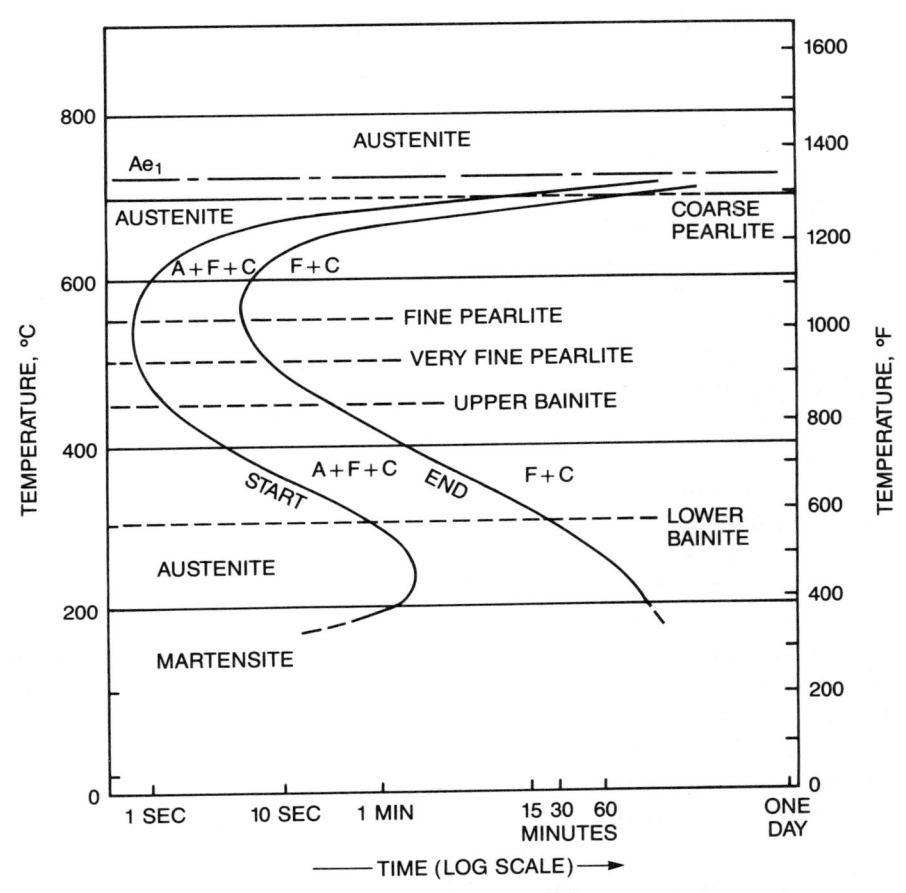

Figure 9.18 — Isothermal transformation diagram for steel of eutectoid composition (0.77% carbon)

Diagram shows time for transformation of austenite to start, and to end at holding temperatures ranging from just below to Ae₁ down to almost room temperature. Austenite is indicated as "A", and undergoes transformation to ferrite "F" and to cementite or carbide "C", and these transformation products form either pearlite or bainite. The transformation to martensite was eventually found athermal.

they found the microstructure in treated specimens to be martensitic. At first, the assumption was made that transformation of austenite to martensite was proceeding with time in a manner similar to the development of pearlite and bainite except that the time delay before start of transformation began to decrease. Therefore, the lower end of the "start" curve was turned toward the left as shown in Figure 9.18, and the "S" curve became the accepted pattern of that early period. However, the assumption concerning martensite formation proved to be incorrect. This early erroneous interpretation of results from isothermal studies deserves telling here because it will emphasize the uniqueness of martensite formation mechanics, and will make clear the need to understand the dependence of martensite formation on temperature of exposure.

Briefly told, when early investigators quenched austenitized steel specimens to a low temperature, held them isothermally for a period, and then brought them to room temperature for microstructural examination, they found the microstructure to be martensite. They incorrectly concluded that the transformation had proceeded at the elevated holding temperature albeit at an accelerated rate. In reality, the martensite formation had initiated in small amount after the temperature passed through a particular level as the specimen was quenched to the selected isothermal treatment temperature, but then further martensite formation was arrested during the entire isothermal holding time. When finally cooled to room temperature, all of the austenite had transformed to martensite, but this transformation had proceeded to completion by successive change in crystalline structure (i.e., fcc to bcc and bct) as the temperature was lowered from the isothermal holding level to room temperature.

Continued research established that the transformation of austenite to martensite begins at a critical point which now is defined as the *martensite start temperature* (M_s). When austenitized steel is quenched to its M_s point and held at this temperature, no delay occurs in the start of martensite formation, but, surprisingly, only a small percentage of the austenite transforms to martensite as the steel is held at that temperature. Further transformation to martensite occurs only when the steel is cooled to lower temperatures. Whenever cooling is interrupted, transformation stops. The amount of martensite formed per degree decrease in temperature is not constant, but starts slowly at the M_s, then increases at a greater rate as temperature is lowered, and finally decreases as the temperature drops sufficiently to bring transformation to martensite to completion. This final temperature is identified as the *martensite finish* point (M_f). Obviously, formation of martensite after reaching the M_s point proceeds only during cooling to lower temperature and is independent of time. This manner of transformation has been termed *athermal*, and its kinetics can be treated mathematically to ascertain the percentage of martensite formed after quenching through the M_s and stopping at a particular temperature level below. The incorrect portrayal of martensite formation by the lower termini of curves in Figure 9.18 should now be recognized. The uniqueness of martensite formation is a very important part of the metallurgical behavior of steels, and it will be given close attention when discussing procedures for welding highly hardenable steels.

Figure 9.19 is an up-to-date IT diagram for a 0.77 percent carbon steel showing the microstructures formed and their hardness when rapidly quenched to temperatures below the Ae_1 and held until isothermal transformation is complete. This diagram also shows the percentage of martensite formed *upon reaching* isothermal levels below the M_s point. Only after cooling below about 75°C (165°F) has the austenitic microstructure transformed essentially completely to martensite. The rate of cooling from the M_s toward room temperature has little influence on the percentage of austenite remaining untransformed at a particular temperature level. Brief mention is made that M_s and the M_f temperatures are altered as the composition of the steel is changed. Also, the matter of retained austenite after cooling below the approximate M_f point is a subject that is more complex than inferred above. The influence of steel composition on the tem-

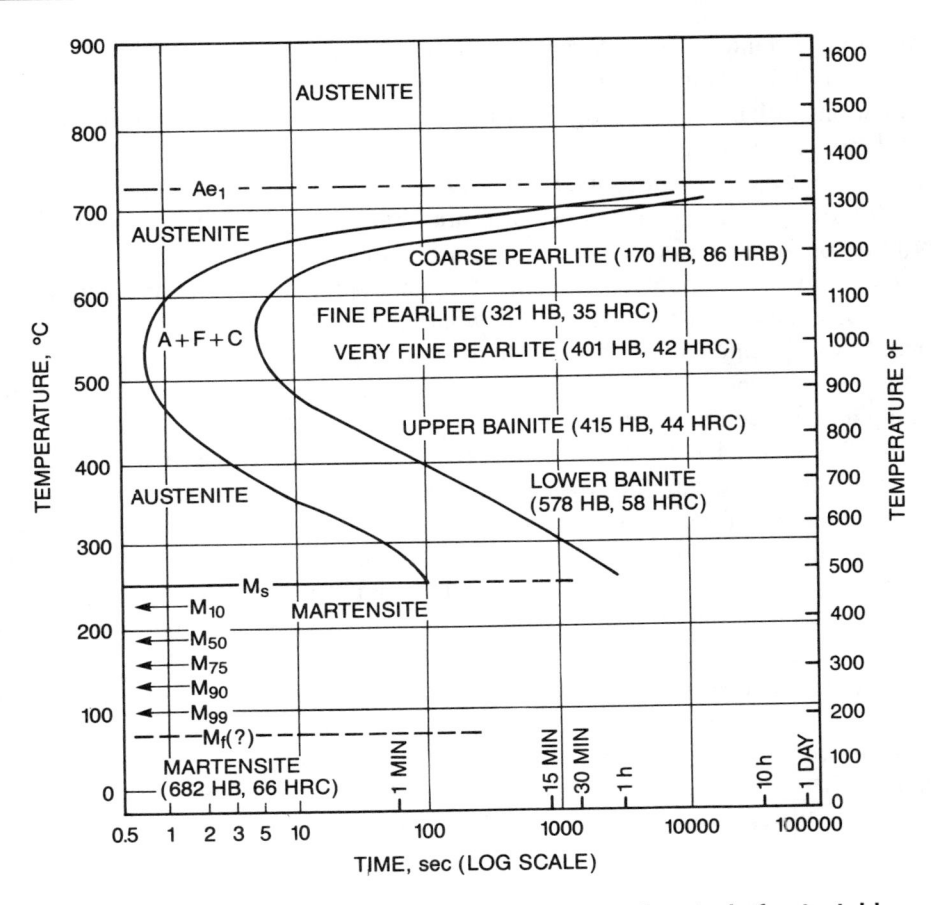

Figure 9.19 — Isothermal transformation diagram for steel of eutectoid composition (0.77% carbon) showing phases, constituents, and hardness values produced by holding at particular isothermal transformation temperatures

Martensite formation does not proceed isothermally, but is athermal as indicated by increasing percentages of transformation when cooled to temperatures below the M_s point. Hardness given for martensite is value obtained after cooling to room temperature.

perature levels of the M_s and M_f points will be discussed shortly when martensite and its formation is examined in detail.

Reappraisal of Microstructures Formed in Steel

Although "bainite" seems to be a concise, appropriate term for the microstructure formed at IT temperatures between those for pearlite and for martensite, bainitic structures actually represent a broad variety of aggregates of ferrite (bcc) and a distribution of precipitated carbide particles. The detailed morphology of the bainites can vary and is dependent on the composition of the steel, particularly its carbon content and the presence of alloying elements that promote hardenability. In addition, the temperature at which the austenite-to-bainite transformation takes place also plays a part in shaping the character of the bainitic microstructure. The temperature factor is so strong that two classifications of bainite are recognized to form toward the extremes of the aforementioned temperature range of 500° down to 200°C (932° to 482°F), and these have been aptly designated *upper bainite*, and *lower bainite*. The approximate temperature regions at which these two kinds of bainite form are indicated in Figure 9.19, but a precise description of the two kinds of microstructure is difficult because of the many variations found in their morphology. In general, upper bainite has a feathery microstructural appearance, and the steel will be somewhat harder and tougher than if pearlite had formed. Lower bainite tends to have a microstructure that appears acicular, but does not have the same morphology as true martensite. Lower bainite is not as hard as martensite, but can be much tougher. The latter property has proved quite suitable for many "hardened" steel articles; hence the commercial utilization of isothermal heat treatment ("austempering") to secure a lower bainite type of microstructure and make use of its good combination of hardness and toughness. Bainitic microstructures formed by IT between the upper and lower temperatures for this kind of transformation product are so varied that systematic classification is not feasible. In fact, bainitic microstructure are quite difficult to interpret by optical metallography. The very high magnification capability of electron microscopy must be relied upon to determine the nature of the ferrite matrix and the distribution of precipitated carbides. Recognition of phases present can be exceedingly difficult when "mixed" microstructures have been produced, and when some of the many forms of bainite are in intimate association with pearlite, martensite, or possibly with retained austenite. The complexity of mixed microstructures is a challenge that arises with weld heat-affected zones formed by continuous cooling which must be examined and interpreted microstructurally.

Upper Bainite

Although feathery in appearance under optical metallography, upper bainite is revealed by electron microscopy to consist of very small lath-like ferrite grains in parallel groups that form plate-shaped sheaves. The laths commonly have an orientation that relates to the parent austenite. The ferritic crystalline structure has a high dislocation population, and the density of dislocations increases with lower IT temperatures. Because of the relatively high temperature at which upper bainite forms, carbon in solid solution in the austenite is able to precipitate readily as cementite (Fe_3C) particles. At the higher temperatures just below the nose of the IT curve, the precipitated cementite may be quite coarse and often forms complete films between the ferrite laths. When this distribution of cementite occurs, the microstructure can have a distinct lamellar appearance. However, the spacing between the lamellae of cementite in bainite usually is greater than in pearlite formed at a similar temperature. Also, the bainite can be recognized by its markedly different etching characteristics, even when viewed by optical metallography. As the IT temperature to form bainite is lowered, the cementite is precipitated in finer particles, and electron microscopy is required to resolve its distribution in the ferritic matrix.

(A) COARSE LAMELLAR PEARLITE FORMED AT 700 °C (1300 °F)

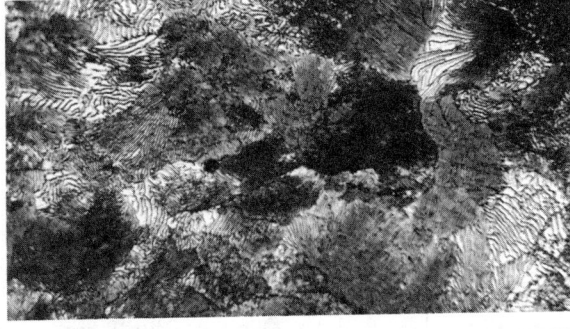

(B) FINE LAMELLAR PEARLITE FORMED AT 600 °C (1100 °F)

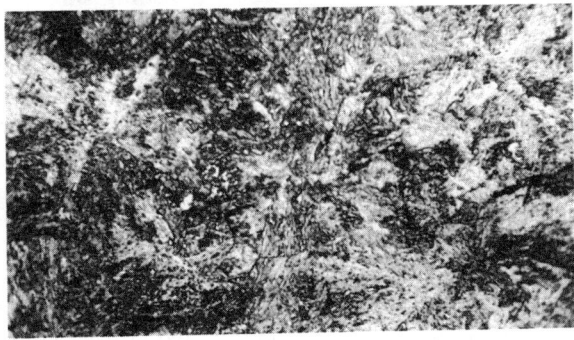

(C) LOWER BAINITE FORMED AT 316 °C (600 °F)

Figure 9.20 — Microstructures produced in eutectoid steel (0.77% carbon) by isothermal transformation at three levels of temperature

Top photomicrograph - coarse lamellar pearlite; middle photomicrograph - fine lamellar pearlite; bottom photomicrograph - lower bainite. (Nital etchant; original magnification: 500X)

Lower Bainite

Lower bainite is formed isothermally at temperature starting about 400°C (752°F) and continuing downward until the M_s point is reached. Lower bainite consists mostly of needlelike ferrite plates that contain a dispersion of extremely small carbide particles. Electron microscopy must be used to resolve the particles, and characteristically the carbides are located *within* the needlelike grains rather than *between* laths or plates as occurs in upper bainite. Also, in lower bainite, the first carbon atoms to precipitate during the fcc-to-bcc transformation form epsilon carbides which have $Fe_{2.4}C$ as their composition. After some amount of ε-carbide forms (depending upon the steel's carbon content), subsequent precipitates have the cementite composition Fe_3C.

Lower bainite is valued for its unique mechanical properties; i.e., good toughness even at strength and hardness levels approaching those of martensite in the same steel. For this reason, certain low-alloy steels have been developed to facilitate the formation of lower bainite. This has been accomplished through judicious alloying and prescribed heat treatment in which cooling rate is of vital importance. When steels of this kind are welded, the character of their prescribed microstructure must be recognized and the procedure for welding, and for possible postweld heat treatment, should be tailored to continue utilizing the lower bainitic microstructure in the completed weldment. Steels of this kind will be discussed in Volume II, Chapter 15.

As a simple demonstration of the value of optical metallography as the first step in appraisal of microstructures, Figure 9.20 presents three photomicrographs showing the microstructure of 0.77 percent carbon steel after rapid quenching of thin section to three different temperatures at which transformation proceeded isothermally. The coarse pearlite [Figure 9.20(A)] formed at 700°C (1300°F) served earlier to illustrate the microstructure formed in this particular steel by the nucleation-and-growth mechanism. The fine pearlite [Figure 9.20(B)] formed at 600°C (1100°F) has a number of nodules where lamellae cannot be seen because of unfavorable orientation to the polished-and-etched surface of the metallographic specimen, or because etching stain has obscured the lamellar character. Nevertheless, a more searching examination, perhaps by SEM, should show that virtually all cementite in the microstructure has been distributed in lamellae of pearlite. The lower bainite shown in Figure 9.20(C) has extremely fine carbide particles dispersed throughout its ferritic matrix, and is devoid of lamellar nodules.

Importance of Critical Cooling Rate

An observation of considerable importance made during studies of isothermal transformation of austenized steel was centered on the rate of cooling above which neither pearlite nor bainite would have the time required for formation, and consequently the too-fast rate of cooling would lead to the formation of martensite. The *critical cooling rate* was mentioned earlier as the "slowest rate at which a given austenitized steel could be cooled and yet prevent the formation of (sic) undesirable transformation products." The unwanted microstructures are, of course, the opinion of steel-heat-treaters aiming for "full hardening," and in many cases this is contrary to aims when welding steel. The first illustration of suppression of decomposition of austenite by excessively rapid cooling was shown in Figure 9.17 where quench cooling delayed transformation to ferrite in a 0.20 percent carbon steel after the rapid temperature drop of 200°C (360°F). The next indication of transformation suppression by cooling rate was shown in Figures 9.18 and 9.19 where the nose of the start-curve showed a delay of almost one second before transformation initiated when rapidly cooling the 0.77 percent carbon steel from 600 down to 500°C (1112 down to 932°F). When the time-temperature path does not enter the nose of the transformation region, cooling of the steel easily can be continued to obtain the formation of martensite. Thusly, the time span ahead of the nose of the start-curve can be a measure of critical cooling rate. Yet, it must be kept in mind that this diagram and most of its plotted results are based on *isothermal* determinations. Continuous cooling, which is more often the case in welding, exerts additional delay, and the effect of continuous cooling will be carefully examined in due course. Nevertheless, isothermal transformation (IT) diagrams frequently are included in the literature because data collection for their plotting is relatively easy. The influence of alloying elements on critical cooling rate has been inferred in earlier chapters (e.g., Chapter 4) when commenting on the hardenability increase promoted by the addition of specific alloying elements. Hardenability equates to the incubation time span before the nose of the transformation start-curve since this indicated how *slowly* an austenitized steel can be cooled and still transform to martensite. Critical cooling rate is the minimum continuous temperature decrease in degrees per second at which a steel can be cooled and still achieve a fully martensitic microstructure after reaching room temperature. Of course, some highly hardenable steels may require cooling to an end-point somewhere below room temperature to ensure passing through the M_f level and thus produce 100 percent martensite (i.e., no retained austenite).

In addition to alloy content, another feature of steels that significantly affects their critical cooling rate is the grain size of the austenite. In Chapters 4 and 5, it was made clear that steels could be manufactured as inherently "fine grained," or "coarse grained." Yet, the temperature to which steel is heated also affects the size of austenite grains. Coarse austenite grains increase the hardenability of steel, that is, they increase the delay time-span before the nose of the transformation start-curve in the region of proeutectoid ferrite and pearlite formation. This is understandable because these microstructural constituents form by nucleation-and-growth, and since nucleation occurs primarily at the austenite grain boundaries, more time is required for nucleation as there are fewer boundaries. Also, longer time is required to complete transformation because of the greater dimensions across which each austenite grain must change. For these reasons, austenite grain size (or, at least, the austenitizing temperature) should be noted in addition to the composition of the steel when evaluating critical cooling rate or hardenability. The influence of grain size is not easily quantified as alloy content. Results with a single steel of modest hardenability provide some insight. This steel had its IT start-curve positioned with the nose at 525°C (977°F), and the following delay times to the nose were exhibited as austenitizing temperature and grain size were changed:

| Austenitizing Temperature | | ASTM | Time to Nose of IT |
°C	°F	Grain Size	Start-Curve, sec
843	1550	8 to 9	1.5
899	1650	7 to 8	2.8
1010	1850	4 to 5	8.5
1093	2000	2 to 3	10.3

Although austenitizing temperature and austenite grain size are shown by the above data to have a pronounced influence when IT delay time is relatively short, this degree of influence may not hold true for other steels with different compositions and having greater hardenability. Nevertheless, austenitizing temperature can exert a significant effect, and one must remember that portions of the HAZ at a weld will be subjected to rather high austenitizing temperatures (as contrasted with the logical, controlled temperatures employed when heating specimens for isothermal transformation studies). Sometimes the real effect of grain size is obscured by homogeneity changes that accompany grain coarsening at a higher austenitizing temperature.

The problem of expressing cooling rates in a manner that can be conveniently correlated with welding phenomena was discussed at length in Chapter 7. This problem again appears when attempting to quantify "critical cooling rate" to obtain a value that can be easily equated to microstructural transformations during cooling. The most practical solution to this problem continues to be hardenability test bars that provide a standardized measurement that is indicative of the ease of martensite formation. Two specimens will be described in Volume II, Chapter 16 — namely, Grossman's Ideal Critical Diameter Round and Jominy's End-Quench Specimen. The latter specimen was used in Chapter 7 to ascertain cooling rates in welds at specific, highly-localized positions in heat-affected zones from hardness data.

Importance of Delay-Time Before Austenite Transformation

Another aspect of austenite transformation clearly seen in isothermal transformation diagrams is the extraordinary length of time (incubation period) before the start of bainite formation. This is a feature which varies with different steels, and although more-or-less characteristic of a given steel composition, the incubation time can be influenced by variables in heating and cooling. The extent of the *bay* positioned in front of the lower bainite region is an especially important feature of a steel for a number of reasons found both in heat treating and in welding.

Steel treaters take note of the bay because its depth dimension at various temperatures indicates the time available to handle steel in the austenitic condition as they perform holding and/or cooling steps of a heat treating operation. Also, a choice can be made whether to allow bainite to form, or to continue cooling (perhaps even at a somewhat slower rate) to form martensite. Some alloy steels can transform to bainite during continuous cooling because the "pearlite nose" of their transformation diagram is further to the right (longer delay-time before austenite starts to transform) because of the alloying elements present.

In welding, the "bay" can have far-reaching involvement. The selection of temperatures for preheating, interpass, reheating, post-weld holding, and cooling endpoint, and the times allowed for these thermal cycles require good knowledge of transformation dynamics to be certain a desired microstructure has actually formed at a particular stage of a fabricating procedure before proceeding to the next step.

Martensite: Implications in Welding

Martensite, and the shear-type transformation by which this microstructure is formed, has been described as being in sharp contrast with transformations that produce pearlite or bainite. Differences in crystal structure, disposition of carbon content, and morphology of microstructure were illustrated in earlier figures, and the unique features of martensite were detailed in Figures 9.12 to 9.15. At this point, certain features of martensite deserve review because of their pertinence when planning procedures for welding steels that can form martensite.

Discussion of martensite in conjunction with isothermal transformation of austenite may seem illogical. After all, martensite formation has no incubation period, nucleation-and-growth does not take place, no diffusion of carbon occurs, and the austenite-to-martensite transformation mechanism does not operate isothermally. Instead, the martensite formation mechanism is athermal and independent of time! Nevertheless, displaying the martensite formation range (M_s to M_f) in the lower portion of IT diagrams has proved to be a very practicable visual form of presentation. The features of martensite to be reviewed here include (1) temperature range for its formation, (2) percentage of martensite formed as temperature decreases over the range of formation, (3) hardness of fully martensitic microstructures as related to composition, and (4) cracking susceptibility.

Temperature Range for Martensite Formation

The range of temperature through which austenite will cool and undergo athermal transformation to martensite occurs over a spread of about 150°C (about 300°F) in most steels. The starting temperature level for martensite formation (the M_s point) is strongly dependent on the steel's chemical composition. Carbon content exerts the most potent influence, and lowers the M_s temperature. Studies have been conducted with relatively pure iron-carbon alloys to establish the potency of carbon alone in lowering the M_s point. Although it is difficult to quench very low carbon alloys at a sufficiently fast cooling rate to form martensite, about 750°C (1382°F) has been judged to be the M_s point for iron containing less than about 0.02 percent carbon. With as little as 0.05 percent carbon present, the M_s temperature is lowered abruptly as can be seen in Figure 9.21, and many investigators have confirmed that the M_s point is steadily lowered as carbon content is increased.

Table 9.1 - Influence of Alloying Elements in Steel on Martensite Formation

I. Qualitative Effects of Alloying Elements on Martensite Start (M_s) Temperature

Alloying Element	Qualitative Effect
Carbon	Significantly Lowers M_s
Manganese	Lowers M_s
Silicon	Virtually no influence
Chromium	Lowers M_s
Nickel	Lowers M_s
Molybdenum	Lowers M_s
Copper	Lowers M_s
Vanadium	Lowers M_s
Tungsten	Lowers M_s
Cobalt	Raises M_s
Aluminum	Raises M_s

II. Empirical Equation for Calculation of M_s Temperature from Composition*

M_s (degrees Celsius) = 550 − 360 (%C) − 40 (%Mn) − 40 (%Cr) − 20 (%Ni) − 30 (%Mo)

This equation appears applicable to steels having chemical composition within the following limits:

Carbon .... 0.1 – 0.6	Silicon .... 0.1 – 0.5	Nickel .... 5.0 max
Manganese ...... 0.2 – 2.0	Chromium .... 3.5 max	Molybdenum ..... 1.0 max

Other alloying elements used in steels affect the M_s point as indicated in Table 9.1. Several investigators have proposed empirical equations for calculating the M_s point from alloy content, and a typical equation is given in Table 9.1. Note that cobalt *raises* the M_s level. This exceptional effect has been utilized in developing high-strength alloy steels that are "self-tempering." These cobalt-containing steels have a martensite formation range at a relatively high level, and atomic mobility at the transformation temperatures allows some carbide precipitation to take place automatically as the austenite transforms to martensite. The net effect is similar to a tempering heat treatment and the martensite so formed has improved toughness.

The trend of influence with most other alloying elements is to *lower* the M_s point; indeed, the entire M_s-to-M_f range usually is lowered. This is not a favorable trend because at lower temperatures carbon is more likely to be retained in solid solution in

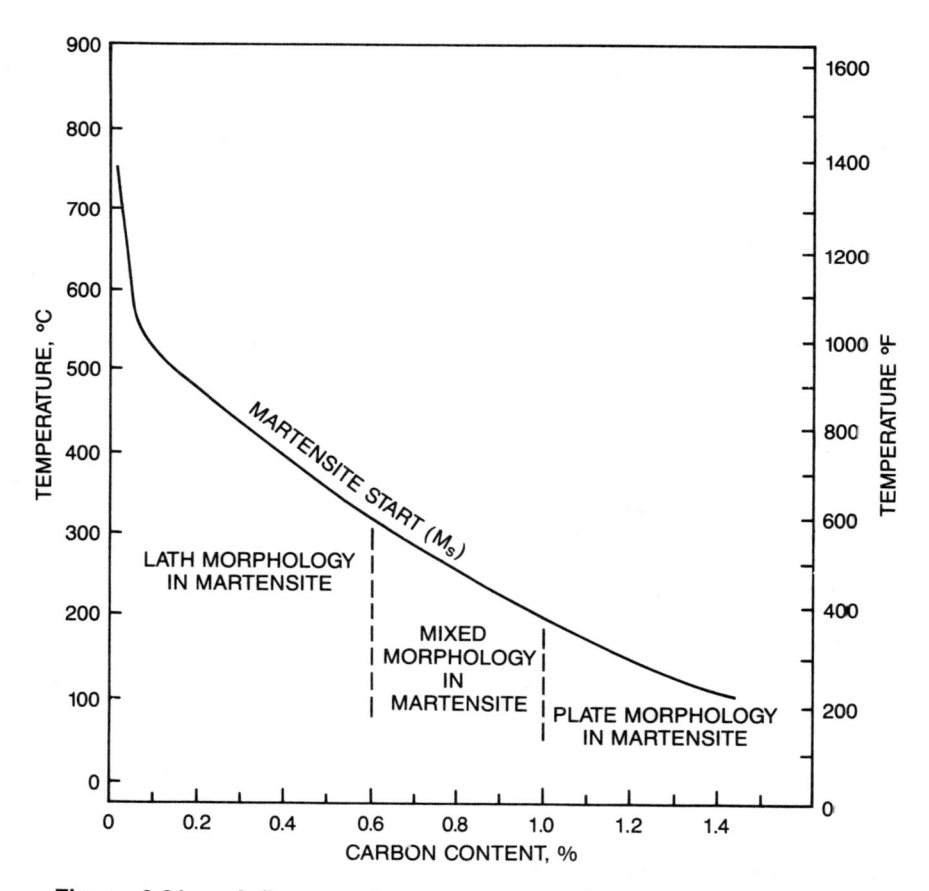

Figure 9.21 — Influence of carbon content in hardenable steels on martensite start (M_s) temperature

A general indication is also given on the diagram of changes in the morphology of the martensitic structure as the carbon content varies.

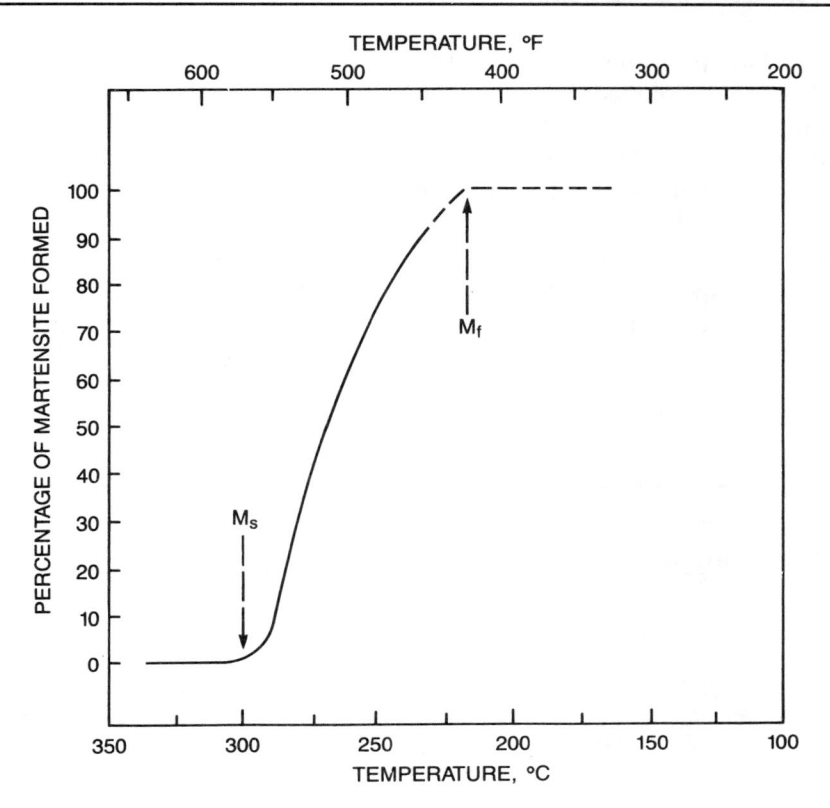

Figure 9.22 — **Martensite formation by athermal transformation in AISI-SAE 2340 (3½% Ni) alloy steel on cooling through the M_s-M_f range**

the bct martensitic microstructure and this contributes to specific volume expansion as the austenite transforms.

Efforts to calculate the M_f point from composition have not been entirely successful. The M_f of many steels has not been established because more discerning methods of microstructural examination continue to find traces of austenite remaining in martensite that had been reported to be 100 percent transformed when examined by earlier x-ray and metallographic methods. More exact knowledge of the actual M_f point can avoid circumstances where retained austenite causes problems in steel with toughness, dimensional stability, and cracking. These difficulties can arise because of the specific volume increase that occurs when *retained* fcc austenite transforms to bct martensite. Even though only a very small percentage of the microstructure may have been retained as austenite, the volume change is forced to take place in relatively brittle martensite at a low temperature where fracture toughness may not be adequate to accommodate the micro-scale volume adjustment wherever a region of austenite transforms.

Quantitative Prediction of Martensite Formation

The amount of martensite that forms as the steel is cooled through its M_s-to-M_f range is important because cooling end-points are not always easily arranged in a welding procedure, particularly if the pertinent M_f point is below room temperature. Figure 9.22 contains a plotted curve of percent martensite formed versus temperature during cooling through the martensite formation range of AISI-SAE 2340 alloy steel. The composition of this steel includes 0.40 percent carbon, 0.80 percent manganese, and 3.50 percent nickel. The equation in Table 9.1 for predicting M_s in steels provides a calculated M_s temperature of 316°C (600°F) for this alloy steel, which is in reasonable agreement with the observed M_s of 300°C (572°F) determined by metallographic examination of specimens. The M_f temperature established metallographically is approximately 218°C (421°F), but the state of 100 percent martensite had not been substantiated by electron microscopy or x-ray diffraction study.

TECHNICAL BRIEF 48:
Martensitic Microstructures

Martensite, so named to honor A. Martens, is the hardest microstructure that can be formed in a carbon or alloy steel. The level of hardness in a fully martensitic microstructure is commensurate with the steel's carbon content; almost regardless of the amounts of other alloying elements present. Consequently, very-low-carbon steel, even in the martensitic condition, will not be very hard (see Figure 9.23). Importantly, with a carbon content of about 0.60 percent, the maximum hardness that can be obtained in steel is roughly 68 HRC (~750 HB); and higher carbon content will not achieve any real increase in maximum obtainable hardness.

A martensitic structure is produced in steel when austenite is continuously cooled at a rate faster than the particular steel's *critical cooling rate*, which is determined by its hardenability potential. With low hardenability, austenitized steel usually must be cooled by quenching to produce martensite. Some alloyed steels with high hardenability will form martensite when their austenitized structure is merely "air cooled." Regions of steel that are austenitized by the localized heat of welding have the potential for forming martensite because cooling rates in welds can be quite fast, but the microstructural outcome also depends on the hardenability of the steel. Furthermore, austenite must be cooled through a range of lower temperature (M_s-M_f) to accomplish full transformation without thermal activation (athermal). A martensitic microstructure may have some constituents and phases intermixed that reduce its expected level of hardness; including pearlite or bainite areas caused by marginally rapid cooling and perhaps some retained austenite due to failure to cool below the M_f point.

As austenite instantly transforms to martensite by its unique shear mechanism, a small increase in volume occurs because the fcc crystalline form is the close-packed arrangement of atoms. Internal stresses are accentuated by the level of carbon in the steel when it becomes martensitic because carbon atoms entrapped in the rapidly-forming ferritic structure force some of the bcc crystals into a bct form which adds to the volume increase.

In general, higher hardness in martensitic microstructures is accompanied by lower ductility and toughness, which in turn increases susceptibility to cracking under many circumstances. Despite its foibles, martensite can be tailored through steel composition and by thermal treatment to be a suitable hard, strong constituent as it originates, or, as is often the case, martensite can be relieved of its shortcomings by thermal tempering. In fact many steel products and articles are heat treated to provide a tempered martensitic microstructure because this is a most effective procedure to obtain highly satisfactory combinations of strength, hardness, ductility and toughness in steel.

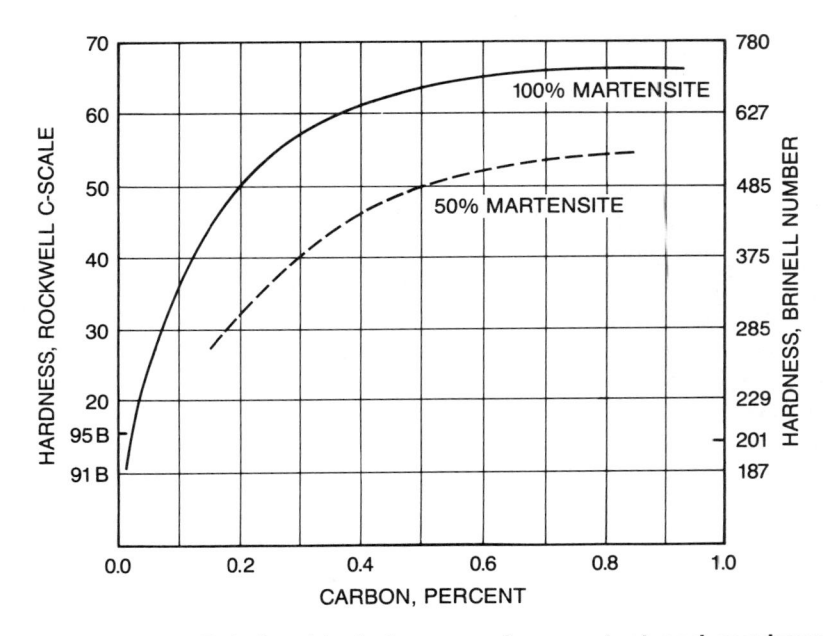

Figure 9.23 — Relationship between carbon content and maximum hardness obtainable in any plain-carbon or alloy steel provided critical cooling rate is exceeded and no significant amount of austenite is retained; i.e., microstructure is 100% martensite

Broken-line curve is hardness when martensite formation is not complete and approximately half of transformed microstructural product is bainite or fine pearlite.

Martensite Hardness Rationale

The hardness of martensite in steels, when methodically investigated many years ago, proved surprising. First, it was found that the maximum hardness that could be obtained in a martensitic microstructure was approximately 68 HRC (745 HB). This level of absolute hardness for steels was found when the carbon content reached approximately 0.6 percent, and higher carbon contents did not significantly increase the maximum obtainable hardness beyond 68 HRC. Equally surprising was finding that the presence of any additional alloying elements did not significantly increase the hardness of martensite. Alloying with other commonly used elements, such as nickel, chromium, etc., increased hardenability, but the hardness of martensite remained quite dependent on its carbon content. Figure 9.23 provides a curve depicting the firm relationship that exists between the maximum obtainable hardness in a completely martensitic microstructure and its carbon content. The curve represents an average of confirming values reported by many investigators on a wide variety of carbon and alloy steels. This is an important fundamental fact. This relationship can be helpful in predicting the approximate maximum hardness that a given steel of known carbon content will attain in its weld HAZ when operating conditions are expected to produce a fast cooling rate that exceeds the critical cooling rate for the steel. (See Technical Brief 48.)

Very low-carbon unalloyed steels, those containing less than 0.10 percent carbon, seldom produce a martensitic microstructure because their critical cooling rate is very high. Yet, when martensite is produced in a very low carbon steel, the toughness can be quite adequate to avoid undue cracking susceptibility and to accommodate service demands. This quality in very low carbon martensite has prompted the recent introduction of steels that contain less than about 0.07 percent and are alloyed with elements to secure strength by mechanisms other than transformation hardening. If the microstructure contains less than 100 percent martensite there will be noticeable decrease in the hardness of the microstructure. As little as five percent of a non-martensitic phase will lower the hardness several Rockwell-C points below the maximum hardness for steels with carbon contents ranging from 0.2 to 0.8 percent. The question sometimes arises whether to regard a steel with less than 100 percent martensite in its microstructure as being "hardened." This question has been resolved in a practical way by considering steel with a minimum of 50 percent martensite as being hardened. In most steels, when the microstructure contains 50 percent or more of martensite, a polished surface usually will etch and remain light in color; whereas a noticeably darker coloration appears when the microstructure contains more than 50 percent of upper bainite or fine pearlite. This color change from a light to a darker etched surface occurs rather abruptly and usually coincides with the half-and-half microstructure morphology. This sharp demarcation helps outline the various heat-affected zones and regions of multi-pass welds.

Martensite Formation Monitoring by Acoustic Emission Signals

Because martensite formation in the weld zone or the HAZ of high-strength, high-hardenability steels, for example, AISI-SAE 4340 Cr-Ni-Mo alloy steel, can be a critical event, much thought has been given to ways of real-time monitoring so that its onset can be detected, and appropriate action can be taken to guard against cold cracking. A number of remedial measures are possible, including (1) immediate application of postweld heating to the weld area to control cooling, (2) avoidance of unnecessary delay before NDE to establish soundness and/or to apply postweld treatment to prevent delayed cold cracking, and (3) alteration of the welding procedure for making further welds in a manner that ensures against cold cracking. Acoustic emission (AE) is a technology that provides a promising method of real-time monitoring that can signal martensite formation, and the occurrence of cold cracking in the weld zone or the HAZ, and can even indicate when a porous weld zone has been made.

The high velocity of transformation of austenite to martensite releases energy that generates elastic waves that propagate far into contiguous sections of the weldment. Sensors can be placed on the weldment at convenient, remote locations to scan all elastic waves emitted during the welding operation and for a period thereafter while cooling takes place. Acoustic emission is directed to a computerized data-acquisition system to identify signal patterns that allow immediate discrimination of the following:

(1) the making of a sound weld,
(2) the making of a porous weld,
(3) the occurrence of martensite formation, and
(4) cold cracking initiation and propagation.

Studies using AE techniques for monitoring welding date back many years; but more recent reports on its application with modern, sophisticated equipment for filtering, gating, amplification and pattern recognition suggest that shop-floor systems may be near-at-hand.

IT Diagrams: Summation of Usefulness

A closing examination is made here of a few typical isothermal transformation diagrams selected from current literature. This brief overview is intended to illustrate how an IT diagram can serve as a helpful guide in dealing with the metallurgical behavior of a steel, and to point out precautions that should normally be observed when taking specific information from a diagram and applying it to a particular welding operation.

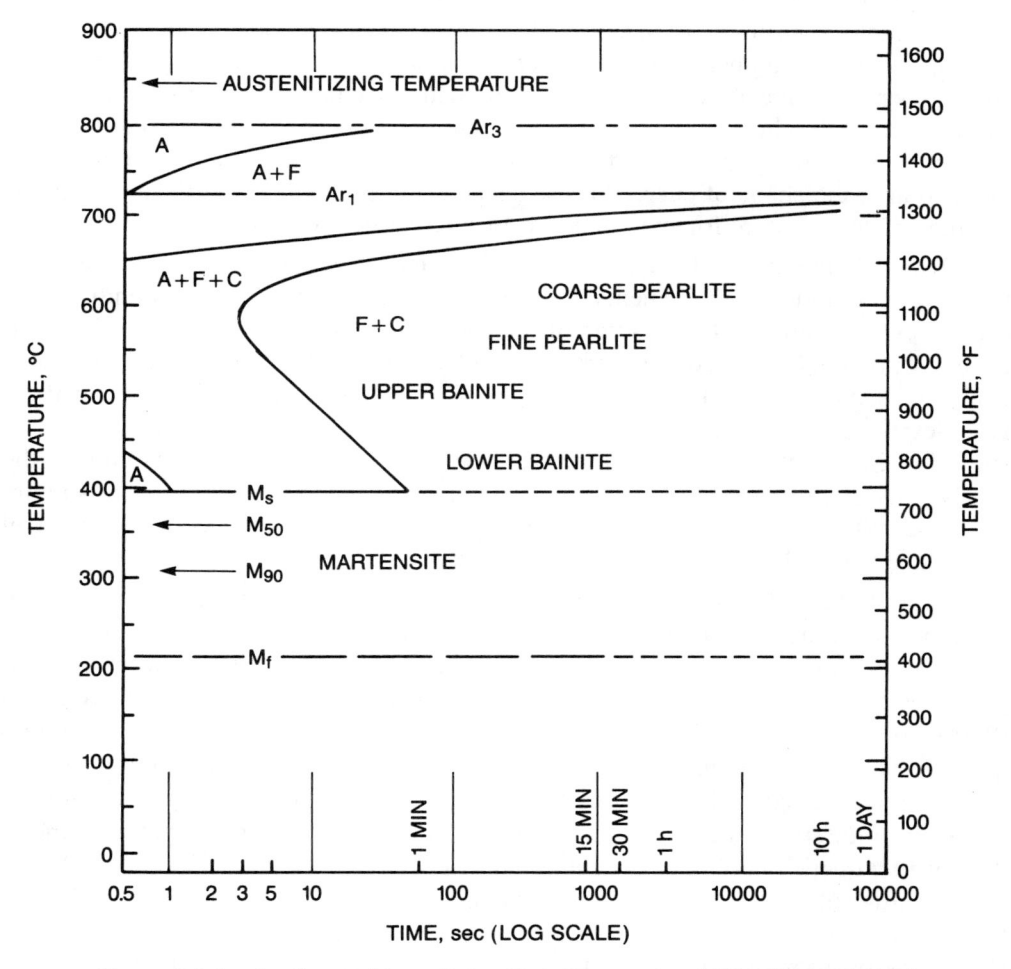

Figure 9.24 — Isothermal transformation diagram for AISI-SAE 1035 steel
This plain steel has a nominal carbon content of 0.35%. The diagram is plotted from data obtained after austenitizing specimens at 850°C (1562°F) and quenching to sub-critical temperatures.

Several popular steels of widely different compositions are represented in Figures 9.24, 9.25, and 9.26 where their respective IT diagrams quickly show the influence of carbon content and of alloying additions with regard to hardenability and martensite formation. More subtle, yet pervasive effects also can be gleaned from these diagrams, and as these are pin-pointed an explanation will be given for weighing them as influential factors when planning proper welding procedures.

The three IT diagrams reviewed here certainly confirm earlier information (see Table 9.1) that the addition of most alloying elements used in steels (except aluminum and cobalt) act to increase hardenability; that is, alloying tends to move the transformation curves of the IT diagrams *to the right*, and downward to some extent. This rightward shift indicates longer delay times when cooling before the austenite starts to transform, and this translates into increased hardenability. The extent to which the IT curves are shifted varies with the different alloying elements. Cobalt moves IT curves *to the left*, thusly, austenite is encouraged to transform earlier and at a somewhat higher temperature. A number of alloying elements, including carbon, manganese, nickel and copper, shift the IT curves to the right without changing their general shape. Certain other alloying elements, particularly those that are strong carbide-formers (e.g., chromium, molybdenum and vanadium), not only shift the IT curves to the right, but also alter the shape of curves. These shape changes portray significant variations in time and temperature for specific kinds of microstructure to form.

Shape-change in the IT curves also can be affected by austenitizing temperature. If strong carbide-forming elements are present in the austenite partly as undissolved carbides, these particles tie up some of the carbon and keep this portion of the hardenability potential out of solution. Consequently, their usual effect in promoting hardenability is nullified. In fact, the heterogeneity created by the undissolved carbide particles *decreases* hardenability because the particles provide additional nucleation sites that assist in starting austenite transformation and hasten completion of the newly-formed microstructure. Of course, increasing the austenitizing temperature could put all carbide particles in solution; and thereafter, during cooling, the steel would display the higher, expected hardenability.

Figure 9.24 is the IT diagram for AISI-SAE 1035 plain-carbon steel. This steel contains only a modest amount of manganese (usually about 0.75 percent) in addition to its carbon content of 0.35 percent to promote hardenability. To construct this diagram, the steel was austenitized at the usual "heat treating" temperature above its A_3 (850°C, or 1562°F) which produced a mixed austenitic grain size consisting of three-quarters No. 2 to 3 and one-quarter No. 7 to 8. When quenched to temperatures below the Ar_1, the IT diagram indicates that the hardenability is so weak that formation of hypoeutectoid ferrite quickly occurs. Formation of pearlite is apt to start almost immediately after quench-cooling to an isothermal temperature of about 650°C (1202°F). To obtain martensite in this steel, the IT start-curve must be displaced rightward to allow time for the cooling austenite to pass a nose on the start-curve without interception. A shift in the position of the curve to the right could be accomplished applying a higher austenitizing temperature to produce a coarser grain size. For the most part, Figure 9.24 indicates that quenching this particular 0.35-percent carbon steel in order to obtain martensite is quite difficult — if not impossible — because of its very high critical cooling rate. Therefore, to secure a fully hardened martensitic microstructure under isothermal study conditions, either the hardenability must be increased, or cooling must be extremely fast in order to bring untransformed austenite into the martensite temperature range. The M_s of 395°C (743°F) found by experimentation with this steel is virtually the same as the 398°C calculated from the equation in Table 9.1.

TECHNICAL BRIEF 49:
Bainitic Microstructures

Bainite is a distinctive array of microstructures in steel that were late in being identified (circa 1930), and were named to honor Edgar C. Bain. Bainitic microstructures in unadulterated classical forms are produced in a controlled manner by isothermal transformation of austenite at temperatures ranging from below the nose of the pearlite formation curve to just above the martensite range. Bainite also forms during continuous cooling, but now the constituent is likely to be interspersed in a mixed microstructure. The presence of bainite affects mechanical properties to some degree, and quite often the end-result is favorable.

When the microstructure of steel is predominantly pearlitic, the inclusion of some bainite will increase hardness, but toughness is likely to improve. When martensite and its normal hardness is being sought, some bainite in the martensitic matrix will reduce hardness to a relatively small extent, but cracking susceptibility probably will be decreased.

Two distinct forms of bainite appear in steel depending on whether the austenite decomposed in the upper, or in the lower part of the temperature range between B_s (i.e., below the pearlite realm) and the M_s.

Upper bainite is mainly an aggregate of ferrite laths formed into parallel groups called "sheaves" which nucleated mostly at prior austenite grain boundaries. Cementite particles usually appear as an interlath precipitate. With higher carbon contents in steel, carbide films between the parallel ferrite laths give the microstructure almost a lamellar appearance: yet, the spacing between the cementite films is larger than that of pearlite lamellae. Also, appearance after etching bainite differs from pearlite. Consequently, upper bainite can be distinguished from lamellar pearlite using optical metallography.

Lower bainite consists mostly of ferrite plates, rather than laths, and the morphology appears feathery; or possibly needle-like if formed at lower temperature just above the M_s. The morphological feature that assists in identification of lower bainite lies in the nature and distribution of carbides, which are more easily examined by electron microscopy. Carbide particles in lower bainite are located within the ferrite plates and are of $Fe_{2.4}C$ composition (ε-carbide). Further heating will form Fe_3C cementite.

Small amounts of bainite are formed in the HAZ of many fusion welds made in steels, but because this constituent is usually unobtrusive particularly note is seldom made of its presence. The inclusion of some bainite is favored by the rates of cooling commonly incurred in the HAZ of many welds. Even though the HAZ microstructure may be predominantly pearlitic, some bainite is likely to be included as suggested by CCT diagrams in Figures 9.27, 9.28 and 9.29-C. When welding steels of high-hardenability (i.e., air hardening types), procedures often include preheating and interpass temperature control, and these precautions will increase the likelihood of having some lower bainite form in the HAZ.

Familiarity with bainitic microstructure is becoming increasingly important. Some newer high-strength low-alloy steels have been tailored in composition to readily form a lower-bainitic microstructure during heat treatment to take advantage of its toughness along with the high level of strength. When these steels are welded, the welding procedure, and any postweld heat treatment, should be planned to re-establish the bainitic microstructure as required in heat-affected zones of the weldment.

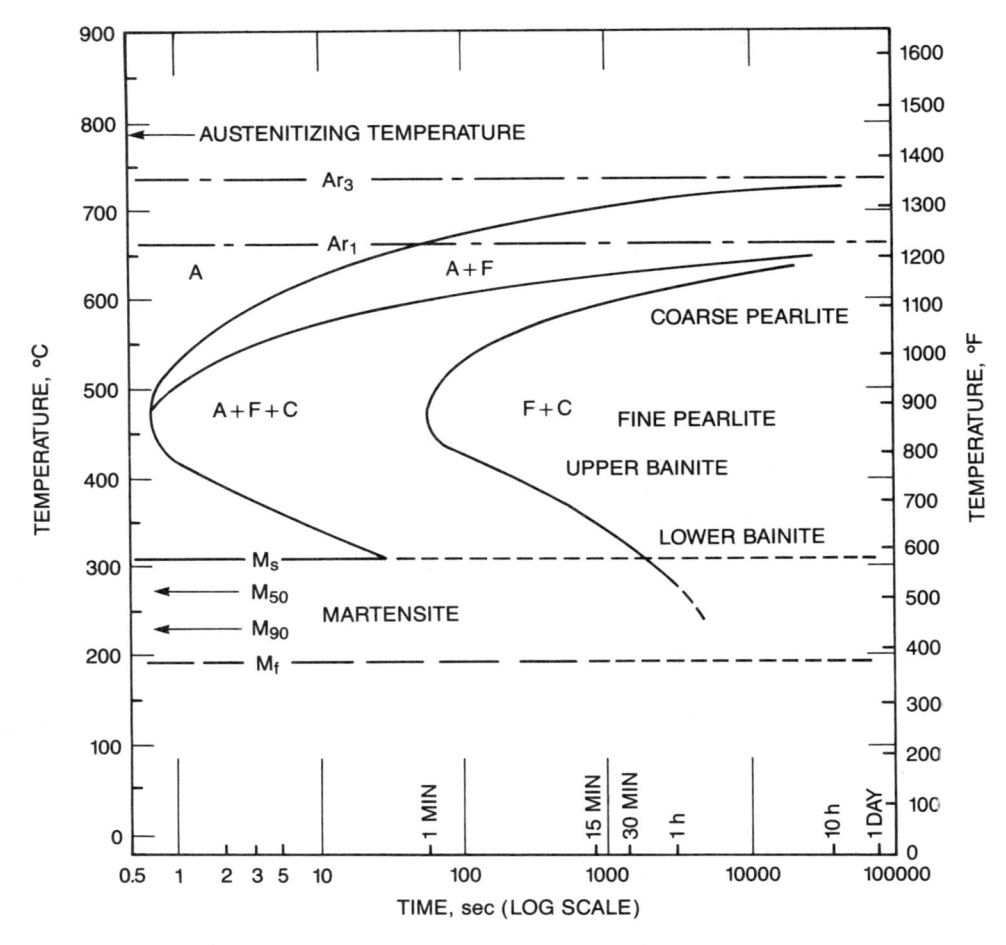

Figure 9.25 — Isothermal transformation diagram for AISI-SAE 2335 alloy steel

This alloy steel contains 3½ percent nickel and a nominal carbon content of 0.35 percent. The diagram is plotted from data obtained after austenitizing specimens at 785 °C (1450 °F) and quenching to sub-critical temperatures.

Figure 9.25 is the IT diagram for AISI-SAE 2335 alloy steel, which contains 3½ percent nickel in addition to 0.35 percent carbon and 0.75 percent manganese. The influence of nickel on hardenability can be clearly seen by comparing the IT start-curve with the previous diagram for AISI-SAE 1035 steel. The IT curves for the nickel-containing steel are displaced to the right and downward so that a time delay of about one-half second appears before the nose of the start-curve at a temperature of about 475°C (887°F). The M_s point also is depressed, and the experimentally determined temperature of 310°C (590°F) is close to the calculated 328°C using the equation in Table 9.1.

Figure 9.26 is the IT diagram for AISI-SAE 4340 (UNS G43400) alloy steel, which is a heat treatable "superstrength" alloy steel that is often welded. The IT curves show the pronounced effects of the 0.40 percent carbon, 0.75 percent manganese, 1.8 percent nickel, 0.80 percent chromium and 0.25 percent molybdenum on hardenability. Approximately three minutes must elapse before any proeutectoid ferrite will start forming in the austenite at the nose of the start-curve at 650°C (1202°F). Consequently, the relatively soft ferrite-and-pearlite microstructure is easily avoided. High strength is achieved by heat treatments that produce either bainite or tempered martensite.

Like many of the more highly-alloyed steels with high hardenability, the AISI-SAE 4340 Ni-Cr-Mo alloy steel transforms under isothermal conditions to bainite with less time delay than just mentioned for ferrite and pearlite. At about 450°C (842°F), upper bainite starts to form after only ten seconds of isothermal holding time. However, complete transformation of the austenite to upper bainite requires a long time, perhaps more than a day at temperature. In the temperature range of 300 to

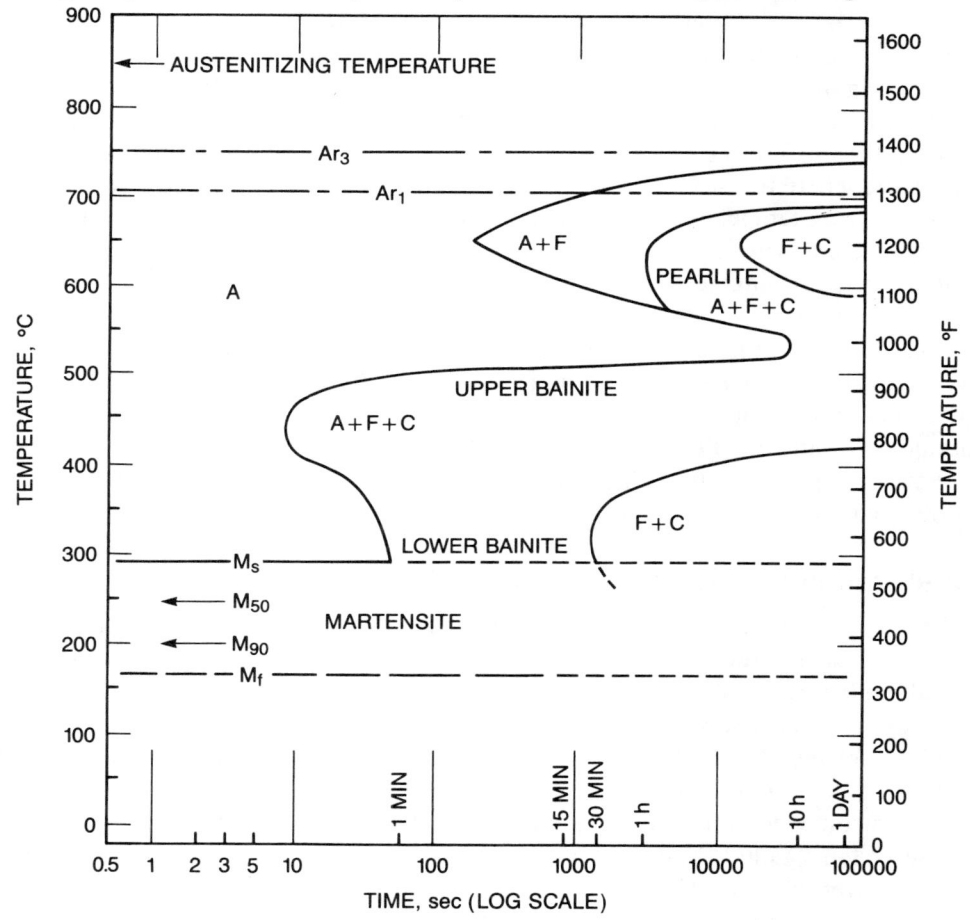

Figure 9.26 — Isothermal transformation diagram for AISI-SAE 4340 alloy steel

This alloy steel, alloyed with nickel, chromium and molybdenum, has a nominal carbon content of 0.40%, and has high hardenability. The diagram is plotted from data obtained after austenitizing specimens at 850°C (1562°F), and quenching to sub-critical temperatures.

400°C (572 to 752°F), the start of transformation to lower (acicular) bainite requires a little more time (almost one minute), but lower bainite formation may be complete in less than one hour. (Refer to Technical Brief 49 for further discussion of Bainitic Microstructures.) The M_s of this particular heat of AISI-SAE 4340 steel is shown to be 295°C (563°F); whereas the equation in Table 9.1 closely predicts 304°C (580°F). The M_f temperature appears to be in the vicinity of 165°C (330°F). The IT diagram for this steel can be very helpful in planning welding operations, because preheating and interpass temperature control are important components of proper welding procedure.

A final perspective is posted here regarding IT diagrams in general, because they usually fall short of realistically predicting microstructural transformations in steels during heat treatment and welding. Personnel with experience in examining microstructures and hardness in heat-affected zones of welded steels know full well that a plain steel containing 0.35 percent carbon indeed *can form* martensite under certain circumstances, contrary to the absence of any delay before a "nose" on the start-curve in Figure 9.24. While the IT diagram is highly instructive regarding the transformation mechanisms by which basic phases and constituents form in steels, the data used to construct IT diagrams were determined using rapid cooling to an arrest temperature where transformation was allowed to proceed isothermally. Furthermore, the austenitizing temperature and time applied to specimens for establishing most IT diagrams are not typical of the thermal cycles generated in the HAZ during welding. Consequently, the singular microstructures produced in the unusually thin IT specimens are not typical of the mixed microstructures often found in real weldments.

Continuous cooling (as contrasted with the rapid quench and isothermal treatment) adds its own influence when austenite transforms. As the next step in gaining an understanding of "simple" welds with typical heat-affected zones and areas formed by continuous cooling, the IT diagram will be re-examined to ascertain whether the many diagrams originally constructed to guide heat treatment of steels can be adjusted to improve their applicability for welding.

Transformation of Austenite During Continuous Cooling

The concept of the IT diagram emerged about 1930, but shortly thereafter the realization grew that certain aspects of the information being conveyed was somewhat misleading. Effort was then directed toward adjusting this highly informative diagram to portray more precisely the transformations during *continuous cooling*. Ten years passed before the first promising method was reported (Grange-Kiefer) for relating reactions on continuous cooling to reactions during isothermal transformation. Other proposed methods for adjusting diagrams soon followed, but it must be recognized that all these systems included assumptions, trial-and-error point placement, and arbitrarily selected reference points. Consequently, the *continuous cooling transformation* (CCT) diagrams that began to appear in the literature still were lacking in precision. Nevertheless, from a welding perspective they were an improvement over the IT portrayals.

Figure 9.27 shows how the IT diagram for 0.77 percent carbon (eutectoid) steel (shown earlier as Figure 9.19) can be adjusted to serve as a CCT diagram. The shift downward-and-to-the-right for curves indicating start and completion of austenite transformation to pearlite during continuous cooling was guided by mathematical treatment (Brick-Gordon-Phillips). Four continuous cooling curves ranging from quenching to slow cooling have been plotted on the diagram as long-dashed lines. As these cooling plots approach the start-curve, the instantaneous cooling rate at those intersecting points has been indicated. The original IT curves are shown as short-dashed lines, whereas the new curves for CCT are solid. Curve "a" for a quenched specimen narrowly exceeded the critical cooling rate for this steel as it passed the nose of the CCT start-curve, and accordingly the austenitized specimen underwent athermal transformation to martensite as it passed through the M_s-M_f temperature range. The heavier lines of curves bearing small solid circles signifies that a transformation was proceeding during passage through the particular range of temperature.

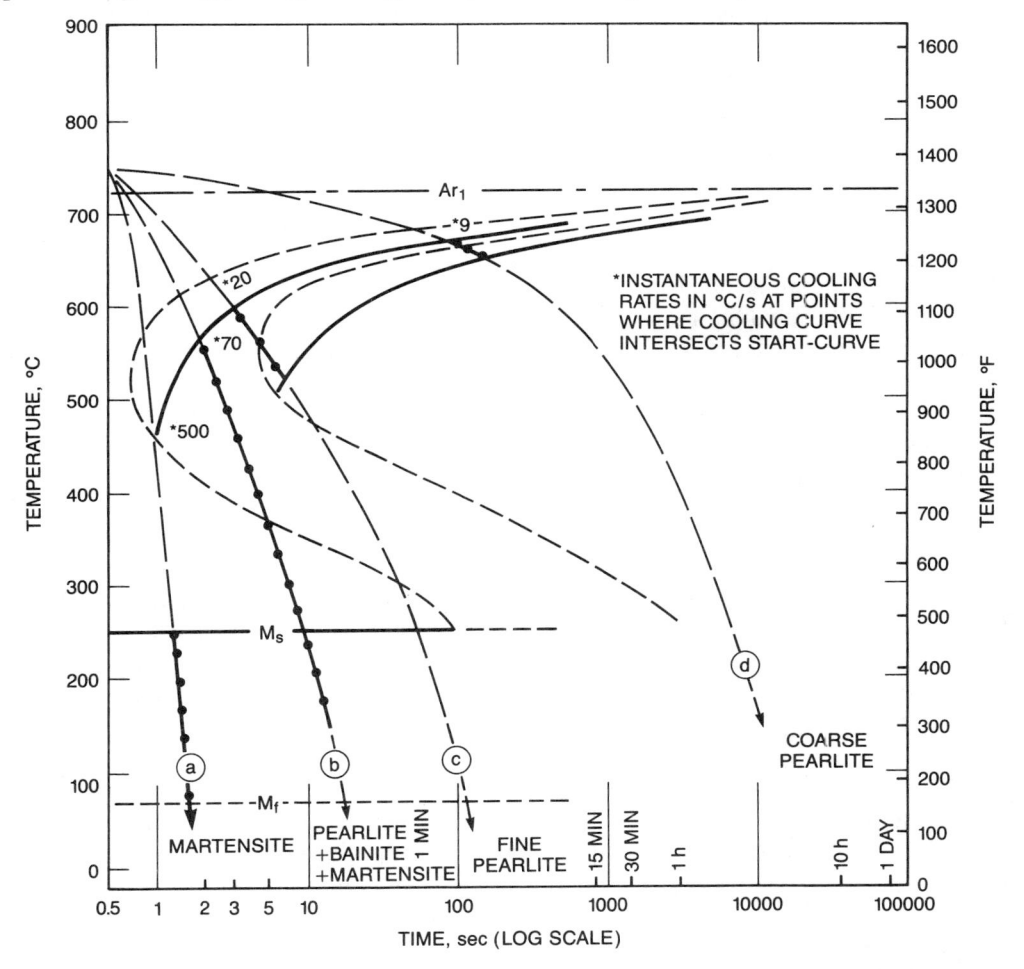

Figure 9.27 — **Combined IT and CCT transformation diagram for steel containing 0.77% carbon (eutectoid composition) showing IT curves (short-dashed lines) for pearlite and bainite formation as compared with transformation during continuous cooling (CCT curves are solid)**

The occurrence of austenite transformation during temperature descent is indicated by those portions of a cooling curve bearing small solid circles on heavier lines.

Curve "b" of Figure 9.27 deserves more detailed discussion. Some investigators have claimed that bainite formation is not possible in a carbon steel during continuous cooling because the plotted curve could not intersect the start-curve *below* its nose and also traverse the upper or the lower bainite regions to allow this phase to form. Yet, study of microstructure in this eutectoid steel shows that continuous cooling at a rate approximately that of curve "b" does not produce complete transformation to pearlite, and that some bainite will form as the steel cools through the temperature range where this phase ordinarily would form under IT conditions. Any remaining austenite in the steel (i.e., not transformed either to pearlite or to bainite) upon reaching the M_s will start athermal transformation to martensite; hence the mixed microstructure of pearlite + bainite + martensite as indicated at the RT terminus of cooling curve "b."

Curve "c" of Figure 9.27 indicates that sufficient time is spent in the temperature range of pearlite formation to undergo complete austenite transformation to fine pearlite. Curve "d" plots CCT that has produced a microstructure consisting entirely of coarse pearlite.

Figure 9.28 is the combined IT and CCT diagram for the AISI-SAE 4340 alloy steel illustrated earlier in Figure 9.26. This diagram appears more complex, but the scheme for indicating transformations, microstructures, etc., is the same as explained for Figure 9.27. This Ni-Cr-Mo alloy steel has a markedly slower critical cooling rate. Cooling rate of about 4°C/s will just clear the nose of the CCT start-curve for bainite formation, and the austenite is shown to continue (via curve "a") to the martensite transformation range. This rate of cooling can be obtained in the relatively thin specimens used simply by cooling them in moving air, which confirms the label "air hardening" often applied to this steel.

When AISI-SAE 4340 alloy steel is cooled more slowly than its critical cooling rate, it transforms to bainite in the temperature range of about 300 to 500°C (572 to 932°F) (curves "b" and "c"). Incidentally, some martensite may be found intermixed with lower bainite (see curve "b"). In order to form a pearlitic microstructure by CCT, a very slow rate of cooling is required, perhaps as slow as 0.006°C/s (curve "e"). This requires furnace cooling over a long period of time in order to drop through the proeutectoid ferrite and pearlite formation regions of the diagram (i.e., about eight hours or longer).

Although the diagram in Figure 9.28 is in close agreement with present knowledge about heat treating AISI-SAE 4340 steel, the diagram probably is less reliable for predicting transformations in a weld HAZ. This view is expressed because the investigators who constructed the diagram used an austenitizing temperature of 800°C (1472°F) which is lower than the 830°C (1526°F) recommended for solution treating this steel. Furthermore, the HAZ of a weld has areas that reach a much higher temperature. This difference in austenitizing temperatures between diagram specimens and actual weld zones would be expected to affect transformation tendencies to a significant degree. Nevertheless, the information in Figure 9.28 appears to be the best currently available.

PREDICTION OF MICROSTRUCTURES IN THE HEAT-AFFECTED ZONES OF WELDS

Discussion of IT and CCT diagrams, as just concluded with the diagram for AISI-SAE 4340 alloy steel, makes clear some of the impediments encountered when attempting to utilize published material intended to assist *heat treating* operations for welding. The first is the lower austenitizing temperature employed that corresponds to proper heat treatment of steel instead of 1150°C (2100°F) which is believed to represent more realistically the maximum temperature of the HAZ during welding. This temperature difference should be reflected to some extent in the cooling cycles exhibited by sections of steel that start their cooling from substantially different temperature levels. Perhaps of greater metallurgical importance is the disparity in austenite grain size that is likely to exist because the weld HAZ usually displays appreciable grain

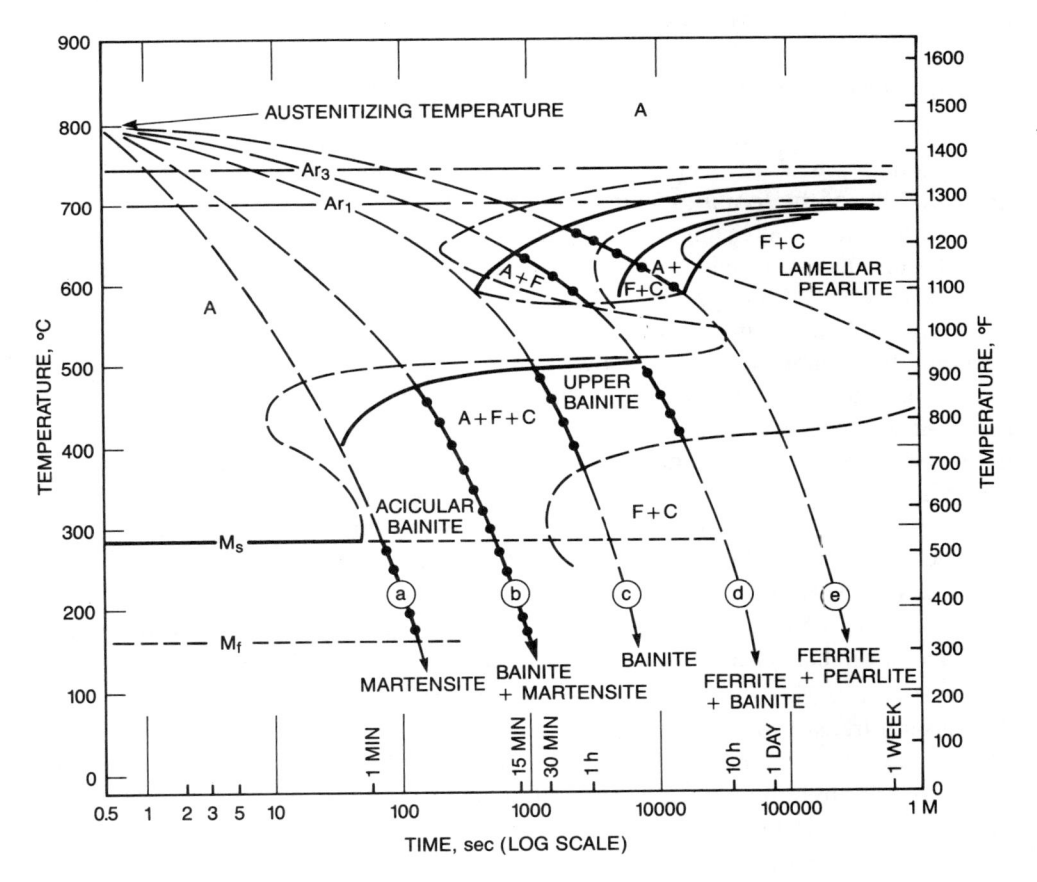

Figure 9.28 — Combined IT and CCT transformation diagram for AISI-SAE 4340 Ni-Cr-Mo alloy steel showing IT curves (short-dashed lines) for pro-eutectoid ferrite, pearlite and bainite formation as compared with transformation during continuous cooling (CCT curves are solid)

The occurrence of austenite transformation during temperature descent is indicated by those portions of a cooling curve bearing small solid circles on heavier lines.

coarsening at the higher temperature. There is a decided difference in time-at-temperature between the thorough soaking of heat treating operations compared to the relatively short-time exposure of the weld HAZ. Nevertheless, the weld HAZ commonly displays evidence of grain growth unless the steel contains alloying elements that effectively minimize grain coarsening at high temperatures. Another dissimilarity between heat-treated specimens and the weld HAZ is the matter of internal elastic and plastic strain. Transformation is known to be initiated sooner and the rate accelerated by elastic strain in the austenitized steel. Strain is more likely to exist in the weld HAZ, and the level of strain present will depend primarily on section thickness and the degree of restraint on the weld during cooling. At the relatively high temperatures where transformation may be initiated by supercooling, the austenitic microstructure will have a relatively low yield strength and cannot sustain a high level of elastic strain. The influence of weld restraint and resulting strain on microstructural transformation in the HAZ is a factor largely unexplored.

Jominy Method of Predicting HAZ Microstructure

In Chapter 7, mention was made of the Jominy end-quench test for hardenability evaluation and its ability to provide a spectrum of cooling rates. Cooling at the water-quenched end of the Jominy specimen is approximately 350°C/second (630°F/sec), and the cooling rate at increasing distance along its length decreases as illustrated in Figure 7.18. Figure 9.29-A provides essential details of the Jominy test. For discussion here, study will be based on hypothetical experiments with AISI-SAE 8630 steel, a Ni-Cr-Mo alloy steel containing 0.30 percent carbon. The literature shows this steel produces a Jominy hardenability curve as shown in Figure 9.29-B; but, of course, this published curve was determined with a Jominy specimen austenitized only at 870°C (1600°F). If the HAZ-simulation temperature of 1150°C (2100°F) had been employed, the curve might have been raised to that shown by the dashed-line, which Jominy testing has shown to be the approximate maximum for this steel. Also, it should be noted that this reported Jominy test performance did not display the "maximum obtainable hardness" for steel containing 0.30 percent carbon at the J-1.5 mm location on the Jominy bar even though the microstructure was reported to be martensitic. Possibly, the expected maximum obtainable hardness of 58 HRC (as per Figure 9.23) might be found on the end-face of the bar where the cooling rate was faster than at the J-1.5 mm location.

In Figure 9.29-C, a CCT diagram for the AISI-SAE 8630 alloy steel has been drawn with transformation start-and-end curves positioned as indicated by published data. Again, these transformation curves are based on specimens austenitized at the lower solution treating temperature regularly employed for heat treating operations. If the HAZ-simulation temperature of 1150°C (2100°F) had been used, the positions of the curves likely would be moved downward and to the right to some extent. Five continuous cooling curves have been plotted on the diagram in Figure 9.29-C. The first three, a, b and c, represent three positions on the Jominy end-quench test bar, J-1.5, J-4 and J-10 mm. The remaining two curves represent "air cooling," and "furnace cooling," rates slower than found in the standard Jominy specimen. Instantaneous cooling rates of these curves as they intersect the 705°C (1300°F) temperature level are listed in Figure 9.29-B; along with hardness values found in the AISI-SAE 8630 steel when cooled at the rates indicated. At the terminus of each curve for an individual specimen cooled with a different cycle, the microstructures and hardness values are shown. This information appears to be generally correct, but it lacks the definition often needed when making critical decisions for welding. For example, when planning the proce-

STANDARD METHOD FOR JOMINY HARDENABILITY TEST

SEE ASTM A 255 OR SAE J406 FOR COMPLETE DETAILS OF SPECIMEN AND PROCEDURE

JOMINY TEST SPECIMEN
SPECIMEN IS MACHINED BAR OF STEEL TO BE TESTED 25.4 mm (1 IN.) DIAMETER ROUND BY 102 mm (4 IN.) LONG WITH FLANGE AT UPPER END AS SHOWN (OR HOOK PROVIDED) FOR SUSPENDING IN VERTICAL POSITION FOR END-QUENCHING.

HEATING
SPECIMEN IS PLACED IN FURNACE ALREADY AT TEMPERATURE SELECTED FOR AUSTENITIZING AND SOAKED ONLY LONG ENOUGH TO ACHIEVE TEMPERATURE UNIFORMITY.

BOTTOM ROUND FACE OF SPECIMEN TO BE PROTECTED AGAINST FORMATION OF HEAVY OXIDE SCALE THAT WOULD INTERFERE WITH HEAT TRANSFER DURING QUENCHING.

END-QUENCHING
APPARATUS CONSISTS OF SQUARE-END TUBE 12.7 mm (½ IN.) ID TO PROJECT STREAM OF WATER VERTICALLY UPWARD AT RATE OF 3.8 LITERS (ONE GALLON) PER MINUTE. THIS FLOW CORRESPONDS TO FREE FLOW HEIGHT OF 63.5 mm (2½ IN.) FROM TUBE END.

WATER TEMPERATURE SHALL BE 16-27 °C (60-80 °F).

A FIXTURE ALLOWS HEATED SPECIMEN TO BE QUICKLY PLACED IN VERTICAL POSITION IN AXIAL ALIGNMENT WITH BOTTOM END-FACE A DISTANCE OF 12.7 mm (½ IN.) ABOVE THE ORIFICE OF THE WATER TUBE.

COOLING RATE FROM WATER END-QUENCH AT EACH INCREMENTAL LOCATION ALONG THE LENGTH OF JOMINY TEST BAR CAN BE OBTAINED FROM FIG. 7.18.

HARDENABILITY CURVE
FOLLOWING QUENCHING TO RT, PARALLEL FLATS 180° APART ARE GROUND ON SPECIMEN 0.38 mm (0.015 in.) DEEP ALONG LENGTH OF CYLINDRICAL SURFACE TO PERMIT PRECISE HARDNESS DETERMINATION.

ROCKWELL-C HARDNESS IS MEASURED ON ONE FLAT SURFACE AT SPECIFIED INCREMENTAL LOCATIONS ALONG LENGTH OF SPECIMEN AND VALUES PLOTTED AS SHOWN IN FIG. 9.29-B.

Figure 9.29-A — Jominy end-quench hardenability test specimen and method

This standardized method is widely used for studying austenite transformation; i.e., hardenability, and is employed further in Figures 9.29B and 9.29C to make predictions on microstructures and hardness likely to be found in the HAZ of welded AISI-SAE 8630 Ni-Cr-Mo alloy steel.

dure for fusion welding an alloy steel, such as AISI-SAE 8630, questions can arise that require the exact relationship between cooling rate and the microstructure (and hardness) expected in the HAZ of a weld made without preheating. These decisions might deal with any number of questions. For example, it may be necessary to decide on the *surety* of continuing fabrication operations of a complex weldment with a given pro-

JOMINY END-QUENCH HARDENABILITY TEST DATA
FOR AISI-SAE 8630 Ni-Cr-Mo ALLOY STEEL

TO SIMULATE WELD HEAT-AFFECTED ZONE EXPOSURE, AUSTENITIZING TEMPERATURE OF 1150 °C (2100 °F) SHOULD BE USED.

JOMINY HARDENABILITY CURVE

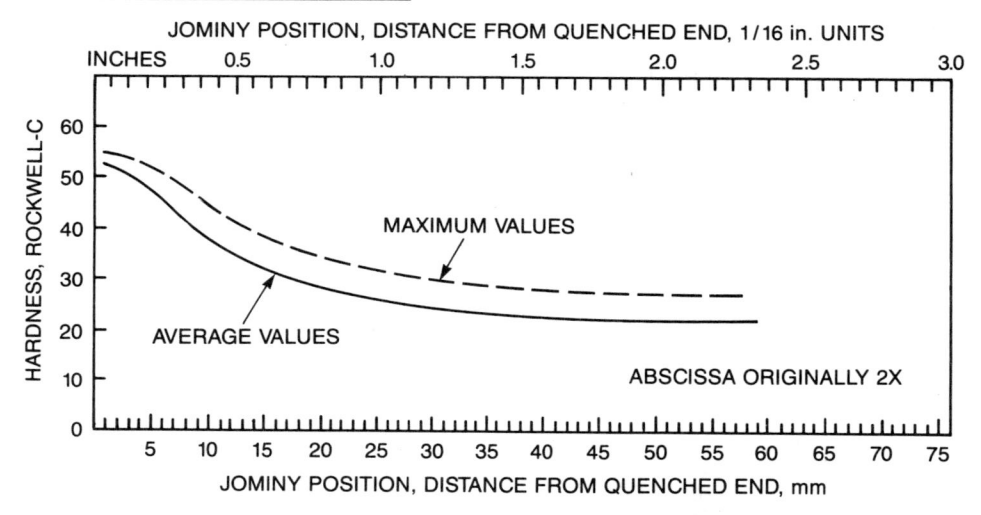

Figure 9.29-B — Jominy hardenability test curves for AISI-SAE 8630 Ni-Cr-Mo alloy steel

The two Jominy curves in the above diagram were taken from literature which reported values for specimens austenitized at regular heat treating temperatures. If an austenitizing temperature of 1150°C (2100°F) were employed to simulate the effective temperature in a weld HAZ, Jominy results would be expected to approach the reported maximum values. Data assembled below includes fixed cooling rates established for specific positions in the standard Jominy test specimen, and for furnace and air cooling. These data have been used in placing five cooling curves applied in Figure 9.29-C.)

Pertinent Data from Jominy Specimen Used for Plotting Cooling Curves on Figure 9.29-C

Curve Identification	Jominy Position Represented mm	1/16ths	Cooling Rate at 705 °C (1300 °F) Temperature Level °C/sec	°F/sec	Hardness (Rockwell-C) Avg.	Max.
a	1.5	1.0	300	540	53	55
b	4.0	2.5	80	144	50	54
c	10.0	6.0	28	50	38	46
d	Air Cooling		2	4	25	29
e	Furnace Cooling		0.05	0.09	20	25

cedure before postweld heat treatment. Consideration of the HAZ microstructure might even constitute a preliminary appraisal of fitness-for-service of the weldment. Such decisions often can be guided by precise knowledge of the microstructure expected to form in the HAZ of welds. Although impressive progress has been made in predicting HAZ cooling rate from calculations based on welding conditions and parameters, prediction of microstructure involves a greater number of influential variables and progress has not been as good.

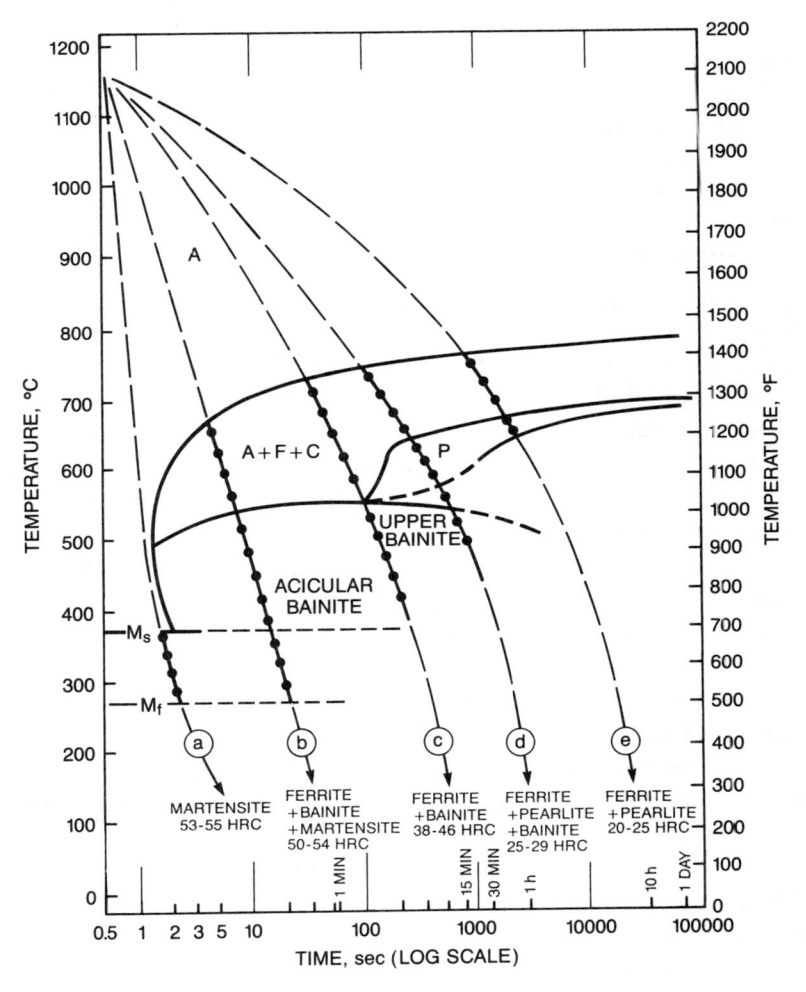

Figure 9.29-C — Constant cooling transformation diagram (CCT) for AISI-SAE 8630 Ni-Cr-Mo alloy steel

Data for plotting five curves (a to e, inclusive) were obtained from the Jominy tests tabulated in Figure 9.29-B. This portrayal of a method for predicting microstructure and hardness of a weld HAZ is only illustrative because Jominy testing has yet to be conducted with a higher austenitizing temperature to simulate the effective temperature to which a HAZ is believed to be exposed during welding (see further explanation in text).

The Jominy end-quench test, with certain modifications to tailor it to simulate welding, has shown promise for predicting the hardness of the HAZ. It may be possible that information regarding microstructure also could be obtained by carrying out *quantitative* metallography at the J-positions on the Jominy test specimen. There is one aspect between Jominy-microstructure and actual weld HAZ-microstructure that requires consideration. A heated bar of a given steel does not appear to have constraints on austenitic grain growth as prevails in the weld HAZ. Consequently, bar specimens usually develop larger grain size for a particular peak temperature. This difference in grain size could make the microstructure of the simulation-specimen different from that of the real weld when comparing points with equal cooling rate. However, if a correction could be introduced for this grain size difference, helpful microstructure data could be obtained from the Jominy specimen and posted on a CCT diagram. Although dilatometry still would be required to establish the temperature levels for transformation curves, quantified microstructural data from the Jominy specimen across the spectrum of cooling rates could confirm regions of specific phases and constituents.

Mathematical Approach to Prejudging HAZ Suitability

A long-sought goal has been the establishment of a reliable mathematical system that would predict the suitability of microstructure in a weld HAZ, especially its susceptibility to cold cracking, as calculated from chemical composition. The need for a quick method of appraisal that did not require time-consuming testing was voiced by users of welding as early as 1930. A number of proposals have been put forth since that time and several have proved truly useful.

Chemical composition of the steel became the center of attention in this effort to utilize mathematical treatment for the following reasons:

(1) Chemical composition of the steel to be welded was either known to lie within the limits set by a procurement specification, or was readily available as reported heat analyses on steel shipment documents, or could be routinely determined by analysis.
(2) Composition had been established as the strongest determinant of the microstructure formed in the HAZ.
(3) Composition had been shown to determine the hardness of the particular microstructure formed in the HAZ as a result of the welding procedure employed.
(4) Composition was known to determine in a general way the toughness of the microstructure in the HAZ, but most importantly, the susceptibility to cold cracking had been found to correlate directly with the kind of microstructure in the HAZ and its level of hardness.

Of course, many aspects of welding procedure and operating conditions were recognized to also influence the HAZ's microstructure, hardness, toughness and cracking susceptibility, but these aspects were regarded as secondary to the steel's composition because the procedure, etc., could usually be altered to accommodate a given composition. Therefore attention was focused early on the composition of the steel to evaluate its "weldability" with respect to cold cracking susceptibility of the HAZ microstructure as produced by the thermal cycle of a stated welding process.

In the early 1940s, several investigative groups began to propose selection guidelines for structural carbon steels based on experience and tests using the SMAW process with the cellulosic covered electrodes which were in common use at the time. Initially, very simple formulas were offered for the steels in gas and oil transmission pipelines. For example, to avoid HAZ cold cracking, a maximum hardness of 30 HRC (285 HB) was reported to be necessary for the HAZ, and this hardness limitation was deemed feasible with the SMAW process if the steel had a carbon content not exceeding 0.35 percent. Shortly thereafter, formulas were introduced for the plain steels that included factors for weighing the influence of manganese and silicon along with that for carbon content. Further work in this direction soon introduced formulas for low-alloy steels. The term *carbon equivalent* (CE) originated during this research because all of the formulas took carbon content as unity and assigned differing terms for other alloying elements present in the steel composition. Presently, a half-dozen formulas with small differences in factors and terms for elements other than carbon appear in the literature. Some of these formulas are also included in certain specifications and codes for welded constructions either as guidance, or as mandatory requirements for regulating steel selection, or for controlling welding procedure.

CE-type formulas are empirical equations that assess steel composition essentially from the standpoint of hardenability. However, some of the recently proposed formulas assign different terms to the alloying elements regularly included as factors, and some include additional minor elements as factors (e.g. boron as included in the P_{cm} equation). The more comprehensive equations take into consideration a number of the welding procedure parameters, and also other operational circumstances now known to influence the form of microstructure in the HAZ and the likelihood of cold cracking occurrence. These additional factors include the following:

(1) heat input in welding and the anticipated cooling rate,
(2) preheating, interpass, and postweld temperature control,
(3) hydrogen potential as judged from consumables and other sources, and the diffusible hydrogen expected to enter the HAZ, and
(4) restraint on the weld as judged from the base metal thickness, or from features of the joint to be welded that increase the level of residual stress in the HAZ.

These additional factors have increased the complexity of mathematical treatment of HAZ suitability to the point where graphics have become a more convenient form of presentation. Basically, the concept of carbon equivalency has merit, and further tailoring of calculations to the welding procedure and other influential conditions of a specific joint to be welded is bound to improve prediction of HAZ microstructure and susceptibility to cold cracking. Detailed discussion of CE-type and P_{cm} formulas, graphs and nomograms will be undertaken in appropriate sections of Volume II, Chapter 13, 14 and 15.

TRANSFORMATIONS IN WELD METAL

Microstructures in steel weld metal are markedly different from those of cast or wrought base metals. Quantitative metallography using both optical and electron microscopy must be employed to analyze the constituents present to rationalize weld metal properties. No radically new phases or compounds are found in weld metal; instead, differences lie mainly in morphology of recognized phases, compounds and nonmetallic inclusions. These differences are promoted by a number of conditions peculiar to welding and weld metal that affect solidification, diffusion, allotropic transformation, and other circumstances to be pointed out shortly.

The variance found in weld metal microstructures has caused some confusion among those communicating on the subject. Therefore efforts have been directed to coining designations and alphabetical symbols for microstructural constituents encountered in steel weld metal. Table 9.2 lists designations and symbols used in the present text to identify specific forms of microstructures found in steel weld metal. Also listed are the many different terms that have appeared in the literature over more

Table 9.2 – Microstructures Found In Steel Weld Metal

Symbol Used in this Text	Description of Micro-Structural Constituent	Other Terms Used in the Literature by Investigators	Figures in this Text Showing Constituent
GF	Grain-boundary ferrite	Allotriomorphic ferrite; Intergranular ferrite; True grain boundary (TGB) ferrite; Proeutectoid ferrite; Primary ferrite (G)	Figures 9.30 and 9.33
AF	Acicular ferrite	Intragranular ferrite plates; Fine intragranular ferrite; Fine bainitic ferrite; Fine grained ferrite; Needle-like ferrite; Coarse acicular ferrite (CAF)	Figures 9.30, 9.31 and 9.33
PF	Polygonal ferrite	Ferrite islands; Primary ferrite (I)	Figure 9.31
LF	Lathe ferrite*		
AC	Ferrite with aligned secondary phases	Ferrite with aligned martensite/austenite/carbides (MAC) Ferrite sideplates growing from grain boundary; allotriomorphs (FSP); Lamellar component; Ferrite with aligned second phase (FS-A); Widmanstätten secondary sideplates	Figure 9.32
NAC	Ferrite with nonaligned second phase	Ferrite plus nonaligned second phase (FS-NA)	Figure 9.33
FC	Ferrite and carbide aggregate	Ferrite-carbide aggregate (including lamellar pearlite)	Figure 9.34
SP	Side-plate microstructure grain boundary nucleated	Side grain boundary (SGB) ferrite	
IP	Side plate microstructure intragranularly nucleated	Ferrite laths intragranularly nucleated	
P	Lamellar pearlite	Ferrite-carbide lamellar aggregate	Figures 9.31 and 9.34
B	Bainite		
M	Martensite		Figure 9.35
A	Austenite	Retained austenite	Figure 9.35

* LF occurs as a predominantly intragranular constituent resembling bainite, and is sometimes found among the AF ferrite and the AC ferrite)

than a decade for the same microstructures. Efforts continue to arrive at internationally-accepted terminology and symbols under the aegis of IIW, but final agreement has yet to be reached.

Greater attention is being given to weld metal and its mechanical properties because of demands for improved and more-consistent toughness. Findings clearly show weld toughness to be controlled by minute details in its microstructure. The challenge to the welding engineer is to become intimately familiar with these details, and to control their occurrence quantitatively.

The microstructure of weld metal is controlled principally by composition and cooling rate, and they wield influence in surprising ways. These subtleties will be reviewed in a preliminary way as "typical" welds are discussed in the present Chapter. However, the many aspects of composition that influence weld metal microstructure are quite complex. Aspects include the levels of minor elements used as deoxidizers and as grain-refining agents, and the amounts of gaseous elements present in the form of nonmetallic inclusions. An in-depth examination of the relationships between weld composition and microstructure will be carried out in Volume II, Chapter 10 on filler metals.

For an overall perspective of transformations in weld metal, a typical low-carbon, plain-steel weld metal should be studied as it solidifies over the temperature range of about 1510 to 1495°C (2750 to 2725°F). This weld metal first forms primary delta ferrite as indicated by the Fe-Fe$_3$C phase diagram (Figure 4.12). As cooling continues, austenite replaces the delta ferrite virtually grain-for-grain over the range of high temperatures where the peritectic reaction takes place. This transformation mechanism retains the columnar grain shape that often develops during solidification. The next transformation occurs as the cooling weld metal reaches the austenite-to-ferrite critical range (below the Ar$_3$ temperature). To gain a useful perspective of microstructures that can be found in this plain-steel weld metal, the effects of different cooling rates must be included. For the moment, the rates to be considered can be described in very general terms; (1) fast rates as produced by low heat input during welding, (2) slow rates that accompany welding with high heat input, and (3) intermediate rates. Later, the cooling rates that prevail when forming particular weld metal microstructures will be expressed quantitatively to establish meaningful correlations.

Again, the familiar problem of how best to express a quantitative cooling rate arises. Some investigators have described their experiments in terms that indicate rate of cooling through the temperature range of 800 down to 500°C (1472 to 932°F), that is, as cooling *rate* in degrees per second through this range, or as *time* in total number of seconds spent cooling through this range. Reasons for selecting this particular temperature range to characterize or portray rate of cooling were given in Chapter 7, but now it must be recognized that Figure 7.6, which made use of Δt_{8-5}, dealt with cooling rate in the base metal HAZ, and not in the weld zone per se. Consequently, the nomograph in Figure 7.6 should not be used to estimate weld metal cooling rate.

Whether a single-constituent microstructure, or a multiplex of constituents form in the weld zone depends upon synergic action between the particular weld composition and the prevailing cooling rate. Ordinarily, cooling rates in the weld zone are regarded as too fast to permit appreciable diffusion. Therefore, diffusion-dependent transformation mechanisms do not operate properly. For this reason, pearlite is not often

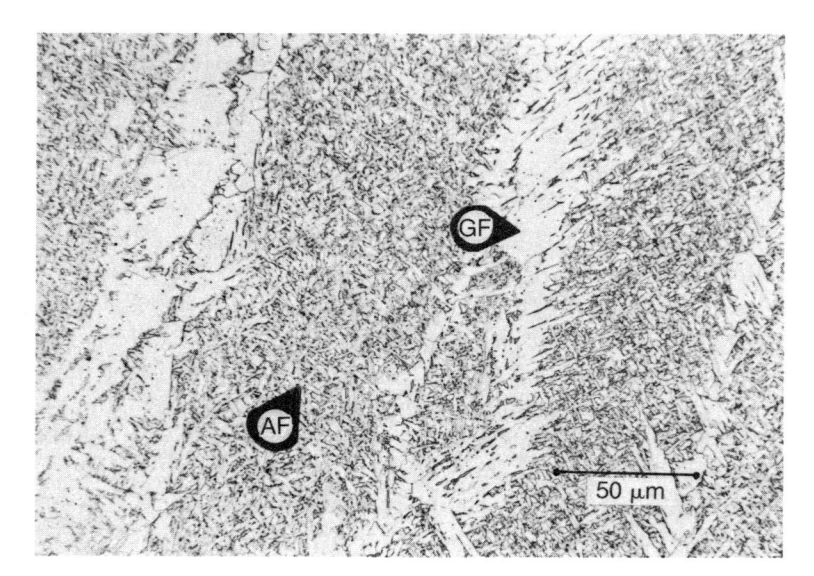

Figure 9.30 — Photomicrograph of steel weld metal

GF in photomicrograph is grain-boundary ferrite which nucleated as proeutectoid phase along the boundaries of columnar austenite grains. AF is acicular ferrite which formed intragranularly as interlocking laths within the intergranularly-veined austenitic microstructure during austenite-to-ferrite transformation below the Ar_3 temperature. The carbide phase is dispersed as tiny particles among the laths of acicular ferrite as the weld metal cools through the Ar_1 temperature. (Nital etchant; original magnification: 500X; courtesy of The Welding Institute, England)

found in as-deposited weld metal. Also, the micro-scale composition heterogeneity called "coring" (described in Chapter 4), which exists in most weld metal after solidification, is not relieved to any appreciable extent during ordinary initial cooling of a fusion weld. There are circumstances in welding, peculiar to many of the fusion processes, that foster unusual forms of proeutectoid and eutectoid transformation products. Proper identification of each constituent in the final microstructure, and measurement or estimates of their volume fraction present in the weld metal morphology are important to carry out, but difficult. These are key steps in correlating microstructure with mechanical properties, and assist in gaining an optimum level of a particular property, such as toughness.

Microstructure typical of weld metal deposited by an arc process and cooled at an intermediate rate as the austenite transforms is illustrated in Figure 9.30. The structure in the photomicrograph is composed of *grain-boundary ferrite* (GF) and *acicular ferrite* (AF). The initial transformation occurred along the boundaries of the columnar austenite grains thus forming a network or "veining" of ferrite throughout the microstructure. This GF constituent can vary from insignificant traces to a complete heavy network of ferrite grains. Ferrite distributed as GF can adversely affect toughness. The interior of the original austenite grains in Figure 9.30 has transformed to acicular ferrite (AF) which is a highly desirable microstructure from the standpoint of toughness. Unfortunately, this form is achieved only when certain composition requirements are satisfied and the cooling rate is suitable for the particular weld-metal composition. Acicular ferrite (AF), seen between the GF veining, consists of very small elongated platelets or laths that form an interlocking pattern. These laths or grains are only about one or two microns (40 to 80 μ in.) in width. The carbon rejected from the austenite during transformation to the AF constituent is present as carbide particles interspersed among the AF laths. Incidentally, the amount of grain-boundary ferrite (GF) present in this microstructure would be expected to have a noticeable adverse influence on the toughness promoted by the acicular ferrite. The distance between the veins of GF ferrite can be seen as about 50 to 75 microns, which actually is the width of the prior columnar austenite grains (i.e., their dimension perpendicular to the long axis of the grain). This has been confirmed as typical of austenite grains in weld metal deposited by SMAW from a four mm size E 7018 electrode.

Figure 9.31 is microstructure produced in the weld zone of a plain-steel ESW weldment. Naturally, the cooling rate was much slower during the ESW process than that for Figure 9.30. *Polygonal ferrite* (PF) has formed in the grain boundaries of the prior austenite, but not in sufficient quantity to form a network. The *acicular ferrite* (AF) present is much coarser because of the slower cooling. Also, there has been greater carbide rejection from the austenite during formation of the acicular ferrite, but electron microscopy would be required at high magnification to compare amounts of

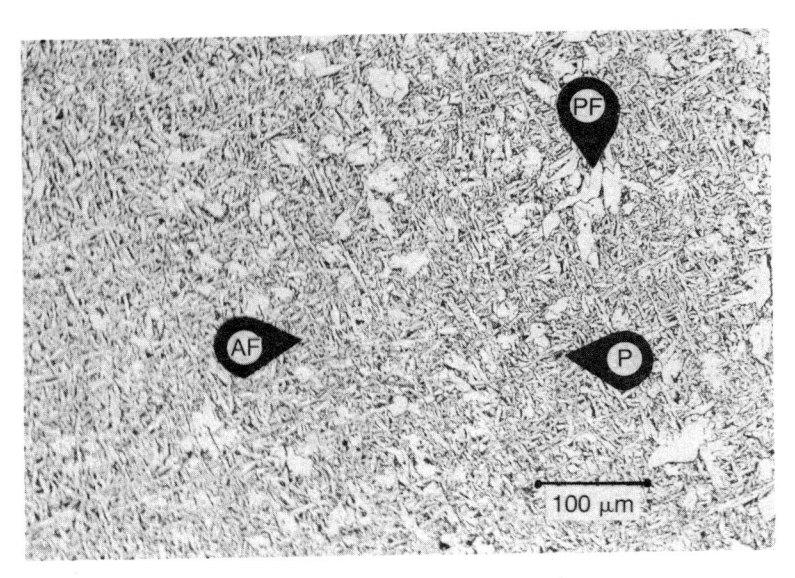

Figure 9.31 — Photomicrograph of steel weld metal

PF is polygonal ferrite which nucleated and formed on prior austenite grain boundaries, but not in sufficient amount to form a network or veining. AF is acicular ferrite which in this ESW weld zone is coarser than that in Figure 9.30 because of much higher heat input during welding. The slower rate of cooling has caused greater carbide precipitation between the laths of acicular ferrite, and also has allowed some very small pearlite (P) colonies to form. (Nital etchant; original magnification: 200X; courtesy of The Welding Institute, England)

this phase in specimens of Figures 9.30 and 9.31. Finally, because of slower cooling and the time available for diffusion, a number of very small *pearlite* (P) colonies have formed among the AF laths.

Figure 9.32 introduces another constituent peculiar to steel weld metal, but which is composed of familiar phases. As indicated by the symbol *AC, ferrite aligned with secondary phases* is present in the microstructure. This constituent develops when proeutectoid ferrite nucleates in austenite grain boundaries at the start of transformation and grows as lamellar-like plates toward the interior of the austenite grains. Very small amounts of martensite and austenite as minute grains can be retained between the ferrite "side plates," but magnification in the order of 1000X is required to reveal these phases. Most of the carbon originally in solution in the austenite is precipitated during a bainitic-like transformation mechanism into needle-shaped areas between the ferrite plates. AC is undesirable in an appreciable volume fraction when good toughness is being sought. The amount of this constituent that forms can vary substantially. As noted in Table 9.2, AC has acquired a number of names from investigative work because of variations in the nature of the secondary phase(s) between the ferrite side plates.

Figure 9.33 illustrates *ferrite with non-aligned secondary phase(s)*, a constituent occasionally observed that has been assigned the symbol *NAC* in the photomicrograph. In this form of microstructure, the ferrite nucleates at austenite grain boundaries (often in close association with the GF form), but progresses as small grains with a high aspect ratio instead of forming the side plates of AC ferrite. Although somewhat similar to acicular ferrite, the grains of NAC are not positioned in the easily-recognized basket-weave or interlocking pattern of AF.

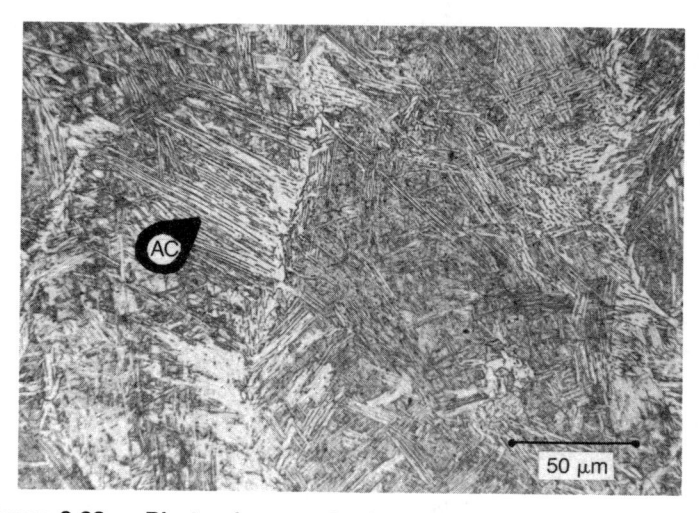

Figure 9.32 — Photomicrograph of steel weld metal

AC is ferrite with aligned secondary phase(s). When this constituent is found to have martensite, austenite and carbides between the lamellae of ferrite side plates, the symbol "MAC" has been used. (Nital etchant; original magnification: 500X; courtesy of The Welding Institute, England)

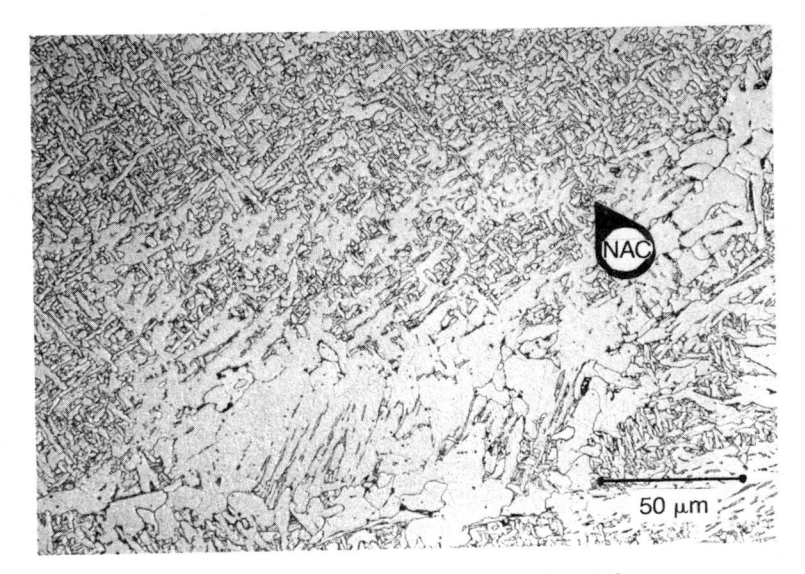

Figure 9.33 — Photomicrograph of steel weld metal
NAC is ferrite with non-aligned secondary phase(s). (Nital etchant; original magnification: 500X; courtesy of The Welding Institute, England)

Figure 9.34 illustrates a multiplex of constituents in weld metal microstructure obtained when the cooling rate is sufficiently slow to permit nucleation of phases favored by the composition, and yet is sufficiently fast to encourage other austenite transformation mechanisms. The relatively coarse microstructure in Figure 9.34 is that of a high heat input weld made by the ESW process. The particular area in the photomicrograph includes four different constituents; three of which were described earlier (AF, GF and P). The fourth constituent, *FC*, a *ferrite and carbide aggregate*, is pointed out for the first time in this text. The FC constituent is closely akin to pearlite inasmuch as both are of eutectoid carbon content. Because the cooling rate when reaching the Ar_1 temperature was a little too fast to permit the growth of lamellae of ferrite and carbide, transformation occurred through a bainitic-type mechanism. The resulting microstructure consists of grains of ferrite with a dispersion of extremely fine carbide particles within the grains. Interestingly, some austenite grains in this particular weld, after being enriched to eutectoid composition, were able to form the classic pearlitic microstructure with very fine lamellae. This array of four constituents in close proximity in one weld zone emphasizes the acute sensitivity of weld metal to circumstances and conditions that influence transformation to microstructural products.

Earlier in this text mention was made of the stirring that occurs in a weld pool melted by an arc which produces a "relatively" homogeneous weld zone mixture from the melted base metal and deposited metal from a consumable electrode (see Chapter 6, *Basic Forms of Arc Welding*). Recent research has shown that weld zones can be heterogeneous in ways that are surprising, and although inappreciable in extent for most cases, they deserve description here because of possible influence on weld metal mechanical properties that may have unexpected consequences. Study of as-

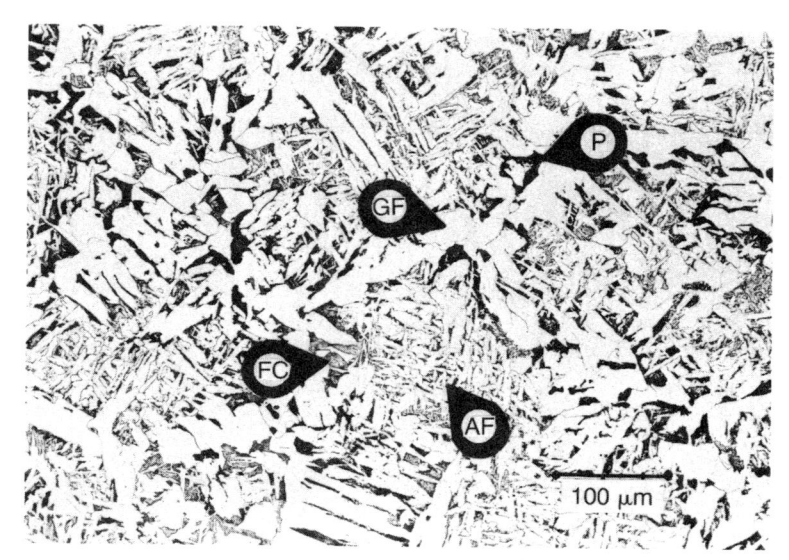

Figure 9.34 — Photomicrograph of steel weld metal with a multiplex of microstructure constituents

P is lamellar pearlite; however, the lamellae cannot be distinguished at this magnification in the photomicrograph, GF is grain boundary ferrite which originated in the boundaries of the prior austenite grains. AF is acicular ferrite. FC is aggregate of ferrite and carbide of eutectoid composition that was not permitted to develop lamellae at the cooling rate that prevailed. (Nital etchant; original magnification: 200X; courtesy of The Welding Institute, England)

deposited single-pass SMAW weld beads has found subtle indications that intrinsic gradients can exist across the section of C-Mn steel weld beads in both microstructure and in mechanical properties.

Metallographic examination of SMAW weld beads during traverse from the fusion line toward the bead's center found the microstructure to gradually change from predominantly grain boundary ferrite (GF) to a mixture of ferrite side plates (IP) and acicular ferrite (AF). As the amount of the two latter forms of finer ferrite increased, a commensurate increase in weld metal hardness occurred. These gradients in microstructure and in mechanical properties are believed to originate because of general coarsening of the austenite grain structure during its epitaxial formation at the fusion line followed by solidification toward the bead center. The influence of stress level in the weld during transformation of the austenite was found to have far less influence on the kind of ferrite microstructure formed than the austenite grain size. Also, and not unexpectedly, transformation of austenite to finer acicular ferrite (AF) was accelerated when the large austenite grains had a large number of very small intragranular nonmetallic inclusions present.

Although the C-Mn steel weld metal composition in this study of homogeneity represented a typical low-hardenability weld metal, the coarser austenite grain structure at the weld bead center increased hardenability just enough to promote transformation to the finer IP and AF ferritic microstructures. Weld beads of low-alloy steel compositions did not exhibit similar gradients in microstructure and mechanical properties because of their higher hardenability and smaller austenite grain size (the latter being attributed to grain boundary pinning by alloying elements and by compounds) which would favor the intragranular formation of finer microstructures, such as acicular ferrite (AF).

This finding regarding intrinsic gradients in C-Mn steel weld metal could be useful when interpreting the results from tension and/or impact testing using specimens trepanned from particular positions in large weld beads. Also, the occurrence of copious amounts of polygonal ferrite, particularly at the grain boundary location (GF), could be important because of the lower strength and fracture toughness possessed by this constituent.

Research using high-resolution analytical techniques and very-high magnification electron metallography has produced findings that clarify some long-standing conceptualizations of steel weld metal microstructures, and they bring to light facts that refute some beliefs held heretofore. For example, cooling rates in weld zones had been viewed as too fast to permit significant diffusion in the solid-state after solidification, especially during the austenite-to-ferrite transformation when cooling through the Ar_3 to Ar_1 critical range. However, recent painstaking work with multipass weld metal from FCAW electrodes that deposited a typical C-Mn weld composition with carbon 0.09 percent, manganese 1.4 percent, silicon 0.6 percent, but with 0.05 percent titanium, and 0.008 percent boron, revealed noteworthy facts concerning (1) diffusion of certain alloying elements during austenite-to-ferrite transformation, (2) micro-scale distribution of these elements in the weld metal microstructure, and (3) retained austenite in otherwise ferritic microstructure.

Sharpened needles were prepared from the FCAW weld metal and first examined at their tip by TEM for the presence of grain boundaries and for small precipates. The tiny volume of weld metal at the needle's point (about 10^{-15} cm^3) then was examined and analyzed with an atom-probe field-ion microscope. The dominant microstructural constituent in this FCAW weld metal was acicular ferrite (AF), but a small amount of

Figure 9.35 — Photomicrograph of steel weld metal

This steel weld metal has been etched with picral solution, which is 4% picric acid in ethyl alcohol. This etchant reveals "retained" phases (in this particular case, austenite and martensite). If nital metallographic etching reagent is employed, the retained phases would be difficult to detect. (Picral etchant; original magnification 700X courtesy of The Welding Institute, England)

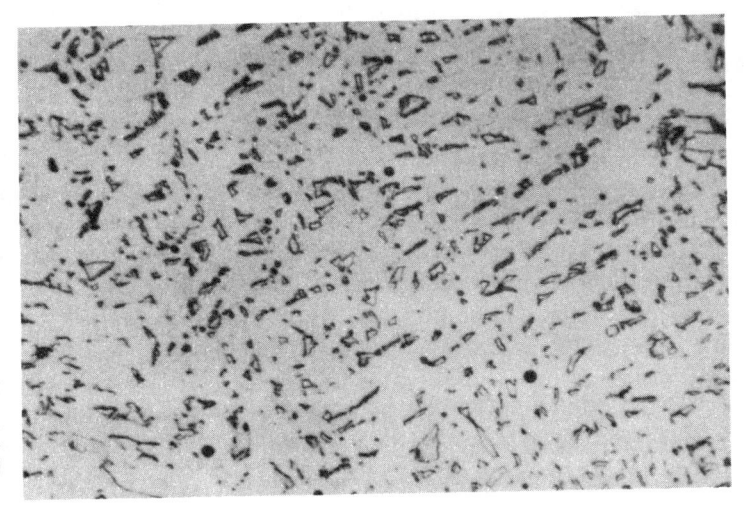

retained austenite was also present. The austenite proved to be surprisingly stabile, and remained untransformed at subzero temperatures as low as -189°C (-340°F).

As progressive evaporation of metal from the needle tip in the ion microscope brought new fields to the surface for analysis, particular attention was directed to concentrations and transitions in composition at grain boundaries. No significant variations in manganese or silicon were found when traversing from ferrite to retained austenite. Although boron has been indicated by other studies to accumulate at the gamma-to-alpha crystallographic interface, no boron was found in this FCAW weld metal study at the γ–α interface, which suggests that diffusion of boron does not occur at temperatures below 615°C (1139°F). Composition profiles on an ultra-high resolution scale were secured for carbon across the retained austenite-ferrite interface. The results gave positive proof that carbon diffused from the acicular ferrite regions into the remaining austenite during the transformation mechanism. Enrichment of austenite to carbon levels as high as seven atomic percent were found in the retained austenite adjacent to the interface with the ferrite, and this enrichment would seem to account for the retention of some austenite in the final weld microstructure. Additional inducement for small regions of austenite to remain untransformed could come from the compressive stress imposed by the surrounding acicular ferrite.

Detection of small amounts of retained austenite in steel weld metal often requires diligent metallographic examination. Very high magnification and special etching technique may be required if the austenite is present as tiny areas. Figure 9.35 illustrates the microstructure of a steel weld metal which, when etched with the commonly-used *nital* solution, probably would appear to consist only of a matrix of ferrite grains and dispersed carbides. Yet, when etched with *picral* solution (four percent picric acid in ethyl alcohol), the presence of considerable retained austenite and martensite is revealed. The volume fraction of these phases in the weld-metal microstructure can either be substantial or only very small amounts. However, their presence, even in small amounts, can significantly affect mechanical properties. Toughness, in particular, can be adversely influenced because these constituents will aid the initiation of cleavage fracture. Thorough examination of steel weld-metal microstructures often requires use of more than one kind of metallographic etching solution to reveal all constituents present.

A fast cooling rate facilitates retention of austenite and martensite, but the presence of alloying elements that promote hardenability in the weld, and are inclined to micro-segregation usually are more likely to play an influential role. Retained austenite and martensite are likely to have increased carbon content, because carbon will be diffusing from the austenite that is undergoing austenite-to-ferrite transformation even in the short time available. Even when only "microphases" are contained in the microstructure, they can exert adverse effects. Martensite is likely to be quite brittle, and any retained austenite can be unstable. There is evidence that these retained microphases can exist in weld metal as isolated islands, or can form films or bands depending on the alloying element(s) involved and the mechanism by which segregation takes place. While traces of brittle microphase can act as cleavage fracture initiators, segregated bands are more damaging because cleavage fracture not only is more easily initiated, but fast propagation is assisted as well.

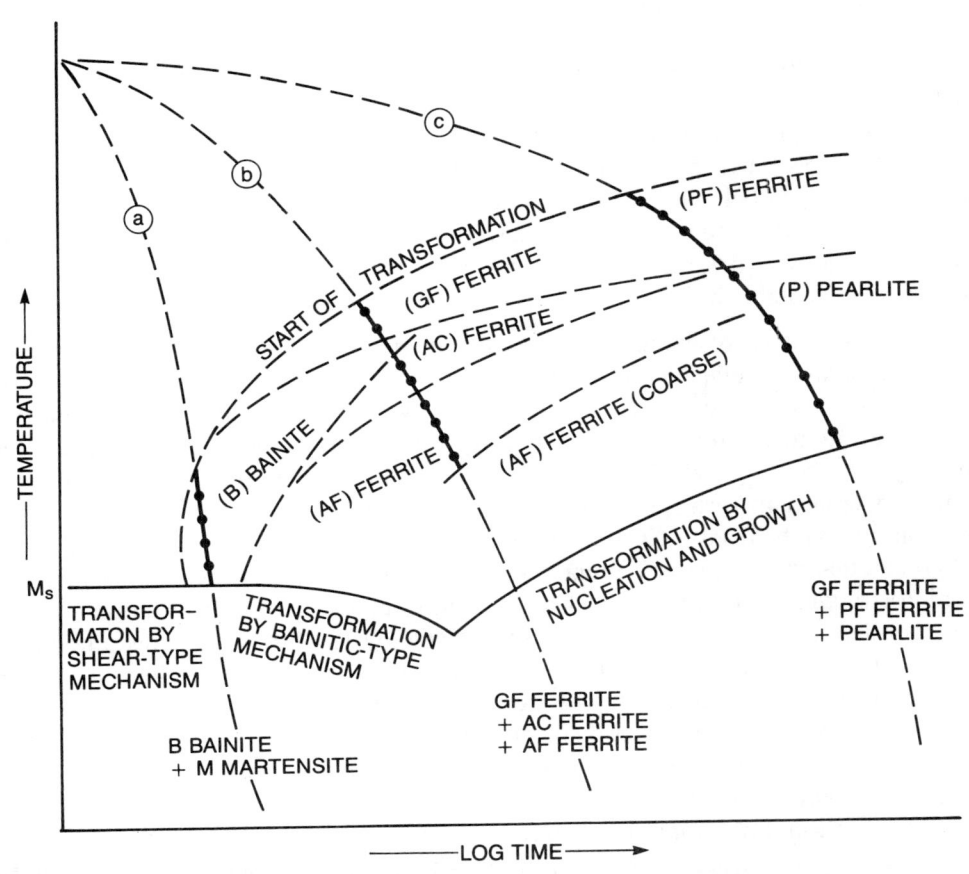

Figure 9.36 — Schematic CCT diagram for a low-carbon, plain steel weld metal

The above diagram includes three cooling curves that may arise in welding. The cooling rates can be pictured as (a) very fast, (b) intermediate, and (c) very slow. Austenite transformation during continuous cooling is indicated by small solid circles on unbroken portions of each curve. The microstructure produced by each cooling rate is noted at the terminus of each curve. (Courtesy of The Welding Institute, England)

Continuous Cooling Transformation Diagrams for Weld Metal

Guidance obtained from continuous cooling transformation (CCT) diagrams for steels (as demonstrated with Figures 9.27, 9.28 and 9.29) encouraged research toward development of similar diagrams for predicting microstructural constituents in weld metal. Diagram preparation for weld metal proved more difficult because investigators had no way of securing very thin specimens of weld metal and closely manipulating their cooling directly after weld-zone solidification to study particular cooling rates and cycles. Nevertheless, studies proceeded either by (1) making welds under a variety of operating parameters and recording the cooling cycle with a thermocouple harpooned into the freezing weld zone, or (2) machining small, thin-wall hollow cylinders from selected weld metals and subjecting the specimens to rapid furnace reheating and controlled cooling cycles to simulate welding. Some research also was conducted using electrical resistance heating of weld metal specimens (on Gleeble equipment) to produce rapid thermal cycling. When reheating techniques were employed on specimens prepared from existing weld metal, the assumption was made that transformation behavior would be much like the weld metal following solidification and continued cooling. These efforts were started about 1975, and initial schematic diagrams proved helpful to portray the relationship between microstructure and cooling rate as established by welding process and operating parameters.

Figure 9.36 is representative of early schematic CCT diagrams that gave general portrayals of weld metals as their austenitic microstructure transformed during cooling to room temperature. It was anticipated that very rapid cooling would produce a weld-metal microstructure of martensite, or of bainite in lean-alloy weld metal, and that very slow cooling would allow the nucleation-and-growth transformation mechanism to operate and form polygonal ferrite and pearlite. However, the most useful information sought from a CCT diagram was the range of cooling rates for a given weld metal that would ensure a preponderance of tough, acicular ferrite in its as-welded microstructure. The schematic CCT diagram in Figure 9.36 indicates that intermediate cooling rates will produce acicular ferrite (AF), but that some amount of grain-boundary ferrite (GF) and aligned ferrite (AC) also may form. The acicular ferrite is believed to form in a temperature range where reconstructive transformations become sluggish and give way to displacing transformations. The latter transformation has been reported to exhibit an "incomplete-reaction phenomenon" inasmuch as the formation of acicular ferrite ceases before the remaining austenite reaches its equilibrium composition. For a long time, researchers believed that a collection of CCT diagrams factually constructed for many compositions of weld metal would eventually allow evaluation of alloying elements for their ability to foster AF-ferrite, and to decrease the volume fractions of GF-ferrite and AC-ferrite with certain cooling rates. This information was expected to indicate types of weld metal with improved toughness, and would possibly guide the cooling-rate control needed for a given weld metal to obtain the best microstructure for optimum toughness.

Much research directed specifically to weld metal has been conducted and reported over the past decade, and some impressive findings concerning acicular ferrite in weld metal microstructure have been reported. These findings will be reviewed shortly. Experimental procedures have been developed using metallography, linear thermal expansion with the dilatometer, and mechanical testing to obtain quantitative data on weld metal transformations. This information has allowed plotting of helpful CCT diagrams like that in Figure 9.37. This example covers weld metal from E7016 electrodes which had the composition shown in the figure. Metallographic results were obtained on weld metal as-deposited, but dilatometry was conducted on specimens machined from multi-layer weld metal and then re-austenitized to simulate the onset of weld cooling. Various techniques were applied to secure controlled cooling rates, which are expressed as "average cooling velocity in °C per second during descent through the 800 to 500°C temperature range" (values in boxes on the Figure), and as "total time in

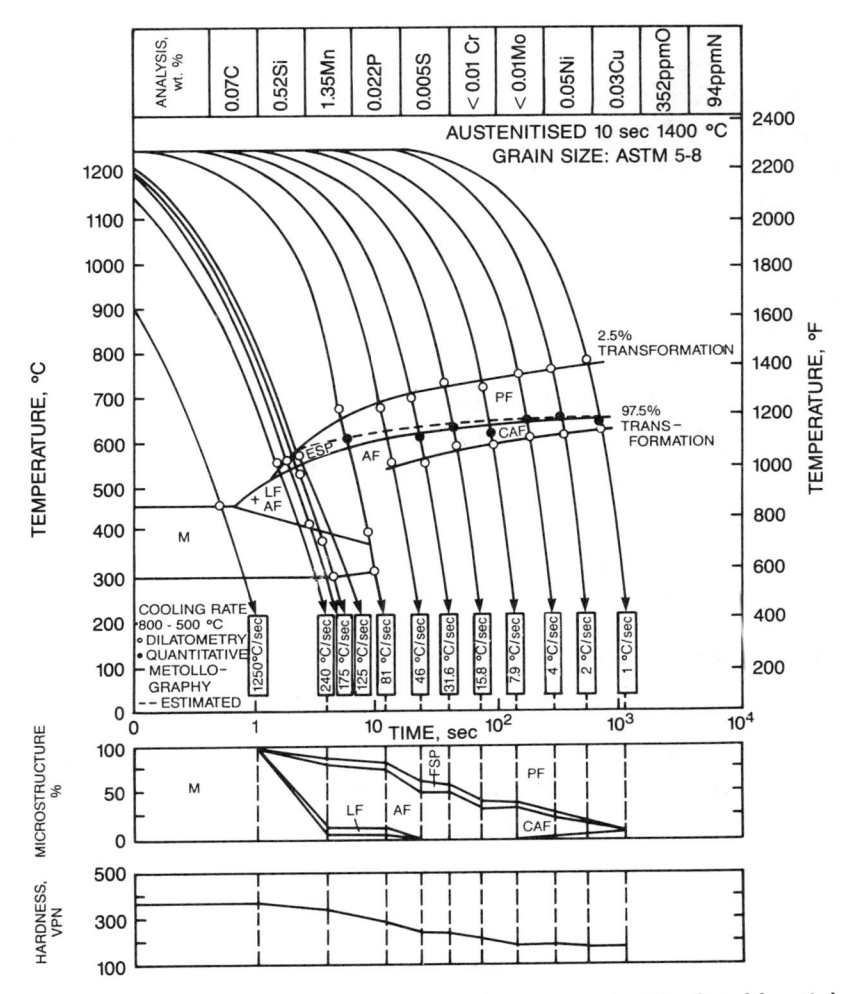

Figure 9.37 — **Factually constructed CCT diagram for steel weld metal deposited by E 7016 SMAW electrodes**

Further details can be found in extensive work reported by Harrison and Farrar as posted in Suggested Reading at the conclusion of this chapter. See Table 9.2 for meaning of symbols denoting microstructural constituents. (Courtesy of The Welding Institute, England)

seconds to cool through the same temperature range" as shown by log scale on the abscissa. Similar CCT diagrams were prepared and published for weld metals with different manganese contents, and for several nickel-containing weld metals. This research enabled the investigators to evaluate many factors that influence the formation of constituents in the microstructure. The dilatometer was selected as the logical tool to measure linear thermal contraction of prepared specimens during cooling to determine the temperature for the start of austenite-to-ferrite transformation as indicated by the familiar sudden expansion during crystallographic change. Unexpectedly, most contraction curves were smooth and gave no real indication of the gamma-to-alpha crystalline transformation. At high cooling rates, the start of martensite formation usually was clearly indicated. Also, at very slow cooling rates, some weld metal signaled the transformation to pearlite. To minimize error in placing transformation start-curves and finish-curves on their CCT diagrams, the investigators used 2½ percent and 97½ percent transformation completion values to plot the curves.

The general effect of increasing cooling rate on transformation in weld metal is much the same as experienced with steel base metals; that is, the time-temperature curves for start and for finish of transformation are moved downward and shifted to the right. This change in position is not as marked as ordinarily experienced with steel base metals simply because the latter are usually compositions that provide greater hardenability. At high cooling rates, most weld metals are apt to form a bainitic microstructure because of limited hardenability, although the more highly alloyed weld metals can transform to martensite. With very slow cooling, most weld metals form polygonal ferrite and either lamellar pearlite or the ferrite and carbide aggregate (FC). As cooling rate is increased, the occurrence of pearlite decreases and the amount of polygonal ferrite is diminished until it is limited to grain-boundary ferrite. The term "allotriomorphic" ferrite sometimes is applied to GF ferrite when present as veins or thin rims because of the ragged or imperfect form of the ferrite grains along the prior-austenite grain boundaries. At intermediate cooling rates, most weld metals undergo intragranular transformations that involve copious amounts of proeutectoid ferrite (because of their usual low carbon contents). As mentioned earlier, the desired transformation product is acicular ferrite, but other products also appear in the microstructure for reasons not yet completely understood. Of course, as the cooling rate becomes slower, the acicular ferrite becomes coarser, but this is not as detrimental to weld toughness as having constituents like GF ferrite and AC ferrite appear in appreciable amounts. Although a microstructure consisting *entirely* of acicular ferrite rarely is formed, a volume fraction of at least 50 percent of the AF product will produce creditable weld toughness.

As appreciation of the value of acicular ferrite in weld metal has grown, study of the AF form has intensified to take further advantage of this constituent when striving to meet growing demands for better weld toughness. Recent research using electron microscopy has been directed to examination of the dislocation density in the crystallographic structure of AF-ferrite. Magnification in the range of about 60 000 to 100 000X was able to resolve individual dislocations, and the dislocation density was estimated to be approximately 10^{14} m^{-2}. Dislocations at this density level were believed to contribute approximately 145 MPa (roughly 20 ksi) to the tensile strength of the weld metal.

Importance of Weld Metal Composition

The composition of weld metal is as important as cooling rate when striving for good as-welded toughness in the structural variety of weld metals. Brief remarks will be made here regarding recent findings on the influence of alloying and residual elements, but detailed discussion will be held in abeyance for Volume II, Chapter 10 on filler metals. Of course, earlier chapters made clear that filler metal alone does not establish weld composition. Regardless of the source elements in weld metal — whether filler metal, base metal, or adjacent material — the following generalities hold true when striving for acicular ferrite in the weld microstructure to ensure good toughness.

Carbon content of the weld metal should not be very low (e.g., not below about 0.07 percent), nor should it be very high, that is, not above about 0.14 percent. For plain-steel weld metal, carbon contents in the range of about 0.08 to 0.12 percent appear most effective. More highly alloyed steel weld metals are yet to be thoroughly studied, but a somewhat lower range of carbon is believed to be desirable. Manganese and nickel have been found quite effective in promoting acicular ferrite and in reducing the amount of polygonal ferrite and sideplate ferrite (SP). While it has been common knowledge that manganese and nickel improve the toughness of steel and weld metals, toughness test values for weld metal now can be correlated with quantitative volume fractions of particular microstructural constituents. Knowledge of these relationships is certain to lead to better weld toughness control. Figure 9.38 is an example of data correlation to ensure toughness in weld metal by determining optimum manganese contents for welds cooled at particular rates. Naturally, slower cooling would call for

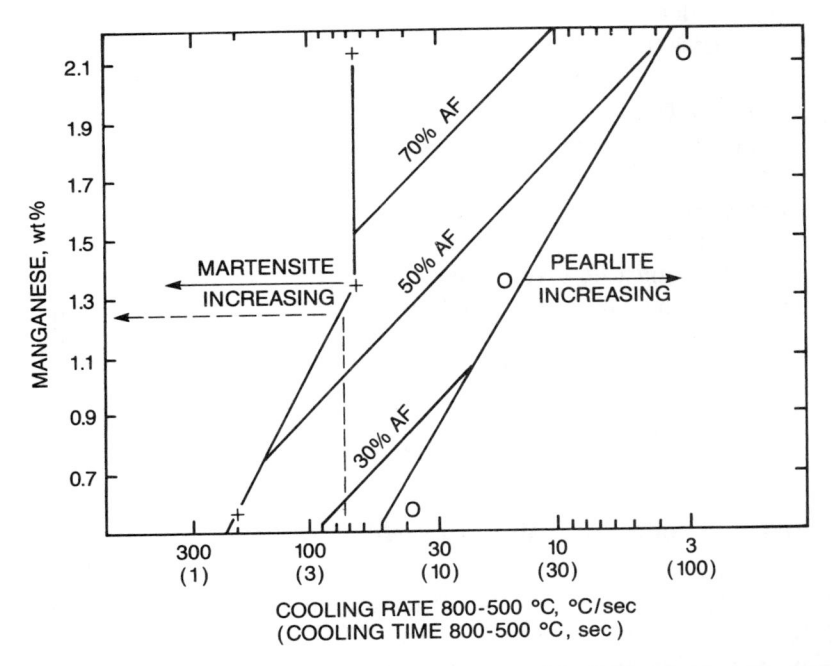

Figure 9.38 — Influence of manganese content in C-Mn steel weld metal over a range of weld cooling rates in producing particular weld zone microstructures

Vertical dashed line from abscissa at about four seconds weld cooling time corresponds to a 20 mm (¾ in.) thick plate welded by SMAW with one kJ/mm (25.4 kJ/in.) heat input. The intercept of the vertical dashed line with the sloping solid line indicates that a manganese content of about 1.3% would be optimum to obtain a maximum percentage of acicular ferrite to ensure good toughness. (Courtesy of The Welding Institute, England)

increased manganese content to provide tough, acicular ferrite in the weld metal microstructure. In Figure 9.38, regimes of three principal forms of microstructure, namely, martensite, acicular ferrite, and pearlite, have been positioned across the scale of cooling rates from left-to-right (fast-to-slow). As cooling rate becomes slower, the manganese contents required to obtain acicular ferrite are indicated by the sloping lines on the left and the right. The connecting lines predict approximate percentages of acicular ferrite to be expected at different levels of manganese. The predictions of this diagram have been found to be in good agreement with the work of other investigators.

Role of Grain Size in Weld Metal

Grain size of the austenitic weld metal microstructure is another feature that influences transformation during cooling, and this feature is related both to composition and to heating-and-cooling conditions. Briefly stated, small austenite grains provide more boundaries for nucleation and this favors formation of grain-boundary (GF) and polygonal (PF) forms of ferrite. Larger austenite grain size increases hardenability, and in weld metal large *austenite* grain size favors the formation of acicular ferrite (AF). While grain size is sufficiently influential to deserve consideration, it is a feature not easily manipulated. Fine grain size is favored by using low weld heat input to produce rapid cooling. Smaller grain size in a *ferritic* microstructure has been recognized to offer better toughness than very coarse grain size. Sometimes, small grain size can be pre-arranged by addition of small amounts of certain elements, such as titanium, to the weld composition. When minute oxides of these elements are dispersed throughout the weld metal, they serve as nucleation sites and not only can promote small grain size, but they also accelerate intragranular transformation. Whether the intragranular transformation product will be acicular ferrite will depend on other influential variables in a given case. For instance, if the weld metal composition includes easily-oxidized elements, consideration must be given to the nature of oxides formed and whether their mode of dispersal in the solid weld metal will influence the grain size of the austenite and subsequent transformation products. This is a metallurgical aspect of weld metal now being given much attention and will be reviewed next. Of course, a larger austenite grain size can be brought about by use of high welding heat input, but toughness is usually lowered in resulting microstructures. Consequently, grain size of the austenite is often a feature of weld metal that does not receive special regulation.

Influence of Nonmetallic Inclusions in Weld Metal

Nonmetallic inclusions in weld metal can be highly influential in the transformation of the primary austenite to the initial ferritic products; and this is especially the case with respect to the amount of acicular (AF) ferrite that forms in the microstructure. The potency of nonmetallic inclusions in weld metal has been somewhat surprising, but is understandable in view of the "peppered" distribution of innumerable, tiny nonmetallics that often exist in weld metal. These micro-particles are usually about 0.5 micron in size, and although larger than the disc-shaped titanium nitride particles of 200 Å found in steel base metal treated with titanium as described in Chapter 4, the weld metal's nonmetallic inclusions can still serve as an effective nucleant to assist all-over initiation of transformation. The inclusions in steel weld metal are usually oxides of one or more of the alloying elements included in the weld zone's composition. While steelmakers have rationalized the chemical composition of non-metallics in cast and wrought steels from the free energy of formation of oxides (ΔF) of elements present, researchers in welding must struggle with (1) several different sources that contribute to the pool of metal that becomes the weld zone, (2) unusually high temperatures at which reactions take place, and (3) the very short times during which reactions occur in the gaseous state of an arc, a laser plasma, or an electron beam, and the molten state before the weld pool undergoes rapid solidification.

About 1975, research began in earnest to examine the role of inclusions in influencing microstructural transformation in the weld zone during cooling, and to determine whether this feature of weld metal could be relied on to promote final microstructures that possessed better toughness. Studies were conducted with both C-Mn and HSLA steels to (1) determine the composition of inclusions in weld metal and the nature of their surface, (2) methodically assess their size and distribution, and (3) correlate nonmetallic inclusion population with the kind of microstructure formed in the weld metal and its toughness. Although much work remains to perfect the efficacy and repeatability of utilizing nonmetallic inclusions in weld metal for microstructural control, the findings to date are helpful in carrying out the following steps:

(1) planning deoxidation of the metal that eventually constitutes the weld zone,
(2) controlling the composition, size, and distribution of inclusions in the weld, and
(3) optimizing weld metal microstructure to obtain best possible strength and toughness.

In the past, the usual recourse in handling the deoxidation of weld metal was simply to make certain that ample amounts of alloying elements (other than carbon) were present that had an affinity for oxygen and would effectively "deoxidize" the steel weld zone and prevent porosity formation through the carbon-oxygen reaction. While manganese and silicon acted as deoxidizers and effectively suppressed porosity the MnO and SiO_2 oxides tended to coalesce into inclusions of appreciable size that sometimes exceeded tolerable limits set by the quality standards for a particular weld. For this reason, a new practice began in which the regular, required manganese and silicon contents were supplemented with relatively small additions of much stronger deoxidizing elements that not only formed smaller oxide particles, but these particles had a lesser tendency to coalesce into large inclusions that might jeopardize compliance with cleanliness or soundness requirements. Elements favored for this kind of deoxidation action included aluminum, calcium, titanium and zirconium. Experience with these stronger deoxidizers augmented by research in this area brought forth findings of a general nature that deserve summation here. Utilization of this technology in formulating the composition of filler metals will be reviewed in greater detail in Volume II, Chapter 10.

Role of Oxygen in Weld Metal

Oxygen is a key ingredient that must be present within a controlled range in the molten weld zone to start the chain of events that leads to a tough weld metal. It has been demonstrated that if the molten weld contains a very low level of available oxygen to form oxides, the weld metal will be likely to have an insufficient number of particles for nucleating the formation of fine, tough acicular ferrite (AF) as the austenitic structure transforms during cooling. Instead, with low oxygen, the weld structure is likely to consist of grain boundary ferrite (GF), side plate ferrite (SP and IP), or if the weld is a HSLA steel composition the final microstructure could be bainitic. The favorable range for the initial, uncombined oxygen in the weld zone is approximately 0.02 to 0.05 percent. When available oxygen is within this range, the volume fraction of micro-size oxide particles that will form with the aid of a strong deoxidizer will be in the vicinity of about 0.3 percent, and this volume has been observed to cause as much as 75 percent of the weld metal microstructure to form as acicular (AF) ferrite. When the oxygen level is below about 0.02 percent, the inclusion population after deoxidation is usually less than 0.2 percent volume fraction, and following austenite transformation very little acicular ferrite is found in the microstructure of C-Mn steel weld metal. At high levels of weld-zone-available oxygen (i.e., 0.06 percent and above) the inclusion volume fraction usually exceeds 0.4 percent, and the transformation products in the final microstructure consist of less than 20 percent acicular (AF) ferrite. Also, the microstructure becomes coarser, and often will contain large amounts of undesirable constituents, such as side plate (SP) ferrite and ferrite with aligned secondary phases (AC).

When neither the steel being welded nor the filler metal provide the desired level of oxygen in the weld zone, certain of the fusion welding processes allow introduction of the needed oxygen. Although the method of addition may be innovative, it must be carefully regulated. For example, gas metal arc welding (GMAW) can make use of oxygen-bearing shielding gas to augment the base and filler metals and achieve the range considered optimum for a particular case. Studies conducted with argon shielding using oxygen additions from one to five percent, and with carbon dioxide additions ranging from 10 to 75 percent found that the oxygen activity or potential of the shielding gas influenced the weld metal oxygen content in a consistent manner; also, higher welding heat input tended to increase the weld metal oxygen level by a small amount for a given oxygen potential in the shielding gas.

Utilization of Nonmetallic Inclusions for Microstructural Control

The use of nonmetallic inclusions in steel weld metal to promote the desirable acicular form of ferrite requires coordination of a number of variables to secure (1) inclusions of a particular kind, (2) within a certain size range, (3) close to a stated volume fraction, and (4) uniformly distributed throughout the solid weld metal. This is quite a challenge to the welding engineer considering the meager guidelines established to date, but when properly coordinated and carried out the improvement in weld metal toughness can be rewarding.

The addition of a deoxidizing agent to suppress weld metal porosity is a relatively simple step which can be accomplished with a single element additions, or with multiple elements. In fact, the practice of adding a multiplicity of deoxidizing elements became entrenched to an extent a long time ago for some filler metals when there was much uncertainty about the influence of deoxidizers on weld metal toughness. This practice seemed to be spurred by the philosophy that if a small dose of a single deoxidizer appeared good for the weld, a larger dosage of several elements should make the weld even better. Now that experience and research findings have clarified some of the reactions and mechanisms by which porosity is suppressed, and the weld structure and toughness are affected, it is possible to be judicious in choice of deoxidizer(s) and the amount(s) to be added for a particular welding operation. For certain fusion welding operations where passes are made with torches or heads operating in tandem, present knowledge can indicate the best sequence of addition when using multiple deoxidizing elements.

Titanium, either as a single deoxidizer, or when included among multiple elements, is now the favored agent to form the nonmetallic nucleant in weld metal to promote acicular ferrite formation and gain better toughness. When added to the extent of approximately 0.03 to 0.05 percent, titanium serves very effectively in the nucleation mechanism. The majority of TiO_2 particles are smaller than the 0.5 micron size mentioned earlier; above which the nonmetallic inclusions lose effectiveness as a nucleant for acicular ferrite formation. Titanium appears to be effective even at weld metal oxygen levels a little below 0.02 percent. This unanticipated effectiveness is attributed to the strong affinity of the element for nitrogen, since titanium nitrides (TiN) have been identified among the dispersed nonmetallic particles in the weld metal.

When titanium is used in combination with another deoxidizing element, for example, aluminum, consideration must be given to their relative affinities for forming oxides. A comparison of free energy of oxide formation (ΔF) of titanium and aluminum will show the latter element as the stronger oxide-former. Consequently, when the two elements are added together into the molten weld zone, aluminum is likely to preempt the available oxygen by forming Al_2O_3 and in so doing will not allow significant formation of TiO_2 particles, which could serve as the more effective nucleant. Competition of this kind should be weighed when fusion welding with two or more filler metals to make a single weld pass, such as done in tandem submerged arc welding. This procedure allows one steel electrode to carry aluminum as a deoxidizer, while another might carry titanium. However, the sequence of multiple deoxidizer usage deserves consideration in planning the procedure because if titanium oxides are wanted as the principal nucleant, aluminum should not precede titanium.

Boron is not especially effective as a deoxidizer because its affinity for oxygen is a little weaker than that of silicon. Nevertheless, when boron is added in combination with titanium for deoxidation, this pair of elements appears to act with synergetic effectiveness in producing the inclusion-acicular ferrite mechanism. Work with the flux cored arc welding process found that an addition of only 0.004 percent boron along with titanium as the principal deoxidizer proved highly effective, and under properly regulated conditions the weld metal microstructure could contain as much as 95 percent acicular ferrite.

Aluminum, when added individually to a C-Mn steel weld metal as a deoxidizer, is not as effective as titanium in promoting the acicular ferrite microstructure even though aluminum forms both oxide and nitride particles. Incidentally, the affinity of aluminum to form nitrides (AlN) is not quite as strong as that of titanium, so less influence is likely to be gained from the formation of AlN particles among the nonmetallic nucleants. Furthermore, aluminum when added individually at too generous a level in the weld metal (possibly above about 0.02 percent) often results in an increase in the number of larger oxide particles, that is, larger than about 0.5 micron as determined by particle analyzing scanning electron microscopy (PASEM). These larger particles not only are less effective than the smaller oxides (i.e., less than 0.4 micron) in nucleating acicular ferrite, but the weld metal microstructure tends to form more grain boundary ferrite (GF) and aligned ferrite (AC) instead of acicular ferrite (AF). Another aspect of deoxidation with aluminum that has drawn cautionary advice from researchers is the tendency for nonmetallic particles to form that are too large for efficient nucleation of AF when the oxygen content is quite high. With oxygen in the weld zone above about 0.06 percent, the nonmetallic inclusions formed by aluminum can have a composition that includes some iron oxide (FeO) in addition to the preponderance of Al_2O_3. These particles appear to trigger the formation of increased amounts of polygonal ferrite (PF), and as a consequence the toughness of the weld metal suffers.

Because much experimentation with deoxidation of weld metal has been cut-and-try, a number of other elements with strong oxide-forming capability, and unusual combinations of multiple elements, have been tried. Only recently, however, has effectiveness been measured against the microstructure formed and its toughness. For example, calcium is a very powerful deoxidizer, but its oxides tend to be larger than one micron. Therefore, these inclusions promote various forms of primary ferrite other than the desirable acicular ferrite. One combination of elements that has shown promise, possibly because of some degree of synergism, is aluminum plus titanium. The optimum level for this combination to produce a significant increase in the proportion of acicular ferrite in weld metal appears to be lower than when the elements are added individually. Minima of 0.006 percent aluminum, and 0.003 percent titanium have been reported to produce a noticeable increase in acicular ferrite.

An extraordinary experiment recently performed that strongly supports the above views of inclusion influence on transformation involved the making of a weld deposit by the submerged arc welding process (SAW) in which the oxygen content was 0.023 percent and the microstructure contained a high proportion of acicular ferrite. This weld deposit was remelted by the laser beam welding process (LBW), and analysis for oxygen showed that only 0.011 percent had been retained. Metallographic examination found that the predominantly acicular ferrite microstructure of the SAW deposit had been replaced in the LBW weld metal by one that was substantially bainitic. This experiment, along with a number of other studies, confirm in a general way the strong influence of oxygen content and minute nonmetallic inclusions on the microstructure formed in weld metal during transformation of the austenite on its initial cooling after solidification. Current studies are revealing evidence that the inclusions taking part in transformation mechanics are polycrystalline and heterogeneous, and that their surface composition may be almost as important as their size. Research along these lines is vital to progress toward improving steel weld metals.

STUDY OF A TYPICAL FUSION WELD IN STEEL

Earlier in this chapter, Figure 9.5 was presented to show the spectrum of zones across a weld in steel plate as created by localized heating. Subsequent text, diagrams and photomicrographs provided information concerning transformations that occurred in steels over the gradient of temperatures and varying cooling rates to which the steel is subjected during welding. Now this information base can be used in examining a so-called "simple" fusion weld.

Figure 9.39 shows a single-pass arc-fusion weld made by SMAW in five mm (0.2 in.) thick C-Mn hot-rolled steel plate containing 0.2 percent carbon. The photograph at the top of the figure is one half of a transverse section polished-and-etched to reveal the weld zone, HAZ, and the base metal. The figure presents a curve of maximum temperatures reached by the weldment during the SMAW operation starting at the left with the molten weld zone at 1675°C (about 3050°F), and moving to the right with decreasing temperatures at increasing distances from the weld center. Because the SMAW process accomplished complete joint penetration and added generous weld face reinforcement, the heat input during welding was substantial. This accentuated microstructural changes and made them easier to illustrate. Alphabetical symbols are used in Figure 9.39 to point out specific locations in the weldment where the microstructures deserve close examination. These microstructures will be shown at higher magnification in a number of photomicrographs to follow, and comments will be provided regarding important details that affect the properties of the weldment. Table 9.3 is provided as a reference to figures in Chapter 9 where additional metallographic portrayal of particular microstructural constituents can be found.

Figure 9.40 is a photomicrograph at 100X of the microstructure in the as-supplied steel plate. The constituents are pearlite and polygonal ferrite. About one-quarter of the microstructure is pearlite, and, according to the multiplication factor of 125 mentioned earlier when reviewing Figure 9.9, the percentage of pearlite appears to confirm the reported carbon content of 0.2 percent. The pearlite in the microstructure of this steel is "banded," which is a distribution often observed in hot-rolled plain steel of hypoeutectoid composition. The lamellae of ferrite and cementite in the pearlite cannot be seen at 100X. A few small, elongated, nonmetallic inclusions (presumably sulfides) can be seen lying in the bands of ferrite. The ASTM grain size of this steel is No. 7 (per Table 2.9). Exposure during welding to only 260°C (500°F) indicated by the temperature distance curve for location "a" produced no change in the microstructure of this steel. Consequently, the base metal in that region has been labeled *unaffected*. The exposure temperature that can cause a detectable microstructural change in 0.2 percent carbon steel will depend on the time at temperature. With the relatively short exposure time during welding, no significant change in microstructure is likely to be found until approximately 620°C (1150°F) is reached or exceeded.

Table 9.3 – Microstructures Found in Cast and Wrought Steels; Including Their Weld Heat-Affected Zones

Symbols	Description of Microstructural Constituent	Figure in Which Microstructure Can Be Found
M	Martensite	Figure 9.13
B	Bainite (lower bainite)	Figure 9.20-C
B	Bainite (upper bainite)	
WF	Widmanstätten pattern	Figure 9.44
FP	Ferrite, polygonal	Figure 9.6
F	Pearlite, lamellar	Figure 9.9
AC	Ferrite, aligned	Figure 9.43
FN	Ferrite, non-aligned	Figure 9.43
AC(MAC)	Aligned ferrite with martensite/austenite/carbides	Figure 9.32

878 Chapter 9

Figure 9.41 shows a subtle change in microstructure which has taken place at the outer edge of the HAZ (location "b") where exposure to a maximum temperature of 650°C (1200°F) occured during welding. An allotropic transformation *did not* occur in this region of the HAZ because this temperature is below the Ac_1. Nevertheless, this exposure did cause cementite in the lamellae of the pearlite to become spheroidized to a noticeable extent. This can be seen at the higher magnification of the photomicro-

① — WELD ZONE (MIXED ZONE)
② — UNMIXED WELD ZONE (SUPERHEAT MELT-BACK)
③ — PARTIALLY-MELTED ZONE
④ — HEAT-AFFECTED ZONE OF BASE METAL
⑤ — UNAFFECTED BASE METAL

CURVE SHOWS MAXIMUM TEMPERATURE REACHED DURING WELDING

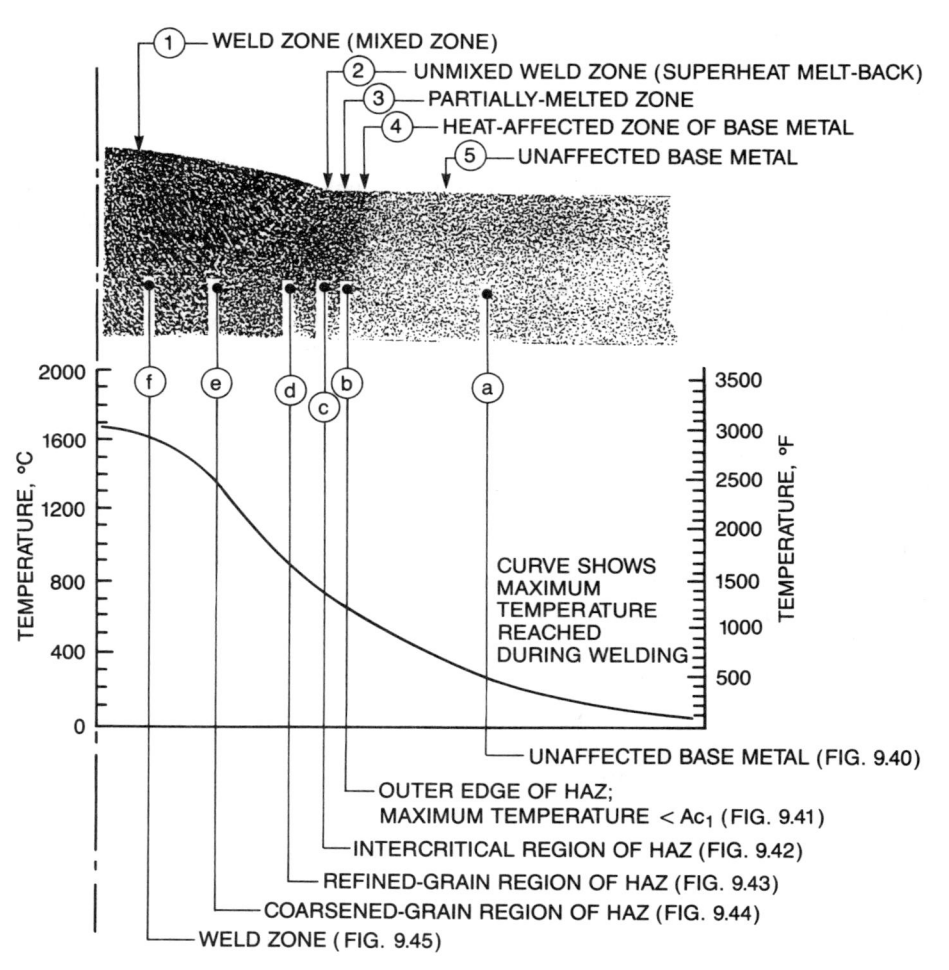

— UNAFFECTED BASE METAL (FIG. 9.40)
— OUTER EDGE OF HAZ; MAXIMUM TEMPERATURE $< Ac_1$ (FIG. 9.41)
— INTERCRITICAL REGION OF HAZ (FIG. 9.42)
— REFINED-GRAIN REGION OF HAZ (FIG. 9.43)
— COARSENED-GRAIN REGION OF HAZ (FIG. 9.44)
— WELD ZONE (FIG. 9.45)

Figure 9.39 — Typical weld made by SMAW showing maximum temperature reached in regions ranging from weld zone to unaffected base metal, and the effects of these temperatures on microstructure

Specimen is transverse etched section of five mm (0.200 in.) thick plate of AISI-SAE 1020 steel with single-square-groove weld. The curve below the photograph shows the temperature gradient from weld zone through the base metal HAZ. The location of five regions of microstructural significance are indicated by arrows placed above the photograph. The nature of microstructures in six locations (letters "a" to "f," inclusive) are pin-pointed and microstructures of this nature can be observed in figures that follow in the text.

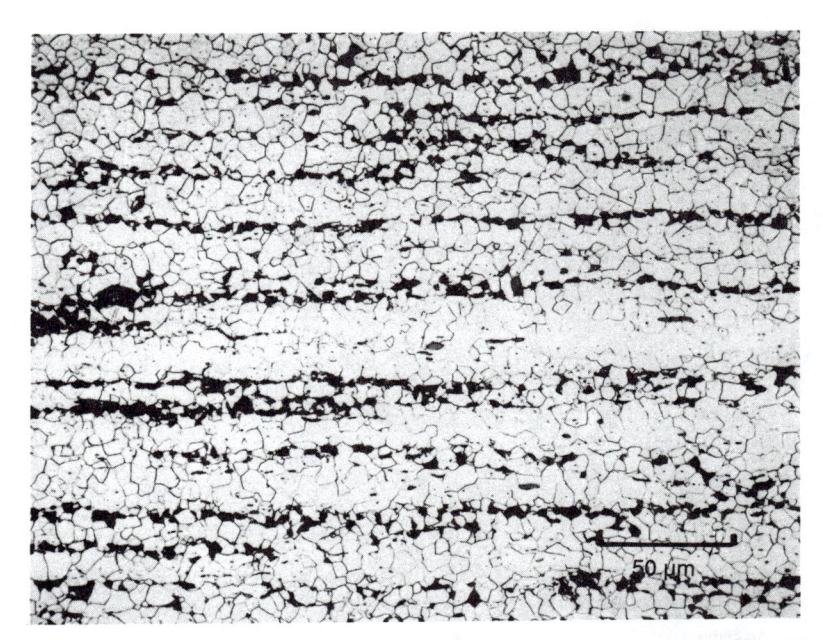

Figure 9.40 — AISI-SAE 1020 steel base metal used to prepare weld specimen in Figure 9.39. Photomicrograph of microstructure at location "a"

This hot rolled plate steel with banded microstructure of pearlite in a matrix of ferrite did not undergo any microstructural change when exposed to the maximum temperature of 260°C (500°F) during welding. (Nital etchant; original magnification: 100X)

graph in Figure 9.41. Although the colonies of pearlite that originally were lamellar have not lost their general outline through diffusion, their platelets of cementite have become reshaped into tiny globules. Also, the "banding" of the constituents remains. This subtle change in microstructure reduces hardness by a small amount, and tensile testing would find this region to have the lowest strength in the weldment. In heat treating language, this region of the HAZ was subjected to a *sub-critical annealing treatment*, albeit a brief one.

Figure 9.42 shows the marked change in microstructure that occurs in the HAZ when heated to a temperature *within* the A_1-A_3 critical range for the steel. Exposure to approximately 760°C (1400°F) during welding produced only partial allotropic transformation; that is, the pearlite was austenitized along with some of the ferrite. This partial "solution treatment" was sufficient to dispel most of the banding. During cooling, this region of the HAZ underwent appreciable grain refinement as grains of the austenitized portion re-transformed to pearlite and ferrite. Those ferrite grains that did not undergo transformation during heating (i.e., the "solution treatment") remained unchanged in size inasmuch as there was not enough strain present to induce recrystallization, and the maximum temperature was not high enough to cause grain growth. Judging from the small volume of untransformed ferrite grains in Figure 9.42, location "c" in Figure 9.39 may have reached a temperature closer to the

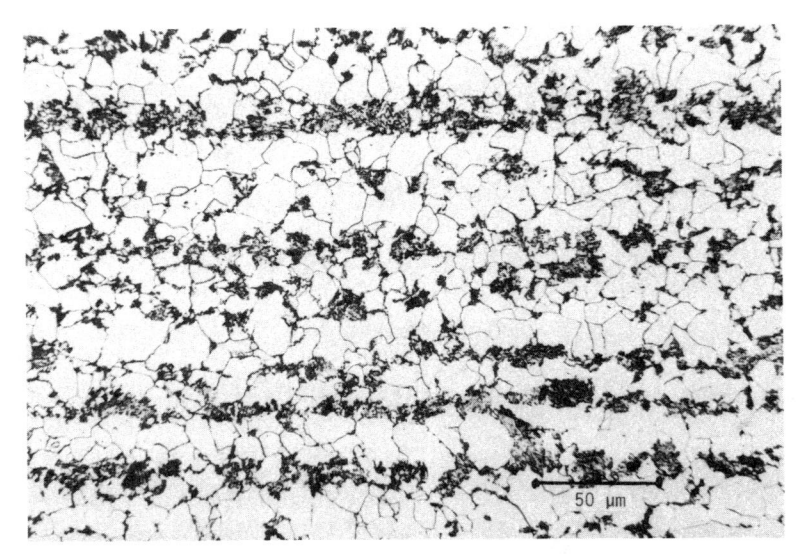

Figure 9.41 — Welded AISI-SAE 1020 steel at location "b" of Figure 9.39, the outer edge of heat-affected zone

Exposure to maximum temperature of 650°C (1200°F) during welding started spheroidization of cementite platelets in the pearlite colonies. (Nital etchant; original magnification: 500X; courtesy of The Welding Institute, England)

Ac$_3$, the than the Ac$_1$. The net effect of this *intercritical heat treatment* is a substantially refined microstructure.

Figure 9.43 illustrates the region (location "d") of the HAZ heated to a temperature just above the Ac$_3$ of the steel's critical range, specifically 927°C (1700°F). The microstructure consists of fine ferrite, both aligned and non-aligned, with extremely fine secondary phases distributed among the grains of ferrite. The secondary phases are difficult to identify, and the metallographer often will indicate *M-A-C* as being dispersed between the ferrite grains. This microstructure represents a favorable kind of heat-effect, and this region of the HAZ often has mechanical properties superior to those of the hot-rolled steel. Unhappily, this beneficially-affected region is fairly narrow, and usually does not extend to the weld zone where good toughness could benefit the fitness of the weld for service.

Figure 9.44 illustrates *unfavorable* treatment of the steel during welding. Location "e" is the HAZ region immediately adjacent to Zone 3 (where partial melting sometimes is observed to have occurred). At location "e," the temperature during welding rose to approximately 1350°C (about 2450°F). When this region is austenitized and reaches this maximum temperature, grain coarsening occurs. The extent of grain growth will depend on the amount of time spent at or near the maximum temperature. If the cooling rate is rather slow, as expected with the high heat input used for the weld in Figure 9.39, transformation of the austenite is likely to produce a potpourri of coarse constituents as shown in Figure 9.44. Pro-eutectoid ferrite appears as polygonal grains forming a network outlining the prior-austenite grains. The intragranular microstructures consist of coarse ferrite which has formed randomly distributed patches of (1) Widmanstätten pattern, (2) aligned ferrite, and (3) non-aligned ferrite. Coarse M-A-C is dispersed between the grains and laths of these intragranular forms of ferrite. Taken in its entirety, this microstructure will not be lacking in strength, but can have poor toughness. The presence of this microstructure at the edge of the weld is unfavorable because this is the location where stresses tend to concentrate, and dynamic loading requires toughness at this particular point in the weldment. A number of steps can be taken to avoid such an undesirable microstructure in the HAZ, and a few incorporated most often in welding procedures will be described shortly.

Weld zone examination will complete the panorama of zones and regions in the weldment of Figure 9.39. In view of the relatively high heat input and slow cooling rate of this single-pass SMAW weld, the weld metal microstructure is not likely to consist of constituents favorable to toughness. Indeed, Figure 9.45 shows the weld zone (location "f") to have substantial polygonal ferrite in the boundaries of the prior-austenite grains and very little acicular ferrite. In fact, much of the intragranular microstructure is ferrite aligned with secondary phases. The toughness of this weld metal probably will be almost as poor as the HAZ at location "e."

There are several additional zones and features in a fusion weld for which microstructural conditions have not been illustrated using Figure 9.39. However, the *fusion boundary*, the *unmixed zone* (identified as "2" in Figure 9.39), and the *partially-melted zone* ((identified as "3") were illustrated earlier in Figures 9.4 and 9.5, and their origins were described in text that accompanied the earlier figures. Recent find-

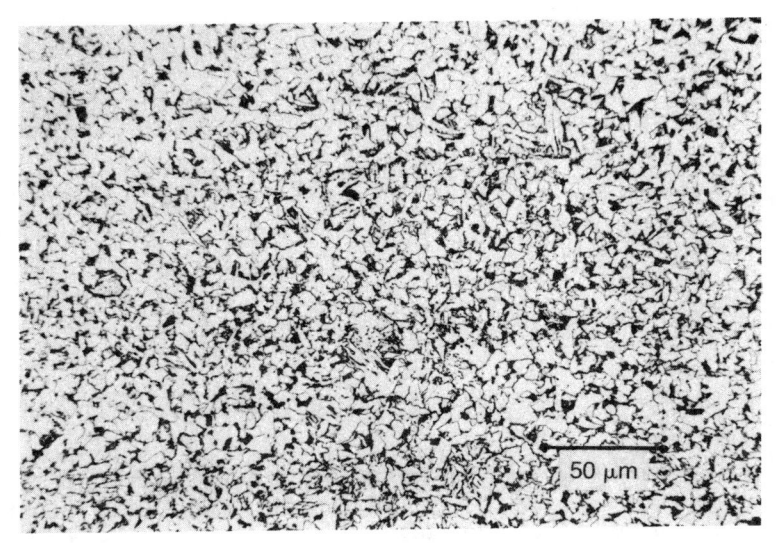

Figure 9.42 — Welded AISI-SAE 1020 steel at location "c" of Figure 9.39, which is within the Ac₁-Ac₃ critical range for this steel

Exposure to a temperature of 700°C (1400°F) represents an intercritical heat treatment and a substantial refinement has occurred in the steel's original microstructure of pearlite-and-ferrite. (Nital etchant; original magnification: 500X; courtesy of The Welding Institute, England)

ings concerning the unmixed melted zone (formed by superheat melt-back, and now commonly identified by the acronym "UNMZ") deserve mention here.

The very narrow UNMZ usually ranges from approximately 0.05 to 2.5 mm (0.002 to 0.10 in.) and is an integral feature of most welds made by a fusion process. The zone consists essentially of base metal composition because the mechanisms that stir the molten weld zone have ceased to operate as imminent solidification of the UNMZ freezes the melted base metal in situ. The microstructure that forms epitaxially in the UNMZ is like that of as-deposited weld metal. Welds made by the SMAW, SAW and ESW processes have been observed to form a wider UNMZ than those made by GMAW, EBW and LBW. Interestingly, the UNMZ tends to be wider near the root of a penetrating weld pass, and is sometimes almost nonexistent at the top surface. This gradation in width along the fusion boundary calls for careful planning in conducting any form of test or examination when studying the UNMZ.

When the weld zone and the base metal are low-carbon steel compositions, the presence of a UNMZ is not likely to influence the performance of the weld in tests or in service. The plain steel weldment in Figure 9.39 displayed nothing significant that deserved illustration; although an UNMZ was present to a limited extent. When dealing with steels that contain larger amounts of alloying elements, especially higher carbon content, the presence of an UNMZ can affect weld performance because of its lower toughness, and increased propensity to crack initiation. Welds subjected to impact loading at very high strain rate have shown degraded fracture toughness in the

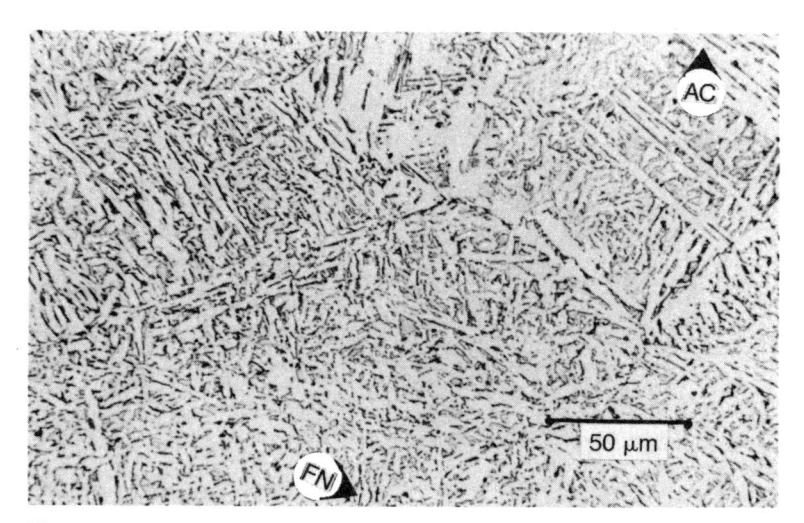

Figure 9.43 — Welded AISI-SAE 1020 steel at location "d" of Figure 9.39, which was heated to a favorable temperature just above the steel's critical range

Exposure to a temperature in the vicinity of 900°C (1652°F) virtually represents a "normalizing heat treatment." A minimum of grain boundary ferrite has formed. The ferrite grains in both the AC and FN constituents are very fine. MAC is dispersed among the ferrite grains. (Nital etchant; original magnification: 500X; courtesy of The Welding Institute, England)

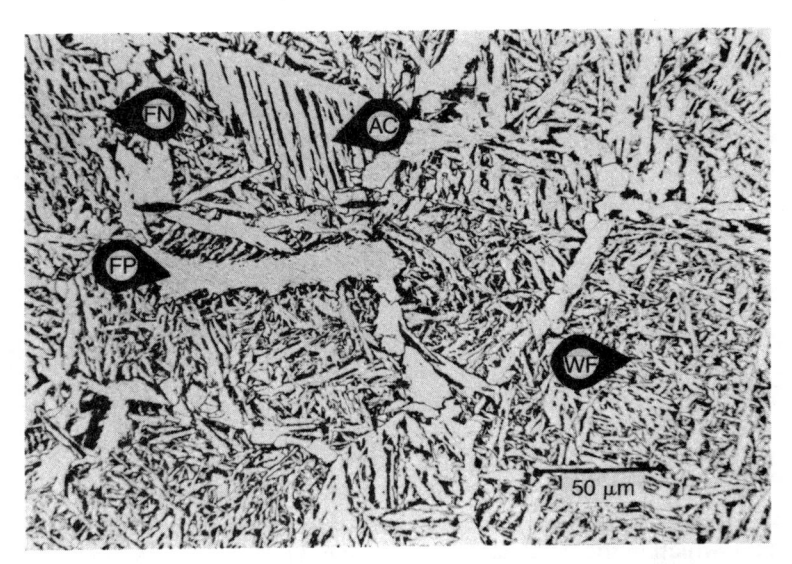

Figure 9.44 — Welded AISI-SAE 1020 steel at location "e" of Figure 9.39, which reached a very high temperature and caused considerable growth in grain size and produced an unfavorable microstructure

This region of the HAZ immediately adjacent to the weld zone was heated to about 1350°C (2450°F) in this particular weld. This severe heat-effect caused marked grain growth in he austenite, which upon cooling transformed into a potpourri of coarse constituents as indicated. The toughness of this region of the HAZ with microstructure of this character will be poor. (Nital etchant; original magnification: 500X; courtesy of The Welding Institute, England)

UNMZ. This matter will be examined in more detail in Volume II, Chapter 15, when the welding of alloy steels for low-temperature service is reviewed.

The partially-melted zone (Location 3 in Figure 9.39) is formed when the base metal composition includes elements that act to depress the liquidus and thusly increase the temperature range between the solidus and the liquidus of the steel. Minor elements (e.g., sulfur) at levels higher than their norm are the most common cause of partial melting (i.e., liquation) in the first increment of the HAZ that reaches the highest temperature. This metallurgical cause-and-effect can easily lead to hot cracking in the partially-melted zone; therefore, more detailed coverage will be relegated to Volume II, Chapter 13 on difficulties and defects.

Making Welds with Good Toughness

In early steel welding, a commonly encountered difficulty was the formation of hardened microstructures both in the weld zone and the HAZ of the base metal. Mainly this was a result of high carbon levels in steels being put into weldments as filler metal and base metal at the time. The remedies regularly considered to alleviate

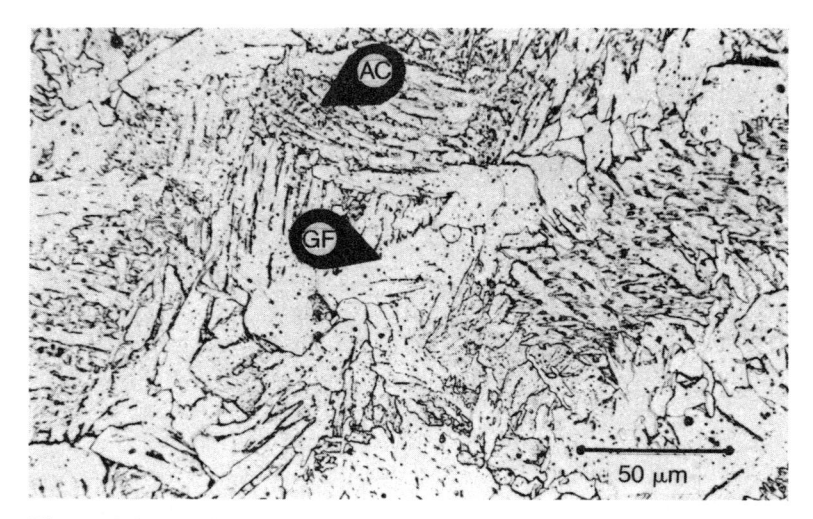

Figure 9.45 — Welded AISI-SAE 1020 steel at location "f" of Figure 9.39, which is the weld zone where the chemical composition consists of a mixture of filler metal and melted base metal.

The microstructure of the weld zone suggests that heat input in making the weld was rather high, and cooling of the completed weld was relatively slow, as evidenced by a massive network of polygonal ferrite grains in the columnar prior-austenite grain boundaries (GF). Much ferrite aligned with secondary phases (AC) is present intragranularly. (Nital etchant; original magnification: 500X; courtesy of The Welding Institute, England)

the low toughness that accompanied hardened microstructures were (1) welding with a high heat input, (2) preheating and maintaining an elevated interpass temperature, and (3) postweld heat treatment. The latter recourse, PWHT, was the least used. Welding with a high heat input reduced the weld cooling rate and this usually resulted in some decrease in hardness, but the improvement in toughness was not always adequate. While the inconvenience and cost of preheating was deplored, general success made preheating a very popular recourse. A philosophic attitude became widespread: "Whenever toughness is in doubt, apply preheat!". Today, there are reasons why high heat input and/or preheating might do more harm than good when fabricating weldments from many modern steels. These reasons will be examined in detail in chapters to follow in Volume II, but they are mentioned at this point in light of the foregoing review of austenite transformation and microstructures in the weldment of Figure 9.39.

Briefly, many steels and filler metals now are made with relatively low carbon contents. This change alone will increase the critical cooling rate of the steel and reduce the likelihood of forming a hardened microstructure (i.e., martensite). Some steels are so low in carbon content that even when a martensitic microstructure is formed, the toughness is acceptable. As carbon content was reduced in modern, weldable steels, additional carefully selected alloying elements were included in the compositions to regain strength by one or more of the hardening mechanisms described in Chapter 4.

> **TECHNICAL BRIEF 50:**
> **Acicular Ferrite in Weld Microstructures**
>
> Acicular ferrite in the microstructure of weld metal in the carbon and low-alloy steels is drawing much attention because of the marked improvement in toughness that this constituent imparts to the weld metal. The acicular form of ferrite consists of laths that grow throughout the decomposing austenite grains to form a fine basketweave or latticework pattern with small carbides and other microphases dispersed between the laths. Essentially, this (AF) microstructure is very fine-grained, is laden with dislocations, and is effectively pinned by secondary phases and grain boundaries in a manner that discourages fracture by the cleavage mechanism. An important part of the basic technology to obtain this AF microstructure is to initiate copious intragranular nucleation of hypoeutectoid ferrite in the austenite and thus encourage highly competitive growth among the numerous ferrite grains to generate the acicular form.
>
> The microstructure of weld metal need not consist entirely of acicular ferrite to benefit toughness by this constituent; in fact, 100 percent acicular ferrite is seldom achieved. A number of other forms of hypoeutectoid ferrite also nucleate, such as larger grains in the prior austenite boundaries, and these non-acicular constituents tend to detract from the toughness of the final microstructure. However, if at least 50 percent is the acicular form, weld toughness will probably be quite acceptable. Beyond 70 percent acicular ferrite, weld toughness appears to decline.
>
> Weld metal compositional features and welding procedure parameters that influence the formation of acicular ferrite in the microstructure have been identified, and rudimentary guidelines have been laid down for promoting the minimum proportion of this constituent wanted to enhance toughness. The number of factors that should be considered requires case-by-case appraisal and coordination between controlling and parametric aspects.
>
> Compositional guidelines for weld metal to obtain acicular ferrite are not particularly difficult to meet. Carbon in the range of about 0.08 to 0.12 percent is preferred for plain-steel weld metals. Elevated manganese contents, and alloying additional of nickel will not only promote acicular ferrite formation, but will tend to reduce the amount of unfavorable forms of microstructural constituents. There is good reason to consider small additions of titanium, boron and zirconium, but not for the usual alloying purposes: instead, these elements can play a vital role in forming minute nonmetallic inclusions that "pepper" the weld metal and serve as nucleation sites throughout the austenite grains for profuse generation of acicular ferrite. An important finding concerning weld zone composition is to have an adequate amount of oxygen available in the molten weld metal to form the needed nonmetallic inclusions. The favorable range for oxygen has been established at about 0.02 to 0.05 percent. This amount of oxygen appears to generate a volume fraction of nonmetallics of approximately 0.3 percent which is the target for the very small (0.5 micron) size inclusions needed to serve as nucleants.
>
> Parameters to consider in planning welding procedures to favor acicular ferrite formation have been explored and documented to a limited extent. In general, the aim is to regulate the cooling rate of the weld metal as its austenitic microstructure undergoes transformation to ferrite. Relatively fast rates of cooling favor acicular ferrite formation, but cooling rate must be coordinated with the hardenability potential of the weld composition. A rate of cooling that is too rapid for a given weld composition can cause an undesirable amount of austenite to be retained, or will form too much lower bainite or martensite.

Alloying elements that enter into hardening by the transformation mechanism were gauged to encourage the formation of bainitic microstructures because these have good strength and are not as low in toughness as martensitic microstructures. Selection of alloying elements for these new steels was guided by two main objectives.

One was centered on the start-curve for the bay beneath the nose of the IT curve. This start-curve was to be displaced farther to the right to facilitate bainite formation. Boron and molybdenum were found especially effective for this purpose. As the second objective, the M_s temperature was to be depressed as little as possible by the alloying. Therefore, whenever cooling rate was sufficiently fast to cause austenite to transform to martensite, the transformation would take place athermally starting at a relatively high temperature where cracking propensity would be minimized, and where some degree of auto-tempering would occur to benefit toughness. These two objectives were to be gained by judicious alloying with multiple elements to take advantage of synergism. Elements that formed carbides which resisted dissolution in austenite were favored (e.g., molybdenum and vanadium). A small addition of boron sometimes was included because it gave a disproportionate increase in hardenability without lowering the M_s temperature significantly. Only modest amounts of the elements in Table 9.1 that lowered the M_s temperature were used (e.g., nickel and chromium). Although cobalt and silicon are unique in their influence to raise the M_s temperature, substantial additions of these two elements had to be made with discretion for reasons of cost, or because of their influence on final mechanical properties.

Selection of alloying elements for weld metal compositions that would provide good toughness usually followed guidelines that were different from those found favorable for base metals. However, one recent finding about low-carbon steel weld metals that makes them compatible with the newer base metals is that the weld metal should not be cooled slowly; this is because less-tough microstructural constituents tend to form rather than the acicular ferrite, which now is recognized to play a key role in pro-

Figure 9.46 — Multi-pass weld in C-Mn steel plate 40 mm (about 1½ in.) thick showing positions of individual weld beads and their heat-affected zones lying both in weld metal and in base metal

Weld beads at top and bottom faces on right-hand side of weld zone in this transverse full-section specimen are in the as-deposited condition. As subsequent weld beads were deposited, those regions of their HAZ that exceeded the Ac_1-Ac_3 critical range have a light color on the etched surface. Those regions that were heated to a lower temperature, that is, below the Ac_1 temperature, underwent tempering and these tend to have darker coloration upon being etched. The horizontal streaks in the plate on the right-hand indicate a small amount of segregation of carbon and/or other residual elements. (Nital etchant; courtesy of The Welding Institute, England)

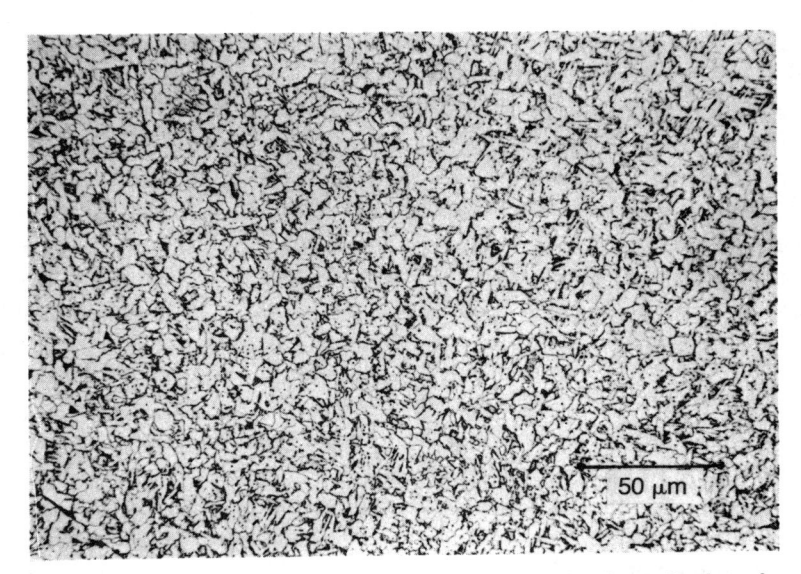

Figure 9.47 — Weld metal in one of the beads deposited early when making the weld in Figure 9.46 showing benefit of multi-pass procedure in refining the microstructure for improved toughness

(Nital etchant; original magnification: 500X; courtesy of The Welding Institute, England)

viding good weld metal toughness. (See Technical Brief 50.) The appropriateness of relatively fast cooling for both base metal HAZ and weld metal of the new, weldable compositions was a help to those concerned with weldment productivity, but the change in metallurgical characteristics required a departure from the earlier philosophy that favored preheating and high weld energy input to secure a relatively slow cooling rate.

As a simple exercise in making a weld intended to have improved toughness, the specimen in Figure 9.46 should be studied closely. This is a complete transverse cross-section that has been polished and etched to reveal individual weld beads and their heat-affected zones. Instead of welding the plate shown in a single pass (as done with SMAW for Figure 9.39), a multi-pass procedure was used. A dozen small weld beads, each deposited with relatively low heat input, fill the double-V-groove joint. With the lower heat input, the width of the HAZ is much narrower than in Figure 9.39. Also, the microstructure of the as-deposited weld metal and the as-welded HAZ are finer (see last weld beads deposited on top and bottom faces at right). Of equal importance for toughness is the refinement and tempering of subsequent weld passes on previously-made passes. Of course, this heat effect involves a gradient of temperatures being applied, which does not produce a uniform microstructure overall. Nevertheless, deposition of each subsequent bead causes the immediately adjacent region of metal to be heated above the Ac_3 critical temperature. The next region is heated to within the Ac_3-Ac_1 critical range, and the more remote regions of this new HAZ are subject to sub-critical annealing, or to tempering. Again, a rapid rate of cooling will occur in these reheated regions because of the relatively low heat input, and this will favor the formation of a fine-grained microstructure.

The benefits of reheating by subsequent weld passes can be easily seen in newly-formed microstructures. Rather than illustrate these one-by-one, as had been done with Figures 9.41 to 9.44, the value of reheating can be gauged by looking at a bead deposited earlier, which now has been reheated above its Ac_3 critical temperature by a subsequent pass and then rapidly cooled. This weld metal initially had a microstructure that likely consisted of intragranular ferrite in the AC or AF forms, and possibly a small amount of ferrite in grain boundaries. Reheating by a subsequent weld pass has produced the microstructure shown in Figure 9.47 consisting of fine polygonal grains of ferrite with a uniform dispersion of minute carbide particles. This microstructure has excellent toughness. Although the entire weld zone of Figure 9.46 does not possess this desirable, tough microstructure, a substantial proportion has been grain-refined, or tempered by subsequent weld passes. Sometimes this multi-pass welding procedure will be carried a step further by depositing an extra bead atop the weld to grain-refine and to temper the beads below. This "temper bead" technique applies excessive reinforcement which possibly may either be ignored or the excess reinforcement can be removed by some non-thermal cutting process thus retaining the benefits from reheating. Despite its heterogeneous microstructure, the multi-bead arc weld in Figure 9.46 with its substantial proportion of fine-grained ferrite (as shown in Figure 9.47) is much tougher than a single pass weld.

The Challenge of Optimizing Welding Procedures

Establishing a satisfactory welding procedure for fabricating a particular weldment is an endeavor that requires surveying and evaluating many aspects before making decisions on how to accommodate influential details. For example, multi-pass welding with low heat input and relatively small weld beads was demonstrated in Figure 9.46 as effective in obtaining a weld with favorable microstructure in a *low-carbon* C-Mn steel. This procedure was effective because it was largely guided by fundamentals covered in preceding chapters of this volume and further improved by findings from recent welding research. The toughness of the weld produced should be adequate for most services. Regrettably, the measures taken in the procedure would not be practicable for the majority of C-Mn steel weldments being fabricated in industry today. The dozen weld passes used to fill the V-grooves would entail excessive costs and fabricating time, and this certainly is not in keeping with the requirement to increase productivity and reduce costs.

If the C-Mn steel plate used for the weldment in Figure 9.46 was higher in carbon content, perhaps about 0.30 percent carbon or above, then the procedure with small beads and low heat input would be risky to carry out because the rapid cooling rate might create a hardened microstructure in the HAZ, which would be susceptible to cracking. This higher carbon steel ordinarily require a procedure that possibly included (1) relatively high heat input, (2) preheating, and (3) low-hydrogen welding conditions. Briefly stated, *compromises* routinely must be made in devising a procedure that will control the metallurgical changes in the steel to ensure fitness-for-service, and yet can be carried out with acceptable costs.

Welding procedures have been promulgated over many years by a number of technical organizations for weldments intended for particular services when fabricated by specific processes. These have been offered as "recommended," or as "prequalified" welding procedures, and they are very instructive and useful documents. Sometimes, codes for construction of equipment and facilities on which public safety is of vital importance will make their welding procedures mandatory. Mention was made in Chapters 1 and 5 of the efforts of AASHTO, API, ASME, AWS, TWI and others in this area. Activity now is underway to make many more "standardized" welding procedures available for a wider variety of steels and welding processes, and to *not link* application of the procedure to weldments for a particular kind of service. These broader documents are being drafted to assist in:

(1) minimizing time required for working document preparation
(2) making clear the requirements in contractual agreements
(3) avoiding oversights in requirements
(4) reducing costs by standardizing all possible aspects of operations
(5) improving consistency of quality in fabrications
(6) providing documentation to satisfy quality assurance requirements, and
(7) adding to a reference library of successful procedures.

As a cautionary note, it must be pointed out that these procedures, although painstakingly-developed, cannot safeguard every detail that might arise. "Fine tuning" the provisions of a procedure often makes use of inner-most knowledge of the steel to be fabricated, and complete familiarity with the welding process to be employed. Sometimes the kind welding equipment to be used and its operating characteristics enter into the final shaping of the procedure. When the pressure of "cost-cutting" is added to procedure planning, optimization becomes a full blown challenge. Here is where "welding metallurgy" serves because it can make resolute appraisal to avoid pitfalls from unwise proposed alterations or procedure deviations to reduce costs. It also assists in technical examination of virtually every detail of a projected weldment. Below is a short-list of principal lineaments in the technology of welding which can be reviewed to gain readiness for study ahead:

Base Metal

(1) Type of iron or steel, including actual analyses of required alloying elements, as well as significant minor and residual elements.
(2) Thickness of base metal, especially variation in dimension.
(3) When steel is a wrought product, the method of working and degree of directionality or anisotropy.
(4) When cast steel, moulding method and size of casting.
(5) Microstructure of base metal as-received before welding.
(6) Mechanical properties of base metal as-received.
(7) Surface character, finish, coating, films, and presence of foreign material.
(8) Cleanliness of steel, internal nonmetallic inclusions.
(9) Soundness of steel, freedom from pipe, laminations, cracks and segregation.

Weld Metal

(1) Autogenous weld metal.
(2) Filler metal, forms and classifications.
(3) Chemical analyses, including minor and residual elements, and especially hydrogen content.

Welding Process and Procedure

(1) Power supply type.
(2) Shielding provided by process, fluxes, gases, other adjuvant materials.
(3) Position of workpieces and weld joint.
(4) Cooling rate of weld as affected by heat input and heat-sink capacity of fixturing.

Weldment Property Testing

(1) Tensile strength.
(2) Hardness and its distribution.
(3) Ductility as judged from tensile elongation and reduction of area.
(4) Toughness as judged from impact tests and fracture mechanics evaluation.
(5) Fatigue crack growth rate determination.
(6) Stress-corrosion cracking threshold.

Nondestructive Examination

(1) Defect detection and measurement.
(2) Fracture mechanics assessment of fitness-for-service.
(3) Fatigue life assessment for prediction of finite service.

In chapters that follow in Volume II, the above list is extended by many additional topics that play technological roles in the modern welding operations. Again, as these topics are singled out, their technology will be examined from a metallurgical viewpoint, and the aim will remain as expressed in the beginning of Volume 1; namely, to advance welding by providing information that can lead to optimization of welding whenever applied. While the present Volume has concentrated on information of a fundamental nature to assist in this regard, Volume II will survey the overall technology of weldment manufacturing. The complex interplay that often takes place between welding and seemingly unrelated manufacturing operations will be singled out whenever it appears, and the nub of the interaction in each case will be spotlighted so that its essential details can be cached as reference for operations to come.

To paraphrase examples of interplay in real-life industrial operations, first consider a case involving hot rolled, pickled finish, low-carbon steel strip that is being fabricated into a high-volume auto component using automated flash welding (FW). On one occasion, the routine of production was suddenly upset by output of components with sub-standard welds. The steel looked bright and clean, and appeared no different from tons of strip consumed earlier. The cause of the welding problem was traced to a change in the acid pickling solution for mill scale removal at the steel processing plant, which increased the electrical resistance of the surface on the strip steel. A substantial change in FW machine settings could have accommodated this particular material, but mass production usually calls for consistency in all qualities important to manufacturing operations - especially those pertinent to welding.

As another example of interplay between remotely associated operations, imagine the consternation among welding personnel as they welded thin-wall cylinders made of dead-soft, annealed and pickled, low-carbon steel using the submerged arc welding process (SAW). As the ends of the cylinders were subjected to a limited cold forming operation, many small cracks developed in the weld HAZ! The cause of this problem originated in the kind of annealing treatment applied to obtain the lowest possible level of hardness in the thin steel (i.e., a spheroidization treatment). The ruinous interplay came into being when the welding travel speed of SAW was increased to approximately 35 mm/s (~80 ipm) to improve productivity. Rapid heating and cooling in the HAZ produced micro-size areas of martensite surrounding partially solution-treated spheroidal carbides in the microstructure. These brittle micro-areas could not tolerate even limited plastic deformation without cracking. Further examples of far-reaching dependence by welding on remote factors could be cited ad infinitum.

As the content of Volume II unfolds, it will be seen that many branches of science and engineering contribute the knowledge for explaining the technology of welding operations, for rationalizing problems, and for devising corrective measures. The fundamental welding metallurgy presented in this Volume I should provide a helpful interdisciplinary base for the more pragmatic study ahead.

SUGGESTED READING

Atlas of Time-Temperature Diagrams for Irons and Steels, edited by G. Vander Voort, ASM International, Materials Park, OH, 1991, ISBN: 0-87170-415-3.

Atlas of Continuous Cooling Transformation Diagrams for Engineering Steels, by M. Atkins, Ibid, 1980.

"A Study of Weld Interface Phenomena in a Low Alloy Steel," by W. F. Savage, E. F. Nippes and E. S. Szekeres, *AWS Welding Journal*, vol 55, no. 9, pp 260s-268s, September 1976.

"The Influence of Cooling Rate and Composition on Weld Metal Microstructures in a C-Mn and a HSLA Steel," by A. G. Glover, et al, *AWS Welding Journal*, vol 56, no. 9, pp 267s-280s, September 1977.

"Ferrite Transformation Characteristics and CCT Diagrams in Weld Metals," by B. G. Kenny, et al, *Metal Construction - The Welding Institute Journal*, vol 17, no. 6, pp 374R-381R, June 1987.

"Microstructural Development and Toughness of C-Mn and C-Mn-Ni Weld Metals," by P. Harrison and R. Farrar, *Metal Construction - The Welding Institute Journal*, vol 19, no 7, pp 392R-399R, July 1987; and vol 19, no. 8, pp 447R-450R, August 1987.

"Microstructure and Impact Toughness of C-Mn Weld Metals," by L. E. Svenson and B. Gretoft, *AWS Welding Journal*, vol 69, no. 12, pp 454-461s, December 1990.

Metallurgy of Arc Welding, I. K. Pokhodnya, et al, published by Riecansky Science publishing Company, ASM International, Materials Park, OH, 1991.

List of Appendixes

APP. NO.	TITLE	PAGE
I	ACRONYMS FOR ORGANIZATIONS	893
II	STANDARD TERMINOLOGY REFERENCES	895
III	SYMBOLS USED IN TEXT AND TABLES	896
IV	ALPHABETS USED IN SCIENTIFIC NOTATION	896
V	ABBREVIATIONS & ALPHABETICAL DESIGNATIONS	897
VI	SI BASE UNITS	901
VII	STRESS CONVERSION: MPa ⇔ ksi	902
VIII	TEMPERATURE CONVERSION: CELSIUS ⇔ FAHRENHEIT	905
IX	THE ELEMENTS: SYMBOLS & PROPERTIES	907
X	THE ELEMENTS: ELECTRON CONFIGURATIONS	913
XI	ELECTRONIC DATABASES & COMPUTER PROGRAMS	919

Appendixes

APPENDIX I
ACRONYMS FOR ORGANIZATIONS

AA	—	Aluminum Association, Washington, DC
AASHTO	—	American Association of State Highway and Transportation Officials, Washington, DC
ABS	—	American Bureau of Shipping, Paramus, NJ
ACI	—	Alloy Casting Institute (Div. of SFSA), Des Plaines, IL
AFS	—	American Foundrymen's Society, Des Plaines, IL
AGA	—	American Gas Association, Arlington, VA
AIME	—	American Institute of Mining, Metallurgical and Petroleum Engineering, New York, NY
AISC	—	American Institute of Steel Construction, Inc., Chicago, IL
AISE	—	Association of Iron and Steel Engineers, Pittsburgh, PA
AISI	—	American Iron and Steel Institute, Washington, DC
AMMRC	—	Army Metals and Materials Research Center, U.S. Army, Watervliet, NY
ANS	—	American Nuclear Society, La Grange Park, IL
ANSI	—	American National Standards Institute, New York, NY
API	—	American Petroleum Institute, Washington, DC
AREA	—	American Railway Engineering Association, Washington, DC
ASM	—	ASM International (formerly, American Society for Metals), Materials Park, OH
ASME	—	American Society of Mechanical Engineers, New York, NY
ASNT	—	American Society for Nondestructive Testing, Columbus, OH
ASTM	—	American Society for Testing and Materials, Philadelphia, PA
ASQC	—	American Society for Quality Control, Milwaukee, WI
AWI	—	American Welding Institute, Knoxville, TN
AWS	—	American Welding Society, Miami, FL
AWWA	—	American Water Works Association, Denver, CO
BMI	—	Battelle Memorial Institute, Columbus, OH
BSI	—	British Standards Institution, London, England
CGA	—	Compressed Gas Association, Arlington, VA
CSA	—	Canadian Standards Association, Rexdale, Ontario, Canada
CWB	—	Canadian Welding Bureau, Toronto, Ontario, Canada

DIN	—	Deutsches Institute fuer Normung, Berlin, Germany
DOD	—	Department of Defense, U.S., Washington, DC
DTRC	—	David Taylor Research Center, U.S. Navy, Annapolis, MD
EPRI	—	Electric Power Research Institute, Palo Alto, CA
EWF	—	European Welding Federation, Dusseldorf, Germany
EWI	—	Edison Welding Institute, Columbus, OH
ITTRI	—	Illinois Institute of Technology Research Institute, Chicago, IL
IIW	—	International Institute of Welding, American Council, Miami, FL
IM	—	Institute of Materials, London, England
ISO	—	International Organization for Standardization, Geneve, Switzerland
JIS	—	Japanese Standards Association, Minato, Tokyo, Japan
MCIC	—	Metals and Ceramics Information Center, Columbus, OH
MIAC	—	Metals Information Analysis Center, West Lafayette, IN
MPC	—	Metal Properties Council, New York, NY
NACE	—	National Association of Corrosion Engineers, Houston, TX
NBBPVI	—	Nat'l Board of Boiler & Pressure Vessel Inspectors, Columbus, OH
NEMA	—	National Electrical Manufacturers' Association, Washington, DC
NF	—	Association Francaise de Normalisation, Paris La Défense, France
NIST	—	National Institute of Standards & Technology (formerly NBS, National Bureau of Standards), Washington, DC
NJC	—	Navy Joining Center, Columbus, OH
NRL	—	Naval Research Laboratory, Washington, DC
NWSA	—	National Welding Supply Association, Philadelphia, PA
PFI	—	Pipe Fabricating Institute, Pittsburgh, PA
RWMA	—	Resistance Welder Manufacturers' Association, Warrendale, PA
SAE	—	Society of Automotive Engineers, Warrendale, PA
SFSA	—	Steel Founder's Society of America, Des Plaines, IL
SME	—	Society of Manufacturing Engineers, Dearborn, MI
SPFA	—	Steel Plate Fabricators Association, Geneva, IL
SRI	—	Southern Research Institute, Birmingham, AL
SRI	—	SRI International, Menlo Park, CA
STI	—	Steel Tank Institute, North Brook, IL
STI	—	Steel Tube Institute, Cleveland, OH
SWRI	—	Southwest Research Institute, San Antonio, TX
TMS	—	The Minerals, Metals, & Materials Society, Warrendale, PA
TWI	—	The Welding Institute, Abington, Cambridge, England
WIC	—	Welding Institute of Canada, Rexdale, Ontario, Canada
WRC	—	Welding Research Council, New York, NY
WSTI	—	Welded Steel Tube Institute, Cleveland, OH

APPENDIX II
STANDARD TERMINOLOGY REFERENCES

Standard terminology has been used in this first Volume whenever possible for description of metallurgical and welding technology. A number of widely available publications that assist this practice to avoid errors in communications are listed below:

Thesaurus of Metallurgical Terms, 10th Edition, ASM International, Materials Park, OH

Mechanical Testing, Methods Of, Standard Definitions of Terms Relating to, ASTM E 6, Book of ASTM Standards, American Society for Testing and Materials, Philadelphia, PA

Metallography, Standard Definitions of Terms Relating to, ASTM E 7 Ibid

Heat Treatment of Metals, Standard Definitions of Terms Relating to, ASTM E 44, Ibid

Fracture Testing, Standard Terminology Relating to, ASTM E 616, Ibid

Corrosion and Corrosion Testing, Standard Definition of Terms Relating to, ASTM G 15, Ibid

Metric Practice Guide for the Welding Industry, AWS A1.1, American Welding Society, Miami, FL

Welding Symbols Charts, AWS 2.1, Ibid

Standard Symbols for Welding, Brazing and Nondestructive Examination, AWS 2.4, Ibid

Standard Welding Terms and Definitions, AWS 3.0, Ibid

APPENDIX III
SYMBOLS USED IN THIS TEXT

Symbol	Meaning
÷	divided by
×	multiplied by; diameters; magnification
·	multiplied by
+	plus; in addition to; positive ion charge
−	minus; negative ion charge
±	maximum deviation; plus-or-minus
=	equals
≡	identically equal to
≠	not equal to
>	greater than
≥	greater than or equal to
≫	much greater than
<	less than
≤	less than or equal to
≪	much less than
~	approximately
≈	approximately equal to
/	per; divided by
√	square root of
%	percent
$	dollars (American)
⇔	either direction

APPENDIX IV
ALPHABETICAL CHARACTERS USED IN SCIENTIFIC NOTATION

Name	Greek	English	Latin
alpha	Α, α	A, a	a (ä)
beta	Β, β	B, b	b
gamma	Γ, γ	G, g	g, n
delta	Δ, δ, ∂	D, d	d
epsilon	Ε, ε	E, e	e
zeta	Ζ, ζ	Z, z	zd, z
eta	Η, η	E, e	ë, ē
theta	Θ, θ, ϑ	Th, th	th
iota	Ι, ι	I, i	i
kappa	Κ, κ	K, k	k
lambda	Λ, λ	L, l	l
mu	Μ, μ	M, m	m
nu	Ν, ν	N, n	n
xi	Ξ, ξ	X, x	x
omicron	Ο, ο	O, o	o
pi	Π, π	P, p	p
rho	Ρ, ρ	R, r	r, hr
sigma	Σ, σ, ς	S, s	s
tau	Τ, τ	T, t	t
upsilon	Υ, υ	U, u	u (ü)
phi	Φ, ϕ	Ph, ph	ph
chi	Χ, χ	Ch, ch	ch
psi	Ψ, ψ	Ps, ps	ps
omega	Ω, ω	W, w	ō

APPENDIX V
ABBREVIATIONS AND ALPHABETICALLY DESIGNATED TERMS

For the many *welding process* designations standardized by AWS see Figures 6.1 and 6.2, and refer to ANSI/AWS A3.0.

Abbreviation or Designation	Meaning in This Text
A	ampere
Å	angstrom unit (10^{-8} cm)
AC	air cool
a	depth of surface crack, half-length of through-thickness crack
ac	alternating current
at.%	atomic per cent
at.wt	atomic weight
atm	atmosphere
avg	average
Ac_1	temperature at which austenite begins to form on heating
Ac_3	temperature at which transformation of ferrite to austenite is completed on heating
Ac_{cm}	in hypereutectoid steel, temperature at which cementite completes its solution in austenite
A_e	equilibrium transformation temperatures in iron-carbon alloys
AI	artificial intelligence
AES	Auger electron spectroscopy
amu	atomic mass unit, ½ the mass of carbon atom of mass 12
b	barn (neutron capture cross-section)
bcc	body-centered cubic
bct	body-centered tetragonal
BOF	basic oxygen steelmaking furnace

Abbreviation or Designation	Meaning in This Text
bp	boiling point temperature at atmospheric pressure
Btu	British thermal unit
B	bainite microstructure
BI	basicity index for fluxes
CAD	computer-aided design
CIM	computer-integrated manufacturing
C_v	Charpy V-notch impact specimen, or test value
°C	Celsius, temperature degree
cal	calorie
cp	chemically pure
cfh	cubic feet per hour
coef	coefficient
cu ft	cubic feet
cu in.	cubic inch
CCD	charge-coupled device
CCT	continuous cooling transformation
CE	carbon equivalent
CRT	cathode ray tube
CTOD	crack tip opening displacement
dc	direct current
deg	degree (angle)
diam	diameter
D_f	fractal dimension, surface roughness characterization
DT	dynamic tear test

Abbreviation or Designation	Meaning in This Text
DWT-NDT	drop-weight test, nil ductility transition
DWT	drop-weight tear test
ε	elastic strain
$\dot{\varepsilon}$	rate of strain
E	elastic modulus (Young's), also a crystallographic plane
eV	electron volt
emf	electromotive force
esu	electrostatic unit
El	elongation
Eq.	equation
EPFM	elastic-plastic fracture mechanics
ESCA	photo-electron spectroscopy
η	viscosity
ft	foot
fcc	face-centered cubic
fpm	feet per minute
fps	feet per second
ft-lb	foot-pound (force)
ΔF	free energy of a metal-oxide system
°F	Fahrenheit, temperature degree
FAX	facsimile transmission
FC	furnace cool
g	gram
gal	gallon
G	Griffith's strain energy release by fracture, also temperature gradient
G_c	Griffith's critical strain energy for unstable fracture
h	hour, also hexagonal
$\hbar$	orbital angular momentum unit
hp	horse power
hcp	hexagonal close-packed
HB	hardness, Brinell
HK	hardness, Knoop
HV	hardness, Vickers
HAZ	heat-affected zone
HF	high frequency (ac)
HRB	hardness, Rockwell, "B" scale
HRC	hardness, Rockwell, "C" scale
Hz	hertz, cycles per second
HSLA	high-strength, low-alloy (steel)
in.	inch
ipm	inches per minute
I	current (electric)
ID	inside diameter
IR	infrared
IT	isothermal transformation
IACS	International Annealed Copper Standard
J	joule
J_{Ic}	parameter for critical threshold of crack extension
J-integral	expression of stress-strain field at crack front
k	thermal transmittance rate
ksi	thousand pounds per square inch
K	crack tip stress intensity factor
$\overset{\circ}{K}$	rate of elastic stress intensity at crack tip
°K	Kelvin, temperature degree
K_c	critical stress intensity under plane-stress conditions
K_I	stress intensity in opening or tensile fracture mode
K_{Ic}	critical value of K_I for unstable crack extension under plane-strain conditions
K_{Ic}	theoretical stress intensity factor for notch-effect
K_{Id}	stress intensity in tensile fracture mode under dynamic loading

Appendixes **899**

Abbreviation or Designation	Meaning in This Text
$K_{I(t)}$	stress intensity in tensile fracture mode when loading rate is between static and dynamic
K_t	theoretical stress-intensity factor for notch-effect
K_{ISCC}	threshold stress intensity for stress corrosion cracking
λ	wave length
lb	pound (avoirdupois weight, 16 ounces)
log	logarithm to base 10
LMC	liquid metal cracking
LSM	laser scanning microscopy
LEFM	linear eleastic fracture mechanics
mp	melting point
max.	maximum
min.	minimum
min	minute
mph	miles per hour
M_f	martensite, transformation finish
M_s	martensite, transformation start
no.	number
N	newton, also Wall neutrality number of SAW flux
μ_N	nuclear magneton
NAA	neutron activation analysis
NDE	nondestructive examination
N_f	fatigue life
NDT	nil ductility transition (drop-weight test)
Nd-YAG	neodynium doped glass rod of yttrium-aluminum-garnet laser
oz	ounce
OB	basicity index for slag from optical spectral radiometric laser reflectance
OD	outside diameter
OH	open hearth steelmaking furnace
OQ	oil quench

Abbreviation or Designation	Meaning in This Text
Ω	ohm (unit of electrical resistance)
ρ	resistivity, electrical, also specific resistance, ohm-cm
pp	pages
ppm	parts per million
psi	pounds per square inch
pH	hydrogen ion concentration
P	pearlite microstructure
PC	personal computer
π	pi, 3.141592
Pa	pascal
PA	positron annihilation technique
PDDF	heat flux or power density distribution function
PTFE	polytetrafluoroethylene
PVC	polyvinyl chloride
μ	Poisson's ratio
$\varnothing$	weld pool trailing acuity
qt	quart
rf	radio frequency (noun)
r-f	radio frequency (adjective)
rms	root mean square
rpm	revolutions per minute
R	resistance (electrical), also solidification front velocity
RT	room temperature (25 °C, or about 75 °F)
RST	rapid solidification technology
°R	Rankine, temperature degree
s	second (time)
sfm	surface feet per minute
sp gr	specific gravity
sp ht	specific heat
SANS	small-angle, neutron-scattering examination
SCC	stress corrosion cracking
SCR	silicon controlled rectifier
SEM	scanning electron microscopy
SHE	standard hydrogen electrode

Abbreviation or Designation	Meaning in This Text
SLAM	scanning laser acoustic microscopy
S	nominal engineering stress
σ	true stress, also electrical conductivity
ε	strain
t	time
T	temperature
T_g	glass transition temperature
T_m	melting temperature
TEM	transmission electron microscopy
TMCP	thermo-mechanical control processing
Θ	angle of twist or rotation per unit length
UNS	Unified Numbering System (ASTM-SAE)
UTS	ultimate tensile strength
UNMZ	unmixed melted zone
v	velocity
V	volt, also volume
VCR	video casette recorder
Vol	volume number of a publication
η	viscosity of a liquid
W	watt, also width of LEFM specimen
WQ	water quench
λ	wave length
yr	year
YP	yield point
YS	yield strength
YAG	yttrium-aluminum-garnet glass rod in laser
z	through-the-thickness direction

APPENDIX VI
SI BASE UNITS

Seven Basic Units of the SI (Metric) System

Quantity	Name of Unit	SI Symbol
amount of substance	mole	mol
electric current	ampere	A
length	meter	m
luminous intensity	camdela	cd
mass	kilogram	kg
thermodynamic temperature	kelvin	K
time	second	s

SI Prefixes; Their Symbols, Names and Factors

SI Symbol	Prefix Name	Exponential Expression	Multiplication Factor
E	exa	10^{18}	1 000 000 000 000 000 000
P	peta	10^{15}	1 000 000 000 000 000
T	tera	10^{12}	1 000 000 000 000
G	giga	10^{9}	1 000 000 000
M	mega	10^{6}	1 000 000
K	kilo	10^{3}	1 000
h	hecto	10^{2}	100
da	deka	10	10
d	deci	10^{-1}	0.1
c	centi	10^{-2}	0.01
m	milli	10^{-3}	0.001
µ	micro	10^{-6}	0.000 001
n	nano	10^{-9}	0.000 000 001
p	pico	10^{-12}	0.000 000 000 001
f	femto	10^{-15}	0.000 000 000 000 001
a	atto	10^{-18}	0.000 000 000 000 000 001

APPENDIX VII
STRESS CONVERSION: SI UNITS ⇔ U.S. CUSTOMARY

MPa (megapascals) ⇔ ksi (thousand pounds per sq. in.)

The special name *pascal* (symbol Pa) was adopted in 1971 for the derived unit N/m². SI prefixes can be applied to the name, pascal, and the stresses commonly handled in calculations concerning steel make MPa (megapascal) the most convenient unit of stress. One ksi is equivalent to 6.894759 MPa.

To use the table below, enter the central (bold-face) column with the stress to be converted. If converting ksi, read the equivalent stress in MPa in the column on the right. If converting MPa, read the equivalent stress in ksi in the column on the left.

ksi		MPa	ksi		MPa	ksi		MPa	ksi		MPa
	0		7.25	**50**	344.7	14.50	**100**	689.5	21.76	**150**	1034
0.15	**1**	6.89	7.40	**51**	351.6	14.65	**101**	696.4	21.90	**151**	1041
0.29	**2**	13.79	7.54	**52**	358.5	14.79	**102**	703.3	22.05	**152**	1048
0.44	**3**	20.68	7.69	**53**	365.4	14.94	**103**	710.2	22.19	**153**	1054
0.58	**4**	27.57	7.83	**54**	372.3	15.08	**104**	717.1	22.34	**154**	1062
0.73	**5**	34.47	7.98	**55**	379.2	15.23	**105**	724.0	22.48	**155**	1069
0.87	**6**	41.37	8.12	**56**	386.1	15.37	**106**	730.8	22.63	**156**	1076
1.02	**7**	48.26	8.27	**57**	393.0	15.52	**107**	737.7	22.77	**157**	1082
1.16	**8**	55.16	8.41	**58**	399.9	15.66	**108**	744.6	22.92	**158**	1089
1.31	**9**	62.05	8.56	**59**	406.8	15.81	**109**	751.5	23.06	**159**	1096
1.45	**10**	68.95	8.70	**60**	413.7	15.95	**110**	758.4	23.21	**160**	1103
1.60	**11**	75.84	8.85	**61**	420.6	16.10	**111**	765.3	23.35	**161**	1110
1.74	**12**	82.74	8.99	**62**	427.5	16.24	**112**	772.2	23.50	**162**	1117
1.89	**13**	89.63	9.14	**63**	434.4	16.39	**113**	779.1	23.64	**163**	1124
2.03	**14**	96.53	9.28	**64**	441.3	16.53	**114**	786.0	23.79	**164**	1131
2.18	**15**	103.40	9.43	**65**	448.2	16.68	**115**	792.9	23.93	**165**	1138
2.32	**16**	110.30	9.57	**66**	455.1	16.82	**116**	799.8	24.08	**166**	1145
2.47	**17**	117.20	9.72	**67**	462.0	16.97	**117**	806.7	24.22	**167**	1151
2.61	**18**	124.10	9.86	**68**	468.8	17.11	**118**	813.6	24.37	**168**	1158
2.76	**19**	131.00	10.01	**69**	475.7	17.26	**119**	820.5	24.51	**169**	1165
2.90	**20**	137.90	10.15	**70**	482.6	17.40	**120**	827.4	24.66	**170**	1172
3.05	**21**	144.80	10.30	**71**	489.5	17.55	**121**	834.3	24.80	**171**	1179
3.19	**22**	151.70	10.44	**72**	496.4	17.69	**122**	481.2	24.95	**172**	1186
3.34	**23**	158.60	10.59	**73**	503.3	17.84	**123**	848.1	25.09	**173**	1193
3.48	**24**	165.50	10.73	**74**	510.2	17.98	**124**	855.0	25.24	**174**	1200
3.63	**25**	172.40	10.88	**75**	517.1	18.13	**125**	861.8	25.38	**175**	1207
3.77	**26**	179.30	11.02	**76**	524.0	18.27	**126**	868.7	25.53	**176**	1213
3.92	**27**	186.20	11.17	**77**	530.9	18.42	**127**	875.6	25.67	**177**	1220
4.06	**28**	193.10	11.31	**78**	537.8	18.56	**128**	882.5	25.82	**178**	1227
4.21	**29**	199.90	11.46	**79**	544.7	18.71	**129**	889.4	25.96	**179**	1234
4.35	**30**	206.80	11.60	**80**	551.6	18.85	**130**	896.3	26.11	**180**	1241
4.50	**31**	213.70	11.75	**81**	558.5	19.00	**131**	903.2	26.25	**181**	1248
4.64	**32**	220.60	11.89	**82**	565.4	19.14	**132**	910.1	26.40	**182**	1255
4.79	**33**	227.50	12.04	**83**	572.3	19.29	**133**	917.0	26.54	**183**	1262
4.93	**34**	234.40	12.18	**84**	579.2	19.44	**134**	923.9	26.69	**184**	1269
5.08	**35**	241.30	12.33	**85**	586.1	19.58	**135**	930.8	26.83	**185**	1276
5.22	**36**	248.20	12.47	**86**	593.0	19.73	**136**	937.7	26.98	**186**	1282
5.37	**37**	255.10	12.62	**87**	599.8	19.87	**137**	944.6	27.12	**187**	1289
5.51	**38**	262.00	12.76	**88**	606.7	20.02	**138**	951.5	27.27	**188**	1296
5.66	**39**	268.90	12.91	**89**	613.6	20.16	**139**	958.4	27.41	**189**	1303
5.80	**40**	275.80	13.05	**90**	620.5	20.31	**140**	965.3	27.56	**190**	1310
5.95	**41**	282.70	13.20	**91**	627.4	20.45	**141**	972.2	27.70	**191**	1317
6.09	**42**	289.60	13.34	**92**	634.3	20.60	**142**	979.1	27.85	**192**	1324
6.24	**43**	296.50	13.49	**93**	641.2	20.74	**143**	986.0	27.99	**193**	1331
6.38	**44**	303.40	13.63	**94**	648.1	20.89	**144**	992.9	28.14	**194**	1338
6.53	**45**	310.30	13.78	**95**	655.0	21.03	**145**	999.7	28.28	**195**	1344
6.67	**46**	317.20	13.92	**96**	661.9	21.18	**146**	1007.0	28.43	**196**	1351
6.82	**47**	324.10	14.07	**97**	668.8	21.32	**147**	1014.0	28.57	**197**	1358
6.96	**48**	331.00	14.21	**98**	675.7	21.47	**148**	1020.0	28.72	**198**	1365
7.11	**49**	337.80	14.36	**99**	682.6	21.61	**149**	1027.0	28.86	**199**	1372
7.25	**50**	344.71	14.50	**100**	689.5	21.76	**150**	1034.0	29.01	**200**	1379

APPENDIX VII
(continued)

ksi		MPa	ksi		MPa	ksi		MPa	ksi		MPa
29.01	200	1379	36.26	250	1724	43.51	300	2068	50.76	350	2413
29.15	201	1386	36.40	251	1731	43.66	301	2075	50.91	351	2420
29.30	202	1393	36.55	252	1737	43.80	302	2082	51.05	352	2427
29.44	203	1400	36.69	253	1744	43.95	303	2089	51.20	353	2434
29.59	204	1407	36.84	254	1751	44.09	304	2096	51.34	354	2441
29.73	205	1413	36.98	255	1758	44.24	305	2103	51.49	355	2448
29.88	206	1420	37.13	256	1765	44.38	306	2110	51.63	356	2455
30.02	207	1427	37.27	257	1772	44.53	307	2117	51.78	357	2461
30.17	208	1434	37.42	258	1779	44.67	308	2124	51.92	358	2468
30.31	209	1441	37.56	259	1786	44.82	309	2130	52.07	359	2475
30.46	210	1448	37.71	260	1793	44.96	310	2137	52.21	360	2482
30.60	211	1455	37.85	261	1800	45.11	311	2144	52.36	361	2489
30.75	212	1462	38.00	262	1806	45.25	312	2151	52.50	362	2496
30.89	213	1469	38.14	263	1813	45.40	313	2158	52.65	363	2503
31.04	214	1475	38.29	264	1820	45.54	314	2165	52.79	364	2510
31.18	215	1482	38.43	265	1827	45.69	315	2172	52.94	365	2517
31.33	216	1489	38.58	266	1834	45.83	316	2179	53.08	366	2523
31.47	217	1496	38.73	267	1841	45.98	317	2186	53.23	367	2530
31.62	218	1503	38.87	268	1848	46.12	318	2193	53.37	368	2537
31.76	219	1510	39.02	269	1855	46.27	319	2199	53.52	369	2544
31.91	220	1517	39.16	270	1862	46.41	320	2206	53.66	370	2551
32.05	221	1524	39.31	271	1868	46.56	321	2213	53.81	371	2558
32.20	222	1531	39.45	272	1875	46.70	322	2220	53.95	372	2565
32.34	223	1538	39.60	273	1882	46.85	323	2227	54.10	373	2572
32.49	224	1544	39.74	274	1889	46.99	324	2234	54.24	374	2579
32.63	225	1551	39.89	275	1896	47.14	325	2241	54.39	375	2585
32.78	226	1558	40.03	276	1903	47.28	326	2248	54.53	376	2592
32.92	227	1565	40.18	277	1910	47.43	327	2255	54.68	377	2599
33.07	228	1572	40.32	278	1917	47.57	328	2261	54.82	378	2606
33.21	229	1579	40.47	279	1924	47.72	329	2268	54.97	379	2613
33.36	230	1586	40.61	280	1931	47.86	330	2275	55.11	380	2620
33.50	231	1593	40.76	281	1937	48.01	331	2282	55.26	381	2627
33.65	232	1600	40.90	282	1944	48.15	332	2289	55.40	382	2634
33.79	233	1606	41.05	283	1951	48.30	333	2296	55.55	383	2641
33.94	234	1613	41.19	284	1958	48.44	334	2303	55.69	384	2648
34.08	235	1620	41.34	285	1965	48.59	335	2310	55.84	385	2654
34.23	236	1627	41.48	286	1972	48.73	336	2317	55.98	386	2661
34.37	237	1634	41.63	287	1979	48.88	337	2324	56.13	387	2668
34.52	238	1641	41.77	288	1986	49.02	338	2330	56.27	388	2675
34.66	239	1648	41.92	289	1993	49.17	339	2337	56.42	389	2682
34.81	240	1655	42.06	290	1999	49.31	340	2344	56.56	390	2689
34.95	241	1662	42.21	291	2006	49.46	341	2351	56.71	391	2696
35.10	242	1669	42.35	292	2013	49.60	342	2358	56.85	392	2703
35.24	243	1675	42.50	293	2020	49.75	343	2365	57.00	393	2710
35.39	244	1682	42.64	294	2027	49.89	344	2372	57.14	394	2717
35.53	245	1689	42.79	295	2034	50.04	345	2379	57.29	395	2723
35.68	246	1696	42.93	296	2041	50.18	346	2386	57.43	396	2730
35.82	247	1703	43.07	297	2048	50.33	347	2392	57.58	397	2737
35.97	248	1710	43.22	298	2055	50.47	348	2399	57.72	398	2744
36.11	249	1717	43.37	299	2062	50.62	349	2406	57.87	399	2751
36.26	250	1724	43.51	300	2068	50.76	350	2413	58.02	400	2758

APPENDIX VII
(continued)

ksi		MPa	ksi		MPa
58.02	**400**	2758	145.04	**1000**	6895
60.92	**420**	2896	174.05	**1200**	8274
63.82	**440**	3034	203.05	**1400**	9652
66.71	**460**	3172	232.06	**1600**	11032
69.62	**480**	3309	261.07	**1800**	12410
72.52	**500**	3447	290.08	**2000**	13789
75.42	**520**	3585	319.08	**2200**	15168
78.32	**540**	3723	348.09	**2400**	16547
81.22	**560**	3861	377.10	**2600**	17926
84.12	**580**	3999	406.11	**2800**	19305
87.02	**600**	4137	435.12	**3000**	20684
89.92	**620**	4275	464.12	**3200**	22063
92.82	**640**	4413	493.13	**3400**	23442
95.72	**660**	4550	522.14	**3600**	24821
98.63	**680**	4688	551.15	**3800**	26200
101.53	**700**	4826	580.16	**4000**	27579
104.43	**720**	4964	609.16	**4200**	28958
107.33	**740**	5102	638.17	**4400**	30337
110.23	**760**	5240	667.18	**4600**	31716
113.13	**780**	5378	696.19	**4800**	33094
116.03	**800**	5516	725.19	**5000**	34474
118.93	**820**	5654	754.20	**5200**	35852
121.83	**840**	5792	783.21	**5400**	37231
124.73	**860**	5929	812.22	**5600**	38610
127.63	**880**	6067	841.22	**5800**	39989
130.54	**900**	6205	870.23	**6000**	41368
133.44	**920**	6343	899.24	**6200**	42747
136.34	**940**	6481	928.25	**6400**	44126
139.24	**960**	6619	957.26	**6600**	45505
142.14	**980**	6757	986.26	**6800**	46884
145.04	**1000**	6895	1015.27	**7000**	48263

APPENDIX VIII
TEMPERATURE CONVERSION: SI UNITS ⟷ U.S. CUSTOMARY

Degrees Celsius ⇔ Degrees Fahrenheit

Despite international acceptance and usage of the Kelvin and the Celsius temperature scales, the Fahrenheit scale continues to be widely used in the USA, hence the conversion table for °C to °F herewith. The term *centigrade* should not be used for temperature because in metric countries this means one hundredth part of the unit of plane angle; i.e., the grade. Between the temperatures of melting ice and boiling water, there are 180° on the Fahrenheit and Rankine scales as compared with 100° on the Celsius and Kelvin scales. The ratio of these numbers is 9:5, therefore, the following equations apply:

$$°C = 5/9 \, (°F - 32), \text{ and, } °F = 9/5 \, °C + 32°$$

To use the tables below, enter the central (bold-face) columns with the number to be converted. If converting Fahrenheit degrees, read the Celsius equivalent in column headed "C" to the left. If converting Celsius degrees, read the Fahrenheit equivalent in the column headed "F" to the right.

C		F	C		F	C		F	C		F	C		F
-273	**-459**		-40.0	**-40**	-40.0	24.4	**76**	168.8	199	**390**	734
-268	**-450**		-34.0	**-30**	-22.0	25.6	**78**	172.4	204	**400**	752
-262	**-440**		-29.0	**-20**	-4.0	26.7	**80**	176.0	210	**410**	770
-257	**-430**		-23.0	**-10**	14.0	27.8	**82**	179.6	216	**420**	788
-251	**-420**		-17.8	**0**	32.0	28.9	**84**	183.2	221	**430**	806
-246	**-410**		-16.7	**2**	35.6	30.0	**86**	186.8	227	**440**	824
-240	**-400**		-15.6	**4**	39.2	31.1	**88**	190.4	232	**450**	842
-234	**-390**		-14.4	**6**	42.8	32.2	**90**	194.0	238	**460**	860
-229	**-380**		-13.3	**8**	46.4	33.3	**92**	197.6	243	**470**	878
-223	**-370**		-12.2	**10**	50.0	34.4	**94**	201.2	249	**480**	896
-218	**-360**		-11.1	**12**	53.6	35.6	**96**	204.8	254	**490**	914
-212	**-350**		-10.0	**14**	57.2	36.7	**98**	208.4	260	**500**	932
-207	**-340**		-8.9	**16**	60.8	37.8	**100**	212.0	266	**510**	950
-201	**-330**		-7.8	**18**	64.4	43.0	**110**	230.0	271	**520**	968
-196	**-320**		-6.7	**20**	68.0	49.0	**120**	248.0	277	**530**	986
-190	**-310**		-5.6	**22**	71.6	54.0	**130**	266.0	282	**540**	1004
-184	**-300**		-4.4	**24**	75.2	60.0	**140**	284.0	288	**550**	1022
-179	**-290**		-3.3	**26**	78.8	66.0	**150**	302.0	293	**560**	1040
-173	**-280**		-2.2	**28**	82.4	71.0	**160**	320.0	299	**570**	1058
-168	**-270**	-454	-1.1	**30**	86.0	77.0	**170**	338.0	304	**580**	1076
-162	**-260**	-436	0.0	**32**	89.6	82.0	**180**	356.0	310	**590**	1094
-157	**-250**	-418	1.1	**34**	93.2	88.0	**190**	374.0	316	**600**	1112
-151	**-240**	-400	2.2	**36**	96.8	93.0	**200**	392.0	321	**610**	1130
-146	**-230**	-382	3.3	**38**	100.4	99.0	**210**	410.0	327	**620**	1148
-140	**-220**	-364	4.4	**40**	104.0	100.0	**212**	414.0	332	**630**	1166
-134	**-210**	-346	5.6	**42**	107.6	104.0	**220**	428.0	338	**640**	1184
-129	**-200**	-328	6.7	**44**	111.2	110.0	**230**	446.0	343	**650**	1202
-123	**-190**	-310	7.8	**46**	114.8	116.0	**240**	464.0	349	**660**	1220
-118	**-180**	-292	8.9	**48**	118.4	121.0	**250**	482.0	354	**670**	1238
-112	**-170**	-274	10.0	**50**	122.0	127.0	**260**	500.0	360	**680**	1256
-107	**-160**	-256	11.1	**52**	125.6	132.0	**270**	518.0	366	**690**	1274
-101	**-150**	-238	12.2	**54**	129.2	138.0	**280**	536.0	371	**700**	1292
-96	**-140**	-220	13.3	**56**	132.8	143.0	**290**	554.0	377	**710**	1310
-90	**-130**	-202	14.4	**58**	136.4	149.0	**300**	572.0	382	**720**	1328
-84	**-120**	-184	15.6	**60**	140.0	154.0	**310**	590.0	388	**730**	1346
-79	**-110**	-166	16.7	**62**	143.6	160.0	**320**	608.0	393	**740**	1364
-73	**-100**	-148	17.8	**64**	147.2	166.0	**330**	626.0	399	**750**	1382
-68	**-90**	-130	18.9	**66**	150.8	171.0	**340**	644.0	404	**760**	1400
-62	**-80**	-112	20.0	**68**	154.4	177.0	**350**	662.0	410	**770**	1418
-57	**-70**	-94	21.1	**70**	158.0	182.0	**360**	680.0	416	**780**	1436
-51	**-60**	-76	22.2	**72**	161.6	186.0	**370**	698.0	421	**790**	1454
-46	**-50**	-58	23.3	**74**	165.2	193.0	**380**	716.0	427	**800**	1472

APPENDIX VIII
(continued)

C	F	C	F	C	F	C	F	C	F
432	**810**	1490	738	**1360**	2480	1043	**1910**	3470	1349	**2460**	4460
438	**820**	1508	743	**1370**	2498	1049	**1920**	3488	1354	**2470**	4478
443	**830**	1526	749	**1380**	2516	1054	**1930**	3506	1360	**2480**	4496
449	**840**	1544	754	**1390**	2534	1060	**1940**	3524	1366	**2490**	4514
454	**850**	1562	760	**1400**	2552	1066	**1950**	3542	1371	**2500**	4532
460	**860**	1580	766	**1410**	2570	1071	**1960**	3560	1377	**2510**	4550
466	**870**	1598	771	**1420**	2588	1077	**1970**	3578	1382	**2520**	4568
471	**880**	1616	777	**1430**	2606	1082	**1980**	3596	1388	**2530**	4586
477	**890**	1634	782	**1440**	2624	1088	**1990**	3614	1393	**2540**	4604
482	**900**	1652	788	**1450**	2642	1093	**2000**	3632	1399	**2550**	4622
488	**910**	1670	793	**1460**	2660	1099	**2010**	3650	1404	**2560**	4640
493	**920**	1688	799	**1470**	2678	1104	**2020**	3668	1410	**2570**	4658
499	**930**	1706	804	**1480**	2696	1110	**2030**	3686	1416	**2580**	4676
504	**940**	1724	810	**1490**	2714	1116	**2040**	3704	1421	**2590**	4694
510	**950**	1742	816	**1500**	2732	1121	**2050**	3722	1427	**2600**	4712
516	**960**	1760	821	**1510**	2750	1127	**2060**	3740	1432	**2610**	4730
521	**970**	1778	827	**1520**	2768	1132	**2070**	3758	1438	**2620**	4748
527	**980**	1796	832	**1530**	2786	1138	**2080**	3776	1443	**2630**	4766
532	**990**	1814	838	**1540**	2804	1143	**2090**	3794	1449	**2640**	4784
538	**1000**	1832	843	**1550**	2822	1149	**2100**	3812	1454	**2650**	4802
543	**1010**	1850	849	**1560**	2840	1154	**2110**	3830	1460	**2660**	4820
549	**1020**	1868	854	**1570**	2858	1160	**2120**	3848	1466	**2670**	4838
554	**1030**	1886	860	**1580**	2876	1166	**2130**	3866	1471	**2680**	4856
560	**1040**	1904	866	**1590**	2894	1171	**2140**	3884	1477	**2690**	4874
566	**1050**	1922	871	**1600**	2912	1177	**2150**	3902	1482	**2700**	4892
571	**1060**	1940	877	**1610**	2930	1182	**2160**	3920	1488	**2710**	4910
577	**1070**	1958	882	**1620**	2948	1188	**2170**	3938	1493	**2720**	4928
582	**1080**	1976	888	**1630**	2966	1193	**2180**	3956	1499	**2730**	4946
588	**1090**	1994	893	**1640**	2984	1199	**2190**	3974	1504	**2740**	4964
593	**1100**	2012	899	**1650**	3002	1204	**2200**	3992	1510	**2750**	4982
599	**1110**	2030	904	**1660**	3020	1210	**2210**	4010	1516	**2760**	5000
604	**1120**	2048	910	**1670**	3038	1216	**2220**	4028	1521	**2770**	5018
610	**1130**	2066	916	**1680**	3056	1221	**2230**	4046	1527	**2780**	5036
616	**1140**	2084	921	**1690**	3074	1227	**2240**	4064	1532	**2790**	5054
621	**1150**	2102	927	**1700**	3092	1232	**2250**	4082	1538	**2800**	5072
627	**1160**	2120	932	**1710**	3110	1238	**2260**	4100	1543	**2810**	5090
632	**1170**	2138	938	**1720**	3128	1243	**2270**	4118	1549	**2820**	5108
638	**1180**	2156	943	**1730**	3146	1249	**2280**	4136	1554	**2830**	5126
643	**1190**	2174	949	**1740**	3164	1254	**2290**	4154	1560	**2840**	5144
649	**1200**	2192	954	**1750**	3182	1260	**2300**	4172	1566	**2850**	5162
654	**1210**	2210	960	**1760**	3200	1266	**2310**	4190	1571	**2860**	5180
660	**1220**	2228	966	**1770**	3218	1271	**2320**	4208	1577	**2870**	5198
666	**1230**	2246	971	**1780**	3236	1277	**2330**	4226	1582	**2880**	5216
671	**1240**	2264	977	**1790**	3254	1282	**2340**	4244	1588	**2890**	5234
677	**1250**	2282	982	**1800**	3272	1288	**2350**	4262	1593	**2900**	5252
682	**1260**	2300	988	**1810**	3290	1293	**2360**	4280	1599	**2910**	5270
688	**1270**	2318	993	**1820**	3308	1299	**2370**	4298	1604	**2920**	5288
693	**1280**	2336	999	**1830**	3326	1304	**2380**	4316	1610	**2930**	5306
699	**1290**	2354	1004	**1840**	3344	1310	**2390**	4334	1616	**2940**	5324
704	**1300**	2372	1010	**1850**	3362	1316	**2400**	4352	1621	**2950**	5342
710	**1310**	2390	1016	**1860**	3380	1321	**2410**	4370	1627	**2960**	5360
716	**1320**	2408	1021	**1870**	3398	1327	**2420**	4388	1632	**2970**	5378
721	**1330**	2426	1027	**1880**	3416	1332	**2430**	4406	1638	**2980**	5396
727	**1340**	2444	1032	**1890**	3434	1338	**2440**	4424	1643	**2990**	5414
732	**1350**	2462	1038	**1900**	3452	1343	**2450**	4442	1649	**3000**	5432

APPENDIX IX
THE ELEMENTS; SYMBOLS AND PROPERTIES

Element	Symbol	Atomic No.	Atomic Weight (a)	Melting Point, °C	Boiling Point, °C	Thermal Conductivity (b)	Electrical Resistivity (c)	Coefficient of Thermal Expansion (d)	Aggregate State at RT (g)	Crystalline Form If Solid at RT (h)
Actinium	Ac	89	227.03	1050	3200	...	...	...	S	...
Aluminum	Al	13	26.98	660	2450	0.53	2.74	23.6	S	fcc
Americium	Am	95	243	994	2607	...	...	...	S	...
Antimony	Sb	51	121.75	631	1750	0.045	41.3	9.0	S	r
Argon	Ar	18	39.95	-189	-186	0.4×10^{-4}	...	...	G	...
Arsenic	As	33	74.92	817	613 (e)	...	33.3	4.7	S	r +
Astatine	At	85	211	302 (f)	337	...	...	...	S	...
Barium	Ba	56	137.34	714	1640	0.04	39.0	...	S	bcc +
Berkelium	Bk	97	249	...	...	...	...	...	S	...
Beryllium	Be	4	9.01	1277	2970	0.35	3.25	11.6	S	hcp +
Bismuth	Bi	83	208.98	271	1560	0.02	116.0	13.3	S	r
Boron	B	5	10.81	2300 (f)	2550 (f)	0.06	1.8×10^{12}	8.3 (f)	S	r
Bromine	Br	35	79.91	-7	58	...	...	...	L	...
Cadmium	Cd	48	112.4	321	765	0.22	6.83	29.8	S	hcp
Calcium	Ca	20	40.08	838	1484	0.30	3.91	22.3	S	fcc +
Californium	Cf	98	252	...	...	...	...	...	S	...
Carbon	C	6	12.01	3727 (e)	4830 (f)	0.36	1375	0.6/4.3	S	h +
Cerium	Ce	58	140.12	804	3470	0.03	81.0	8.0	S	fcc +
Cesium	Cs	55	132.92	29	669	0.08	20.0	97.0	S	bcc
Chlorine	Cl	17	35.45	-101	-35	0.2×10^{-4}	...	...	G	...
Chromium	Cr	24	51.99	1875	2665	0.22	12.9	6.2	S	bcc +
Cobalt	Co	27	58.93	1495	2870	0.24	6.2	13.8	S	hcp +

APPENDIX IX
(continued)

Element	Symbol	Atomic No.	Atomic Weight (a)	Melting Point, °C	Boiling Point, °C	Thermal Conductivity (b)	Electrical Resistivity (c)	Coefficient of Thermal Expansion (d)	Aggregate State at RT (g)	Crystalline Form If Solid at RT (h)
Copper	Cu	29	63.54	1083	2595	0.96	1.7	16.5	S	fcc
Curium	Cm	96	245	1340	...	...	...	...	S	
Dysprosium	Dy	66	162.5	1409	2335	0.02	90.0	9.0	S	hcp
Einsteinium	Es	99	254	...	...	...	...	...	S	
Erbium	Er	68	167.26	1522	2510	0.03	81.0	9.0	S	hcp
Europium	Eu	63	151.96	826	1597	0.03	89.0	26.0	S	bcc
Fermium	Fm	100	257	...	...	...	...	...	S	
Fluorine	F	9	18.99	-220	1188	0.7×10^{-4}	...	...	G	
Francium	Fr	87	233	27 (f)	...	...	...	...	S	
Gadolinium	Gd	64	157.25	1312	3233	0.02	140.5	4.0	S	hcp
Gallium	Ga	31	69.72	30	2237	0.08	14.8	18.0	S	o
Germanium	Ge	32	72.59	937	2830	0.15	450.0	5.8	S	dc
Gold	Au	79	196.98	1064	3080	0.76	2.4	14.2	S	fcc
Hafnium	Hf	72	178.49	2222	4602	0.05	30.6	519.0	S	hcp +
Hahnium	Ha	105	260	...	...	...	...	...	S	
Helium	He	2	4	-270	-269	3.6×10^{-4}	...	...	G	
Holmium	Ho	67	164.93	1461	2720	0.04	87.0	11.2	S	hcp
Hydrogen	H	1	1.01	-259	-253	3.4×10^{-4}	...	...	G	
Indium	In	49	114.82	156	2080	0.06	8.37	33.0	S	fct
Iodine	I	53	126.9	114	183	1.04×10^{-3}	1.3×10^{15}	93.0	S	o
Iridium	Ir	77	192.2	2410	4130	0.35	5.3	6.8	S	fcc
Iron	Fe	26	55.85	1538	2870	0.18	9.7	11.8	S	bcc +

APPENDIX IX
(continued)

Element	Symbol	Atomic No.	Atomic Weight (a)	Melting Point, °C	Boiling Point, °C	Thermal Conductivity (u)	Electrical Resistivity (c)	Coefficient of Thermal Expansion (d)	Aggregate State at RT (g)	Crystalline Form If Solid at RT (h)
Krypton	Kr	36	83.8	-157	-152	2.26×10^{-5}	...	...	G	
Lanthanum	La	57	138.91	920	3470	0.33	5.7	5.0	S	h
Lawrencium	Lr	103	260	...	...	...	...	...	S	
Lead	Pb	82	207.19	327	1725	0.08	20.6	29.3	S	fcc
Lithium	Li	3	6.94	181	1330	0.2	8.6	56.0	S	bcc
Lutetium	Lu	71	174.97	1652	3315	0.04	53.0	...	S	hcp
Magnesium	Mg	12	24.31	650	1107	0.37	4.4	27.1	S	hcp
Manganese	Mn	25	54.94	1245	2150	0.02	139.0	22.0	S	c +
Mendelevium	Md	101	258	...	...	...	...	...	S	
Mercury	Hg	80	200.59	-38	357	0.02	98.4	...	L	
Molybdenum	Mo	42	95.94	2610	5560	0.34	5.2	4.9	S	bcc
Neodymium	Nd	60	144.24	1019	3180	0.04	64.0	6.0	S	h +
Neon	Ne	10	20.18	-249	-246	1.1×10^{-4}	...	...	G	
Neptunium	Np	93	237	637	2730	0.02	118.0	...	S	
Nickel	Ni	28	58.71	1453	2732	0.22	6.8	13.3	S	fcc
Niobium	Nb	41	92.91	2468	4927	0.13	20.0	7.3	S	bcc
Nitrogen	N	7	14.01	-210	-196	0.6×10^{-4}	...	...	G	
Nobelium	No	102	259	...	...	...	...	...	S	
Osmium	Os	76	190.2	3045	5027	...	9.5	4.6	S	hcp
Oxygen	O	8	15.99	-219	-183	0.6×10^{-4}	...	...	G	
Palladium	Pd	46	106.4	1552	3980	0.17	10.8	11.8	S	fcc
Phosphorus	P	15	30.97	45	280	0.6×10^{-3}	1×10^{17}	125	S	c +

APPENDIX IX
(continued)

Element	Symbol	Atomic No.	Atomic Weight (a)	Melting Point, °C	Boiling Point, °C	Thermal Conductivity (b)	Electrical Resistivity (c)	Coefficient of Thermal Expansion (d)	Aggregate State at RT (g)	Crystalline Form If Solid at RT (h)
Platinum	Pt	78	195.09	1769	4530	0.17	10.6	8.9	S	fcc
Plutonium	Pu	94	239	640	3235	0.02	141.4	55.0	S	mc +
Polonium	Po	84	210	254	...	...	46.0	...	S	mc
Potassium	K	19	39.1	64	760	0.24	6.2	83.0	S	bcc
Praseodymium	Pr	59	140.91	919	3020	0.03	68.0	4.0	S	h +
Promethium	Pm	61	145	1027 (f)	2460 (f)	0.04	75.0 (f)	...	S	
Protactinium	Pa	91	231	1230 (f)	...	...	...	...	S	
Radium	Ra	88	226.05	700	1140	0.04	...	...	S	
Radon	Rn	86	222	-71	-62 (f)	0.09×10^{-4}	...	...	G	
Rhenium	Re	75	186.2	3180	5900	0.13	19.3	6.7	S	hcp
Rhodium	Rh	45	102.91	1966	4500	0.3	4.5	8.3	S	fcc +
Rubidium	Rb	37	85.47	39	688	0.14	12.5	90.0	S	bcc
Ruthenium	Ru	44	101.07	2500	4900	0.28	7.6	9.1	S	hcp +
Rutherfordium	Rf	104	259	...	...	...	...	...	S	
Samarium	Sm	62	150.35	1072	1630	0.03	88.0	...	S	r +
Scandium	Sc	21	44.96	1539	2830	0.04	61.0	...	S	hcp
Selenium	Se	34	78.96	217	685	7.2×10^{-3}	10^6	37.0	S	h +
Silicon	Si	14	28.09	1410	2680	0.34	4×10^6	2.8/7.3	S	dc
Silver	Ag	47	107.87	961	2210	1.02	1.6	19.7	S	fcc
Sodium	Na	11	22.99	98	892	0.34	4.2	71.0	S	bcc
Strontium	Sr	38	87.62	768	1380	0.08	23	...	S	fcc +
Sulfur	S	16	32.06	119	2445	0.6×10^{-3}	2×10^{23}	64.0	S	o

APPENDIX IX
(continued)

Element	Symbol	Atomic No.	Atomic Weight (a)	Melting Point, °C	Boiling Point, °C	Thermal Conductivity (b)	Electrical Resistivity (c)	Coefficient of Thermal Expansion (d)	Aggregate State at RT (g)	Crystalline Form If Solid at RT (h)
Tantalum	Ta	73	180.95	2996	5425	0.14	13.5	6.5	S	bcc
Technetium	Tc	43	99	2130 (f)	4877	0.12	18.5	7.0	S	hcp
Tellurium	Te	52	127.6	450	990	0.01	2×10^5	16.8	S	h
Terbium	Tb	65	158.92	1356	2530	0.02	111.0	7.0	S	hcp
Thallium	Tl	81	204.37	303	1457	0.11	18.0	28.0	S	hcp +
Thorium	Th	90	232.04	1750	3850	0.13	13.0	12.5	S	fcc
Thulium	Tm	69	168.93	1545	1720 (f)	0.04	79.0	13.0	S	hcp
Tin	Sn	50	118.69	232	2270	0.16	11.0	21.0	S	t
Titanium	Ti	22	47.9	1668	3260	0.13	42.0	8.4	S	hcp +
Tungsten	W	74	183.92	3410	5930	0.4	5.6	4.6	S	bcc +
Uranium	U	92	238.03	1132	3818	0.07	30.0	6.8/14.1	S	o +
Vanadium	V	23	50.94	1900	3400	0.07	25.4	8.3	S	bcc
Xenon	Xe	54	131.3	-112	-108	1.4×10^5	...	...	G	
Ytterbium	Yb	70	173.04	824	1530	0.08	29.0	25.0	S	fcc +
Yttrium	Y	39	88.91	1509	3030	0.04	57.0	...	S	hcp +
Zinc	Zn	30	65.37	420	906	0.28	5.9	39.7	S	hcp
Zirconium	Zr	40	91.22	1852	3580	0.21	40.0	5.8	S	hcp +
(Un-named)		106	263							
(Un-named)		107	258							
(Un-named)		109	266							

APPENDIX IX
(continued)

(a) Atomic weight based upon 12 as the exact atomic weight of carbon -12

(b) Thermal conductivity at about room temperature (20 °C) in cal/sq cm/cm/sec/deg C

(c) Electrical resistivity (p) in microhm−cm

(d) Coefficient of linear thermal expansion at about room temperature (20 °C) in μm/m · °K

(e) Sublimes

(f) Estimated

(g) State of aggregate, if known, at about room temperature (20 °C); G − gas, L − liquid, S − solid

(h) Crystalline form, if solid, at about room temperature (20 °C) as indicated below:

bcc	-	body-centered cubic	hcp	-	hexagonal close-packed
bct	-	body-centered tetragonal	mc	-	monoclinic
fcc	-	face-centered cubic	o	-	orthorhombic
dc	-	diamond cubic	r	-	rhombohedral
c	-	cubic	t	-	tetragonal
h	-	hexagonal	+	-	Other crystalline forms are known through allotropic transformation at certain temperatures, or as determined by method of producing the aggregate

APPENDIX X
THE ELEMENTS; ELECTRON CONFIGURATIONS

Atomic No. (Z)	Element	K(1) 1(2) s(3)	L 2 s	L 2 p	M 3 s	M 3 p	M 3 d	N 4 s
1	H	1						
2	He	2						
3	Li	2	1					
4	Be	2	2					
5	B	2	2	1				
6	C	2	2	2				
7	N	2	2	3				
8	O	2	2	4				
9	F	2	2	5				
10	Ne	2	2	6				
11	Na	2	2	6	1			
12	Mg	2	2	6	2			
13	Al	2	2	6	2	1		
14	Si	2	2	6	2	2		
15	P	2	2	6	2	3		
16	S	2	2	6	2	4		
17	Cl	2	2	6	2	5		
18	Ar	2	2	6	2	6		
19	K	2	2	6	2	6		1

APPENDIX X
(continued)

Atomic No. (Z)	Element	K(1) 1(2) s(3)	L 2 s	L 2 p	M 3 s	M 3 p	M 3 d	N 4 s	N 4 p	N 4 d	N 4 f	O 5 s	O 5 p	O 5 d	O 5 f	P 6 s	P 6 p	P 6 d	P 6 f	Q 7 s	Q 7 p	Q 7 d	Q 7 f
20	Ca	2	2	6	2	6		2															
21	Sc	2	2	6	2	6	1	2															
22	Ti	2	2	6	2	6	2	2															
23	V	2	2	6	2	6	3	2															
24	Cr	2	2	6	2	6	5	1															
25	Mn	2	2	6	2	6	5	2															
26	Fe	2	2	6	2	6	6	2															
27	Co	2	2	6	2	6	7	2															
28	Ni	2	2	6	2	6	8	2															
29	Cu	2	2	6	2	6	10	1															
30	Zn	2	2	6	2	6	10	2															
31	Ga	2	2	6	2	6	10	2	1														
32	Ge	2	2	6	2	6	10	2	2														
33	As	2	2	6	2	6	10	2	3														
34	Se	2	2	6	2	6	10	2	4														
35	Br	2	2	6	2	6	10	2	5														
36	Kr	2	2	6	2	6	10	2	6														
37	Rb	2	2	6	2	6	10	2	6			1											
38	Sr	2	2	6	2	6	10	2	6			2											

APPENDIX X
(continued)

Atomic No. (Z)	Element	K(1) 1(2) s(3)	L 2 s	L 2 p	M 3 s	M 3 p	M 3 d	N 4 s	N 4 p	N 4 d	N 4 f	O 5 s	O 5 p	O 5 d	O 5 f	P 6 s	P 6 p	P 6 d	P 6 f	Q 7 s	Q 7 p	Q 7 d	Q 7 f
39	Y	2	2	6	2	6	10	2	6	1	⋮	2											
40	Zr	2	2	6	2	6	10	2	6	2	⋮	2											
41	Nb	2	2	6	2	6	10	2	6	4	⋮	1											
42	Mo	2	2	6	2	6	10	2	6	5	⋮	1											
43	Tc	2	2	6	2	6	10	2	6	6	⋮	1											
44	Ru	2	2	6	2	6	10	2	6	7	⋮	1											
45	Rh	2	2	6	2	6	10	2	6	8	⋮	1											
46	Pd	2	2	6	2	6	10	2	6	10	⋮												
47	Ag	2	2	6	2	6	10	2	6	10	⋮	1											
48	Cd	2	2	6	2	6	10	2	6	10	⋮	2											
49	In	2	2	6	2	6	10	2	6	10	⋮	2	1										
50	Sn	2	2	6	2	6	10	2	6	10	⋮	2	2										
51	Sb	2	2	6	2	6	10	2	6	10	⋮	2	3										
52	Te	2	2	6	2	6	10	2	6	10	⋮	2	4										
53	I	2	2	6	2	6	10	2	6	10	⋮	2	5										
54	Xe	2	2	6	2	6	10	2	6	10	⋮	2	6										
55	Cs	2	2	6	2	6	10	2	6	10	⋮	2	6		⋮	1							
56	Ba	2	2	6	2	6	10	2	6	10	⋮	2	6		⋮	2							
57	La	2	2	6	2	6	10	2	6	10	⋮	2	6	1	⋮	2							

APPENDIX X
(continued)

Atomic No. (Z)	Element	K(1) 1(2) s(3)	L 2 s	L 2 p	M 3 s	M 3 p	M 3 d	N 4 s	N 4 p	N 4 d	N 4 f	O 5 s	O 5 p	O 5 d	O 5 f	P 6 s	P 6 p	P 6 d	P 6 f	Q 7 s	Q 7 p	Q 7 d	Q 7 f
58	Ce	2	2	6	2	6	10	2	6	10	2	2	6	.	.	2							
59	Pr	2	2	6	2	6	10	2	6	10	3	2	6	.	.	2							
60	Nd	2	2	6	2	6	10	2	6	10	4	2	6	.	.	2							
61	Pm	2	2	6	2	6	10	2	6	10	5	2	6	.	.	2							
62	Sm	2	2	6	2	6	10	2	6	10	6	2	6	.	.	2							
63	Eu	2	2	6	2	6	10	2	6	10	7	2	6	.	.	2							
64	Gd	2	2	6	2	6	10	2	6	10	7	2	6	1	.	2							
65	Tb	2	2	6	2	6	10	2	6	10	9	2	6	.	.	2							
66	Dy	2	2	6	2	6	10	2	6	10	10	2	6	.	.	2							
67	Ho	2	2	6	2	6	10	2	6	10	11	2	6	.	.	2							
68	Er	2	2	6	2	6	10	2	6	10	12	2	6	.	.	2							
69	Tm	2	2	6	2	6	10	2	6	10	13	2	6	.	.	2							
70	Yb	2	2	6	2	6	10	2	6	10	14	2	6	.	.	2							
71	Lu	2	2	6	2	6	10	2	6	10	14	2	6	1	.	2							
72	Hf	2	2	6	2	6	10	2	6	10	14	2	6	2	.	2							
73	Ta	2	2	6	2	6	10	2	6	10	14	2	6	3	.	2							
74	W	2	2	6	2	6	10	2	6	10	14	2	6	4	.	2							
75	Re	2	2	6	2	6	10	2	6	10	14	2	6	5	.	2							
76	Os	2	2	6	2	6	10	2	6	10	14	2	6	6	.	2							

APPENDIX X
(continued)

Atomic No. (Z)	Element	K(1) 1(2) s(3)	L 2 s	L 2 p	M 3 s	M 3 p	M 3 d	N 4 s	N 4 p	N 4 d	N 4 f	O 5 s	O 5 p	O 5 d	O 5 f	P 6 s	P 6 p	P 6 d	P 6 f	Q 7 s	Q 7 p	Q 7 d	Q 7 f
77	Ir	2	2	6	2	6	10	2	6	10	14	2	6	7	:	2							
78	Pt	2	2	6	2	6	10	2	6	10	14	2	6	9	:	1							
79	Au	2	2	6	2	6	10	2	6	10	14	2	6	10	:	1							
80	Hg	2	2	6	2	6	10	2	6	10	14	2	6	10	:	2							
81	Tl	2	2	6	2	6	10	2	6	10	14	2	6	10	:	2	1						
82	Pb	2	2	6	2	6	10	2	6	10	14	2	6	10	:	2	2						
83	Bi	2	2	6	2	6	10	2	6	10	14	2	6	10	:	2	3						
84	Po	2	2	6	2	6	10	2	6	10	14	2	6	10	:	2	4						
85	At	2	2	6	2	6	10	2	6	10	14	2	6	10	:	2	5						
86	Rn	2	2	6	2	6	10	2	6	10	14	2	6	10	:	2	6						
87	Fr	2	2	6	2	6	10	2	6	10	14	2	6	10	:	2	6	:	:	1			
88	Ra	2	2	6	2	6	10	2	6	10	14	2	6	10	:	2	6	:	:	2			
89	Ac	2	2	6	2	6	10	2	6	10	14	2	6	10	:	2	6	1	:	2			
90	Th	2	2	6	2	6	10	2	6	10	14	2	6	10	2	2	6	2	:	2			
91	Pa	2	2	6	2	6	10	2	6	10	14	2	6	10	3	2	6	1	:	2			
92	U	2	2	6	2	6	10	2	6	10	14	2	6	10	4	2	6	1	:	2			
93	Np	2	2	6	2	6	10	2	6	10	14	2	6	10	6	2	6	1	:	2			
94	Pu	2	2	6	2	6	10	2	6	10	14	2	6	10	6	2	6	:	:	2			
95	Am	2	2	6	2	6	10	2	6	10	14	2	6	10	7	2	6	:	:	2			

APPENDIX X
(continued)

Atomic No. (Z)	Element	K(1) 1(2) s(3)	L 2 s	L 2 p	M 3 s	M 3 p	M 3 d	N 4 s	N 4 p	N 4 d	N 4 f	O 5 s	O 5 p	O 5 d	O 5 f	P 6 s	P 6 p	P 6 d	P 6 f	Q 7 s	Q 7 p	Q 7 d	Q 7 f
96	Cm	2	2	6	2	6	10	2	6	10	14	2	6	10	7	2	6	1	..	2			
97	Bk	2	2	6	2	6	10	2	6	10	14	2	6	10	9	2	6		..	2			
98	Cf	2	2	6	2	6	10	2	6	10	14	2	6	10	10	2	6		..	2			
99	Es	2	2	6	2	6	10	2	6	10	14	2	6	10	11	2	6		..	2			
100	Fm	2	2	6	2	6	10	2	6	10	14	2	6	10	12	2	6		..	2			
101	Md	2	2	6	2	6	10	2	6	10	14	2	6	10	13	2	6		..	2			
102	No	2	2	6	2	6	10	2	6	10	14	2	6	10	14	2	6		..	2			
103	Lr	2	2	6	2	6	10	2	6	10	14	2	6	10	14	2	6	1		2			
104	Rf	2	2	6	2	6	10	2	6	10	14	2	6	10	14	2	6	2		2			
105	Ha																						
106	Un-named																						
107	Un-named																						
108	Un-named																						
109	Un-named																						
Maximum total number of electrons that can be held on main orbital shell		2	8		18			32				32											

(1) Alphabetical designations commonly used for main orbital electron shells
(2) Numerical designations used for main orbital electron shells
(3) Subshell alphabetical designations

APPENDIX XI
ELECTRONIC DATABASES & COMPUTER PROGRAMS

Resources listed below were current at time this Volume was published, but rapid developments may have eliminated some and new versions of others may have been issued; nevertheless, these examples illustrate the broad scope of technical information pertinent to welding that can be quickly accessed selectively. Attention must be given to electronic operating systems employed to ensure compatibility between the chosen resource and the computer (i.e., pc (IBM), UNIX, APPLE, etc.).

DATABASES AND INFORMATION SERVICES

These examples are stored on a mainframe platform by the originating organization, and selected information can be retrieved by various means of communication in a number of forms as arranged with the provider.

WELDASEARCH, TWI

More than 110 000 citations from, worldwide literature dating back to 1967 on welding, brazing, and soldering; including virtually every topic pertaining to welded design, metallurgy, fracture mechanics, fatigue of metal components, corrosion, automation, quality control, patents, standards, etc., are stored on a mainframe at Abington, England. Most non-English material has between translated into English, and every entry has been abstracted in English by TWI staff using keywords from the IIW *International Welding Thesaurus.* Online searching can be accomplished through Orbit Search Service under the file name "WELD." Selected information can be obtained from TWI as magnetic tape, diskettes, hard copy, abstracts, reference lists, or special forms as requested.

METADEX, ASM

This database dates back to 1966, and presently holds more than 500 000 citations from literature published in 49 countries in 24 languages. This resource is stored in the USA at Metals Park, OH, and is available as a "Metals Information" service offered jointly by ASM and MI. Online searching can be accomplished through Orbit Search Service where the file name is "METADEX." However, access to this resource also can be gained through services located in France, Germany and Italy. The systems holds most of the articles in the originating language, but the majority of the resource is in English. An abstract has been entered for each citation keyworded according to ASM's *Thesaurus of Metallurgical Terms,* which presently is in the 10th Edition, 1992. All aspects of metals and alloys are covered in this resource; including material designations, composition limits, metallurgy, mechanical; properties, physical characteristics, corrosion, testing, quality control, etc. A wide variety of services are available both online and offline to provide selected information in a form that suits the needs of the requestor.

Mat. DB, ASM

This database is a collection of highly organized properties data for engineered materials; including metals, plastics, composites and ceramics. After incoming information from international literature is reviewed, the data are loaded in a three-component system, identified as (1) Mat, DB, (2) Mat. DB Data, and (3) EnPlot. The systems is designed for PC (IBM) access, and both digital data and stored graphics can be transmitted for terminal display or printing. Diskettes are provided which contain files on specific groups of materials, and the files have been presorted into folders for convenient review.

APPENDIX XI
(continued)

Metals and Ceramics Information Center, BMI

This is a bibliographic database operated by Battelle's Columbus Laboratories for the DOD in the USA. Since its origin in 1955, several chartered information centers of DOD have been merged into this single center to cover both metals and ceramics. The MCIC collects, evaluates and disseminates timely information on the characteristics and utilization of advanced metals and ceramics to all sectors of government and industry.

MCIC also prepares summaries, state-of-the-art reports, and handbooks. The BMI staff provides technical advice and assistance. This center is the largest of its kind in the world.

COMPUTER PROGRAMS

Below are examples of computer programs that assist in many of the phases of welding procedure planning, process selection, design calculations, qualification requirements, code compliance testing, etc. A typical software package consists of a diskette and an instruction manual for program installation and operation. New information and changes in requirements are usually accommodated by an issue of an updated version of software, or the originating organization may also offer online connection for searching current requirements, and for interactive consultation.

WELDING DATABASES & SOFTWARE, TWI

A series of software packages, which are available also through EWI, are listed below to illustrate some of the TWI programs:

PROGRAM TITLE	OPERATIONAL FUNCTION
WELDARCQUEST	- for arc welding procedures
WELDCONQUEST	- for weld filler metals
WELDSPEC	- for storing and retrieving welding procedures
WELDSPEC Utilities	- for management of additional graphics
MAGDATA	- for guidelines on welding mild steel by GMAW
PREHEAT	- for welding steels without HAZ hydrogen cracking
WELDFRACQUEST	- for fracture test results
WELDFATQUEST	- for fatigue test results
PARENT/MATERIAL LOG	- for base metal properties
UNITCALC	- for conversion of welding engineering units
WELDCOST	- for cost of arc welded fabrications
WELDERQUAL	- for storage of welder qualification records
WELDVOL	- for calculation of weld sizes and filler metal needs

WELDBEST 1.32, EWI

This computer program is user-friendly to plan welding procedures for the GMAW and the FCAW processes in conformance with AWS D1.1, *Structural Welding Code-Steel*. Procedures for welds made by the SMAW, SAW and GTAW processes also can be planned from the guidance included in the program.

APPENDIX XI
(continued)

WELDSELECTOR, AWI

Information is provided in this program to plan complete procedures for welding approximately 900 steels as classified by ASTM. Guidelines are included for filler metal selection for the SMAW, GMAW and FCAW processes.

WELDSYMPLE, AWI

This program generates the proper symbol to be used on drawings for indicating specific welds to be made according to ANSI/AWS A2.4, *Standard Symbols for Welding, Brazing and Nondestructive Examination*.

CORRAL D1 WPS/PQR, AWI

This database program facilitates storing, searching and retrieving welding procedure specifications and procedure qualification records as would be carried out when conforming to AWS D1.1. A graphics package to assist in drawing welds to meet requirements and to record details is also included. The program provides an online code referencing system to selected portions of the ANSI/AWS D1.1 code.

TURBO-IX, C-Spec

This program was produced by C-Spec, Pleasant Hill, CA. It assists in review of welding procedure specifications to determine compliance with ASME Section IX, *Boiler and Pressure Vessel Code*. A database incorporated in the software includes (1) text of Section IX, (2) 1400 ASME/ASTM base metals, (3) 1100 AWS/SFA filler metals, (4) 1000 cross-referenced interpretations, (5) qualified thickness ranges, (6) specified mechanical properties, and (7) required tests. In addition to updates, technical support is provided online.

WELDING ENGINEERS HELPER, AWS

Software can be provided to cover welding procedure specification and procedure qualification for AWS Codes D1.1, D1.2, D1.3, D1.5 and D9.1. Included in the programs are standard weld joint design, document revision security, automatic welder qualification renewal, etc.

* * * * * * * * *

In addition to software for use in desktop or laptop computers, as illustrated by the above examples, a variety of notebook or handheld computers are available that have a fixed internal, foundation operating system. One example of a very small, pocketsize welding data computer is POCKETWELDER II available from AWS.

The above information on databases, software and computers is provided to illustrate the kind of resources that have become available. No warranty or guarantee as to suitability or reliability of the items listed is made or implied by their inclusion in this table.

About the Index:

Welding Metallurgy, Volume 1, was written to cover as much fundamental information as possible concerning the welding of carbon and alloy steels. The necessity for brevity and conciseness, however, precludes the inclusion of everything pertinent to the subject. Therefore, considerable attention was given to the use of key words throughout the text. Key words are derived from standardized terms, as well as from the currently used verbiage in research and in industry for describing important topics. This feature of the book should facilitate the finding of specific information by the reader; also, when a particular subject could be given only brief mention because of space limitations, the key words employed can assist in locating other references that provide further coverage. The following index, for the most part, consists of key words employed throughout the text in the discussion of significant topics. This index also recognizes, as much as possible, the idiom and jargon which still persists in the welding community. Such terms are included solely for the purpose of enhancing the usefulness of this book as a reference to busy welding personnel in industry.

Index

A

Abbreviations
    Table V 897
Absorption of gases 374, 386, 708, 710
Acetylene 465, 582
Acicular ferrite 860, 885
Acid steelmaking 368
Acoustic microscopy 122
Acronyms for organizations
    Table I 893
Actinides 40
Adams' peak temperature equation 667
Adaptive control of welding 5
Agglomerated SAW fluxes 749
Aggregates of atoms 41
Aerospace Material Specifications (AMS) 426
Aircraft quality steel 431
Air carbon arc cutting (CAC-A) 471, 639
AISI system for generic sheet steels 423
AISI/SAE steel composition specifications 418
    carbon and alloy steel sheet 421
Alkali metals 40
Alkaline earth metals 40
Allotropy in crystalline structure 46, 75, 103
Alloying
    AISI-SAE alloy steel specifications 419
    alloy steels defined 411
    element transfer during SAW 758
    element transfer during SMAW 729
    elements in iron 295
    immiscibility in liquid state 296
Alnico magnet alloy 271
Alpha iron 75, 321, 820
Alpha particles 27, 37
Alphabetical characters in text
    Table IV 896
Alternating current
    for arc processes 455, 470, 481, 490
    balanced sine wave form 455
Alumina as solidification nuclei 375, 875

Aluminum
    alloying element in iron 348
    coating on steel 409, 530
    deoxidizing agent for steel 375
    grain refining additive 376
    weld metal usage 875
American Iron and Steel Institute (AISI) 418
American Society of Mechanical Engineers
    (ASME) 426
    Boiler and Pressure Vessel Code 427
    Pressure Piping Code 428
American Petroleum Institute (API) 424
Amer. Soc. for Testing & Materials (ASTM) 420
    book of annual ASTM standards 422
    grain size standards of ASTM 98
American Welding Society (AWS) 429
Amorphous metal 47
Analyzers for chemical elements
    atom probe microanalyzer 118
    Auger electron spectroscope (AES) 120
    electron-probe microanalyzer 117
    ion-probe microanalyzer 118
    photo-electron spectroscopy (ESCA) 121
Angstrom unit (Å, 10^{-8} cm) 18
Angular momentum of electrons 22
Anisotropy in metal properties 135
Annealing 408
Annihilation energy 32
Anode 453
Arc, electric, for cutting 637
    air carbon arc cutting (CAC-A) 471, 639
    gas metal arc (GMAC) 638
    gas tungsten arc (GTAC) 638
    oxygen (AOC) 639
    plasma arc (PAC) 640
Arc, electric, for welding (AW) 448, 467
    alternating current (ac) 455, 470
    atomic hydrogen (AHW) 471, 518
    bare metal electrode (BMAW) 455
    carbon electrode (CAW) 471, 519
    constant arc voltage 469
    constriction of arc 452, 511, 641

covered metal electrode (SMAW) 429, 472
direct current (dc)....................468, 470
drooping arc voltage 469
efficiency of energy transfer 687
electrical metal-explosion welding 541
electrogas process (EGW) 547, 780
fall space......................................453
flux-cored metal electrode (FCAW) 501, 738
heat input equations 458
initiation of arc........................455, 480
magnetically impelled arc 515
metal transfer through arc 456, 474, 489
742, 773
oxide removal by DCEP in GMAW.......454, 490
pinch-effect 456, 492, 495, 497
power equation 458
power sources 467, 492, 495, 497
reverse polarity (DCEP).............................454
stabilizing agents 725
straight polarity (DCEN)....................453
stud welding process (SW).................477
submerged arc process (SAW) 505
tungsten electrode (GTAW)......................481
work function, anode and cathode 453, 485
Arc, electric, for spraying....................630
Arcing during electron beam welding.........555, 557
Argon
argon-oxygen decarburization (AOD)..........396
shielding for welding...................482, 490, 500
512, 575
Atomic hydrogen arc welding (AHW).........471, 518
Atmospheres for shielding 717, 739, 767
781
Atom 18
construction....................................32
diameter................................38, 311
mass units (amu)........................31, 35
nucleus (nucleon).............................26
number (Z) 33
valency....................................35, 315
Auger electron spectroscopy (AES)...................120
Austenite
continuous cooling transformation 848
crystallographic structure 819
elements favoring austenite................319, 327
isothermal transformation behavior 824, 827
836
grain size 376, 835
retained austenite 838, 865
Austenitic manganese steel 413
Autogeneous fusion welding 787

Automation of welding..5

B

Backing for welding by SAW.......................511
Bainite 831, 845
formation during cont. cooling 848, 850
isothermal transformation 828
Banding in steel microstructure........................879
Barn (b), neutron fluence unit 30
Base metal
HAZ from fusion welding 801
HAZ from solid-state welding.....................802
Basic steelmaking
determined by furnace refractories..............368
oxygen processes (BOF) 380
Basicity index for SAW fluxes...........368, 506, 753
computed BI from flux components.............754
optical basicity assessment 756
Beach marks on fatigue fracture......................178
Bessemer steelmaking..369
Beta iron 327
Beta rays 37
Binary alloy phase diagram.........................299, 304
Biological corrosion 281
Blast furnace 362
Blistering from hydrogen 289
Blowholes in weld...8
Blue brittleness..240
Body-centered cubic crystals 62
Bohr's electron concept..23
Boiling point 49
of metals 266
Bollman specimen for TEM...............................111
Boron
alloying element in iron.................................354
neutron capture cross-section 251
use in steel weld metal 875
Bravais crystal lattices.............................51, 55
Boyle's law for gases 50
Brazing 620
atmospheres 622
diffusion controlled......................448, 623
filler metals...623
fluxes..621
master chart of processes 449
surface preparation........................622
Breaking strength of metals 143
Bright annealing 408
Brinell hardness test (HB) 150

British Standard 5762 for CTOD 206
Brittle fracture 126, 157, 159, 163, 168
    loading rate effect 171
    stress axiality effect 166
    temperature effect 164
    testing procedures 190, 197
Brittleness in metals
    blue brittleness 240
    hot (red) shortness 341
    hydrogen induced 289
    phosphorus induced 338
Buildup (see surfacing) 628
Burger's vectors 90
Burning of steel
    overheating 809
    oxygen cutting (OC) 634, 637
Buttering (see surfacing) 628

C

C-curve for isothermal transformation 828
Calcium
    deoxidizer in steel 721
    desulfurizing agent in steelmaking 388
    treatment of sulfide inclusions 118
Capacitor discharge welding (CDW) 477
    percussion welding (PEW) 514
    stud arc welding (SW) 477
Capped steel 373
Carbides in steel 335
Carbon, alloying element in steel 329
Carbon arc
    cutting processes (CAC) (CAC-A) 638, 639
    shielding from air 722
    welding process (CAW) 722
Carbon dioxide
    carbon dioxide laser 463, 568
    evolution from molten steel 710
    shielding for welding 718, 769
Carbon equivalency formulas 857
Carbon steels 410
    AISI-SAE steel specifications 418
    classification of categories 361
Carbon/oxygen reaction in molten steel 710
Carburization
    by air carbon arc cutting (CAC-A) 639
    by oxyfuel welding flame (OFW) 582
Cast iron 366
    cutting by oxygen (OC) processes 635
    types; gray, malleable, white 366

    welding behavior 366
Cathode 450
Cathode ray tube (CRT) 25, 123
Cathodic etching 454, 490
Caustic embrittlement of steel 292
Cellular grain formation 792, 794
    sub-structure in grains 70
    weld metal microstructure 794
Cellulosic cover'g, SMAW electrodes 476, 727
    732
Cementite, a microstructural phase 329, 806
    cementite eutectic 334
Ceria in GTAW electrodes 482
Cerium, use in steels 388
Charles' law for gases 49
Charpy impact test 156, 221
    C_v values for metals at low temperature 221
    improving the value of the C_v test 227
    transitions in C_v test parameters 222
Chemical flux cutting (FOC) 636
Chemical heat generation 465
Chemical properties of metals 134, 275
Chevron fracture pattern 162
Chisel test of resistance spot welds 523
Chlorine, shielding gas additive 767, 771
Chromium, alloy in steel 345
Cladding (also see surfacing) 632
Cleanliness of steel
    internal 436
    surface 410, 535, 890
Cleavage fracture 157, 160
Cleaning oxidized surface by DCEP arc 454, 490
Closed-loop control in welding 5
Coatings on steel
    fusion welding effects 410
    resistance welding effects 526
Cobalt, alloy in steel 319, 355, 837
Coextrusion welding process (CEW) 591
Coherent precipitates 324
Cohesive strength of metals 174
Cold welding (CW) (USW) 613, 616
Cold-shut defect in GMAW-S 494
Cold working
    effects on metals 88, 248
    finishing operations 407
Columbium (niobium), alloy in steel 347
Columnar grains in weld metal 792
Combined IT and CCT diagrams 848
Combustion velocity of gases 465, 662

Computer programs
    Table XI..919
Conduction of heat 655
Conductivity
    electrical...269
    thermal 261
Constant voltage power source 468
Constitutional diagram..296
Constitutional supercooling........................791, 794
Constricted arc.................................452, 511, 641
Consumable electrode arc welding............455, 471
Consumable guide in ESW 544
Consumable insert in GTAW 478
Continuous casting of steel...............398, 400, 404
Continuous coating steel strip............................409
Continuous cooling transformation diagram (CCT)
    for base metal HAZ..............................848
    for weld zone 866, 869
Contraction coefficient, thermal..........................266
Conversion tables (SI units ⇔ U. S. Customary)
    stress (MPa ⇔ ksi), Table VII 902
    temperature (°C ⇔ °F), Table VIII 905
Cooling curves for metals and alloys............65, 299
Cooling rates in welds, as affected by
    bead shape and size 687
    cooling end point.............................696
    heat input 677
    temperature distribution 668
    temperature gradient 668, 669
    travel speed 686
    workpiece temperature.........681, 685, 698
Cooling rate/hardness/microstructure in steel....693
Coordination numbers for crystal systems...........52
Copper
    alloy in steel............................340, 343
    backing bar for SAW......................511
    dams to retain weld melt 543, 780
    plating on filler metal.......................499
Coring in microstructures 302
Corrosion of metals 275
    forms of attack 275, 279
    gases, hot 282
    metals, molten................................282
    stress-corrosion cracking................284
    susceptibility of weldments 283
Cottrell atmospheres 89
Covered electrodes (SMAW) 472, 724
Cracking
   9
    arrest 154, 162, 164, 229, 251
    cold...164

critical extension force 196
critical size 197
hot...507, 544, 792, 800
initiation 157
propagation 157
stable crack growth.......................................158
subcritical extension 197, 200, 206
Crack tip opening displacement (CTOD) 205
Crater in arc welds 470
Crevice corrosion 280
Creep of metals
    at elevated temperature.......................242, 243
    creep brittle response 245
Critical cold work, effect on recrystallization 100
Critical cooling rate for hardening steel 834
Critical resolved shear stress...............................78
Critical temperatures for steels 331
Cross-wire weld'g resistance process (PW) 533
Cryogenic service.....................................164, 413
Crystals (also see grains) 45, 56, 58, 61
    Axes 51
    covalent form 59
    crystallites 70
    fundamentals 50
    growth in melt 64
    hydrogen bond...59
    inert gas form...58
    ionic form 58
    lattice forms...52
    metallic..59
    plasticity 77
    systems...52
Crystalline structure of metals......................61, 74
Cupola 366
Curie point of iron....................................273, 332
Cutting processes, thermal (TC) 633
Cyclic stress (fatigue)..174

D

Databases
    Table XI...919
Decarburization
    argon-oxygen steelmaking............................396
    carbon/oxygen reaction................710, 740, 759
    thin steel during furnace treatment...............408
Deformation welding
    cold (CW) (USW)..................................613, 616
    hot (UW) (HPW) (ROW) (CEW) 533, 587
                                   589, 590, 591
Delta iron 331

Index

Dendrites in weld metal................................66, 792
Density of metals..66
Deoxidation of steel..712, 721
Design curve, CTOD assessment.........................208
Design of weldments..11
Desulfurization during steelmaking............388, 408
Deuterium..35
Dew point for water vapor....................................783
Diamond surface film on steel.............................633
Diffusion
    bonding..591
    brazing...622
    welding (DFW)....................................591, 612
Dilatometric behavior of metals...........................814
Dilution in hardfacing...631
Dimples in fracture texture..................................126
Direct current arc welding..............................468, 470
Direct reduced iron from ore (DR) (DRI)............365
Directionality in properties of metals.................135
Dislocations in crystalline lattice....................70, 84
    edge dislocations...86
    Frank-Read sources.....................................89
    point defects..70, 85
    screw dislocations.......................................88
    stacking faults..88
Dissimilar metal welds......................................4, 437
Dispersion hardening..323
Domains, magnetic..271
Drag rods..475
DRI, direct reduction of iron................................365
Drooping arc voltage...468
Drop weight tear test (DWTT)..............................230
Drop weight test (DWT-NDT)................................229
Ductility..147
    ductile fracture...................................126, 155
    special tests..149
Duplex grain size measurement...........................99
Dynamic fracture toughness (K_{Id})......................198
Dynamic tear test (DT)...235

E

Elasticity in metals...137
    elastic limit..141
    elastic modulus...139
Elastic-plastic fracture mechanics (EPFM) 188, 201
Electric arc...448, 519
    cutting processes (AC).............................637
    spraying processes (ASP) (THSP) (EASP) (PSP)...630
    steelmaking processes............................389
    welding processes........448, 455, 471, 472, 489, 501, 505, 511, 517
Electric induction furnace steelmaking..............389
Electric resistance heating..................................461
Electrical metal-explosion welding.....................541
Electrical properties of metals............................268
    conductivity...268
    high-frequency conductance...................535
    Ohm's law..269
    Peltier effect..271
    resistivity...269
    superconductivity.....................................269
    thermoelectric (Seebeck) effect...............271
    Thompson effect..271
Electrochemical corrosion...................................276
Electrode extension in FCAW...............................504
Electrodeposition welding....................................619
Electrodes, arc welding
    bare metal-arc (BMAW)............................472
    flux cored (FCAW).............................501, 738
    gas metal-arc (GMAW).............................489
    nonconsumable (GTAW) (CAW).........478, 722
    shielded metal-arc (SMAW).....................472
    submerged arc (SAW)..............................505
Electrodes, resistance welding
    high-frequency processes.......................535
    seam welding..534
    spot welding.......................................522, 527, 529
Electrogas, welding process (EGW)............547, 780
Electromagnetic radiation....................................461
    spectrum of wavelengths........................462
    laser beams...463
Electromagnetic stirring of weld pool.........472, 863
Electromotive force values, metals....................278
Electron...20, 22
    compounds..314
    configurations in atoms....................315, 913
    covalent bonding..59
    microscopes (SEM) (TEM) (STEM)......106, 108, 112
    probe microanalyzer....................10, 18, 117
    shells of atoms....................................23, 120
    spectroscopy (Auger, AES) (ESCA)..........120
    theory for metal crystals....................59, 317
    valence electrons......................................315
    volts, electron (eV).....................................24
    wave length..23, 25
Electron beam...459
    cutting (EBC)...643

Index

detection of defects 554
heat generation equation 460
steelmaking furnace 395
weld defects 554
welding in vacuum (EBW-HV) 548
welding sans vacuum (EBW-NV) 459, 565
Electronegativiity of elements 314
Electronic databases and computer programs
    Table XI 919
Electroslag
    remelting steel (ESR) 391
    welding fluxes 765
    welding process (ESW) 542
Electropostive elements 314
Electrostatic bonding 617
Elementary particles 20
Elements, chemical 32, 39
    atomic numbers (Z) 33
    atomic weight 34
    electron configurations 317, 913
    mass number (A) 34
    periodic table 42
    properties; Table IX 907
    symbols 42
    valency 35, 310, 315
Elevated temp. mechanical properties 165, 238
    creep rupture 242
    Larson-Miller parameter 247
    relaxation testing 246
    service regimes defined 238
    stress-rupture 245
    tensile properties, short time 241
    thermal shock resistance 246
Elongation in tension test 148
Emissivity 263
Endurance limit, fatigue 181
Energy output during welding, equations
    arc welding 458
    electric resistance welding 461
    electron beam welding 460
Energy transition, impact testing 222
Enthalpy 656
Entropy 656
Epitaxial grain formation 69, 788, 795
Equicohesive temperature 165
Equilibrium diagrams 298
Etching metallographic specimens 93
Eutectic type phase diagrams 304
Eutectic/diffusion brazing 448, 623

Eutectoid microstructure in steel 333
Expansion coefficient, thermal 267, 814
Explosion testing of weldments 171, 227
Explosion welding processes
    chemical explosion (EXW) 607
    electrical metal-explosion 541

F

Face-centered cubic metal crystals 62
Fahrenheit ⇔ Celsius temperature conversion
    Table VIII 905
Fatigue of metal 174
    appearance of failure 12, 128, 178
    constant lifetime fatigue diagram 185
    counting cumulative damage 185
    endurance limit 181
    fatigue limit 181
    fatigue ratio 183
    fatigue strength 181
    fractals 128
    fracture mech. assessment of fatigue 188, 215
    Goodman fatigue diagram, original form 184
    initiation of fatigue failure 176
    propagation of fatigue fracture 176
    stress concentration at notches 176
    striae in fatigue fracture 128, 176, 178
    thermal fatigue 247
    variable loading 183, 185
Feather pattern in fractures 161
Feed-back control of welding 5
Ferrite, a microstructural constituent 804
    acicular ferrite in weld metal 860, 885
    crystallographic structure 820
Ferrite-forming alloy elements in iron 319, 327
Ferrimagnetism 273
Ferromagnetism 273
Fiber optics usage in welding 702
Field ion microscopy 18
Firebox quality steel plate 435
Fission of atoms 30
Fitness-for-service assessment 188
Flame spraying (FLSP) 630
Flange quality steel plate 435
Flash welding, electrical (FW) 537
Flashover during EBW 554
Flow strength, metals 174
Fluidity
    molten slag 714
    molten metal 264

Index

Flux cored arc welding (FCAW) 501, 738
    gas shielded electrodes for 503, 739, 779
    self-shielded electrodes for...................503, 741
    flux formulas for FCAW 740
Fluxes 716
    arc welding (AW).................473, 501, 723, 779
    brazing (B)..621, 624, 782
    electroslag welding (ESW)..........................765
    flux, oxygen cutting (FOC).........................636
    forge welding (FOW)....................................590
    soldering (S)..784
    submerged arc welding (SAW)............506, 744
Foil butt seam resistance welding....................532
Forge welding (FOW) 590
Foundry operations 397
Fractals treatment
    for fracture assessment (D_F) 128
Fractography 123
Fracture appearance assessment 91
    brittle 125, 157
    cleavage 125, 157, 159
    dimples...126, 157
    ductile..126, 157
    fractal dimensioning (D_F) 128
    initiation 157
    intergranular 162
    propagation 157
    shear 158
Fracture in metals..156
    assessment from appearance 91
    assessment by fracture mechanics 188
    axiality of stresses 155, 166
    brittle type 125, 159
    characteristic forms...157
    chevron markings...162
    cleavage type 125, 157, 160
    cleavage tongues...161
    correlation of toughness test results (IIW)....236
    crack arrest 164, 229
    crack plane..193
    crack surface displacement modes 193
    crack tip opening displacem't test (CTOD) ...205
    critical stress intensity factor (K_C)................191
    drop-weight test (DWT-NDT).........................229
    ductile type fracture 158
    dynamic tear test (DT) 235
    elastic-plastic fracture mech. (EPFM)...........201
    explosion bulge test 227
    fatigue fracture 174

feather markings on fracture surface............161
fractals (D_F) for fracture assessment............128
general yielding fracture mechanics 201
Griffith's theory of fracture 189
herringbone markings on fractures................161
initiation mechanics 157, 222
intergranular fracture 157, 162
J-integral test method (J_{Ic})...........................211
J-R curve determination................................214
linear elastic fracture mechanics (LEFM).....188
mechanics of fracture 188
microvoids on fracture surface 130, 159
mixed fracture modes 163
plane-strain conditions (K_{Ic}) 195
plane-strain dynamic conditions (K_{Id}) 198
plane-stress conditions (K_c)........................195
plastic collapse...201
plastic zone 193
propagation..157, 160
quasicleavage 162
resistance curve test method (R-curve) 210
river pattern on fracture surface 161
rock candy fracture appearance 163
shear type fracture..157
strain rate effects 171
stress distribution effects 196
stress intensity factor (K) 190
stretch zone in fracture morphology.............206
temperature effect...164
tests for brittle fracture propensity 190
toughness as a weld quality 431
transition in fracture appearance..........217, 225
transition in energy absorption 222, 224
Fracture mechanics toughness assessment
    linear elastic analysis (LEFM)......................188
    elastic-plastic analysis (EPFM)....................201
Frank-Read sources of dislocations....................89
Free-machining steels...341
Free energy of oxide formation (ΔF) 721
Freezing (solidification) of weld metal..............792
Friction welding processes (FRW) 593
    co-extrusion 607
    direct-drive, rotational 595
    hydropillar..607
    inertia-drive, rotational 594
    linear reciprocating.......................................605
    materials suitable for FRW 595
    mechanical properties of FRW welds...........598
    orbital friction welding 605

930 Index

post-rotational twist effects 599
radial friction welding 602
rotational friction welding 593
stored-energy (inertia-drive) FRW 594
surfacing by friction welding 602
underwater friction welding 599
Fuel gases for weld'g & thermal cutt'g 582, 584
635
Fused SAW fluxes 748
Fusion
atomic, thermonuclear 32
latent heat, metals 264
Fusion welding 441, 445, 787
Adaptive control systems 702
diagrams for parameter prediction 699
mathematical modeling 699
master chart, fusion welding processes 447
parametric tolerance for SAW 699
typical weld in steel dissected 876

G

Galvanic relationship of metals 278
Galvanized steel (also see zinc coated) 409, 528
Gamma iron 326
Gamma rays 37, 461
x-rays 549
Gas content of steels 380, 394, 741, 761
Gas for shielding brazing 782
Gas metal arc cutting (GMAC) 638
Gas metal arc welding (GMAW) 489
axial spray transfer 490
difficulties and defects 490
globular transfer 491, 775
metal transfer modes 489, 772
narrow grove welding 497
oxide cleaning by DCEP 490
pulsed spray transfer in GMAW-P 494
shielding gases 772
short circuiting transfer in GMAW-S 492
spray transfer 490
synergic pulse transfer in GMAW-P 497
Gas tungsten arc cutting (GTAC) 638
Gas tungsten arc welding (GTAW) 478
electrodes 481
gases for shielding 478, 482, 771
joint penetration consistency 483
Marangoni convection 486
minor element effects 485
polarity effects 481
power sources 480

slag patch formation 479
Gas welding, oxyfuel (OFW) 582
G_c value for fracture toughness 189
General yielding fracture mechanics 201
Ghost bands in steel microstructure 339
Gibbs phase rule 310
Glass 47
glass transition temperature (T_g) 47
Gleeble
HAZ simulator using resistance heating 676
Goodman fatigue diagram 184
Graphite
in cast iron 329, 366
in steel 329, 331
Grains 68
ASTM size number methods 96
boundaries 70, 87
cross-section 94
formation 68
growth at high temperature 101
orientation 101, 135
recrystallization 100
refinement 72, 100, 103
single crystal 77
size and shape 71, 94
weld zone 788
Gravity arc welding 475
Gray cast iron 366
Greek alphabet for scientific notation
Table IV 896
Griffith's fracture theory 189
Gun discharge during EBW 557
Gurney on fatigue 186

H

Hadfield manganese steel 338, 413
Hadron 21
Half-life of radioactive elements 37
Halogens 41, 43
Hardenability of steel 336, 693, 816
Jominy end-quench test 693, 852
maximum obtainable hardness in steels 840
Hardfacing processes (see surfacing) 628
Hardness/microstructure/weld cooling rate 693
Hardness testing 149
Brinell (HB) 150
conversion of test values 153
dynamic hardness tests 152
file-hardness 153
indentation hardness tests 150

Index

Knoop microhardness 151
microhardness tests 151
Rockwell hardness (HR) 151
Scleroscope hardness (HS) 152
scratch test for hardness 153
static indentation hardness tests 150
ultrasonic hardness test 152
Vickers hardness (HV) 150
Heat, defined 653
Heat flow in welding 654
Heat generation for welding 448, 661
chemical 465, 662
electric arc 448, 662
electrical resistance 60, 663
electromagnetic radiation 461, 663
electron beam 459, 663
equations 657
mechanical 466, 662
rate of heating 659
source characterization 658
Heat-affected zones (HAZ) 656, 657
from power beam glazing 61, 571, 633
from thermal cutting 635, 643
from welding, single pass 801
from welding, multiple passes 887
prediction of cooling rates 677
prediction of microstructures 851, 856
Heat of fusion, latent 264
Heat-resisting steels 415
(also see high-alloy steels)
Heat sink 677
Heat treatment of steel 408
Helium, use for alloying 356
arc welding processes 768
EBW-NV 549, 565, 781
EGW 780
GMAW 489, 777
GTAW 482, 771
LBW 575, 781
PAW 780
Hematite, iron ore 362
Herringbone pattern in fractures 161
Heyn grain size determination 98
Hexagonal metal crystals; 62
High-alloy steels, definition 413
High-frequency electric current
arc starting 480
heat generation 461
welding processes 535
High-strength low-alloy steels (HSLA) 412

High-temperature mechanical properties 238
H-iron ore reduction process 366
Hole drilling by EBC 643
Holloman, regarding fracture 189
Hooke's law 139, 167
Hot cracking in welds by SAW 507
Hot isostatic pressing (HIP) of castings 397
Hot pressure welding (HPW) 589
Hot wire welding technique 478, 581
Hot working effects 248
Hot working operations 404, 406, 437
Hume-Rothery alloying principles 310, 313, 315
Hydrogen 35
addition to GTAW shielding 482, 770
atomic-hydrogen arc welding 471, 518
blistering in steel 289
content of alloy steels 380, 387
content of weld metal 476, 498, 501, 511, 763
content of SMAW electrode coverings 732
diffusible hydrogen in weld metal 736
hydrogen-assisted corrosion 289
hydrogen embrittlement (HE) 289, 498
hydrogen induced cracking (HIC) 289, 770
hydrogen in SAW process 511, 763
hydrogen stress cracking (HSC) 289
hydrogen sulfide SCC 288
low-hydrogen SMAW electrode cover'gs 733
shielding for welding 718, 770
use in oxyfuel gas welding (OHW) 465
Hydrogen bond in crystals 59
Hypereutectoid steel 331, 808
Hypersonic thermal spraying 630
Hypoeutectoid steel 333, 334, 808

I

Immiscibility in molten alloys 296
Impact testing
at low temperatures 220
improving the value of C_v test 227
lateral expansion in C_v tested specimen 224
mechanics of fracture 226
Imperfections in crystalline structure 71
Impurity substructure 71
Inclusions, nonmetallic 93
analyses 118, 130
in microvoids of fractures 130
utilization as nucleant 874
Incoherent precipitates 323

Incomplete fusion
    in EBW weld 552, 561
    in FCAW weld 504
    in GMAW weld 499
    in LBW weld 572
Induction, electric 461
    brazing (IB) 622
    furnace melting 389
    fusion welding (IW) 542, 592
    hot pressure welding (HPW) 589
Inert gases 40
    crystals 56
Infrared radiation 461
Ingot steelmaking practice 398
In-process regulation of arc welding 6
Instantaneous cooling rate 683
Intercritical heating of steel 813, 879
Intergranular corrosive attack 280
Intergranular fracture 162
Intermetallic compounds 316
Internal soundness, weldability factor 435
International Institute of Welding (IIW)
    classificat'n, GMAW metal transfer modes ... 773
    classification of SAW fluxes 754
    correlat'n, fracture toughness test results 236
    welding thesaurus 14
Interplay between factors in welding 890
Interstitial solid solution 77, 310
Interstitial-free steel (I-F) 357
Invertor power source for arc welding 497
Ion implantation 631
Ionic bond in crystals 58
Ionization 36, 452
    potential of elements 37
Ion-probe analyzer 118
Iron
    allotropy 103
    as an alloy matrix 326, 357
    iron-iron carbide phase diagram 329
Iron carbide 335, 806
Iron ores 362
Iron powder
    addition of SAW flux 509
    electrode covering addition (SMAW) 737
Irradiation by neutrons, influence of 249
    boron 251
    cobalt 251
    copper 256
    helium 250
    microstructure 254
    nickel 256
    nitrogen 257
    phosphorus 256
    sulfur 257
    vanadium 257
Irwin's fracture theory 190
Isobar 34
Isothermal transformation diagram (IT) 827, 843
Isotope 33
Isotropy of metal properties 135
I-T and CCT combined diagrams 850
Izod impact test 156

J

Jeffries grain size procedure 98
Jominy end-quench hardenability test 693
    HAZ microstructural forecasting 851
    Jominy hardening factor 696
Joules, heat units 653
J-integral fracture test method 211

K

Kaldo steelmaking process 383
Kelvin temperature scale (°K) 654
Keyhole in fusion welding processes 513, 551
    572
Kikuchi lines 110
Killed steel 373
Kinematic viscosity 714
Kinetic energy 45
Knoop microhardness test 151
KR iron ore reduction process 366

L

Lack-of-fusion defect in GMAW-S 494
Ladle analyses of steel 400
Ladle degassing of molten steel 387
Ladle refining in steelmaking 385
Lamellar pearlite 807, 817, 826
Lamellar tearing 136
Laminations in hot rolled steel sections 399
Lance-bubbling-equilibrium
    steelmaking technique 385
Lanthana in GTAW electrodes 482
Lanthanides 40
Larmor precession 22
Larson-Miller creep and rupture parameters 247
Laser 463, 568
    excimer type 463, 570

difficulties and defects 572
gas type 464, 568
heat generation mechanics 464
plasma generation 464, 569
scanning microscopy (LSM) 94, 96, 130
shielding gas effects 575
solid-state type 463
surfacing 631
welding techniques 568, 571, 576
Latent heat of fusion 264
Lateral expansion in C_v tested specimen 224
Lattice, crystalline 50
imperfections 71, 84, 88
parameters 60
Laue photographs of x-ray diffraction 18
Laves phases 314
L-D steelmaking process 381
Lead
alloying element in steel 297, 355
lead-coated steel 409, 527
Ledeburite microstructure 333
Lepton 21
Level-crossing counting, fatigue cycles 187
Lever law of alloy solidification 301
Liberty ships 3
Lineage structure 71
Linear-elastic fracture mechs. (LEFM) 188, 190
198
Linz-Donawitz steelmaking process 381
Liquidation during incipient melting 881
Liquid metal cracking (LMC) 282
Liquid state 48
Liquidus 296
Lobes of resistance welding parameters 529
Lorentz forces, electromotive stirring of melt 485
Low-alloy steels 412
Lower bainite microstructure 831
Low-hydrogen electrode cover'gs for SMAW 733
Low-temperature
mechanical properties of steels 218
service temperature ranges defined 413

M

Macroetching 436
Madelung energy in ionic crystals 59
Magnesium, use in desulphurizing 365
Magnetic deflection during EBW 562
Magnetic particle examination 436
Magnetic properties 21, 270
coercive force 272

magnetostriction 274
permeability 273
remanence 272
Seebeck effect 271
Magnetically impelled arc weld'g processes 515
Magnetite iron ore 362
Malleable cast iron 366
Manganese 337
alloying element in steels 337
carbide in steel 338
killed steel production 375
sulfide former 314, 337, 341
Manganese steel, austenitic, Hadfield 338, 413
Marangoni convection 486
Martensite 828, 840
athermal transformation 839
crystallographic structure 820
hardness rationale 841
implications in welding 836
lath microstructure 822
maximum obtainable hardness 840
microstructural appearance 809, 816, 822
monitoring formation by AE signals 842
M_s calculation from steel composition 837
M_s to M_f formation range 837
plate microstructure 822
self-tempering 837
twinning in microstructure 822
Mass number (A) for atoms of elements 33
McQuaid-Ehn austenite grain size test 376
Mechanical heat generation 466
explosion (EXW) 467
friction (FRW) 466
Mechanical properties 10, 133, 136
after plastic work 248
alteration by alloying 319
anisotrophy 135
at elevated temperatures 238
at low temperatures 218, 221
directionality 135
elastic limit 141
fatigue strength 181
proportional limit 141
strength, tensile 144
structure sensitivity 133
toughness 154
weldability factor 434
yield point 143
yield strength 143

Mechanically mixed SAW fluxes 749
Melting
    mechanics 47
    temperature points and ranges, materials 264
Mendeleyev's periodic table 39
Meshlength theory of work hardening 90
Mesnager impact test 155
Meson 21
METADEX information retrieval system
    Table XI 919
Metal arc welding
    bare electrode process (BMAW) 723
    covered electrode process (SMAW) 724
    electrode covering compositions 725
    element recovery during SMAW 729
Metal powder cutting (POC) 636
Metallizing by spraying 630
Metallographic grain size determination 94
Metallography 10, 92
    acoustic 122
    electron 106
    field force 18
    ion probe 18, 115
    optical 17, 92
    scanning laser acoustic (SLAM) 123
    thermal wave 121
    tunnel effect (STM) 116
    quantitative 105, 855
Metallurgy 6
    physical 7
    process 7
Metalloid 342
Microbiological corrosion of metals 281
Microhardness tests 151
    Knoop 151
    Ultrasonic 152
    Vickers 151
Microprobe analyzer 18, 117
Microstructure 17
    microsegregation 163, 302
    transformations during cooling 816
    transformations during heating 810
    transformations in weld metal 857
    weldability factor 435
Microvoids 130
    appearance in fractures 126, 130
    coalescence 126, 130, 159
Microwave energy 461
    for plasma welding and cutting 464

Midrex iron ore reduction process 366
Miller indices 56
Miner's rule for fatigue 187
Modulus of elasticity, Young's 139
Mohs scale of scratch hardness 153
Moisture in SMAW electrode coverings 733
Molybdenum, alloy in steel 346
Multi-pass weld microstructures 887

N
Narrow groove welding w/ GMAW process 497
Natural gas, methane 635
Neutron 20, 28
    activation analysis (NAA) 32
    capture cross section of atoms 27, 250
    delayed neutrons 30
    diffraction by residual stress 28
    fast neutrons 29
    fluence rate 31
    irradiation effects, surveillance (RT_{NDT}) 31
    magnetic moment 28
    neutrino 28
    prompt neutrons 30
    secondary neutrons 30
    small-angle scattering (SANS) exam. 123
    thermal neutrons 29
    velocity 31
Nick-break fracture assessment 91
Nickel, alloy in steel 346
Niobium (columbium), alloy in steel 347
Nitrogen 349
    alloy in steel 350
    content of alloy steels 380, 394
    content of weld metals 743, 762
    fixation in steels 349, 370
    ion implantation 632
    shielding for welding 718
Noble metals 41
Nonconsumable electrode arc weld'g 450, 471
Nonmetallic inclusions 93, 130, 436
    a welding quality 436
    exogenous types 436
    in hot rolled steel 118
    in weld heat-affected zones 796, 801
    in weld metal 872, 874
    indigenous types 436
Nontransferred plasma arc 511
Notches
    ductility influence 147

Index 935

notched tensile strength.................................147
stress intensity factor (K_t) 169
toughness influence................................155, 164
Nuclear fusion 32
Nugget
arc weld zone 658, 686, 691
resistance welds 522, 703
Nuclei of atoms (nucleons)....................................26
Nu-iron ore reduction process.............................366

O

Off-gas (OG-BOF) steelmaking process............383
Offset yield strength 143
Ohm's law for electrical resistance 269
Olsen ductility test 149
Open hearth steelmaking process 370
Optical basicity
OB method of slag evaluation 756
Orbit, search services 14
Orbital angular momentum....................................23
Orbital electron shells, atom..................................35
Order-disorder hardening....................................321
Orientation of crystalline structure in grains.........68
Orowan, fracture theory 190
Overaging, precipitates 323
Overhead position arc welding 456, 475, 726
Overlaying by welding 632
Oxidation
of iron 708
of molten steel 710
of solid steel...282, 713
prevention during welding.............................714
Oxide removal during GMAW with DCEP 490
Oxidizing acids 277
Oxidizing OAW welding flame............................582
Oxyacetylene welding process (OAW) 582
Oxyfuel gas cutting (OFC) 635
Oxyfuel gas welding (OFW) 582
Oxygen
content of steels 374, 380, 386
solubility in molten steel........................710, 761
reference atom..33
Oxygen arc cutting (AOC)...................................639
Oxygen cutting (OC) 634, 637
Oxygen, laser beam, cutting (LBC-O)................650
Oxygen steelmaking processes 380
Ozone 771

P

Painted surfaces on steel....................................409
Parafin as neutron moderator 29
Parametric tolerance envelopes.........................699
resistance spot welding 529
submerged arc welding..................................701
Partially-melted HAZ 800, 801
Peak and trough counting, fatigue cycles..........187
Peak weld temperature equation, Adams' 667
Pearlite 817
isothermal formation 826
microstructural configuration........................807
Peel test for resistance spot welds 523
Peening 8, 248
Penetrators in flash welds...................................540
Percussion welding process (PEW)...................514
Peritectic in iron-carbon alloys 305
Permeability, magnetic...273
pH scale of acidity and basicity...........................277
Phase diagrams 299
Photo-electron spectroscopy (ESCA) 121
Photon, electromagnetic energy 23, 37, 461
Physical metallurgy, definition................................7
Physical properties......................................133, 258
atomic weight 34
boiling point..266
coefficient of expansion 266
density...258
electrical conductivity....................................269
electrical resistivity 269
gases, properties 717, 766
heat of fusion 264
heat of vaporization 266
ionization potential 36
melting temperature.......................................264
specific heat 260
strength-weight ratio of materials 259
structure sensitivity..133
surface tension of liquids......................265, 715
thermal conductivity 261
thermal expansion..266
thermal properties..259
thermionic work function 268
viscosity of liquids..................................264, 715
Phosphorus, alloy in iron and steel....................338
Pig iron 361
Pitting type corrosion...279
Planar mode grain growth in weld metal 792
Planimetric grain size procedure..........................98

Plasma arc
    cutting (PAC) 640
    jet formation 456, 485
    overlay coating 514
    spraying (PSP) 514
    welding (PAW) 511, 780
Plasmared iron ore reduction process 366
Plastic instability and collapse 201
Plasticity in metal crystals 77, 142
Point, a dislocation defect 70, 85
Poisson's ratio for elastic behavior 139, 168
Polarity in arc welding 451
Polygonized structure 71
Polymorphism 46, 306
Porosity in welds
    EBW process 566
    GMAW process 498
    GTAW process 480
    LBW process 572
    OAW process 584
    SAW process 763
Positrons 26
Pourbaix diagrams for corrosion 277
Powder metallurgy 356, 367
Power sources for arc welding 467
    GMAW-P (pulsed) process 494
    GMAW-S (short circuiting transfer) 492
    GTAW process 480
Precipitation hardening 322
Precoated steel sheet
    painted, metallic coated 409
Preheating 697, 884
Pressure gas welding (PGW) 585
Pressure piping, ASME code 428
Pressure vessels and boilers, ASME code 426
Procedure optimization for welding 888
Process metallurgy, definition 7
Projection welding, electrical resistance (PW) ... 532
Propane 770
    oxyfuel heat source 635
    shielding for welding 770, 767
Proportional limit, elastic 141
Protective atmospheres 717
Proton 20, 27
Pulsed spray welding (GMAW-P) 494

Q

Quality characterization of steels 420
Qualities of steel important to welding 431
Quantitative metallography 105, 856

Quantum tunneling of electrons 116
Quelle-basic (Q-BOP) steelmaking process 383
Quenched and tempered treatment for steel 408

R

Radiation damage in steel 249
Radiant energy 461, 463
    heat transfer mechanisms 263, 464, 655
Radioactivity 37
Rainflow counting, fatigue cycles 187
Range counting, fatigue cycles 187
Rapid solidification technology (RST) 47
Rare earth metals 40
Ratchet marks from fatigue 176
R-curve fracture toughness test method 210
Real-time welding control 6
Recovery of alloying elements
    during fusion welding 729, 739, 746, 758
Recrystallization
    cold worked metal 100
    temperature limits 100, 165
Recycled SAW fluxes 750
Red shortness 240
Reducing acids 277
Reducing fluxes 717
Reducing gases 622, 717
Reduction of area, tension test 149
Relaxation testing at elevated temperature 246
Repair welding 439
Residual elements in iron and steel 358
Residual stresses 28, 86, 199, 202
                                                     284, 289
Resistance welding processes 520
    adaptive control 703
    continuous upset seam process 534
    electrical metal-explosion welding 541
    electroslag welding (ESW) 542
    flash welding (FW) 537
    heat generation 461
    high-frequency ac seam welding 535
    induction welding (IW) 542
    pipe and tube welding 534
    projection welding (PW) 532
    seam welding (RSEW) 531
    spot welding (RSW) 10, 522
    upset welding (UW) 533
Resistance curve toughness test (R-curve) 210
Resistivity, electrical 269
Resonant tunneling effect by electrons 26
Resources, welding technology 12

Rimmed steel 372
Ripples on fusion weld surfaces 797
River pattern in fractures 161
R-N iron ore reduction process 365
Robertson arrest test for toughness.................. 164
Rocket motor casings... 3
Rockwell hardness test (HR) 151
Roll spot resistance welding 531
Roll welding (ROW).. 590
Rolling mills for steelmaking.............................. 405
Rosenthal's thermal equations........................... 689
Rotary forging operation..................................... 405
Ruby laser 463
Rupture strength, elevated temperature 245

S

SAE-AISI steel specifications............................ 418
SAE-ASTM Unified Numbering System............ 416
Scaling of steel... 713, 890
Scanning electron microscopy (SEM)........ 112, 124
Scanning laser acoustic microscopy
    (SLAM)... 123
Scanning transmission electron microscopy
    (STEM)... 114
Scanning tunneling microscopy (STM) 116
Scarfing of steel sections by OFC 636
Scleroscope hardness test................................ 152
Scratch hardness test 153
Seam welding, electrical resistance (RSEW)
    (RSEW-HF) (RSEW-I)................. 531, 535, 536
Seebeck thermoelectric effect........................... 271
Selenium, alloy in steel 340
Semiconductor... 44
Semikilled steel 379
Semimetal... 45
Shear fracture in metal...................................... 157
    shear lips... 159
Shelf regions in toughness testing..... 204, 222, 253
Shielded metal arc welding (SMAW) 472
Shielding from air 714, 722
    atmospheres 482, 717
    deoxidizers 721
    fluxes..716
    gases 717, 766
    liquid blankets 722
    protective surface alloys 721
    slags..714
    vacuum.. 718

Ships, fractures in service
    Liberty and Victory designs 3
Short circuiting transfer in GMAW-S.................. 492
SI base units and prefixes
    Table VI... 901
Silicon
    alloy in iron... 342
    killed steel production 374
Single crystal... 69, 77
Slag
    inclusions in metal............... 476, 511, 714, 728
        743, 795
    patches on weld surface........................... 486
    removal from melt during steelmaking 385
    shielding molten weld metal 476, 714
Slenderness ratio in tension test specimens 148
Slip in metal crystals 78
Small-angle neutro-scattering from
    crystal structure examination (SANS)........... 123
SMP steelmaking process................................. 384
S-N fatigue testing diagrams............................. 182
Soldering (S) 624
    filler metals.. 624
    fluxes.. 627, 784
    master chart of processes 449
Solid solution hardening.................................... 320
Solid state... 45, 50
Solid-state welding (SSW) 441, 444, 587
    adaptive control systems 703
    master chart of processes 446
Solidification of metals.. 64
    liquidus and solidus 299
    volume changes.. 66
Solidification in steelmaking...................... 353, 437
    continuous casting 400
    ingots 69, 398
Solidification of weld metal................................ 788
    epitaxial... 788, 792
    modes of primary structure........................ 794
    nucleation 66, 353, 437, 788, 874
Solubility in alloys
    liquid state... 296
    solid state..................................... 76, 302, 310
Specific heat.. 260
Spiking during EBW 553
Spin of elementary particles................................ 21
Spot welding, electrical resistance (RSW)......... 522
Spray transfer in GMAW 490
Spraying, metal 630
Stacking fault dislocation.............................. 70, 88

Index

Stainless steels (also see high-alloy steels) 414
Standard terminology references
    Table II 895
Steel composition specifications 416
Step-down machined test, cleanliness...............436
Stirring weld pool by electromagnetic forces 472
Stora-Kaldo steelmaking process 383
Strain hardening 78
Strain rate, effects 155, 172
Strategic-Udy iron ore reduction process 366
Strength of metals 133
    at low temperatures 219, 238
Strength-weight ratio of materials 259
Stress
    axiality 155, 166
    cyclic (fatigue) 174
    distribution around cracks.............................196
    gradient 169
    intensity factor (K_t) 169
    multiaxial 167, 170
    rate of application, effects............................171
    residual 28, 86, 199
                                     202, 284, 289
Stress conversion, SI/U.S. Customary
    Table VII...902
Stress corrosion cracking....................................284
Stress-rupture testing at high temperatures 245
Stress-strain diagrams for tension tests.......141,142
Stretch zone in fracture morphology 206
Structure sensitivity of properties.......................133
Stud arc welding process (SW)..........................477
Sublimation, heat of 57
Submerged arc welding process (SAW)............505
    electrodes and filler metals..........................509
    fluxes 506, 508, 744, 752
    parametric tolerances envelopes.................699
    weld cracking 507
Substitutional solid solution.........................77, 310
Sub-zero service 164
Sulfide stress corrosion cracking (SSC) 288
Sulfur 340
    alloy in steel..340, 411
    desulfurization of solid sheet steel 408
    fluxing effect in fusion welding.............766, 771
    removal during steelmaking 363, 364, 388
    sulfide inclusions in steel..............118, 130, 337
                                               340, 434
Super alloys (also see heat resisting steels) 415
Superconductivity, electrical................................269
Superficial Rockwell hardness test (N scale).....151
Superheat meltback zone (UNMZ) 472, 799, 881
Superlattices...321
Surface defects in steel product forms 431
Surface tension
    of molten metals and slags...................264, 715
Surfacing
    by thermal spraying 630
    by welding...628
Symbols for welding
    Table II 895
Symbols used in this text
    Table III 896

T

Tabs extending from start and finish of welds ...470
Taconite, iron ore 362
Tailored blanks of steel sheet 438
Tellurium, alloy in steel.......................................356
Temperature conversion °C ⇔ °F
    Table VIII..905
Temperature, defined 654
    numerical modeling.......................................666
Temperature effects on properties 164, 218, 220
Temperature in welding......................................653
    cooling rates of heated zones 670, 677
    distribution in welded plates 668
    effective maximum heat-affected zone.........689
    gradients 656, 668
    multipass effects 887
    OSR/LR measurement 703
    peak temperatures in welds.........................667
    pre-weld temperature effect 681, 697, 884
    temper colors as indicant.............................675
    travel speed effects......................................686
Temper bead technique 888
Temper colors on steel.......................................675
Tempering heat treatment..................................325
Tensile properties......................................144, 154
Terminology, references
    Table II 895
Ternary phase diagrams 307
Thermal conductivity 261
Thermal cutting processes (TC).........................633
Thermal expansion and contraction...................266
Thermal fatigue 247
Thermal flow...654
Thermal neutron cross-section...........................250
Thermal properties 259
    boiling point of molten metal........................266
    conductivity 261

emittance of radiant heat 263
expansion and contraction 266
heat of fusion 264
heat of vaporization 266
melting point or range 264
specific heat 260
surface tension of molten metal 264
thermionic work function 268
viscosity of molten metal 264
Thermal shock 246
Thermal spraying (THSP) 628, 630
Thermal wave microscopy 121
Thermionic work function 268
Thermite welding process (TW) 7, 466, 585
Thermoelectric current flow during EBW 562
Thermographic temperature measurement 6
Thermo-mechanical control process
(T-MCP) 406, 437
Thermonuclear fusion 32
Thesaurus of welding terms (IIW) 14
Thoria in GTAW electrodes 482
Time-temperature transformation diagrams
(IT) (CCT) 827, 848
Tin, allotropic transformation 75
Titania type covering for SMAW electrodes 732
Titanium
alloy in steel 352
SAW process additions 764
weld metal usage 875
Tool steels 415
Toughness criteria
correlation of test data (IIW) 236
elongation, tensile 148
energy absorption, impact 221
fracture appearance 154, 225
impact testing 155
reduction of area, tensile 149
plastic deformation, impact 225
welds in steel 883
z-direction properties 136, 170
Tracking joints during welding
EBW process 565
GMAW process 499, 500
LBW process 572
Transactinides 40
Transferred plasma arc 511
Transformation hardening 325, 816
Transition current in GMAW process 489
Transition metals 41

Transition temperature
energy absorption 222
fracture appearance 225
plastic deformation capacity 224
Transition welds between dissimilar steels 4
Transmission electron microscopy
(TEM) 108, 124
Tritium 35
True stress-strain tension test diagram 146
Tungsten
alloy in steel 355
electrodes for GTAW process 481
Tunnel-effect microscopy 116
Twinning in metal crystals 81

U
Ultimate tensile strength 144
Ultrasonic soldering 784
Ultrasonic welding process (USW) 613
Ultraviolet radiation 461, 462
Undercooling during metal solidification 72, 104
Undercooled liquids (glass) 47
Undercooling of transformat'ns in solid state 104
Undercutting in fusion welds 771
Underwater friction welding (FRW) 599
Unified Numbering System, UNS
for AISI-SAE steels 416
Units of cracking susceptibility (UCS)
in SAW 507
Unmixed weld zone
of base metal (UNMZ) 472, 799, 881
Upper bainite microstructure 832
Upset welding process (UW)
electrical resistance 533
Upturned fiber in flash welds (FW) 540
Uranium, alloy in steel 356

V
Vacuum 718
consumable-electrode remelting (VAR) 393
degassing of steel 386
deoxidation of steel 379
diffusion bonding & welding (DFW) 591, 612
effect on arc 720
electron beam welding environment 548
induction furnace steelmaking (VIM) 392
protection against oxidation 718
Valency
atom 35, 315
ionic binding forces 59
valency bond of crystals 60

Vanadium, alloy in steel 348
Van Der Waals forces 58
Vaporization, latent heat 49, 265
Veining as sub-boundaries...................71
Vickers hardness test (HV) 150
Victory ships 3
Viscosity of molten metal and slag 264
Voltage of current for arc welding 467

W

Wall neutrality number for flux (N) 747
Water vapor leakage in GMAW 499
Wavelength
    electromagnetic radiation.............................461
    electrons..23
    neutrons 29
Weathering steels 276, 340
Weld backing, for
    electron beam hole drilling...........................645
    root of joints by SAW 511
Weld filler metals (AWS) 429
Weld surface patterns (ripples) 797
Weld overlay..632
Weld pool 472, 476, 483, 500
    depression 484, 485
    humping 484
    oscillation 484
    penetration 484, 486
    Marangoni convection.................................486
    motion 485
Weld zone 797
    acicular ferrite importance 868
    CCT diagram for microstructure 867
    composition homogeneity 472
    cracking, hot, SAW 507
    grain size role 872
    hydrogen content................476, 498, 501, 511
        736, 763
    interface with base metal............................795
    microstructural heterogeneity 863
    microstructures 858, 866
    mixed zone..797
    multi-pass weld 887
    nonmetallic inclusions.........................872, 874
    optimizing weld zone composition...............870
    oxygen content..873
    retained austenite 865
    unmixed weld zone (UNMZ)...............799, 881
Weldability 432
    carbon equivalency formulas......................857
    qualities of steel important to welding 431

WELDASEARCH information retrieval system 13
Welding simulation, by
    furnace heating 675
    resistance heating.......................................676
Welding and allied processes
    AWS master charts.....................................441
Welding metallurgy, definition 6
Welding procedures, optimization 888
Welding thesaurus, international (IIW)................14
Wells wide plate test 202
White cast iron..336
Widmanstätten pattern
    in microstructure 248, 809
Work hardening 248
Wrought iron...361, 367

X

x-rays
    diffraction camera for crystal structure 17
    radiation generation during EBW 460, 549
        565
    spectrographic analysis, photo-electron 121

Y

Yield strength 143
Young's modulus of elasticity............................139
Yttrium aluminum garnet neodymium laser
    (YAG) 463

Z

z-direction mech. properties of steel..........136, 170
Zener, regarding fracture 189
Zinc
    arc welding coated steel 409
    atom construction..19
    coated steels.......................................409, 527
    resistance spot welding 526
Zirconia in GTAW electrode 481
Zirconium, deoxidizer in steel 721